风景园林理论·方法·技术系列丛书

西安建筑科技大学风景园林系　主编

城市空间数据分析方法

——基础试验

包瑞清　[厄瓜多尔]亚历克西斯·阿里亚斯·贝当古　著

(Alexis Arias Betancourt)

中国建筑工业出版社

图书在版编目（CIP）数据

城市空间数据分析方法：基础试验 / 包瑞清，（厄瓜）亚历克西斯·阿里亚斯·贝当古（Alexis Arias Betancourt）著. -- 北京：中国建筑工业出版社，2024. 12. --（风景园林理论·方法·技术系列丛书）. -- ISBN 978-7-112-30513-1

Ⅰ. TU984.11-39

中国国家版本馆CIP数据核字第2024RL0453号

本书中城市空间数据分析方法理论研究框架系统化的建立，正是试图明确跨学科关联知识广度和深度上认知，促进分科知识融通发展为完整的知识体系，强化城市空间数据分析方法发展的深度，推动研究者从工具的使用者转变为工具的创造者，摆脱传统软件工具的束缚，增加城市空间数据分析方法的创造性。本书内容共4章，包括：准备；理论工具；基础试验；附：Python基础核心，为城市空间数据分析方法的基础试验部分。本书的读者群体主要是风景园林、城乡规划、建筑学专业，以数字化规划设计、大数据、机器学习和深度学习、智能化规划设计、城市空间数据分析处理为目的，相关的高等院校、规划设计院等的技术人员、科研人员以及相关专业人士、学者和学生等。

责任编辑：王华月
责任校对：赵　菲

风景园林理论·方法·技术系列丛书
西安建筑科技大学风景园林系　主编

城市空间数据分析方法——基础试验

包瑞清　[厄瓜多尔] 亚历克西斯·阿里亚斯·贝当古（Alexis Arias Betancourt）　著

*

中国建筑工业出版社出版、发行（北京海淀三里河路9号）
各地新华书店、建筑书店经销
北京锋尚制版有限公司制版
廊坊市海涛印刷有限公司印刷

*

开本：787毫米×1092毫米　1/16　印张：44¾　字数：1534千字
2025年3月第一版　2025年3月第一次印刷
定价：**198.00**元
ISBN 978-7-112-30513-1
（43116）

作者:

包瑞清

任教于西安建筑科技大学建筑学院，从事数字化设计相关研究。已出版《参数化逻辑构建过程》《参数模型构建》《编程景观》《学习PYTHON—做个有编程能力的设计师》《ArcGIS下的Python编程》和《折叠的程序》等专著。

亚历克西斯 · 阿里亚斯 · 贝当古（Alexis Arias Betancourt）

为伊利诺伊理工大学博士学者，专注于可持续交通基础设施研究，主要侧重于普遍存在无人驾驶技术的应用，发表有无人驾驶城市：自动驾驶汽车及导航安全在城市的可能性等研究论文。

Alexis Arias, is a Ph.D scholar affiliated with the Illinois Institute of Technology. His research pursuits center around sustainable transportation infrastructure, with a primary emphasis on the application of autonomous and ubiquitous technologies. Alexis Arias has authored and co-authored publications such as “Driverless City: The Urban Possibilities of Autonomous Vehicles and Navigation Safety.”

技术审稿人:

程宏

目前，在韩国光云大学攻读计算机工程博士学位。他的研究兴趣包括机器学习、深度学习、通信领域路径损耗建模和基于地理信息的图生图方法等。同时，也正在向大模型微调训练领域拓展更多的可能性。

贡献者:

数字营造学社：王育辉、刘航宇、张旭阳、柴金玉、戴礽祁、许保平、赵丽播、张卜予、王垚

资助:

国家重点研发计划资助（城市蓝绿空间生态涵养关键技术研究与示范（2022YFC3802603）——城市蓝绿空间规划与生态系统服务功能优化技术）。

总序

风景园林学是综合运用科学与艺术的手段，研究、规划、设计、管理自然和建成环境的应用型学科，以协调人与自然之间的关系为宗旨，保护和恢复自然环境，营造健康优美人居环境为目标。风景园林学研究人类居住的户外空间环境，其研究内容涉及户外自然和人工境域，是综合考虑气候、地形、水系、植物、场地容积、视景、交通、构筑物和居所等因素在内的景观区域的规划、设计、建设、保护和管理。风景园林的研究工作服务于社会发展过程中人们对于优美人居环境以及健康良好自然环境的需要，旨在解决人居环境建设中人与自然之间的矛盾和问题，诸如国家公园与自然保护地体系建设中的矛盾与问题、棕地修复中的技术与困难、气候变化背景下的城市生态环境问题、城市双修中的技术问题、新区建设以及城市更新中的景观需求与矛盾等。当前的生态文明建设和乡村振兴战略为风景园林的研究提供了更为广阔的舞台和更为迫切的社会需求，这既是风景园林学科的重大机遇，同时也给学科自身的发展带来巨大挑战。

西安建筑科技大学的历史可追溯至始建于1895年的北洋大学，从梁思成先生在东北大学开办建筑系到1956年全国高等院校院系调整、学校整体搬迁西安，由东北工学院、西北工学院、青岛工学院和苏南工业专科学校的土木、建筑、市政系（科）整建制合并而成，积淀了我国近代高等教育史上最早的一批土木、建筑、市政类学科精华，形成当时的西安建筑工程学院及建筑系。我校风景园林学科的发展正是根植于这样历史深厚的建筑类教育土壤。1956年至1980年代并校初期，开设园林课程，参与重大实践项目，考察探索地方园林风格；1980年代至2003年，招收风景园林方向硕士和博士研究生，搭建研究团队，确立以中国地景文化为代表的西部园林理论思想；2002年至今，从“景观专门化”到“风景园林”新专业，再到“风景园林学”新学科独立发展，形成地域性风景园林理论方法与实践的特色和优势。从开办专业到2011年风景园林一级学科成立以来，我院汇集了一批从事风景园林教学与研究的优秀中青年学者，这批中青年学者学缘背景丰富、年龄结构合理、研究领域全面、研究方向多元，已经成长为我校风景园林教学与科研的骨干力量。

“风景园林理论·方法·技术系列丛书”便是各位中青年学者多年研究成果的汇总，选题涉及黄土高原聚落场地雨洪管控、城市开放空间形态模式与数据分析、城郊乡村景观转型、传统山水景象空间模式、城市高密度区小微绿地更新营造、城市风环境与绿地空间协同规划、城市夜景、大遗址景观、城市街道微气候、地域农宅新模式、城市绿色生态系统服务以及朱鹮栖息生境保护等内容。在这些作者中：

杨建辉长期致力于地域性规划设计方法以及传统生态智慧的研究，建构了晋陕黄土高原沟壑型聚落场地适地性雨洪管控体系和场地规划设计模式与方法。

刘恺希对空间哲学与前沿方法应用有着强烈的兴趣，提出了“物质空间表象-内在动力机制”的研究范型，总结出四类形态模式并提出系统建构的方法。

包瑞清长期热衷于数字化、智能化规划设计方法研究，通过构建基础实验和专项研究的数据分析代码实现途径，形成了城市空间数据分析方法体系。

吴雷对西部地区乡村景观规划与设计研究充满兴趣，提出了未来城乡关系变革中西安都市区城郊乡村景观的空间异化转型策略。

董芦笛长期致力于中国传统园林及风景区规划设计研究，聚焦人居环境生态智慧，提出了“象思维”的空间模式建构方法，建构了传统山水景象空间基本空间单元模式和体系。

李莉华热心于探索西北城市高密度区绿地更新设计方法，从场地生境融合公众需求的角度研究了城市既有小微绿地更新营造的策略。

薛立尧专注于以西安为代表的我国北方城市绿地系统的生态耦合机制研究及规划方法创新，尤其在绿地与风环境因子的耦合规划建设方面取得了一定的进展。

孙婷长期致力于城市夜景规划与景观照明的设计与研究工作，研究了昼、夜光环境下街道空间景观构成特征及关系，提出了“双面街景”的设计模式。

段婷热心于文化遗产的保护工作，挖掘和再现了西汉帝陵空间格局的历史图景，揭示了其内在结构的组织规律，初步构建了西汉帝陵大遗址空间展示策略。

樊亚妮的研究聚焦于微气候与户外空间活动及空间形态的关联性，建立了户外空间相对热感觉评价方法，构建了基于微气候调控的城市街道空间设计模式。

沈葆菊对“遗址–绿地”的空间融合研究充满兴趣，阐述了遗址绿地与城市空间的耦合关系，提出了遗址绿地对城市空间的影响机制及城市设计策略。

孙自然长期致力于乡土景观与乡土建筑的研究，将传统建筑中绿色营建智慧经验进行当代转译，为今天乡村振兴服务。

王丁冉对数字技术与生态规划设计研究充满热情，基于多尺度生态系统服务供需测度，响应精细化城市更新，构建了绿色空间优化的技术框架。

赵红斌长期致力于朱鹮栖息生境保护与修复规划研究，基于栖息地生境的具体问题，分别从不同生境尺度，探讨朱鹮栖息地的保护与修复规划设计方法。

近年来本人作为西安建筑科技大学建筑学院的院长，目睹了上述中青年教师群体从科研的入门者逐渐成长为学科骨干的曲折历程。他（她）们在各自或擅长或热爱的领域潜心研究，努力开拓，积极进取，十年磨一剑，终于积淀而成的这套“风景园林理论·方法·技术系列丛书”，是对我校风景园林学科研究工作阶段性的、较为全面的总结。这套丛书的出版，是我校风景园林学科发展的里程碑，这批中青年学者，必将成为我国风景园林学科队伍中的骨干，未来必将为我国风景园林事业的进步贡献积极的力量。

值“风景园林理论·方法·技术系列丛书”出版之际，谨表祝贺，以为序。

刘加平

中国工程院院士，西安建筑科技大学教授

序言

Ron Henderson FASLA

We must leverage the information and insights that we gain from computing to plan, build, and adapt cities for health, safety, equity, and beauty, among other values. Otherwise, we squander the cultural and ecological promise that new and emerging technologies could help attain. Autonomous vehicles are one such computing-based technology – with the opportunity for traffic-related injuries and deaths to trend toward zero, with greater land use equity for people in streets, and with travel efficiencies that give people time back to their lives, among others. It was within the context of autonomous vehicles, and The Driverless City Project at Illinois Institute of Technology in Chicago, that Bao Ruiqing led an investigation of urban spatial data analytics to amplify and elevate our research. It is joyful to see some of this work in this book.

The Driverless City Project is an interdisciplinary initiative initiated in 2015 and funded by, among others, the National Science Foundation. It has involved landscape architects, roboticists, navigation engineers, law professionals, urban planners, transportation planners, social scientists, political scientists, and civil engineers. In 2018, I extended an invitation to Bao Ruiqing, an expert specializing in integrated research pertaining to urban spatial data analysis methods, to join the Driverless City Project as a visiting scholar working alongside our team on his research. Concurrently, and with guidance from Bao Ruiqing, graduate research assistant and PhD candidate Alexis Arias Bettancourt explored urban spatial data analysis methods to enrich our research with theoretical models of analytics.

In conjunction with the overarching Driverless City Project, Bao and Arias conducted experimental analyses exploring diverse facets of urban spatial structures in the city of Chicago. This research encompassed the evaluation of land cover type connectivity based on remote sensing (RS) imagery, scrutiny of the network structure within urban open spaces, examination of building height distribution, spatiotemporal data analysis, parametric models, as well as the examination of patterns and the generation of landmark points within the framework of driverless city studies. Over the span of approximately two years, the project successfully executed a total of approximately 20 exploratory experiments.

In this book, which has reached fruition about two and a half years after Bao Ruiqing's term as visiting scholar, I can witness at least four valuable accomplishments in this rigorous work:

1) Theoretical methods can be successfully intergrated into urban spatial data analysis.
2) Theoretical methods explored in Python code achieve operability and practical significance.
3) Theoretical knowledge that is interactive, based on Python code, can be utilized to understand complex formulas and scenarios.
4) Explorations that utilize multi-source data with different formats and contents provide resiliency and a rich diversity of scenarios for both data processing and analytical results.

Urban spatial data analysis methods integrates many theoretical models across disciplines to promote interactive exploration of data analysis. It is book that will continue to influence the expansion of analytics to plan and design great cities.

Ron Henderson FASLA is Professor of Landscape Architecture + Urbanism at Illinois Institute of Technology, inaugural Director of Research at The Alphawood Arboretum at Illinois Institute of Technolgy, and founding principal of LIRIO Landscape Architecture.

前言

网络的持续建设升级，数据的可获取，机器学习和深度学习的快速发展，理论方法的不断更新迭代，可实现数据处理、分析和理论方法等Python编程语言扩展库的急速增加和开源，全球知识共享，计算机硬件性能的提升，使得对城市机制的理解，空间模式的发现都可以尝试从数据中去探寻，这为城市空间数据分析方法的探索提供了契机。笔者在2017年末时开始思考究竟有哪些知识可以在专业领域中发挥作用，并且用Python自行书写代码工具而不受制于既有的软件仅能提供有限计算功能，及无法即时更新最新理论方法的限制，且能从根本上来学习、理解这些理论方法，从而以此为根基将其用于专业领域相关问题的分析，甚至为自行探索新的方法、算法提供可能性。同时，将对理论方法的学习、解释和代码的实现相结合，通过具体观察计算过程数值的变化有效的理解理论方法。

对城市空间数据分析方法的探索和写作基本经历了三个阶段：

第一个阶段从2019年1月至2020年10月，随机的进行大量（大于50次）的探索性试验，例如城市色彩、POI数据分析、SIR模型、数据库试验、景观质量视觉评估、试验用网络应用平台的部署、生活圈、城市热环境、视域分析、时空数据和无人驾驶城市等，详细内容的地址为：https://github.com/richieBao/python-urbanPlanning。

第二个阶段从2020年10月至2023年1月，将第一阶段进行的大量试验归类并增加新的基础内容为基础试验部分，同时思考开始专项研究部分。基础试验部分包括编程统计学及数据统计分析、编程线性代数及遥感影像数据处理、编程微积分及SIR空间传播模型、计算机视觉及城市街道空间、机器学习实验、深度学习实验、时空数据分析、复杂网络（图论）、智能体模型（ABM）和点云数据处理与内存管理等部分。对于专项研究部分是希望能够通过具体的多个案例来解释数据分析的具体流程，初步分为城市空间：形态-结构-功能、城市生态：自然-生物-小气候和城市智能：微控制器-传感器等3个部分，并尝试开展了10个试验。同时，结合高校课程，完成了Python基础核心部分内容。该阶段具体内容的地址为：https://richiebao.github.io/USDA_CH_final。

第三个阶段从2023年1月至2023年12月，因为第二个阶段的专项研究部分开展的并不理想，主要原因是以数据分析过程演示案例并不会给读者，甚至作者自身带来太多有价值的知识和信息。此外，大量代码占据了版面。就上述原因，做出了两个决定，一是以专项探索的方式，集成具有价值的理论方法，形成城市空间数据分析方法理论集成研究框架，除基础试验外，包括标记距离、权重决策、更新策略、模式生成、推理学习、时空序列、维度空间、尺度效应、强化学习和复杂网络等10个专题；二是建立USDA工具包，通过调用工具包中的工具减少代码量，并形成可用的科学工具。该阶段形成了最终稿，在线的更新地址为：https://richiebao.github.io/USDA_CH_endup。

虽然集成了大量理论方法于本书，但这只是城市空间数据分析方法的一隅。在未来的探索里，除了基于本书集成的方法理论和USDA工具包开展纵深向更深入的研究外；也继续横向拓展不断增加新的理论方法并将其融入至工具包中；此外不断调整更新已有内容和工具包，修正错误并使其趋于完善、合理。

理论方法的集成涉及跨专业多学科领域，期待读者批评指正以不断修正完善。作者邮箱：richiebao@outlook.com。

2023年11月18日于西安

指南

1．面向的读者

《城市空间数据分析方法》可用于的研究专业方向包括：城市规划、风景园林、建筑、生态、地理信息等专业。人群为高校在校学生和研究者、企事业的规划设计师等。

2．电脑的配置与库的安装

电脑的配置

本书涉及深度学习。为了增加运算速度，用到GPU。关于选择哪种GPU可以查看相关信息，或者使用Colaborator（CoLab，由Google推出的云端深度学习计算平台）。作者所使用的配置为（仅作参考）：

```
OMEN by HP Laptop 15
Processor - Intel(R) Core(TM) i9-9880H CPU @2.30GHz
Installed RAM - 64.0 GB (63.9 GB usable)
System type - 64-bit operating system, x64-based processor

Display adapter - NVIDA GeForce RTX 2080 with Max-Q Design
Copy to clipboardErrorCopied
```

建议内存最低 32.0GB，最好 64.0GB 以上或更高。如果内存较低，则考虑分批读取数据处理后，再合并等方法。善用内存管理，数据读写方式，例如用 HDF5[①] 等格式。

城市空间数据分析通常涉及海量数据处理，例如芝加哥城三维激光雷达数据约有1.41TB。如果使用的是笔记本电脑，最好配置容量大的外置硬盘。虽然会影响到读写速度，但是避免了对笔记本自身存储空间的占用。

库的安装

Anaconda集成环境[②]，使得Python库的安装，尤其处理库之间的依赖关系变得轻松。虽然Anaconda提供了非常友好的库管理环境，但还是会有个别库之间存在冲突，有可能造成库安装失败，甚至已经安装的库也可能出现问题无法调入，无形之中增加了学习的时间成本。因此，在开始本书代码前，不要在一个环境之下一气安装所有需要的库，而应是跟随代码的编写，由提示再行逐个安装。

同时，对于某一项目工程，最好仅建立支持本项目的新环境，不安装未使用到的库，使集成环境轻便，在网络环境部署时也不会冗余，减少出错几率。对于自行练习或者不是很繁重的项目任务，只要环境能够支持，代码运行不崩溃就可以。

3．如何学习

善用`print()`

`print()`是Python语言中使用最为频繁的语句。在代码编写、调试过程中，不断地用该语句来查看数据的结构、值和变化，变量所代表的内容，监测程序进程及显示计算结果等。通过`print()`实时查看数据反馈，才可以知道是否实现了代码编写的目的，并做出反馈。善用`print()`是用Python写代码的基础。

充分利用搜索引擎

代码的世界摸不到边际，不可能记住所有方法函数；而且，代码库在不断地更新、完善，记住的不一定是对的。学习代码关键是养成写代码的习惯，训练写代码的思维，最终借助这个工具来解决实际的问题。那些无以计数的库、涉及的函数、方法、属性等内容，通常是在写代码过程中，充分利用搜索引擎和对应库的官网在线文档来查找，其中一个不错的问答平台是StackOverflow[3]。经常用到的方法函数会不知不觉的被记住，或有意识的来记忆经常用到的方法函数。即使对使用过的方法仅存有蛛丝马迹的印象，通常也可以依据这些线索快速地再行搜索寻找答案。

由案例来学习

对方法函数再多的解释，不如一个简单的案例示范来得清晰、明了。通常，查看某一个方法函数及属性的功用，最好的途径是直接搜索案例来查看数据结构的变化，往往不需要阅读相关解释文字就可以从数据的前后变化中获悉该方法的作用，再辅助阅读文字说明进一步深入理解或者对比查看仍旧不明晰地方的解释。

库的学习

城市空间数据分析方法在各类实验中使用有大量的相关库，例如[4]:

- 数据结构：NumPy，Pandas；
- 科学计算：SciPy，SymPy；
- 统计推断：statsmodels；
- 机器学习：scikit-learn；
- 深度学习：PyTorch，TensorFLow，TorchVision，segmentation-models-pytorch，Kornia，TorchGeo，Transformers，pytorch-lightning；
- GIS工具：GDAL，GeoPandas，rasterstats，rasterio，EarthPy，PySAL，Shapely，Fiona，rio-tiler，rioxarray，Xarray；
- 复杂网络：NetworkX，igraph；
- 点云处理：PDAL，PCL，open3D；
- 影像视觉：scikit-image，OpenCV，Pillow；
- 数据可视化：matplotlib，Plotly，seaborn，bokeh，VTK，gradio；
- Web应用：flask；
- 数据库：sqlite3；
- GUI：tkinter，PySide6，PyQt，Kivy，Pygame；
- 其他类：re，pathlib，itertools，urllib。

很多都为大体量库，单独学习每个库都会花费一定的时间。阅读该书时，如果应用到的库之前没有接触过。一种方式是，提前快速的学习各个库官方提供的教程（tutorial），配合手册（manual）快速掌握库的主要结构，再继续阅读本书实验；另一种方式是，为了不打断阅读的连贯性，可以只查找用到该库相关方法部分的内容。

一般每个库的内容都会非常多，通常不需要一一学习。当用到时，再根据需要有针对性和有目的性地查阅。但是，也有些库是需要系统的学习，例如scikit-learn、PyTorch、PySide6等重量型的库。

用代码学习数学公式

在科学中，公式是用符号表达信息的一种简明方式。数学公式通常使用特定逻辑语言的符号和形成规则构建事物间的联系，以更好地理解事物的本质和内涵。当用公式代替文字阐述方法时，可以更直观地表达内在逻辑和变化规律。然而很多时候，公式及文字阐述让人费解，读不懂作者所要表达的含义和目的，尤其包含有大量公式，而给出的案例也都只是“白纸黑字”的文字论述时。这是因为很难去调整参数再实验作者的案例。但是，用Python等语言进行数据分析，可以用`print()`查看每一步数据的变化，尤其对于不易理解的地方，通过比较前后的数据变化，容易发现内在的逻辑，理解用公式和文字描述不易阐明的地方。这也是为什么本书的所有阐述都基于代码。当公式和文字阐释不易理解的时候，只要逐行地运行代码，往往能明了。同时，尽可能地将数据及其变化以图表数据可视化的方式表达，显现事物中隐藏的联系，让问题更容易浮出水面。

书中公式保留了各参考文献中的符号表述和对应的含义。

避免重复造轮子

“避免重复造轮子”是程序员的一个“座右铭”。当然，每个人所持的观点不尽相同。但是可以肯定的是，没有必要从零开始搭建所有项目内容，这如同再写Python的所有库，甚至Python核心内容的东西。例如，scikit-learn、PyTorch集成了大量的模型算法，通常直接调用，不会再重复写这些代码。本书阐述过程中，会有意识地对最为基本的部分按照计算公式重写代码，目的是认识方法的内在逻辑，而实际的项目则无须如此。

对于那些绕弯的代码和理解算法

读代码不是读代码本身，而是读代码解决问题的逻辑策略。有些代码比较简单，例如数组的加减乘除；但是有些代码往往“不知其所以然”，却解决了某一问题。那么这里面可能就有一些非常巧妙的处理手段，例如梯度下降法、主成分分析、生成对抗网络，以及各类算法等。很多时候解决问题的方法越巧妙，逻辑方法越不同于常规，因此阅读起来可能不那么容易。对于理解这些绕弯的代码，除了要不断查看数据变化，甚至要结合可交互图表动态分析，也可以网络搜索寻找更清晰易懂的解释方式，或者自行寻找到易于理解的途径，找到作者所发现的方法的内在逻辑。

库中的方法通常是解决某一问题算法逻辑的实现，一般而言直接调用该方法，避免重复造轮子，这毋庸置疑。但是，有时很多算法能够启发我们，从而对某些问题提出新的解决策略，因此有意识的去理解一些算法是必要的。

对于“千行+”复杂代码的学习

能够一下子看懂百千行代码的可能性不大，尤其这些代码以类的形式出现时，各类参数、变量，甚至函数自身在各个函数下不断的跳跃，乃至分支、合并，不断产生新的变量。第1阶段，捋清楚初始化的条件及初始化参数传输的路径，是首先需要明确的事情，这个过程的难易视代码的复杂程度决定。一旦清楚了代码运行的流程，即语句执行的顺序（有顺序，也可能并行），可以将其记录下来，方便进一步的分析；进而开始第2阶段的代码分析，因为已经清楚了流程，那么就从开始逐行的运行代码，打印数据，理解每一行代码所要解决的是哪些问题。这个过程与第1阶段的捋清代码运行流程可以同时进行。边梳理、边运行行查看数据，边分析理解代码解决问题的逻辑。逐行分析代码时，往往会遇到那些绕弯的代码（算法）。如果一时无法理解，在不影响后续代码运行前提下，可以先绕过，直接获取用于运行后续代码行的数据；最后，再重点分析这些绕弯的代码。如果无法绕过，那么就只能花费精力先来解决这个问题。

参考文献的推荐

城市空间数据分析方法是多学科的融合，这包括作为分析工具的Python编程语言知识及城乡规划、建筑和风景园林等专业知识，以及数学（统计学、微积分、线性代数）、数据库、地理信息系统、生态学等大类和其下更为繁杂的小类细分等知识。虽然涉猎内容众多，但实际上只是在解决某一问题时用到所需要的相关知识。当阅读该书时，如果某处的知识点不是很明白，可以查找相关文献补充阅读。在每一部分，也提供了所参考的文献。必要时，可以翻看理解或印证。

书中对应的知识点均详细地给出了参考文献，方便对应知识点的溯源。

关于英语和翻译

Python编程语言使用英文书写；同时，本书集成的大部分理论方法来源的期刊论文和论著基本为英文文献；再者，很多英文的同一词汇翻译为中文时往往有多种翻译结果，造成术语不统一，因此本书书写过程中保留了大量关键术语的英文词汇，以避免混淆，甚至产生的歧义，而且有助于读者以英文搜索相关文献。并且，迁移代码中的英文释义也有选择性地给予了保留。

开源和数据

形成本书各阶段的代码均已开源，包括：

python-urbanPlanning：https://github.com/richieBao/python-urbanPlanning；

USDA_CH_final：https://github.com/richieBao/USDA_CH_final；

USDA_CH_endup：https://github.com/richieBao/USDA_CH_endup；

USDA_special_study：https://github.com/richieBao/USDA_special_study；

caDesign_ExperimentPlatform：https://github.com/richieBao/caDesign_ExperimentPlatform；

USDA_dashboard：https://github.com/richieBao/USDA_dashboard；

USDA_PyPI：https://github.com/richieBao/USDA_PyPI。

读者可以自行下载查看或者运行。

对于试验中的数据，均使用的为开源数据且给出了来源地址；而且，因为部分数据量较大，因此目前并未提供单独统一下载地址。为了方便读者试验，未来应会寻找适合服务器提供下载，可以从USDA_CH_endup地址获取未来的更新信息。

基础试验部分和部分专项探讨部分，书写代码时将调用的模块放置于函数的内部，用于明确定义函数所要调用的模块。在实际的项目任务中，调用模块一般统一置顶，特此说明。

注释（Notes）：

① Hierarchical Data Format，HDF是一组文件格式（HDF4，HDF5），用于大数据的存储和组织。其中h5py包（https://docs.h5py.org/en/stable/）是HDF5二进制数据格式的一个Python接口。

② Anaconda，用于科学计算（数据科学、机器学习应用、大数据处理、预测分析等）的Python和R编程语言的环境管理器，旨在简化包管理和部署（https://www.anaconda.com/）。

③ Stack Overflow，是一个面向专业程序员和爱好者，关于计算机编程广泛主题的问答网站（https://stackoverflow.com/）。

④ 例举的相关库，印刷版本未给出网页链接时，可以通过搜索引擎检索。

目录

1 准备

2 理论工具

3 基础试验

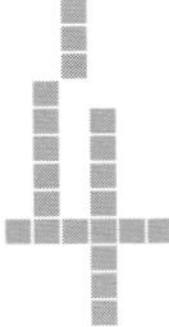

附：Python 基础核心

1

准备

1.1 代码的整洁之道

书写代码时，能够有意识地保持代码的整洁，养成一种代码书写的习惯，本身就是一种智慧。

1.1.1 代码的整洁之道

每一篇代码都应该是一篇蕴含着作者解决某一问题逻辑思考的“散文”，即使不懂代码的读者也能够行云流水般的阅读，从而有个大概的认知，知道作者想要解决什么问题，以及大概是怎么解决的。因此，有必要提及代码的整洁之道，梳理一些关键事项，以备参考。

● **变量的命名应能反映变量本身的意义**

代码书写时，为了图一时方便，往往不注重变量名的命名，而是随意起名，例如代码行中充斥着x、y、i、j、k等单个变量名，及jh、ik等不反映变量名意义的字母组合。日后需要返回来应用已书写的代码或者迁移代码于新项目时，无法快速理解变量的意义，使得代码可读性差，从而会浪费更多时间再度理解，这也为他人迁移应用该代码形成了一定的阻碍。

变量名命名时，建议的形式包括三种：

一是单个词，例如markers=['.','+','o','^','x','s']，这是应用matplotlib图表库打印图表时，定义图表标记类型所定义的类型列表，其名字能够反映所要表述变量的意义。

二是字母组合，其一示例为landmarkPts=[Point(coordi[0],coordi[1]) for coordi in np.stack((landmarks[0], landmarks[1]), axis=-1)]中的landmarkPts，或者不引入缩写，为landmarkPoints。可以直接理解该变量名为地标点，形式为字母组合时后一字母首字母大写，称为驼形（Camel-case）；其二是，landmark_pts或者landmark_points形式，即多个字母组合时中间以短下画线隔离，称为蛇形（underscore-separated，Snake-case）。之后，字母的首字母不须大写。

● **给出关键的注释**

“烂笔头”的重要性无需质疑。虽然好的变量名能够一定程度上解释语句的意义，但是解决问题的逻辑通常需要给与注释（使用# 或者```note```），尤其记录解决问题时思维碰撞的关键想法。例如，对函数功能的注释：

```python
def recursive_factorial(n):
    '''
    阶乘计算。

    Parameters
    ----------
    n : int
        阶乘计算的末尾值.

    Returns
    -------
    int
        阶乘计算结果.

    '''
    if n == 1:
        return 1
    else:
        return n * recursive_factorial(n - 1)
```

及对关键语句的解释：

```python
idx = tempDf[tempDf['cluster'] == -1].index # 删除字段 'cluster' 值为 -1 对应的行，即未形成聚
类的独立 OSM 点数据 tempDf.drop(idx,inplace=True)
```

如果以上示例中不给出注释，则很难立刻理解出这些代码书写的意图，往往必须花费较大精力来推断，甚至不得不阅读整段代码。当然，要以对函数和变量适合的命名为先，再注释补充。不应该是草草命名后，再用大量的注释来解释，本末倒置。

● **尽量保持定义的一个函数仅作一件事情，及最小的迁移代价**

每一段函数代码都是解决某一问题的一种方法，很多时候在同一项目或不同项目中，这种

方法会不断被重复调用，因此该函数应该能够以最少的代价来迁移，无须或者很少的改动代码就可以调用。

一是，一个函数最好仅做一件事情，保持代码具有更好的易读性和可迁移性。这个并不难以理解。如果该函数可以同时做多件事情，当其他项目仅需要该函数的部分功能时，事情就会变得复杂起来。

二是，有时在单独的函数内部包含了全局变量，迁移后会提示全局变量未定义的情况。因此建议函数定义时，尽量通过函数参数传递变量。

三是，在单独函数命名，及函数内变量命名时，尽量保持名称通用性，避免迁移后的名称只能反映当初项目的内容，例如函数`ffill(arr)`表示向前填充数组中的空数据。如果函数名及参数名改为具体的`landmarks_ffill(landmarks_arr)`，迁移后的变量可能不是`landmarks`，就会容易引起歧义，造成代码阅读上的干扰。

- **复杂的程序要学会使用类和分文件——系统的思维**

代码可以在不知不觉中变得复杂起来，少则千行，多则万行。那么，所有代码位于一个文件（.py）中，包含有数十个单独的函数，代码的管理就会成为问题。因此学会应用类的面向对象编程（Object-oriented programming，OOP）来组织相关属性和方法，并用多个子包（Subpackage）和分文件（Module，模块）来切分代码，在相关文件中只是引用，例如`import driverlessCityProject_spatialPointsPattern_association_basic as basic`，调入模块并取别名为`basic`来使用代码。那么代码的结构就会比较清晰，也避免了处于同一文件中，不容易查找的弊病；同时，方便后续代码打包，发布到PyPI[1]，分享工具。

类和分文件只是系统思维表现的一种手段，对于项目所有代码的合理组织，确定数据的流动走向、前后关联层级的配置、数据库的使用、整体结构的把握，才能够保证代码的稳健性（robustness）。

\# 《代码的整洁之道 (Clean Code: A Handbook of Agile Software Craftsmanship》这本阐述代码哲理的书，不管是对刚刚进入代码领域，还是早已浸淫代码多年的程序员，通读或偶尔翻阅，都会受益良多。

① Python Package Idnex, PyPI，为Python编程语言的软件库，可以安装发布Python库（包）或者发布包文件（https://pypi.org/）。

1.1.2 代码的风格

\# 参考引用：Google 开源项目风格指南[2]，并根据该书情况稍作调整。

② Google 开源项目风格指南（https://zh-google-styleguide.readthedocs.io/en/latest/google-python-styleguide/python_style_rules/#section-2）；同时少量参考“Python PEP-8编码风格指南中文版”（https://alvin.red/2017/10/07/python-pep-8/）。

- **风格是为了代码具有更好的可读性；**
- **风格同一模块和同一函数保持一致，同一项目保持一致，同风格指南或共识性规范尽量保持一致；**
- **缩进风格统一，示例：**

```
# 同开始分界符（左括号）对齐（Aligned with opening delimiter）
foo = long_function_name(var_one, var_two,
                         var_three, var_four)

# 续行多缩进一级以同其他代码区别
def long_function_name(
        var_one, var_two, var_three,
        var_four):
    print(var_one)

# 悬挂缩进需要多缩进一级（space hanging indent; nothing on first line）
foo = long_function_name(
    var_one, var_two,
    var_three, var_four)

# 与字典中的开始分隔符对齐（Aligned with opening delimiter in a dictionary）
foo = {
     long_dictionary_key: value1 + value2,
     ...
}
```

- **行长度，示例**

```
# Python 会将圆括号，中括号和花括号中的行隐式的连接起来
Yes: foo_bar(self, width, height, color='black',
             design=None, x='foo',emphasis=None,
```

```
         highlight=0)
if (width == 0 and height == 0 and
    color == 'red' and emphasis == 'strong'):

# 如果一个文本字符串在一行放不下，可以使用圆括号来实现隐式行连接
x = ('This will build a very long long '
     'long long long long long long string')

# 在注释中，如果必要，将长的 URL 放在一行上
# http://www.example.com/us/developer/documentation/api/content/v2.0/csv_
file_name_extension_full_specification.html
```

- 序列元素尾部逗号

```
Yes:   golomb3 = [0, 1, 3]
Yes:   golomb4 = [
         0,
         1,
         4,
         6,
         ]
No:    golomb4 = [
         0,
         1,
         4,
         6
         ]
```

- 空行：顶级定义之间空两行，方法定义之间空一行

顶级定义之间空两行，比如函数或者类定义。方法定义，类定义与第一个方法之间，都应该空一行。函数或方法中，根据具体情况确定空一行，还是空两行。

- 空格

括号内不要有空格。

```
Yes: spam(ham[1], {eggs: 2}, [])
No:  spam( ham[ 1 ], { eggs: 2 }, [ ] )
```

不要在逗号、分号、冒号前面加空格，但应该在它们后面加（除了在行尾）。

```
Yes: if x == 4:
         print(x, y)
     x, y = y, x
No:  if x == 4 :
         print(x , y)
     x , y = y , x
```

参数列表，索引或切片的左括号前不应加空格。

```
Yes: spam(1)
no: spam (1)
Yes: dict['key'] = list[index]
No:  dict ['key'] = list [index]
```

在二元操作符两边都加上一个空格，比如赋值（=），比较（==, <, >, !=, <>, <=, >=, in, not in, is, is not），布尔（and, or, not）。至于算术操作符两边的空格该如何使用，需要具体情况判断，不过两侧务必要保持一致。

```
Yes: x == 1
No:  x<1
```

当 = 用于指示关键字参数或默认参数值时，不要在其两侧使用空格。但若存在类型注释的时候，需要在 = 周围使用空格。

```
Yes: def complex(real, imag=0.0): return magic(r=real, i=imag)
Yes: def complex(real, imag: float = 0.0): return Magic(r=real, i=imag)
No: def complex(real, imag = 0.0): return magic(r = real, i = imag)
No: def complex(real, imag: float=0.0): return Magic(r = real, i = imag)
```

不要用空格来垂直对齐多行间的标记，因为这会成为维护的负担（适用于:, #, =等）：

```
Yes:
    foo = 1000  # 注释
```

```
    long_name = 2  # 不应该对齐注释
    dictionary = {
        "foo": 1,
        "long_name": 2,
        }
No:
    foo       = 1000  # 注释
    long_name = 2     # 不应该对齐注释
    dictionary = {
        "foo"      : 1,
        "long_name": 2,
        }
```

- **注释**

函数部分注释按Spyder[①]提供的方法，例如：

```
def recursive_factorial(n):
    '''
    阶乘计算。

    Parameters
    ----------
    n : int
        阶乘计算的末尾值 .

    Returns
    -------
    int
        阶乘计算结果 .

    '''
    if n==1:
        return 1
    else:
        return n*recursive_factorial(n-1)
```

#之后空一格

```
Yes:
    # the next element is i+1
No:
    #the next element is i+1
```

无须注释模块中的所有函数：

1. 公共的API需要注释；
2. 在代码的安全性、清晰性和灵活性上进行权衡是否注释；
3. 对于容易出现类型相关错误的代码进行注释；
4. 难以理解的代码请进行注释；
5. 若代码中的类型已经稳定，可以进行注释。对于一份成熟的代码，多数情况下，即使注释了所有的函数，也不会丧失太多的灵活性。

- **字符串**

%格式化字符前后空一格。

```
Yes:
    x = '%s, %s!' % (imperative, expletive)
No:
    x = '%s, %s!'%(imperative, expletive)
```

避免在循环中用+和+=操作符来累加字符串。由于字符串是不可变的，这样做会创建不必要的临时对象，并且导致二次方而不是线性的运行时间。作为替代方案，可以将每个子串加入列表，然后在循环结束后用 .join连接列表（也可以将每个子串写入一个cStringIO.StringIO缓存中）。

```
Yes: items = ['<table>']
    for last_name, first_name in employee_list:
        items.append('<tr><td>%s, %s</td></tr>' % (last_name, first_name))
    items.append('</table>')
    employee_table = ''.join(items)
No: employee_table = '<table>'
    for last_name, first_name in employee_list:
```

① Spyder，免费开源的Python交互式解释器，由科学家、工程师和数据分析师设计并为其服务。具有科学软件包的数据分析探索，交互执行，深度检验和数据可视化等能力，包括高级编辑、分析，调试和剖析等功能（https://www.spyder-ide.org/）。

```
        employee_table += '<tr><td>%s, %s</td></tr>' % (last_name, first_name)
    employee_table += '</table>'
```

在同一个文件中，保持使用字符串引号的一致性。使用单引号'或者双引号"之一用以引用字符串，并在同一文件中沿用。在字符串内，可以使用另外一种引号，以避免在字符串中使用。

```
Yes:
    Python('Why are you hiding your eyes?')
    Gollum("I'm scared of lint errors.")
    Narrator('"Good!" thought a happy Python reviewer.')
No:
    Python("Why are you hiding your eyes?")
    Gollum('The lint. It burns. It burns us.')
    Gollum("Always the great lint. Watching. Watching.")
```

为多行字符串使用三重双引号"""而非三重单引号'''。当且仅当项目中使用单引号'来引用字符串时，才可能会使用三重'''为非文档字符串的多行字符串来标识引用。文档字符串必须使用三重双引号"""。多行字符串不应随着代码其他部分缩进的调整而发生位置移动。如果需要避免在字符串中嵌入额外的空间，可以使用串联的单行字符串或者使用textwrap.dedent()来删除每行多余的空间。

```
No:
long_string = """This is pretty ugly.
Don't do this.
"""
Yes:
long_string = """This is fine if your use case can accept
    extraneous leading spaces."""
Yes:
long_string = ("And this is fine if you cannot accept\n"
    + "extraneous leading spaces.")
Yes:
long_string = ("And this too is fine if you cannot\n"
    "accept extraneous leading spaces.")
Yes:
import textwrap
long_string = textwrap.dedent("""\
    This is also fine, because textwrap.dedent()
    will collapse common leading spaces in each line.""")
```

- 文件和sockets

除文件外，sockets或其他类似文件的对象在没有必要的情况下打开，会有许多副作用，例如：

1. 它们可能会消耗有限的系统资源，如文件描述符。如果这些资源在使用后没有及时归还系统，那么用于处理这些对象的代码会将资源消耗殆尽；
2. 持有文件将会阻止对于文件的其他操作，诸如移动、删除之类；
3. 仅仅是从逻辑上关闭文件和sockets，那么它们仍然可能会被其共享的程序在无意中进行读或者写操作。只有当它们真正被关闭后，对于它们尝试进行读或者写操作将会抛出异常，并使得问题快速显现出来。

而且，幻想当文件对象析构时，文件和sockets会自动关闭，试图将文件对象的生命周期和文件的状态绑定在一起的想法，都是不现实的。因为有如下原因：

1. 没有任何方法可以确保运行环境会真正的执行文件的析构。不同的Python实现采用不同的内存管理技术，比如延时垃圾处理机制。延时垃圾处理机制可能会导致对象生命周期被任意无限制地延长。
2. 对于文件意外的引用，会导致对于文件的持有时间超出预期（比如对于异常的跟踪，包含有全局变量等）。

推荐使用 with语句 以管理文件：

```
with open("hello.txt") as hello_file:
    for line in hello_file:
        print(line)
```

对于不支持使用with语句的类似文件的对象，使用contextlib.closing()：

```
import contextlib
with contextlib.closing(urllib.urlopen("http://www.python.org/")) as front_page:
```

```
for line in front_page:
    print(line)
```

- **导入格式（采用部分Google开源项目风格指南）**

导入应该按照从最通用到最不通用的顺序分组：

```
from __future__ import absolute_import # __future__ 导入
import sys # 标准库导入
import tensorflow as tf # 第三方库导入
from otherproject.ai import mind # 本地代码子包导入
```

每种分组中，应该根据每个模块的完整包路径按字典序排序，忽略大小写。

```
import collections
import queue
import sys
from absl import app
from absl import flags
import bs4
import cryptography
import tensorflow as tf
from book.genres import scifi
from myproject.backend import huxley
from myproject.backend.hgwells import time_machine
from myproject.backend.state_machine import main_loop
from otherproject.ai import body
from otherproject.ai import mind
from otherproject.ai import soul

# 旧风格的代码可能存在下面这些导入：
# from myproject.backend.hgwells import time_machine
# from myproject.backend.state_machine import main_loop
```

导入代码统一位置，1，位于模块文件开始行；2，位于函数内开始行。一次性在模块开始行调入，可以避免函数内重复调入。但如果导入的模块体量较大，且存在库之间的冲突，如果一次性在模块开始行导入，代码无法顺利运行，此时可以尝试在函数内开始行调入（不推荐）。因此，具体调入位置，根据具体情况比较确定。

- **命名**

模块名写法: module_name；包名写法: package_name；类名: ClassName；方法名: method_name；异常名: ExceptionName；函数名: function_name；全局常量名: GLOBAL_CONSTANT_NAME；全局变量名: global_var_name；实例名: instance_var_name；函数参数名: function_parameter_name；局部变量名: local_var_name。函数名，变量名和文件名应该是描述性的，尽量避免缩写，特别要避免使用非项目人员不清楚难以理解的缩写，不要通过删除单词中的字母来进行缩写。始终使用 .py作为文件后缀名，不要用破折号。

应该避免的名称：

1. 单字符名称，除了计数器和迭代器，作为`try/except`中异常声明的`e`，作为`with`语句中文件句柄的 `f`；
2. 包/模块名中的连字符(-)；
3. 双下画线开头并结尾的名称（Python保留, 例如`__init__`）。

命名约定：

1. 所谓“内部（Internal）”表示仅模块内可用，或者在类内是保护或私有的；
2. 用单下画线（_）开头表示模块变量或函数是受保护的（protected）的（使用`from module import *`时不会包含）；
3. 用双下画线（__）开头的实例变量或方法表示类内私有；
4. 将相关的类和顶级函数放在同一个模块里，不像Java，没必要限制一个类一个模块；
5. 对类名使用大写字母开头的单词（如`CapWords`, 即Pascal风格），但是模块名应该用小写加下画线的方式（如lower_with_under.py）。尽管已经有很多现存的模块使用类似于CapWords.py这样的命名，但现在已经不鼓励这样做，因为如果模块名碰巧和类名一致，会让人困扰。

文件名：

所有Python脚本文件都应该以.py为后缀名且不包含 -。若是需要一个无后缀名的可执行文件，可以使用软链接或者包含 exec “$0.py” “$@” 的bash脚本。

也可以进一步参考Python之父Guido推荐的规范（表1.1–1）：

表1.1–1 Guido 推荐的Python 代码书写风格

Type	Public	Internal
Modules（模块）	lower_with_under	_lower_with_under
Packages（包）	lower_with_under	
Classes（类）	CapWords	_CapWords
Exceptions（异常）	CapWords	
Functions（函数）	lower_with_under()	_lower_with_under()
Global/Class Constants（全局/类常量）	CAPS_WITH_UNDER	_CAPS_WITH_UNDER
Global/Class Variables（全局/类变量）	lower_with_under	_lower_with_under
Instance Variables（实例变量）	lower_with_under	_lower_with_under (protected) or __lower_with_under (private)
Method Names（方法名）	lower_with_under()	_lower_with_under() (protected) or __lower_with_under() (private)
Function/Method Parameters（函数/方法参数）	lower_with_under	
Local Variables（局部变量）	lower_with_under	

- Main

\# 即使是一个打算被用作脚本的文件，也应该是可导入的。并且简单的导入不应该导致这个脚本的主功能（main functionality）被执行，这是一种副作用。主功能应该放在一个 main() 函数中。

在Python中，pydoc及单元测试要求模块必须是可导入的。代码应该在执行主程序前总是检查 `if __name__ == '__main__'`。这样，当模块被导入时，主程序就不会被执行。

若使用absl模块，则使用 `app.run`：

```
from absl import app
...
def main(argv):
    # process non-flag arguments
    ...
if __name__ == '__main__':
    app.run(main)
```

否则，使用：

```
def main():
    ...
if __name__ == '__main__':
    main()
```

所有的顶级代码在模块导入时都会被执行。要小心不要去调用函数、创建对象，或者执行那些不应该在使用pydoc时执行的操作。

- **类型注释（略）**

类型注释具体解释查看Google 开源项目风格指南。

目前，已经有很多可以自动格式化代码风格（例如Python Enhancement Proposals，PEP编码规范）的编辑器、转换工具或者基于解释器的插件，这大幅度提升了代码编辑的效率及其可读性，例如在线的格式化工具Code Beutiful-Python Formatter①，以及Visual Studio Code（VSC）的扩展插件 autopep8等。

① Python Beautifier Online，在线格式化Python代码工具，有助于保存和共享代码。（https://code-beautify.org/python-formatter-beautifier）。

参考文献（References）：

[1] Martin, R. (2008). Clean Code: A Handbook of Agile Software Craftsmanship. Prentice Hall.

[2] Google 开源项目风格指南（https://zh-google-styleguide.readthedocs.io/en/latest/google-python-styleguide/python_style_rules/#section-2）.

[3] Python PEP-8编码风格指南中文版（https://alvin.red/2017/10/07/python-pep-8/）.

1.2 Python解释器和笔记，与GitHub代码托管

1.2.1 Python解释器和笔记

用代码来解决专业的问题时，代码只是解决问题的工具，而本质是发现解决问题的方法。“工欲善其事，必先利其器”。作为工具，除了Python本身内置的各类方法外，各类库所提供的方法更是数不胜数。为了方便库的安装和管理，就数据分析类，本书主要使用Anaconda环境管理器（Python包管理器）；当涉及用Flask[①]等扩展库构建Web应用时，则推荐使用PyCharm。

为方便交流，本书在阐述“基础实验”时，主要使用Anaconda下的JupyterLab[②]，基于网页单元式的集成开发环境（Integrated development environment，IDE）。JupyterLab包括Code格式的代码交互式解释器；Markdown[③]格式的文本编辑器；和保持原始字符Raw格式的方式。集成有文本编辑器的代码交互式解释器，可以边书写代码、边记录笔记，不仅方便个人注释代码，同时方便代码分享，便于他人学习或查阅。在解决专项问题时，通常会包含大量代码，需要架构合理的文件结构，以包、子包和模块的方式管理代码，方便代码迁移和发布。因此，在“专项研究”部分使用Spyder交互式解释器书写代码，而不再建议使用JupyterLab的模式。

书写代码时，有时需要快速的浏览代码和数据，可以辅助使用Notepad++文本编辑器。NotePad++支持众多各类编程语言语法，并能够高亮显示查看；同时，支持多种编码、多国语言编写功能，以及一些拓展了文本编辑能力的有用工具。

另一经常使用到的源码编辑器是Visual Studio Code（VS Code），支持代码调试、语法高亮、智能代码提示、代码重构和嵌入Git[④]等功能。本书用docsify[⑤]部署网页版时，使用VS Code实现。

每个人都会有自己做笔记的习惯，作代码的笔记，与我们常规使用Microsoft OneNote[⑥]等工具有所不同，要能高亮显示代码格式，最好能运行查看结果，因此需要结合自身情况来选择笔记工具，使得学习代码这件事事半功倍，如表1.2-1所示。

表1.2-1 用于书写Python代码的解释器比较

序号	解释器名称	免费与否	推荐说明	官方网址
1	Python 官方	免费	不推荐，但轻便，可以安装，偶尔用于随手简单代码的验证	https://www.python.org
2	Anaconda	个人版（Individual Edition）免费；团队版（Team Edition）和企业版（Enterprise Edition）付费	集成了众多科学包及其依赖项，强烈推荐，因为自行解决库依赖项是件令人十分苦恼的事情。其中，包含Spyder和JupyterLab为本书所使用	https://www.anaconda.com
3	Visual Studio Code(VS Code)	免费	推荐使用，用于查看书写代码非常方便，并支持多种语言格式	https://code.visualstudio.com
4	PyCharm	付费	一般推荐，通常只在部署网页时使用，本书“实验用网络应用平台的部署部分”用到该平台	https://www.jetbrains.com/pycharm
5	Notepad++	免费	仅用于查看代码、文本编辑，推荐使用，轻便的纯文本代码编辑器，可以用于代码查看，尤其兼容众多文件格式，当有些文件乱码时，不妨尝试用此编辑器	https://notepad-plus-plus.org

① Flask，使用Python编写的轻量级Web应用框架（https://flask.palletsprojects.com/），因为不需要特定的工具或依赖库，也称为微框架（microframework）。

② JupyterLab，基于网页单元式的集成开发环境（https://jupyterlab.readthedocs.io/en/stable/index.html）。

③ Markdown，轻量级标记语言，支持图片、图表和数学公式等，且易读易写，广泛用于撰写帮助文档（https://www.markdownguide.org/）。

④ Git，是一个开源的分布式版本控制系统，可以有效、高速地处理从很小到很大的项目版本管理（https://git-scm.com/）。VS Code集成了源码控制管理（Source control management，SCM），包括开箱即用的Git支持（https://code.visualstudio.com/docs/sourcecontrol/overview）。

⑤ docsify，通过智能的加载和解析Markdown文件，即时显示生成一个网站（https://docsify.js.org）。可以通过配置index.html文件，部署到GitHub页面上，发布网站。

⑥ Microsoft OneNote，为数字记录笔记，支持笔记、绘图、屏幕截图和音频等功能的记事程序。

1.2.2 GitHub代码托管

本书书写过程中使用GitHub托管代码和用VS Code书写Markdown说明文档。在本地和云端

同步是直接应用GitHub提供的GitHub Desktop[①]来推送（push）和拉取（pull）代码及相关文档。有时也会直接应用StackEdit[②]等工具书写推送Markdown文档到云端。亦可以直接在GitHub上在线编辑。

GitHub是一个使用Git进行软件开发和版本控制的互联网托管服务平台。它为每个项目提供Git的分布式版本控制及访问控制、错误跟踪、软件功能请求、任务管理、持续集成和维基（wikis）[③]。截至2022年6月，GitHub报告有8300万名开发者和超过2亿个代码仓库，包括至少2800万个公共代码仓库。

在学习本书或者建立自己的代码开发项目进行数据分析，开发应用（Application, App），布局网站等任何应用编程语言的代码工程，都强烈推荐使用GitHub代码托管平台，①方便不断增加的个人代码仓库的管理；②避免本地代码丢失；③方便云端多人或者团队协作；④可以配置代码仓库为私有，也可以配置为公共开源，易于代码分享传播；⑤可以应用GitHub网页功能，将个人的代码仓库发布为网页形式，通常用于代码仓库的说明等。

① GitHub Desktop，可以用图形用户界面（Graphical User Interface，GUI）而不是命令行或网页浏览器与GitHub互动。GitHub Desktop可以完成大多数Git命令，例如推送和拉取、克隆远程仓库、对更改内容以可视化确认等（https://docs.github.com/en/desktop）。

② StackEdit，Markdown编辑器，可以将本地工作区的任何文件与云端的Google Drive、Dropbox和GitHub等账户中的文件同步。同步机制每分钟都会在后台进行，实现文件的下载、合并和上传修改文件等功能（https://stackedit.io/）。

③ wikis（维基），为超文本出版物，使用网络浏览器协作编辑和管理。一个典型的wiki包括项目主题、领域等多个页面，可以向公众开放，也可以仅限于在一个组织内部使用，以维护其知识库。

1.3 数据库与数据分析基本流程组织

当数据量开始膨胀，常规存储数据方式的简单文件形式，虽然逻辑简单，但可扩展性差，不能有效解决数据的完整性、一致性及安全性等一系列问题，由此产生了数据管理系统（Database Management System，DBMS），即数据库（database）。数据库是按照一定规则保存数据，能给予多个用户共享，具有尽可能小的冗余度，与应用程序彼此独立，并能通过查询取回程序检索所需数据的数据存储和管理方式。数据库有多种类型，例如与分布处理技术结合产生的分布式数据库，与并行处理技术结合产生的并行数据库，特定领域的地理数据库、空间数据库等。Web最常使用基于关系模型的数据库，即关系型数据库（Relational Database Management System，RDBMS），或称为SQL（Structured Query Language）数据库，使用结构化查询语言操作。与之相反的是最近流行的文档数据库和键-值对数据库，即NoSQL数据库。其中，关系型数据库把数据存储在表中，表的列（colunm）为字段（field），每一字段为"样本"的一个属性，行（row）为每一"样本"数据，包含一个或多个属性。常用的关系型数据库有MySQL（其替代品包括MariaDB等）、SQLite和PostgreSQL。

在城市空间数据分析方法研究中，主要使用SQLite①和PostgreSQL②两个数据库。当涉及地理空间信息数据，需要配置投影坐标系统及在QGIS③中读取地理信息建立地图时，使用PostgreSQL；否则，一般使用轻量型的数据库SQLite。

1.3.1 SQLite数据库

SQLite是一个用C编程语言编写的SQL数据库引擎，具有小型、快速、自包含（self-contained）、高可靠性、功能齐全等特点。已有超过1万亿（1×10^{12}）SQLite数据库在活跃的使用。其文档格式稳定、跨平台，向后兼容，同时其开发人员保证到2050年一直保持这种格式。

对SQLite关系型数据库的操作，包含通过SQLite命令执行（SQL语句），通过Python等语言执行（大多数数据库引擎都有对应的Python包）。对于Python，使用两个库：一个是sqlite3④操作SQLite数据库的库；另一是SQLAlchemy（flask_sqlalchemy）库⑤（数据库抽象层代码包，可以直接处理高等级的Python对象，而不用关注表、文档或查询语言等数据库实体）。当然，Pandas等库也提供了直接读写数据库对应的方法，进一步简化了对数据库的操作。

SQLite数据库应用途径，引用《漫画数据库》中的数据，结合代码实现阐释。同时，使用DB Browser for SQLite（DB4S）⑥辅助查看、管理SQLite数据库。

1.3.1.1 查看版本

```
%%cmd
sqlite3 version
--------------------------------------------------------------------------------
Microsoft Windows [Version 10.0.22000.856]
(c) Microsoft Corporation. All rights reserved.
(USDAlab) C:\Users\richie\omen_richiebao\omen_github\USDA_CH_endup\USDA\
notebook>sqlite3 version
(USDAlab) C:\Users\richie\omen_richiebao\omen_github\USDA_CH_endup\USDA\
notebook>
--------------------------------------------------------------------------------
import sqlalchemy
sqlalchemy.__version__
--------------------------------------------------------------------------------
'1.4.32'
```

1.3.1.2 根据漫画数据库中的销售数据集录入数据

该数据包括4个表，分别是销售表sales_table，包含的字段（列）有报表编码idx，日期date和出口国编码exporting_country_ID；出口国表exporting_country_table，包含的字段有出口国编码exporting_country_ID和出口国名称exporting_country_name；销售明细表sale_details_table，包含的字段有报表编码idx，商品编码commodity_code和

① SQLite，不是一个独立的应用程序，而是用C编程语言编写的数据库引擎，可以将其嵌入到APP中，因此属于嵌入式数据库系列（https://www.sqlite.org/index.html）。

② PostgreSQL，也称为Postgres，一个自由开源的关系型数据库管理系统，强调可扩展性和SQL兼容性（https://www.postgresql.org/）。

③ QGIS，为一个免费开源的跨平台桌面地理信息系统（geographic information system，GIS）应用程序，支持查看、编辑、打印和分析地理空间数据（https://www.qgis.org/en/site/）。

④ sqlite3，是一个C语言库，提供了一个基于磁盘的轻量级数据库，不需要单独的服务器进程，并允许使用SQL查询语言的非标准变体访问数据库。一些应用程序可以使用SQLite进行内部数据存储或建立一个应用程序的原型，然后将代码移植到一个更大的数据库，例如PostgreSQL或Oracle等（https://docs.python.org/3/library/sqlite3.html）。

⑤ SQLAlchemy，是Python SQL工具包和对象关系映射器，为应用程序开发人员提供了SQL的全部功能和灵活性（https://www.sqlalchemy.org/）。

⑥ DB Browser for SQLite (DB4S)，是一个高质量、可视化、开源的工具，用于创建、设计和编辑与SQLite兼容的数据库文件（https://sqlitebrowser.org/）。

数量number；商品表commodity_table，包含的字段有商品编码commodity_code和商品名称commodity_name。

数据录入时的数据结构为字典，然后转换为Pandas的DataFrame数据结构。注意，时间格式采用datetime库的datetime方法格式化时间。

```
import pandas as pd
from datetime import datetime

# 定义字典类型的假设数据
sales_dic = {'idx': [1101, 1102, 1103, 1104, 1105],
             'date': [datetime(2020, 3, 5), datetime(2020, 3, 7), datetime(2020, 3, 
8), datetime(2020, 3, 10), datetime(2020, 3, 12)],
             "exporting_country_ID": [12, 23, 25, 12, 25]}
exporting_country_dic = {"exporting_country_ID": [12, 23, 25],
                         'exporting_country_name': ['kenya', 'brazil', 'peru']}
sale_details_dic = {'idx': [1101, 1101, 1102, 1103, 1104, 1105, 1105],
                    'commodity_code': [101, 102, 103, 104, 101, 103, 104],
                    'number': [1100, 300, 1700, 500, 2500, 2000, 700]}
commodity_dic = {'commodity_code': [101, 102, 103, 104],
                 'commodity_name': ['muskmelon', 'strawberry', 'apple', 'lemon']}

# 为方便数据管理，将字典格式数据转换为 Pandas 的 DataFrame 格式
sales_table = pd.DataFrame.from_dict(sales_dic)
exporting_country_table = pd.DataFrame.from_dict(exporting_country_dic)
sale_details_table = pd.DataFrame.from_dict(sale_details_dic)
commodity_table = pd.DataFrame.from_dict(commodity_dic)
print("-"*50, "sales_table", "\n", sales_table)
print("-"*50, "exporting_country_table", "\n", exporting_country_table)
print("-"*50, "sale_details_table", "\n", sale_details_table)
print("-"*50, "commodity_table", "\n", commodity_table)
-------------------------------------------------- sales_table
    idx       date  exporting_country_ID
0  1101 2020-03-05                    12
1  1102 2020-03-07                    23
2  1103 2020-03-08                    25
3  1104 2020-03-10                    12
4  1105 2020-03-12                    25
-------------------------------------------------- exporting_country_table
   exporting_country_ID exporting_country_name
0                    12                  kenya
1                    23                 brazil
2                    25                   peru
-------------------------------------------------- sale_details_table
    idx  commodity_code  number
0  1101             101    1100
1  1101             102     300
2  1102             103    1700
3  1103             104     500
4  1104             101    2500
5  1105             103    2000
6  1105             104     700
-------------------------------------------------- commodity_table
   commodity_code commodity_name
0             101      muskmelon
1             102     strawberry
2             103          apple
3             104          lemon
```

1.3.1.3 创建数据库(链接)

首先,调用函数库sqlalchemy,使用from sqlalchemy import create_engine调入create_engine方法;

在当前目录下创建数据库,例如使用engine=create_engine('sqlite:///x.sqlite')语句;

在相对或绝对路径创建数据库,例如使用engine=create_engine('sqlite:///./data/fruits.sqlite')或engine=create_engine('sqlite:///absolute/data/fruits.sqlite')语句;

如果创建内存数据库,例如使用engine=create_engine('sqlite://')或engine=create_engine('sqlite:///:memory:', echo=True)语句。

Unix、Max及Window系统的文件路径分隔符可能不同。如果出现异常,可以尝试在/或\切换,同时注意\也是转义符号,因此可能需要写成\\。

```
from sqlalchemy import create_engine
db_fp = r'./database/fruits.sqlite'
engine = create_engine('sqlite:///'+'\\\\'.join(db_fp.split('\\')),echo=True)
```

执行create_engine语句,只是建立了数据库链接,只有向其写入表数据(或者对数据库执行任务,例如engine.connect())等操作,才会在硬盘指定路径下找到该文件。如果没有,则建立新数据库文件。如果存在同名数据库,重复执行此语句,只执行数据库链接操作。

```
connection=engine.connect()
connection.close()
```

1.3.1.4 向数据库中写入表及数据

1. 写入方法-pandas.DataFrame.to_sql()

其中,参数if_exists,可以选择fail-为默认值。如果表存在,则返回异常;replace-先删除已经存在的表,再重新插入新表;append-向存在的表中追加行。

Pandas方法不需要建立模型(表结构)。数据库模型的表述方法通常易于与机器学习模型,或者算法模型的说法混淆,因此为了便于区分,这里用表结构代替模型的说法。Pandas可以根据DataFrame格式数据信息,尤其包含有自动生成的数据类型,直接在数据库中建立对应数据格式的表,不需要自行预先定义表结构。但是,在应用程序中调入表中数据时,又往往需要调用表结构来读取数据库信息,例如Flask Web框架(参看Flask部分阐述)等。因此,可以用DB4S来查看刚刚建立的SQLite数据库及写入的表,可以看到表结构。根据表结构的数据类型信息,再手工建立表结构。表结构通常以类(class)的形式定义。因为手工定义相对比较繁琐,尤其字段比较多,不容易确定数据类型时,可以使用数据库逆向工程的方法,例如使用sqlacodegen库生成数据库表结构。

```
def df2SQLite(db_fp, df, table_name, method='fail'):
    '''
    function - 把 pandas DataFrame 格式数据写入数据库(同时创建表)

    Paras:
        db_fp - 数据库链接; string
        df - 待写入数据库的 DataFrame 格式数据; DataFrame
        table_name - 表名称; string
        method - 写入方法, 'fail', 'replace' 或 'append'; string

    Returns:
        None
    '''
    from sqlalchemy import create_engine
    engine = create_engine(
        'sqlite:///'+'\\\\'.join(db_fp.split('\\')), echo=True)
    try:
        df.to_sql(table_name, con=engine, if_exists="%s" % method)
        if method == 'replace':
            print("_"*10, 'the %s table has been overwritten...' % table_name)
        elif method == 'append':
```

```
                print("_"*10, 'the %s table has been appended...' % table_name)
            else:
                print("_"*10, 'the %s table has been written......' % table_name)
        except:
            print("_"*10, 'the %s table has been existed......' % table_name)
df2SQLite(db_fp, sales_table, 'sales', 'fail')
--------------------------------------------------------------------------------
2022-10-15 10:57:10,965 INFO sqlalchemy.engine.Engine PRAGMA main.table_
info("sales")
...
CREATE TABLE sales (
    "index" BIGINT,
    idx BIGINT,
    date DATETIME,
    "exporting_country_ID" BIGINT
)
...
2022-10-15 10:57:10,990 INFO sqlalchemy.engine.Engine COMMIT
__________ the sales table has been written......
```

'sales'表已经写入数据库，如果配置method='fail'，再次写入时，则返回异常，即提示表已经存在。

```
df2SQLite(db_fp,sales_table,'sales','fail')
--------------------------------------------------------------------------------
2022-10-15 10:58:19,755 INFO sqlalchemy.engine.Engine PRAGMA main.table_
info("sales")
2022-10-15 10:58:19,756 INFO sqlalchemy.engine.Engine [raw sql] ()
__________ the sales table has been existed......
```

- 由sqlacodegen库生成SQLite数据库中'sales'表结构。对于sqlacodegen方法可以在命令行中输入sqlacodegen --help查看。生成的表结构写入到指定的文件中（下述代码写入到了sales_table_structure.py文件下）。

sqlacodegen 库的安装通常用 pip install sqlacodegen

```
%%cmd
sqlacodegen sqlite:///./database/fruits.sqlite --tables sales --outfile sales_
table_structure.py
--------------------------------------------------------------------------------
Microsoft Windows [Version 10.0.22000.856]
(c) Microsoft Corporation. All rights reserved.
(USDAlab) C:\Users\richie\omen_richiebao\omen_github\USDA_CH_endup\USDA\
notebook>sqlacodegen sqlite:///./database/fruits.sqlite --tables sales --outfile
sales_table_structure.py
(USDAlab) C:\Users\richie\omen_richiebao\omen_github\USDA_CH_endup\USDA\
notebook>
```

打开存储有表结构的sales_table_structure.py文件，内容如下：

```
# coding: utf-8
from sqlalchemy import BigInteger, Column, DateTime, MetaData, Table
from sqlalchemy.ext.declarative import declarative_base
metadata = MetaData()
t_sales = Table(
    'sales', metadata,
    Column('index', BigInteger, index=True),
    Column('idx', BigInteger),
    Column('date', DateTime),
    Column('exporting_country_ID', BigInteger)
)
```

2. 写入方法-sqlalchemy创建表结构及写入表

定义的表结构需要继承declarative_base()映射类。完成表结构的定义后，执行BASE.metadata.create_all(engine, checkfirst=True)写入表结构，注意此时并未写入数据。

```
from sqlalchemy.ext.declarative import declarative_base
import sqlalchemy as db

BASE = declarative_base()  # 基本映射类，需要自定义的表结构继承

class exporting_country(BASE):
    __tablename__ = 'exporting_country'
```

```
    index = db.Column(db.Integer, primary_key=True,
                      autoincrement=True)  # 自动生成的索引列
    exporting_country_ID = db.Column(db.Integer)
    exporting_country_name = db.Column(db.Text)

    def __repr__(self):  #用于表结构打印时输出的字符串，亦可以不用写
        return '<exporting_country %r>' % self.exporting_country_ID

exporting_country.__table__  # 查看表结构
--------------------------------------------------------------------------------
Table('exporting_country', MetaData(), Column('index', Integer(), table=<exporting_
country>, primary_key=True, nullable=False), Column('exporting_country_ID', Integer(),
table=<exporting_country>), Column('exporting_country_name', Text(), table=<exporting_
country>), schema=None)
```

checkfirst=True，用于检查该表是否存在。如果存在则不建立，默认为True。可以增加tables=[Base.metadata.tables['exporting_country']]参数指定创建哪些表，或者直接使用exporting_country.__table__.create(engine, checkfirst=True)方法。

```
BASE.metadata.create_all(engine, checkfirst=True)
--------------------------------------------------------------------------------
2022-10-15 11:00:02,381 INFO sqlalchemy.engine.Engine BEGIN (implicit)
...
CREATE TABLE exporting_country (
    "index" INTEGER NOT NULL,
    "exporting_country_ID" INTEGER,
    exporting_country_name TEXT,
    PRIMARY KEY ("index")
)
...
2022-10-15 11:00:02,394 INFO sqlalchemy.engine.Engine COMMIT
```

将数据写入到新定义的表中。使用session.add_all方法可以一次性写入多组数据，但是需要将其转换为对应的格式。

```
from sqlalchemy.orm import sessionmaker

SESSION = sessionmaker(bind=engine)  #建立会话链接
session = SESSION()  #实例化

def zip_dic_tableSQLite(dic, table_model):
    '''
    function - 按字典的键，成对匹配，返回用于写入 SQLite 数据库的列表

    Paras:
    dic - 字典格式数据；dict
    table_model - 表结构（模型）。数据将写入到该表中；Class
    '''
    keys = list(dic.keys())
    vals = dic.values()
    vals_zip = list(zip(*list(vals)))
    return [table_model(**{k: i for k, i in zip(keys, v)}) for v in vals_zip]
exporting_country_table_model = zip_dic_tableSQLite(
    exporting_country_dic, exporting_country)
session.add_all(exporting_country_table_model)
session.commit()
--------------------------------------------------------------------------------
2022-10-15 11:00:15,709 INFO sqlalchemy.engine.Engine BEGIN (implicit)
2022-10-15 11:00:15,712 INFO sqlalchemy.engine.Engine INSERT INTO exporting_
country ("exporting_country_ID", exporting_country_name) VALUES (?, ?)
...
2022-10-15 11:00:15,722 INFO sqlalchemy.engine.Engine COMMIT
```

提取一组数据kenya（为类的形式），通过__dict__方式可以查看该类的属性值，包括待写入数据库中表的字段和对应值。

```
exporting_country_table_model
--------------------------------------------------------------------------------
2022-10-15 11:01:14,818 INFO sqlalchemy.engine.Engine BEGIN (implicit)
...
2022-10-15 11:01:14,827 INFO sqlalchemy.engine.Engine SELECT exporting_
country."index" AS exporting_country_index, exporting_country."exporting_country_
ID" AS "exporting_country_exporting_country_ID", exporting_country.exporting_
```

```
country_name AS exporting_country_exporting_country_name
FROM exporting_country
WHERE exporting_country."index" = ?
2022-10-15 11:01:14,828 INFO sqlalchemy.engine.Engine [cached since 0.006631s
ago] (3,)
[<exporting_country 12>, <exporting_country 23>, <exporting_country 25>]
--------------------------------------------------------------------------------
kenya = exporting_country_table_model[0]
print(kenya.__dict__)
--------------------------------------------------------------------------------
{'_sa_instance_state': <sqlalchemy.orm.state.InstanceState object at
0x0000021F3716BC70>, 'index': 1, 'exporting_country_name': 'kenya', 'exporting_
country_ID': 12}
```

将剩下的两组数据用Pandas写入SQLite数据库的方法写入。同时，应用sqlacodegen库生成对应的数据库表结构。

```
df2SQLite(db_fp,sale_details_table,table_name='sale_details')
df2SQLite(db_fp,commodity_table,table_name='commodity')
--------------------------------------------------------------------------------
2022-10-15 11:06:05,909 INFO sqlalchemy.engine.Engine PRAGMA main.table_
info("sale_details")
...
2022-10-15 11:06:05,980 INFO sqlalchemy.engine.Engine COMMIT
__________ the commodity table has been written......
```

使用sqlacodegen库分别生成对应的数据库表结构。

```
%%cmd
sqlacodegen sqlite:///./database/fruits.sqlite --tables sale_details --outfile
sale_details_table_structure.py
sqlacodegen sqlite:///./database/fruits.sqlite --tables commodity --outfile
commodity_table_structure.py
--------------------------------------------------------------------------------
Microsoft Windows [Version 10.0.22000.856]
...
(USDAlab) C:\Users\richie\omen_richiebao\omen_github\USDA_CH_endup\USDA\
notebook>sqlacodegen sqlite:///./database/fruits.sqlite --tables commodity
--outfile commodity_table_structure.py
...
```

打开存储有表结构的sale_details_table_structure.py文件，内容如下：

```
# coding: utf-8
from sqlalchemy import BigInteger, Column, MetaData, Table
from sqlalchemy.ext.declarative import declarative_base
metadata = MetaData()
t_sale_details = Table(
    'sale_details', metadata,
    Column('index', BigInteger, index=True),
    Column('idx', BigInteger),
    Column('commodity_code', BigInteger),
    Column('number', BigInteger)
)
```

打开存储有表结构的commodity_table_structure.py文件，内容如下：

```
# coding: utf-8
from sqlalchemy import BigInteger, Column, MetaData, Table, Text
from sqlalchemy.ext.declarative import declarative_base
metadata = MetaData()
t_commodity = Table(
    'commodity', metadata,
    Column('index', BigInteger, index=True),
    Column('commodity_code', BigInteger),
    Column('commodity_name', Text)
)
```

1.3.1.5 查询、增删和修改数据库

1. 查询数据库

- 使用session.query()方法

使用类定义的表结构，在应用session.query()读取数据库时返回的是一个对象'<exporting_country 12>'，需要给定字段读取具体的值。

读取的方法有多种，可以自行搜索查询。

```
exporting_country_query = session.query(exporting_country).filter_by(exporting_
country_ID=12).first() #.all() 将读取所有匹配，.first() 仅返回首个匹配对象

print("_"*50)
print(exporting_country_query)
print(exporting_country_query.exporting_country_name,exporting_country_query.
exporting_country_ID)
--------------------------------------------------------------------------------
2022-10-15 17:30:19,221 INFO sqlalchemy.engine.Engine SELECT exporting_
country."index" AS exporting_country_index, exporting_country."exporting_country_
ID" AS "exporting_country_exporting_country_ID", exporting_country.exporting_
country_name AS exporting_country_exporting_country_name
FROM exporting_country
WHERE exporting_country."exporting_country_ID" = ?
 LIMIT ? OFFSET ?
2022-10-15 17:30:19,222 INFO sqlalchemy.engine.Engine [cached since 2.283e+04s
ago] (12, 1, 0)
--------------------------------------------------------------------------------
<exporting_country 12>
kenya 12
```

使用sqlacodegen库生成数据库表结构，是使用sqlalchemy.Table定义。在应用session.query()读取数据库时，返回的是一个元组，顺序包含所有字段的值。

```
# coding: utf-8
from sqlalchemy import BigInteger, Column, DateTime, MetaData, Table
metadata = MetaData()
t_sales = Table(
    'sales', metadata,
    Column('index', BigInteger, index=True),
    Column('idx', BigInteger),
    Column('date', DateTime),
    Column('exporting_country_ID', BigInteger)
)
sales_query = session.query(t_sales).filter_by(idx=1101).first()
print("_"*50)
print(sales_query)
--------------------------------------------------------------------------------
...
2022-10-15 11:11:10,127 INFO sqlalchemy.engine.Engine [generated in 0.00057s]
(1101, 1, 0)
--------------------------------------------------------------------------------
(0, 1101, datetime.datetime(2020, 3, 5, 0, 0), 12)
```

- 使用Pandas库提供的方法

应用Pandas读取数据库，相对sqlite3和SQLAlchemy库而言较为简单，不需要配置表结构，能直接读取。读取结果见表1.3-1。

表1.3-1 用Pandas直接读取数据库

	type	name	tbl_name	rootpage	sql
0	table	sales	sales	2	CREATE TABLE sales (\n\t "index" BIGINT, \n\tid...
1	index	ix_sales_index	sales	3	CREATE INDEX ix_sales_index ON sales ("index")
2	table	exporting_country	exporting_country	4	CREATE TABLE exporting_country (\n\t "index" IN...
3	table	sale_details	sale_details	5	CREATE TABLE sale_details (\n\t "index" BIGINT,...
4	index	ix_sale_details_index	sale_details	6	CREATE INDEX ix_sale_details_index ON sale_det...
5	table	commodity	commodity	7	CREATE TABLE commodity (\n\t "index" BIGINT, \n...
6	index	ix_commodity_index	commodity	8	CREATE INDEX ix_commodity_index ON commodity (...

```
import sqlite3
import pandas as pd
db_fp = r'./database/fruits.sqlite'
conn = sqlite3.connect(db_fp)
df_sqlite = pd.read_sql('select * from sqlite_master',con=conn) # pd.read_sql将读取数据库结构(database structure)信息
df_sqlite
```

用pd.read_sql_table()从数据库中读取指定的表1.3-2。

```
df_sales = pd.read_sql_table('sales', 'sqlite:///./database/fruits.sqlite')
df_sales
```

表1.3-2 用Pandas从数据库中读取表‘sales’

	index	idx	date	exporting_country_ID
0	0	1101	2020-03-05	12
1	1	1102	2020-03-07	23
2	2	1103	2020-03-08	25
3	3	1104	2020-03-10	12
4	4	1105	2020-03-12	25

pd.read_sql_query()将根据SQL query 或 SQLAlchemy Selectable查询语句读取特定的值，例如读取select idx和exporting_country_ID，读取结果见表1.3-3。

```
df_sales_query = pd.read_sql_query('select idx,exporting_country_ID from sales', con=conn)
df_sales_query
```

表1.3-3 用Pandas从数据库中读取表‘sales’中的指定字段

	idx	exporting_country_ID
0	1101	12
1	1102	23
2	1103	25
3	1104	12
4	1105	25

2．增-删数据库

● 向已有表中增加数据

sqlacodegen库生成数据库表结构并运行，'sales'表则被存储于metadata元数据中。如果再定义一个类，同样指向这个表，则需要配置'extend_existing': True。表示在已有列基础上进行扩展，即sqlalchemy允许类是表的子集（一个表可以指向多个表结构的类）。

```
metadata.tables
-----------------------------------------------------------------------------
FacadeDict({'sales': Table('sales', MetaData(), Column('index', BigInteger(), table=<sales>), Column('idx', BigInteger(), table=<sales>), Column('date', DateTime(), table=<sales>), Column('exporting_country_ID', BigInteger(), table=<sales>), schema=None)})
-----------------------------------------------------------------------------
from sqlalchemy.orm import sessionmaker
class sales(BASE):
    __tablename__ = 'sales'
    __table_args__ = {'extend_existing': True}

    # 因为该sales类是在执行t_sales之后定义，只能是在原有表上扩展，无法修改原表结构属性，因此index字段并不会实现自动增加的属性。需要手动增加index字段值
    index = db.Column(db.Integer, primary_key=True, autoincrement=True)
    idx = db.Column(db.Integer)
    date = db.Column(db.DateTime)
```

```
    exporting_country_ID = db.Column(db.Integer)

SESSION = sessionmaker(bind=engine)  # 建立会话链接
session = SESSION()  # 实例化

new_sale = sales(index=5, idx=1106, date=datetime(
    2020, 12, 18), exporting_country_ID=25)
session.add(new_sale)
session.commit()
--------------------------------------------------------------------------------
2022-10-15 17:39:40,423 INFO sqlalchemy.engine.Engine BEGIN (implicit)
...
2022-10-15 17:39:40,430 INFO sqlalchemy.engine.Engine COMMIT
```

从表中读取新增加的数据。

```
del_sale = session.query(sales).filter_by(idx=1106).first() # 如果该行中有值为空，
例如在增加该行数据时未定义写入 index=5 字段，该语句返回值会为空。如允许出现空值，在定义表结构时
需要配置
nullabley = True
print("_"*50)
print(del_sale.exporting_country_ID,del_sale.date)
--------------------------------------------------------------------------------
2022-10-15 17:41:57,556 INFO sqlalchemy.engine.Engine BEGIN (implicit)
...
2022-10-15 17:41:57,559 INFO sqlalchemy.engine.Engine [generated in 0.00078s]
(1106, 1, 0)
--------------------------------------------------------------------------------
25 2020-12-18 00:00:00
```

- 从表中删除已有的数据

```
session.delete(del_sale)
session.commit()
--------------------------------------------------------------------------------
2022-10-15 17:41:59,500 INFO sqlalchemy.engine.Engine DELETE FROM sales WHERE
sales."index" = ?
2022-10-15 17:41:59,501 INFO sqlalchemy.engine.Engine [generated in 0.00152s]
(5,)
2022-10-15 17:41:59,504 INFO sqlalchemy.engine.Engine COMMIT
```

3. 修改数据库

```
mod_sale=session.query(sales).filter_by(idx=1105).first()
mod_sale.exporting_country_ID = 23 # 修改字段值
session.commit()
--------------------------------------------------------------------------------
2022-10-15 17:42:23,038 INFO sqlalchemy.engine.Engine BEGIN (implicit)
...
2022-10-15 17:42:23,044 INFO sqlalchemy.engine.Engine [generated in 0.00066s]
(23, 4)
2022-10-15 17:42:23,046 INFO sqlalchemy.engine.Engine COMMIT
```

1.3.2 SQLite数据库的表间关系（多表关联）

上文建立表结构时，并未配置表间关联，各个表是独立的，如果想通过一个表的数据字段查询另一个表的字段内容就比较困难，例如想根据销售数量，查询对应的商品名称时，是无法直接在'commodity'商品表中直接查询商品名称的，需要先根据待查询的销售数量例如300，在'sale_details'销售明细表里找到对应的commodity_code商品编码为102。根据这个商品编码，再在'commodity'商品表找到对应的商品名称为'strawberry'。因为这个过程很繁琐，尤其数据库结构和表结构再复杂时，这个问题会更凸显，因此需要建立表间的联系。

SQLite的表间关系配置，可以包括1对多，多对1，1对1和多对多，SQLAlchemy给出表结构配置的方法Relationship Configuration[①]，可以根据其阐述进行配置。配置时，参数'back_populates'定义反向引用，用于建立双向关系，例如销售明细表->商品表，均包括relationship()语句，显示定义关系属性。如果使用参数'backref'添加反向引用，会自动在另一侧建立关系属性，为'back_populates'的简化形式。参数uselist=True（默认值）时，为1对多关系，如果配置1对1时，需要将其配置为uselist=False。在配置表关系

① Relationship Configuration，SQLAlchemy中relationship()函数的用法（https://docs.sqlalchemy.org/en/14/orm/relationships.html）。

时，为了能够清晰易读，通常以表名作为变量名，例如销售明细表->商品表，销售明细表为父表（parent），商品表为子表（child），父表中语句commodity=relationship('commodity',uselist=False,back_populates="sale_details")以子表为变量名，而子表中语句sale_details=relationship('sale_details',back_populates="commodity")以父表为变量名，这样可以更清晰地表述表之间的关系。

可以使用sqlacodegen库生成数据库表结构，往往应用于类似Pandas写入数据库而没有定义表结构的情况下，这是一种逆向工程。下述已经定义4个表结构，那么则可以使用逆向工程反馈表的内容和表之间的关系，例如使用Visual Paradigm①反馈有下表关系（图1.3-1），即统一建模语言（Unified Modeling Language，UML）。可以清晰、直观地读出表结构和表间关系，其中'sales'是'exporting_country'的父表，连接的关键字段（ForeignKey）是'exporting_country_ID'；同时'sales'是'sale_details'的子表，联系的关键字段是'idx'，其他的关系以此类推，一目了然。

① Visual Paradigm，为统一建模语言（Unified Modeling Language，UML），是设计、分析和管理工具套件，可推动IT项目开发。除了支持建模，还可以生成报告和生成代码，并可以依据代码生成逆向工程图表（https://www.visual-paradigm.com/）。

```
from sqlalchemy import create_engine
db_fp = r'./database/fruits_relational.sqlite'
engine = create_engine('sqlite:///'+'\\\\'.join(db_fp.split('\\')),echo=True)
from sqlalchemy.ext.declarative import declarative_base
import sqlalchemy as db
from sqlalchemy.orm import relationship
from sqlalchemy.schema import ForeignKey

BASE = declarative_base()  # 基本映射类，需要自定义的表结构继承

# 销售明细表
class sale_details(BASE):
    __tablename__ = 'sale_details'
    index = db.Column(db.Integer, primary_key=True, autoincrement=True)
    commodity_code = db.Column(db.Integer)
    number = db.Column(db.Integer)
    idx = db.Column(db.Integer)
    sales = relationship('sales', uselist=False, back_populates="sale_details")
    commodity = relationship('commodity', uselist=False,
                             back_populates="sale_details")

# 销售表
class sales(BASE):
    __tablename__ = 'sales'
    index = db.Column(db.Integer, primary_key=True,
                      autoincrement=True)  # 自动生成的索引列
    date = db.Column(db.DateTime)
    exporting_country_ID = db.Column(db.Integer)
    exporting_country = relationship(
        'exporting_country', uselist=False, back_populates="sales")
    idx = db.Column(db.Integer, ForeignKey('sale_details.idx'))
    sale_details = relationship('sale_details', back_populates="sales")

# 出口国表
class exporting_country(BASE):
    __tablename__ = 'exporting_country'
```

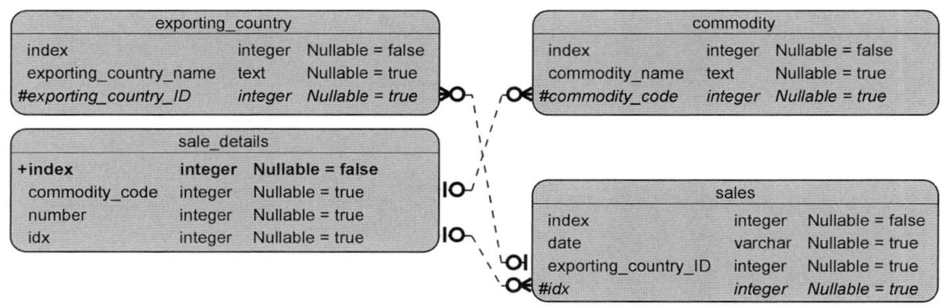

图1.3-1 数据库表间结构

```
    index = db.Column(db.Integer, primary_key=True, autoincrement=True)
    exporting_country_name = db.Column(db.Text)
    exporting_country_ID = db.Column(
        db.Integer, db.ForeignKey('sales.exporting_country_ID'))
    sales = relationship('sales', back_populates="exporting_country")

# 商品表
class commodity(BASE):
    __tablename__ = 'commodity'
    index = db.Column(db.Integer, primary_key=True, autoincrement=True)
    commodity_name = db.Column(db.Text)
    commodity_code = db.Column(
        db.Integer, ForeignKey('sale_details.commodity_code'))
    sale_details = relationship('sale_details', back_populates="commodity")

BASE.metadata.create_all(engine, checkfirst=True)  # 将所有表结构写入数据库
--------------------------------------------------------------------------------
2022-10-15 18:08:14,506 INFO sqlalchemy.engine.Engine BEGIN (implicit)
...
2022-10-15 18:08:14,545 INFO sqlalchemy.engine.Engine [no key 0.00102s] ()
2022-10-15 18:08:14,553 INFO sqlalchemy.engine.Engine COMMIT
```

将数据写入各个表。

```
from sqlalchemy.orm import sessionmaker

SESSION = sessionmaker(bind=engine)  # 建立会话链接
session = SESSION()  # 实例化

sales_ = zip_dic_tableSQLite(sales_dic, sales)
exporting_country_ = zip_dic_tableSQLite(
    exporting_country_dic, exporting_country)
sale_details_ = zip_dic_tableSQLite(sale_details_dic, sale_details)
commodity_ = zip_dic_tableSQLite(commodity_dic, commodity)
session.add_all(sales_)
session.add_all(exporting_country_)
session.add_all(sale_details_)
session.add_all(commodity_)
session.commit()
--------------------------------------------------------------------------------
2022-10-15 18:08:38,919 INFO sqlalchemy.engine.Engine BEGIN (implicit)
...
2022-10-15 18:08:39,001 INFO sqlalchemy.engine.Engine COMMIT
```

通过正向引用或者反向引用轻松的获取关联表中对应的数据。例如由商品销售数量找到对应的商品名称。

```
sale_details_info = session.query(sale_details).filter_by(number=300).first()
commodity_info = sale_details_info.commodity
commodity_name = commodity_info.commodity_name
print("_"*50)
print("销量 number=300 的商品名为 :%s"%commodity_name)
--------------------------------------------------------------------------------
2022-10-15 18:08:55,687 INFO sqlalchemy.engine.Engine BEGIN (implicit)
...
2022-10-15 18:08:55,696 INFO sqlalchemy.engine.Engine [generated in 0.00101s]
(102,)
--------------------------------------------------------------------------------
销量 number=300 的商品名为 :strawberry
```

- 使用DB Browser for SQLite(DB4S)查看数据库（图1.3–2）

1.3.3 PostgreSQL数据库

PostgreSQL[①]是一个强大开源的对象关系数据库系统（open source object-relational database system）。经过30多年的发展，其在可靠性、特征的健壮性和性能方面赢得了很高的声誉。同时，因为PostgreSQL可以存储具有投影坐标系统的地理空间数据，在QGIS[②]等地理信息系统工具平台下可以直接从PostgreSQL（PostGIS）中读入与显示数据，建立地图，弥补了Python在地图表达上的不足，而又可以充分利用Python的数据处理能力。

① PostgreSQL，也称为Postgres，自由开源的关系型数据库管理系统（relational database management system，RDBMS）（https://www.postgresql.org/）。

② QGIS，在Windows、Mac、Linux、BSD和移动设备上创建、编辑、可视化、分析和发布地理空间信息（https://qgis.org/en/site/）。

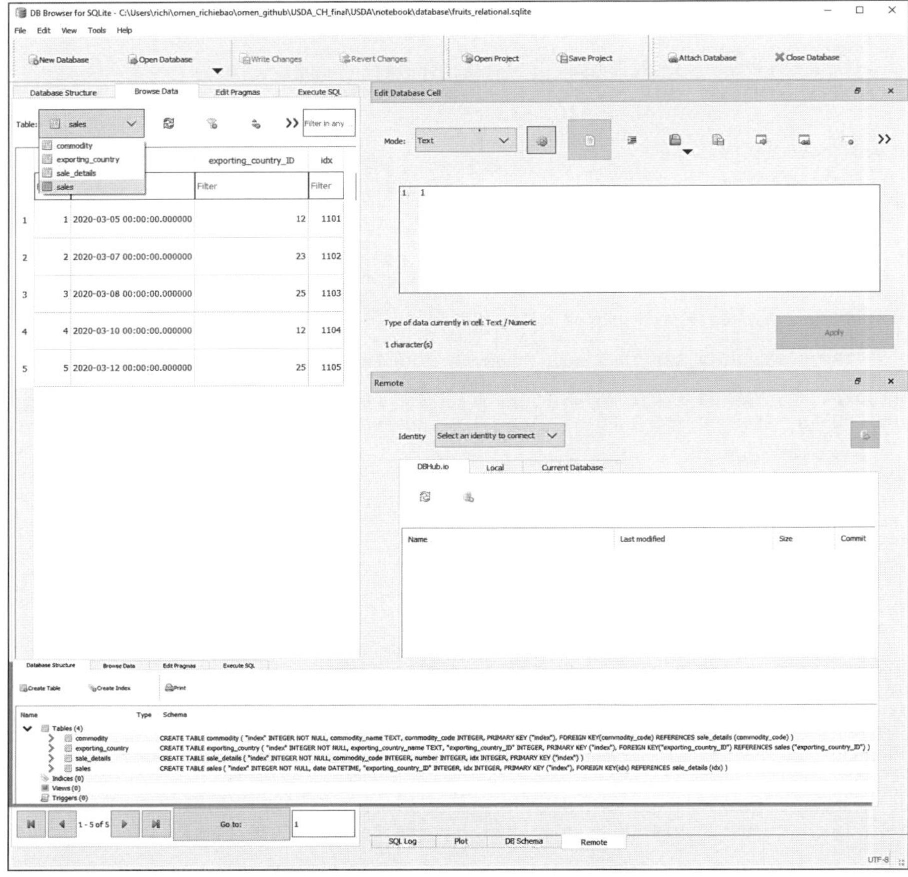

图1.3-2　用DB Browser for SQLite查看数据库

通常，使用开源的pgAdmin①工具查看管理PostgreSQL数据库。

用Array of Things(AoT) 城市环境传感器②数据演示Python数据处理、写入和读取PostgreSQL数据库，及使用QGIS调入数据库中的数据，建立地图的方法（数据下载地址：https://www.mcs.anl.gov/research/projects/waggle/downloads/datasets/index.php）。

2019年10月，AoT团队基于已有研究成功申请了国家科学基金（美）的资助，创建新的软硬件基础设施构建城市环境传感器网络，并开源了获取的实时传感器数据。关于数据的详细说明，可以查看数据包中的说明文档，详细解释了各字段的含义；同时，给出了所使用传感器的型号和详细说明链接，这对于研究城市环境下的局地小气候具有重要价值。

① pgAdmin，为最流行和功能丰富的PostgreSQL开源管理和开发平台（https://www.pgadmin.org/）。

② Array of Things（AoT）城市环境传感器，以收集城市环境、基础设施和活动的实时数据供研究和公共使用（https://arrayofthings.github.io/）。

1.3.3.1　读取nodes数据，并转换为GeoDataFrame格式

nodes.csv数据文件，包括所有布置于城市的传感器位置节点编号、坐标（wgs84）、地址等信息。使用经纬度，通过Shapely库建立地理空间点后，用Geopandas库，给定坐标系统wgs84的epsg编号4326，转换为GeoDataFrame格式数据，方便地理信息数据的存储、分析和写入PostgreSQL数据库。

```
import pandas as pd
import geopandas as gpd
from shapely.geometry import Point
AoT_nodes_fp = ’./data/AoT_Chicago.complete.2021-12-20/nodes.csv’
AoT_nodes_df = pd.read_csv(AoT_nodes_fp,sep=”,”,header=0)
epsg_wgs84 = 4326
AoT_nodes_df[“geometry”] = AoT_nodes_df.apply(lambda row:Point(row.lon,row.
```

```
lat),axis=1) # 使用 shapely 库建立几何点数据
AoT_nodes_gdf = gpd.GeoDataFrame(AoT_nodes_df,crs=epsg_
wgs84)
print("crs{}{}".format("-"*10,AoT_nodes_gdf.crs))
AoT_nodes_gdf
----------------------------------------------------------
```

输出 AoT 传感器位置节点信息（表 1.3-4）。

```
crs----------epsg:4326
126 rows × 10 columns
```

表1.3-4 AoT传感器位置节点信息

	node_id	project_id	vsn	address	lat	lon	description	start_timestamp	end_timestamp	geometry
0	001e0610ba46	AoT_Chicago	004	State St & Jackson Blvd Chicago IL	41.878377	-87.627678	AoT Chicago (S) [C]	2017/10/09 00:00:00	NaN	POINT (-87.62768 41.87838)
…	…	…	…	…	…	…	…	…	…	…
125	001e061144be	AoT_Chicago	890	UChicago, Regenstine Chicago IL	41.792543	-87.600008	AoT Chicago (S) [C] {UChicago}	2018/03/15 00:00:00	NaN	POINT (-87.60001 41.79254)

从Chicago Data Portal[①]中搜索下载行政区划数据，读取后与nodes数据叠合显示，方便定位传感器在城市中的位置。

```
Chicago_community_areas_fp = './data/ChicagoCommunityAreas/
ChicagoCommunityAreas.shp'
Chicago_community_areas = gpd.read_file(Chicago_community_
areas_fp)
print("crs<Chicago_community_areas>:{}".format(Chicago_
community_areas.crs))
Chicago_community_areas
----------------------------------------------------------
crs<Chicago_community_areas>:GEOGCS["WGS84(DD)",
DATUM["WGS84", SPHEROID["WGS84", 6378137, 298.257223563]],
PRIMEM["Greenwich", 0], UNIT["degree", 0.0174532925199433],
AXIS["Longitude", EAST], AXIS["Latitude", NORTH]]
```

输出芝加哥城行政区划信息（表 1.3-5）。

```
77 rows × 10 columns
----------------------------------------------------------
import matplotlib.pyplot as plt
fig, ax = plt.subplots(figsize=(10, 15))
Chicago_community_areas.plot(ax=ax, color='white',
edgecolor='black')
AoT_nodes_gdf.plot(ax=ax, markersize=20,
column='description',
                    legend=True, cmap='tab20c', legend_
kwds={'loc': 'lower left'})
plt.show()
----------------------------------------------------------
```

输出 芝加哥城 AoT 传感器位置分布地图（图 1.3-3）。

1.3.3.2 GeoDataFrame数据读写PostgreSQL数据库

读写GeoDataFrame数据于PostgreSQL数据库，分别定义gpd2postSQL和postSQL2gpd函数，可以放置于自定义的.py文件下，例如本书定义的util_database.py文件，方便日后调用。需要注意，建立数据库时，首先本地安装PostgreSQL，再使用安装的pgAdmin工具建立数据库，例如本例建立数据库名为'AoT'，用户名为'postgres'，密码为'123456'。同时，要在pgAdmin的Query Tool下执行CREATE EXTENSION postgis;命令，从而可以存储具有坐标系统的地理几何对象；否则，将GeoDataFrame数据写入PostgreSQL数据库时，会提示错误。

```
def gpd2postSQL(gdf, table_name, **kwargs):
    '''
    function - 将 GeoDataFrame 格式数据写入 PostgreSQL 数据库

    Paras:
        gdf - GeoDataFrame 格式数据，含 geometry 字段（几何对象，
点、线和面，数据值对应定义的坐标系统）；GeoDataFrame
        table_name - 写入数据库中的表名；string
        **kwargs - 连接数据库相关信息，包括 myusername（数据库的
用户名），mypassword（用户密钥），mydatabase（数据库名）；string

    Returns:
        None
    '''
    from sqlalchemy import create_engine
    engine = create_engine("postgresql://
```

① Chicago Data Portal，为芝加哥城开放数据门户，可免费下载数据用于相关分析，其中许多数据集每天至少更新一次或数次（https://data.cityofchicago.org/）。

表1.3-5 芝加哥城行政区划信息

	area	area_num_1	area_numbe	comarea	comarea_id	community	perimeter	shape_area	shape_len	geometry
0	0.0	35	35	0.0	0.0	DOUGLAS	0.0	4.600462e+07	31027.054510	POLYGON ((-87.60914 41.84469, -87.60915 41.844...
...	...	...	...	...	...	...	...	...	...	...
76	0.0	9	9	0.0	0.0	EDISON PARK	0.0	3.163631e+07	25937.226841	POLYGON ((-87.80676 42.00084, -87.80676 42.000...

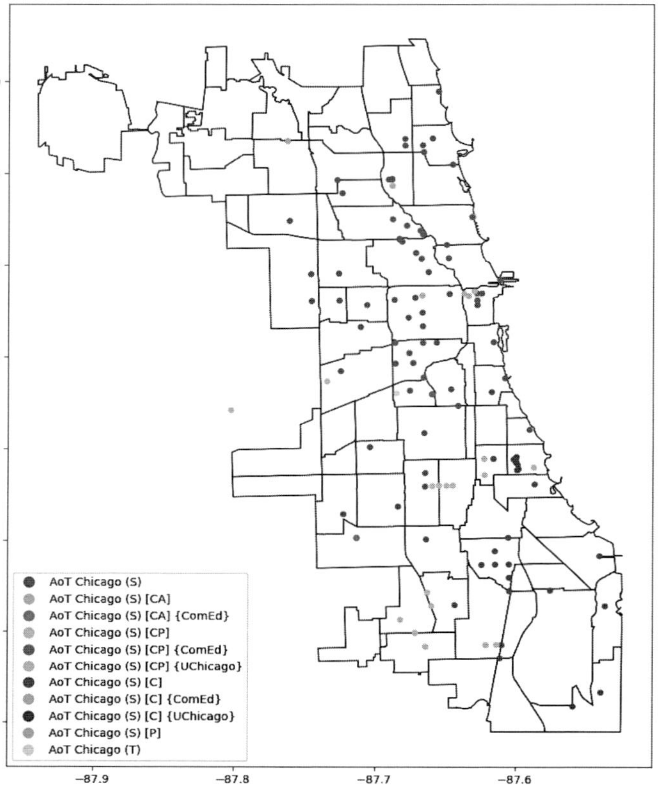

图1.3-3 芝加哥城AoT传感器位置分布

```
{myusername}:{mypassword}@localhost:5432/{mydatabase}".format(
        myusername=kwargs['myusername'], mypassword=kwargs['my
password'], mydatabase=kwargs['mydatabase']))
    gdf.to_postgis(table_name, con=engine, if_exists='replace',
index=False,)
    print("_"*50)
    print('The GeoDataFrame has been written to the PostgreSQL
database.The table name is {}.'.format(table_name))
def postSQL2gpd(table_name, geom_col='geometry', **kwargs):
    '''
    function - 读取 PostgreSQL 数据库中的表为 GeoDataFrame 格式数据

    Paras:
        table_name - 待读取数据库中的表名; string
        geom_col='geometry' - 几何对象，常规默认字段为 'geometry';
string
        **kwargs - 连接数据库相关信息,包括myusername(数据库的用户名),
mypassword(用户密钥), mydatabase(数据库名); string
    Returns:
        读取的表数据; GeoDataFrame
    '''
    from sqlalchemy import create_engine
    import geopandas as gpd
    engine = create_engine("postgresql://
{myusername}:{mypassword}@localhost:5432/{mydatabase}".format(
        myusername=kwargs['myusername'], mypassword=kwargs['my
password'], mydatabase=kwargs['mydatabase']))
    gdf = gpd.read_postgis(table_name, con=engine, geom_
col=geom_col)
    print("_"*50)
    print('The data has been read from PostgreSQL database.
The table name is {}.'.format(table_name))
```

```
    return gdf
gpd2postSQL(AoT_nodes_gdf, table_name='AoT_nodes',
            myusername='postgres', mypassword='123456', mydatabase='AoT')
gpd2postSQL(Chicago_community_areas, table_name='Chicago_community_areas',
            myusername='postgres', mypassword='123456', mydatabase='AoT')
-------------------------------------------------------------------------------
-------------------------------------------------------------------------------
The GeoDataFrame has been written to the PostgreSQL database.The table name is
AoT_nodes.
-------------------------------------------------------------------------------
The GeoDataFrame has been written to the PostgreSQL database.The table name is
Chicago_community_areas.
```

为方便后期QGIS调用建立地图，可以直接定义投影后再写入数据库。

```
epsg_Chicago = 32616
gpd2postSQL(AoT_nodes_gdf.to_crs(epsg_Chicago),table_name='AoT_nodes_sername='po
stgres',mypassword='123456',mydatabase='AoT')
gpd2postSQL(Chicago_community_areas.to_crs(epsg_Chicago),table_name='Chicago_
community_areas_sername='postgres',mypassword='123456',mydatabase='AoT')
-------------------------------------------------------------------------------
-------------------------------------------------------------------------------
The GeoDataFrame has been written to the PostgreSQL database.The table name is
AoT_nodes_prj.
-------------------------------------------------------------------------------
The GeoDataFrame has been written to the PostgreSQL database.The table name is
Chicago_community_areas_prj.
```

在建立一个研究项目时，通常将基本的数据写入数据库后，再从数据库中调用对应的表读入数据，进行后续的分析。

```
AoT_nodes_gdf = postSQL2gpd(table_name='AoT_nodes',geom_sername='postgres',mypas
sword='123456',mydatabase='AoT')
Chicago_community_areas = postSQL2gpd(table_name='Chicago_community_areas',geom_
sername='postgres',mypassword='123456',mydatabase='AoT')
-------------------------------------------------------------------------------
-------------------------------------------------------------------------------
The data has been read from PostgreSQL database. The table name is AoT_nodes.
-------------------------------------------------------------------------------
The data has been read from PostgreSQL database. The table name is Chicago_
community_areas.
```

打开pgAdmin工具，可以查看写入的数据（图1.3–4）。

1.3.3.3 读取data传感器记录的数据与初步处理

截止December 20 2021 19:18:35 CS时，data数据约有39.5GB，并且为单独一个csv格式文件。如果内存量较小，则需要分批读入处理再写入数据库。这里，仅示范读取2018/01/01一天所记录的数据（表1.3–6），并计算每小时的温度均值（对应处理'value_hrf'字段），再将其对应'node_id'字段，与AoT_nodes_gdf变量合并后，写入数据库。

```
AoT_data_fp = r"F:\data\AoT_Chicago.complete.2021-12-20\data"
chunksize = 10**6
for chunk in pd.read_csv(AoT_data_fp, chunksize=chunksize):
    AoT_data_part = chunk
    break
AoT_data_20180101 = AoT_data_part[(AoT_data_part['timestamp'] >= '2018/01/01
00:00:00')
                                   & (AoT_data_part['timestamp'] <= '2018/01/01
23:59:59')]
print("parameter-Sensor parameter that was measured:{}\n{}".format(
    AoT_data_20180101.parameter.unique(), "_"*50))
print("sensor-Sensor that was measured:{}\n{}".
format(AoT_data_20180101.sensor.unique(), "_"*50))
AoT_data_20180101
-------------------------------------------------------------------------------
parameter-Sensor parameter that was measured:['temperature' 'id' 'concentration'
'pressure' 'humidity' 'ir_intensity'
 'uv_intensity' 'visible_light_intensity' 'intensity']
-------------------------------------------------------------------------------
sensor-Sensor that was measured:['at0' 'at1' 'at2' 'at3' 'chemsense' 'co' 'h2s' 'lps25h'
'no2' 'o3'
```

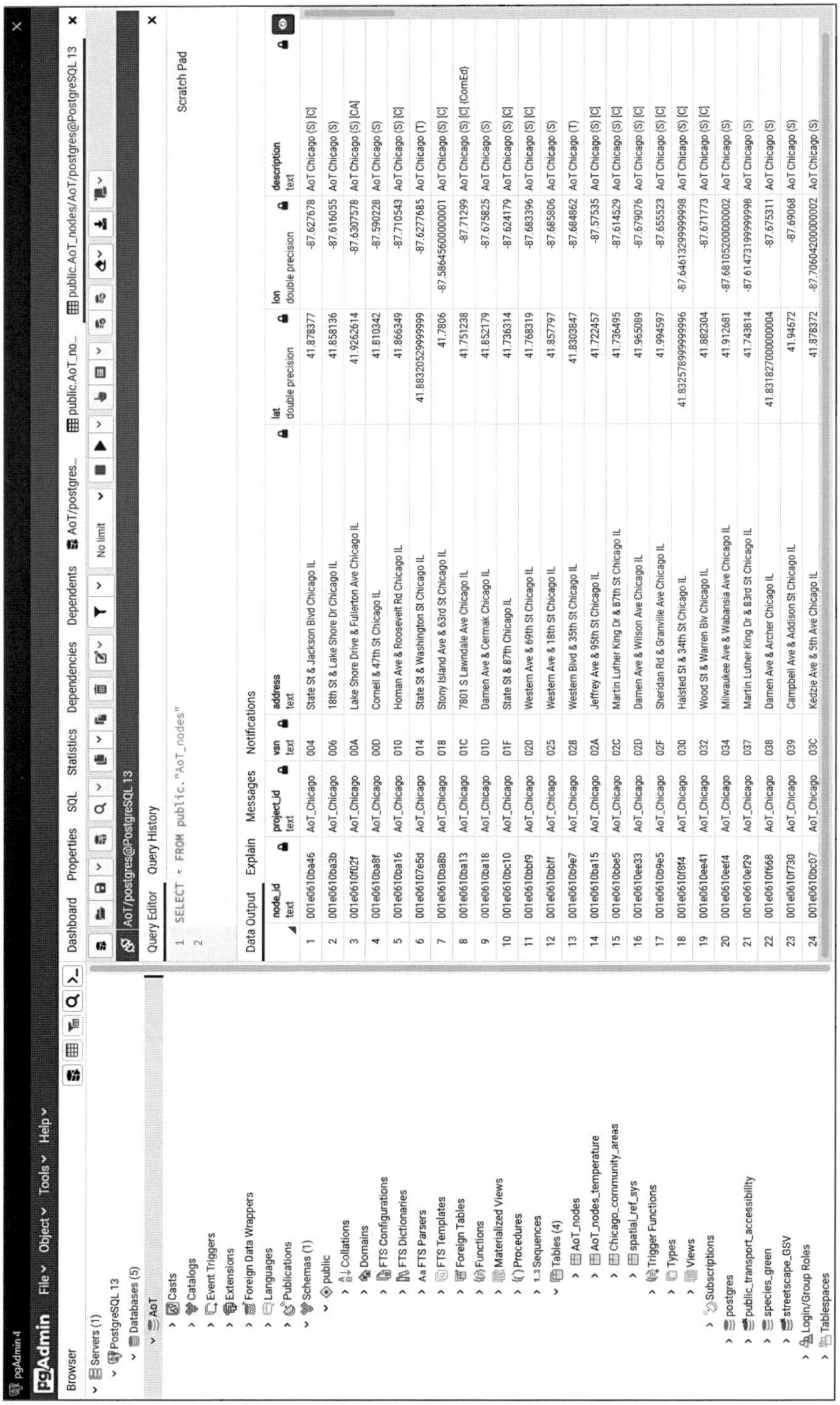

```
SELECT * FROM public."AoT_nodes"
```

	node_id text	project_id text	vsn text	address text	lat double precision	lon double precision	description text
1	001e0610ba46	AoT_Chicago	004	State St & Jackson Blvd Chicago IL	41.878377	-87.627678	AoT Chicago (S) [C]
2	001e0610ba3b	AoT_Chicago	006	18th St & Lake Shore Dr Chicago IL	41.858136	-87.616055	AoT Chicago (S)
3	001e0610f02f	AoT_Chicago	00A	Lake Shore Drive & Fullerton Ave Chicago IL	41.9262614	-87.6307578	AoT Chicago (S) [CA]
4	001e0610ba8f	AoT_Chicago	00D	Cornell & 47th St Chicago IL	41.810342	-87.590228	AoT Chicago (S)
5	001e0610ba16	AoT_Chicago	010	Homan Ave & Roosevelt Rd Chicago IL	41.866349	-87.710543	AoT Chicago (S) [C]
6	001e06107e5d	AoT_Chicago	014	State St & Washington St Chicago IL	41.88320529999999	-87.6277685	AoT Chicago (T)
7	001e0610ba8b	AoT_Chicago	018	Stony Island Ave & 63rd St Chicago IL	41.7806	-87.58645600000001	AoT Chicago (S) [C]
8	001e0610ba13	AoT_Chicago	01C	7801 S Lawndale Ave Chicago IL	41.751238	-87.71299	AoT Chicago (S) [C] {ComEd}
9	001e0610ba18	AoT_Chicago	01D	Damen Ave & Cermak Chicago IL	41.852179	-87.675825	AoT Chicago (S)
10	001e0610bc10	AoT_Chicago	01F	State St & 87th Chicago IL	41.736314	-87.624179	AoT Chicago (S) [C]
11	001e0610bbf9	AoT_Chicago	020	Western Ave & 69th St Chicago IL	41.768319	-87.683396	AoT Chicago (S) [C]
12	001e0610bbff	AoT_Chicago	025	Western Ave & 18th St Chicago IL	41.857797	-87.685806	AoT Chicago (S)
13	001e0610b9e7	AoT_Chicago	028	Western Blvd & 35th St Chicago IL	41.8303847	-87.684862	AoT Chicago (T)
14	001e0610ba15	AoT_Chicago	02A	Jeffrey Ave & 95th St Chicago IL	41.722457	-87.57535	AoT Chicago (S) [C]
15	001e0610bbe5	AoT_Chicago	02C	Martin Luther King Dr & 87th St Chicago IL	41.736495	-87.614529	AoT Chicago (S) [C]
16	001e0610ee33	AoT_Chicago	02D	Damen Ave & Wilson Ave Chicago IL	41.965089	-87.679076	AoT Chicago (S) [C]
17	001e0610b9e5	AoT_Chicago	02F	Sheridan Rd & Granville Ave Chicago IL	41.994597	-87.655523	AoT Chicago (S) [C]
18	001e0610f8f4	AoT_Chicago	030	Halsted St & 34th St Chicago IL	41.832578999999996	-87.64613299999998	AoT Chicago (S) [C]
19	001e0610ee41	AoT_Chicago	032	Wood St & Warren Blv Chicago IL	41.882304	-87.671773	AoT Chicago (S) [C]
20	001e0610eef4	AoT_Chicago	034	Milwaukee Ave & Wabansia Ave Chicago IL	41.912681	-87.68105200000002	AoT Chicago (S)
21	001e0610ef29	AoT_Chicago	037	Martin Luther King Dr & 83rd St Chicago IL	41.743814	-87.61473199999998	AoT Chicago (S)
22	001e0610f668	AoT_Chicago	038	Damen Ave & Archer Chicago IL	41.831827000000004	-87.675311	AoT Chicago (S)
23	001e0610f730	AoT_Chicago	039	Campbell Ave & Addison St Chicago IL	41.94672	-87.69068	AoT Chicago (S)
24	001e0610bc07	AoT_Chicago	03C	Kedzie Ave & 5th Ave Chicago IL	41.878372	-87.70604200000002	AoT Chicago (S)

图1.3-4 用pgAdmin工具查看PostgreSQL数据库

表1.3-6 芝加哥城AoT data文件数据

	timestamp	node_id	subsystem	sensor	parameter	value_raw	value_hrf
0	2018/01/01 00:00:06	001e0610e532	chemsense	at0	temperature	-1106	-11.06
...	...	...	...	...	...	...	...
769631	2018/01/01 23:59:59	001e0610e540	metsense	tsl250rd	intensity	2	0.101
769632	2018/01/01 23:59:59	001e0610e540	metsense	tsys01	temperature	NaN	-18.47

```
 'oxidizing_gases' 'reducing_gases' 'sht25' 'si1145' 'so2' 'apds_9006_020'
 'hih6130' 'ml8511' 'mlx75305' 'tmp421' 'tsl250rd' 'tsl260rd' 'bmp180'
 'hih4030' 'htu21d' 'metsense' 'pr103j2' 'spv1840lr5h_b' 'tmp112' 'tsys01']
------------------------------------------------------------------------
769633 rows × 7 columns
```

读取的data数据，各个字段数据为字符串类型，应用`pd.to_numeric`方法将`'value_hrf'`数值字段（已转换的各传感器测量值）转换为数值类型（表1.3–7）。

```
from tqdm import tqdm
AoT_data_20180101_temperature = AoT_data_20180101[(
    AoT_data_20180101.parameter == 'temperature')]  # &(AoT_data_20180101.
sensor=='at2')
print("列数据类型：\n{}".format(AoT_data_20180101_temperature.dtypes))
print("_"*50)
print("列名（数据类型为字符串 -str-object）:{}".format(
    AoT_data_20180101_temperature.columns[AoT_data_20180101_temperature.dtypes.
eq('object')]))
columns_dtypeEQstr = ['value_raw', 'value_hrf']
AoT_data_20180101_temperature[columns_dtypeEQstr] = AoT_data_20180101_
temperature[columns_dtypeEQstr].apply(
    pd.to_numeric, errors='coerce', axis=1)
node_id_unqiue = AoT_data_20180101_temperature.node_id.unique()
print("node_id-ID of node which did the measurement:{}\n{}".format(node_id_
unqiue, "_"*50))
AoT_data_20180101_temperature_grouped = AoT_data_20180101_temperature.groupby(
    AoT_data_20180101_temperature.node_id)
value_raw_dic = {}
for n_id in tqdm(node_id_unqiue):
    sub_df = AoT_data_20180101_temperature_grouped.get_group(n_id)
    sub_df.set_index(pd.to_datetime(sub_df["timestamp"]), inplace=True)
    value_raw_dic[n_id] = sub_df.groupby(sub_df.index.hour)[
        ['value_hrf']].mean()['value_hrf']
value_raw_df = pd.DataFrame.from_dict(value_raw_dic, orient='columns')
value_raw_df
------------------------------------------------------------------------
timestamp    object
node_id      object
subsystem    object
sensor       object
parameter    object
value_raw    object
value_hrf    object
dtype: object
------------------------------------------------------------------------
列名（数据类型为字符串 -str-object）:Index(['timestamp', 'node_id', 'subsystem',
'sensor', 'parameter', 'value_raw',
       'value_hrf'],
      dtype='object')
node_id-ID of node which did the masurement:['001e0610e532' '001e0610bc07'
'001e0610ef27' '001e0610e540'...'001e0610eef4']
------------------------------------------------------------------------

100%|██████████| 10/10 [00:00<00:00, 61.43it/s]
------------------------------------------------------------------------
AoT_nodes_temperature_gdf = pd.merge(AoT_nodes_gdf,value_raw_df.T.reset_index().
rename(columns={'index':'node_id'}),on="node_id")
AoT_nodes_temperature_gdf
------------------------------------------------------------------------
```

输出芝加哥城行政区划信息和AoT data测量数值合并后的空间地理信息数据（表1.3–8）。

表1.3-7　芝加哥城AoT测量值转换为数值型

	001e0610e532	001e0610bc07	001e0610ef27	001e0610e540	001e0610ee61	001e0610fb4c	001e0610ba18	001e0610ba3b	001e0610ba57	001e0610eef4
timestamp										
0	-9.495390	-9.917552	-10.226980	-8.861876	-9.366752	-8.686524	-9.450483	-7.710883	-9.393620	NaN
...										
23	-14.141487	-15.405905	-15.061102	-14.382365	-13.325823	-12.918876	-13.744962	-13.581819	-13.886686	-13.877945

表1.3-8　芝加哥城AoT data 测量数值空间地理信息数据

	node_id	project_id	vsn	address	lat	lon	description	start_timestamp	end_timestamp	geometry	...	22	23
0	001e0610ba3b	AoT_Chicago	006	18th St & Lake Shore Dr Chicago IL	41.858136	-87.616055	AoT Chicago (S)	2017/08/08 00:00:00	NaN	POINT (-87.61606 41.85814)	...	-12.155292	-13.581819
...													
9	001e0610e540	AoT_Chicago	05A	Fort Dearborn Dr & 31st St Chicago IL	41.838618	-87.607817	AoT Chicago (S) [C]	2017/11/29 00:00:00	NaN	POINT (-87.60782 41.83862)	...	-13.413810	-14.382365

```
10 rows × 34 columns
gpd2postSQL(AoT_nodes_temperature_gdf, table_name='AoT_nodes_temperature',
            myusername='postgres', mypassword='123456', mydatabase='AoT')
gpd2postSQL(AoT_nodes_temperature_gdf.to_crs(epsg_Chicago), table_name='AoT_
nodes_temperature_prj',
            myusername='postgres', mypassword='123456', mydatabase='AoT')
--------------------------------------------------------------------------------
has been written to into the PostSQL database...
--------------------------------------------------------------------------------
has been written to into the PostSQL database...
```

1.3.3.4 QGIS读取PostgreSQL数据库

虽然Python中的GeoPandas，以及其他图表库可以直接打印地图，但不是很方便地处理地图的细节表达，尤其将其用于论文发表、专著或者其他传播用途时。应用QGIS来构建地图，可以直接从PostgreSQL数据库中读取表（在PostGIS下建立数据库连接），和Python基本无缝结合。这可以充分利用Python的数据处理能力和QGIS的地图表达能力，如图1.3-5所示。

1.3.3.5 计算样本总长度（附）

data数据为单独的一个csv格式文件，可以通过下述代码来计算总样本数，即行数。也可以读取指定范围的部分行（表1.3-9）。但是，读取部分行时，仍旧需要耗费一定时间来略过需要忽略的行。通常，可以分批处理后，分别写入数据库或存储为单独的文件再读取处理。

表1.3-9 指定范围读取部分行数据

	timestamp	node_id	subsystem	sensor	parameter	value_raw	value_hrf
0	2018/05/15 08:00:25	001e06113a24	lightsense	hmc5883l	magnetic_field_y	564	512.727
...	...	...	...	...	...	...	...
999997	2018/05/15 11:12:05	001e0610f703	lightsense	apds_9006_020	intensity	34	2.733
999998	2018/05/15 11:12:05	001e0610f703	lightsense	hih6130	humidity	23629	44.22

```
from tqdm.auto import tqdm
count = 0
for chunk in tqdm(pd.read_csv(AoT_data_fp,chunksize=chunksize)):
    count += 1 # 样本分组数
    last_len = len(chunk)  # 最后一组的样本数量
data_length = (count*chunksize+last_len-chunksize) # 数据行（样本）总长度
print("数据行(样本)总长度={}".format(data_length))
--------------------------------------------------------------------------------
0it [00:00, ?it/s]

数据行(样本)总长度=573074785

rows_diff = data_length-chunksize
AoT_data_lastChunck = pd.read_csv(AoT_data_fp,skiprows=range(1,rows_
diff),nrows=chunksize-1)
AoT_data_lastChunck
--------------------------------------------------------------------------------
999999 rows × 7 columns
```

1.3.4 数据分析基本流程组织

开始一个以数据分析为主的研究课题（项目），通常建立一个单独的文件夹。在该文件夹下，放置该项目的所有代码，以及相关的数据、图表等内容。一般定义的子文件夹包括：

1. data - 存放原始的数据
2. data_processed - 存放处理的过程数据
3. database - 放置数据库
4. graph - 存放图表（一般由Python代码直接输出）
5. imgs - 存放一般的图像

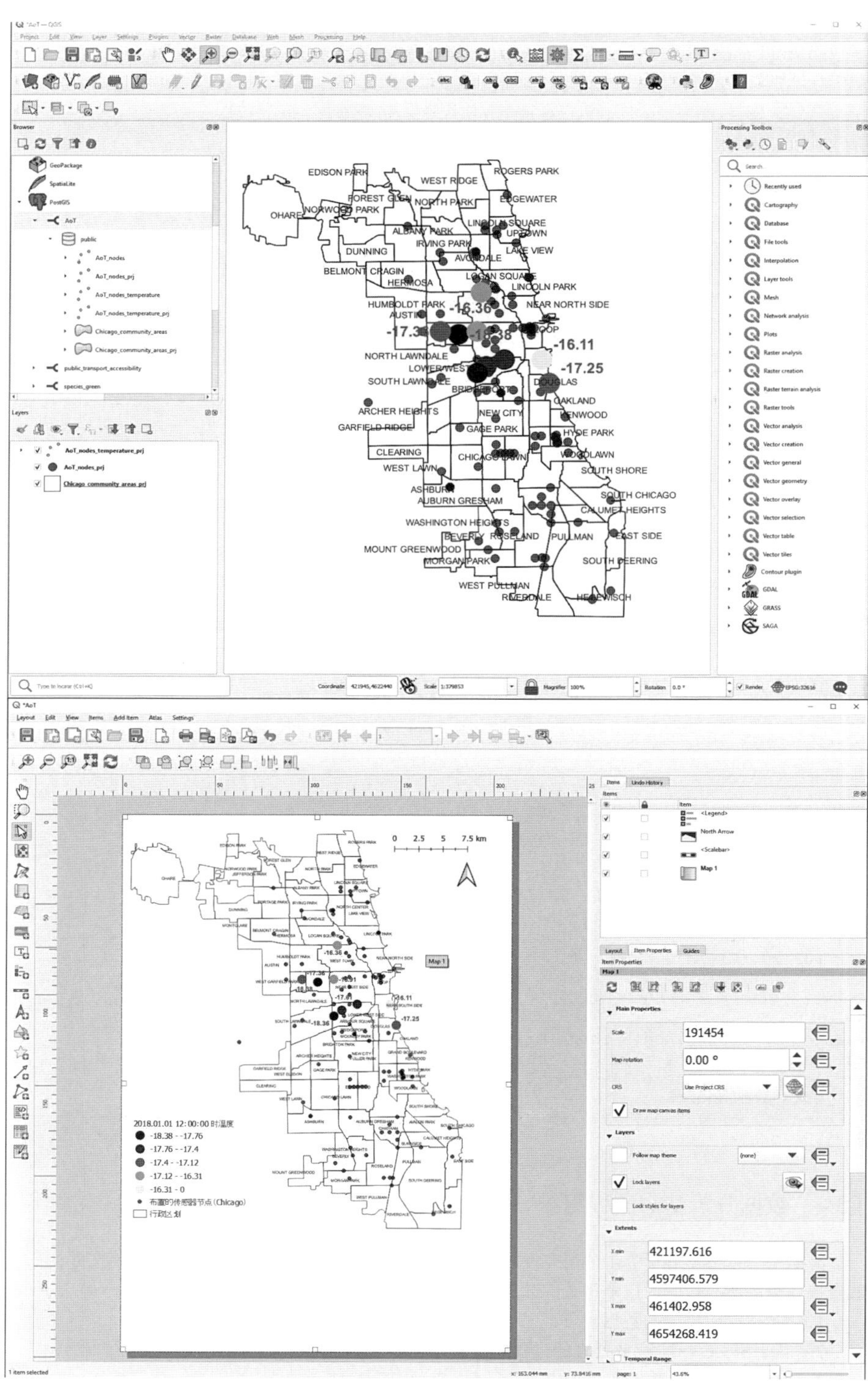

图1.3-5 用QGIS查看PostgreSQL数据库中的数据并布局打印地图

6. map - 放置地图文件（例如QGIS）

7. model - 存储训练好的模型（例如，机器/深度学习、网络模型等）

文件夹命名可以参考上述，亦可自行灵活调整。因为代码工程量会随着分析内容的深入不断增加，为了防止代码丢失，有必要将其推送（push）到GitHub代码托管平台，或国内相关的代码托管平台上。当代码更新时，可以推送更新云端仓库（repository）。可以使用GitHub的桌面版（GitHub Desktop）操作，具体方法可以查看官网。

通过子文件夹的结构，可以明了数据分析基本流程组织。从data下读取原始数据（如果数据文件较大，也会存储于外置硬盘中）；处理的过程数据则放置于data_processed中；为了方便数据的管理和读写，优先选择将数据写入数据库。SQLite为单独的文件，可以直接放置于database中，而postgreSQL是直接写入默认的安装路径下，在单独一个项目完结后，可以备份导出数据库；分析过程图表（用于说明分析结果，或过程数据分析描述，往往为书写科研论文的重要部分或用以报告）放置于graph下；其他非Python直接输出的图表或图像，存放于imgs下；QGIS读取数据库构建的地图放在map下；训练的模型则存储于model子文件中；.py的代码则直接位于根目录下，与子文件夹并列，这是为了方便直接读写数据。例如`db_fp = r'./database/fruits.sqlite'`和`AoT_nodes_fp = './data/AoT_Chicago.complete.2021-12-20/nodes.csv'`等，使用相对路径比较简单明了并便于代码迁移。如果有特殊需要，再建立存储代码的子文件夹。

为了方便调用，常用自定义或者迁移的代码工具，通常存储于单独的.py文件下，例如本书的util_database.py（用于数据库操作的代码函数），util_misc.py（包括显示文件的结构等杂项代码类或函数）等。

下述迁移的代码可以查看文件夹的结构，并打印了本书代码工程项目的阶段性文件夹内容结构。

```
from pathlib import Path
class DisplayablePath(object):
    '''
    class - 返回指定路径下所有文件夹及其下文件的结构。代码未改动，迁移于'https://
stackoverflow.com/questions/9727673/list-directory-tree-structure-in-python'
    '''
    display_filename_prefix_middle = '├──'
    display_filename_prefix_last = '└──'
    display_parent_prefix_middle = '    '
    display_parent_prefix_last = '│   '
    def __init__(self, path, parent_path, is_last):
        from pathlib import Path
        self.path = Path(str(path))
        self.parent = parent_path
        self.is_last = is_last
        if self.parent:
            self.depth = self.parent.depth + 1
        else:
            self.depth = 0
    @property
    def displayname(self):
        if self.path.is_dir():
            return self.path.name + '/'
        return self.path.name
    @classmethod
    def make_tree(cls, root, parent=None, is_last=False, criteria=None):
        from pathlib import Path
        root = Path(str(root))
        criteria = criteria or cls._default_criteria
        displayable_root = cls(root, parent, is_last)
        yield displayable_root
        children = sorted(list(path
                               for path in root.iterdir()
                               if criteria(path)),
                          key=lambda s: str(s).lower())
        count = 1
        for path in children:
            is_last = count == len(children)
            if path.is_dir():
                yield from cls.make_tree(path,
 parent=displayable_root,
                                         is_last=is_last,
```

```
        criteria=criteria)
                else:
                    yield cls(path, displayable_root, is_last)
                count += 1
        @classmethod
        def _default_criteria(cls, path):
            return True
        @property
        def displayname(self):
            if self.path.is_dir():
                return self.path.name + '/'
            return self.path.name
        def displayable(self):
            if self.parent is None:
                return self.displayname
            _filename_prefix = (self.display_filename_prefix_last
                                if self.is_last
                                else self.display_filename_prefix_middle)
            parts = ['{!s} {!s}'.format(_filename_prefix,
                                        self.displayname)]
            parent = self.parent
            while parent and parent.parent is not None:
                parts.append(self.display_parent_prefix_middle
                             if parent.is_last
                             else self.display_parent_prefix_last)
                parent = parent.parent
            return ''.join(reversed(parts))
app_root = r'C:\Users\richi\omen_richiebao\omen_github\USDA_CH_final\USDA\
notebook'
paths = DisplayablePath.make_tree(Path(app_root))
for path in paths:
    print(path.displayable())
--------------------------------------------------------------------------------
notebook/
├── .ipynb_checkpoints/
│   ├── 1.3_ 数据库与数据分析基本流程组织 -checkpoint.ipynb
│   └── 2.1.1 数据 POI 与描述性统计和正态分布 -checkpoint.ipynb
├── 1.3_ 数据库与数据分析基本流程组织 .ipynb
├── 2.1.1 数据 POI 与描述性统计和正态分布 .ipynb
├── __pycache__/
│   ├── coordinate_transformation.cpython-38.pyc
│   └── util_database.cpython-38.pyc
├── commodity_table_structure.py
├── coordinate_transformation.py
├── data/
│   ├── AoT_Chicago.complete.2021-12-20/
│   │   ├── nodes.csv
│   │   ├── offsets.csv
│   │   ├── provenance.csv
│   │   ├── README.md
│   │   └── sensors.csv
│   ├── Chicago Community Areas/
│   │   ├── Chicago Community Areas.dbf
│   │   ├── Chicago Community Areas.prj
│   │   ├── Chicago Community Areas.shp
│   │   └── Chicago Community Areas.shx
│   ├── poi_csv.csv
│   └── poi_json.json
├── data_processed/
├── database/
│   ├── AoT.1.3.sql
│   ├── fruits.sqlite
│   └── fruits_relational.sqlite
├── graph/
├── imgs/
│   ├── 1_3_01.png
│   ├── 1_3_02.jpg
│   ├── 1_3_03.jpg
│   ├── 1_3_04.jpg
│   ├── 1_3_04.psd
│   ├── 1_3_05.jpg
│   └── 1_3_06.jpg
├── map/
│   └── AoT.qgz
```

```
├── model/
├── sale_details_table_structure.py
├── sales_table_structure.py
├── util_database.py
└── util_misc.py
```

参考文献（References）:

[1] Miguel Grinberg (2018). Flask Web Development: Developing Web Applications with Python. O'Reilly Media.（中文版：Miguel Grinberg．安道译．Flask Web开发：基于Python的Web应用开发实战[M]．人民邮电出版社，2018.）

[2] 高桥麻奈著，崔建锋译．株式会社TREND-PRO漫画制作．漫画数据库[M]．北京：科学出版社．2010．5．

2

理论工具

2.1 城市空间数据分析方法理论集成研究框架

2.1.1 城市空间数据分析方法理论集成研究框架

数据分析是就所分析研究的内容和目的，使用与之相关的单一数据或多源数据，以统计学方法为基础，涉及应用线性代数、微积分、概率论、图论、计算机视觉、机器学习、深度学习等交叉学科的方法或算法模型解释挖掘数据、寻找模式、总结规律，从数据中发现有用的信息、得出结论以支持决策。数据分析方法在各个领域中具有广泛的应用，对城乡规划、风景园林等领域，结合地理信息系统提出有城市空间数据分析，以数据分析的方式解决城市相关问题。

数据分析是以数据为源动力，多学科方法理论综合应用的过程。因此，城市空间数据分析方法理论并无固定的框架，而是以解决所研究的内容为目的，适合选择一种或多种方法，乃至依据已有方法理论提出新的方法或算法模型。整合集成城市空间数据分析方法理论的方式，则变得复杂多样。为了使得集成的方法理论易于实现且方便应用于实际的研究过程、以数据科学领域最常用的编程语言Python为媒介，充分利用可结合方法理论的开源代码，按照基础试验和专项探索的方式集成可操作的城市空间数据分析方法理论。

基础试验包括知识间具有潜在迭代路径的统计学、线性代数、微积分、计算机视觉、机器学习和深度学习所包含的基础核心内容。专项探索是集成可用于城市空间数据分析能够构成独立框架的方法理论，包括标记距离、权重决策、更新策略、模式生成、推理学习、时空序列、维度空间、尺度效应、强化学习和复杂网络10个专题。在集成不同学科领域，可服务于城市空间数据分析的方法理论时，所有的知识均是可代码实现，并开展探索性试验。表2.1–1列出了所集成的方法理论，并分为一级、二级分类和计算分析内容，以及探索性试验。

表2.1–1 城市空间数据分析方法理论集成研究框架

一级分类	二级分类	计算分析内容	探索性试验或示例（Python 代码实现）
基础试验			
统计学	描述性统计	1. 频数（次数）分布表和直方图； 2. 均值、中位数、箱型图、标准差与标准计分（z_score）	1. 示例数据的描述性统计； 2. 百度地图POI数据的描述性统计
	正态分布与概率密度函数	1. 偏度与峰度； 2. 检验数据集是否服从正态分布； 3. 异常值处理； 4. 给定特定值计算概率，及找到给定概率的值	百度地图POI数据分类的正态分布与概率密度函数相关计算
	核密度估计	1. 单变量（一维数组）的核密度估计； 2. 多变量（多维数组）的核密度估计	开放街道地图（OpenStreetMap，OSM）点分类的核密度估计
	基本统计量	1. 标准误； 2. 中心极限定理； 3. t 分布 (Student's t-distribution); 4. 统计显著性； 5. 效应量； 6. 置信区间	示例数据的基本统计量计算
	相关性	1. 相关系数（数值数据和数值数据）； 2. 相关比（数值数据和分类数据）； 3. 克莱姆相关系数（分类数据和分类数据）	1. 示例数据的相关性计算; 2. 公共健康数据的相关性分析
	卡方分布与独立性检验	1. 卡方分布； 2.（卡方）独立性检验； 3. 协方差估计	示例数据的卡方检验与协方差估计

续表

一级分类	二级分类	计算分析内容	探索性试验或示例（Python 代码实现）
	回归模型（含机器学习-回归）	1. 反函数； 2. 指数函数与自然对数函数； 3. 回归与微分； 4. 矩阵； 5. 简单线性回归求解与显著性检验（F检验）、置信区间估计、精度计算； 6. 多元线性回归求解与显著性检验、置信区间估计、精度计算； 7. K-近邻模型（k-nearest neighbors algorithm，k-NN）； 8. 平均绝对误差（mean absolute error，MAE）和均方误差（mean squared error，MSE）； 9. 多项式回归； 10. 正则化； 11. 梯度下降法（Gradient Descent）	1. 示例数据的线性回归计算； 2. 公共健康数据中经济条件数据与疾病数据的简单线性回归建模； 3. 公共健康数据的多元回归、k-NN和多项式回归建模
线性代数		1. 矩阵； 2. 向量； 3. 线性映射； 4. 特征值和特征向量	bildstein_station1（城市点云数据）的线性变换
		主成分分析（Principal components analysis，PCA）	Landsat遥感影像波段的PCA降维与RGB显示
		1. NDVI 指数； 2. 采样精度计算	（建立简单工具）建立对遥感影像波段的采样工具
微积分		1. 导数与微分； 2. 积分； 3. 泰勒展开式； 4. 偏微分	示例数据的微积分相关计算
	卷积与SIR空间传播模型	1. 一维卷积与曲线分割； 2. 二维卷积与图像特征提取； 3. SIR 传播模型； 4. SIR 空间传播模型	根据土地覆盖配置空间阻力值建立SIR空间传播模型
计算机视觉	图像特征提取	1. SIFT（Scale-Invariant Feature Transform）特征检测和描述算法； 2. Star 特征检测器提取图像特征； 3. 特征匹配	用特征检测提取KITTI数据集（城市影像）特征，通过特征匹配探索动态街景视觉感知
	超像素级分割方法	1. Felzenszwalb 超像素级分割方法； 2. Quickshift 超像素级分割方法	Sentinel-2影像超像素级分割下高空分辨率特征尺度界定
机器学习	聚类	1. K-Means 聚类； 2. 多种聚类算法比较	用调研图像聚类图像主题色，分析城市环境色彩
	图像分类器与识别器	1. 视觉词袋与构建图像映射特征； 2. 决策树（Decision trees）与随机森林（Random forests）； 3. 信息增益（Information gain）； 4. 基尼不纯度（Gini impurity）； 5. 交叉验证（cross_val_score）； 6. 图像分类器与识别器	（建立简单工具）图像分类器与识别器的机器学习模型网络实验应用平台部署
深度学习	多层感知机（多层神经网络）	1. 张量（tensor）计算； 2. 微积分-链式法则； 3. 激活函数（activation function）； 4. 前向与后向传播（propagation）； 5. 自动求梯度（gradient）； 6. 多层神经网络； 7. runx.logx-深度学习试验管理	示例数据的多层神经网络构建
	逻辑回归二分类与SoftMax回归多分类	1. 伯努利分布； 2. 似然函数； 3. 极大似然估计； 4. 逻辑回归（Logistic Regression，LR）； 5. SoftMax回归多分类	示例数据（图像数据集Fashion-MNIST）的SoftMax回归多分类建模
	卷积神经网络（Convolutional neural network，CNN）	1. CNN，及其卷积层、池化层和填充； 2. tensorboard可视化； 3. 可视化卷积层/卷积核； 4. VGG网络	示例数据的CNN建模

续表

一级分类	二级分类	计算分析内容	探索性试验或示例（Python 代码实现）
	对象检测与实例分割	1. 对象检测； 2. 对象实例分割与对象统计； 3. 交并比（Intersection over Union，IoU）； 4. Faster R-CNN ResNet-50 FPN 图像语义分割模型	1. PennFudan和COCO数据集行人检测； 2. KITTI数据集行人检测与人流量估算； 3. KITTI数据集对象实例分割与统计和关联网络结构
	图像分割	1. 开放神经网络交换 —— ONNX（Open Neural Network，Exchage）； 2. Netron 网络可视化工具； 3. DUC（Dense Upsampling Convolution）图像分割； 4. VGG16 卷积神经网络； 5. SegNet 图像分割模型； 6. 图像数据增强变换； 7. 图像分割模型库； 8. Unet 网络模型（resnet34 编码器）图像分割	1. cityscapes数据集DUC图像语义分割下城市空间要素组成，时空量度，绿视率和均衡度计算； 2. sentinel-2影像的无监督土地分类——聚类方法； 3. 用VGGNet模型预测图像内容对象； 4. ISPRS遥感影像SegNet图像分割模型解译； 5. NAIP航拍影像解译（图像分割）
其他		天空视域因子（Sky View Factor，SVF）计算	基于数字表面模型（Digital Surface Model，DSM）计算 SVF
专项探讨			
标记距离		1. 地图分类算法； 2. 模式标记特征（Pattern Signature）： a. 类/簇大小直方图（连通域标签）； b. 共现关系； c. 3 层级分解。 3. 距离度量（Distance meatrics）； 4. 模式发现： a. 模式级聚——距离矩阵与层级聚类； b. 模式搜索； c. 模式监测； d. 模式分割	1. 对NLCD（National Land Cover Database）美国土地覆盖类型数据集执行模式发现； 2. 地表温度（Land Surface Temperature，LST）冷热点样方标记土地覆盖类型模式标记特征
权重决策		多准则决策法，Multiple Criteria Decision-Making（MCDM）： a. 熵值权重法； b. 理想解法（TOPSIS）； c. AHP、F-AHP、ARAS、BWM（基于GWO元启发式算法）、DEMATEL、IDOCRIW（基于GA元启发式算法）、ELECTRE-I、WASPAS等MCDM算法	1. 对AoT城市环境传感器测量数据中污染气体浓度数据测量点污染浓度综合指数评估（应用 TOPSIS算法）； 2. LST（降温）与绿地和建设用地矛盾的匹配评估（WASPAS算法评估样方单元综合评价指数）
更新策略		1. 一维度权重决策； 2. 复杂网络更新（遗传算法）； 3. 二维度布局优化——空间决策支持系统； 4. 元启发式算法（Meta-Heuristic Algorithm）： a. 粒子群优化算法（Particle Swarm Optimization, PSO）； b. 布谷鸟搜索算法（Cuckoo Search，CS）； c. 萤火虫算法（Firefly Algorithm，FA）	1. 一维度权重决策解决不同建设内容分配有限不同土地资源条件问题； 2. 二维度布局优化——空间决策支持系统复现
模式生成		1. 聚类模式特征分析； 2. 生成对抗网络（Generative Adversarial Networks，GAN）的模式生成； 3. SytleGAN； 4. 条件对抗网络（Conditional Adversarial Networks，cGAN）	1. 多源数据的区域统计与数据合并； 2. 多源数据聚类模式组成结构； 3. LST冷热点舒适区域与地表覆盖类型的簇分布统计； 4.（建立简单工具）WGAN模型中G网络各层数据演化情况计算分析工具； 5. StyleGAN模型生成NAIP航拍影像数据； 6. NAIP遥感影像和土地覆盖类型之间的翻译转化； 7.（建立简单工具）遥感影像和土地覆盖类型互译工具构建； 8. 遥感影像缺失区域的补全

续表

一级分类	二级分类	计算分析内容	探索性试验或示例（Python 代码实现）
推理学习	概率论与概率图	1．概率论； 2．概率图——贝叶斯网络： a. 基于概率模型的分类器； b. 离散型贝叶斯网络——表征（Representation）与推理（Inference）； c. 连续型贝叶斯网络； d. 模型学习：参数估计；结构学习	1．示例数据概率论与概率图相关计算； 2．用生境威胁、生境类型、碳储存、生境质量和生境风险5个生态系统服务指数关系的简化模型建立贝叶斯网络
	生态系统服务价值（InVEST模型）	1．InVEST和NWV生态系统服务计算工具； 2．生态系统服务价值中的碳储存和封存、生境质量、作物授粉和作物生产（InVEST）	生态系统服务价值的贝叶斯网络构建与推理
时空序列	时间序列分析	白噪声检验、平稳性检验和周期性模式判断	城市环境传感器AoT时序数据的白噪声检验、平稳性检验和周期性模式判断
	空间自相关分析	1．空间权重； 2．全局空间自相关； 3．局部空间自相关； 4．地理轮廓（Geosilhouettes）	公共健康数据的空间自相关分析
	空间动力学	1．空间马尔可夫链： a．经典离散（时间）马尔可夫链； b．空间马尔科夫（Spatial Markov）； c．局部空间自相关马尔科夫（LISA Markov）； d．流动性测量。 2．空间 Kendall's Tau： a．经典肯德尔等级相关系数（Kendall's Tau）； b．空间Kendall's Tau	1．犯罪率年统计量的经典和空间离散型马尔可夫链计算； 2．空气质量指数（Air quality index，AQI）的空间 Kendall's Tau 计算
	不平等性和空间隔离	1．不平等性（基尼系数和洛伦兹曲线）； 2．空间隔离： a．空间隔离指数； b．组间比较与作用（决定）因素成分分解和贡献度； c．空间隔离指数推断； d．局部空间隔离指数； e．多类空间隔离指数	芝加哥城土地利用样方统计下开放空间的不平等性和空间隔离相关计算
	从自然语言处理到视觉模型	1．RNN（Recurrent Neural Networks）； 2．LSTM（long short-term memory）； 3．Word2Vec和Seq2Seq； 4．Transformer——自然语言处理（Natural Language Processing，NLP）； 5．GPT； 6．Transformers 的视觉模型	1．百度全景静态图视域景观指数计算； 2．视域景观指数不同邻里尺度最优簇数选择和指数贡献度； 3．街道行业分类（POI）服务空间组成结构； 4．特征指数的LSTM时空预测模型
维度空间		1．流形学习（Manifold learning）； 2．MDS、KPCA、Isomap、LLE、t-SNE和UMAP等流形学习算法	1．街道空间多维景观指数的MDS降维试验； 2．街道空间单一景观指数的KPCA映射试验； 3．街道视域全景语义分割图的Isomap降维和嵌入空间的主成分分布； 4．街道POI行业类别空间分布结构； 5．街道空间多维景观指数的UMAP空间特征区段划分
尺度效应	不同尺度作用结果的比较分析方法	1．标记距离的尺度变化矩阵； 2．分类面积尺度变化曲线； 3．常规统计； 4．半变异（方差）函数（semi-variograms）	1．NAIP航拍影像解译（图像分割）为土地覆盖数据； 2．土地覆盖数据标记距离的尺度变化矩阵和分类面积尺度变化曲线等不同尺度作用结果比较分析； 3．用半变异函数寻找人口数量分布数据尺度效应的幅度变化关系

续表

一级分类	二级分类	计算分析内容	探索性试验或示例（Python 代码实现）
	空间相互作用模型（Spatial Interaction Modelling，SIM）	1. SIM 族； 2. 广义线性模型（GLM）与泊松对数线性回归； 3. SIM 模型拟合统计	绿地公园供给与人口需求关系的SIM参数估计
强化学习	智能体模型（Agent-Based Models，ABM）	NetLogo，Mesa 和 Repast4Py 模拟系统比较	
	强化学习	1. 表格型求解方法； 2. Gymnasium 环境构建； 3. 深度强化学习（Deep RL）； 4. 多智能体强化学习（MARL）	1. 动物的运动及其与土地覆盖类型关系的试验； 2. 考虑土地覆盖类型多智能体的移动模拟
复杂网络（图论）		1. 图论； 2. 图属性的基本度量； 3. 图嵌入和图神经网络	1. 构建土地利用类型的样方式复杂网络进行图属性基本度量计算； 2. 公共交通复合网络构建与公园潜在服务人口压力； 3. 植物调查样方物种的图嵌入； 4. 植物调查样方物种间的关联

2.1.2 数据分类和处理

数据是数据分析的源动力，在应用方法理论进行所研究内容的数据分析前，通常需要对数据进行预处理，这包括数据的检索获取、清洗、格式转化、提取、集成，及写入数据库，从数据库中读写数据等。为了实现完成广泛类型数据的预处理任务，在城市空间数据分析方法理论集成时，尽量使用具有不同格式和内容的多源数据作为开展探索性试验的数据源（表2.1–2），包括的数据类型主要有地理空间矢量数据和栅格数据、文本数据、影像等。根据数据内容可以分类有兴趣点、点云数据、遥感数据、航拍影像、土地利用和覆盖、地表温度、城市环境传感器测量数据、空气质量指数、数字高程、建筑高度、夜间灯光、城市影像、全景静态图、观测鸟分布、动物跟踪数据库、人口分布、GPS轨迹数据、物种分布及丰度、公共健康数据、犯罪事件数据集，以及城市的基础数据，例如路网、公交和地铁线路站点、行政边界等。

各类数据处理完全使用Python编程语言实现，这与城市空间数据分析方法理论集成的Python代码实现保持一致，从而充分利用Python生态系统，构建名为USDA的工具包和科学计算软件工具，弹性集成数据处理技术、数据分析方法理论，建立方法理论可实现计算的工具。

表2.1–2 数据分类和预处理

数据分类	数据名称	数据来源	数据预处理内容
兴趣点（points of interest，POI）	百度地图POI数据	百度地图开放平台	1. 检索； 2. CSV转DF格式数据； 3. DF转GDF地理空间格式数据； 4. 数据可视化
	开放街道地图（OpenStreetMap，OSM）	OpenStreetMap	1. 检索； 2. 读取、转换OSM数据为SHP格式
点云数据	bildstein_station1（城市点云数据）	semantic3d	读取和可视化
	伊利诺伊州LAS格式激光雷达数据	伊利诺斯州草原地质调查研究所（Illinois state geological survey - prairie research institute）	1. 查看点云数据信息； 2. 点云数据可视化； 3. 由点云建立数字表面模型（Digital Surface Model，DSM）与分类栅格； 4. 建筑高度提取

续表

数据分类	数据名称	数据来源	数据预处理内容
遥感影像	Landsat	美国地质调查局（United States Geological Survey，USGS-earthexplorer）	1. 元数据读取； 2. 波段的合成显示
	Sentinel-2	哥白尼数据空间生态系统（Copernicus Data Space Ecosystem）	1. 以Web Mercator方式显示Sentinel-2的一个波段； 2. Sentinel-2 波段合成显示； 3. 元数据读取； 4. 影像裁切
	ISPRS 遥感图像数据集	International Society for Photogrammetry and Remote Sensing	读取与建立样本数据集
航拍影像	NAIP 航拍影像。（美）国家农业图像项目（National Agriculture Imagery Program，NAIP）	美国地质调查局（United States Geological Survey，USGS-earthexplorer）	1. 数据检索； 2. 建立样本数据集
土地覆盖	NLCD（National Land Cover Database）美国土地覆盖类型数据集	美国地质调查局（United States Geological Survey，USGS）与多分辨率土地特征联盟（Multi-Resolution Land Characteristics Consortium，MRLC）	数据检索与建立样本数据集
	高精度芝加哥土地覆盖数据	ArcGIS-Chicago Regional Land Cover	读取与可视化
	MCD12Q1_v006土地覆盖类型数据集	美国国家航空和航天局（National Aeronautics and Space Administration，NASA）	1. 数据检索和地图打印； 2. 批量处理和保存MCD12Q1_v006 各年数据； 3. 按采样点提取数据
	ESA WorldCover	欧空局（European Space Agency，ESA）	
土地利用数据	芝加哥城土地利用	芝加哥都市规划局（Chicago Metropolitan Agency for Planning，CMAP）	1. SHP格式转GDF格式与名称编码； 2. 矢量SHP格式土地利用数据转栅格（raster）； 3. 土地利用地图打印
地表温度（Land Surface Temperature，LST）	中分辨率成像光谱仪（Moderate Resolution Imaging Spectroradiometer，MODIS）的地表温度和辐射率（Emissivity）产品（MYD21A1D v006）	美国地质调查局（United States Geological Survey）	数据检索与建立样本数据集
城市环境传感器	Array of Things，AoT（分类测量内容为气体类、颗粒物、热环境、噪声、光、电磁场和惯性测量等）	西北-阿贡科学与工程研究所（Northwestern-Argonne Institute for Science and Engineering，NAISE）领导	1. 数据清洗（数值有效区间，数据精度处理）； 2. 按时间范围提取数据； 3. 分类数据
空气质量指数（Air quality index，AQI）	空气质量指数数据集（Air Quality Index Dataset）	Kaggle	读取与可视化
数字高程	数字高程模型（digital elevation model，DEM）	美国地质调查局（United States Geological Survey，USGS-earthexplorer）	读取与可视化
建筑高度（层数）数据	芝加哥城建筑轮廓数据，含有层数字段	芝加哥数据门户（Chicago Data Portal，CDP）	读取与可视化

续表

数据分类	数据名称	数据来源	数据预处理内容
夜间灯光数据	Annual VNL V2	地球观测组（Earth Observation Group）	
城市影像	KITTI 数据集（无人驾驶场景下计算机视觉算法评测数据集）	The KITTI Vision Benchmark Suite	KITTI文件信息读取和可视化
	PennFudan数据集	Penn-Fudan Database for Pedestrian Detection and Segmentation	读取与可视化
	COCO数据集	COCO	读取与可视化
	Cityscapes数据集	Cityscapes Dataset	读取与可视化
	ImageNet数据集	ImageNet	读取与可视化
全景静态图	百度全景静态图	百度地图开放平台	1. 全景图像素级语义分割； 2. 语义分割图的投影变换
	调研图像	用手机应用记录调研路径	1. 图像数据读取与可视化； 2. 读取图像KML调研路径数据与可视化； 3. Exif（Exchangeable image file format）可交换图像格式数据读取； 4. RGB色彩的三维图示
观测鸟分布数据	eBird Basic Dataset（EBD）	eBird	读取与转化为GDF格式数据，以及可视化
动物跟踪数据库	森林公园生活实验室（Forest Park Living Lab，FPL）动物运动跟踪数据	Movebank	读取与可视化
人口分布数据	WorldPop	WorldPop	读取与可视化
GPS 轨迹数据	GeoLife	微软亚洲研究院（Microsoft Research Asia）	1. 合并GeoLife数据集； 2. 合并GPS轨迹和标签； 3. 交通工具时速统计
物种分布及丰度	Countryside Survey 1978 vegetation plot data数据集	英国生态与水文中心（Centre for Ecology & Hydrology）	读取为DF格式数据
其他	公共健康数据（公共健康指标）	芝加哥数据门户（Chicago Data Portal，CDP）	合并社区边界数据，转CSV至GDF
	芝加哥市发生报告的犯罪事件数据集	芝加哥数据门户（Chicago Data Portal，CDP）	时空数据（面板数据），按照给定的分组，时间长度，数值计算方法重采样数值列
数据类型缩写	DataFrame（DF）； GeoDataFrame（GDF）		

2.2 USDA工具包和科学计算工具构建

2.2.1 USDA 工具包的基本情况

构建城市空间数据分析方法理论集成研究框架时的代码书写有两种方式：一种是，用Python及其相关库完全书写，基础试验部分基本全部采用该种方式；另一种是，构建USDA工具包，通过调用USDA工具完成书写，从而大幅度减少代码量，而更多关注于理论方法本身，专项探索部分大部分采用该种方式。

USDA工具包源码托管于GitHub仓库USDA_PyPI，地址为：https://github.com/richieBao/USDA_PyPI。

USDA工具包存储于Python Package Index（PyPI），一个Python编程语言的软件存储库，索引名为usda，地址为：https://pypi.org/project/usda。通过`pip install usda`方式安装。如果指定版本则为`pip install usda==0.0.30`。

USDA工具包的说明文档为城市空间数据分析方法-USDA库手册，地址为：https://richiebao.github.io/USDA_PyPI。

截至本书完成时，USDA工具包的版本为usda 0.0.30。

因为USDA工具包主要用于阐述城市空间数据分析集成的理论方法，涉及的内容广泛庞杂，用到的依赖库数量较多且依赖库之间可能存在兼容性等问题，因此USDA工具包不强制安装相关依赖库；并且，不同工具往往是独立的，方便直接调用或者直接复制所用源码于具体的计算过程。

2.2.2 USDA 工具包的内容（0.0.30 版本）

USDA工具包在集成城市空间数据分析方法理论过程中逐步形成，且将不断调整和更新。具体的模块及相关简要介绍如下：

- data_process：常用数据（集）的预处理和信息读取，及常用数据处理工具；
- datasets：实验性数据集，以及数据集检索工具；
- database：数据库读写工具；
- utils 和 tools：通用类一般工具集；
- data_visualization 和 data_visual：数据可视化工具；
- stats：统计类工具；
- geodata_process 和 geodata_process_opt：地理信息数据处理工具（栅格数据和矢量数据等）；
- models：模型类；
- maths：基本数学计算类；
- indices：指数类；
- pattern_signature：标记距离与模式；
- weight：多准则决策法（权重决策）；
- meta_heuristics：元启发式算法；
- network：图与复杂网络；
- net：深度学习网络；
- migrated_project：迁移的代码库，包括 invest（生态系统服务计算工具），pass_panoseg（全景图语义分割），pix2pix（cGAN 通用框架），RL_an_introduction（强化学习），stylegan（生成对抗网络）等；
- imgs_process：图像处理；
- pgm：概率图（概率论）；
- rl：强化学习；
- mpe_realworld：自定义多智能体强化学习（MARL）环境示例；
- demo：说明类演示图表；
- manifold：流形学习（维度空间）；

- pano_projection_transformation：全景语义分割图的投影变换。

USDA 工具包及其说明文档尚不完善，这将需要更长的时间来调整更新。但是，当前 0.0.30 版本作为本书的配套工具，可以辅助读者完成书中演示的内容。如果因为依赖库冲突等问题，可以直接对应复制迁移 USDA 源码进行试验。

2.2.3 USDA 科学计算软件工具构建

为了进一步拓展城市空间数据分析方法理论集成研究框架的应用途径，当USDA工具包初步完成后，基于该Python工具包，应用基于Qt[①]的PySide6[②]开发科学计算软件工具，主要在Qt Creater 集成开发环境（integrated development environment，IDE）中完成。

USDA软件的功能架构基本同该书专项讨论部分的章节组成，包括标记距离、权重决策、更新策略、模式生成、推理学习、时空序列、维度空间、尺度效应、强化学习和复杂网络10个专题，并增加有简单的地图显示，所用地图打印模式，常用数据处理等功能。在软件开发过程中，会根据需要增补或调整内容。例如，在更新策略部分，引入景观指数计算，并开发基于景观指数计算的二维度布局优化探索等。

USDA软件开发过程的测试版可以从coding-x.tech（https://coding-x.tech）下的USDA软件工具及下载页面中获取。软件的界面见图2.2-1。为了方便使用者熟悉软件使用，尤其参数的配置，在开始页面增加了使用手册（也可以从网站获取），并链接了该书的在线版本（非最终出版社编辑版本）。

① Qt 为软件开发工具，创建软件应用程序或嵌入式设备所需的所有工具，为从规划和设计到开发、测试和面向未来的产品。

② PySide6为Qt项目的官方Python开发模块，其提供了对完整Qt 6.0+框架的访问，且为开源项目。

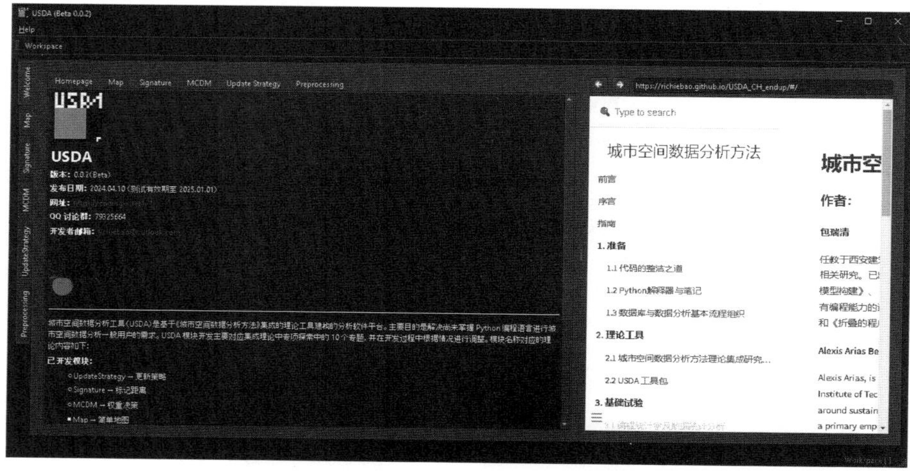

图2.2-1 USDA软件工具界面

基础试验

3-A 编程统计学及数据统计分析

3.1 POI数据与描述性统计和正态分布

3.1.1 单个分类POI数据检索与地理空间点地图

在城市空间分析中，涉及两个方面的数据，一类是物质空间相关数据，进一步可以分为二维平面数据（例如遥感影像，城市地图等）和三维空间数据（例如雷达数据，含高空扫描和地面扫描，及一般意义上各类格式的城市三维模型等）；另一类则是物质空间承载的各类属性数据，包括反映二维平面数据的属性（与二维地理位置结合，可以划分到二维平面数据当中），例如用地类型、自然资源分布、地址名称等；和反映人类、动物及无形物质各类活动性质的数据，例如人们的活动轨迹（出租车和共享单车车行轨迹、夜间灯光、手机基站定位用户数量等）、动物迁徙路径和各类小气候测量指标变化等。

百度地图开放平台[①]提供地图相关的功能与服务，其Web服务API为开发者提供http/https接口。开发者可以通过http/https形式发起检索请求，获取返回JSON（JavaScript Object Notation）或XML（Extensible Markup Language）格式的检索数据。其中，兴趣点（POI，points of interest）[②]数据，目前包括21个大类和无数小类，是反映物质空间承载人类活动的属性数据（业态分布），如表3.1–1所示。

表 3.1–1 百度地图POI分类

一级行业分类	二级行业分类
美食	中餐厅、外国餐厅、小吃快餐店、蛋糕甜品店、咖啡厅、茶座、酒吧、其他
酒店	星级酒店、快捷酒店、公寓式酒店、民宿、其他
购物	购物中心、百货商场、超市、便利店、家居建材、家电数码、商铺、市场、其他
生活服务	通讯营业厅、邮局、物流公司、售票处、洗衣店、图文快印店、照相馆、房产中介机构、公用事业、维修点、家政服务、殡葬服务、彩票销售点、宠物服务、报刊亭、公共厕所、步骑行专用道驿站、其他
丽人	美容、美发、美甲、美体、其他
旅游景点	公园、动物园、植物园、游乐园、博物馆、水族馆、海滨浴场、文物古迹、教堂、风景区、景点、寺庙、其他
休闲娱乐	度假村、农家院、电影院、KTV、剧院、歌舞厅、网吧、游戏场所、洗浴按摩、休闲广场、其他
运动健身	体育场馆、极限运动场所、健身中心、其他
教育培训	高等院校、中学、小学、幼儿园、成人教育、亲子教育、特殊教育学校、留学中介机构、科研机构、培训机构、图书馆、科技馆、其他
文化传媒	新闻出版、广播电视、艺术团体、美术馆、展览馆、文化宫、其他
医疗	综合医院、专科医院、诊所、药店、体检机构、疗养院、急救中心、疾控中心、医疗器械、医疗保健、其他
汽车服务	汽车销售、汽车维修、汽车美容、汽车配件、汽车租赁、汽车检测场、其他
交通设施	飞机场、火车站、地铁站、地铁线路、长途汽车站、公交车站、公交线路、港口、停车场、加油加气站、服务区、收费站、桥、充电站、路侧停车位、普通停车位、接送点、其他
金融	银行、ATM、信用社、投资理财、典当行、其他
房地产	写字楼、住宅区、宿舍、内部楼栋、其他
公司企业	公司、园区、农林园艺、厂矿、其他
政府机构	中央机构、各级政府、行政单位、公检法机构、涉外机构、党派团体、福利机构、政治教育机构、社会团体、民主党派、居民委员会、其他
出入口	高速公路出口、高速公路入口、机场出口、机场入口、车站出口、车站入口、门（备注：建筑物和建筑物群的门）、停车场出入口、自行车高速出口、自行车高速入口、自行车高速出入口、其他
自然地物	岛屿、山峰、水系、其他
行政地标	省、省级城市、地级市、区县、商圈、乡镇、村庄、其他
门址	门址点、其他

引自：百度地图开放平台更新于 2020 年 5 月 11 日。

① 百度地图开放平台（API），是为开发者免费提供的一套基于百度地图服务的应用接口，包括JavaScript API、Web服务API、Android SDK、iOS SDK、定位SDK、车联网API、LBS云等多种开发工具与服务，提供基本地图展现、搜索、定位、逆/地理编码、路线规划、LBS云存储与检索等功能，适用于PC端、移动端、服务器等多种设备，多种操作系统下的地图应用开发（https://lbsyun.baidu.com/）。

② 兴趣点（POI，points of interest），是一个具有一定意义的特定点位置，例如地球上代表菲尔铁塔（Eiffel Tower）的位置，或是火星上代表最高峰奥林匹斯山（Olympus Mons）的位置。在提及酒店、露营地、加油站或汽车导航系统中任何其他类别都会使用这个术语。

3.1.1.1 单个分类检索

检索数据，需要查看百度提供的检索方法，根据要求来配置对应参数实现下载，具体查看其服务文档[①]。本次实验对应配置检索语句为，矩形区域检索:http://api.map.baidu.com/place/v2/search?query=银行unds=39.915,116.404,39.975,116.414&output=json&ak={您的密钥} //GET请求，百度地图例举的语句仅包含个别请求参数，实际上提供的行政区划区域检索请求参数约15个，可以根据对数据的下载需求来确定使用哪些请求参数。检索服务包括圆形区域检索、地点详情检索和多边形区域检索服务等。此次矩形区域检索（多边形区域检索）的请求参数配置说明如下：

① 服务文档，接口功能介绍，含参数说明（https://lbsyun.baidu.com/index.php?title=webapi/guide/webservice-placeapi）。

```
query_dic={
    'query':检索关键字。圆形区域检索和矩形区域内检索支持多个关键字并集检索，不同关键字间以"$"符号分隔，最多支持10个关键字检索，如:"银行$酒店"。如果需要按POI分类进行检索，将分类通过query参数进行设置，例如 query='旅游景点'
    'page_size': 单次召回POI数量，默认为10条记录，最大返回20条。多关键字检索时，返回的记录数为"关键字个数*page_size"，例如 page_size='20'
    'page_num':分页页码，默认为0, 0代表第一页， 1代表第二页，以此类推。常与page_size搭配使用，例如 page_num='0'
    'scope':检索结果详细程度。取值为1或空，则返回基本信息；取值为2，返回检索POI详细信息，例如 scope='2'
    'bounds':检索矩形区域，为左下角和右上角坐标，多组坐标间以","分隔， 例如 str(leftBottomCoordi[1]) + ',' + str(leftBottomCoordi[0]) + ','+str(rightTopCoordi[1]) + ',' + str(rightTopCoordi[0]),
    'output':输出格式为json或者xml，例如output='json'
    'ak':开发者的访问密钥，必填项。v2之前该属性为key,
}
```

如需下载数据，请求参数需要申请访问密钥'ak'（注意，需要注册登录）。

首先，配置基本参数，最终要实现的目的是将下载后的数据分别存储为CSV（Comma-Separated Values，逗号（字符）分隔值，以纯文本形式存储表格数据）和JSON（JavaScript Object Notation，轻量级数据交换格式，层次结构简洁清晰）数据格式，因此调入json和csv库，辅助读写对应格式的文件。因为要通过网页地址来检索数据，因此调入urllib库实现相应的URL（Uniform Resource Locator，统一资源定位符，即网络地址）处理。对文件路径的管理则可以使用os及pathlib库。

#注意，多边形区域检索目前为高级权限，如有需求，需要在百度地图开放平台上提交工单咨询。

- 多边形区域检索（矩形区域检索）

```python
import os

data_path='./data' #配置数据存储位置
#定义存储文件名
poi_fn_csv = os.path.join(data_path,'poi_csv.csv')
poi_fn_json = os.path.join(data_path,'poi_json.json')
```

配置请求参数，注意其中page_num参数为页数递增，初始参数时，配置页数范围为page_num_range=range(20)。保存的文件类型直接由定义的存储文件名的后缀名确定。由于百度API的限制，检索区域内最多返回的POI数据量有限制，造成下载疏漏，因此如果下载区域很大时，最好是将其切分为数个矩形逐一下载，通过增加一个配置参数partition实现检索区域的切分次数，如果设置为2，则切分矩形为4份检索区域分别下载。

注意在配置坐标时，使用百度地图坐标拾取系统[②]。

② 百度地图坐标拾取系统，支持POI点坐标显示、复制；地址精确/模糊查询；坐标鼠标跟随显示；支持坐标查询（http://api.map.baidu.com/lbsapi/getpoint/index.html）。

```python
bound_34.186027],'rightTop':[109.129275,34.382171]}
page_num_range = range(20)
partition = 4
query_dic = {
    'query':'旅游景点',
    'page_size':'20',
    'scope':2,
    'ak':'2Zh7jNunzIzKoWx59ucjHLlZ63oI9St0',
}
```

通过定义一个函数实现百度地图开放平台下，基于矩形区域检索爬取POI数据，方便日后相关项目的代码迁移或者调用，增加代码融合的力度。因此，需要谨慎确定输入参数，避免在调用时还需调整函数内的变量，造成不必要的错误。由百度地图下载数据，其经纬度坐标为百度坐标系，因此需要对其进行转换，调入转换代码文件coordinate_transformation.py（从该书配套GitHub仓库下载）。

在坐标转换过程中，调用了两个转换函数，bd09togcj02(bd_lon, bd_lat)和gcj02towgs84(lng,lat)。当运行检索函数时，需要将坐标转换文件置于与该.ipynb文件（或者读者新建立包含检索函数的 .py 文件）同一文件夹下，以备调用。

```
def baiduPOI_dataCrawler(query_dic, bound_coordinate, partition, page_num_range, poi_fn_list=False):
    '''
    function - 百度地图开放平台 POI 数据检索--多边形区域检索（矩形区域检索）方式

    Params:
        query_dic - 请求参数配置字典，详细参考上文或者百度服务文档; dict
        bound_coordinate - 以字典形式配置下载区域; dict
        partition - 检索区域切分次数; int
        page_num_range - 配置页数范围; range()
        poi_fn_list=False - 定义的存储文件名列表; list

    Returns:
        None
    '''
    import coordinate_transformation as cc
    import urllib
    import json
    import csv
    import pathlib

    urlRoot = 'http://api.map.baidu.com/place/v2/search?'  #数据下载网址，查询百度地图服务文档
    #切分检索区域
    if bound_coordinate:
        xDis = (bound_coordinate['rightTop'][0] -
                bound_coordinate['leftBottom'][0])/partition
        yDis = (bound_coordinate['rightTop'][1] -
                bound_coordinate['leftBottom'][1])/partition
    #判断是否要写入文件
    if poi_fn_list:
        for file_path in poi_fn_list:
            fP = pathlib.Path(file_path)
            if fP.suffix == '.csv':
                poi_csv = open(fP, 'w', encoding='utf-8')
                csv_writer = csv.writer(poi_csv)
            elif fP.suffix == '.json':
                poi_json = open(fP, 'w', encoding='utf-8')
    num = 0
    jsonDS = []  #存储读取的数据，用于.json格式数据的保存
  #循环切分的检索区域，逐区下载数据
    print("Start downloading data...")
    for i in range(partition):
        for j in range(partition):
            leftBottomCoordi = [bound_coordinate['leftBottom']
                                [0]+i*xDis, bound_coordinate['leftBottom']
[1]+j*yDis]
            rightTopCoordi = [bound_coordinate['leftBottom'][0] +
                              (i+1)*xDis, bound_coordinate['leftBottom']
[1]+(j+1)*yDis]
            for p in page_num_range:
                #更新请求参数
                query_dic.update({'page_num': str(p),
                                  'bounds': str(leftBottomCoordi[1]) + ',' +
str(leftBottomCoordi[0]) + ',' +
                                  str(rightTopCoordi[1]) +
                                  ',' + str(rightTopCoordi[0]),
                                  'output': 'json',
                                  })
                url = urlRoot+urllib.parse.urlencode(query_dic)
                data = urllib.request.urlopen(url)
                responseOfLoad = json.loads(data.read())
                # print(url,responseOfLoad.get("message"))
                if responseOfLoad.get("message") == 'ok':
                    results = responseOfLoad.get("results")
                    for row in range(len(results)):
```

```
                        subData = results[row]
                        baidu_coordinateSystem = [subData.get('location').get(
                            'lng'), subData.get('location').get('lat')]  #获取百
度坐标系
                        Mars_coordinateSystem = cc.bd09togcj02(
                            baidu_coordinateSystem[0], baidu_coordinateSystem[1])
#百度坐标系 --> 火星坐标系
                        WGS84_coordinateSystem = cc.gcj02towgs84(
                            Mars_coordinateSystem[0], Mars_coordinateSystem[1])
#火星坐标系 -->WGS84

                        #更新坐标
                        subData['location']['lat'] = WGS84_coordinateSystem[1]
                        subData['detail_info']['lat'] = WGS84_coordinateSystem[1]
                        subData['location']['lng'] = WGS84_coordinateSystem[0]
                        subData['detail_info']['lng'] = WGS84_coordinateSystem[0]
                        if csv_writer:
                            csv_writer.writerow([subData])  #逐行写入 .csv 文件
                        jsonDS.append(subData)
            num += 1
            print("No."+str(num)+" was written to the .csv file.")
    if poi_json:
        json.dump(jsonDS, poi_json)
        poi_json.write('\n')
        poi_json.close()
    if poi_csv:
        poi_csv.close()
    print("The download is complete.")
baiduPOI_dataCrawler(query_dic, bound_coordinate, partition,
                     page_num_range, poi_fn_list=[poi_fn_csv, poi_fn_json])
--------------------------------------------------------------------------------
Start downloading data...
No.1 was written to the .csv file.
No.2 was written to the .csv file.
...
No.15 was written to the .csv file.
No.16 was written to the .csv file.
The download is complete.
```

- 圆形区域检索（百度地图开放平台没有限制该方法）

可设置圆心和半径，检索圆形区域内的地点信息。官方给出的检索方式为：https://api.map.baidu.com/place/v2/search?query=银行&location=39.915,116.404&radius=2000&output=xml&ak=您的密钥 //GET请求

```
def baiduPOI_dataCrawler_circle(query_dic, poi_save_path, page_num_range):
    '''
    function - 百度地图开放平台 POI 数据检索--圆形区域检索方式

    Params:
        query_dic - 请求参数配置字典，详细参考上文或者百度服务文档；dict
        poi_save_path - 存储文件路径；string
        page_num_range - 配置页数范围；range()

    Returns:
        None
    '''
    import coordinate_transformation as cc
    import urllib
    import json
    import csv
    import os
    import pathlib
    from tqdm import tqdm

    urlRoot = 'http://api.map.baidu.com/place/v2/search?'  #数据下载网址，查询百度地
图服务文档
    poi_json = open(poi_save_path, 'w', encoding='utf-8')
    jsonDS = []  #存储读取的数据，用于 .json 格式数据的保存
    for p in tqdm(page_num_range):
        #更新请求参数
        query_dic.update({'page_num': str(p)})
        url = urlRoot+urllib.parse.urlencode(query_dic)
        data = urllib.request.urlopen(url)
```

```
        responseOfLoad = json.loads(data.read())
        if responseOfLoad.get("message") == 'ok':
            results = responseOfLoad.get("results")
            for row in range(len(results)):
                subData = results[row]
                baidu_coordinateSystem = [subData.get('location').get(
                    'lng'), subData.get('location').get('lat')]  #获取百度坐标系
                Mars_coordinateSystem = cc.bd09togcj02(
                    baidu_coordinateSystem[0], baidu_coordinateSystem[1]) #百度坐
标系 --> 火星坐标系
                WGS84_coordinateSystem = cc.gcj02towgs84(
                    Mars_coordinateSystem[0], Mars_coordinateSystem[1])  #火星坐
标系 -->WGS84

                #更新坐标
                subData['location']['lat'] = WGS84_coordinateSystem[1]
                subData['detail_info']['lat'] = WGS84_coordinateSystem[1]
                subData['location']['lng'] = WGS84_coordinateSystem[0]
                subData['detail_info']['lng'] = WGS84_coordinateSystem[0]
                jsonDS.append(subData)
    if poi_json:
        json.dump(jsonDS, poi_json)
        poi_json.write('\n')
        poi_json.close()
    print("The download is complete.")
page_num_range = range(20)
query_dic = {
    'location': '34.265708,108.953431',
    'radius': 1000,
    'query': '旅游景点',
    'page_size': '20',
    'scope': 2,
    'output': 'json',
    'ak': 'YuN8HxzYhGNfNLGX0FVo3NU3NOrgSNdF'
}
poi_save_path = './data/poi_circle.json'
baiduPOI_dataCrawler_circle(query_dic, poi_save_path, page_num_range)
-------------------------------------------------------------------------------
100%|██████████| 20/20 [00:05<00:00,  3.82it/s]
The download is complete.
-------------------------------------------------------------------------------
import json
with open(poi_save_path) as f:
    poi_circle = json.load(f)
print(poi_circle[0])
-------------------------------------------------------------------------------
{'name': '西安城墙', 'location': {'lat': 34.25321785828024, 'lng':
108.94356940023889}, 'address': '陕西省西安市碑林区南大街', 'province': '陕西省',
'city': '西安市', 'area': '碑林区', 'street_id': '6e7e1320b113d8e3bea82230',
'telephone': '(029)87272792', 'detail': 1, 'uid': '6e7e1320b113d8e3bea82230',
'detail_info': {'distance': 874, 'tag': '旅游景点;文物古迹', 'navi_location':
{'lng': 108.95260787133, 'lat': 34.259100951615}, 'type': 'scope', 'detail_url':
'http://api.map.baidu.com/place/e1320b113d8e3bea82230&output=html&source=place
api_v2', 'overall_rating': '4.3', 'comment_num': '200', 'shop_hours': '08:00-
22:00', 'children': [], 'lat': 34.25321785828024, 'lng': 108.94356940023889}}
    #行政区划区域检索、地点详情检索等方法可以查看服务文档。
```

3.1.1.2 将CSV格式的POI数据转换为Pandas的DataFrame格式数据

读取已经保存的poi_csv.csv文件，因为在文件保存时，使用的是csv和json库，因此读取时仍旧使用对应的库。经常使用的Pandas库中也有`read_csv()`和`read_json()`等方法。在将数据存储为CSV格式文件，因为保存的数据格式可能存在差异，因此Pandas读取CSV或者JSON文件最好是其自身所存储的文件，数据格式能够对应。如果直接读取上述保存的POI的.csv文件，则会出现错误。只有知道数据格式，才能够有目的地提取数据，读取每一行的数据格式如下：

```
{
    "name": "昆明池遗址",
    "location": {
        "lat": 34.210991,
        "lng": 108.779778
    },
```

```
    "address": " 西安市长安区昆明池七夕公园内 ",
    "province": " 陕西省 ",
    "city": " 西安市 ",
    "area": " 长安区 ",
    "detail": 1,
    "uid": "c7332cd7fbcc0d82ebe582d9",
    "detail_info": {
        "tag": " 旅游景点 ; 景点 ",
        "navi_location": {
            "lng": 108.7812626866,
            "lat": 34.217484892966
        },
        "type": "scope",
        "detail_url": "http://api.map.baidu.com/place/cc0d82ebe582d9&amp;output=
html&amp;source=placeapi_v2",
        "overall_rating": "4.3",
        "comment_num": "77",
        "children": []
    }
}
```

读取.csv数据之后，直接使用CSV格式数据来提取和分析数据并不方便。最为常用的数据格式是NumPy库提供的数组（array）和Pandas提供的DataFrame及Series。其中，NumPy的数据组织形式更倾向于科学计算，为数阵形式，每一数组为同一数据类型；而Pandas的DataFrame与地理信息数据中的属性表类似，其列（column）可以理解为属性字段（field），每一列的数据类型相同，因此一个DataFrame可以包含多种数据类型。就POI的CSV格式数据，将其转换为DataFrame的数据格式是最好的。进一步而言，在城市空间数据分析方法中，更多的数据是基于地理空间位置，因此对于具有地理属性的数据，最好能够以地理信息系统常用的数据格式来处理。在处理为DataFrame数据格式之后，则进一步应用GeoPandas等地理信息库来处理地理空间信息数据。

在数据格式转换时可能会出现错误，例如原始数据可能存在错误的数据格式：

```
{'name': ' 励进海升酒店 - 多功能厅 ', 'location': {'lat': 34.218525, 'lng':
108.891524}, 'address': ' 西安市高新区沣惠南路 34 号励进海升酒店 4 层 ', 'province': ' 陕西
省 ', 'city': ' 西安市 ', 'a{'name': ' 红蚂蚁少儿创意美术馆 '", " 'location': {'lat':
34.306666", " 'lng': 108.822465}", " 'address': ' 陕西省西安市未央区后围寨启航佳
苑 B 区 3 层商铺 '", " 'province': ' 陕西省 '", " 'city': ' 西安市 '", " 'area': ' 未央
区 '", " 'telephone': '18209227178", "15229372642'", " 'detail': 1", " 'uid':
'e3fd730bb528b40015c9050c'", " 'detail_info': {'tag': ' 文化传媒 ; 美术馆 '", " 'type':
'scope'", " 'detail_url': 'http://api.map.baidu.com/place/b40015c9050c&amp;outpu
t=html&amp;source=placeapi_v2'", " 'overall_rating': '0.0'", ' \'children\': []}}
```

其中，'a{'name': '红蚂蚁少儿创意美术馆'",为多余的部分，并不符合任何格式语法，因此在数据处理时，需要对此做出反应。通常使用try/except语句，最好返回可以检索的信息，加以处理，避免数据损失。

```
def csv2df(poi_fn_csv):
    '''
    function- 转换 CSV 格式的 POI 数据为 pandas 的 DataFrame

    Params:
        poi_fn_csv - 存储有 POI 数据的 CSV 格式文件路径

    Returns:
        poi_df - DataFrame(pandas)
    '''
    import pandas as pd
  # benedict 库是 dict 的子类，支持键列表（keylist）/ 键路径（keypath）。用该库的 flatten 方法展平嵌套字典，
用于 DataFrame 数据结构
    from benedict import benedict
    import csv
    n = 0
    with open(poi_fn_csv, newline='', encoding='utf-8') as csvfile:
        poi_reader = csv.reader(csvfile)
        poi_dict = {}
        poiExceptions_dict = {}
        for row in poi_reader:
            if row:
                try:
                    # 用 eval 方法，将字符串字典" {}" 转换为字典 {}
```

```
                row_benedict = benedict(eval(row[0]))
                flatten_dict = row_benedict.flatten(
                    separator='_')  #展平嵌套字典
                poi_dict[n] = flatten_dict
            except:
                print("incorrect format of data_row
number:%s" % n)
                poiExceptions_dict[n] = row
        n += 1
        # if n==5:break # 因为循环次数比较多，在调试代码时，可以设置
停止的条件，节省时间与方便数据查看
    poi_df = pd.concat([pd.DataFrame(poi_dict[d_k].values(),
index=poi_dict[d_k].keys(
    ), columns=[d_k]).T for d_k in poi_dict.keys()],
sort=True, axis=0)
    print("_"*50)
    for col in poi_df.columns:
        try:
            poi_df[col] = pd.to_numeric(poi_df[col])
        except:
            print("%s data type is not converted..." % (col))
    print("_"*50)
    print(".csv to DataFrame completed!")
    return poi_df
```

几乎Python的所有数据类型，列表、字典、集合和类等都可以用pickle来序列化存储，Pandas提供了写入`pandas.DataFrame.to_pickle`，读取`pandas.DataFrame.read_pickle`的方法，因此为了避免每次将CSV转换为DataFrame，可以将Pandas类型的文件按Pandas提供的方法加以存储，或者写入数据库。

```
poi_fn_csv = './data/poi_csv.csv'
poi_df = csv2df(poi_fn_csv)
poi_df.to_pickle('./data/poi_df.pkl')
print("_"*50)
print(poi_df.columns)
poi_df.head()
-----------------------------------------------------------
incorrect format of data_row number:4
incorrect format of data_row number:28
incorrect format of data_row number:36
-----------------------------------------------------------
address data type is not converted...
...
.csv to DataFrame completed!
-----------------------------------------------------------
Index(['address', 'area', 'city', 'detail',...
       'street_id', 'telephone', 'uid'],
      dtype='object')
```

输出POI数据信息（表3.1-2）。

表3.1-2 百度地图POI数据信息

	address	area	city	...	detail_info_tag	detail_info_type	location_lat	location_lng	name	province	...	telephone	uid
0	西安市长安区昆明池七夕公园内	长安区	西安市	...	旅游景点;景点	scope	34.206767	108.768459	昆明池遗址	陕西省	...	NaN	c7332cd7fbcc0d82ebe582d9
...													
10	陕西省西安市雁塔区西宝疏导线关中驾校南200米	雁塔区	西安市	...	旅游景点;其他	scope	34.224172	108.815592	大明怀远将军园	陕西省	...	NaN	8ed2e7b406ac382b4bcc4a18

```
5 rows × 29 columns
```

3.1.1.3 将数据格式为DataFrame的POI数据转换为GeoPandas地理空间数据GeoDataFrame

可以使用GeoPandas库将Pandas的DataFrame和Series数据转换为具有地理意义的GeoDataFrame和GeoSeries地理数据。GeoPandas基于Pandas库，因此在数据结构上二者之间最大的区别是GeoPandas有一个列（column）名为`'geometry'`，用来存储几何数据，例如点数据`POINT (163.85316 -17.31631)`，单个多边形`POLYGON ((33.90371 -0.95000, 31.86617 -1.02736...))`，多边形集合`MULTIPOLYGON (((120.83390 12.70450, 120.32344...)))`等。其中，对几何数据的表达是使用shapely库实现。大部分SHP地理信息矢量数据，在Python语言中，通常使用该库来建立几何对象。同时，GeoDataFrame对象的建立，需要配置坐标系统。这是地理信

息数据显著的一个标志，对于坐标系统可以查看Spatial Reference[①]。

GeoDataFrame可以直接通过.plot()方法显示地理空间信息数据（图3.1–1）。

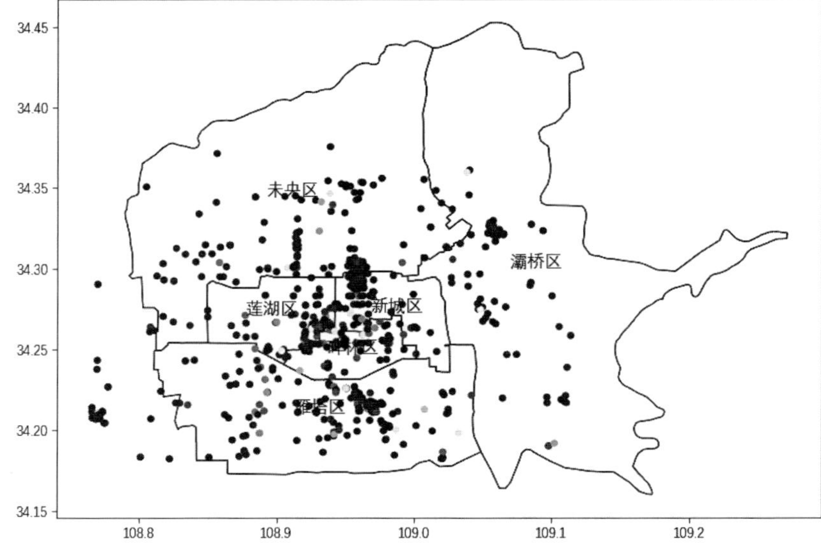

图3.1–1　POI位置点空间分布

```
import pandas as pd
import geopandas as gpd
from shapely.geometry import Point
import matplotlib.pyplot as plt
from pylab import mpl
poi_df = pd.read_pickle('./data/poi_df.pkl')  # 读取已经保存的 .pkl(pickle) 数据格式的 POI
poi_2gdf = poi_df.copy(deep=True)
poi_2gdf['geometry'] = poi_2gdf.apply(lambda row:Point(row.location_lng,row.
location_lat),axis=1)
epsg_wgs84=4326  # 配置坐标系统，参考：https://spatialreference.org/
poi_gdf = gpd.GeoDataFrame(poi_2gdf,crs=epsg_wgs84)
xian_region_gdf = gpd.read_file('./data/xian_region/xian_region.shp')
mpl.rcParams['font.sans-serif']=['DengXian']  # 解决中文字符乱码问题
fig, ax = plt.subplots(figsize=(10,15))
xian_region_plot = xian_region_gdf.plot(ax=ax,color='white', edgecolor='black')
xian_region_gdf.apply(lambda row: xian_region_plot.annotate(text=row.NAME,
xy=row.geometry.centroid.coords[0], ha='center',fontsize=13),axis=1) # 标注
poi_gdf.plot(ax=ax,column='detail_info_comment_num',markersize=20)
plt.show()
```

- 使用Plotly库建立地图

GeoPandas库提供的地图显示类型功能有限，但却便捷，通常用于数据的查看。当需要打印的地图具有一定质量，表达更多信息，甚至能够交互操作时，可以使用Plotly[②]图表库实现（通常使用Plotly Python[③]）。下述地图打印结果的底图使用了mapbox[④]提供的地图数据。要使用其功能，首先需要注册，并获取访问许可（access token）。

使用Plotly库建立地图（图3.1–2），不需要将DataFrame转换为GeoDataFrame。

```
import plotly.express as px
poi_gdf.detail_info_price = poi_gdf.detail_info_price.fillna(0)  # Pandas 库的方法同样适用于 GeoPandas 库，例如对 `nan` 位置填充指定数值
mapbox_token = 'pk.mFvIiwiYSI6ImNrYjB3N2NyMzBlMG8yc254dTRzNnMyeHMifQ.
QT7MdjQKs9Y6OtaJaJAn0A'
px.set_mapbox_access_token(mapbox_token)
fig=px.scatter_mapbox(poi_gdf,
                      lat=poi_gdf.location_lat,
                      lon=poi_gdf.location_lng,
```

① Spatial Reference，空间参考系统（spatial reference system，SRS)或坐标参考系统（coordinate reference system，CRS）是用于精确测量地球表面位置，用坐标表示的一个体系。是坐标系和解析几何的抽象数学在地理空间的应用。例如，一个CRS坐标表示为：Universal Transverse Mercator WGS 84 Zone 16，包括选择地球椭圆体、水平基准、地图投影，原点和计量单位等（https://spatialreference.org/）。

② Plotly，是一个通过数据应用实现数据驱动决策的实践者，构建有支持多种编程语言的图表库（https://plotly.com/）。

③ Plotly Python，Plotly的Python图形库可以制作互动的、具有出版质量的图表。例如制作线图（line plots）、散点图（scatter plots）、面积图（area charts）、条形图（bar charts）、误差条（error bars）、箱形图（box plots）、柱状图（histograms）、热图（heatmaps）、子图（subplots）、多轴图（multiple-axes）、极地图（polar charts）和气泡图（bubble charts）等（https://plotly.com/python/）。

④ mapbox，用于移动和网络应用的位置数据平台，提供定制在线地图的供应商（https://www.mapbox.com/）。

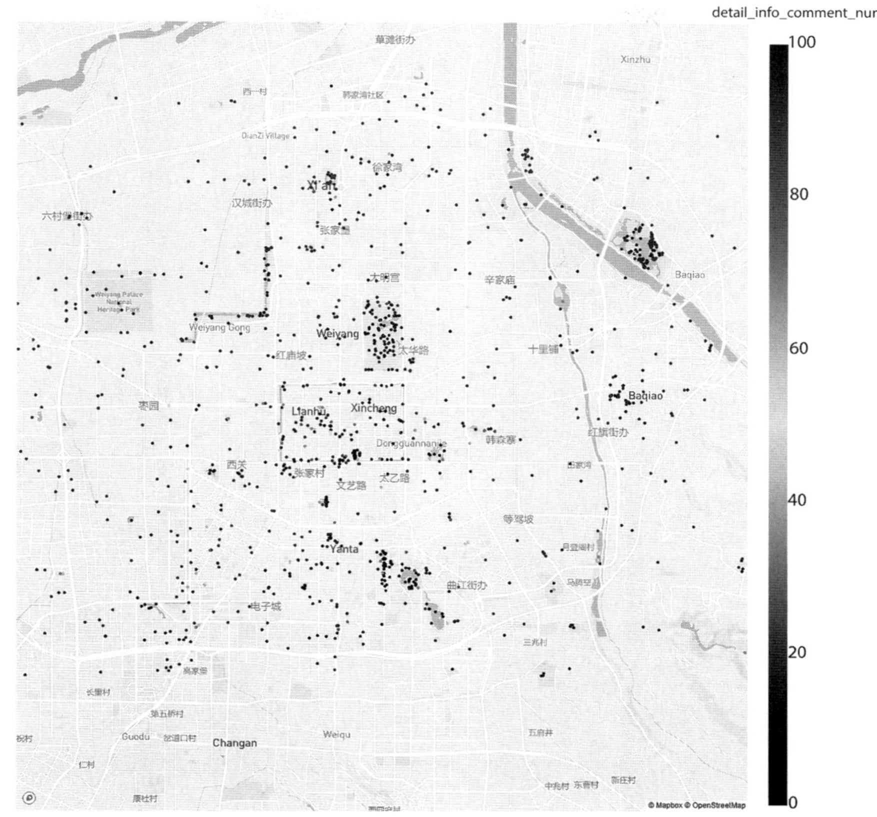

图3.1-2　用Plotly库打印POI 地图

```
                        color="detail_info_comment_num",
                        color_continuous_scale=px.colors.cyclical.IceFire,
                        size_max=15,
                        zoom=10)  #亦可以选择列，通过 size="" 配置增加显示信息
fig.show()
fig.write_html('./graph/POI_singlClassi.html') #保存地图
```

3.1.2　多个分类POI数据检索

3.1.2.1　多个分类POI检索

前文定义了两个函数，百度地图开放平台POI数据检索baiduPOI_dataCrawler()和转换CSV格式的POI数据为Pandas的DataFrame格式数据csv2df()。为了方便应用建立的函数工具，使用Anaconda的Spyder创建一个新的文件（模块）为util_A.py，将上述两个函数置于其中，同时包括函数所使用的库。对于所包括的库，为方便日后函数迁移和打包自定义Python库发布，明确每个函数所调用的库，将对应调用库的语句分别置于各个函数内部。util_A.py与调用该工具的代码文件置于同一文件夹下。调入语句如下：

```
import util_A
```

根据百度地图一级行业分类，建立映射字典，用于多个分类POI的数据检索。可以根据数据分析的需求选择行业分类，因为在进一步分析中不需要的分类包括出入口、自然地物、行政地标和门址等，因此在映射字典中未包含上述分类。

```
poi_classificationName={
        " 美食 ":"delicacy",
        " 酒店 ":"hotel",
        " 购物 ":"shopping",
        " 生活服务 ":"lifeService",
        " 丽人 ":"beauty",
        " 旅游景点 ":"spot",
        " 休闲娱乐 ":"entertainment",
        " 运动健身 ":"sports",
        " 教育培训 ":"education",
        " 文化传媒 ":"media",
        " 医疗 ":"medicalTreatment",
        " 汽车服务 ":"carService",
        " 交通设施 ":"trafficFacilities",
        " 金融 ":"finance",
        " 房地产 ":"realEstate",
        " 公司企业 ":"corporation",
        " 政府机构 ":"government"
        }
```

配置基本参数。上文配置用到query_dic。此次批量下载将所有参数在循环函数外以字典形式单独给出，方便调用。而query_dic字典参数在批量下载函数内配置。

```
query_dic={
    'query':' 旅游景点 ',
    'page_size':'20',
    'scope':2,
    'ak':'2Zh7jNunzIzKoWx59ucjHLlZ63oI9St0',
}
poi_config_para={
    'data_path':'./data/poi_batchCrawler/',  # 配置数据存储位置
    'bound_.186027],'rightTop':[109.129275,34.382171]},  # 使用百度地图坐标拾取系统 
http://api.map.baidu.com/lbsapi/getpoint/index.html
    'page_num_range':range(20),
    'partition':6,
    'page_size':'20',
    'scope':2,
    'ak':'2Zh7jNunzIzKoWx59ucjHLlZ63oI9St0', # 需要改为读者自行申请的开发者访问密钥
}
```

建立批量下载的循环函数，依据给出的poi_classificationName字典键值逐次调用单个分类POI检索函数baiduPOI_dataCrawler()下载POI数据。检索过程中，可以将每一次小批量下载数据存储在同一变量下，待全部下载完后一次性存储。但是，这种一次性存储的方式并不推荐：其一，网络有时并不稳定，可能造成下载中断，那么已下载的数据未存储，造成数据丢失，并需要重新下载；其二，有时数据量很大，如果都存储在一个变量下，可能造成内存溢出。

```
def baiduPOI_batchCrawler(poi_config_para):
    '''
    function - 百度地图开放平台 POI 数据批量检索，
               需要调用单个分类 POI 检索函数 baiduPOI_dataCrawler(query_dic,bound_
coordinate,partition,page_num_range,poi_fn_list=False)
    Params:
        poi_config_para - 参数配置，包含：
            'data_path'（配置数据存储位置），
            'bound_coordinate'（矩形区域检索下、右上经纬度坐标），
            'page_num_range'（配置页数范围），
            'partition'（检索区域切分次数），
            'page_size'（单次召回 POI 数量），
            'scope'（检索结果详细程度），
            'ak'（开发者的访问密钥）

    Returns:
        None
    '''
    import os
    import util_A
    for idx, (poi_ClassiName, poi_classMapping) in enumerate(poi_
classificationName.items()):
        print(str(idx+16)+"_"+poi_ClassiName)
        poi_subFileName = "poi_"+str(idx+16)+"_"+poi_classMapping
        data_path = poi_config_para['data_path']
```

```
        poi_fn_csv = os.path.join(data_path, poi_subFileName+'.csv')
        poi_fn_json = os.path.join(data_path, poi_subFileName+'.json')
        query_dic = {
            'query': poi_ClassiName,
            'page_size': poi_config_para['page_size'],
            'scope': poi_config_para['scope'],
            'ak': poi_config_para['ak']
        }
        bound_coordinate = poi_config_para['bound_coordinate']
        partition = poi_config_para['partition']
        page_num_range = poi_config_para['page_num_range']
        util_A.baiduPOI_dataCrawler(
            query_dic, bound_coordinate, partition, page_num_range, poi_fn_
list=[poi_fn_csv, poi_fn_json])
baiduPOI_batchCrawler(poi_config_para)
--------------------------------------------------------------------------------
16_美食
Start downloading data...
No.1 was written to the .csv file.
...
No.35 was written to the .csv file.
No.36 was written to the .csv file.
The download is complete.
```

3.1.2.2 批量转换CSV格式数据为GeoDataFrame格式

在单个分类部分定义了CSV格式数据到GeoDataFrame数据的转换方法csv2df()。基于已有代码，该部分将达到两个目的：一是定义单独函数实现CSV格式数据批量转换为GeoDataFame格式数据并存储为.pkl文件；二是批量读取存储为.pkl文件的GeoDataFrame格式数据，并根据需要提取信息存储在单一变量下，再存储为.pkl文件。当读取所有数据于单一变量时，内存需要满足要求。如果有内存溢出，则需考虑是否根据内存情况调整每次读取的数据量。

- **定义提取文件夹下所有文件路径的函数**

因为要批量下载POI数据为多个.csv文件及.json文件，因此在批量处理这些数据时，第一件事是要提取所有文件的路径。定义返回所有指定后缀名文件路径的函数为常用的函数之一，在之后的很多实验中，都需要调用该函数，因此可以将其保存在util_misc.py文件中，方便日后调用。文件夹下通常包括子文件夹，需要使用os.walk()方法遍历目录，给出条件语句判断是否存在子文件夹。如果存在，则需要返回该文件夹下的文件路径。

```
def filePath_extraction(dirpath, fileType):
    '''
    funciton  - 以所在文件夹路径为键，值为包含该文件夹下所有文件名的列表。文件类型可以自行定义
    Params:
        dirpath - 根目录，存储所有待读取的文件；string
        fileType - 待读取文件的类型；list(string)
    Returns:
        filePath_Info - 文件路径字典，文件夹路径为键，文件夹下的文件名列表为值；dict
    '''
    import os
    filePath_Info = {}
    i = 0
    # os.walk() 遍历目录，使用 help(os.walk) 查看返回值解释
    for dirpath, dirNames, fileNames in os.walk(dirpath):
        i += 1
        if fileNames:  # 仅当文件夹中有文件时才提取
            tempList = [f for f in fileNames if f.split('.')[-1] in fileType]
            if tempList:  # 剔除文件名列表为空的情况，即文件夹下存在不为指定文件类型的文件时，
上一步列表会返回空列表 []
                filePath_Info.setdefault(dirpath, tempList)
    return filePath_Info
dirpath = './data/poi_batchCrawler/'
fileType = ["csv"]
poi_paths = filePath_extraction(dirpath, fileType)
print(poi_paths)
--------------------------------------------------------------------------------
{'./data/poi_batchCrawler/': ['poi_0_delicacy.csv', 'poi_10_medicalTreatment.
csv', 'poi_11_carService.csv', 'poi_12_trafficFacilities.csv', 'poi_13_finance.
csv', 'poi_14_realEstate.csv', 'poi_15_corporation.csv', 'poi_16_government.
csv', 'poi_1_hotel.csv', 'poi_2_shopping.csv', 'poi_3_lifeService.csv', 'poi_4_
```

```
beauty.csv', 'poi_5_spot.csv', 'poi_6_entertainment.csv', 'poi_7_sports.csv',
'poi_8_education.csv', 'poi_9_media.csv']}
```

- CSV格式POI数据批量转换为GeoDataFrame格式数据

```
Index(['address', 'area', 'city', 'detail', 'detail_info_checkin_num','detail_
info_children', 'detail_info_comment_num',
       'detail_info_detail_url', 'detail_info_facility_rating','detail_info_
favorite_num', 'detail_info_hygiene_rating',
       'detail_info_image_num', 'detail_info_indoor_floor','detail_info_navi_
location_lat', 'detail_info_navi_location_lng',
       'detail_info_overall_rating', 'detail_info_price','detail_info_service_
rating', 'detail_info_tag', 'detail_info_type',
       'location_lat', 'location_lng', 'name', 'province', 'street_
id','telephone', 'uid'],dtype='object')
```

上述为POI数据字段，可以用来确定提取的字段名。除了增加循环语句循环POI的.csv文件，逐一转换为Pandas的DataFrame格式，再进一步转换为GeoPandas的GeoDataFrame格式之外，其他的条件同前文所述。用GeoDataFrame.plot()方法初步查看地理空间信息数据。

在文件保存部分，可以有多种保存格式，GeoPandas提供了Shapefile、GeoJSON和GeoPackage三种方式；也可以使用pickle库保存数据，或存入数据库。用GeoPandas提供的保存格式再读取后不再包含多重索引，而pickle格式则保持。

转换为SHP格式文件时，在QGIS等桌面GIS平台下打开时会出现两个问题：一是，如果列（字段）名称过长，转化为属性表的字段名会被字段压缩修改，往往不能反映字段的意义，因此需要置换列名称；二是，用POI一级行业分类名作为index时，列中并不包含该字段，转化为SHP文件时也不包含该字段，因此需要将index转换为列，再存储为.shp文件。

```
def poi_csv2GeoDF_batch(poi_paths, fields_extraction, save_path):
    '''
    funciton - CSV 格式 POI 数据批量转换为 GeoDataFrame 格式数据，需要调用转换 CSV 格式的 POI
数据为 Pandas 的 DataFrame 函数 csv2df(poi_fn_csv)

    Params:
        poi_paths - 文件夹路径为键，值为包含该文件夹下所有文件名列表的字典；dict
        fields_extraction - 配置需要提取的字段；list(string)
        save_path - 存储数据格式及保存路径的字典；string

    Returns:
        poisInAll_gdf - 提取给定字段的 POI 数据；GeoDataFrame（GeoPandas）
    '''
    import os
    import pathlib
    import util_A
    import pandas as pd
    import geopandas as gpd
    from shapely.geometry import Point
    from pyproj import CRS
    poi_df_dic = {}
    i = 0
    for key in poi_paths:
        for val in poi_paths[key]:
            poi_csvPath = os.path.join(key, val)
            poi_df = util_A.csv2df(poi_csvPath)
            # 注释掉 csv2df() 函数内部的 print("%s data type is not converted..." %(col)) 语句，以 pass 替代，
减少提示内容，避免干扰
            print(val)
            poi_df_path = pathlib.Path(val)
            poi_df_dic[poi_df_path.stem] = poi_df
            i += 1
    poi_df_concat = pd.concat(
        poi_df_dic.values(), keys=poi_df_dic.keys(), sort=True)
    # print(poi_df_concat.loc[['poi_0_delicacy'],:]) # 提取 index 为 'poi_0_delicacy' 的行，验证结果
    poi_fieldsExtraction = poi_df_concat.loc[:, fields_extraction]
    poi_geoDF = poi_fieldsExtraction.copy(deep=True)
    poi_geoDF['geometry'] = poi_geoDF.apply(
        lambda row: Point(row.location_lng, row.location_lat), axis=1)
    crs_4326 = CRS('epsg:4326') # 配置坐标系统，参考：https://spatialreference.org/
    poisInAll_gdf = gpd.GeoDataFrame(poi_geoDF, crs=crs_4326)
    poisInAll_gdf.to_pickle(save_path['pkl'])
    poisInAll_gdf.to_file(save_path['geojson'],
```

```
                                    driver='GeoJSON', encoding='utf-8')
    # 不指定 level 参数，例如 Level=0，会把多重索引中的所有索引转换为列
    poisInAll_gdf2shp = poisInAll_gdf.reset_index()
    poisInAll_gdf2shp.rename(columns={
        'location_lat': 'lat', 'location_lng': 'lng',
        'detail_info_tag': 'tag', 'detail_info_overall_rating': 'rating',
'detail_info_price': 'price'}, inplace=True)
    poisInAll_gdf2shp.to_file(save_path['shp'], encoding='utf-8')
    return poisInAll_gdf
fields_extraction = ['name', 'location_lat', 'location_lng',
                     'detail_info_tag', 'detail_info_overall_rating',
                     'detail_info_price']  # 配置需要提取的字段，即列（columns）
# 分别存储为 GeoJSON、Shapefile 和 pickle 三种数据格式
save_path = {'geojson': './data/poisInAll/poisInAll_gdf.geojson',
             'shp': './data/poisInAll/poisInAll_gdf.shp',
             'pkl': './data/poisInAll/poisInAll_gdf.pkl'}
poisInAll_gdf = poi_csv2GeoDF_batch(poi_paths, fields_extraction, save_path)
--------------------------------------------------------------------------------
--------------------------------------------------------------------------------
.csv to DataFrame completed!
poi_0_delicacy.csv
--------------------------------------------------------------------------------
...
--------------------------------------------------------------------------------
.csv to DataFrame completed!
poi_9_media.csv
--------------------------------------------------------------------------------
import matplotlib.pyplot as plt
from mpl_toolkits.axes_grid1 import make_axes_locatable
import geopandas as gpd
from pylab import mpl

mpl.rcParams['font.sans-serif'] = ['DengXian']  # 解决中文字符乱码问题

xian_region_gdf = gpd.read_file('./data/xian_region/xian_region.shp')
fig, ax = plt.subplots(figsize=(15, 20))
xian_region_plot = xian_region_gdf.plot(
    ax=ax, color='white', edgecolor='black')
# 标注
xian_region_gdf.apply(lambda row: xian_region_plot.annotate(text=row.NAME,
xy=row.geometry.centroid.coords[0],
ha='center',
fontsize=13),
                      axis=1)
divider = make_axes_locatable(ax)
cax = divider.append_axes("right", size="5%", pad=0.1)
poisInAll_gdf.loc[['poi_0_delicacy'], :].plot(ax=ax,
column='detail_info_overall_rating',
markersize=10,
cmap='cividis',
legend=True,
cax=cax)  # 提取 index 为'poi_0_delicacy'的行查看结果
plt.show()
```

输出图3.1-3为GeoPandas直接打印的地图。

在 QGIS 开源桌面 GIS 平台下打开保存的 .shp 数据（图 3.1-4）。虽然所有工作基本都是在 Python 中完成，但是有些工作与其他平台相互联系，需要以这些平台为辅助。在辅助平台选择上，尽可能使用具有广泛应用的开源软件。在地理信息系统（Geographic Information Systems，GIS）中，主要使用的集成式平台有 QGIS 和 ArcGIS。

3.1.2.3 使用Plotly库建立地图

用颜色表示POI一级分类，用大小表示rating字段。

```
import geopandas as gpd
import plotly.express as px

poi_gdf=gpd.read_file('./data/poisInAll/poisInAll_gdf.shp')  # 读取存储的 .shp 格式文件
poi_gdf.rating=poi_gdf.rating.fillna(0) # Pandas 库的方法同样适用于 GeoPandas 库，例如对
`nan` 位置填充指定数值
mapbox_token='pk.mFvIiwiYSI6ImNrYjB3N2NyMzBlMG8yc254dTRzNnMyeHMifQ.
```

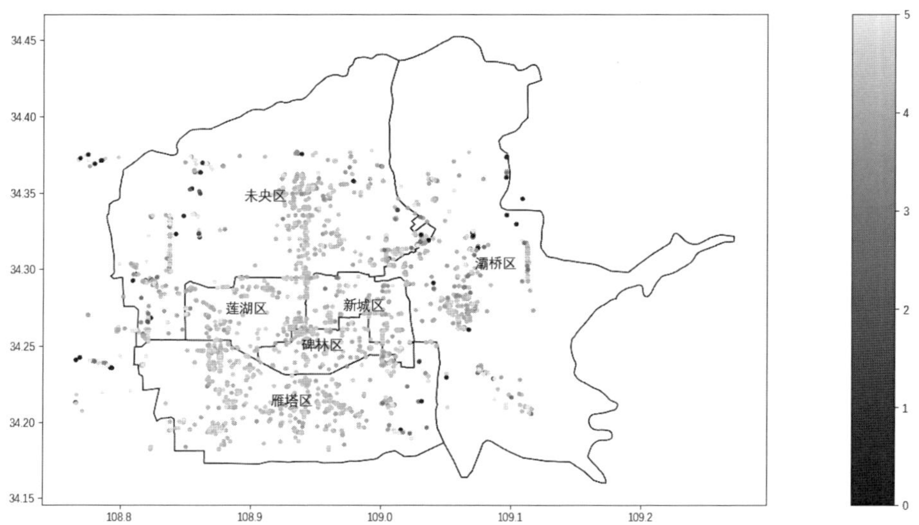

图3.1-3 用 GeoPandas输出POI地图

```
QT7MdjQKs9Y6OtaJaJAn0A'
px.set_mapbox_access_token(mapbox_token)
fig=px.scatter_mapbox(poi_gdf,lat=poi_gdf.lat, lon=poi_gdf.lng,
                      color="level_0",size='rating',
                      color_continuous_scale=px.colors.cyclical.IceFire, size_
max=5, zoom=10) #亦可以选择列，通过 size=" " 配置增加显示信息
fig.show()
```

用Plotly库建立地图（图3.1-5）。

3.1.3 描述性统计图表

3.1.3.1 读取与查看数据

- 读取已经保存的.pkl数据。通过.plot()确认读取的数据是否正常（图3.1-6），或者直接使用poi_gdf.head()查看数据。

```
import pandas as pd
poi_gdf = pd.read_pickle('./data/poisInAll/poisInAll_gdf.pkl')
poi_gdf.plot(marker=".",markersize=5,column='detail_info_overall_
rating',figsize=(10,15)) #只有不设置 columns 参数时，可以使用 color=' green' 参数
print(poi_gdf.columns) #查看列名称
--------------------------------------------------------------------------------
Index(['name', 'location_lat', 'location_lng', 'detail_info_tag','detail_info_
overall_rating', 'detail_info_price', 'geometry'],
      dtype='object')
```

3.1.3.2 用Plotly表格显示DataFrame格式数据

print()是查看数据最为主要的方式，主要用于代码调试。当需要展示数据时，对于DataFrame格式的数据可以使用Plotly转换为表格形式。因为POI数据有万行之多，仅显示每一级行业分类的前两行内容。为方便调用，将该功能定义为一个函数ploly_table()。提取数据时，因为数据格式是多重索引的DataFrame，因此使用pandas.IndexSlice()函数辅助执行多重索引切分。因为用Plotly显示表3.1-3时，如果为多重索引会显示错误，因此需要df.reset_index()重置索引。Plotly也不能够显示'geometry'几何对象，在列提取时需要移除该列。

如果在Spyder解释器中打印Plotly图表，通常要增加如下代码：

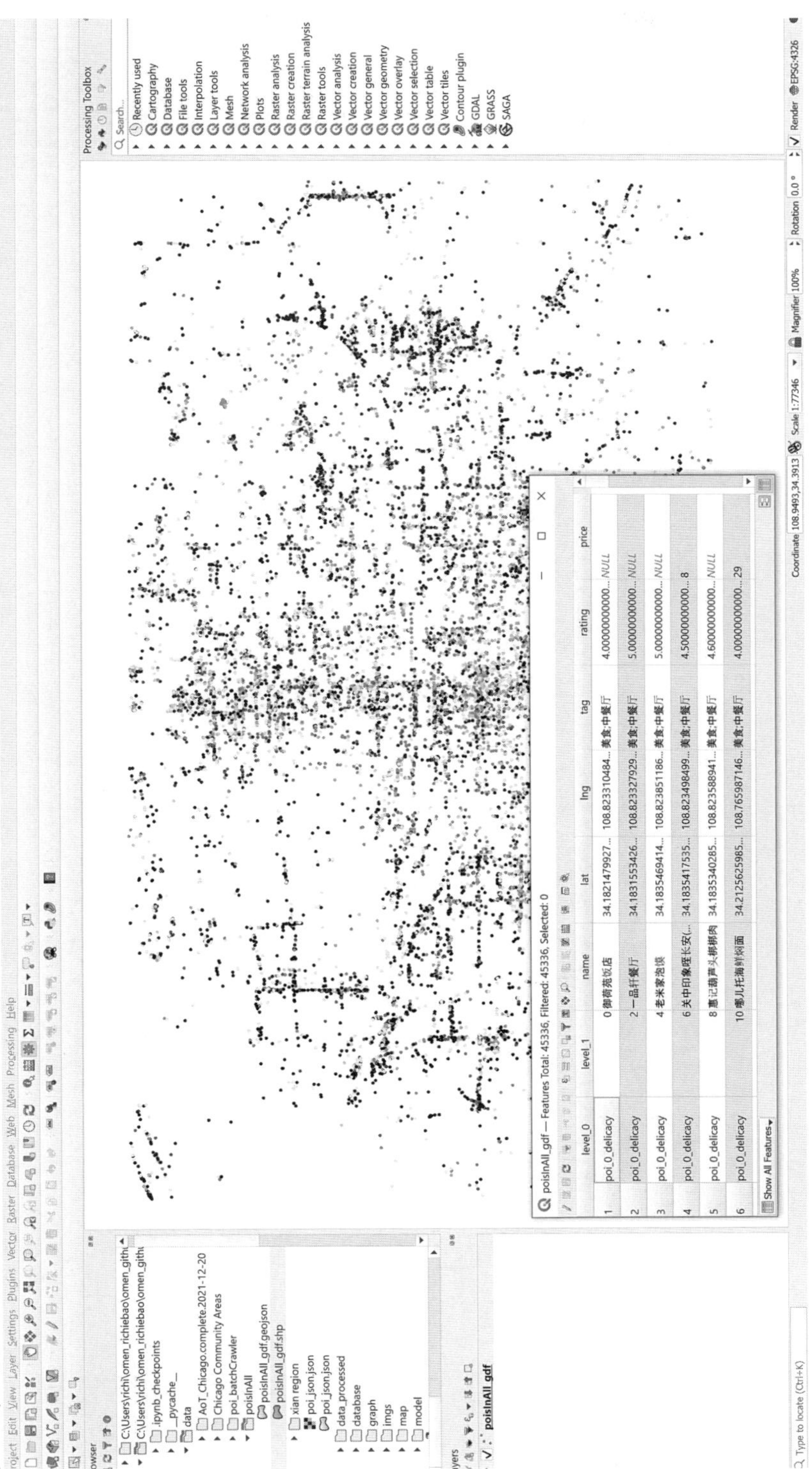

图3.1-4 用QGIS查看POI地图

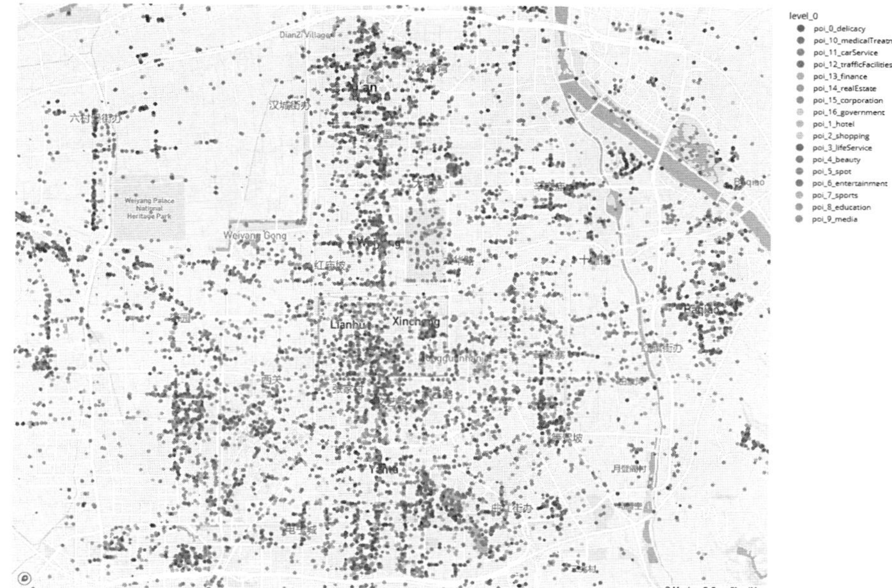

图3.1-5　用Plotly库输出POI地图

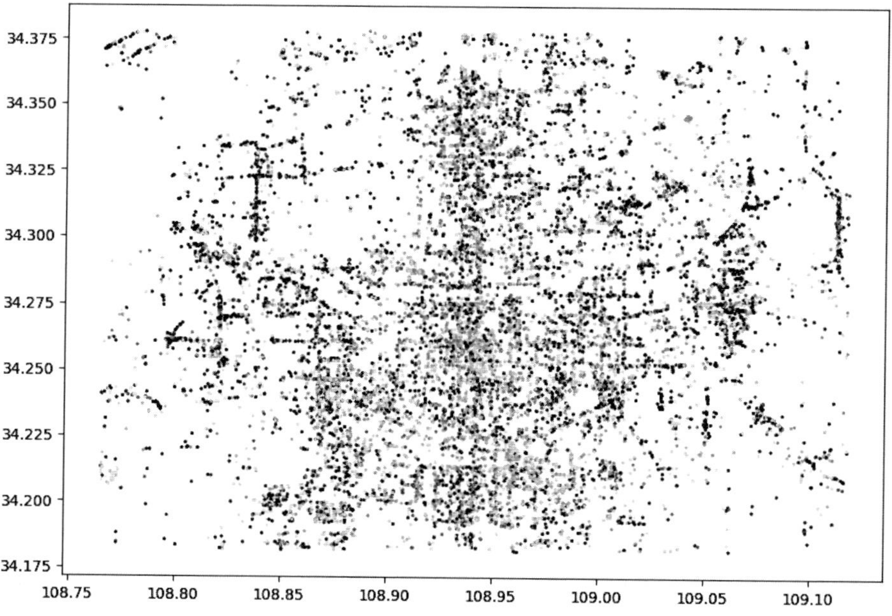

图3.1-6　打印保存后读取的数据

表 3.1-3 用Plotly打印DataFrame格式数据

level_0	name	location_lat	location_lng	detail_info_tag	detail_info_overall_rating	detail_info_price
poi_0_delicacy	御荷苑饭店	34.182147992713084	108.82331048461374	美食;中餐厅	4	null
poi_0_delicacy	一品轩餐厅	34.183155342604536	108.82332792933936	美食;中餐厅	5	null
poi_10_medicalTreatment	书香镇街文化中心(细柳店)	34.18220346751758	108.82198680408702	教育培训;其他	0	null
poi_10_medicalTreatment	镐京骨科	34.2436857774664	108.78120928654668	医疗;诊所	0	null
poi_11_carService	西安大诚行二手车(中心店)	34.23884428449097	108.82408017416553	汽车服务;汽车销售	5	null
poi_11_carService	西安真诚汽车销售服务有限公司	34.24215755245006	108.80386091108363	汽车服务;汽车销售	0	null
poi_12_trafficFacilities	昆明池-七夕公园-停车场	34.212900293344596	108.77087024084808	交通设施;停车场	null	null

```
import plotly.io as pio
pio.renderers.default = 'browser'
def plotly_table(df, column_extraction):
    '''
    funciton - 使用Plotly，以表格形式显示DataFrame格式数据

    Params:
        df - 输入的DataFrame或者GeoDataFrame; [Geo]
DataFrame
        column_extraction - 提取的字段（列）; list(string)

    Returns:
        None
    '''
    import plotly.graph_objects as go
    import pandas as pd
    fig = go.Figure(data=[go.Table(
        header=dict(values=column_extraction,
                    fill_color='paleturquoise',
                    align='left'),
        cells=dict(values=df[column_extraction].values.
T.tolist(),  #values参数值为按列的嵌套列表，因此需要使用参数.T反
转数组
                   fill_color='lavender',
                   align='left'))
    ])
    fig.show()
df = poi_gdf.loc[pd.IndexSlice[:, :2], :]
df = df.reset_index()
column_extraction = ['level_0', 'name', 'location_lat',
'location_lng',
                     'detail_info_tag', 'detail_info_
overall_rating', 'detail_info_price']
plotly_table(df, column_extraction)
```

3.1.3.3 描述性统计

本小节内容主要参考《漫画数据库》《漫画统计学》及维基百科（Wikipedia）。其将枯燥的知识以漫画的形式讲出来，并结合实际的案例由简入繁，使枯燥的学习变得有趣起来。欧姆社学习漫画系列和众多以漫画及图示方式讲解知识的优秀图书都值得推荐。但有利有弊，大部分的漫画图书往往以基础知识为主，深入的研究还是要搜索科学文献和相关论著。同时，漫画形式可以引起读者的兴趣，但是因为穿插故事情节，知识点不易定位，核心的知识内容相对分散，阅读上也要花费更多的时间。因此，想学习一门知识，以哪种形式入手，需要根据个人的情况确定。

描述性统计分析是对调查总体所有变量的有关数据做统计性描述，了解各变量内的观察值集中与分散的情况。

- 表示集中趋势（集中量数）的有平均数、中位数、众数、几何平均数、调和平均数等；
- 表示离散程度（变异量数）的有极差（全距）、平均差、标准差、相对差、四分差等。数据的次数分配情况，往往会呈现正态分布；
- 数据的频数（次数）分配情况，往往会呈现正态分布。为了表示测量数据与正态分布偏离的情况，会使用偏度、峰度这两种统计数据；
- 为了解个别观察值在整体中所占的位置，会需要将观察值转换为相对量数，如百分等级、标准分数、四分位数等。

通常，在描述性统计中将数据图表化，以直观的方式了解整体资料数据的分布情况，例如直方图、散点图、饼图、折现图、箱形图等。

- 数据种类

通常，数据可以分为两类：不可测量的数据称为分类数据；可测量的数据称为数值数据。上述图表中，'level_0'、'detail_info_tag'字段均为分类数据，'location_lat'、'location_lng'、'detail_info_overall_rating'、'detail_info_price'均为数值数据，而'name'字段则是数据的索引名称。

- 数值数据的描述性统计
- 频数（次数）分布表和直方图

建立简单数据示例，数据来源于《漫画统计学》中“美味拉面畅销前50”上刊载的拉面馆的拉面价格。一般，对于只有一组（一个特征、一个自变量/解释变量）的数据通常使用pandas.Series()建立Series格式数据，但是后续分析会加入新的数据，因此仍旧建立DataFrame格式数据。使用df.describe()可以初步查看主要统计值。

```
ramen_price = pd.DataFrame([700, 850, 600, 650, 980, 750, 500, 890, 880, 700,
890, 720, 680, 650, 790, 670, 680, 900, 880, 720, 850, 700, 780, 850, 750,
                        780, 590, 650, 580, 750, 800, 550, 750, 700, 600,
800, 800, 880, 790, 790, 780, 600, 690, 680, 650, 890, 930, 650, 777, 700],
columns=["price"])
print(ramen_price.describe())
--------------------------------------------------------------------------------
            price
count   50.000000
mean   743.340000
std    108.261891
min    500.000000
25%    672.500000
50%    750.000000
75%    800.000000
max    980.000000
```

因为有些价格是相同的，一般可以直接使用上述ramen_price数据直接计算频数。但是，很多时候相同的数据并不多，并且希望分析内容为数值区段间的比较，分析才更具有意义，因此转换为相对量数，以100间隔为一级。范围则根据数据的最大值和最小值来确定，见表3.1-4。

```
bins = range(500,1000+100,100) #配置分割区间（组距）
ramen_price['price_bins'] = pd.cut(x=ramen_price.price,bins=bins,right=False)  #
参数 right=False 指定为包含左边值，不包括右边值
ramenPrice_bins = ramen_price.sort_values(by=['price']) #按照分割区间排序
ramenPrice_bins.set_index(['price_bins',ramenPrice_bins.index],drop=False,
inplace=True) #以 price_bins 和原索引值设置多重索引，同时配置 drop=False 参数保留原列
ramenPrice_bins.head(10)
```

表 3.1-4 拉面价格频数分布表

		price	price_bins
price_bins			
[500, 600)	6	500	[500, 600)
	31	550	[500, 600)
	28	580	[500, 600)
	26	590	[500, 600)
[600, 700)	34	600	[600, 700)
	41	600	[600, 700)
	2	600	[600, 700)
	44	650	[600, 700)
	3	650	[600, 700)
	27	650	[600, 700)

数值区段间的频数计算，见表3.1-5。

```
ramenPriceBins_frequency = ramenPrice_bins.price_bins.value_counts()
ramenPriceBins_relativeFrequency = ramenPrice_bins.price_bins.value_
counts(normalize=True)  # 参数 normalize=True 将计算相对频数（次数）
ramenPriceBins_freqANDrelFreq = pd.DataFrame({'fre':ramenPriceBins_frequency,
'relFre':ramenPriceBins_relativeFrequency})
ramenPriceBins_freqANDrelFreq
```

组中值计算，见表3.1-6。

```
ramenPriceBins_median = ramenPrice_bins.groupby(level=0).median(numeric_
only=True)
ramenPriceBins_median.rename(columns={'price':'median'},inplace=True)
ramenPriceBins_median
```

表 3.1-5　拉面价格数值区段间频数

	fre	relFre
[700, 800)	18	0.36
[600, 700)	13	0.26
[800, 900)	12	0.24
[500, 600)	4	0.08
[900, 1000)	3	0.06

表 3.1-6　拉面价格组中值

	median
price_bins	
[500, 600)	565.0
[600, 700)	650.0
[700, 800)	750.0
[800, 900)	865.0
[900, 1000)	930.0

合并分割区间、频数计算和组中值的DataFrame格式数据，见表3.1-7。

```
ramen_fre = ramenPriceBins_freqANDrelFreq.join(ramenPriceBins_median).sort_
index().reset_index()  # 在合并时会自动匹配 index
ramen_fre
```

表 3.1-7　合并分割区间、频数和组中值

	index	fre	relFre	median
0	[500, 600)	4	0.08	565.0
1	[600, 700)	13	0.26	650.0
2	[700, 800)	18	0.36	750.0
3	[800, 900)	12	0.24	865.0
4	[900, 1000)	3	0.06	930.0

计算频数比例，即各个区间频数占总数的百分比，能够更清晰地比较之间的差异大小，见表3.1-8。配合经常使用的`df.apply()`和`lambda`匿名函数，能够巧妙地以一种简洁的方式书写代码。

```
ramen_fre['fre_percent%'] = ramen_fre.apply(lambda row:row['fre']/ramen_fre.fre.
sum()*100,axis=1)
ramen_fre
```

表 3.1-8　区间频数占比

	index	fre	relFre	median	fre_percent%
0	[500, 600)	4	0.08	565.0	8.0
1	[600, 700)	13	0.26	650.0	26.0
2	[700, 800)	18	0.36	750.0	36.0
3	[800, 900)	12	0.24	865.0	24.0
4	[900, 1000)	3	0.06	930.0	6.0

拉面价格直方图（图3.1-7）。

Pandas自身就带有不少图表打印的功能（基于Matplotlib库），不必过多地调整数据自身结构，就可迅速地预览。但是，不像Plotly具有图表交互功能。

```
ramen_fre.loc[:,['fre','index']].
plot.bar(x='index',rot=0,figsi
ze=(5,5))
```

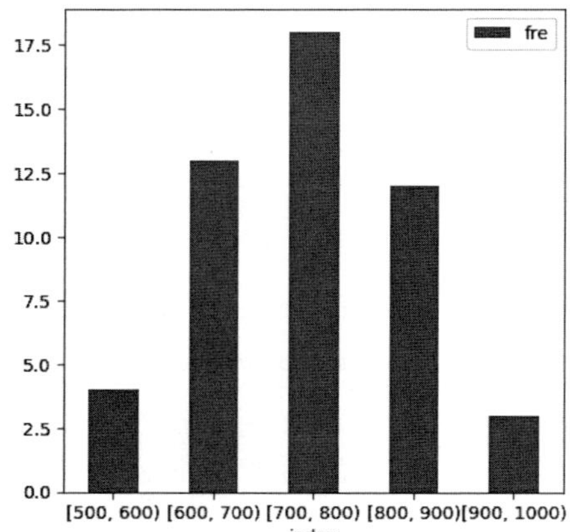

图3.1-7　拉面价格直方图

有了上述简单数据的示例，再返回到POI实验数据，可以直接迁移上述代码，略作调整后分析POI一级分类美食'poi_0_delicacy'的价格总体分布情况（表3.1-9）。将上述的所有分析代码置于一个函数中frequency_bins()，函数的主要功能是计算DataFrame数据格式下，指定组距，一列数据的频数分布。将上述零散的代码纳入到一个函数中，需要注意几点事宜。一是，尽可能让变量名具有普适性，例如原变量名ramenPrice_bins在函数中更改为df_bins，因为该函数同样可以计算'detail_info_overall_rating'字段的频数分布情况；二是，公用的常用变量尽量在开始配置，例如column_name和column_bins_name，这样避免不断地用原始的语句，例如重复使用df.columns[0]+'_bins'，从而导致代码可读性较差；再者，基本不可能直接迁移上述代码于函数中使用，需要逐句或者逐段的迁移测试，例如原代码ramenPrice_bins.price_bins，在函数中改为df_bins[column_bins_name]，因为列名是以变量名形式存储，无法直接用.的方式读取数据。最后，需要注意函数返回值的灵活性，例如并未在函数内部定义图表打印，而是返回DataFrame格式的数据。因为打印的方式比较多样化，可以仅打印一列数据的柱状图，或者多列，以及图表的形式也会多样化。因此，这部分工作放置于函数外处理，增加灵活性。

```
def frequency_bins(df, bins, field):
    '''
    function - 频数分布计算

    Params:
        df - 单列（数值类型）的DataFrame数据；DataFrame(Pandas)
        bins - 配置分割区间（组距）；range()，例如：range(0,600+50,50)
        field - 字段名；string

    Returns:
        df_fre - 统计结果字段包含：index（为bins）、fre、relFre、median和fre_
percent%；DataFrame
    '''
    import pandas as pd

    #A-组织数据
    column_name = df.columns[0]
    column_bins_name = df.columns[0]+'_bins'
    #参数right=False指定为包含左边值，不包括右边值
    df[column_bins_name] = pd.cut(x=df[column_name], bins=bins, right=False)
    df_bins = df.sort_values(by=[column_name])  #按照分割区间排序
    #以price_bins和原索引值设置多重索引，同时配置drop=False参数保留原列
    df_bins.set_index([column_bins_name, df_bins.index],
                      drop=False, inplace=True)

    #B-频数计算
    dfBins_frequency = df_bins[column_bins_name].value_counts()  #dropna=False
    #参数normalize=True将计算相对频数(次数)

    dfBins_relativeFrequency = df_bins[column_bins_name].value_counts(
        normalize=True)
    dfBins_freqANDrelFreq = pd.DataFrame(
        {'fre': dfBins_frequency, 'relFre': dfBins_relativeFrequency})
```

```
    # C- 组中值计算
    df_bins[field] = df_bins[field].astype(float)
    dfBins_median = df_bins.groupby(level=0).median(numeric_only=True)
    dfBins_median.rename(columns={column_name: 'median'}, inplace=True)

    # D- 合并分割区间、频数计算和组中值的 DataFrame 格式数据。
    df_fre = dfBins_freqANDrelFreq.join(
        dfBins_median).sort_index().reset_index()  # 在合并时会自动匹配 index

    # E- 计算频数比例
    df_fre['fre_percent%'] = df_fre.apply(
        lambda row: row['fre']/df_fre.fre.sum()*100, axis=1)
    return df_fre
pd.options.mode.chained_assignment = None
bins = range(0, 600+50, 50)  # 配置分割区间（组距）
delicacy_price_df = poi_gdf[["detail_info_price"]]
delicacy_price_df["detail_info_price"] = delicacy_price_df["detail_info_price"].
apply(
    pd.to_numeric, errors='coerce')  # 将 str(object) 数据类型转换为数值类型
poiPrice_fre_50 = frequency_bins(
    delicacy_price_df.dropna(), bins, "detail_info_price")
poiPrice_fre_50
```

表 3.1-9 POI美食价格频数计算

	index	fre	relFre	median	fre_percent%
0	[0, 50)	2824	0.445918	20.0	44.591821
1	[50, 100)	2030	0.320543	73.0	32.054319
2	[100, 150)	689	0.108795	115.0	10.879520
3	[150, 200)	286	0.045160	170.0	4.516027
4	[200, 250)	164	0.025896	223.0	2.589610
5	[250, 300)	145	0.022896	268.0	2.289594
6	[300, 350)	71	0.011211	321.0	1.121112
7	[350, 400)	45	0.007106	373.0	0.710564
8	[400, 450)	25	0.003948	423.0	0.394758
9	[450, 500)	21	0.003316	475.0	0.331596
10	[500, 550)	21	0.003316	522.0	0.331596
11	[550, 600)	12	0.001895	573.5	0.189484

计算POI美食价格直方图（图3.1–8）。

```
poiPrice_fre_50.loc[:,['fre','index']].plot.bar(x='index',rot=0,figsize=(15,5))
```

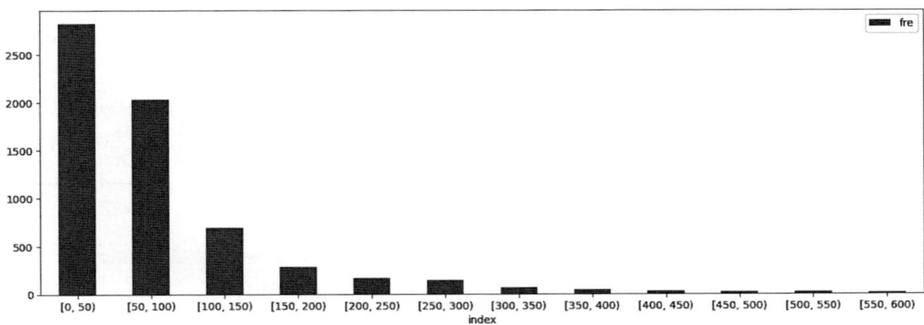

图3.1–8 POI美食价格直方图

调整组距，查看总体频数分布情况（图3.1-9）。

```
import matplotlib.pyplot as plt

bins = list(range(0,300 + 5,5)) + [600]  #通过df.describe() 查看数据后，发现 72% 的价格位于 72 元之下，结合上图柱状图，重新配置组距，尽可能地显示数据变化的趋势。
poiPrice_fre_5 = frequency_bins(delicacy_price_df,bins,"detail_info_price")
poiPrice_fre_5.loc[:,['fre','index']].plot.bar(x='index',rot=0,figsize=(30,5))
_ = plt.xticks(rotation=90)
```

- **集中量数与变异量数**

建立简单数据示例，数据来源于《漫画统计学》保龄球大赛的结果（表3.1-10）。首先建立嵌套字典，然后将其转换为多重索引的DataFrame数据。

```
bowlingContest_scores_dic = {'A_team': {'Barney': 86, 'Harold': 73, 'Chris': 124, 'Neil': 111, 'Tony': 90, 'Simon': 38},
                             'B_team': {'Jo': 84, 'Dina': 71, 'Graham': 103, 'Joe': 85, 'Alan': 90, 'Billy': 89},
                             'C_team': {'Gordon': 229, 'Wade': 77, 'Cliff': 59, 'Arthur': 95, 'David': 70, 'Charles': 88}
                             }

# 可以逐步拆解来查看每一步的数据结构，结合搜索相关方法解释，可以理解每一步的作用。例如，df.stack() 是由列返回多重索引的DataFrame, 具体解释可以查看官方案例
bowlingContest_scores = pd.DataFrame.from_dict(
    bowlingContest_scores_dic, orient='index').stack().to_frame(name='score')
# 使用 print() 或者直接在每一 JupyterLab 的 Cell 最后给出要查看的变量名，都可以查看数据，只是可能显示的模式略有差异。但是，建议使用 print() 查看。因为代码迁移时，单独变量的出现可能会造成代码运行错误
bowlingContest_scores
```

表 3.1-10　保龄球大赛结果数据

		score
A_team	Barney	86.0
	Harold	73.0
	Chris	124.0
	Neil	111.0
	Tony	90.0
	Simon	38.0
B_team	Jo	84.0
	Dina	71.0
	Graham	103.0
	Joe	85.0
	Alan	90.0
	Billy	89.0
C_team	Gordon	229.0
	Wade	77.0
	Cliff	59.0
	Arthur	95.0
	David	70.0
	Charles	88.0

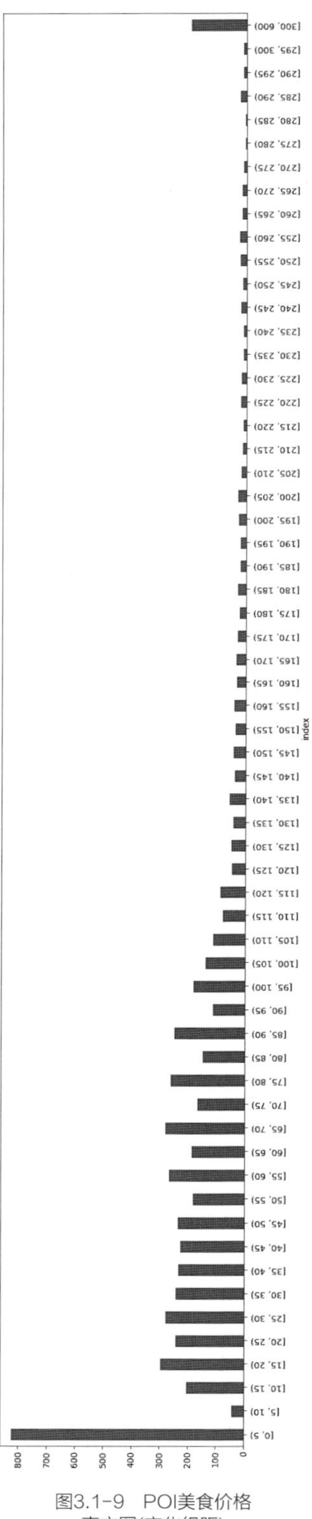

图3.1-9　POI美食价格直方图(变化组距)

求每一队的均值（算数平均数），见表3.1-11。

```
bowlingContest_mean = bowlingContest_scores.groupby(level=0).mean()
bowlingContest_mean
```

求每一队的中位数（表3.1-12）。C队的均值最高，其原因不是每一个队员的成绩都高，而是Gordon获得了229分，远超其他队员的得分。因此，求中位数更为适合。

```
bowlingContest_median = bowlingContest_scores.groupby(level=0).median()
bowlingContest_median
```

表 3.1-11 保龄球大赛结果均值

	score
A_team	87.0
B_team	87.0
C_team	103.0

表 3.1-12 保龄球大赛结果中位数

	score
A_team	88.0
B_team	87.0
C_team	82.5

箱形图（Box plot），又称盒须图、盒式（状）图或箱线图，一种用作显示一组数据分散情况的统计图。显示的一组数据包括最大值、最小值、中位数和上下四分位数，因此使用箱形图较之单一的数值而言，可以更清晰地观察数据的分布情况。如图3.1-10（引自：Wikipedia）所示。

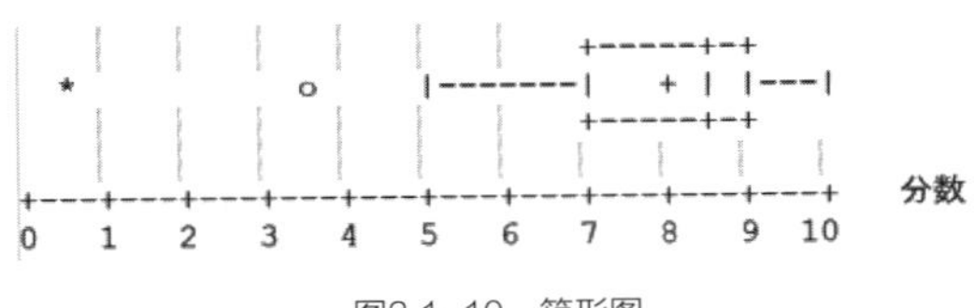

图3.1-10 箱形图

这组数据显示：最小值（minimum）=5；下四分位数（Q1）=7；中位数（Med，即Q2）=8.5；上四分位数（Q3）=9；最大值（maximum）=10；平均值=8；四分位间距(interquartile range)=(Q3–Q2)=2（即ΔQ）。使用Pandas自带的plot功能打印箱形图（图3.1-11），查看各队分数的分布情况。

```
bowlingContest_scores_transpose = bowlingContest_scores.stack().unstack(level=0)
boxplot = bowlingContest_scores_transpose.boxplot(column=['A_team', 'B_team', 'C_team'])
```

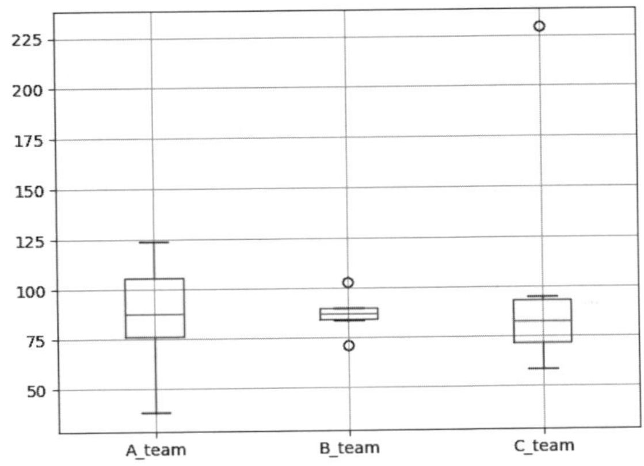

图3.1-11 保龄球大赛结果各队箱形图

Plotly库提供的箱形图可以互动显示具体的数值，具有相对更强的图示能力（图3.1–12）。

```
import plotly.express as px
fig = px.box(bowlingContest_scores.xs('C_team',level=0), y="score")
fig.update_layout(
    autosize=False,
    width=500,
    height=500,
    title='C_team'
    )
fig.show()
```

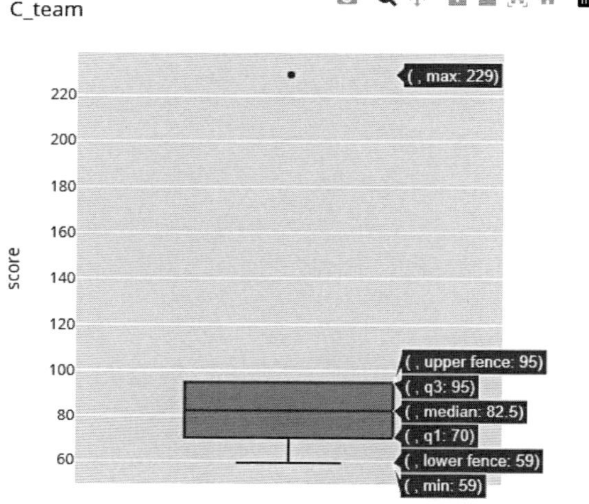

图3.1–12 用Plotly库打印可交互的保龄球大赛结果各队箱形图

求标准差，又称准偏差或均方差（Standard Deviation，SD），数学符号通常用 σ（sigma），用于测量一组数值的离散程度，公式：$SD=\sqrt{\frac{1}{N}\sum_{i=1}^{N}(x_i-\mu)^2}$。其中，$\mu$ 为平均值。虽然A_team和B_team具有相同的均值，但是数值的分布情况迥然不同，通过计算标准差比较离散程度（表3.1–13），标准差越小，代表数据的离散程度越小；反之，标准差越大，离散程度越大。

上述公式为整体标准差，本书的实验数据通常为全部数据样本，而不是抽样，因此使用整体标准差。计算样本的标准差，公式为：$SD=\sqrt{\frac{1}{N-1}\sum_{i=1}^{N}(x_i-\mu)^2}$

```
bowlingContest_std = bowlingContest_scores.groupby(level=0).std()
bowlingContest_std
```

表 3.1–13 保龄球大赛结果各队标准差

	score
A_team	30.172835
B_team	10.373042
C_team	63.033325

返回到POI实验数据，因为一级行业分类中美食部分的'detial_info_tag'包含餐厅的子分类，使用箱形图显示子分类评分'detail_info_overall_rating'的数值分布情况，并计算标准差（表3.1–14）。

```
delicacy = poi_gdf.xs('poi_0_delicacy',level=0)
delicacy_rating = delicacy[['detail_info_tag','detail_info_overall_rating']]
```

```
delicacy_rating
```

表 3.1-14 POI 美食二级分类评分

	detail_info_tag	detail_info_overall_rating
0	美食;中餐厅	4.0
2	美食;中餐厅	5.0
...	...	...
6594	美食;小吃快餐店	NaN
6596	美食;小吃快餐店	NaN
6598	美食;茶座	NaN

```
3300 rows × 2 columns
```

查看餐厅类型。并移除错误的分类数据，例如'教育培训;其他'。同时，可以调整子分类的名称，例如由'美食;中餐厅'修改为'中餐厅'，其中使用了`df.applay()`方法。最后，将其映射为英文字符，在打印时也可以避免显示错误（表3.1-15）。如果显示中文字符错误，需要增加相应的处理代码。

`pd.options.mode.chained_assignment=None`方法为Pandas的警告异常处理，选项包括：`None`，忽略警告；`warn`，打印警告消息；`raise`，引发异常。

```
pd.options.mode.chained_assignment = None
print(delicacy_rating.detail_info_tag.unique())
print("_"*50)
delicacy_rating["clean_bool"] = delicacy_rating.detail_info_tag.apply(
    lambda row: row.split(";")[0] == "美食" if isinstance(row, str) else False)
delicacy_rating_clean = delicacy_rating[delicacy_rating.clean_bool].
drop(columns=[
"clean_bool"])
print(delicacy_rating_clean.detail_info_tag.unique())
print("_"*50)

# 定义一个函数，传入 df.apply() 函数处理字符串

def str_row(row):
    if type(row) == str:
        row_ = row.split(';')[-1]
    else:
```

表 3.1-15 POI 美食二级分类评分数据预处理

	detail_info_overall_rating
detail_info_tag	
CakeANDdessert_shop	0.375599
Chinese_restaurant	0.703622
Foreign_restaurant	0.265208
Snake_bar	0.826895
bar	0.589164
cafe	0.273044
delicacy	0.529150
others	0.751135
teahouse	0.678449

```
            row_ = 'nan'  # 原数据类型为 nan，通过 type(row) 查看后为 float 数据类型，此时将其转换为字符串
    return row_
delicacy_rating_clean.loc[:, ["detail_info_tag"]
                          ] = delicacy_rating_clean["detail_info_tag"].apply(str_row)
print(delicacy_rating_clean.detail_info_tag.unique())
print("_"*50)

tag_mapping = {'中餐厅': 'Chinese_restaurant',
               '小吃快餐店': 'Snake_bar',
               'nan': 'nan',
               '其他': 'others',
               '外国餐厅': 'Foreign_restaurant',
               '蛋糕甜品店': 'CakeANDdessert_shop',
               '咖啡厅': 'cafe',
               '茶座': 'teahouse',
               '酒吧': 'bar',
               '美食': 'delicacy',
               '公司': 'company',
               '商铺': 'store',
               '洗浴按摩': 'massage',
               '超市': 'supermarket',
               '快捷酒店': 'budgetHotel',
               '园区': 'Park'}
delicacy_rating_clean.loc[:, ["detail_info_tag"]
                          ] = delicacy_rating_clean["detail_info_tag"].replace(tag_mapping)
print(delicacy_rating_clean.detail_info_tag.unique())
---------------------------------------------------------------------------
['美食;中餐厅' '美食;其他' '美食;小吃快餐店' '美食;咖啡厅' '美食;蛋糕甜品店' '美食;茶座' '房地产;其他'
 '美食;外国餐厅' '美食' '公司企业;公司' '购物;商铺' nan '休闲娱乐;洗浴按摩' '美食;酒吧' '交通设施;其他'
 '公司企业;其他' '购物;超市' '酒店;快捷酒店' '教育培训;其他' '酒店;其他' '公司企业;园区']
---------------------------------------------------------------------------
['美食;中餐厅' '美食;其他' '美食;小吃快餐店' '美食;咖啡厅' '美食;蛋糕甜品店' '美食;茶座' '美食;外国餐厅' '美食'
 '美食;酒吧']
---------------------------------------------------------------------------
['中餐厅' '其他' '小吃快餐店' '咖啡厅' '蛋糕甜品店' '茶座' '外国餐厅' '美食' '酒吧']
---------------------------------------------------------------------------
['Chinese_restaurant' 'others' 'Snake_bar' 'cafe' 'CakeANDdessert_shop'
 'teahouse' 'Foreign_restaurant' 'delicacy' 'bar']
---------------------------------------------------------------------------
```

输出美食二级分类评分的箱形图（图3.1-13）。

```
delicacy_rating_clean.boxplot(column=['detail_info_overall_rating'],by=['detail_info_tag'],figsize=(25,8))
delicacy_rating_clean_std=delicacy_rating_clean.set_index(['detail_info_tag']).groupby(level=0).std()
delicacy_rating_clean_std
```

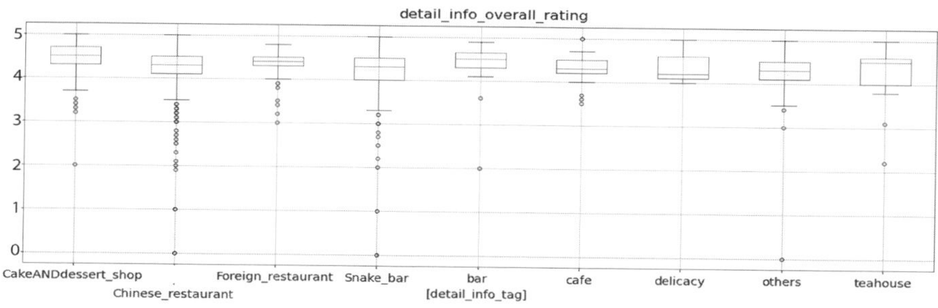

图3.1-13　POI 美食二级分类评分箱形图

- **标准计分（分数）**

建立简单数据示例（表3.1-16），数据来源于《漫画统计学》中的考试成绩数据。在这个案例中，虽然Mason和Reece分别在English和Chinese科目中，而history和biology科目中具有相同的分数。但是，因为标准差，即离散程度不同，所表示的重要程度亦不一样。标准差越小，离散程度越小，则数值每一单位的变化都会影响最终排名，每一分都很重要。也可以理解为标准差小时，落后的同学相对容易追上成绩在前的同学；但是标准差大时，落后的同学相对不容易追上排名靠前的同学。

```
test_score_dic = {"English": {"Mason": 90, "Reece": 81, 'A': 73, 'B': 97, 'C': 85, 'D': 60, 'E': 74, 'F': 64, 'G': 72, 'H': 67, 'I': 87, 'J': 78, 'K': 85, 'L': 96, 'M': 77, 'N': 100, 'O': 92, 'P': 86},
                  "Chinese": {"Mason": 71, "Reece": 90, 'A': 79, 'B': 70, 'C': 67, 'D': 66, 'E': 60, 'F': 83, 'G': 57, 'H': 85, 'I': 93, 'J': 89, 'K': 78, 'L': 74, 'M': 65, 'N': 78, 'O': 53, 'P': 80},
                  "history": {"Mason": 73, "Reece": 61, 'A': 74, 'B': 47, 'C': 49, 'D': 87, 'E': 69, 'F': 65, 'G': 36, 'H': 7, 'I': 53, 'J': 100, 'K': 57, 'L': 45, 'M': 56, 'N': 34, 'O': 37, 'P': 70},
                  "biology": {"Mason": 59, "Reece": 73, 'A': 47, 'B': 38, 'C': 63, 'D': 56, 'E': 75, 'F': 53, 'G': 80, 'H': 50, 'I': 41, 'J': 62, 'K': 44, 'L': 26, 'M': 91, 'N': 35, 'O': 53, 'P': 68},
                  }
test_score = pd.DataFrame.from_dict(test_score_dic)
test_score.head()
```

表3.1-16　考试成绩数据

	English	Chinese	history	biology
Mason	90	71	73	59
Reece	81	90	61	73
A	73	79	74	47
B	97	70	47	38
C	85	67	49	63

求标准计分（Standard Score），又称z-score，即Z-分数，或标准化值（表3.1-17）。z-score代表原始数值和平均值之间的距离，并以标准差为单位计算，即z-score是从感兴趣的点到均值之间有多少个标准差。这样，就可以在不同组数据间比较某一数值的重要程度。公式为：$z=(x-\mu)/\sigma$，其中，$\sigma \neq 0$；并且，x是需要被标准化的原始分数；μ是平均值；σ是标准差。

标准计分的特征：

1. 无论作为变量的取值范围是多少（例如，科目满分为100分还是150分），其标准计分的平均数势必为0，而其标准差势必为1；
2. 无论作为变量的单位是什么（例如分、公斤和小时等），其标准计分的平均数势必为0，而其标准差势必为1。

```
from scipy.stats import zscore
test_Zscore = test_score.apply(zscore)
test_Zscore.head()
```

表3.1-17　考试成绩标准计分

	English	Chinese	history	biology
Mason	0.770054	−0.296174	0.780635	0.162355
Reece	−0.029617	1.392020	0.207107	1.014719
A	−0.740436	0.414644	0.828429	−0.568242
B	1.392020	−0.385027	−0.462008	−1.116191
C	0.325792	−0.651584	−0.366420	0.405887

其中，Mason在English科目中的标准计分为0.77，在整体分布中位于平均分之上0.71个标准差的地位；而Reece在Chinese中的标准计分为1.39，在整体分布中位于平均分之上1.39个标准差的地位，即Reece获得的每一分值价值高于Mason所获取的每一分值。

返回到POI实验数据，分别计算美食部分总体评分'detail_info_overall_rating'和价格'detail_info_price'的标准计分（表3.1-18）。

```
pd.options.mode.chained_assignment = None
delicacy = poi_gdf.xs('poi_0_delicacy',level=0)
delicacy_dropna = delicacy.dropna(subset=['detail_info_overall_rating', 'detail_info_price'])
delicacy_dropna[['detail_info_overall_rating', 'detail_info_price']] = delicacy_dropna[['detail_info_overall_rating', 'detail_info_price']].astype(float)
delicacy_Zscore = delicacy_dropna[['detail_info_overall_rating', 'detail_info_price']].apply(zscore).join(delicacy["name"])
delicacy_Zscore
```

表3.1-18 POI美食评分和价格的标准计分

	detail_info_overall_rating	detail_info_price	name
6	0.500851	−0.903030	关中印象咥长安（创汇店）
10	−0.597102	−0.493965	哪儿托海鲜焖面
22	1.598804	−0.493965	赛百味（昆明池店）
...	...	...	...
6490	−0.377511	0.051455	窑村猪蹄坊总店（纺渭路店）
6518	0.500851	−0.727716	食膳坊中式快餐

2286 rows × 3 columns

通过计算z_score，比较某一饭店的价格和评分的重要性。其意义是，是否价格越接近均值，对应的评分越高于（低于）均值。但是，单独看单一饭店的数据很难判断是否存在这样的一种关系，因此可以打印曲线（图3.1-14），观察曲线的变化规律。使用df.rolling()方法平滑数据后绘制曲线观察数据。

```
import matplotlib.pyplot as plt

delicacy_Zscore_selection = delicacy_Zscore.select_dtypes(include="number") #通过.select_dtypes()方法，提取数值列
delicacy_Zscore_selection.rolling(50, win_type='triang').sum().plot.line(figsize=(25,8))
plt.axhline(y=0,color='gray',linestyle='--')
plt.show()
```

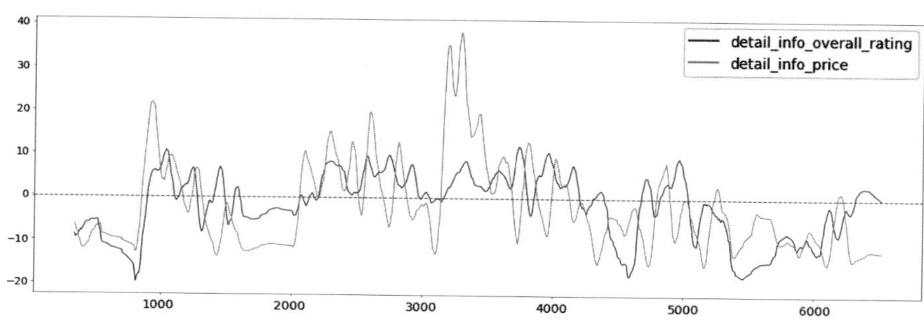

图3.1-14 POI美食评分和价格折线图

3.1.4 正态分布与概率密度函数，异常值处理

3.1.4.1 正态分布

在开始概率密度函数之前，认识下正态分布，又称常态分布、高斯分布（normal distribution/Gaussian distribution），因为正态分布的形状，正态分布常见的名称为钟形曲线（bell curve）。用NumPy库中的numpy.random.normal()方法生成满足指定平均值和标准差的一维数据（也可以用SciPy库stats类中提供的norm方法），并打印直方图和对应由概率密度函数计算的分布曲线。图3.1–15中，y轴表示随机生成数值的频数（因为配置了density=Ture，返回的频数为标准化后的结果）；x轴为随机生成的数据集数值分布，因为设置了plt.hist()中bins=30，将数值分为30等分，计算每一等分（频数宽度）的频数。曲线的顶点为均值、中位数，频数为最高，对应的值为0，该值也为众数。可以从新定义平均数mu参数的值，曲线顶点也会随之变动。从众数向两侧移动，曲线高度下降，表示这些值出现的情形逐渐减少，频数降低。在统计学上图3.1–15可以文字描述为：x服从平均值为0，标准差为0.1的正态分布。

依据正态分布的形状，其三个基本性质为：一，它是对称的，意味着左右部分互为镜像；二，均值、中位数和众数处于同一位置，并且在分布的中心，即钟形的顶点。表现为曲线中心最高，首尾两端向下倾斜，呈单峰状；三，正态分布是渐近的，意味着分布的左尾和右尾永远不会触及底线，即x轴。

正态分布具有重要意义，自然界与人类社会中经常出现正态分布的各类数据，例如经济中的收入水平、人的智商（IQ）分数和考试成绩、自然界中受大量微小随机扰动影响的事物等。因此，能够依据这一现象推断出现某种情形的准确概率。同时，需要注意，正态分布是统计学中所谓的理论分布，很少有数值严格服从正态分布；而是近似于，也有可能相差较远。违背正态分布假设，则依据正态分布假设所计算的概率结果将不再有效。

对于统计学的知识，通常是依据已出版的专著或教材为依据进行解释，并根据阐述的内容，以Python语言为工具，分析数据的变化。从而，可以直接应用代码来解决对应的问题，并能够通过具体的数据分析更深入地理解统计学的相关知识。

```
# 下述案例参考 SciPy.org 中 `numpy.random.normal` 案例
import math
import numpy as np
import matplotlib.pyplot as plt

# 依据分布配置参数生成数据 (Draw samples from the distribution)
mu, sigma = 0, 0.1  # 配置平均值和标准差 (mean and standard deviation)
s = np.random.normal(mu, sigma, 1000)
# 验证生成的数据是否满足配置参数 (Verify the mean and the variance)
print("mean<0.01:", abs(mu-np.mean(s)) < 0.01)
print("sigmag variance<0.01", abs(sigma - np.std(s, ddof=1)) < 0.01)
count, bins, ignored = plt.hist(s, bins=30, density=True)
```

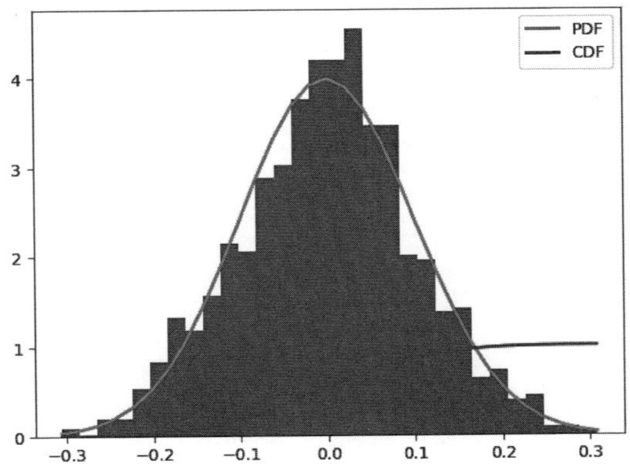

图3.1–15 标准正态分布

```
PDF = 1/(sigma * np.sqrt(2 * np.pi)) * np.exp(- (bins - mu)
                                   ** 2 / (2 * sigma**2))  #y 方向的值
计算公式即为概率密度函数
plt.plot(bins, PDF, linewidth=2, color='r', label='PDF')

CDF = PDF.cumsum()  #计算累积分布函数
CDF /= CDF[-1]  #通过除以最大值，将 CDF 数值分布调整在 [0,1] 的区间
plt.plot(bins, CDF, linewidth=2, color='g', label='CDF')
plt.legend()
plt.show()
------------------------------------------------------------------
mean<0.01: True
sigmag variance<0.01 True
```

3.1.4.2 概率密度函数（Probability density function，PDF）

当直方图的组距无限缩小至极限后，能够拟合出一条曲线，计算这个分布曲线的公式为概率密度函数：$f(x)=\frac{1}{\sigma\times\sqrt{2\pi}}\times \mathrm{e}^{-\frac{1}{2}[\frac{x-\mu}{\sigma}]^2}$。式中，$\sigma$为标准差；$\mu$为平均值；e为自然对数的底，其值大约为2.7182...。在上述程序中，计算概率密度函数时，并未计算每个数值，而是使用`plt.hist()`返回值`bins`替代，为每一组距的左边沿和右边沿，这里划分了30份，因此首位数为第1个频数宽度的左边沿，末位数为最后一个频数宽度的右边沿，而中间的所有为左右边沿重叠。概率密度函数的积分，为累积分布函数（cumulative distribution function，CDF），可以用`numpy.cumsum()`计算，为给定轴上数组元素的累积和。

图表打印的库主要包括Matplotlib、Plotly（含dash）、Bokeh、seaborn等①。具体选择哪个打印图表，没有固定的标准，通常根据数据图表打印的目的、由哪个库能够满足要求，以及个人习惯用哪个库来确定。在上述使用Matplotlib库打印概率密度函数曲线时，是自行计算，而Seaborn提供了`seaborn.histplot()`方法，指定`bins`参数后可以直接获取上述结果。`bins`参数的配置，如果为一个整数值，则是划分同宽度频数宽度（bin）的数量；如果是列表，则为频数宽度的边缘，例如`[1,2,3,4]`，表示`[[1,2),[2,3),[3,4]]`的频数宽度列表，`[`代表包含左边沿数据，`]`代表包含右边沿数据，而(，和)则是分别代表不包含左或右边沿数据。同Pandas的`pandas.core.indexes.range.RangeIndex`的RangeIndex数据格式。

① Matplotlib，https://matplotlib.org/；Plotly（含dash），https://plotly.com/；Bokeh，https://docs.bokeh.org/en/latest/index.html；seaborn，https://seaborn.pydata.org/。

返回到POI实验数据，直接计算打印POI美食数据的价格概率密度函数的值（纵轴），并连为曲线（分布曲线）。为了进一步观察，组距不断缩小时，与曲线的拟合程度，定义一个循环打印多个连续组距变化的直方图（图3.1-16），来观察直方图的变化情况。

```
import pandas as pd
import seaborn as sns
import matplotlib.pyplot as plt
import math

#读取已经存储为 .pkl 格式的 POI 数据，其中包括 geometry 字段，为 GeoDataFrame 地理信息数据，可以通过
poi_gpd.plot() 迅速查看数据。
poi_gdf = pd.read_pickle('./data/poisInAll/poisInAll_gdf.pkl')
delicacy_price = poi_gdf.xs(
    'poi_0_delicacy', level=0).detail_info_price  #提取美食价格数据
delicacy_price_df = delicacy_price.to_frame(name='price').astype(float)
```

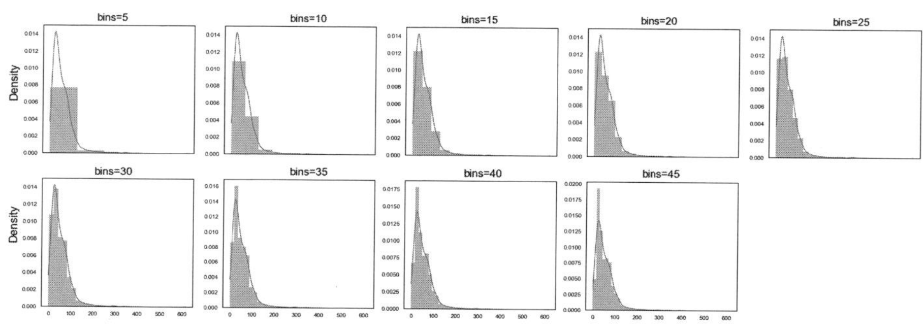

图3.1-16 连续变化组距的直方图

```
sns.set(style="white", palette="muted", color_codes=True)
bins = list(range(0, 50, 5))[1:]
bin_num = len(bins)
ncol = 5
nrow = math.ceil(bin_num/ncol)
fig, axs = plt.subplots(ncols=ncol, nrows=nrow, figsize=(30, 10), sharex=True)
ax = [(row, col) for row in range(nrow) for col in range(ncol)]
i = 0
for bin in bins:
    sns.histplot(delicacy_price_df.price, bins=bin, color="r",
                 ax=axs[ax[i]], kde=True, stat="density", linewidth=0).
set(title="bins=%d" % bin)
    i += 1
```

3.1.4.3 偏度与峰度

描述正态分布的特征（或概率密度函数的分布曲线）有两个概念，一个是偏度（skew），另一个是峰度（kurtosis）。

由SciPy库skewnorm方法建立具有正偏态和负偏态属性的数据，并由skew方法计算偏度值，其公式为：$skewness=\frac{3(\ \mu-median)}{\sigma}$。式中，$\mu$为均值；$\sigma$为标准差。偏度值为负，即负偏态，则概率密度函数左侧尾部比右侧长；偏度值为正，即正偏态，概率密度函数右侧尾部比左侧长。

由NumPy建立具有尖峰分布（瘦尾）和扁峰分布（厚尾）属性的数据，配置绝对值参数均为0，变化标准差，使用SciPy库kurtosis方法计算峰度值（峰度值计算有多个公式，不同软件平台公式也会有所差异），结果如图3.1–17所示。当平均值为0、标准差为1时的正态分布为标准正态分布，概率密度函数公式可以简化为：$f(x)=\frac{1}{\sqrt{2\pi}}\times \mathrm{e}^{-\frac{x^2}{2}}$。

在用NumPy生成数据集时，间接使用了方差σ^2的参数，其概率密度函数公式可以调整为：$f(x|\mu,\sigma^2)=\frac{1}{\sqrt{2\pi\sigma^2}}\times \mathrm{e}^{-\frac{(x-\mu)^2}{2\sigma^2}}$，$\sigma^2$为方差：$\sigma^2=\frac{1}{N}\sum_{i=1}^{N}(x_i-\mu)^2$

```
import math
import numpy as np
from scipy.stats import kurtosis
from scipy.stats import skewnorm
from scipy.stats import skew
import matplotlib.pyplot as plt

#建立具有正偏态和负偏态属性的数据
skew_list = [7, -7]
skewNorm_list = [skewnorm.rvs(
    a, scale=1, size=1000, random_state=None) for a in skew_list]
skewness = [skew(d) for d in skewNorm_list]
print('skewness for data:', skewness)  #验证偏度值
```

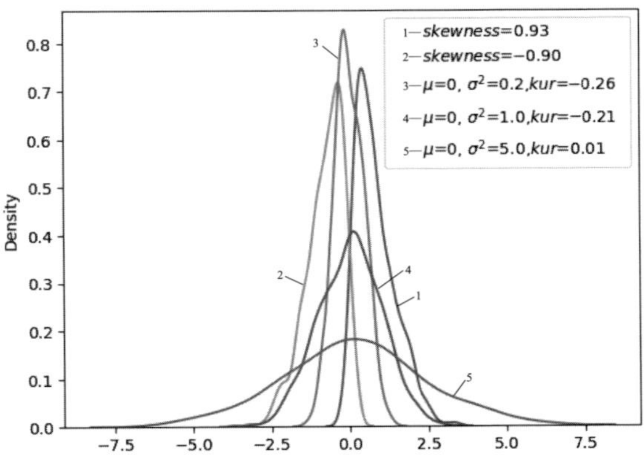

图3.1–17 尖峰分布和扁峰分布

```
#建立具有尖峰分布和扁峰分布属性的数据
mu_list = [0, 0, 0,]
variance = [0.2, 1.0, 5.0,]
normalDistr_paras = zip(mu_list, [math.sqrt(sig2)
                        for sig2 in variance]) #配置多个平均值和标准差
s_list = [np.random.normal(para[0], para[1], 1000)
          for para in normalDistr_paras]
kur = [kurtosis(s, fisher=True) for s in s_list]
print("kurtosis for data:", kur)
i = 0
for skew_data in skewNorm_list:
    sns.kdeplot(skew_data, label="skewness=%.2f" % skewness[i])
    i += 1
n = 0
for s in s_list:
    sns.kdeplot(s, label="μ=%s, $σ^2$=%s,kur=%.2f" %
                (mu_list[n], variance[n], kur[n]))
    n += 1
plt.legend()
_ = plt.plot()
--------------------------------------------------------------------------------
skewness for data: [0.9722605767531536, -0.8485166469879731]
kurtosis for data: [0.1385811357392006, 0.012989429359109739,
-0.11650103191067807]
```

返回POI实验数据，计算POI美食数据价格的偏度和峰度。偏度计算结果值为正，为正偏态，概率密度函数右侧尾部比左侧长。说明多数价格较低，少数较高价格将分布的尾巴托向另一端。峰度为尖峰分布。

```
delicacy_price_df_clean = delicacy_price_df.dropna()
print("skewness:%.2f, kurtosis:%.2f"%(skew(delicacy_price_df_clean.
price),kurtosis(delicacy_price_df_clean.price,fisher=True)))
--------------------------------------------------------------------------------
skewness:4.18, kurtosis:29.84
```

3.1.4.4 检验数据集是否服从正态分布

首先，计算标准计分，标准化价格数据集后其平均值为0，标准差为1，绘制概率密度函数分布曲线（理论值）。同时，叠加满足上述条件，由NumPy随机生成，满足正态分布数据集的分布曲线（观察值），可以比较二者的差异（图3.1-18）。因为实验数据（美食价格）是正偏态，可以观察到位于均值左侧的数据有所差异，而右侧趋势基本吻合。同时，具有较高的峰度值，高于标准正态分布曲线。

```
import pandas as pd
def comparisonOFdistribution(df, field, bins=100):
    '''
    funciton-数据集z-score概率密度函数分布曲线（即观察值/实验值 observed/empirical
data）与标准正态分布（即理论值 theoretical set）比较
```

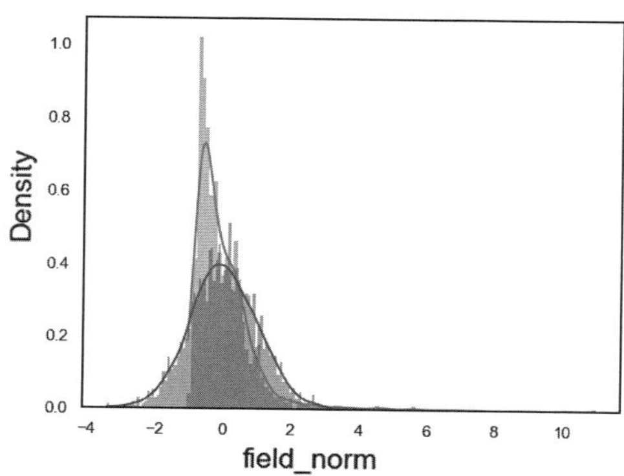

图3.1-18 POI美食价格分布与标准正态分布

```
    Params:
        df - 包含待分析数据集的DataFrame格式类型数据; DataFrame(Pandas)
        field - 指定分析数据数据(DataFrame格式)的列名; string
        bins - 指定频数宽度,为单一整数代表频数宽度(bin)的数量;或者列表,代表多个频数宽度
的列表。The default is 100; int;list(int)

    Returns:
        None
    '''
    import pandas as pd
    import numpy as np
    import seaborn as sns
    df_field_mean = df[field].mean()
    df_field_std = df[field].std()
    print("mean:%.2f, SD:%.2f" % (df_field_mean, df_field_std))

    # 标准化价格(标准计分,z-score),或者使用 `from scipy.stats import zscore` 方法
    df['field_norm'] = df[field].apply(
        lambda row: (row-df_field_mean)/df_field_std)

    # 验证 z-score,标准化后的均值必为 0, 标准差必为 1.0
    df_fieldNorm_mean = df['field_norm'].mean()
    df_fieldNorm_std = df['field_norm'].std()
    print("norm_mean:%.2f, norm_SD:%.2f" %
          (df_fieldNorm_mean, df_fieldNorm_std))
    sns.histplot(df['field_norm'], bins=bins, kde=True,
                 stat="density", linewidth=0, color='r')
    s = np.random.normal(0, 1, len(df[field]))
    sns.histplot(s, bins=bins, kde=True,
                 stat="density", linewidth=0, color='b')
pd.options.mode.chained_assignment = None
comparisonOFdistribution(delicacy_price_df_clean, 'price', bins=100)
--------------------------------------------------------------------------------
mean:54.19, SD:51.22
norm_mean:0.00, norm_SD:1.00
```

- **异常值处理**

从直方图或箱形图中,可以观察到偏度的差异,有少数部分较高的价格出现,估计为异常值。

处理异常值,先建立简单数据来理解异常值,并找到相应的方法。使用*How to Detect and Handle Outliers*中提供的方法,其公式为:$MAD = median_i\{|x_i - \widetilde{x}|\}$,式中MAD(the median of the absolute deviation about the median /Median Absolute Deviation)为绝对中位差,其中$\widetilde{x}$为中位数;$M_i = \frac{0.6745(x_i - \widetilde{x})}{MAD}$,式中$M_i$为修正的z-score(modified z-score),$\widetilde{x}$为中位数。

由计算结果(图3.1-19),确定该公式能有效地判断异常值。

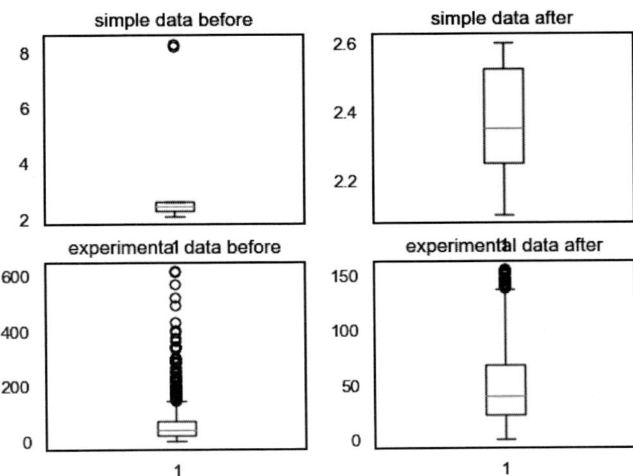

图3.1-19 数据的异常值处理

```
import numpy as np
import matplotlib.pyplot as plt
outlier_data = np.array([2.1, 2.6, 2.4, 2.5, 2.3, 2.1,
                        2.3, 2.6, 8.2, 8.3]) #建立简单的数据，便于观察
ax1 = plt.subplot(221)
ax1.margins(0.05)
ax1.boxplot(outlier_data)
ax1.set_title('simple data before')
def is_outlier(data, threshold=3.5):
    '''
    function-判断异常值

    Params:
        data - 待分析的数据，列表或者一维数组；list/array
        threshold - 判断是否为异常值的边界条件，The default is 3.5；float

    Returns
        is_outlier_bool - 判断异常值后的布尔值列表；list(bool)
        data[~is_outlier_bool] - 移除异常值后的数值列表；list
    '''
    import numpy as np
    MAD = np.median(abs(data-np.median(data)))
    modified_ZScore = 0.6745*(data-np.median(data))/MAD
    is_outlier_bool = abs(modified_ZScore) > threshold
    return is_outlier_bool, data[~is_outlier_bool]
is_outlier_bool, data_clean = is_outlier(outlier_data, threshold=3.5)
print(is_outlier_bool, data_clean)
plt.rcParams["figure.figsize"] = (10, 10)
ax2 = plt.subplot(222)
ax2.margins(0.05)
ax2.boxplot(data_clean)
ax2.set_title('simple data after')
_, delicacyPrice_outliersDrop = is_outlier(
    delicacy_price_df_clean.price, threshold=3.5)
print("原始数据描述：", delicacy_price_df_clean.price.describe(), "\n")
print("-"*50)
print("异常值处理后数据描述：", delicacyPrice_outliersDrop.describe(), "\n")

ax3 = plt.subplot(223)
ax3.margins(0.05)
ax3.boxplot(delicacy_price_df_clean.price)
ax3.set_title('experimental data before')
ax3 = plt.subplot(224)
ax3.margins(0.05)
ax3.boxplot(delicacyPrice_outliersDrop)
ax3.set_title('experimental data after')
plt.show()
--------------------------------------------------------------------------------
[False False False False False False False False  True  True] [2.1 2.6 2.4 2.5 2.3
2.1 2.3 2.6]
原始数据描述：count    2303.000000
mean       54.186062
std        51.215379
min         0.000000
25%        22.000000
50%        41.000000
75%        72.000000
max       617.000000
Name: price, dtype: float64
--------------------------------------------------------------------------------
异常值处理后数据描述：count    2232.000000
mean       47.695116
std        30.958886
min         0.000000
25%        22.000000
50%        40.000000
75%        68.000000
max       155.000000
Name: price, dtype: float64
plt.rcParams["figure.figsize"] = (5,5)
comparisonOFdistribution(pd.DataFrame(delicacyPrice_outliersDrop,columns=['price
']),'price',bins=100)
--------------------------------------------------------------------------------
mean:47.70, SD:30.96
```

```
norm_mean:0.00, norm_SD:1.00
```

如果将较高的价格视为异常值移除，调用定义的comparisonOFdistribution(df,field,bins=100)函数打印概率密度函数（图3.1-20），比较标准正态分布可以发现右侧较高值的部分发生了变化，其他部分未发生明显变化。那么，如何检验数据集是否服从正态分布？SciPy库提供有多种正态性检验的方法，分别为kstest、shapiro、normaltest和anderson。计算结果显示的p-value基本为0，即p-value<0.05，拒绝原假设。表明清理异常值后的美食价格数据集仍不服从正态分布，为非正态数据集。只有服从正态分布的数据，可以计算从一个总体中选取特定值或区间的概率。本次美食价格数据集为正偏和尖峰分布，为非正态数据集，正态分布的概率不能很好地适用于该类数据集。

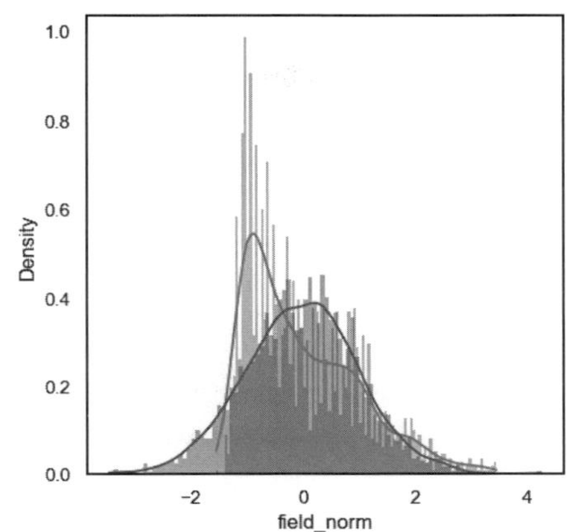

图3.1-20　经异常值处理后的概率密度分布

SciPy检验数据集是否服从正态分布的返回值，第1个为统计量，该值越接近0表明数据和标准正态分布拟合得越好；第2个为p-value值，如果该值大于0.05的显著性水平，接受原假设，判断样本的总体服从正态分布。

```
from scipy import stats
kstest_test = stats.kstest(delicacyPrice_outliersDrop,cdf='norm')
print("original data:",kstest_test)
z_score_kstest_test = stats.kstest(stats.zscore(delicacyPrice_
outliersDrop),cdf='norm')
print('z_score:',z_score_kstest_test)

# 官方文档注释，当 N>5000 时，只有统计量正确，但是 p-value 值不一定正确。本次实验 `len(delicacyPrice_
outliersDrop)` 数据量为 770，可以使用
shapiro_test=stats.shapiro(delicacyPrice_outliersDrop)
print("original data - shapiroResults(statistic=%f,pvalue=%f)"%(shapiro_test))
normaltest_test=stats.normaltest(delicacyPrice_outliersDrop,axis=None)
print("original data:",normaltest_test)
anderson_test = stats.anderson(delicacyPrice_outliersDrop,dist='norm')
print("original data:",anderson_test)
--------------------------------------------------------------------------
original data: KstestResult(statistic=0.999551684674593, pvalue=0.0)
z_score: KstestResult(statistic=0.11464960111854455,
pvalue=5.153632014592587e-26)
original data - shapiroResults(statistic=0.918487,pvalue=0.000000)
original data: NormaltestResult(statistic=232.93794469002853,
pvalue=2.6191959927510468e-51)
original data: AndersonResult(statistic=53.19820994109705, critical_
values=array([0.575, 0.655, 0.786, 0.916, 1.09 ]), significance_level=array([15. ,
10. ,  5. ,  2.5,  1. ]))
```

3.1.4.5 给定特定值计算概率，及找到给定概率的值

如果数据集服从正态分布（图3.1-21），则可以直接使用SciPy库stats类中提供的norm方法进行计算。下述案例使用norm.cdf()、norm.sf()和norm.ppf()分别计算小于或等于特定值、大于或等于特定值或者找到给定概率的值进行计算。

```
from scipy.stats import norm
import matplotlib.pyplot as plt
import numpy as np
fig, ax = plt.subplots(1, 1)
mean, var, skew, kurt = norm.stats(moments='mvsk')
print('mean, var, skew, kurt=',(mean, var, skew, kurt))  # 验证符合标准正态分布的相关
统计量
x = np.linspace(norm.ppf(0.01),norm.ppf(0.99), 100)  # norm.ppf 百分比点函数 - Percent
```

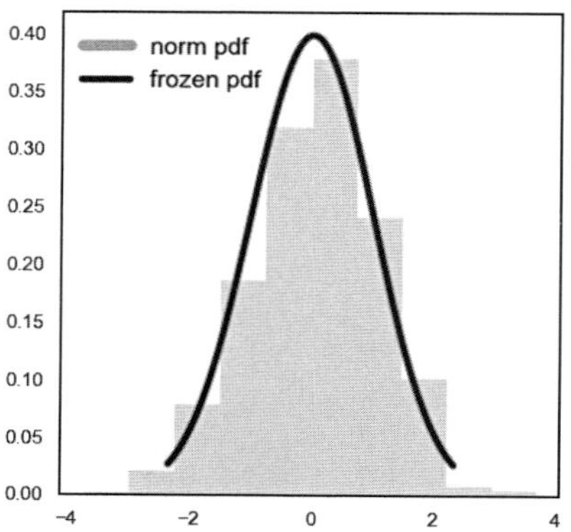

图3.1-21 符合正态分布的数据集

```
point function (inverse of cdf — percentiles)
ax.plot(x, norm.pdf(x), 'r-', lw=5, alpha=0.6, label='norm pdf')  # norm.pdf 为概率密度函数
rv=norm()  # 固定形状（偏度和峰度）、位置 loc（平均值）和比例 scale（标准差）参数，即指定固定值
ax.plot(x, rv.pdf(x), 'k-', lw=3, label='frozen pdf')  # 固定 / "冻结" 分布（frozen distribution）

vals = norm.ppf([0.001, 0.5, 0.999]) # 返回概率为 0.1%、50% 和 99.9% 的值，默认 loc=0,scale=1
print("验证累积分布函数 CDF 返回值与其 PPF 返回值是否相等或近似：",np.allclose([0.001, 0.5, 0.999], norm.cdf(vals)))

r = norm.rvs(size=1000) # 指定数据集大小
ax.hist(r, density=True, histtype='stepfilled', alpha=0.2)
ax.legend(loc='best', frameon=False)
plt.show()
--------------------------------------------------------------------------------
mean, var, skew, kurt= (array(0.), array(1.), array(0.), array(0.))
验证累积分布函数 CDF 返回值与其 PPF 返回值是否相等或近似： True
```

```
_ # 如果需要计算概率，则定义固定分布的数据集。
print("用 .cdf 计算值小于或等于 113 的概率为：",norm.cdf(113,100,12))  # pdf(x, loc=0, scale=1) 配置 Loc(均值) 和 scale(标准差)
print("用 .sf 计算值大于或等于 113 的概率为：",norm.sf(113,100,12))
print("可以观察到 .cdf（<=113）概率结果 +.sf(>=113) 概率结果为：",norm.cdf(113,100,12)+norm.sf(113,100,12))
print("用 .ppf 找到给定概率值为 98% 的数值为：",norm.ppf(0.86066975,100,12))
--------------------------------------------------------------------------------
用 .cdf 计算值小于或等于 113 的概率为： 0.8606697525503779
用 .sf 计算值大于或等于 113 待概率为： 0.13933024744962208
可以观察到 .cdf（<=113）概率结果 +.sf(>=113) 概率结果为： 1.0
用 .ppf 找到给定概率值为 98% 的数值为： 112.99999986204986
```

在传统的概率计算中，已知z-score（标准计分），可以通过查表的方式来获取对应的概率值，这种方式已经很少使用。为了直观地观察概率值，由曲线、横轴和通过对应概率密度函数值的垂直线围合的面积即为概率值，通过绘制图表可以更清晰地进行观察（图3.1–22）。

```
def probability_graph(x_i, x_min, x_max, x_s=-9999, left=True, step=0.001, subplot_num=221, loc=0, scale=1):
    '''
    function - 正态分布概率计算及图形表述

    Paras:
        x_i - 待预测概率的值；float
        x_min - 数据集区间最小值；float
        x_max - 数据集区间最大值；float
        x_s - 第 2 个带预测概率的值，其值大于 x_i 值。The default is -9999; float
        left - 是否计算小于或等于，或者大于或等于指定值的概率。The default is True; bool
        step - 数据集区间的步幅。The default is 0.001; float
        subplot_num - 打印子图的序号，例如 221 中，第一个 2 代表列，第二个 2 代表行，第三个是子图的序号，即总共 2 行 2 列总共 4 个子图，1 为第一个子图。The default is 221; int
        loc - 即均值。The default is 0; float
        scale - 标准差。The default is 1; float
```

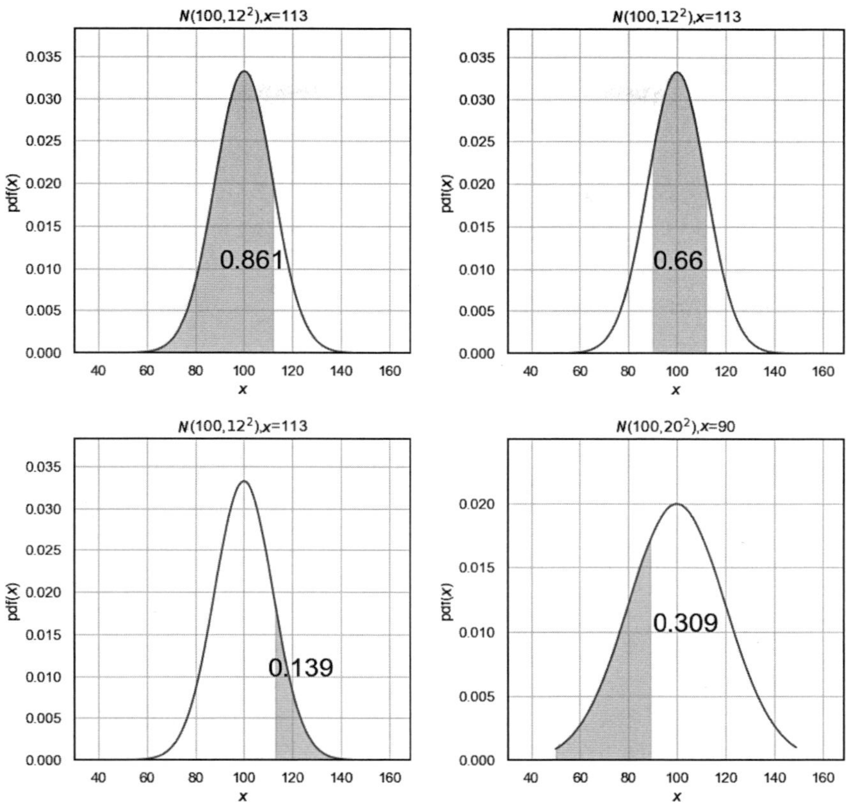

图3.1-22 概率值计算

```
    Returns:
        None
    '''
    import math
    import numpy as np
    from scipy.stats import norm
    x = np.arange(x_min, x_max, step)
    ax = plt.subplot(subplot_num)
    ax.margins(0.2)
    ax.plot(x, norm.pdf(x, loc=loc, scale=scale))
    ax.set_title('N(%s,$%s^2$),x=%s' % (loc, scale, x_i))
    ax.set_xlabel('x')
    ax.set_ylabel('pdf(x)')
    ax.grid(True)
    if x_s == -9999:
        if left:
            px = np.arange(x_min, x_i, step)
            ax.text(loc-loc/10, 0.01,
                    round(norm.cdf(x_i, loc=loc, scale=scale), 3), fontsize=20)
        else:
            px = np.arange(x_i, x_max, step)
            ax.text(loc+loc/10, 0.01, round(1-norm.cdf(x_i,
                    loc=loc, scale=scale), 3), fontsize=20)
    else:
        px = np.arange(x_s, x_i, step)
        ax.text(loc-loc/10, 0.01, round(norm.cdf(x_i, loc=loc, scale=scale) -
                norm.cdf(x_s, loc=loc, scale=scale), 2), fontsize=20)
    ax.set_ylim(0, norm.pdf(loc, loc=loc, scale=scale)+0.005)
    ax.fill_between(px, norm.pdf(px, loc=loc, scale=scale),
                    alpha=0.5, color='g')
plt.figure(figsize=(10, 10))
probability_graph(x_i=113, x_min=50, x_max=150, step=1,
```

```
                     subplot_num=221, loc=100, scale=12)
probability_graph(x_i=113, x_min=50, x_max=150, step=1,
                     left=False, subplot_num=223, loc=100, scale=12)
probability_graph(x_i=113, x_min=50, x_max=150, x_s=90,
                     step=1, subplot_num=222, loc=100, scale=12)
probability_graph(x_i=90, x_min=50, x_max=150, step=1,
                     subplot_num=224, loc=100, scale=20)
plt.show()
```

参考文献（References）：

[1]（日）高桥麻奈著，崔建锋译．株式会社TREND-PRO漫画制作．漫画数据库[M]．北京：科学出版社，2010．5.
[2]（日）高桥信著，陈刚译．株式会社TREND-PRO漫画制作．漫画统计学[M]．北京：科学出版社，2019．8.
[3] Timothy C.Urdan（蒂莫西·C·厄丹），彭志文译．白话统计学[M]．北京：中国人民大学出版社，2013．12．第3版．<Timothy C. Urdan (2022). Statistics in Plain English. Routledge.>.
[4] Boris Iglewicz and David Hoaglin (1993), “Volume 16: How to Detect and Handle Outliers”, The ASQC Basic References in Quality Control: Statistical Techniques, Edward F. Mykytka, Ph.D., Editor.

3.2 OSM数据与核密度估计

3.2.1 OpenStreetMap（OSM）数据处理

OpenStreetMap（OSM）①。“欢迎访问OpenStreetMap，这是个为全世界创建和分发免费地理数据的项目。我们之所以这么做，是因为人们认为作为免费的地图，在使用方面实际上有法律或者技术上的限制，阻碍了人们以创作性的、多产的或意想不到的方式使用它们”。OSM提供了世界各地的道路、小径、咖啡馆、火车站等诸多地理信息数据，是用于城市空间方面研究的宝贵数据财富。OSM数据的下载通过查看官网信息，一般来讲有两个途径：一个是直接下载显示窗口范围的数据，或输入坐标自定义范围。但是，这种方式下载的数据量有限制，如果范围过大，则无法下载。另一种方式是直接从资源库下载，不同资源下载的方式也有所不同，需要根据研究目的来确定。同时，OSM提供了多年历史数据，这为城市空间变化的研究提供了数据支持。本书使用的数据从Geofabrik②库中下载，因为分析的目标区域为芝加哥城，下载了llinois-latest-free.shp.zip③压缩文件340MB，解压后3.73GB。根据下载后点数据的分布情况，为了保持点连续的区域，增加下载wisconsin-latest.osm.bz2④数据文件324MB，解压后3.73GB。可以使用QGIS初步查看数据（图3.2-1）。包含的数据层有：lines，multilinestrings，multipolygons，other_relations 和points。

下图同时打开了 Ilinois 和 wisconsin 的点数据，内部红色半透明区域为芝加哥城行政范围⑤，数据来源于 Chicago Data Portal⑥。外部红色虚线为实验数据确定的初步边界。最小的黑色矩形是用于代码调试提取的小规模数据范围。

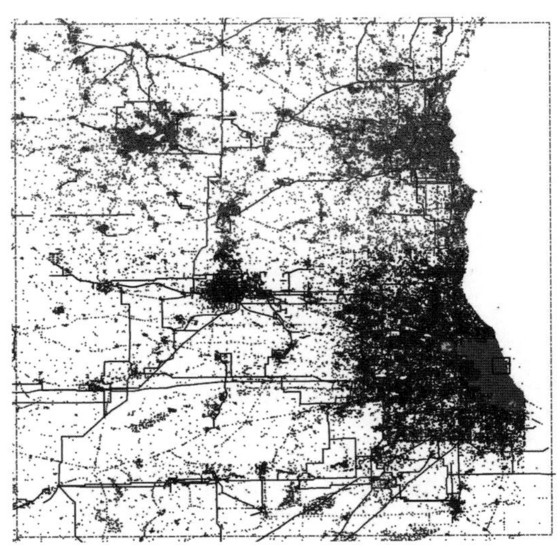

图3.2-1 不同芝加哥城OSM数据

3.2.1.1 OSM原始数据处理

OSM数据处理包括合并两个区域数据及裁切，或者先分别裁切再合并。可以根据电脑内存需求和处理速度确定前后顺序。裁切OSM数据，这里选用的裁切方法是使用osmosis⑦命令行工具，非常适合处理大数据文件，裁切和更新数据。同时，可以参考Manipulating Data with Osmosis⑧。查询osmosis给出的案例，寻找应用Polygon提取数据的代码为：`osmosis --read-xml file="planet-latest.osm" --bounding-polygon file="country2pts.txt" --write-xml file="germany.osm"`，涉及三个参数：原始OSM数据；裁切边界Polygon（TXT数据格式），该处的Polygon为osmosis格式的Polygon格式数据，需要编写转换代码；输出路径。

① OpenStreetMap（OSM），为开源世界地图，可以免费下载。地图数据托管服务由UCL、Fastly、Bytemark Hosting和其他合作伙伴支持（https://www.openstreetmap.org。

② Geofabrik，Geofabrik是一家位于德国Karlsruhe的咨询和软件开发公司，专门从事OpenStreetMap服务（https://download.geofabrik.de/north-america/us.html）。

③ llinois-latest-free.shp.zip，Illinois区域OpenStreetMap数据（https://download.geofabrik.de/north-america/us/illinois.html）。

④ wisconsin-latest.osm.bz2，wisconsin区域OpenStreetMap数据（https://download.geofabrik.de/north-america/us/wisconsin.html）。

⑤ 芝加哥城行政范围（https://data.cityofchicago.org/Facilities-Geographic-Boundaries/Boundaries-City/ewy2-6yfk）。

⑥ Chicago Data Portal，为芝加哥城开放数据门户，可免费下载数据用于相关分析。其中，许多数据集每天至少更新一次或数次（https://data.cityofchicago.org/）。

⑦ osmosis，用于处理OSM数据的命令行工具。该工具包含的组件（例如，读写数据库和文件，获取和应用数据源的变化，进行数据分类等）可以自由组合，以执行更大的操作（https://wiki.openstreetmap.org/wiki/Osmosis）。

⑧ Manipulating Data with Osmosis，用于操作和处理原始.osm数据，并通常处理大的数据文件，将OSM文件分割成较小的部分，以及应用变化集更新现有文件（https://learnosm.org/en/osm-data/osmosis/）。

目视粗略判断点集聚的范围，在QGIS中绘制常规的Polygon边界，如上图外红色虚线。首先，编写Polygon到osmosis格式的Polygon代码，查询其数据格式为：

```
australia_v
first_area
     0.1446693E+03    -0.3826255E+02
     0.1446627E+03    -0.3825661E+02
     ...
     0.1446758E+03    -0.3826229E+02
     0.1446693E+03    -0.3826255E+02
END
second_area
     0.1422436E+03    -0.3839315E+02
     0.1422496E+03    -0.3839070E+02
     ...
     0.1422420E+03    -0.3839857E+02
     0.1422436E+03    -0.3839315E+02
END
END
```

定义转换格式函数shpPolygon2OsmosisTxt()，调用了osgeo类，该类包含于GDAL[①]库中。GDAL是一个用于栅格（raster）和矢量（vector）地理空间数据格式的开源转换库，包含大量格式驱动。大多数Python的地理空间数据处理库通常基于GDAL编写，是最为基础的库。GeoPandas库基于Fiona库，方便用户对地理空间数据的处理，而Fiona包含链接到GDAL的扩展模块。通常，在使用Python地理空间数据库时，并没有使用哪个库的限制。目前，最常用处理地理空间数据的库是GeoPandas，但是有时这些库不能满足所有要求，因此需要调用GDAL来处理。查看GDAL帮助信息，可以浏览GDAL/OGR Cookbook![②]和GDAL documentation[③]。

① GDAL，一些用于编程和操作GDAL地理空间数据抽象库的工具，实际上包括两个库——GDAL库和OGR库，分别用于处理地理空间的栅格数据和矢量数据（https://pypi.org/project/GDAL/）。

② GDAL/OGR Cookbook，关于如何使用Python GDAL/OGR API的简单代码片断（http://pcjericks.github.io/py-gdalogr-cookbook/index.html）。

③ GDAL documentation，GDAL文档（https://gdal.org/）。

```
def shpPolygon2OsmosisTxt(shape_polygon_fp, osmosis_txt_fp):
    '''
    function - 转换 shape 的 polygon 为 osmium 的 polygon 数据格式（.txt），用于 .osm 地图数据的裁切
    Params:
        shape_polygon_fp - 输入 shape 地理数据格式的 polygon 文件路径；string
        osmosis_txt_fp - 输出为 osmosis 格式的 polygon 数据格式 .txt 文件路径；string
    Returns:
        None
    '''
    from osgeo import ogr  # osgeo 包含在 GDAL 库中
    # GDAL 能够处理众多地理数据格式，此时调入了 ESRI Shapefile 数据格式驱动
    driver = ogr.GetDriverByName('ESRI Shapefile')
    infile = driver.Open(shape_polygon_fp)  # 打开 .shp 文件
    layer = infile.GetLayer()  # 读取层
    f = open(osmosis_txt_fp, "w")
    f.write("osmosis polygon\nfirst_area\n")
    for feature in layer:
        feature_shape_polygon = feature.GetGeometryRef()
        print(feature_shape_polygon)  # 为 polygon
        firsts_area_linearring = feature_shape_polygon.GetGeometryRef(
            0)  # polygon 不包含嵌套，为单独的形状
        print(firsts_area_linearring)  # 为 linearRing
        area_vertices = firsts_area_linearring.GetPointCount()  # 提取 linearRing 对象的点数量
        for vertex in range(area_vertices):  # 循环点，并向文件中写入点坐标
            lon, lat, z = firsts_area_linearring.GetPoint(vertex)
            f.write("%s  %s\n" % (lon, lat))
    f.write("END\nEND")
    f.close()
# 转换初步实验边界
shape_polygon_fp = './data/OSMBoundary/OSMBoundary.shp'
osmosis_txt_fp = './data/OSMBoundary.txt'
shpPolygon2OsmosisTxt(shape_polygon_fp, osmosis_txt_fp)
# 转换代码调试小批量数据边界
shape_polygon_small_fp = './data/OSMBoundary_small/OSMBoundary_small.shp'
osmosis_txt_small_fp = './data/OSMBoundary_small.txt'
shpPolygon2OsmosisTxt(shape_polygon_small_fp, osmosis_txt_small_fp)
----------------------------------------------------------------------------------
POLYGON ((-90.0850881031402 40.9968994947319,-90.0850881031402
43.6657936592248,-87.383039973871 43.6657936592248,-87.383039973871
40.9968994947319,-90.0850881031402 40.9968994947319))
LINEARRING (-90.0850881031402 40.9968994947319,-90.0850881031402
```

```
43.6657936592248,-87.383039973871 43.6657936592248,-87.383039973871
40.9968994947319,-90.0850881031402 40.9968994947319)
POLYGON ((-87.6807286451907 41.8373927809521,-87.6807286451907
41.9214101975252,-87.5941157249019 41.9214101975252,-87.5941157249019
41.8373927809521,-87.6807286451907 41.8373927809521))
LINEARRING (-87.6807286451907 41.8373927809521,-87.6807286451907
41.9214101975252,-87.5941157249019 41.9214101975252,-87.5941157249019
41.8373927809521,-87.6807286451907 41.8373927809521)
```

执行转换后，写入.txt格式文件的实际实验边界Polygon如下：

```
osmosis polygon
first_area
-90.08508810314017  40.99689949473193
-90.08508810314017  43.66579365922478
-87.38303997387102  43.66579365922478
-87.38303997387102  40.99689949473193
-90.08508810314017  40.99689949473193
END
END
```

用于调试小批量数据提取的.txt边界：

```
osmosis polygon
first_area
-87.68072864519071  41.83739278095207
-87.68072864519071  41.92141019752525
-87.59411572490187  41.92141019752525
-87.59411572490187  41.83739278095207
-87.68072864519071  41.83739278095207
END
END
```

osmosis也提供了多个.osm地理空间数据的合并方法，合并示例代码为`osmosis --rx 1.osm --rx 2.osm --rx 3.osm --merge --wx merged.osm`。首先，执行合并操作，针对该实验的osmosis合并代码为`osmosis --rx "F:/GitHubBigData/illinois-latest.osm" --rx "F:/GitHubBigData/wisconsin-latest.osm"--merge --wx "F:/GitHubBigData/illinois-wisconsin.osm"`，合并后的文件大小为7.57GB。再执行裁切命令`osmosis --read-xml file="F:\GitHubBigData\illinois-wisconsin.osm" --bounding-polygon file="C:\Users\richi\omen_richiebao\omen_github\USDA_CH_final\USDA\notebook\data\OSMBoundary.txt" --write-xml file="F:\GitHubBigData\osm_clip.osm"`，裁切后的文件大小为3.80GB。可以再通过QGIS查看数据是否已经按照预期合并裁切完毕。用于代码调试的小规模数据提取直接裁切`osmosis --read-xml file="F:\GitHubBigData\illinois-wisconsin.osm" --bounding-polygon file="C:\Users\richi\omen_richiebao\omen_github\USDA_CH_final\USDA\notebook\data\OSMBoundary_small.txt" --write-xml file="F:\GitHubBigData\osm_small_clip.osm"`。

osmosis 命令行，于 Windows 系统的命令行终端中执行，建议在 Windows PowerShell[①] 终端中执行代码。具体文件路径须根据个人情况调整。

OSM使用附加在基本数据结构上的标签（tag）来表示地面上的物理特征（feature），例如道路或者建筑物等。在QGIS中，打开属性表3.2–1，可以查看各要素属于的标签。对于具体标签的内容，可以查看Map Features[②]，下面仅列出主要标签的分类（表3.2–2）。

3.2.1.2 读取、转换OSM数据

用Python读取OSM数据仍然使用osmcode.org提供的工具Pyosmium[③]。Pyosmium用于处理不同格式的OSM文件，内核为c++的osmium库，能够有效、快速地处理OSM数据。上述处理后的OSM数据文件osm_clip.osm为3.80GB，如果一开始就使用较大的数据来编写程序，花费的时间成本可能较高，因此可以给予更少的数据编写、调试，达到预期效果后，再使用待要分析的大文件数据。

① Windows PowerShell，是一个跨平台的任务自动化解决方案，由一个命令行外壳（shell）、一种脚本语言和一个配置管理框架（framework）组成（https://learn.microsoft.com/en-us/powershell/）。

② Map Features，使用附加在基本数据结构（nodes、ways、relations）上的标签来表示地面上的物理特征（例如，道路和建筑）。每个标签都描述了该node、way和relation所表示特征的一个地理属性（https://wiki.openstreetmap.org/wiki/Map_features）。

③ Pyosmium，是一个处理不同OSM文件格式的库，为一个C++库 的osmium封装器，允许快速、有效、连续地处理OpenStreetMap数据（https://docs.osmcode.org/pyosmium/latest/）。

表3.2-1 在QGIS中打开OSM属性表

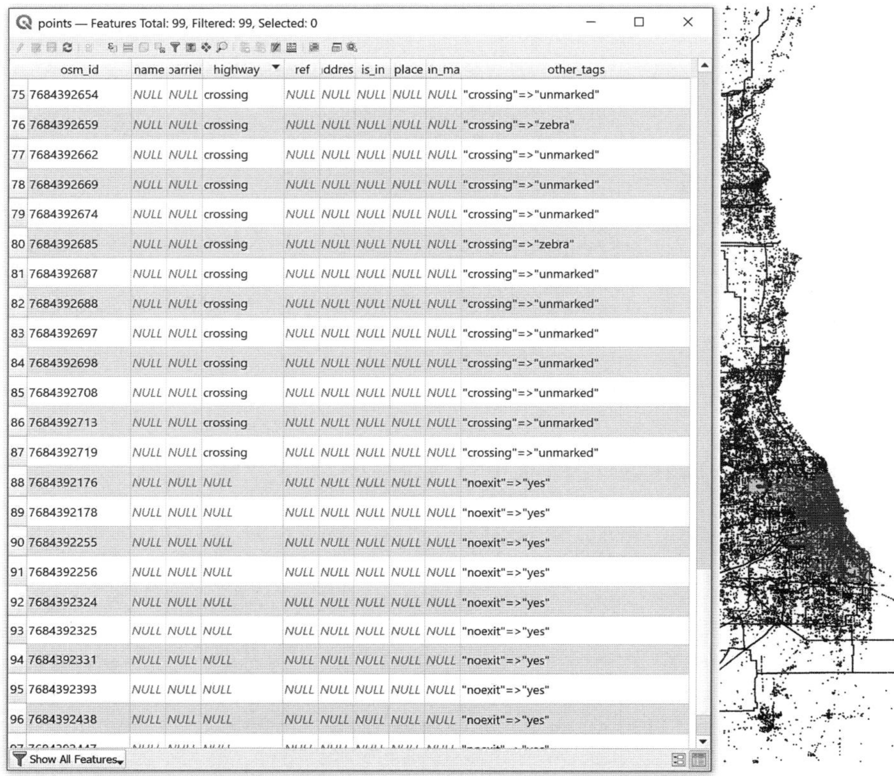

points — Features Total: 99, Filtered: 99, Selected: 0

	osm_id	name	arrie	highway	ref	ddres	is_in	place	n_ma	other_tags
75	7684392654	NULL	NULL	crossing	NULL	NULL	NULL	NULL	NULL	"crossing"=>"unmarked"
76	7684392659	NULL	NULL	crossing	NULL	NULL	NULL	NULL	NULL	"crossing"=>"zebra"
77	7684392662	NULL	NULL	crossing	NULL	NULL	NULL	NULL	NULL	"crossing"=>"unmarked"
78	7684392669	NULL	NULL	crossing	NULL	NULL	NULL	NULL	NULL	"crossing"=>"unmarked"
79	7684392674	NULL	NULL	crossing	NULL	NULL	NULL	NULL	NULL	"crossing"=>"unmarked"
80	7684392685	NULL	NULL	crossing	NULL	NULL	NULL	NULL	NULL	"crossing"=>"zebra"
81	7684392687	NULL	NULL	crossing	NULL	NULL	NULL	NULL	NULL	"crossing"=>"unmarked"
82	7684392688	NULL	NULL	crossing	NULL	NULL	NULL	NULL	NULL	"crossing"=>"unmarked"
83	7684392697	NULL	NULL	crossing	NULL	NULL	NULL	NULL	NULL	"crossing"=>"unmarked"
84	7684392698	NULL	NULL	crossing	NULL	NULL	NULL	NULL	NULL	"crossing"=>"unmarked"
85	7684392708	NULL	NULL	crossing	NULL	NULL	NULL	NULL	NULL	"crossing"=>"unmarked"
86	7684392713	NULL	NULL	crossing	NULL	NULL	NULL	NULL	NULL	"crossing"=>"unmarked"
87	7684392719	NULL	NULL	crossing	NULL	NULL	NULL	NULL	NULL	"crossing"=>"unmarked"
88	7684392176	NULL	NULL	NULL	NULL	NULL	NULL	NULL	NULL	"noexit"=>"yes"
89	7684392178	NULL	NULL	NULL	NULL	NULL	NULL	NULL	NULL	"noexit"=>"yes"
90	7684392255	NULL	NULL	NULL	NULL	NULL	NULL	NULL	NULL	"noexit"=>"yes"
91	7684392256	NULL	NULL	NULL	NULL	NULL	NULL	NULL	NULL	"noexit"=>"yes"
92	7684392324	NULL	NULL	NULL	NULL	NULL	NULL	NULL	NULL	"noexit"=>"yes"
93	7684392325	NULL	NULL	NULL	NULL	NULL	NULL	NULL	NULL	"noexit"=>"yes"
94	7684392331	NULL	NULL	NULL	NULL	NULL	NULL	NULL	NULL	"noexit"=>"yes"
95	7684392393	NULL	NULL	NULL	NULL	NULL	NULL	NULL	NULL	"noexit"=>"yes"
96	7684392438	NULL	NULL	NULL	NULL	NULL	NULL	NULL	NULL	"noexit"=>"yes"

Show All Features

表3.2-2 OSM数据主要标签

序号	一级标签	二级标签
1	Aerialway	
2	Aeroway	
3	Amenity	Sustenance,Education, Transportation, Financial, Healthcare, Entertainment, Arts & Culture, Others
4	Barrier	Linear barriers, Access control on highways
5	Boundary	Attributes
6	Building	Accommodation, Commercial, Religious,Civic/Amenity, Agricultural/Plant production, Sports,Storage,Cars, Power/Technical buildings,Other Buildings, Additional Attributes
7	Craft	
8	Emergency	Medical Rescue,Firefighters, Lifeguards, Assembly point,Other Structure
9	Geological	
10	Highway	Roads, Link roads, Special road types, Paths,Lifecycle, Attributes,Other highway features
11	Historic	
12	Landuse	Common Landuse Key Values - Developed land, Common Landuse Key Values - Rural and agricultural land, Other Landuse Key Values
13	Leisure	
14	Man_made	
15	Military	
16	Natural	Vegetation or surface related, Water related, Landform related

续表

序号	一级标签	二级标签
17	Office	
18	Place	Administratively declared places, Populated settlements, urban,Populated settlements, urban and rural, Other places
19	Power	Public Transport
20	Public Transport	
21	Railway	Tracks, Additional features, Stations and Stops, Other railways
22	Route	
23	Shop	Food, beverages, General store, department store, mall, Clothing, shoes, accessories, Discount store, charity, Health and beauty, Do-it-yourself, household, building materials, gardening, Furniture and interior, Electronics, Outdoors and sport, vehicles, Art, music, hobbies, Stationery, gifts, books, newspapers, Others
24	Sport	
25	Telecom	
26	Tourism	
27	Waterway	Natural watercourses, Man-made waterways, Facilities,Barriers on waterways,Other features on waterways
附加属性 (Additional properties)		
1	Addresses	Tags for individual houses, For countries using hamlet, subdistrict, district, province, state, Tags for interpolation ways
2	Annotation	
3	Name	
4	Properties	
5	References	
6	Restrictions	

osmosis 工具是由 OSM 及其开源社区成员建立的 osmcode.org 开发的工具。

Pyosmium GitHub 仓库及安装详见：https://github.com/osmcode/pyosmium

编写读取.osm数据，需要对OSM的数据结构有所了解，从而能够清晰地提取所需要的值。元素（elements）[①]是OSM物理世界概念数据模型的基本组成部分，包括节点nodes、路径或区域ways、关系relations及标签tag，见表3.2–3。

① 元素（elements），是OpenStreetMap物理世界概念数据模型的基本组成部分，包括nodes、ways和relations（https://wiki.openstreetmap.org/wiki/Elements）。

表3.2–3 OSM 数据元素

元素（elements）	图标	解释	对位shape地理空间数据（vector矢量）
node		由经纬度坐标定义的地理空间点	point
way		由点（20～2000个）构成的路径以及闭合的区域，包含open way、closed way和area	polyline,polygon
relation		记录两个或多个元素之间关系的多用途数据结构，关系可以有不同的含义，其意义由对应的标签定义	
tag	k=v	nodes、ways和relations都可以有描述其意义的标签，一个标签由键key:值value组成，Key必须是唯一的	字段 field

- 示例

Node

```
<node id="25496583" lat="51.5173639" lon="-0.140043" version="1"
changeset="203496" user="80n" uid="1238" visible="true" timestamp="2007-01-
28T11:40:26Z">
    <tag k="highway" v="traffic_signals"/>
</node>
```

way

简单的路径或区域（simple way）

```
  <way id="5090250" visible="true" timestamp="2009-01-19T19:07:25Z" version="8"
changeset="816806" user="Blumpsy" uid="64226">
    <nd ref="822403"/>
    <nd ref="21533912"/>
    <nd ref="821601"/>
    <nd ref="21533910"/>
    <nd ref="135791608"/>
    <nd ref="333725784"/>
    <nd ref="333725781"/>
    <nd ref="333725774"/>
    <nd ref="333725776"/>
    <nd ref="823771"/>
    <tag k="highway" v="residential"/>
    <tag k="name" v="Clipstone Street"/>
    <tag k="oneway" v="yes"/>
  </way>
```

多边形区域集合（multipolygon area），见图 3.2–2。

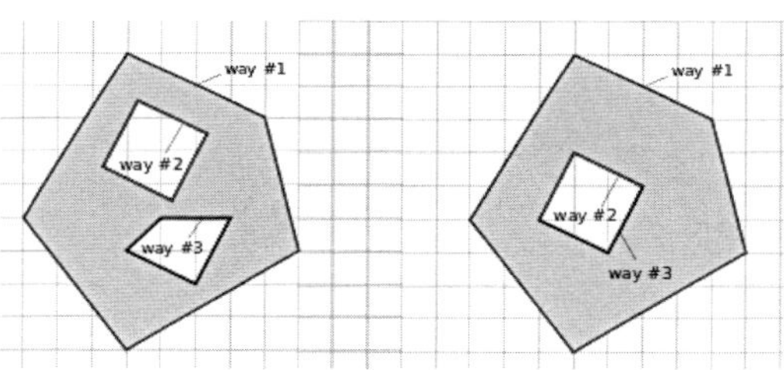

图3.2–2　多边形区域集合

```
  <relation id="12" timestamp="2008-12-21T19:31:43Z" user="kevjs1982"
uid="84075">
    <member type="way" ref="2878061" role="outer"/> <!-- picture ref="1" -->
    <member type="way" ref="8125153" role="inner"/> <!-- picture ref="2" -->
    <member type="way" ref="8125154" role="inner"/> <!-- picture ref="3" -->
    <member type="way" ref="3811966" role=""/> <!-- empty role produces
        a warning; avoid this; most software works around it by computing
        a role, which is more expensive than having one set explicitly;
        not shown in the sample pictures to the right -->
    <tag k="type" v="multipolygon"/>
  </relation>
```

- 元素的属性

了解OSM基本的数据类型、结构和属性（表3.2–4），通过继承osmium的类.SimpleHandler，用.apply_file方法传入.osm文件，并定义所要提取的元素类型，给出该元素类型的属性提取对应的属性值。在地理空间数据分析中，通常比较关键的属性包括：元素修改的最后时间（.timestamp），标签（tags,<tag.k,tag.v>），生成的几何对象（geometry<point,linestring,multipolygon>）。下述函数分别提取了node、way（area）对象的属性和几何对象；同时，将其转换为GeoDataFrame的数据格式，并存储为GPKG的格式数据，方便日后调用，尤其是大批量数据。因为涉及大批量数据，因此调入datatime时间模块，观察所用时间，帮助调试代码。

表3.2-4 OSM 数据类型

名称	值类型	解释
id	integer (64-bit)	用于表示元素
user	character string	最后修改对象的用户名
uid	integer	最后修改对象的用户ID
timestamp	W3C standard date and time	最后修改时间
visible	"true" or "false"	数据库中的对象是否被删除
version	integer	版本控制
changeset	integer	创建或更新对象时使用的变更集编号

```
import datetime
import osmium as osm
class osmHandler(osm.SimpleHandler):
    '''
    class-通过继承osmium类 class osmium.SimpleHandler读取.osm数据.
    '''
    def __init__(self):
        osm.SimpleHandler.__init__(self)
        self.osm_node = []
        self.osm_way = []
        self.osm_area = []
    def node(self, n):
        import pandas as pd
        import shapely.wkb as wkblib
        wkbfab = osm.geom.WKBFactory()
        wkb = wkbfab.create_point(n)
        point = wkblib.loads(wkb, hex=True)
        self.osm_node.append([
            'node',
            point,
            n.id,
            n.version,
            n.visible,
            pd.Timestamp(n.timestamp),
            n.uid,
            n.user,
            n.changeset,
            len(n.tags),
            {tag.k: tag.v for tag in n.tags},
        ])
    def way(self, w):
        import pandas as pd
        import shapely.wkb as wkblib
        wkbfab = osm.geom.WKBFactory()
        try:
            wkb = wkbfab.create_linestring(w)
            linestring = wkblib.loads(wkb, hex=True)
            self.osm_way.append([
                'way',
                linestring,
                w.id,
                w.version,
                w.visible,
                pd.Timestamp(w.timestamp),
                w.uid,
                w.user,
                w.changeset,
                len(w.tags),
                {tag.k: tag.v for tag in w.tags},
            ])
        except:
            pass
    def area(self, a):
        import pandas as pd
        import shapely.wkb as wkblib
        wkbfab = osm.geom.WKBFactory()
        try:
            wkb = wkbfab.create_multipolygon(a)
```

```
                multipolygon = wkblib.loads(wkb, hex=True)
                self.osm_area.append([
                    'area',
                    multipolygon,
                    a.id,
                    a.version,
                    a.visible,
                    pd.Timestamp(a.timestamp),
                    a.uid,
                    a.user,
                    a.changeset,
                    len(a.tags),
                    {tag.k: tag.v for tag in a.tags},
                ])
            except:
                pass
a_T = datetime.datetime.now()
print("start time:", a_T)
# osm_Chicago_fp=r"F:\data\osm_small_clip.osm"  # 待读取的 .osm 数据路径，用提取的小范围数据调试代码
osm_Chicago_fp = r"F:\data\osm_clip.osm"  # 用小批量数据调试完之后，计算实际的实验数据
osm_handler = osmHandler()  # 实例化类 osmHandler()
# 调用 class osmium.SimpleHandler 的 apply_file 方法
osm_handler.apply_file(osm_Chicago_fp, locations=True)
b_T = datetime.datetime.now()
print("end time:", b_T)
duration = (b_T-a_T).seconds/60
print("Total time spend:%.2f minutes" % duration)
--------------------------------------------------------------------------------
start time: 2021-12-24 11:00:18.357860
end time: 2021-12-24 11:34:11.350215
Total time spend:33.87 minutes
```

当读取全部OSM元素数据后，定义保存函数。如果是小批量数据，通常可以一起保存。但是，本次实验数据有3.80GB，将读取后的数据转换为GeoDataFrame数据格式保存，将花费较多时间。因此将OSM元素逐个转换保存。同时，注意到在小批量调试时，保存node为GeoJSON格式文件其大小为104MB，而保存为GPKG仅有52.3MB。因此，对于实验数据的保存，这里选择后者。

```
def save_osm(osm_handler, osm_type, save_path=r"./data/", fileType="GPKG"):
    '''
    function - 根据条件逐个保存读取的 osm 数据（node, way and area）
    Params:
        osm_handler - osm 返回的 node,way 和 area 数据，配套类 osmHandler(osm.
SimpleHandler) 实现；Class
        osm_type - 要保存的 osm 元素类型，包括 "node", "way" 和 "area"; string
        save_path - 保存路径。The default is "./data/"; string
        fileType - 保存的数据类型,包括 "shp", "GeoJSON", "GPKG"。The default is "GPKG";
string
    Returns:
        osm_node_gdf - OSM 的 node 类；GeoDataFrame(GeoPandas)
        osm_way_gdf - OSM 的 way 类；GeoDataFrame(GeoPandas)
        osm_area_gdf - OSM 的 area 类；GeoDataFrame(GeoPandas)
    '''
    import geopandas as gpd
    import os
    import datetime
    a_T = datetime.datetime.now()
    print("start time:", a_T)
    def duration(a_T):
        b_T = datetime.datetime.now()
        print("end time:", b_T)
        duration = (b_T-a_T).seconds/60
        print("Total time spend:%.2f minutes" % duration)
    def save_gdf(osm_node_gdf, fileType, osm_type):
        if fileType == "GeoJSON":
            osm_node_gdf.to_file(os.path.join(
                save_path, "osm_%s.geojson" % osm_type), driver='GeoJSON')
        elif fileType == "GPKG":
            osm_node_gdf.to_file(os.path.join(
                save_path, "osm_%s.gpkg" % osm_type), driver='GPKG')
        elif fileType == "shp":
            osm_node_gdf.to_file(os.path.join(
                save_path, "osm_%s.shp" % osm_type))
```

```
    epsg_wgs84 = 4326  #配置坐标系统，参考：https://spatialreference.org/
    osm_columns = ['type', 'geometry', 'id', 'version', 'visible', 'ts', 'uid', 'user',
                   'changeet', 'tagLen', 'tags']
    if osm_type == "node":
        osm_node_gdf = gpd.GeoDataFrame(
            osm_handler.osm_node, columns=osm_columns, crs=epsg_wgs84)
        save_gdf(osm_node_gdf, fileType, osm_type)
        duration(a_T)
        return osm_node_gdf
    elif osm_type == "way":
        osm_way_gdf = gpd.GeoDataFrame(
            osm_handler.osm_way, columns=osm_columns, crs=epsg_wgs84)
        save_gdf(osm_way_gdf, fileType, osm_type)
        duration(a_T)
        return osm_way_gdf
    elif osm_type == "area":
        osm_area_gdf = gpd.GeoDataFrame(
            osm_handler.osm_area, columns=osm_columns, crs=epsg_wgs84)
        save_gdf(osm_area_gdf, fileType, osm_type)
        duration(a_T)
        return osm_area_gdf
node_gdf=save_osm(osm_handler,osm_type="node",save_path="./data/OSM_
processed",fileType="GPKG")
--------------------------------------------------------------------------------
start time: 2021-12-24 11:35:34.746935
end time: 2021-12-24 13:03:08.123318
Total time spend:87.55 minutes
--------------------------------------------------------------------------------
way_gdf=save_osm(osm_handler,osm_type="way",save_path=r"./data/OSM_
processed",fileType="GPKG")
--------------------------------------------------------------------------------
start time: 2021-12-24 14:16:00.736027
end time: 2021-12-24 14:28:42.828275
Total time spend:12.70 minutes
--------------------------------------------------------------------------------
area_gdf=save_osm(osm_handler,osm_type="area",save_path=r"./data/OSM_
processed",fileType="GPKG")
--------------------------------------------------------------------------------
start time: 2021-12-24 14:29:30.864597
end time: 2021-12-24 14:39:25.170465
Total time spend:9.90 minutes
```

存储过程中，该部分实验数据node元素存储时间约87分钟，存储的GPKG文件大小为3.10GB，way和area元素则存储时间相对较短，存储文件较小。因为已经转换为GeoDataFrame格式地理空间数据，因此可以直接`.plot()`查看数据分布情况，初步判断是否读取和转换正确。下述代码用来测试读取的时间，其中way（图3.2–3）和area（图3.2–4）读取时间较短，而node（图3.2–5）元素读取时间较长。

```
import geopandas as gpd
def start_time():
    '''
    function-计算当前时间
    '''
    import datetime
    start_time = datetime.datetime.now()
    print("start time:", start_time)
    return start_time
def duration(start_time):
    '''
    function-计算持续时间
    Params:
    start_time - 开始时间；datatime
    '''
    import datetime
    end_time = datetime.datetime.now()
    print("end time:", end_time)
    duration = (end_time-start_time).seconds/60
    print("Total time spend:%.2f minutes" % duration)
start_time=start_time()
read_way_gdf = gpd.read_file("./data/OSM_processed/osm_way.gpkg")
duration(start_time)
--------------------------------------------------------------------------------
```

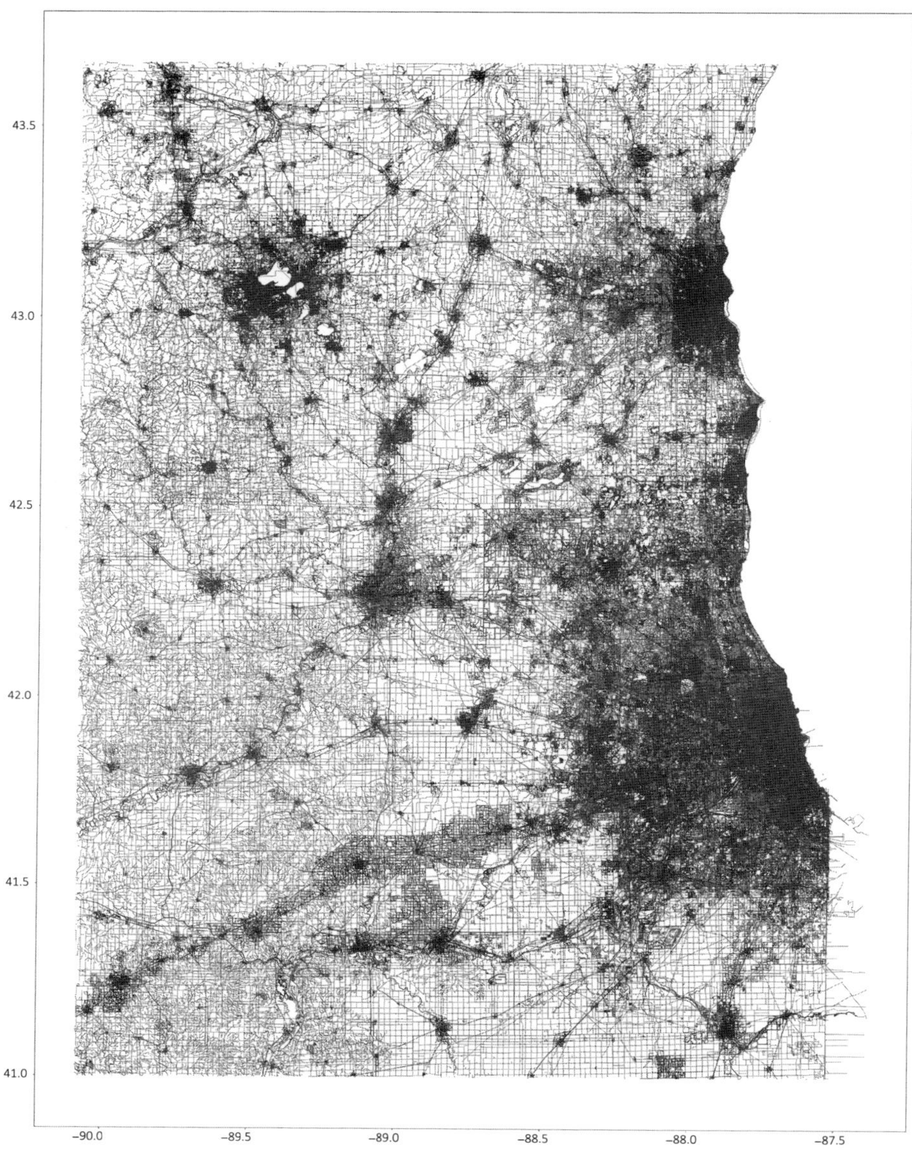

图3.2-3 OSM的Way元素

```
start time: 2021-12-24 17:14:44.041892
end time: 2021-12-24 17:17:46.256983
Total time spend:3.03 minutes
    # 因为数据量较大，直接gdf.plot()会花费较长时间
import matplotlib.pyplot as plt
Chicago_boundary_city_fp = './data/Chicago boundaries_city/Chicago boundaries_
city.shp'
Chicago_boundary_city = gpd.read_file(Chicago_boundary_city_fp)
fig, ax = plt.subplots(figsize=(40,40))
read_way_gdf.plot(ax=ax,linewidth=0.5,zorder=1)
Chicago_boundary_city.plot(ax=ax,color='none', edgecolor='red',alpha=1,linewidth
=5,zorder=2)
plt.show()
```

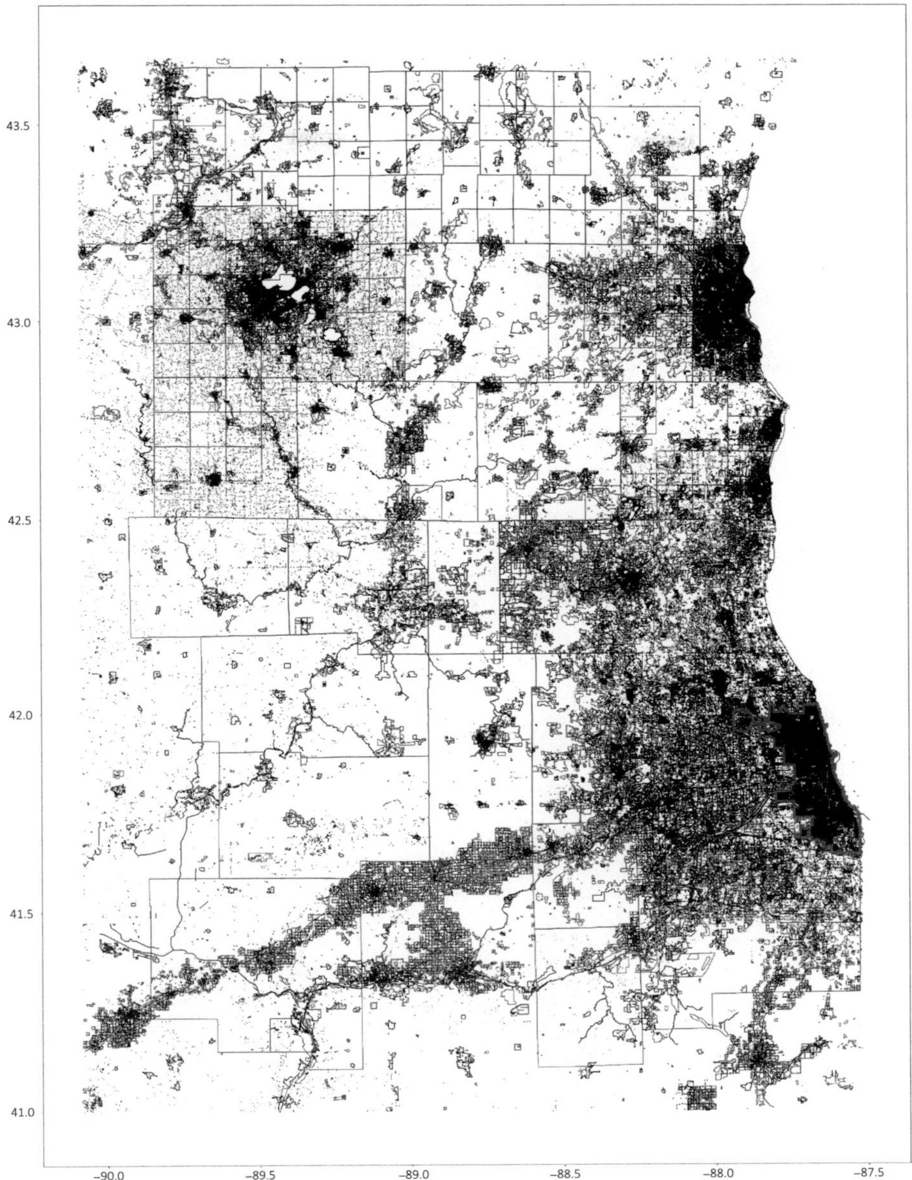

图3.2-4 OSM的Area元素

```
del read_way_gdf #如果内存有限，可以使用 del 删除不再使用的变量，节约内存
----------------------------------------------------------------------------
start_time = start_time()
read_area_gdf = gpd.read_file(“./data/OSM_processed/osm_area.gpkg”)
duration(start_time)
----------------------------------------------------------------------------
start time: 2021-12-24 17:10:05.802124
end time: 2021-12-24 17:12:52.787002
Total time spend:2.77 minutes
----------------------------------------------------------------------------
fig, ax = plt.subplots(figsize=(40,40))
read_area_gdf.plot(ax=ax,color='none', edgecolor = 'black',linewidth=0.5,zorder=1)
Chicago_boundary_city.plot(ax=ax,color='none', edgecolor='red',alpha=1,linewidth
```

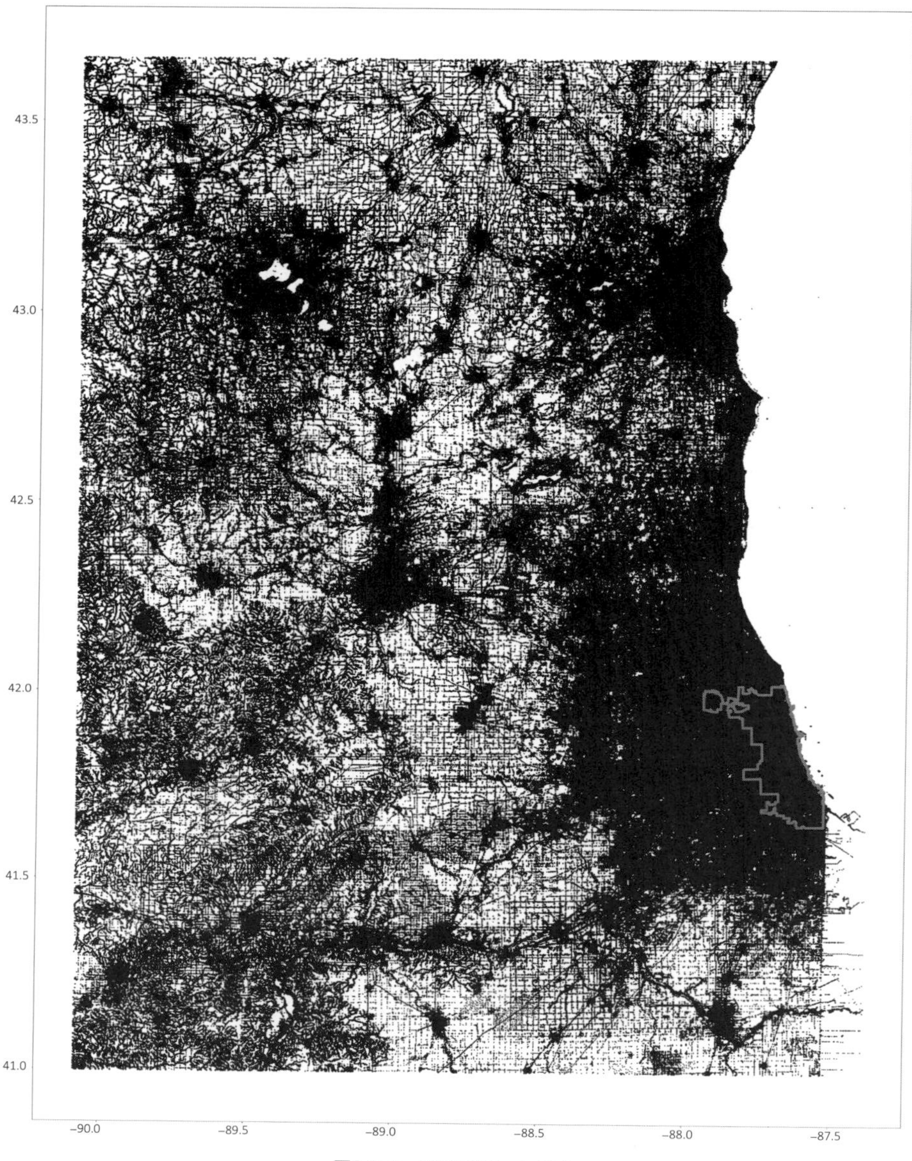

图3.2-5 OSM的Node元素

```
=5,zorder=2)
plt.show()
start_time = start_time()
read_node_gdf = gpd.read_file(“./data/OSM_processed/osm_node.gpkg”)
duration(start_time)
--------------------------------------------------------------------------------
start time: 2021-12-24 17:43:04.319218
end time: 2021-12-24 18:13:09.240528
Total time spend:30.07 minutes
--------------------------------------------------------------------------------
fig, ax = plt.subplots(figsize=(40,40))
read_node_gdf.plot(ax=ax,color='black',markersize=1)
Chicago_boundary_city.plot(ax=ax,color='none', edgecolor='red',alpha=1,linewidth
=5,zorder=2)
plt.show()
```

3.2.2 核密度估计与地理空间点密度分布

3.2.2.1 单变量（一维数组）的核密度估计

在正态分布一节中阐述有，当直方图的组距无限缩小至极限后，能够拟合出一条曲线，计算这个分布曲线的公式为概率密度函数（Probability Density Function，PDF）。用于估计概率密度函数的非参数方法就是核密度估计。核密度估计（Kernel density estimation，KDE）是一个基本的数据平滑问题（fundamental data smoothing problem），例如对不平滑的直方图平滑，给定一个核K，并指定带宽（bandwidth），其值为正数，核密度估计定义为：$\hat{f}_n(x)=\frac{1}{nh}\sum_{i=1}^{n}K(\frac{x-x_i}{h})$。式中，$K$为核函数，$h$为带宽，核函数有多个，例举其中高斯核为：$\frac{1}{\sqrt{2\pi}}\mathrm{e}^{-\frac{1}{2}x^2}$，将其带入核密度估计公式结果为：$\hat{f}_n(x)=\frac{1}{\sqrt{2\pi}nh}\sum_{i=1}^{n}\mathrm{e}^{-\frac{(x-x_i)^2}{2h^2}}$。

在下述代码中绘制了三条曲线（图3.2-6），红色粗线为概率密度函数；两条细线均为核密度估计（高斯核），只是蓝色线是依据核密度公式直接编写代码，并设置带宽h=0.4；绿色线则是使用SciPy库下的`stats.gaussian_kde()`方法计算高斯核密度估计。

非参数统计（Nonparametric Statistics）是统计的一个分支，但不是完全基于参数化的概率分布，例如通过参数均值和方差（或标准差）定义一个正态分布，非参数统计基于自由分布（distribution-free）或指定的分布但未给分布参数，例如当处理 PDF 的一般情况时，不能像正态分布那样给定参数进行分类。其基本思想是在尽可能少的假定时，利用数据对一个未知量做出推断，通常意味着利用具有无穷维的统计模型。

对于核密度估计名词中密度一词可以形象理解为下图中橄榄绿小竖线的分布密度。

```
import numpy as np
from scipy import stats
import matplotlib.pyplot as plt
import math

# 等分概率为 0.1% 到 99.9% 之间的数值。如果不给参数 loc 和 scale，则默认为标准正态分布，即 loc=0, scale=1
x = np.linspace(stats.norm.ppf(0.001, loc=0, scale=1),
                stats.norm.ppf(0.999, loc=0, scale=1), 100)
pdf = stats.norm.pdf(x)
plt.figure(figsize=(25, 5))
plt.plot(x, pdf, 'r-', lw=5, alpha=0.6, label='norm_pdf')
random_variates = stats.norm.rvs(loc=0, scale=1, size=500)
count, bins, ignored = plt.hist(
    random_variates, bins=100, density=True, histtype='stepfilled', alpha=0.2)
plt.eventplot(random_variates, color='y', linelengths=0.03,
              lineoffsets=0.025)  # 给定位置画出对应的短线

rVar_sort = np.sort(random_variates)
h = 0.4  # 带宽（bandwidth,bw）
n = len(rVar_sort)
kde_Gaussian = [sum(math.exp(-1*math.pow(vi-vj, 2)/(2*math.pow(h, 2)))
                    for vj in rVar_sort)/(h*n*math.sqrt(2*math.pi)) for vi in
rVar_sort]  # 将上述高斯核密度估计公式转换为代码
plt.plot(rVar_sort, kde_Gaussian, 'b-', lw=2,
        alpha=0.6, label='kde_formula,h=%s' % h)
scipyStatsGaussian_kde = stats.gaussian_kde(random_variates)
plt.plot(bins, scipyStatsGaussian_kde(bins), 'g-', lw=2,
        alpha=0.6, label='scipyStatsGaussian_kde')
plt.legend()
plt.show()
```

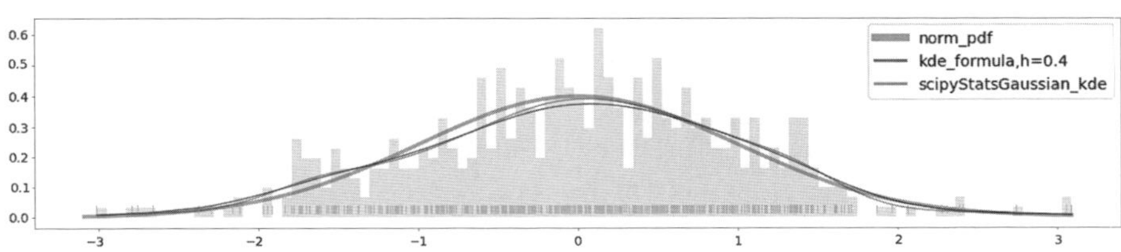

图3.2-6 核密度估计（单变量）

带宽（bandwidth）影响光滑的程度，下述实验设置不同的值，观察核密度曲线的变化情况（图3.2–7）。关于最适宜的带宽推断，*Bonus algorithm for large scale stochastic nonlinear programming problems*第3章，*Probability Density Function and Kernel Density Estimation*一节中提到一种方法，可以参考。

```
bws = np.arange(0.1, 1, 0.2)
colors_kde = ['C{}'.format(i) for i in range(len(bws))]
# maplotlib，指定颜色
i = 0
plt.figure(figsize=(25, 5))
for h in bws:
    kde_Gaussian = [sum(math.exp(-1*math.pow(vi-vj, 2)/(2*math.pow(h, 2)))
for vj in rVar_sort)/(h*n*math.sqrt(2*math.pi)) for vi in rVar_sort]  #将上述高斯核密度估计公式转换为代码
    plt.plot(rVar_sort, kde_Gaussian,
             color=colors_kde[i], lw=2, alpha=0.6, label='kde_formula,h=%.2f' % h)
    i += 1
plt.legend()
plt.show()
```

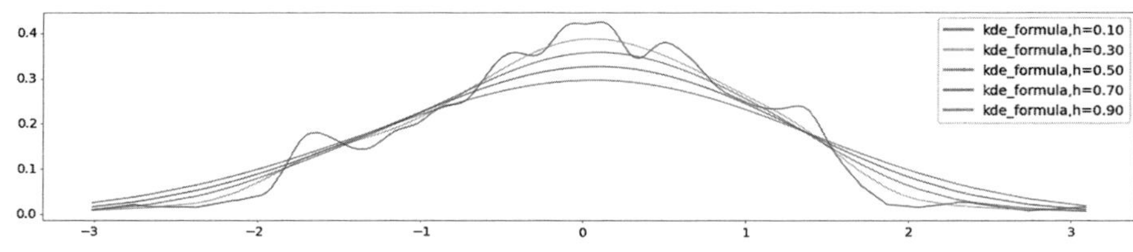

图3.2–7　不同带宽核密度曲线

3.2.2.2　多变量（多维数组）的核密度估计

上文叙述了OSM数据处理的方法，并简要地概述了OSM地理空间数据集的结构。OSM的node标签分类丰富，针对不同的问题可以提取不同的标签用于分析。因为标签中的便利设施（amenity）与人们的日常生活栖息相关，包括生计、教育、交通、金融、医疗、娱乐、艺术和文化等，更进一步的分类则多达100多类。使用GeoPandas的`gdf.clip(polygon)`方法进一步裁切数据到目标区域保存，读取后再提取标签（tags）为amenity的空间点数据（表3.2–5），计算核密度估计，查看其分布情况。

```
import util_misc
import geopandas as gpd
start_time = util_misc.start_time() # start_time() 和 duration(start_time) 函数放置于 util_misc.py 文件中，方便调用
read_node_gdf = gpd.read_file("./data/OSM_processed/osm_node.gpkg")
util_misc.duration(start_time)
--------------------------------------------------------------------------------
start time: 2021-12-24 21:11:29.190573
end time: 2021-12-24 21:40:36.262947
Total time spend:29.12 minutes
--------------------------------------------------------------------------------
read_node_gdf.head()
```

- 给定边界（SHP格式，例如用于物种-鸟观测分析的边界），裁切OSM的node数据。

```
chicago_species_study_area_wgs84_fp = './data/Chicago study area_species/
chicago_species_study_area_wgs84.shp'
chicago_species_study_area_wgs84 = gpd.read_file(chicago_species_study_area_
wgs84_fp)
start_time = util_misc.start_time()
chicago_species_node_gdf = read_node_gdf.clip(chicago_species_study_area_wgs84)
util_misc.duration(start_time)
chicago_species_node_gdf.to_file('./data/OSM_processed/osm_node_4speciesStudy.
gpkg',driver='GPKG')
--------------------------------------------------------------------------------
start time: 2021-12-24 21:49:56.113615
```

```
end time: 2021-12-24 22:07:42.284118
Total time spend:17.77 minutes
-------------------------------------------------------
import geopandas as gpd
osm_node_4speciesStudy = gpd.read_file('G:/data/OSM_
processed/osm_node_4speciesStudy.gpkg')
amenity_poi = osm_node_4speciesStudy[osm_node_4speciesStudy.
tags.apply(lambda row: "amenity" in eval(row).keys())] # 提取
标 tags 列，含标签 amenity 的所有行
print("Amenity extraction finished!")
-------------------------------------------------------
Amenity extraction finished!
```

提取仅含amenity的行后，数据大幅度减少（图3.2–8）。

```
print(
    "the overal data number:",osm_node_4speciesStudy.
shape,'\n',"the amenity data number:",amenity_poi.
shape,'\n',
    )
import matplotlib.pyplot as plt
Chicago_boundary_city_fp = './data/Chicago_boundaries_city/
Chicago_boundaries_city.shp'
Chicago_boundary_city = gpd.read_file(Chicago_boundary_
city_fp)
fig, ax = plt.subplots(figsize=(15,15))
amenity_poi.plot(ax=ax,marker=".",markersize=5,zorder=1)
Chicago_boundary_city.plot(ax=ax,color='none', edgecolor='r
```

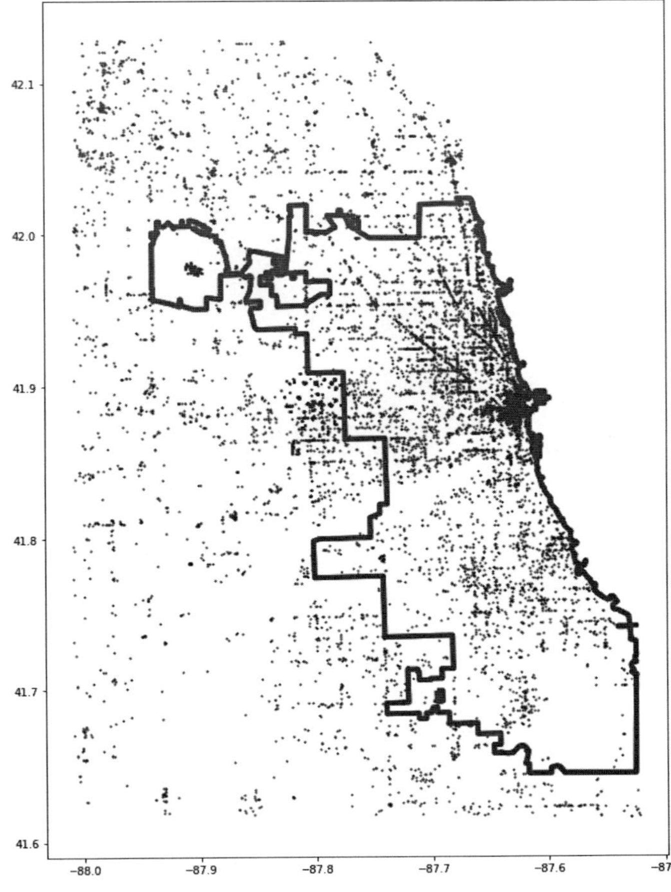

图3.2–8 设施（amenity）分布

表3.2–5 node标签

	type	id	version	ts	tags	geometry
0	node	219850	55	2018-02-20T05:50:28	{ "highway" : "motorway_junction" , "ref" : "276C" }	POINT (-87.91012 41.75859)
1	node	219851	48	2018-02-20T05:50:29	{ "highway" : "motorway_junction" , "ref" : "277A" }	POINT (-87.90764 41.75931)
3	node	219968	12	2015-08-04T05:38:49	{ "ref" : "73B" , "highway" : "motorway_junction" }	POINT (-87.92464 43.05606)

```
ed',alpha=1,linewidth=5,zorder=2)
plt.show()
----------------------------------------------------------------
the overal data number: (7124076, 11)
the amenity data number: (12606, 11)
```

- **核密度估计-scipy.stats.gaussian_kde()方法**

核密度估计可以平滑多维数据，例如热力图的制作是基于核密度估计的二维平滑，以OSM数据为例，直接使用scipy.stats.gaussian_kde()计算点分布的核密度。

```
import pandas as pd
import numpy as np
from scipy import stats
pd.options.mode.chained_assignment = None
poi_coordinates = np.array([amenity_poi.geometry.x,amenity_poi.
geometry.y])
amenity_kernel = stats.gaussian_kde(poi_coordinates) #核密度估计
amenity_poi['amenityKDE'] = amenity_kernel(poi_coordinates)
```

建立地图，打印显示核密度分布（图3.2-9）及其值（表3.2-6）。

```
# mapbox 为连接至境外服务器，国内有可能无法显示地图

import plotly.express as px
mapbox_token = 'pk.YmFvIiwiYSI6ImNrYjB3N2NyMzBlMG8yc254dTRzNnMye
HMifQ.QT7MdjQKs9Y6OtaJaJAn0A'
px.set_mapbox_access_token(mapbox_token)
fig = px.scatter_mapbox(amenity_poi, lat=amenity_poi.geometry.y,
lon=amenity_poi.geometry.x,
                        color='amenityKDE', color_continuous_
scale=px.colors.sequential.PuBuGn, size_max=15, zoom=10)
fig.update_layout(autosize=False, width=800, height=800,)
fig.show()
amenity_poi.head()
```

表3.2-6 OSM设施（amenity）的核密度值

	type	id	ts	tags	geometry	amenityKDE
51751	node	354150593	2009-08-09T20:12:06	{"ele": "185", "name": "Wacker Elementary Scho...	POINT (-87.64847 41.71624)	3.342937
73722	node	354128001	2009-03-01T01:48:33	{"ele": "179", "name": "Our Lady School", "ame...	POINT (-87.56699 41.71559)	1.879984
73753	node	354247611	2010-02-17T23:04:22	{"ele": "179", "name": "Our Lady Gate of Heave...	POINT (-87.56671 41.71587)	1.900003

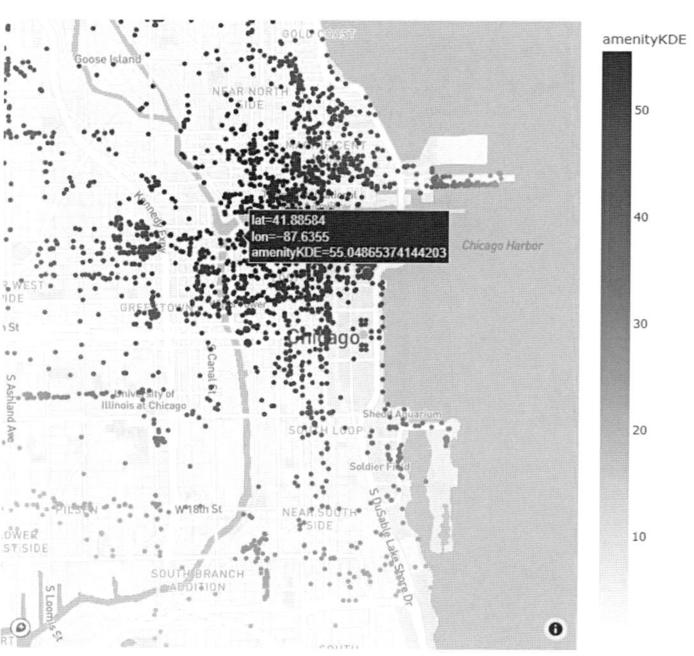

图3.2-9 OSM设施（amenity）的核密度分布

提取amenity字段，查看分类，并单独计算restaurant的核密度（图3.2-10）。

```
amenity_poi['amenity'] = amenity_poi.tags.apply(lambda row:eval(row)["amenity"])
print("服务设施子类：\n",amenity_poi.amenity.unique())
服务设施子类：
['school' 'place_of_worship' 'fountain' 'fuel' 'bank' 'parking' 'restaurant' 'fire_
station' 'bar' 'bureau_de_change' 'ice_cream' 'public_building' 'fast_food' 'courthouse'
'bicycle_rental' 'pharmacy' 'atm' 'clinic' 'social_facility' 'drinking_water' 'vending_
machine'
 'money_transfer' 'dentist' 'cafe' 'bicycle_parking' 'bench' 'theatre' 'waste_basket'
'recycling' 'parking_entrance' 'social_centre' 'arts_centre' 'car_wash' 'post_box'
'bicycle_repair_station' 'cinema' 'pub' 'car_sharing' 'shower' 'community_centre'
'car_rental' 'toilets' 'veterinary' 'post_office' 'marketplace' 'boat_rental' 'photo_
booth' 'charging_station' 'taxi' 'food_court' 'police' 'loading_dock' 'library' 'venue'
'nightclub' 'hospital' 'dojo' 'shelter' 'ferry_terminal'
 'doctors' 'college' 'studio' 'embassy' 'university' 'events_venue' 'clock'
'retail' 'internet_cafe' 'driving_school' 'health_center' 'parking_space'
'research_institute' 'bus_station' 'kindergarten' 'townhall' 'community_center'
'Auto Repair Shop' 'public_bookcase'
 'public_bath' 'device_charging_station' 'book_return' 'animal_boarding'
'childcare' 'music_school' 'dancing_school' 'Community Center' 'dental' 'grave_
yard' 'telephone' 'language_school' 'prep_school' 'spa' 'gym'
 'animal_shelter' 'payment_centre' 'salon' 'dance_theater' 'brewery' 'water_
point' 'bbq' 'gymnasium' 'waste_disposal' 'music_venue' 'ATM' 'swimming_pool'
'payment_terminal' 'child_care']
--------------------------------------------------------------------------------
restaurant_df = amenity_poi.loc[amenity_poi.amenity=='restaurant']
restaurant_coordinates = np.array([restaurant_df.geometry.x,restaurant_df.
geometry.y])
restaurant_kernel = stats.gaussian_kde(restaurant_coordinates)
restaurant_df['restaurant_kde'] = restaurant_kernel(restaurant_coordinates)
import plotly.express as px
mapbox_token = 'pk.YmFvIiwiYSI6ImNrYjB3N2NyMzBlMG8yc254dTRzNnMyeHMifQ.
QT7MdjQKs9Y6OtaJaJAn0A'
px.set_mapbox_access_token(mapbox_token)
fig = px.scatter_mapbox(restaurant_df,lat=restaurant_df.geometry.y,
```

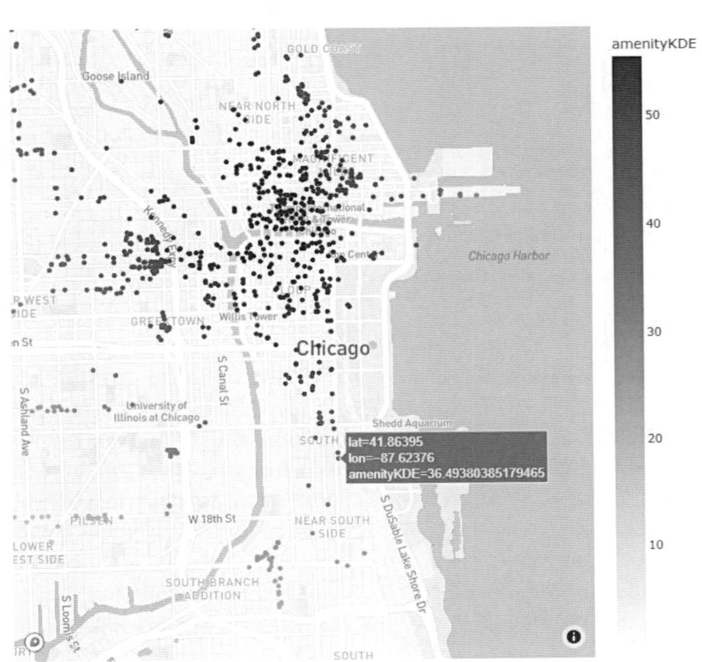

图3.2-10 餐厅（restaurant）的核密度分布

```
lon=restaurant_df.geometry.x,color='amenityKDE',color_continuous_scale=px.
colors.sequential.PuBuGn,size_max=15,zoom=10)
fig.update_layout(autosize=False,width=800,height=800,)
fig.show()
```

● 核密度估计结果转换为地理栅格数据

在地图表达上，上述二维地理空间点的核密度估计最终的计算值落在了点位置本身。如果希望能够以栅格的形式显示估计值，可以将其转换为栅格数据，即为地理空间点数据转栅格数据。在下述的方法中，提供了两种转换方式，一种方式是将存储有核密度估计值的GeoDataFrame格式数据先使用gdf.to_file()存储为SHP格式的点数据，然后定义SHP转栅格的函数；第二种方式是定义一个函数，能够直接计算核密度估计并同时直接存储为栅格数据。

EPSG编号体系

在这两种方式中均需要定义GeoDataFrame格式数据的坐标投影，并提取信息用于使用GDAL库提取或定义坐标投影，因此必须在GeoPandas库和GDAL库之间有可以互相转换的共同坐标体系，EPSG编号体系是比较好的选择。EPSG（European Petroleum Survey Group）最早建立了该编码体系，其中最为重要的编码包括：

EPSG:4326 - 即WGS84，广泛应用于地图和导航系统，在地理空间数据分析中，这个地理坐标系统是最为基础表征位置数据的坐标系统，通常各类地理空间的数据信息都以WGS84为基本的坐标系统，而后可以在不同数据类型或者平台，以及根据分析目的的不同，尤其实际地理位置的变化来配置投影。

EPSG:3857 - 伪墨卡托投影，也被称为球体墨卡托，用于显示许多基于网页的地图工具，包括Google地图和OpenStreetMap等。

关于EPSG编码可以查看：Spatial Reference①和epsg.io。②

本次实验输出栅格的坐标投影系统为EPSG:32616，即WGS 84 / UTM zone 16N，对应芝加哥城Landsat遥感影像所使用的坐标投影系统。通常，在分析某一城市空间问题时，可能用到很多类型的、不同来源的数据，尽量以最基本的WGS84作为数据存储坐标系统之外，数据的显示往往需要对应区域的坐标系统，从而优化显示结果。

① Spatial Reference，可以用于查找EPSG编号和CRS坐标系（https://spatialreference.org/）。

② epsg.io，世界各地坐标系（Coordinate Systems Worldwide），用于查找EPSG编号（https://epsg.io/）。

● SHP格式地理空间点数据转栅格数据

在定义字段名时，如果写入 .shp 格式数据后，字段名有可能被裁切，例如如果定义字段名为"amenity_kde"，那么用 GeoPandas 存储为 .shp 后，字段名可能被裁切为"amenity_kd"。如果读取该数据，不注意字段名的变化，可能带来不易察觉的错误。

```
import geopandas as gpd
from pyproj import CRS
import os
print("original projection:",amenity_poi.crs)
amenity_poi_copy = amenity_poi.copy(deep=True)
amenity_poi_copy = amenity_poi.to_crs(CRS("EPSG:32616"))  # EPSG:32616 - WGS 84 / UTM
zone 16N - Projected
print("re-projecting:",amenity_poi_copy.crs)
amenity_kde_fn = './data_processed/amenity_kde/amenity_kde.shp'
amenity_poi_copy.to_file(amenity_kde_fn)
--------------------------------------------------------------------------
original projection: epsg:4326
re-projecting: EPSG:32616
```

定义pts2raster()函数的核心是GDAL库提供的gdal.RasterizeLayer()方法，可以将读取的.shp点层属性字段的值写入对应位置的栅格，避免编写对位栅格单元位置和字段值数组的代码。在栅格投影定义上直接提取.shp点数据的坐标投影系统，代码位置于向栅格单元写入数据之后。如果位于之前，会出现坐标投影错误。在栅格定义时，空值通常设置为-9999。GDAL提供栅格定义的方式为先获取栅格驱动gdal.GetDriverByName('GTiff')，再建立.Create(raster_path, x_res, y_res, 1, gdal.GDT_Float64)。并配置地理变换target_ds.SetGeoTransform((x_min, cellSize, 0, y_max, 0, -cellSize))，通过读取栅格波段band=target_ds.GetRasterBand(1)，向其写入值band.WriteArray()完成栅格的定义。地理变换中，因为地图通常向上为北向，因此第3个和第5个参数通常配置为0。

```
def pts2raster(pts_shp, raster_path, cellSize, field_name=False):
    '''
    function - 将 .shp 格式的点数据转换为 .tif 栅格 (raster)
               将点数据写入为 raster 数据。使用 raster.SetGeoTransform，栅格化数据。参考
GDAL 官网代码
    Params:
        pts_shp - .shp 格式点数据文件路径；SHP 点数据
        raster_path - 保存的栅格文件路径；string
        cellSize - 栅格单元大小；int
        field_name - 写入栅格的 .shp 点数据属性字段；string
    Returns:
        返回读取已经保存的栅格数据；array
    '''
    from osgeo import gdal, ogr
    # 定义空值（没有数据）的栅格数值
    NoData_value = -9999
    # 打开 .shp 点数据，并返回地理区域范围
    source_ds = ogr.Open(pts_shp)
    source_layer = source_ds.GetLayer()
    x_min, x_max, y_min, y_max = source_layer.GetExtent()
   # 使用 GDAL 库建立栅格
    x_res = int((x_max - x_min) / cellSize)
    y_res = int((y_max - y_min) / cellSize)
    # create(filename,x_size,y_size,band_count,data_type,creation_options)。gdal 的数据类型 gdal.GDT_
Float64,gdal.GDT_Int32...
    target_ds = gdal.GetDriverByName('GTiff').Create(
        raster_path, x_res, y_res, 1, gdal.GDT_Float64)
    target_ds.SetGeoTransform((x_min, cellSize, 0, y_max, 0, -cellSize))
    outband = target_ds.GetRasterBand(1)
    outband.SetNoDataValue(NoData_value)
    # 向栅格层中写入数据
    if field_name:
        gdal.RasterizeLayer(target_ds, [1], source_layer, options=[
                            "ATTRIBUTE={0}".format(field_name)])
    else:
        gdal.RasterizeLayer(target_ds, [1], source_layer, burn_values=[-1])
    # 配置投影坐标系统
    spatialRef = source_layer.GetSpatialRef()
    target_ds.SetProjection(spatialRef.ExportToWkt())
    outband.FlushCache()
    return gdal.Open(raster_path).ReadAsArray()
amenity_kde_fn = './data_processed/amenity_kde/amenity_kde.shp'
raster_path = './data_processed/amenity_epsg32616.tif'
cellSize = 300
field_name = 'amenityKDE'
poiRaster_array = pts2raster(amenity_kde_fn, raster_path, cellSize, field_name)
print("conversion completed!")
--------------------------------------------------------------------------------
conversion completed!
```

Python中，处理栅格数据的库可以使用rasterio，相比GDAL能够大幅度减少代码书写量，方法相较也更便捷。用该库读取栅格数据的相关信息，并读取数据为数组（array），打印栅格查看数据（图3.2-11）。

如果环境中已经安装了 GeoPandas，再安装 rasterio（conda install -c conda-forge rasterio）后运行代码如果提示错误（库之间可能存在冲突），可以新建环境安装 rasterio，在新建环境下执行下述代码。

```
import rasterio
raster_path = './data_processed/amenity_epsg32616.tif'
dataset = rasterio.open(raster_path)
print(
    "band count:",dataset.count,'\n', # 查看栅格波段数量
    "columns wide:",dataset.width,'\n', # 查看栅格宽度
    "rows hight:",dataset.height,'\n', # 查看栅格高度
    "dataset's index and data type:",{i: dtype for i, dtype in zip(dataset.
indexes, dataset.dtypes)},'\n',# 查看波段及其数据类型
    "bounds:",dataset.bounds,'\n', # 查看外接矩形边界左下角与右上角坐标
    "geospatial transform:",dataset.transform,'\n', # 数据集的地理空间变换
    "lower right corner:",dataset.transform*(dataset.width,dataset.height),'\n',
# 计算外接矩形边界右下角坐标
    "crs:",dataset.crs,'\n', # 地理坐标投影系统
    "band's index number:",dataset.indexes,'\n' # 栅格层（波段）索引
    )
```

```
band1 = dataset.read(1) #读取栅格波段数据为数组
print(band1)
--------------------------------------------------------------------------------
band count: 1
columns wide: 134
rows hight: 190
dataset's index and data type: {1: 'float64'}
bounds: BoundingBox(left=416035.55653533165, bottom=4607564.55138286,
right=456235.55653533165, top=4664564.55138286)
geospatial transform: | 300.00, 0.00, 416035.56|
| 0.00,-300.00, 4664564.55|
| 0.00, 0.00, 1.00|
lower right corner: (456235.55653533165, 4607564.55138286)
crs: PROJCS["WGS 84 / UTM zone 16N",GEOGCS["WGS 84",DATUM["WGS_1984",SPHEROI
D["WGS G","4326"]],PROJECTION["Transverse_Mercator"],PARAMETER["latitude_of_
origin",0],PARAMETER["central_meridian",-87],PARAMETER["scale_factor",0.9996],PA
RAMETER["false_easting",500000],PARAMETER["false_Northing",NORTH],AUTHORITY["EPS
G","32616"]]
band's index number: (1,)
[[-9999. -9999. -9999. ... -9999. -9999. -9999.]
 [-9999. -9999. -9999. ... -9999. -9999. -9999.]
 [-9999. -9999. -9999. ... -9999. -9999. -9999.]
 ...
 [-9999. -9999. -9999. ... -9999. -9999. -9999.]
 [-9999. -9999. -9999. ... -9999. -9999. -9999.]
 [-9999. -9999. -9999. ... -9999. -9999. -9999.]]
from rasterio.plot import show
import matplotlib.pyplot as plt
plt.figure(figsize=(15,10))
show((dataset,1),cmap='Greens')
plt.show()
```

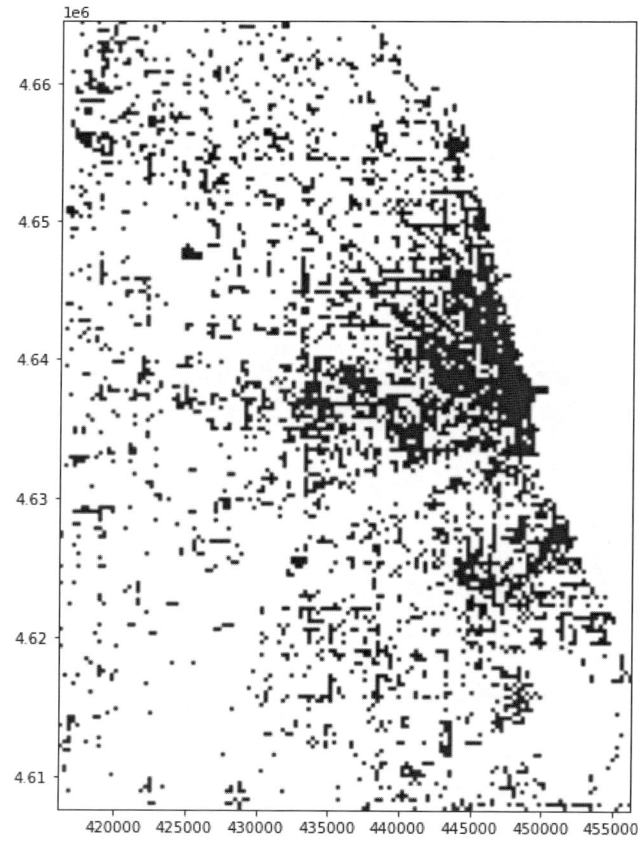

图3.2-11 核密度分布的栅格数据类型

rasterio库支持的颜色（基本同matplotlib库，可以在该库查看具体名称对应的颜色带）：

```
'Accent', 'Accent_r', 'Blues', 'Blues_r', 'BrBG', 'BrBG_r', 'BuGn', 'BuGn_r',
'BuPu', 'BuPu_r', 'CMRmap', 'CMRmap_r', 'Dark2', 'Dark2_r', 'GnBu', 'GnBu_r',
'Greens', 'Greens_r', 'Greys', 'Greys_r', 'OrRd', 'OrRd_r', 'Oranges', 'Oranges_r',
'PRGn', 'PRGn_r', 'Paired', 'Paired_r', 'Pastel1', 'Pastel1_r', 'Pastel2', 'Pastel2_
r', 'PiYG', 'PiYG_r', 'PuBu', 'PuBuGn', 'PuBuGn_r', 'PuBu_r', 'PuOr', 'PuOr_r',
'PuRd', 'PuRd_r', 'Purples', 'Purples_r', 'RdBu', 'RdBu_r', 'RdGy', 'RdGy_r', 'RdPu',
'RdPu_r', 'RdYlBu', 'RdYlBu_r', 'RdYlGn', 'RdYlGn_r', 'Reds', 'Reds_r', 'Set1',
'Set1_r', 'Set2', 'Set2_r', 'Set3', 'Set3_r', 'Spectral', 'Spectral_r', 'Wistia',
'Wistia_r', 'YlGn', 'YlGnBu', 'YlGnBu_r', 'YlGn_r', 'YlOrBr', 'YlOrBr_r', 'YlOrRd',
'YlOrRd_r', 'afmhot', 'afmhot_r', 'autumn', 'autumn_r', 'binary', 'binary_r', 'bone',
'bone_r', 'brg', 'brg_r', 'bwr', 'bwr_r', 'cividis', 'cividis_r', 'cool', 'cool_
r', 'coolwarm', 'coolwarm_r', 'copper', 'copper_r', 'cubehelix', 'cubehelix_r',
'flag', 'flag_r', 'gist_earth', 'gist_earth_r', 'gist_gray', 'gist_gray_r', 'gist_
heat', 'gist_heat_r', 'gist_ncar', 'gist_ncar_r', 'gist_rainbow', 'gist_rainbow_
r', 'gist_stern', 'gist_stern_r', 'gist_yarg', 'gist_yarg_r', 'gnuplot', 'gnuplot2',
'gnuplot2_r', 'gnuplot_r', 'gray', 'gray_r', 'hot', 'hot_r', 'hsv', 'hsv_r',
'inferno', 'inferno_r', 'jet', 'jet_r', 'magma', 'magma_r', 'nipy_spectral', 'nipy_
spectral_r', 'ocean', 'ocean_r', 'pink', 'pink_r', 'plasma', 'plasma_r', 'prism',
'prism_r', 'rainbow', 'rainbow_r', 'seismic', 'seismic_r', 'spring', 'spring_r',
'summer', 'summer_r', 'tab10', 'tab10_r', 'tab20', 'tab20_r', 'tab20b', 'tab20b_r',
'tab20c', 'tab20c_r', 'terrain', 'terrain_r', 'twilight', 'twilight_r', 'twilight_
shifted', 'twilight_shifted_r', 'viridis', 'viridis_r', 'winter', 'winter_r'
```

在QGIS中打开栅格数据查看结果（图3.2-12）。该部分的栅格化过程，开始配置空值`outband.SetNoDataValue(-9999)`。后续向栅格写入数据时，仅含有值的位置替换原空值，因此所看到的栅格空值部分透明。

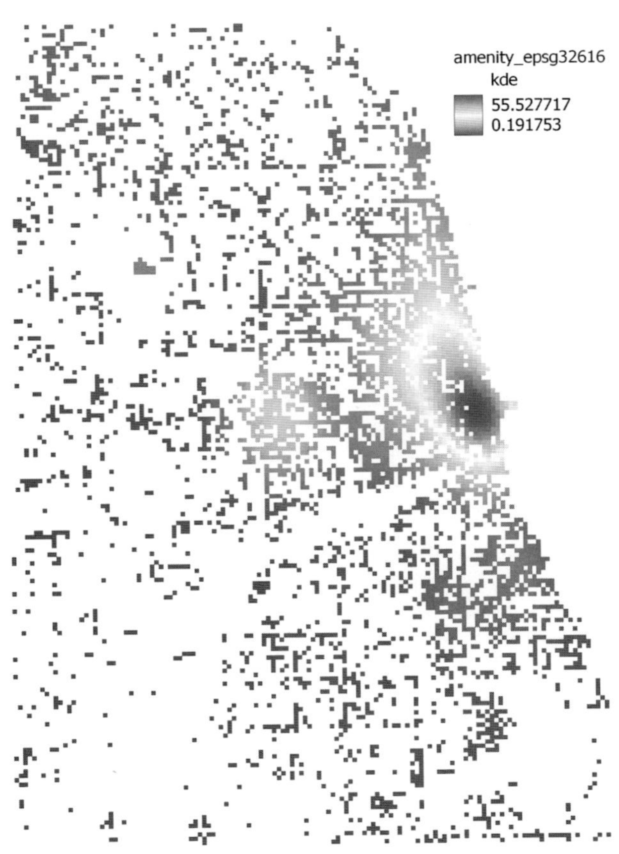

图3.2-12 QGIS中查看核密度分布的栅格数据

● 给定GeoDataFrame格式的地理空间点数据，计算核密度估计并存储为栅格数据

下述函数ptsKDE_geoDF2raster()的定义将核密度估计置于函数之内，可以传入GeoDataFrame格式地理空间点数据（pts_gdf），给定保存位置（raster_path），栅格单元大小（cellSize），以及调整核密度估计值比例缩放因子（scale）直接获取最后的核密度估计值栅格图（热力图）图3.2-13。这种计算方法能够减少中间步骤，不需先转换为SHP格式的点数据后再存储为栅格数据。但是这种将多个步骤置于一个函数中的计算方式，会将“单步”的计算时间拉长，因此最好在函数内通过print()函数打印相关完成信息，避免大批量数据计算时，不知道程序完成的进度，无法确定程序是否仍在正常运行，或者已经完成，甚至已经中断。

GDAL库提供建立栅格，向栅格单元写入数值的方法是outband.WriteArray()，传入的参数为数组，这个数组对应着栅格的位置，因此计算完核密度估计之后，定义提取估计值的位置坐标需要重新进行定义，而不能直接用点坐标。在位置（positions）定义上借助np.meshgrid()实现，同时需要注意上述定义的positions位置提取估计值，其顺序是逆反的，由下往上逐行读取，这个通常符合对图片像素的定义顺序；在地理栅格数据中，通常是由上往下写入，因此用np.flip(Z,0)翻转数组，将最后一行提为正数第一行，倒数第二行为正数第二行，以此类推。

```
def ptsKDE_geoDF2raster(pts_geoDF, raster_path, cellSize, scale):
    '''
    function - 计算 GeoDaraFrame 格式的点数据核密度估计，并转换为栅格数据
    Params:
        pts_geoDF - GeoDaraFrame 格式的点数据；GeoDataFrame(GeoPandas)
        raster_path - 保存的栅格文件路径；string
        cellSize - 栅格单元大小；int
        scale - 缩放核密度估计值；int/float
```

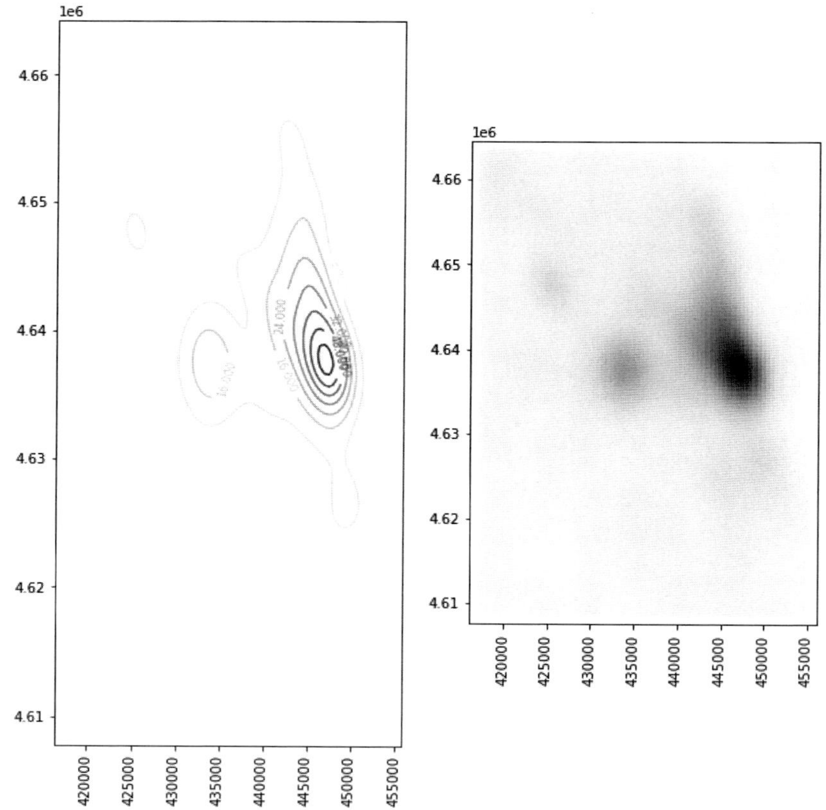

图3.2-13 核密度估计的栅格数据结果

```
    Returns:
        返回读取已经保存的核密度估计栅格数据; array
    '''
    from osgeo import gdal, ogr, osr
    import numpy as np
    from scipy import stats
    #定义空值(没有数据)的栅格数值
    NoData_value = -9999
    x_min, y_min, x_max, y_max = pts_geoDF.geometry.total_bounds
    #使用 GDAL 库建立栅格
    x_res = int((x_max - x_min) / cellSize)
    y_res = int((y_max - y_min) / cellSize)
    target_ds = gdal.GetDriverByName('GTiff').Create(
        raster_path, x_res, y_res, 1, gdal.GDT_Float64)
    target_ds.SetGeoTransform((x_min, cellSize, 0, y_max, 0, -cellSize))
    outband = target_ds.GetRasterBand(1)
    outband.SetNoDataValue(NoData_value)
    #配置投影坐标系统
    spatialRef = osr.SpatialReference()
    epsg = int(pts_geoDF.crs.srs.split(":")[-1])
    spatialRef.ImportFromEPSG(epsg)
    target_ds.SetProjection(spatialRef.ExportToWkt())
    #向栅格层中写入数据
    X, Y = np.meshgrid(np.linspace(x_min, x_max, x_res), np.linspace(
        y_min, y_max, y_res))  #用于定义提取核密度估计值的栅格单元坐标数组
    positions = np.vstack([X.ravel(), Y.ravel()])
    values = np.vstack([pts_geoDF.geometry.x, pts_geoDF.geometry.y])
    print("Start calculating kde...")
    kernel = stats.gaussian_kde(values)
    Z = np.reshape(kernel(positions).T, X.shape)
    print("Finish calculating kde!")
    # print(values)
    outband.WriteArray(np.flip(Z, 0)*scale)  #需要翻转数组, 写栅格单元
    outband.FlushCache()
    print("conversion completed!")
    return gdal.Open(raster_path).ReadAsArray()
raster_path_gpd = './data_processed/amenity_kde_gdf.tif'
cellSize = 500  #cellSize 值越小, 需要计算的时间越长。开始调试时, 可以尝试将其调大, 以节约计算时间
scale = 10**10  #相当于 math.pow(10,10)
poiRasterGeoDF_array = ptsKDE_geoDF2raster(
    amenity_poi_copy, raster_path_gpd, cellSize, scale)
------------------------------------------------------------------------------
Start calculating kde...
Finish calculating kde!
conversion completed!
------------------------------------------------------------------------------
import rasterio
from rasterio.plot import show
import matplotlib.pyplot as plt
dataset_gpd = rasterio.open(raster_path_gpd)
fig, axs = plt.subplots(1,2,figsize=(10, 10))
show((dataset_gpd,1),contour=True,cmap='Greens',ax=axs[0]) #开启等高线模式
show((dataset_gpd,1),cmap='Greens',ax=axs[1])
axs[0].tick_params(axis='x', rotation=90)
axs[1].tick_params(axis='x', rotation=90)
plt.show()
```

参考文献(References):

[1] Urmila Diwekar, A. (2015). BONUS Algorithm for Large Scale Stochastic Nonlinear Programming Problems. Springer.

[2] [美]Larry Wasserman, 吴喜之译. 现代非参数统计[M]. 北京: 科学出版社, 2008. 5. (Larry Wasserman(2006). All Of Nonparametric Statistics. Springer.)

3.3 基本统计量与公共健康数据的相关性分析

3.3.1 基本统计量：标准误、中心极限定理、t分布、统计显著性、效应量和置信区间

3.3.1.1 标准误

定义：标准误是某一统计量（例如均值、两个均值只差、相关系数等）抽样分布（sampling distribution）的标准差（即样本均值的标准差，而不是样本的标准差），度量了从同一总体中抽取相同容量样本的预期随机差异。下述代码中，从服从平均值为30、标准差为5的正态分布中，随机提取2000个样本（sample），各个样本容量为1000，并计算每一样本的均值，查看这2000个样本的均值分布（图3.3–1），即均值抽样分布（sampling distribution of the mean）。普通分布具有均值和标准差。在均值的抽样分布中，均值则称为均值的期望值（expected value of the mean），这是因为样本均值的最佳猜测与总体均值一致。下述代码中，计算了均值抽样分布的均值几乎与总体（population）一样，证明了这一点。均值抽样分布的标准差则称为标准误。标准差是分布中的单个取值与分布均值之间的平均差异或平均离差，均值的标准误提供了同样的信息。只是单个样本均值与其期望值之间的平均差异，可以理解为对样本均值代表实际总体均值的确信程度，有助于确定样本统计值（例如样本均值）与总体参数（例如总体均值）之间的差异是否有意义。同时，计算了均值抽样分布标准计分的K-S test正态性检验，每次代码运行产生的数据是符合上述条件下的随机值，其p-value是变化的，但通常高于0.5，即>0.05，因此可以判断均值抽样分布符合正态分布。

均值标准误的计算公式：

$\sigma_{\overline{x}}=\frac{\sigma}{\sqrt{n}}$ 或 $S_{\overline{x}}=\frac{s}{\sqrt{n}}$，式中$\sigma$为总体标准差，$s$为标准差的样本估计，$n$为样本容量。因为通常$\sigma$总体的标准差未知，因此用标准差的样本估计，计算标准误。

```
import numpy as np
from scipy import stats
import seaborn as sns
import math
sns.set()
mu, sigma = 30, 5
sample_size = 1000
sample_mus = np.array([np.random.normal(mu, sigma, sample_size).mean() for i in
range(
    2000)])  # 从服从平均值为 30，标准差为 5 的正态分布中，随机提取 2000 个样本，每个样本容量为
1000 的样本，并计算每一样本的均值
```

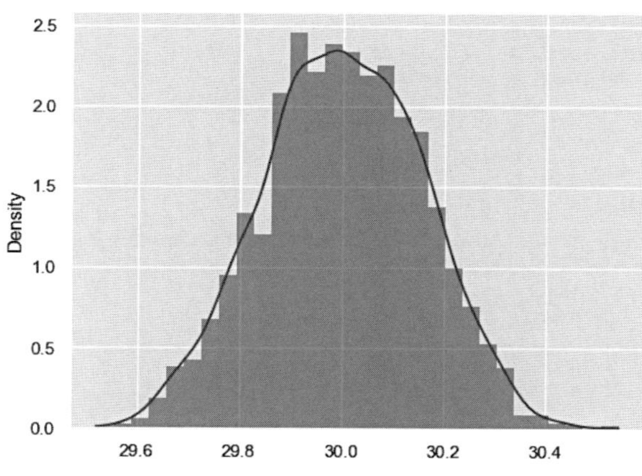

图3.3–1　2000个样本的均值分布

```
bins = 30
sns.histplot(sample_mus, bins=bins, kde=True,
             stat="density", linewidth=0)  # 查看 2000 个样本均值分布
print("sample_mus mu:%.2f,sigma:%.2f" % (sample_mus.mean(), sample_mus.std()))
kstest_test = stats.kstest(stats.zscore(sample_mus), cdf='norm')
print("sampling distribution of the mean - K-S test statistic:%.2f,p-value:%.2f"
% kstest_test)
print("standard error of mean:", sample_mus.std() /
      math.sqrt(sample_size))  # 计算标准误
print("standard error of mean_scipy.stats.sem():",
      stats.sem(sample_mus, ddof=0))  # 使用 scipy.stats 库直接计算标准误
--------------------------------------------------------------------------------
sample_mus mu:30.00,sigma:0.16
sampling distribution of the mean - K-S test statistic:0.02,p-value:0.41
standard error of mean: 0.0049183131322453714
standard error of mean_scipy.stats.sem(): 0.0034777725678095507
```

单个样本的容量越大，就越接近于总体，均值抽样分布的均值越趋近于总体均值，而其标准差即标准误越小，表明样本均值的统计量（或计算其他的统计量）能够代表该总体的统计量，对总体的估计也就越准确。利用下述代码分析上述观点，计算多个不用样本容量均值抽样分布的标准误。随着样本容量的增加，标准误迅速地减小后逐步趋于缓和（图3.3–2），说明单个样本容量越大，越接近总体，对总体的估计越准确。

```
import matplotlib.pyplot as plt
sample_mu_list = []
sample_sigma_list = []
sample_size_list = list(range(10, 2000, 20))
for sample_size in sample_size_list:
    sample_size_mu = np.array(
        [np.random.normal(mu, sigma, sample_size).mean() for i in range(2000)])
    sample_mu_list.append(sample_size_mu.mean())
    sample_sigma_list.append(sample_size_mu.std())
ax = sns.lineplot(x=sample_size_list, y=sample_sigma_list)
ax.set(xlabel='sample_size', ylabel='sample_sigma')
plt.show()
```

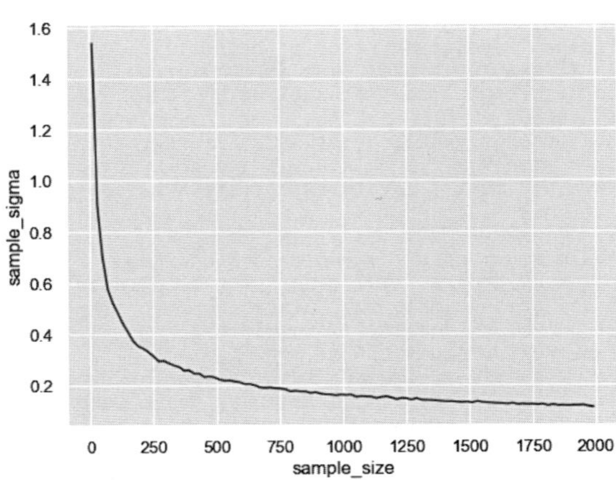

图3.3–2　2000个样本的均值分布

3.3.1.2　中心极限定理

中心极限定义（central limit theorem）表明，只要样本容量足够大，即使样本取值的总体分布不是正态分布，均值的抽样分布也服从正态分布。或理解为不管样本总体服从什么分布，当样本数量足够大时，样本的均值以正态分布的形式围绕总体均值波动。为验证中心极限，使用百度POI数据，提取美食的价格数据。在正态分布与概率密度函数一节，已经通过正态性检验确定了该美食价格数据集为非正态分布，下述代码重现了K-S test正态性检验，p-value=5.153632014592587e-26，小于0.05，拒绝原假设，即不符合正态分布。从该数据集中随机抽取单个样本容量为350，总共2000个样本，计算均值抽样分布（图3.3–3）后的K-S test正态性

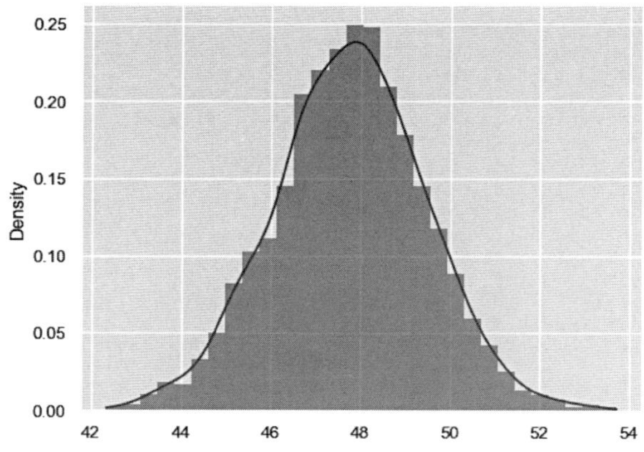

图3.3-3　POI美食价格数据的均值抽样分布

检验，其值p-value通常大于0.5。表明均值抽样分布为正态分布，满足中心极限定义。

很多统计量都依赖于从正态分布中得到概率，而中心极限定理确定了足够多的样本均值服从正态分布（即使总体分布不是正态的），即均值抽样分布为正态分布，因此这一定理使得众多统计量的概率计算成为可能。

```
import util_misc
import pandas as pd
# 读取已经存储为 .pkl 格式的 POI 数据，其中包括 geometry 字段，为 GeoDataFrame 地理信息数据，可以通过 poi_gpd.plot() 迅速查看数据。
poi_gpd = pd.read_pickle('./data/poisInAll/poisInAll_gdf.pkl')
delicacy_price = poi_gpd.xs(
    'poi_0_delicacy', level=0).detail_info_price  # 提取美食价格数据
delicacy_price_df = delicacy_price.to_frame(name='price').astype(float)
delicacy_price_df_clean = delicacy_price_df.dropna()
_, delicacyPrice_outliersDrop = util_misc.is_outlier(
    delicacy_price_df_clean.price, threshold=3.5)  # 将异常值处理函数放置于 util_misc.py 文件中，直接调用。见异常值处理一节
delicacy_price_array_dropna = delicacyPrice_outliersDrop.to_numpy().reshape(-1)
print("original data mean:%.2f" % delicacy_price_array_dropna.mean())
print("original data K-S test:", stats.kstest(stats.zscore(delicacy_price_array_dropna),
      cdf='norm'))  # 计算标准化后的 K-S test 正态性检验
delicacy_price_sample_mus = np.array([np.random.choice(
    delicacy_price_array_dropna, 350).mean() for i in range(2000)])
sns.histplot(delicacy_price_sample_mus, bins=bins,
             kde=True, stat="density", linewidth=0)
print("sample_mus mu:%.2f,sigma:%.2f" %
      (delicacy_price_sample_mus.mean(), delicacy_price_sample_mus.std()))
kstest_test = stats.kstest(stats.zscore(delicacy_price_sample_mus), cdf='norm')
print("sample_mus K-S test statistic:%.2f,p-value:%.2f" % kstest_test)
--------------------------------------------------------------------------------
original data mean:47.70
original data K-S test: KstestResult(statistic=0.11464960111854455,
pvalue=5.153632014592587e-26)
sample_mus mu:47.76,sigma:1.69
sample_mus K-S test statistic:0.01,p-value:0.76
```

3.3.1.3　t分布 (Student's t-distribution)

如果要使用正态分布，通过z-score（标准计分）获得精确概率，至少需要满足两个条件：一个是，σ总体标准差已知；另一个是，大样本（即样本容量足够大）。如果上述条件均不满足，则需要考虑样本容量，这就用到t分布。t分布的形状受样本容量影响，大样本条件下，t分布与正态分布的形状几乎完全一样，而随样本容量减少，t分布的形状变得中间更平坦、两端更粗厚，即均值周围的取值变得更少，而远离均值，位于分布尾部的取值变得更多（图3.3-4）。以SciPy库scipy.stats.t[①]给出的案例来观察t分布，其代码方法与正态分布基本相同，可以互相比较查

① scipy.stats.t，a Student's T continuous random variable（https://docs.scipy.org/doc/scipy-0.13.0/reference/generated/scipy.stats.t.html）。

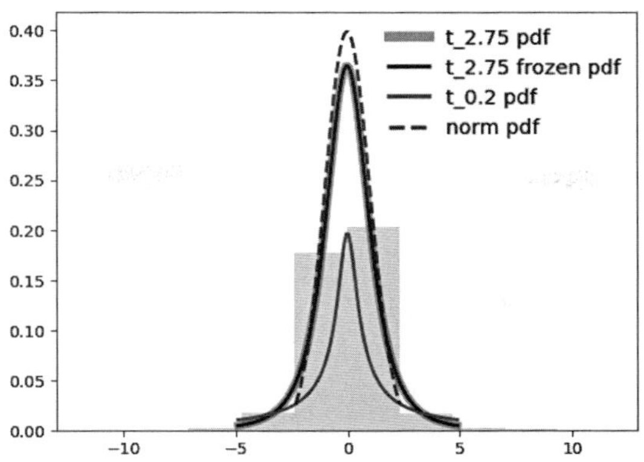

图3.3-4 POI美食价格数据的均值抽样分布

看之间的差异。

t分布中，增加了一个新的参数自由度（degree of freedom，df），通常用符号 ν 来表示，当以样本的统计量来估计总体的参数时，对N个随机样本而言，其自由度为N–1。

数据集为正态分布，给定一个取值，计算该取值的z-score后，就可以获得该取值的概率。而在t分布下，要获取对应取值的概率，则需要计算t值（t统计量），其公式定义为：$t=\frac{\overline{X}-\mu}{S_{\overline{x}}}$。式中，$\mu$为总体均值，$\overline{X}$为样本均值，$S_{\overline{x}}$为标准误的样本估计。知道$t$值，及自由度（样本容量$N$-1），可以查t值表得到该取值的概率。当然，目前直接使用SciPy等库计算，无须再对应t值查表。

```
from scipy.stats import norm
from scipy.stats import t
import matplotlib.pyplot as plt
import numpy as np
fig, ax = plt.subplots(1, 1)
df = 2.75  # 配置自由度
mean, var, skew, kurt = t.stats(df, moments='mvsk')  # 查看服从 t 分布的相关统计量
print('mean, var, skew, kurt=', (mean, var, skew, kurt))
x = np.linspace(t.ppf(0.01, df), t.ppf(0.99, df),100)  # 获取服从自由度 df，概率位于 1% 到 99% 的 100 个取值
ax.plot(x, t.pdf(x, df), 'r-', lw=5, alpha=0.6, label='t_%.2f pdf' % df)
rv = t(df)  # 指定固定自由度的随机变量（random variable）序列
ax.plot(x, rv.pdf(x), 'k-', lw=2, label='t_%.2f frozen pdf' % df)
rv_df_02 = t(0.2)  # 指定固定自由度为 0.2 的随机变量序列
ax.plot(x, rv_df_02.pdf(x), 'g-', lw=2, label='t_0.2 pdf')
vals = t.ppf([0.001, 0.5, 0.999], df)  # 返回概率为 0.1%、50% 和 99.9% 的取值
print("验证累积分布函数 CDF 返回值与其 PPF 返回值是否相等或近似：", np.allclose(
    [0.001, 0.5, 0.999], t.cdf(vals, df)))
rv_1000 = t.rvs(df, size=1000)  # 获取服从自由度 df 的 1000 个随机取值
ax.hist(rv_1000, density=True, histtype='stepfilled', alpha=0.2)
# 比较标准正态分布
x_nd = np.linspace(norm.ppf(0.01), norm.ppf(0.99), 100)
ax.plot(x_nd, norm.pdf(x_nd), 'b--', lw=2, alpha=0.8, label='norm pdf')
ax.legend(loc='best', frameon=False)
plt.show()
--------------------------------------------------------------------------------
mean, var, skew, kurt= (array(0.), array(3.66666667), array(nan), array(inf))
验证累积分布函数 CDF 返回值与其 PPF 返回值是否相等或近似： True
# 如果需要计算服从 t 分布的概率则指定 df，及 loc（均值，默认为 0）和 scale（标准差，默认为 1）
print("用 .cdf 计算值小于或等于 113 的概率为：",t.cdf(113,df=999,loc=100,scale=12))
print("用 .sf 计算值大于或等于 113 待概率为：",t.sf(113,df=999,loc=100,scale=12))
print("可以观察到 .cdf（<=113）概率结果 +.sf(>=113) 概率结果为：",t.cdf(113,df=999,loc=
100,scale=12)+t.sf(113,df=999,loc=100,scale=12))
print("用 .ppf 找到给定概率值为 98% 的数值为：",t.ppf(0.81766,df=999,loc=100,scale=12))
--------------------------------------------------------------------------------
用 .cdf 计算值小于或等于 113 的概率为： 0.8605390547558804
用 .sf 计算值大于或等于 113 待概率为： 0.13946094524411956
可以观察到 .cdf(<=113) 概率结果 +.sf(>=113) 概率结果为： 1.0
用 .ppf 找到给定概率值为 98% 的数值为： 110.88276310459135
```

3.3.1.4 统计显著性

使用SciPy库建立服从均值为100、标准差为12的正态分布。绘制样本容量为1000的单个样本的随机取值分布（图3.3–5）；同时，绘制样本容量为1000，样本数为2000的均值抽样分布（图3.3–6）。计算2000个样本均值抽样分布的标准差，将其作为绘制t分布（图3.3–7）的标准差，均值保持不变为100，自由度为1999（即样本容量为2000，等于均值抽样分布的样本数量）。观察均值抽样分布与对应的t分布（同均值和标准差，及自由度为均值抽样分布的样本数），曲线形状基本保持一致，再一次验证中心极限定理。在这样本容量为1000的2000个样本中，可以依据t分布估计样本均值概率小于0.05（5%）的取值为 99.387，大于0.95（95%）的取值为100.641。就此，我们能够估计取值小于99.387或者大于100.641的样本均值出现的概率（机率）小于5%（0.05）。

如果我们不是根据概率获取对应的取值（即样本均值），而是估计取值的概率，例如估计样本均值小于或者等于99.0的概率为多少，计算结果为0.00439。这个概率值是否能确定样本统计量（例如样本均值）与总体参数（例如总体均值）之间某种差异（例如样本均值为99.0，而总体均值为100，100–99.0=1.0的差异），是否仅仅由随机抽样误差或偶然因素导致？惯例是取概率*P*值p-value=0.05作为一个水平的界限。这一水平就为，当原假设为真时所得到的样本观察结果或更极端结果出现的概率，其*P*值很小，说明原假设发生的概率很小，根据小概率原理，0.00439<0.05，有理由拒绝原假设。*P*值越小，拒绝原假设的理由越充分，就是说样本均值为99.0与总体均值之间的差异不是偶然的，得出结论认为99.0的样本均值实际上异于总计均值。从该样本中观察到的相关情况，不能代表总体中的实际现象。

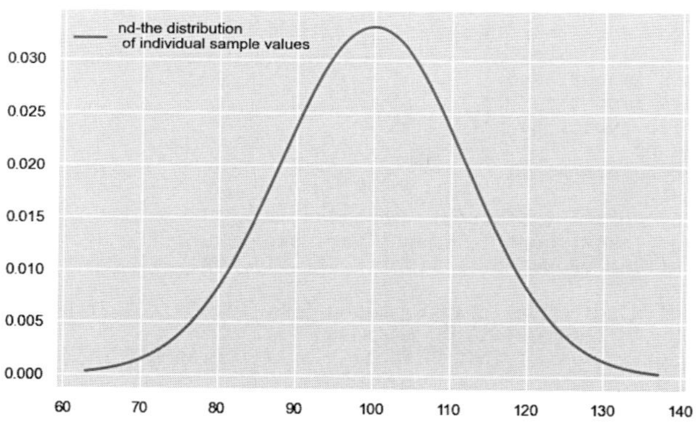

图3.3–5 样本容量为1000单个样本的随机取值分布

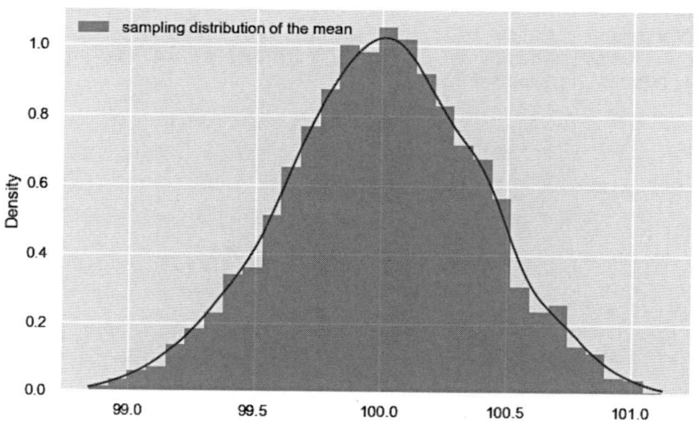

图3.3–6 样本容量为1000，样本数为2000的均值抽样分布

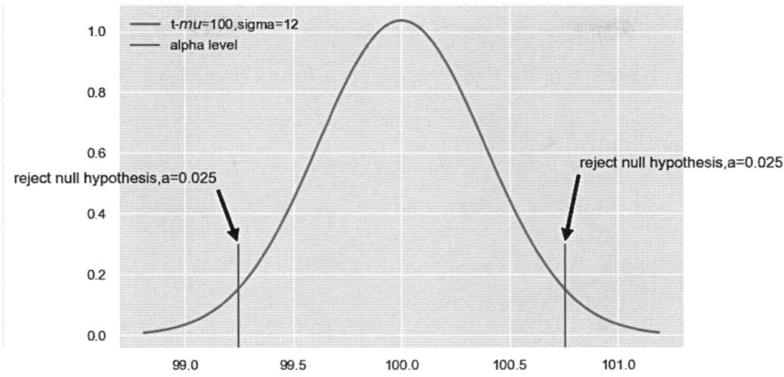

图3.3-7 对应 2000个样本均值抽样分布标准差的 t分布

对于上述过程，可转换为假设检验的推断过程，提出一种假设并确立一种准则，用于决定保留或拒绝假设。假设样本均值不异于总体均值为原假设即零假设（null hypothesis，H_0），零假设一般指总体中的效应不存在（总体均值与样本均值不会不同），符号表示为：$H_0:\mu=\overline{X}$，其中μ为总体均值，$\overline{X}$为样本均值。如果假设样本均值异于总体均值，则为对立假设（一种替代假设），符号表示为：$H_A:\mu\neq\overline{X}$。样本均值趋向于等同总体均值，但是因为毕竟不是总体，之间总会存在差异。引起这个差异的原因就是随机抽样误差或偶然因素，这正是均值抽样分布呈正态性的内在原因。其p-value<=0.05的概率对应着显著性水平（α，alpha level），例如样本均值99.0的概率为0.00439，小于显著性水平α，拒绝零假设，样本均值异于总体均值，其结果是统计显著的（指在原假设为真的条件下，用于检验的样本统计量的值落在了拒绝域内，做出了拒绝原假设的决定）。显著性水平常用值除了0.05，还有0.01。

上述计算过程中，计算有概率小于0.025和大于0.975的两种情况。如果同时包括两种情况，则为双尾检测；仅有一种情况，但是概率需要小于0.05或者大于0.95的情况下，为单尾检测。

```
import matplotlib.pyplot as plt
df, loc, scale, sample_size = 1999, 100, 12, 1000
# 绘制样本容量为 1000 的单个样本的随机取值分布
x = np.linspace(norm.ppf(0.001, loc=loc, scale=scale),norm.ppf(0.999, loc=loc,
                scale=scale), sample_size)
fig, axs = plt.subplots(1, 3, figsize=(27, 5))
axs[0].plot(x, norm.pdf(x, loc=loc, scale=scale), 'r-', lw=2, alpha=0.8,
           label='nd-the distribution \n of individual sample values')
axs[0].legend(loc='upper left', frameon=False)
# 绘制样本容量为 1000、样本数为 2000 的均值抽样分布
samples = np.array(
    [norm.rvs(loc=loc, scale=scale, size=sample_size).mean() for i in
range(df+1)])
bins = 30
sns.histplot(samples, bins=bins,
             ax=axs[1], label='sampling distribution of the mean', kde=True,
stat="density", linewidth=0)
axs[1].legend(loc='upper left', frameon=False)
# 计算 2000 个样本均值抽样分布的标准差，作为绘制 t 分布的标准差，均值保持不变为 100，自由度为 1999（即样本容量为 2000，等于均值抽样分布的样本数量）
samples_std = samples.std()
t_x = np.linspace(t.ppf(0.001, df=df, loc=loc, scale=samples_std), t.ppf(
    0.999, df=df, loc=loc, scale=samples_std), df+1)  # 获取服从自由度 df，概率位于
1% ~ 99% 的 100 个取值
t_rv = t(df=df, loc=loc, scale=samples_std)  # 指定固定自由度
axs[2].plot(t_x, t_rv.pdf(t_x), 'g-', lw=2,
           label='t-$mu$=%d,sigma=%d' % (loc, scale))
# 显著性水平为 0.05 时，曲线区间绘制
pValue_5percent = t.ppf(0.025, df=df, loc=loc, scale=samples_std)
pValue_95percent = t.ppf(0.975, df=df, loc=loc, scale=samples_std)
axs[2].axvline(pValue_5percent, 0, 0.3, c='gray', label='alpha level')
axs[2].axvline(pValue_95percent, 0, 0.3, c='gray')
axs[2].annotate('reject null hypothesis,a=%.3f' % 0.025, xy=(pValue_5percent, 0.3),
```

```
xytext=(
    pValue_5percent-0.1, 0.5), arrowprops=dict(facecolor='black', shrink=0.01),
horizontalalignment='right',)
axs[2].annotate('reject null hypothesis,a=%.3f' % 0.025, xy=(pValue_95percent,
0.3), xytext=(
    pValue_95percent+0.07, 0.55), arrowprops=dict(facecolor='black',
shrink=0.01), horizontalalignment='left',)
axs[2].legend(loc='upper left', frameon=False)
plt.show()
print("用.ppf找到给定概率值为5%的数值为：", pValue_5percent)
print("用.ppf找到给定概率值为95%的数值为：", pValue_95percent)
print("用.cdf计算值小于或等于99的概率为：", t.cdf(99.0, df=df, loc=loc,
scale=samples_std))
用.ppf找到给定概率值为5%的数值为： 99.24538217329646
用.ppf找到给定概率值为95%的数值为： 100.75461782670354
用.cdf计算值小于或等于99的概率为： 0.004710985065683475
```

3.3.1.5 效应量

效应量（effect size）是量化现象强度的数值。其绝对值越大，表示效应越强，也就是现象越明显。对于均值抽样分布，效应量的计算公式为：$d=\frac{\overline{X}-\mu}{s}$，式中$d$为效应量，$\overline{X}$为样本均值，$\mu$为总体均值，$s$为标准差的样本估计。效应量所代表的就是以标准差为单位所度量的差异，这个与标准计分（z-score）非常相似，标准计分是以标准差为单位，从感兴趣的点到均值之间有多少个标准差。关于效应量是否有意义时，所建议的效应量值的区间也所有差异，关键是检验什么以及所持观点，一般而言效应量小于0.20算小，在0.25～0.75之间算中等，超过0.80算大。对于样本均值99的效应量计算结果为–2.583。其绝对值大于0.8，表明样本均值99到总体均值100之间所差的100–99=1的差异。按照标准差为单位，效应量显著，即样本均值99远离总体均值100。

```
print("样本均值99的效应量：",(99-100)/samples_std)
--------------------------------------------------------------------------------
样本均值99的效应量： -2.5988670698883993
```

3.3.1.6 置信区间

一个概率样本的置信区间（Confidence interval，CI），是对产生这个样本的总体的参数分布（Parametric Distribution）中的某一个位置参数值，以区间的形式给出估计。对于现实中均值抽样分布，实际上总体的均值（总体参数的实际值）并不知道，我们所拥有的只是样本数据；而通过样本数据是可以估计总体均值，给出总体均值分布的区间，这个区间即为置信区间。计算公式为：$CI_{95}=\overline{X}\pm(t_{95})(s_{\overline{X}})$ 和 $CI_{99}=\overline{X}\pm(t_{99})(s_{\overline{X}})$。式中，$CI_{95}$为95%置信区间；$CI_{99}$为99%置信区间；$\overline{X}$为样本均值；$s_{\overline{X}}$为标准误；$t_{95}$为给定自由度条件下，$a$水平为0.05双尾检验所对应的t值；$t_{99}$为给定自由度条件下，$a$水平为0.01双尾检验所对应的t值。

对于置信区间的计算，可以直接使用SciPy库。

```
print("0.05_置信区间：",stats.t.interval(0.95, len(samples)-1, loc=np.
mean(samples), scale=stats.sem(samples)))
print("0.01_置信区间：",stats.t.interval(0.99, len(samples)-1, loc=np.
mean(samples), scale=stats.sem(samples)))
--------------------------------------------------------------------------------
0.05_置信区间： (99.98972207782829, 100.02347805302809)
0.01_置信区间： (99.98441087421692, 100.02878925663946)
```

3.3.2 相关性

如果要判断两个变量之间是否相互联系，则需要计算相关系数。在变量之间相关性分析时，涉及三种情况（表3.3–1），包括数值数据与数值数据、数值数据与分类数据，以及分类数据与分类数据。因此，不同的变量类型之间的相关性计算指标有所差异：

表3.3-1 相关性

数据类型	指标	值的范围	计算公式	说明
数值数据和数值数据	相关系数	–1 ~ 1	$r=\frac{S_{xy}}{\sqrt{S_{xx}\cdot S_{yy}}}$ $=\frac{\sum_{i=1}^{n}(x_i-\overline{x})(y_i-\overline{y})}{\sqrt{\sum_{i=1}^{n}(x_i-\overline{x})^2\cdot\sum_{i=1}^{n}(y_i-\overline{y})^2}}\in[-1,1]$	式中S_{xx}叫作x的方差（Variance），S_{yy}叫作y的方差，S_{xy}叫作x和y的协方差（Covariance）
数值数据和分类数据	相关比	0 ~ 1	$WSS=\sum_{k=1}^{K}S_{x^k x^i}$ $=\sum_{k=1}^{K}\sum_{i=1}^{n_i}(x_i^k-\overline{x^k})^2$ $BSS=\sum_{k=1}^{K}n_k(\overline{x^k}-\overline{x})^2$ $cr=\frac{BSS}{BSS+WSS}$	设共有n个数值数据，它们被分成了K个类别，$x_1^1,x_2^1,\cdots,x_{n_1}^1$；$x_1^2,x_2^2,\cdots,x_{n_2}^2,\cdots;x_1^K,x_2^K,\cdots,x_{n_K}^K$。式中，$n_K$代表第$K$个类别的数值的个数，记$\overline{x^k}$为第$K$个类别数值的均值，$\overline{x}$为所有数据的均值。WSS（within sum of squares）为组内变异，BSS（between sum of squares）为组间变异
分类数据和分类数据	克莱姆相关系数	0 ~ 1	$\chi^2=\sum_{i,j}\frac{(n_{ij}-\frac{n_i n_j}{n})^2}{\frac{n_i n_j}{n}}$ $V=\sqrt{\frac{\chi^2/n}{min(k-1,r-1)}}$	设有两个类别变量A和B，观测样本总数为n，对于$i=1,\cdots,r;j=1,\cdots,k$，$n_{ij}$为$(A_i,B_j)$的观测次数（频数），$k$为观察次数表列，$r$为其行。$\frac{n_i n_j}{n}$为计算期望次数

对于数值数据和数值数据之间的相关性分析，使用相关系数计算（通常为Pearson's r，即皮尔森相关系数 r，为线性相关）。先建立简单数据集，数据来源于《漫画统计学》询问10名20多岁女性的化妆品费和置装费。待分析的两个变量为x和y值，打印散点图3.3–8观察点的分布是否存在规律，可以比较明显地观察到点的分布似乎沿一条隐藏的倾斜直线分布，可以初步判断化妆品费和置装费之间存在关联。

皮尔森相关系数，又称积差相关系数、积矩相关系数（Pearson product-moment correlation coefficient，PPMCC 或 PCCs）

```
import pandas as pd
dressUp_cost = {'name': ["miss_A", "miss_B", "miss_C", "miss_D", "miss_E", "miss_
F", "miss_G", "miss_H", "miss_I", "miss_J"],
                "cosmetics_fee": [3000, 5000, 12000, 2000, 7000, 15000, 5000,
6000, 8000, 10000],
                "clothes_fee": [7000, 8000, 25000, 5000, 12000, 30000, 10000,
15000, 20000, 18000]
                }
dressUp_cost_df = pd.DataFrame.from_dict(dressUp_cost)
dressUp_cost_df.plot.scatter(x='cosmetics_fee', y='clothes_fee', c='DarkBlue')
plt.show()
```

使用上述公式计算相关系数（表3.3–2）。

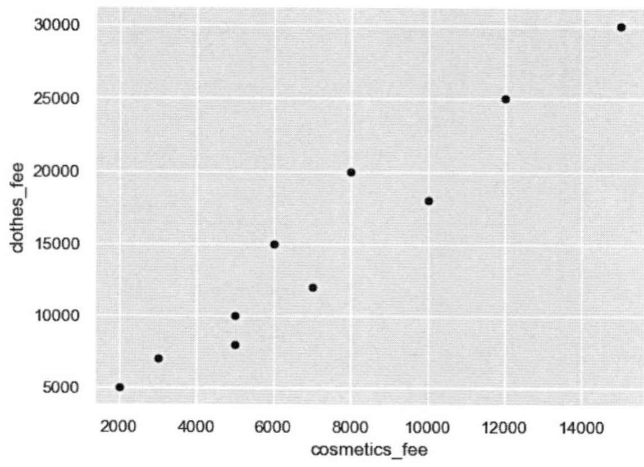

图3.3–8 化妆费和置装费散点图

表3.3-2 相关系数逐步计算过程

	name	cosmetics_fee	clothes_fee	xSx_mean	ySy_mean	square_xSx_mean	square_ySy_mean	xSx_meanMySy_mean
0	miss_A	3000	7000	-4300.0	-8000.0	18490000.0	64000000.0	34400000.0
1	miss_B	5000	8000	-2300.0	-7000.0	5290000.0	49000000.0	16100000.0
2	miss_C	12000	25000	4700.0	10000.0	22090000.0	100000000.0	47000000.0
3	miss_D	2000	5000	-5300.0	-10000.0	28090000.0	100000000.0	53000000.0

```
#人工计算的过程有助于理解概念及公式。

import math
dressUp_cost_df["xSx_mean"] = dressUp_cost_
df.cosmetics_fee.apply(lambda row: row-dressUp_cost_
df.cosmetics_fee.mean())
dressUp_cost_df["ySy_mean"] = dressUp_cost_df.clothes_
fee.apply(lambda row: row-dressUp_cost_df.clothes_fee.
mean())
dressUp_cost_df["square_xSx_mean"] = dressUp_cost_
df.xSx_mean.apply(lambda row: math.pow(row,2))
dressUp_cost_df["square_ySy_mean"] = dressUp_cost_
df.ySy_mean.apply(lambda row: math.pow(row,2))
dressUp_cost_df["xSx_meanMySy_mean"] = dressUp_cost_
df.xSx_mean*dressUp_cost_df.ySy_mean
r=dressUp_cost_df.xSx_meanMySy_mean.sum()/math.
sqrt(dressUp_cost_df.square_xSx_mean.sum()*dressUp_
cost_df.square_ySy_mean.sum())
print("Pearson's r:",r)
dressUp_cost_df
-------------------------------------------------------
Pearson's r: 0.968019612860768
```

使用SciPy库直接计算，其结果保持一致。

```
from scipy import stats
r_ = stats.pearsonr(dressUp_cost_df.cosmetics_
fee,dressUp_cost_df.clothes_fee)
print(
    "pearson's r:",r_[0],"\n",
    "p_value:",r_[1]
     )
-------------------------------------------------------
pearson's r: 0.968019612860768 p_value:
4.402991448166131e-06
```

相关系数的强弱并没有严格的规定，通常需要根据具体的应用背景和目的确定，例如复杂多变因素影响的变量，0.9的值则是相当高的。对于相关系数值的意义，不同的文献给出的参考意义也有所差别，这里给出《漫画统计学》里的划分（表3.3–3）。

表3.3–3 相关系数强弱细分（《漫画统计学》）

相关系数的绝对值	若细分	若大体上划分
1.0 ~ 0.9	非常强	相关
0.9 ~ 0.7	有点强	相关
0.7 ~ 0.5	有点弱	相关
未满0.5	非常弱	不相关

以及，Wikipedia给出的参考（表3.3–4）。

表3.3–4 相关系数强弱细分（维基百科）

相关性	负	正
无	−0.09 ~ 0.0	0.0 ~ 0.09
弱	−0.3 ~ −0.1	0.1 ~ 0.3
中	−0.5 ~ −0.3	0.3 ~ 0.5
强	−1.0 ~ −0.5	0.5 ~ 1.0

若相关系数接近±1，则相关性越强；如果接近0，则相关性越弱。如果为正，则为正相关；如果为负，则为负相关。

通过计算相关系数，获得结果r=0.968，说明抽样的10名20多岁女性在化妆品费和置装费这两个变量间存在强相关性，那么该值是否代表其抽样的总体（所有女性）中两个变量（化妆品费和置装费）间也存在非常强的相关关系？为了确认这个现象，需要检验相关系数是否统计显著。提出零假设，代表总体下两个变量（化妆品费和置装费）之间完全无关，即总体的相关系数为0。通常，用t分布来检验相关系数是否统计显著，进行t检验。其t值的计算公式为：$t=r\sqrt{\frac{N-2}{1-r^2}}$，式中自由度为$N$–2，即样本对象数减去2，即$N$–2=10–2=8。使用t分布的累计分布函数`t.cdf()`或者`t.sf()`计算P值，因为双尾检查，因此再乘以2，其计算结果与`stats.pearsonr()`计算相关系数时给出的P值保持一致。因为p_value=4.402991448104743e-06<0.05，因此拒绝原假设，即拒绝总体下两个变量（化妆品费和置装费）之间完全无关。就是说，总体下两个变量是相关的。因此，所计算的相关系数值为0.968，说明了总体所有女性在化妆品费和置装费的消费上是强关联的。如果一个女性花费较多的钱在化妆品费上，那么她花费在置装费的钱也相对较多，反之亦然。

```
t_value = r*math.sqrt((10-2)/(1-math.pow(r,2)))
print("p-value_cdf:",(1-stats.t.cdf(t_value,10-2))*2)
print("p-value_sf:",stats.t.sf(t_value,10-2)*2)
--------------------------------------------------------------------------------
p-value_cdf: 4.402991448104743e-06 p-value_sf: 4.402991448166121e-06
```

3.3.3 卡方分布与独立性检验

3.3.3.1 卡方分布（Chi-Square Distribution，χ^2-distribution）

若k个随机变量$Z_1,\ldots\ldots,Z_k$，是相互独立，符合标准正态分布的随机变量（数学期望为0，方差为1），则随机变量Z的平方和$X=\sum_{i=1}^{k}x_i^2$被称为服从自由度为k的卡方分布，记作$X\sim x^2(k)$或$X\sim x_{\mathrm{k}}^2$。卡方分布的概率密度函数（即计算卡方分布曲线的公式）为：$f_{\mathrm{k}}(x)=\frac{\frac{1}{2}^{\frac{k}{2}}}{\Gamma(\frac{k}{2})}x^{\frac{k}{2}-1}\mathrm{e}^{\frac{-x}{2}}$，式中$x\geqslant 0$，当$x\leqslant 0$时$f_{\mathrm{k}}(x)=0$。$\Gamma$代表Gamma函数。在Python中，绘制卡方分布，仍旧与正态分布、t分布一样，直接由SciPy库完成。

因为卡方分布表述的是多个事件（随机变量）的机率，每个事件符合标准正态分布，而标准正态分布表为记录对应横轴刻度的机率表，卡方分布表则是记录对应机率的横轴刻度表。

对于卡方分布的理解，可以结合比较均值抽样分布，两者具有类似的逻辑。可以表述为从同一总体中抽取相同容量样本平方和的分布（图3.3–9），假设从服从平均值为30、标准差为5的正态分布中，随机提取2000个样本（事件，或随机变量，即Z_k），各个样本的容量为1000，计算每一样本的平方和，观察这2000个样本的卡方分布情况。从下述实验打印结果来看，x^2趋近服从自由度为（2000–1）的x^2分布。即可以通过一个检验统计量（平方和）来比较期望结果和实际结果之间的差别，然后得出观察频数极值的发生概率。因此，以特定概率分布为某种情况建模时，事件长期结果较为稳定，能够清晰地进行把握。但是，如果期望与事实存在差异时，则可以应用卡方分布判断偏差是正常的小幅度波动还是建模上的错误。一是，可以检验一组给定数据与指定分布的吻合程度；二是，可以检验两个变量的独立性，即变量之间是否存在某种关系。

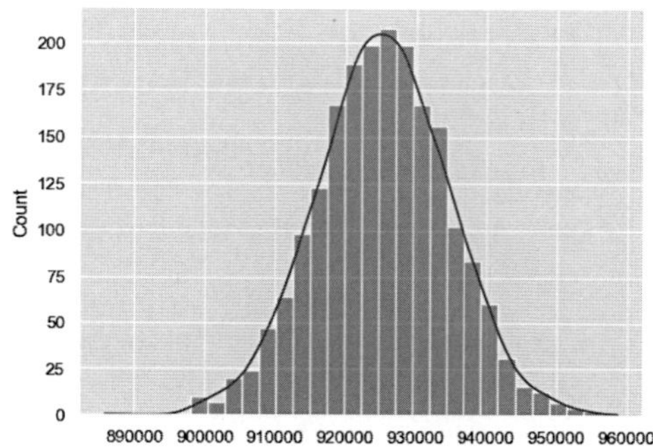

图3.3–9　2000个样本（一个样本容量1000）的平方和分布

- Γ函数

在数学中，Γ函数，也称为伽马函数（Gamma函数），是阶乘函数在实数与复数域上的扩展。如果n为正整数，则：$\Gamma(n)=(n-1)!$，即正整数的阶乘；对于实数部分为正的复数z，伽马函数定义为：$\Gamma(z)=\int_0^\infty t^{z-1}e^{-t}dt$。发现$\Gamma$函数的起因是数列插值问题，即找到一个光滑的曲线连接那些由$y=(x-1)!$所给定的点(x,y)，并要求x为正整数。但是，如果x由正整数拓展到实数，即可以计算$2!,3!,...$，那么是否可以计算$2.5!$，并绘制$(n,n!)$的平滑曲线？而Γ函数正是借由微积分的积分与极限表达阶乘。

伽马（Gamma）分布，假设$X_1,X_2,...,X_n$为连续发生事件的等候时间，且这n次等候时间为独立的，那么这n此等候时间之和$Y(Y=X_1,X_2,...,X_n)$服从伽马分布，即$Y\sim Gamma(\alpha,\beta)$。式中，$\alpha=n,\beta=\gamma$，$\alpha$是伽马分布中的母数，称为形状参数；$\beta$为尺度参数；$\gamma$是连续发生事件的平均发生频率。指数分布是伽马分布$\alpha=1$的特殊情况。

令 $X\sim\Gamma(\alpha,\beta)$，且 $\lambda=\beta$（即 $X\sim\Gamma(\alpha,\gamma)$），则伽马分布的概率密度函数为：$f(x)=\frac{x^{a-1}\gamma^a e^{-\gamma x}}{\Gamma(a)},x>0$，式中伽马函数的特征为：

$$\begin{cases}\Gamma(a)=(a-1)! & if\ a\ is\ \mathbb{Z}^+\\ \Gamma(a)=(a-1)\Gamma(a-1) & if\ a\ is\ \mathbb{R}^+\\ \Gamma(\frac{1}{2}=\sqrt{\pi}) & \end{cases}$$

```
import numpy as np
from scipy import stats
import seaborn as sns
import math
sns.set()
mu,sigma = 30,5
sample_size = 1000
sample_square = np.array([sum(np.square(np.random.normal(mu, sigma, sample_
size))) for i in range(2000)]) # 从服从平均值为 30、标准差为 5 的正态分布中，随机提取 2000 个
样本，每个样本容量为 1000 的样本，并计算每一样本的均值
bins = 30
sns.histplot(sample_square,bins=bins,kde=True,legend=False) # 查看 2000 个样本平方和的分布
```

使用SciPy库计算打印卡方分布（图3.3–10）及伽马分布（图3.3–11）。

```
from scipy.stats import gamma
import matplotlib.pyplot as plt
from scipy.stats import chi2
import numpy as np
fig, axs = plt.subplots(1, 2, figsize=(18/1.5, 8/1.5))
# A- 卡方分布
df = 55
mean, var, skew, kurt = chi2.stats(df, moments='mvsk')
# 打印卡方分布的概率密度函数
x = np.linspace(chi2.ppf(0.01, df), chi2.ppf(0.99, df), 100)
axs[0].plot(x, chi2.pdf(x, df), 'r-', lw=5, alpha=0.6, label='chi2 pdf_55')
df_lst = [20, 30, 40, 80, 100]
fmts = ['b-', 'g-', 'y-', 'm-', 'c-']
for i in range(len(df_lst)):
    axs[0].plot(x, chi2.pdf(x, df_lst[i]), fmts[i], lw=3,
                alpha=0.6, label='chi2 pdf_%d' % df_lst[i])
# 固定分布 (Alternatively, freeze the distribution and display the frozen pdf)
rv = chi2(df)
axs[0].plot(x, rv.pdf(x), 'k-', lw=2, label='frozen pdf_55')
vals = chi2.ppf([0.001, 0.5, 0.999], df)
print("Chi_2_Check accuracy of cdf and ppf:", np.allclose(
    [0.001, 0.5, 0.999], chi2.cdf(vals, df)))
r = chi2.rvs(df, size=1000)
axs[0].hist(r, density=True, histtype='stepfilled', alpha=0.2)
axs[0].legend(loc='best', frameon=False)
# B-Gamma 分布
a = 1.99323054838
mean_, var_, skew_, kurt_ = gamma.stats(a, moments='mvsk')
# 打印 Gamma 分布的概率密度函数
x_ = np.linspace(gamma.ppf(0.01, a), gamma.ppf(0.99, a), 100)
axs[1].plot(x_, gamma.pdf(x_, a), 'r-', lw=5, alpha=0.6, label='gamma pdf')
rv_ = gamma(a)
axs[1].plot(x_, rv_.pdf(x_), 'k-', lw=2, label='frozen pdf')
```

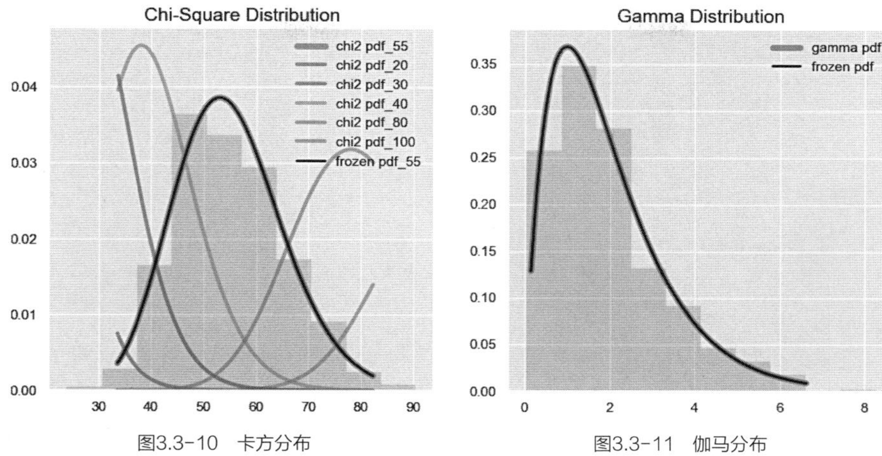

图3.3-10 卡方分布　　图3.3-11 伽马分布

```
vals = gamma.ppf([0.001, 0.5, 0.999], a)
print("Gamma_Check accuracy of cdf and ppf:", np.allclose(
    [0.001, 0.5, 0.999], gamma.cdf(vals, a)))
r_ = gamma.rvs(a, size=1000)
axs[1].hist(r_, density=True, histtype='stepfilled', alpha=0.2)
axs[1].legend(loc='best', frameon=False)
axs[0].set_title(r'Chi-Square Distribution', fontsize=15)
axs[1].set_title(r'Gamma Distribution', fontsize=15)
plt.show()
-------------------------------------------------------------------
Chi_2_Check accuracy of cdf and ppf: True
Gamma_Check accuracy of cdf and ppf: True
```

3.3.3.2 （卡方）独立性检验

卡方检验（Chi-Squared Test，或x^2 Test），是假设检验的一种，一种非参数假设检验，主要用于类别/分类变量（类别变量就是取值为离散值的变量，例如性别即为一个类别变量，有男女两类，又或者国籍、学科、植物等的类别等）。在没有其他限制条件或说明时，卡方检验一般指代的是皮尔森卡方（Pearson）检验。1900年，Pearson发表了著名的x^2检验的论文，假设实验中从总体随机取样得到的n个观察值被划分为k个互斥的分类中，这样每个分类都有一个对应实际的观察次数（或观测频数，observed frequencies）$x_i(i=1,2,\dots,k)$。对实验中各个观察值落入第i个分类的概率p_i的分布提出零假设，获得对应所有第i分类的理论期望次数（或预期频数，expected frequencies）及限制条件，$\sum_{i=1}^{k} p_i=1, and\ \sum_{i=1}^{k} m_i=\sum_{i=1}^{k} x_i=n$。在上述零假设成立以及$n$趋向$\infty$时，以下统计量的极限分布趋向$x^2$分布，$X^2=\sum_{i=1}^{k}\frac{(x_i-m_i)^2}{m_i}=\sum_{i=1}^{k}\frac{x_i^2}{m_i}-n$。$X^2$值的计算公式通常表示为：$X^2=\sum(\frac{(O-E)^2}{E})$。其中，$O$为各个单元格（对列联表而言）的观测值（观测频数），E为各个单元格的预期值（预期频数）。零假设中所有分类的理论期望次数m_i均为足够大且已知的情况，同时假设各分类的实际观察次数x_i均服从正态分布，得出样本容量n足够大时，x^2趋近服从自由度为$(k-1)$的x^2分布。

通常，将用于卡方检验的数据以表格的形式给出并依据表格进行计算，这个表格即为列联表（contingency tabel），如表3.3–6所示。以《白话统计学》性别与专业的修订数据（表3.3–5）为例，利用表格每一单元格中的观测频数，以及行、列和整个样本的合计频数，计算每个单元格的预期频数。男女两行中每一单元格的预期值都相等，是因为样本中的男女生人数相等。并根据上述X2值的计算公式计算X2，其和为：0.19+0.19+4.17+4.17+7.5+7.5=23.72。

表3.3–5 性别与专业数据

性别/专业	心理学	英语	生物学	行合计
男生	35	50	15	100
女生	30	25	45	100
列合计	65	75	60	200

表3.3-6 列联表

性别/专业	心理学	英语	生物学
男生	观测频数：35 预期频数：$\frac{100\times65}{200}=32.5$ x^2值：$\frac{(35-32.5)^2}{32.5}=0.19$	观测频数：50 预期频数：$\frac{100\times75}{200}=37.5$ x^2值：$\frac{(50-37.5)^2}{37.5}=4.17$	观测频数：15 预期频数：$\frac{100\times60}{200}=30$ x^2值：$\frac{(15-30)^2}{30}=7.5$
女生	观测频数：30 预期频数 $\frac{100\times65}{200}=32.5$ x^2值：$\frac{(30-32.5)^2}{32.5}=0.19$	观测频数：25 预期频数：$\frac{100\times75}{200}=37.5$ x^2值：$\frac{(25-37.5)^2}{37.5}=4.17$	观测频数：45 预期频数：$\frac{100\times60}{200}=30$ x^2值：$\frac{(45-30)^2}{30}=7.5$

注意到x^2值较大是因为男女生在选择英语或生物专业时存在相对较大的差异。而心理学专业的观测值和预期值之差相对较小，对整体x^2值的贡献不大。获得观测的x^2值，则需要查表（或程序）查找临界x^2值，其自由度$df=(R-1)(C-1)=(2-1)\times(3-1)=2$，使用SciPy的`chi2.ppf(q=1-0.05,df=2)`计算可得0.05的$\alpha$水平，自由度为2的条件下临界$x^2$值为5.99，而观测的$x^2$值为23.72，所以可以得出结论，男女生在专业选择上存在统计显著的差异。而因为观测的x^2值足够大，在0.001的显著性水平上`chi2.ppf(q=1-0.001,df=2)`（临界值为13.815510557964274），也是统计显著的（即$p<0.001$）。

```
print("p<0.05,df=2,Chi-Squared=%.3f"%chi2.ppf(q=1-0.05,df=2))
print("p<0.001,df=2,Chi-Squared=%.3f"%chi2.ppf(q=1-0.001,df=2))
------------------------------------------------------------------------
p<0.05,df=2,Chi-Squared=5.991
p<0.001,df=2,Chi-Squared=13.816
```

使用SciPy的`chi2_contingency`方法计算列联表，其计算结果与手工计算结果保持一致。

```
from scipy.stats import chi2_contingency
import numpy as np
schoolboy = (35,50,15)
schoolgirl = (30,25,45)
statistical_data=np.array([schoolboy,schoolgirl])
chi2_results = chi2_contingency(statistical_data)
print("卡方值：%.3f \n P值：%.10f \n 自由度：%d \n 对应预期频数：期望值):\n %s"%chi2_
results)
------------------------------------------------------------------------
卡方值：23.718
P值：0.0000070748
自由度 :2
对应预期频数（期望值）： [[32.5 37.5 30. ] [32.5 37.5 30. ]]
```

3.3.3.3 协方差估计（Covariance Estimators）

统计学上常用的统计量包括平均值、方差、标准差等。平均值描述了样本集合的中间点；方差描述了一组数据与其平均值的偏离程度，方差越小，数据越集中；方差越大，数据越离散；标准差描述了样本集中各个样本点到均值的距离的平均值，同方差类似，描述数据集的集聚离散程度。这些统计量是针对一维数组，到处理高维时，用到协方差，度量多个随机变量关系的统计量，结果均为正则正相关，都为负则负相关，均趋近于0，则不相关。协方差是计算不同特征之间的统计量，不是不同样本之间的统计量。同时，协方差的大小，除了和变量之间的相关程度有关，也与变量本身的方差大小有关，因此引入相关系数，移除变量本身的影响。在协方差计算时，可以使用协方差（矩阵）计算公式（查看方差和协方差部分）。而有时，并不使用全部的样本数据计算协方差矩阵，而是利用部分样本数据计算，这是就需要考虑样本计算得到的协方差矩阵是否与总体的协方差矩阵相同和近似。大多数情况下，估计总体的协方差矩阵必须在样本的性质［大小（size）、结构（structure）、同质性（homogeneity）］对估计质量有很大影响下进行。在`sklearn.covariance`模块中，则提供了多个健壮的协方差估计算法（表3.3-7），列表如下。

表3.3-7 协方差估计算法

协方差估计方法	解释
covariance.EmpiricalCovariance(*[, …])	最大似然协方差估计（Maximum likelihood covariance estimator）
covariance.EllipticEnvelope(*[, …])	用于检测高斯分布数据集中异常值的对象（An object for detecting outliers in a Gaussian distributed dataset.）
covariance.GraphicalLasso([alpha, mode, …])	带L1惩罚估计量的稀疏逆协方差估计（Sparse inverse covariance estimation with an l1-penalized estimator.）
covariance.GraphicalLassoCV(*[, alphas, …])	稀疏逆协方差w/交叉验证l1惩罚的选择（Sparse inverse covariance w/ cross-validated choice of the l1 penalty.）
covariance.LedoitWolf(*[, store_precision, …])	LedoitWolf估计量（LedoitWolf Estimator）
covariance.MinCovDet(*[, store_precision, …])	最小协方差行列式（MCD）：协方差的稳健估计（Minimum Covariance Determinant（MCD）: robust estimator of covariance.）
covariance.OAS(*[, store_precision, …])	Oracle逼近收缩估计（Oracle Approximating Shrinkage Estimator）
covariance.ShrunkCovariance(*[, …])	协方差缩水（shrinkage）估计（Covariance estimator with shrinkage）
covariance.empirical_covariance(X, *[, …])	计算最大似然协方差估计量（Computes the Maximum likelihood covariance estimator）
covariance.graphical_lasso(emp_cov, alpha, *)	l1-惩罚项协方差估计量（l1-penalized covariance estimator）
covariance.ledoit_wolf(X, *[, …])	估计缩水的Ledoit-Wolf协方差矩阵（Estimates the shrunk Ledoit-Wolf covariance matrix.）
covariance.oas(X, *[, assume_centered])	用Oracle近似缩水算法估计协方差（Estimate covariance with the Oracle Approximating Shrinkage algorithm.）
covariance.shrunk_covariance(emp_cov[, …])	计算在对角线上缩水的协方差矩阵（Calculates a covariance matrix shrunk on the diagonal）

下述假设了一个协方差矩阵，并根据该协方差矩阵生产一组数据集，分布使用了sklearn.covariance提供的GraphicalLassoCV、EmpiricalCovariance、MinCovDet及NumPy库提供的np.cov()方法进行计算比较观察。其结果相近，向真实假设的协方差矩阵值靠近。

```
import numpy as np
from sklearn.covariance import GraphicalLassoCV, EmpiricalCovariance, MinCovDet
#假设的协方差矩阵，包含4个特征量
true_cov = np.array([[0.8, 0.0, 0.2, 0.0],
                     [0.0, 0.4, 0.0, 0.0],
                      [0.2, 0.0, 0.3, 0.1],
                      [0.0, 0.0, 0.1, 0.7]])
np.random.seed(0)
#生成满足假设协方差矩阵的特征值
X = np.random.multivariate_normal(mean=[0, 0, 0, 0], cov=true_cov, size=200)
#A-使用GraphicalLassoCV方法
cov = GraphicalLassoCV().fit(X)
print("A-GraphicalLassoCV algorithm:\n{},estimated location(the estimated
mean):{}".format(
    np.around(GraphicalLassoCV().fit(X).covariance_, decimals=3), np.around(cov.
location_, decimals=3)))
#B-EmpiricalCovariance
print("A-EmpiricalCovariance algorithm:\n{}".format(
    np.around(EmpiricalCovariance().fit(X).covariance_, decimals=3)))
#C-MinCovDet
print("A-MinCovDet:\n{}".format(np.around(MinCovDet().fit(X).covariance_,
decimals=3)))
#D-np.cov
print("A-np.cov:\n{}".format(np.around(np.cov(X.T), decimals=3)))
--------------------------------------------------------------------------------
A-GraphicalLassoCV algorithm:
[[0.816 0.051 0.22  0.017]
 [0.051 0.364 0.018 0.036]
```

```
 [0.22  0.018 0.322 0.094]
 [0.017 0.036 0.094 0.69 ]],estimated location(the estimated mean):[0.073 0.04
0.038 0.143]
A-EmpiricalCovariance algorithm:
[[0.816 0.059 0.228 0.009]
 [0.059 0.364 0.025 0.044]
 [0.228 0.025 0.322 0.103]
 [0.009 0.044 0.103 0.69 ]]
A-MinCovDet:
[[ 0.741 -0.005  0.162  0.089]
 [-0.005  0.305  0.024  0.061]
 [ 0.162  0.024  0.237  0.117]
 [ 0.089  0.061  0.117  0.55 ]]
A-np.cov:
[[0.82  0.059 0.229 0.009]
 [0.059 0.366 0.025 0.044]
 [0.229 0.025 0.324 0.103]
 [0.009 0.044 0.103 0.694]]
```

3.3.4 公共健康数据的地理空间分布与相关性分析

3.3.4.1 公共健康数据的地理空间分布

公共健康数据（表3.3-8）为芝加哥社区选定的公共健康（卫生）指标（public health indicator）。该数据集包括27项重要的公共健康指标，这些指标为比率、百分比或者出生率、死亡率、传染病、铅中毒及经济状况相关的指标。该数据集是按照社区进行统计，在字段中给出了社区名，因此可以将数据按照社区名配对到各个社区范围的地理空间数据上。芝加哥社区边界及公共健康数据均来源于Chicago Data Portal，CDP[①]。关于数据的细节描述，可以参考CDP提供的文件。

① Chicago Data Portal，CDP，为芝加哥城开放数据门户，可免费下载数据用于相关分析。其中，许多数据集每天至少更新一次或数次（https://data.cityofchicago.org/）。

在 JupyterLab 中打印数据的方式，包括 `print()` 和直接将要显示数据的变量放置于单个单元格（cell）两种方式。但是，前者显示为固定宽度，不能按照屏幕宽度自动缩放；后者虽为 100% 宽度自动缩放显示，但是如果要将 .ipynb 的 Jupyter 文件输出为 .md 的 Mardown 文件，可能会产生乱码。因此，可以使用 `from IPython.display import HTML`，将要显示的数据转换为 HTML 格式数据后，会解决上述两种问题。

```
import pandas as pd
dataFp_dic = {
    "ublic_Health_Statistics_byCommunityArea_fp": r'./data/Public_Health_
Statistics_Selected_public_health_indicators_by_Chicago_community_area.csv',
    "Boundaries_Community_Areas_current": r'./data/ChicagoCommunityAreas/
ChicagoCommunityAreas.shp',
}
pubicHealth_Statistic = pd.read_csv(
    dataFp_dic["ublic_Health_Statistics_byCommunityArea_fp"])
# 中英对照表（字段映射表）
PubicHealth_Statistic_columns = {
    'Community Area': '社区',
    'Community Area Name': '社区名',
    'Birth Rate': '出生率', 'General Fertility Rate': '一般生育率', 'Low Birth Weight':
'低出生体重', 'Prenatal Care Beginning in First Trimester': '产前3个月护理', 'Preterm
Births': '早产','Teen Birth Rate': '青少年生育率', 'Assault (Homicide)': '攻击（杀
人）', 'Breast cancer in females': '女性乳腺癌', 'Cancer (All Sites)': '癌症', 'Colorectal
Cancer': '结肠直肠癌', 'Diabetes-related': '糖尿病相关', 'Firearm-related': '枪支相关',
'Infant Mortality Rate': '婴儿死亡率','Lung Cancer': '肺癌','Prostate Cancer in
Males': '男性前列腺癌','Stroke (Cerebrovascular Disease)': '中风(脑血管疾病)', 'Childhood
Blood Lead Level Screening': '儿童血铅水平检查', 'Childhood Lead Poisoning': '儿童铅中毒',
'Gonorrhea in Females': '女性淋病', 'Gonorrhea in Males': '男性淋病', 'Tuberculosis':
'肺结核', 'Below Poverty Level': '贫困水平以下','Crowded Housing': '拥挤的住房',
'Dependency': '依赖', 'No High School Diploma': '没有高中文凭', 'Per Capita Income': '人
均收入', 'Unemployment': '失业',
    }
def print_html(df, row_numbers=5):
    '''
    function - 在 Jupyter 中打印 DataFrame 格式数据为 HTML
    Params:
        df - 需要打印的 DataFrame 或 GeoDataFrame 格式数据; DataFrame
        row_numbers - 打印的行数，如果为正，从开始打印如果为负，从末尾打印; int
    Returns:
        转换后的 HTML 格式数据;
    '''
```

表3.3-8 芝加哥城公共健康数据

Community Area	Community Area Name	Birth Rate	General Fertility Rate	Low Birth Weight	Prenatal Care Beginning in First Trimester	Preterm Births	Teen Birth Rate	Assault (Homicide)	Breast cancer in females
1	Rogers Park	16.4	62	11	73	11.2	40.8	7.7	23.3
2	West Ridge	17.3	83.3	8.1	71.1	8.3	29.9	5.8	20.2
3	Uptown	13.1	50.5	8.3	77.7	10.3	35.1	5.4	21.3
4	Lincoln Square	17.1	61	8.1	80.5	9.7	38.4	5	21.7
5	North Center	22.4	76.2	9.1	80.4	9.8	8.4	1	16.6
6	Lake View	13.5	38.7	6.3	79.1	8.1	15.8	1.4	20.1
Cancer(All Sites)	**Colorectal Cancer**	**Diabetes-related**	**Firearm-related**	**Infant Mortality Rate**	**Lung Cancer**	**Prostate Cancer in Males**	**Stroke (Cerebrovascular Disease)**	**Childhood Blood Lead Level Screening**	**Childhood Lead Poisoning**
176.9	25.3	77.1	5.2	6.4	36.7	21.7	33.7	364.7	0.5
155.9	17.3	60.5	3.7	5.1	36	14.2	34.7	331.4	1
183.3	20.5	80	4.6	6.5	50.5	25.2	41.7	353.7	0.5
153.2	8.6	55.4	6.1	3.8	43.1	27.6	36.9	273.3	0.4
152.1	26.1	49.8	1	2.7	42.4	15.1	41.6	178.1	0.9
126.9	13	38.5	1.8	2.2	32.5	17	24.4	179.2	0.4
Gonorrhea in Females	**Gonorrhea in Males**	**Tuberculosis**	**Below Poverty Level**	**Crowded Housing**	**Dependency**	**No High School Diploma**	**Per Capita Income**	**Unemployment**	
322.5	423.3	11.4	22.7	7.9	28.8	18.1	23714	7.5	
141	205.7	8.9	15.1	7	38.3	19.6	21375	7.9	
170.8	468.7	13.6	22.7	4.6	22.2	13.6	32355	7.7	
98.8	195.5	8.5	9.5	3.1	25.6	12.5	35503	6.8	
85.4	188.6	1.9	7.1	0.2	25.5	5.4	51615	4.5	
81.8	357.6	3.2	10.5	1.2	16.5	2.9	58227	4.7	

```
    from IPython.display import HTML
    if row_numbers > 0:
        return HTML(df.head(row_numbers).to_html())
    else:
        return HTML(df.tail(abs(row_numbers)).to_html())
print_html(pubicHealth_Statistic, 6)
```

在数据合并时，公共健康数据以"Community Area"为对位列，社区边界数据以"area_numbe"为对位列，使用df.merge()方法完成融合。打印两组数据（图3.3-12，图3.3-13），其中肺癌（Lung Cancer）为每100000人所占的比例。GeoDataFrame格式数据在打印时如果需要在图上每个Polygon上标注数据增加地图的信息量，可以按照下述代码处理。

```
import geopandas as gpd
from mpl_toolkits.axes_grid1 import make_axes_locatable
import matplotlib.pyplot as plt
community_area = gpd.read_file(
    dataFp_dic["Boundaries_Community_Areas_current"])
print(community_area.dtypes)
community_area.area_numbe = community_area.area_numbe.astype('int64')
print("_"*50)
print("boundaries_community.area_numbe dtype:",
      community_area.area_numbe.dtypes)
pubicHealth_gpd = community_area.merge(
    pubicHealth_Statistic, left_on='area_numbe', right_on='Community Area')
fig, axs = plt.subplots(1, 2, figsize=(40, 40))
# 打印第 1 组数据
divider = make_axes_locatable(axs[0])
cax_0 = divider.append_axes("left", size="5%", pad=0.1)  # 配置图例参数
# 如果提示如下错误，则需要按照信息安装对应的库 ,ImportError: The descartes package is required for plotting polygons in geopandas. You can install it using 'conda install -c conda-forge descartes' or 'pip install descartes'.
```

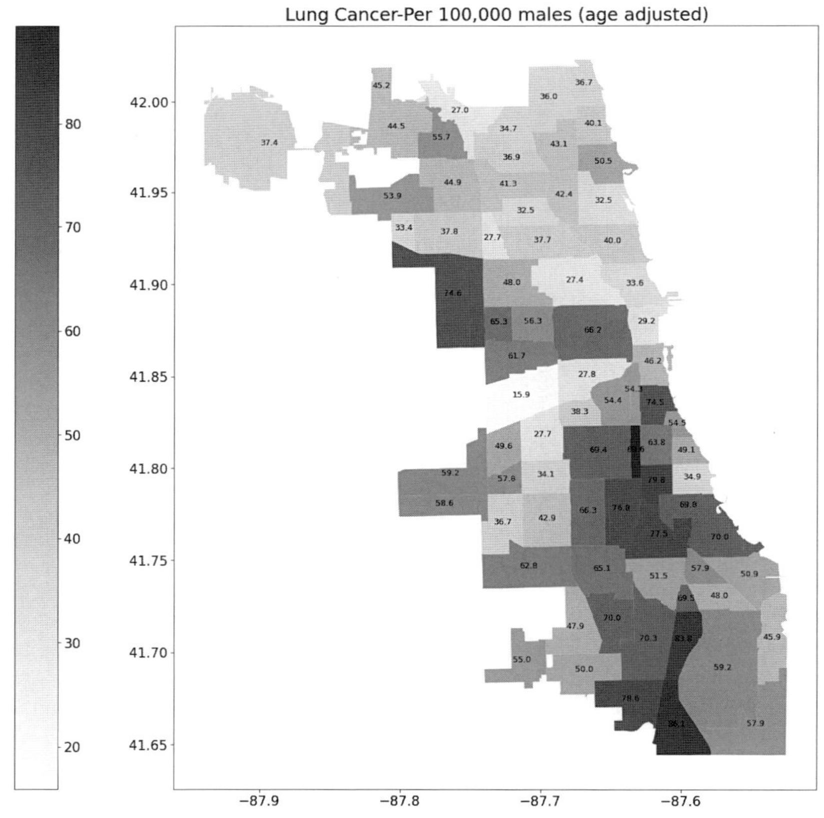

图3.3-12　肺癌患者社区统计

```
pubicHealth_gpd.plot(column='Lung Cancer',
                    ax=axs[0], cax=cax_0, legend=True, cmap='OrRd')
axs[0].set_title("Lung Cancer-Per 100,000 males (age adjusted)", fontsize=20)
pubicHealth_gpd.apply(lambda x: axs[0].annotate(
    text=x["Lung Cancer"], xy=x.geometry.centroid.coords[0], ha='center'), axis=1)
# 增加标注
axs[0].tick_params(axis='both', labelsize=15)
cb_ax_0 = axs[0].figure.axes[2]
cb_ax_0.tick_params(labelsize=15)
# 打印第 2 组数据
divider = make_axes_locatable(axs[1])
cax_1 = divider.append_axes("right", size="5%", pad=0.1)
pubicHealth_gpd.plot(column='Per Capita Income', ax=axs[1], cax=cax_1,
legend=True, cmap='OrRd')
axs[1].set_title("Per Capita Income-2011 inflation-adjusted dollars", fontsize=20)
pubicHealth_gpd.apply(lambda x: axs[1].annotate(
    text=x["Community Area Name"], xy=x.geometry.centroid.coords[0], ha='center'),
axis=1)
axs[1].tick_params(axis='both', labelsize=15)
cb_ax_1 = axs[1].figure.axes[3]
cb_ax_1.tick_params(labelsize=15)
plt.show()
```

```
--------------------------------------------------------------------------
area            float64
area_num_1       object
area_numbe       object
comarea         float64
comarea_id      float64
community        object
perimeter       float64
shape_area      float64
shape_len       float64
geometry       geometry
dtype: object
--------------------------------------------------------------------------
```

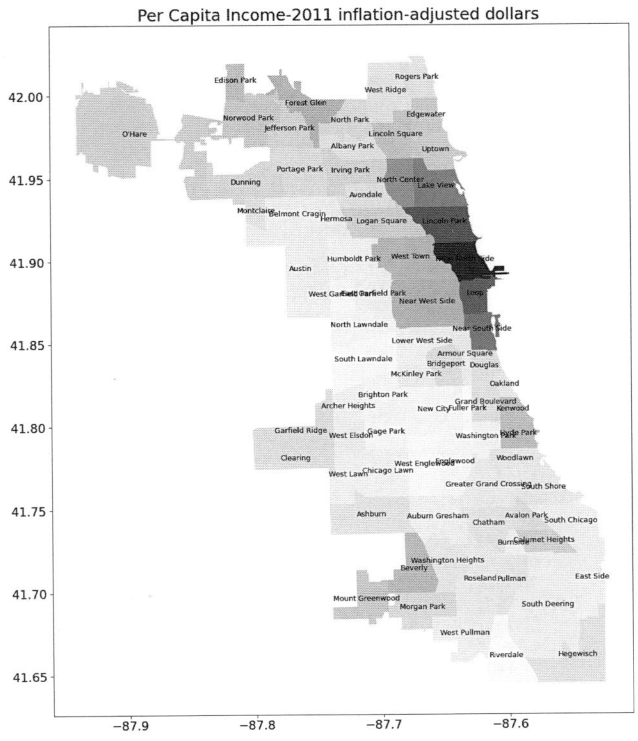

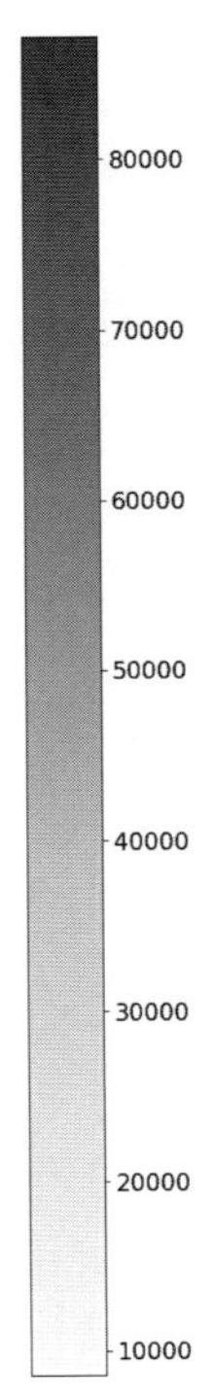

图3.3-13 人均收入社区统计

```
boundaries_community.area_numbe dtype: int64
```

3.3.4.2 公共健康数据的相关性分析

观察图3.3-13，似乎肺癌的发生比例与收入多少有关联，收入高的区域发生肺癌的机率低；相反，收入低的区域，发生肺癌的机率高。粗略的观察只能够给出一个初步不是十分确定的判断结果。如果需要以数据分析的方式证明之间是否存在关联，则需要作相关性分析（表3.3-9）。

```
pubicHealth_Statistic_mapping = {
     'Community Area': 'CommunityArea',
     'Community Area Name': 'CommunityArea_Name',
     'Birth Rate': 'Birth_Rate',
     'General Fertility Rate': 'General_FertilityRate',
     'Low Birth Weight': 'Low_BirthWeight',
     'Prenatal Care Beginning in First Trimester':
'PrenatalCareBeginning_inFirstTrimester',
     'Preterm Births': 'Preterm_Births',
     'Teen Birth Rate': 'TeenBirth_Rate',
     'Assault (Homicide)': 'Assault_Homicide',
     'Breast cancer in females': 'BreastCancer_
infemales',
     'Cancer (All Sites)': 'Cancer_AllSites',
     'Colorectal Cancer': 'Colorectal_Cancer',
     'Diabetes-related': 'Diabetes_related',
     'Firearm-related': 'Firearm_related',
     'Infant Mortality Rate': 'InfantMortality_Rate',
     'Lung Cancer': 'Lung_Cancer',
     'Prostate Cancer in Males': 'ProstateCancer_
inMales',
     'Stroke (Cerebrovascular Disease)': 'Stroke_
CerebrovascularDisease',
     'Childhood Blood Lead Level Screening':
'ChildhoodBloodLeadLevel_Screening',
     'Childhood Lead Poisoning': 'ChildhoodLead_
Poisoning',
     'Gonorrhea in Females': 'Gonorrhea_inFemales',
     'Gonorrhea in Males': 'Gonorrhea_inMales',
     'Tuberculosis': 'Tuberculosis',
     'Below Poverty Level': 'BelowPoverty_Level',
     'Crowded Housing': 'Crowded_Housing',
     'Dependency': 'Dependency',
     'No High School Diploma': 'NoHighSchool_Diploma',
     'Per Capita Income': 'PerCapita_Income',
     'Unemployment': 'Unemployment',
}
pubicHealth_rename = pubicHealth_gpd.rename(
    columns=pubicHealth_Statistic_mapping)
pubicHealth_extract_columns = [
     'Birth_Rate', 'General_FertilityRate', 'Low_
BirthWeight',
     'PrenatalCareBeginning_inFirstTrimester', 'Preterm_
Births',
     'TeenBirth_Rate', 'Assault_Homicide', 'BreastCancer_
infemales',
     'Cancer_AllSites', 'Colorectal_Cancer', 'Diabetes_
related',
     'Firearm_related', 'InfantMortality_Rate', 'Lung_
Cancer',
     'ProstateCancer_inMales', 'Stroke_
CerebrovascularDisease',
     'ChildhoodBloodLeadLevel_Screening', 'ChildhoodLead_
Poisoning',
     'Gonorrhea_inFemales', 'Gonorrhea_inMales',
'Tuberculosis',
     'BelowPoverty_Level', 'Crowded_Housing',
'Dependency',
     'NoHighSchool_Diploma', 'PerCapita_Income',
'Unemployment'
]
```

表3.3-9 公共健康数据两两相关系数

	Birth_Rate	General_FertilityRate	Low_BirthWeight	…	ChildhoodLead_Poisoning	Gonorrhea_inFemales	Tuberculosis	BelowPoverty_Level	Crowded_Housing
Birth_Rate	1	0.810334	0.108179		0.210016	0.126671	0.279755	0.249764	0.603219
General_FertilityRate	0.810334	1	0.142189		0.303624	0.250984	0.144972	0.173846	0.655826
Low_BirthWeight	0.108179	0.142189	1		0.388319	0.737727	0.125231	0.681049	-0.080532
PrenatalCareBeginning_inFirstTrimester	-0.178847	-0.134292	-0.532546		-0.516579	-0.653985	-0.270147	-0.516145	-0.074951
Preterm_Births	0.004334	0.122235	0.8431		0.379146	0.800491	0.001197	0.550906	-0.140244

```
pubicHealth_extract = pubicHealth_rename[pubicHealth_extract_columns].select_
dtypes(
    include=np.number)
publicHealth_correlation = pubicHealth_extract.corr()
print_html(publicHealth_correlation)
```

通过上述相关性计算，已经计算了两两之间的相关系数。为了更清晰的查看相关系数，可以同时打印散点图3.3–14（通常在计算相关系数前打印散点图，查看两个变量间的线性关系），或查看相关系数变化的热力图3.3–15。

```
import plotly.graph_objects as go
fig = go.Figure(data=go.Splom(dimensions=[dict(label=k,values=list(v.values()))
for k,v in pubicHealth_extract.to_dict().items()],diagonal_visible=False, # 移除对
角线上的子图
                ))
fig.update_layout(title='公共健康数据',dragmode='select',width=1800,height=1800,ho
vermode='closest',)
fig.show()
import seaborn as sns
plt.figure(figsize=(20, 20))
```

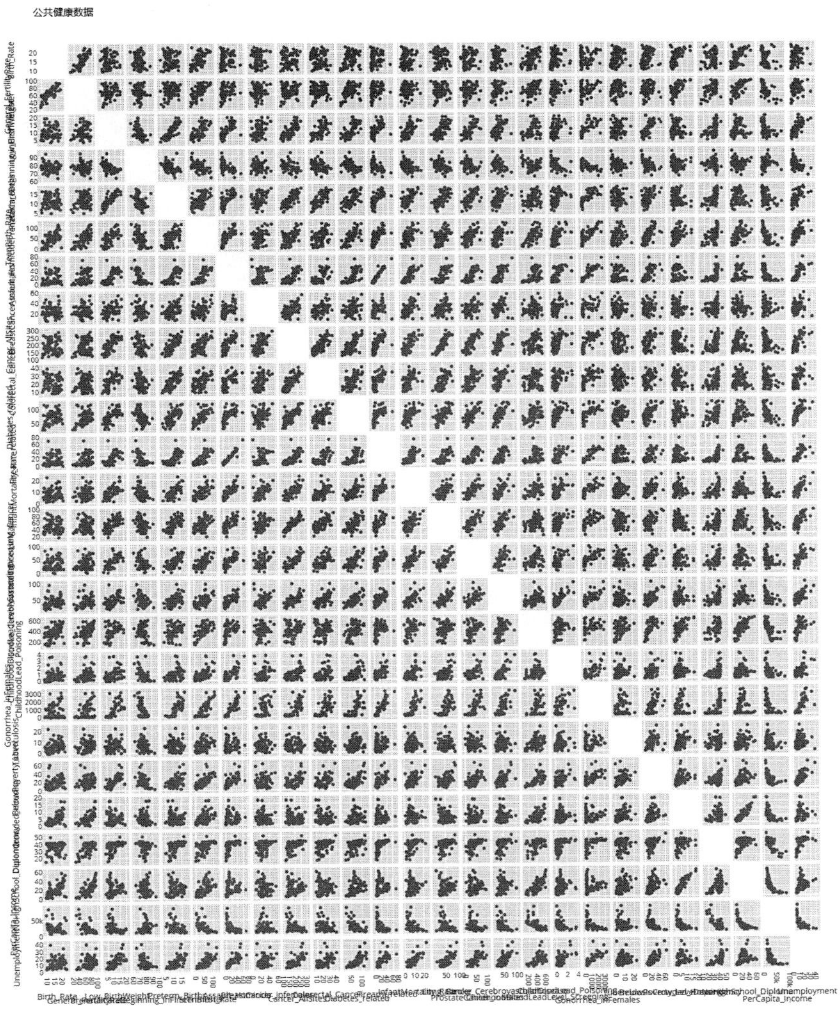

图3.3–14 公共健康数据两两散点图

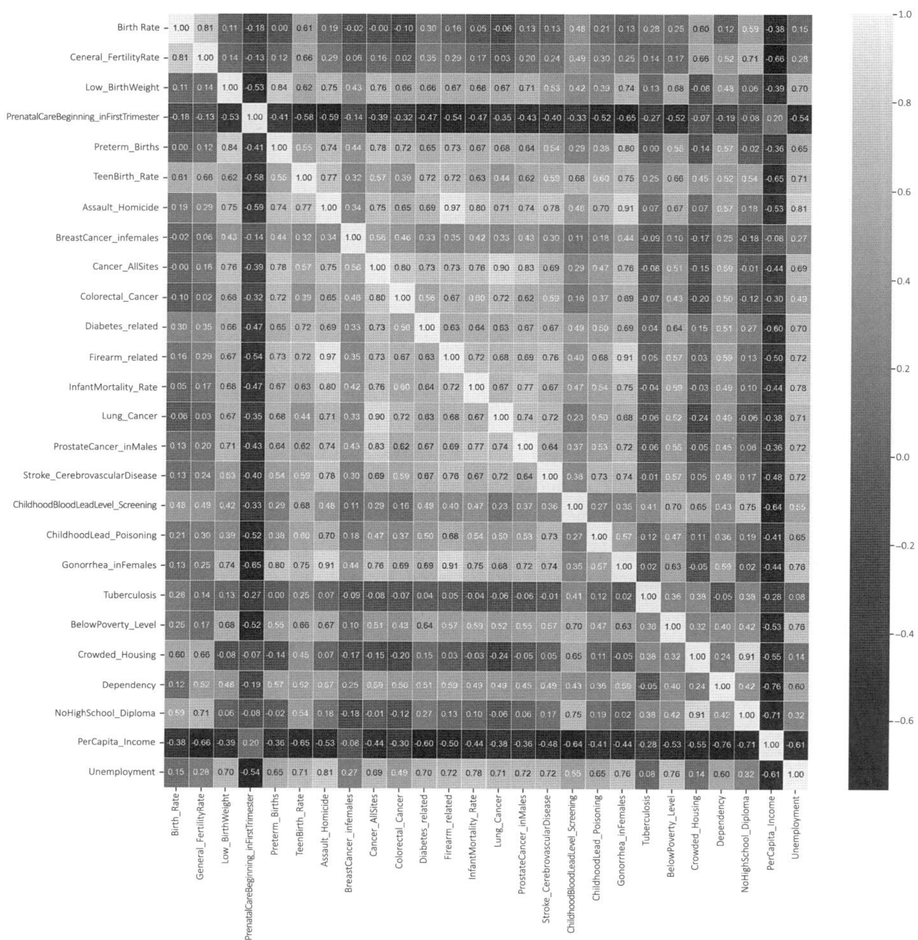

图3.3-15　公共健康数据两两相关系数变化热力图

```
sns.heatmap(publicHealth_correlation,annot=True, fmt=".2f", linewidths=.5,)
plt.show()
```

不管是散点图还是热力图，以及相关系数矩阵，在观察多个两两之间相关性时，都不是很容易观察所关注的对象。因此，可以根据分析的目的将需要关注的变量提取出来，只关注该变量与其他所有变量的相关系数，如图3.3-16所示。

```
import numpy as np
economic_factors = ['BelowPoverty_Level', 'Crowded_Housing', 'Dependency',
'NoHighSchool_Diploma', 'PerCapita_Income', 'Unemployment']
publicHealth_indicator = publicHealth_correlation[economic_factors]
publicHealth_indicator_columns = publicHealth_indicator.T.columns.to_numpy()
plt.rcdefaults()
plt.rcParams.update({'font.size': 9})
nrows = 3
ncols = 2
fig, axs = plt.subplots(nrows=nrows, ncols=ncols, figsize=(10*2, 10*2))
y_pos = np.arange(len(publicHealth_indicator_columns))
i = 0
for idx in [(row, col) for row in range(nrows) for col in range(ncols)]:
    axs[idx].barh(
        y_pos, publicHealth_indicator[economic_factors[i]].to_numpy(),
align='center')
    axs[idx].set_yticks(y_pos)
    axs[idx].set_yticklabels(publicHealth_indicator_columns)
```

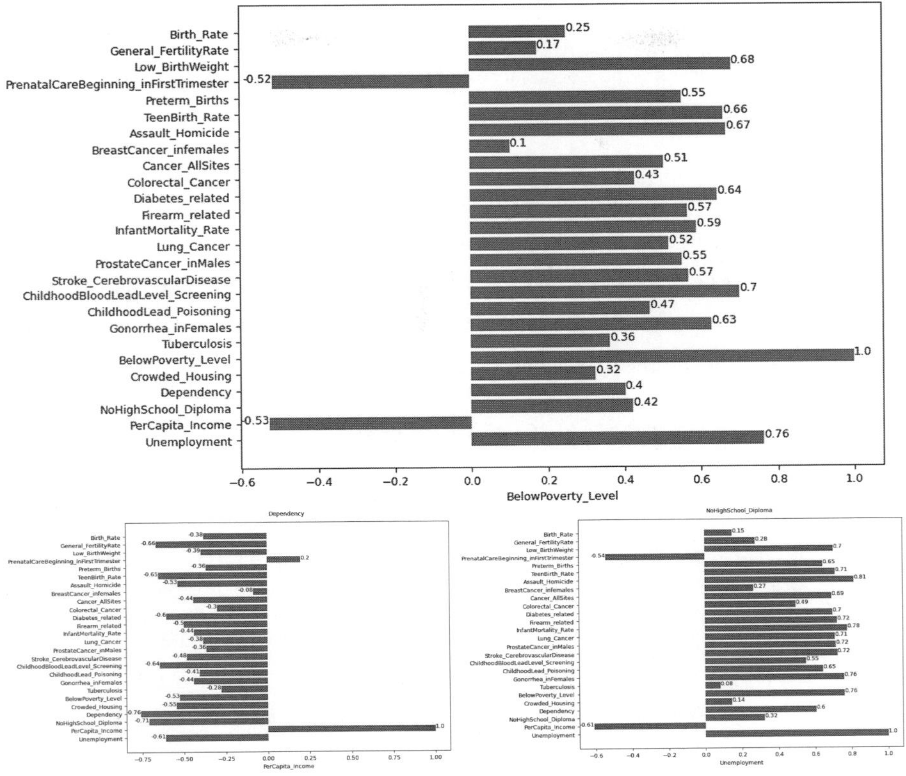

图3.3-16 公共健康数据两两相关系数（关注单个变量）

```
    axs[idx].invert_yaxis()
    axs[idx].set_xlabel(economic_factors[i])
    for index, value in enumerate(publicHealth_indicator[economic_factors[i]].
to_numpy()):
        if value >= 0:
            axs[idx].text(value, index, str(round(value, 2)),horizontalalignment='left')
        else:
            axs[idx].text(value, index, str(round(value, 2)),horizontalalignment='right')
    i += 1
plt.subplots_adjust(top=0.92, bottom=0.08, left=0.10,
                    right=0.95, hspace=0.1, wspace=0.4)
plt.show()
```

虽然计算了两两公共健康指标之间的相关系数，但这并不代表该样本所计算得出的相关性能够反映总体是否也存在样本所得出的现象。这时，需要关注假设检验计算得出的P值（表3.3-10），将P值不满足显著性水平为0.05的两两变量剔除掉，所得到的结果即为能够反映总体相关性的部分。首先，用所关注的变量作为索引（index），为所有的经济变量`economic_factors=['BelowPoverty_Level', 'Crowded_Housing', 'Dependency', 'NoHighSchool_Diploma', 'PerCapita_Income', 'Unemployment']`，方便后续的数据分析。

```
from scipy import stats
disease_columns = [
    'Birth_Rate', 'General_FertilityRate', 'Low_BirthWeight', 'PrenatalCareBeginning_
inFirstTrimester', 'Preterm_Births', 'TeenBirth_Rate', 'Assault_Homicide',
'BreastCancer_infemales', 'Cancer_AllSites', 'Colorectal_Cancer', 'Diabetes_related',
'Firearm_related', 'InfantMortality_Rate', 'Lung_Cancer', 'ProstateCancer_inMales',
'Stroke_CerebrovascularDisease', 'ChildhoodBloodLeadLevel_Screening', 'ChildhoodLead_
Poisoning', 'Gonorrhea_inFemales', 'Tuberculosis', ]
pubicHealth_pearsonr = {}
```

```
for factor in economic_factors:
    disease_temp = {}
    for disease in disease_columns:
        desease_series = pd.to_numeric(pubicHealth_extract[disease],
errors='coerce')
        factor_series = pd.to_numeric(pubicHealth_extract[factor],
errors='coerce')
        mask = pd.notna(desease_series) & pd.notna(factor_series)
        disease_temp[disease] = stats.pearsonr(pd.to_numeric(factor_series[mask],
errors='ignore'), pd.to_numeric(desease_series[mask], errors='ignore'))
    pubicHealth_pearsonr[factor] = disease_temp
pubicHealth_pearsonr_df = pd.DataFrame.from_dict(
    pubicHealth_pearsonr, orient='index').stack().to_frame(name='corr_pV')
pubicHealth_pearsonr_df['correlation'] = pubicHealth_pearsonr_df.corr_pV.apply(lambda
row: row[0])
pubicHealth_pearsonr_df['p_value'] = pubicHealth_pearsonr_df.corr_pV.apply(lambda
row: row[1])
print_html(pubicHealth_pearsonr_df)
```

表3.3-10 相关系数的显著性水平

		corr_pV	correlation	p_value
BelowPoverty_Level	Birth_Rate	(0.24976406537114698, 0.02847574386002521 6)	0.249764	2.847574e-02
	General_FertilityRate	(0.17384625982353086, 0.13051281178487212)	0.173846	1.305128e-01
	Low_BirthWeight	(0.6810485083790098, 9.384671983490509e-12)	0.681049	9.384672e-12
	PrenatalCareBeginning_inFirstTrimester	(-0.5161448655124206, 1.5503295893976403e-06)	-0.516145	1.550330e-06
	Preterm_Births	(0.5509056684724956, 2.086303329322895e-07)	0.550906	2.086303e-07

以相关系数的值为横坐标，以相关系数对应的P值为纵坐标，以0.05作为显著性水平，打印散点图3.3-17，可以较为直观地查看哪些两两公共健康指标的相关系数能够反映或者不能够反映总体的现象。

```
pubicHealth_pearsonr_resetIdx = pubicHealth_pearsonr_df.reset_index()
significance_level = 0.05
fig = plt.figure(figsize=(30, 10))
ax = fig.add_subplot(111, facecolor='#FFFFCC')
X = pubicHealth_pearsonr_resetIdx.correlation
Y = pubicHealth_pearsonr_resetIdx.p_value
ax.plot(X, Y, 'o')
ax.axhline(y=significance_level, color='r',
           linestyle='--', label='significance_level')
i = 0
for x, y in zip(X, Y):
    label = pubicHealth_pearsonr_resetIdx.loc[i].level_0 + \ ' : '+pubicHealth_
pearsonr_resetIdx.loc[i].level_1
    plt.annotate(label, (x, y), textcoords="offset points", xytext=(0, 10),
                 ha='center', rotation=90)
    i += 1
plt.annotate('alpha level=%.2f' % significance_level, xy=(-0.7, significance_
level), xytext=(-0.75, 0.15), arrowprops=dict(facecolor='black', shrink=0.01),
horizontalalignment='right', size=15)
plt.xlabel("correlation", fontsize=20)
plt.ylabel("p-value", fontsize=20)
plt.tick_params(axis='both', labelsize=15)
plt.show()
```

提取P值小于等于0.05的所有行（表3.3-11），并以柱状图的形式打印相关系数结果（图3.3-18）。

```
pubicHealth_pearsonr_alpha = pubicHealth_pearsonr_df[pubicHealth_pearsonr_df.p_
value<=0.05]
print_html(pubicHealth_pearsonr_alpha)
```

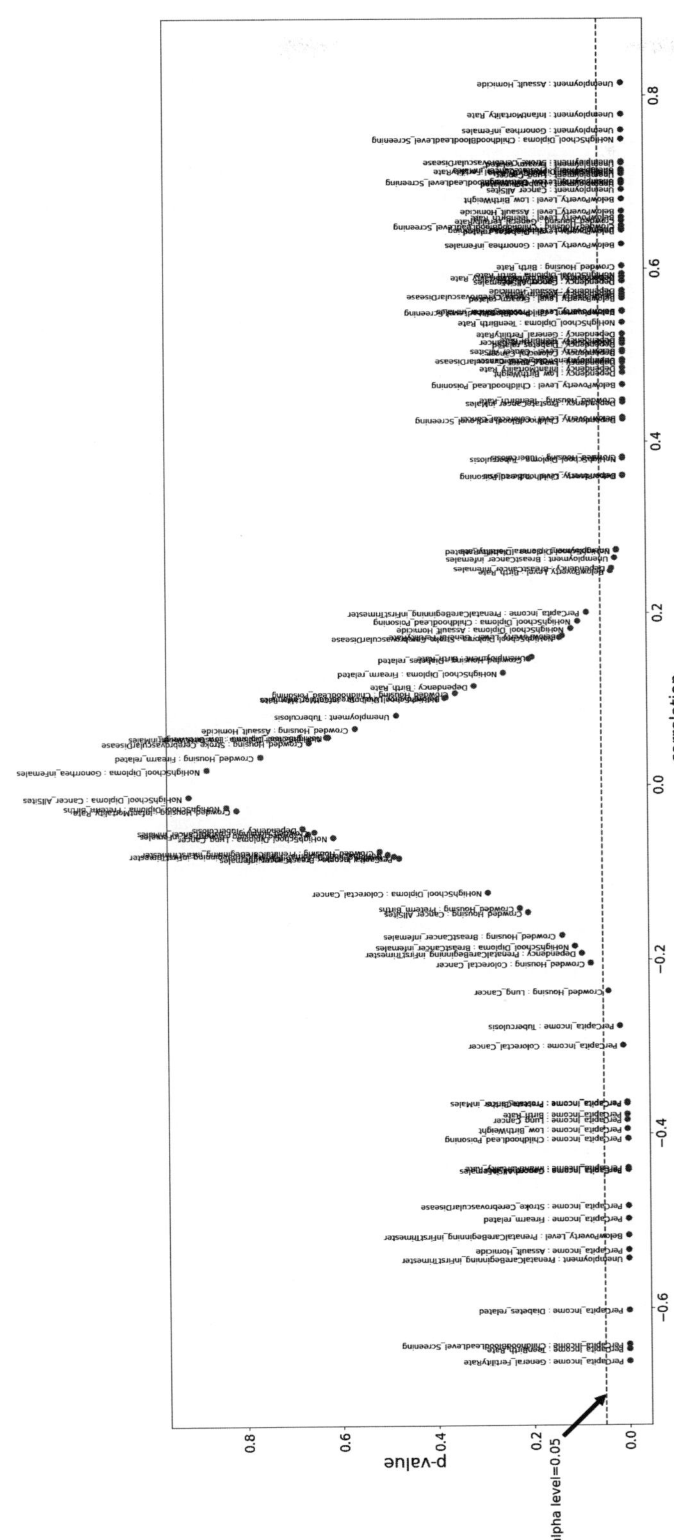

图3.3-17 相关系数与显著性水平

表3.3-11 *P*值小于等于0.05的行

		corr_pV	correlation	p_value
BelowPoverty_Level	Birth_Rate	(0.24976406537114698, 0.028475743860025216)	0.249764	2.847574e-02
	Low_BirthWeight	(0.6810485083790098, 9.384671983490509e-12)	0.681049	9.384672e-12
	PrenatalCareBeginning_inFirstTrimester	(-0.5161448655124206, 1.5503295893976403e-06)	-0.516145	1.550330e-06
	Preterm_Births	(0.5509056684724956, 2.086303329322895e-07)	0.550906	2.086303e-07
	TeenBirth_Rate	(0.6600382763346119, 6.598083934767818e-11)	0.660038	6.598084e-11

图3.3-18 以柱状图表述相关系数

```
import plotly.graph_objects as go
import random
def generate_colors():
    '''
    function - 生成颜色列表或者字典
    Returns:
        hex_colors_only - 16 进制颜色值列表; list
        hex_colors_dic - 颜色名称: 16 进制颜色值; dict
        rgb_colors_dic - 颜色名称: (r,g,b); dict
    '''
    import matplotlib
    hex_colors_dic = {}
    rgb_colors_dic = {}
    hex_colors_only = []
    for name, hex in matplotlib.colors.cnames.
items():
        hex_colors_only.append(hex)
        hex_colors_dic[name] = hex
        rgb_colors_dic[name] = matplotlib.colors.
to_rgb(hex)
    return hex_colors_only, hex_colors_dic, rgb_
colors_dic
generate_colors, _, _ = generate_colors()
economic_factors = ['BelowPoverty_Level', 'Crowded_
Housing', 'Dependency','NoHighSchool_Diploma',
'PerCapita_Income', 'Unemployment']
fig = go.Figure()
for idx, data in pubicHealth_pearsonr_alpha.
groupby(level=0):
    fig.add_trace(go.Bar(x=disease_columns,
y=data.correlation,name=idx,marker_color=random.
choice(generate_colors)))
fig.update_layout(title='Public health indicators_
correlation,p-value<0.05',xaxis_tickfont_size=14,y
axis=dict(title='correlation)',titlefont_size=16,
tickfont_size=14,),
    barmode='group',
    bargap=0.1,  # 相邻位置坐标条之间的间隙。
    bargroupgap=0.5  # 相同位置坐标条之间的间距。
)
fig.show()
```

参考文献（References）：

[1] Timothy C. Urdan（蒂莫西·C·厄丹），彭志文译. 白话统计学[M]. 北京：中国人民大学出版社，2013，12. 第3版. <Timothy C. Urdan (2022). Statistics in Plain English. Routledge.>.

[2]（日）高桥 信著，陈刚译. 株式会社TREND-PRO漫画制作. 漫画统计学[M]. 北京：科学出版社，2019. 8.

[3] Dawn Griffiths (2008). Head First Statistics. O'Reilly Media.

3.4 简单回归与多元回归

在相关性分析部分计算了两两之间的相关系数，但在相关系数计算中并没有区分自变量（independent variable）或预测变量（predictor variable）亦或解释变量，与因变量(dependent variable)或结果变量（outcome variable）亦或效标变量（criterion variable）；而回归分析除了可以反映变量间关系的性质和强度，也可以根据自变量预测因变量，进行统计推断，其预测的准确程度取决于相关的程度。相关性越强，预测就越准确。根据自变量的数量，可以将回归分为简单回归和多元回归。多元回归包括两个或两个以上的自变量及一个因变量，那么这多个自变量作为一个整体与因变量有多大关系；作为个体，各个自变量与因变量的关系强度如何；同时，自变量之间的相对强度又如何，以及自变量之间是否存在交互效应等。回答上述问题后，基本就解释了自变量和因变量之间的关系。

回归涉及的内容异常丰富，回归的类型丰富多彩。在实际应用回归来预测因变量时，并不会从最基础的开始一步步计算，而是直接使用Python中集成的库。这时，就涉及了著名的机器学习库scikit-learn（sklearn）[①]库。该库包含了数据的预处理、降维、分类模型、聚类模型和回归模型，以及模型选择等内容，能够辅助处理众多城市空间数据分析问题。同时，也可以使用statsmodels[②]库等。scikit-learn库倾向于预测，而statsmodels库倾向于统计推断。但是，如果一开始就使用集成的模型，对于理解回归乃至任何统计学的知识点或者数据分析的算法都是不利的，往往囫囵吞枣、一知半解，因此亦然一步步，借助Python语言来把这个最为基础部分的脉络梳理清楚。

3.4.1 预备知识

3.4.1.1 反函数（inverse function）

如果函数$f(x)$有$y=f(x)$，f是y关于自变量x的函数；如果存在一个函数g，使得，$g(y)=g(f(x))=x$，则g是x关于自变量y的函数，称g为f的反函数。若一函数有反函数，此函数便是可逆的。可记作函数f和它的反函数f^{-1}。假设函数$f=2x+1$，则$x=(f/2-1/2)$，替换x为f^{-1}，f为x，则结果为反函数$f^{-1}=x/2-1/2$。在下述代码表述假设函数及其反函数时，使用了SymPy[③]代数计算库，轻量型的SymPy的语法方式保持了Python语言本身语法形式特点，使得数学公示的表述和计算上清晰、简便。打印f（图3.4–1）和它的反函数f^{-1}（图3.4–2）。可以看到，两者的相对横纵坐标发生了置换。求一个函数的反函数，当前有很多在线自动转换的平台，可以搜索输入公式获得其反函数的公式。例如，Symbolab[④]，如果$f(x)=\frac{x^2+x+1}{x}$，则其反函数为$g(x)=\frac{-1+x+\sqrt{x^2-2x-3}}{2}$，及$g(x)=\frac{-1+x-\sqrt{x^2-2x-3}}{2}$。

```
import sympy
from sympy import init_printing, pprint, sqrt
import numpy as np
import seaborn as sns
import matplotlib.pyplot as plt

init_printing()  # sympy 提供有多种公式打印模式

# 示例 -A
# 定义字符
x = sympy.Symbol('x')

# 定义表达式
f = 2*x+1  # 函数 fx
g = x/2-1/2  # fx 的反函数

# 转换表达式为等价的 numpy 函数实现数值计算
x_array = np.arange(-5, 10)
f_ = sympy.lambdify(x, f, "numpy")
g_ = sympy.lambdify(x, g, "numpy")

# 求解函数并绘制图表
```

① scikit-learn，也称为sklearn，针对Python编程语言的免费软件机器学习库（https://scikit-learn.org/stable/index.html）。

② statsmodels，为许多不同统计模型的估计提供了Python类和函数，进行统计测试和统计数据探索。每个估计器都返回一个包含广泛结果的统计列表，这些结果与现有的统计包进行了测试，确保其正确性（https://www.statsmodels.org/stable/index.html）。

③ SymPy，用于符号数学（symbolic mathematics）的Python库（https://docs.sympy.org/latest/index.html）。

④ Symbolab，为一个在线数学教学工具，允许用户使用数学符号和科学符号及文本来学习、练习和发现数学知识，例如代数、三角和微积分。同时，提供了丰富的智能计算器，包括：方程（equations）、同步方程（simultaneous equations）、不等式（inequalities）、积分（integrals）、导数（derivatives）、极限（limits）、切线（tangent line）、三角方程（trigonometric equations）、函数（functions）等（https://www.symbolab.com/）。

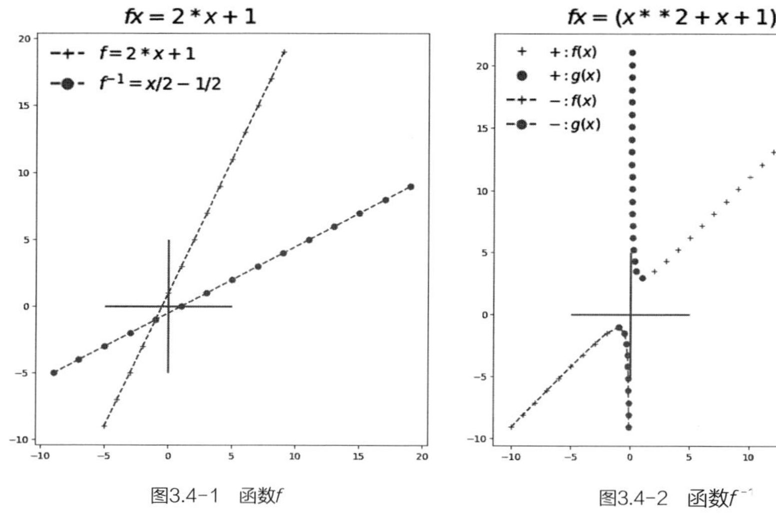

图3.4-1　函数f　　图3.4-2　函数f^{-1}

```
fig, axs = plt.subplots(1, 2, figsize=(16, 8))
axs[0].plot(x_array, f_(x_array), '+--', color='tab:blue', label='$f=2*x+1$')
axs[0].plot(f_(x_array), g_(f_(x_array)), 'o--',
            color='tab:red', label='$f^{-1}=x/2-1/2$')
axs[0].set_title('$fx=2*x+1$')
axs[0].legend(loc='upper left', frameon=False)
axs[0].hlines(y=0, xmin=-5, xmax=5, lw=2, color='gray')
axs[0].vlines(x=0, ymin=-5, ymax=5, lw=2, color='gray')

# 示例 -B
f_B = (x**2+x+1)/x
print("使用 pprint 方式打印公式：")
pprint(f_B, use_unicode=True)  # 使用 pprint 方式打印公式
g_B_negative = (-1+x+sqrt(x**2-2*x-3))/2
g_B_positive = (-1+x-sqrt(x**2-2*x-3))/2
f_B_ = sympy.lambdify(x, f_B, "numpy")
g_B_positive_ = sympy.lambdify(x, g_B_positive, "numpy")
g_B_negative_ = sympy.lambdify(x, g_B_negative, "numpy")
x_B_array = np.arange(-10, 21)
x_B_array_positive = x_B_array[x_B_array > 0]
axs[1].plot(x_B_array_positive, f_B_(x_B_array_positive),
            '+', color='tab:blue', label='$+:f(x)$')
axs[1].plot(g_B_positive_(f_B_(x_B_array_positive)), f_B_(
    x_B_array_positive), 'o', color='tab:red', label='$+:g(x)$')
x_B_array_negative = x_B_array[x_B_array < 0]
axs[1].plot(x_B_array_negative, f_B_(x_B_array_negative),
            '+--', color='tab:blue', label='$-:f(x)$')
axs[1].plot(g_B_negative_(f_B_(x_B_array_negative)), f_B_(
    x_B_array_negative), 'o--', color='tab:red', label='$-:g(x)$')
axs[1].hlines(y=0, xmin=-5, xmax=5, lw=2, color='gray')
axs[1].vlines(x=0, ymin=-5, ymax=5, lw=2, color='gray')
axs[1].legend(loc='upper left', frameon=False)
axs[1].set_title('$fx=(x**2+x+1)/x$')
plt.show()
print("JupyterLab 直接输出公式 :g_B_negative=")
g_B_negative  # 用 JupyterLab 直接输出公式
--------------------------------------------------------------------------------
```

```
使用 pprint 方式打印公式：
 2
x  + x + 1
──────────
    x
JupyterLab 直接输出公式 :g_B_negative=
```

$$\frac{x}{2}+\frac{\sqrt{x^2-2x-3}}{2}-\frac{1}{2}$$

3.4.1.2 指数函数与自然对数函数

指数函数（Exponential function）是形式为b^x的数学函数。其中，b是底数（或称基数，base），而x是指数（index/exponent），如图3.4–3和图3.4–4所示。

对数（logarithm）是幂运算的逆运算。假如$x=b^y$，则有$y=\log_b x$，其中b是对数的底（或称基数），而y就是x对于底数b，x的对数。典型的底数有e、10或2，如图3.4–5所示。

自然对数（Natural logarithm）为以数学常e为底数的对数函数，标记为$\ln x$或$\log_e x$，其反函数为指数函数e^x。

● 指数函数与对数函数的性质：

1. $(e^a)^b=e^{ab}$
2. $\frac{e^a}{e^b}=e^{a-b}$
3. $a=\log(e^a)$
4. $\log(a^b)=b\times(\log a)$
5. $\log a+\log b=\log(a\times b)$

```
import math
from sympy import ln, log, Eq
x = sympy.Symbol('x')
f_exp = 2**x
f_exp_ = sympy.lambdify(x, f_exp, "numpy")
fig, axs = plt.subplots(1, 3, figsize=(25, 8))
exp_x = np.arange(-10, 10, 0.5, dtype=float)
axs[0].plot(exp_x, f_exp_(exp_x), label="f(x)=2**x")
axs[0].legend(loc='upper left', frameon=False)
f_exp_e = math.e**x
f_exp_e_ = sympy.lambdify(x, f_exp_e, "numpy")
axs[1].plot(exp_x, f_exp_e_(exp_x), label="f(x)=e**x", color='r')
axs[1].legend(loc='upper left', frameon=False)
log_x = np.arange(1, 20, dtype=float)
f_ln = ln(x)
f_ln_ = sympy.lambdify(x, f_ln, "numpy")
axs[2].plot(log_x, f_ln_(log_x), label='base=e')
f_log_2 = log(x, 2)
f_log_2_ = sympy.lambdify(x, f_log_2, "numpy")
axs[2].plot(log_x, f_log_2_(log_x), label="base=2")
f_log_10 = log(x, 10)
f_log_10_ = sympy.lambdify(x, f_log_10, "numpy")
axs[2].plot(log_x, f_log_10_(log_x), label="base=10")
axs[2].legend(loc='upper left', frameon=False)
plt.show()
```

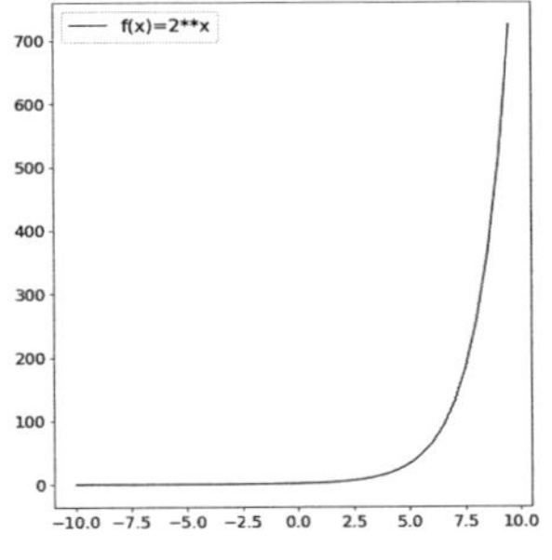

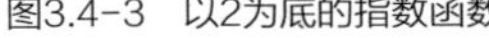

图3.4–3 以2为底的指数函数

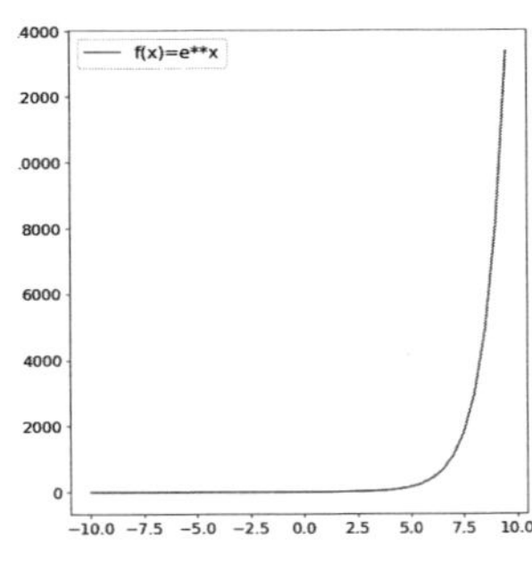

图3.4–4 以e为底的指数函数

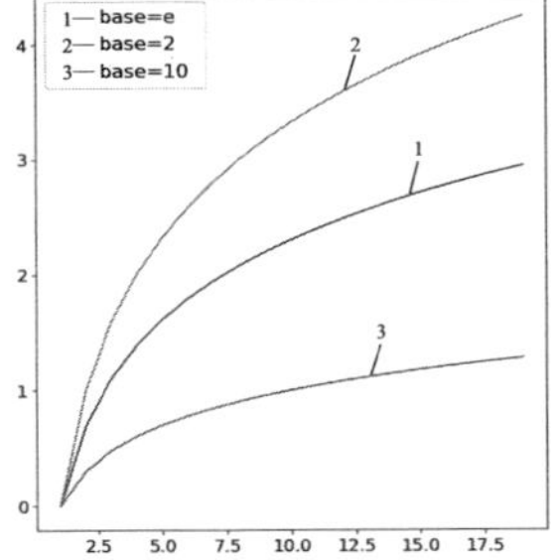

图3.4–5 以e、2和10为底的对数函数

3.4.1.3 回归与微分

首先，根据《漫画统计学之回归分析》中美羽的年龄和身高数据建立数据集，实现计算年龄和身高的相关系数，结果p-value<0.05，即pearson's r=0.942的相关系数能够说明年龄和身高直接存在强相关关系。既然二者之间存在相关性，就可以建立回归方程。在下述代码中，给出了三种回归模型（方程），如图3.4–6所示。一种是《漫画统计学之回归分析》

给出的$f(x)=-\frac{326.6}{x}+173.3$方程；另外两种是直接使用sklearn库Linear Models线性模型中的LinearRegression线性回归，和基于LinearRegression的Polynomial regression多项式回归。关于sklearn的语法规则，可以参考官方网站scikit-learn给出的指南。sklearn的语法结构秉承了Python自身语法特点，具有很强的易读性，代码编写流畅、自然。三种回归模型中，以多项式回归拟合得最好，《漫画统计学之回归分析》中给出的公式次之，而简单、粗暴的简单线性回归因为呈现线性，与真实值近似对数函数曲线的形状相异。

微分是对函数的局部变化率的一种线性描（图3.4–7），可以近似的描述当函数自变量的取值足够小时的改变时，函数的值是怎样变化的。$y=-\frac{326.6}{x}+173.3$关于x求微分（图3.4–8），即是求x岁到x岁之后极短时间内，身高的年平均增长量（自变量以年为单位），$\frac{(-\frac{326.6}{x+\triangle}+173.3)-(-\frac{326.6}{x}+173.3)}{\triangle}=\frac{326.6}{x^2}$。对于微分的计算，可以直接使用SymPy提供的`diff`工具计算，计算结果记作$\frac{\mathrm{d}y}{\mathrm{d}x}=\frac{\mathrm{d}f}{\mathrm{d}x}=y'=f'=\frac{326.6}{x^2}$。

常用函数求微分示例：

1. $y=x$，关于x进行微分，$\frac{\mathrm{d}y}{\mathrm{d}x}=1$
2. $y=x^2$，关于x进行微分，$\frac{\mathrm{d}y}{\mathrm{d}x}=2x$
3. $y=\frac{1}{x}$，关于x进行微分，$\frac{\mathrm{d}y}{\mathrm{d}x}=-x^{-2}$
4. $y=\frac{1}{x^2}$，关于x进行微分，$\frac{\mathrm{d}y}{\mathrm{d}x}=-2x^{-3}$
5. $y=(5x-7)^2$，关于x进行微分，$\frac{\mathrm{d}y}{\mathrm{d}x}=2(5x-7)\times 5$
6. $y=(ax+b)^n$，关于x进行微分，$\frac{\mathrm{d}y}{\mathrm{d}x}=n(ax+b)^{n-1}\times a$
7. $y=\mathrm{e}^x$，关于x进行微分，$\frac{\mathrm{d}y}{\mathrm{d}x}=\mathrm{e}^x$
8. $y=\log x$，关于x进行微分，$\frac{\mathrm{d}y}{\mathrm{d}x}=\frac{1}{x}$
9. $y=\log(ax+b)$，关于x进行微分，$\frac{\mathrm{d}y}{\mathrm{d}x}=\frac{1}{ax+b}\times a$
10. $y=\log(1+\mathrm{e}a^{x+b})$，关于$x$进行微分，$\frac{\mathrm{d}y}{\mathrm{d}x}=\frac{1}{1+\mathrm{e}^{ax+b}}\times a\mathrm{e}^{ax+b}$

在代码的领域里直接用SymPy的`diff`方法，或其他库提供的方法计算。

微分可以阅读该书微积分基础的代码表述部分

```
from sympy import diff
from sympy import Symbol
from sklearn.pipeline import Pipeline
from sklearn.preprocessing import PolynomialFeatures
from sklearn.linear_model import LinearRegression
import pandas as pd
from scipy import stats
emma_statureAge = {"age": list(range(4, 20)), "stature": [
    100.1, 107.2, 114.1, 121.7, 126.8, 130.9, 137.5, 143.2, 149.4, 151.6, 154.0,
154.6, 155.0, 155.1, 155.3, 155.7]}
emma_statureAge_df = pd.DataFrame(emma_statureAge)
r_ = stats.pearsonr(emma_statureAge_df.age, emma_statureAge_df.stature)
print(
    "pearson's r:", r_[0], "\n",
    "p_value:", r_[1]
)
```

原始数据散点图

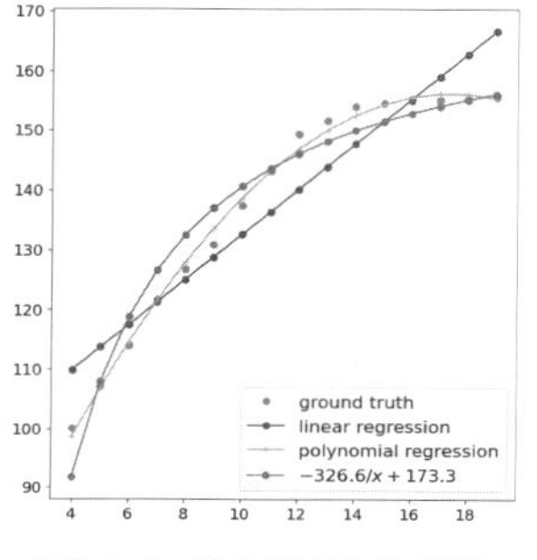

图3.4–6　建立的3种回归模型

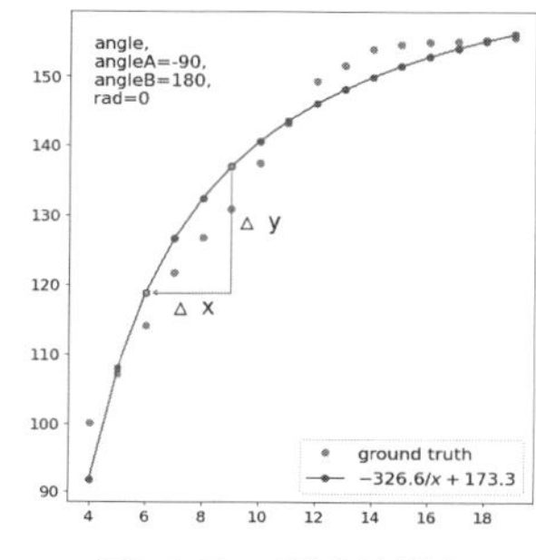

图3.4–7　对微分的描述

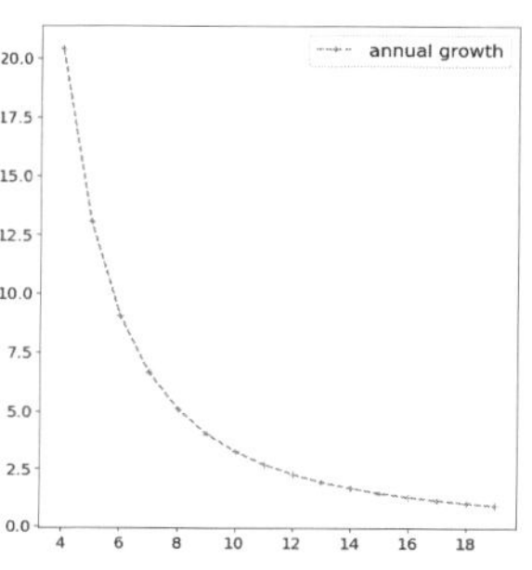

图3.4–8　微分（函数的局部变化率）

```
fig, axs = plt.subplots(1, 3, figsize=(25, 8))
axs[0].plot(emma_statureAge_df.age, emma_statureAge_df.stature,
            'o', label='ground truth', color='r')

# A - 使用 sklearn 库 sklearn.linear_model.LinearRegression(), Ordinary least squares Linear Regression- 普通最小二乘线性回归，获取回归方程
X = emma_statureAge_df.age.to_numpy().reshape(-1, 1)
y = emma_statureAge_df.stature.to_numpy()

# 拟合模型
LR = LinearRegression().fit(X, y)

# 模型参数
print("slop:%.2f,intercept:%.2f" % (LR.coef_, LR.intercept_))
print(LR.get_params())

# 模型预测
axs[0].plot(emma_statureAge_df.age, LR.predict(X),
            'o-', label='linear regression')

# B - 多项式回归 Polynomial regression
model = Pipeline([('poly', PolynomialFeatures(degree=2)),
                  ('linear', LinearRegression(fit_intercept=False))])
reg = model.fit(X, y)
axs[0].plot(emma_statureAge_df.age, reg.predict(X),
            '+-', label='polynomial regression')

# C - 使用'漫画统计学之回归分析'给出的公式
x = Symbol('x')
f_emma = -326.6/x+173.3
f_emma_ = sympy.lambdify(x, f_emma, "numpy")
axs[0].plot(emma_statureAge_df.age, f_emma_(
    emma_statureAge_df.age), 'o-', label='$-326.6/x+173.3$')
axs[1].plot(emma_statureAge_df.age, emma_statureAge_df.stature,
            'o', label='ground truth', color='r')
axs[1].plot(emma_statureAge_df.age, f_emma_(
    emma_statureAge_df.age), 'o-', label='$-326.6/x+173.3$')
def demo_con_style(a_coordi, b_coordi, ax, connectionstyle):
    '''
    function - 在 matplotlib 的子图中绘制连接线。参考： matplotlib 官网 Connectionstyle 
Demo
    Params:
        a_coordi - a 点的 x, y 坐标；tuple
        b_coordi - b 点的 x, y 坐标；tuple
        ax - 子图；ax(plot)
        connectionstyle - 连接线的形式；string
    Returns:
        None
    '''
    x1, y1 = a_coordi[0], a_coordi[1]
    x2, y2 = b_coordi[0], b_coordi[1]
    ax.plot([x1, x2], [y1, y2], ".")
    ax.annotate("",
                xy=(x1, y1), xycoords='data',
                xytext=(x2, y2), textcoords='data',
                arrowprops=dict(arrowstyle="->", color="0.5",
                                shrinkA=5, shrinkB=5,
                                patchA=None, patchB=None,
    connectionstyle=connectionstyle,
                                ),
                )
    ax.text(.05, .95, connectionstyle.replace(",", ",\n"),
            transform=ax.transAxes, ha="left", va="top")
dx = 3
demo_con_style((6, f_emma.evalf(subs={x: 6})), (6+dx, f_emma.evalf(
    subs={x: 6+dx})), axs[1], "angle,angleA=-90,angleB=180,rad=0")
axs[1].text(7, f_emma.evalf(subs={x: 6})-3, "△ x", family="monospace", size=20)
axs[1].text(9.3, f_emma.evalf(subs={x: 9.3}) -
            10, "△ y", family="monospace", size=20)

# 用 sympy 提供的 diff 方法求微分
print("f_emma=-326.6/x+173.3 关于 x 求微分：")
pprint(diff(f_emma), use_unicode=True)
diff_f_emma_ = sympy.lambdify(x, diff(f_emma), "numpy")
```

```
axs[2].plot(emma_statureAge_df.age, diff_f_emma_(
    emma_statureAge_df.age), '+--', label='annual growth', color='r')
axs[2].legend(loc='upper right', frameon=False)
axs[1].legend(loc='lower right', frameon=False)
axs[0].legend(loc='upper left', frameon=False)
plt.show()
--------------------------------------------------------------------------------
pearson's r: 0.9422225583501309
 p_value: 4.943118398567093e-08
slop:3.78,intercept:94.82
{'copy_X': True, 'fit_intercept': True, 'n_jobs': None, 'normalize': 'deprecated',
'positive': False}
f_emma=-326.6/x+173.3 关于 x 求微分:
326.6
─────────
   2
  x
```

3.4.1.4 矩阵

一个$m \times n$的矩阵是一个由m行（row）、n列（column）元素排列成的矩形阵列。矩阵里的元素可以是数字、符号或数学式。例如：$\begin{bmatrix} 1 & 9 & -13 \\ 20 & 5 & -6 \end{bmatrix}$，如果$\begin{cases} x_1 + 2x_2 = -1 \\ 3x_1 + 4x_2 = 5 \end{cases}$可以写作：$\begin{bmatrix} 1 & 2 \\ 3 & 4 \end{bmatrix} \begin{bmatrix} x_1 \\ x_2 \end{bmatrix} = \begin{bmatrix} -1 \\ 5 \end{bmatrix}$，而如果$\begin{cases} x_1 + 2x_2 \\ 3x_1 + 4x_2 \end{cases}$，可以写作：$\begin{bmatrix} 1 & 2 \\ 3 & 4 \end{bmatrix} \begin{bmatrix} x_1 \\ x_2 \end{bmatrix}$。

矩阵的操作和运算可以直接应用SymPy库的Matrices部分方法，或者其他库。更多的内容需要参考官方教程，此处不再赘述。

```
#矩阵可以阅读该书线性代数基础的代码表述部分

from sympy import Matrix,init_printing,pprint
init_printing()
M_a = Matrix([[1, -1], [3, 4], [0, 2]])
pprint(M_a)
--------------------------------------------------------------------------------
⎡1  -1⎤
⎢     ⎥
⎢3   4⎥
⎢     ⎥
⎣0   2⎦
```

3.4.2 简单线性回归

在统计学中，线性回归（linear regression）是利用称为线性回归方程的最小平方函数对一个或多个自变量和因变量之间关系进行建模的一种回归分析。这种函数是一个或多个称为回归系数的模型参数的线性组合。只有一个自变量的情况，称为简单线性回归（simple linear regression）；大于一个自变量的情况，叫多元回归（multivariable linear regression）。

- （一元）回归分析的流程:

1. 数据可视化：画出自变量和因变量的散点图，并求解相关系数，确定自变量因变量之间是否可以用（一元）线性回归方程$y=Ax+b$解释；
2. 求解回归方程中A和b的值；
3. 确认回归方程的精度；
4. 进行回归系数的检验；
5. 计算总体回归$y=Ax+b$的估计；
6. 进行预测。

3.4.2.1 建立数据集并计算相关系数

使用《漫画统计学之回归分析》中最高温度（℃）与冰红茶销售量（杯）的数据（表3.4–1），首先建立基于DataFrame格式的数据集。该数据集使用了时间戳（timestamp）作为索引。

```
import pandas as pd
import util_misc
from scipy import stats
dt = pd.date_range('2020-07-22', periods=14, freq='D')
dt_temperature_Sales":[77,62,93,84,59,64,80,75,58,91,51,73,65,84]}
iceTea_df = pd.DataFrame(dt_temperature_iceTeaSales).set_index("dt")
util_misc.print_html(iceTea_df,14)
```

表3.4–1 温度与冰红茶销售量

	temperature	iceTeaSales
dt		
2020-07-22	29	77
2020-07-23	28	62
2020-07-24	34	93
2020-07-25	31	84
2020-07-26	25	59
2020-07-27	29	64
2020-07-28	32	80
2020-07-29	31	75
2020-07-30	24	58
2020-07-31	33	91
2020-08-01	25	51
2020-08-02	31	73
2020-08-03	26	65
2020-08-04	30	84

打印温度与销量之间的散点图（图3.4–9）并计算其相关系数，确定两者之间存在关联。其p-value=7.661412804450245e-06，小于0.05的显著性水平，确定pearson's r=0.90，能够表明两者之间是强相关性，适合采用线性拟合模型求解两者之间的关联公式。

```
from scipy import stats
r_=stats.pearsonr(iceTea_df["temperature"],iceTea_df["iceTeaSales"])
print(
    "pearson's r:",r_[0],
    "\np-value:",r_[1]
```

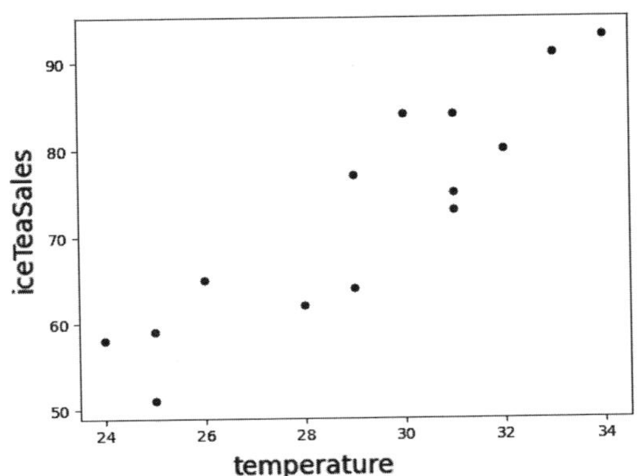

图3.4–9 冰红茶销售量与温度散点图

```
    )
iceTea_df.plot.
ter(x="temperature",y="iceTeaSales",c="DarkBLue")
plt.show
--------------------------------------------------------------------------
pearson's r: 0.9069229780508894
p-value: 7.661412804450245e-06
```

3.4.2.2 求解回归方程

求解回归方程使用了两种方法：一种是逐步计算的方式（表3.4–2）；另一种是直接使用sklearn库的LinearRegression模型。逐步计算的方式可以更深入地理解回归模型。熟悉基本计算过程后，直接应用sklearn机器学习库中的模型，也会对各种参数的配置有一个比较清晰的了解。

求解回归方程即是使所有真实值与预测值之差的和为最小，求出a和b，就是所有变量残差（residual）的平方（s_residual）的和（S_residual）为最小。因为温度与销量为线性相关，因此使用一元一次方程式：$y=ax+b$，x为自变量温度，y为因变量销量，a和b为回归系数（参数），分别称为斜率（slop）和截距（intercept）。求解a和b的过程，可以使用最小二乘法（least squares method），又称最小平方法。通过最小化误差的平方（残差平方和），寻找数据的最佳函数匹配。残差平方和为：

$$(-34a-b+93)^2+(-33a-b+91)^2+(-32a-b+80)^2+(-31a-b+73)^2+\\(-31a-b+75)^2+(-31a-b+84)^2+(-30a-b+84)^2+(-29a-b+64)^2+\\(-29a-b+77)^2+(-28a-b+62)^2+(-26a-b+65)^2+(-25a-b+51)^2+\\(-25a-b+59)^2+(-24a-b+58)^2$$

先对a和b分别求微分$\frac{df}{da}$和$\frac{df}{db}$，是$\triangle a$，即a在横轴上的增量，及$\triangle b$，即b在横轴上的增量趋近于无穷小。无限接近a和b时，因变量的变化量，这个因变量就是残差平方和的值。残差平方和的值是由a和b确定的。当a和b取不同的值时，残差平方和的值随之变化（图3.4–10）；当残差平方和的值为0时，说明由自变量温度所有值通过回归方程预测的销量，与真实值的差值之和为0；单个温度值通过回归模型预测的销量与真实值之差则趋于0。在实际计算中，手工推算时，对残差平方和关于a和b求微分，是对公式进行整理，最终获得求解回归方程回归系数的公式为：$a=\frac{S_{xy}}{S_{xx}}$，式中S_{xy}即变量SS_xy是x和y的离差积，S_{xx}即变量SS_x是x的离差平方和。求得a后，可以根据推导公式：$b=\overline{y}-\overline{x}a$计算$b$，式中$\overline{x}$和$\overline{y}$分别为$x$和$y$的均值，分别对应iceTea_df.temperature.mean()和iceTea_df.iceTeaSales.mean()。

在Python语言中，使用相关库则可以避免上述繁琐的手工推导过程，在逐步计算中，使用SymPy库约简残差平方和公式为：$12020\cdot a^2+816\cdot a\cdot b-60188\cdot a+14\cdot b^2-2032\cdot b+75936$，并直接分别对$a$和$b$微分，获得结果为：$\frac{df}{da}=24040\cdot a+816\cdot b-60188$和$\frac{df}{db}=816\cdot a+28\cdot b-2032$，另两者分别为0，使用SymPy库的solve求解二元一次方程组，计算获取a和b值（图3.4–11）。

最后，使用sklearn库的LinearRegression模型求解决回归模型，仅需要几行代码，所得结果与上述同（图3.4–12）。可以用sklearn返回的参数，建立回归方程公式。但是，在实际的应用中并不会这么做，而是直接应用以变量形式代表的回归模型直接预测值。

```
from sklearn.linear_model import LinearRegression
from sklearn import preprocessing
import math
import sympy
from sympy import diff, Eq, solveset, solve, simplify, pprint
import matplotlib.pyplot as plt
from matplotlib import cm
import numpy as np

# 原始数据散点图
fig, axs = plt.subplots(1, 3, figsize=(25, 8), dpi=300)
axs[0].plot(iceTea_df.temperature, iceTea_df.iceTeaSales,
            'o', label='ground truth', color='r')
axs[0].set(xlabel='temperature', ylabel='ice tea sales')

# A - 使用‘最小二乘法’逐步计算
# 1 - 求出 x 和 y 的离差及离差平方和
iceTea_df["x_deviation"] = iceTea_df.temperature.apply(
    lambda row: row-iceTea_df.temperature.mean())
```

表3.4-2 逐步求解回归方程

	temperature	iceTeaSales	x_deviation	y_deviation	S_x_deviation	S_y_deviation	S_xy_deviation	prediciton	residual	s_residual
dt										
2020-07-22	29	77	-0.142857	4.428571	0.020408	19.612245	-0.632653	29*a + b	-29*a - b + 77	(-29*a - b + 77)**2
2020-07-23	28	62	-1.142857	-10.571429	1.306122	111.755102	12.081633	28*a + b	-28*a - b + 62	(-28*a - b + 62)**2
2020-07-24	34	93	4.857143	20.428571	23.591837	417.326531	99.224490	34*a + b	-34*a - b + 93	(-34*a - b + 93)**2
2020-07-25	31	84	1.857143	11.428571	3.448980	130.612245	21.224490	31*a + b	-31*a - b + 84	(-31*a - b + 84)**2
2020-07-26	25	59	-4.142857	-13.571429	17.163265	184.183673	56.224490	25*a + b	-25*a - b + 59	(-25*a - b + 59)**2
2020-07-27	29	64	-0.142857	-8.571429	0.020408	73.469388	1.224490	29*a + b	-29*a - b + 64	(-29*a - b + 64)**2
2020-07-28	32	80	2.857143	7.428571	8.163265	55.183673	21.224490	32*a + b	-32*a - b + 80	(-32*a - b + 80)**2
2020-07-29	31	75	1.857143	2.428571	3.448980	5.897959	4.510204	31*a + b	-31*a - b + 75	(-31*a - b + 75)**2
2020-07-30	24	58	-5.142857	-14.571429	26.448980	212.326531	74.938776	24*a + b	-24*a - b + 58	(-24*a - b + 58)**2
2020-07-31	33	91	3.857143	18.428571	14.877551	339.612245	71.081633	33*a + b	-33*a - b + 91	(-33*a - b + 91)**2
2020-08-01	25	51	-4.142857	-21.571429	17.163265	465.326531	89.367347	25*a + b	-25*a - b + 51	(-25*a - b + 51)**2
2020-08-02	31	73	1.857143	0.428571	3.448980	0.183673	0.795918	31*a + b	-31*a - b + 73	(-31*a - b + 73)**2
2020-08-03	26	65	-3.142857	-7.571429	9.877551	57.326531	23.795918	26*a + b	-26*a - b + 65	(-26*a - b + 65)**2
2020-08-04	30	84	0.857143	11.428571	0.734694	130.612245	9.795918	30*a + b	-30*a - b + 84	(-30*a - b + 84)**2

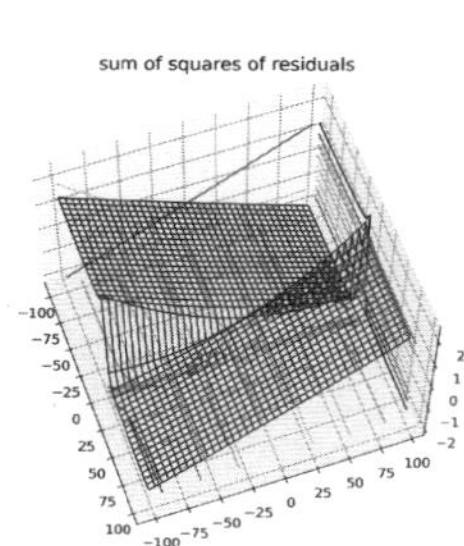

图3.4-10 回归方程的残差平方和

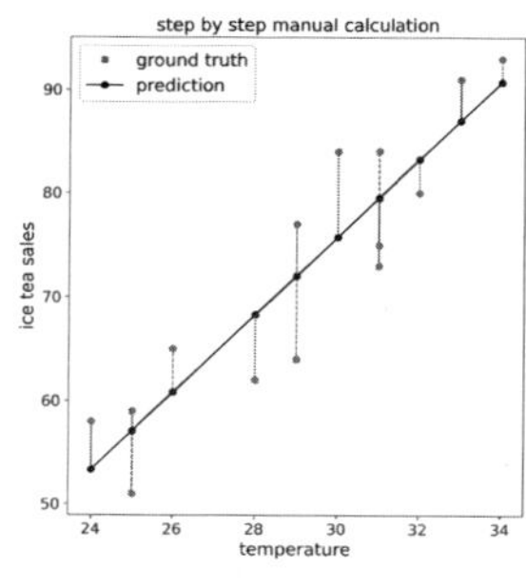

图3.4-11 回归方程逐步推导结果

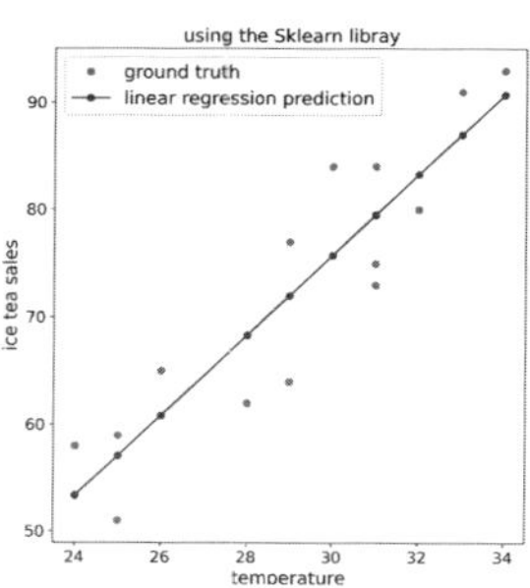

图3.4-12 由Sklearn库计算回归方程

```
iceTea_df["y_deviation"] = iceTea_df.iceTeaSales.apply(
    lambda row: row-iceTea_df.iceTeaSales.mean())
iceTea_df["S_x_deviation"] = iceTea_df.temperature.apply(
    lambda row: math.pow(row-iceTea_df.temperature.mean(), 2))
iceTea_df["S_y_deviation"] = iceTea_df.iceTeaSales.apply(
    lambda row: math.pow(row-iceTea_df.iceTeaSales.mean(), 2))
SS_x = iceTea_df["S_x_deviation"].sum()
SS_y = iceTea_df["S_y_deviation"].sum()

#2-求出x和y的离差积及其和
iceTea_df["S_xy_deviation"] = iceTea_df.apply(lambda row: (
    row["temperature"]-iceTea_df.temperature.mean())*(row["iceTeaSales"]-iceTea_
df.iceTeaSales.mean()), axis=1)
SS_xy = iceTea_df["S_xy_deviation"].sum()

#3-运算过程
a, b = sympy.symbols('a b')
iceTea_df["prediciton"] = iceTea_df.temperature.apply(lambda row: a*row+b)
iceTea_df["residual"] = iceTea_df.apply(
    lambda row: row.iceTeaSales-(a*row.temperature+b), axis=1)
iceTea_df["s_residual"] = iceTea_df.apply(lambda row: (
    row.iceTeaSales-(a*row.temperature+b))**2, axis=1)
S_residual = iceTea_df["s_residual"].sum()
S_residual_simplify = simplify(S_residual)
print("S_residual simplification(Binary quadratic equation):")
pprint(S_residual_simplify)  #残差平方和为一个二元二次函数
print("_"*50)

#打印残差平方和图形
S_residual_simplif_ = sympy.lambdify([a, b], S_residual_simplify, "numpy")
a_ = np.arange(-100, 100, 5)
a_3d = np.repeat(a_[:, np.newaxis], a_.shape[0], axis=1).T
b_ = np.arange(-100, 100, 5)
b_3d = np.repeat(b_[:, np.newaxis], b_.shape[0], axis=1)
z = S_residual_simplif_(a_3d, b_3d)
z_scaled = preprocessing.scale(z)  #标准化z值，同from scipy.stats import zscore方法

axs[1] = fig.add_subplot(1, 3, 2, projection='3d')
axs[1].plot_wireframe(a_3d, b_3d, z_scaled)
axs[1].contour(a_3d, b_3d, z_scaled, zdir='z', offset=-2, cmap=cm.coolwarm)
axs[1].contour(a_3d, b_3d, z_scaled, zdir='x', offset=-100, cmap=cm.coolwarm)
axs[1].contour(a_3d, b_3d, z_scaled, zdir='y', offset=100, cmap=cm.coolwarm)

#4-对残差平方和S_residual关于a和b求微分，并使其为0
diff_S_residual_a = diff(S_residual, a)
diff_S_residual_b = diff(S_residual, b)
print("diff_S_residual_a=",)
pprint(diff_S_residual_a)
print("\n")
print("diff_S_residual_b=",)
pprint(diff_S_residual_b)

#对a和b的微分公式构造二元一次方程组，即diff_S_residual_a=0，diff_S_residual_b=0
Eq_residual_a = Eq(diff_S_residual_a, 0)
Eq_residual_b = Eq(diff_S_residual_b, 0)
```

```
# 设置二元一次方差组中的变量为 a 和 b，并求解
slop_intercept = solve((Eq_residual_a, Eq_residual_b), (a, b))
print("_"*50)
print("slop and intercept:\n")
pprint(slop_intercept)
slop = slop_intercept[a]
intercept = slop_intercept[b]

# 用求解回归方程回归系数的推导公式之间计算斜率 slop 和截距 intercept
print("_"*50)
slop_ = SS_xy/SS_x
print("derivation formula to calculate the slop=", slop_)
intercept_ = iceTea_df.iceTeaSales.mean()-iceTea_df.temperature.mean()*slop_
print("derivation formula to calculate the intercept=", intercept_)
print("_"*50)

# 5 - 建立简单线性回归方程
x = sympy.Symbol('x')
fx = slop*x+intercept
print("linear regression_fx=:\n")
pprint(fx)
fx_ = sympy.lambdify(x, fx, "numpy")

# 在残差平方和图形上标出 a,b 的位置
axs[1].text(slop, intercept, -1.7, "a/b", color="red", size=20)
axs[1].scatter(slop, intercept, -2, color="red", s=80)
axs[1].view_init(60, 340)  # 可以旋转图形的角度，方便观察

# 6 - 绘制简单线性回归方程的图形
axs[0].plot(iceTea_df.temperature, fx_(iceTea_df.temperature),
            'o-', label='prediction', color='blue')

# 绘制真实值与预测值之间的连线
i = 0
for t in iceTea_df.temperature:
    axs[0].arrow(t, iceTea_df.iceTeaSales[i], t-t, fx_(t)-iceTea_
df.iceTeaSales[i],
                 head_width=0.1, head_length=0.1, color="gray", linestyle="--")
    i += 1

# B - 使用 sklearn 库 sklearn.linear_model.LinearRegression()，Ordinary least squares Linear Regression- 普通最小二
乘线性回归，获取回归方程
X, y = iceTea_df.temperature.to_numpy(
).reshape(-1, 1), iceTea_df.iceTeaSales.to_numpy()

# 拟合模型
LR = LinearRegression().fit(X, y)
# 模型参数
print("_"*50)
print("Sklearn slop:%.2f,intercept:%.2f" % (LR.coef_, LR.intercept_))
# 模型预测
axs[2].plot(iceTea_df.temperature, iceTea_df.iceTeaSales,
            'o', label='ground truth', color='r')
axs[2].plot(X, LR.predict(X), 'o-', label='linear regression prediction')
axs[2].set(xlabel='temperature', ylabel='ice tea sales')

axs[0].legend(loc='upper left', frameon=False)
axs[2].legend(loc='upper left', frameon=False)

axs[0].set_title('step by step manual calculation')
axs[1].set_title('sum of squares of residuals')
axs[2].set_title('using the Sklearn libray')
plt.show()
util_misc.print_html(iceTea_df, 14)
```

```
--------------------------------------------------------------------------------
S_residual simplification(Binary quadratic equation):
       2                              2
12020·a  + 816·a·b - 60188·a + 14·b  - 2032·b + 75936
--------------------------------------------------
diff_S_residual_a=
24040·a + 816·b - 60188
diff_S_residual_b= 816·a + 28·b - 2032
--------------------------------------------------
slop and intercept:
```

```
⎧   1697      -8254 ⎫
⎨a: ────, b: ───── ⎬
⎩   454        227  ⎭
-------------------------------------------------------------------
derivation formula to calculate the slop= 3.7378854625550666
derivation formula to calculate the intercept= -36.361233480176224
-------------------------------------------------------------------
linear regression_fx=:

1697·x   8254
────── - ────
 454      227
-------------------------------------------------------------------
Sklearn slop:3.74,intercept:-36.36
```

3.4.2.3 确认回归方程的精度

回归方程（模型）的精度可以通过计算判断系数确认。判断系数（也称决定系数，coefficient of determination），记为R^2或r^2，用于度量因变量的变异中可由自变量解释部分所占的比例，以此来判断回归模型的解释力，来表示实测值（图表中的点）与回归方程拟合的程度。对于简单线性回归而言，判断系数为样本相关系数R的平方。其复（重）相关系数计算公式为：$R=\frac{\sum_{i=1}^{n}(y_i-\overline{y})^2(\widehat{y}_i-\overline{\widehat{y}})^2}{\sqrt{(\sum_{i=1}^{n}(y_i-\overline{y})^2)(\sum_{i=1}^{n}(\widehat{y}_i-\overline{\widehat{y}})^2)}}$。式中，$y$为观测值，$\overline{y}$为观测值的均值，$\widehat{y}$为预测值，$\overline{\widehat{y}}$为预测值的均值。而判定系数$R^2$，则为重相关系数的平方。判定系数的取值在0到1，其值越接近于1，回归方程的精度越高。第二种计算公式为：$R^2=1-\frac{SS_{\text{res}}}{SS_{\text{tot}}}=1-\frac{\sum_{i=1}^{n}e_i^2}{SS_{\text{tot}}}=1-\frac{\sum_{i=1}^{n}(y_i-\widehat{y}_i)^2}{\sum_{i=1}^{n}(y_i-\overline{y})^2}$。其中，$SS_{\text{res}}$为残差平方和，$SS_{\text{tot}}$为观测值离差平方和（总平方和，或总的离差平方和），e_i为残差，y_i为观测值，$\widehat{y}$为预测值，$\overline{y}$为观测值均值。第三种是直接使用sklearn库提供的`r2_score`方法直接计算。

根据计算结果，第1、2、3种方法结果一致。在后续的实验中，直接使用sklearn提供的方法进行计算。

```python
from sklearn.metrics import r2_score
def coefficient_of_determination(observed_vals, predicted_vals):
    '''
    function - 回归方程的决定系数

    Params:
        observed_vals - 观测值（实测值）; list(float)
        predicted_vals - 预测值; list(float)

    Returns:
        R_square_a - 决定系数，由观测值和预测值计算获得; float
        R_square_b - 决定系数，由残差平方和和总平方和计算获得; float
    '''
    import pandas as pd
    import numpy as np
    import math
    vals_df = pd.DataFrame({'obs': observed_vals, 'pre': predicted_vals})
    # 观测值的离差平方和(总平方和，或总的离差平方和)
    obs_mean = vals_df.obs.mean()
    SS_tot = vals_df.obs.apply(lambda row: (row-obs_mean)**2).sum()
    # 预测值的离差平方和
    pre_mean = vals_df.pre.mean()
    SS_reg = vals_df.pre.apply(lambda row: (row-pre_mean)**2).sum()
    # 观测值和预测值的离差积和
    SS_obs_pre = vals_df.apply(lambda row: (
        row.obs-obs_mean)*(row.pre-pre_mean), axis=1).sum()

    # 残差平方和
    SS_res = vals_df.apply(lambda row: (row.obs-row.pre)**2, axis=1).sum()

    # 判断系数
    R_square_a = (SS_obs_pre/math.sqrt(SS_tot*SS_reg))**2
    R_square_b = 1-SS_res/SS_tot
    return R_square_a, R_square_b
R_square_a, R_square_b = coefficient_of_determination(
```

```
    iceTea_df.iceTeaSales.to_list(), fx_(iceTea_df.temperature).to_list())
print("R_square_a=%.5f,R_square_b=%.5f" % (R_square_a, R_square_b))
R_square_ = r2_score(iceTea_df.iceTeaSales.to_list(),
                     fx_(iceTea_df.temperature).to_list())
print("using sklearn libray to calculate r2_score=", R_square_)
--------------------------------------------------------------------------------
R_square_a=0.82251,R_square_b=0.82251
using sklearn libray to calculate r2_score= 0.8225092881166944
```

3.4.2.4 回归系数的检验（回归显著性检验）| F分布与方差分析

F-分布（F-distribution）图3.4–13是两个服从卡方分布的独立随机变量各除以其自由度后的比值的抽样分布，是一种非对称分布，且位置不可互换。F-分布是一种连续概率分布，广泛应用于似然比率检验，特别是方差分析（Analysis of variance, ANOVA，或变异数分析）中。对于F-分布的阐释，使用`scipy.stats.f`的官方案例。函数方法基本同正态分布和t分布。

如果随机变量X有参数d_1和d_2的F-分布，写作$X \sim F(d_1, d_2)$，那么对于实数$x \geqslant 0$，X的概率密度函数（pdf）为：$f(x;d_1,d_2)=\frac{\sqrt{\frac{((d_1x))^{d_1}d_2^{d_2}}{(d_1x+d_2)^{d_1+d_2}}}}{xB(\frac{d_1}{2},\frac{d_2}{2})}=\frac{1}{B(\frac{d_1}{2},\frac{d_2}{2})}(\frac{d_1}{d_2})^{\frac{d_1}{2}}x^{\frac{d_1}{2}-1}(1+\frac{d_1}{d_2}x)^{-\frac{d_1+d_2}{2}}$。式中，$B$是B函数。在很多应用中，参数$d_1$和$d_2$是正整数；但是，对于这些参数为正实数时，也有定义。一个 F-分布的随机变量是两个卡方分布变量除以自由度的比率：$\frac{U_1/d_1}{U_2/d_2}=\frac{U_1/U_2}{d_1/d_2}$。式中，$U_1$和$U_2$呈卡方分布，它们的自由度（degree of freedom）分别是d_1和d_2；并且，U_1和U_2是相互独立的。

```
from scipy.stats import f
import matplotlib.pyplot as plt
fig, ax = plt.subplots(1, 1)

dfn, dfd = 29, 18  # 分别是第 1 自由度，即分子中卡方分布的自由度，和第 2 自由度，即分母中卡方分布的自由度
mean, var, skew, kurt = f.stats(dfn, dfd, moments='mvsk')
print("mean=%f, var=%f, skew=%f, kurt=%f" % (mean, var, skew, kurt))

# 打印概率密度函数 (probability density function,pdf)
x = np.linspace(f.ppf(0.01, dfn, dfd), f.ppf(0.99, dfn, dfd),
                100)  # 取服从自由度 dfn 和 dfd，位于 1% ~ 99% 的 100 个取值
ax.plot(x, f.pdf(x, dfn, dfd), '-', lw=5, alpha=0.6, label='f pdf')

# 固定分布形状，即固定自由度
rv = f(dfn, dfd)
ax.plot(x, rv.pdf(x), 'k-', lw=2, label='frozen pdf')
vals = f.ppf([0.001, 0.5, 0.999], dfn, dfd)
print(" 验证累计分布函数 CDF 返回值与其 PPF 返回值是否相等或近似：", np.allclose(
    [0.001, 0.5, 0.999], f.cdf(vals, dfn, dfd)))

# 生成服从 F- 分布的随机数，并打印直方图
r = f.rvs(dfn, dfd, size=1000)
ax.hist(r, density=True, histtype='stepfilled', alpha=0.2)
ax.legend(loc='best', frameon=False)
plt.show()
----------------------------------------------------------------
mean=1.125000, var=0.280557, skew=1.806568, kurt=7.074636
```

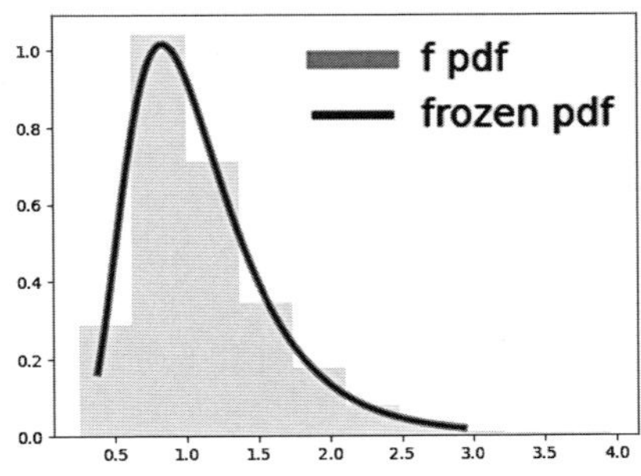

图3.4–13 由Sklearn库计算回归方程

验证累计分布函数 CDF 返回值与其 PPF 返回值是否相等或近似：`True`

- $SS_{\text{tot}} = SS_{\text{reg}} + SS_{\text{res}}$，即总平方和=回归平方和+残差平方和

公式为：$SS_{\text{tot}} = \sum_{i=1}^{n}(y_i - \overline{y})^2 = \sum_{i=1}^{n}(\widehat{y}_i - \overline{y})^2 + \sum_{i=1}^{n}(y_i - \widehat{y}_i)^2$，式中$SS_{\text{reg}}$为回归平方和，其他同上。回归平方和$SS_{\text{reg}} = \sum_{i=1}^{n}(\widehat{y}_i - \overline{y})^2$，是预测值（回归值）与观测值（真实值、实测值）均值之差的平方和。该统计量反映了自变量$x_1, x_2, \ldots, x_m$的变化引起的$y(y_k(k=1,2,\ldots,n))$的波动，其自由度为$df_{\text{reg}} = m$。其中，m为自变量的个数，温度与销量求解的一元一次线性方程只有一个自变量，因此其自由度为1，即只有这一个因素可以自由变化；残差平方和$SS_{\text{res}} = \sum_{i=1}^{n}(y_i - \widehat{y}_i)^2$，是观测值与预测值之差的平方和，残差的存在是由实验误差及其他因素引起的，其自由度为$df_{\text{res}} = n - m - 1$。其中，$n$为样本数量，即对应的$y$的取值数量。总的离差平方和$SS_{\text{tot}}$的自由度为$n-1$。

观测值（样本）通常是给定的，因此总的离差平方和是固定的，构成总的离差平方和的因素为回归平方和和残差平方和，分别代表所求得的回归方程，或实验误差和其他因素引起y值的变化。当残差平方和越小（就是实验误差和其他因素影响小），则回归平方和越大，则说明所求得的回归方程的预测值越准确。

自由度的再讨论

在统计学中，自由度（defree of freedom, df）是指当以样本的统计量估计总体的参数时，样本中独立或能自由变化的数据的个数，称为该统计量的自由度。范例：

1. 若存在两个变量x和y，如果$y = x + c$，其中c为常量，则其自由度为 1，因为实际上只有x才能真正的自由变化，y会被x取值的不同所限制。
2. 估计总体的平均数μ时，由于样本中n个数都是相互独立的，任一个尚未抽出的数都不受已抽出任何数值的影响，所以自由度为n。
3. 估计总体的方差σ^2时所使用的统计量是样本的方差s^2，而s^2必须用到样本平均数$\overline{x}$来计算，$\overline{x}$在抽样完成后已确定，所以大小为n的样本中只要$n-1$个数确定，第n个数就只有一个能使样本符合$\overline{x}$的数值。也就是说，样本中只有$n-1$个数可以自由变化，只要确定了这$n-1$个数，方差也就确定了。这里，平均数$\overline{x}$就相当于一个限制条件，由于加了这个限制条件，样本方差s^2的自由度为$n-1$。
4. 统计模型的自由度等于可自由取值的自变量的个数。如在回归方程中，如果共有p个参数需要估计，则其中包括了$p-1$个自变量（与截距对应的自变量是常量），因此该回归方程的自由度为$p-1$。

无偏估计（unbiased estimator）在统计学中，一个总体的标准差通常是由总体中随机抽取的样本的估计，样本标准差的定义为：$s = \sqrt{\frac{\sum_{i=1}^{n}(x_i - \overline{x})^2}{n-1}}$，其中$x_1, x_2, \ldots, x_n$为样本，样本容量为$n$，$\overline{x}$为样本均值。使用$n-1$替代$n$，被称为 Bessel's correction（贝塞尔矫正），纠正了总体方差估计中的偏差（总体方差估计是使用随机抽取的样本的估计，不等于总体方差），以及总体标准差估计中的部分偏差，但不是全部偏差。因为偏差取决于特定的分布，不可能找到对所有总体分布无偏的标准偏差的估计。

- 方差分析（或变异数分析，Analysis of variance，ANOVA）

上述总平方和=回归平方和+残差平方和分析，实际上是在分析因变量（总平方和，即总的离差平方和）与影响因变量变化的两个因素（或称为两个类别），即回归平方和及残差平方和的关系探索。这个过程即称为方差分析。求解上述回归方程前，温度与销量的关系不一定是线性的，可能存在两种情况：一种是不管温度（x）取什么值，销量（y）都在一条水平线上下波动；二是，温度和销量存在除线性外其他类型的关系，例如非线性等。

对于上述所求得回归方程$f_{\text{x}} = ax + b = \frac{1697}{454}x - \frac{8254}{227}$（样本回归模型），对于总体而言，$F_{\text{x}} = Ax + B$（总体回归模型），斜率$A$约为$a(A \sim a)$，截距$B$约为$b$（$B \sim b$），$\sigma^2 = \frac{SS_{\text{res}}}{n-2}$（无偏估计量，残差平方和有$n-2$个自由度，这是因为两个自由度与得到预测值的估计值$A$和$B$相关），$\sigma^2$的平方根有时称为回归标准误差（由残差平方和求得$\sigma^2$的推导过程，可以参考《线性回归分析导论》（Introduction to linear regression analysis），简单线性回归部分）。

总体回归方程$F_{\text{x}} = Ax + B$非常重要的特例是，$H_0: A = 0, H_1: A \neq 0$，原假设意味$x$和$y$之间不存在线性关系，$x$对解释$y$的方差几乎是无用的；如果拒绝原假设，而接受备择假设，意味x对解释y的方差是有用的，可能意味线性模型是合适的，但是也可能存在需要用高阶多项式拟合的非线性模型。对于回归显著性检验可以使用t统计量，也可以使用方差分析。回归系数检验的F统计量为：$F_0 = \frac{SS_{\text{reg}}/df_{\text{reg}}}{SS_{\text{res}}/df_{\text{res}}}$，式中$SS_{\text{reg}}$为回归平方和，自由度$df_{\text{reg}} = m$为1，$SS_{\text{res}}$为残差平方

和，其自由度$df_{\mathrm{res}}=n-m-1$为$14-1-1=12$。如果原假设成立，那么检验统计量就服从第1自由度$m=1$，第2自由度$n-m-1=12$的F分布。p-value=0.000008，小于显著性水平0.05，拒绝原假设，备择假设成立。

```
def ANOVA(observed_vals, predicted_vals, df_reg, df_res):
    '''
    function - 简单线性回归方程 - 回归显著性检验（回归系数检验）

    Params:
        observed_vals - 观测值（实测值）; list(float)
        predicted_vals - 预测值; list(float)

    Returns:
        None
    '''
    import pandas as pd
    import numpy as np
    import math
    from scipy.stats import f
    vals_df = pd.DataFrame({'obs': observed_vals, 'pre': predicted_vals})
    # 观测值的离差平方和(总平方和，或总的离差平方和)
    obs_mean = vals_df.obs.mean()
    SS_tot = vals_df.obs.apply(lambda row: (row-obs_mean)**2).sum()

    # 残差平方和
    SS_res = vals_df.apply(lambda row: (row.obs-row.pre)**2, axis=1).sum()

    # 回归平方和
    SS_reg = vals_df.pre.apply(lambda row: (row-obs_mean)**2).sum()

    print("总平方和 =%.6f, 回归平方和 =%.6f, 残差平方和 =%.6f" % (SS_tot, SS_reg, SS_
res))
    print("总平方和 = 回归平方和 + 残差平方和: SS_tot=SS_reg+SS_res=%.6f+%.6f=%.6f" %
          (SS_reg, SS_res, SS_reg+SS_res))
    Fz = (SS_reg/df_reg)/(SS_res/df_res)
    print("F- 分布统计量 =%.6f;p-value=%.6f" % (Fz, f.sf(Fz, df_reg, df_res)))

ANOVA(iceTea_df.iceTeaSales.to_list(), fx_(
    iceTea_df.temperature).to_list(), df_reg=1, df_res=12)
-------------------------------------------------------------
总平方和 =2203.428571, 回归平方和 =1812.340466, 残差平方和 =391.088106
总平方和 = 回归平方和 + 残差平方和: SS_tot=SS_reg+SS_res=1812.340466+391.088106=2203.428571
F- 分布统计量 =55.609172;p-value=0.000008
```

利用F检验对回归方程进行显著性检验的方法就是方程分析，将上述过程可以归结为一个方程分析表3.4–3，从而更容易理清脉络。

表3.4–3 方程分析表

统计量	平方和	自由度	方差	方差比
回归	$SS_{\mathrm{reg}}=\sum_{i=1}^{n}(\widehat{y}_i-\overline{y})^2$	$df_{\mathrm{reg}}=m$	$SS_{\mathrm{reg}}/df_{\mathrm{reg}}$	$F_0=\frac{SS_{\mathrm{reg}}/df_{\mathrm{reg}}}{SS_{\mathrm{res}}/df_{\mathrm{res}}}$
残差	$SS_{\mathrm{res}}=\sum_{i=1}^{n}(y_i-\widehat{y}_i)^2$	$df_{\mathrm{res}}=n-m-1$	$SS_{\mathrm{res}}/df_{\mathrm{res}}$	
总体	$SS_{\mathrm{tot}}=\sum_{i=1}^{n}(y_i-\overline{y})^2$	$df_{\mathrm{tot}}=n-1$		

式中，n为样本数量，m为自变量的个数。

3.4.2.5 总体回归$Ax+b$的估计——置信区间估计

对于温度与销量的回归模型，温度为任意值时，所对应的销量不是一个固定的值，而是服从平均值为$Ax+B$（总体回归）、标准差为σ的正态分布，因此在给定置信度（95%，99%等），总体回归$Ax+B$（即预测值）一定会在某个值以上、某个值以下的区间中，计算任意温度所对应销量的置信区间（图3.4–14），是由预测值加减一个区间，该区间的计算公式为：$\sqrt{F(1,n-m-1;0.05)\times(\frac{1}{n}+\frac{(x_i-\overline{x})^2}{S_{\mathrm{xx}}})\times\frac{SS_{\mathrm{res}}}{n-2}}$。其中，$n$为样本个数，$m$为自变量$x$（温度）的个数，$x_i$为自变量（温度）样本取值，$\overline{x}$为样本均值，$S_{\mathrm{xx}}$为自变量$x$（温度）样本的离差平方和，

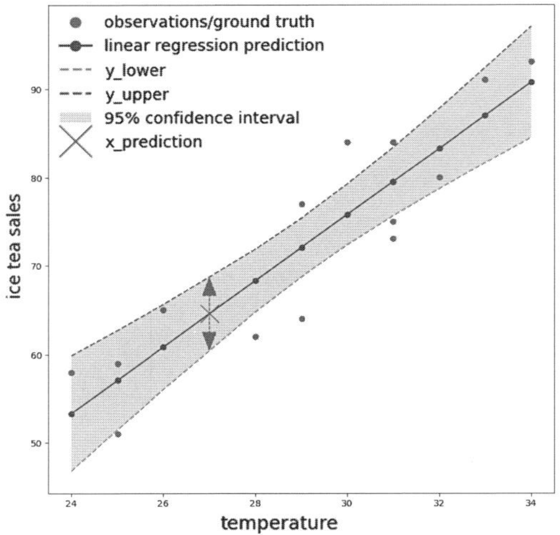

图3.4-14 置信区间

SS_{res}为残差平方和。

```
def confidenceInterval_estimator_LR(x, sample_num, X, y, model, confidence=0.05):
    '''
    function - 简单线性回归置信区间估计，及预测区间

    Params:
        x - 自变量取值；float
        sample_num - 样本数量；int
        X - 样本数据集 - 自变量；list(float)
        y - 样本数据集 - 因变量；list(float)
        model -使用 sklearn 获取的线性回归模型；model
        confidence -  置信度。The default is 0.05" ；float

    Returns:

       CI - 置信区间；list(float)
    '''
    import numpy as np
    import math
    from scipy.stats import f
    import matplotlib.pyplot as plt
    X_ = X.reshape(-1)
    X_mu = X_.mean()
    s_xx = (X_-X_mu)**2
    S_xx = s_xx.sum()
    ss_res = (y-LR.predict(X))**2
    SS_res = ss_res.sum()
    # dfn=1, dfd=sample_num-2
    probability_val = f.ppf(q=1-confidence, dfn=1, dfd=sample_num-2)
    CI = [math.sqrt(probability_val*(1/sample_num+(x-X_mu) **
                2/S_xx)*SS_res/(sample_num-2)) for x in X_]
    y_pre = LR.predict(X)
    fig, ax = plt.subplots(figsize=(10, 10))
    ax.plot(X_, y, 'o', label='observations/ground truth', color='r')
    ax.plot(X_, y_pre, 'o-', label='linear regression prediction')
    ax.plot(X_, y_pre-CI, '--', label='y_lower')
    ax.plot(X_, y_pre+CI, '--', label='y_upper')
    ax.fill_between(X_, y_pre-CI, y_pre+CI, alpha=0.2,
                label='95% confidence interval')

    # 给定值的预测区间
    x_ci = math.sqrt(probability_val*(1/sample_num+(x-X_mu)
                 ** 2/S_xx)*SS_res/(sample_num-2))
    x_pre = LR.predict(np.array([x]).reshape(-1, 1))[0]
    x_lower = x_pre-x_ci
    x_upper = x_pre+x_ci
```

```
    print("x prediction=%.6f;confidence interval=[%.6f,%.6f]" % (
        x_pre, x_lower, x_upper))
    ax.plot(x, x_pre, 'x', label='x_prediction', color='r', markersize=20)
    ax.arrow(x, x_pre, 0, x_upper-x_pre, head_width=0.3, head_length=2,
             color="gray", linestyle="--", length_includes_head=True)
    ax.arrow(x, x_pre, 0, x_lower-x_pre, head_width=0.3, head_length=2,
             color="gray", linestyle="--", length_includes_head=True)
    ax.set(xlabel='temperature', ylabel='ice tea sales')
    ax.legend(loc='upper left', frameon=False)
    plt.show()
    return CI
sample_num = 14
confidence = 0.05
iceTea_df_sort = iceTea_df.sort_values(by=['temperature'])
X, y = iceTea_df_sort.temperature.to_numpy(
).reshape(-1, 1), iceTea_df_sort.iceTeaSales.to_numpy()
CI = confidenceInterval_estimator_LR(27, sample_num, X, y, LR, confidence)
---------------------------------------------------------------------------
x prediction=64.561674;confidence interval=[60.496215,68.627133]
```

3.4.2.6 预测区间

给定特定值例如温度为31，则预测值为79.51，但是实际值不一定为该值，而是在置信度（置信水平，置信系数）为95%，对应的置信区间［66.060470,92.965962］［66.060470,92.965962］内浮动，这个区间称为预测区间。

3.4.3 多元线性回归

包含多于一个回归变量的回归模型，称为多元回归模型。如果为线性，则为多元线性回归（multivariable linear regression）。实际问题处理上，尤其大数据，会涉及很多自变量，例如sklearn机器学习库经典的鸢尾花（Iris）数据集包含的自变量有花萼长度、花萼宽度、花瓣长度和花瓣宽度（Sepal Length, Sepal Width, Petal Length and Petal Width），总共4个。其因变量为鸢尾花的种类，如果要根据自变量与因变量建立回归模型，则需要5个参数。

3.4.3.1 建立数据集

在应用Python语言解析该部分内容时，仍然使用比较简单的数据集，用《漫画统计学之回归分析》中店铺的数据集（表3.4–4），自变量包括店铺的面积（m^2）、最近的车站距离（m），因变量为月营业额（万元）。

```
import pandas as pd
import util_misc
from scipy import stats
store_info = {"location": ['Ill.', 'Ky.', 'Lowa.', 'Wis.', 'MIch.', 'Neb.',
'Ark.', 'R.I.', 'N.H.', 'N.J.'], "area": [10, 8, 8, 5, 7, 8, 7, 9, 6, 9],
              "distance_to_nearestStation": [80, 0, 200, 200, 300, 230, 40, 0,
330, 180], "monthly_turnover": [469, 366, 371, 208, 246, 297, 363, 436, 198,
364]}
storeInfo_df = pd.DataFrame(store_info)
util_misc.print_html(storeInfo_df, 10)
```

表3.4–4 店铺的数据

	location	area	distance_to_nearestStation	monthly_turnover
0	Ill.	10	80	469
1	Ky.	8	0	366
2	Lowa.	8	200	371
3	Wis.	5	200	208
4	MIch.	7	300	246
5	Neb.	8	230	297
6	Ark.	7	40	363
7	R.I.	9	0	436

续表

	location	area	distance_to_nearestStation	monthly_turnover
8	N.H.	6	330	198
9	N.J.	9	180	364

3.4.3.2 相关性分析

为了判断依据上述数据是否具有建立多元线性回归模型的意义，同样需要进行相关性分析。因为涉及变量增加，需要计算两两之间的相关系数及对应的p-value（表3.4–5）。为了方便日后对此种类型数据的相关性分析，建立correlationAnalysis_multivarialbe()函数。自变量与因变量之间的相关系数反映了自变量所能解释因变量的程度，其相关系数分别为0.8924, –0.7751，两个自变量均与因变量具有较强的相关关系，能够解释因变量，可以建立回归模型；同时，自变量之间的相关关系，可以初步判断自变量之间是否存在多重共线性，即自变量之间存在精确相关关系或高度相关关系，使得模型估计失真，或者难以估计准确。根据计算结果，两个自变量之间的相关系数为–0.4922，但是对应p-value=0.1485，即拒绝原假设，说明两个自变量之间不存在线性相关关系。因此，同时使用这两个自变量解释因变量，初步判断不会使回归模型失真。为了观察变量间的关系，同时打印了店铺数据两两变量的散点图3.4–15。

```
import seaborn as sns
import matplotlib.pyplot as plt
sns.pairplot(storeInfo_df)
plt.show()
def correlationAnalysis_multivarialbe(df):
    '''
    function - DataFrame 数据格式，成组计算 pearsonr 相关系数

    Params:
        df - DataFrame 格式数据集；DataFrame(float)

    Returns:
        p_values - P 值；DataFrame(float)
        correlation - 相关系数；DataFame(float)
    '''
    from scipy.stats import pearsonr
    import pandas as pd
    df = df.dropna()._get_numeric_data()
    df_cols = pd.DataFrame(columns=df.columns)
    p_values = df_cols.transpose().join(df_cols, how='outer')
    correlation = df_cols.transpose().join(df_cols, how='outer')
    for r in df.columns:
        for c in df.columns:
            p_values[r][c] = round(pearsonr(df[r], df[c])[1], 4)
            correlation[r][c] = round(pearsonr(df[r], df[c])[0], 4)
    return p_values, correlation
p_values, correlation = correlationAnalysis_multivarialbe(storeInfo_df)
print("p_values:")
print(p_values)
print("_"*78)
print("correlation:")
print(correlation)
```

表3.4–5 店铺数据变量两两相关系数

p_values:			
	area	**distance_to_nearestStation**	**monthly_turnover**
area	0.0	0.1485	0.0005
distance_to_nearestStatio	0.1485	0.0	0.0084
monthly_turnover	0.0005		
correlation:			
area	1.0	-0.4922	0.8924
distance_to_nearestStation	-0.4922	1.0	-0.7751
monthly_turnover	0.8924	-0.7751	1.0

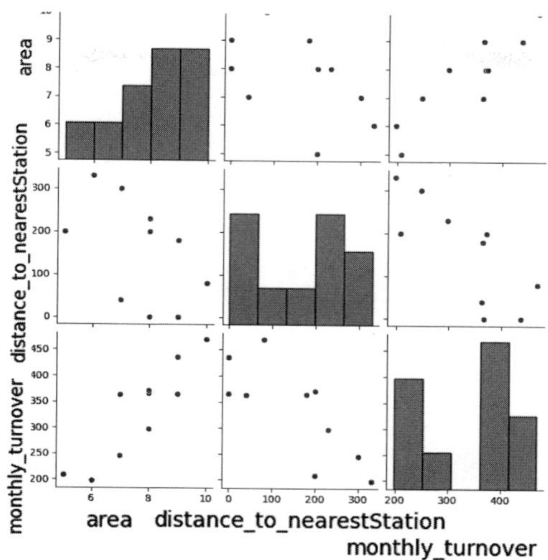

图3.4-15 店铺数据两两变量的散点图

因为总共涉及了3个变量，可以使用Plotly库提供的三元图（Ternary Plot），查看两个自变量与一个因变量之间的分布关系。可能变量之间数值的取值范围相差较大，在三元图打印时，某些变量的值可能全部贴近图形边缘，无法清晰地表述变量间的关系，因此使用：$\frac{x_i-\bar{x}}{x_{\max}-x_{\min}}$方法分别标准化各个变量。从图3.4-16可得店铺面积（颜色表示面积）逐渐增加，月营业额逐渐增加（点的大小表示月营业额数值大小）；而最近的车站距离逐步减小时，月营业额逐渐增加。

```
import plotly.express as px
pd.options.mode.chained_assignment = None
columns = ['area', 'distance_to_nearestStation', 'monthly_turnover']
storeInfo_plot = storeInfo_df[columns]
normalize_df = storeInfo_plot.T.apply(lambda row: (
    row-row.min())/(row.max()-row.min()), axis=1).T
normalize_df["location"] = storeInfo_df.location
fig = px.scatter_ternary(normalize_df, a="monthly_turnover", b="area",
c="distance_to_nearestStation", hover_name="location",
                         color="area", size="monthly_turnover", size_max=8)
fig.show()
```

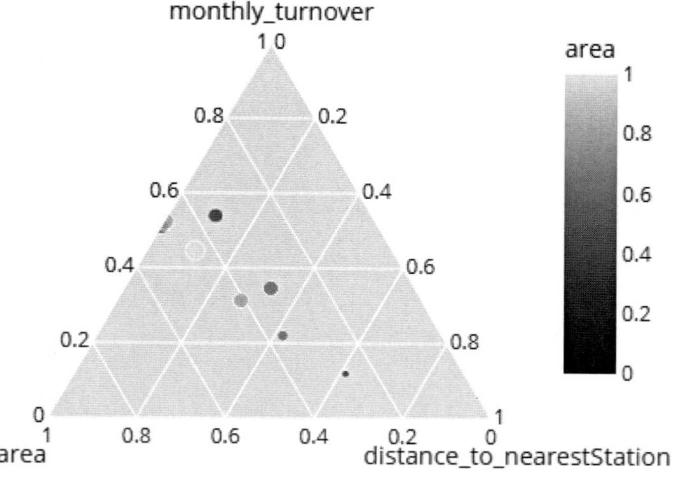

图3.4-16 月营业额、最近的车站距离和店铺面积的三元图

3.4.3.3 求解多元回归方程

求解多元回归方程的方法与简单线性回归求解方式相类似，使用最小二乘法对偏回归系数进行求解。求解过程中，使用了三种方法：一是，使用SymPy分别对残差平方和SS_{res}的$a1$、$a2$和b求微分，当各自微分的值等于0时，所反映的残差平方和为0，即观测值和预测值差值的平方和为0，而单个观测值与对应的预测值之间的差值趋于0；二是，使用矩阵计算的方式求解参数，其计算公式为：$\widehat{A}=(X'X)^{-1}X'Y$，其中$X=\begin{bmatrix}1&10&80\\1&8&0\\1&8&200\\1&5&200\\1&7&300\\1&8&230\\1&7&40\\1&9&0\\1&6&330\\1&9&180\end{bmatrix}$，$X'=\begin{bmatrix}1&1&1&1&1&1&1&1&1&1\\10&8&8&5&7&8&7&9&6&9\\80&0&200&200&300&230&40&0&330&180\end{bmatrix}$即$X$的转置，也可记作$X^{\text{T}},X^{\text{tr}}$等，$Y=\begin{bmatrix}469\\366\\371\\208\\246\\297\\363\\436\\198\\364\end{bmatrix}$。对于一个矩阵$X$，其逆矩阵（inverse matrix）为$X^{-1}$。使用矩阵的计算方法时，仍然是使用SymPy库，该库提供了建立矩阵和矩阵计算的功能。最后一种求解多元线性回归方程的方式是直接使用`sklearn.linear_model.LinearRegression`计算，并获得回归模型。

偏回归系数（partial regression coefficient），是多元回归问题出现的一个特殊性质。设自变量$x_1,x_2,\ldots,x_m$，与因变量y具有线性关系，有$y=a_1x_1+a_2x_2+\ldots+a_nx_n+b$，则$a_1,a_2,\ldots,a_n$为相对于各自变量的偏回归系数，表示当其他的各自变量都保持一定时，指定的某一自变量每变动一个单位，因变量y增加或减少的数值。

- 对矩阵表示的求解参数公式$\widehat{A}=(X'X)^{-1}X'Y$再理解：

多元线性回归模型公式：$y=a_1x_1+a_2x_2+\ldots+a_nx_n+b$，$b$首先可以被看做每一个样本中固定的一个自变量$x_0$，再引入一个可以学习的$a_0$，即$y=a_0x_0+a_1x_1+a_2x_2+\ldots+a_nx_n$。为进一步方便表示和学习，设公式中的$x_0$为不需要学习的值，即$x_0=1$。此时，$a_0$值的意义则代表原公式中的偏置$b$，因此多元线性回归模型公式可简单表示为：$Y=AX$，对于所有样本$X=\begin{bmatrix}X_1\\X_2\\\vdots\\X_n\end{bmatrix}$和$Y=\begin{bmatrix}Y_1\\Y_2\\\vdots\\Y_n\end{bmatrix}$，其矩阵的表示方式为：$Y=AX=\begin{bmatrix}Y_1\\Y_2\\\vdots\\Y_n\end{bmatrix}=[A_1,A_2,\ldots,A_n]\begin{bmatrix}X_1\\X_2\\\vdots\\X_n\end{bmatrix}=\begin{bmatrix}A_1X_1\\A_2X_2\\\vdots\\A_nX_n\end{bmatrix}$。因为矩阵不能相除，因此不能直接对$Y=AX$两边同时除以$X$，以求取$A$。但是，可以两边同时乘以$X$的逆矩阵$X'$避免除法（矩阵乘以自身的逆矩阵结果为1）。同时，只有方阵才可能可逆，而样本的数量是无法控制的，因此用X乘以其转置产生一个可以求逆的方阵$X'X$。完整推导过程如下：

$Y = AX$

首先，右乘X'，得到$YX' = \hat{A}XX'$

然后，右乘$(XX')^{-1}$，得到$YX'(XX')^{-1} = \hat{A}XX'(XX')^{-1}$

因为$XX'(XX')^{-1} = 1$，所以化简如下：

$YX'(XX')^{-1} = \hat{A}$

即$\widehat{A} = (X'X)^{-1}X'Y$

```
from sklearn.linear_model import LinearRegression
import sympy
import math
from sympy import diff, Eq, solveset, solve, simplify, pprint, Matrix
a1, a2, b = sympy.symbols('a1 a2 b')
# 计算残差平方和
storeInfo_df["ss_res"] = storeInfo_df.apply(lambda row: (
    row.monthly_turnover-(row.area*a1+row.distance_to_nearestStation*a2+b))**2,
axis=1)
util_misc.print_html(storeInfo_df, 10)
SS_res = storeInfo_df["ss_res"].sum()

# A- 使用 sympy 求解多元回归方程
# 对残差平方和 SS_res 关于 a1, a1 和 b 求微分, 并使微分值为 0
diff_SSres_a1 = diff(SS_res, a1)
diff_SSres_a2 = diff(SS_res, a2)
diff_SSres_b = diff(SS_res, b)

# 当微分值为 0 时, 解方程组, 获得 a1、a2 和 b 的值
Eq_residual_a1 = Eq(diff_SSres_a1, 0)  # 设所求 a1 微分为 0
Eq_residual_a2 = Eq(diff_SSres_a2, 0)  # 设所求 a2 微分为 0
Eq_residual_b = Eq(diff_SSres_b, 0)  # 设所求 a2 微分为 0
slop_intercept = solve((Eq_residual_a1, Eq_residual_a2,
                        Eq_residual_b), (a1, a2, b))  # 计算三元一次方程组
print("diff_a1,a2 and b:\n")
pprint(slop_intercept)
print("_"*50)

# B - 使用矩阵 ( 基于 sympy ) 求解多元回归方程
if 'one' not in storeInfo_df.columns:
    X_m = Matrix(storeInfo_df.insert(loc=1, column='one', value=1)
                 [['one', 'area', 'distance_to_nearestStation']])
else:
    X_m = Matrix(storeInfo_df[['one', 'area', 'distance_to_nearestStation']])
y_m = Matrix(storeInfo_df.monthly_turnover)

parameters_reg = (X_m.T*X_m)**-1*X_m.T*y_m  # 注意在矩阵计算时, 矩阵相乘不能任意变化位置
print("matrix_a1,a2 and b:\n")
pprint(parameters_reg)

# C - 使用 sklearn 求解多元回归方程
# B - 使用 sklearn 库 sklearn.linear_model.LinearRegression(), Ordinary least squares Linear Regression- 普通最小二乘线性回归, 获取回归方程
X = storeInfo_df[['area', 'distance_to_nearestStation']].to_numpy()
y = storeInfo_df['monthly_turnover'].to_numpy()

# 拟合模型
LR_multivariate = LinearRegression().fit(X, y)
# 模型参数
print("_"*50)
print("Sklearn a1=%.2f,a2=%.2f,b=%.2f" % (LR_multivariate.coef_[0],
      LR_multivariate.coef_[1], LR_multivariate.intercept_))

# 建立回归方程
x1, x2 = sympy.symbols('x1,x2')
fx_m = slop_intercept[a1]*x1+slop_intercept[a2]*x2+slop_intercept[b]
print("linear regression_fx=:\n")
pprint(fx_m)
fx_m = sympy.lambdify([x1, x2], fx_m, "numpy")
--------------------------------------------------------------------------------
diff_a1,a2 and b:
```

$$\left\{a_1: \frac{4073344}{98121},\ a_2: \frac{-44597}{130828},\ b: \frac{6409648}{98121}\right\}$$

```
-------------------------------------------------
matrix_a1,a2 and b:
```

$$\left[\begin{matrix}\frac{6409648}{98121}\\\frac{4073344}{98121}\\-\frac{44597}{130828}\end{matrix}\right]$$

```
-------------------------------------------------
Sklearn a1=41.51,a2=-0.34,b=65.32
linear regression_fx=:
```

$$\frac{4073344 \cdot x_1}{98121} - \frac{44597 \cdot x_2}{130828} + \frac{6409648}{98121}$$

```
# 可以将矩阵打印为 Latex 格式的数学表达式，方便在 markdown 中表述，避免自行输入。

from sympy import latex
print(latex(X_m.T))
------------------------------------------------------------------------------
\left[\begin{matrix}1 & 1 & 1 & 1 & 1 & 1 & 1 & 1 & 1 & 1\\10 & 8 & 8 & 5
& 7 & 8 & 7 & 9 & 6 & 9\\80 & 0 & 200 & 200 & 300 & 230 & 40 & 0 & 330 &
180\end{matrix}\right]
```

同样，使用三元图打印两个自变量，以及预测值之间的图3.4-17，观察变量之间的关系。

```
import plotly.express as px
pd.options.mode.chained_assignment = None
storeInfo_df['pre'] = LR_multivariate.predict(X)
columns = ['area', 'distance_to_nearestStation', 'monthly_turnover', 'pre']
storeInfo_plot = storeInfo_df[columns]
normalize_df = storeInfo_plot.T.apply(lambda row: (
    row-row.min())/(row.max()-row.min()), axis=1).T
normalize_df["location"] = storeInfo_df.location
fig = px.scatter_ternary(normalize_df, a="pre", b="area", c="distance_to_
nearestStation", hover_name="location",
                         color="area", size="pre", size_max=8,
                         )
fig.show()
```

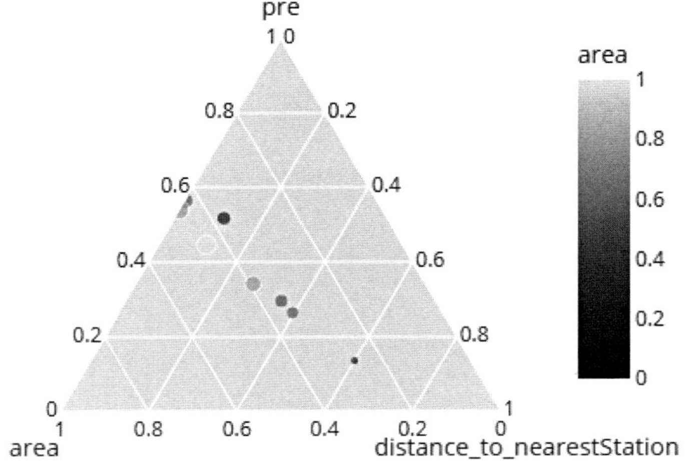

图3.4-17 最近的车站距离、店铺面积和预测值（月营业额）的三元图

3.4.3.4 确认多元回归方程的精度

非修正自由度的判定系数计算同简单线性回归，将定义的计算函数coefficient_of_determination放置于util_A.py文件（模块）中，直接调用计算；同时，也使用sklearn提供的r2_score计算。其计算结果约为0.94，表示实测值与回归方程的预测值拟合程度的指标较高，能够比较好地根据店铺面积和最近车站距离预测月营业额。

```
# 计算复相关系数 R
from sklearn.metrics import r2_score
import util_A
R_square_a, R_square_b = util_A.coefficient_of_determination(
    storeInfo_df.monthly_turnover.to_list(), storeInfo_df.pre.to_list())
print("R_square_a=%.5f,R_square_b=%.5f" % (R_square_a, R_square_b))
R_square_ = r2_score(storeInfo_df.monthly_turnover.to_list(),
                     storeInfo_df.pre.to_list())
print("using sklearn libray to calculate r2_score=", R_square_)
---------------------------------------------------------------------------
R_square_a=0.94524,R_square_b=0.94524
using sklearn libray to calculate r2_score= 0.945235852681711
```

- 修正自由度的判定系数

直接使用判定系数时，其自变量的数量越多，判定系数的值越高，但是并不是每一个自变量都是有效的，因此通常使用修正自由度的判定系数，其公式为：$R^2=1-\frac{SS_{\text{res}}}{n_{\text{s}}-n_{\text{v}}-1}/\frac{SS_{\text{tot}}}{n_{\text{s}}-1}$，其中$n_{\text{s}}$为样本个数，$n_{\text{v}}$为自变量个数，$SS_{\text{res}}$为残差平方和，$SS_{\text{tot}}$为总的离差平方和。

```
def coefficient_of_determination_correction(observed_vals, predicted_vals,
independent_variable_n):
    '''
    function - 回归方程修正自由度的判定系数

    Params:
        observed_vals - 观测值（实测值）；list(float)
        predicted_vals - 预测值；list(float)
        independent_variable_n - 自变量个数；int

    Returns:
        R_square_correction - 正自由度的判定系数；float
    '''
    import pandas as pd
    import numpy as np
    import math

    vals_df = pd.DataFrame({'obs': observed_vals, 'pre': predicted_vals})
    # 观测值的离差平方和(总平方和，或总的离差平方和)
    obs_mean = vals_df.obs.mean()
    SS_tot = vals_df.obs.apply(lambda row: (row-obs_mean)**2).sum()

    # 残差平方和
    SS_res = vals_df.apply(lambda row: (row.obs-row.pre)**2, axis=1).sum()

    # 判断系数
    sample_n = len(observed_vals)
    R_square_correction = 1 - \
        (SS_res/(sample_n-independent_variable_n-1))/(SS_tot/(sample_n-1))
    return R_square_correction
R_square_correction = coefficient_of_determination_correction(
    storeInfo_df.monthly_turnover.to_list(), storeInfo_df.pre.to_list(), 2)
print("修正自由度的判定系数=", R_square_correction)
---------------------------------------------------------------------------
修正自由度的判定系数= 0.929588953447914
```

3.4.3.5 回归显著性检验

在简单回归模型中的回归系数检验，只需要给定$H_0:A=0,H_1:A\neq 0$。但是在多元回归中，就总体而言$F_{\text{x}}=A_1x_1+A_2x_2+B$，其中$A_1\sim a_1,A_2\sim a_2,B\sim b$，$\sim$为约为。包括$A_1$和$A_2$两个偏相关系数，因此可以分为两种情况，一种是全面讨论偏回归系数的检验，原假设：$A_1=A_2=0$，备择假设：$A_1=A_2=0$不成立，即以下任意一组关系成立，$A_1\neq 0$且$A_2\neq 0$，

$A_1 \neq 0$且$A_2=0$，或$A_1=0$且$A_2 \neq 0$。另一种是分别讨论偏回归系数的检验，例如原假设：$A_1=0$，备择假设：$A_1 \neq 0$。在这两种方式中，检验统计量是不同的，对于全面检验，其统计量为：$F_0=\frac{SS_{\text{tot}}-SS_{\text{res}}}{n_{\text{v}}} / \frac{SS_{\text{res}}}{n_{\text{s}}-n_{\text{v}}-1}$，式中$SS_{\text{tot}}$为总平方和$SS_{\text{res}}$为残差平方和，$n_{\text{s}}$为样本个数，$n_{\text{v}}$为自变量个数；对于单个回归系数的检验，其统计量为：$F_0=\frac{a_1^2}{C_{jj}} / \frac{SS_{\text{res}}}{n_{\text{s}}-n_{\text{v}}-1}$，式中$C_{jj}$为$(X'X)^{-1}$对角线相交位置的值，即$(X'X)^{-1}=\begin{bmatrix} \frac{511351}{98121} & -\frac{55781}{98121} & -\frac{1539}{327070} \\ \frac{55781}{98121} & \frac{6442}{98121} & \frac{66}{163535} \\ \frac{1539}{327070} & \frac{66}{163535} & \frac{67}{6541400} \end{bmatrix}$，对角线的值为$\frac{6442}{98121}$。

对于全部回归系数的总体检验，及单个回归系数的检验，其结果p-value均小于0.05，意味着所求得的多元线性回归模型是合适的。

```
def ANOVA_multivarialbe(observed_vals, predicted_vals, independent_variable_n, a_i, X):
    '''
    function - 多元线性回归方程 - 回归显著性检验（回归系数检验），全部回归系数的总体检验，以及单个回归系数的检验

    Paras:
        observed_vals - 观测值（实测值）; list(float)
        predicted_vals - 预测值; list(list)
        independent_variable_n - 自变量个数; int
        a_i - 偏相关系数列表; list(float)
        X - 样本数据集 _ 自变量; array(numpy)

    Returns:
        None
    '''
    import pandas as pd
    import numpy as np
    import math
    from scipy.stats import f
    from sympy import Matrix, pprint
    vals_df = pd.DataFrame({'obs': observed_vals, 'pre': predicted_vals})
    # 总平方和，或总的离差平方和
    obs_mean = vals_df.obs.mean()
    SS_tot = vals_df.obs.apply(lambda row: (row-obs_mean)**2).sum()

    # 残差平方和
    SS_res = vals_df.apply(lambda row: (row.obs-row.pre)**2, axis=1).sum()

    # 回归平方和
    SS_reg = vals_df.pre.apply(lambda row: (row-obs_mean)**2).sum()

    # 样本个数
    n_s = len(observed_vals)
    dfn = independent_variable_n
    dfd = n_s-independent_variable_n-1

    # 计算全部回归系数的总体检验统计量
    F_total = ((SS_tot-SS_res)/dfn)/(SS_res/dfd)
    print("F- 分布统计量 _total=%.6f;p-value=%.6f" %
          (F_total, f.sf(F_total, dfn, dfd)))

    # 逐个计算单个回归系数的检验统计量
    X = np.insert(X, 0, 1, 1)
    X_m = Matrix(X)
    M_inverse = (X_m.T*X_m)**-1
    C_jj = M_inverse.row(1).col(1)[0]
    pprint(C_jj)
    F_ai_list = []
    i = 0
    for a in a_i:
        F_ai = (a**2/C_jj)/(SS_res/dfd)
        F_ai_list.append(F_ai)
        print("a%d=%.6f 时, F- 分布统计量 _=%.6f;p-value=%.6f" %
              (i, a, F_ai, f.sf(F_total, 1, dfd)))
        i += 1
a1_, a2_ = LR_multivariate.coef_[0], LR_multivariate.coef_[1]
X = storeInfo_df[['area', 'distance_to_nearestStation']].to_numpy()
```

```
ANOVA_multivarialbe(storeInfo_df.monthly_turnover.to_list(),
                    storeInfo_df.pre.to_list(), 2, a_i=[a1_, a2_], X=X)
--------------------------------------------------------------------------------
F-分布统计量_total=60.410426;p-value=0.000038
 6442
————————
98121
a0=41.513478时，F-分布统计量_=44.032010;p-value=0.000110
a1=-0.340883时，F-分布统计量_=0.002969;p-value=0.000110
```

3.4.3.6 总体回归$A_1X_1+A_2X_2+\ldots+A_nX_n+B$的估计——置信区间

多元线性回归模型的预测值置信区间估计使用了两种计算方式，一是，自定义函数逐步计算，其计算公式为：$\sqrt{F(1,n_s-n_v-1;0.05)\times(\frac{1}{n_s}+\frac{D^2}{n_s-1})\times\frac{SS_{res}}{n_s-n_v-1}}$，其中$n_s$为样本个数，$n_v$为自变量个数，$D^2$为马氏距离（Mahalanobis distance）的平方，SS_{res}为残差平方和；D^2马氏距离的平方计算公式为：先求$S=\begin{bmatrix}S_{11} & S_{12} & \ldots & S_{1p}\\ S_{21} & S_{22} & \ldots & S_{2p}\\ \vdots & \vdots & \ddots & \vdots\\ S_{p1} & S_{p2} & \ldots & S_{pp}\end{bmatrix}$的逆矩阵$S^{-1}$，式中，$S_{22}$代表第2个自变量的离差平方和，$S_{25}$代表第2个自变量和第5个自变量的离差积和，$S_{25}$与$S_{52}$是相等的，以此类推；然后，根据$S^{-1}$，求取马氏距离的平方公式为：

$$D^2=[(x_1-\overline{x_1})(x_1-\overline{x_1})S^{11}+(x_1-\overline{x_1})(x_2-\overline{x_2})S^{12}]+\ldots+(x_1-\overline{x_1})(x_p-\overline{x_p})S^{1p}$$
$$+(x_2-\overline{x_2})(x_1-\overline{x_1})S^{21}+(x_2-\overline{x_2})(x_2-\overline{x_2})S^{12}]+\ldots+(x_2-\overline{x_2})(x_p-\overline{x_p})S^{2p}$$
$$\ldots+(x_p-\overline{x_p})(x_1-\overline{x_1})S^{p1}+(x_p-\overline{x_p})(x_2-\overline{x_2})S^{12}]+\ldots+(x_p-\overline{x_p})(x_p-\overline{x_p})S^{pp}(n_s-1),$$

其中n_s为样本个数。

二是，使用statsmodels的`statsmodels.regression.linear_model.OLS`普通最小二乘法（Ordinary Least Squares，OLS）求得多元线性回归方程，其语法结构与sklearn基本相同。所求的回归模型包含有置信区间的属性，可以通过`dt=res.get_prediction(X).summary_frame(alpha=0.05)`的方式提取。可以打印statsmodels计算所得回归模型的概要（summary），见表3.4-6，比较求解回归方程的偏回归系数和截距（coef_const/area/distance_to_nearestStation），以及确认多元回归方程的精度R-squared（R^2）和修正自由度的判定系数Adj. R-squared，和回归显著性检验全面讨论偏回归系数的检验F-分布统计量F-statistic，对应p-value/Prob (F-statistic)，全部相等，互相印证了所使用的方法是否保持一致。

表3.4-6 statsmodels计算所得回归模型的概要

OLS Regression Results

==

Dep. Variable:	monthly_turnover	R-squared:	0.945
Model:	OLS	Adj. R-squared:	0.93
Method:	Least Squares	F-statistic:	60.41
Date:	Mon, 24 Oct 2022	Prob (F-statistic):	3.84E-05
Time:	13:22:21	Log-Likelihood:	-44.358
No. Observations:	10	AIC:	94.72
Df Residuals:	7	BIC:	95.62
Df Model:	2		
Covariance Type:	nonrobust		

==

	coef	std err	t	P>|t|	[0.025	0.975]
const	65.3239	55.738	1.172	0.28	-66.476	197.124
area	41.5135	6.256	6.636	0	26.72	56.307
distance_to_nearestStation	-0.3409	0.078	-4.362	0.003	-0.526	-0.156

==

续表

Omnibus:	0.883	Durbin-Watson:	3.44
Prob(Omnibus):	0.643	Jarque-Bera (JB):	0.448
Skew:	0.479	Prob(JB):	0.799
Kurtosis:	2.603	Cond. No.	1.40E+03

```
==============================================================================

Notes:
[1] Standard Errors assume that the covariance matrix of the errors is correctly specified.
[2] The condition number is large, 1.4e+03. This might indicate that there are
strong multicollinearity or other numerical problems.
Index(['area', 'distance_to_nearestStation'], dtype='object')
4173.006119994701
```

对于两种方法在预测变量置信区间比较上，分别打印了各自的三维分布图3.4-18，其结果显示两者的图形保持一致，即通过statsmodels求解多元回归方程与逐步计算所得结果保持一致。

- 马氏距离（Mahalanobis distance）

马氏距离表示数据的协方差矩阵，有效计算两个未知样本集相似度的方法。与欧式距离（Euclidean distance）不同的是它考虑各种特性之间的联系（例如，身高和体重是有关联的），并且是尺度无关的（scale-invariant，例如去掉单位），独立于测量尺度。计算公式如上所述，也可以简化表示为，对于一个均值为$\vec{\mu}=(\mu_1,\mu_2,\mu_3,\ldots,\mu_{\mathrm{N}})^{\mathrm{T}}$（即为各个自变量的均值）的多变量（多个自变量）的矩阵，$\vec{x}=(x_1,x_2,x_3,\ldots,x_{\mathrm{N}})^{\mathrm{T}}$，其马氏距离为$D_{\mathrm{M}}(\vec{x})=\sqrt{(\vec{x}-\vec{\mu})^{\mathrm{T}}S^{-1}(\vec{x}-\vec{\mu})}$。

```
# statsmodels 供了一些类和函数，用于估计许多不同的统计模型，以及执行统计测试和统计数据研究。每个估计器都有一个广泛的结果统计信息列表，可以用以查看相关信息，以确保所求得的估计器（模型）的准确性、正确性。

import numpy as np
import statsmodels.api as sm
import warnings
warnings.filterwarnings("ignore")

# 使用 statsmodels 库求解回归方程，与获得预测值的置信区间
storeInfo_df_sort = storeInfo_df.sort_values(by=['area'])
X = storeInfo_df_sort[['area', 'distance_to_nearestStation']]
# 因为在上述逐步计算或者使用 Sklearn 求解回归方程过程中，多元回归方程均增加了常量截距的参数，因此此处增加一个常量 adding a constant
X = sm.add_constant(X)
y = storeInfo_df_sort['monthly_turnover']
```

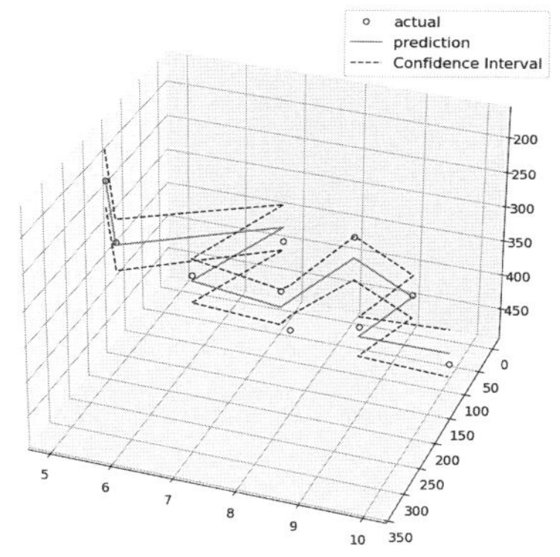

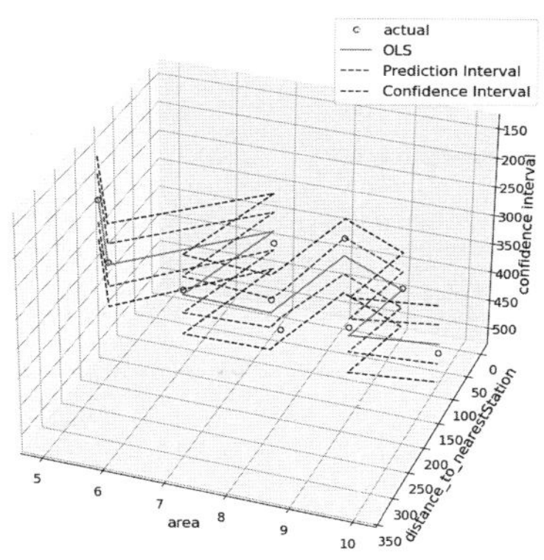

图3.4-18 逐步计算（左）与 statsmodels库计算（右）结果预测变量置信区间比较

```
mod = sm.OLS(y, X)  #构建最小二乘模型
res = mod.fit()  #拟合模型（Fit model）
print(res.summary())
dt = res.get_prediction(X).summary_frame(alpha=0.05)
y_prd = dt['mean']
yprd_ci_lower = dt['obs_ci_lower']
yprd_ci_upper = dt['obs_ci_upper']
ym_ci_lower = dt['mean_ci_lower']
ym_ci_upper = dt['mean_ci_upper']

#逐步计算
def confidenceInterval_estimator_LR_multivariable(X, y, model, confidence=0.05):
    '''
    function - 多元线性回归置信区间估计及预测区间

    Params:
        X - 样本自变量；DataFrame 数据格式
        y - 样本因变量；list(float)
        model - 多元回归模型；model
        confidence - 置信度，The default is 0.05; float

    return:
        CI- 预测值的置信区间；list(float)
    '''
    import pandas as pd
    from sympy import Matrix, pprint
    import numpy as np

    #根据指定数目，划分列表的函数
    def chunks(lst, n):
        for i in range(0, len(lst), n):
            yield lst[i:i + n]

    X_deepCopy = X.copy(deep=True)  #不进行深度拷贝，如果传入的参数变量X发生了改变，则
该函数外部的变量值也会发生改变
    columns = X_deepCopy.columns
    n_v = len(columns)
    n_s = len(y)

    #求S，用于马氏距离的计算
    SD = []
    SD_name = []
    for col_i in columns:
        i = 0
        for col_j in columns:
            SD_column_name = col_i+'S'+str(i)
            SD_name.append(SD_column_name)
            if col_i == col_j:
                X_deepCopy[SD_column_name] = X_deepCopy.apply(
                    lambda row: (row[col_i]-X_deepCopy[col_j].mean())**2,
axis=1)
                SD.append(X_deepCopy[SD_column_name].sum())
            else:
                X_deepCopy[SD_column_name] = X_deepCopy.apply(lambda row: (
                    row[col_i]-X_deepCopy[col_i].mean())*(row[col_j]-X_
deepCopy[col_j].mean()), axis=1)
                SD.append(X_deepCopy[SD_column_name].sum())
            i += 1
    M = Matrix(list(chunks(SD, n_v)))

    #求S的逆矩阵
    M_invert = M**-1
    # pprint(M_invert)
    M_invert_list = [M_invert.row(row).col(col)[0]
                     for row in range(n_v) for col in range(n_v)]
    X_mu = [X_deepCopy[col].mean() for col in columns]

    #求马氏距离的平方
    SD_array = X_deepCopy[SD_name].to_numpy()
    D_square_list = [sum([x*y for x, y in zip(SD_selection,
                         M_invert_list)])*(n_s-1) for SD_selection in SD_array]

    #计算CI- 预测值的置信区间
    print(columns)
```

```
    ss_res = (y-model.predict(X_deepCopy[columns].to_numpy()))**2
    SS_res = ss_res.sum()
    print(SS_res)
    probability_val = f.ppf(q=1-confidence, dfn=1, dfd=n_s-n_v-1)
    CI = [math.sqrt(probability_val*(1/n_s+D_square/(n_s-1)) *
                SS_res/(n_s-n_v-1)) for D_square in D_square_list]
    return CI
X_ = storeInfo_df_sort[['area', 'distance_to_nearestStation']]
y_ = storeInfo_df_sort['monthly_turnover']
CI = confidenceInterval_estimator_LR_multivariable(
    X_, y_, LR_multivariate, confidence=0.05)

# 打印图表
fig, axs = plt.subplots(1, 2, figsize=(25, 11))
x_ = X.area
y_ = X.distance_to_nearestStation

# 由自定义函数，逐步计算获得的置信区间
axs[0] = fig.add_subplot(1, 2, 1, projection='3d')
axs[0].plot(x_, y_, y, linestyle="None", marker="o",
            markerfacecolor="None", color="black", label="actual")
X_array = X_.to_numpy()
LR_pre = LR_multivariate.predict(X_array)
axs[0].plot(x_, y_, LR_pre, color="red", label="prediction")
axs[0].plot(x_, y_, LR_pre+CI, color="darkgreen",
            linestyle="--", label="Confidence Interval")
axs[0].plot(x_, y_, LR_pre-CI, color="darkgreen", linestyle="--")

# 由 statsmodels 库计算所得的置信区间
axs[1] = fig.add_subplot(1, 2, 2, projection='3d')
axs[1].plot(x_, y_, y, linestyle="None", marker="o",
            markerfacecolor="None", color="black", label="actual")
axs[1].plot(x_, y_, y_prd, color="red", label="OLS")
axs[1].plot(x_, y_, yprd_ci_lower, color="blue",
            linestyle="--", label="Prediction Interval")
axs[1].plot(x_, y_, yprd_ci_upper, color="blue", linestyle="--")
axs[1].plot(x_, y_, ym_ci_lower, color="darkgreen",
            linestyle="--", label="Confidence Interval")
axs[1].plot(x_, y_, ym_ci_upper, color="darkgreen", linestyle="--")

axs[1].view_init(210, 250)  # 可以旋转图形的角度，方便观察
axs[1].set_xlabel('area')
axs[1].set_ylabel('distance_to_nearestStation')
axs[1].set_zlabel('confidence interval')
axs[0].legend()
axs[1].legend()
axs[0].view_init(210, 250)  # 可以旋转图形的角度，方便观察
axs[1].view_init(210, 250)  # 可以旋转图形的角度，方便观察
plt.show()
-----------------------------------------------------------------
```

参考文献（References）：

[1] [日]高桥信著作，Inoue Iroha，株式会社 TREND-PRO漫画制作，张仲恒译. 漫画统计学之回归分析[M]. 北京：科学出版社，2009. 08.

[2] Timothy C.Urdan（蒂莫西・C・厄丹），彭志文译. 白话统计学[M]. 中国人民大学出版社. 2013，12. 第3版. <Timothy C. Urdan (2022). Statistics in Plain English. Routledge.>.

[3] Douglas C.Montgomery，Elizabeth A.Peck，G.Geoffrey Vining著. 王辰勇译. 线性回归分析导论[M]. 北京：机械工业出版社，2016. 04（第5版）. <Douglas C. Montgomery, G. (2012). Introduction to Linear Regression Analysis. Wiley..>

3.5 公共健康数据的线性回归与梯度下降法

公共健康数据可以分为三类，分别为地理信息数据、疾病数据和经济条件数据。在公共健康数据中，将经济条件数据视为自变量，疾病数据视为因变量。通常自变量为因，因变量为果。自变量是可以改变的因素，而因变量为不能改变的因素。公共健康数据中英对照表（字段映射表）：

```
PubicHealth_Statistic_columns={
'geographic information':
{'Community Area':'社区',
'Community Area Name':'社区名',},
                              'disease':
{'natality':{'Birth Rate':'出生率',
'General Fertility Rate':'一般生育率',
'Low Birth Weight':'低出生体重',
'Prenatal Care Beginning in First Trimester':'产前3个月护理',
'Preterm Births':'早产',
'Teen Birth Rate':'青少年生育率',},
'mortality':{'Assault (Homicide)':'攻击（杀人）',
'Breast cancer in females':'女性乳腺癌',
'Cancer (All Sites)':'癌症',
'Colorectal Cancer':'结肠直肠癌',
'Diabetes-related':'糖尿病相关',
'Firearm-related':'枪支相关',
'Infant Mortality Rate':'婴儿死亡率',
'Lung Cancer':'肺癌',
'Prostate Cancer in Males':'男性前列腺癌',
'Stroke (Cerebrovascular Disease)':'中风(脑血管疾病)',},
'lead':{'Childhood Blood Lead Level Screening':'儿童血铅水平检查',
'Childhood Lead Poisoning':'儿童铅中毒',},
'infectious':{'Gonorrhea in Females':'女性淋病',
'Gonorrhea in Males':'男性淋病',
'Tuberculosis':'肺结核',},
'economic condition':
{'Below Poverty Level':'贫困水平以下',
'Crowded Housing':'拥挤的住房',
'Dependency':'依赖',
'No High School Diploma':'没有高中文凭',
'Per Capita Income':'人均收入',
'Unemployment':'失业',},
                              }
```

在实际的回归模型应用中，使用scikit-learn（sklearn）[①]库为主。对于有些描述也做相应的变化，自变量为解释变量（explanatory variable）即特征（features）或属性（attributes），其值为特征向量，通常用X表示特征向量（vector）数组（array）或矩阵；因变量为响应变量（response variable），其值通常用y表示（为目标值，target value）；机器学习（machine learning）的算法（algorithms）和模型（models），被称为估计器（estimator），不过，算法、模型和估计器的叫法有时会混淆使用。

① scikit-learn，也称为sklearn，针对Python编程语言的免费软件机器学习库（https://scikit-learn.org/stable/index.html）。

首先，读取公共健康数据为DataFrame数据格式（表3.5-1），因为sklearn的数据处理过程，以NumPy的数组形式为主，需要在两者之间进行转换。

#通常用大写字母表示矩阵（大于或等于2维的数组），用小写字母表示向量。

```
import pandas as pd
import geopandas as gpd
dataFp_dic = {
    "ublic_Health_Statistics_byCommunityArea_fp": r'./data/Public_Health_
Statistics_Selected_public_health_indicators_by_Chicago_community_area.csv',
    "Boundaries_Community_Areas_current": r'./data/ChicagoCommunityAreas/
ChicagoCommunityAreas.shp',
}
pubicHealth_Statistic = pd.read_csv(
    dataFp_dic["ublic_Health_Statistics_byCommunityArea_fp"])
community_area = gpd.read_file(
    dataFp_dic["Boundaries_Community_Areas_current"])
community_area.area_numbe = community_area.area_numbe.astype('int64')
pubicHealth_gpd = community_area.merge(
```

```
    pubicHealth_Statistic, left_on='area_numbe', right_
on='Community Area')
pubicHealth_gpd.head()
```

表3.5-1 公共健康数据

	area_num_1	area_numbe	community	shape_area	geometry	...	Gonorrhea in Females	Gonorrhea in Males	Tuberculosis	Below Poverty Level	Crowded Housing	Dependency	No High School Diploma	Per Capita Income	Unemployment
0	35	35	DOUGLAS	4.60E+07	POLYGON ((-87.60914 41.84469, -87.60915 41.844...	...	1063.3	727.4	4.2	26.1	1.6	31	16.9	23098	16.7
1	36	36	OAKLAND	1.69E+07	POLYGON ((-87.59215 41.81693, -87.59231 41.816...	...	1655.4	1629.3	6.7	38.1	3.5	40.5	17.6	19312	26.6

3.5.1 公共健康数据的简单线性回归

在回归前，需要做相关性分析，确定解释变量和响应变量是否存在相关性。返回到“公共健康数据的相关性分析”部分查看已经计算的结果，在经济条件数据和疾病数据中选取“Per Capita Income”（人均收入）和“Childhood Blood Lead Level Screening”（儿童血铅水平检查），其相关系数为-0.64，呈现线性负相关关系。在回归部分阐述过求取简单线性回归方程的方法为普通最小二乘法（Ordinary Lease Squares，OLS）或线性最小二乘，即通过最小化残差平方和，对回归系数求微分并另微分结果为0，解方程组得回归系数值，构建简单线性回归方程，或模型、估计器。如果模型预测的响应变量都接近观测值，那么模型就是拟合的，用残差平方和来衡量模型拟合性的方法称为残差平方和（RSS）代价函数。代价函数也被称为损失函数，用于定义和衡量一个模型的误差。其公式同回归部分的残差平方和：$SS_{\text{res}}=\sum_{i=1}^{n}(y_i-\widehat{y_i})^2$或$SS_{\text{res}}=\sum_{i=1}^{n}(y_i-f(x_i))^2$，式中$y_i$为观测值，$\widehat{y_i}$和$f(x_i)$均为预测值（回归值），$f(x_i)$即为所求得的估计器。因此，对残差平方和回归系数求微分的过程就是对损失函数求极小值找到模型参数值的过程，两者只是表述上的不同。

在*Mastering Machine Learning with scikit-learn*中，作者给出了另一种求解简单线性回归系数的方法，计算斜率的公式为：$\beta=\frac{cov(x,y)}{var(x)}=\frac{\sum_{i=1}^{n}(x_i-\overline{x})(y_i-\overline{y})}{n-1}/\frac{\sum_{i=1}^{n}(x_i-\overline{x})^2}{n-1}=\frac{\sum_{i=1}^{n}(x_i-\overline{x})(y_i-\overline{y})}{(x_i-\overline{x})^2}$。其中，$var(x)$为解释变量的方差，$n$为训练数据的总量；$cov(x,y)$为协方差，$x_i$表示训练数据集中第$i$个$x$的值，$\overline{x}$为解释变量的均值，$y_i$为第$i$个$y$的值，$\overline{y}$为响应变量的均值。求得$\beta$之后，再由公式：$\alpha=\overline{y}-\beta\overline{x}$求得$\alpha$截距。计算结果与使用sklearn的LinearRegression（OLS）方法保持一致。

#方差（variance）用来衡量一组值的偏离程度，通常标记为σ^2，即标准差的平方，或方差的平方根即为标准差。当集合中的所有数值都相等时，其方差为0。方差描述了一个随机变量离其期望值的距离，通常期望值为该组值的均值。

协方差（covariance）用来衡量两个变量的联合变化程度。方差则是协方差的一种特殊情况，即变量与自身的协方差。协方差表示两个变量总体的误差，这与只表示一个变量误差的方差不同。如果两个变量的变化趋势一致，即其中一个大于自身的期望值，另一个也大于其自身的期望值，则两个变量之间的协方差为正值；如果两个变量的变化趋势相反，即其中一个大于自身期望值，而另一个小于自身期望值，则两个变量之间的协方差为负值；当变量所有值都趋近于各自的期望值时，协方差趋近于0。两个变量统计独立时，协方差为0。但是协方差为0，不仅包括统计独立一种情况，如果两个变量间没有线性关系，二者之间的协方差为0，其线性无关但不一定是相对独立。

- 训练（数据）集，测试（数据）集和验证（数据）集

通常，在训练模型前将数据集划分为训练数据集，测试数据集，也经常增加有验证数据集。训练数据集用于模型的训练，测试数据集依据衡量标准用来评估模型性能，测试数据集不能包含训练数据集中的数据，否则很难评估算法是否真地从训练数据集中学到了泛化能力，还是只是简单地记住了训练的样本。一个能够很好泛

化的模型可以有效地对新数据进行预测。如果只是记住了训练数据集中解释变量和响应变量之间的关系，则称为过拟合。对于过拟合，通常用正则化的方法加以处理。验证数据集常用来微调超参数的变量，超参数通常由人为调整配置，控制算法如何从训练数据中学习。各自部分划分没有固定的比例，通常训练集占50%～75%，测试集占10%～25%，余下的为验证集。

决定系数的计算结果为0.475165，其分数并没有过0.5，所训练的简单线性回归模型并不能根据解释变量很好地预测响应变量（图3.5-1）。事实上，在现实的世界中，很少用到简单的线性回归模型，数据的复杂性往往需要借助更有利的算法来解决实际问题。但是，简单线性回归的逐步计算方式的阐述，可以对回归模型有个比较清晰的理解。

```
from sklearn.linear_model import LinearRegression
import numpy as np
import matplotlib.pyplot as plt
from sklearn.model_selection import train_test_split
ax1 = pubicHealth_Statistic.plot.scatter(
    x='Per Capita Income', y='Childhood Blood Lead Level Screening',
c='DarkBlue', figsize=(8, 8), label='ground truth')
data_IncomeLead = pubicHealth_Statistic[[
    'Per Capita Income', 'Childhood Blood Lead Level Screening']].dropna()
# sklearn 求解模型需要移除数据集中的空值
X = data_IncomeLead['Per Capita Income'].to_numpy(
).reshape(-1, 1)  # 将特征值数据格转换为 numpy 的特征向量矩阵
# 将目标值数据格式转换为 numpy 格式的向量
y = data_IncomeLead['Childhood Blood Lead Level Screening'].to_numpy()
X_train, X_test, y_train, y_test = train_test_split(
    X, y, test_size=0.33, random_state=42)

# 构建与拟合模型
LR = LinearRegression().fit(X_train, y_train)
# 模型参数
print("_"*50)
print("sklearn slop:%.6f,intercept(b):%.6f" % (LR.coef_, LR.intercept_))
ax1.plot(X_train, LR.predict(X_train), 'o-',
         label='linear regression', color='r', markersize=4)
ax1.set(xlabel='Per Capita Income',
        ylabel='Childhood Blood Lead Level Screening')

# 逐步计算回归系数
# 用 nu.cov() 求方程及协方差，注意返回值，如果是对两个变量求协方差，返回值为各自变量的方差及两个变量的协方差的矩阵
income_variance = np.cov(X_train.T)
income_lead_covariance = np.cov(X_train.T, y_train)
print("income_variance=%.6f,income_lead_covariance=%.6f" %
```

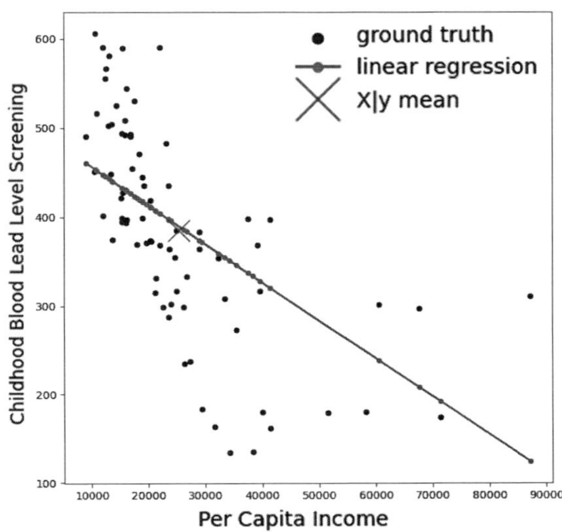

图3.5-1　由人均收入预测儿童血铅水平检查结果的简单线性回归模型

```
        (income_variance, income_lead_covariance[0, 1]))
beta = income_lead_covariance[0, 1]/income_variance
alpha = y_train.mean()-beta*X_train.T.mean()
print("beta=%.6f,alpha=%.6f" % (beta, alpha))

ax1.plot(X.T.mean(), y.mean(), 'x', label='X|y mean',
         color='green', markersize=20)
ax1.legend(loc='upper right', frameon=False)
plt.show()

# 简单线性回归方程 - 回归显著性检验（回归系数检验）
print("决定系数, coefficient of determination, r_squared=%.6f" %
      LR.score(X_test, y_test))
-------------------------------------------------
sklearn slop:-0.004310,intercept(b):499.029224
income_variance=266830859.633061,income_lead_covariance=-1150133.371184
beta=-0.004310,alpha=499.029224
决定系数, coefficient of determination, r_squared=0.475165
```

3.5.2 公共健康数据的K-近邻模型（k-nearest neighbors algorithm，k-NN）

3.5.2.1 k-NN

k-NN是一种用于回归任务和分类任务的简单模型。在2、3维空间中，可以将解释变量与响应变量映射到可以观察的空间中，这个能够定义数据集中所有成员之间距离的特征空间就是度量空间。但是，如果解释变量数量比较多，就很难映射变量到可以观察的2、3维空间中，而是更高的空间维度。所有变量的成员都可以在这个度量空间中表示。k-NN的原理就是找到邻近的样本，在分类任务中是看这些样本对应占据多数的响应变量的类别；而回归任务中，则是计算邻近样本响应变量的均值（图3.5–2）。在k-NN中有个需要人为控制的参数k，即超参k是用于指定近邻的个数，不同的k值训练出模型的决定系数不同。在下述代码中定义了一个k值的区间，通过循环k值，计算比较决定系数找到决定系数最大时的k值。计算结果为当k=5时，决定系数r_squared=0.675250。用k-NN算法求得的回归模型较之简单线性回归模型的预测能力有很大改善（图3.5–3），表明儿童血铅水平检查（Childhood Blood Lead Level Screening）变量的方差很大比例上可以被模型解释。

k-NN中成员之间的距离（或理解为两点之间或多点之间的距离），通常使用欧式（欧几里得）距离（Euclidean Distance），对于n维空间中两个点$x_1(x_{11}, x_{12}, \ldots, x_{1n})$和$x_2(x_{21}, x_{22}, \ldots, x_{2n})$间的欧式距离公式为：$d(x,y) := \sqrt{(x_1 - y_1)^2 + (x_2 - y_2)^2 + \ldots + (x_n - y_n)^2} = \sqrt{\sum_{i=1}^{n}(x_i - y_i)^2}$，二维空间中的欧式距离公式即为直角三角形的勾股定理：$\rho = \sqrt{(x_2 - x_1)^2 + (y_2 - y_1)^2}$。

同时，sklearn提供了权重（weight）参数，这对于一些受城市空间地理位置影响的变量分析

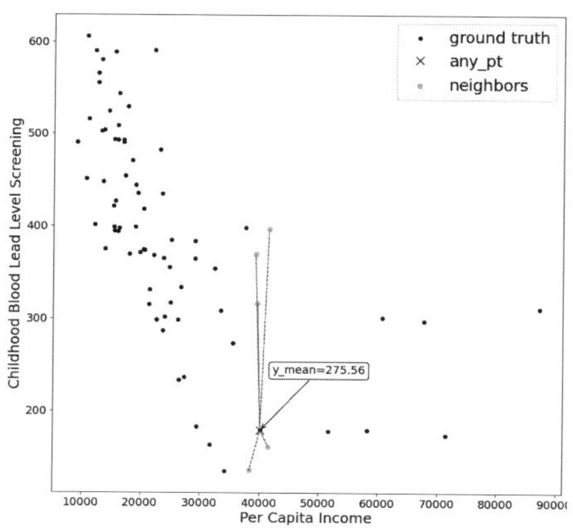

图3.5–2 k-NN邻域点和成员之间的距离

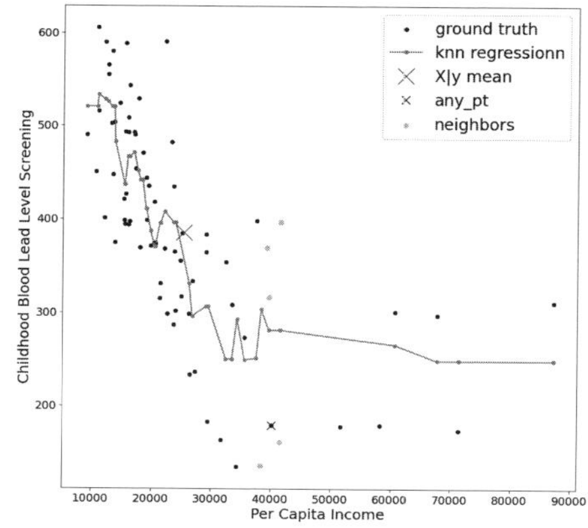

图3.5–3 由人均收入预测儿童血铅水平检查结果的k-NN模型

尤其有用，而PySAL[①]库在空间权重部分有大量可以直接应用的方法。在权重参数配置上，包含三个可选参数，分别为unifom，每个邻域内所有点的权重是相等的；distance，权重为距离的倒数，即越接近查询点的邻居点权重越高，越远的则越低；callable，用户定义权重。对于本次人均收入与儿童血铅水平检查之间的模型建立，因为并未涉及空间地理位置的影响分析，因此参数配置为默认，即uniform模式。

① PySAL，支持空间数据科学计算的开源项目（https://pysal.org/）。

```
# 使用 skearn 库训练 k-NN
from matplotlib.text import OffsetFrom
from pointpats import PointPattern
from sklearn.neighbors import KNeighborsRegressor
import math
fig, axs = plt.subplots(1, 2, figsize=(25, 11))
k_neighbors = range(1, 15, 1)
r_squared_temp = 0
for k in k_neighbors:
    knn = KNeighborsRegressor(n_neighbors=k)
    knn.fit(X_train, y_train)
    if r_squared_temp < knn.score(X_test, y_test):  # knn- 回归显著性检验（回归系数检验）
        r_squared_temp = knn.score(X_test, y_test)
        k_temp = k
knn = KNeighborsRegressor(n_neighbors=k_temp).fit(X_train, y_train)
print("在区间 %s，最大的 r_squared=%.6f，对应的 k=%d" %
      (k_neighbors, knn.score(X_test, y_test), k_temp))
pubicHealth_Statistic.plot.scatter(
    x='Per Capita Income', y='Childhood Blood Lead Level Screening',
c='DarkBlue', label='ground truth', ax=axs[0])
X_train_sort = np.sort(X_train, axis=0)
axs[0].plot(X_train_sort, knn.predict(X_train_sort), 'o-',
            label='knn regressionn', color='r', markersize=4)
axs[0].set(xlabel='Per Capita Income',
           ylabel='Childhood Blood Lead Level Screening')
axs[0].plot(X.T.mean(), y.mean(), 'x', label='X|y mean',
            color='green', markersize=20)

# 关于 K- 近邻
Xy = data_IncomeLead.to_numpy()
# A - 自定义返回 K- 近邻点索引的函数

def k_neighbors_entire(xy, k=3):
    '''
    function - 返回指定邻近数目的最近点坐标

    Params:
        xy - 点坐标二维数组，例如
            array([[23714. ,   364.7],
                   [21375. ,   331.4],
                   [32355. ,   353.7],
                   [35503. ,   273.3]])
        k - 指定邻近数目；int

    return:
        neighbors - 返回各个点索引，及各个点所有指定数目邻近点索引；list(tuple)
    '''
    import numpy as np
    neighbors = [(s, np.sqrt(np.sum((xy-xy[s])**2, axis=1)).argsort()[1:k+1])
                 for s in range(xy.shape[0])]
    return neighbors
neighbors = k_neighbors_entire(Xy, k=5)
any_pt_idx = 70
any_pt = np.take(Xy, neighbors[any_pt_idx][0], axis=0)
neighbor_pts = np.take(Xy, neighbors[any_pt_idx][1], axis=0)

axs[0].plot(any_pt[0], any_pt[1], 'x', label='any_pt',
            color='black', markersize=10)
axs[0].scatter(neighbor_pts[:, 0], neighbor_pts[:, 1],
               c='orange', label='neighbors')

# B - 用 PySAL 库下的 pointpats 求得近邻点
pubicHealth_Statistic.plot.scatter(
    x='Per Capita Income', y='Childhood Blood Lead Level Screening',
c='DarkBlue', label='ground truth', ax=axs[1])
Xy = data_IncomeLead.to_numpy()
```

```
pp = PointPattern(Xy)
pp_neighbor = pp.knn(5)
any_pt = np.take(Xy, any_pt_idx, axis=0)
neighbor_pts_idx = np.take(pp_neighbor[0], any_pt_idx, axis=0)
neighbor_pts = np.take(Xy, neighbor_pts_idx, axis=0)
axs[1].plot(any_pt[0], any_pt[1], 'x', label='any_pt',
            color='black', markersize=10)
axs[1].scatter(neighbor_pts[:, 0], neighbor_pts[:, 1],
               c='orange', label='neighbors')
for coodi in neighbor_pts:
    axs[1].arrow(any_pt[0], any_pt[1], coodi[0]-any_pt[0], coodi[1]-any_pt[1],
head_width=1,
                 head_length=1, color="gray", linestyle="--", length_includes_
head=True)
axs[1].annotate("y_mean=%s" % neighbor_pts[:, 1].mean(), xy=any_pt,
xycoords="data", xytext=(
    50000, 250), va="top", ha="center", bbox=dict(boxstyle="round", fc="w"), arr
owprops=dict(arrowstyle="->"))
axs[0].legend(loc='upper right', frameon=False)
axs[1].legend(loc='upper right', frameon=False)
plt.show()
--------------------------------------------------------------------------------
在区间 range(1, 15), 最大的 r_squared=0.675250, 对应的 k=5
```

3.5.2.2 平均绝对误差（mean absolute error，MAE）和均方误差（mean squared error，MSE）

平均绝对误差（MAE），又被称为L_1范数损失（L1_Loss），是预测结果误差绝对值的均值，一种回归损失函数，其公式为：$MAE(y,\widehat{y})=\frac{1}{n_{\text{samples}}}\sum_{i=0}^{n_{\text{samples}}-1}|y_i-\widehat{y_i}|$。式中，$\widehat{y_i}$为预测值，$y_i$为观测值，范围为$[0,\infty]$。$L_1$范数损失函数MAE，虽然能够较好的衡量回归模型的好坏，但是绝对值的存在导致函数不光滑。虽然对于什么样的输入值，都有稳定的梯度，不会导致梯度爆炸问题，具有较为稳健型的解，但是在中心点是折点，不能求导，不方便求解。

均方误差（MSE），又称均方偏差，比起MAE来说是一种更为常用的指标，也称为L_2范数损失（L2_Loss），是最为常用的损失函数，即为在回归部分阐述的残差平方和（SS_{res}）再除以样本容量，公式为：$MSE(y,\widehat{y})=\frac{1}{n_{\text{samples}}}\sum_{i=0}^{n_{\text{samples}}-1}(y_i-\widehat{y_i})^2$。$L_2$范数损失MSE，各点连续光滑，方便求导，具有较为稳定的解。但是，不是特别稳健。当函数的输入值距离中心值较远时，使用梯度下降法求解时梯度很大，导致在神经网络训练过程中网络权重的大幅度更新，使得网络变得不稳定。极端情况下，权值可能溢出，即梯度爆炸。

从下述的图形中，通过打印MAE和MSE的函数图3.5-4，更容易观察到函数的变化。回归部分对残差平方和的a、b求微分，并令其结果等于0，求得回归方程。扩展到sklearn机器学习部分，则是对损失函数求解，这样能够更容易理解什么是损失函数。注意下述代码建立的简单线性回归方程为：$y=ax$，并没有加截距b，以便简化运算。

范数（norm），是具有“长度”概念的函数。在线性代数、泛函分析及相关的数学领域，是一个函数，为向量空间内的所有向量赋予非零的正长度或大小。例如，一个二维度的欧式几何空间R^2就有欧式范数。在这个向量空间的元素(x_i,y_i)，常常在笛卡尔坐标系统被画成一个从原点出发的箭号，每一个向量的欧式范数就是箭号的长度。或者，理解为向量空间（即度量空间）中的向量都是有大小，度量这个大小的方式就是用范数来度量，不同的范数都可以来度量这个大小，其范数的定义为：向量的范数是一个函数$||x||$，满足非负性$||x||>=0$，齐次性$||cx||=|c|||x||$，三角不等式$||x+y||<=||x||+||y||$。

在阐述 k-NN 时，谈到度量空间，对于 2、3 维的向量可以画出几何图形，但是对于多维、高维，超出三维空间时，则引入范数的概念，从而可以定义任意维度两个向量的距离。常用向量的范数有：

1. L_1范数：$||x||$为x向量各个元素绝对值之和，公式为：$||x||_1=\sum_i|x_i|=|x_1|+|x_2|+\ldots+|x_i|$
2. L_2范数：$||x||$为x向量各个元素平方和的 1/2 次方，公式为：$||x||_2=\sqrt{(\sum_i x_i^2)}=\sqrt{x_1^2+x_2^2+\ldots+x_i^2}$
3. L_p范数：$||x||$为x向量各个元素绝对值p次方和的$1/p$次方，公式为：$||x||_\text{p}=(\sum_i|x_i|^\text{p})^{1/\text{p}}$
4. L^∞范数：$||x||$为x向量各个元素绝对值最大那个元素的绝对值。

```
import pandas as pd
import sympy
from sklearn.metrics import mean_absolute_error, mean_squared_error, r2_score
print("coefficient of determination=%s" %
      r2_score(y_test, knn.predict(X_test)))
```

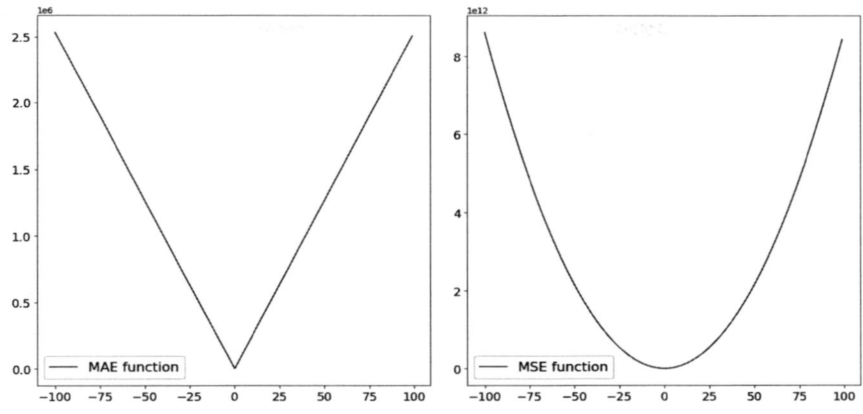

图3.5-4 MAE（左）和MSE（右）函数图形

```
print("mean absolute error=%s" %
      mean_absolute_error(y_test, knn.predict(X_test)))
print("mean squared error=%s" %
      mean_squared_error(y_test, knn.predict(X_test)))

# 打印 MAE 函数图形
fig, axs = plt.subplots(1, 2, figsize=(18, 8))
a = sympy.symbols('a')
data_IncomeLead_copy = data_IncomeLead.copy(deep=True)
data_IncomeLead_copy.rename(columns={
                                "Per Capita Income": "PCI", "Childhood Blood Lead
Level Screening": "CBLLS"}, inplace=True)
data_IncomeLead_copy["residual"] = data_IncomeLead_copy.apply(
    lambda row: row.CBLLS-(a*row.PCI), axis=1)
data_IncomeLead_copy["abs_residual"] = data_IncomeLead_copy.residual.apply(
    lambda row: abs(row))
n_s = data_IncomeLead_copy.shape[0]
MAE = data_IncomeLead_copy.abs_residual.sum()/n_s
MAE_ = sympy.lambdify([a], MAE, "numpy")
a_val = np.arange(-100, 100, 1)  # 假设的 a 值

axs[0].plot(a_val, MAE_(a_val), '-', label='MAE function')

# 打印 MSE 函数图形
data_IncomeLead_copy["residual_squared"] = data_IncomeLead_copy.residual.apply(
    lambda row: row**2)
MSE = data_IncomeLead_copy.residual_squared.sum()/n_s
MSE_ = sympy.lambdify([a], MSE, "numpy")
axs[1].plot(a_val, MSE_(a_val), '-', label='MSE function')
axs[0].legend(loc='upper right', frameon=False)
axs[1].legend(loc='upper right', frameon=False)
plt.show()
--------------------------------------------------------------------------------
coefficient of determination=0.675249990123995
mean absolute error=59.75384615384613
mean squared error=4912.952276923076
```

3.5.2.3 特征值的比例缩放（标准化）

人均收入的特征值范围在［9016, 87163］区间，单位为美元；儿童血铅水平检查特征值范围在［133.6, 605.9］区间，单位为每1,000名儿童（0～6岁）。两个特征向量单位不同，取值范围相差也很大，必然会影响模型训练，如果将其范围缩放到相同的取值范围下，学习算法会比较好地运行。特征值的缩放可以使用sklearn提供的`sklearn.preprocessing.StandardScaler`方法，其计算公式为：$z=(x-\mu)/s$，其中μ为训练样本的均值，s为训练样本的标准差。

计算结果显示，通过特征值的标准化后其决定系数为0.72，大于特征值处理前的值0.68，模型预测能力得以再次提升。

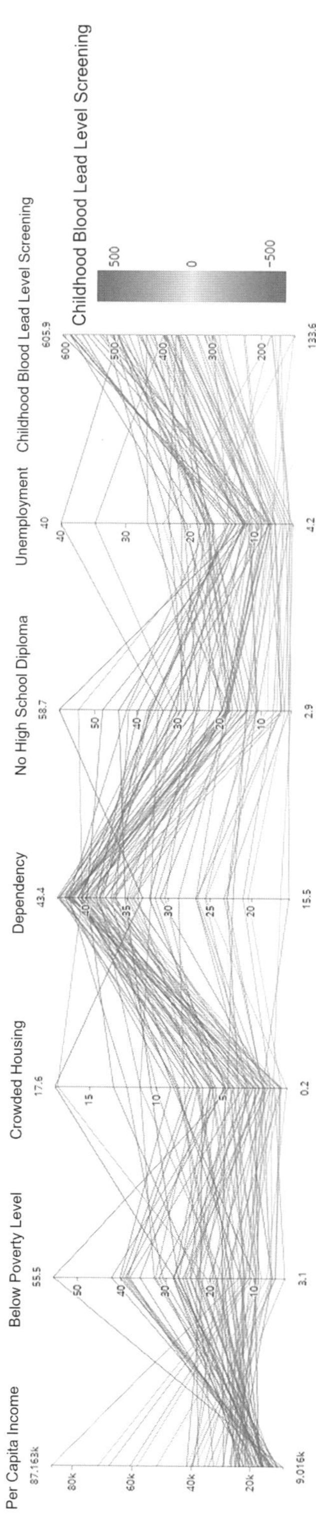

图3.5-5　平行坐标图

```
from sklearn.metrics import mean_absolute_
error, mean_squared_error, r2_score
from numpy.polynomial import polyutils as pu
from sklearn.preprocessing import
StandardScaler
SS = StandardScaler()
X_train_scaled = SS.fit_transform(X_train)
X_test_scaled = SS.fit_transform(X_test)
k_neighbors = range(1, 15, 1)
r_squared_temp = 0
for k in k_neighbors:
    knn_ = KNeighborsRegressor(n_neighbors=k)
    knn_.fit(X_train_scaled, y_train)
    if r_squared_temp < knn_.score(X_test_
scaled, y_test):  #knn-回归显著性检验(回归系数检验)
        r_squared_temp = knn_.score(X_test_
scaled, y_test)
        k_temp = k
knn_ = KNeighborsRegressor(n_neighbors=k_
temp).fit(X_train_scaled, y_train)
print("coefficient of determination=%s" %
      r2_score(y_test, knn_.predict(X_test_
scaled)))
print("mean absolute error=%s" %
      mean_absolute_error(y_test, knn_.
predict(X_test_scaled)))
print("mean squared error=%s" % mean_squared_
error(
    y_test, knn_.predict(X_test_scaled)))
Per_Capita_Income_domain = pu.getdomain(X.
reshape(-1))
Childhood_Blood_Lead_Level_Screening_domain
= pu.getdomain(y)
print("Per Capita Income domain=%s" % Per_
Capita_Income_domain)
print("Childhood Blood Lead Level Screening
domain=%s" %
      Childhood_Blood_Lead_Level_Screening_
domain)
----------------------------------------
coefficient of determinati
on=0.7150373630356484
mean absolute error=56.32587412587413
mean squared error=4311.032466624287
Per Capita Income domain=[ 9016. 87163.]
Childhood Blood Lead Level Screening
domain=[133.6 605.9]
```

3.5.3 公共健康数据的多元回归

3.5.3.1 多元线性回归

超过3个的多个解释变w量（多维或高维数组），仍然有很多方法打印图表观察数据的关联情况，例如使用平行坐标图（parallel coordinates plot），响应变量为Childhood Blood Lead Level Screening，其他的为解释变量。通过折线的变化趋势，能够初步判断各个解释变量之间，以及与响应变量之间是正相关还是负相关，而折线的分布密度则表明各个变量的数值分布情况（图3.5-5）。

```
import numpy as np
import pandas as pd
import matplotlib.pyplot as plt
from sklearn.model_selection import train_
test_split
from sklearn.preprocessing import
StandardScaler
```

```
from sklearn.linear_model import LinearRegression
import plotly.express as px
columns = ['Per Capita Income', 'Below Poverty Level', 'Crowded Housing', 'Dependency',
           'No High School Diploma', 'Unemployment', 'Childhood Blood Lead Level
Screening']
data_Income = pubicHealth_Statistic[columns].dropna()  # sklearn 求解模型需要移除数据集
中的空值

fig = px.parallel_coordinates(data_Income, labels=columns, color_continuous_
scale=px.colors.diverging.Tealrose,
                              color_continuous_midpoint=2, color='Childhood
Blood Lead Level Screening')
fig.show()
```

由“Per Capita Income”一个特征训练回归模型的决定系数为0.48，由多个特征，此次为6个解释变量，获得的决定系数为0.76，较之单一特征有大幅度的提升，增加的解释变量提升了模型的性能。同时，比较k-NN模型。当k=4时，6个特征作为解释变量，其决定系数为0.82。而一个特征时为0.72，模型的性能也得以大幅度提升，表明增加的解释变量对模型的预测是有贡献的。

```
from sklearn.neighbors import KNeighborsRegressor
X = data_Income[['Per Capita Income', 'Below Poverty Level', 'Crowded Housing',
'Dependency',
                 'No High School Diploma', 'Unemployment']].to_numpy()  #将特征值
数据格转换为 numpy 的特征向量矩阵
#将目标值数据格式转换为 numpy 格式的向量
y = data_Income['Childhood Blood Lead Level Screening'].to_numpy()
SS = StandardScaler()
X_train, X_test, y_train, y_test = train_test_split(
    X, y, test_size=0.33, random_state=42)
X_train_scaled = SS.fit_transform(X_train)
X_test_scaled = SS.fit_transform(X_test)
LR_m = LinearRegression()
LR_m.fit(X_train_scaled, y_train)
print("Linear Regression - Accuracy on test data: {:.2f}".format(
    LR_m.score(X_test_scaled, y_test)))
y_pred = LR_m.predict(X_train_scaled)

#使用 k-NN
k_neighbors = range(1, 15, 1)
r_squared_temp = 0
for k in k_neighbors:
    knn = KNeighborsRegressor(n_neighbors=k)
    knn.fit(X_train_scaled, y_train)
    if r_squared_temp < knn.score(X_test_scaled, y_test):  #knn- 回归显著性检验（回归
系数检验）
        r_squared_temp = knn.score(X_test_scaled, y_test)
        k_temp = k
knn = KNeighborsRegressor(n_neighbors=k_temp).fit(X_train_scaled, y_train)
print("k-NN 在区间 %s，最大的 r_squared=%.6f，对应的 k=%d" %
      (k_neighbors, knn.score(X_test_scaled, y_test), k_temp))
--------------------------------------------------------------------------------
Linear Regression - Accuracy on test data: 0.76
k-NN 在区间 range(1, 15)，最大的 r_squared=0.815978，对应的 k=4
```

3.5.3.2 多项式回归

真实的世界，很多因果关系并不是线性的。对于非线性的解释变量和响应变量之间的关系建模，可以有多种途径。这里，用多项式回归的方法，该方法在微分部分用于拟合模型曲线，得到的结果较之线性模型有很大提升。首先，打印响应变量“Childhood Blood Lead Level Screening”与其他所有经济条件解释变量的散点图3.5–6，观察发现与“Per Capita Income”特征趋于线性，而与其他的特征似乎呈现一定的弧度。

```
import plotly.express as px
fig=px.scatter_matrix(data_Income)
fig.update_layout(
    autosize=True,
    width=1800,
    height=1800,
    )
fig.show()
```

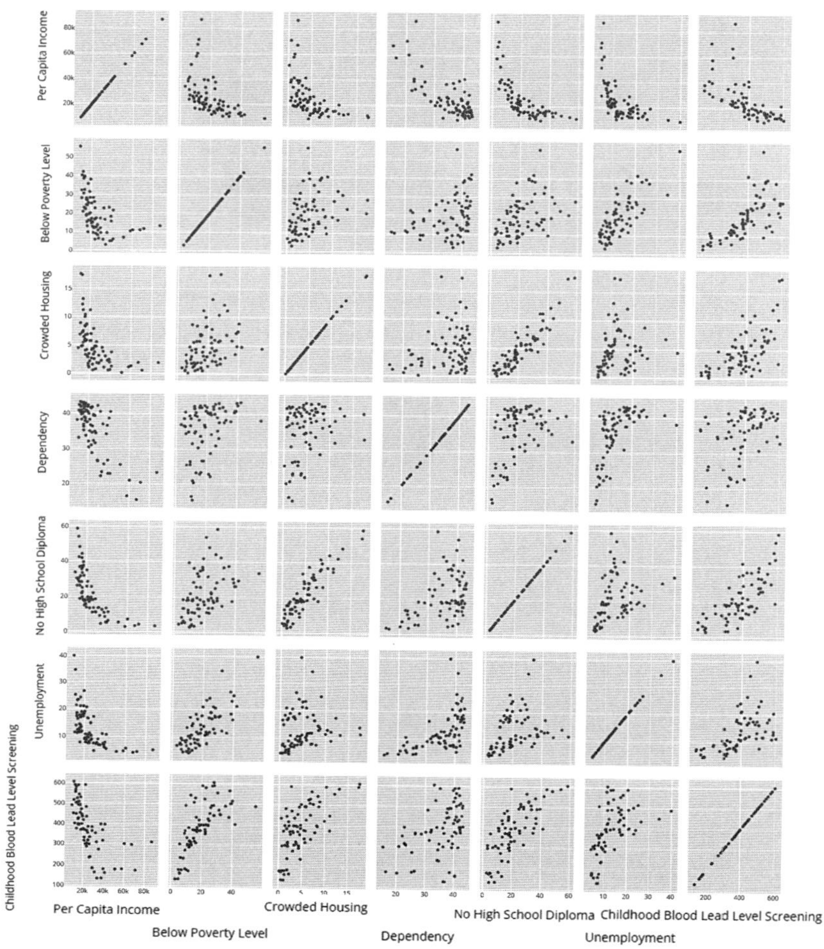

图3.5-6 公共健康数据变量两两散点图

为了方便观察拟合曲线，只选择Below Poverty Level一个解释变量，响应变量依旧为Childhood Blood Lead Level Screening。定义PolynomialFeatures_regularization()函数，用于多项式回归degree次数选择对回归结果影响的比较和讨论。包含对不同正则化方法的调用，由参数regularization配置，包括“linear”“Ridge”和“LASSO”三种方式，分别代表不采用正则化，采取L2正则化（又称岭回归，Ridge Regression）和L1正则化（又称LASSO）。本小节只对不加正则化的多项式回归方法进行讨论，即配置regularization='linear'，下一节则讨论两种正则化方法对回归结果的影响。PolynomialFeatures方法的传入参数degree配置多项式特征的次数，默认为2。同时，循环不同的degree值，绘制对应的拟合曲线（图3.5-7），并通过比较决定系数的值获得值最高时的degree值。计算结果为当degree=4时，决定系数值0.81为最高。

```
X = data_Income['Below Poverty Level'].to_numpy(
).reshape(-1, 1)  #将特征值数据格转换为 numpy 的特征向量矩阵
#将目标值数据格式转换为 numpy 格式的向量
y = data_Income['Childhood Blood Lead Level Screening'].to_numpy()
def PolynomialFeatures_regularization(X, y, regularization='linear'):
    '''
    function - 多项式回归 degree 次数选择，及正则化

    Params:
        X - 解释变量；array
```

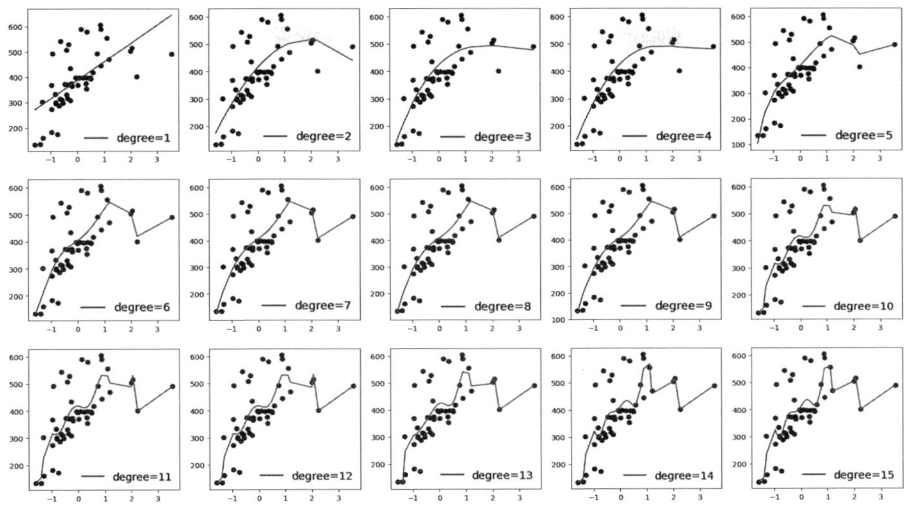

图3.5-7 多项式回归不同degree次数拟合曲线

```
        y - 响应变量; array
        regularization - 正则化方法，为'linear'时，不进行正则化，正则化方法为'Ridge'
和'LASSO'; string

    Returns:
        reg - model
    '''
    from sklearn.preprocessing import PolynomialFeatures
    from sklearn.linear_model import LinearRegression
    from sklearn.pipeline import Pipeline
    from sklearn.preprocessing import StandardScaler
    from sklearn.linear_model import Ridge
    from sklearn.linear_model import Lasso
    X_train, X_test, y_train, y_test = train_test_split(
        X, y, test_size=0.33, random_state=42)
    SS = StandardScaler()
    X_train_scaled = SS.fit_transform(X_train)
    X_test_scaled = SS.fit_transform(X_test)
    degrees = np.arange(1, 16, 1)
    fig_row = 3
    fig_col = degrees.shape[0]//fig_row
    fig, axs = plt.subplots(fig_row, fig_col, figsize=(21, 12))
    r_squared_temp = 0
    p = [(r, c) for r in range(fig_row) for c in range(fig_col)]
    i = 0
    for d in degrees:
        if regularization == 'linear':
            model = Pipeline([('poly', PolynomialFeatures(degree=d)),
                              ('regular', LinearRegression(fit_
intercept=False))])
        elif regularization == 'Ridge':
            model = Pipeline([('poly', PolynomialFeatures(degree=d)),
                              ('regular', Ridge())])
        elif regularization == 'LASSO':
            model = Pipeline([('poly', PolynomialFeatures(degree=d)),
                              ('regular', Lasso())])
        reg = model.fit(X_train_scaled, y_train)
        x_ = X_train_scaled.reshape(-1)
        print("训练数据集的-r_squared=%.6f，测试数据集的-r_squared=%.6f，对应的
degree=%d" %
              (reg.score(X_train_scaled, y_train), reg.score(X_test_scaled, y_
test), d))
        print("系数：", reg['regular'].coef_)
        print("_"*50)
        X_train_scaled_sort = np.sort(X_train_scaled, axis=0)
        axs[p[i][0]][p[i][1]].scatter(
```

```
            X_train_scaled.reshape(-1), y_train, c='black')
        axs[p[i][0]][p[i][1]].plot(X_train_scaled_sort.reshape(-1),
                                   reg.predict(X_train_scaled_sort),
label='degree=%s' % d)
        axs[p[i][0]][p[i][1]].legend(loc='lower right', frameon=False)

        if r_squared_temp < reg.score(X_test_scaled, y_test): #knn-回归显著性检验(回
归系数检验)
            r_squared_temp = reg.score(X_test_scaled, y_test)
            d_temp = d
        i += 1
    plt.show()
    model = Pipeline([('poly', PolynomialFeatures(degree=d_temp)),
                      ('linear', LinearRegression(fit_intercept=False))])
    reg = model.fit(X_train_scaled, y_train)
    print("_"*50)
    print("在区间%s,最大的r_squared=%.6f,对应的degree=%d" %
          (degrees, reg.score(X_test_scaled, y_test), d_temp))
    return reg
reg = PolynomialFeatures_regularization(X, y, regularization='linear')
```

```
训练数据集的-r_squared=0.383777,测试数据集的-r_squared=0.649537,对应的degree=1
系数: [387.116        72.44793532]
________________________________________________________________________________
...
________________________________________________________________________________
训练数据集的-r_squared=0.604060,测试数据集的-r_squared=-3.340680,对应的degree=15
系数: [ 4.11556610e+02  1.61553839e+02  8.85250923e+01 -1.38796631e+03
 -1.12016163e+03  5.09282046e+03  2.32199849e+03 -7.12596446e+03
 -1.28518030e+03  4.64340360e+03 -2.16517056e+02 -1.40585554e+03
  3.29106662e+02  1.41310366e+02 -6.37668285e+01  6.73200011e+00]
在区间[ 1  2  3  4  5  6  7  8  9 10 11 12 13 14 15],最大的r_squared=0.812928,对
应的degree=4
```

打印多项式回归模型（图3.5–8）。

```
reg
```

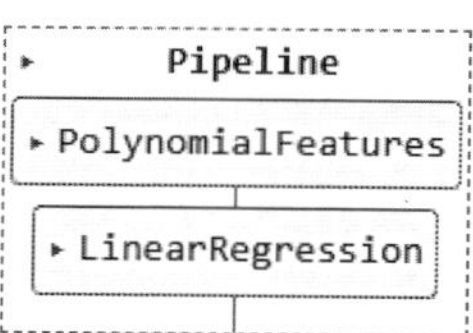

图3.5–8　多项式回归模型

多个特征作为解释变量输入时，多项式回归决定系数最高为0.76，degree为1，即为线性回归模型。说明对于该数据集而言，当下述6个变量作为解释变量时，线性回归模型的预测精度要高于多项式回归。

```
from sklearn.pipeline import Pipeline
from sklearn.preprocessing import PolynomialFeatures
X_ = data_Income[['Per Capita Income', 'Below Poverty Level', 'Crowded Housing',
'Dependency',
                  'No High School Diploma', 'Unemployment',]].to_numpy() #将特征
值数据格转换为numpy的特征向量矩阵
#将目标值数据格式转换为numpy格式的向量
y_ = data_Income['Childhood Blood Lead Level Screening'].to_numpy()
X_train, X_test, y_train, y_test = train_test_split(
    X_, y_, test_size=0.33, random_state=42)
SS = StandardScaler()
X_train_scaled = SS.fit_transform(X_train)
X_test_scaled = SS.fit_transform(X_test)
degrees = np.arange(1, 16, 1)
r_squared_temp = 0
for d in degrees:
    model = Pipeline([('poly', PolynomialFeatures(degree=d)),
```

```
                                ('linear', LinearRegression(fit_intercept=False))])
        reg = model.fit(X_train_scaled, y_train)
        x_ = X_train_scaled.reshape(-1)

        if r_squared_temp < reg.score(X_test_scaled, y_test):  # knn- 回归显著性检验（回归
系数检验）
            r_squared_temp = reg.score(X_test_scaled, y_test)
            d_temp = d

model = Pipeline([('poly', PolynomialFeatures(degree=d_temp)),
                  ('linear', LinearRegression(fit_intercept=False))])
reg = model.fit(X_train_scaled, y_train)
print("在区间 %s, 最大的 r_squared=%.6f, 对应的 degree=%d" %
      (degrees, reg.score(X_test_scaled, y_test), d_temp))
--------------------------------------------------------------------------
在区间 [ 1  2  3  4  5  6  7  8  9 10 11 12 13 14 15], 最大的 r_squared=0.764838, 对
应的 degree=1
```

3.5.3.3 正则化

多项式回归中，计算决定系数时，在训练数据集和测试数据集上同时进行。由其结果观察到，当degree即变量次数（阶数）增加时，训练数据集上的决定系数是增加的。但是，测试数据集上所获得的决定系数并不与训练数据集的变化保持一致。这个问题称为过拟合。因为求得的单个参数数值可能非常大，当模型面对全新数据时就会产生很大波动，是模型含有巨大方差误差的问题。这往往是由于样本的特征很多，样本的容量却很少，模型就容易陷入过拟合。本次实验的特征数为6，样本容量为76，样本容量相对较少。解决过拟合的方法，一种是减少特征数量，或增加样本容量；另一种是正则化。

正则化是一个能用于防止过拟合的技巧集合，sklearn提供了Ridge回归方法，即L_2正则化，和LASSO回归，即L_1正则化的方法。对于多元线性回归模型：$y = a_0 + a_1x_1 + a_2x_2 + \ldots + a_\mathrm{n}x_\mathrm{n}$，其一个样本$X^{(i)}$的预测值为：$\widehat{y}^{(i)} = a_0 + a_1X_1^{(i)} + a_2X_2^{(i)} + \ldots + a_\mathrm{n}X_\mathrm{n}^{(i)}$，模型最终要求解参数$A = (a_0, a_1, \ldots, a_\mathrm{n})^T$，使得均方误差MSE尽可能小。但是通过上述实验，从所获取的系数中观察到，有些系数比较大，对于测试数据集预测值可能将会有很大波动，因此为了模型的泛化能力，对参数a加以限制，岭回归通过增加系数的L_2范数来修改RSS损失函数，其公式为：$RSS_\text{ridge} = \sum_{i=1}^{n}(y_i - x_i^T a_i)^2 + \lambda\sum_{i=1}^{p} a_i^2$。注意，前文的手动实现里，$a_0$是被放到最后一项，且值为1，即$a_0 = 1$，这样是为了和$y = ax + b$的形式呼应。

通过岭回归正则化的方式，打印计算结果及图3.5-9，可以发现训练数据集上的决定系数已经不具有明显随degree参数增加而增加的趋势，相对比较稳定。从打印的图形中也可以看到，拟合

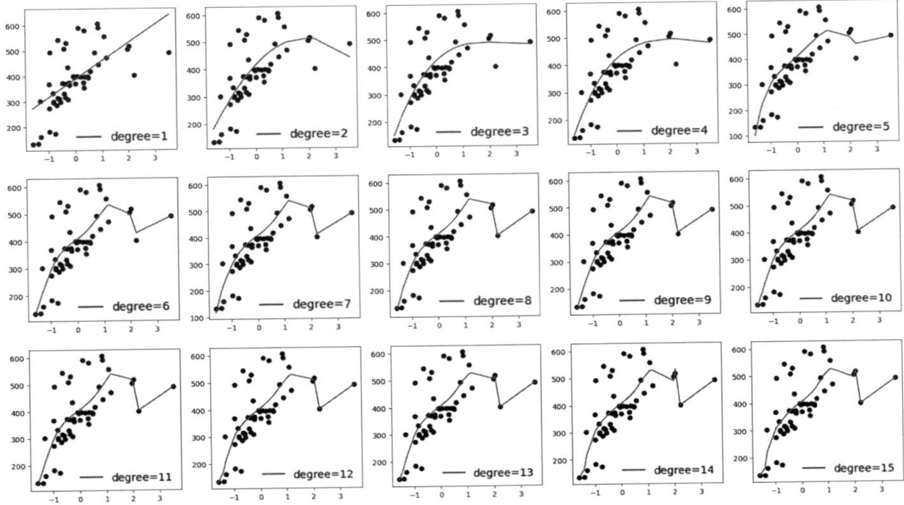

图3.5-9　岭回归正则化不同degree次数拟合曲线

曲线较之未作正则化之前平滑。最高的决定系数基本没有变化，为0.812620，但是degree的值由4变为3。

同时，可以修正Ridge的参数alpha正则化的强度，来进一步优化。值越大，正则化越强。当alpha参数足够大，损失函数中起作用的几乎为正则项，曲线会成为与x轴平行的直线。其默认值为1.0。

```
reg=PolynomialFeatures_regularization(X,y,regularization='Ridge')
--------------------------------------------------------------------------------
训练数据集的-r_squared=0.383629,测试数据集的-r_squared=0.643434,对应的degree=1
系数：[ 0.          71.02738756]
--------------------------------------------------------------------------------
…
--------------------------------------------------------------------------------
训练数据集的-r_squared=0.569994,测试数据集的-r_squared=0.725188,对应的degree=15
系数：[  0.          72.49854347  -0.11330674  50.1999039   10.14026572
   5.68532456   1.07771445 -22.12227329  -5.86763851 -14.14479811
   1.71433656  18.83966612  -5.26612263  -3.85954692   1.86774214
  -0.21511299]
在区间[ 1  2  3  4  5  6  7  8  9 10 11 12 13 14 15],最大的r_squared=0.812620,对应的degree=3
```

LASSO（Least Absolute Shrinkage and Selection Operator Regression）算法则通过对损失函数增加L_1范数来惩罚系数（图3.5-10），公式为：$RSS_{\text{lasso}}=\sum_{i=1}^{n}(y_i-x_i^T a_i)^2+\lambda\sum_{i=1}^{p}|a_i|$，LASSO的特性使得部分$a$变为0，可以作为特征选择用，系数为0的特征说明该解释变量对模型精度的提升几乎不起作用。与Ridge在计算速度比较上，对于特征比较多的数据集，可以考虑用LASSO；否则，使用Ridge，因为Ridge相对准确些。

```
from warnings import filterwarnings
filterwarnings('ignore')
reg_lasso = PolynomialFeatures_regularization(X, y, regularization='LASSO')
--------------------------------------------------------------------------------
训练数据集的-r_squared=0.383704,测试数据集的-r_squared=0.645268,对应的degree=1
系数：[ 0.          71.44793532]
--------------------------------------------------------------------------------
...
--------------------------------------------------------------------------------
训练数据集的-r_squared=0.555024,测试数据集的-r_squared=0.788065,对应的degree=15
系数：[ 0.00000000e+00  9.66281005e+01  5.39448929e+00  8.89912644e+00
 -1.60563919e+01  1.27744308e+00 -5.78371064e-02  6.53875403e-02
  6.34304681e-03  2.41707918e-03  3.70114582e-04  8.65078199e-05
```

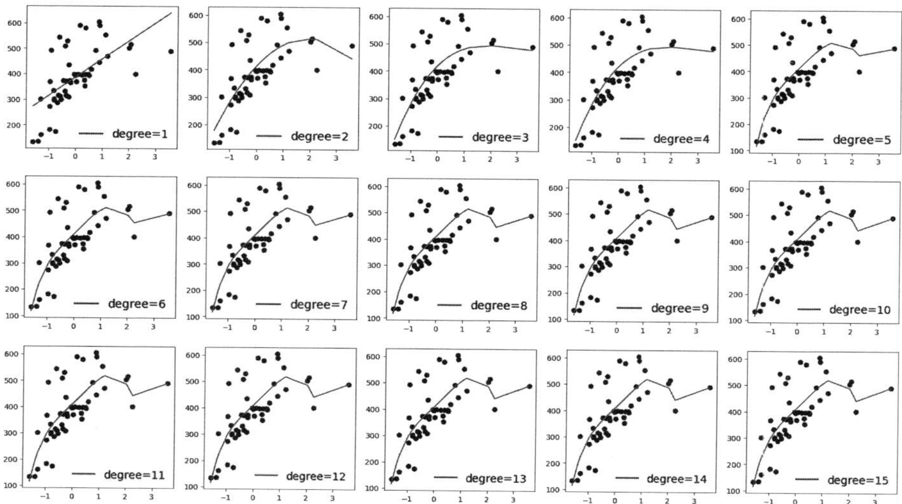

图3.5-10　LASSO回归不同degree次数拟合曲线

```
  1.53329695e-05  3.09976035e-06  5.69537369e-07  1.08937732e-07]
------------------------------------------------------------------------
在区间[ 1  2  3  4  5  6  7  8  9 10 11 12 13 14 15]，最大的 r_squared=0.812928，对应的 degree=4。
```

3.5.4 梯度下降法（Gradient Descent）

在回归部分使用$\widehat{A}=(X'X)^{-1}X'Y$矩阵计算的方法求解模型参数值（即回归系数），其中矩阵求逆的计算较复杂，同时有些情况则无法求逆，因此引入另一种估计模型参数最优值的方法——梯度下降法。当下，对于梯度下降法的解释文献繁多，主要包括通过形象的图式进行说明，使用公式推导过程等。仅有图式的说明只能大概的理解方法的过程，却对真正计算的细节无从理解；对于纯粹的公式推导，没有实例，只能空凭象形，无法落实。Udacity在线课程有一篇文章*Gradient Descent - Problem of Hiking Down a Mountain*对梯度下降法的解释包括了图示、推导及实例，解释得透彻、明白。因此，以该文章为主导，来一步步地理解梯度下降法，这对于机器学习及深度学习有很大的帮助。

对于梯度下降法最为通用的描述是下到山谷处，找到最低点，即损失函数的最小值；在下山的过程中，每一步试图找到该步通往下山路最陡路径，找到给定点的梯度，梯度的方向就是函数变化最快的方向；朝梯度反向，就能让函数值下降最快。然后，不断反复这个过程，最终到达山谷。快速、准确地达到山谷，需要衡量每一步在寻找最陡方向时测量的距离。如果步子过大，可能错过最终山谷点；如果步子过小，则增加计算的时长，并可能陷入局部最低点。因此，每一步的跨度与当前地形的陡峭程度成比例。如果很陡，就迈大步；如不很平缓，就走小步。梯度下降法是估计函数局部最小值的优化算法。

3.5.4.1 梯度（与微分和导数）

函数图像中，对切线斜率的理解就是导数，导数是函数图像在某点处的斜率，即纵坐标增量（$\triangle y$）和横坐标增量（$\triangle x$），在$\triangle x\mapsto 0$时的比值。其一般定义为：设有定义域和取值都在实数域中的函数$y=f(x)$，若$f(x)$在点x_0的某个邻域内有定义，则当自变量x在x_0处取得增量$\triangle x$（点$x_0+\triangle x$仍在该邻域内）时，相应的y取得增量$\triangle y=f(x_0+\triangle x)-f(x_0)$，如果$\triangle x\mapsto 0$时，$\triangle y$与$\triangle x$之比的极限存在，则称函数$y=f(x)$在点$x_0$处可导，并称这个极限为函数$y=f(x)$在点$x_0$处的导数，记为$f(x_0)$，即：$f(x_0)=\lim_{\triangle x\to 0}\frac{\triangle y}{\triangle x}=\lim_{\triangle x\to 0}\frac{f(x_0+\triangle x)-f(x_0)}{\triangle x}$，也可记作$y(x_0)$，$\frac{\mathrm{d}y}{\mathrm{d}x}|_{x=x_0}$，$\frac{\mathrm{d}y}{\mathrm{d}x}(x_0)$，$\frac{\mathrm{d}f}{\mathrm{d}x}|_{x=x_0}$等。

微分是对函数的局部变化率的一种线性描述。可以近似地描述当函数自变量的取值足够小的改变时，函数的值是怎样变化的。其定义为：设函数$y=f(x)$在某个了邻域内有定义，对于邻域内一点x_0，当x_0变动到附近的$x_0+\triangle x$（也在邻域内）时，如果函数的增量$\triangle y=f(x_0+\triangle x)-f(x_0)$可表示为$\triangle y=A\triangle x+o(\triangle x)$。其中，$A$是不依赖于$\triangle x$的常数，$o(\triangle x)$是比$\triangle x$高阶的无穷小，那么称函数$f(x)$在点$x_0$是可微的，且$A\triangle x$称作函数在点$x_0$相应于自变量增量$\triangle x$的微分，记作$\mathrm{d}y$，即$\mathrm{d}y=A\triangle x$，$\mathrm{d}y$是$\triangle y$的线性主部。通常，把自变量$x$的增量$\triangle x$称为自变量的微分，记作$\mathrm{d}x$，即$\mathrm{d}x=\triangle x$。

微分和导数是两个不同的概念。但是，对于一元函数来说，可微与可导是完全等价的。可微的函数，其微分等于导数乘以自变量的微分$\mathrm{d}x$，即函数的微分与自变量的微分之上商等于该函数的导数，因此导数也叫微商。函数$y=f(x)$的微分又可记作$\mathrm{d}y=f(x)\mathrm{d}x$。

梯度实际上是多元函数导数（或微分）的推广，用θ作为函数$J(\theta_1,\theta_2,\theta_3)$的变量，其$J(\Theta)=0.55-(5\theta_1+2\theta_2-12\theta_3)$，则$\triangle J(\Theta)=\langle\frac{\partial J}{\partial\theta_1},\frac{\partial J}{\partial\theta_2},\frac{\partial J}{\partial\theta_3}\rangle=\langle -5,-2,12\rangle$。其中，$\triangle$作为梯度的一种符号，$\partial$符号用于表示偏微分（部分的意思），即梯度就是分别对每个变量求偏微分；同时，梯度用$\langle\rangle$括起，表示梯度为一个向量。

梯度是微积分一个重要的概念。在单变量的函数中，梯度就是函数的微分（或对函数求导），代表函数在某个给定点切线的斜率；在多变量函数中，梯度是一个方向（向量的方向）。梯度的方向指出了函数给定点上升最快的方向，而反方向是函数在给定点下降最快的方向。

#导数是对含有一个自变量函数(一元)求导；偏导数是对含有多个自变量函数(多元)中的一个自变量求导。偏微分与偏导数类似，是对含有多个自变量函数（多元）中的一个自变量微分。

3.5.4.2 梯度下降算法数学解释

公式为：$\Theta^1=\Theta^0-\alpha\nabla J(\Theta)$。其中，$J$是关于 Θ的函数，当前所处的位置为Θ^0，要从这个点走到$J(\Theta)$的最小值点Θ^1，方向为梯度的反向，并用α（学习率或者步长，Learning rate or step size）超参数，控制每一步的距离，避免步子过大错过最低点。当越趋近于最低点，学习率的控制越重要。梯度是上升最快的方向，如果往下走，则需要在梯度前加负号。

- **单变量函数的梯度下降**

定义函数为：$J(x)=x^2$，对该函数x微分，结果为：$J'=2x$，因此梯度下降公式为：$x_{\text{next}}=x_{\text{current}}-\alpha*2x$。同时，给定了迭代次数（iteration），通过打印图3.5-11可以方便地查看每一迭代梯度下降的幅度。

```
import sympy
import matplotlib.pyplot as plt
import numpy as np

# 定义单变量的函数，并绘制曲线
x = sympy.symbols('x')
J = 1*x**2
J_ = sympy.lambdify(x, J, "numpy")
fig, axs = plt.subplots(1, 2, figsize=(25, 12), dpi=300)
x_val = np.arange(-1.2, 1.3, 0.1)
axs[0].plot(x_val, J_(x_val), label='J function')
axs[1].plot(x_val, J_(x_val), label='J function')

# 函数微分
dy = sympy.diff(J)
dy_ = sympy.lambdify(x, dy, "math")

# 初始化
x_0 = 1  # 初始化起点
a = 0.1  # 配置学习率
iteration = 15  # 初始化迭代次数

axs[0].scatter(x_0, J_(x_0), label='starting point')
axs[1].scatter(x_0, J_(x_0), label='starting point')

# 根据梯度下降公式迭代计算
for i in range(iteration):
    if i == 0:
        x_next = x_0-a*dy_(x_0)
    x_next = x_next-a*dy_(x_next)
    axs[0].scatter(x_next, J_(x_next), label='epoch=%d' % i)

# 调整学习率，比较梯度下降速度
a_ = 0.2
for i in range(iteration):
    if i == 0:
```

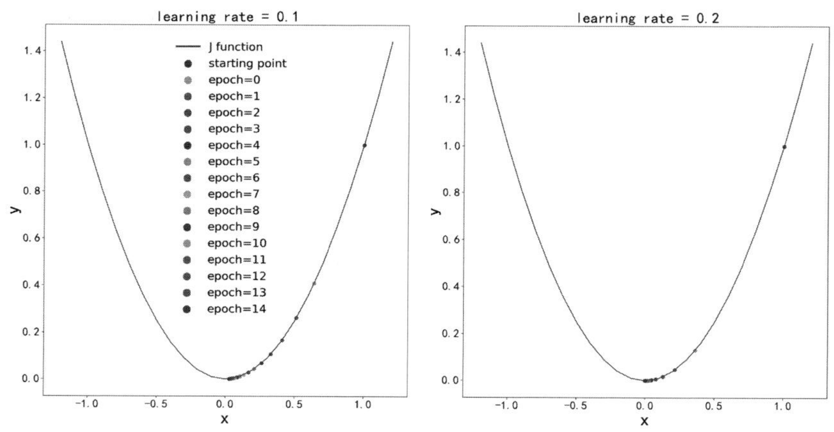

图3.5-11 不同学习率单变量函数梯度下降幅度

```
        x_next = x_0-a_*dy_(x_0)
    x_next = x_next-a_*dy_(x_next)
    axs[1].scatter(x_next, J_(x_next), label='epoch=%d' % i)
axs[0].set(xlabel='x', ylabel='y')
axs[1].set(xlabel='x', ylabel='y')
axs[0].legend(loc='lower right', frameon=False)
axs[1].legend(loc='lower right', frameon=False)
axs[0].set_title("learning rate = 0.1")
axs[1].set_title("learning rate = 0.2")
plt.show()
```

上一部分代码是给定了回归系数的函数，其系数为1，通过梯度下降算法来查看梯度下降的趋势变化。需要注意的是，上述定义的$J(x)=x^2$函数，是代表损失函数［可以写成$J(\beta)=\beta^2$，β是回归系数］，不是模型（例如回归方程）。是由最初对损失函数求微分，并令结果为0，求回归系数，调整为在损失函数曲线上先求某点微分，找到该点的下降方向和大小（向量），并乘以学习率（调整下降速度）；然后，根据该向量移到下一个点，以此类推，直至找到下降趋势（梯度变化）趋近于0的位置。这个位置就是所要求的模型系数。

根据上述表述完成下述代码，即先定义模型，这个模型是回归模型，为了和上述损失函数的模型区别开来，定义其为：$y=\omega x^2$，并假设$\omega=3$来建立数据集，解释变量X和响应变量y。定义模型的函数为`model_quadraticLinear(w,x)`，使用SymPy库辅助表达和计算。有了模型之后，就可以根据模型和数据集计算损失函数，这里用MSE作为损失函数，计算结果为：

$MSE=1.75435200000001(0.333333333333333w-1)^2+0.431208(0.333333333333333w-1)^2$。

然后，定义梯度下降函数，就是对MSE的ω求微分，加入学习率后计算结果为：$G=0.0728520000000003w-0.218556000000001$。准备好了这三个函数，就可以开始训练模型，顺序依次为：

1. 定义函数；
2. 定义损失函数；
3. 定义梯度下降；
4. 指定$\omega=5$初始值，用损失函数计算残差平方和即MSE；
5. 比较MSE是否满足预设的精度'accuracy=1e-5'，如果不满足则开始循环；
6. 由梯度下降公式计算下一个趋近于0的点，并再计算MSE，并比较MSE是否满足要求，周而复始；
7. 直至'L<accuracy'，达到要求，跳出循环，此时'w_next'即为模型系数ω的值。

计算结果为'w=3.008625'，约为3，正是最初用于生成数据集所假设的值。

```
#定义数据集，服从函数 y= 3*x**2，方便比较计算结果
import sympy
from sympy import pprint
import numpy as np
x = sympy.symbols('x')
y = 3*x**2
y_ = sympy.lambdify(x, y, "numpy")
X = np.arange(-1.2, 1.3, 0.1)
y = y_(X)
n_size = X.shape[0]

#初始化
a = 0.15  #配置学习率
accuracy = 1e-5  #给出精度
w_ = 5  #随机初始化系数

#定义模型
def model_quadraticLinear(w, x):
    '''定义一元二次方程，不含截距 b'''
    return w*x**2

#定义损失函数
def Loss_MSE(model, X, y):
    '''用均方误差（MSE）作为损失函数'''
    model_ = sympy.lambdify(x, model, "numpy")
    loss = (model_(X)-y)**2
    return loss.sum()/n_size/2
```

```
#定义梯度下降函数，是对损失函数求梯度
def gradientDescent(loss, a, w):
    '''定义梯度下降函数，即对模型变量微分'''
    return a*sympy.diff(loss, w)

#训练模型
def train(X, y, a, w_, accuracy):
    '''根据精度值，训练模型'''
    x, w = sympy.symbols(['x', 'w'])
    model = model_quadraticLinear(w, x)
    print("定义函数：")
    pprint(model)
    loss = Loss_MSE(model, X, y)
    print("定义损失函数：")
    pprint(loss)
    grad = gradientDescent(loss, a, w)
    print("定义梯度下降：")
    pprint(grad)
    print("_"*50)
    grad_ = sympy.lambdify(w, grad, "math")
    w_next = w_-grad_(w_)
    loss_ = sympy.lambdify(w, loss, "math")
    L = loss_(w_next)
    i = 0
    print("迭代梯度下降，直至由损失函数计算的结果小于预设的值，w即为权重值（回归方程的系数）")
    while not L < accuracy:
        w_next = w_next-grad_(w_next)
        L = loss_(w_next)
        if i % 10 == 0:
            print("iteration:%d,Loss=%.6f,w=%.6f" % (i, L, w_next))
        i += 1
        # if i%100:break
    return w_next
w_next = train(X, y, a, w_, accuracy)
--------------------------------------------------------------------------------
定义函数：
w·x²
定义损失函数：
1.75435200000001·(0.333333333333333·w - 1)²  + 0.431208·(0.333333333333333·w - 1)²

定义梯度下降：
0.0728520000000003·w - 0.218556000000001
__________________________________________________
迭代梯度下降，直至由损失函数计算的结果小于预设的值，w即为权重值（回归方程的系数）
iteration:0,Loss=0.717755,w=4.719207
iteration:10,Loss=0.158109,w=3.806898
iteration:20,Loss=0.034829,w=3.378712
iteration:30,Loss=0.007672,w=3.177746
iteration:40,Loss=0.001690,w=3.083424
iteration:50,Loss=0.000372,w=3.039154
iteration:60,Loss=0.000082,w=3.018377
iteration:70,Loss=0.000018,w=3.008625
```

- **多变量函数的梯度下降**

多变量函数的梯度下降与单变量函数的梯度下降类似，只是在求梯度时是分别对各个变量求梯度，两者之间不互相干扰。计算结果如下（图3.5-12）：

```
import sympy
import matplotlib.pyplot as plt
from matplotlib import cm
import numpy as np

#定义单变量的函数，并绘制曲线
x1, x2 = sympy.symbols(['x1', 'x2'])
J = x1**2+x2**2
J_ = sympy.lambdify([x1, x2], J, "numpy")
x1_val = np.arange(-5, 5, 0.1)
x2_val = np.arange(-5, 5, 0.1)
x1_mesh, x2_mesh = np.meshgrid(x1_val, x2_val)
y_mesh = J_(x1_mesh, x2_mesh)
fig, axs = plt.subplots(1, 2, figsize=(25, 12))
axs[0] = fig.add_subplot(1, 2, 1, projection='3d')
surf = axs[0].plot_surface(x1_mesh, x2_mesh, y_mesh,
```

```
                           cmap=cm.coolwarm, linewidth=0, antialiased=False,
alpha=0.5,)
axs[1] = fig.add_subplot(1, 2, 2, projection='3d')
surf = axs[1].plot_surface(x1_mesh, x2_mesh, y_mesh,
                           cmap=cm.coolwarm, linewidth=0, antialiased=False,
alpha=0.5,)
# fig.colorbar(surf, shrink=0.5, aspect=5)

# 函数 x1,x2 微分
dx1 = sympy.diff(J, x1)
dx2 = sympy.diff(J, x2)
dx1_ = sympy.lambdify(x1, dx1, "math")
dx2_ = sympy.lambdify(x2, dx2, "math")

# 初始化
x1_0 = 4  # 初始化 x1 起点
x2_0 = 4  # 初始化 x2 起点
iteration = 15  # 初始化迭代次数
a = 0.1  # 配置学习率

axs[0].scatter(x1_0, x2_0, J_(x1_0, x2_0),
               label='starting point', c='black', s=80)
axs[1].scatter(x1_0, x2_0, J_(x1_0, x2_0),
               label='starting point', c='black', s=80)

# 根据梯度下降公式迭代计算
for i in range(iteration):
    if i == 0:
        x1_next = x1_0-a*dx1_(x1_0)
        x2_next = x2_0-a*dx2_(x2_0)
    x1_next = x1_next-a*dx1_(x1_next)
    x2_next = x2_next-a*dx2_(x2_next)
    axs[0].scatter(x1_next, x2_next, J_(x1_next, x2_next),
                   label='epoch=%d' % i, s=80)

# 调整学习率，比较梯度下降速度
a_ = 0.2
for i in range(iteration):
    if i == 0:
        x1_next = x1_0-a_*dx1_(x1_0)
        x2_next = x2_0-a_*dx2_(x2_0)
    x1_next = x1_next-a_*dx1_(x1_next)
    x2_next = x2_next-a_*dx2_(x2_next)
    axs[1].scatter(x1_next, x2_next, J_(x1_next, x2_next),
                   label='epoch=%d' % i, s=80)
axs[0].set(xlabel='x', ylabel='y', zlabel='z')
axs[1].set(xlabel='x', ylabel='y')
axs[0].legend(loc='lower right', frameon=False)
axs[1].legend(loc='lower right', frameon=False)
```

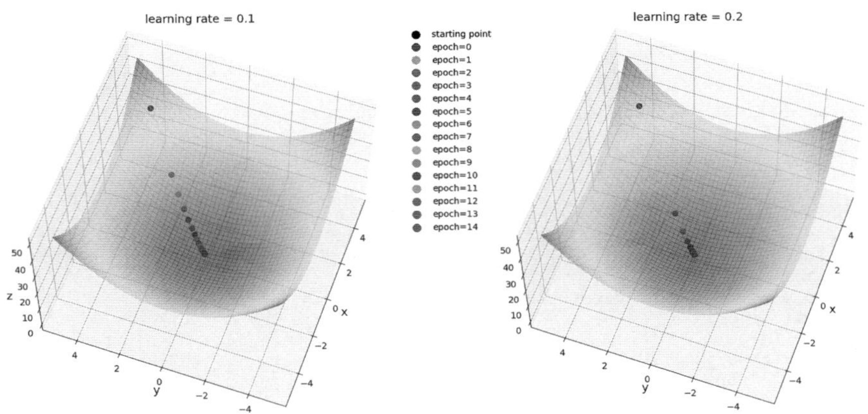

图3.5-12 不同学习率多变量函数梯度下降幅度

```
axs[0].view_init(60, 200)  #可以旋转图形的角度，方便观察
axs[1].view_init(60, 200)
axs[0].set_title("learning rate = 0.1")
axs[1].set_title("learning rate = 0.2")
plt.show()
```

用梯度下降方法求解二元函数模型，其过程基本同上述求解一元二次函数模型。注意在下述求解过程中，学习率的配置对计算结果有较大影响。可以尝试不同的学习率，观察所求回归系数的变化。计算结果在$\alpha = 0.01$的条件下，$w = 3.96$，$v = 4.04$，与假设的值3，5还是有段距离；也可以打印图3.5-13，观察真实平面与训练所得模型的平面之间的差距。可以正则化，即增加惩罚项L_2或L_1尝试改进。不过，这已经能够对梯度下降算法有个比较清晰的理解，这是后续应用sklearn机器学习库和PyTorrch[①]深度学习库非常重要的基础。

同时，也应用了sklearn库提供的SGDRegressor（Stochastic Gradient Descent）随机梯度下降方法训练该数据，其$w = 2.9999586$，$v = 4.99993523$，即约为3和5，与原假设的系数值一样。

① PyTorch，一个基于Torch库的开源机器学习框架，用于计算机视觉和自然语言处理等应用（https://pytorch.org/）。

```
#定义数据集，服从函数 y= 3*x**2，方便比较计算结果
from sklearn.model_selection import train_test_split
from sklearn.linear_model import SGDRegressor
import sympy
from sympy import pprint
import numpy as np
x1, x2 = sympy.symbols(['x1', 'x2'])
y = 3*x1+5*x2
y_ = sympy.lambdify([x1, x2], y, "numpy")
X1_val = np.arange(-5, 5, 0.1)
X2_val = np.arange(-5, 5, 0.1)
y_val = y_(X1_val, X2_val)
n_size = y_val.shape[0]

#初始化
a = 0.01  #配置学习率
accuracy = 1e-10  #给出精度
w_ = 5  #随机初始化系数，对应 x1 系数
v_ = 5  #随机初始化系数，对应 x2 系数

#定义模型
def model_quadraticLinear(w, v, x1, x2):
    '''定义二元一次方程，不含截距 b'''
    return w*x1+v*x2

#定义损失函数
def Loss_MSE(model, X1, X2, y):
    '''用均方误差（MSE）作为损失函数'''
    model_ = sympy.lambdify([x1, x2], model, "numpy")
    loss = (model_(x1=X1, x2=X2)-y)**2
    return loss.sum()/n_size/2

#定义梯度下降函数，是对损失函数求梯度
def gradientDescent(loss, a, w, v):
    '''定义梯度下降函数，即对模型变量微分'''
    return a*sympy.diff(loss, w), a*sympy.diff(loss, v)

#训练模型
def train(X1_val, X2_val, y_val, a, w_, v_, accuracy):
    '''根据精度值，训练模型'''
    w, v = sympy.symbols(['w', 'v'])
    model = model_quadraticLinear(w, v, x1, x2)
    print("定义函数：")
    pprint(model)
    loss = Loss_MSE(model, X1_val, X2_val, y_val)
    print("定义损失函数：")
    pprint(loss)
    grad_w, grad_v = gradientDescent(loss, a, w, v)
    print("定义梯度下降：")
    pprint(grad_w)
    pprint(grad_v)
    print("_"*50)
    gradw_ = sympy.lambdify([v, w], grad_w, "math")
    gradv_ = sympy.lambdify([v, w], grad_v, "math")
    w_next = w_-gradw_(v_, w_)
    v_next = v_-gradv_(v_, w_)
```

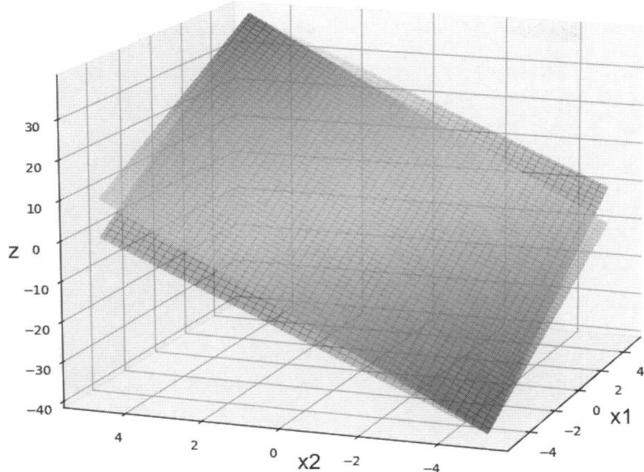

图3.5-13 真实平面与训练所得模型的预测平面

```
    loss_ = sympy.lambdify([v, w], loss, "math")
    L = loss_(w=w_next, v=v_next)
    i = 0
    print("迭代梯度下降，直至由损失函数计算的结果小于预设的值，w,v 即为权重值（回归方程的系数）")
    while not L < accuracy:
        w_next = w_next-gradw_(w=w_next, v=v_next)
        v_next = v_next-gradv_(v=v_next, w=w_next)
        L = loss_(w=w_next, v=v_next)
        if i % 10 == 0:
            print("iteration:%d,Loss=%.6f,w=%.6f,v=%.6f" %
                  (i, L, w_next, v_next))
        i += 1
    return w_next, v_next
w_next, v_next = train(X1_val, X2_val, y_val, a, w_, v_, accuracy)
fx = w_next*x1+v_next*x2
fx_ = sympy.lambdify([x1, x2], fx, "numpy")
fig, ax = plt.subplots(figsize=(10, 10))
ax = fig.add_subplot(projection='3d')
x1_mesh, x2_mesh = np.meshgrid(X1_val, X2_val)
y_mesh = y_(x1_mesh, x2_mesh)
y_pre_mesh = fx_(x1_mesh, x2_mesh)
surf = ax.plot_surface(x1_mesh, x2_mesh, y_mesh, cmap=cm.coolwarm,
                       linewidth=0, antialiased=False, alpha=0.5,)
surf = ax.plot_surface(x1_mesh, x2_mesh, y_pre_mesh,
                       cmap=cm.ocean, linewidth=0, antialiased=False, alpha=0.2,)
# fig.colorbar(surf, shrink=0.5, aspect=5)

# 用 sklearn 库提供的 SGDRegressor（ Stochastic Gradient Descent ）随机梯度下降方法训练
X_ = np.stack((x1_mesh.flatten(), x2_mesh.flatten())).T
y_ = y_mesh.flatten()
X_train, X_test, y_train, y_test = train_test_split(
    X_, y_, test_size=0.33, random_state=42)
SDGreg = SGDRegressor(loss='squared_loss')  # 配置损失函数，其正则化，即惩罚项默认为 L2
SDGreg.fit(X_train, y_train)
print("_"*50)
print("Sklearn SGDRegressor test set r-squared score%s" %
      SDGreg.score(X_test, y_test))
print("Sklearn SGDRegressor coef_:", SDGreg.coef_)
ax.set(xlabel='x1', ylabel='x2', zlabel='z')
ax.view_init(13, 200)  # 可以旋转图形的角度，方便观察
plt.show()
--------------------------------------------------------------------------
```

定义函数：

$v \cdot x_2 + w \cdot x_1$

定义损失函数：

$137.360000000001 \cdot (-0.125 \cdot v - 0.125 \cdot w + 1)^2 + 129.359999999998 \cdot (0.125 \cdot v + 0.125 \cdot w - 1)^2$

定义梯度下降：

```
0.0833499999999994·v + 0.0833499999999994·w - 0.666799999999996
0.0833499999999994·v + 0.0833499999999994·w - 0.666799999999996
--------------------------------------------------
迭代梯度下降，直至由损失函数计算的结果小于预设的值，w,v 即为权重值（回归方程的系数）
iteration:0,Loss=8.172455,w=4.694389,v=4.705967
iteration:10,Loss=0.251475,w=4.091926,v=4.153720
iteration:20,Loss=0.007738,w=3.986244,v=4.056846
iteration:30,Loss=0.000238,w=3.967706,v=4.039853
iteration:40,Loss=0.000007,w=3.964454,v=4.036872
iteration:50,Loss=0.000000,w=3.963883,v=4.036349
iteration:60,Loss=0.000000,w=3.963783,v=4.036258
iteration:70,Loss=0.000000,w=3.963766,v=4.036241
--------------------------------------------------
Sklearn SGDRegressor test set r-squared score0.9999999998894413
Sklearn SGDRegressor coef_: [2.99996858 4.99994742]
```

3.5.4.3 用Sklearn库提供的SGDRegressor（Stochastic Gradient Descent）随机梯度下降方法训练公共健康数据

使用SGDRegressor随机梯度下降训练多元回归模型，参数设置中损失函数配置为'squared_loss'，惩罚项（penalty）默认为L2（具体信息可以从Sklearn官网获取）。计算结果中决定系数为0.76，较多项式回归偏小。计算获取6个系数，对应6个特征。

```
from sklearn.linear_model import SGDRegressor
from sklearn.model_selection import train_test_split
from sklearn.pipeline import make_pipeline
X_ = data_Income[['Per Capita Income', 'Below Poverty Level', 'Crowded Housing', 'Dependency',
                  'No High School Diploma', 'Unemployment',]].to_numpy()  #将特征值数据格转换为 numpy 的特征向量矩阵
#将目标值数据格式转换为 numpy 格式的向量
y_ = data_Income['Childhood Blood Lead Level Screening'].to_numpy()
X_train, X_test, y_train, y_test = train_test_split(
    X_, y_, test_size=0.33, random_state=42)
SDGreg = make_pipeline(StandardScaler(),
                       SGDRegressor(loss='squared_loss',
                                    max_iter=1000,
                                    tol=1e-3))
SDGreg.fit(X_train, y_train)
print("_"*50)
print("Sklearn SGDRegressor test set r-squared score%s" %
      SDGreg.score(X_test, y_test))
print("Sklearn SGDRegressor coef_:", SDGreg[1].coef_)
--------------------------------------------------------------------------------
Sklearn SGDRegressor test set r-squared score0.7622218232194027
Sklearn SGDRegressor coef_: [ 3.09516933 36.26810545 27.85043992  5.83100125
42.48193716 15.37895748]
```

图3.5-14打印查看测试数据集观测值和预测值的分布情况。从结果来看，预测值基本趋向于观测值，结合决定系数为0.76，可以判断训练的模型具有较好的预测表现能力。

```
y_pred=SDGreg.predict(X_test)
x_ax = range(len(y_test))
plt.plot(x_ax, y_test, label="original")
plt.plot(x_ax, y_pred, label="predicted")
plt.title("Childhood Blood Lead Level Screening test and predicted data")
plt.xlabel('X-axis')
plt.ylabel('Y-axis')
plt.legend(loc='best',fancybox=True, shadow=True)
plt.grid(True)
plt.show()
```

#关于损失函数，代价函数 / 成本函数

损失函数（Loss Function），针对单个样本，衡量单个样本的预测值$\widehat{y}^{(i)}$和观测值$y^{(i)}$之间的差距；

代价函数 / 成本函数（Cost Function），针对多个样本，衡量多个样本的预测值$\sum_{i=1}^{n}\widehat{y}^{(i)}$和观测值$\sum_{i=1}^{n}y^{(i)}$之间的差距；

实际上，对于这三者的划分在实际相关文献使用上并没有体现出来，往往混淆，因此也不做特殊界定。

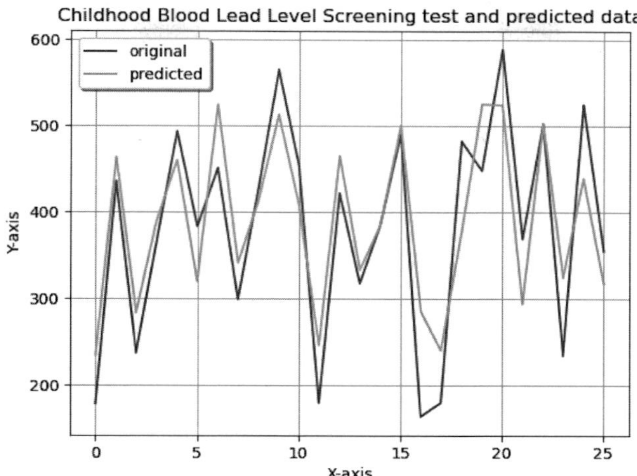

图3.5-14 使用 SGDRegressor训练多元回归模型的预测值与观测值的比较

参考文献（References）：

[1] Cavin Hackeling．张浩然译．scikit-learning机器学习[M]．北京：人民邮电出版社，2019．2．(Gavin Hackeling (2017). Mastering Machine Learning with scikit-learn: Apply effective learning algorithms to real-world problems using scikit-learn. Packt Publishing.)

[2] Gradient Descent - Problem of Hiking Down a Mountain. Udacity (https://jums.club/pdf/Gradient_Descent.pdf).

3-B 编程线性代数及遥感影像数据处理

3.6 编程线性代数

如果想对某一解决问题的方法有比较清晰的理解，数学知识则很难规避，线性代数则是其中之一。回归部分方程式的矩阵求解，以及特征向量与降维（例如，主成分分析，Principal component analysis，PCA），空间变换等内容都需要基础的线性代数知识。因此以《漫画线性代数》结构为主要思路，结合*Gilbert Strang. Introduction to linear algbra*及相关参考文献，借助Python编程语言，Python图表打印可视化，将主要的知识点串联起来，作为用到线性代数知识解决相关问题的基础。主要使用的库为SymPy（Matrices-linear algebra）[①]。

① SymPy，一个用于符号数学的Python库，目的是构建一个全功能的计算机代数系数（computer algebra system，CAS）；同时，保持代码尽可能地简单，便于理解和扩展（https://www.sympy.org/）。

线性代数（linear algebra）是关于向量空间和线性映射的一个数学分支，包括对线、面和子空间的研究。同时，也涉及所有的向量空间的一般性质。笼统地说，是使用向量和矩阵的数学空间表达方式，将m维空间（世界）与n维空间（世界）联系起来的数学分支。实际上，三维软件平台对大量几何对象空间变换的操作就是对线性代数的应用。

3.6.1 矩阵

形如$\begin{bmatrix} a_{11} & a_{12} & \dots & a_{1n} \\ a_{21} & a_{22} & \dots & a_{2n} \\ \vdots & \vdots & \ddots & \vdots \\ a_{m1} & a_{m2} & \dots & a_{mn} \end{bmatrix}$的数据集合即为$m$行$n$列，$m \times n$矩阵，这同于NumPy的数组形式，也可以用Pandas的DataFrame格式表达。在统计学上，就可以通过矩阵方式对样本特征表述，尤其具有多个特征的样本。而m可以称为行标，n则称为列标，组合起来就代表了矩阵中的每个元素。对于形如$\begin{bmatrix} a_{11} & a_{12} & \dots & a_{1n} \\ a_{21} & a_{22} & \dots & a_{2n} \\ \vdots & \vdots & \ddots & \vdots \\ a_{n1} & a_{n2} & \dots & a_{nn} \end{bmatrix}$，即$m = n$的矩阵为$n$阶方阵，对角线上的元素称为对角元素。在相关性分析中，所计算两两相关性的数组结果，就是方阵。其对角元素皆为1，为自身相关性计算结果。

3.6.1.1 矩阵的运算

1. 加（减）法（addition，subtraction ）

$m \times n$矩阵A和B的和（差）：$A \pm B$为一个$m \times n$矩阵，其中每个元素是A和B相应元素的和（差），$(A \pm B)_{i,j} = A_{i,j} \pm B_{i,j}$，例如$\begin{bmatrix} 1 & 3 & 1 \\ 1 & 0 & 0 \end{bmatrix} + \begin{bmatrix} 0 & 0 & 5 \\ 7 & 5 & 0 \end{bmatrix} = \begin{bmatrix} 1+0 & 3+0 & 1+5 \\ 1+7 & 0+5 & 0+0 \end{bmatrix} = \begin{bmatrix} 1 & 3 & 6 \\ 8 & 5 & 0 \end{bmatrix}$

2. 数乘（倍数，multiple）

标量c与矩阵A的数乘，cA的每个元素是A的相应元素与c的乘积，$(cA)_{i,j} = c \cdot A_{i,j}$，例如

$$2 \cdot \begin{bmatrix} 1 & 3 & 1 \\ 1 & 0 & 0 \end{bmatrix} = \begin{bmatrix} 2 \cdot 1 & 2 \cdot 3 & 2 \cdot 1 \\ 2 \cdot 1 & 2 \cdot 0 & 2 \cdot 0 \end{bmatrix} = \begin{bmatrix} 2 & 6 & 2 \\ 2 & 0 & 0 \end{bmatrix}$$

3. 乘法（Multiplication）

两个矩阵的乘法仅当第1个矩阵A的列数（column）和另一个矩阵B的行数（row）相等时，才能定义。如果A是$m \times n$矩阵，B是$n \times p$，它们的乘积AB是一个$m \times p$矩阵，$[AB]_{i,j} = A_{i,1}B_{1,j} + A_{i,2}B_{2,j} + \ldots + A_{i,n}B_{n,j} = \sum_{r=1}^{n} A_{i,r}B_{r,j}$，例如

$$\begin{bmatrix} 1 & 0 & 2 \\ -1 & 3 & 1 \end{bmatrix} \times \begin{bmatrix} 3 & 1 \\ 2 & 1 \\ 1 & 0 \end{bmatrix} = \begin{bmatrix} (1 \times 3 & 0 \times 2 & 2 \times 1) & (1 \times 1 & 0 \times 1 & 2 \times 0) \\ (-1 \times 3 & 3 \times 2 & 1 \times 1) & (-1 \times 1 & 3 \times 1 & 1 \times 0) \end{bmatrix} = \begin{bmatrix} 5 & 1 \\ 4 & 2 \end{bmatrix}$$

4. 幂（exponentiation）

相等于矩阵乘法，例如$\begin{bmatrix}1 & 2\\3 & 4\end{bmatrix}^3=\begin{bmatrix}1 & 2\\3 & 4\end{bmatrix}\begin{bmatrix}1 & 2\\3 & 4\end{bmatrix}\begin{bmatrix}1 & 2\\3 & 4\end{bmatrix}$

$=\begin{bmatrix}1\times1+2\times3 & 1\times2+2\times4\\3\times1+4\times2 & 3\times2+4\times4\end{bmatrix}\begin{bmatrix}1 & 2\\3 & 4\end{bmatrix}$

$=\begin{bmatrix}7 & 10\\15 & 22\end{bmatrix}\begin{bmatrix}1 & 2\\3 & 4\end{bmatrix}=\begin{bmatrix}7\times1+10\times3 & 7\times2+10\times4\\15\times1+22\times3 & 15\times2+22\times4\end{bmatrix}=\begin{bmatrix}37 & 54\\81 & 118\end{bmatrix}$

- **线性方程组**

矩阵乘法的一个基本应用是在线性方程组上，线性方程组是数学方程组的一种，它符合以下的形式：$\begin{cases}a_{1,1}x_1+a_{1,2}x_2+\ldots+a_{1,n}x_n=b_1\\a_{2,1}x_1+a_{2,2}x_2+\ldots+a_{2,n}x_n=b_2\\\vdots\\a_{m,1}x_1+a_{m,2}x_2+\ldots+a_{m,n}x_n=b_m\end{cases}$，式中$a_{1,1}a_{1,2}$及$b_1b_2$等是已知的常数，而$x_1x_2$等则是要求的未知数。如果用线性代数中的概念来表达，则线性方程组可写为：$Ax=b$。其中，A是$m\times n$矩阵，x是含有n个元素列向量，b是含有m个元素列向量。

$$A=\begin{bmatrix}a_{1,1} & a_{1,2} & \ldots & a_{1,n}\\a_{2,1} & a_{2,2} & \ldots & a_{2,n}\\\vdots & \vdots & \ddots & \vdots\\a_{m,1} & a_{m,2} & \ldots & a_{m,n}\end{bmatrix},\ x=\begin{bmatrix}x_1\\x_2\\\vdots\\x_n\end{bmatrix},\ b=\begin{bmatrix}b_1\\b_2\\\vdots\\b_m\end{bmatrix}$$

这是线性方程组的另一种记录方法。已知矩阵A和向量b的情况下，求得未知向量x是线性代数的基本问题。

在多元回归部分，阐述过使用矩阵的方式求解回归模型，对于样本包含多个特征（解释变量）即n列，多个样本（样本容量）即m行，常规表达为$y=\alpha+\beta_1x_1+\beta_2x_2+\ldots+\beta_nx_n$，其中$x_1,x_2,\ldots,x_n$为$n$个特征，每个特征下实际上包含有多个样本实例（$m$行），即$\begin{cases}Y_1=\alpha+\beta_1x_{11}+\beta_2x_{12}+\ldots+\beta_nx_{1n}\\Y_2=\alpha+\beta_1x_{21}+\beta_2x_{22}+\ldots+\beta_nx_{2n}\\\vdots\\Y_n=\alpha+\beta_1x_{m1}+\beta_2x_{m2}+\ldots+\beta_nx_{mn}\end{cases}$，表达为矩阵形式为

$\begin{bmatrix}Y_1\\Y_2\\\vdots\\Y_n\end{bmatrix}=\begin{bmatrix}\alpha+\beta_1x_{11}+\beta_2x_{12}+\ldots+\beta_nx_{1n}\\\alpha+\beta_1x_{21}+\beta_2x_{22}+\ldots+\beta_nx_{2n}\\\vdots\\\alpha+\beta_1x_{m1}+\beta_2x_{m2}+\ldots+\beta_nx_{mn}\end{bmatrix}$，可以进一步简化为，

$\begin{bmatrix}Y_1\\Y_2\\\vdots\\Y_n\end{bmatrix}=\begin{bmatrix}1+x_{11}+x_{12}+\ldots+x_{1n}\\1+x_{21}+x_{22}+\ldots+x_{2n}\\\vdots\\1+x_{m1}+x_{m2}+\ldots+x_{mn}\end{bmatrix}\times\begin{bmatrix}\alpha\\\beta_1\\\beta_2\\\vdots\\\beta_n\end{bmatrix}$，实际上就是应用矩阵的乘法。

```
import sympy
from sympy import Matrix, pprint
A = Matrix([[1, 3, 1], [1, 0, 0]])
B = Matrix([[0, 0, 5], [7, 5, 0]])
print("A 矩阵 =")
pprint(A)
print("B 矩阵 =")
pprint(B)
print("-"*50)

print(" 矩阵加法：A+B=")
pprint(A+B)
print("_"*50)

print(" 数乘：2*A=")
pprint(2*A)
```

```
print("_"*50)
C = Matrix([[1, 0, 2], [-1, 3, 1]])
D = Matrix([[3, 1], [2, 1], [1, 0]])
print("C 矩阵 =")
pprint(C)
print("D 矩阵 =")
pprint(D)
print("-"*50)
print(" 乘法: C*D=")
pprint(C*D)
print("_"*50)
E = Matrix([[1, 2], [3, 4]])
print("E 矩阵 =")
pprint(E)
print("-"*50)
print(" 幂: E**3")
pprint(E**3)
print("_"*50)
```

```
--------------------------------------------------------------------------------
A 矩阵 =

⎡1 3 1⎤
⎢     ⎥
⎣1 0 0⎦
B 矩阵 =

⎡0 0 5⎤
⎢     ⎥
⎣7 5 0⎦
--------------------------------------------------------------------------------
矩阵加法：A+B=

⎡1 3 6⎤
⎢     ⎥
⎣8 5 0⎦
--------------------------------------------------------------------------------
数乘：2*A=

⎡2 6 2⎤
⎢     ⎥
⎣2 0 0⎦
--------------------------------------------------------------------------------
C 矩阵 =

⎡1  0 2⎤
⎢      ⎥
⎣-1 3 1⎦
D 矩阵 =

⎡3 1⎤
⎢   ⎥
⎢2 1⎥
⎢   ⎥
⎣1 0⎦
--------------------------------------------------------------------------------
乘法：C*D=

⎡5 1⎤
⎢   ⎥
⎣4 2⎦
--------------------------------------------------------------------------------
E 矩阵 =

⎡1 2⎤
⎢   ⎥
⎣3 4⎦
--------------------------------------------------------------------------------
幂：E**3

⎡37 54 ⎤
⎢      ⎥
⎣81 118⎦
```

3.6.1.2 特殊矩阵

1．零矩阵

所有元素均为0的矩阵，例如，$\begin{bmatrix} 0 & 0 \\ 0 & 0 \end{bmatrix}$。

2．转置矩阵（transpose）

是指将 $m \times n$ 矩阵 $A = \begin{bmatrix} a_{1,1} & a_{1,2} & \cdots & a_{1,n} \\ a_{2,1} & a_{2,2} & \cdots & a_{2,n} \\ \vdots & \vdots & \ddots & \vdots \\ a_{m,1} & a_{m,2} & \cdots & a_{m,n} \end{bmatrix}$，交换行列后得到的 $n \times m$，$\begin{bmatrix} a_{1,1} & a_{2,1} & \cdots & a_{m,1} \\ a_{1,2} & a_{2,2} & \cdots & a_{m,2} \\ \vdots & \vdots & \ddots & \vdots \\ a_{1,n} & a_{2,n} & \cdots & a_{m,n} \end{bmatrix}$，可以用$A^{\mathrm{T}}$和$A$表示，例如$3 \times 2$的矩阵$\begin{bmatrix} 1 & 2 \\ 3 & 4 \\ 5 & 6 \end{bmatrix}$的转置矩阵为$2 \times 3$矩阵$\begin{bmatrix} 1 & 3 & 5 \\ 2 & 4 & 6 \end{bmatrix}$。

3．对称矩阵（symmetric matrix）

以对角元素为中心线对称的n阶方阵，例如$\begin{bmatrix} 1 & 5 & 6 & 7 \\ 5 & 2 & 8 & 9 \\ 6 & 8 & 3 & 10 \\ 7 & 9 & 10 & 4 \end{bmatrix}$，对称矩阵与其转置矩阵完全相同。

4．(5.)，上三角矩阵和下三角矩阵（triangular matrix）

如$\begin{bmatrix} 1 & 5 & 6 & 7 \\ 0 & 2 & 8 & 9 \\ 0 & 0 & 3 & 10 \\ 0 & 0 & 0 & 4 \end{bmatrix}$，对角元素左下角的所有元素均为0的$n$阶矩阵。

如$\begin{bmatrix} 1 & 0 & 0 & 0 \\ 5 & 2 & 0 & 0 \\ 6 & 8 & 3 & 0 \\ 7 & 9 & 10 & 4 \end{bmatrix}$，对角元素右上角的所有元素均为0的$n$阶矩阵。

6．对角矩阵 (diagonal matrix)

如$\begin{bmatrix} 1 & 0 & 0 & 0 \\ 0 & 2 & 0 & 0 \\ 0 & 0 & 3 & 0 \\ 0 & 0 & 0 & 4 \end{bmatrix}$，对角元素以外的元素均为0的$n$阶矩阵，可表示为$\mathrm{diag}(1,2,3,4)$。

对角矩阵的p次幂，等于对角元素的p次幂，公式为：

$$\begin{bmatrix} a_{1,1} & 0 & 0 & 0 \\ 0 & a_{2,2} & 0 & 0 \\ 0 & 0 & a_{3,3} & 0 \\ 0 & 0 & 0 & a_{n,n} \end{bmatrix}^p = \begin{bmatrix} a_{1,1}^p & 0 & 0 & 0 \\ 0 & a_{2,2}^p & 0 & 0 \\ 0 & 0 & a_{3,3}^p & 0 \\ 0 & 0 & 0 & a_{n,n}^p \end{bmatrix}，例如\begin{bmatrix} 2 & 0 \\ 0 & 3 \end{bmatrix}^2 = \begin{bmatrix} 2^2 & 0 \\ 0 & 3^2 \end{bmatrix} = \begin{bmatrix} 4 & 0 \\ 0 & 9 \end{bmatrix}$$

7．单位矩阵（identity matrix）

如$\begin{bmatrix} 1 & 0 & 0 & 0 \\ 0 & 1 & 0 & 0 \\ 0 & 0 & 1 & 0 \\ 0 & 0 & 0 & 1 \end{bmatrix}$，对角元素均为1，对角元素以外的其他元素全部为0的n阶方阵，即$\mathrm{diag}(1,1,\ldots,1)$。单位矩阵与任何矩阵相乘，都对这个矩阵没有任何影响。

8．逆矩阵（inverse matrix）

又称反矩阵。在线性代数中，给定一个n阶方阵A，若存在n阶方阵B，使得$AB = BA = I_{\mathrm{n}}$，其中I_{n}为n阶单位矩阵，则称A是可逆的，且B是A的逆矩阵，记作A^{-1}。只有方阵$n \times n$才可能有逆矩阵。若方阵A的逆矩阵存在，则称A为非奇异方阵或可逆矩阵。逆矩阵的求法有代数余子式法（不实用）、消元法等，对解法感兴趣的可以参看《漫画线性代数》。在代码的世界里，直接使用SymPy库提供的方法。

并不是所有的方阵都有逆矩阵，可以使用行列式指标determinant(det)，用SymPy库的det()方法计算判断。如果$det(A) \neq 0$，则矩阵A可逆。

```
print("零矩阵：")
pprint(sympy.zeros(2))

print("转置矩阵：")
F = Matrix([[1, 2], [3, 4], [5, 6]])
pprint(F.T)

print("对称矩阵的转置矩阵：")
G = Matrix([[1, 5, 6, 7], [5, 2, 8, 9], [6, 8, 3, 10], [7, 9, 10, 4]])
pprint(G.T)

print("对角矩阵的 2 次幂：")
H = Matrix([[2, 0], [0, 3]])
pprint(H**2)

print("单位矩阵：")
pprint(sympy.eye(4))

print("单位矩阵与任何矩阵相乘，都对这个矩阵没有任何影响：")
pprint(sympy.eye(4)*G)

print("求解逆矩阵，并乘以自身：")
print("使用 -1 次方计算逆矩阵：")
G_inverse = G**-1
pprint(G*G_inverse)
print("使用 .inv() 计算逆矩阵：")
G_inverse_ = G.inv()
pprint(G*G_inverse_)

print("判断方阵是否有逆矩阵：")
print(G.det())
print(G_inverse.det())
---------------------------------------------------------------------------
```

零矩阵：

$$\begin{bmatrix} 0 & 0 \\ 0 & 0 \end{bmatrix}$$

转置矩阵：

$$\begin{bmatrix} 1 & 3 & 5 \\ 2 & 4 & 6 \end{bmatrix}$$

对称矩阵的转置矩阵：

$$\begin{bmatrix} 1 & 5 & 6 & 7 \\ 5 & 2 & 8 & 9 \\ 6 & 8 & 3 & 10 \\ 7 & 9 & 10 & 4 \end{bmatrix}$$

对角矩阵的 2 次幂：

$$\begin{bmatrix} 4 & 0 \\ 0 & 9 \end{bmatrix}$$

单位矩阵：

$$\begin{bmatrix} 1 & 0 & 0 & 0 \\ 0 & 1 & 0 & 0 \\ 0 & 0 & 1 & 0 \\ 0 & 0 & 0 & 1 \end{bmatrix}$$

单位矩阵与任何矩阵相乘，都对这个矩阵没有任何影响：

$$\begin{bmatrix}1 & 5 & 6 & 7\\5 & 2 & 8 & 9\\6 & 8 & 3 & 10\\7 & 9 & 10 & 4\end{bmatrix}$$

求解逆矩阵，并乘以自身：
使用 -1 次方计算逆矩阵：

$$\begin{bmatrix}1 & 0 & 0 & 0\\0 & 1 & 0 & 0\\0 & 0 & 1 & 0\\0 & 0 & 0 & 1\end{bmatrix}$$

使用 .inv() 计算逆矩阵：

$$\begin{bmatrix}1 & 0 & 0 & 0\\0 & 1 & 0 & 0\\0 & 0 & 1 & 0\\0 & 0 & 0 & 1\end{bmatrix}$$

判断方阵是否有逆矩阵：
-3123
-1/3123

3.6.2 向量（euclidean vector）

3.6.2.1 向量概念、表达及运算

一般，同时满足具有大小和方向两个性质的几何对象，即可认为是向量。其代数表示为，在指定了一个坐标系之后，用一个向量在该坐标系下的坐标来表示该向量，兼具符号的抽象性和几何形象性，具有最高的实用性，广泛应用于定量分析的情景。对于自由向量，将向量的起点平移到坐标原点后，向量就可以用一个坐标系下的一个点来表示，该点的坐标值即向量的终点坐标

设一个向量$\vec{a}$，有坐标系S。在S中定义好若干个特殊的基本向量（称为基向量Base Vector，各个基向量共同组成该坐标系下的基底）$\vec{e_1}, \vec{e_2}, \ldots, \vec{e_n}$之后，则向量在各个基向量下的投影即为对应的坐标系，各个投影值组成了该向量在该坐标系S下可唯一表示的有序数组（即坐标），且与向量的终点一一对应。换而言之，其他的向量只须通过将这些基本向量拉伸后，再按照平行四边形法则进行向量加法即可表示（通常被称为“用基底线性表出一个向量”，即该向量是基向量的某种线性组合），即：$\vec{a} = a_1\vec{e_1}, a_2\vec{e_2}, \ldots, a_n\vec{e_n}$，其中$\vec{a_1}, \vec{a_2}, \ldots, \vec{a_n}$是$\vec{a}$分别在$\vec{e_1}, \vec{e_2} \ldots, \vec{e_n}$下对应的投影。当基底已知，可直接省略各基向量的符号，类似于坐标系上的点，直接用坐标表示为$\vec{a} = (a_1, a_2, \ldots, a_n)$。在矩阵运算中，向量更多的被写成类似于矩阵的列向量或行向量。在线性代数中所指的向量，通常默认为列向量。如一个向量$\vec{a} = (a, b, c)$可以写成：$\vec{a} = \begin{bmatrix}a\\b\\c\end{bmatrix}$，$\vec{a} = \begin{bmatrix}a & b & c\end{bmatrix}$，其中第一个为列向量写法，第二个为行向量写法。$n$维列向量可视为$n \times 1$矩阵，$n$维行向量可视为$1 \times n$矩阵。

在常见的三维空间直角坐标系$\bigcirc xyz$（三维笛卡尔坐标系）里，基本向量就是行轴（$\bigcirc x$）、竖轴（$\bigcirc y$）、及纵轴（$\bigcirc z$）为方向的三个长度为1的单位向量$\vec{i}, \vec{j}, \vec{k}$，即基向量。这三个向量取好之后，其他的向量就可以透过三元数组来表示，因为它们可以表示成一定倍数的三个基本向量的总和。比如说一个标示为（2,1,3）的向量就是2个向量$\vec{i}$加上1个向量$\vec{j}$加上3个向量$\vec{k}$得到的向量，即：$(a, b, c) = a\vec{i} + b\vec{j} + c\vec{k}$。因此，向量本质上讲即为矩阵，向量的计算同于矩阵的计算。

SymPy库对向量计算有两种方式：一种是完全使用矩阵，因为二者计算相同，例如和、差、倍数和积；另一种是该库提供了专门针对向量的类：“Vector”，通过Vector提供的方法可以实现向量的表达及向量的基本运算。向量的主要目的是要解决空间几何问题，因此对于向量的理解要

将空间几何图形与表达式结合起来，才能够有效地理解前因后果。在了解基本的向量概念、坐标空间、运算之后，如果要用代码来表达和计算向量，有必要阅读SymPy对应的Vector①及Matrices（linear algebra）②两个部分，那么再阅读下述代码将会相对轻松。

为了方便使用matplotlib打印三维空间下向量，定义了vector_plot_3d函数。首先，定义了名称为'C'的三维坐标系统，获取其基向量'i,j,k'，用基向量构建向量'v1'；同时，通过提取该向量在三个轴向上的系数建立v1向量在各轴上的分量（投影向量），各轴的投影向量之和即为v1向量，打印三维空间上的向量图形（图3.6-1），观察向量间的关系。注意，向量空间起始位置的确定，vector_k的起始点，以vector_i和vector_j之和的向量作为起点，而vector_j以vector_i作为起点。

① Vector，为SymPy库模块，该模块提供了与三维直角坐标系有关的基本矢量数学和微分计算的工具（https://docs.sympy.org/dev/modules/vector/index.html）。

② Matrices（linear algebra），为线性代数模块，并使设计的代码结构尽可能简单（https://docs.sympy.org/latest/modules/matrices/matrices.html）。

- 把$n \times 1$向量$\begin{bmatrix} a_1 \\ a_2 \\ \vdots \\ a_n \end{bmatrix}$，所有分量构成的集合表示为$R^n$。

```
import matplotlib.pyplot as plt
from sympy.vector.coordsysrect import CoordSys3D
from sympy.vector.vector import Vector, BaseVector
from sympy.vector import Vector
def vector_plot_3d(ax_3d, C, origin_vector, vector, color='r', label='vector', arrow_length_ratio=0.1):
    '''
    funciton - 转换SymPy的vector及Matrix数据格式为matplotlib可以打印的数据格式

    Params:
        ax_3d - matplotlib的3d格式子图；ax(matplotlib)
        C - /coordinate_system - SymPy下定义的坐标系；CoordSys3D()
        origin_vector - 如果是固定向量，给定向量的起点（使用向量，即表示从坐标原点所指向的位置）；如果是自由向量，起点设置为坐标原点；vector(SymPy)
        vector - 所要打印的向量；vector(SymPy)
        color - 向量色彩，The default is 'r'；string
        label - 向量标签，The default is 'vector'；string
        arrow_length_ratio - 向量箭头大小，The default is 0.1；float
    Returns:
        None
    '''
    origin_vector_matrix = origin_vector.to_matrix(C)
    x = origin_vector_matrix.row(0)[0]
    y = origin_vector_matrix.row(1)[0]
    z = origin_vector_matrix.row(2)[0]
    vector_matrix = vector.to_matrix(C)
    u = vector_matrix.row(0)[0]
    v = vector_matrix.row(1)[0]
```

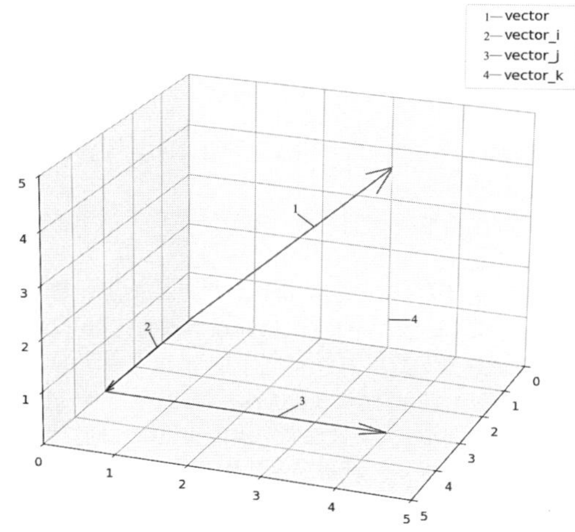

图3.6-1　三维空间上的向量

```
    w = vector_matrix.row(2)[0]
    ax_3d.quiver(x, y, z, u, v, w, color=color, label=label,
                 arrow_length_ratio=arrow_length_ratio)
fig, ax = plt.subplots(figsize=(12, 12))
ax = fig.add_subplot(projection='3d')

#定义坐标系统，及打印向量 v1=3*i+4*j+5*k
C = CoordSys3D('C')
i, j, k = C.base_vectors()
v1 = 3*i+4*j+5*k
v1_origin = Vector.zero
vector_plot_3d(ax, C, v1_origin, v1, color='r',
               label='vector', arrow_length_ratio=0.1)

#打印向量 v1=3*i+4*j+5*k 在轴上的投影
v1_i = v1.coeff(i)*i
vector_plot_3d(ax, C, v1_origin, v1_i, color='b',
               label='vector_i', arrow_length_ratio=0.1)
v1_j = v1.coeff(j)*j
vector_plot_3d(ax, C, v1_i, v1_j, color='g',
               label='vector_j', arrow_length_ratio=0.1)
v1_k = v1.coeff(k)*k
vector_plot_3d(ax, C, v1_i+v1_j, v1_k, color='yellow',
               label='vector_k', arrow_length_ratio=0.1)
ax.set_xlim3d(0, 5)
ax.set_ylim3d(0, 5)
ax.set_zlim3d(0, 5)
ax.legend()
ax.view_init(20, 20)  #可以旋转图形的角度，方便观察
plt.show()
```

3.6.2.2 线性无关

在线性代数里，向量空间的一组元素中，若没有向量，可用有限个其他向量的线性组合所表示，则称为线性无关（linearly independent）或线性独立［例如，（1，0，0），（0，1，0）和（0，0，1）］；反之，称为线性相关（linearly dependent）［例如（2，-1，1），（1，0，1）和（3，-1，2），因为第3个是前两个的和］。即假设V在域K上的向量空间，如果从域K中有非全零的元素$a_1, a_2, \ldots, a_n$，使得$a_1v_1 + a_2v_2 + \ldots + a_nv_n = 0$，或表示为，$\sum_{i=1}^{n} a_iv_i = 0$，其中$v_1, v_2, \ldots, v_n$是$V$的向量，称它们为线性相关，其中右边的0，为$\vec{0}$即0向量（vector），而不是0标量（scalar）；如果K中不存在这样的元素，那么$v_1, v_2, \ldots, v_n$线性无关。对线性无关可以给出更直接的定义，向量$v_1, v_2, \ldots, v_n$线性无关，当且仅当它们满足以下条件：如果$a_1, a_2, \ldots, a_n$是K的元素，适合：$a_1v_1 + a_2v_2 + \ldots + a_nv_n = 0$，那么对所有$i = 1, 2, \ldots, n$都有$a_i = 0$。

在V中的一个无限集，如果它任何一个有限子集都是线性无关，那么原来的无限集也是线性无关。线性相关性是线性代数的重要概念，因为线性无关的一组向量可以生成一个向量空间，而这组向量则是这个向量空间的基。

首先，通过`sympy.vector`的`CoordSys3D`方法建立三维坐标系（可以直接提取单位向量$\vec{i}$、$\vec{j}$、$\vec{k}$），依托该坐标系建立（V）向量集合，包括v2、v3、v4、v5，均由单位向量的倍数建立。如果倍数等于0，例如`v2=a*1*N.i+a*0*N.j`，也保持了0的存在，以保持各个方向上的一致性，便于观察。可以通过打印各个单独向量，查看变量在SymPy中的表现形式。该向量集合是线性相关，可以给出除了$a = b = c = d = 0$外，其他a、b、c、d的值，即有非全零的元素。例如，`a_,b_,c_,d_=1,2,0,-1`，或者`a_,b_,c_,d_=1,-3,-1,2`，均也满足$a_1v_1 + a_2v_2 + \ldots + a_nv_n = 0$，上述的a、b、c、d即为$a_1, a_2, \ldots, a_n$。因此，可以打印图形，通过定义`move_alongVectors`函数实现，可以看到第1，2个图形（图3.6–2左，中），形成闭合平面二维图形（回到起点）。同样，在三维空间中，定义了向量集合v6、v7、v8、v9，因为线性相关，如第3个图形（图3.6–2右），形成空间闭合的折线（回到起点）。

判断一个向量集合V，是否线性无关，可以使用`matrix.rref()`将表示向量的矩阵转换为行阶梯形矩阵（Row Echelon Form）。如果返回的简化行阶梯形矩阵末尾行存在全零值，则表明该向量数据集为线性相关。例如，使用`matrix.to_matrix(C)`方法将v2、v3、v4、v5等向量转换为矩阵表达形式，并通过`matrix.col_insert()`方法合并为矩阵，使用`matrix.`

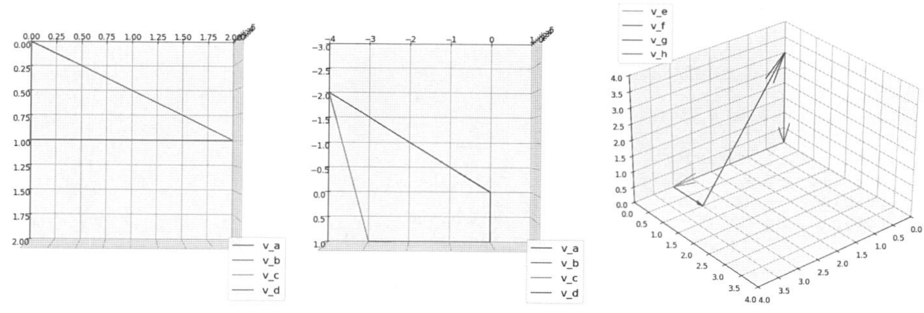

图3.6-2 线性相关向量集合（形成空间闭合折线）

rref()计算。对于数据集v2、v3、v4、v5，其简化行阶梯形矩阵计算结果为$\begin{bmatrix}1&0&0\\0&1&0\\0&0&1\\0&0&0\end{bmatrix}$，末尾包括全零行，因此为线性相关。同时，返回了主元位置的列表为（0，1，2）。而v10、v11组成的向量集合则末尾不含全零行，因此为线性无关，无法通过各个向量构建闭合的空间折线，则可以生成向量空间，这组向量就是这个向量空间的基。对于R_{m}的任意元素向量$\begin{bmatrix}y_1\\y_2\\\vdots\\y_{\mathrm{m}}\end{bmatrix}$，当$\begin{bmatrix}y_1\\y_2\\\vdots\\y_{\mathrm{m}}\end{bmatrix}=c_1\begin{bmatrix}a_{11}\\a_{21}\\\vdots\\a_{\mathrm{m}1}\end{bmatrix}+c_2\begin{bmatrix}a_{12}\\a_{22}\\\vdots\\a_{\mathrm{m}2}\end{bmatrix}+\ldots+c_{\mathrm{n}}\begin{bmatrix}a_{1\mathrm{n}}\\a_{2\mathrm{n}}\\\vdots\\a_{\mathrm{mn}}\end{bmatrix}$的解$c_1,c_2,\ldots,c_{\mathrm{n}}$均为零时，则把集合$\left\{\begin{bmatrix}a_{11}\\a_{21}\\\vdots\\a_{\mathrm{m}1}\end{bmatrix},\begin{bmatrix}a_{12}\\a_{22}\\\vdots\\a_{\mathrm{m}2}\end{bmatrix},\ldots,\begin{bmatrix}a_{1\mathrm{n}}\\a_{2\mathrm{n}}\\\vdots\\a_{\mathrm{mn}}\end{bmatrix}\right\}$叫作基，即为了表示$R_{\mathrm{m}}$任意元素所必需的最少向量构成的集合。

- 阶梯型矩阵

线性代数中，一个矩阵如果符合下列条件，则称为行阶梯型矩阵（Row Echelon Form）：

1. 所有非零行（矩阵的行至少有一个非零元素）在所有全零行的上面，即全零行都在矩阵的底部；
2. 非零行的首项系数（leading coefficient），也称主元，即最左边的首个非零元素，严格地比上面行的首项系数更靠右（有些版本会要求非零行的首项系数必须是1）；
3. 首项系数所在列，在该首项系数下面的元素都是零（前两条的推论）。

例如：$\begin{bmatrix}1&a_1&a_2|&b_1\\0&2&a_3|&b_2\\0&0&1|&b_3\end{bmatrix}$。

增广矩阵，又称广置矩阵，是在线性代数中系数矩阵的右边填上线性方程组等号右边的常数列得到的矩阵。例如，方程$AX=B$的系数矩阵为A，它的增广矩阵为$A|B$。方程组唯一确定增广矩阵，通过增广矩阵的初等行变换可用于判断对应线性方程组是否有解，以及化简原方程组的解。其竖线可以省略。

化简后的行阶梯型矩阵（reduced row echelon form，或译为简约行梯形式），也称作行规范形矩阵（row canonical form）。如果满足额外的条件：每个首项系数是1，且是其所在列的唯一的非零元素，例如，$\begin{bmatrix}1 & 0 & a_1 & 0 & \mid & b_1\\0 & 1 & 0 & 0 & \mid & b_2\\0 & 0 & 0 & 1 & \mid & b_3\end{bmatrix}$，注意化简后的行阶梯型矩阵的左部分（系数部分）不意味着总是单位阵。

通过有限步的行初等变换，任何矩阵可以变换到行阶梯形。由于行初等变换保持了矩阵的行空间，因此行阶梯型矩阵的行空间与变换前的原矩阵的行空间相同（这也是在使用`sympy.rref()`计算化简后的行阶梯型矩阵（reduced row echelon form），需要将形状为$\begin{bmatrix}a_1i & a_2i & a_3i\\a_1j & a_2j & a_3j\\\vdots & \vdots & \vdots\\a_1k & a_2k & a_3k\end{bmatrix}$的向量集合，转置矩阵变换为$\begin{bmatrix}a_1i & a_1j & a_1k\\a_2i & a_2j & a_2k\\\vdots & \vdots & \vdots\\a_3i & a_3j & a_3k\end{bmatrix}$，保持矩阵行为$a_n\vec{i}, a_n\vec{j}, a_n\vec{k}$）。行阶梯形的结果并不是唯一的。例如，行阶梯型乘以一个标量系数仍然是行阶梯形。但是，可以证明一个矩阵的化简后的行阶梯形是唯一的。

```
from sympy.vector import CoordSys3D
# 相当于 a,b=sympy.symbols([ "a" ," b" ])
from sympy.abc import a, b, c, d, e, f, g, h, o, p
from sympy.vector import Vector
from sympy import pprint, Eq, solve
import matplotlib.pyplot as plt
import matplotlib.colors as mcolors
fig, axs = plt.subplots(1, 3, figsize=(30, 10))
axs[0] = fig.add_subplot(1, 3, 1, projection='3d')
axs[1] = fig.add_subplot(1, 3, 2, projection='3d')
axs[2] = fig.add_subplot(1, 3, 3, projection='3d')

# A - 2 维度，绘制 v2+v3+v4+v5=0 的解，线性相关，有多个解
N = CoordSys3D('N')
v2 = a*1*N.i+a*0*N.j
v3 = b*0*N.i+b*1*N.j
v4 = c*3*N.i+c*1*N.j
v5 = d*1*N.i+d*2*N.j
v0 = Vector.zero
def move_alongVectors(vector_list, coeffi_list, C, ax,):
    '''
    function - 给定向量及对应系数，沿向量绘制
    Params:
        vector_list - 向量列表，按移动顺序；list(vector)
        coeffi_list - 向量的系数，例如：list(tuple(symbol,float))
        C - SymPy 下定义的坐标系；CoordSys3D()
        ax - 子图；ax(matplotlib)
    Returns:
        None
    '''
    import random
    import sympy
    # mcolors.BASE_COLORS, mcolors.TABLEAU_COLORS,mcolors.CSS4_COLORS
    colors = [color[0] for color in mcolors.TABLEAU_COLORS.items()]
    colors__random_selection = random.sample(colors, len(vector_list)-1)
    v_accumulation = []
    v_accumulation.append(vector_list[0])
    # 每个向量绘制以之前所有向量之和为起点
    for expr in vector_list[1:]:
        v_accumulation.append(expr+v_accumulation[-1])
    v_accumulation = v_accumulation[:-1]
    for i in range(1, len(vector_list)):
        vector_plot_3d(ax, C, v_accumulation[i-1].subs(coeffi_list), vector_
list[i].subs(
            coeffi_list), color=colors__random_selection[i-1], label='v_%s' %
coeffi_list[i-1][0], arrow_length_ratio=0.2)

# v2+v3+v4+v5=0，向量之和为 0 的解，解 -1
```

```
vector_list = [v0, v2, v3, v4, v5]
a_, b_, c_, d_ = 1, 2, 0, -1
coeffi_list = [(a, a_), (b, b_), (c, c_), (d, d_)]
move_alongVectors(vector_list, coeffi_list, N, axs[0],)

# v2+v3+v4+v5=0，向量之和为 0 的解，解 -2
vector_list = [v0, v2, v3, v4, v5]
a_, b_, c_, d_ = 1, -3, -1, 2
coeffi_list = [(a, a_), (b, b_), (c, c_), (d, d_)]
move_alongVectors(vector_list, coeffi_list, N, axs[1],)

# B - 3 维度，绘制 v6+v7+v8+v9=0 的解，线性相关
M = CoordSys3D('M')
v6 = e*1*M.i+e*0*M.j+e*0*M.k
v7 = f*0*M.i+f*1*M.j+f*0*M.k
v8 = g*3*M.i+g*1*M.j+g*-3*M.k
v9 = h*0*M.i+h*0*M.k+h*3*M.k
v0 = Vector.zero
vector_list = [v0, v6, v7, v8, v9]
e_, f_, g_, h_ = 3, 1, -1, -1
coeffi_list = [(e, e_), (f, f_), (g, g_), (h, h_)]
move_alongVectors(vector_list, coeffi_list, M, axs[2],)

# C - 向量转换为矩阵，判断线性无关
# C-1 - 对于向量集合 v2,v3,v4,v5
v_2345_matrix = v2.to_matrix(N)
for v in [v3, v4, v5]:
    v_temp = v.to_matrix(N)
    v_2345_matrix = v_2345_matrix.col_insert(-1, v_temp)
print("v_2345_matrix:")
pprint(v_2345_matrix)
print("v_2345_matrix.T rref:")
pprint(v_2345_matrix.T.rref())

# C-2 - 对于向量集合 v6,v7,v8,v9
print("_"*50)
v_6789_matrix = v6.to_matrix(M)
for v in [v7, v8, v9]:
    v_temp = v.to_matrix(M)
    v_6789_matrix = v_6789_matrix.col_insert(-1, v_temp)
print("v_6789_matrix:")
pprint(v_6789_matrix)
print("v_6789_matrix.T rref:")
pprint(v_6789_matrix.T.rref())

# C-3 对于向量集合 v10,v11
print("_"*50)
C = CoordSys3D('C')
v10 = o*1*C.i+o*0*C.j
v11 = p*0*C.i+p*1*C.j
v_10_11_matrix = v10.to_matrix(C).col_insert(-1, v11.to_matrix(C))
print("v_10_11_matrix:")
pprint(v_10_11_matrix)
print("v_10_11_matrix.T rref:")
pprint(v_10_11_matrix.T.rref())
axs[0].set_xlim3d(0, 2)
axs[0].set_ylim3d(0, 2)
axs[0].set_zlim3d(0, 5)
axs[1].set_xlim3d(-3, 1)
axs[1].set_ylim3d(-4, 1)
axs[1].set_zlim3d(0, 5)
axs[2].set_xlim3d(0, 4)
axs[2].set_ylim3d(0, 4)
axs[2].set_zlim3d(0, 4)
axs[0].legend()
axs[1].legend()
axs[2].legend()
axs[0].view_init(90, 0)  # 可以旋转图形的角度，方便观察
axs[1].view_init(90, 0)
axs[2].view_init(30, 50)
plt.show()
--------------------------------------------------------------------------------
v_2345_matrix:
```

$$\begin{bmatrix} 0 & 3\cdot c & d & a \\ b & c & 2\cdot d & 0 \\ 0 & 0 & 0 & 0 \end{bmatrix}$$

v_2345_matrix.T rref:

$$\left(\begin{bmatrix} 1 & 0 & 0 \\ 0 & 1 & 0 \\ 0 & 0 & 0 \\ 0 & 0 & 0 \end{bmatrix}, (0, 1)\right)$$

----

v_6789_matrix:

$$\begin{bmatrix} 0 & 3\cdot g & 0 & e \\ f & g & 0 & 0 \\ 0 & -3\cdot g & 3\cdot h & 0 \end{bmatrix}$$

v_6789_matrix.T rref:

$$\left(\begin{bmatrix} 1 & 0 & 0 \\ 0 & 1 & 0 \\ 0 & 0 & 1 \\ 0 & 0 & 0 \end{bmatrix}, (0, 1, 2)\right)$$

----

v_10_11_matrix:

$$\begin{bmatrix} 0 & o \\ p & 0 \\ 0 & 0 \end{bmatrix}$$

v_10_11_matrix.T rref:

$$\left(\begin{bmatrix} 1 & 0 & 0 \\ 0 & 1 & 0 \end{bmatrix}, (0, 1)\right)$$

上述代码是以sympy.vector提供的向量方式建立向量数据集，能够比较直观地理解向量空间及向量在空间中的运算。下述代码则直接建立矩阵模式的向量系数矩阵并进行相关计算，包括求得行阶梯形矩阵，简化的行阶梯形矩阵，以及求解系数，从而可以判断该向量集V_C_matrix为线性相关，有非全零解，可以在空间中构建回到起点闭合的图形。

```
from sympy import Matrix, solve_linear_system
C = CoordSys3D('C')
V_C_matrix = Matrix([[1, 4, 2, -3], [7, 10, -4, -1], [-2, 1, 5, -4]])
print("V_C_matrix=")
pprint(V_C_matrix)
print("-"*50)
lambda_1, lambda_2, lambda_3 = sympy.symbols(
    ['lambda_1', 'lambda_2', 'lambda_3'])
coeffi_expr = V_C_matrix.T*Matrix([lambda_1, lambda_2, lambda_3])
print("向量集合（线性方程组）系数矩阵：")
pprint(coeffi_expr)
print("echelon_form:")
pprint(coeffi_expr.echelon_form())
print("reduced row echelon form:")
pprint(coeffi_expr.rref())

#解线性方程组
print("解线性方程组（矩阵模式），sympy提供的几种方式：")
pprint(solve(coeffi_expr, (lambda_1, lambda_2, lambda_3)))
pprint(solve(coeffi_expr, set=True))
```

```
pprint(solve_linear_system(V_C_matrix.T, lambda_1, lambda_2,))
```

V_C_matrix=

$$\begin{bmatrix}1 & 4 & 2 & -3\\7 & 10 & -4 & -1\\-2 & 1 & 5 & -4\end{bmatrix}$$

向量集合（线性方程组）系数矩阵：

$$\begin{bmatrix}\lambda1 + 7\cdot\lambda2 - 2\cdot\lambda3\\4\cdot\lambda1 + 10\cdot\lambda2 + \lambda3\\2\cdot\lambda1 - 4\cdot\lambda2 + 5\cdot\lambda3\\-3\cdot\lambda1 - \lambda2 - 4\cdot\lambda3\end{bmatrix}$$

echelon_form:

$$\begin{bmatrix}\lambda1 + 7\cdot\lambda2 - 2\cdot\lambda3\\0\\0\\0\end{bmatrix}$$

reduced row echelon form:

$$\left(\begin{bmatrix}1\\0\\0\\0\end{bmatrix}, (0,)\right)$$

解线性方程组（矩阵模式），sympy 提供的几种方式：

$$\left\{\lambda1: \frac{-3\cdot\lambda3}{2}, \lambda2: \frac{\lambda3}{2}\right\}$$

$$\left([\lambda1, \lambda2], \left\{\frac{-3\cdot\lambda3}{2}, \frac{\lambda3}{2}\right\}\right)$$

{λ1: 3/2, λ2: -1/2}

3.6.2.3 维数

假设c为任意实数，若R_{m}的子集W满足下述三个条件：1. W的任意元素的c倍也是W的元素；2. W的任意元素的和也是W的元素；3. 零向量$\vec{0}$在W中。即满足这三个条件时，1. 如果$\begin{bmatrix}a_{1i}\\a_{2i}\\\vdots\\a_{\mathrm{m}i}\end{bmatrix}\in W$，那么$c\begin{bmatrix}a_{1i}\\a_{2i}\\\vdots\\a_{\mathrm{m}i}\end{bmatrix}\in W$；2. 如果$\begin{bmatrix}a_{1i}\\a_{2i}\\\vdots\\a_{\mathrm{m}i}\end{bmatrix}\in W$并且$\begin{bmatrix}a_{1j}\\a_{2j}\\\vdots\\a_{\mathrm{m}j}\end{bmatrix}\in W$，那么$\begin{bmatrix}a_{1i}\\a_{2i}\\\vdots\\a_{\mathrm{m}i}\end{bmatrix}+\begin{bmatrix}a_{1j}\\a_{2j}\\\vdots\\a_{\mathrm{m}j}\end{bmatrix}\in W$；同时$\vec{0}\in W$，则把$W$叫作$R_{\mathrm{m}}$的线性子空间，简称子空间。在理解向量子空间时可以先从2、3维空间思考，如果乘以一个倍数，实际上只是对向量的缩放，而对于向量的和也只是延着多个向量的行进，这些计算都存在于一个空间维度中，进而可以帮助理解拓展到大于3个的维度（无法类似2，3维度直接观察），例如“通过原点的直线”“通过原点的平面”等等描述。

W线性无关的元素，即基元素，$\left\{\begin{bmatrix}a_{11}\\a_{21}\\\vdots\\a_{m1}\end{bmatrix}, \begin{bmatrix}a_{12}\\a_{22}\\\vdots\\a_{m2}\end{bmatrix}, \dots, \begin{bmatrix}a_{1n}\\a_{2n}\\\vdots\\a_{mn}\end{bmatrix}\right\}$，其中基元素的个数$n$叫作

子空间W的维度，一般表示为$dimW$（dim为dimension缩写）。

例如W是R^3的子空间，向量vector_a=0*C.i+3*C.j+1*C.k，即$\begin{bmatrix}0\\3\\1\end{bmatrix}$，与向量vector_b=0*C.i+1*C.j+2*C.k，即$\begin{bmatrix}0\\1\\2\end{bmatrix}$是$W$的线性无关（通过pprint(v_a_b.T.rref())查看，不含全零行）的元素。显然，$\left\{c_1\begin{bmatrix}0\\3\\1\end{bmatrix}+c_2\begin{bmatrix}0\\1\\2\end{bmatrix}\right\}$，$c_1,c_2$为任意实数等式成立，因此集合$\left\{\begin{bmatrix}0\\3\\1\end{bmatrix},\begin{bmatrix}0\\1\\2\end{bmatrix}\right\}$是子空间$W$的基，子空间的维数为2（图3.6-3）。

```
from matplotlib.patches import PathPatch, Rectangle
import mpl_toolkits.mplot3d.art3d as art3d
fig, ax = plt.subplots(figsize=(12, 12))
ax = fig.add_subplot(projection='3d')
p = Rectangle((-5, -5), 10, 10, color='red', alpha=0.1)
ax.add_patch(p)
art3d.pathpatch_2d_to_3d(p, z=0, zdir="x",)
C = CoordSys3D('C')
vector_0 = Vector.zero
vector_1 = 5*C.j
vector_plot_3d(ax, C, vector_0, vector_1, color='gray',
               label='vector_j', arrow_length_ratio=0.1)
vector_2 = 5*C.k
vector_plot_3d(ax, C, vector_0, vector_2, color='gray',
               label='vector_k', arrow_length_ratio=0.1)
vector_a = 0*C.i+3*C.j+1*C.k
vector_b = 0*C.i+1*C.j+2*C.k
vector_plot_3d(ax, C, vector_0, vector_a, color='green',
               label='vector_a', arrow_length_ratio=0.1)
vector_plot_3d(ax, C, vector_0, vector_b, color='olive',
               label='vector_b', arrow_length_ratio=0.1)
ax.set_xlim3d(-5, 5)
ax.set_ylim3d(-5, 5)
ax.set_zlim3d(-5, 5)
ax.legend()
ax.view_init(20, 20)  #可以旋转图形的角度，方便观察
plt.show()
```

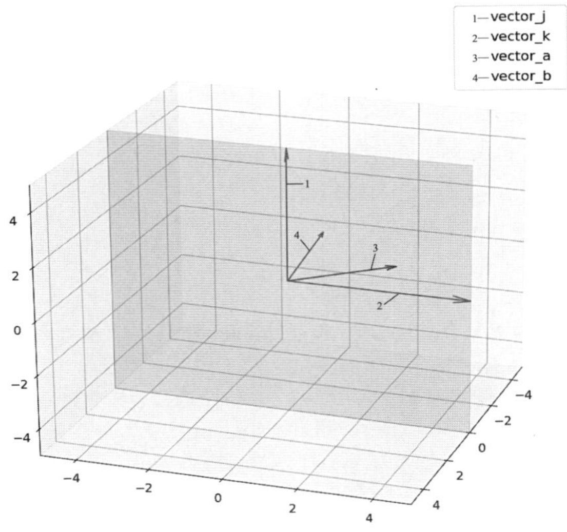

图3.6-3　线性无关与维数

```
v_a_b = vector_a.to_matrix(C).col_insert(-1, vector_b.to_matrix(C))
print("v_a_b.T rref:")
pprint(v_a_b.T.rref())
v_a_b.T rref:
```

$$\left(\begin{bmatrix}0 & 1 & 0\\ 0 & 0 & 1\end{bmatrix}, (1, 2)\right)$$

3.6.3 线性映射

3.6.3.1 定义，线性映射的矩阵计算方法

假设$\begin{bmatrix}x_{1i}\\ x_{2i}\\ \vdots\\ x_{ni}\end{bmatrix}$和$\begin{bmatrix}x_{1j}\\ x_{2j}\\ \vdots\\ x_{nj}\end{bmatrix}$为$R^n$的任意元素，$f$为从$R^n$到$R^m$的映射。当映射$f$满足以下两个条件时，则称映射$f$是从$R^n$到$R^m$线性映射。

1. $f\left\{\begin{bmatrix}x_{1i}\\ x_{2i}\\ \vdots\\ x_{ni}\end{bmatrix}\right\}+f\left\{\begin{bmatrix}x_{1j}\\ x_{2j}\\ \vdots\\ x_{nj}\end{bmatrix}\right\}$与$f\left\{\begin{bmatrix}x_{1i}+x_{1j}\\ x_{2i}+x_{2j}\\ \vdots\\ x_{ni}+x_{nj}\end{bmatrix}\right\}$相等；
2. $cf\left\{\begin{bmatrix}x_{1i}\\ x_{2i}\\ \vdots\\ x_{ni}\end{bmatrix}\right\}$与$f\left\{c\begin{bmatrix}x_{1i}\\ x_{2i}\\ \vdots\\ x_{ni}\end{bmatrix}\right\}$相等。

从R^n到R^m线性映射，可以被称为线性变换或一次变换。为了方便理解线性变换，可以将映射f理解为一个函数（变换矩阵），输入一个向量，经过f作用后，而后输出一个向量的过程，即向量发生了运动。就是，当$\begin{bmatrix}x_1\\ x_2\\ \vdots\\ x_n\end{bmatrix}$（输入）通过$m\times n$矩阵$\begin{bmatrix}a_{11} & a_{12} & \dots & a_{1n}\\ a_{21} & a_{22} & \dots & a_{2n}\\ \vdots & \vdots & \ddots & \vdots\\ a_{m1} & a_{m2} & \dots & a_{mn}\end{bmatrix}$对应的从$R^n$到$R^m$的线性映射，形成的像是$\begin{bmatrix}y_1\\ y_2\\ \vdots\\ y_n\end{bmatrix}$（输出），其矩阵计算公式表示为：

$$f\left\{\begin{bmatrix}x_1\\ x_2\\ \vdots\\ x_n\end{bmatrix}\right\}=\begin{bmatrix}a_{11} & a_{12} & \dots & a_{1n}\\ a_{21} & a_{22} & \dots & a_{2n}\\ \vdots & \vdots & \ddots & \vdots\\ a_{m1} & a_{m2} & \dots & a_{mn}\end{bmatrix}\begin{bmatrix}x_1\\ x_2\\ \vdots\\ x_n\end{bmatrix}$$

$$=\begin{bmatrix}a_{11} & a_{12} & \dots & a_{1n}\\ a_{21} & a_{22} & \dots & a_{2n}\\ \vdots & \vdots & \ddots & \vdots\\ a_{m1} & a_{m2} & \dots & a_{mn}\end{bmatrix}\left\{x_1\begin{bmatrix}1\\ 0\\ \vdots\\ 0\end{bmatrix}+x_2\begin{bmatrix}0\\ 1\\ \vdots\\ 0\end{bmatrix}+\dots+x_n\begin{bmatrix}0\\ 0\\ \vdots\\ 1\end{bmatrix}\right\}。$$

对像而言，即假设x_i是集合X的元素，把通过映射f与x_i对应的集合Y的元素，称为x_i通过映射f形成的像。

定义三维的C向量空间，单位向量为$\vec{i},\vec{j},\vec{k}$，定义v_1向量的系数为a1,a2,a3=-1,2,0，其中0可以省略，即为v_1=a1*i+a2*j=-1*i+2*j向量。给定新的向量集合v_2，v_3，通过计算简化的行阶梯形矩阵，判断为向量无关，因此将该数据集作为基，可以建立新的向量空间，并由v_2，v_3的基构建变换矩阵，v_1的系数矩阵乘以变换矩阵计算得新向量空间下的v_1在相对于C（原）向量空间下的向量v_1_N（图3.6–4）。

```
from sympy.vector import matrix_to_vector
fig, ax = plt.subplots(figsize=(12, 12))
ax = fig.add_subplot(projection='3d')
a1, a2 = sympy.symbols(["a1", "a2",])
a1 = -1
a2 = 2
C = CoordSys3D('C')
i, j, k = C.base_vectors()
v_0 = Vector.zero
v_1 = a1*i+a2*j
vector_plot_3d(ax, C, v_0, v_1, color='red',
               label='v_1', arrow_length_ratio=0.1)
v_2 = -1*i+3*j
vector_plot_3d(ax, C, v_0, v_2, color='blue',
               label='v_2', arrow_length_ratio=0.1)
v_3 = 2*i+0*j
vector_plot_3d(ax, C, v_0, v_3, color='blue',
               label='v_3', arrow_length_ratio=0.1)
def vector2matrix_rref(v_list, C):
    '''
    function - 将向量集合转换为向量矩阵，并计算简化的行阶梯形矩阵
    Params:
        v_list - 向量列表；list(vector)
        C - sympy 定义的坐标系统；CoordSys3D()

    Returns:
        v_matrix.T - 转换后的向量矩阵，即线性变换矩阵
    '''
    v_matrix = v_list[0].to_matrix(C)
    for v in v_list[1:]:
        v_temp = v.to_matrix(C)
        v_matrix = v_matrix.col_insert(-1, v_temp)
    print("_"*50)
    pprint(v_matrix.T.rref())
    return v_matrix.T

print("v_2,v_3 简化的行阶梯形矩阵 :")
# 通过 rref() 返回简化的行阶梯形矩阵，判断向量集合 v_2,v_3 为向量无关。返回线性变换矩阵
v_2_3 = vector2matrix_rref([v_2, v_3], C)

# 由向量无关的向量集合 v_2,v_3 生成新的向量空间，相对于原向量空间下 i,j,k 单位向量，i,j,k 新的位置
v_1_N_matrix = Matrix([a1, a2]).T*v_2_3  # 根据变换矩阵，计算系数同为 a1,a2 向量的位置
v_1_N = matrix_to_vector(v_1_N_matrix, C)  # 向量转矩阵
vector_plot_3d(ax, C, v_0, v_1_N, color='orange',
               label='v_1_N', arrow_length_ratio=0.1)  # 绘制新位置向量
```

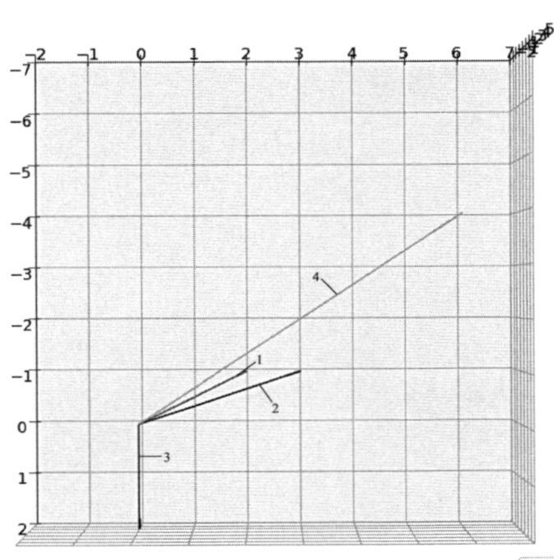

图3.6-4　线性映射的矩阵计算结果

```
ax.set_xlim3d(-7, 2)
ax.set_ylim3d(-2, 7)
ax.set_zlim3d(-2, 5)
ax.legend()
ax.view_init(90, 0)  # 可以旋转图形的角度，方便观察
plt.show()
---------------------------------------------------------------------------------
v_2,v_3 简化的行阶梯形矩阵：
---------------------------------------------------
```

$$\left(\begin{bmatrix}1 & 0 & 0\\0 & 1 & 0\end{bmatrix}, (0, 1)\right)$$

3.6.3.2 特殊的线性映射（线性变换）

线性映射的矩阵计算方法就是将待变换的对象（向量集合），通过乘以变换矩阵，就可获得同一向量空间下新的对象。使用SymPy方便数据量比较小的矩阵运算，方便观察矩阵形式。但是，如果数据量大，则使用NumPy库提供的矩阵计算方法。在下述案例中，从semantic3d[①]下载城市点云数据“bildstein_station1”，并对其进行了降采样，以加快计算速度。对于点云数据的处理，可参考点云部分。

① semantic3d，为大规模点云分类基准数据（http://semantic3d.net/）。

点云数据读取后，包括的信息有点坐标和点颜色。点坐标可以理解为在单位向量为$\vec{i},\vec{j},\vec{k}$的C向量空间下的向量。如果提供相应的变换矩阵，就可以对原点云数据加以空间上的变换，例如移动、旋转、缩放、透视、镜像、切变等。缩放对应的矩阵为：$\begin{bmatrix}scale & 0 & 0\\0 & scale & 0\\0 & 0 & scale\end{bmatrix}$；延z轴（$\vec{k}$）旋转对应的矩阵为：$\begin{bmatrix}\cos(\theta) & -\sin(\theta) & 0\\\sin(\theta) & \cos(\theta) & 0\\0 & 0 & 1\end{bmatrix}$；给定点S(x,y,z)的透视为：$\frac{1}{-s_z}\begin{bmatrix}s_z & 0 & s_x & 0\\0 & -s_z & s_y & 0\\0 & 0 & 1 & 0\\0 & 0 & 1 & -s_z\end{bmatrix}$，透视变换中使用了齐次坐标（homogeneous coordinates）（图3.6-5）。

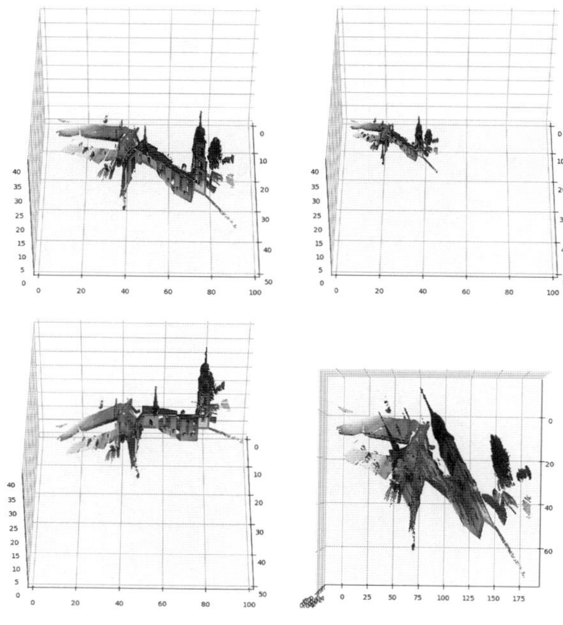

图3.6-5　线性映射（左上：原始点云数据；右上：缩放；左下：旋转；右下：透视）

```
import open3d as o3d
import numpy as np
import pandas as pd
import matplotlib
cloud_pts_fp = "./data/bildstein_station1.txt"
cloud_pts = pd.read_csv(cloud_pts_fp)
# cloud_pts['hex']=cloud_pts.apply(lambda row:'%02x%02x%02x' % (int(row.
r/255),int(row.g/255),int(row.b/255)),axis=1)
print("data reading completed")
cloud_pts['hex'] = cloud_pts.apply(lambda row: matplotlib.colors.to_hex(
    [row.r/255, row.g/255, row.b/255]), axis=1)  #将色彩 rgb 格式转换为 hex 格式，用于
matplotlib 图形打印
--------------------------------------------------------------------------------
data reading completed
--------------------------------------------------------------------------------
import matplotlib.pyplot as plt
import matplotlib
import sympy
import math
from sympy import Matrix, pprint
import numpy as np
pts = cloud_pts[["x", "y", "z"]].to_numpy()
fig, axs = plt.subplots(ncols=2, nrows=2, figsize=(20, 20))
axs[0, 0] = fig.add_subplot(2, 2, 1, projection='3d')
axs[0, 1] = fig.add_subplot(2, 2, 2, projection='3d')
axs[1, 0] = fig.add_subplot(2, 2, 3, projection='3d')
axs[1, 1] = fig.add_subplot(2, 2, 4, projection='3d')
axs[0, 0].scatter(cloud_pts.x, cloud_pts.y,
                  cloud_pts.z, c=cloud_pts.hex, s=0.1)
#A - 缩放
scale = 0.5
f_scale = np.array([[scale, 0, 0], [0, scale, 0], [0, 0, scale]])
pts_scale = np.matmul(pts, f_scale)
axs[0, 1].scatter(pts_scale[:, 0], pts_scale[:, 1],
                  pts_scale[:, 2], c=cloud_pts.hex, s=0.1)

#B - 旋转
angle = math.radians(-20)  #度转换为弧度
f_rotate = np.array([[math.cos(angle), -math.sin(angle), 0],
                     [math.sin(angle), math.cos(angle), 0], [0, 0, 1]])
pts_rotate = np.matmul(pts, f_rotate)
axs[1, 0].scatter(pts_rotate[:, 0], pts_rotate[:, 1],
                  pts_rotate[:, 2], c=cloud_pts.hex, s=0.1)

#C - 透视
s1, s2, s3 = 2, 2, 2
f_persp = np.array([[-s3, 0, 0, 0], [0, -s3, 0, 0],
                    [s1, s2, 0, 1], [0, 0, 0, -s3]])*1/(1-s3)
pts_homogeneousCoordinates = np.hstack(
    (pts, np.ones(pts.shape[0]).reshape(-1, 1)))
pts_persp = np.matmul(pts_homogeneousCoordinates, f_persp)
axs[1, 1].scatter(pts_persp[:, 0], pts_persp[:, 1],
                  pts_persp[:, 2], c=cloud_pts.hex, s=0.1)
axs[0, 0].set_xlim3d(0, 50)
axs[0, 0].set_ylim3d(0, 100)
axs[0, 0].set_zlim3d(0, 40)
axs[0, 1].set_xlim3d(0, 50)
axs[0, 1].set_ylim3d(0, 100)
axs[0, 1].set_zlim3d(0, 40)
axs[1, 0].set_xlim3d(0, 50)
axs[1, 0].set_ylim3d(0, 100)
axs[1, 0].set_zlim3d(0, 40)
axs[0, 0].view_init(45, 0)  #可以旋转图形的角度，方便观察
axs[0, 1].view_init(45, 0)
axs[1, 0].view_init(45, 0)
axs[1, 1].view_init(90, 0)
plt.show()
```

3.6.3.3 秩

f 为从R^n到R^m的映射，被映射到零元素（$\vec{0}$）的全体元素的集合，即集合

$\left\{\begin{bmatrix}x_1\\x_2\\\vdots\\x_n\end{bmatrix}\middle|\begin{bmatrix}0\\0\\\vdots\\0\end{bmatrix}=\begin{bmatrix}a_{11}&a_{12}&\cdots&a_{1n}\\a_{21}&a_{22}&\cdots&a_{2n}\\\vdots&\vdots&\ddots&\vdots\\a_{m1}&a_{m2}&\cdots&a_{mn}\end{bmatrix}\begin{bmatrix}x_1\\x_2\\\vdots\\x_n\end{bmatrix}\right\}$称为映射 f 的核，一般表示为 $K_{er}f$。为了与“映射 f 的核”相呼应，则把“映射 f 的值域”，即集合 $f\left\{\begin{bmatrix}y_1\\y_2\\\vdots\\y_m\end{bmatrix}\middle|\begin{bmatrix}y_1\\y_2\\\vdots\\y_m\end{bmatrix}=\begin{bmatrix}a_{11}&a_{12}&\cdots&a_{1n}\\a_{21}&a_{22}&\cdots&a_{2n}\\\vdots&\vdots&\ddots&\vdots\\a_{m1}&a_{m2}&\cdots&a_{mn}\end{bmatrix}\begin{bmatrix}x_1\\x_2\\\vdots\\x_n\end{bmatrix}\right\}$称为“映射$f$的像空间”，一般表示为$I_mf$。

$K_{er}f$ 是 R^n 的子空间，I_mf 是 R^m 的子空间。在 $K_{er}f$ 与 I_mf 之间，有“维数公式”：$n-dimK_{er}f=dimI_mf$。

把向量$\begin{bmatrix}a_{11}\\a_{21}\\\vdots\\a_{m1}\end{bmatrix},\begin{bmatrix}a_{12}\\a_{22}\\\vdots\\a_{m2}\end{bmatrix},\dots,\begin{bmatrix}a_{1n}\\a_{2n}\\\vdots\\a_{mn}\end{bmatrix}$中线性无关的向量的个数，即$R^m$子空间$I_mf$的维数称为 $m\times n$ 矩阵$\begin{bmatrix}a_{11}&a_{12}&\cdots&a_{1n}\\a_{21}&a_{22}&\cdots&a_{2n}\\\vdots&\vdots&\ddots&\vdots\\a_{m1}&a_{m2}&\cdots&a_{mn}\end{bmatrix}$的秩，一般表示为 $rank\begin{bmatrix}a_{11}&a_{12}&\cdots&a_{1n}\\a_{21}&a_{22}&\cdots&a_{2n}\\\vdots&\vdots&\ddots&\vdots\\a_{m1}&a_{m2}&\cdots&a_{mn}\end{bmatrix}$，或 $r(A),rank(A),rk(A)$。

SymPy计算秩（rank）的方法为`matrix.rank()`。

3.6.4 特征值和特征向量

在特殊的线性映射部分，假设给定单位向量为$\vec{i},\vec{j},\vec{k}$，即$\begin{bmatrix}1&0&0\\0&1&0\\0&0&1\end{bmatrix}$的$C$向量空间，通过变换矩阵$\begin{bmatrix}scale1&0&0\\0&scale2&0\\0&0&scale3\end{bmatrix}$实现了对原向量集（点云数据）的缩放。为方便解释，如果仅考虑点云数据中的一个点，$c_1,c_2,c_3=x,y,z=39,73,22$，则该点在$C$向量空间的向量为 $C_1i+C_2j+C_3k=C_1\begin{bmatrix}1\\0\\0\end{bmatrix}+C_2\begin{bmatrix}0\\1\\0\end{bmatrix}+C_3\begin{bmatrix}0\\0\\1\end{bmatrix}$。进行线性映射和公式变换，

$$\begin{bmatrix}scale1&0&0\\0&scale2&0\\0&0&scale3\end{bmatrix}\left\{C_1\begin{bmatrix}1\\0\\0\end{bmatrix}+C_2\begin{bmatrix}0\\1\\0\end{bmatrix}+C_3\begin{bmatrix}0\\0\\1\end{bmatrix}\right\}$$

$$=C_1\begin{bmatrix}scale1&0&0\\0&scale2&0\\0&0&scale3\end{bmatrix}\begin{bmatrix}1\\0\\0\end{bmatrix}+C_2\begin{bmatrix}scale1&0&0\\0&scale2&0\\0&0&scale3\end{bmatrix}\begin{bmatrix}0\\1\\0\end{bmatrix}$$

$$+C_3\begin{bmatrix}scale1&0&0\\0&scale2&0\\0&0&scale3\end{bmatrix}\begin{bmatrix}0\\0\\1\end{bmatrix}=C_1\begin{bmatrix}scale1\\0\\0\end{bmatrix}+C_1\begin{bmatrix}0\\scale2\\0\end{bmatrix}+C_1\begin{bmatrix}0\\0\\scale3\end{bmatrix}$$

$=C_1\left\{scale1\begin{bmatrix}1\\0\\0\end{bmatrix}\right\}+C_2\left\{scale2\begin{bmatrix}0\\1\\0\end{bmatrix}\right\}+C_3\left\{scale3\begin{bmatrix}0\\0\\1\end{bmatrix}\right\}$，变换后的公式保持了$C$向量空间下的单位向量$\vec{i},\vec{j},\vec{k}$不变，则$scale1,scale2,scale3$为变换矩阵的特征值，而$scale1$对应的$\begin{bmatrix}1\\0\\0\end{bmatrix}$，$scale2$

对应的$\begin{bmatrix}0\\1\\0\end{bmatrix}$，$scale3$对应的$\begin{bmatrix}0\\0\\1\end{bmatrix}$，为其特征向量。

当$\begin{bmatrix}x_1\\x_2\\\vdots\\x_n\end{bmatrix}$（输入）通过$m\times n$矩阵$\begin{bmatrix}a_{11}&a_{12}&\cdots&a_{1n}\\a_{21}&a_{22}&\cdots&a_{2n}\\\vdots&\vdots&\ddots&\vdots\\a_{m1}&a_{m2}&\cdots&a_{mn}\end{bmatrix}$对应的从$R^n$到$R^m$的线性映射$f$，形成的像是$\begin{bmatrix}y_1\\y_2\\\vdots\\y_n\end{bmatrix}=\lambda\begin{bmatrix}x_1\\x_2\\\vdots\\x_n\end{bmatrix}$（输出），则把$lambda$叫作方阵$\begin{bmatrix}a_{11}&a_{12}&\cdots&a_{1n}\\a_{21}&a_{22}&\cdots&a_{2n}\\\vdots&\vdots&\ddots&\vdots\\a_{m1}&a_{m2}&\cdots&a_{mn}\end{bmatrix}$的特征值（eigenvalue），把$\begin{bmatrix}x_1\\x_2\\\vdots\\x_n\end{bmatrix}$叫作与特征值λ对应的特征向量（eigenvector）。此外，零向量不能解释为特征向量。

SymPy库提供了`matrix.eigenvals()`，`matrix.eigenvects()`方法直接计算特征值和特征向量。

```
scale1, scale2, scale3 = sympy.symbols(['scale1', 'scale2', 'scale3'])
M0 = Matrix(3, 3, [1, 0, 0, 0, 1, 0, 0, 0, 1])
print("original matrix")
pprint(M0)
M = Matrix(3, 3, [scale1*1, 0, 0, 0, scale2*1, 0, 0, 0, scale3*1])
print("modified scaled matrix")
pprint(M)
print("eigen values:")
pprint(M.eigenvals())
print("eigen vectors:")
pprint(M.eigenvects())
------------------------------------------------------------------------
original matrix
```

$$\begin{bmatrix}1&0&0\\0&1&0\\0&0&1\end{bmatrix}$$

```
modified scaled matrix
```

$$\begin{bmatrix}scale1&0&0\\0&scale2&0\\0&0&scale3\end{bmatrix}$$

```
eigen values:
{scale₁: 1, scale₂: 1, scale₃: 1}
eigen vectors:
```

$$\left[\left(scale1, 1, \left[\begin{bmatrix}1\\0\\0\end{bmatrix}\right]\right), \left(scale2, 1, \left[\begin{bmatrix}0\\1\\0\end{bmatrix}\right]\right), \left(scale3, 1, \left[\begin{bmatrix}0\\0\\1\end{bmatrix}\right]\right)\right]$$

参考文献（References）：

[1] [日]高桥 信著作，Inoue Iroha，株式会社TREND-PRO漫画制作，腾永红译. 漫画线性代数［M］. 北京：科学出版社，2009. 08.

[2] Gilbert Strang (2016). Introduction to Linear Algebra. Wellesley-Cambridge Press.

3.7 主成成分分析与Landsat遥感影像

3.7.1 Landsat地理信息数据读取、裁切、融合和打印显示

从1970年代，地球资源卫星（The Landsat series of satellites）几乎不间断地提供全球范围地球表面中分辨率多光谱遥感影像数据（图3.7–1）。不间断的数据记录，增强了应用数据分析的潜力，可以帮助各领域的研究者在人口不断增长、城市不断发展下，关注粮食、水、森林等自然资源，应用其无以伦比的影像质量、覆盖率、细节价值分析城市的发展与自然平衡的问题（美国地质调查局 United States Geological Survey，USGS-earthexplorer[①]）。

下述图 3.7–1 为 Mermaid[②] 图表代码生成。

① 美国地质调查局United States Geological Survey，USGS-earthexplorer，EarthExplor（EE）为美国地质调查局（USGS）档案中的地球科学数据提供在线检索，浏览显示，元数据（metadata）输出和数据下（https://earthexplorer.usgs.gov/）。

② Mermaid，是一个基于JavaScript的图表工具，使用Markdown-inspired文本定义和渲染器来创建和修改复杂的图表（https://mermaid-js.github.io/mermaid）。

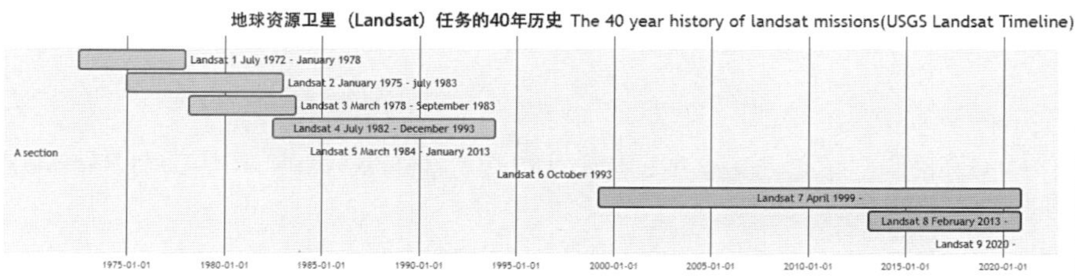

图3.7–1　Landsat 历史

Landsat陆地资源卫星的数据为多光谱，通过安装在环绕地球轨道的太空卫星上的平台收集。各个卫星传感器所获遥感影像数据的波段情况有所差异，表3.7–1为Landsat 8各波段的情况。

表3.7–1　Landsat8波段信息

波段（band）	波长（wavelength range）/nanometers	空间分辨率（spatial resolution）/m	光谱宽度（spetral width）/nm	用途
Band 1- Coastal aerosol	430 - 450	30	20	沿海和气溶胶研究
Band 2- Blue 蓝色波段	450 - 510	30	60	水深绘图，从植被中区分土壤，针叶林植被中区分落叶
Band 3- Green 绿色波段	530 - 590	30	60	强调植被生长高峰，用于评估植被活力
Band 4- Red 红色波段	640 - 670	30	30	处于叶绿素吸收区，用于观测道路、裸露土壤、植被种类等
Band 5- Near Infrared(NIR) 近红外波段	850 - 880	30	30	强调生物量和海岸线
Band 6- Shortwave Infrared 1(SWIR1) 短波红外波段1	1570 - 1650	30	80	区分土壤和植被含水量，并能穿透薄云
Band 7- Shortwave Infrared 2(SWIR2) 短波红外波段2	2110 - 2290	30	180	增加的波段，区分土壤和植被含水量，并能穿透薄云
Band 8- Panchromatic 全色波段	500 - 680	15	180	15m分辨率，更清晰的图像清晰度
Band 9- Cirrus clouds 卷云波段	1360 - 1380	30	20	改进了对卷云污染的检测
Band 10- Thermal Infrared 1(TIRS_1) 热红外波段1	10600 - 11190	100	590	100m分辨率，热红外图，可估计的土壤湿度
Band 11- Thermal Infrared 2(TIRS_2) 热红外波段2	11500 - 12510	100	1010	100m分辨率，增加的热红外图，可估计的土壤湿度

Landsat系列遥感影像在城市研究中作用举足轻重，在不同的研究领域中均需要使用该数据分析。例如，从20世纪70年代至今城市及自然环境的演化，用地类型的变化，反演地表温度研究城市热环境、绿地变化与生态影响、水资源情况等，不计其数。除了Landsate中分辨率卫星，还有高空分辨率低于1m的高分辨率遥感影像，探究城市的细微变化。尤其包含有近红外波段的高分辨率影像，可以在微尺度上深入植被研究，以及拓展相关研究。

- Landsat文件名所包含的信息

Landsat文件名格式：LXSS_LLLL_PPPRRR_YYYYMMDD_yyyymmdd_CC_TX，例如：LC08_L1TP_023031_20180310_20180320_01_T1，其文件名包含的信息如表3.7-2所示。

表3.7-2 Landsat文件名信息解释

序号	占位符	代表含义	备注	案例中的值
1	L	卫星名	Landsat	L
2	X	传感器	“C”代表OLI/TIRS组合，“O”代表OLI，“T”代表TIRS，“E”代表ETM+，“T”代表TM，“M”代表MSS	C
3	SS	卫星编号	截至2021年，编号为1～9	08
4	LLLL①	处理矫正水平（Processing correction level），由地面控制点、数字高程和传感器收集的数据确定	“L1TP”（Precision Terrain）为1级精度地形矫正产品，如果参考数据不充足，则处理为系统的、地形矫正的L1GT（Systematic Terrain）和L1GS（Systematic）产品	L1TP
5	PPP	WRS path世界参考系统（the Worldwide Reference），是Landsat卫星数据的全球定位系统	可以指定path，row查询世界上任何部分的卫星图像	023
6	RRR	WRS row	\	031
7	YYYYMMDD	获取年份	\	20180310
8	yyyymmdd	处理年份	\	20180320
9	CC	采集编号	（01，02）	01
10	TX②	采集类别	“RT”为实时，“T1”为层1（Tier1）包含最高数据质量；“T2”为层2，因为缺失数据精度较T1低	T1

对于遥感影像的处理，即为对raster（栅格数据）的处理，可以使用rasterio③、EarthPy④、GDAL⑤等库读写，定义或重定义投影坐标系统，裁切与融合，数据处理及可视化（图3.7-2）等。

```
import earthpy.spatial as es
import earthpy.plot as ep
import earthpy.mask as em
import matplotlib.pyplot as plt
from glob import glob
import os
import re
plt.rcParams.update({'font.size': 20})
Landsat_fp = {
    "w_180310": r"F:\data\Landsat\LC08_L1TP_023031_20180310_20180320_01_T1",
#冬季
    "s_190820": r"F:\data\Landsat\LC08_L1TP_023031_20190804_20190820_01_T1",
#夏季
    "a_191018": r"F:\data\Landsat\LC08_L1TP_023031_20191007_20191018_01_T1"
#秋季
}
w_landsat = glob(os.path.join(Landsat_fp["w_180310"], "*_B[0-9]*.tif"))
band_name = {'B1': 'coastal_aerosol', 'B2': 'blue', 'B3': 'green', 'B4': 'red', 'B5': 'near_infrared',
             'B6': 'SWIR_1', 'B7': 'SWIR_2', 'B8': 'panchromatic', 'B9': 'cirrus', 'B10': 'TIRS_1', 'B11': 'TIRS_2'}
def fp_sort(fp_list, str_pattern, prefix=""):
    '''
    function - 按照文件名中的数字排序文件列表
```

① LLLL，Landsat Levels of Processing（https://www.usgs.gov/landsat-missions/landsat-levels-processing）。

② TX，Landsat Collection（https://www.usgs.gov/landsat-missions/landsat-collection-1?qt-science_support_page_related_con=1#qt-science_support_page_related_con）。

③ rasterio，地理信息系统（Geographic information systems，GIS）使用GeoTIFF和其他格式组织和存储栅格数据（raster datasets），如卫星影像和地形模型等。rasterio读写这些文件格式，并提供基于NumPy N-维数组和GeoJSON的Python API（https://rasterio.readthedocs.io/en/latest）。

④ EarthPy，是一个Python包，使得使用开源工具绘制和处理空间栅格和矢量数据变得更加容易。EarthPy依赖于专注矢量数据的GeoPandas库和方便输入输出栅格数据的rasterio库，并需要matplotlib库绘图，其目标是让科学家更容易地处理空间数据（https://earthpy.readthedocs.io/en/latest）。

⑤ GDAL，一些用于编程和操作GDAL地理空间数据抽象库的工具，实际上包括两个库GDAL库和OGR库，分布用于处理地理空间的栅格数据和矢量数据（https://gdal.org/）。

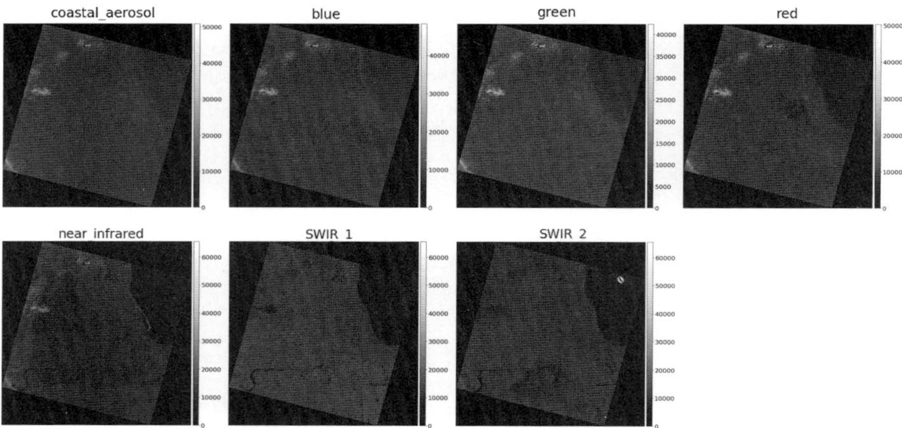

图3.7-2 打印Landsat8各波段

```
    Params:
        fp_list - 文件列表；list(string)
        str_pattern - 字符串匹配模式，用于提取文件名中的数字；re.compile()
        prefix - 字典数据格式的键名前缀；string

    Returns:
        fn_sort - 返回排序后的列表；list(string)
        fn_dict - 返回字典；dict
    '''
    fn_num_extraction = [(int(re.findall(str_pattern, fn)[0]), fn)
                         for fn in fp_list]
    fn_sort = sorted(fn_num_extraction)
    fn_dict = dict([("%s" % prefix+str(i[0]), i[1]) for i in fn_sort])
    return fn_sort, fn_dict
str_pattern = re.compile(r'B(.*?)[.]', re.S)
fn_sort, fn_dict = fp_sort(w_landsat, str_pattern, prefix="B")

#打印遥感影像波段
array_stack, meta_data = es.stack([fn[1] for fn in fn_sort][:7], nodata=-9999)
ep.plot_bands(array_stack, title=list(band_name.values())
              [:7], cols=7, cbar=True, figsize=(10*7, 10))
plt.show()
```

上述通过给定Landsat文件夹，用glob库获取指定字符串匹配模式的所有文件路径，即Landsat各个波段的路径名，然后使用EarthPy库.stack叠合层为一个变量。注意，只有空间分辨率相同的层可以叠合，因此叠合了前7个波段，并使用该库打印显示。下述代码则定义了LandsatMTL_info(fp)函数，给定Landsat所在根目录，读取元文件，即_MTL.txt文件。通过读取元文件获取相关影像信息，并获取各个波段的路径。

```
def LandsatMTL_info(fp):
    '''
    function - 读取 landsat *_MTL.txt 文件，提取需要的信息

    Paras:
        fp - Landsat 文件根目录；string

    return:
        band_fp_dic - 返回各个波段的路径字典；dict
        Landsat_para - 返回 Landsat 参数；dict
    '''
    import os
    import re
    fps = [os.path.join(root, file) for root, dirs, files in os.walk(fp)
           for file in files]  #提取文件夹下所有文件的路径
    MTLPattern = re.compile(r'_MTL.txt', re.S)  #匹配对象模式，提取 _MTL.txt 遥感影像的
元数据文件
```

```
    MTLFn = [fn for fn in fps if re.findall(MTLPattern, fn)][0]
    with open(MTLFn, 'r') as f:  #读取所有元数据文件信息
        MTLText = f.read()
    bandFn_Pattern = re.compile(
        r'FILE_NAME_BAND_[0-9]\d* = "(.*?)"\n', re.S)  #Landsat 波段文件
    band_fn = re.findall(bandFn_Pattern, MTLText)
    band_fp = [[(re.findall(r'B[0-9]\d*', fn)[0], re.findall(r'.*?%s$' % fn, f)[0])
                for f in fps if re.findall(r'.*?%s$' % fn, f)] for fn in band_
fn]  #(文件名, 文件路径)
    band_fp_dic = {i[0][0]: i[0][1] for i in band_fp}
    #需要数据的提取标签/根据需要读取元数据信息
    values_fields = ["RADIANCE_ADD_BAND_10",
                     "RADIANCE_ADD_BAND_11",
                     "RADIANCE_MULT_BAND_10",
                     "RADIANCE_MULT_BAND_11",
                     "K1_CONSTANT_BAND_10",
                     "K2_CONSTANT_BAND_10",
                     "K1_CONSTANT_BAND_11",
                     "K2_CONSTANT_BAND_11",
                     "DATE_ACQUIRED",
                     "SCENE_CENTER_TIME",
                     "MAP_PROJECTION",
                     "DATUM",
                     "UTM_ZONE"]
    Landsat_para = {field: re.findall(re.compile(
        r'%s = "*(.*?)"*\n' % field), MTLText)[0] for field in values_fields}  #(参
数名, 参数值)
    return band_fp_dic, Landsat_para  #返回所有波段路径和需要的参数值

band_fp_dic, Landsat_para = LandsatMTL_info(Landsat_fp["w_180310"])
print(band_fp_dic)
print(Landsat_para)
-------------------------------------------------------------------------------
{'B1': 'F:\\data\\Landsat\\LC08_L1TP_023031_20180310_20180320_01_T1\\LC08_L1TP_0
23031_20180310_20180320_01_T1_B1.TIF', 'B2': 'F:\\data\\Landsat\\LC08_L1TP_02303
1_20180310_20180320_01_T1\\LC08_L1TP_023031_20180310_20180320_01_T1_B2.TIF',...
}
{'RADIANCE_ADD_BAND_10': '0.10000', 'RADIANCE_ADD_BAND_11': '0.10000', 'RADIANCE_
MULT_BAND_10': '3.3420E-04', 'RADIANCE_MULT_BAND_11': '3.3420E-04', 'K1_CONSTANT_
BAND_10': '774.8853', 'K2_CONSTANT_BAND_10': '1321.0789', 'K1_CONSTANT_BAND_11':
'480.8883', 'K2_CONSTANT_BAND_11': '1201.1442', 'DATE_ACQUIRED': '2018-03-10', 'SCENE_
CENTER_TIME': '16:34:42.6511940Z', 'MAP_PROJECTION': 'UTM', 'DATUM': 'WGS84', 'UTM_
ZONE': '16'}
```

从元文件获取的信息中，能够得知最基本的信息，包括投影坐标系统，为WGS84，UTM_ZONE 16。如果要对影像裁切，需要保持用于裁切的边界SHP文件与Landsat影像保持相同的投影。可以用rasterio库读取一个Landsat波段，直接获取该波段的CRS格式的投影坐标系统来转化SHP边界的坐标系统。EarthPy库提供的裁切工具.crop_all可以一次性裁切多个波段（即栅格文件）。

```
import geopandas as gpd
from pyproj import CRS
import rasterio as rio
shape_polygon_fp = r'.\data\LandsatChicago_boundary\LandsatChicago_boundary.shp'
crop_bound = gpd.read_file(shape_polygon_fp)

#获取 Landsat 遥感影像的投影坐标系统，用于 .shp 格式的裁切边界，保持投影一致，裁切正确
with rio.open(band_fp_dic['B1']) as raster_crs:
    crop_raster_profile = raster_crs.profile
    crop_bound_utm16N = crop_bound.to_crs(crop_raster_profile["crs"])
print(crop_bound_utm16N.crs)
workspace = r"F:\data\Landsat\data_processed"
output_dir = os.path.join(workspace, r"w_180310")
w_180310_band_paths_list = es.crop_all(band_fp_dic.values(
), output_dir, crop_bound_utm16N, overwrite=True)  #对所有波段 band 执行裁切
print("cropping finished ...")
-------------------------------------------------------------------------------
PROJCS["WGS 84 / UTM zone 16N",
       GEOGCS["WGS 84",
       DATUM["WGS_1984",
       SPHEROID["WGS 84",6378137,298.257223563,
       AUTHORITY["EPSG","7030"]],
```

```
        AUTHORITY["EPSG","6326"]],
        PRIMEM["Greenwich",0,
        AUTHORITY["EPSG","8901"]],
        UNIT["degree",0.0174532925199433,
        AUTHORITY["EPSG","9122"]],
        AUTHORITY["EPSG","4326"]],
        PROJECTION["Transverse_Mercator"],
        PARAMETER["latitude_of_origin",0],
        PARAMETER["central_meridian",-87],
        PARAMETER["scale_factor",0.9996],
        PARAMETER["false_easting",500000],
        PARAMETER["false_northing",0],
        UNIT["metre",1,
        AUTHORITY["EPSG","9001"]],
        AXIS["Easting",EAST],
        AXIS["Northing",NORTH],
        AUTHORITY["EPSG","32616"]]
cropping finished ...
```

将裁切后的文件堆叠在一个变量下，可以直接使用索引的方式读取任何一层，方便对多光谱遥感影像的管理。

```
w_180310_stack_fp = os.path.join(workspace, r"w_180310_stack.tif")
w_180310_array, w_180310_raster_prof = es.stack(
    w_180310_band_paths_list[:7], out_path=w_180310_stack_fp)
print("stacking finished...")
--------------------------------------------------------------------------------
stacking finished...
```

将不同的波段合成显示（图3.7–3），可以有不同的显示结果，方便不同目的性的观察，通常如表3.7–3所示。

图3.7–3　自然真彩色（左）；农业用波段合成（右）

表3.7–3　合成波段

RGB通道编号	波段名	主要用途
4、3、2	Red、Green、Blue	自然真彩色
7、6、4	SWIR2、SWIR1、Red	城市
5、4、3	NIR、Red、Green	标准假彩色图像，植被
6、5、2	SWIR1、NIR、Blue	农业
7、6、5	SWIR2、SWIR1、NIR	穿透大气层
5、6、2	NIR、SWIR1、Blue	健康植被
5、6、4	NIR、SWIR1、Red	陆地/水
7、5、3	SWIR2、NIR、Green	移除大气影响的自然表面
7、5、4	SWIR2、NIR、Red	短波红外
6、5、4	SWIR1、NIR、Red	植被分析

```
from rasterio.plot import plotting_extent
extent = plotting_extent(w_180310_array[0],  w_180310_raster_prof["transform"])
plt.rcParams.update({'font.size': 12})
fig, axs = plt.subplots(1, 2, figsize=(25, 12))
ep.plot_rgb(
    w_180310_array,
    rgb=[3, 2, 1],  # 对应波段文件为：B4,B3,B2
    stretch=True,
    extent=extent,
    str_clip=0.5,
    title="w_180310_B4,B3,B2",
    ax=axs[0]
)
ep.plot_rgb(
    w_180310_array,
    rgb=[5, 4, 1],  # 对应波段文件为：B6,B5,B2
    stretch=True,
    extent=extent,
    str_clip=0.5,
    title="w_180310_B6,B5,B2",
    ax=axs[1]
)
plt.show()
```

3.7.2 主成分分析（Principal components analysis，PCA）

3.7.2.1 主成分分析解析

除了全色之外的所有波段，是否有一种方式可以将10个波段的信息降维为1个，并保留最大的信息量，然后再结合相关波段共同打印RGB图像？主成分分析（principal component analysis，PCA）即为其中一种降低数据维度的技术，可以用于压缩数据，减少内存使用和减负CPU/GPU的处理能力，探索数据集并将高维数据在二三维度上显示观察。

理解该部分的知识，建议先阅读“编程线性代数”部分内容。

主成分分析是一种统计分析，简化数据集的方法。利用正交变换对一系列可能相关的变量观测值进行线性变换，从而投影为一系列线性无关变量的值，这些不相关变量称为主成分（Principal Componnets）。主成分分析经常用于减少数据集的维数，并保留数据集中对方差贡献最大的特征。主成分分析依赖所给数据，数据的准确性对分析结果影响很大。这里，为了方便阐述PCA，自行生成了一个随机数据集X，包含两个特征，即两组解释变量。为了使得数据两个特征具有一定的线性关系，对该数据集乘以一个变换矩阵`transformation= [[-0.6, -0.6], [-0.3, -0.9]]`（参看线性映射部分）。

这些特征点的分布（分别由一个特征作为x坐标，另一个作为y坐标），将其绘制于二维的笛卡尔坐标系（Cartesian coordinate system，也称直角坐标系）。目的是试图找到一条直线（新的坐标轴），再将所有点投影到该直线上，从而构建一个新属性。这个属性将由w_1x+w_2y线性组合定义，每条线对应w_1和w_2的特定值。那么，在下述旋转的直线中，哪条直线的投影点能够保留原特征数据集的最大信息量？首先，这条线上投影点的差异应该最大化，即反应分布的方差最大。从图3.7-4的变化中，可以观察到新特征即红色投影点（重建于原本的两个特征）方差的变化。当直线旋转到蓝色轴`v_n_j`时，方差达到最大，从特征点的分布上也能够容易观察到这个变化趋势。这个投影点当其方差大于到第2个轴的投影时为第1主成分，否则为第2主成分。整体的重建误差通过特征点到相应投影点的距离均方根来衡量，而两个旋转的轴永远是垂直的。投影点到特征点的距离与投影点到原点的距离正好与特征点到原点的距离构成直角三角形，因此对于重建误差的测量可以转换为投影点方差的测量，即投影点方差越高，误差越低。

使得投影点方差为最大的新坐标轴是由特征点的协方差求得，$\begin{bmatrix}(x(i)|x(i)) & (y(j)|x(i))\\(x(i)|y(j)) & (y(j)|y(j))\end{bmatrix}=\begin{bmatrix}0.36372533 & 0.55326727\\0.55326727 & 1.10198785\end{bmatrix}$，其中$x(i)|x(i)$及$y(j)|y(j)$是各个特征自身的方差；而$y(j)|x(i)$和$x(i)|y(j)$则是特征之间的协方差，其值相同。所要寻找新的坐标系是使得特征点到新坐标系其中一个轴的投影的方差为最大，相对到其他轴投影方差则次第减小，则说明轴之间的方差即$y(j)|x(i)$和$x(i)|y(j)$趋近于0，其他轴的投影点方差也相对较小。

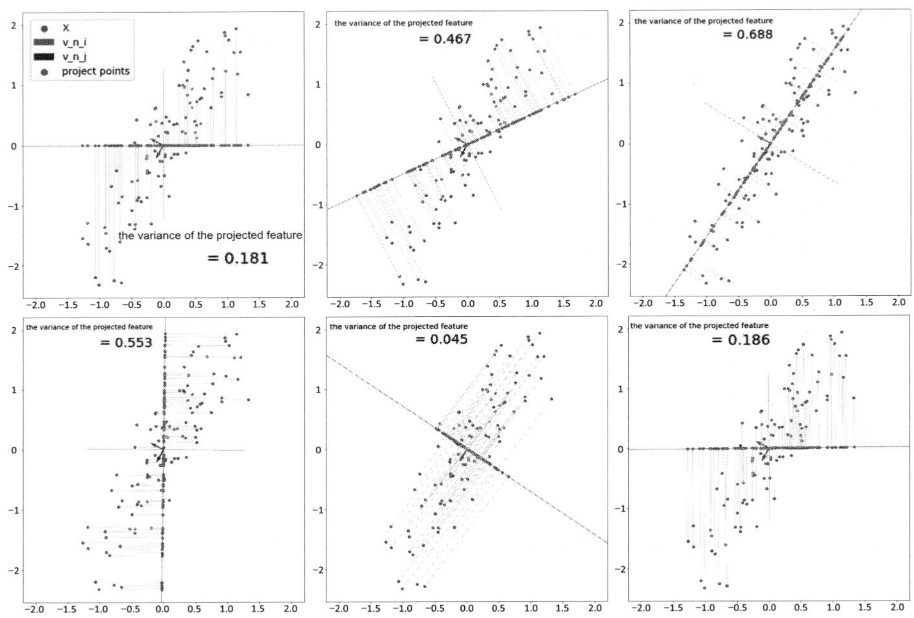

图3.7-4　PCA 正交变换投影点方差变化

$\begin{bmatrix}(x(i)|x(i)) & (y(j)|x(i))\\(x(i)|y(j)) & (y(j)|y(j))\end{bmatrix}=\begin{bmatrix}3.11020298\text{e}-04 & -4.96686164\text{e}-18\\-4.96686164\text{e}-18 & 2.73202237\text{e}+00\end{bmatrix}$，从计算的结果也能观察到这一特点。其中，只有$y(j)|y(j)$的方差最大，其他的均趋于0。轴之间的协方差基本为0，基本不存在相关性；即当协方差为0时，表示两个特征基本完全独立。而为了让协方差为0，在确定第2个轴（基）时只能在与第一个轴（基）正交方向上选择，因此新坐标系的两个轴是正交垂直的。

由上述表述可以进一步推广到一般维度上的降维问题，将一组N维特征向量降到K维，其中$0<K<N$，选择K个单位（模为1）的正交基（新的坐标系或向量空间），使得原始数据变换到这组基上后，各个特征间协方差为0；而特征的方差则尽可能大，在正交的约束下取最大的K各方差对应的基（轴）。

由原始特征值的协方差方阵计算特征向量和特征值，所计算的单位特征向量就是新坐标轴的方向，而特征值是坐标轴的长度（相对于原始对应坐标轴单位的倍数）。对于为什么协方差矩阵的特征向量和特征值的求得就为新的坐标系，可以理解为，假设原始特征的坐标系的轴表示为：$\vec{i}(x),\vec{j}(y)$，单位长度为1，所有特征向量可以表示为$\begin{cases}c_{11}i+c_{12}j\\c_{21}i+c_{22}j\\\cdots\\c_{n1}i+c_{n2}j\end{cases}$，如果将方差的值认为是能反映该类特征分布的一个代表，则所有的特征向量可以简化表示为s_1i+s_2j，那么协方差矩阵可以表示为：$\begin{bmatrix}s_1i & s\_ji\\s\_ij & s_2j\end{bmatrix}$。原坐标系（向量空间）可以通过单位向量矩阵表示为：$\begin{bmatrix}1*i & 0*j\\0*i & 1*j\end{bmatrix}=\begin{bmatrix}i & 0\\0 & j\end{bmatrix}$，新的坐标系可以表示为：$\begin{bmatrix}w_1i & 0\\0 & w_2j\end{bmatrix}$。可以观察到，协方差矩阵与新坐标系形式相似。对协方差矩阵对角化（即除了对角线外的其他元素化为0）后，就可以获得新坐标系的轴（基），对角化就是求特征向量和特征值。

- **方差和协方差**

在方差和协方差部分，已经给出了方差和协方差的解释。方差（Variance）用于衡量一组值的分布，是每个值和均值平方差的平均，公式为：$s^2=\frac{\sum_{i=1}^{n}(X_i-\overline{X})^2}{n-1}$；协方差（Covariance）用于行列两个变量之间分布相关性程度的方法，其公式为：$cov(X,Y)=\frac{\sum_{i=1}^{n}(X_i-\overline{x})(Y_i-\overline{y})}{n-1}$。

如果协方差为0，则变量不相关；如果非0，正负号表示正负相关。正相关时，一个变量随另一个变量的增加而增加；为负相关时，一个变量随其均值增加时，另一个变量相对于其均值减小。协方差矩阵（离差矩阵，或方差-协方差矩阵）描述了一个数据集中每一对维度数变量的协方差。如果元素(i,j)表示数据i^{th}维和j^{th}维的协方差，则一个三维协方差矩阵公式如下：$C=\begin{bmatrix}cov(x_1,x_1) & cov(x_1,x_2) & cov(x_1,x_3)\\cov(x_2,x_1) & cov(x_2,x_2) & cov(x_2,x_3)\\cov(x_3,x_1) & cov(x_3,x_2) & cov(x_3,x_3)\end{bmatrix}$。使用经典的iris数据集，拓展协方差二维到4维数据，每一维度为一个特征。协方差的计算直接使用NumPy、Pandas等库。

- **再议特征值和特征向量**

对于公式$A\xi=\lambda\xi$，或表述为$A\vec{v}=\lambda\vec{v}$，其中在A变换矩阵的作用下，向量ξ或 $\vec{v}$仅仅在尺度上变为原来的λ倍。称向量ξ或 $\vec{v}$是A的一个特征向量，λ是对应的特征值。即一个矩阵乘以它的特征向量，等于对这个特征向量作缩放。特征向量和特征值只能由方阵衍生，同时并非所有的方阵都有特征向量和特征值。如果一个矩阵有特征向量和特征值，它的每一个维度上都有一个特征向量和特征值。一个矩阵的主成分是它的协方差矩阵的特征向量，对应最大特征值的特征向量是第1个主成分，对应第2大特征值的特征向量是第2个主成分，以此类推。

```
%matplotlib inline
import numpy as np
from numpy import linalg as LA
from sympy.vector.coordsysrect import CoordSys3D
from sympy.vector.vector import Vector, BaseVector
from sympy import pprint
import matplotlib.pyplot as plt
import matplotlib.animation
from IPython.display import HTML
fig, axs = plt.subplots(1, 2, figsize=(25, 12))
np.random.seed(42)

# 定义向量空间 C
C = CoordSys3D('C')
i, j, k = C.base_vectors()
v1_origin = Vector.zero

# 建立随机数据集
n_samples = 100
X = np.random.randn(100, 2)
transformation = [[-0.6, -0.6], [-0.3, -0.9]]
X = np.dot(X, transformation)
axs[0].scatter(X[:, 0], X[:, 1], label='X')
axs[1].scatter(X[:, 0], X[:, 1], label='X')

# 计算特征值和特征向量
w, v = LA.eig(np.cov(X.T))
print("covariance:\n", np.cov(X.T))
print("eigenvalues:\n", w)
print("eigenvectors:\n", v)

# 特征值乘以特征向量和特征值，获得样本新的特征
new_X = np.matmul(X, w*v)
print("new build fearure covariance:\n", np.cov(new_X.T))

# 根据特征向量建立新坐标轴
v_n_i = v[0][0]*i+v[1][0]*j
v_n_j = v[0][1]*i+v[1][1]*j
def vector_plot_2d(ax_2d, C, origin_vector, vector, color='r', label='vector', 
width=0.022):
    '''
    funciton - 转换 SymPy 的 vector 及 Matrix 数据格式为 matplotlib 可以打印的数据格式

    Params:
        ax_2d - matplotlib 的 2d 格式子图：ax(matplotlib)
        C - /coordinate_system - SymPy 下定义的坐标系；CoordSys3D()
        origin_vector - 如果是固定向量，给定向量的起点（使用向量，即表示从坐标原点所指向的
```

```
位置）；如果是自由向量，起点设置为坐标原点；vector(SymPy)
        vector - 所要打印的向量；vector(SymPy)
        color - 向量色彩，The default is 'r'；string
        label - 向量标签，The default is 'vector'；string
        arrow_length_ratio - 向量箭头大小，The default is 0.022；float
    Returns:
        None
    '''
    origin_vector_matrix = origin_vector.to_matrix(C)
    x = origin_vector_matrix.row(0)[0]
    y = origin_vector_matrix.row(1)[0]
    vector_matrix = vector.to_matrix(C)
    u = vector_matrix.row(0)[0]
    v = vector_matrix.row(1)[0]
    ax_2d.quiver(float(x), float(y), float(u), float(v),
                 color=color, label=label, width=width)
vector_plot_2d(axs[0], C, v1_origin, v_n_i,
               color='r', label='v_n_i', width=0.005)
vector_plot_2d(axs[0], C, v1_origin, v_n_j,
               color='b', label='v_n_j', width=0.005)
vector_plot_2d(axs[1], C, v1_origin, v_n_i,
               color='r', label='v_n_i', width=0.005)
vector_plot_2d(axs[1], C, v1_origin, v_n_j,
               color='b', label='v_n_j', width=0.005)

#绘制旋转的坐标轴
def circle_lines(center, radius, division):
    '''
    function - 给定圆心，半径，划分份数，计算所有直径的首尾坐标
    Params:
        center - 圆心，例如(0,0)；tuple
        radius - 半径，float
        division - 划分份数；int

    Returns:
        xy - 首坐标数组；array
        xy_ -尾坐标数组；array
        xy_head_tail - 收尾坐标数组；array
    '''
    import math
    import numpy as np
    angles = np.linspace(0, 2*np.pi, division)
    x = np.cos(angles)*radius
    y = np.sin(angles)*radius
    xy = np.array(list(zip(x, y)))
    xy = xy+center
    x_ = -x
    y_ = -y
    xy_ = np.array(list(zip(x_, y_)))
    xy_ = xy_+center
    xy_head_tail = np.concatenate((xy, xy_), axis=1)
    return xy, xy_, xy_head_tail
center = (0, 0)
radius = 5
division = 360
_, _, xy_head_tail = circle_lines(center, radius, division)

#点到直线的投影
def point_Proj2Line(line_endpts, point):
    '''
    function - 计算二维点到直线上的投影

    Params:
        line_endpts - 直线首尾点坐标，例如((2,0),(-2,0))；tuple/list
        point - 待要投影的点，例如[-0.11453985,  1.23781631]；tuple/list

    Returns:
        P - 投影点；tuple/list
    '''
    import numpy as np
    pts = np.array(point)
    Line_s = np.array(line_endpts[0])
    Line_e = np.array(line_endpts[1])
    n = Line_s - Line_e
```

```
    n_ = n/np.linalg.norm(n, 2)
    P = Line_e + n_*np.dot(pts - Line_e, n_)
    return P
line, = axs[0].plot([2, -2], [0, 0], color='gray', linewidth=1)  #旋转的轴 1 初始化
line_v, = axs[0].plot([0, 0], [2, -2], color='silver',
                      linewidth=1)  #旋转的轴 2（垂直轴 1）初始化

pts_proj_initial = np.array(
    [point_Proj2Line(((2, 0), (-2, 0)), p) for p in X])  #计算投影点
pts_proj = axs[0].scatter(
    pts_proj_initial[:, 0], pts_proj_initial[:, 1], c='r', label='project
points')  #投影点初始化

var_text = axs[0].text(0.05, 0.8, '', transform=axs[0].transAxes)  #打印文字初始化
var_template = 'the variance of the projected feature = %.3f'
v_lines = axs[0].plot([pts_proj_initial[:, 0], X[:, 0]], [
                      pts_proj_initial[:, 1], X[:, 1]], color='palegoldenrod',
linewidth=1)  #特征点和投影点垂直线初始化

#子图更新
def update(xy_head_tail):
    '''
    function - matplolib 动画更新方法
    '''
    x, y, x_, y_ = xy_head_tail[0], xy_head_tail[1], xy_head_tail[2], xy_head_
tail[3]
    pts_p = np.array([point_Proj2Line(((x, y), (x_, y_)), p) for p in X])
    line.set_data([x, x_], [y, y_])
    line_v.set_data([-y/4, -y_/4], [x/4, x_/4])
    pts_proj.set_offsets(pts_p)
    pts_proj_var = np.var(pts_p)
    var_text.set_text(var_template % (pts_proj_var))
    for i in range(pts_p.shape[0]):
        v_lines[i].set_data([pts_p[i][0], X[i][0]], [pts_p[i][1], X[i][1]])

    return line,  #需要保留 “,”

axs[0].legend()
axs[1].legend()
anima = matplotlib.animation.FuncAnimation(
    fig, update, frames=xy_head_tail, blit=True, interval=100)
HTML(anima.to_html5_video())  # ffmpeg 库的安装 conda install -c conda-forge ffmpeg
-----------------------------------------------------------------------------------
covariance:
 [[0.36372533 0.55326727]
 [0.55326727 1.10198785]]
eigenvalues:
 [0.06775316 1.39796002]
eigenvectors:
 [[-0.88175912 -0.47169996]
 [ 0.47169996 -0.88175912]]
new build fearure covariance:
 [[ 3.11020298e-04 -4.96686164e-18]
 [-4.96686164e-18  2.73202237e+00]]
#保存动画为 .gif 文件
from matplotlib.animation import FuncAnimation, PillowWriter
writer = PillowWriter(fps=25)
anima.save("./graph/pca.gif", writer=writer)
```

3.7.2.2 三维度PCA

上述阐述PCA时，从二维度入手，这样比较方便观察。下述则以机器学习经典鸢尾花（iris）数据集为例，提取其中的三个特征用于说明三个维度下特征值的PCA降维。基本过程同两个维度，只是在图表分析上结合使用三维图形（图3.7–5），方便观察。对于更高的维度，则可以借助对2，3维的理解，想象构建方便理解的高纬度空间。目前，对于高纬度空间的图表研究也有很多研究成果，可以尝试借助其中的方法来可视化高维空间。同时，计算了特征的协方差矩阵图3.7–6。

```
import numpy as np
import pandas as pd
import plotly.express as px
import matplotlib.pyplot as plt
```

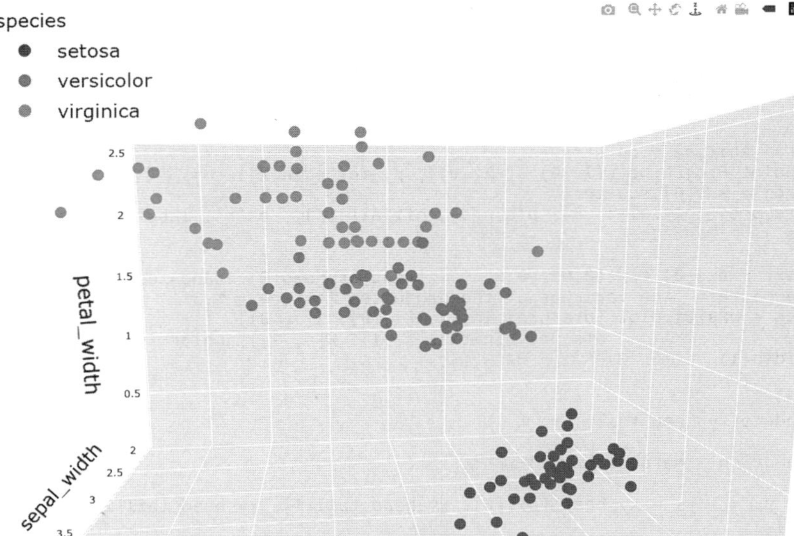

图3.7-5　鸢尾花数据集特征（选择3个）在3维空间中的分布

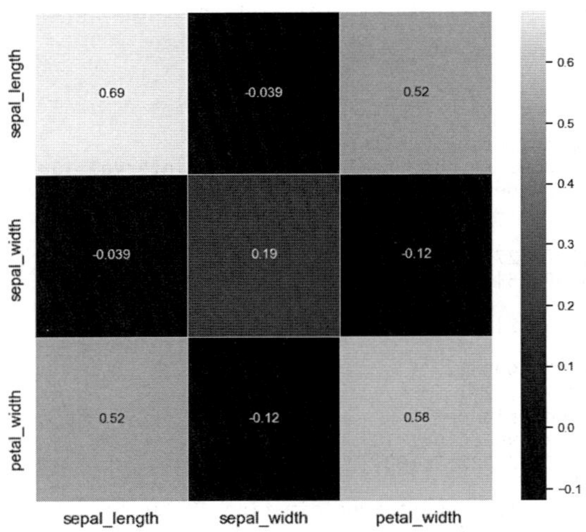

图3.7-6　鸢尾花数据集特征（选择3个）协方差矩阵

```
import seaborn as sns
sns.set()
iris_df = px.data.iris()
feature_selection = ["sepal_length", "sepal_width", "petal_width", "species"]
iris_selection = iris_df[feature_selection]
print(iris_selection)
print("-"*50)
# 先从 3 维度讲起，方便打印观察，[ "sepal_length" ," sepal_width" ," petal_length" ," petal_width" ]
COV_iris = iris_selection[["sepal_length", "sepal_width", "petal_width"]].cov()
print("iris covariance")
print(COV_iris)
f, ax = plt.subplots(figsize=(9, 8))
sns.heatmap(COV_iris, annot=True, linewidths=.5, ax=ax)
fig = px.scatter_3d(iris_df, x='sepal_length', y='sepal_width', z='petal_width',
```

```
                    color='species', width=1200, height=800)  # symbol=' species'
fig.show()
--------------------------------------------------------------------------------
     sepal_length  sepal_width  petal_width    species
0             5.1          3.5          0.2     setosa
1             4.9          3.0          0.2     setosa
..            ...          ...          ...        ...
148           6.2          3.4          2.3  virginica
149           5.9          3.0          1.8  virginica
[150 rows x 4 columns]
--------------------------------------------------
iris covariance
              sepal_length  sepal_width  petal_width
sepal_length      0.685694    -0.039268     0.516904
sepal_width      -0.039268     0.188004    -0.117981
petal_width       0.516904    -0.117981     0.582414
```

在三维空间打印特征点和对应的投影点，以及各自的坐标系（图3.7–7）。第1主成分延着N_i轴分布，具有较大的方差；而另外两个轴的特征值相对第一个轴值很小，即在2，3轴的特征点对应的投影点的方差很小。因此，第一主成分很好的保留了iris特征值的属性，达到了降维的目的。

```
import pandas as pd
from numpy import linalg as LA
import util_A
from sympy.vector.coordsysrect import CoordSys3D
from sympy.vector.vector import Vector, BaseVector
from sympy import pprint
COV_iris_array = COV_iris.to_numpy()
w, v = LA.eig(COV_iris_array)
print("eigenvalues:\n", w)
print("eigenvectors_normalized:\n", v)
iris_3Features = iris_df[feature_selection]
pd.options.mode.chained_assignment = None
species_colors = {'setosa': 'red', 'versicolor': 'blue', 'virginica': 'yellow'}
iris_3Features['color'] = iris_3Features.species.apply(
    lambda row: species_colors[row])
fig, axs = plt.subplots(1, 2, figsize=(25, 12))
axs[0] = fig.add_subplot(1, 2, 1, projection='3d')
axs[1] = fig.add_subplot(1, 2, 2, projection='3d')
axs[0].scatter(iris_3Features.sepal_length, iris_3Features.sepal_width,
               iris_3Features.petal_width, c=iris_3Features.color,)
C = CoordSys3D('C')
i, j, k = C.base_vectors()
v1_origin = Vector.zero
util_A.vector_plot_3d(axs[0], C, v1_origin, 2*i,
                      color='salmon', label='C_i', arrow_length_ratio=0.1)
util_A.vector_plot_3d(axs[0], C, v1_origin, 2*j,
                      color='maroon', label='C_j', arrow_length_ratio=0.1)
util_A.vector_plot_3d(axs[0], C, v1_origin, 2*k,
                      color='sandybrown', label='C_k', arrow_length_ratio=0.1)

# 单位特征向量 * 特征值（倍数）
N_i_ = w[0]*(v[0][0]*i+v[0][1]*j+v[0][2]*k)
N_j_ = w[1]*(v[1][0]*i+v[1][1]*j+v[1][2]*k)
N_k_ = w[2]*(v[2][0]*i+v[2][1]*j+v[2][2]*k)

# 单位特征向量
N_i = v[0][0]*i+v[0][1]*j+v[0][2]*k
N_j = v[1][0]*i+v[1][1]*j+v[1][2]*k
N_k = v[2][0]*i+v[2][1]*j+v[2][2]*k
util_A.vector_plot_3d(axs[0], C, v1_origin, N_i, color='forestgreen',
                      label='N_i:1st eigenvector', arrow_length_ratio=0.1)
util_A.vector_plot_3d(axs[0], C, v1_origin, N_j, color='limegreen',
                      label='N_j:2nd eigenvector', arrow_length_ratio=0.1)
util_A.vector_plot_3d(axs[0], C, v1_origin, N_k, color='mediumaquamarine',
                      label='N_k:3rd eigenvector', arrow_length_ratio=0.1)
util_A.vector_plot_3d(axs[1], C, v1_origin, N_i_, color='forestgreen',
                      label='N_i:1st eigenvector', arrow_length_ratio=0.1)
util_A.vector_plot_3d(axs[1], C, v1_origin, N_j_, color='limegreen',
                      label='N_j:2nd eigenvector', arrow_length_ratio=0.1)
util_A.vector_plot_3d(axs[1], C, v1_origin, N_k_, color='mediumaquamarine',
                      label='N_k:3rd eigenvector', arrow_length_ratio=0.1)
```

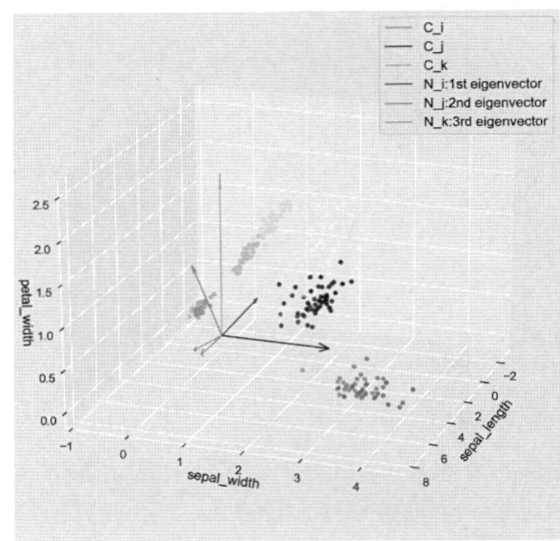

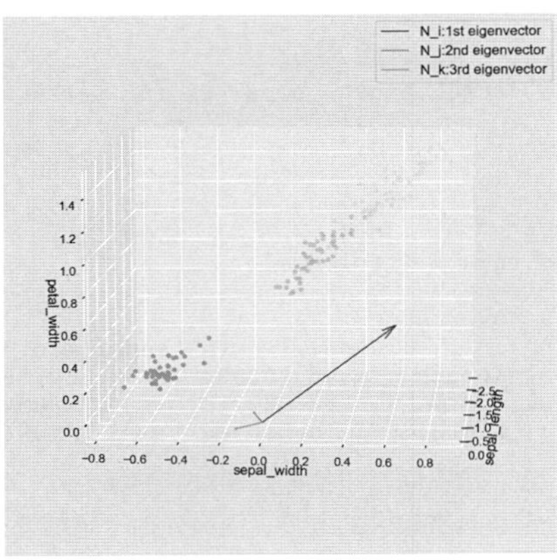

图3.7-7 笛卡尔三维坐标系和PCA向量空间及特征点分布（左）；PCA向量空间下的特征点分布（右）

```
# 特征值映射到 N_i, N_j,N_k 的向量空间 N
# 构建变换矩阵
N_f_matrix = N_i_.to_matrix(C)
for v in [N_j_, N_k_]:
    v_temp = v.to_matrix(C)
    N_f_matrix = N_f_matrix.col_insert(-1, v_temp)
print("变换矩阵：")
pprint(N_f_matrix)
iris_mapping_N = np.matmul(
    iris_df[feature_selection[:3]].to_numpy(), N_f_matrix.T).astype(float)  #线性映射
species_colors_ = {'setosa': 'coral',
                   'versicolor': 'lightblue', 'virginica': 'wheat'}
iris_3Features['color_'] = iris_3Features.species.apply(
    lambda row: species_colors_[row])
axs[0].scatter(iris_mapping_N[:, 0], iris_mapping_N[:, 1],
               iris_mapping_N[:, 2], c=iris_3Features.color_,)
axs[1].scatter(iris_mapping_N[:, 0], iris_mapping_N[:, 1],
               iris_mapping_N[:, 2], c=iris_3Features.color_,)
axs[0].legend()
axs[1].legend()
axs[0].set(xlabel='sepal_length', ylabel='sepal_width', zlabel='petal_width')
axs[1].set(xlabel='sepal_length', ylabel='sepal_width', zlabel='petal_width')
axs[1].legend(loc='lower left', frameon=False)
axs[0].view_init(20, 20)
axs[1].view_init(10, 5)
plt.show()
-------------------------------------------------------------------------------
eigenvalues:
 [1.1655692  0.07718794 0.21335471]
eigenvectors_normalized:
 [[-0.73255285  0.53278011  0.42368818]
 [ 0.11049104 -0.52110099  0.84631288]
 [-0.671683   -0.66678266 -0.32286659]]
变换矩阵：
```

$$\begin{bmatrix} 0.00852857583562477 & -0.143306730789405 & -0.85384104677628 \\ -0.0402227130979333 & -0.14226122114419 & 0.620992083779818 \\ 0.0653251494721517 & -0.0688851077741475 & 0.493837898598557 \end{bmatrix}$$

在三维空间中，打印鸢尾花数据集特征PCA变换后的特征值（图3.7–8）。

```
iris_df_N = pd.DataFrame(np.hstack((iris_mapping_N, iris_df.species.to_numpy(
).reshape(-1, 1))), columns=["1st_eigenvector", "2nd_eigenvector", "3rd_
eigenvector", "species"])
fig = px.scatter_3d(iris_df_N, x='1st_eigenvector', y='2nd_eigenvector',
                    z='3rd_eigenvector', color='species', width=1200, height=800)
# symbol='species'
fig.show()
```

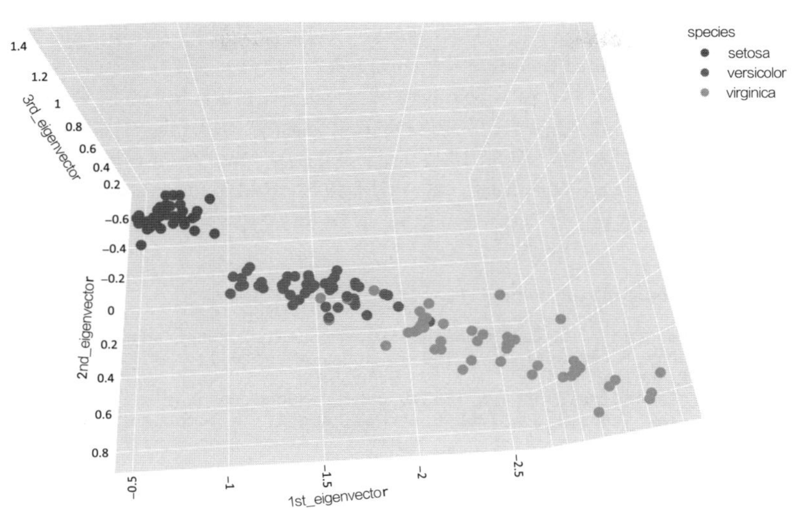

图3.7-8 鸢尾花数据集特征（选择3个）PCA变换后在3维空间中的分布

仅保留第1，2个主成分，通过二维图表查看特征点（值）的分布情况（图3.7-9）。因为具有较大的方差，各个特征值对应的响应变量（iris的种类）能够很好地分离。同样，仅打印第1个主成分（图3.7-10），线性分布的结果也可以很好地分离响应变量。

```
# A - 降维 - 保留第 1，2 主成分
fig, axs = plt.subplots(1, 2, figsize=(25, 12))
iris_PCA_1st_2nd = np.matmul(iris_df[feature_selection[:3]].to_numpy(
), np.array(N_f_matrix).astype(float)[0:2].T).astype(float)  # 降维第 1，2 主成分
axs[0].scatter(iris_PCA_1st_2nd[:, 0], iris_PCA_1st_2nd[:, 1],
               c=iris_3Features.color_)
axs[0].set(xlabel='1st_eigenvector', ylabel='2nd_eigenvector',)
iris_PCA_1st = np.matmul(iris_df[feature_selection[:3]].to_numpy(), np.array(
    N_f_matrix).astype(float)[0:1].T).astype(float)  # 降维第 1 主成分
axs[1].scatter(iris_PCA_1st, np.zeros(
    iris_PCA_1st.shape), c=iris_3Features.color_)
axs[1].set(xlabel='1st_eigenvector', ylabel='y',)
plt.show()
```

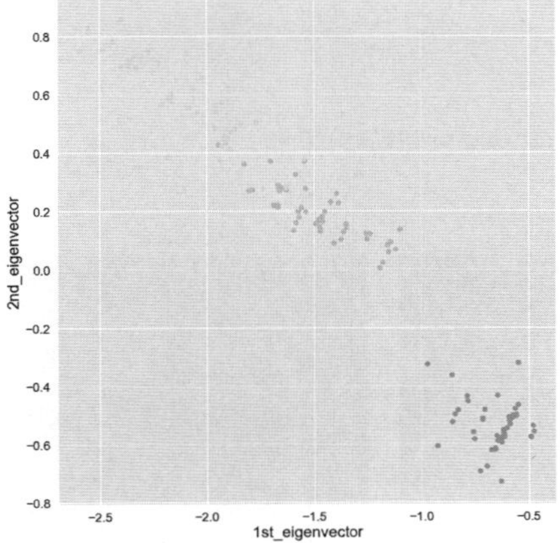

图3.7-9 鸢尾花数据集特征PCA变换后第1、2主成分

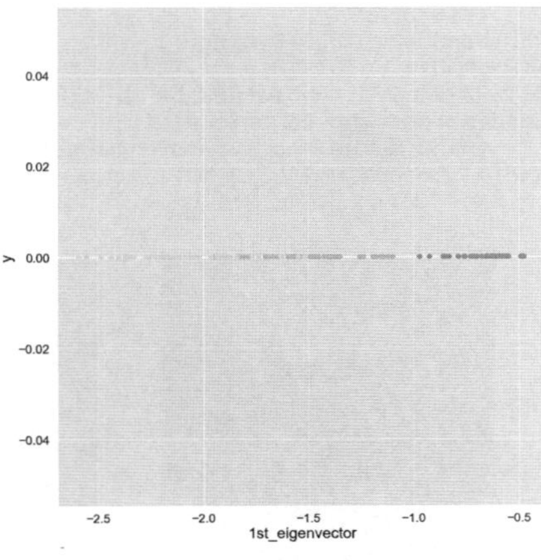

图3.7-10 鸢尾花数据集特征PCA变换后第1主成分

3.7.2.3 Landsat遥感影像波段的PCA降维与RGB显示

降维Landate的前7个波段为1个，打印比较PCA_1（主成分分析第1主成分）(图3.7–11左)，归一化植被指数（Normalized Difference Vegetation Index，NDVI）(图3.7–11中)，及［PCA_1，NDVI，band_5］组合（图3.7–11右）。NDVI能够很好地区分出植被区域。当与PCA和band5组合时，水体区域能够比较好地区分开来。对于何种波段组合或波段间的计算，能够较好地反映哪种地物。除了在上述开始引入了各个波段的主要功能，也有很多计算指数辅助解译。例如，NDVI，NDWI（Normalized Difference Water Index，归一化水指数），NDBI（Normalized Difference Built-up Index，归一化建筑指数）等，其公式分别为：$NDVI=\frac{NIR-Red}{NIR+Red}$，$NDWI=\frac{NIR-SWIR}{NIR+SWIR}$，$NDBI=\frac{SWIR-NIR}{SWIR+NIR}$。

```
from sklearn.decomposition import PCA
n_components = 1
w_180310_band_reduced = PCA(n_components=n_components).fit_transform(
    w_180310_array.T.reshape(-1, w_180310_array.shape[0]))
print("PCA finished.")
w_180310_band_reduced_reshape = w_180310_band_reduced.reshape(
    w_180310_array.shape[2], w_180310_array.shape[1], n_components).T
--------------------------------------------------------------------------------
PCA finished.
--------------------------------------------------------------------------------
from rasterio.plot import plotting_extent
band_merge = np.concatenate(
    (w_180310_array, w_180310_band_reduced_reshape), axis=0)
extent = plotting_extent(w_180310_array[0],  w_180310_raster_prof["transform"])
plt.rcParams.update({'font.size': 12})
fig, axs = plt.subplots(1, 3, figsize=(25, 12))
axs[0].imshow(w_180310_band_reduced_reshape[0], cmap='flag')  # 'prism'
axs[0].set_title("PCA_1")

# 计算 NDVI
def NDVI(RED_band, NIR_band):
    '''
    function - 计算 NDVI 指数

    Params:
        RED_band - 红色波段; array
        NIR_band - 近红外波段; array

    Returns:
        NDVI - NDVI 指数值; array
    '''
    RED_band = np.ma.masked_where(NIR_band+RED_band == 0, RED_band)
    NDVI = (NIR_band-RED_band)/(NIR_band+RED_band)
    NDVI = NDVI.filled(-9999)
    print("NDVI"+"_min:%f,max:%f" % (NDVI.min(), NDVI.max()))
    return NDVI
RED_band = w_180310_array[3]
NIR_band = w_180310_array[4]
w_180310_NDVI = NDVI(RED_band, NIR_band)
band_merge = np.concatenate(
    (band_merge, np.expand_dims(w_180310_NDVI, axis=0)), axis=0)
axs[1].imshow(w_180310_NDVI, cmap='flag')
```

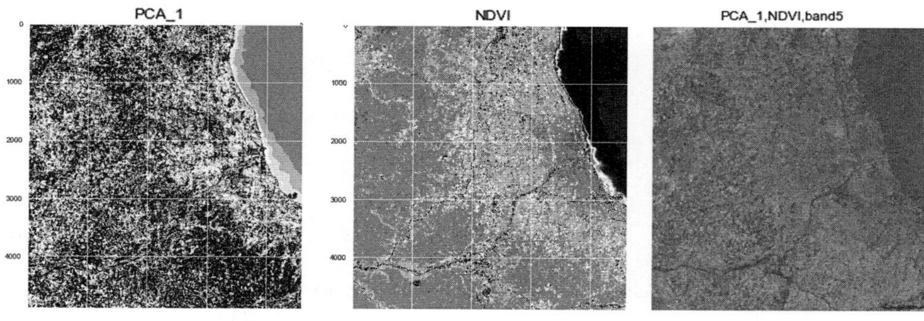

图3.7–11 Landate PCA第1主成分（左）；NDVI（中）；PCA_1，NDVI和band_5组合（右）

```
axs[1].set_title("NDVI")
ep.plot_rgb(
    band_merge,
    rgb=[-1, -2, 4],
    stretch=True,
    extent=extent,
    str_clip=0.5,
    title="PCA_1,NDVI,band5",
    ax=axs[2]
)
plt.show()
--------------------------------------------------------------------------
!_min:0.000000,max:15.654902
```

3.7.3 数字高程（Digital Elevation）（附）

数字高程模型（digital elevation model，DEM）数据也可以从美国地质调查局（United States Geological Survey，USGS-earthexplorer）下载。DEM的数据格式为栅格（raster，.tif）格式，处理方法同Landsat遥感影像的处理方式，所有栅格数据的处理模式是基本相同的（图3.7-12）。

```
import os
import numpy as np
import matplotlib.pyplot as plt
import earthpy as et
import earthpy.spatial as es
import earthpy.plot as ep
import rasterio as rio
from rasterio.warp import calculate_default_transform, reproject, Resampling
DE_Chicago_fp = r"F:\data\Landsat\GMTED2010N30W090_075\30n090w_20101117_gmted_
mea075.tif"
DEChicago_reprojecting_savePath = os.path.join(workspace, r"DE_Chicago.tif")
# 投影转换 DEM，.tif 栅格数据
dst_crs = crop_raster_profile["crs"]
with rio.open(DE_Chicago_fp) as src:
    transform, width, height = calculate_default_transform(
        src.crs, dst_crs, src.width, src.height, *src.bounds)
    kwargs = src.meta.copy()
    kwargs.update({
        'crs': dst_crs,
        'transform': transform,
        'width': width,
        'height': height
    })
    with rio.open(DEChicago_reprojecting_savePath, 'w', **kwargs) as dst:
        for i in range(1, src.count + 1):
            reproject(
                source=rio.band(src, i),
```

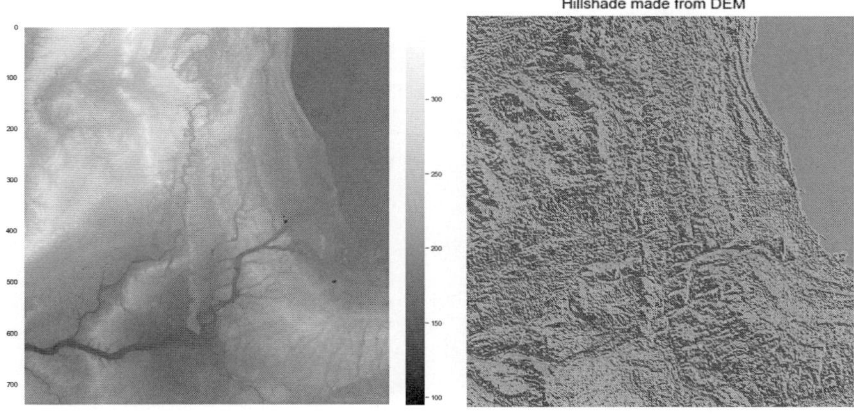

图3.7-12 DEM数据（左）；DEM数据的山体阴影方式显示（右）

```
                    destination=rio.band(dst, i),
                    src_transform=src.transform,
                    src_crs=src.crs,
                    dst_transform=transform,
                    dst_crs=dst_crs,
                    resampling=Resampling.nearest)
print("reprojecting finished ...")
--------------------------------------------------------------------------------
reprojecting finished ...
--------------------------------------------------------------------------------
# 裁切 DEM .tif 数据
DE_Chicago = es.crop_all([DEChicago_reprojecting_savePath],
                         workspace, crop_bound_utm16N, overwrite=True)  # 对所有波段 band 执行裁切
print("cropping finished ...")
--------------------------------------------------------------------------------
cropping finished ...
--------------------------------------------------------------------------------
DE_Chicago_tif = rio.open(DE_Chicago[0])

# DEM 打印
fig, axs = plt.subplots(1, 2, figsize=(25, 12))
dem = axs[0].imshow(DE_Chicago_tif.read(1), cmap='pink')
plt.colorbar(dem, fraction=0.0485, pad=0.04, ax=axs[0])

# 增加山体阴影
hillshade = es.hillshade(DE_Chicago_tif.read(1), azimuth=210, altitude=10)
ep.plot_bands(hillshade, cbar=False, title="Hillshade made from DEM",
              figsize=(10, 6), ax=axs[1], cmap='terrain')
axs[1].imshow(hillshade, cmap="Greys", alpha=0.5)
axs[0].grid(False)
plt.show()
```

参考文献（References）：

[1] Cavin Hackeling．张浩然译．scikit-learning机器学习[M]．北京：人民邮电出版社．2019．2.（Cavin Hackeling. Mastering Machine Learning with scikit-learn[M]. Packt Publishing Ltd. July 2017. Second published.）

[2] [日]高桥信著作，Inoue Iroha，株式会社TREND-PRO漫画制作，腾永红译．漫画线性代数[M]．北京：科学出版社，2009．08．

[3] PCA解读，Making sense of principal component analysis, eigenvectors & eigenvalues (https://stats.stackexchange.com/questions/2691/making-sense-of-principal-component-analysis-eigenvectors-eigenvalues/140579#140579).

3.8 基于NDVI指数解译影像与建立采样工具

如果只是分析城市的绿地、裸地和水体，不涉及更精细的土地覆盖分类，例如灌丛、草地、裸地、居民地、园地、耕地、河流，湖泊等，可以利用Landsat系列遥感影像通过NDVI（归一化植被指数）、NDWI（归一化水体指数）和NDBI（归一化建筑指数）等手段提取绿地（可进一步细分耕地和林地）、裸地和水体等。基于NDVI的遥感影像解译过程如下：

1. 首先，读取能够反映不同季节绿地情况的Landsat不同季节影像，根据研究的目的范围裁切，裁切的边界可以在QGIS中完成。
2. 然后，计算不同季节的NDVI。
3. 再通过使用交互的Plotly图表，分析NDVI取值范围，判断不同土地覆盖阈值范围，解译影像。
4. 如果要判断解译的精度，需要给出采样，随机提取点的真实土地覆盖类型，这个过程是一个需要手工操作的过程。
5. Python内嵌库tkinter①的图形用户界面（Graphical User Interface，GUI）能够方便帮助快速地建立交互操作平台，完成采样工作。
6. 最后，计算混淆矩阵和百分比精度，判断解译的精度。

① tkinter，是Tcl/Tk GUI工具包的标准Python接口（https://docs.python.org/3/library/tkinter.html）。

3.8.1 影像数据处理

为了方便影像的处理，通常将一些常用的工具构建为函数，例如下述的影像裁切函数raster_clip()。影像处理最需要关心的是坐标投影系统，通常Landsat影像都包含对应的投影系统，例如“UTM, DATUM: WGS84, UTM_ZONE16”，因此最好直接统一为该坐标投影系统，不建议转换为其他投影。对于其他地理文件而言，如果是栅格数据通常需要转换投影；对于SHP的矢量地理文件，一般保持为WGS84，即“EPSG: 4326”，通常不含投影，方便用于不同坐标投影系统平台，以及作为中转的数据格式类型。因此，在QGIS等平台下建立裁切边界时，仅保持坐标系为WGS84，不配置投影。在定义的裁切函数中，根据读取的Landsat影像投影系统，再进行定义，方便数据的处理。

影像数据一般都很大，因此影像数据的处理，尤其高分辨率的影像会花费更长时间，有必要将处理后的数据即刻保存到本地磁盘下。当需要时，可以直接从硬盘中读取，避免再次花费时间计算。

```
import os
import util_A
workspace = r"F:\data\Landsat\data_processed"
Landsat_fp = {
    "w_180310": r"F:\data\Landsat\LC08_L1TP_023031_20180310_20180320_01_T1",  #冬季
    "s_190820": r"F:\data\Landsat\LC08_L1TP_023031_20190804_20190820_01_T1",  #夏季
    "a_191018": r"F:\data\Landsat\LC08_L1TP_023031_20191007_20191018_01_T1"  #秋季
}
w_180310_band_fp_dic, w_180310_Landsat_para = util_A.LandsatMTL_info(
    Landsat_fp["w_180310"])  # LandsatMTL_info(fp) 函数，在 Landsat 遥感影像处理部分阐述，将其
放置于 util_A.py 文件中后调用
s_190820_band_fp_dic, s_190820_Landsat_para = util_A.LandsatMTL_info(
    Landsat_fp["s_190820"])
a_191018_band_fp_dic, a_191018_Landsat_para = util_A.LandsatMTL_info(
    Landsat_fp["a_191018"])
print("w_180310-MAP_PROJECTION:%s,DATUM:%s,UTM_ZONE%s" %
      (w_180310_Landsat_para['MAP_PROJECTION'], w_180310_Landsat_para['DATUM'],
w_180310_Landsat_para['UTM_ZONE']))
print("s_190820-MAP_PROJECTION:%s,DATUM:%s,UTM_ZONE%s" %
      (s_190820_Landsat_para['MAP_PROJECTION'], s_190820_Landsat_para['DATUM'],
s_190820_Landsat_para['UTM_ZONE']))
print("a_191018-MAP_PROJECTION:%s,DATUM:%s,UTM_ZONE%s" %
      (a_191018_Landsat_para['MAP_PROJECTION'], a_191018_Landsat_para['DATUM'],
a_191018_Landsat_para['UTM_ZONE']))
def raster_clip(raster_fp, clip_boundary_fp, save_path):
    '''
```

```
    function - 给定裁切边界，批量裁切栅格数据

    Params:
        raster_fp - 待裁切的栅格数据文件路径（.tif）；string
        clip_boundary - 用于裁切的边界（.shp，WGS84，无投影），与栅格具有相同的坐标投
影系统；string

    Returns:
        rasterClipped_pathList - 裁切后的文件路径列表；list(string)
    '''
    import earthpy.spatial as es
    import geopandas as gpd
    from pyproj import CRS
    import rasterio as rio
    clip_bound = gpd.read_file(clip_boundary_fp)
    with rio.open(raster_fp[0]) as raster_crs:
        raster_profile = raster_crs.profile
        clip_bound_proj = clip_bound.to_crs(raster_profile["crs"])
    rasterClipped_pathList = es.crop_all(
        raster_fp, save_path, clip_bound_proj, overwrite=True)  # 对所有波段 band 执
行裁切
    print("clipping finished.")
    return rasterClipped_pathList
clip_boundary_fp = './data/LandsatChicago_boundary/LandsatChicago_boundary.shp'
save_path = os.path.join(workspace, "s_190820")
s_190820_clipped_fp = raster_clip(
    list(s_190820_band_fp_dic.values()), clip_boundary_fp, save_path)
--------------------------------------------------------------------------------
w_180310-MAP_PROJECTION:UTM,DATUM:WGS84,UTM_ZONE16
s_190820-MAP_PROJECTION:UTM,DATUM:WGS84,UTM_ZONE16
a_191018-MAP_PROJECTION:UTM,DATUM:WGS84,UTM_ZONE16
clipping finished.
```

通过定义的裁切函数，直接计算冬季和秋季影像。

```
save_path = os.path.join(workspace,"w_180310")
w_180310_clipped_fp = raster_clip(list(w_180310_band_fp_dic.values()),clip_
boundary_fp,save_path)
save_path = os.path.join(workspace,"a_191018")
a_191018_clipped_fp=raster_clip(list(a_191018_band_fp_dic.values()),clip_
boundary_fp,save_path)
--------------------------------------------------------------------------------
clipping finished.
clipping finished.
```

从美国地质调查局下载的Landsat影像，各个波段是单独的文件。为了避免每次读取单个文件，使用EarthPy库的.stack方法将所有波段放置于一个文件下（数组形式），方便数据处理。

```
import earthpy.spatial as es
w_180310_array, w_180310_raster_prof = es.stack(
    w_180310_clipped_fp[:7], out_path=os.path.join(workspace, r"w_180310_stack.
tif"))
print("stacking_1 finished...")
s_190820_array, s_190820_raster_prof = es.stack(
    s_190820_clipped_fp[:7], out_path=os.path.join(workspace, r"s_190820_stack.
tif"))
print("stacking_2 finished...")
a_191018_array, a_191018_raster_prof = es.stack(
    a_191018_clipped_fp[:7], out_path=os.path.join(workspace, r"a_191018_stack.
tif"))
print("stacking_3 finished...")
--------------------------------------------------------------------------------
stacking_1 finished...
stacking_2 finished...
stacking_3 finished...
```

只有将数据显示出来，才能更好地判断地理空间数据处理的结果是否正确。建立w_180310_array影像波段显示函数，查看影像。可以通过band_num输入参数，确定合成显示的波段。

```
def bands_show(img_stack_list, band_num):
    '''
    function - 指定波段，同时显示多个遥感影像

    Params:
```

```
        img_stack_list - 影像列表; list(array)
        band_num - 显示的层列表; list(int)
    '''
    import matplotlib.pyplot as plt
    from rasterio.plot import plotting_extent
    import earthpy.plot as ep
    def variable_name(var):
        '''
        function - 将变量名转换为字符串

        Params:
            var - 变量名
        '''
        return [tpl[0] for tpl in filter(lambda x: var is x[1], globals().
items())][0]
    plt.rcParams.update({'font.size': 12})
    img_num = len(img_stack_list)
    fig, axs = plt.subplots(1, img_num, figsize=(12*img_num, 12))
    i = 0
    for img in img_stack_list:
        ep.plot_rgb(
            img,
            rgb=band_num,
            stretch=True,
            str_clip=0.5,
            title="%s" % variable_name(img),
            ax=axs[i]
        )
        i += 1
    plt.show()
img_stack_list = [w_180310_array, s_190820_array, a_191018_array]
band_num = [3, 2, 1]
bands_show(img_stack_list, band_num)
```

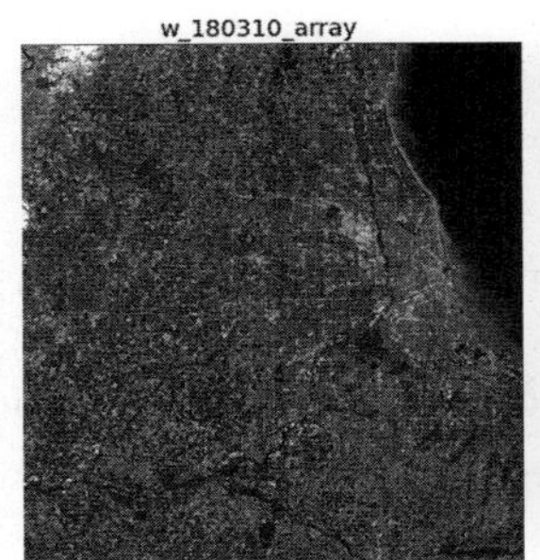

图3.8-1 合成显示波段

从上述直接显示的合成波段影像（图3.8–1、图3.8–2）来看，显示效果偏暗，不利于细节的观察。借助scikit-image[①]图像处理库中的exposure方法，可以拉伸图像，调亮与增强对比度。处理时，需要注意要逐波段的处理，然后再合并。

① scikit-image，是一个用于图像处理的算法集合（https://scikit-image.org/）。

```
def image_exposure(img_bands, percentile=(2, 98)):
    '''
    function - 拉伸图像 contract stretching

    Params:
        img_bands - landsat stack后的波段; array
        percentile - 百分位数, The default is (2,98); tuple
    Returns:
        img_bands_exposure - 返回拉伸后的影像; array
    '''
    from skimage import exposure
    import numpy as np
    bands_temp = []
    for band in img_bands:
        p2, p98 = np.percentile(band, (2, 98))
        bands_temp.append(exposure.rescale_intensity(band, in_range=(p2, p98)))
```

```
    img_bands_exposure = np.concatenate(
        [np.expand_dims(b, axis=0) for b in bands_temp], axis=0)
    print("exposure finished.")
    return img_bands_exposure
w_180310_exposure = image_exposure(w_180310_array)
s_190820_exposure = image_exposure(s_190820_array)
a_191018_exposure = image_exposure(a_191018_array)
--------------------------------------------------------------------------------
exposure finished.
exposure finished.
exposure finished.
```

显示处理后的遥感影像，对比原显示效果，可见有明显的提升。

```
img_stack_list = [w_180310_exposure,s_190820_exposure,a_191018_exposure]
band_num = [3,2,1]
bands_show(img_stack_list,band_num)
```

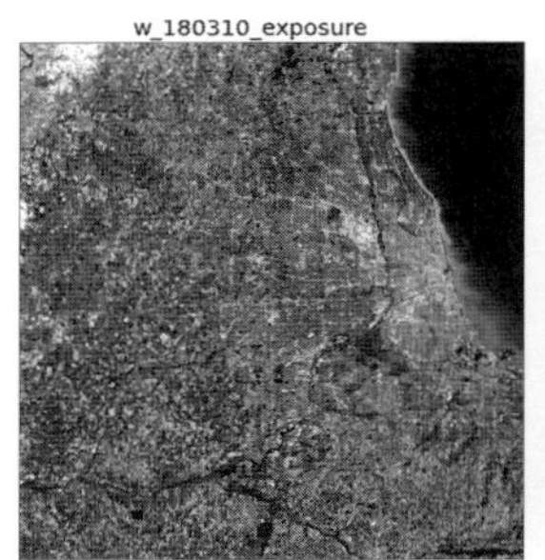

图3.8-2　合成显示波段(曝光处理后)

3.8.2　计算NDVI，交互图像与解译

调用阐述影像波段计算指数时定义的NDVI计算函数NDVI(RED_band,NIR_band)计算NDVI归一化植被指数。同时，查看计算结果的最大和最小值，即值的区间。可以看到，s_190820_NDVI数据（夏季）明显有错误。通过查看上述图像，也能够发现下载的夏季Landsat影像在城区中有很多云，初步判断可能是造成数据异常的原因。

```
import util_A
import rasterio as rio
import os
workspace = r"F:\data\Landsat\data_processed"
w_180310 = rio.open(os.path.join(workspace, r"w_180310_stack.tif"))
s_190820 = rio.open(os.path.join(workspace, r"s_190820_stack.tif"))
a_191018 = rio.open(os.path.join(workspace, r"a_191018_stack.tif"))
w_180310_NDVI = util_A.NDVI(w_180310.read(4), w_180310.read(5))
s_190820_NDVI = util_A.NDVI(s_190820.read(4), s_190820.read(5))
a_191018_NDVI = util_A.NDVI(a_191018.read(4), a_191018.read(5))
--------------------------------------------------------------------------------
NDVI_min:0.000000,max:15.654902
NDVI_min:-9999.000000,max:8483.000000
NDVI_min:0.000000,max:15.768559
```

因为NDVI的数据为一个维度数组，可以直接使用异常值处理部分定义的is_outlier (data,threshold=3.5)函数处理异常值，并打印NDVI图3.8-3。查看处理结果，能够较为清晰地看到NDVI对绿地植被的识别，可以较好地区分水体和裸地。

```
import matplotlib.pyplot as plt
import earthpy.plot as ep
fig, axs = plt.subplots(1, 3, figsize=(30, 12))
ep.plot_bands(w_180310_NDVI, cbar=False,
              title="w_180310_NDVI", ax=axs[0], cmap='flag')
```

#对异常值采取了原地取代的方式，因此如果多次运行异常值检测，需要重新计算 NDVI，保持原始数据不变

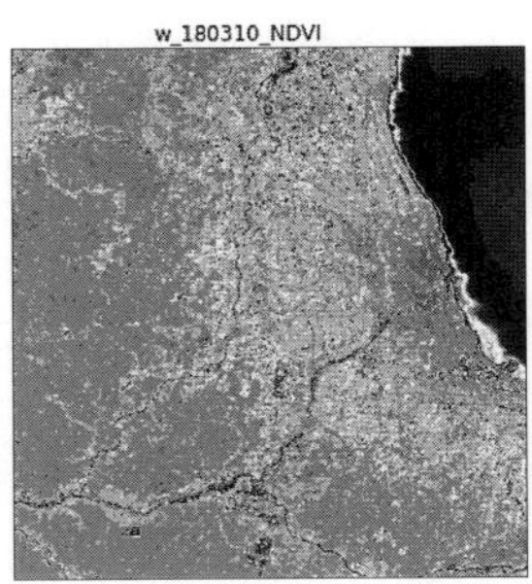

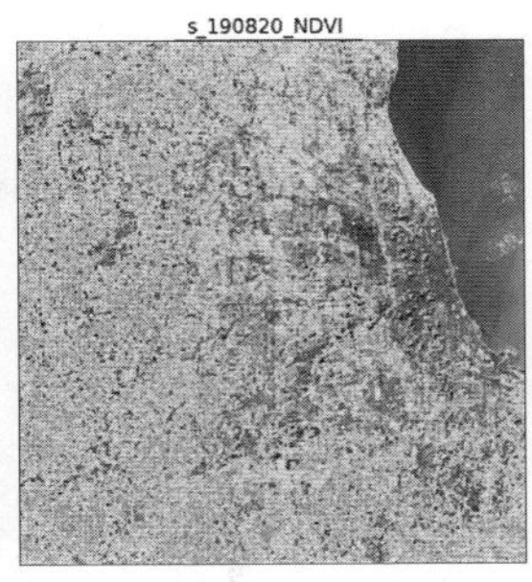

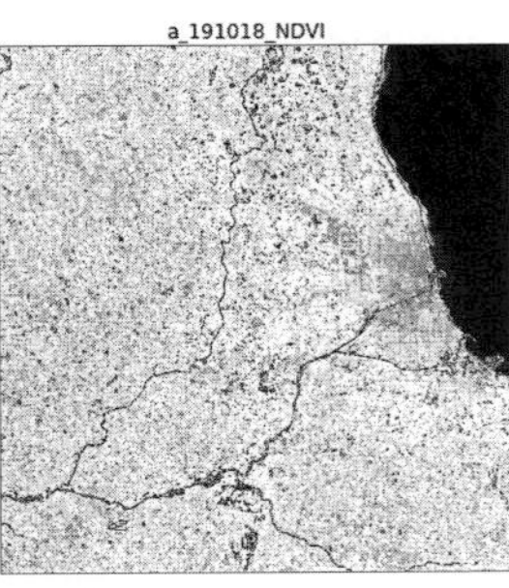

图3.8-3 不同季节NDVI计算结果

```
s_190820_NDVI = util_A.NDVI(s_190820.read(4), s_190820.read(1))
is_outlier_bool, _ = util_A.is_outlier(s_190820_NDVI, threshold=3)
s_190820_NDVI[is_outlier_bool] = 0  #原地取代
ep.plot_bands(s_190820_NDVI, cbar=False,
              title="s_190820_NDVI", ax=axs[1], cmap='terrain')
ep.plot_bands(a_191018_NDVI, cbar=False,
              title="a_191018_NDVI", ax=axs[2], cmap='flag')
plt.show()
-------------------------------------------------------------------------------
NDVI_min:0.000000,max:64105.000000
```

直接处理形式为（4900，4604）的NDVI数组，包含4900×4604多个数据。如果计算机硬件条件可以，保证在所能够承担的计算时间长度下，可以不用压缩数据；否则，在不影响最终计算结果对后续分析的影响，可以适当地压缩影像，直接使用scikit-image图像处理库中的rescale方法；同时，也调入了resize，downscale_local_mean方法。一般计算的方法是求取局部均值。

打印夏季影像计算所得的NDVI，按阈值显示颜色的图3.8-4，方便查看不同阈值区间的区域范围。便于确定阈值，从而解译土地用地分类。

```
import plotly.express as px
from skimage.transform import rescale, resize, downscale_local_mean
s_190820_NDVI_rescaled = rescale(s_190820_NDVI, 0.1, anti_aliasing=False)
fig = px.imshow(img=s_190820_NDVI_rescaled, zmin=s_190820_NDVI_rescaled.min(
), zmax=s_190820_NDVI_rescaled.max(), width=800, height=800, color_continuous_
scale=px.colors.sequential.speed)
fig.show()
```

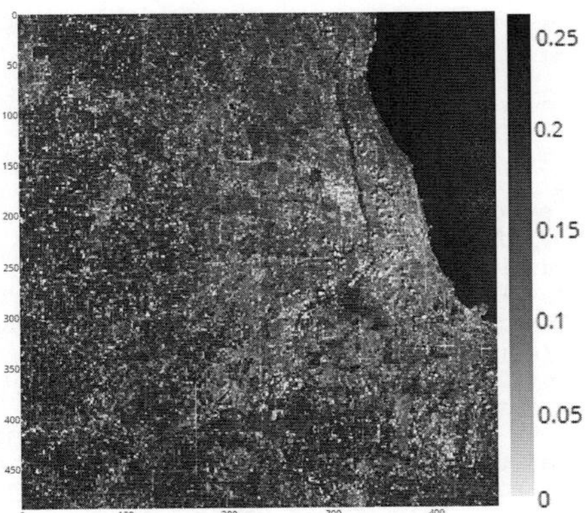

图3.8-4 NDVI（夏季）显示颜色阈值

计算的NDVI值是一个无量纲、维度为1的数组，值通常为浮点型小数。在使用matplotlib，Plotly，seaborn和bokeh等图表库时，一般会使用RGB，RGBA，十六进制形式，以及浮点型等表示颜色的数据格式。对于计算的NDVI，定义函数data_division()，将其按照分类的阈值转换为RGB的颜色格式方便图像打印（图3.8-5）。因为NDVI本身不是颜色数据，首先根据分类阈值用.percentile方法计算百分位数，然后根据百分位数使用.digitize方法，返回数值对应区间的索引（整数）。由区间数量即唯一的索引数定义不同的位于0～255的随机整数（每个值对应三个随机整数值），代表颜色。

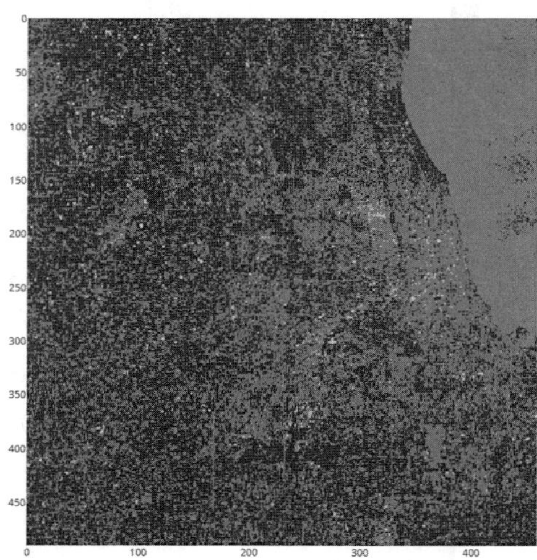

图3.8-5　将NDVI按分类阈值转化为RGB的颜色格式

```
from skimage.transform import rescale, resize, downscale_local_mean
def data_division(data, division, right=True):
    '''
    function - 将数据按照给定的百分数划分，并给定固定的值，整数值或 RGB 色彩值

    Params:
        data - 待划分的 numpy 数组; array
        division - 百分数列表，例如 [0,35,85]; list
        right - bool,optional. The default is True.
                Indicating whether the intervals include the right or the left
bin edge.
                Default behavior is (right==False) indicating that the interval
does not include the right edge.
                The left bin end is open in this case, i.e., bins[i-1] <= x <
bins[i] is the default behavior for monotonically increasing bins.
    Returns:
        data_digitize - 返回整数值; array(int)
        data_rgb - 返回 RGB，颜色值; array
    '''
    import numpy as np
    import pandas as pd
    percentile = np.percentile(data, np.array(division))
    data_digitize = np.digitize(data, percentile, right)
    unique_digitize = np.unique(data_digitize)
    random_color_dict = [{k: np.random.randint(
        low=0, high=255, size=1)for k in unique_digitize} for i in range(3)]
    data_color = [pd.DataFrame(data_digitize).replace(
        random_color_dict[i]).to_numpy() for i in range(3)]
    data_rgb = np.concatenate([np.expand_dims(i, axis=-1)
                               for i in data_color], axis=-1)
    return data_digitize, data_rgb
w_180310_NDVI_rescaled = rescale(w_180310_NDVI, 0.1, anti_aliasing=False)
s_190820_NDVI_rescaled = rescale(s_190820_NDVI, 0.1, anti_aliasing=False)
a_191018_NDVI_rescaled = rescale(a_191018_NDVI, 0.1, anti_aliasing=False)
print("rescale finished.")
--------------------------------------------------------------------------------
rescale finished.
--------------------------------------------------------------------------------
division = [0, 35, 85]
_, s_190820_NDVI_RGB = data_division(s_190820_NDVI_rescaled, division)
fig = px.imshow(img=s_190820_NDVI_RGB, zmin=s_190820_NDVI_RGB.min(),
```

```
zmax=s_190820_NDVI_RGB.max(
), width=800, height=800, color_continuous_scale=px.colors.sequential.speed)
fig.show()
```

为辅助解译，确定阈值，可以通过直方图（图3.8–6频数分布）查看NDVI的数据分布情况。因为地物的数量不同，某些转折或断裂的位置点可能代表不同的地物。

```
fig, axs = plt.subplots(1, 3, figsize=(20, 6))
count, bins, ignored = axs[0].hist(
    w_180310_NDVI.flatten(), bins=100, density=True)
count, bins, ignored = axs[1].hist(
    s_190820_NDVI.flatten(), bins=100, density=True)
count, bins, ignored = axs[2].hist(
    a_191018_NDVI.flatten(), bins=100, density=True)
plt.show()
```

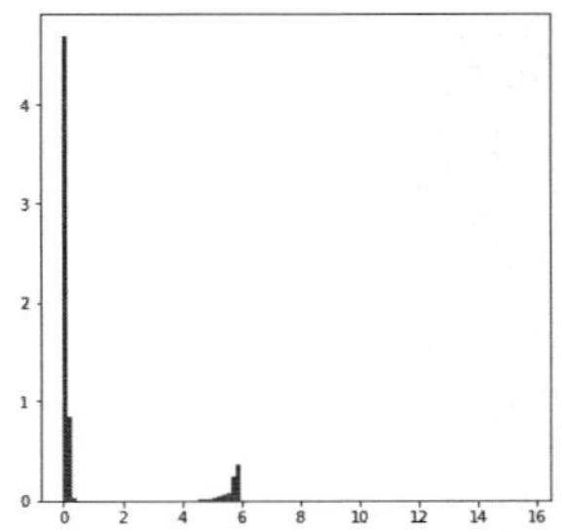
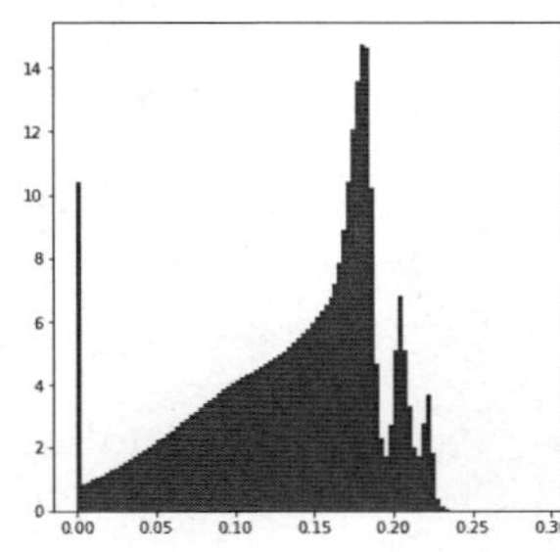
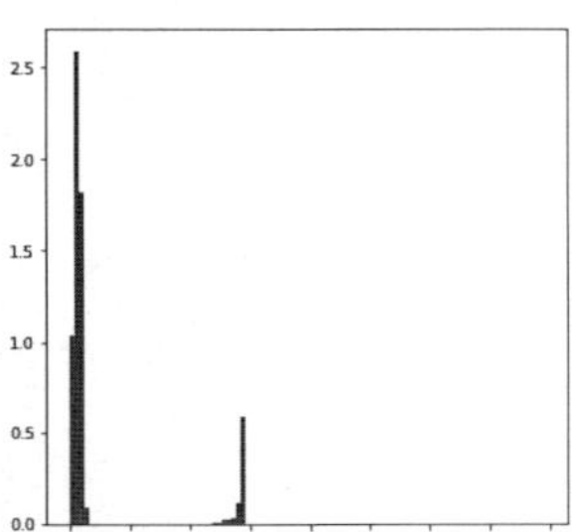

图3.8–6　3个季节NDVI值频数分布

虽然上述代码可以打印，按阈值显示NDVI图像，但是不能够交互操作，不能通过不断地调整阈值区间，即刻地查看阈值图像，从而判断适合于分类地物的阈值。Plotly图表工具提供了简单的交互式方法，可以通过增加滑条窗口工具（widgets.IntSlier）及下拉栏（widgets.Dropdown）实现交互。这里，定义了三个滑动条来定义阈值区间（图3.8–7），一个下拉栏定义可选择的三个不同季节的NDVI数据，调用上述定义的data_division()函数，按阈值区间配置颜色显示图像。通过不断调整阈值，与波段3，2，1组合的真彩色图像比较，判断三个季节土地用地分类阈值区间为［0, 35, 85］。注意，通常不同影像的NDVI分类阈值可能不同。

w_180310_NDVI_rescaled，［0, 35, 85］；s_190820_NDVI_rescaled，［0, 35, 85］；a_191018_NDVI_rescaled，［0, 35, 85］。

```
def percentile_slider(season_dic):
    '''
    function - 多个栅格数据，给定百分比，交互观察

    Params:
        season_dic - 多个栅格字典，键为自定义键名，值为读取的栅格数据(array)，
例如{"w_180310": w_180310_NDVI_rescaled,"s_190820": s_190820_NDVI_
rescaled,"a_191018": a_191018_NDVI_rescaled}; dict
    Returns:
        None
    '''
    import numpy as np
    import pandas as pd
    import plotly.graph_objects as go
    from ipywidgets import widgets
    from IPython.display import display
    p_1_slider = widgets.IntSlider(
        min=0, max=100, value=10, step=1, description="percentile_1")
    p_2_slider = widgets.IntSlider(
        min=0, max=100, value=30, step=1, description="percentile_2")
    p_3_slider = widgets.IntSlider(
        min=0, max=100, value=50, step=1, description="percentile_3")
    season_keys = list(season_dic.keys())
    season = widgets.Dropdown(
        description='season',
        value=season_keys[0],
```

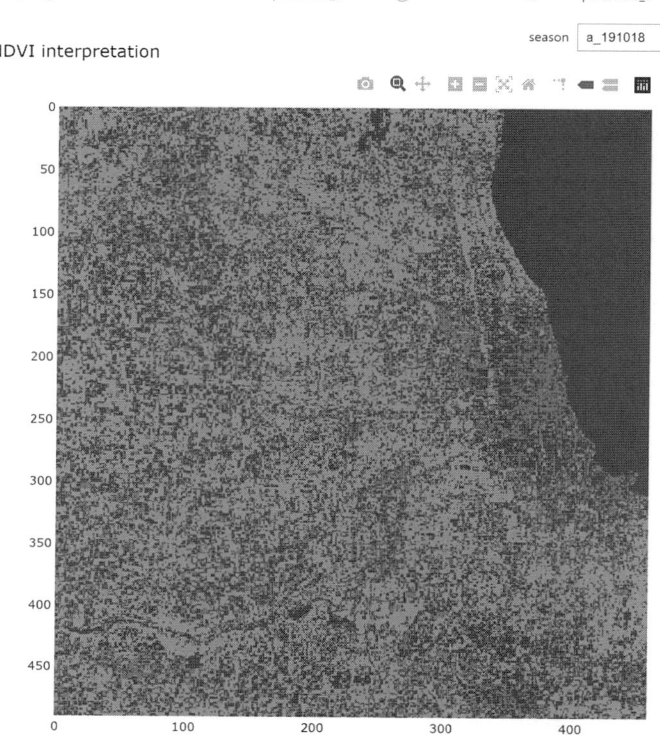

图3.8-7 交互式调整观察NDVI阈值

```
        options=season_keys
    )
    season_val = season_dic[season_keys[0]]
    _, img = data_division(season_val, division=[10, 30, 50], right=True)
    trace1 = go.Image(z=img)
    g = go.FigureWidget(data=[trace1,],
                        layout=go.Layout(
        title=dict(
            text='NDVI interpretation'
        ),
        width=800,
        height=800
    ))
    def validate():
        if season.value in season_keys:
            return True
        else:
            return False
    def response(change):
        if validate():
            division = [p_1_slider.value, p_2_slider.value, p_3_slider.value]
            _, img_ = data_division(
                season_dic[season.value], division, right=True)
            with g.batch_update():
                g.data[0].z = img_
    p_1_slider.observe(response, names="value")
    p_2_slider.observe(response, names="value")
    p_3_slider.observe(response, names="value")
    season.observe(response, names="value")
    container = widgets.HBox([p_1_slider, p_2_slider, p_3_slider, season])
    box = widgets.VBox([container, g])
    display(box)
season_dic = {"w_180310": w_180310_NDVI_rescaled,
```

```
            "s_190820": s_190820_NDVI_rescaled, "a_191018": a_191018_NDVI_
rescaled}
percentile_slider(season_dic)
```

确定［0, 35, 85］的阈值区间后，就可计算所有季节的NDVI土地用地分类。

```
division=[0,35,85]
w_interpretation,_ = data_division(w_180310_NDVI_rescaled,division,right=True)
s_interpretation,_ = data_division(s_190820_NDVI_rescaled,division,right=True)
a_interpretation,_ = data_division(a_191018_NDVI_rescaled,division,right=True)
```

获得三个不同季节NDVI的解译结果后（图3.8-8左1～3），根据冬季大部分农田无种植，秋季部分农田收割，夏季大部分农田生长；以及落叶树木夏季茂盛、常绿树木四季常青的特点，可以解译出农田、常绿、落叶和裸地、水体等土地覆盖类型（图3.8-8右1），这里直接将只要各季节为绿色的就划分为绿地（.logical_or），值为2；而水体中湖泊的阈值区分较好，河流的则不是很清晰，确定只有都为水体才为水体的判断（.logical_and），值为3；其余的则为裸地，值为0和1。

```
import plotly.express as px
from plotly.subplots import make_subplots
import plotly.graph_objects as go
import numpy as np
fig = make_subplots(rows=1, cols=4, shared_xaxes=False, subplot_titles=(
    'w_interpretation', 's_interpretation', 'a_interpretation', 'green_water_
bareLand'))
fig1 = px.imshow(img=np.flip(w_interpretation, 0), zmin=w_interpretation.min(),
zmax=w_interpretation.max(),
width=800, height=800, color_continuous_scale=px.colors.sequential.deep,
title="winter")
fig2 = px.imshow(img=np.flip(s_interpretation, 0), zmin=s_interpretation.min(
), zmax=s_interpretation.max(), width=800, height=800, color_continuous_scale=px.
colors.sequential.haline)
fig3 = px.imshow(img=np.flip(a_interpretation, 0), zmin=a_interpretation.min(
), zmax=a_interpretation.max(), width=800, height=800, color_continuous_scale=px.
colors.sequential.haline)
green = np.logical_or(w_interpretation == 2, s_interpretation ==
                      2, a_interpretation == 2)  # 只要为绿地（值为 2），就为 2
water = np.logical_and(w_interpretation == 3, s_interpretation ==
                       3, a_interpretation == 3)  # 只有都为水体（值为 3），才为 3
green_v = np.where(green == True, 2, 0)
water_v = np.where(water == True, 3, 0)
green_water_bareLand = green_v+water_v
fig4 = px.imshow(img=np.flip(green_water_bareLand, 0), zmin=green_water_bareLand.
min(
), zmax=green_water_bareLand.max(), width=800, height=800, color_continuous_
scale=px.colors.sequential.haline)
trace1 = fig1['data'][0]
trace2 = fig2['data'][0]
trace3 = fig3['data'][0]
trace4 = fig4['data'][0]
fig.add_trace(trace1, row=1, col=1)
fig.add_trace(trace2, row=1, col=2)
fig.add_trace(trace3, row=1, col=3)
fig.add_trace(trace4, row=1, col=4)
fig.update_layout(height=500, width=1600, title_text="interpretation")
fig.show()
```

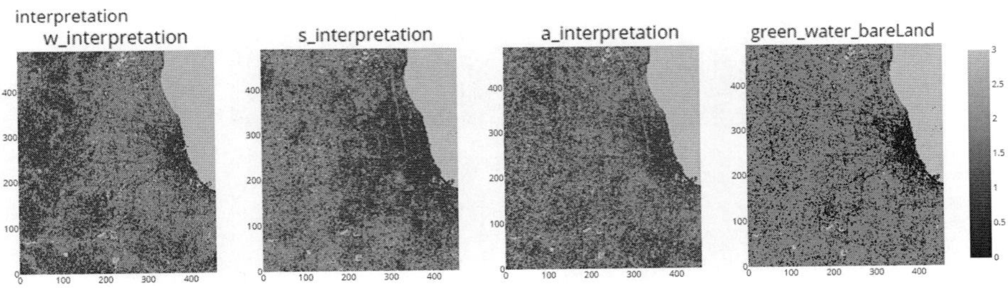

图3.8-8 按NDVI阈值解译土地覆盖类型，分别为冬季（左1）；夏季（左2）；秋季（左3）和综合结果（右1）

3.8.3 采样交互操作平台的建立，与精度计算

3.8.3.1 使用tkinter建立采样交互操作平台

对于采样的工作，可以借助于QGIS等平台，建立SHP点数据格式，随机生成一定数量的点，由目视确定每个点的土地覆盖类型；或者直接手工点取，然后将该采样数据在Python下读取，再进行后续的精度分析。该方法需要数据在不同平台下转换，操作稍许烦琐。此处，通过tkinter自行建立可交互的GUI采样工具。采样的参照底图选择了秋季的影像（图3.8-9），将该影像单独保存，方便调用。

```
import matplotlib.pyplot as plt
import earthpy.plot as ep
from skimage.transform import rescale, resize, downscale_local_mean
import os
import numpy as np
a_191018_exposure_rescaled = np.concatenate([np.expand_dims(rescale(
    b, 0.1, anti_aliasing=False), axis=0) for b in a_191018_exposure], axis=0)
fig, ax = plt.subplots(figsize=(12, 12))
ep.plot_rgb(
    a_191018_exposure_rescaled,
    rgb=[3, 2, 1],
    stretch=True,
    # extent=extent,
    str_clip=0.5,
    title="a_191018_exposure_rescaled",
    ax=ax
)
plt.show()
print("a_191018_exposure_rescaled shape:", a_191018_exposure_rescaled.shape)
save_path = "./data"
np.save(os.path.join(save_path, 'a_191018_exposure_rescaled.npy'),
        a_191018_exposure_rescaled)
a_191018_exposure_rescaled shape: (7, 490, 460)
```

图3.8-9 秋季自然真彩色合成显示

- **样本的大小**

使用抽样方法进行精度估计，样本越小，产生的总体估计量误差就越大。应用时，需要计算在允许的误差范围内，一组样本应包含的最少样本个数。其公式为：$n=\frac{pqz^2}{d^2}$。其中，p和q分别为解译图判读正确和错误的百分比，或表示为$p(1-p)$；z为对应于置信水平的双侧临界值；d

误差允许范围。一般情况下，先假定解译精度为90%，置信水平为95%。通过scipy.stats的norm工具计算置信水平为95%时，双侧临界值为1.95996，带入公式可求得样本数约为138。

```
from scipy.stats import norm
import math
val = norm.ppf((1+0.95)/2)
n = val**2*0.9*0.1/0.05**2
print("样本数量为：",n)
--------------------------------------------------------------------------------
样本数量为： 138.2925175449885
```

- **采样交互平台的代码结构**

完成一个能够处理一个或者多个任务的综合性代码，一般要结合类来梳理代码结构。如果仅依靠函数，虽然一样能完成任务，但是代码结构松散，不方便代码的编写及查看。开始一个任务前，最好先捋清楚所要实现的功能，例如采样GUI工具需要实现如下功能：

1. 影像数据处理；
2. 显示背景影像，可缩放；
3. 选择分类，在图像中采样；
4. 显示当前采样分类状态；
5. 计算采样位置坐标；
6. Python控制台信息打印。

对于一个不大的功能开发，例如可能一开始只是想实现一个采样的工具，一些细节也许考虑得并不清晰。开始任务后，随着代码编写的深入，可以再作调整。不过，如果一开始，尤其功能复杂的工具开发，就需要将可能的问题尽量考虑清晰，尤其细节上的实现及综合结构，那么会提升代码编写的效率。避免由于考虑不周，使得代码不断调整，甚至造成重新编写。当然，即使考虑得再清晰，问题仍旧会层出不穷，因为代码本身就是一个不断调试的过程。

写代码的过程要不断地调试，而读代码则要捋清楚代码的结构，尤其各项功能实现的先后顺序，以及之间的数据关联。代码的结构图3.8-10，能够很好地帮助我们识别代码的结构。如果没有这个结构图，则需要一步步从头运行程序，结合print()打印需要查看的变量数据，推导出整个代码的流程。首先，确定“GUI包含功能”，由所要实现的功能出发，捋清楚脉络，按照实现的顺序，由a、b、c、d、e字母标识，可以沿着字母标识顺序，查看所调用的类和函数。这个图表并没有给出对应的具体代码行，不过根据这个顺序已经能够把握住整个代码的结构流程。对应的代码行，根据给出的对应函数可以很方便地找到。

这个随手工具的开发，有两个关键点需要注意。一个是，图像缩放功能；另一个是，采样点坐标变换。对于第一个问题，迁移了*stack overflow-tkinter canvas zoom+ move/pan*[①]给出的代码，因为原代码是直接读取图像文件，而这里需要读取数组结构的遥感影像波段文件。因此，需要增加图像处理的部分代码，并修改源代码适应增加的部分。源代码图像缩放的核心部分不需要做出调整，包括滚动条配置，左键拖动配置和鼠标滚动缩放部分，而最为核心的则是图像显示函数show_image(self, event=None)部分。因为图像移动和缩放引发了画布（canvas）的比例缩放，滚动条（_scroll_x/__scroll_y）和鼠标拖动（_move_from/__move_to）引发的是移动，而鼠标滑轮（__wheel）引发了画布的缩放。缩放需要控制图像的大小。当缩小到一定程度，图像不再缩小。这个由图像缩放比例因子（imscale）参数控制，初始值为1.0。而滚轮每滚动一次的缩放比例，则由delta参数控制。可以尝试修改不同的参数值，观察图像变化情况。因为画布的移动缩放，图像显示需要对应做出调整，读取当前图像位置（bbox1），画布可视区域位置（bbox2），计算滚动条区域（bbox），确定图像是部分还是整个位于画布可视区域。如果部分，则计算可视区域图像的平铺坐标（x_1,y_1,x_2,y_2），并裁切图像；否则，为整个图像。图像随画布缩放self.canvas.scale('all', x, y, scale, scale)，即缩放画布及画布上所有物件。对于该部分的理解，最好的方法是运行代码、打印相关变量、查看数据变化。

① stack overflow-tkinter canvas zoom+ move/pan，（https://stackoverflow.com/questions/41656176/tkinter-canvas-zoom-move-pan）。

因为图像移动缩放，引发了第2个问题，采样点坐标的变换。采样的基本思想是，确定所要采样的类别，由3个单选按钮控制。点击鼠标触发click_xy_collection()函数，绘制点（实际上是圆形），并按照类别保存点的索引至字典xy。采样点的坐标是由当前画布确定，画布是移动和缩放的，采样点的坐标随画布的变化而变化，那么如何获得对应实际图像大小（图像是数组，由行列表示，那么一个像素的坐标可以表示为(x_i,y_j)，注意图像x_i为列，y_j为行，与numpy

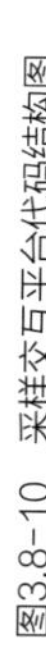

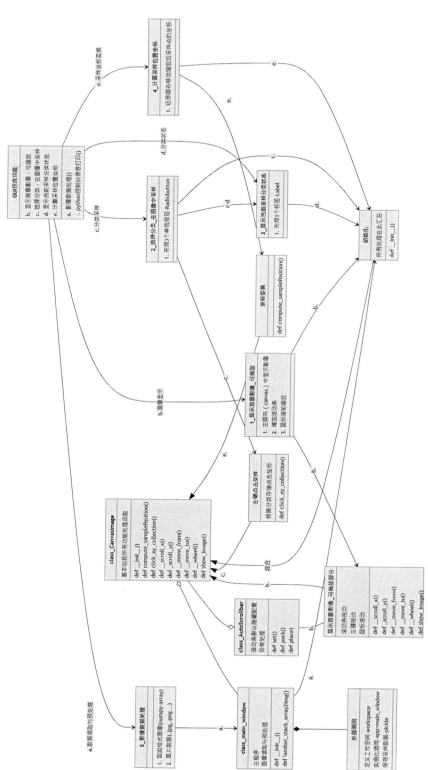

图3.8-10 采样交互平台代码结构图

数组正好相反，即NumPy数组x_i为行，y_j为列），采样点的坐标随图像缩放变换实际上正是“编程线性代数”部分阐述的内容，通过直接线性变换就可以返回到原始坐标。要实现线性变换，需要获得比例缩放因子，即`scale`参数。为了确定该参数，也可以在画布未移动缩放时，自动生成两个点，计算该坐标，即为实际的坐标，以其为参照。这两个点随后，会随其他采样点跟着画布移动缩放，坐标值也发生了变换。在新的画布空间下，获取当前坐标，计算这两个点前后的距离比值，即为缩放比例。其中，与`scale`参数保持一致。因此，可以将`1/scale`作为线性变换的比例因子，返回缩放前的状态。采用点坐标变换，通过点击按钮（`button_computePosition`）调用函数`compute_samplePosition()`（GUI中显示的文本为，calculate sampling position）实现。线性变换用`np.matmul()`计算，即两个矩阵之积。最好将计算的采样点坐标根据给定的文件位置保存，用于后续的精度分析。完成后的采样交互操作平台见图3.8-11。

注意，tkinter编写的GUI不能在Jupyter（Lab）下运行，需要在spyder等解释器下打开运行。可以新建.py文件，将下述代码复制于该文件，再运行。如果自行编写基于tkinter开发的GUI工具，也需要在spyder等解释器下编写调试。完成的代码文件名为“image_pixel_sampling_zoom.py”。

```
import math
import os
import random
import warnings
import tkinter as tk
import numpy as np
from tkinter import ttk
```

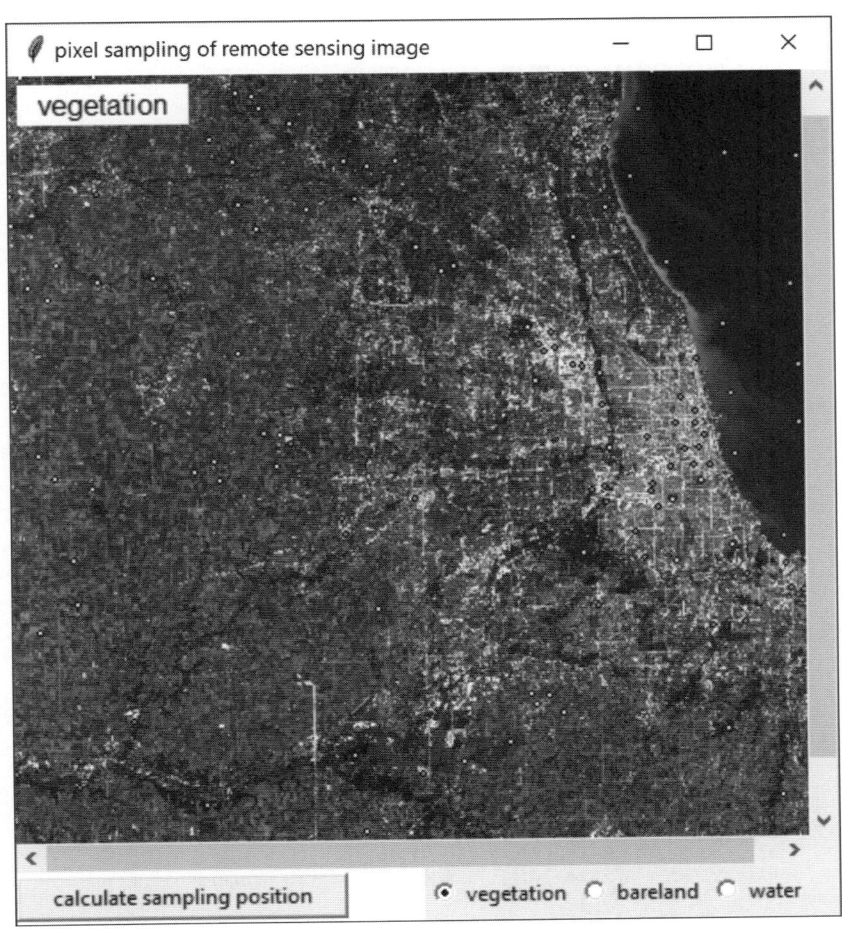

图3.8-11 采样交互平台

```
from PIL import Image, ImageTk
from skimage.util import img_as_ubyte
class AutoScrollbar(ttk.Scrollbar):
    '''滚动条默认时隐藏'''

    def set(self, low, high):
        if float(low) <= 0 and float(high) >= 1.0:
            self.grid_remove()
        else:
            self.grid()
            ttk.Scrollbar.set(self, low, high)
    def pack(self, **kw):
        raise tk.TclError("Cannot use pack with the widget" +
                          self.__class__.__name__)
    def place(self, **kw):
        raise tk.TclError("Cannot use pack with the widget" +
                          self.__class__.__name__)
class CanvasImage(ttk.Frame):
    '''显示图像，可缩放'''

    def __init__(self, mainframe, img):
        '''初始化Frame框架'''
        ttk.Frame.__init__(self, master=mainframe)
        self.master.title("pixel sampling of remote sensing image")
        self.img = img
        self.master.geometry('%dx%d' % self.img.size)
        self.width, self.height = self.img.size

        #增加水平、垂直滚动条
        hbar = AutoScrollbar(self.master, orient='horizontal')
        vbar = AutoScrollbar(self.master, orient='vertical')
        hbar.grid(row=1, column=0, columnspan=4, sticky='we')
        vbar.grid(row=0, column=4, sticky='ns')
        #创建画布并绑定滚动条
        self.canvas = tk.Canvas(self.master, highlightthickness=0,
xscrollcommand=hbar.set,
                                yscrollcommand=vbar.set, width=self.width,
height=self.height)
        self.canvas.config(scrollregion=self.canvas.bbox('all'))
        self.canvas.grid(row=0, column=0, columnspan=4, sticky='nswe')
        self.canvas.update()  #更新画布
        hbar.configure(command=self.__scroll_x)  #绑定滚动条于画布
        vbar.configure(command=self.__scroll_y)

        self.master.rowconfigure(0, weight=1)  #使得画布（显示图像）可扩展
        self.master.columnconfigure(0, weight=1)

        #于画布绑定事件（events）
        self.canvas.bind(
            '<Configure>', lambda event: self.show_image())  #调整画布大小
        self.canvas.bind('<ButtonPress-1>', self.__move_from)  #原画布位置
        self.canvas.bind('<B1-Motion>', self.__move_to)  #移动画布到新的位置
        #Windows 和 MacOS 下缩放，不适用于 Linux
        self.canvas.bind('<MouseWheel>', self.__wheel)
        self.canvas.bind('<Button-5>', self.__wheel)  #Linux 下，向下滚动缩放
        self.canvas.bind('<Button-4>',   self.__wheel)  #Linux 下，向上滚动缩放
        #处理空闲状态下的击键，因为太多击键，会使得性能低的电脑运行缓慢
        self.canvas.bind('<Key>', lambda event: self.canvas.after_idle(
            self.__keystroke, event))

        self.imscale = 1.0  #图像缩放比例
        self.delta = 1.2  #滑轮，画布缩放量级

        #将图像置于矩形容器中，宽高等于图像的大小

        self.container = self.canvas.create_rectangle(
            0, 0, self.width, self.height, width=0)
        self.show_image()
        self.xy = {"water": [], "vegetation": [], "bareland": []}
        self.canvas.bind('<Button-1>', self.click_xy_collection)
        self.xy_rec = {"water": [], "vegetation": [], "bareland": []}
        #配置按钮，用于选择样本，以及计算样本位置
        button_frame = tk.Frame(
            self.master, bg='white', width=5000, height=30, pady=3).grid(row=2,
```

```
sticky='NW')
        button_computePosition = tk.Button(button_frame, text='calculate sampling
position', fg='black',
                                           width=25, height=1, command=self.
compute_samplePosition).grid(row=2, column=0, sticky='w')
        self.info_class = tk.StringVar(value='empty')
        button_green = tk.Radiobutton(button_frame, text="vegetation",
variable=self.info_class,
                                      value='vegetation').grid(row=2, column=1,
sticky='w')
        button_bareland = tk.Radiobutton(
            button_frame, text="bareland", variable=self.info_class,
value='bareland').grid(row=2, column=2, sticky='w')
        button_water = tk.Radiobutton(
            button_frame, text="water", variable=self.info_class, value='water').
grid(row=2, column=3, sticky='w')
        self.info = tk.Label(self.master, bg='white', textvariable=self.info_
class, fg='black', text='empty', font=(
            'Arial', 12), width=10, height=1).grid(row=0, padx=5, pady=5,
sticky='nw')
        self.scale_ = 1

        #绘制一个参考点
        self.ref_pts = [self.canvas.create_oval((0, 0, 1.5, 1.5), fill='white'),
self.canvas.create_oval(
            (self.width, self.height, self.width-0.5, self.height-0.5),
fill='white')]
        self.ref_coordi = {'ref_pts': [((self.canvas.coords(i)[2]+self.canvas.
coords(
            i)[0])/2, (self.canvas.coords(i)[3]+self.canvas.coords(i)[1])/2) for
i in self.ref_pts]}
        self.sample_coordi_recover = {}
    def compute_samplePosition(self):
        self.xy_rec.update({'ref_pts': self.ref_pts})
        # print(self.xy_rec)
        sample_coordi = {key: [((self.canvas.coords(i)[2]+self.canvas.coords(i)
[0])/2, (self.canvas.coords(
            i)[3]+self.canvas.coords(i)[1])/2) for i in self.xy_rec[key]] for
key in self.xy_rec.keys()}
        print("+"*50)
        print("sample coordi:", sample_coordi)
        print("_"*50)
        print(self.ref_coordi)
        print("image size:", self.width, self.height)
        print("_"*50)
        def distance(p1, p2): return math.sqrt(
            ((p1[0]-p2[0])**2)+((p1[1]-p2[1])**2))
        scale_byDistance = distance(sample_coordi['ref_pts'][0], sample_
coordi['ref_pts'][1])/distance(
            self.ref_coordi['ref_pts'][0], self.ref_coordi['ref_pts'][1])
        print("scale_byDistance:", scale_byDistance)
        print("scale_by_self.scale_:", self.scale_)

        #缩放回原始坐标系
        f_scale = np.array([[1/scale_byDistance, 0], [0, 1/scale_byDistance]])
        sample_coordi_recover = {key: np.matmul(np.array(
            sample_coordi[key]), f_scale) for key in sample_coordi.keys() if
sample_coordi[key] != []}
        print("sample_coordi_recove", sample_coordi_recover)
        relative_coordi = np.array(sample_coordi_recover['ref_pts'][0])-1.5/2
        sample_coordi_recover = {
            key: sample_coordi_recover[key]-relative_coordi for key in sample_
coordi_recover.keys()}
        print("sample_coordi_recove", sample_coordi_recover)
        self.sample_coordi_recover = sample_coordi_recover
    def click_xy_collection(self, event):
        multiple = self.imscale
        length = 1.5*multiple  #根据图像缩放比例的变化调节所绘制矩形的大小，保持大小一致

        def event2canvas(e, c): return (c.canvasx(e.x), c.canvasy(e.y))
        # cx,cy=event2canvas(event,self.canvas)
        cx, cy = event2canvas(event, self.canvas)
        print(cx, cy)
        if self.info_class.get() == 'vegetation':
```

```
                self.xy["vegetation"].append((cx, cy))
                rec = self.canvas.create_oval(
                    (cx-length, cy-length, cx+length, cy+length), fill='yellow')
                self.xy_rec["vegetation"].append(rec)
            elif self.info_class.get() == 'bareland':
                self.xy["bareland"].append((cx, cy))
                rec = self.canvas.create_oval(
                    (cx-length, cy-length, cx+length, cy+length), fill='red')
                self.xy_rec["bareland"].append(rec)
            elif self.info_class.get() == 'water':
                self.xy["water"].append((cx, cy))
                rec = self.canvas.create_oval(
                    (cx-length, cy-length, cx+length, cy+length), fill='aquamarine')
                self.xy_rec["water"].append(rec)
            print("_"*50)
            print("sampling count", {
                  key: len(self.xy_rec[key]) for key in self.xy_rec.keys()})
            print("total:", sum([len(self.xy_rec[key])
                  for key in self.xy_rec.keys()]))
        def __scroll_x(self, *args, **kwargs):
            ''' 水平滚动画布，并重画图像 '''
            self.canvas.xview(*args, **kwargs)  #滚动水平条
            self.show_image()  #重画图像

        def __scroll_y(self, *args, **kwargs):
            """ 垂直滚动画布，并重画图像 """
            self.canvas.yview(*args, **kwargs)  #垂直滚动
            self.show_image()  #重画图像

        def __move_from(self, event):
            ''' 鼠标滚动，前一坐标 '''
            self.canvas.scan_mark(event.x, event.y)

        def __move_to(self, event):
            ''' 鼠标滚动，下一坐标 '''
            self.canvas.scan_dragto(event.x, event.y, gain=1)
            self.show_image()  #重画图像

        def __wheel(self, event):
            ''' 鼠标滚轮缩放 '''
            x = self.canvas.canvasx(event.x)
            y = self.canvas.canvasy(event.y)
            bbox = self.canvas.bbox(self.container)  #图像区域
            if bbox[0] < x < bbox[2] and bbox[1] < y < bbox[3]:
                pass  #鼠标如果在图像区域内部
            else:
                return  #只有鼠标在图像内才可以滚动缩放
            scale = 1.0
            # 响应 Linux (event.num) 或 Windows (event.delta) 滚轮事件
            if event.num == 5 or event.delta == -120:  #向下滚动
                i = min(self.width, self.height)
                if int(i * self.imscale) < 30:
                    return  #图像小于 30 pixels
                self.imscale /= self.delta
                scale /= self.delta
            if event.num == 4 or event.delta == 120:  #向上滚动
                i = min(self.canvas.winfo_width(), self.canvas.winfo_height())
                if i < self.imscale:
                    return  #如果 1 个像素大于可视图像区域
                self.imscale *= self.delta
                scale *= self.delta
            self.canvas.scale('all', x, y, scale, scale)  #缩放画布上的所有对象
            self.show_image()
            self.scale_ = scale*self.scale_
        def show_image(self, event=None):
            ''' 在画布上显示图像 '''
            bbox1 = self.canvas.bbox(self.container)  #获得图像区域
            #在 bbox1 的两侧移除 1 个像素的移动
            bbox1 = (bbox1[0] + 1, bbox1[1] + 1, bbox1[2] - 1, bbox1[3] - 1)
            bbox2 = (self.canvas.canvasx(0),  #获得画布上的可见区域
                     self.canvas.canvasy(0),
                     self.canvas.canvasx(self.canvas.winfo_width()),
                     self.canvas.canvasy(self.canvas.winfo_height()))
            bbox = [min(bbox1[0], bbox2[0]), min(bbox1[1], bbox2[1]),  #获取滚动区域框
```

```
                max(bbox1[2], bbox2[2]), max(bbox1[3], bbox2[3])]
        if bbox[0] == bbox2[0] and bbox[2] == bbox2[2]:  # 整个图像在可见区域
            bbox[0] = bbox1[0]
            bbox[2] = bbox1[2]
        if bbox[1] == bbox2[1] and bbox[3] == bbox2[3]:  # 整个图像在可见区域
            bbox[1] = bbox1[1]
            bbox[3] = bbox1[3]
        self.canvas.configure(scrollregion=bbox)  # 设置滚动区域
        x1 = max(bbox2[0] - bbox1[0], 0)  # 得到图像平铺的坐标 (x1, y1, x2, y2)
        y1 = max(bbox2[1] - bbox1[1], 0)
        x2 = min(bbox2[2], bbox1[2]) - bbox1[0]
        y2 = min(bbox2[3], bbox1[3]) - bbox1[1]
        if int(x2 - x1) > 0 and int(y2 - y1) > 0:  # 显示图像，如果它在可见的区域
            x = min(int(x2 / self.imscale), self.width)   # 有时大于 1 个像素 ...
            y = min(int(y2 / self.imscale), self.height)  # ... 有时不是
            image = self.img.crop(
                (int(x1 / self.imscale), int(y1 / self.imscale), x, y))
            imagetk = ImageTk.PhotoImage(
                image.resize((int(x2 - x1), int(y2 - y1))))
            imageid = self.canvas.create_image(max(bbox2[0], bbox1[0]), max(
                bbox2[1], bbox1[1]), anchor='nw', image=imagetk)
            self.canvas.lower(imageid)  # 将图像设置为背景
            # 保持一个额外的引用来防止垃圾收集
            self.canvas.imagetk = imagetk
class main_window:
    ''' 主窗口类 '''

    def __init__(self, mainframe, rgb_band, img_path=0, landsat_stack=0,):
        ''' 读取图像 '''
        if img_path:
            self.img_path = img_path
            self.__image = Image.open(self.img_path)
        if rgb_band:
            self.rgb_band = rgb_band
        if type(landsat_stack) is np.ndarray:
            self.landsat_stack = landsat_stack
            self.__image = self.landsat_stack_array2img(
                self.landsat_stack, self.rgb_band)
        self.MW = CanvasImage(mainframe, self.__image)
    def landsat_stack_array2img(self, landsat_stack, rgb_band):
        r, g, b = self.rgb_band
        landsat_stack_rgb = np.dstack(
            (landsat_stack[r], landsat_stack[g], landsat_stack[b]))  # 合并三个波段
        # 使用 skimage 提供的方法，将 float 等浮点型色彩，转换为 0-255 整型
        landsat_stack_rgb_255 = img_as_ubyte(landsat_stack_rgb)
        landsat_image = Image.fromarray(landsat_stack_rgb_255)
        return landsat_image
if __name__ == "__main__":
    # img_path=r'C:\Users\richi\Pictures\n.png'
    workspace = "./data"
    img_fp = os.path.join(workspace, 'a_191018_exposure_rescaled.npy')
    landsat_stack = np.load(img_fp)
    rgb_band = [3, 2, 1]
    mainframe = tk.Tk()
    # img_path=img_path,landsat_stack=landsat_stack,
    app = main_window(mainframe, rgb_band=rgb_band,
                      landsat_stack=landsat_stack)
    # app=main_window(mainframe, rgb_band=rgb_band,img_path=img_path)
    mainframe.mainloop()

    # 保存采样点
    import pickle as pkl
    with open(os.path.join(workspace, 'sampling_position.pkl'), 'wb') as handle:
        pkl.dump(app.MW.sample_coordi_recover, handle)
```

3.8.3.2 分类精度计算

混淆矩阵（confusion matrix），每一行（列）代表一个类的实例预测，而每一列（行）代表一个实际的类的实例，是一种特殊的、具有两个维度（实际值和预测值）的列联表（contingency table）；并且，两个维度中都有着一样的类别的集合。例如：

$$\begin{bmatrix} & & \text{Category of prediction} & & \\ & & \mathit{baraland} & \mathit{vegetation} & \mathit{water} \\ \text{Actual class} & \mathit{bareland} & 51 & 2 & 0 \\ & \mathit{vegetation} & 10 & 54 & 0 \\ & \mathit{water} & 0 & 2 & 19 \end{bmatrix},$$

可以解读为，总共138个样本（采样点），真实裸地为53个，误判为绿地的2个；真实绿地为64个，误判为裸地的10个；真实水体为21个，误判为绿地的2个。混淆矩阵的计算使用sklearn库提供的confusion_matrix方法计算。

通过计算混淆矩阵，分析获知绿地中误判为裸地的相对较多，一方面是在解译过程中，%35的阈值界限可以适当调大，使部分绿地划分到裸地类别；另一方面，可能是在采样过程中，采样点设置的比较大，同时覆盖了绿地和裸地，不能确定最终坐标是落于绿地还是落于裸地，造成采样上的错误。此时，可以调小采样点直径，更精确地定位。

除了计算混淆矩阵，同时计算百分比精度，即正确的分类占总样本数的比值。

```
from sklearn.metrics import confusion_matrix
import pickle as pkl
import os
import pandas as pd
workspace = r"./data"
with open(os.path.join(workspace, 'sampling_position_138.pkl'), 'rb') as handle:
    sampling_position = pkl.load(handle)
sampling_position_int = {key: sampling_position[key].astype(
    int) for key in sampling_position.keys() if key != 'ref_pts'}
i = 0
sampling_position_int_ = {}
for key in sampling_position_int.keys():
    for j in sampling_position_int[key]:
        sampling_position_int_.update({i: [j.tolist(), key]})
        i += 1
sampling_position_df = pd.DataFrame.from_dict(sampling_position_int_, columns=[
                                              'coordi', 'category'],
orient='index')  # 转换为 pandas 的 DataFrame 数据格式，方便数据处理
sampling_position_df['interpretation'] = [
    green_water_bareLand[coordi[1]][coordi[0]] for coordi in sampling_position_
df.coordi]
interpretation_mapping = {1: "bareland",
                          2: "vegetation", 3: "water", 0: "bareland"}
sampling_position_df.interpretation = sampling_position_df.interpretation.
replace(
    interpretation_mapping)
sampling_position_df.to_pickle(os.path.join(
    workspace, 'sampling_position_df.pkl'))
precision_confusionMatrix = confusion_matrix(
    sampling_position_df.category, sampling_position_df.interpretation)
precision_percent = np.sum(
    precision_confusionMatrix.diagonal())/len(sampling_position_df)
print("precision - confusion Matrix:\n", precision_confusionMatrix)
print("precision - percent:", precision_percent)
--------------------------------------------------------------------------------
precision - confusion Matrix:
 [[50  2  1]
 [ 8 54  2]
 [ 0  2 19]]
precision - percent: 0.8913043478260869
```

参考文献（References）：

[1] 吴健平. 遥感影像解译精度的分析[J]. 遥感信息，1992（2）：17-18.

3-C 编程微积分及SIR空间传播模型

3.9 编程微积分

微积分（Calculus），是研究极限、微分、积分和无穷级数等的一个数学分支。本质来讲，微积分是一门研究变化的学问。在本书中，多处涉及微积分的知识，例如阐述回归部分对残差平方和关于回归系数求微分，另微分结果为0解方程组得回归系数值，构建回归方程；在梯度下降法中，梯度就是分别对每个变量求偏微分；本部分SIR传播模型的阐述中则通过建立易感人群、恢复人群和受感人群的微分方程建立SIR传播模型。可见，微积分在数据分析中具有重要的作用，因此有必要以代码的途径结合图表表述阐释微积分的基础知识，为相关数据分析预备。主要使用SymPy（Calculus）[①]库解释微积分。

3.9.1 导数（Derivative）与微分（Differentiation）

- **导数**

导数是用来分析变化的，即曲线（函数图像）在某点处的斜率，表示倾斜的程度。对于直线函数求导，会得到直线的斜率；对曲线函数求导，则能得到各点的斜率（即瞬间斜率）。下述代码使用SymPy库的`diff`方法计算了曲线上采样点各处的斜率，具体的过程是先建立曲线图形的函数表达式为：$y=\sin(2\pi x)$，由`diff`方法关于x求微分结果为：$y(x_0)=2\pi\cos(2\pi x)$，通过该微分方程，给定任意一点的横坐标，就可以计算获得曲线对应点的斜率。为了清晰地表述采样点各处斜率的变化情况，由导数（斜率值）`derivative_value`，采样点横坐标`sample_x`，采样点纵坐标`sample_y`。假设采样点横坐标的固定变化值为$delta_x=0.1$，计算$sample_x+delta_x$处的纵坐标，从而绘制各点处的切线。为了能够清晰地看到各个采样点处导数的倾斜程度，即变化趋势的强弱，对齐所有切线原点于横坐标上，保持各点横轴变化量不变，计算各个结束点的纵坐标。通过向量长度的变化，可以确定各点变化趋势的大小；通过向量方向的变化，可以确定各点变化趋势的走势。

- **极限**

上述计算斜率的方法是直接使用了SymPy库，为了更清晰地理解计算的过程，需要首先了解什么是极限。极限可以描述一个序列的指标越来越大时，数列中元素的性质变化趋势；也可以描述函数的自变量接近某一个值时，相对应函数值变化的趋势。例如，对于数列（sequence）$a_n=\frac{1}{n}$，随着n的增大，a_n从0的右侧越来约接近0，于是可以认为0是这个序列的极限。对于函数的极限，可以假设$f(x)$是一个实函数，c是一个实数，那么$\lim_{x\to c}f(x)=L$。表示当x充分靠近c时，$f(x)$可以任意地靠近L。即x趋向c时，函数$f(x)$的极限是L。用数学算式表示极限，例如$\lim_{n\to 1}(1-n)$表示使n无限接近1时，$1-n$无限接近1-1，即无限接近0。又如，$\lim_{n\to 1}\frac{n^2-3n+2}{n-1}=\lim_{n\to 1}\frac{(n-1)(n-2)}{n-1}=\lim_{n\to 1}(n-2)$，即$n$无限接近1时，$\lim_{n\to 1}\frac{n^2-3n+2}{n-1}$无限接近2。

如果要求函数$f(x)$图形点A的斜率，点A的坐标为$(x_0,f(x_0))$，将点A向右移动$\triangle x$，即横向长度差，则纵向长度差为$f(x_0+\triangle x)-f(x_0)$，过点$A$的斜率，即$f(x)$在$x_0$处的导数（导函数）为：$f'(x_0)$，即：$f'(x_0)=\lim_{\triangle x\to 0}\frac{\triangle y}{\triangle x}=\lim_{\triangle x\to 0}\frac{f(x_0+\triangle x)-f(x_0)}{\triangle x}$，也可记作$y'(x_0),\frac{\mathrm{d}y}{\mathrm{d}x}|_{x=x_0},\frac{\mathrm{d}y}{\mathrm{d}x}(x_0),\frac{\mathrm{d}f}{\mathrm{d}x}|_{x=x_0}$等，读作“对$y(f(x))$关于$x$求导”，$d$是derivative（导数）的第一个字母。$\frac{\mathrm{d}}{\mathrm{d}x}$表示一个整体，是“求关于$x$的导数”的求导计算。

已知曲线图形的函数表达式为$y=\sin(2\pi x)$，在SymPy下建立表达式，根据上述求导极限方程可以得到函数在点x_0处的求导表达式（导函数），即下述代码中变量`limit_x_0_expr`为$\frac{\sin(\pi(2d+1.6))-\sin(1.6\pi)}{d}$。其中，$d$为横向长度差$\triangle x$，当$\triangle x\mapsto 0$时的极限值即为导数/斜率，计算结果为$2.0\pi\cos(1.6\pi)$。这与使用SymPy的`diff`方法求得的关于$x$求导方程$2.0\pi\cos(2\pi x)$，代入$x_0=0.8$的结果一致。计算结果和解释见图3.9-1、图3.9-2。

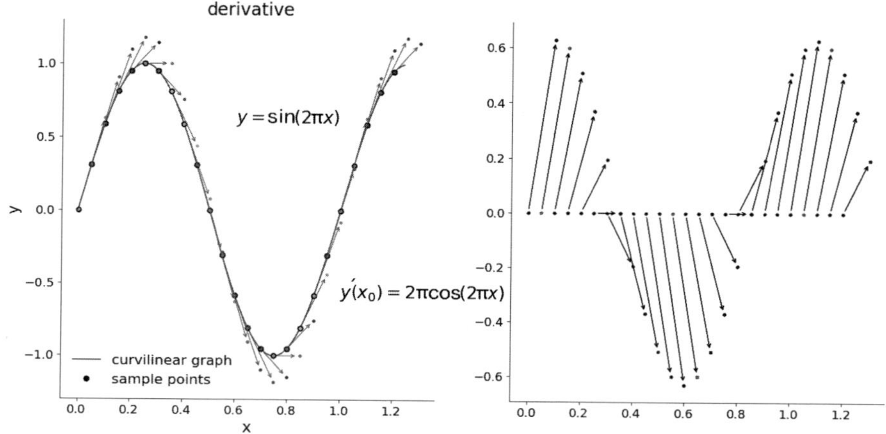

图3.9-1 各个采样点上的斜率（左）；各个采样点上的斜率比较（右）

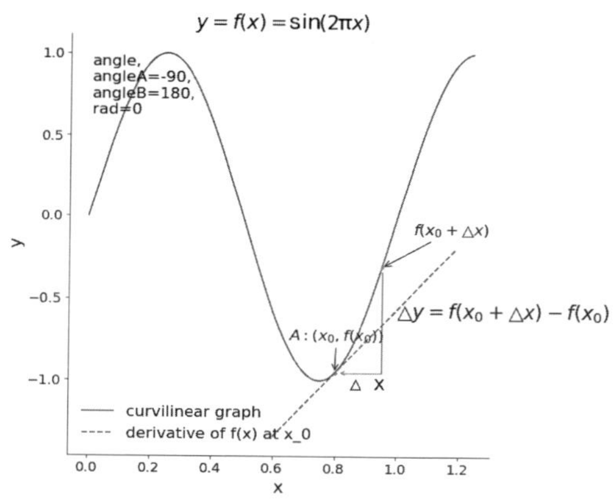

图3.9-2 曲线上给定采样点处斜率的解释

- **误差率**

获得$f(x)$导函数，可以求任一点的斜率，例如在$x_0 = 0.8$点处斜率为$2.0\pi\cos(1.6\pi)$。欲建立该点处的切线方程，需要求得截距，根据$g(x) = ax + b$，其中a为斜率，b为截距，则有$b = ax - g(x)$。此时，$g(x) = f(x)$，而斜率a已知，$x = x_0 = 0.8$。求得截距，则可建立该点的切线方程$g(x)$。误差率则是以x为起点进行变化时，$f(x)$和$g(x)$值之间的差异，占x的变化量的百分比，即$= \frac{(fg)}{(x)}$。离x_0越近，误差率越小。所谓近似成一次函数，就是令原函数的误差率局部为0的情况。所以，在讨论局部性质时，可以用一次函数替代原函数进而推导出正确的结论。

- **求导的基本公式**

1. $p' = 0(p)$;
2. $(px)' = pp$;
3. $(af(x))' = af'(x)$ # 常系数微分
4. $\{f(x) + g(x)\}' = f'(x) + g'(x)$; # 和的微分
5. $(x^n)' = nx^{n-1}$; # 幂函数的导数
6. $\{f(x)g(x)\}' = f(x)'g(x) + f(x)g(x)'$; # 积的微分
7. $\{\frac{g(x)}{f(x)}\}' = \frac{g'(x)f(x)-g(x)f'(x)}{f(x)^2}$ # 商的微分
8. $\{g(f(x))\}' = g'(f(x))f'(x)$ # 复合函数的微分

9. $\{f^{-1}(x)\}' = \frac{1}{f'(x)}$ #反函数的微分方程

- **微分**

对于微分的理解，可以拓展为函数图像中，某点切线的斜率和函数的变化率。微分是对函数的局部变化率的一种线性描述，其可以近似的描述当函数自变量的取值足够小的改变时，函数的值是怎样变化的。

- **由"微分=0"可知极值**

极大点和极小点是函数增减性发生变化的地方，对研究函数的性质来说很重要。极大点、极小点常常会变成最大点、最小点，是求解某些（最优解）问题时十分关键的点。极值条件：$y = f(x)$在$x = a$处为极大点或极小点，则有$f'(a) = 0$。即求极大点和极小点，只须找到满足$f'(a) = 0$的a即可。

增减性的判断条件：当$f'(a) > 0$时，所近似的一次函数在$x = a$处呈现递增的趋势，因此可知$f(x)$同样呈现递增趋势；同样，当$f'(a) < 0$时，$f(x)$处于下降的状态，既不在顶端也不在谷底。

- **平均值定理**

对于$a, b(a < b)$来说，存在一个ζ，当$a < \zeta < b$时，满足$f(b) = f'(\zeta)(b - a) + f(a)$。

```
import util_A
import numpy as np
import matplotlib.pyplot as plt
import sympy
from sympy import diff, pprint, limit

x = sympy.symbols('x')
curvilinear_expr = sympy.sin(2*sympy.pi*x)  #定义曲线函数

#A-使用 SymPy 库 diff 方法求导
derivative_curvilinear_expr = diff(
    curvilinear_expr, x)  #curvilinear_expr 关于 x 求微分 / 导数方程
print("curvilinear_expr 关于 x 求微分 / 导数方程 :")
pprint(derivative_curvilinear_expr, use_unicode=True)
curvilinear_expr_ = sympy.lambdify(x, curvilinear_expr, "numpy")
derivative_expr = sympy.lambdify(x, derivative_curvilinear_expr, "numpy")
t = np.arange(0.0, 1.25, 0.01)
y = curvilinear_expr_(t)
fig, axs = plt.subplots(1, 3, figsize=(26, 8))
axs[0].plot(t, y, label="curvilinear graph")
axs[0].set_title(r'derivative', fontsize=20)
axs[0].text(1, -0.6, r'$y\'(x_{0}) =2 \pi cos(2 \pi x)$', fontsize=20)
axs[0].text(0.6, 0.6, r'$y=sin(2 \pi x)$', fontsize=20)
axs[0].set_xlabel('x')
axs[0].set_ylabel('y')
axs[0].spines['right'].set_visible(False)
axs[0].spines['top'].set_visible(False)

#采样原点
sample_x = t[::5]
sample_y = curvilinear_expr_(sample_x)
axs[0].plot(sample_x, sample_y, 'o', label='sample points', color='black')

#采样终点
derivative_value = derivative_expr(sample_x)  #求各个采样点的导数（斜率）
delta_x = 0.1  #x 向变化量
sample_x_endPts = sample_x+delta_x
sample_y_endPts = derivative_value*delta_x+sample_y
def demo_con_style_multiple(a_coordi, b_coordi, ax, connectionstyle):
    '''
    function - 在 matplotlib 的子图中绘制多个连接线
    reference: matplotlib 官网 Connectionstyle Demo: https://matplotlib.
org/3.3.2/gallery/userdemo/connectionstyle_demo.html#sphx-glr-gallery-userdemo-
connectionstyle-demo-py
    Params:
        a_coordi - 起始点的 x, y 坐标; tuple
        b_coordi - 结束点的 x, y 坐标; tuple
        ax - 子图; ax(plot)
        connectionstyle - 连接线的形式; string

    Returns:
        None
    '''
```

```
    x1, y1 = a_coordi[0], a_coordi[1]
    x2, y2 = b_coordi[0], b_coordi[1]
    ax.plot([x1, x2], [y1, y2], ".")
    for i in range(len(x1)):
        ax.annotate("",
                    xy=(x1[i], y1[i]), xycoords='data',
                    xytext=(x2[i], y2[i]),
textcoords='data',
                    arrowprops=dict(arrowstyle="<-",
color="0.5",
                                    shrinkA=5, shrinkB=5,
                                    patchA=None, patchB=None,
                                    connectionstyle=connectionstyle,
                                    ),
                    )
demo_con_style_multiple(
    (sample_x, sample_y), (sample_x_endPts, sample_y_endPts), axs[0],
"arc3,rad=0.")
demo_con_style_multiple((sample_x, sample_y*0), (sample_x_endPts,
                        sample_y_endPts-sample_y), axs[1], "arc3,rad=0.")

#B-使用极限方法求导
axs[2].set_title(r'$y=f(x)=sin(2 \pi x)$', fontsize=20)
axs[2].plot(t, y, label="curvilinear graph")
dx = 0.15
x_0 = 0.8
util_A.demo_con_style((x_0, curvilinear_expr_(
    x_0)), (x_0+dx, curvilinear_expr_(x_0+dx)), axs[2], "angle,angleA=-
90,angleB=180,rad=0")
axs[2].text(x_0+0.05, curvilinear_expr_(x_0)-0.1,
            "Δ x", family="monospace", size=20)
axs[2].text(x_0+0.2, curvilinear_expr_(x_0+dx)-0.3,
            r"$Δ y=f(x_{0}+Δ x)-f(x_{0})$", family="monospace", size=20)
color = 'blue'
axs[2].annotate(r'$A:(x_{0},f(x_{0}))$', xy=(x_0, curvilinear_expr_(x_0)),
xycoords='data', xytext=(x_0-0.15, curvilinear_expr_(
    x_0)+0.2), textcoords='data', weight='bold', color=color, arrowprops=dict(ar
rowstyle='->', connectionstyle="arc3", color=color))
axs[2].annotate(r'$f(x_{0}+Δ x)$', xy=(x_0+dx, curvilinear_expr_(x_0+dx)),
xycoords='data', xytext=(x_0+dx+0.1, curvilinear_expr_(
    x_0+dx)+0.2), textcoords='data', weight='bold', color=color, arrowprops=dict
(arrowstyle='->', connectionstyle="arc3", color=color))
axs[2].set_xlabel('x')
axs[2].set_ylabel('y')
axs[2].spines['right'].set_visible(False)
axs[2].spines['top'].set_visible(False)
d = sympy.symbols('d')
limit_x_0_expr = (curvilinear_expr.subs(x, x_0+d) -
                  curvilinear_expr.subs(x, x_0))/d  #函数f(x)在点x_0处的极限方程
print("f(x)在x_0处的求导方程:")
pprint(limit_x_0_expr)
limit_x_0 = limit(limit_x_0_expr, d, 0)
print(r"f(x)在x_0处的导数/斜率为:")
pprint(limit_x_0)
t_ = np.arange(0.6, 1.2, 0.01)
intercept = curvilinear_expr_(x_0)-limit_x_0*x_0
#limit_x_0*t_+intercept即为x_0处的切线方程
axs[2].plot(t_, limit_x_0*t_+intercept, '--r',
            label="derivative of f(x) at x_0")

#C-(x_0)误差率
gx = limit_x_0*x+intercept
x_1 = x_0+dx
err = (curvilinear_expr.subs(x, x_1)-gx.subs(x, x_1))/dx
print("x_0点导函数，在x_1点的误差率：%.2f" % err)

axs[0].legend(loc='lower left', frameon=False)
axs[2].legend(loc='lower left', frameon=False)
plt.show()
--------------------------------------------------------------------------
curvilinear_expr 关于x求微分/导数方程:
2·π·cos(2·π·x)
f(x)在x_0处的求导方程:
```

```
sin(π·(2·d + 1.6)) - sin(1.6·π)
───────────────────────────────────
                d
f(x) 在 x_0 处的导数 / 斜率为：
  π    √5·π
- ─ + ─────
  2     2
x_0 点导函数，在 x_1 点的误差率：2.34
```

3.9.2 积分（Integrate）

积分是导数的逆运算（针对计算方式而言），利用积分可以求出变化的规律和不规整图形的面积。积分和导数通常配套使用，合称为微积分。积分通常分为定积分和不定积分两种。对于定积分，给定一个正实值函数$f(x)$，$f(x)$在一个实数区间$[a,b]$上的定积分为$\int_a^b f(x)\mathrm{d}x$，可以在数值上理解为在o_{xy}坐标平面上，由曲线$x,f(x)(x\in[a,b])$，直线$x=ax=b$及x轴围成的曲边梯形的面积值（一种确定的实数值），如图3.9-3所示。$f(x)$的不定积分（或原函数）是指任何满足导数是函数$f(x)$的函数$F(x)$。一个函数$f(x)$的不定积分不是唯一的：只要$F(x)$是$f(x)$的不定积分，那么与之相差一个常数的函数$F(x)+C$也是f的不定积分。$\int_a^b f(x)\mathrm{d}x$读作求函数$f(x)$关于x的积分，$\int_a^b f(x)\mathrm{d}y$读作求函数$f(x)$关于y的积分，因为积分是导数的逆运算。因此，可以理解为"关于x求导得到$f(x)$的原函数即为积分"。因此，对于表述"计算求导后得$f(x)$的函数""求$f(x)$的不定积分""求$f(x)$的原函数"，这三种表达方式意思相同。

$f(x)$是基础函数，$f'(x)=\lim_{\triangle x\to 0}\frac{\triangle y}{\triangle x}=\lim_{\triangle x\to 0}\frac{f(x+\triangle x)-f(x)}{\triangle x}=\frac{\mathrm{d}f(x)}{\mathrm{d}x}$是$f(x)$导函数的各种表述，$F(x)=\int f'(x)\mathrm{d}x$，为$f'(x)$的不定积分，即原函数［如果确定了常数$C$，即为$f(x)$］。通常，$\int_a^b f(x)\mathrm{d}x=F(x)|_a^b$表示定积分，$\int f(x)\mathrm{d}x=F(x)$表示不定积分。

- **不定积分、定积分和面积**

$\int f(x)\mathrm{d}x$实际上表示将$f(x)\times\mathrm{d}x$进行$\int$（积分），而$\int$是"summation（合计）"的开头字母的变形，表示对$f(x)\times\mathrm{d}x$的合计之意。$f(x)$是"与x对应的y轴坐标"，$\mathrm{d}x$表示延x轴的最小增量。因此$f(x)\times\mathrm{d}x$就是变化横轴增量$\mathrm{d}x$下矩形的面积。当对所有位于区间$[a,b]$下变化增量为$\mathrm{d}x$的矩形面积求积分（合计）后（宽度极限小的长方形的集合），即为区间为$[a,b]$的横轴与曲线围合的面积。

- **区分求积法**

对于函数$f(x)$，给定区间$[a,b]$，假设进行n次分割，长方形从左向右依次为$x_1,x_2,x_3,\ldots,x_{\mathrm{k}},\ldots,x_{\mathrm{n}}$，单个矩形的面积为$\frac{b-a}{n}\times f(x_{\mathrm{k}})$，将全部的长方形面积加起来为

$S_{\mathrm{a-b}}=\frac{b-a}{n}\times f(x_1)+\frac{b-a}{n}\times f(x_2)+\ldots+\frac{b-a}{n}\times f(x_{\mathrm{n}})=\frac{b-a}{n}\{f(x_1)+f(x_2)+\ldots+f(x_{\mathrm{n}})\}$

$=\lim_{n\to\infty}\frac{b-a}{n}\{f(x_1)+f(x_2)+\ldots+f(x_{\mathrm{n}})\}=\lim_{n\to\infty}\frac{b-a}{n}\sum_{k=1}^{n}f(x_{\mathrm{k}})$

（或$\lim_{n\to\infty}\frac{b-a}{n}\sum_{k=0}^{n-1}f(x_{\mathrm{k}})$）。

下述代码在使用区分求积法时，给定的函数为$f'(x)=x^2$（是函数$f(x)=\frac{x^3}{3}$的导函数），已知区间为$[a,b]$，依据上述公式则有$S_{\mathrm{a-b}}=\lim_{n\to\infty}\frac{b-a}{n}\sum_{k=0}^{n-1}f(x_k)=\lim_{n\to\infty}\frac{b-a}{n}\sum_{k=0}^{n-1}(a+k\times\frac{b-a}{n})^2$。使用极限计算求和公式时，需要使用`doit()`方法计算不被默认计算的对象（极限、积分、求和及乘积等）；否则，不能计算极限。

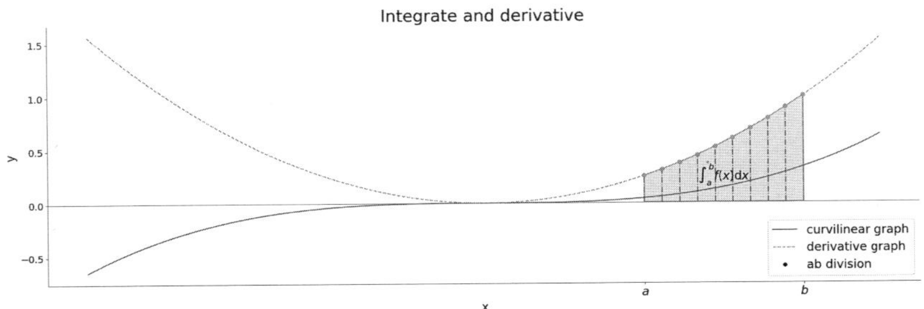

图3.9-3　定积分图解

定积分求给定区间的面积，直接使用SymPy提供的integrate方法，给定区间计算结果约为0.29，与区分求积法计算结果相同。

- 换元积分公式

对于$f(x)$，将变量x替换为一个关于变量y的函数，即$x=g(y)$时，对于$f(x)$的定积分$S=\int_a^b f(x)\mathrm{d}x$的值，用$y$表示为：$\int_a^b f(x)\mathrm{d}x=\int_\alpha^\beta f(g(y))g'(y)\mathrm{d}y$。

```
import numpy as np
import matplotlib.pyplot as plt
from matplotlib.patches import Polygon
import sympy
from sympy import diff, pprint, integrate, oo, Sum, limit  #oo为正无穷

x, n, k = sympy.symbols('x n k')
curvilinear_expr = x**3/3  #定义曲线函数

derivative_curvilinear_expr = diff(
    curvilinear_expr, x)  #curvilinear_expr关于x求微分/导数方程
print("曲线函数curvilinear_expr为:")
pprint(curvilinear_expr)
print("curvilinear_expr 关于x求微分/导数，导函数为:")
pprint(derivative_curvilinear_expr, use_unicode=True)
integrate_derivative_curvilinear_expr = integrate(
    derivative_curvilinear_expr, x)
print("curvilinear_expr导函数的积分:")
pprint(integrate_derivative_curvilinear_expr)
curvilinear_expr_ = sympy.lambdify(x, curvilinear_expr, "numpy")
derivative_expr = sympy.lambdify(x, derivative_curvilinear_expr, "numpy")
t = np.arange(-1.25, 1.25, 0.01)
y = curvilinear_expr_(t)
fig = plt.figure(figsize=(26, 8))
ax = fig.add_subplot(111)
ax.plot(t, y, label="curvilinear graph")
ax.plot(t, derivative_expr(t), '--c', label="derivative graph")
ax.set_title(r'Integrate and derivative', fontsize=20)
ax.set_xlabel('x')
ax.set_ylabel('y')
ax.spines['right'].set_visible(False)
ax.spines['top'].set_visible(False)
ax.axhline(0, color='black', linewidth=0.5)

a, b = 0.5, 1.0  #定义区间
ix = np.linspace(a, b, 10)
iy = derivative_expr(ix)
ax.plot(ix, iy, 'o', label='ab division', color='black')
verts = [(a, 0)]+list(zip(ix, iy))+[(b, 0)]
poly = Polygon(verts, facecolor='0.9', edgecolor='0.5')  #绘制面积区域
ax.add_patch(poly)
plt.text(0.5 * (a + b), 0.2,
         r"$\int_a^b f(x)\mathrm{d}x$", horizontalalignment='center',
fontsize=20)
ax.set_xticks((a, b))
ax.set_xticklabels(('$a$', '$b$'))
ax.stem(ix, iy, linefmt='-.')

#A-使用区分求积法求取面积
Sum_ab = (b-a)/n*Sum(derivative_curvilinear_expr.subs(x,
a+k*(b-a)/n), (k, 0, n-1))  #面积求和公式
print("所有长方形面积之和的公式: \n")
pprint(Sum_ab)
print("doit():\n")
pprint(Sum_ab.doit())
S_ab = limit(Sum_ab.doit(), n, oo)
print("区分求积法计算的面积=", S_ab)

#B-使用定积分求面积(函数)
S_ab_integrate = integrate(derivative_curvilinear_expr, (x, a, b))
print("定积分计算的面积=", S_ab_integrate)

ax.legend(loc='lower left', frameon=False)
plt.xticks(fontsize=20)
plt.show()
-----------------------------------------------------------------
```

```
曲线函数 curvilinear_expr 为：
 3
x
──
3
curvilinear_expr 关于 x 求微分 / 导数，导函数为 :
 2
x
curvilinear_expr 导函数的积分：
 3
x
──
3
所有长方形面积之和的公式：
     n - 1
      ____
      ╲
       ╲                 2
        ╲        ⎛k     ⎞
0.5⋅    ╱  0.25⋅⎜─ + 1⎟
       ╱         ⎝n     ⎠
      ╱
      ‾‾‾‾
     k = 0
───────────────────────────────────────
               n
doit():
      ⎛                ⎛ 2    ⎞        ⎛ 3    2    ⎞⎞
      ⎜                ⎜n    n⎟        ⎜n    n    n⎟⎟
      ⎜          0.5⋅⎜── - ─⎟  0.25⋅⎜── - ── + ─⎟⎟
      ⎜                ⎝2    2⎠        ⎝3    2    6⎠⎟
0.5⋅⎜0.25⋅n + ──────────────── + ────────────────⎟
      ⎜                  n                  2     ⎟
      ⎝                                    n      ⎠
────────────────────────────────────────────────────────
                          n
区分求积法计算的面积 = 7/24
定积分计算的面积 = 0.291666666666667
```

3.9.3 泰勒展开式（Taylor expansion）

上述阐释误差率时，在曲线局部使用一次函数替代（近似）曲线，例如对于函数$f(x)$，令$p=f'(a)q=f(a)$，则在距$x=a$很近的地方，能够将$f(x)$近似为$f(x)\sim q+p(x-a)$。使用一次函数其误差率相对较高，如果近似为二次函数或者三次函数（图3.9–4），是否可以降低误差率？泰勒展开就是将复杂的函数改写成多项式。

一般函数$f(x)$（要能够无限次的进行微分），则可以表示成如下形式，$f(x)=a_0+a_1x+a_2x^2+a_3x^3+\ldots+a_nx^n+\ldots$，右边称为$f(x)$的泰勒展开。这个公式在包含$x=0$的某个限制区域内，才意味着函数$f(x)$同无限次多项式时完全一致的；然而，一旦超出这个无限制区域，右边将会变成一个无法确定的数。例如，对于函数$f(x)=\frac{1}{1-x}$，有$(f(x)=)\frac{1}{1-x}=1+x+x^2+x^3+x^4+\ldots$，令$x=0.1$，有$f(0.1)=\frac{1}{1-0.1}=\frac{1}{0.9}=\frac{10}{9}=1.111\ldots$，及右边为$1+0.1+0.1^2+0.1^3+0.1^4+\ldots=1.1111111\ldots$，因此左右相等。但是，如果令$x=2$时，左右则不会相等。对于上述函数，只有满足$-1<x<1$的$x$才成立。

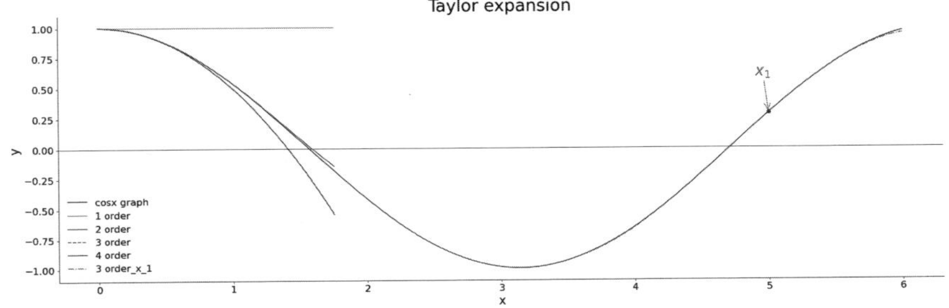

图3.9–4　泰勒展开多阶近似结果比较

- **泰勒展示的求解方法——确定系数**

对于$f(x) = a_0 + a_1x + a_2x^2 + a_3x^3 + ... + a_nx^n + ...$ 式（1）

首先，带入$x = 0$，由$f(0) = a_0$,可知0次常数项a_0为$f(0)$。 ------（A）

然后，对式（1）进行微分，$f'(x) = a_1 + 2a_2x + 3a_3x^2 + ... + na_nx^{n-1} + ...$ 式（2）

将$x = 0$带入式（2），由$f'(0) = a_1$，可知一次系数a_1为$f'(0)$。 ------（B）

继续对式（2）进行微分，$f''(x) = 2a_2 + 6a_3x + ... + n(n-1)a_nx^{n-2} + ...$ 式（3）

代入$x = 0$，可知二次系数a_2为$\frac{1}{2}f''(0)$。 ------（C）

对式（3）进行微分，$ff'''(x) = 6a_3 + ... + n(n-1)(n-2)a_nx^{n-3} + ...$,

由此可知，三次系数a_3为$\frac{1}{6}ff'''(0)$。

持续进行这种运算，n次微分后，就应该得到，$f^{(n)}(x) = n(n-1)... \times 2 \times 1a_n + ...$，其中$f^{(n)}(x)$表示$n$次微分后的$f(x)$。

由此可知，n次系数a_n为$\frac{1}{n!}f^{(n)}(0)$。$n!$，读作“n的阶乘”，它表示$n \times (n-1) \times (n-2) \times ... \times 2 \times 1$。

对$f(x)$进行泰勒展开，便有

$$f(x) = f(0) + \frac{1}{1!}f'(0)x + \frac{1}{2!}f''(0)x^2 + \frac{1}{3!}ff'''(0)x^3 + ... + \frac{1}{n!}f^{(n)}(0)x^n + ...$$

上述公式中，

$f(0)$ <------0次的常数项，即$a_0 = f(0)$ ------（A）

$f'(0)x$ <------1次项，即$a_1 = f'(0)$ ------（B）

$\frac{1}{2!}f''(0)x^2$<------2次项，即$a_2 = \frac{1}{2!}f''(0)$ ------（C）

$\frac{1}{3!}ff'''(0)x^3$<------3次项，即 $a_3 = \frac{1}{3!}ff'''(0)$ ------（D）

泰勒展开，不一定非要从$x = 0$的地方开始，也可以从$(x_0, f(x_0))$处开始。此时，只需要将0替换为x_0，展开方法同上，得：

$$f(x) = f(x_0) + \frac{1}{1!}f'(x_0)(x - x_0) + \frac{1}{2!}f''(x_0)(x - x_0)^2 + \frac{1}{3!}ff'''(x_0)(x - x_0)^3 + ... + \frac{1}{n!}f^{(n)}(x_0)(x - x_0)^n + ...$$

- **误差**

$R_n(x) = \frac{f^{(n+1)}(\xi)}{(n+1)!}(x - x_0)^{n+1}$（推导过程略）

```
import numpy as np
import matplotlib.pyplot as plt
import sympy
from sympy import pprint, solve, diff, factorial
x, a_1, b_1, x_i = sympy.symbols('x a_1 b_1 x_i')

#定义原函数
cos_curve = sympy.cos(x)
cos_curve_ = sympy.lambdify(x, cos_curve, "numpy")
#定义区间
a, b = 0, 6
ix = np.linspace(a, b, 100)
fig = plt.figure(figsize=(26, 8))
ax = fig.add_subplot(111)
ax.plot(ix, cos_curve_(ix), label="cosx graph")

#A-x=0 位置点近似多项式
x_0 = 0
#近似曲线系数计算
a_0 = cos_curve.subs(x, x_0)
a_1 = diff(cos_curve, x).subs(x, x_0)/factorial(1)
a_2 = diff(cos_curve, x, x).subs(x, x_0)/factorial(2)
a_3 = diff(cos_curve, x, x, x).subs(x, x_0)/factorial(3)
a_4 = diff(cos_curve, x, x, x, x).subs(x, x_0)/factorial(4)
ix_ = ix[:30]
#1 阶近似
f_1 = a_1*x+a_0
print("1 阶函数：", f_1)
ax.plot(ix_, [1]*len(ix_), label="1 order")
```

```
#2 阶近似
f_2 = a_2*x**2+a_1*x+a_0
f_2_ = sympy.lambdify(x, f_2, "numpy")
ax.plot(ix_, f_2_(ix_), label="2 order")
print("2 阶函数：")
pprint(f_2)

#3 阶近似
f_3 = a_3*x**3+a_2*x**2+a_1*x+a_0
f_3_ = sympy.lambdify(x, f_3, "numpy")
ax.plot(ix_, f_3_(ix_), '--', label="3 order")
print("3 阶函数：")
pprint(f_3)

#4 阶近似
f_4 = a_4*x**4+a_3*x**3+a_2*x**2+a_1*x+a_0
f_4_ = sympy.lambdify(x, f_4, "numpy")
ax.plot(ix_, f_4_(ix_), label="4 order")
print("4 阶函数：")
pprint(f_4)

#B-x 任一点近似多项式 (3 阶为例)
x_1 = 5
f_x = cos_curve.subs(x, x_i)+diff(cos_curve, x).subs(x, x_i)*(x-x_i)/
factorial(1)+diff(cos_curve, x, x).subs(
    x, x_i)*(x-x_i)**2/factorial(2)+diff(cos_curve, x, x, x).subs(x, x_i)*(x-x_
i)**3/factorial(3) # 近似多项式
f_x_1 = f_x.subs(x_i, x_1)
f_x_1_ = sympy.lambdify(x, f_x_1, "numpy")
ax.plot(ix[60:], f_x_1_(ix[60:]), label="3 order_x_1", c='red', ls='-.')
ax.plot(x_1, cos_curve.subs(x, x_1), 'o')
ax.annotate(r'$x_1$', xy=(x_1, cos_curve.subs(x, x_1)), xycoords='data',
xytext=(x_1-0.1, cos_curve.subs(x, x_1)+0.3),
            textcoords='data', weight='bold', color='red', arrowprops=dict(arrow
style='->', connectionstyle="arc3", color='red'), fontsize=25)

#误差项（3 阶）
xi = x_1+0.25
c, d = x_1-0.5, x_1+0.5
error = diff(cos_curve, x, x, x, x).subs(x, xi)*(d-c)**4/factorial(4)
print("3 阶多项式区间 [%.2f,%.2f] 内位置点 %.2f 的误差为：%.2f" % (c, d, xi, error))

ax.set_title(r'Taylor expansion', fontsize=20)
ax.set_xlabel('x')
ax.set_ylabel('y')
ax.spines['right'].set_visible(False)
ax.spines['top'].set_visible(False)
ax.axhline(0, color='black', linewidth=0.5)
ax.legend(loc='lower left', frameon=False)
plt.xticks(fontsize=13)
plt.yticks(fontsize=13)
plt.show()
-----------------------------------------------------------------------------
1 阶函数：1
2 阶函数：
     2
    x
1 - ──
    2
3 阶函数：
     2
    x
1 - ──
    2
4 阶函数：
 4    2
x    x
── - ── + 1
24   2
3 阶多项式区间 [4.50,5.50] 内位置点 5.25 的误差为：0.02
```

3.9.4 偏微分（偏导数Partial derivative）

- **偏微分**

函数$z=f(x,y)$，在某个邻域内的所有点(x,y)都可以关于x进行偏微分时，在点(x,y)处，关于x的偏微分系数$f_x(x,y)$所对应的函数$(x,y)\mapsto f_x(x,y)$被称为$z=f(x,y)$关于x的偏导数。可表示为：$f_x, f_x(x,y), \frac{\partial f}{\partial x}, \frac{\partial z}{\partial x}$。

同样，在这个邻域内的所有点(x,y)都可以关于y进行偏微分时，所对应的$(x,y)\mapsto f_y(x,y)$被称为$z=f(x,y)$关于y的偏导数。可表示为：$f_y, f_y(x,y), \frac{\partial f}{\partial y}, \frac{\partial z}{\partial y}$。求偏导数的过程叫作偏微分。

偏微分计算直接使用SymPy库的`diff`方法，示例计算结果如图3.9-5所示。

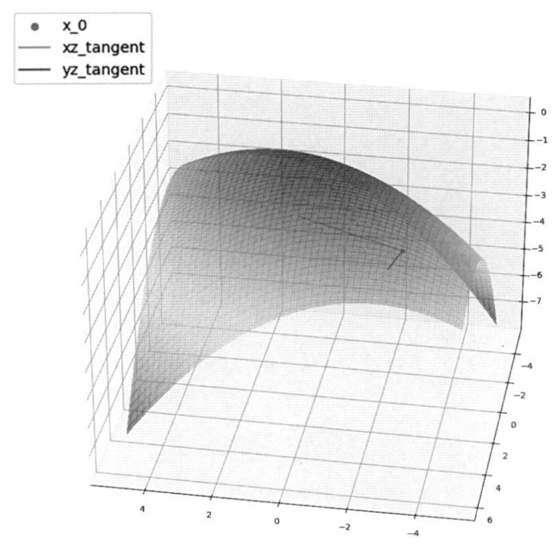

图3.9-5　偏微分计算结果

- **全微分**

由$z=f(x,y)$在$(x,y)=(a,b)$处的近似一次函数可知

$f(x,y)\sim f_x(a,b)(x-a)+f_y(a,b)(x-b)+f(a,b)$，可以将其改写为：

$$f(x,y)-f(a,b)\sim \frac{\partial f}{\partial x}(a,b)(x-a)+\frac{\partial f}{\partial y}(a,b)(x-b) \qquad \text{式(1)}$$

$f(x,y)-f(a,b)$意味着，当点(a,b)向(x,y)变化时，高度$z(=f(x,y))$的增量，效仿一元函数的情况写作$\triangle z$。另外，$x-a$为$\triangle x$，$y-b$为$\triangle y$。

此时，式（1）可以写作

$$\triangle z\sim \frac{\partial z}{\partial x}\triangle x+\frac{\partial z}{\partial y}\triangle y \qquad \text{式(2)}$$

$(x\sim a, y\sim b)$时，这个式子意味着：对于函数$z=f(x,y)$，当x由a增加了$\triangle x$，y由b增加了$\triangle y$后，z就相应增加了$\frac{\partial z}{\partial x}\triangle x+\frac{\partial z}{\partial y}\triangle y$。

$\frac{\partial z}{\partial x}\triangle x$表示"$y$固定在$b$时$x$方向上的增量"，$\frac{\partial z}{\partial y}\triangle y$表示"$x$固定在$a$时$y$方向上的增量"。

说明"$z(=f(x,y))$"的增量可以分解为x方向上的增量与y方向上的增量之和。

将式（2）作理想化（瞬时化）处理了，得，

$$\mathrm{d}z=\frac{\partial z}{\partial x}\mathrm{d}x+\frac{\partial z}{\partial y}\mathrm{d}y \qquad \text{式(3)}$$

或者，$\mathrm{d}f=f_x\mathrm{d}x+f_y\mathrm{d}y$　　式（4）

式（3）（4）被称为全微分公式。即，（曲面高度的增量）=（x方向上的微分系数）×（x方向上的增量）+（y方向上的微分系数）×（y方向上的增量）

- **链式法则公式(Chain rule)**

当$z=f(x,y), x=a(t), y=b(t)$时，$\frac{\mathrm{d}z}{\mathrm{d}t}=\frac{\partial f}{\partial x}\frac{\mathrm{d}a}{\mathrm{d}t}+\frac{\partial f}{\partial y}\frac{\mathrm{d}b}{\mathrm{d}t}$。（推导过程略）

#（偏）微分在机器学习领域中广泛应用，具体可以参看"梯度下降法"部分，对寻找极值的解释。

```
import numpy as np
import matplotlib.pyplot as plt
from matplotlib import cm
import sympy
from sympy import pprint, diff
```

```
x, y = sympy.symbols('x y')
f = -(x**2+x*y+y**2)/10
f_ = sympy.lambdify([x, y], f, "numpy")
fig = plt.figure(figsize=(12, 12))
ax = fig.add_subplot(111, projection='3d')
x_i = np.arange(-5, 5, 0.1)
y_i = np.arange(-5, 5, 0.1)
x_mesh, y_mesh = np.meshgrid(x_i, y_i)
mesh = f_(x_mesh, y_mesh)
surf = ax.plot_surface(x_mesh, y_mesh, mesh, cmap=cm.coolwarm,
                       linewidth=0, antialiased=False, alpha=0.5,)

# 偏微分
px = diff(f, x)
py = diff(f, y)
print(" 偏微分 x：∂f/∂x=")
pprint(px)
print(" 偏微分 y：∂f/∂y=")
pprint(py)
x_0, y_0 = -3, 3
z_0 = f.subs([(x, x_0), (y, y_0)])
print(x_0, y_0, z_0)
ax.scatter(x_0, y_0, float(z_0), marker="o", color="red", label="x_0")

# 平行于 xz 面，绘制点（x_0,y_0）的切线。关于 x 的偏导数
xz_dx = 3
xz_dz = px.subs([(x, x_0), (y, y_0)])*xz_dx
ax.plot((x_0, x_0+xz_dx), (y_0, y_0), (z_0, z_0+xz_dz), label="xz_tangent")

# 平行于 yz 面，绘制点（x_0,y_0）的切线。关于 y 的偏导数
yz_dy = 3
yz_dz = py.subs([(x, x_0), (y, y_0)])*yz_dy
ax.plot((x_0, x_0), (y_0, y_0+yz_dy), (z_0, z_0+xz_dz), label="yz_tangent")

ax.view_init(30, 100)  # 可以旋转图形的角度，方便观察
ax.legend()
plt.show()
--------------------------------------------------------------------------------
偏微分 x：∂f/∂x=
   x    y
 - ─ - ──
   5   10
偏微分 y：∂f/∂y=
    x    y
 - ── - ─
   10   5
-3 3 -9/10
```

参考文献（References）:

[1] [日]石山平，大上丈彦著．李巧丽译．7天搞定微积分[M]．海口：南海出版公司，2010．8.

[2] [日]小岛宽之著，十神真漫画绘制，株式会社BECOM漫画制作，张仲恒译．漫画微积分[M]．北京：科学出版社，2009．8.

3.10 卷积与SIR空间传播模型

从生物体角度研究的生态学家发现，区别生物体生活在哪里与生物体在做什么是很有用的。生物体的栖息地（habitat）是指它们生活的地方。栖息地的特征要用清晰的物理特征来描述，这常常是植物或动物主要的生活型。例如，森林栖息地、荒漠栖息地和珊瑚栖息地等。因为栖息地概念强调了生物所暴露的条件，为生物环境保护及栖息条件营造提供参考。生物的生态位（niche）代表生物能够耐受的条件范围，它们的生活方式，即在生态学系统中的作用。每个物种都具有一个独特的生态位。生物的独特形态和生理特征决定它们的耐受条件，例如如何取食、逃避天敌等。没有一种生物具备在地球上任何条件下存活的能力。在空间变化上，不同地方的环境条件是有区别的，气候、地形和土壤类型的变化导致大尺度的异质性（从数米到几百公里）。小尺度上的异质性一般是由于植物结构、动物活动和土壤内含物引起的。特定的空间尺度变化可能对于一种动物重要，而对于另一种不重要。当一个个体穿行于空间不断变化着的环境时，个体移动越快，空间变化的尺度越小，就更快地遇到新的环境条件。地形和地质能够改变其他方面相同气候区域内的局部环境。例如，多山区域，坡度倾斜和阳光方向影响着土壤的温度和湿度，这同样适用于城市区域。建筑的高度及分布形成了城市的多样局地微气候环境。

虽然植物对于生活在哪里具有相对小的选择性，但是仍然向着高浓度的土壤矿物质等适合生长的环境中趋向。而动物则能够自主地在环境中四处移动，并选择其栖息地。即使在栖息地内，温度、湿度、盐度和其他因子也有明显不同，能够进一步区分位微栖息地（microhabitats）或微环境（microenvironments）。例如，荒漠中灌木丛下的树荫比暴露在直射阳光下的地区更凉快、潮湿。由于动物生活在各式各样的和变化的环境中，它们经常需要做出如何决策的行动。这些决策中很多涉及食物，到哪里寻找食物，在某些板块栖息地中进食多长时间，吃哪一种食物类型等。动物期望获取选择获利最大的行为，即以最适摄食（optimal foraging）理论试图阐释，例如中心摄食、鸟类饲喂巢穴中的子女时，以巢穴位中心，在远处自由搜索食物；风险-敏感摄食，取决于个体能够搜索食物的速率，以及这个地区的相对安全性；猎物选择，每种类型的食物基于其营养和能量，以及处理难度和毒素的潜在危险，形成固有的价值；以及混合食物，以满足所必需的营养。

涉及地理空间因素，除了满足适宜的栖息地（生态位）及生物的行动决策，还有个体移动可以保持种群的空间连接。种群生物学家把种群内的个体移动称为散布（dispersal），指的是个体在各亚种群之间的移动。在规划领域，为了保护某一物种，常用的方法是生物栖息地的适宜性评价；或综合考虑生物、水安全、历史价值、地价、风景价值、游憩价值、居住价值，计算成本栅格来规划城市布局、道路选线等内容。上述方法广泛应用于规划、风景园林、生态规划等领域，而SIR（Susceptible Infected Recovered Model）[①]模型是一种传播模型，典型地应用于传染病传播路径模拟中，也包括估计如何传播、感染总数、流行病持续时间和相关参数等。

因为SIR模型所涉及易感染着的空间阻力分布与规划中适宜性评价的成本栅格如出一辙，而感染源及病毒的蔓延类似于生态学中物种的迁移；同时，SIR模型也是最为简单的分室/分区模型（compartmental models）之一，许多模型都是这种基本形式的衍生物。因此，为了进一步拓展成本栅格（cost raster）已有研究成果，尝试将其引入规划领域。

在SIR模型实现过程中，使用到卷积的方法。卷积也是深度学习中卷积神经网络的基础，亦是图像处理或特征提取的基础。

① SIR（Susceptible Infected Recovered Model），最简单的分室模型（compartmental model）之一，许多模型是这种基本形式的衍生物（https://en.wikipedia.org/wiki/Compartmental_models_in_epidemiology#The_SIR_model）。

3.10.1 卷积

在泛函分析中，卷积（又称叠积或褶积，convolution），是透过两个函数f和g生成第3个函数的一种数学算子。如果把一个函数看作区间的指示函数（系统响应函数，g），另一个为待处理的输入函数（输入信号函数，f），卷积可以看作“滑动平均”的拓展。设$f(x)$、$g(x)$是$\mathbb{R}$上两个可积函数，作积分：$\int_{-\infty}^{\infty} f(\tau)g(x-\tau)\mathrm{d}\tau$。可以证明，关于几乎所有的$x \in (-\infty,\infty)$，上述积分是存在的。这样，随着$x$的不同取值，这个积分就定义了一个新的函数$h(x)$，称为$f$与$g$的卷积，记为$h(x)=(f*g)(x)$。对于上述公式的理解，需要具备积分方面的知识；并需要注意，为方便理解，将指示函数和输入函数看作一个时间区段上的曲线函数，固定$f(\tau)$在T时间点上（横坐标表示为时间点），翻转$g(\tau)$为$g(-\tau)$，并平移对齐$f(\tau)$的时间点，结果为$g(T-\tau)$。

对于卷积计算的解释，可以通过图3.10–1理解。其中，需要注意的一个关键点是T时刻时，例如T=20时，20时刻是最新进入的信号，之前的时刻越小，信号进入的时间越久。例如，0时刻时的信号已经历经了20个时间点；1时刻，经历了19个时间点。而指示函数则在0时刻时作用于输入函数的新进入信号即20时刻时的值，指示函数1时刻作用于输入函数19时刻的值。以此类推，正好是一个反向的过程，也因此出现了对响应函数进行翻转、移动等操作。实际上，如果固定指示函数，而随时间逐步地移动输入函数的值进入，则不会有上述操作，也更容易理解。

#泛函分析（Functional Analysis）研究主要对象是函数构成的函数空间。使用泛函一词作为表述，代表作用于函数的函数。这意味着，一个函数的参数是函数。

```
import numpy as np
import sympy
from sympy import pprint, Piecewise
import matplotlib.pyplot as plt
t, t_ = sympy.symbols('t t_')

''' 定义指示函数 '''
e = sympy.E
# 参考 Hyperbolic tangent function 即双曲正切函数  y=tanh x
g_t = -1*((e**t-e**(-t))/(e**t+e**(-t)))+1
g_t_ = sympy.lambdify(t, g_t, "numpy")

''' 定义输入函数 '''
f_t = 1*sympy.sin(4*np.pi*0.05*t_)+2.5  # +np.random.randn(1)
f_t_ = sympy.lambdify(t_, f_t, "numpy")

''' 定义时间段 '''
T = 20  #时间点
t_bucket = np.linspace(0, T, T+1, endpoint=True)

''' 绘制图形 '''
# 函数原始位置
fig, axs = plt.subplots(1, 3, figsize=(30, 3))
axs[0].plot(t_bucket, g_t_(t_bucket), 'o', c='r', label='g_t')
axs[0].plot(t_bucket, f_t_(t_bucket), 'o', c='b', label='f_t')
axs[0].plot([t_bucket, np.flip(t_bucket)], [
            g_t_(t_bucket), np.flip(f_t_(t_bucket))], '--', c='gray')
axs[0].set_title('original position')

# 翻转响应函数
axs[1].plot(-t_bucket, g_t_(t_bucket), 'o', c='r', label='g_t')
axs[1].plot(t_bucket, f_t_(t_bucket), 'o', c='b', label='f_t')
axs[1].plot([np.flip(-t_bucket), t_bucket],
            [np.flip(g_t_(t_bucket)), f_t_(t_bucket)], '--', c='gray')
axs[1].set_title('flip g(t)')

# 移动响应函数
axs[2].plot(-t_bucket+T, g_t_(t_bucket), 'o', c='r', label='g_t')
axs[2].plot(t_bucket, f_t_(t_bucket), 'o', c='b', label='f_t')
axs[2].plot([np.flip(-t_bucket+T), t_bucket],
            [np.flip(g_t_(t_bucket)), f_t_(t_bucket)], '--', c='gray')
axs[2].set_title('move g(t)')
axs[0].legend(loc='upper right', frameon=False)
axs[1].legend(loc='upper right', frameon=False)
axs[2].legend(loc='upper right', frameon=False)
plt.show()
```

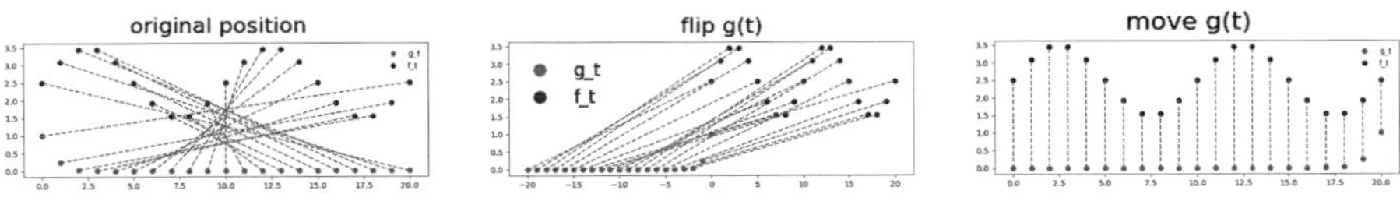

图3.10–1 指示函数与输入函数，原始位置（左）；翻转（中）与移动响应函数（右）

3.10.1.1 一维卷积与曲线分割

对于一维卷积给出一个简单的例子（表3.10–1）：指示函数的一组结果［1, 1, 1］，输入函数的一组结果［0, 0, 1, 1, 1, 1, 1, 0, 0］，为了方便计算，固定输入函数（通常固定指示函数，而滑动输入函数），滑动指示函数，对应位置求积，最后求和，所有时刻的结果即为卷积结果。通过使用np.convolve(mode='full')，计算结果同为array(［0, 0, 1, 2, 3, 3, 3, 2, 1, 0, 0］)。

表3.10–1 一维卷积过程示例

momentc/step	step-0	step-1	step-2	step-3	step-4	step-5	step-6	step-7	step-8	step-sum
t-0	0	0	1	1	1	1	1	0	0	*0*
t-1	**1***0	-*0	-*1	-*1	-*1	-*1	-*1	-*0	-*0	*0*
t-2	**1***0	**1***0	-*1	-*1	-*1	-*1	-*1	-*0	-*0	*0*
t-3	**1***0	**1***0	**1***1	-*1	-*1	-*1	-*1	-*0	-*0	*1*
t-4	-*0	**1***0	**1***1	**1***1	-*1	-*1	-*1	-*0	-*0	*2*
t-5	-*0	-*0	**1***1	**1***1	**1***1	-*1	-*1	-*0	-*0	*3*
t-6	-*0	-*0	-*1	**1***1	**1***1	**1***1	-*1	-*0	-*0	*3*
t-7	-*0	-*0	-*1	-*1	**1***1	**1***1	**1***1	-*0	-*0	*3*
t-8	-*0	-*0	-*1	-*1	-*1	**1***1	**1***1	**1***0	-*0	*2*
t-9	-*0	-*0	-*1	-*1	-*1	-*1	**1***1	**1***0	**1***0	*1*
t-10	-*0	-*0	-*1	-*1	-*1	-*1	-*1	**1***0	**1***0	*0*
t-11	-*0	-*0	-*1	-*1	-*1	-*1	-*1	-*0	**1***0	*0*
t-12	-*0	-*0	-*1	-*1	-*1	-*1	-*1	-*0	-*0	*0*

基于上述的例子，更容易从信号分析的角度来理解卷积，因此对应指示函数（$g(x)$）名称为系统响应函数（或简称为响应函数），对应输入函数（$f(x)$）为输入信号函数（或简称为信号函数），定义一个可以给定响应函数和信号函数，输出动态卷积的类。在类的定义过程中，使用matplotlib库中给出的animation方法定义动态图表，并通过继承animation.TimedAnimation父类，实现多个子图的动画。类的配置实现了三个图表功能：第一个是定义的响应函数，响应函数通常为固定的时间段一段变化的曲线，用于作用于信号函数，增强、或特定变化信号函数的值；第2个子图是动态的信号函数，信号函数随时间而变化，可以理解为信号在不同时间上的“位置”变化，例如图中的-4时刻，信号位于子图的左侧，即开始进入。而在4时刻，信号位于子图的右侧，即开始出去。或理解为信号随时间的流逝；第3个子图则为计算的卷积，响应函数是固定的，整个作用域，即有值的区间均作用于指定时间区段信号函数的每一时刻，因此卷积的位置变化随信号函数而变化，但是并不完全对齐。

类的输入参数包括G_T_fun()响应函数；F_T_fun()信号函数；以及t时间的开始（s）与结束（t）点。时间段的步幅step用于配置帧（frame），步幅越小，时间点更新越小，时间精度越高。linespace为时间段的划分，用于图表的x轴坐标，以及信号函数和响应函数的输入参数；mode参数是对应np.convolve(mode='')，其模式包括same、full和valid，该类的定义中未处理valid模式。

所定义的类中，F_T_fun输入参数，即信号函数的定义比较特殊，包括一个函数的输入参数timing（f_t_sym=self.f_t_(i)），用于保持信号随时间变化值的移动，应用到sympy库提供的Piecewise多段函数的方法；以及函数内部所定义sympy公式的输入参数（f_t_val=f_t_sym(self.t)）。

```
%matplotlib inline
from IPython.display import HTML
import numpy as np
from scipy import signal
import matplotlib.pyplot as plt
import matplotlib.animation as animation
import matplotlib.pyplot as plt
```

```
import math
import sympy
from sympy import pprint, Piecewise
class dim1_convolution_SubplotAnimation(animation.TimedAnimation):
    '''
    function - 一维卷积动画解析，可以自定义系统函数和信号函数

    Params:
        G_T_fun - 系统响应函数；func
        F_T_fun - 输入信号函数；func
        t={"s":-10,"e":10,'step':1,'linespace':1000} - 时间开始点、结束点、帧的步幅、
时间段细分；dict
        mode='same' - numpy 库提供的 convolve 卷积方法的卷积模式；string
    '''
    def __init__(self, G_T_fun, F_T_fun, t={"s": -10, "e": 10, 'step': 1, 'linespace':
1000}, mode='same'):
        self.mode = mode
        self.start = t['s']
        self.end = t['e']
        self.step = t['step']
        self.linespace = t['linespace']
        self.t = np.linspace(self.start, self.end,
                             self.linespace, endpoint=True)
        fig, axs = plt.subplots(1, 3, figsize=(24, 3))
        # 定义 g(t)，系统响应函数
        self.g_t_ = G_T_fun
        g_t_val = self.g_t_(self.t)
        self.g_t_graph, = axs[0].plot(
            self.t, g_t_val, '--', label='g(t)', color='r')

        # 定义 f(t)，输入信号函数
        self.f_t_ = F_T_fun
        self.f_t_graph, = axs[1].plot([], [], '-', label='f(t)', color='b')

        # 卷积（动态 - 随时间变化）
        self.convolution_graph, = axs[2].plot(
            [], [], '-', label='1D convolution', color='g')
        axs[0].set_title('g_t')
        axs[0].legend(loc='lower left', frameon=False)
        axs[0].set_xlim(self.start, self.end)
        axs[0].set_ylim(-1.2, 1.2)
        axs[1].set_title('f_t')
        axs[1].legend(loc='lower left', frameon=False)
        axs[1].set_xlim(self.start, self.end)
        axs[1].set_ylim(-1.2, 1.2)
        axs[2].set_title('1D convolution')
        axs[2].legend(loc='lower left', frameon=False)
        axs[2].set_xlim(self.start, self.end)
        axs[2].set_ylim(-1.2*100, 1.2*100)
        plt.tight_layout()
        animation.TimedAnimation.__init__(
            self, fig, interval=500, blit=True)  # interval 配置更新速度

    # 更新图形
    def _draw_frame(self, framedata):
        i = framedata
        f_t_sym = self.f_t_(i)  # 1- 先输入外部定义的 F_T_fun 函数的输入参数
        f_t_val = f_t_sym(self.t)  # 2- 再定义 F_T_fun 函数内部由 sympy 定义的公式的输入参数

        self.f_t_graph.set_data(self.t, f_t_val)
        g_t_val = self.g_t_(self.t)
        g_t_val = g_t_val[~np.isnan(g_t_val)]  # 移除空值，仅保留用于卷积部分的数据

        if self.mode == 'same':
            conv = np.convolve(g_t_val, f_t_val, 'same')  # self.g_t_(t)
            self.convolution_graph.set_data(self.t, conv)
        elif self.mode == 'full':
            conv_ = np.convolve(g_t_val, f_t_val, 'full')  # self.g_t_(t)
            t_diff = math.ceil((len(conv_)-len(self.t))/2)
            conv = conv_[t_diff:-t_diff+1]
            self.convolution_graph.set_data(self.t, conv)
        else:
            print("please define the mode value--'full' or 'same' ")
```

```
    # 配置帧 frames
    def new_frame_seq(self):
        return iter(np.arange(self.start, self.end, self.step))

    # 初始化图形
    def _init_draw(self):
        graphs = [self.f_t_graph, self.convolution_graph,]
        for G in graphs:
            G.set_data([], [])
```

- **一个简单的响应函数与信号函数的定义实现**

响应函数是时间长度为1、值为1的函数。信号函数为给定时间点（timing），以及该时间点后长度为1时间段内值为1、其他时间值为0的函数（实际上，是保持每一时间点持续的产生一段信号）。将该响应函数作用于信号函数，结果为值先增加数倍后落回的三角形坡度（图3.10-2）。

```
def G_T_type_1():
    '''
    function - 定义系统响应函数 . 类型 -1

    Returns:
        g_t_ - sympy 定义的函数
    '''
    import sympy
    from sympy import pprint, Piecewise
    t, t_ = sympy.symbols('t t_')
    g_t = 1
    # 定义位分段函数，系统响应函数在区间 [0,1] 之间作用。
    g_t_piecewise = Piecewise((g_t, (t >= 0) & (t <= 1)))
    g_t_ = sympy.lambdify(t, g_t_piecewise, "numpy")
    return g_t_
def F_T_type_1(timing):
    '''
    function - 定义输入信号函数，类型 -1

    Returns:
        函数计算公式
    '''
    import sympy
```

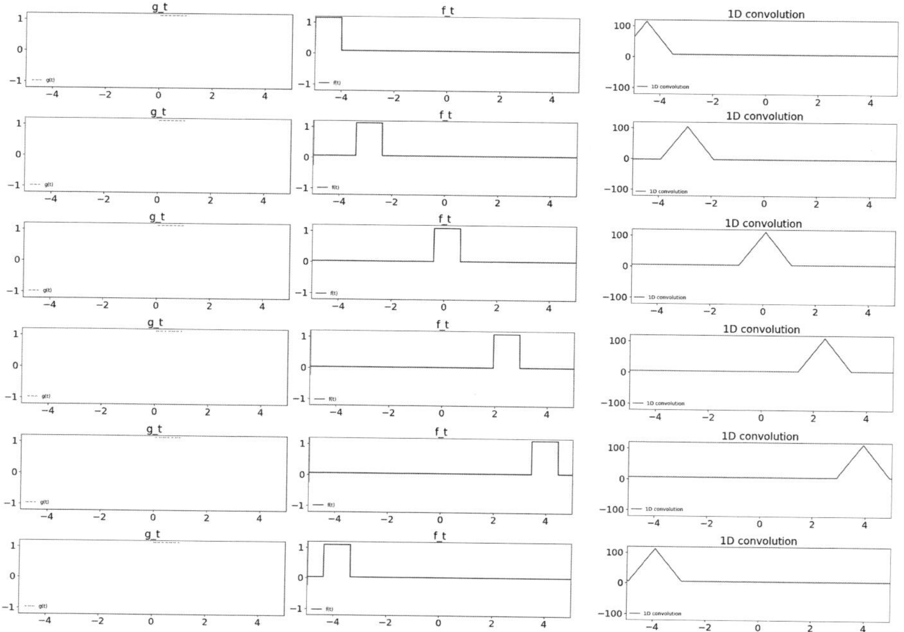

图3.10-2　响应函数（g_t）作用于信号函数（f_t）的过程示例

```
    from sympy import pprint, Piecewise
    t, t_ = sympy.symbols('t t_')
    f_t = 1
    # 定义位分段函数，系统响应函数在区间 [0,1] 之间作用
    f_t_piecewise = Piecewise((f_t, (t > timing) & (t < timing+1)), (0, True))
    f_t_ = sympy.lambdify(t, f_t_piecewise, "numpy")
    return f_t_

G_T_fun = G_T_type_1()  # 系统响应函数
F_T_fun = F_T_type_1  # 输入信号函数
t = {"s": -5, "e": 5, 'step': 0.1, 'linespace': 1000, 'frame': 1}  # 时间参数配置
ani = dim1_convolution_SubplotAnimation(
    G_T_fun, F_T_fun, t=t, mode='same')  # mode:'full','same'
HTML(ani.to_html5_video())  # conda install -c conda-forge ffmpeg
# 保存动画为 .gif 文件
from matplotlib.animation import FuncAnimation, PillowWriter
writer = PillowWriter(fps=25)
ani.save(r"./imgs/2_3_2/convolution_a.gif", writer=writer)
```

- **一个稍微复杂一些的响应函数与信号函数**

这里借助tanh双曲正切函数的一段作为响应函数（图3.10-3），值快速下降到趋于平缓逐渐降低的一个过程。而信号函数是参考给定时间点上，两段时间区间上值为1，其他时候值为0的函数。卷积后的结果在上升和下降段，均产生一定的弧度曲张（凸形爬升，凹形回落）。

```
def G_T_type_2():
    '''
    function - 定义系统响应函数 . 类型 -2

    Returns:
        g_t_ - sympy 定义的函数
    '''
    import sympy
    from sympy import pprint, Piecewise
    t, t_ = sympy.symbols('t t_')
    e = sympy.E
    # 参考 Hyperbolic tangent function 即双曲正切函数  y=tanh x
    g_t = -1*((e**t-e**(-t))/(e**t+e**(-t)))+1
```

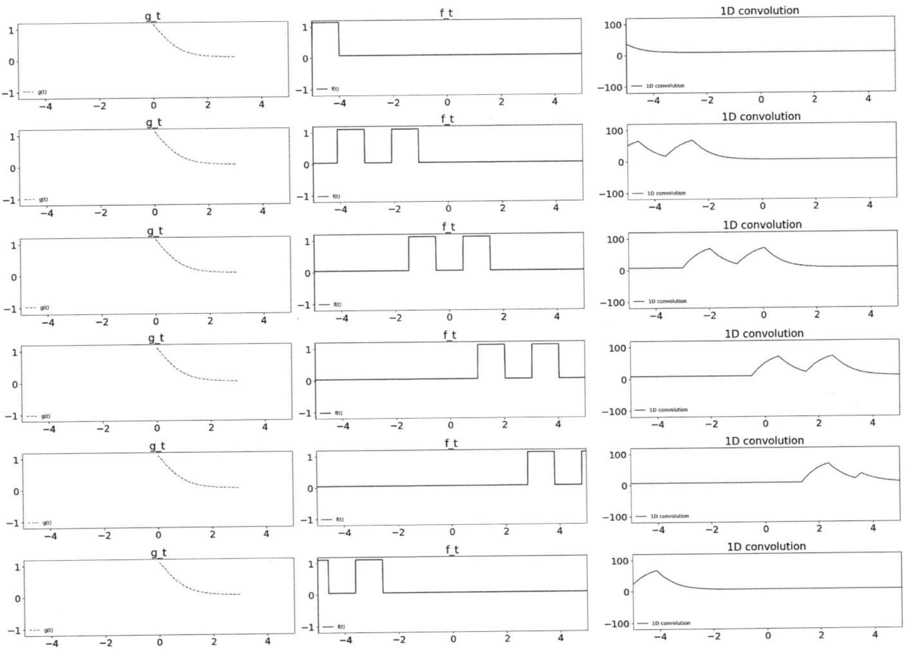

图3.10-3 以（tanh）双曲正切函数作为响应函数的卷积过程示例

```
    #定义位分段函数，系统响应函数在区间[0,3]之间作用
    g_t_piecewise = Piecewise((g_t, (t >= 0) & (t < 3)))
    g_t_ = sympy.lambdify(t, g_t_piecewise, "numpy")
    return g_t_
def F_T_type_2(timing):
    '''
    function - 定义输入信号函数，类型-2

    Returns:
        函数计算公式
    '''
    import sympy
    from sympy import pprint, Piecewise
    t, t_ = sympy.symbols('t t_')
    f_t = 1
    f_t_piecewise = Piecewise((f_t, (t > timing) & (
        t < timing+1)), (f_t, (t > timing-2) & (t < timing-1)), (0, True)) #定义
位分段函数，系统响应函数在区间[0,1]之间作用
    f_t_ = sympy.lambdify(t, f_t_piecewise, "numpy")
    return f_t_
G_T_fun = G_T_type_2()
F_T_fun = F_T_type_2
t = {"s": -5, "e": 5, 'step': 0.1, 'linespace': 1000, 'frame': 1} #时间参数配置
ani = dim1_convolution_SubplotAnimation(
    G_T_fun, F_T_fun, t=t, mode='same')  #mode:'full','same'
HTML(ani.to_html5_video())
#保存动画为.gif文件
from matplotlib.animation import FuncAnimation, PillowWriter
writer = PillowWriter(fps=25)
ani.save(r"./imgs/2_3_2/convolution_b.gif", writer=writer)
```

- **基于1维卷积曲线跳变点识别切分**

给定指示函数，可以对输入函数施加影响，即如果给定一个卷积核（convolution kernel，指示函数的计算值），作用于一维数组（输入函数的计算值），可以变化输入数据达到预期的结果。例如，可以使用一维卷积的方法由跳变点分割曲线。这里使用的卷积核为$[-1,2,-1]$。在该实验分析中，数据使用了无人驾驶城市项目①的PHMI模拟车载激光雷达导航评估值（图3.10-4），当$PHMI>10^{-5}$时，说明从激光雷达扫描点云中提取的特征点可以很好地导航无人车，否则存在风险。对于实验的输入数据，也可以自行随机生成一维数组（一组值列表）。因为PHMI的计算最初由MatLab完成，生成了对应图表保存为.fig的MatLab图表文件。可以用`scipy.io`库提供的工具加载该数据，提取对应的值。

① Driverless City Project, Illinois Institute of Technolog, IIT,((https://www.thedriverlesscityproject.org/))。

#注意，MatLab的图表文件.fig，可能因为操作系统不同，MatLab版本不同，提取数据的代码可能对应做出调整。下述提取.fig数据的方法不能提取所有类型的.fig文件。

```
data_PHMI_fp = r'./data/04-10-2020_312LM_PHMI.fig'
def read_MatLabFig_type_A(matLabFig_fp, plot=True):
    '''
    function - 读取MatLab的图表数据，类型-A

    Params:
        matLabFig_fp - MatLab的图表数据文件路径；string

    Returns:
        fig_dic - 返回图表数据，(X,Y,Z)
    '''
    from scipy.io import loadmat
    import matplotlib.pyplot as plt
    matlab_fig = loadmat(matLabFig_fp, squeeze_me=True, struct_as_record=False)
    fig_dic = {}  #提取MatLab的.fig值
    ax1 = [c for c in matlab_fig['hgS_070000'].children if c.type == 'axes']
```

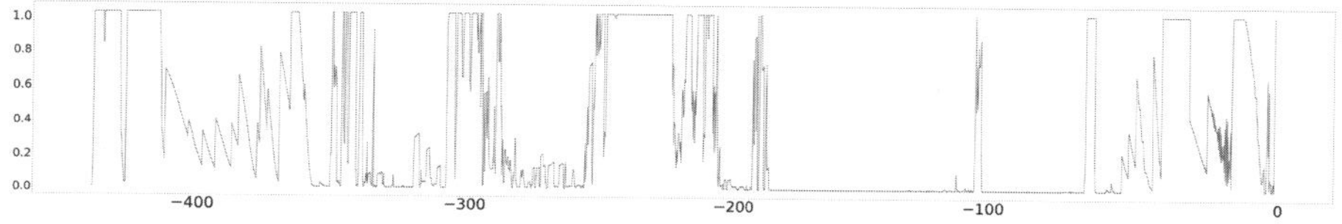

图3.10-4 无人驾驶城市项目的PHMI模拟车载激光雷达导航评估值

```
    if (len(ax1) > 0):
        ax1 = ax1[0]
    i = 0
    for line in ax1.children:
        try:
            X = line.properties.XData
            Y = line.properties.YData
            Z = line.properties.ZData
            fig_dic[i] = (X, Y, Z)
        except:
            pass
        i += 1
    if plot == True:
        fig = plt.figure(figsize=(130, 20))
        markers = ['.', '+', 'o', '', '', '']
        colors = ['#7f7f7f', '#d62728', '#1f77b4', '', '', '']
        linewidths = [2, 10, 10, 0, 0, 0]
        plt.plot(fig_dic[1][1], fig_dic[1][2], marker=markers[0],
                 color=colors[0],
        linewidth=linewidths[0])
        plt.tick_params(axis='both', labelsize=40)
        plt.show()
    return fig_dic
```

由跳变点分割曲线的基本思路是:

1. 计算卷积，获取反映跳变点变化的特征值；
2. 由标准计分标准化卷积值，方便设置阈值；
3. 返回满足阈值的索引值；
4. 定义根据索引，分割列表的函数，根据满足阈值的索引值分割列表，包括分割卷积值、分割原始数据和分割索引值等；
5. 返回分割结果，打印显示结果图3.10-5。

```
import matplotlib.pyplot as plt
def curve_segmentation_1DConvolution(data, threshold=1):
    '''
    function - 应用一维卷积，根据跳变点分割数据

    Params:
        data - 待处理的一维度数据; list/array

    Returns:
        data_seg - 列表分割字典，"dataIdx_jump"- 分割索引值，"data_jump"- 分割原始数据，
"conv_jump"- 分割卷积结果
    '''
    import numpy as np
    from scipy import stats
    def lst_index_split(lst, args):
        '''
        function - 根据索引，分割列表

        transfer:https://codereview.stackexchange.com/questions/47868/splitting-
a-list-by-indexes/47877
        '''
        if args:
            args = (0,) + tuple(data+1 for data in args) + (len(lst)+1,)
        seg_list = []
        for start, end in zip(args, args[1:]):
            seg_list.append(lst[start:end])
        return seg_list
    data = data.tolist()
    kernel_conv = [-1, 2, -1]  #定义卷积核，即指示函数
```

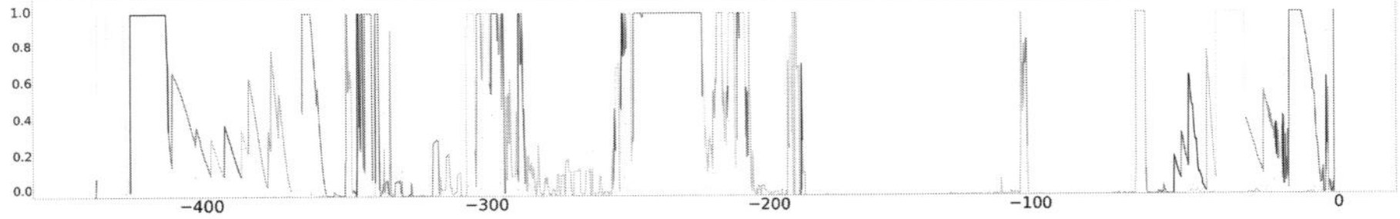

图3.10-5 跳变点分割曲线结果

```
    result_conv = np.convolve(data, kernel_conv, 'same')
    #标准化，方便确定阈值，根据阈值切分列表
    z = np.abs(stats.zscore(result_conv))  #标准计分 - 绝对值
    z_ = stats.zscore(result_conv)  #标准计分

    threshold = threshold
    breakPts = np.where(z > threshold)  #返回满足阈值的索引值
    breakPts_ = np.where(z_ < -threshold)

    #根据满足阈值的索引值，切分列表
    conv_jump = lst_index_split(
        result_conv.tolist(), breakPts_[0].tolist())  #分割卷积结果
    data_jump = lst_index_split(data, breakPts_[0].tolist())  #分割原始数据
    dataIdx_jump = lst_index_split(
        list(range(len(data))), breakPts_[0].tolist())  #分割索引值
    data_seg = {"dataIdx_jump": dataIdx_jump,
                "data_jump": data_jump, "conv_jump": conv_jump}
    return data_seg
p_X = PHMI_dic[1][1]
p_Y = PHMI_dic[1][2]
p_Y_seg = curve_segmentation_1DConvolution(p_Y)

'''展平列表函数'''
def flatten_lst(lst): return [m for n_lst in lst for m in flatten_lst(
    n_lst)] if type(lst) is list else [lst]

#打印分割结果
plt.figure(figsize=(130, 20))
plt.scatter(p_X, [abs(v) for v in flatten_lst(p_Y_seg["conv_jump"])], s=1)
def nestedlst_insert(nestedlst):
    '''
    function - 嵌套列表，子列表前后插值

    Params:
        nestedlst - 嵌套列表；list

    Returns:
        nestedlst - 分割后的列表；list
    '''
    for idx in range(len(nestedlst)-1):
        nestedlst[idx+1].insert(0, nestedlst[idx][-1])
    nestedlst.insert(0, nestedlst[0])
    return nestedlst
def uniqueish_color():
    import matplotlib.pyplot as plt
    import numpy as np
    '''
    function - 使用matplotlib提供的方法随机返回浮点型RGB
    '''
    return plt.cm.gist_ncar(np.random.random())
data_jump = p_Y_seg["data_jump"]
data_jump = nestedlst_insert(data_jump)
dataIdx_jump = p_Y_seg['dataIdx_jump']
dataIdx_jump = nestedlst_insert(dataIdx_jump)
p_X_seg = [[p_X[idx] for idx in g] for g in dataIdx_jump]
for val, idx in zip(data_jump, p_X_seg):
    plt.plot(idx, val, color=uniqueish_color(), linewidth=5.0)
plt.ylim(0, 1.1)
plt.tick_params(axis='both', labelsize=40)
plt.show()
```

3.10.1.2 二维卷积与图像特征提取

一维卷积是卷积核滑动，对应位置求积后计算和。二维卷积同样，此时卷积核的维度为2，通常为奇数。在二维平面上滑动，对应位置上求积后计算和。使用的方法为SciPy库提供的 `scipy.signal import convolve2d`。其中，同样有mode参数（`full,valid,same`），并有边界处理参数（`fill,wrap,symm`）。在下述的测试代码中，调入了一个图像（图3.10-6左），该图像RGB值中，G和B值均为0，只是变化R值，并仅有红（255）和黑（0）两类值。通过给定卷积核，计算卷积（图3.10-6右），观察前后的数据变化，确认是卷积核滑动，对应位置乘积求和替换卷积核中心对应位置的图像值。

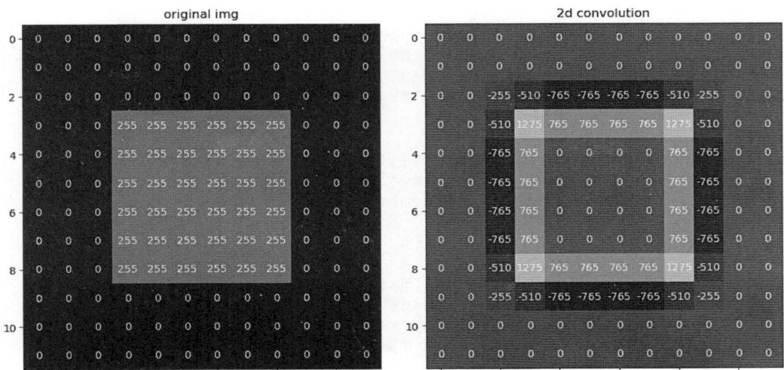

图3.10-6 自定义数据（左）；及其卷积（右）

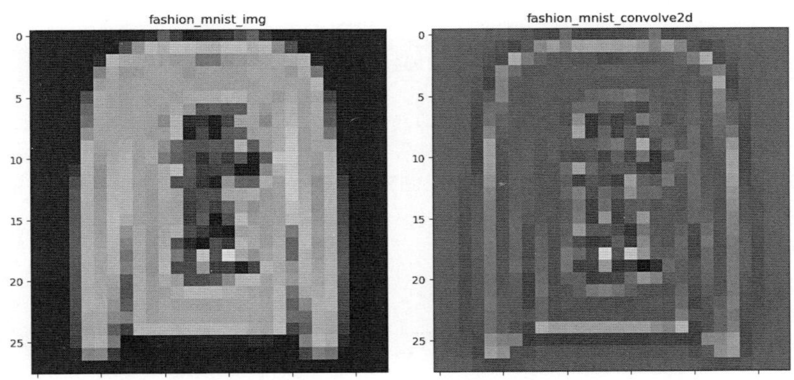

图3.10-7 Fashion_MNIST数据（左）；及其卷积（右）

同样，使用Fashion_MNIST①数据集中的图像（图3.10-7）测试卷积，边缘检测卷积核也能够探测出图像变化的边界位置。

Fashion_MNIST 数据集是深度学习的基础数据集，可以替代已经被广泛，无数次重复使用似乎不再新鲜的 MNIST② 手写数据集。

① Fashion_MNIST，是Zalando论文图像数据集，由60,000个样本的训练集和10,000个样本的测试集组成。每个样本是一个28×28的灰度图像，与10个类别的标签相关。（https://www.kaggle.com/datasets/zalando-research/fashionmnist）。

② MNIST，是一个手写数字数据集，由60,000个样本组成的训练集，及一个由10,000个样本组成的测试集。这些数字的大小已被标准化，并在一个固定大小的图像中居中。（http://yann.lecun.com/exdb/mnist/）。

```
from tensorflow import keras
import matplotlib.pyplot as plt
from skimage import io
import numpy as np
from skimage.morphology import square
from scipy.signal import convolve2d
img_12Pix_fp = r'./data/12mul12Pixel.bmp'
fig, axs = plt.subplots(1, 4, figsize=(28, 7))

#A - 子图 1，原始图像，以及 RGB 值中的 R 值
img_12Pix = io.imread(img_12Pix_fp)
struc_square = square(12)
axs[0].imshow(img_12Pix, cmap="Paired", vmin=0, vmax=12)
for i in range(struc_square.shape[0]):
    for j in range(struc_square.shape[1]):
        axs[0].text(j, i, img_12Pix[:, :, 0][i, j],
                    ha="center", va="center", color="w")
axs[0].set_title('original img')

#B - 子图 2，二维卷积，卷积核可以探测边界
kernel_edge = np.array([[-1, -1, -1],
                        [-1, 8, -1],
                        [-1, -1, -1]])  # 边缘检测卷积核 / 滤波器。统一称为卷积核
```

```
#有时，图像默认值为'int8'。如果不修改数据类型，convolve 计算结果会出错
img_12Pix_int32 = img_12Pix[..., 0].astype(np.int32)
print('Verify that the dimensions are the same:',
      img_12Pix_int32.shape, kernel_edge.shape)  #仅计算 R 值
img_12Pix_convolve2d = convolve2d(img_12Pix_int32, kernel_edge, mode='same')
axs[1].imshow(img_12Pix_convolve2d)
for i in range(struc_square.shape[0]):
    for j in range(struc_square.shape[1]):
        axs[1].text(j, i, img_12Pix_convolve2d[i, j],
                    ha="center", va="center", color="w")
axs[1].set_title('2d convolution')

#C-使用 fashion_mnist 数据集中的图像，实验卷积
fashion_mnist = keras.datasets.fashion_mnist
(train_images, train_labels), (test_images,
                               test_labels) = fashion_mnist.load_data()
fashion_mnist_img = train_images[900]  #随机提取一个图像
axs[2].imshow(fashion_mnist_img)
axs[2].set_title('fashion_mnist_img')

#D- fashion_mnist 数据集中随机一个图像的卷积
fashion_mnist_convolve2d = convolve2d(
    fashion_mnist_img, kernel_edge, mode='same')
axs[3].imshow(fashion_mnist_convolve2d)
axs[3].set_title('fashion_mnist_convolve2d')

plt.show()
--------------------------------------------------------------------------------
Verify that the dimensions are the same: (12, 12) (3, 3)
```

图像处理中广泛使用二维卷积处理图像达到特殊的效果，或者提取图像的特征。下述列出图像处理主要使用的卷积核[①]:

① 图像处理主要使用的卷积核，Lode's Computer Graphics Tutorial（https://lodev.org/cgtutor/filtering.html）。

1. 同一性（identity）: $\begin{bmatrix}0&0&0\\0&1&0\\0&0&0\end{bmatrix}$

2. 锐化（sharpen）: $\begin{bmatrix}0&-1&0\\-1&5&-1\\0&-1&0\end{bmatrix}$, $\begin{bmatrix}-1&-1&-1\\-1&9&-1\\-1&-1&-1\end{bmatrix}$, $\begin{bmatrix}1&1&1\\1&-7&1\\1&1&1\end{bmatrix}$, $\begin{bmatrix}-k&-k&-k\\-k&8k+1&-k\\-k&-k&-k\end{bmatrix}$,

$\begin{bmatrix}-1&-1&-1&-1&-1\\-1&2&2&2&-1\\-1&2&8&2&-1\\-1&2&2&2&-1\\1&-1&-1&-1&-1\end{bmatrix}$

3. 边缘检测（edge detection）: $\begin{bmatrix}-1&-1&-1\\-1&8&-1\\-1&-1&-1\end{bmatrix}$, $\begin{bmatrix}0&1&0\\1&-4&1\\0&1&0\end{bmatrix}$, $\begin{bmatrix}0&0&0&0&0\\0&0&0&0&0\\-1&-1&2&0&0\\0&0&0&0&0\\0&0&0&0&0\end{bmatrix}$,

$\begin{bmatrix}0&0&-1&0&0\\0&0&-1&0&0\\0&0&4&0&0\\0&0&-1&0&0\\0&0&-1&0&0\end{bmatrix}$, $\begin{bmatrix}-1&0&0&0&0\\0&-2&0&0&0\\0&0&6&0&0\\0&0&0&-2&0\\0&0&0&0&-1\end{bmatrix}$

4. 浮雕（embossing filter）: $\begin{bmatrix}-1&-1&0\\-1&0&1\\0&1&1\end{bmatrix}$, $\begin{bmatrix}-1&-1&-1&-1&0\\-1&-1&-1&0&1\\-1&-1&0&1&1\\-1&0&1&1&1\\0&1&1&1&1\end{bmatrix}$

5. 模糊（blur）: 均值模糊（box filter, averaging）$\begin{bmatrix}1&1&1\\1&1&1\\1&1&1\end{bmatrix}\times\frac{1}{9}$, $\begin{bmatrix}0.&0.2&0.\\0.2&0.&0.2\\0.&0.2&0.\end{bmatrix}$,

$\begin{bmatrix} 0&0&1&0&0\\ 0&1&1&1&0\\ 1&1&1&1&1\\ 0&1&1&1&0\\ 0&0&1&0&0 \end{bmatrix}$；高斯模糊（gaussian blur,approximation）$\begin{bmatrix} 1&2&1\\ 2&4&2\\ 1&2&1 \end{bmatrix} \times \frac{1}{16}$；运动模糊（motion blur）

$$\begin{bmatrix} 1&0&0&0&0&0&0&0&0\\ 0&1&0&0&0&0&0&0&0\\ 0&0&1&0&0&0&0&0&0\\ 0&0&0&1&0&0&0&0&0\\ 0&0&0&0&1&0&0&0&0\\ 0&0&0&0&0&1&0&0&0\\ 0&0&0&0&0&0&1&0&0\\ 0&0&0&0&0&0&0&1&0\\ 0&0&0&0&0&0&0&0&1 \end{bmatrix}$$

3.10.2 SIR传播模型

开始数学模型的建立，先确定自变量（independent variables）和因变量（dependent variables）。自变量为时间t，以d为单位。考虑两组相关的因变量。

第一组因变量计算每一组（归类）中的人数，均为时间的函数。$\begin{cases} S = S(t) & susceptible \\ I = I(t) & infected \\ R = R(t) & recovered \end{cases}$。

第二组因变量表示这三种类型中每一种占总人口数的比例，假设N为总人口数，$\begin{cases} s(t) = S(t)/N & susceptible \\ i(t) = I(t)/N & infected \\ r(t) = R(t)/N & recovered \end{cases}$，因此$s(t)+i(t)+r(t)=1$。使用总体计数可能看起来更自然。但是，如果使用分数替代，则会计算更简单。这两组因变量是成比例的，所以具有病毒传播的相同信息。

- 因为忽略了出生率和移民等改变总体人口数据的因素，“易感人群”中的个体被感染，则会进入“受感人群”。而“受感人群”中的个体恢复（或死亡），则进入“恢复人群”。假设$S(t)$的时间变化率，“易感人群”的人口数量取决于已经感染的人数，以及易感染者和被感染者接触的数量。特别是，假设每个被感染的人每天有固定数量（b）的接触者，这足以传播疾病。同时，并不是所有的接触者都是“易感人群”。如果假设人群是均匀混合的，接触者中易感染者的比例是$s(t)$。那么，平均而言，每个感染者每天产生$bs(t)$个新的被感染者（由于“易感人群”众多，而“受感人群”相对较少，可以忽略一些较为棘手的计算情况，例如一个易感个体同一天遭遇多个受感个体）。
- 同时假设在任何给定的一天内，受感人群有固定比例（k）的人恢复（或死亡），进入“恢复人群”。

易感人群微分方程：根据上述假设，可以得到，$\frac{dS}{dt} = -bs(t)I(t)$。为了方便理解，可以假设总人口数$N=10$。当前时刻$t$下，易感人数$S=4$（则，$s=4/10$），受感人数$I=3$（则，$i=3/10$），恢复人数$R=3$（则，$r=3/10$）。在变化时间$\triangle t$下，因为参数$b$为“受感人群”中每个人的固定数量接触者。假设$b=2$，那么对于整个“受感人群”的接触者为$bI(t)=2\times 3=6$。因为并不是所有接触者都是“易感人群”，假设了人群为均匀混合，则接触者中易感染者的比例是$s(t)=4/10$。因此，$\triangle S = bI(t)s(t) = 2\times 3\times 4/10 = 2.4$，因为“易感人群”中的个体接触了“受感人群”中的个体而转换为“受感人群”中新的个体。“易感人群”中个体的数量是减少的，因此要加上符号。最终，当变化时间$\triangle t$后（未考虑I到R的转换），$S=1.6 I=5.4 R=3$。依据上述微分方程，根据总体计数和分数（或百分比）替代是成比例的，因此最终得到易感人群$\triangle t$变化下单位微分方程为：$\frac{ds}{dt} = -bs(t)i(t)$。

恢复人群微分方程：根据上述假设，可以得到，$\frac{dr}{dt} = ki(t)$，其中k是“受感人群”到“恢复人群的转换比例”，同样应用上述假设，如果$k=0.2$，则$\triangle R = 0.2*3 = 0.6$，最终，

$S = 1.6I = 4.8, R = 3.6$。

受感人群微分方程： 因为 $s(t)+i(t)+r(t)=1$，可以得到 $\frac{\mathrm{d}s}{\mathrm{d}t}+\frac{\mathrm{d}i}{\mathrm{d}t}+\frac{\mathrm{d}r}{\mathrm{d}t}=1$，因此 $\frac{\mathrm{d}i}{\mathrm{d}t}=bs(t)i(t)-ki(t)$。

最终的SIR传播模型为：$\begin{cases}\frac{\mathrm{d}s}{\mathrm{d}t}=-bs(t)i(t)\\ \frac{\mathrm{d}i}{\mathrm{d}t}=bs(t)i(t)-ki(t)\\ \frac{\mathrm{d}r}{\mathrm{d}t}=ki(t)\end{cases}$。定义SIR传播模型微分方程函数`SIR_deriv()`，模拟结果如图3.10–8所示。

对于 SIR 模型的解释参考来自于 MAA[①]（Mathematical Association of America），The SIR Model for Spread of Disease - The Differential Equation Model[②③]<作者：David Smith and Lang Moore>; 代码参考 The SIR epidemic model。

① MAA，Mathematical Association of America，美国数学协会是世界上最大的数学家、学生和爱好者的社区，其使命是促进人们对数学的理解和数学对人类世界的影响（https://www.maa.org/）。

② The SIR Model for Spread of Disease - The Differential Equation Model，Author(s): David Smith and Lang Moore（https://www.maa.org/press/periodicals/loci/joma/the-sir-model-for-spread-of-disease-the-differential-equation-model）。

③ The SIR epidemic model，对疾病在人群中传播的一个简单数学描述模型（https://scipython.com/book/chapter-8-scipy/additional-examples/the-sir-epidemic-model/）。

```
#定义 SIR 模型微分方程函数
import numpy as np
def SIR_deriv(y, t, N, beta, gamma, plot=False):
    '''
    function - 定义 SIR 传播模型微分方程

    Params:
        y - S,I,R 初始化值（例如，人口数）; tuple
        t - 时间序列; list
        N - 总人口数; int
        beta - 易感人群到受感人群转化比例; float
        gamma - 受感人群到恢复人群转换比例; float

    Rreturns:
        SIR_array - S, I, R 数量; array
    '''
    import numpy as np
    from scipy.integrate import odeint
    import matplotlib.pyplot as plt
    def deriv(y, t, N, beta, gamma):
        S, I, R = y
        dSdt = -beta*S*I/N
        dIdt = beta*S*I/N-gamma*I
        dRdt = gamma*I
        return dSdt, dIdt, dRdt
    deriv_integration = odeint(deriv, y, t, args=(N, beta, gamma))
    S, I, R = deriv_integration.T
    SIR_array = np.stack([S, I, R])
    if plot == True:
        fig = plt.figure(facecolor='w', figsize=(12, 6))
        ax = fig.add_subplot(111, facecolor='#dddddd', axisbelow=True)
        ax.plot(t, S/N, 'b', alpha=0.5, lw=2, label='Susceptible')
        ax.plot(t, I/N, 'r', alpha=0.5, lw=2, label='Infected')
        ax.plot(t, R/N, 'g', alpha=0.5, lw=2, label='Recovered')
        ax.set_label('Time/days')
```

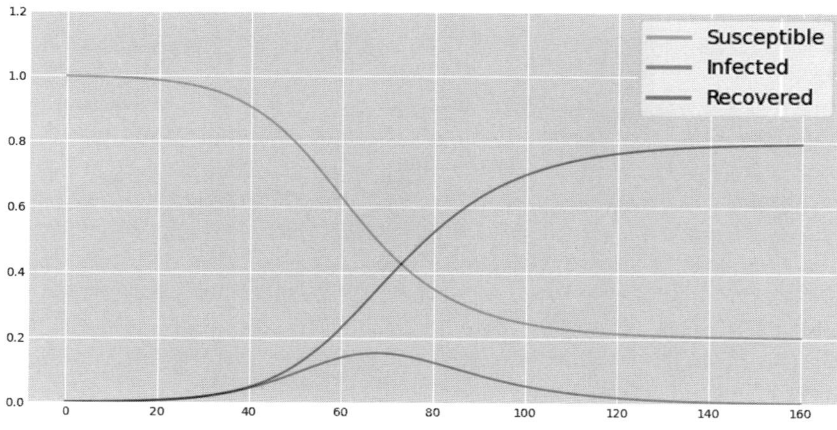

图3.10–8 SIR模拟结果

```
        ax.set_ylim(0, 1.2)
        ax.yaxis.set_tick_params(length=0)
        ax.xaxis.set_tick_params(length=0)
        ax.grid(visible=True, which='major', c='w', lw=1, ls='-')
        legend = ax.legend()
        legend.get_frame().set_alpha(0.5)
        for spine in ('top', 'right', 'bottom', 'left'):
            ax.spines[spine].set_visible(False)
        plt.show()
    return SIR_array

# 参数配置
N = 1000  # 总人口数
I_0, R_0 = 1, 0  # 初始化受感人群，及恢复人群的人口数
S_0 = N-I_0-R_0  # 有受感人群和恢复人群，计算得易感人群人口数
beta, gamma = 0.2, 1./10  # 配置参数 b( 即 beta) 和 k( 即 gamma)
t = np.linspace(0, 160, 160)  # 配置时间序列

y_0 = S_0, I_0, R_0
SIR_array = SIR_deriv(y_0, t, N, beta, gamma, plot=True)
```

3.10.3 SIR空间传播模型

3.10.3.1 卷积扩散

如果配置卷积核为$\begin{bmatrix} 0.5 & 1 & 0.5 \\ 1 & -6 & 1 \\ 0.5 & 1 & 0.5 \end{bmatrix}$，假设存在一个传播（传染）源，如下述程序配置一个栅格（.bmp图像，图3.10-9），其值除了源为1（R值）外，其他的单元均为0。每次卷积，值都会以源为中心，向四周扩散；并且，四角的值（绝对值）最小，水平垂直向边缘值（绝对值）稍大，越向内部值（绝对值）越大，形成了一个逐步扩散的趋势，并且具有强弱。即越趋向于传播源，其绝对值越大。同时，达到一定阶段后，扩散开始逐步消失，恢复为所有单元值为0的阶段。观察卷积扩散，可以通过打印每次卷积后的值查看，也可以记录每次的结果最终存储为.gif文件，动态地观察图像的变化。

```
def img_struc_show(img_fp, val='R', figsize=(7, 7)):
    '''
    function - 显示图像以及颜色 R 值，或 G,B 值
```

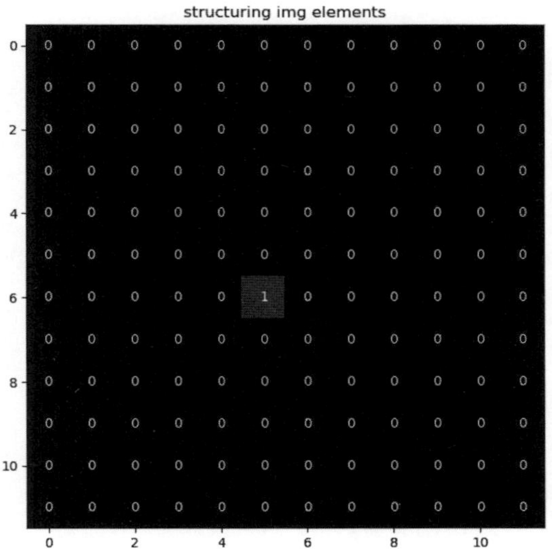

图3.10-9 假设值为1的传播源

```
    Params:
        img_fp - 输入图像文件路径; string
        val - 选择显示值, R, G, 或 B, The default is 'R'; string
        figsize - 配置图像显示大小, The default is (7,7); tuple

    Returns:
        None
    '''
    from skimage import io
    from skimage.morphology import square, rectangle
    import matplotlib.pyplot as plt
    img = io.imread(img_fp)
    shape = img.shape
    struc_square = rectangle(shape[0], shape[1])
    fig, ax = plt.subplots(figsize=figsize)
    ax.imshow(img, cmap="Paired", vmin=0, vmax=12)
    for i in range(struc_square.shape[0]):
        for j in range(struc_square.shape[1]):
            if val == 'R':
                ax.text(j, i, img[:, :, 0][i, j],
                        ha="center", va="center", color="w")
            elif val == 'G':
                ax.text(j, i, img[:, :, 1][i, j],
                        ha="center", va="center", color="w")
            elif val == 'B':
                ax.text(j, i, img[:, :, 2][i, j],
                        ha="center", va="center", color="w")
    ax.set_title('structuring img elements')
    plt.show
img_12Pix_1red_fp = r'./data/12mul12Pixel_1red.bmp'
img_struc_show(img_fp=img_12Pix_1red_fp, val='R')
```

扩散是一个随时间变化的动态过程，为了记录每一次扩散的结果，形成动态的变化显示，使用MoviePy①记录每次图像变化的结果，并存储为.gif文件，方便查看。卷积扩散主要是更新SIR变量值，其初始值配置为图像的R颜色值，除了源配置为1外，其他单元的值（R）均为0，这样可以比较清晰的观察卷积扩散的过程。卷积的计算使用SciPy库提供的convolve方法。

① MoviePy，是一个用于视频编辑的Pyton模块，可用于基本操作（如剪辑、连接、标题插入）、视频合成（又称非线性编辑）、视频处理或创建高级效果等。其可以读写最常见的视频格式，包括GIF格式（https://zulko.github.io/moviepy/）。

```
class Convolution_diffusion_img:
    '''
    class - 定义基于 SIR 模型的二维卷积扩散

    Params:
        img_path - 图像文件路径; string
        save_path - 保持的 .gif 文件路径; string
        hours_per_second - 扩散时间长度; int
        dt - 时间记录值，开始值; int
        fps - 配置 moviepy，write_gif 写入 GIF 每秒帧数; int
    '''
    def __init__(self, img_path, save_path, hours_per_second, dt, fps):
        from skimage import io
        import numpy as np
        self.save_path = save_path
        self.hours_per_second = hours_per_second
        self.dt = dt
        self.fps = fps
        img = io.imread(img_path)
        # 在配置 SIR 数组时，为三维度，是为了对接后续物种散步程序对 SIR 的配置
        SIR = np.zeros((1, img.shape[0], img.shape[1]), dtype=np.int32)
        SIR[0] = img.T[0]  # 将图像的 RGB 中 R 通道值赋值给 SIR
        self.world = {'SIR': SIR, 't': 0}
        self.dispersion_kernel = np.array([[0.5, 1, 0.5],
                                          [1, -6, 1],
                                           [0.5, 1, 0.5]])  # SIR 模型卷积核

    def make_frame(self, t):
        ''' 返回每一步卷积的数据到 VideoClip 中 '''
        while self.world['t'] < self.hours_per_second*t:
            self.update(self.world)
        if self.world['t'] < 6:
            print(self.world['SIR'][0])
        return self.world['SIR'][0]
    def update(self, world):
        ''' 更新数组，即基于前一步卷积结果的每一步卷积 '''
```

```
        disperse = self.dispersion(world['SIR'], self.dispersion_kernel)
        world['SIR'] = disperse
        world['t'] += dt  #记录时间，用于循环终止条件

    def dispersion(self, SIR, dispersion_kernel):
        '''卷积扩散'''
        import numpy as np
        from scipy.ndimage.filters import convolve

        #注意卷积核与待卷积数组的维度
        return np.array([convolve(SIR[0], self.dispersion_kernel,
mode='constant', cval=0.0)])
    def execute_(self):
        '''执行程序'''
        import moviepy.editor as mpy
        self.animation = mpy.VideoClip(
            self.make_frame, duration=1)  # duration=1
        self.animation.write_gif(self.save_path, self.fps)

img_12Pix_fp = r'./data/12mul12Pixel_1red.bmp'  #图像文件路径
SIRSave_fp = r'./data/12mul12Pixel_1red_SIR.gif'
hours_per_second = 20
dt = 1  #时间记录值，开始值
fps = 15  #配置 moviepy，write_gif 写入 GIF 每秒帧数
convDiff_img = Convolution_diffusion_img(
    img_path=img_12Pix_fp, save_path=SIRSave_fp, hours_per_second=hours_per_
second, dt=dt, fps=fps)
convDiff_img.execute_()
--------------------------------------------------------------------------------
[[0 0 0 0 0 0 0 0 0 0 0 0]
 [0 0 0 0 0 0 0 0 0 0 0 0]
 [0 0 0 0 0 0 0 0 0 0 0 0]
 [0 0 0 0 0 0 0 0 0 0 0 0]
 [0 0 0 0 0 0 0 0 0 0 0 0]
 [0 0 0 0 0 0 1 0 0 0 0 0]
 [0 0 0 0 0 0 0 0 0 0 0 0]
 [0 0 0 0 0 0 0 0 0 0 0 0]
 [0 0 0 0 0 0 0 0 0 0 0 0]
 [0 0 0 0 0 0 0 0 0 0 0 0]
 [0 0 0 0 0 0 0 0 0 0 0 0]
 [0 0 0 0 0 0 0 0 0 0 0 0]]
MoviePy - Building file ./data/12mul12Pixel_1red_SIR.gif with imageio.
[[0 0 0 0 0 0 0 0 0 0 0 0]
 [0 0 0 0 0 0 0 0 0 0 0 0]
 [0 0 0 0 0 0 0 0 0 0 0 0]
 [0 0 0 0 0 0 0 0 0 0 0 0]
 [0 0 0 0 0 0 0 0 0 0 0 0]
 [0 0 0 0 0 0 1 0 0 0 0 0]
 [0 0 0 0 0 0 0 0 0 0 0 0]
 [0 0 0 0 0 0 0 0 0 0 0 0]
 [0 0 0 0 0 0 0 0 0 0 0 0]
 [0 0 0 0 0 0 0 0 0 0 0 0]
 [0 0 0 0 0 0 0 0 0 0 0 0]
 [0 0 0 0 0 0 0 0 0 0 0 0]]
[[  0   0   0   0   0   0   0   0   0   0   0   0]
 [  0   0   0   0   0   0   0   0   0   0   0   0]
 [  0   0   0   0   0   0   0   0   0   0   0   0]
 [  0   0   0   0   0   0   1   0   0   0   0   0]
 [  0   0   0   0   0  -1 -11  -1   0   0   0   0]
 [  0   0   0   0   1 -11  40 -11   1   0   0   0]
 [  0   0   0   0   0  -1 -11  -1   0   0   0   0]
 [  0   0   0   0   0   0   1   0   0   0   0   0]
 [  0   0   0   0   0   0   0   0   0   0   0   0]
 [  0   0   0   0   0   0   0   0   0   0   0   0]
 [  0   0   0   0   0   0   0   0   0   0   0   0]
 [  0   0   0   0   0   0   0   0   0   0   0   0]]
[[    0    0    0    0    0    0    0    0    0    0    0    0]
 [    0    0    0    0    0    0    0    0    0    0    0    0]
 [    0    0    0    0    0    0    1    0    0    0    0    0]
 [    0    0    0    0    0   -5  -18   -5    0    0    0    0]
 [    0    0    0    0   -5    5   94    5   -5    0    0    0]
 [    0    0    0    1  -18   94 -286   94  -18    1    0    0]
 [    0    0    0    0   -5    5   94    5   -5    0    0    0]
 [    0    0    0    0    0   -5  -18   -5    0    0    0    0]
```

```
 [   0    0    0    0    0    0    1    0    0    0    0    0]
 [   0    0    0    0    0    0    0    0    0    0    0    0]
 [   0    0    0    0    0    0    0    0    0    0    0    0]
 [   0    0    0    0    0    0    0    0    0    0    0    0]]
[[   0    0    0    0    0    0    0    0    0    0    0    0]
 [   0    0    0    0    0    0    1    0    0    0    0    0]
 [   0    0    0    0   -2  -13  -29  -13   -2    0    0    0]
 [   0    0    0   -2   -7   62  198   62   -7   -2    0    0]
 [   0    0    0  -13   62  -13 -769  -13   62  -13    0    0]
 [   0    0    1  -29  198 -769 2102 -769  198  -29    1    0]
 [   0    0    0  -13   62  -13 -769  -13   62  -13    0    0]
 [   0    0    0   -2   -7   62  198   62   -7   -2    0    0]
 [   0    0    0    0   -2  -13  -29  -13   -2    0    0    0]
 [   0    0    0    0    0    0    1    0    0    0    0    0]
 [   0    0    0    0    0    0    0    0    0    0    0    0]
 [   0    0    0    0    0    0    0    0    0    0    0    0]]
```

为了方便查看动态变化的.gif文件（图3.10-10），定义animated_gif_show()函数。对图像的读取和处理使用了Pillow[①]库。

① Pillow，是由Fredrik Lundh和贡献者编写的Python图像处理库（https://pillow.readthedocs.io/en/stable/）。

```
def animated_gif_show(gif_fp, figsize=(8, 8)):
    '''
    function - 读入 .gif，并动态显示

    Params:
        gif_fp - GIF 文件路径; string
        figsize - 图表大小，The default is (8,8); tuple

    Returns:
        HTML
    '''
    from PIL import Image, ImageSequence
    import numpy as np
    import matplotlib.pyplot as plt
    import matplotlib.animation as animation
    from IPython.display import HTML
    gif = Image.open(gif_fp, 'r')
    frames = [np.array(frame.getdata(), dtype=np.uint8).reshape(
        gif.size[0], gif.size[1], -1) for frame in ImageSequence.Iterator(gif)]
    fig = plt.figure(figsize=figsize)
    imgs = [(plt.imshow(img, animated=True),) for img in frames]
    anim = animation.ArtistAnimation(
        fig, imgs, interval=300, repeat_delay=3000, blit=True)
    return HTML(anim.to_html5_video())
gif_fp = r'./data/12mul12Pixel_1red_SIR.gif'
animated_gif_show(gif_fp, figsize=(8, 8))
```

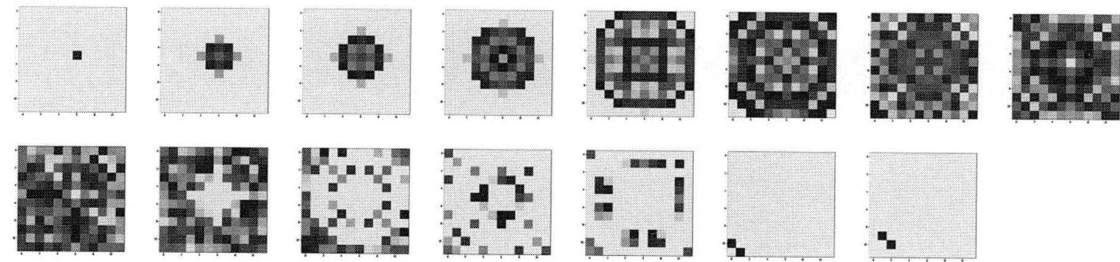

图3.10-10　卷积扩散过程（原为GIF文件）

3.10.3.2　SIR空间传播模型

SIR传播模型是给定了总体人口数N及S、I、R的初始值，和β（beta）、γ（gamma）转换系数，计算S、I、R人口数量的变化。SIR传播模型不具有空间属性，因此引入了卷积扩散，可以实现一个源向四周扩散的变化过程；并且，始终以源的空间位置强度最大，并向四周逐步减弱。这样，通过结合SIR模型和卷积扩散，可以实现SIR的空间传播模型。

可以这样理解SIR空间传播模型的过程，对于空间分布（即栅格，每个栅格单元有一个值或

多个值，即存在多层栅格）配置有三个对应的栅格层（'SIR'），分别对应S、I和R值的空间分布。例如，对于S而言，其初始值对应了用地类型对于物种散步的影响，即成本栅格或者是空间阻力值，如物种容易在林地、农田中迁移散步；而建筑和道路则会阻挡其散步，通过配置不同大小的值可以反映物种能够散步的强度（注意，有时可以用大的值表示阻力值小，而小的值表示阻力值大，具体需要根据算法的方式确定，或调整为符合习惯的数值大小表达）；对于对应的I层栅格而言，初始值需要设定源，可以为一个或者多个。多个源可以是分散的，也可以是聚集为片区的，源也可以根据源自身的强度配置代表不同强度的值，但是除了源之外的所有值均为0；对于对应的R层，因为在开始时间点，并没有恢复者（或死亡）案例，因此所有值设置为0。

配置好SIR空间分布的栅格层后，计算可以理解为纵向的SIR模型传播，以及水平向的卷积扩散过程的结合。在纵向上，每一个空间点（即一个位置的单元栅格，每个位置有S，I和R 3 个层的栅格）都有一定数量的人口数（即空间阻力值）。每一个位置对应的S，I和R单元栅格的SIR传播模型计算过程与上述解释的SIR模型传播过程是相同的。只是在一开始的时候，只有I层对应源的位置有非0值，因此源的栅格位置是有纵向（S-->I-->R）的传播。但是，为0的位置，SIR模型计算公式变化结果为0，即没有纵向上的传播；除了纵向SIR传播，各层的水平向发生着卷积扩散，对于S 、I和R三个栅格层，水平向的扩散速度是不同的。通过变量dispersion_rates扩散系数确定强度，三个值分别对应S、I和R的3个栅格层。因为I层是给了值非0的源，因此会发生水平扩散，新扩散的区域在SIR模型计算时，纵向上就会发生传播变化。而在配置扩散系数时，对应的S层的扩散系数为0，即S栅格层水平向上并不会发生卷积扩散，卷积扩散仅发生在I和R层，这与S层为“易感人群”但并传播；而I层“受感人群”和R层“恢复人群或死亡”是病毒携带者是可以传染扩散的过程相一致。每一时间步，由纵向SIR传播和水平向卷积扩散共同作用，因此每一时刻的SIR空间栅格值变化为world['SIR'] += dt*(deriv_infect+disperse)，其中deriv_infect为纵向SIR的传播结果，disperse为水平向的扩散结果，通过求和作为两者共同作用的结果。同时，乘以dt时间参数，开始时，时间参数比较小，而随时间的流逝，时间参数的值越来越大，那么各层的绝对值大小也会增加得更快，即传播强度会增加。

同样，应用MoviePy库记录每一时刻的变化，如图3.10-11所示。

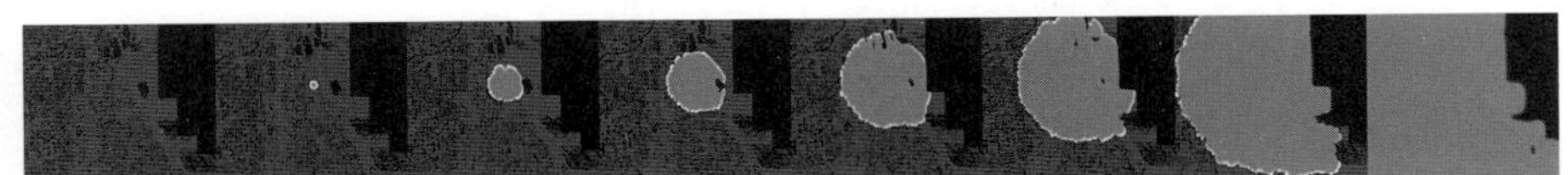

图3.10-11 SIR空间传播模型过程示例（原为GIF文件）

- **确定源的位置**

首先，读取分类数据，因为不同分类数据的标识各异，通常无法正常显示为颜色方便地物的辨别，因此建立las_classi_colorName分类值到颜色的映射字典，将其转换为不同颜色的表达。根据颜色辨别地物，方便观察位置信息。位置信息的获取，使用skimage库提供的ImageViewer方法。

分类数据为点云数据处理部分的分类数据，因为SIR的空间传播模型计算中，卷积扩散计算过程比较耗时，因此需要压缩数据量，并确保分类值的正确。压缩数据时，skimage.transform提供了rescale，resize，和downscale_local_mean等方法，但是没有提供给定一个区域（block），返回区域内频数最大值的方式；skimage.measure中的block_reduce方法，仅有numpy.sum，numpy.min，numpy.max，numpy.mean和 numpy.median返回值。因此，自定义函数downsampling_blockFreqency，实现给定二维数组和区域大小，以区域内出现次数最多的值为该区域的返回值，实现降采样。

```
import sys
from skimage.measure import block_reduce
import numpy as np
from matplotlib import colors
import pandas as pd
from skimage.transform import rescale
import skimage.viewer
```

```
import skimage.io
def downsampling_blockFreqency(array_2d, blocksize=[10, 10]):
    '''
    fuction - 降采样二维数组，根据每一 block 内值得频数最大值，即最多出现得值为每一 block 的采样值

    Params:
        array_2d - 待降采样的二维数组；array(2d)
        blocksize - block 大小，即每一采用的范围，The default is [10,10]；tuple

    Returns:
        downsample - 降采样结果；array
    '''
    import numpy as np
    from statistics import multimode
    from tqdm import tqdm
    shape = array_2d.shape
    row, col = blocksize
    row_p, row_overlap = divmod(shape[1], row)  # divmod(a,b) 方法为除法取整，以及 a 对 b 的余数
    col_p, col_overlap = divmod(shape[0], col)
    print("row_num:", row_p, "col_num:", col_p)
    array_extraction = array_2d[:col_p*col, :row_p*row]  # 移除多余部分，规范数组，使其正好切分均匀
    print("array extraction shape:", array_extraction.shape,
          "original array shape:", array_2d.shape)
    h_splitArray = np.hsplit(array_extraction, row_p)
    v_splitArray = [np.vsplit(subArray, col_p) for subArray in h_splitArray]
    blockFrenq_list = []
    for h in tqdm(v_splitArray):
        temp = []
        for b in h:
            blockFrenq = multimode(b.flatten())[0]
            temp.append(blockFrenq)
        blockFrenq_list.append(temp)
    downsample = np.array(blockFrenq_list).swapaxes(0, 1)
    return downsample
classi_fp = r'F:\data\classification_mosaic.tif'
mosaic_classi_array = skimage.io.imread(classi_fp)
mosaic_classi_array_rescaled = downsampling_blockFreqency(
    mosaic_classi_array, blocksize=(20, 20))
las_classi_colorName = {0: 'black', 1: 'white', 2: 'beige', 3: 'palegreen', 4: 'lime', 5: 'green', 6: 'tomato', 7: 'silver', 8: 'grey', 9: 'lightskyblue',
                        10: 'purple', 11: 'slategray', 12: 'grey', 13: 'cadetblue', 14: 'lightsteelblue', 15: 'brown', 16: 'indianred', 17: 'darkkhaki', 18: 'azure', 9999: 'white'}
las_classi_colorRGB = pd.DataFrame({key: colors.hex2color(
    colors.cnames[las_classi_colorName[key]]) for key in las_classi_colorName.keys()})
classi_array_color = [pd.DataFrame(mosaic_classi_array_rescaled).replace(
    las_classi_colorRGB.iloc[idx]).to_numpy() for idx in las_classi_colorRGB.index]
classi_img = np.stack(classi_array_color).swapaxes(0, 2).swapaxes(0, 1)
print("finished img preprocessing...")
--------------------------------------------------------------------------------
...
row_num: 1250 col_num: 1125
array extraction shape: (22500, 25000) original array shape: (22501, 25001)
100%|██████████| 1250/1250 [01:19<00:00, 15.74it/s]
finished img preprocessing...
--------------------------------------------------------------------------------
viewer=skimage.viewer.ImageViewer(classi_img)
viewer.show()
------------------------------------------------------------------------------[]
```

- 定义SIR的空间传播模型类

```
class SIR_spatialPropagating:
    '''
    funciton - SIR 的空间传播模型

    Params:
        classi_array - 分类数据（.tif，或者其他图像类型），或者其他可用于成本计算的数据类型
        cost_mapping - 分类数据对应的成本值映射字典
```

```
        beta - beta 值，确定 S-->I 的转换率
        gamma - gamma 值，确定 I-->R 的转换率
        dispersion_rates - SIR 三层栅格各自对应的卷积扩散率
        dt - 时间更新速度
        hours_per_second - 扩散时间长度 / 终止值（条件）
        duration - moviepy 参数配置，持续时长
        fps - moviepy 参数配置，每秒帧数
        SIR_gif_savePath - SIR 空间传播计算结果 .gif 文件保存路径
    '''
    def __init__(self, classi_array, cost_mapping, start_pt=[10, 10], beta=0.3,
gamma=0.1, dispersion_rates=[0, 0.07, 0.03], dt=1.0, hours_per_second=7*24,
duration=12, fps=15, SIR_gif_savePath=r'./SIR_sp.gif'):
        from sklearn.preprocessing import MinMaxScaler

        # 将分类栅格，按照成本映射字典，转换为成本栅格（配置空间阻力）
        for idx, (identity, cost_value) in enumerate(cost_mapping.items()):
            classi_array[classi_array == cost_value[0]] = cost_value[1]
        self.mms = MinMaxScaler()
        normalize_costArray = self.mms.fit_transform(classi_array)  # 标准化成本栅格

        # 配置 SIR 模型初始值，将 S 设置为空间阻力值
        SIR = np.zeros(
            (3, normalize_costArray.shape[0], normalize_costArray.shape[1]),
dtype=float)
        SIR[0] = normalize_costArray

        # 配置 SIR 模型中 I 的初始值。1，可以从设置的 1 个或多个点开始；2，可以将森林部分直接设置
为 I 有值，而其他部分保持 0。
        # start_pt=int(0.7*normalize_costArray.shape[0]), int(0.2*normalize_costArray.shape[1]) # 根据行列拾取
点位置
        # print("起始点：",start_pt)
        start_pt = start_pt
        SIR[1, start_pt[0], start_pt[1]] = 0.8  # 配置起始点位置值

        # 配置转换系数，以及卷积核
        self.beta = beta  # β 值
        self.gamma = gamma  # γ 值
        self.dispersion_rates = dispersion_rates  # 扩散系数
        dispersion_kernelA = np.array([[0.5, 1, 0.5],
                                       [1, -6, 1],
                                       [0.5, 1, 0.5]])  # 卷积核_类型 A
        dispersion_kernelB = np.array([[0, 1, 0],
                                       [1, 1, 1],
                                       [0, 1, 0]])  # 卷积核_类型 B

        self.dispersion_kernel = dispersion_kernelA  # 卷积核
        self.dt = dt  # 时间记录值，开始值
        self.hours_per_second = hours_per_second  # 终止值（条件）
        self.world = {'SIR': SIR, 't': 0}  # 建立字典，方便数据更新

        # moviepy 配置
        self.duration = duration
        self.fps = fps

        # 保存路径
        self.SIR_gif_savePath = SIR_gif_savePath
    def deriv(self, SIR, beta, gamma):
        '''SIR 模型'''
        S, I, R = SIR
        dSdt = -1*beta*I*S
        dRdt = gamma*I
        dIdt = beta*I*S-gamma*I
        return np.array([dSdt, dIdt, dRdt])
    def dispersion(self, SIR, dispersion_kernel, dispersion_rates):
        ''' 卷积扩散 '''
        from scipy.ndimage.filters import convolve
        return np.array([convolve(e, dispersion_kernel, cval=0)*r for (e, r) in
zip(SIR, dispersion_rates)])
    def update(self, world):
        ''' 执行 SIR 模型和卷积，更新 world 字典 '''
        deriv_infect = self.deriv(world['SIR'], self.beta, self.gamma)
        disperse = self.dispersion(
            world['SIR'], self.dispersion_kernel, self.dispersion_rates)
        world['SIR'] += self.dt*(deriv_infect+disperse)
```

```
            world['t'] += self.dt
    def world_to_npimage(self, world):
        ''' 将模拟计算的值转换到 [0,255]RGB 色域空间 '''
        coefs = np.array([2, 20, 25]).reshape((3, 1, 1))
        SIR_coefs = coefs*world['SIR']
        accentuated_world = 255*SIR_coefs
        # 调整数组格式为用于图片显示的（x,y,3）形式
        image = accentuated_world[::-1].swapaxes(0, 2).swapaxes(0, 1)
        return np.minimum(255, image)
    def make_frame(self, t):
        ''' 返回每一步的 SIR 和卷积综合蔓延结果 '''
        while self.world['t'] < self.hours_per_second*t:
            self.update(self.world)
        return self.world_to_npimage(self.world)
    def execute(self):
        ''' 执行程序 '''
        import moviepy.editor as mpy
        animation = mpy.VideoClip(
            self.make_frame, duration=self.duration)
        animation.write_gif(self.SIR_gif_savePath, fps=self.fps)
```

空间阻力值的配置，需要根据具体的研究对象做出调整。也可以增加新的多个条件，通过栅格计算后作为输入的一个条件栅格。cost_mapping成本映射字典，键为文字标识，值为一个元组，第一个值为分类值，第二个值为空间阻力值。

```
# 成本栅格（数组）
import util_misc
classi_array = mosaic_classi_array_rescaled

# 配置用地类型的成本值（空间阻力值）
cost_H = 250
cost_M = 125
cost_L = 50
cost_Z = 0
cost_mapping = {
    'never classified': (0, cost_Z),
    'unassigned': (1, cost_Z),
    'ground': (2, cost_M),
    'low vegetation': (3, cost_H),
    'medium vegetation': (4, cost_H),
    'high vegetation': (5, cost_H),
    'building': (6, cost_Z),
    'low point': (7, cost_Z),
    'reserved': (8, cost_M),
    'water': (9, cost_M),
    'rail': (10, cost_L),
    'road surface': (11, cost_L),
    'reserved': (12, cost_M),
    'wire-guard(shield)': (13, cost_M),
    'wire-conductor(phase)': (14, cost_M),
    'transimission': (15, cost_M),
    'wire-structure connector(insulator)': (16, cost_M),
    'bridge deck': (17, cost_L),
    'high noise': (18, cost_Z),
    'null': (9999, cost_Z)
}
s_t = util_misc.start_time()
# 参数配置
start_pt = [418, 640]  # [3724,3415]
beta = 0.3
gamma = 0.1
dispersion_rates = [0, 0.07, 0.03]  # S 层卷积扩散为 0，I 层卷积扩散为 0.07，R 层卷积扩散为 0.03
dt = 1.0
hours_per_second = 30*24
duration = 12
fps = 15
SIR_gif_savePath = r"./imgs/SIR_sp.gif"
SIR_sp = SIR_spatialPropagating(classi_array=classi_array, cost_mapping=cost_
mapping, start_pt=start_pt, beta=beta, gamma=gamma,
                                dispersion_rates=dispersion_rates, dt=dt, hours_
per_second=hours_per_second, duration=duration, fps=fps, SIR_gif_savePath=SIR_
gif_savePath)
SIR_sp.execute()
```

```
util_misc.duration(s_t)
--------------------------------------------------------------------------------
start time: 2022-01-06 18:55:30.966194
MoviePy - Building file ./imgs/SIR_sp.gif with imageio.
end time: 2022-01-06 19:17:01.796161
Total time spend:21.50 minutes
```

参考文献（References）:

[1] Robert E.Ricklefs著. 孙儒泳等译. 生态学[M]. 北京：高等教育出版社，2004. 7第5版.（Robert E.Ricklefs. The economy of nature[M]. New York: W. H. Freeman and Company,2008, sixth edition.）——非常值得推荐的读物（教材），图文并茂

3-D 计算机视觉及城市街道空间

3.11 图像特征提取与动态街景视觉感知

城市的规划与设计笼统来说，是在平衡可见事物（建筑、道路、植被、车辆…）与不可见事物（能量、感知、关系、结构…），是理解不可见来规划可见，可见事物则影射着城市的结构、肌理、物质信息流、生活方式、城市感知、自然自身运行的发展等。图像（照片，影像）是对可见事物的捕捉，是否可以从图像的信息中捕捉到不可见事物呢？目前，计算机视觉的发展，可以解构图像信息，这为规划设计行业注入了新的技术手段。为方便计算机视觉的理解，可以粗略分为两个层级：一是，基于OpenCV[①]库提供的各类算法对图像解析，计算图像处理，图像特征提取，对象检测等；二是，应用深度学习实现图像的语义分割、图像特征迁移、对抗生成等。这些方法对规划行业有着重要的影响，百度、Google的街景图像，以及无人驾驶项目带来的大量序列图像（连接有GPS，惯性测量单元IMU（Inertial Measurement Unit）等传感器信息数据），以及社交网络的图像（通常包括GPS信息），都推动着计算机视觉在规划领域潜在的应用前景。

OpenCV是计算机视觉开源库，包括有数百种计算机视觉算法。对应的OpenCV-Python[②]库包括其核心功能、图像处理、视频分析、相机标定与三维重建、对象检测和特征提取等。

① OpenCV，是一个开源的计算机视觉和机器学习软件库。该库有超过2500种优化算法，其中包括一套全面的经典和最先进的计算机视觉和机器学习算法。这些算法可用于检测和识别人脸、识别物体、对视频中的人类行为进行分类、跟踪摄像机运动、跟踪移动物体、提取物体的三维模型、从立体摄像机中产生三维点云、将图像拼接起来以产生整个场景的高分辨率图像、从图像数据库中寻找相似的图像、从使用闪光灯拍摄的图像中去除红眼、跟踪眼睛运动、识别景物并建立标记以与增强现实叠加等。（https://opencv.org/）。

② OpenCV-Python，OpenCV的Python实现库（https://docs.opencv.org/4.x/d6/d00/tutorial_py_root.html）。

3.11.1 OpenCV读取图像与图像处理示例

OpenCV的具体内容可以查看官网手册。图像处理部分，下述代码示例了图像的读取和图像模糊、边缘检测、角点检测及建立图像金字塔显示等内容（图3.11–1），粗略了解图像处理的一些基本内容。对于该部分，也可以结合到卷积部分的知识阐述。其中，建立图像金字塔显示对于应用GIS等平台处理地理信息图像时并不陌生，大数据量的遥感影像显示并不是一次加载所有单元数据；而是根据所需要的显示效果，建立一系列不同分辨率的图像。这些图像叠合在一起，最高分辨率的位于底层，最低的则位于顶部，类似金字塔。因此，图像的集合被称为图像金字塔，从而提升了显示的速度。

角点检测时，为了能够清晰地显示主要的角点内容，通过放大菱角标记和定义显示阈值实

图3.11–1 OpenCV图像处理示例

现。注意角点检测在官方的归类中，是将其纳入到特征提取的部分。

```
# cv2 即 OpenCV-Python 库，通常需要安装 pip install opencv-python 和 pip install opencv-contrib-python。

import cv2 as cv
import sys
img_1_fp = r'./data/cv_IMG_0405.JPG'
img_2_fp = r'./data/cv_IMG_0407.JPG'
img_1 = cv.imread(cv.samples.findFile(img_1_fp))
img_2 = cv.imread(cv.samples.findFile(img_2_fp))
if img_1 is None:
    sys.exit("could not read the image")
cv.imshow("display window", img_1)  # 图像将在一个新打开的窗口中显示
k = cv.waitKey(0)
```

使用cv.imshow会打开一个新的窗口显示图像内容。因为cv.imread读取图像为数组格式，因此可以使用matplotlib库打印。只是需要注意，OpenCV读取的图像为BGR，需要将其转换为RGB；否则，显示的颜色会不正确。

```
import matplotlib.pyplot as plt
import numpy as np
import copy
fig, axs = plt.subplots(ncols=2, nrows=3, figsize=(30, 15))

# 01- 原始图像
# 注意 OpenCV 读取图像，其格式为 BGR，需要将其调整为 RGB 再用 matplotlib 库打印
img_2_RGB = cv.cvtColor(img_2, cv.COLOR_BGR2RGB)
axs[0][0].imshow(img_2_RGB)
axs[0][0].set_title("original image-shape=(684, 1920, 3)")

# 02-averaging 均值模糊
axs[0][1].imshow(cv.blur(img_2_RGB, (15, 15)))
axs[0][1].set_title("averaging blure")

# 03-Canny 边缘检测
axs[1][0].imshow(cv.Canny(img_2_RGB, 200, 300))
axs[1][0].set_title("Canny edge detection")

# 04-Harris 角点检测
# 将图像转换为灰度，为每一像素位置 1 个值，可理解为图像的强度 (颜色，易受光照影响，难以提供关键信息，故将图像进行灰度化，同时也可以加快特征提取的速度)。并强制转换为浮点值，用于棱角检测。
img_2_gray = np.float32(cv.cvtColor(img_2, cv.COLOR_BGR2GRAY))
img_2_harris = cv.cornerHarris(img_2_gray, 7, 5, 0.04)  # 哈里斯角检测器
img_2_harris_dilate = cv.dilate(img_2_harris, np.ones((1, 1)))  # 放大棱角标记
img_2_copy = copy.deepcopy(img_2_RGB)
img_2_copy[img_2_harris_dilate > 0.01 *
           img_2_harris_dilate.max()] = [40, 75, 236]  # 定义阈值，显示重要的棱
axs[1][1].imshow(img_2_copy)
axs[1][1].set_title("Harris corner detector")

# 05- 建立图像显示金字塔 -pyrDown-1 次
axs[2][0].imshow(cv.pyrDown(img_2_RGB))
axs[2][0].set_title("image pyramids-shape=(342, 960, 3)")

# 06- 建立图像显示金字塔 -pyrDown-3 次 -pyrUp-1 次
axs[2][1].imshow(cv.pyrUp(cv.pyrDown(cv.pyrDown(cv.pyrDown(img_2_RGB)))))
axs[2][1].set_title("image pyramids-shape=(172, 480, 3)")
plt.show()
```

3.11.2 图像特征提取

3.11.2.1 SIFT（Scale-Invariant Feature Transform）算法关键点阐述

- A – 高斯分布与高斯模糊（高斯卷积核）

高斯分布即正态分布，可以查看正态分布与概率密度函数部分，其概率密度函数为：$f(x)=\frac{1}{\sigma\times\sqrt{2\pi}}\times \mathrm{e}^{-\frac{1}{2}[\frac{x-\mu}{\sigma}]^2}$。式中$\sigma$为标准差；$\mu$为平均值；e为自然对数的底。因为使用二维高斯分布作为卷积核，中心点对应着原点，因此$\mu=0$，公式可以简化为：$f(x)=\frac{1}{\sigma\sqrt{2\pi}}\mathrm{e}^{\frac{-x^2}{2\sigma^2}}$，对应的2维高斯函数为：$G(x,y)=\frac{1}{2\pi\sigma^2}\mathrm{e}^{\frac{-(x^2+y^2)}{2\sigma^2}}$。代码中的计算直接调用SciPy库，分别使用norm和

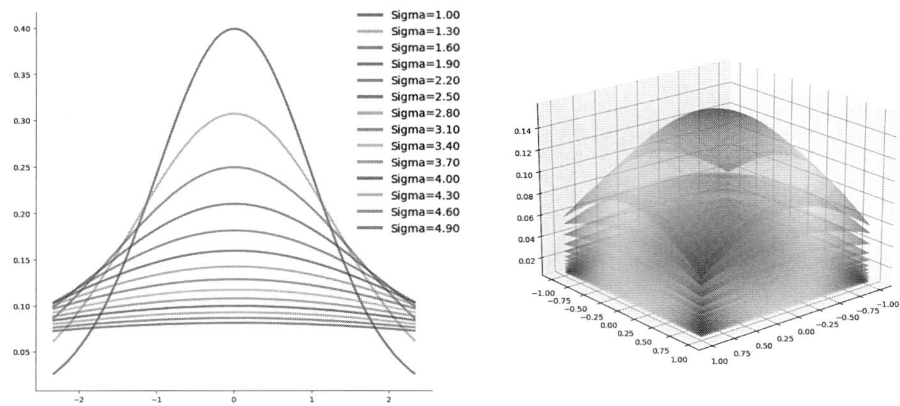

图3.11-2　1维高斯分布（左）；2维高斯分布（右）

multivariate_normal方法实现1维和2维高斯函数的计算（图3.11–2）。从打印的结果查看2维的高斯分布，以中心点为最大值，向四周逐渐降低，其分布的幅度由σ控制。σ较小时，中心点最高，曲线（曲面）较陡；反之，σ值较大时，中心点较低。曲线（曲面）趋于缓和。

高斯模糊（Gaussian blur）是以二维高斯分布为卷积核（权重值）对图像（或二维数组）进行卷积操作，即卷积计算为每个像素值与周围相邻像素值的加权平均。原始像素值对应着卷积核的中心，具有最大的权重值，相邻像素随距离中心点的距离增加而降低。高斯卷积核可以在模糊的同时，较好地保留图像的边缘效果。高斯卷积核做归一化处理，确保在［0，1］区间。例如，下述代码中假设了4种典型的矩阵（数组）类型，均匀型、凸型、凹型和偏斜型；以及假设高斯权重卷积核，通过计算卷积（权重加权平均）得到结果为：均匀型值为100，凸型值为138.46，凹型值为61.54，偏斜型值为107.69。均匀型为衡量的标准，可见凹凸型都较大的偏离均匀型值，而偏斜型则靠近均匀型，因此通过这个典型的案例，可以清楚地理解高斯卷积核可以计算目标像素与周边像素值之间的差异程度。

```
from matplotlib import cm
from scipy.stats import multivariate_normal
import numpy as np
from scipy.stats import norm
import matplotlib.pyplot as plt
fig = plt.figure(figsize=(20, 9))
ax1 = fig.add_subplot(121)
ax2 = fig.add_subplot(122, projection='3d')

#A- 一维高斯分布（即正态分布）
x = np.linspace(norm.ppf(0.01), norm.ppf(0.99), 100)
for scale in np.arange(1, 5, 0.3):
    ax1.plot(x, norm.pdf(x, loc=0, scale=scale), '-', lw=3, alpha=0.6,
             label='Sigma=%.2f' % scale)  #位置 loc（平均值）和比例 scale（标准差）参数

#B- 二维高斯分布
#x,y=np.meshgrid(np.linspace(-1,1,10), np.linspace(-1,1,10))
x_, y = np.mgrid[-1.0:1.0:30j, -1.0:1.0:30j]
xy = np.column_stack([x_.flat, y.flat])
for scale in np.arange(1, 5, 0.3):
    mu = np.array([0.0, 0.0])
    sigma = np.array([scale, scale])
    covariance = np.diag(sigma**2)
    z = multivariate_normal.pdf(xy, mean=mu, cov=covariance)
    z = z.reshape(x_.shape)
    ax2.plot_surface(x_, y, z, cmap=cm.coolwarm, alpha=0.4)
ax1.legend(loc='upper left', frameon=False)
ax1.spines['right'].set_visible(False)
ax1.spines['top'].set_visible(False)
ax2.view_init(20, 50)
plt.show()
matrix_even = np.array(
```

```
    [[100, 100, 100], [100, 100, 100], [100, 100, 100]])  #均匀型
matrix_convex = np.array(
    [[100, 100, 100], [100, 200, 100], [100, 100, 100]])  #凸型
matrix_concave = np.array(
    [[100, 100, 100], [100, 0, 100], [100, 100, 100]])  #凹型
matrix_skew = np.array(
    [[200, 100, 100], [100, 100, 100], [100, 100, 100]])  #偏斜型

kernal = np.array([[0.1, 0.1, 0.1], [0.1, 0.5, 0.1], [0.1, 0.1, 0.1]])
even_weightedAverage = np.average(matrix_even, weights=kernal)
convex_weightedAverage = np.average(matrix_convex, weights=kernal)
concave_weightedAverage = np.average(matrix_concave, weights=kernal)
skew_weightedAverage = np.average(matrix_skew, weights=kernal)
print("even_WA=%.2f,convex_WA=%.2f,concave_WA=%.2f,skew_WA=%.2f" % (
    even_weightedAverage, convex_weightedAverage, concave_weightedAverage, skew_
weightedAverage))
--------------------------------------------------------------------------------
even_WA=100.00,convex_WA=138.46,concave_WA=61.54,skew_WA=107.69
```

OpenCV可以增加滑动条，实时调整数据，与显示更新后的图像，方便观察参数变化对计算结果的影响。下述实现了给定卷积核大小和σ两个参数，应用OpenCV库提供的GaussianBlur函数实现图像高斯模糊的过程（图3.11-3）。

```
def Gaussion_blur(img_fp):
    '''
    function - 应用 OpenCV 计算高斯模糊，并给定滑动条调节参数

    Params:
        img_fp - 图像路径；string

    Returns:
        None
    '''
    import cv2 as cv

    #回调函数
    def Gaussian_blur_size(GBlur_size):  #高斯核（卷积核大小），值越大，图像越模糊
        global KSIZE
        KSIZE = GBlur_size * 2 + 3
        print("changes in kernel size:", KSIZE, SIGMA)
        dst = cv.GaussianBlur(img, (KSIZE, KSIZE), SIGMA, KSIZE)
        cv.imshow(window_name, dst)

    def Gaussian_blur_Sigma(GBlur_sigma):  # σ(sigma) 设置，值越大，图像越模糊
        global SIGMA
        SIGMA = GBlur_sigma/10.0
        print("changes in sigma:", KSIZE, SIGMA)
        dst = cv.GaussianBlur(img, (KSIZE, KSIZE), SIGMA, KSIZE)
        cv.imshow(window_name, dst)

    #全局变量
    GBlur_size = 1
    GBlur_sigma = 15
```

图3.11-3 图像高斯模糊（滑动条调整参数）

```
    global KSIZE
    global SIGMA
    KSIZE = 1
    SIGMA = 15
    max_value = 300
    max_type = 6
    window_name = "Gaussian Blur"
    trackbar_size = "Size*2+3"
    trackbar_sigema = "Sigma/10"

    #读入图片，模式为灰度图，创建窗口
    img = cv.imread(img_fp, 0)
    cv.namedWindow(window_name)

    #创建滑动条
    cv.createTrackbar(trackbar_size, window_name, GBlur_size,
                      max_type, Gaussian_blur_size)
    cv.createTrackbar(trackbar_sigema, window_name,
                      GBlur_sigma, max_value, Gaussian_blur_Sigma)

    #初始化
    Gaussian_blur_size(GBlur_size)
    Gaussian_blur_Sigma(GBlur_sigma)
    if cv.waitKey(0) == 27:
        cv.destroyAllWindows()
img_2_fp = r'./data/cv_IMG_0407.JPG'
Gaussion_blur(img_2_fp)
----------------------------------------------------------------------------------
changes in kernel size: 5 15
changes in sigma: 5 1.5
changes in kernel size: 5 1.5
changes in sigma: 5 1.5
```

● B－尺度空间（scale space）与高斯差分金字塔

从高斯分布（函数）到高斯模糊可知，在固定卷积核大小时，不同的σ值，因为产生权重值的分布陡峭程度不同，图像的模糊程度会对应变化。可理解为基于变化权重值的图像拉伸，从图像整体上来讲就是各个像素值与各周边像素值的差异程度。而σ值的变化带来差异程度分布即加权均值分布的差异，这个差异变化是固定卷积核大小下尺度空间的纵深向空间。当σ设置值较大时，差异程度分布趋向于均值，反映图像的概貌特征；设置较小时，差异程度分布明显，强化了目标像素与周边像素的差异性，对应于图像的细节特征。如果对图像连续降采样，每一个降采样都是前一图像的1/4，即长宽分别缩小一倍。这一连续降采样获得的多个图像，就是尺度空间的水平向空间。降采样类似于遥感影像的空间分辨率，如果分辨率为1m，即遥感影像的每一单元（cell）大小为1m，可以分辨出建筑、道路、车辆甚至行人的信息。但是，如果降采样到空间分辨率为30m时，则较大的建筑、农田、林地可以分辨出。而小于30m的建筑，车辆和行人等信息是不能分辨的。降采样的水平向空间实际上反映了不同大小地物存在的尺度，或理解为不同的尺度反映不同大小的地物信息，结合到图像上，即为图像的不同内容存在不同的水平向尺度。将上述阐述对应到D.Lowe的论文明确其相应的定义，可以确定降采样的水平向空间为八度（octave）的定义，每一个八度即为连续降采样的图像之一，可以称为组。而对每一降采样（octave）图像进行高斯卷积核计算，这些不同参数下的卷积结果即为每一八度的层。组（降采样-八度，octave）反映图像内容（地物）的水平向尺度（空间分辨率），构成了高斯金字塔的尺度空间。

D.Lowe将图像的尺度空间定义为$L(x,y,\sigma)=G(x,y,\sigma)*I(x,y)$。式中，$G(x,y,\sigma)$为变化尺度的高斯函数（2维），$I(x,y)$为原图像（像素位置），$*$表示卷积运算。

为在连续的尺度空间下检测到特征点，建立高斯差分金字塔（difference-of-Gaussian，DOG），其高斯差分算子为$D(x,y,\sigma)=(G(x,y,k\sigma)-G(x,y,\sigma))*I(x,y)=L(x,y,k\sigma)-L(x,y,\sigma)$。即在图3.11-4左每一组（octave）下每相邻的两层图像相减，得到右侧的图像。层之间的差值可以理解为，相邻σ值的变化反映目标像素与周围像素差异程度表现在层之间的变化程度。而层间较大的程度变化，代表着尺度空间纵深向目标像素点与周边像素点差异程度表现最为突出所在的层，即该层反映了目标像素的特征。如果同时考虑到水平向空间，假设5m的空间分辨率为车辆的水平向尺度空间的界定边界（一个组），即如果大于5m，车辆信息被淹没在更大尺度的地物中，例如车辆被分配到绿地空间里。要想在5m的水平向尺度下，将车辆提取出来，必须比较车辆所在像素与周边像素的差异。如果差异程度较高，则可以提取，但是必须寻找到体现差异程度最大的

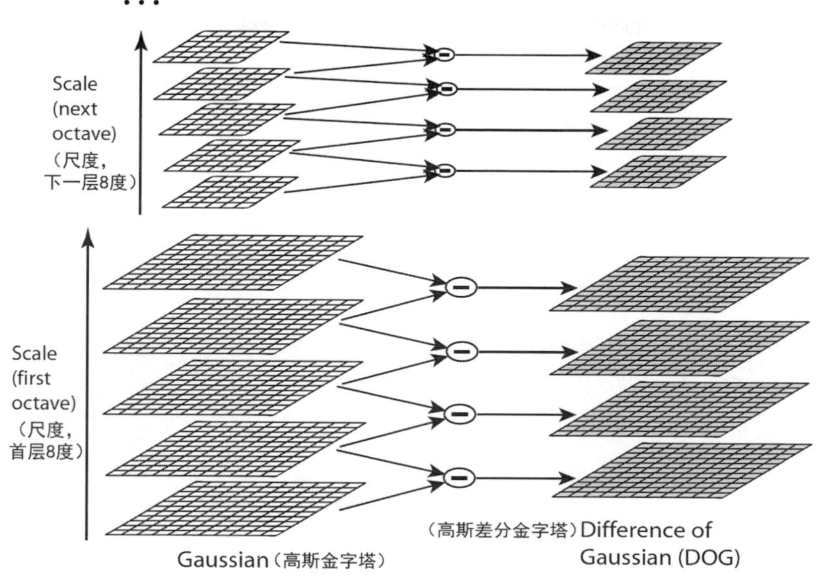

图3.11-4 高斯金字塔和高斯差分金字塔（引自参考文献[1]）

纵深向尺度空间的层。

- C－空间关键点（极值点）检测

已获得DOG，为寻找极值点，需要比较中间的检测点与其同纵深向尺度空间的8个相邻点和上下相邻纵深向尺度层的18个点，共26个点进行比较（图3.11-5）。每组中的纵深向尺度空间，设置连续的σ值，可以产生无数连续的层。对于一个目标像素而言，就是需要不断比较连续的DOG中的相邻层，从而获得无数个离散的局部的极值点。离散的空间极值点并不是真正的极值点，因此可以对尺度空间DOG函数通过泰勒展开式（具体可参看微积分部分的泰勒展开式）拟合函数（二次方程）为：$D(X)=D+\frac{\partial D^{\mathrm{T}}}{\partial X}X+\frac{1}{2}X^{\mathrm{T}}\frac{\partial^2 D}{\partial X^2}X$，其中$X=(x,y,\sigma)^{\mathrm{T}}$。对$X$求导，并令方程等于0，求得极值点（关键点）的位置$\widehat{X}=\frac{\partial^2 D^{-1}}{\partial X^2}\frac{\partial D}{\partial X}$。

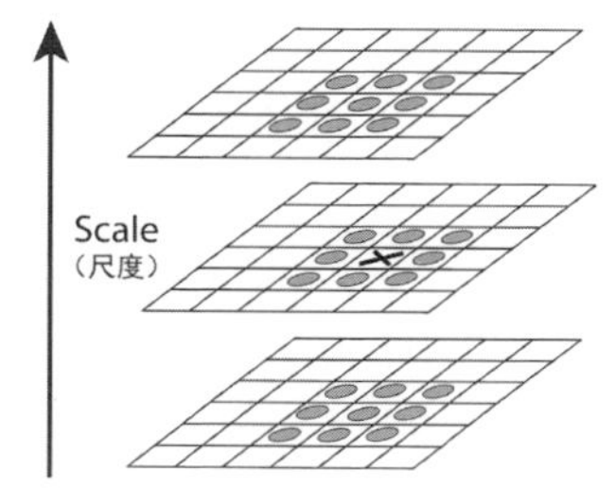

图3.11-5 邻域点（引自参考文献[1]）

- D－关键点方向（用一组向量描述关键点，特征向量）

计算像素点的幅度（梯度大小，gradient magnitude）$m(x,y)$和方向（orientation）$\theta(x,y)$，使用像素的位置差异预先计算。例如，像素的位置标识：$\begin{bmatrix}(-1,1) & (0,1) & (1,1)\\(-1,0) & (0,0) & (1,0)\\(-1,-1) & (0,-1) & (1,-1))\end{bmatrix}$，以关键点为中心，以$3\times1.5\sigma$为半径的同一纵向尺度空间的邻域内计算，$m(x,y)=\sqrt{(L(x+1,y)-L(x-1,y)^2+(L(x,y+1)-L(x,y-1)^2}$，$\theta(x,y)=\tan^{-1}\frac{L(x,y+1)-L(x,y-1)}{L(x+1,y)-L(x-1,y)}$，$L$为关键点所在的尺度空间值，即对应的高斯模糊图像。完成关键点的梯度大小和方向计算后，统计频数，将360°的方向划分为36个等分（bin），频数最大值代表关键点的主方向。

在关键点位置周围区域内（指定半径的圆形区域，或关键点周围16×16个像素点），首先计算每个图像样本点的梯度大小和方向（高斯卷积核加权），创建关键点描述子（符）。然后，这些样本累积成4×4次区域内方向直方图［每个关键点有16个子块（sub-block），即有16个直方图，每个

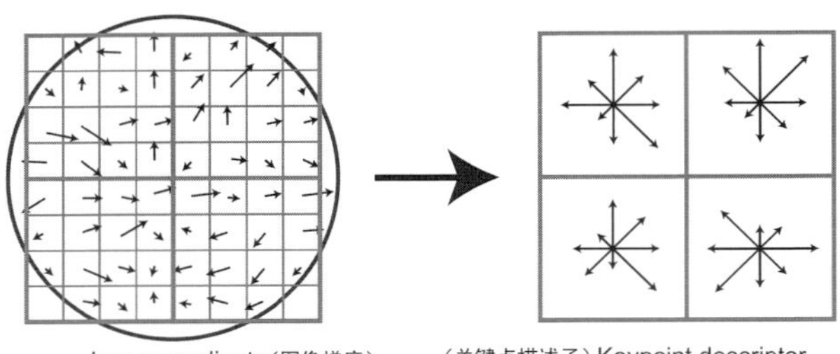

图3.11-6　关键点描述子（引自参考文献[1]）

图3.11-7　SIFT方法实现尺度不变特征转换关键点信息

块为4 × 4的像素点，图3.11–6仅以8 × 8邻域为例]。每个箭头的长度对应区域内该方向附近的梯度大小之和，将所有方向特征相邻连接起来，够成16 × 8 = 128维的特征向量。至此，每个关键点有三个信息$(x, y, \sigma, \theta, magnitude, region)$，即位置，纵深向空间尺度、方向和大小，以及邻域。

使用OpenCV的SIFT_create()方法实现尺度不变特征转换（图3.11–7），所返回的关键点信息对应位置点坐标、邻域（直径），水平向空间尺度（即octave）、梯度大小（响应程度）和方向，对图像分类（–1为未设置）。OpenCV的SIFT计算配置邻域大小同样为16 × 16，划分为4 × 4大小的16个子块，总共128份（bin）。

```
def SIFT_detection(img_fp, save=False):
    '''
    function - 尺度不变特征变换 (scale invariant feature transform, SIFT) 特征点检测

    Params:
        img_fp - 图像文件路径; string
        save- 是否保存特征结果图像。 The default is False; bool

    Returns:
        None
    '''
    import cv2 as cv
    import numpy as np
    import matplotlib.pyplot as plt
    img = cv.imread(img_fp)
    img_gray = cv.cvtColor(img, cv.COLOR_BGR2GRAY)
    sift = cv.SIFT_create()  # SIFT 特征实例化 cv.xfeatures2d.SIFT_create()
    key_points = sift.detect(img_gray, None)  # 提取 SIFT 特征关键点 detector

    # 示例打印关键点数据
    for k in key_points[:5]:
        print("关键点点坐标 :%s, 直径 :%.3f, 金字塔层 :%d, 响应程度 :%.3f, 分类 :%d, 方向 :%d" %
              (k.pt, k.size, k.octave, k.response, k.class_id, k.angle))
        """
        关键点信息包含:
        k.pt 关键点点的坐标（图像像素位置）
        k.size 该点范围的大小（直径）
```

```
        k.octave 从高斯金字塔的哪一层提取得到的数据
        k.response 响应程度，代表该点强壮大小，即角点的程度。角点：极值点，某方面属性特别突出
的点（最大或最小）。
        k.class_id 对图像进行分类时，可以用 class_id 对每个特征点进行区分，未设置时为 -1
        k.angle 角度，关键点的方向。SIFT 算法通过对邻域做梯度运算，求得该方向。-1 为初始值
        """
    print("_"*50)
    # 提取 SIFT 调整描述子 -descriptor，返回的列表长度为 2，第 1 组数据为关键点，第 2 组数据为描述子（关
键点周围对其有贡献的像素点）
    descriptor = sift.compute(img_gray, key_points)
    print("key_points 数据类型 :%s,descriptor 数据类型 :%s" %
         (type(key_points), type(descriptor)))
    print(" 关键点：")
    print(descriptor[0][:1])  # 关键点
    print(" 描述子：")
    print(descriptor[1][:1])  # 描述子
    print(" 描述子 shape:", descriptor[1].shape)
    cv.drawKeypoints(img, key_points, img,
                     flags=cv.DRAW_MATCHES_FLAGS_DRAW_RICH_KEYPOINTS)
    if save:
        cv.imshow('sift features', img)
        cv.imwrite('./data/sift_features.jpg', img)  # 保存图像
        cv.waitKey()
    else:
        fig, ax = plt.subplots(figsize=(30, 15))
        ax.imshow(cv.cvtColor(img, cv.COLOR_BGR2RGB))
        plt.show()
SIFT_detection(img_fp=img_2_fp)
```

```
-----------------------------------------------------------------
关键点坐标 :(2.318413734436035, 498.3979187011719), 直径 :1.818, 金字塔层 :852479, 响应
程度 :0.019, 分类 :-1, 方向 :300
...
关键点坐标 :(3.029824733734131, 606.0113525390625), 直径 :2.157, 金字塔层 :13238783, 响
应程度 :0.042, 分类 :-1, 方向 :61
-------------------------------------------------
key_points 数据类型 :<class 'tuple'>,descriptor 数据类型 :<class 'tuple'>
关键点：
(< cv2.KeyPoint 00000255E1C765A0>,)
描述子：
[[  0.   0.   0.   0.   0.   0.   0.   0.   0.   0.   0.   0.   0.   4.  14.   3.
8.   0.   0.   0.   0.   9.  43.  48.  30.   2.   0.   0.   0.  14.  32.  24.  12.
11.   0.   0.   8.  23.   2.   1. 129.  39.   0.   0.   3.  21.  81.  43.  24.
0.   1.   2.  66.  65. 129.  74.   8.   5.  23.   7.  27.  20.  33.  23.  42.  30.
15.  13.  10.  55.  37.  23. 129.  18.   1.   6.   7.  26.  55. 129.  56.   6.
4. 121. 129.  28.   8.  14.   1.   6.  93. 106.  60.   2.   0.   1. 129.  91.
20.  15.   2.   1.  12.  53.  73.   4.   1.   6.  28.  24.  47. 103.   4.   1.
1. 111. 129.  38.   3.   4.   3.   2.  17.  80.  55.   4.   0.   1.]]
描述子 shape: (19810, 128)
```

OpenCV提供有多种类型的特征检测和描述算法，如果SIFT计算速度不够快，可以尝试使用SURF（Speeded-Up Robust Features），或者BRIEF（Binary Robust Independent Elementary Feature）算法。如果需要实时性，例如机器人领域里的SLAM（simultaneous localization and mapping）配合使用，则可以应用FAST（Features from accelerated segment test）算法。下述为Star算法实现（图3.11-8）。每种算法都有各自的优缺点，需要针对分析目标进行比较确定。

图3.11-8 Star方法实现尺度不变特征转换关键点信息

```
def STAR_detection(img_fp, save=False):
    '''
    function - 使用Star特征检测器提取图像特征
    Params:
        img_fp - 图像文件路径
        save- 是否保存特征结果图像。 The default is False; bool

    Returns:
        None
    '''
    import cv2 as cv
    import numpy as np
    import copy
    import matplotlib.pyplot as plt
    img = cv.imread(img_fp)
    star = cv.xfeatures2d.StarDetector_create()
    key_point = star.detect(img)
    cv.drawKeypoints(img, key_point, img,
                     flags=cv.DRAW_MATCHES_FLAGS_DRAW_RICH_KEYPOINTS)
    if save:
        cv.imshow('star features', img_copy)
        cv.imwrite('./data/star_features.jpg', img)  #保存图像
        cv.waitKey()
    else:
        fig, ax = plt.subplots(figsize=(30, 15))
        ax.imshow(cv.cvtColor(img, cv.COLOR_BGR2RGB))
        plt.show()
STAR_detection(img_fp=img_2_fp)
```

3.11.2.2 特征匹配

通过特征检测器可以获取图像关键点的描述子，那么包含同一事物（具有同样的特征）的两幅图像就可以通过描述子匹配。OpenCV提供了三种方式[①]：一种是基于ORB（Oriented FAST and Rotated BRIEF）描述子的蛮力匹配（Brute-Force Matching with ORB Descriptors），结果如图3.11–9所示；再者，为使用SIFT描述子和比率检测蛮力匹配（Brute-Force Matching with SIFT Descriptors and Ratio Test），结果如图3.11–10所示；以及，基于FLANN的匹配器（FLANN based Matcher），结果如图3.11–11所示。

① OpenCV提供了三种方式特征匹配（https://docs.opencv.org/4.x/dc/dc3/tutorial_py_matcher.html）。

图3.11-9 ORB描述子的蛮力匹配结果

图3.11-10 SIFT描述子和比率检测蛮力匹配结果

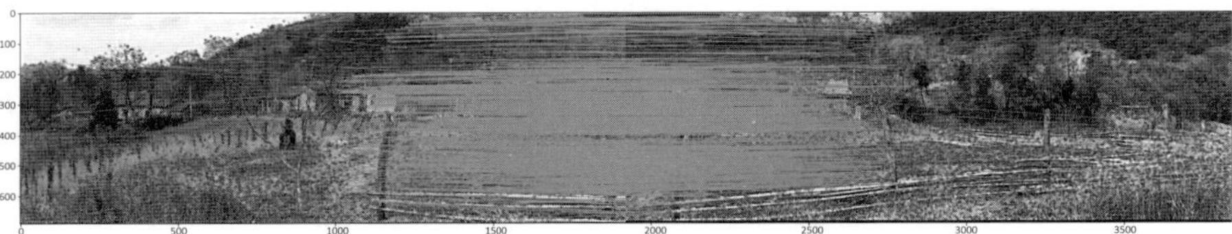

图3.11-11 基于FLANN的匹配器匹配结果

```
    #注意，有些算法已申请专利。

def feature_matching(img_1_fp, img_2_fp, index_params=None, method='FLANN'):
    '''
    function - OpenCV 根据图像特征匹配图像。迁移官网的三种方法，1-ORB 描述子蛮力匹配
（Brute-Force Matching with ORB Descriptors）；2 —使用 SIFT 描述子和比率检测蛮力匹配
（Brute-Force Matching with SIFT Descriptors and Ratio Test）；3- 基于 FLANN 的匹配器
（FLANN based Matcher）
    Params:
        img_1 - 待匹配图像 1 路径；string
        img_2 - 待匹配图像 2 路径；string
        method - 参数为：'ORB','SIFT','FLANN'。The default is 'FLANN'; string

    Returns:
        None
    '''
    import numpy as np
    import cv2 as cv
    import matplotlib.pyplot as plt
    plt.figure(figsize=(30, 15))
    img1 = cv.imread(img_1_fp, cv.IMREAD_GRAYSCALE)  #读取图像 1
    img2 = cv.imread(img_2_fp, cv.IMREAD_GRAYSCALE)  #读取图像 2
    if method == 'ORB':
        #初始化 ORB 探测器
        orb = cv.ORB_create()
        #用 ORB 查找关键点和描述符
        kp1, des1 = orb.detectAndCompute(img1, None)
        kp2, des2 = orb.detectAndCompute(img2, None)

        #建立 BFMatcher 对象
        bf = cv.BFMatcher(cv.NORM_HAMMING, crossCheck=True)
        #匹配描述子.
        matches = bf.match(des1, des2)
        #按距离排序
        matches = sorted(matches, key=lambda x: x.distance)
        #绘制前 10 个匹配.
        img3 = cv.drawMatches(
            img1, kp1, img2, kp2, matches[:10], None, flags=cv.DrawMatchesFlags_
NOT_DRAW_SINGLE_POINTS)
        plt.imshow(img3), plt.show()
    if method == 'SIFT':
        #初始化 SIFT 探测器
        sift = cv.SIFT_create()
        #用 SIFT 查找关键点和描述符
        kp1, des1 = sift.detectAndCompute(img1, None)
        kp2, des2 = sift.detectAndCompute(img2, None)
        #带有默认参数的 BFMatcher
        bf = cv.BFMatcher()
        matches = bf.knnMatch(des1, des2, k=2)
        #应用比率测试
        good = []
        for m, n in matches:
            if m.distance < 0.75*n.distance:
                good.append([m])
        #绘制满足要求的匹配.
        img3 = cv.drawMatchesKnn(img1, kp1, img2, kp2, good[0:int(
            1*len(good)):int(0.1*len(good))], None, flags=cv.DrawMatchesFlags_
NOT_DRAW_SINGLE_POINTS)
        plt.imshow(img3), plt.show()
    if method == 'FLANN':
        #初始化 SIFT 探测器
        sift = cv.SIFT_create()
        #用 SIFT 查找关键点和描述符
        kp1, des1 = sift.detectAndCompute(img1, None)
        kp2, des2 = sift.detectAndCompute(img2, None)
        search_params = dict(checks=50)  #或者传入空字典
        flann = cv.FlannBasedMatcher(index_params, search_params)
        matches = flann.knnMatch(des1, des2, k=2)
        #只需要绘制好的匹配，所以创建一个掩码
        matchesMask = [[0, 0] for i in range(len(matches))]
        #根据 Lowe 的论文的比率测试
        for i, (m, n) in enumerate(matches):
            if m.distance < 0.7*n.distance:
                matchesMask[i] = [1, 0]
```

```
        draw_params = dict(matchColor=(0, 255, 0),
                           singlePointColor=(255, 0, 0),
                           matchesMask=matchesMask,
                           flags=cv.DrawMatchesFlags_DEFAULT)
        img3 = cv.drawMatchesKnn(img1, kp1, img2, kp2,
                                 matches, None, **draw_params)
        plt.imshow(img3,), plt.show()
feature_matching(img_1_fp, img_2_fp, method='ORB',)
feature_matching(img_1_fp,img_2_fp,method='SIFT',)
# FLANN 参数
FLANN_INDEX_KDTREE = 1
index_params = dict(algorithm = FLANN_INDEX_KDTREE, trees = 5)
feature_matching(img_1_fp,img_2_fp,index_params=index_params,method='FLANN',)
```

3.11.3 动态街景视觉感知

上述解释尺度空间是从算法数理逻辑层面进行的解释。如果从感性上来理解，不同尺度的图像，可以近似理解为人眼观察事物从远及近的过程。这个过程中，可以认为尺度空间满足视觉不变性，即图像的分析不受图像灰度、对比度的影响，并满足平移不变性、尺度不变性、欧几里得不变性和仿射不变性。因此，可以借助计算机视觉分析技术来分析研究城市街道景观的感知变化。为了保持街道景观分析数据的统一性，使用KITTI数据为例，这样可以保持图像的连续性、拍摄高度及视角的不变性，提高数据分析结果的可靠性。

3.11.3.1 KITTI数据集

KITTI数据集（The KITTI Vision Benchmark Suite）[①]是用于无人驾驶场景下计算机视觉算法评测数据集，连续记录有车行路线下的连续城市景观图像，以及GPS定位等信息。在这里，并不是用KITTI数据集研究无人驾驶，但是无人驾驶项目可以为城市规划带来用于研究城市的数据。随着无人驾驶技术的发展，以及实际应用的推广，该部分的数据信息量将会成倍增加。此次实验使用2011_09_29_drive_0071 (4.1 GB)[②]标识数据，为城市内的街巷空间。

查看“oxts/dataformat.txt”文件，可以获知“oxts/data”下文件的数据格式，信息罗列如表3.11–1所示。并定义KITTI_info()函数，读取查看数据，见表3.11–2。

① The KITTI Vision Benchmark Suite，利用无人驾驶平台Annieway开发具有挑战性的现实世界计算机视觉基准（computer vision benchmarks）(https://www.cvlibs.net/datasets/kitti/index.php)

② 2011_09_29_drive_0071 (4.1 GB)，数据下载地址（https://www.cvlibs.net/datasets/kitti/raw_data.php）。

表3.11–1 KITTI数据集数据格式信息

(@ARTICLE{Geiger2013IJRR, author = {Andreas Geiger and Philip Lenz and Christoph Stiller and Raquel Urtasun}, title = {Vision meets Robotics: The KITTI Dataset}, journal = {International Journal of Robotics Research (IJRR)}, year = {2013}})

Column1	Column2
lat	latitude of the oxts-unit (deg)
lon	longitude of the oxts-unit (deg)
alt	altitude of the oxts-unit (m)
roll	roll angle (rad), 0 = level, positive = left side up, range: -pi .. +pi
pitch	pitch angle (rad), 0 = level, positive = front down, range: -pi/2 .. +pi/2
yaw	heading (rad), 0 = east, positive = counter clockwise, range: -pi .. +pi
vn	velocity towards north (m/s)
ve	velocity towards east (m/s)
vf	forward velocity, i.e. parallel to earth-surface (m/s)
vl	leftward velocity, i.e. parallel to earth-surface (m/s)
vu	upward velocity, i.e. perpendicular to earth-surface (m/s)
ax	acceleration in x, i.e. in direction of vehicle front (m/s^2)
ay	acceleration in y, i.e. in direction of vehicle left (m/s^2)
ay	acceleration in z, i.e. in direction of vehicle top (m/s^2)
af	forward acceleration (m/s^2)
al	leftward acceleration (m/s^2)
au	upward acceleration (m/s^2)

续表

Column1	Column2
wx	angular rate around x (rad/s)
wy	angular rate around y (rad/s)
wz	angular rate around z (rad/s)
wf	angular rate around forward axis (rad/s)
wl	angular rate around leftward axis (rad/s)
wu	angular rate around upward axis (rad/s)
pos_accuracy	velocity accuracy (north/east in m)
vel_accuracy	velocity accuracy (north/east in m/s)
navstat	navigation status (see navstat_to_string)
numsats	number of satellites tracked by primary GPS receiver
posmode	position mode of primary GPS receiver (see gps_mode_to_string)
velmode	velocity mode of primary GPS receiver (see gps_mode_to_string)
orimode	orientation mode of primary GPS receiver (see gps_mode_to_string)

```
import util_misc
def KITTI_info(KITTI_info_fp, timestamps_fp):
    '''
    function - 读取 KITTI 文件信息，1- 包括经纬度，惯性导航系统信息等的 .txt 文件，2- 包含时间
戳的 .txt 文件

    Params:
        KITTI_info_fp - 数据文件路径；string
        timestamps_fp - 时间戳文件路径；string

    Returns:
        drive_info - 返回数据；DataFrame
    '''
    import pandas as pd
    import util_misc
    import os
    drive_fp = util_misc.filePath_extraction(KITTI_info_fp, ['txt'])
    ''' 展平列表函数 '''
    def flatten_lst(lst): return [m for n_lst in lst for m in flatten_lst(
        n_lst)] if type(lst) is list else [lst]
    drive_fp_list = flatten_lst(
        [[os.path.join(k, f) for f in drive_fp[k]] for k, v in drive_fp.items()])
    columns = ["lat", "lon", "alt", "roll", "pitch", "yaw", "vn", "ve", "vf", "vl",
"vu", "ax", "ay", "ay", "af", "al", "au",
                "wx", "wy", "wz", "wf", "wl", "wu", "pos_accuracy", "vel_
accuracy", "navstat", "numsats", "posmode", "velmode", "orimode"]
    drive_info = pd.concat([pd.read_csv(item, delimiter=' ', header=None)
                            for item in drive_fp_list], axis=0)
    drive_info.columns = columns
    drive_info = drive_info.reset_index()
    timestamps = pd.read_csv(timestamps_fp, header=None)
    timestamps.columns = ['timestamps_']
    drive_info = pd.concat([drive_info, timestamps], axis=1, sort=False)
    # drive_29_0071_info.index=pd.to_datetime(drive_29_0071_info["timestamps_"]) # 用时间戳作为行 (row)
索引
    return drive_info
KITTI_info_fp = r'F:\data\2011_09_29_drive_0071_sync\oxts\data'
timestamps_fp = r'F:\data\2011_09_29_drive_0071_sync\image_03\timestamps.txt'
drive_29_0071_info = KITTI_info(KITTI_info_fp, timestamps_fp)
util_misc.print_html(drive_29_0071_info)
```

通过提取的数据，应用Plotly库打印包含地图信息的图3.11-12，观察数据在现实世界里具体的位置。

```
def plotly_scatterMapbox(df, **kwargs):
    '''
    function - 使用 plotly 的 go.Scattermapbox 方法，在地图上显示点及其连线，坐标为经纬度

    Paras:
        df - DataFrame 格式数据，含经纬度；DataFrame
```

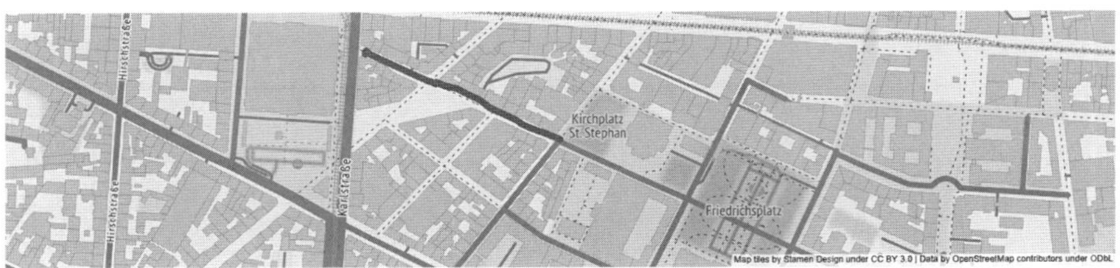

图3.11-12 KITTI数据位置地图

```
        field - 'lon':df 的 longitude 列名, 'lat': 为 df 的
latitude 列名, 'center_lon': 地图显示中心精经度定位, "center_lat":
地图显示中心维度定位, "zoom": 为地图缩放; string
    '''
    import pandas as pd
    import plotly.graph_objects as go
    field = {'lon': 'lon', 'lat': 'lat', "center_lon":
8.398104,
              "center_lat": 49.008645, "zoom": 16}
    field.update(kwargs)
    fig = go.Figure(go.Scattermapbox(mode="markers",
lat=df[field['lat']], lon=df[field['lon']], marker={
                         'size': 10}))  # 亦可以选择列, 通过 size="" 配
置增加显示信息 ; mode=" markers+lines"
    fig.update_layout(
        margin={'l': 0, 't': 0, 'b': 0, 'r': 0},
        mapbox={
            'center': {'lon': 10, 'lat': 10},
            'style': "stamen-terrain",
            'center': {'lon': field['center_lon'], 'lat':
field['center_lat']},
            'zoom': 16})
    fig.show()
drive_29_0071_info.sort_values(by="timestamps_",
inplace=True)
# print(drive_29_0071_info["timestamps_"].head(50))
plotly_scatterMapbox(drive_29_0071_info)
```

动态查看连续记录的影像，如图3.11-13所示。

```
from IPython.display import HTML
import matplotlib.image as mpimg
import matplotlib.animation as animation
import matplotlib.pyplot as plt
import os
drive_29_0071_img_fp = util_misc.filePath_extraction(
    r'F:\data\2011_09_29_drive_0071_sync\image_03\data',
['png'])
drive_29_0071_img_fp_list = util_misc.flatten_lst([[os.path.
join(
    k, f) for f in drive_29_0071_img_fp[k]] for k, v in
drive_29_0071_img_fp.items()])
drive_29_0071_img_fp_list.sort()
fig = plt.figure(figsize=(20, 10))
ims = [[plt.imshow(mpimg.imread(f), animated=True)]
       for f in drive_29_0071_img_fp_list[:200]]
print("finished reading the imgs.")
ani = animation.ArtistAnimation(
    fig, ims, interval=100, blit=True, repeat_delay=1000)
# interval = 50
# ani.save(r'./imgs/drive_29_0071_imgs.mp4')
HTML(ani.to_html5_video())
-------------------------------------------------------------
finished reading the imgs.
```

图像匹配可以返回两幅图像中特征点基本相似的关键点。对于无人驾驶拍摄的连续影像而言，假设确定一个固定位置，即选择该位置上的一张影像，将其分别与之后的所有影像进行图像匹配，返回关键点，计算每一对的特征点匹配数量。这个特征点匹配数量的变化体现了当前位

表3.11-2 KITTI文件信息

	index	lat	lon	alt	roll	pitch	yaw	…	pos_accuracy	vel_accuracy	numsats	posmode	velmode	orimode	timestamps_
0	0	49.00865	8.398092	112.923836	0.028517	0.033114	2.638139	…	0.265887	0.031113	10	4	4	6	55:00.0
1	0	49.008777	8.397611	112.555534	0.02484	0.020511	2.685955	…	0.592571	0.102616	6	4	0	0	55:00.1
2	0	49.009162	8.396541	111.875397	0.05351	0.009328	2.637656	…	5.410355	0.436904	5	4	0	0	55:00.2
3	0	49.008962	8.397075	112.19326	0.034518	0.016115	2.720819	…	1.296095	0.160181	3	0	0	0	55:00.3
4	0	49.009505	8.395251	116.018288	0.027009	0.009783	2.68212	…	1.212669	0.103966	5	5	5	0	55:00.4

图3.11-13 连续影像（部分）

置的图像与之后顺序影像相似的程度。因为后一帧（后一位置）影像包含前一影像的一部分，当距离越近，两者匹配返回的特征点匹配数量越多；反之，离当前位置越远，匹配的数量越少，而且这个过程基本上是逐渐减少的。因此，可以由上述计算推测一条街道视觉感知变化的情况，找到感知消失的距离；同时，也可以比较不同街道感知变化的差异；甚至，可以比较不同街道感知的相似度。

代码实现上主要包括三个部分：第一个是定义批量特征提取及返回匹配点数量的类DynamicStreetView_visualPerception；第二个是匹配点数量是随距离增加而逐渐降低的，需要找到降低后，基本不再变化的位置点，即感知消失的位置，定义类MovingAverage_inflection；最后，计算这个感知消失的距离，定义函数vanishing_position_length()。

- **A－返回特征匹配点数量**

返回特征点匹配数量的类，是使用Star方法提取关键点，这个方法可以降低使用SIFT产生的噪声；然后，再用SIFT根据Star提取的关键点来返回描述子；最后，利用描述子进行图像匹配，返回特征点匹配的数量。

```
class DynamicStreetView_visualPerception:
    '''
    class - 应用 Star 提取图像关键点，结合 SIFT 获得描述子，根据特征匹配分析特征变化（视觉感知变化），即动态街景视觉感知

    Params:
        imgs_fp - 图像路径列表；list(string)
        knnMatch_ratio - 图像匹配比例，默认为 0.75；float
    '''
    def __init__(self, imgs_fp, knnMatch_ratio=0.75):
        self.knnMatch_ratio = knnMatch_ratio
        self.imgs_fp = imgs_fp
    def kp_descriptor(self, img_fp):
        '''
        function - 提取关键点和获取描述子
        '''
        import cv2 as cv
        img = cv.imread(img_fp)
        star_detector = cv.xfeatures2d.StarDetector_create()
        key_points = star_detector.detect(img)  # 应用处理 Star 特征检测相关函数，返回检测出的特征关键点
        img_gray = cv.cvtColor(img, cv.COLOR_BGR2GRAY)  # 将图像转为灰度
        kp, des = cv.xfeatures2d.SIFT_create().compute(
            img_gray, key_points)  # SIFT 特征提取器提取特征
        return kp, des
    def feature_matching(self, des_1, des_2, kp_1=None, kp_2=None):
        '''
        function - 图像匹配
        '''
        import cv2 as cv
        bf = cv.BFMatcher()
        matches = bf.knnMatch(des_1, des_2, k=2)
        '''
        可以由匹配 matches 返回关键点（train,query）的位置索引，train 图像的索引，以及描述子之间的距离
        DMatch.distance - 描述符之间的距离。越低越好 .
        DMatch.trainIdx - 训练（train）描述子中描述子的索引
        DMatch.queryIdx - 查询（query）描述子中描述子的索引
        DMatch.imgIdx - 训练图像的索引 .
        '''
        '''
        if kp_1 !=None and kp_2 != None:
            kp1_list=[kp_1[mat[0].queryIdx].pt for mat in matches]
```

```
            kp2_list=[kp_2[mat[0].trainIdx].pt for mat in matches]
            des_distance=[(mat[0].distance,mat[1].distance) for mat in matches]
            print(des_distance[:5])
        '''
        good = []
        for m, n in matches:
            if m.distance < self.knnMatch_ratio*n.distance:
                good.append(m)
        return good
    def sequence_statistics(self):
        '''
        function - 序列图像匹配计算，每一位置图像与后续所有位置匹配分析
        '''
        from tqdm import tqdm
        des_list = []
        print("计算关键点和描述子...")
        for f in tqdm(self.imgs_fp):
            _, des = self.kp_descriptor(f)
            des_list.append(des)
        matches_sequence = {}
        print("计算序列图像匹配数...")
        for i in tqdm(range(len(des_list)-1)):
            matches_temp = []
            for j_des in des_list[i:]:
                matches_temp.append(self.feature_matching(des_list[i], j_des))
            matches_sequence[i] = matches_temp
        matches_num = {k: [len(v) for v in val]
                       for k, val in matches_sequence.items()}
        return matches_num
dsv_vp = DynamicStreetView_visualPerception(drive_29_0071_img_fp_list)
# kp1,des1=dsv_vp.kp_descriptor(drive_29_0071_img_fp_list[10])
# kp2,des2=dsv_vp.kp_descriptor(drive_29_0071_img_fp_list[50])
# dsv_vp.feature_matching(des1,des2,kp1,kp2)
matches_num = dsv_vp.sequence_statistics()
```

```
计算关键点和描述子...
100%|██████████| 1059/1059 [01:27<00:00, 12.15it/s]
计算序列图像匹配数...
100%|██████████| 1058/1058 [38:58<00:00,  2.21s/it]
```

- B－寻找拐点，及跳变稳定的区域（视觉感知变化的位置，以及消失的位置）

寻找拐点，即寻找变化最快的位置（图3.11-14）。在该点之前，特征点匹配的数量快速减低，该点之后变化开始缓慢。可以理解为一开始视觉信息快速的流失，因为此时不同影像重叠的信息明确，但随着距离增加，明确重叠的信息在减少，因此特征点数量变化也会变得缓慢。下述以第一张影像为位置点，分析其与之后所有的影像，可以得知拐点为12。因为位置图索引为0，因此12为索引为13影像所在的位置。

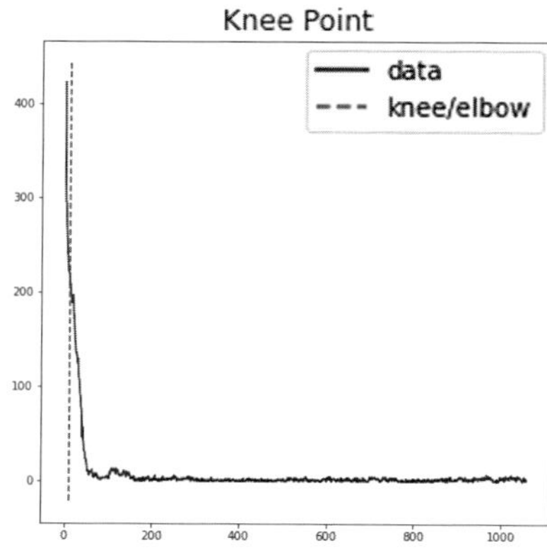

图3.11-14 特征点匹配数量曲线和拐点计算

```
def knee_lineGraph(x, y):
    '''
    function - 绘制折线图及其拐点。需要调用kneed库的KneeLocator，及DataGenerator文件

    Paras:
        x - 横坐标，用于横轴标签
        y - 纵坐标，用于计算拐点
    '''
    import matplotlib.pyplot as plt
    from data_generator import DataGenerator
    from knee_locator import KneeLocator

    #如果调整图表样式，需调整knee_locator文件中的plot_knee()函数相关参数
    kneedle = KneeLocator(x, y, curve='convex', direction='decreasing')
    print('曲线拐点（凸）：', round(kneedle.knee, 3))
    print('曲线拐点（凹）：', round(kneedle.elbow, 3))
    kneedle.plot_knee(figsize=(8, 8))
idx = 0
x = range(idx,idx+len(matches_num[idx]))
y = matches_num[idx]
knee_lineGraph(x,y)
-----------------------------------------------------------------
曲线拐点（凸）： 12
曲线拐点（凹）： 12
```

上述方法找到的拐点，是曲线变化最快的点。但是，需要确定在哪个位置，视觉感知的联系降为最低（即特征点匹配的数量基本不再降低）的位置。这个位置是曲线变化基本为0的位置，就是计算每一点与前一点的差值。差值越小，说明变化越小。如果这个差值变化基本保持不变，就说明已经找到了这个位置点，即保证差值的差值与前一差值保持相等。例如，如果有一组数据为［20, 10, 8, 1, 1, 1］，做第一次差值（取绝对值），结果为［10, 2, 7, 0, 0］；做差值的差值，结果为［8, 5, 7, 0］。那么，如果`diff(x) == diff(diff(x))`，即差值的差值第3个数7，与第一次差值的第3个数7相等。说明该位置之后曲线基本保持水平一致，不再变化。同时，满足`diff(x) != 0`。如果为0，即为数据相等。此时，曲线已经保持基本不变。

下述图3.11-15的绘制，对曲线做了平滑处理，降低噪声，以便找到跳变基本为0的位置点。同时，给出了置信区间，标注异常点，便于观察数据平滑后的变化。平滑后的曲线尽量要在置信区间内，虽然在曲线降低的区段出现异常点，但是因为不在跳变稳定的区段，所以这个异常点错误对结果并没有影响。

```
import numpy as np
import pandas as pd
class MovingAverage_inflection:
    '''
    class - 曲线（数据）平滑，与寻找曲线水平和纵向的斜率变化点

    Params:
        series - pandas 的Series格式数据
        window - 滑动窗口大小，值越大，平滑程度越大
        plot_intervals - 是否打印置信区间，某人为False
```

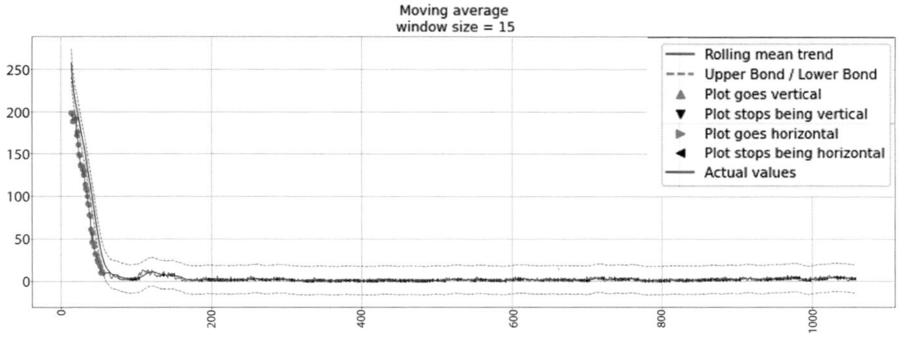

图3.11-15 寻找曲线变化基本为0的位置

```
        scale - 偏差比例，默认为1.96,
        plot_anomalies - 是否打印异常值，默认为False,
        figsize - 打印窗口大小，默认为(15,5),
        threshold - 拐点阈值，默认为0
    '''
    def __init__(self, series, window, plot_intervals=False, scale=1.96, plot_
anomalies=False, figsize=(15, 5), threshold=0):
        self.series = series
        self.window = window
        self.plot_intervals = plot_intervals
        self.scale = scale
        self.plot_anomalies = plot_anomalies
        self.figsize = figsize
        self.threshold = threshold
        self.rolling_mean = self.movingAverage()
    def masks(self, vec):
        '''
        function - 寻找曲线水平和纵向的斜率变化，参考 https://stackoverflow.com/
questions/47342447/find-locations-on-a-curve-where-the-slope-changes
        '''
        d = np.diff(vec)
        dd = np.diff(d)

        # 根据参数 vec 确定曲线超向垂直或水平位置的掩码
        to_mask = ((d[:-1] != self.threshold) & (d[:-1] == -dd-self.threshold))
        # 根据参数 vec 确定曲线所来自垂直或水平位置的掩码
        from_mask = ((d[1:] != self.threshold) & (d[1:] == dd-self.threshold))
        return to_mask, from_mask
    def apply_mask(self, mask, x, y):
        return x[1:-1][mask], y[1:-1][mask]
    def knee_elbow(self):
        '''
        function - 返回拐点的起末位置
        '''
        x_r = np.array(self.rolling_mean.index)
        y_r = np.array(self.rolling_mean)
        to_vert_mask, from_vert_mask = self.masks(x_r)
        to_horiz_mask, from_horiz_mask = self.masks(y_r)
        to_vert_t, to_vert_v = self.apply_mask(to_vert_mask, x_r, y_r)
        from_vert_t, from_vert_v = self.apply_mask(from_vert_mask, x_r, y_r)
        to_horiz_t, to_horiz_v = self.apply_mask(to_horiz_mask, x_r, y_r)
        from_horiz_t, from_horiz_v = self.apply_mask(from_horiz_mask, x_r, y_r)
        return x_r, y_r, to_vert_t, to_vert_v, from_vert_t, from_vert_v, to_
horiz_t, to_horiz_v, from_horiz_t, from_horiz_v
    def movingAverage(self):
        rolling_mean = self.series.rolling(window=self.window).mean()
        return rolling_mean
    def plot_movingAverage(self, inflection=False):
        """
        function - 打印移动平衡 / 滑动窗口，及拐点
        """
        import numpy as np
        from sklearn.metrics import median_absolute_error, mean_absolute_error
        import matplotlib.pyplot as plt
        plt.figure(figsize=self.figsize)
        plt.title("Moving average\n window size = {}".format(self.window))
        plt.plot(self.rolling_mean, "g", label="Rolling mean trend")

        # 打印置信区间（Plot confidence intervals for smoothed values）
        if self.plot_intervals:
            mae = mean_absolute_error(
                self.series[self.window:], self.rolling_mean[self.window:])
            deviation = np.std(
                self.series[self.window:] - self.rolling_mean[self.window:])
            lower_bond = self.rolling_mean - (mae + self.scale * deviation)
            upper_bond = self.rolling_mean + (mae + self.scale * deviation)
            plt.plot(upper_bond, "r--", label="Upper Bond / Lower Bond")
            plt.plot(lower_bond, "r--")

            # 显示异常值（Having the intervals, find abnormal values）
            if self.plot_anomalies:
                anomalies = pd.DataFrame(
                    index=self.series.index, columns=self.series.to_frame().
columns)
```

```
                anomalies[self.series < lower_bond] = self.series[self.series <
lower_bond].to_frame()
                anomalies[self.series > upper_bond] = self.series[self.series >
upper_bond].to_frame()
                plt.plot(anomalies, "ro", markersize=10)
        if inflection:
            x_r, y_r, to_vert_t, to_vert_v, from_vert_t, from_vert_v, to_horiz_t,
to_horiz_v, from_horiz_t, from_horiz_v = self.knee_elbow()
            plt.plot(x_r, y_r, 'b-')
            plt.plot(to_vert_t, to_vert_v, 'r^', label='Plot goes vertical')
            plt.plot(from_vert_t, from_vert_v, 'kv',
                     label='Plot stops being vertical')
            plt.plot(to_horiz_t, to_horiz_v, 'r>',
                     label='Plot goes horizontal')
            plt.plot(from_horiz_t, from_horiz_v, 'k<',
                     label='Plot stops being horizontal')
        plt.plot(self.series[self.window:], label="Actual values")
        plt.legend(loc="upper right")
        plt.grid(True)
        plt.xticks(rotation='vertical')
        plt.show()
idx = 0
x = np.array(range(idx, idx+len(matches_num[idx])))
y = np.array(matches_num[idx])
y_ = pd.Series(y, index=x)
MAI = MovingAverage_inflection(y_, window=15, plot_intervals=True,
                                    scale=1.96, plot_anomalies=True, figsize=(15*2,
5*2), threshold=0)
MAI.plot_movingAverage(inflection=True)
```

- C - 计算感知消失的距离

定位到第一个跳变为0的位置，就是曲线变化基本平缓的位置，获取其索引值。然后，找出观察位置与第一个跳变为0位置之间所有图像的坐标，定义其为地理空间数据GeoDataFrame的数据格式（表3.11-3）。因为研究区域在德国，因此找到德国的EPSG编号，用于投影；然后，计算路径的长度，并统计相关量。由结果可知，视觉感知消失距离的均值约为29m，16～46之间的数量约占到62%。

表3.11-3 观察位置与第一个跳变为0位置之间所有图像的坐标GDF格式数据

	start_idx	end_idx	geometry	length
0	0	91	LINESTRING (934871.288 6276329.185,...)	11310.68584
...	...	...	...	...
4	4	82	LINESTRING (934555.096 6276474.385, ...)	9474.584042

```
def vanishing_position_length(matches_num, coordi_df, epsg, **kwargs):
    '''
    function - 计算图像匹配特征点几乎无关联的距离，即对特定位置视觉随距离远去而感知消失的距离

    Params:
        matches_num - 由类 dynamicStreetView_visualPerception 计算的特征关键点匹配数量
        coordi_df - 包含经纬度的 DataFrame，其列名为：lon,lat
        **kwargs - 同类 movingAverage_inflection 配置参数
    '''
    from shapely.geometry import Point, LineString, shape
    import geopandas as gpd
    import pyproj
    import numpy as np
    MAI_paras = {'window': 15, 'plot_intervals': True, 'scale': 1.96,
                 'plot_anomalies': True, 'figsize': (15*2, 5*2), 'threshold': 0}
    MAI_paras.update(kwargs)
    vanishing_position = {}
    for idx in range(len(matches_num)):
        x = np.array(range(idx, idx+len(matches_num[idx])))
        y = np.array(matches_num[idx])
        y_ = pd.Series(y, index=x)
        MAI = movingAverage_inflection(y_,
                                       window=MAI_paras['window'],
                                       plot_intervals=MAI_paras['plot_
```

```
intervals'],
                                        scale=MAI_paras['scale'],
                                        plot_anomalies=MAI_paras['plot_
anomalies'],
                                        figsize=MAI_paras['figsize'],
                                        threshold=MAI_paras['threshold'])
        _, _, _, _, from_vert_t, _, _, _, from_horiz_t, _ = MAI.knee_elbow()
        if np.any(from_horiz_t != None):
            vanishing_position[idx] = (idx, from_horiz_t[0])
        else:
            vanishing_position[idx] = (idx, idx)
    vanishing_position_df = pd.DataFrame.from_dict(
        vanishing_position, orient='index', columns=['start_idx', 'end_idx'])
    vanishing_position_df['geometry'] = vanishing_position_df.apply(lambda idx:
LineString(
        coordi_df[idx.start_idx:idx.end_idx][['lon', 'lat']].apply(lambda row:
Point(row.lon, row.lat), axis=1).values.tolist()), axis=1)
    crs_4326 = 4326
    vanishing_position_gdf = gpd.GeoDataFrame(
        vanishing_position_df, geometry='geometry', crs=crs_4326)
    crs_ = pyproj.CRS(epsg)
    vanishing_position_gdf_reproj = vanishing_position_gdf.to_crs(crs_)
    vanishing_position_gdf_reproj['length'] = vanishing_position_gdf_reproj.
geometry.length
    return vanishing_position_gdf_reproj
coordi_df = drive_29_0071_info
vanishing_gpd = vanishing_position_length(
    matches_num, coordi_df, epsg="EPSG:3857", threshold=0)
print("感知消失距离统计：", "_"*50, "\n")
print(vanishing_gpd[vanishing_gpd["length"] > 1].length.describe())
print("频数统计：", "_"*50, "\n")
print(vanishing_gpd[vanishing_gpd["length"] > 1]
      ["length"].value_counts(bins=5))
util_misc.print_html(vanishing_gpd)
----------------------------------------------------------------------------
感知消失距离统计：__________________________________________________
count     1000.000000
mean      9472.084030
std       3682.111894
min       1686.790065
25%       6785.848838
50%       9019.067193
75%      11666.403897
max      25517.460098
dtype: float64
频数统计：__________________________________________________
(6452.924, 11219.058]     513
(11219.058, 15985.192]    221
(1662.958, 6452.924]      202
(15985.192, 20751.326]     52
(20751.326, 25517.46]      12
Name: length, dtype: int64
```

查看一个观测点感知消失位置的影像，如图3.11-16所示。

```
import matplotlib.pyplot as plt
import matplotlib.image as mpimg
fig, axs = plt.subplots(ncols=2,nrows=1,figsize =(30, 15))
starting_idx = vanishing_gpd.iloc[0].start_idx
ending_idx = vanishing_gpd.iloc[0].end_idx
axs[0].imshow(mpimg.imread(drive_29_0071_img_fp_list[starting_idx]))
axs[0].set_title("starting position ")
axs[1].imshow(mpimg.imread(drive_29_0071_img_fp_list[ending_idx]))
```

图3.11-16 查看一个观测点及其感知消失位置的影像

```
axs[1].set_title("ending position ")
plt.show()
```

查看上述一个观测点及其感知消失位置影像的特征匹配（图3.11-17）。

```
FLANN_INDEX_KDTREE = 1
im_1,im_2 = drive_29_0071_img_fp_list[starting_idx],drive_29_0071_img_fp_
list[ending_idx]
index_params = dict(algorithm=FLANN_INDEX_KDTREE, trees=5)
feature_matching(im_1,im_2,index_params=index_params,method='FLANN',)
```

- D－比较两个街区的变化

图3.11-17 查看一个观测点及其感知消失位置影像的特征匹配

与上述方法相同，选取了自然景观偏多的另一个街区。比较计算结果如表3.11-4所示。

表3.11-4 两个街区视觉感知消失距离比较

统计量	街区_26_0009	街区_29_0071
count	983	388
mean	29.147147	64.207892
std	15.663416	32.833346
min	1.034268	1.065737
25%	18.190853	42.206708
50%	29.592061	61.216303
75%	40.271488	84.249092
max	76.957715	190.549947

从计算的结果可以观察到，不同的街区视觉感知消失的距离不同。城市内的街道小巷视觉感知消失的距离较之较为开阔自然植被相对较多的街道要小，标准差也偏小，即分布的离散程度要小。也就是说，在变化丰富的小巷中行走时，视觉感知变化会比较快，有琳琅满目的感觉；但是，在开阔区域行走时，视觉感知变化就会很慢，这里是慢了约2倍的距离。当然，对于同是小巷而言，不同的设计，这个视觉变化也会不同。例如，推断景观单一的小巷，视觉感知变化应该比较慢，注意这里并没有数据计算验证。

比较街区位置地图见图3.11-18，视觉感知消失距离的计算同前文，此处不再赘述。

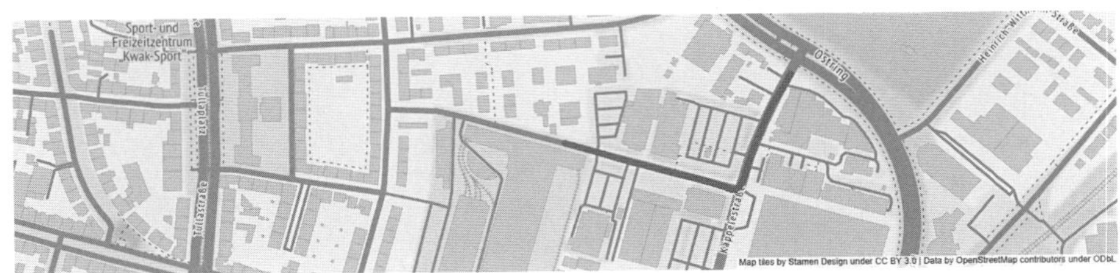

图3.11-18 比较街区位置地图

```
KITTI_info_fp = r'F:\data\2011_09_26_drive_0009_sync\oxts\data'
timestamps_fp = r'F:\data\2011_09_26_drive_0009_sync\image_03\timestamps.txt'
drive_26_0009_info = KITTI_info(KITTI_info_fp,timestamps_fp)
drive_26_0009_info.sort_values(by="timestamps_",inplace=True)
# util_misc.print_html(drive_26_0009_info)
drive_26_0009_info.sort_values(by="timestamps_",inplace=True)
plotly_scatterMapbox(drive_26_0009_info,center_lon=8.437134,center_
lat=49.009348)
```

参考文献（References）:

[1] Lowe, D. G. (2004). Distinctive Image Features from Scale-Invariant Keypoints. International Journal of Computer Vision, 60(2), 91–110. doi:10.1023/b:visi.0000029664.99615.94 10.1023/b:visi.0000029664.99615.94

3.12 Sentinel-2及超像素级分割下高空分辨率特征尺度界定

3.12.1 Sentinel-2 遥感影像

Sentinel-2[①]为高分辨率多光谱成像卫星，为2A和2B两颗卫星，分别于2015-06-23和2017-03-07日发射升空。每颗卫星重访周期为10d，两者为每5d完成一次对地球赤道地区的完整成像。卫星寿命为7.25年，其携带的多光谱器（MultiSpectral Instrument，MSI），覆盖13个光谱波段，地面分辨率分别为10m、20m和60m。数据可以从Copernicus Open Access Hub[②]处下载。具体的波段解释与Landsat-8比较，如表3.12–1所示。

表3.12–1 Sentinel–2与 Landsat–8影像波段比较

Landset-8			Sentinel-2		
Band（波段）	Wavelenght（nm）(波长)	Resolution（m）（分辨率）	Band（波段）	Wavelength（nm）(波长)	Resolution（m）（分辨率）
1（Coastal）	430-450	30	1（Coastal）	433-453	60
2（Blue）	450-515	30	2（Blue）	458-523	10
3（Green）	525-600	30	3（Green）	543-578	10
4（Red）	630-680	30	4（Red）	650-680	10
			5（red Edge）	698-713	20
			6（red Edge）	733-748	20
			7（red Edge）	773-793	20
5（NIR）	845-885	30	8（NIR）	785-900	10
			9（Water vapor）	935-955	60
			10（SWIR-Cirrus）	1360-1390	60
6（SWIR-1）	1560-1660	30	11（SWIR-1）	1565-1655	20
7（SWIR-2）	2100-2300	30	12（SWIR-2）	2100-2280	20
8（PAN）	503-676	15			

Sentinel-2与Landset-8最大的区别除了各个波段的分辨率不同外，还有在近红外波段NIR与红色波段之间细分了Red Edge红边波段，这对检测植被健康信息非常有效。

Sentinel-2产品级别可以划分为：

- Level-0，原始数据；
- Level-1A，包含元信息的几何粗矫正产品；
- Level-1B，辐射率产品，嵌入经GCP优化的几何模型，但未进行相应的几何校正；
- Level-1C，经正射校正和亚像元级几何精校正后的大气表观反射率产品；
- Level-2A，由Level-1C产品经过大气校正的大气底层反射率数据（Bottom Of Atmosphere（BOA）reflectance images derived from the associated Level-1C products）。

在由Level-1C生成Level-2A（即经辐射定标和大气校正），可以使用欧空局（European Space Agency，ESA）发布的Sen2Cor[③]工具，在Windows下的Command Prompt（CMD）终端下安装（执行`L2A_Process--help`命令），并执行`L2A_Process`+数据位置+参数（可选）。ESA发布的产品中混合有标识为“MSIL1C”的Level-1C产品，标识为“MSIL2A”的Level-2A产品，需要注意区分，Sen2Cor工具只对Level-1C产品执行大气校正产生Level-2A产品。

rio_tiler 库[④]是 rasterio[⑤]的插件（plugin），用于从栅格数据集读取网页地图瓦片（web map tiles）。

3.12.1.1 以Web Mercator方式显示Sentinel-2的一个波段

```
import rio_tiler
help(rio_tiler)
-----------------------------------------------------------------------------
Help on package rio_tiler:
NAME
```

① Sentinel-2 遥感影像，Sentinel-2由两颗极地轨道卫星组成，被置于同一个太阳同步轨道上，彼此相距180°。它的目的是监测陆地表面状况的变化，其宽幅扫描（290公里）和高重访时间（在赤道上用一颗卫星10天，在无云条件下用两颗卫星5天，在中纬度地区为2-3天）将支持监测地球表面的变化（https://sentinels.copernicus.eu/web/sentinel/missions/sentinel-2）。

② Copernicus Open Access Hub，提供完整、免费和开放的Sentinel-1、Sentinel-2、Sentinel-3和Sentinel-5P的用户产品（https://scihub.copernicus.eu/dhus/#/home）。

③ Sen2Cor，是一个用于Sentinel-2 2A级产品生成和格式化的处理器，对顶部大气层1C级输入数据进行大气、地形和卷云校正。Sen2Cor创建大气底层、可选择的地形和卷云校正的反射率图像，及气溶胶光学厚度、水汽、场景分类图以及云和雪概率的质量指标。其输出产品格式等同于1C级用户产品：PEG 2000图像，三种不同的分辨率：60m、20m和10m（http://step.esa.int/main/snap-supported-plugins/sen2cor/）。

④ rio_tiler库，最初被设计用来从大型栅格数据源中创建滑动地图瓦片（slippy map tiles），并在网络地图上动态渲染这些瓦片。从rio-tiler v2.0开始，则增加了许多辅助方法，从Rasterio/GDAL支持的任何栅格源中读取数据和元数据。这包括通过HTTP、AWS S3、谷歌云存储的本地和远程文件等（https://cogeotiff.github.io/rio-tiler/）。

⑤ rasterio，地理信息系统（Geographic information systems，GIS）使用GeoTIFF和其他格式组织和存储栅格数据（raster datasets），如卫星影像和地形模型等。rasterio读写这些文件格式，并提供基于NumPy N-维数组和GeoJSON的Python API（https://rasterio.readthedocs.io/en/latest/）。

```
    rio_tiler - rio-tiler.
PACKAGE CONTENTS
    cmap_data (package)
    colormap
    constants
    errors
    expression
    io (package)
    logger
    models
    mosaic (package)
    profiles
    reader
    tasks
    types
    utils
VERSION
    3.1.6
FILE
    c:\users\richi\anaconda3\envs\usda\lib\site-packages\rio_tiler\__init__.py
```

Web墨卡托投影（Web Mercator）是墨卡托投影的一种变体，是Web地图应用的事实标准。自2005年Google地图采用该投影之后，几乎所有的在线地图提供商都使用这一标准，包括Google map、Mapbox、Bing map、OpenStreetMap、MapQuest和Esri等[①]。其正式的EPSG标识符是EPSG:3857。

① Google map（https://www.google.com/maps；Mapbox（https://www.mapbox.com/）；Bing map（https://www.bing.com/maps）；OpenStreetMap（https://www.openstreetmap.org；MapQuest（https://www.mapquest.com/）和Esri（https://www.esri.com/en-us/home）。

几个世纪以来，人们一直在使用坐标系统和地图投影将地球的形状转换成可用的平面地图。而世界地图很大，不能直接在电脑上显示，所以引出快速浏览和缩放地图的机制，地图瓦片（map tiles）。将世界划分为很多小方块，每个小方块都有固定的地理面积和规模。这样，可以在不加载整个地图的情况下浏览其中的一小部分。这涉及几种表示方法：大地坐标、投影系统、像素坐标和金子塔瓦片，以及他们之间的相互转换。

1. 度（Degrees），用于Geodetic coordinates WGS84（EPSG:4326）坐标系统。使用1984年定义的世界大地测量系统（World Geodetic System），GPS设备用于定义地球位置的经纬度坐标。
2. 米（Meters），用于Projected coordinates Spherical Mercator（EPSG:3857）坐标系统。全球投影坐标（Global projected coordinates ），用于GIS，A Web Map Tile Service（WM（T）S）服务的栅格瓦片（raster tile）生成。
3. 像素（Pixels），用于Screen coordinates XY pixels at zoom坐标系统。影像金子塔每一层（each level of the pyramid）的特定缩放像素坐标。顶级（zoom=0）通常有256×256像素，下一等级为512×512。带有屏幕的设备（电脑、手机）等在定义的缩放级别计算像素坐标，并确定应该从服务器加载的区域用于可视屏幕。
4. 瓦片（Tiles），用于Tile coordinates Tile Map Service (ZXY)坐标系统。影像金子塔中指定缩放级别下（zoom level）瓦片的索引，即x轴和y轴的位置/索引。每一级别下所有瓦片都有相同的尺寸，通常为256×256像素。就是由粗到细不同分辨率的影像集合。其底部为图像的高分辨率表示，为原始图像，瓦片数应与原始图像的大小同；顶部为低分辨率的近似影像，最顶层只有1个瓦片，而后为4、16等。

球面墨卡托投影金字塔的分辨率和比例列于表3.12-2。

表3.12-2 球面墨卡托投影金字塔的分辨率和比例列表

Zoom level（缩放级别）	**Resolution（meters/pixel）**（分辨率）	**Map Scale（at 96 dpi）**（地图比例尺）	**Width and Height of map（pixels）**（地图的宽度和高度）
0	156,543.03	1：591,658,710.90	256
1	78,271.52	1：295,829,355.45	512
2	39,135.76	1：147,914,677.73	1,024
3	19,567.88	1：73,957,338.86	2,048
4	9,783.94	1：36,978,669.43	4,096
5	4,891.97	1：18,489,334.72	8,192
6	2,445.98	1：9,244,667.36	16,384
7	1,222.99	1：4,622,333.68	32,768

续表

Zoom level（缩放级别）	Resolution（meters/pixel）（分辨率）	Map Scale（at 96 dpi）（地图比例尺）	Width and Height of map（pixels）（地图的宽度和高度）
8	611.4962263	1：2,311,166.84	65,536
9	305.7481131	1：1,155,583.42	131,072
10	152.8740566	1：577,791.71	262,144
11	76.43702829	1：288,895.85	524,288
12	38.21851414	1：144,447.93	1,048,576
13	19.10925707	1：72,223.96	2,097,152
14	9.554728536	1：36,111.98	4,194,304
15	4.777314268	1：18,055.99	8,388,608
16	2.388657133	1：9,028.00	16,777,216
17	1.194328566	1：4,514.00	33,554,432
18	0.597164263	1：2,257.00	67,108,864
19	0.298582142	1：1,128.50	134,217,728
20	0.149291071	10：24.2	268,435,456
21	0.074645535	05：42.1	536,870,912
22	0.037322768	03：21.1	1,073,741,824
23	0.018661384	02：10.5	2,147,483,648

对于金字塔瓦片和坐标之间的转换，可以查看Tiles à la Google Maps[①]，其中给出了转换的源代码。下述迁移的函数deg2num()可以将经纬度坐标转换为指定zoom level缩放级别下，金子塔中瓦片的坐标。其输入参数为经纬度坐标值和缩放级别。在使用rio_tiler库COGReader方法读取栅格文件后，可以通过src.dataset、src.tms.identifier、src.minzoom、src.maxzoom、src.bounds、src.crs、src.geographic_bounds和src.colormap等方式提取读取的栅格的属性值信息。

① Tiles à la Google Maps，坐标、瓦片界限和投影（Coordinates，Tile Bounds and Projection）。了解可缩放地图如何工作，什么是坐标系，以及如何在它们之间转换（https://www.maptiler.com/google-maps-coordinates-tile-bounds-projection）。

```
def deg2num(lat_deg, lon_deg, zoom):
    '''
    code migration
    function - 将经纬度坐标转换为指定 zoom level 缩放级别下，金子塔中瓦片的坐标。

    Params:
        lat_deg - 纬度；float
        lon_deg - 经度；float
        zoom - 缩放级别；int

    Returns:
        xtile - 金子塔瓦片 x 坐标；int
        ytile - 金子塔瓦片 y 坐标；int
    '''
    import math
    lat_rad = math.radians(lat_deg)
    n = 2.0 ** zoom
    xtile = int((lon_deg + 180.0) / 360.0 * n)
    ytile = int((1.0 - math.log(math.tan(lat_rad) +
                (1 / math.cos(lat_rad))) / math.pi) / 2.0 * n)
    return (xtile, ytile)
def centroid(bounds):
    '''
    code migration
    function - 根据获取的地图边界坐标 [ 左下角经度，左下角纬度，右上角经度，右上角维度 ] 计算中
心点坐标

    Params:
        bounds - [ 左下角经度，左下角纬度，右上角经度，右上角维度 ]；numerical

    Returns:
        lat - 边界中心点维度；float
        lng - 边界中心点经度；float
    '''
    lat = (bounds[1] + bounds[3]) / 2
```

```
    lng = (bounds[0] + bounds[2]) / 2
    return lat, lng
```

用src.bounds提取栅格边界（左下角经度，左下角纬度，右上角经度，右上角维度）时，为投影坐标；而用src.geographic_bounds提取的边界为经纬度坐标。下述从COG中用src.tile()方法读取网络地图瓦片（Web Map tile），需要输入的参数tile_x (int)，tile_y (int)和tile_z (int)为瓦片的水平、垂直和缩放级别索引，tilesize (int, optional)为输出图像的大小。打印Sentinel-2的一个波段瓦片影像如图3.12-1所示。

```
from rio_tiler.io import COGReader
from skimage import exposure
import numpy as np
scene_id = r'G:\data\S2B_MSIL2A_20200709T163839_N0214_R126_T16TDM_20200709T211044.SAFE\GRANULE\L2A_T16TDM_A017455_20200709T164859\IMG_DATA\R10m\T16TDM_20200709T163839_B04_10m.jp2'
z = 9  #需要调整不同的缩放级别，查看显示结果。如果缩放级别过大，影像则会模糊，无法查看细节；如果缩放比例来过大，数据量增加，则会增加计算时长
with COGReader(scene_id) as src:
    # print(help(image))
    print('影像边界坐标：', src.geographic_bounds)
    #指定缩放级别，转换影像中心点的经纬度坐标为金子塔瓦片坐标
    x, y = deg2num(*centroid(src.geographic_bounds), z)
    print("影像中心点瓦片索引：", x_, y_)
    img = src.tile(x, y, z, tilesize=512)  #tilesize参数为瓦片大小，默认值为256
tile = img.data
#将颜色维度移动到最后一个轴
tile = np.transpose(tile, (1, 2, 0))
#重新调整强度，适合视觉观察
low, high = np.percentile(tile, (1, 97))
tile = exposure.rescale_intensity(tile, in_range=(low, high)) / 65535
```

图3.12-1　Sentinel-2的一个波段瓦片影像

```
#该瓦片仅显示整个图像的一个子集
print("瓦片的形状：", tile.shape)
--------------------------------------------------------------------------------
影像边界坐标：(-88.21648331497754, 41.45751520585338, -86.88130582637692,
42.4526911672117)
影像中心点瓦片索引：647168 148
瓦片的形状：(512, 512, 1)
import matplotlib.pyplot as plt
plt.figure(figsize=(10,10))
plt.imshow(tile)
plt.axis("off")
plt.show()
```

3.12.1.2 Sentinel-2波段合成显示

波段合成显示的波段组合与Landsat部分阐述基本相同，例如“4_Red、3_Green 、2_Blue”波段组合为自然真彩色，如图3.12-2所示。

```
import os
def Sentinel2_bandsComposite_show(RGB_bands, zoom=10, tilesize=512, figsize=(10,
10)):
    '''
    function - Sentinel-2 波段合成显示。需要 deg2num(lat_deg, lon_deg, zoom) 和
centroid(bounds) 函数

    Params:
        RGB_bands - 波段文件路径名字典，例如 {"R":path_R,"G":path_G,"B":path_B}; dict
        zoom - zoom level 缩放级别。The defalut is 10; int
        tilesize - 瓦片大小。The default is 512; int
        figsize- 打印图表大小。The default is (10,10); tuple

    Returns:
        None
    '''
```

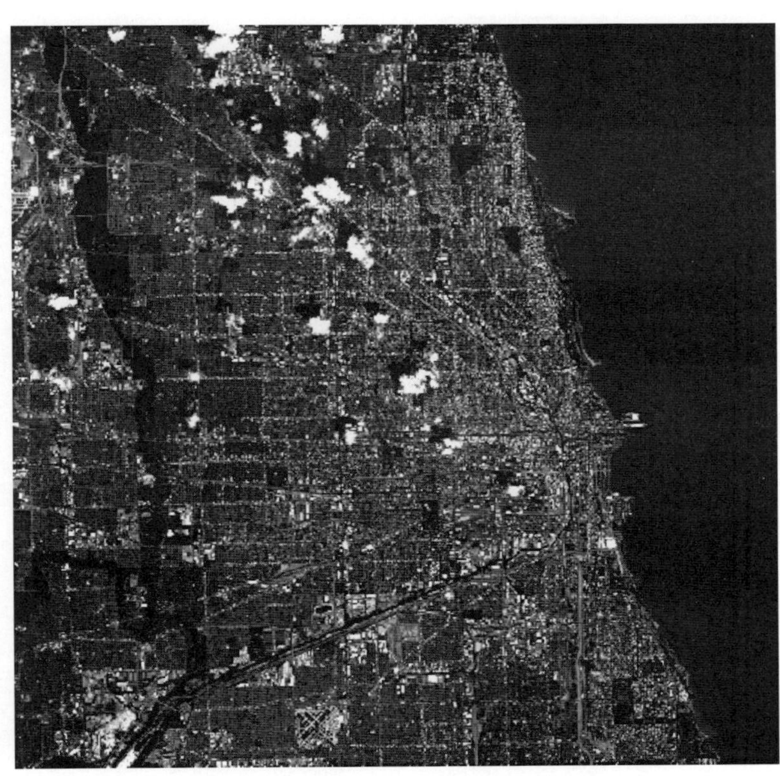

图3.12-2 Sentinel-2自然真彩色

```
    %matplotlib inline
    import matplotlib.pyplot as plt
    import math
    import os
    import numpy as np
    from rio_tiler.io import COGReader
    from skimage import exposure
    from rasterio.plot import show
    B_band = RGB_bands["B"]
    G_band = RGB_bands["G"]
    R_band = RGB_bands["R"]
    def band_reader(band):
        with COGReader(band) as image:
            bounds = image.geographic_bounds
            print('影像边界坐标：', bounds)
            x, y = deg2num(*centroid(bounds), zoom)
            print("影像中心点瓦片索引：", x, y)
            img = image.tile(x, y, zoom, tilesize=tilesize)
            return img.data
    tile_RGB_list = [np.squeeze(band_reader(band))
                     for band in RGB_bands.values()]
    tile_RGB_array = np.array(tile_RGB_list).transpose(1, 2, 0)
    p2, p98 = np.percentile(tile_RGB_array, (2, 98))
    image = exposure.rescale_intensity(
        tile_RGB_array, in_range=(p2, p98)) / 65535
    plt.figure(figsize=(10, 10))
    plt.imshow(image)
    plt.axis("off")
    plt.show()
sentinel2_root = r"G:\data\S2B_MSIL2A_20200709T163839_N0214_R126_T16TDM_20200709T211044.SAFE\GRANULE\L2A_T16TDM_A017455_20200709T164859\IMG_DATA\R10m"
RGB_bands = {
    "R": os.path.join(sentinel2_root, 'T16TDM_20200709T163839_B04_10m.jp2'),
    "G": os.path.join(sentinel2_root, 'T16TDM_20200709T163839_B03_10m.jp2'),
    "B": os.path.join(sentinel2_root, 'T16TDM_20200709T163839_B02_10m.jp2'), }
Sentinel2_bandsComposite_show(RGB_bands)
--------------------------------------------------------------------------------
影像边界坐标：(-88.21648331497754, 41.45751520585338, -86.88130582637692, 42.4526911672117)
影像中心点瓦片索引：262 380
```

3.12.2 超像素级分割下高空分辨率特征尺度界定

在景观生态学中，斑块—廊道—基质模型是构成景观空间结构，描述景观空间异质性的一个基本模式。其中，斑块是景观格局中的基本组成单元，是指不同于周围背景，相对均质的非线性区域。自然界各种等级系统都普遍存在时间和空间的斑块化，反映系统内部和系统间的相似性或相异性。不同斑块的大小、形状、边界性质及斑块的距离等空间分布特征构成了不同的生态带，形成了生态系统的差异，调节生态过程。廊道是不同于景观基质的现状或带状的景观要素，例如河流廊道、生态廊道等。其中，生态廊道又称野生动物生态廊道或绿色廊道，是指用于连接因人类活动或构筑物而被隔开的野生动物种群生境的区域。生态廊道有利于野生动物的迁移扩散，提高生境间的连接，促进濒危物种不同群间的基因交流，降低种群灭绝风险。基质则是景观中面积最大，连接性最好的景观要素类型。斑廊基景观空间结构的提出为城市格局规划提供了依据，在宏观尺度上给出了保护自然生物的空间形式。那么，对于一个区域，如何自然界定斑块、廊道和基质的区域？或者，即使是一个可以肉眼辨识的斑块，这个斑块自身也是呈现变化的，可以表现在地物的变化，例如可见的不同林地、不同物种的农田等。或者，不可见的地表温度变化、物质流动等。那么，又如何细分斑块的空间区域，挖掘斑块变化区域的流动方向或子区域？

一方面，需要能够反映地物变化的信息数据，例如遥感影像的各个波段对不同地物的探测，Sentinel-2影像中新增加的5、6、7波段（red edge），可以有效地监测植被健康信息；或衍生数据，如反演的地表温度，以及NDVI等反映植被分布的指数，NDWI反映水体分布的指数，NDBI反映建城区分布的指数等。另一方面，在分析这些数据时，可以介入超像素级分割的概念，探索由像素（或空间点数据）局部分组形成的区域。这类似于聚类的方法，将具有同一或近似属性的区域优先聚集即分割，分割区域的变化根据所提供反应不同内容的数据确定，例如探索植被分布的

NDVI则优先聚集植被指数临近的区域。也可以组合波段，例如red、green和blue波段组合更倾向于优先聚集同一地物，例如建筑区域、林地区域等。

超像素级分割是一种语义分割，是计算机视觉的基本方法，可以更加精准地执行地物分割、探测和分类等深度学习任务。这一方法也同样为景观、生态专业探索地物变化和地物之间的关系提供一种新的策略。scikit-image[①]提供了四种分割的算法。在下述实验中，计算了Felzenszwalb和Quickshift两种方法。Felzenszwalb方法在分割图像时，虽然逐级增加scale参数大小，但是分割的图像并不是上级区域覆盖下级区域；而Quickshift方法，则基本是逐级覆盖的，因此选用Quickshift方法，指定逐级增加的kernel_size参数，获取不同深度分割的结果。通过计算逐级覆盖的分割类别频数统计及方差等指数，试图找到研究区域不同深度分割下区域间的关联，以及区域的差异性程度。

① scikit-image，分割和超级像素算法的比较（Comparison of segmentation and superpixel algorithms）（https://scikit-image.org/docs/dev/auto_examples/segmentation/plot_segmentations.html#sphx-glr-auto-examples-segmentation-plot-segmentations-py）。

- 01 - 读取所下载的Sentinel-2影像的元文件，获取影像波段路径。

Sentinel-2影像的信息均记录于下载文件夹下的“MTD_MSIL2A.xml”中，因此可以从该文件获取各个波段的路径。该文件给出的路径为相对于影像文件夹的相对路径。

```
def Sentinel2_bandFNs(MTD_MSIL2A_fn):
    '''
    funciton - 获取 sentinel-2 波段文件路径 ， 和打印主要信息

    Params:
        MTD_MSIL2A_fn - MTD_MSIL2A 文件路径 ; string

    Returns:
        band_fns_list - 波段相对路径列表 ; list(string)
        band_fns_dict - 波段路径为值 , 反应波段信息的字段为键的字典 ; dict
    '''
    import xml.etree.ElementTree as ET
    Sentinel2_tree = ET.parse(MTD_MSIL2A_fn)
    Sentinel2_root = Sentinel2_tree.getroot()
    print("GENERATION_TIME:{}\nPRODUCT_TYPE:{}\nPROCESSING_LEVEL:{}".
format(Sentinel2_root[0][0].find('GENERATION_TIME').text,

Sentinel2_root[0][0].find(

'PRODUCT_TYPE').text,

Sentinel2_root[0][0].find(

'PROCESSING_LEVEL').text
    ))
    print("MTD_MSIL2A.xml 文件父结构 :")
    for child in Sentinel2_root:
        print(child.tag, "-", child.attrib)
    print("_"*50)
    # [elem.text for elem in Sentinel2_root[0][0][11][0][0].iter()]
    band_fns_list = [elem.text for elem in Sentinel2_root.iter('IMAGE_FILE')]
    band_fns_dict = {f.split('_')[-2]+'_'+f.split('_')
                     [-1]: f+'.jp2' for f in band_fns_list}
    print('获取 sentinel-2 波段文件路径 :\n', band_fns_dict)

    return band_fns_list, band_fns_dict
MTD_MSIL2A_fn = r'G:\data\S2B_MSIL2A_20200709T163839_N0214_R126_
T16TDM_20200709T211044.SAFE\MTD_MSIL2A.xml'
band_fns_list, band_fns_dict = Sentinel2_bandFNs(MTD_MSIL2A_fn)
--------------------------------------------------------------------------------
GENERATION_TIME:2020-07-09T21:10:44.000000Z
PRODUCT_TYPE:S2MSI2A
PROCESSING_LEVEL:Level-2A
MTD_MSIL2A.xml 文件父结构 :
{https://psd-14.sentinel2.eo.esa.int/PSD/User_Product_Level-2A.xsd}General_Info
- {}
{https://psd-14.sentinel2.eo.esa.int/PSD/User_Product_Level-2A.xsd}Geometric_
Info - {}
{https://psd-14.sentinel2.eo.esa.int/PSD/User_Product_Level-2A.xsd}Auxiliary_
Data_Info - {}
{https://psd-14.sentinel2.eo.esa.int/PSD/User_Product_Level-2A.xsd}Quality_
Indicators_Info - {}
__________________________________________________
获取 sentinel-2 波段文件路径 :
```

```
{'B02_10m': 'GRANULE/L2A_T16TDM_A017455_20200709T164859/IMG_DATA/R10m/
T16TDM_20200709T163839_B02_10m.jp2', 'B03_10m': 'GRANULE/L2A_T16TDM_
A017455_20200709T164859/IMG_DATA/R10m/T16TDM_20200709T163839_B03_10m.jp2',
...'SCL_60m': 'GRANULE/L2A_T16TDM_A017455_20200709T164859/IMG_DATA/R60m/
T16TDM_20200709T163839_SCL_60m.jp2'}
```

- 02 - 裁切到研究区域。裁切边界由QGIS绘制。

```
import util_A
import os
raster_fp = [os.path.join(r"G:\data\S2B_MSIL2A_20200709T163839_N0214_R126_
T16TDM_20200709T211044.SAFE",
                          band_fns_dict[k]) for k in band_fns_dict.keys() if
k.split("_")[-1] == "20m"]
clip_boundary_fp = r'.\data\superPixel_boundary\superPixel_boundary.shp'
save_path = r'G:\data_processed\RSi\crop_20'
util_A.raster_clip(raster_fp, clip_boundary_fp, save_path)
--------------------------------------------------------------------------------
finished clipping.
['G:\\data\\RSi\\crop_20\\T16TDM_20200709T163839_B02_20m_crop.jp2',
...
 'G:\\data\\RSi\\crop_20\\T16TDM_20200709T163839_SCL_20m_crop.jp2']
```

- 03 - 读取裁切后的影像，显示查看（图3.12–3）。实验中仅分析了red、blue和green波段组合。可以再深入分析red edge波段对植物的分割，以及计算NDVI、NDWI、NDBI，或反演地表温度来进一步研究不同信息数据的分割结果。

```
import glob
import earthpy.spatial as es
import earthpy.plot as ep
import matplotlib.pyplot as plt
save_path = r'G:\data_processed\RSi\crop_20'
croppedImgs_fns = glob.glob(save_path+"/*.jp2")
croppedBands_fnsDict = {
    f.split('_')[-3]+'_'+f.split('_')[-2]: f for f in croppedImgs_fns}
bands_selection_ = ['B02_20m', 'B03_20m', 'B04_20m', 'B05_20m', 'B06_20m', 'B07_20m',
                    'B8A_20m', 'B11_20m', 'B12_20m']  # 'TCI_20m', 'AOT_20m', 'WVP_20m',
'SCL_20m'
cropped_stack_bands = [croppedBands_fnsDict[b] for b in bands_selection_]
cropped_array_stack, _ = es.stack(cropped_stack_bands)
ep.plot_bands(cropped_array_stack, title=bands_selection_,
              cols=cropped_array_stack.shape[0], cbar=True, figsize=(20, 10))
plt.show()
```

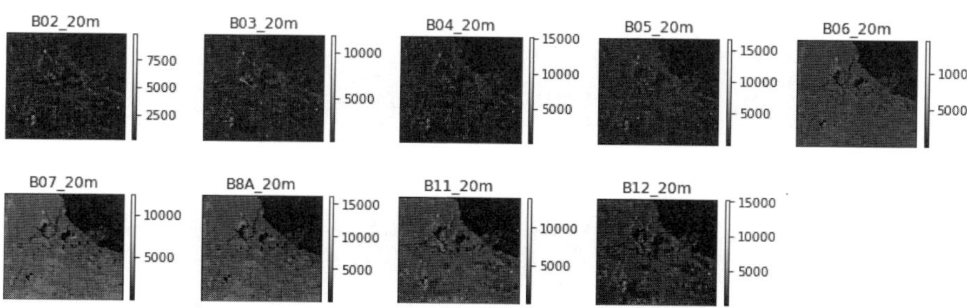

图3.12–3　查看影像波段

- 04 - Felzenszwalb超像素级分割方法，结果如图3.12–4所示。

```
import matplotlib.pyplot as plt
import numpy as np
from skimage.color import rgb2gray
from skimage.filters import sobel
from skimage.segmentation import felzenszwalb, slic, quickshift, watershed
from skimage.segmentation import mark_boundaries
from skimage.util import img_as_float
import pickle
img = cropped_array_stack[[2, 1, 0]]
```

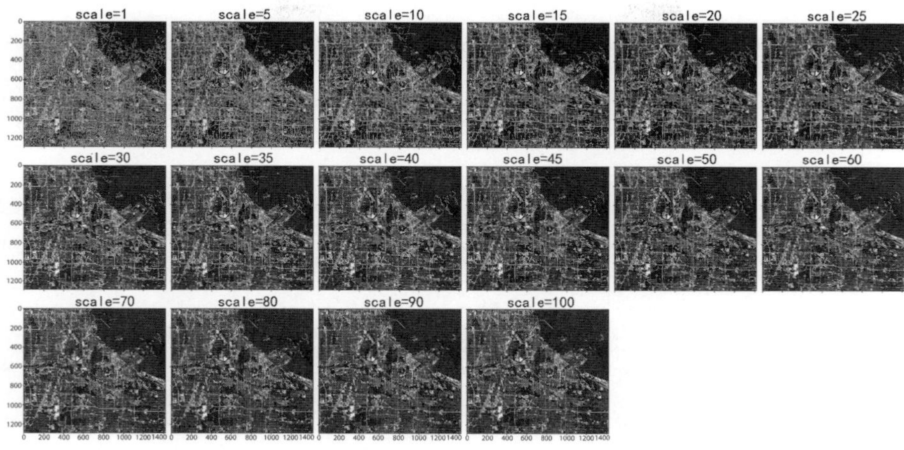

图3.12-4 给定多个scale参数值的felzenszwalb方法分割结果

```
img = img.transpose(1, 2, 0)
def superpixel_segmentation_Felzenszwalb(img, scale_list, sigma=0.5, min_
size=50):
    '''
    function - 超像素分割，skimage 库 felzenszwalb 方法。给定 scale 参数列表，批量计算

    Params:
        img - 读取的遥感影像、图像；ndarray
        scale_list - 分割比例列表；list(float)
        sigma - 用于预处理高斯核的宽度（标准差）. The default is 0.5; float
        min_size - 最小组成大小。使用后处理强制执行 . The default is 50; int

    Returns:
        分割结果。表示分割标签的整数掩码；ndarray
    '''
    import numpy as np
    from skimage.segmentation import felzenszwalb
    # conda install -c conda-forge tqdm ;conda install -c conda-forge ipywidgets
    from tqdm import tqdm
    segments = [felzenszwalb(img, scale=s, sigma=sigma,
                             min_size=min_size) for s in tqdm(scale_list)]
    return np.stack(segments)
scale_list = [1, 5, 10, 15, 20, 25, 30, 35, 40, 45, 50, 60, 70, 80, 90, 100]
segs = superpixel_segmentation_Felzenszwalb(img, scale_list)
with open('./data_processed/segs_superpixel.pkl', 'wb') as f:
    pickle.dump(segs, f)
--------------------------------------------------------------------------------
100%|████████████████| 16/16 [00:55<00:00,  3.49s/it]
```

- 05 - 显示分割图像，分割边界。

```
import math
from skimage import exposure
scale_list = [1, 5, 10, 15, 20, 25, 30, 35, 40, 45, 50, 60, 70, 80, 90, 100]
with open('./data_processed/segs_superpixel.pkl', 'rb') as f:
    segs = pickle.load(f)
p2, p98 = np.percentile(img, (2, 98))
img_ = exposure.rescale_intensity(img, in_range=(p2, p98)) / 65535
def markBoundaries_layoutShow(segs_array, img, columns, titles, prefix,
figsize=(15, 10)):
    '''
    function - 给定包含多个图像分割的一个数组，排布显示分割图像边界。

    Paras:
        segs_array - 多个图像分割数组；ndarray
        img - 底图；ndarray
        columns - 列数；int
```

```
        titles - 子图标题；string
        figsize - 图表大小。The default is (15,10); tuple

    Returns:
        None
    '''
    import math
    import os
    import matplotlib.pyplot as plt
    from PIL import Image
    from skimage.segmentation import mark_boundaries
    rows = math.ceil(segs_array.shape[0]/columns)
    fig, axes = plt.subplots(rows, columns, sharex=True,
                             sharey=True, figsize=figsize)   #布局多个子图，每个子图
显示一幅图像
    ax = axes.flatten()  #降至1维，便于循环操作子图
    for i in range(segs_array.shape[0]):
        ax[i].imshow(mark_boundaries(img, segs_array[i]))  #显示图像
        ax[i].set_title("{}={}".format(prefix, titles[i]))
    invisible_num = rows*columns-len(segs_array)
    if invisible_num > 0:
        for i in range(invisible_num):
            ax.flat[-(i+1)].set_visible(False)
    fig.tight_layout()  #自动调整子图参数，使之填充整个图像区域
    fig.suptitle("segs show", fontsize=14, fontweight='bold', y=1.02)
    plt.show()
columns = 6
markBoundaries_layoutShow(
    segs, img_, columns, scale_list, 'scale', figsize=(30, 20))
```

- 06 - Quickshift 超像素级分割方法及显示，结果如图3.12-5所示。

```
def superpixel_segmentation_quickshift(img, kernel_sizes, ratio=0.5):
    '''
    function - 超像素分割，skimage 库 quickshift 方法。给定 kernel_size 参数列表，批量计算

    Params:
        img - 输入图像。对应颜色通道的轴可以通过 channel_axis 参数指定；ndarray
        kernel_sizes - 高斯核的大小用于平滑样本密度。高值意味着得到更少的聚类簇；float,
optional
        ratio - 平衡色彩空间的邻近性和图像空间的邻近性。数值越高，色彩空间的权重越大 . The
default is 0.5; float, optional, between 0 and 1

    Returns:
        表示分割标签的整数掩码 .
    '''
    import numpy as np
    from skimage.segmentation import quickshift
    # conda install -c conda-forge tqdm ;conda install -c conda-forge ipywidgets
    from tqdm import tqdm
    segments = [quickshift(img, kernel_size=k, ratio=ratio)
```

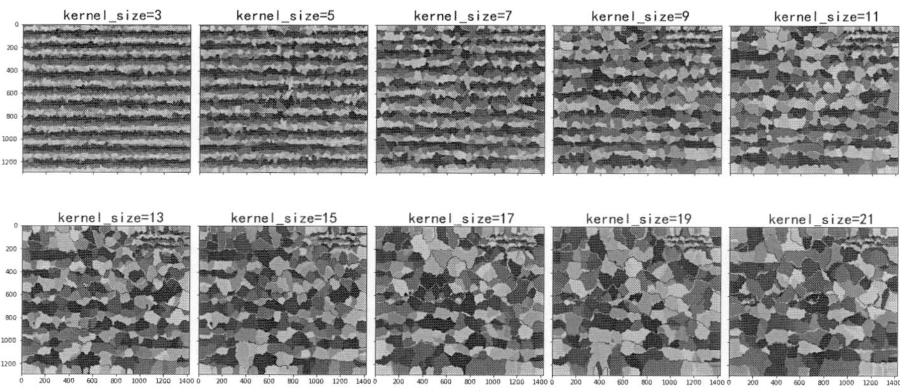

图3.12-5 Quickshift超像素级分割结果

```
                for k in tqdm(kernel_sizes)]
    return np.stack(segments)
kernel_sizes = [3, 5, 7, 9, 11, 13, 15, 17, 19, 21]
segs_quickshift = superpixel_segmentation_quickshift(img, kernel_sizes)
with open('./data_processed/segs_superpixel_quickshift.pkl', 'wb') as f:
    pickle.dump(segs_quickshift, f)
```

```
100%|██████████| 10/10 [32:29<00:00, 194.94s/it]
```

```
def segMasks_layoutShow(segs_array, columns, titles, prefix, cmap='prism',
figsize=(20, 10)):
    '''
    function - 给定包含多个图像分割的一个数组，排布显示分割图像掩码。

    Paras:
        segs_array - 多个图像分割数组；ndarray
        columns - 列数；int
        titles - 子图标题；string
        figsize - 图表大小。The default is (20,10)；tuple(int)
    '''
    import math
    import os
    import matplotlib.pyplot as plt
    from PIL import Image
    rows = math.ceil(segs_array.shape[0]/columns)
    fig, axes = plt.subplots(rows, columns, sharex=True,
                             sharey=True, figsize=figsize)   #布局多个子图，每个子图
显示一幅图像
    ax = axes.flatten()  #降至1维，便于循环操作子图
    for i in range(segs_array.shape[0]):
        ax[i].imshow(segs_array[i], cmap=cmap)  #显示图像
        ax[i].set_title("{}={}".format(prefix, titles[i]))
    invisible_num = rows*columns-len(segs_array)
    if invisible_num > 0:
        for i in range(invisible_num):
            ax.flat[-(i+1)].set_visible(False)
    fig.tight_layout()  #自动调整子图参数，使其填充整个图像区域
    fig.suptitle("segs show", fontsize=14, fontweight='bold', y=1.02)
    plt.show()
columns = 5
segMasks_layoutShow(segs_quickshift, columns, kernel_sizes, 'kernel_size')
```

- 07 - 多尺度超像素级分割结果叠合频数统计，包括各个层级与其之后所有层级间的计算。

```
def multiSegs_stackStatistics(segs, save_fp):
    '''
    function - 多尺度超像素级分割结果叠合频数统计

    Params:
        segs - 超级像素分割结果。表示分割标签的整数掩码；ndarray(int)
        save_fp - 保存路径名（pickle）；string

    Returns:
        stack_statistics - 统计结果字典；dict
    '''
    from scipy.ndimage import label
    from tqdm import tqdm
    segs = list(reversed(segs))
    stack_statistics = {}
    for i in tqdm(range(len(segs)-1)):
        labels = np.unique(segs[i])
        coords = [np.column_stack(np.where(segs[i] == k)) for k in labels]
        i_j = {}
        for j in range(i+1, len(segs)):
            j_k = {}
            for k in range(len(coords)):
                covered_elements = [segs[j][x, y]
                                    for x, y in zip(*coords[k].T)]
                freq = list(
                    zip(np.unique(covered_elements, return_counts=True)))
                j_k[k] = freq
            i_j[j] = j_k
        stack_statistics[i] = i_j
    with open(save_fp, 'wb') as f:
```

```
        pickle.dump(stack_statistics, f)
    return stack_statistics
stack_statistics = multiSegs_stackStatistics(
    segs_quickshift, './data_processed/multiSegs_stackStatistics.pkl')
----------------------------------------------------------------
100%|██████████| 9/9 [01:09<00:00,  7.78s/it]
```

● 08 - 读取保存的分割层级叠合频数统计文件，提取卷积核即`kernel_size`最大的层级与之后所有层级的频数统计，转换为DataFrame数据格式（表3.12-3）。

```
from tqdm import tqdm
import pickle
with open('./data_processed/multiSegs_stackStatistics.pkl', 'rb') as f:
    stack_statistics = pickle.load(f)
segsOverlay_0_num = {k: [stack_statistics[0][k][i][0][0].shape[0]
                         for i in stack_statistics[0][k].keys()] for k in
tqdm(stack_statistics[0].keys())}
--------------------------------------------------------------------------------
100%|██████████| 9/9 [00:00<00:00, 3008.11it/s]
--------------------------------------------------------------------------------
import pandas as pd
segsOverlay_0_num_df = pd.DataFrame.from_dict(segsOverlay_0_num)
segsOverlay_0_num_df
```

表3.12-3 分割层级叠合频数统计

	1	**2**	**3**	**4**	**5**	**6**	**7**	**8**	**9**
0	2	3	5	3	3	5	5	7	12
1	2	2	2	2	3	5	6	7	11
2	6	8	10	12	12	16	17	20	30
3	3	3	6	6	7	7	8	7	8
4	2	4	5	7	8	12	10	11	15
...	...	...	...	...	...	...	...	...	...
434	3	3	3	4	5	4	9	15	19
435	4	2	3	2	3	3	3	6	8
436	2	2	3	1	3	4	3	6	10
437	3	3	4	5	6	9	8	9	15
438	3	4	4	4	4	5	6	8	14

```
439 rows × 9 columns
```

● 09 - 显示所有对应深度层级的频数变化（图3.12-6）。

```
import plotly.express as px
x = list(segsOverlay_0_num_df.index)
y = list(segsOverlay_0_num_df.columns)
fig = px.scatter(segsOverlay_0_num_df, x=x, y=y,
                 # hover_data=[],
                 title='id_info_df'
                )
fig.show()
```

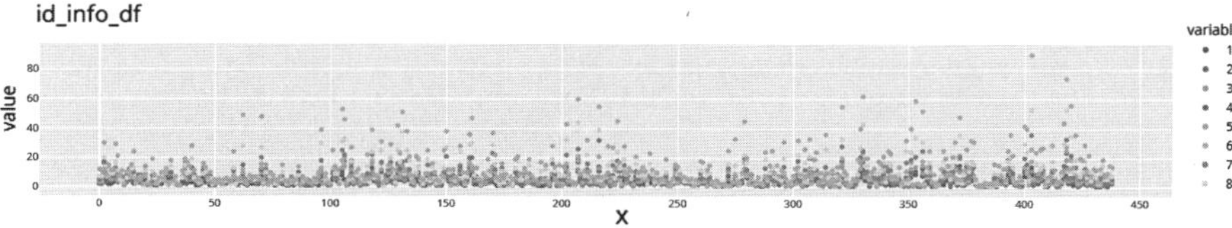

图3.12-6 所有对应深度层级的频数变化

● 10 - 计算对应所有层级，父级分割（kernel_size最大的层级）每一分割类，在各个深度层级上对应覆盖分割类数量的方差统计。可以分析父级每一个分割区域内深度层级下破碎（父级分割区域内子层分割的种类数量）的变化情况。如果值越大，往往父级分割区域内的“斑块”破碎化程度比较高，即区域内具有明显的异质性（差异性）；如果方差值越小，则说明父级分割区域内“斑块”属性基本近似，区域同质性。

将计算结果叠合到分割图上，以颜色显示方差变化（图3.12-7），方便对应地理空间位置，并观察邻域间的情况。

```
from skimage import exposure
from mpl_toolkits.axes_grid1.inset_locator import inset_axes
import matplotlib.pyplot as plt
from skimage.measure import regionprops
import numpy as np
var = segsOverlay_0_num_df.var(axis=1)
var_dict = var.to_dict()
seg_old = np.copy(segs_quickshift[-1])
seg_new = np.copy(seg_old).astype(float)
for old, new in var_dict.items():
    seg_new[seg_old == old] = new
regions = regionprops(segs_quickshift[-1])
seg_centroids = {}
for props in regions:
    seg_centroids[props.label] = props.centroid
x, y = zip(*seg_centroids.values())
labels = seg_centroids.keys()
p2, p98 = np.percentile(img, (2, 98))
img_ = exposure.rescale_intensity(img, in_range=(p2, p98)) / 65535
fig, ax = plt.subplots(1, 1, frameon=False, figsize=(15, 15))
im1 = ax.imshow(mark_boundaries(img_, segs_quickshift[-1]))
im2 = ax.imshow(seg_new, cmap='terrain', alpha=.35)
for k, coordi in seg_centroids.items():
    label = ax.text(x=coordi[1], y=coordi[0], s=k,
                    ha='center', va='center', color='white')
axins = inset_axes(ax,
                   width="5%",
                   height="50%",
                   loc='lower left',
```

图3.12-7 深度层级分割方差

```
                    bbox_to_anchor=(1.05, 0., 1, 1),
                    bbox_transform=ax.transAxes,
                    borderpad=0,
                    )
fig.colorbar(im2, cax=axins)
plt.show()
```

参考文献（References）:

[1] Robert E.Ricklefs著. 孙儒泳等译. 生态学[M]. 北京：高等教育出版社，2004.7第5版.（Robert E. Ricklefs. The economy of nature[M]. New York: W. H. Freeman and Company, 2008, sixth edition.）

[2] Felzenszwalb, P. F., & Huttenlocher, D. P. (2004). Efficient Graph-Based Image Segmentation. International Journal of Computer Vision, 59(2), 167–181. doi:10.1023/b:visi.0000022288.19776.77 10.1023/b:visi.0000022288.19776.77.

3-E 机器学习试验

3.13 聚类与城市色彩

3.13.1 调研图像

3.13.1.1 用手机App记录调研路径

如果区域调研的位置精度要求不是很高，可以在手机的应用（Application，App）中搜索GPS追踪（tracker）用于调研路线的记录。不同的应用存储的文件格式可能不同，例如本例中将调研路线存储为KML格式文件。KML全称Keyhole Markup Language，是基于XML（eXtensible Markup Language，可扩展标记语言）语法标准的一种标记语言（markup language），采用标记结构，含有嵌套的元素和属性。KML通常应用于Google地球相关软件中，例如Google Earth、Google Map、Google Maps for mobile等，用于显示数据（包括点、线、面、多边形、多面体及模型等）。KML文件可以用文本编辑器打开，例如Notepad++。下述列举了GPS跟踪KML文件开头部分的内容：

```
<?xml version="1.0" encoding="UTF-8"?>
<kml xmlns="http://www.opengis.net/kml/2.2" xmlns:gx="http://www.google.com/kml/ext/2.2" xmlns:kml="http://www.opengis.net/kml/2.2" xmlns:atom="http://www.w3.org/2005/Atom">
<Document>
<name>default_20170720081441</name>
<open>1</open>
<description>线路开始时间：2017-07-20 08:14:41，结束时间：2017-07-20 20:53:03，线路长度：197801。由 GPS 工具箱导出。</description>
<Style id="yellowLineGreenPoly" >
    <LineStyle>
    <color>7f00ffff</color>
    <width>4</width>
    </LineStyle>
    <PolyStyle>
    <color>7f00ff00</color>
    </PolyStyle>
    </Style>
<Folder>
<name> 线路标记点 </name>
    <Placemark>
        <name> 线路起点 </name>
        <description><![CDATA[2017-07-20 08:14:41]]></description>
        <Point>
        <coordinates>120.132007,30.300508,9.7</coordinates>
        </Point>
        <markerStyle>-1</markerStyle>
    </Placemark>
    <Placemark>
        <name> 线路追踪路径 </name>
        <visibility>1</visibility>
        <description>GPS 工具箱导出数据 </description>
        <styleUrl>#yellowLineGreenPoly</styleUrl>
        <LineString>
        <tessellate>1</tessellate>
        <coordinates>
        120.130187,30.211812,18.3
        120.130298,30.211757,19.5
        120.130243,30.211673,20.3
        120.13012,30.211692,20.5
        120.130095,30.21169,20.5
        </coordinates>
        </LineString>
    </Placemark>
    <Placemark>
        <name></name>
        <description><![CDATA[<img src="20170720091655_30.21169-120.130095-
```

```
20.5_.jpg" width="250"/>2017-07-20 09:16:53]]></description>
        <Point>
        <coordinates>120.130095,30.21169,20.5</coordinates>
        </Point>
        <markerStyle>0</markerStyle>
    </Placemark>
```

该文件中，记录有文件名、开始与结束时间、线路长度、地标（placemark）名、描述（description）及点坐标（coordinates）。通常，有用的信息为地标点坐标，对坐标的提取在下述的代码中采用了两种方式：一种是自定义函数；再者直接使用GeoPandas库实现。自定义函数可以根据提取数据的要求，更为精准地提取，返回的数据格式也更自由。提取的数据保留了地标名与地标点坐标的对应关系。注意上述KML文件中形式为<name></name>的坐标位置，通常为在该位置拍摄有对应的照片。不同的App记录的GPS信息不同，需要根据具体情况调整代码以便提取正确的信息。表3.13–1为提取结果示例。

表3.13–1　KML信息提取结果示例

	Name	Description	geometry
0	线路起点	2017-07-20 08:14:41	POINT Z (120.13201 30.30051 9.70000)
1	线路追踪路径	GPS工具箱导出数据	LINESTRING Z (120.13019 30.21181 18.30000, 120.13030 30.21176 19.50000, 120.13024 30.21167 20.30000, 120.13012 30.21169 20.50000, 120.13009 30.21169 20.50000)
2		<img src=" 20170720091655_30.21169-120.130095-20.5_.jpg" width=" 250" />2017-07-20 09:16:53	POINT Z (120.13009 30.21169 20.50000)

```
import fiona
from fiona.drvsupport import supported_drivers
import geopandas as gpd
import util_misc
import re
import os
surveyPath_kml_fn = util_misc.filePath_extraction(
    r'./data/default_20170720081441', ["kml"])  # .kml和.jpg文件在同一文件夹下，读取.kml
文件

# A-自定义读取.kml文件坐标信息函数

def kml_coordiExtraction(kml_pathDict):
    '''
    function - 提取.kml文件中的坐标信息

    Params:
        kml_pathDict - .kml文件路径字典。文件夹名为键，值为包含该文件夹下所有文件名的列表。
使用filePath_extraction()函数提取。

    Returns:
        kml_coordi_dict - 返回坐标信息；dict
    '''
    kml_CoordiInfo = {}
    '''正则表达式函数，将字符串转换为模式对象.号匹配除换行符之外的任何字符串，但只匹配一个字母，
增加*?字符代表匹配前面表达式的0个或多个副本，并匹配尽可能少的副本'''
    pattern_coodi = re.compile('<coordinates>(.*?)</coordinates>')
    pattern_name = re.compile('<name>(.*?)</name>')
    count = 0
    kml_coordi_dict = {}
    for key in kml_pathDict.keys():
        temp_dict = {}
        for val in kml_pathDict[key]:
            f = open(os.path.join(key, val), 'r',
                     encoding='UTF-8')  # .kml文件中含有中文
            content = f.read().replace('\n', ' ')  # 移除换行，从而可以根据模式对象提取
```

```
标识符间的内容，同时忽略换行
            name_info = pattern_name.findall(content)
            coordi_info = pattern_coodi.findall(content)
            coordi_info_processing = [coordi.strip(
                ' ').split('\t\t') for coordi in coordi_info]
            #名称中包含了文件名 <name>default_20170720081441</name> 和文件夹名 <name> 线路标记点
</name>。位于文头。
            print("名称数量：%d，坐标列表数量：%d" %
                  (len(name_info), len(coordi_info_processing)))
            name_info_id = [name_info[2:][n]+'_ID_' +
                            str(n) for n in range(len(name_info[2:]))]  #名称有重
名，用 ID 标识
            name_coordi = dict(zip(name_info_id, coordi_info_processing))
            for k in name_coordi.keys():
                temp = []
                for coordi in name_coordi[k]:
                    coordi_split = coordi.split(',')
                    #提取的坐标值字符，可能不正确，不能转换为浮点数，因此通过异常处理
                    try:
                        one_coordi = [float(i) for i in coordi_split]
                        if len(one_coordi) == 3:  #可能提取的坐标值除了经纬度和高程，
会出现多余或者少于 3 的情况，判断后将其忽略
                            temp.append(one_coordi)
                    except ValueError:
                        count = +1
                temp_dict[k] = temp
            kml_coordi_dict[os.path.join(key, val)] = temp_dict
            print("kml_坐标字典键：", kml_coordi_dict.keys())
    f.close()
    return kml_coordi_dict
kml_coordi = kml_coordiExtraction(surveyPath_kml_fn)

#B-使用 Geopandas 库提取
supported_drivers['KML'] = 'rw'  # Enable fiona driver
surveyPath_kml_fn = './data/default_20170720081441/default_20170720081441.kml'
kml_coordi_ = gpd.read_file(surveyPath_kml_fn, driver='KML')
util_misc.print_html(kml_coordi_)
--------------------------------------------------------------------------------
名称数量：81，坐标列表数量：79
kml_坐标字典键：dict_keys(['./data/default_20170720081441\\default_20170720081441.
kml'])
```

提取的地标GPS坐标有可能存在异常值（离群值），这里剔除异常值的方法采用“异常值处理”部分所定义的`is_outlier()`函数处理，可以选取经度、纬度或高程作为异常值的判断。打印的图3.13-1中左图为原始值，存在异常值；右图为异常值处理后的结果。

```
import numpy as np
import matplotlib.pyplot as plt
kml_coordi_lst = []
for key_1 in kml_coordi.keys():
    for key_2 in kml_coordi[key_1]:
```

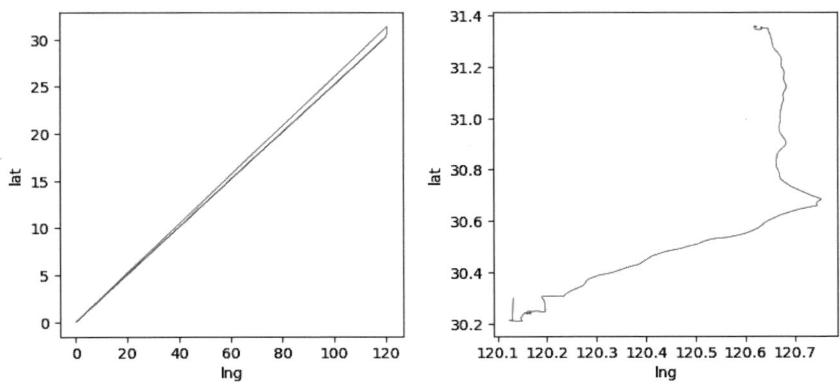

图3.13-1 存在异常值的KML路径（左）；处理异常值后的KML路径（右）

```
        for coordi in kml_coordi[key_1][key_2]:
            kml_coordi_lst.append(coordi)
kml_coordi_array = np.array(kml_coordi_lst)
x_kml = kml_coordi_array[:, 0]
is_outlier_bool, _ = util_misc.is_outlier(x_kml, threshold=3.5)
kml_coordi_clean = kml_coordi_array[~is_outlier_bool]
fig = plt.figure(figsize=(18/2, 8/2))
# 例如”111”为 1×1 的格网，第 1 个子图；”234”为 2×3 的格网，第 4 个子图。
ax_1 = fig.add_subplot(121)
ax_1.plot(kml_coordi_array[:, 0],
          kml_coordi_array[:, 1], ‘r-’, lw=0.5, markersize=5)
ax_2 = fig.add_subplot(122)
ax_2.plot(kml_coordi_clean[:, 0],
          kml_coordi_clean[:, 1], ‘r-’, lw=0.5, markersize=5)
ax_1.set_xlabel(‘lng’)
ax_1.set_ylabel(‘lat’)
ax_2.set_xlabel(‘lng’)
ax_2.set_ylabel(‘lat’)
plt.show()
```

使用Matplotlib库打印图表可以快速地查看数据信息，但是对于地理信息数据的表达通常不是很清楚。这里，仍然使用Plotly库调用地图打印查看信息（图3.13–2），色彩标识了高程数据的变化。

```
import pandas as pd
import plotly.express as px
kml_coordi_clean_df = pd.DataFrame(data=kml_coordi_clean, columns=[
                                   “lon”, “lat”, “elevation”])
mapbox_token = ‘pk.eyJ...An0A’
px.set_mapbox_access_token(mapbox_token)
fig = px.scatter_mapbox(kml_coordi_clean_df, lat=kml_coordi_clean_df.lat,
```

图3.13–2　打印KML高程信息和地图

```
lon=kml_coordi_clean_df.lon, color=”elevation”,
                        color_continuous_scale=px.colors.cyclical.IceFire, size_
max=15, zoom=10)  # 亦可以选择列，通过 size=” ” 配置增加显示信息
fig.show()
```

● 图像显示

为方便查看图像或排布图像（图3.13–3）用于研究文章的图表说明，定义图像排布显示的函数。图像处理过程，例如打开、调整图像大小，均使用PIL图像处理库，在调整大小时要保持图像的R、G、B三个通道不变。

```
def imgs_layoutShow(imgs_root, imgsFn_lst, columns, scale, figsize=(15, 10)):
    '''
    function - 显示一个文件夹下所有图片，便于查看。

    Params:
        imgs_root - 图像所在根目录；string
        imgsFn_lst - 图像名列表；list(string)
        columns - 列数；int

    Returns:
        None
    '''
    import math
    import os
    import matplotlib.pyplot as plt
    from PIL import Image
    rows = math.ceil(len(imgsFn_lst)/columns)
```

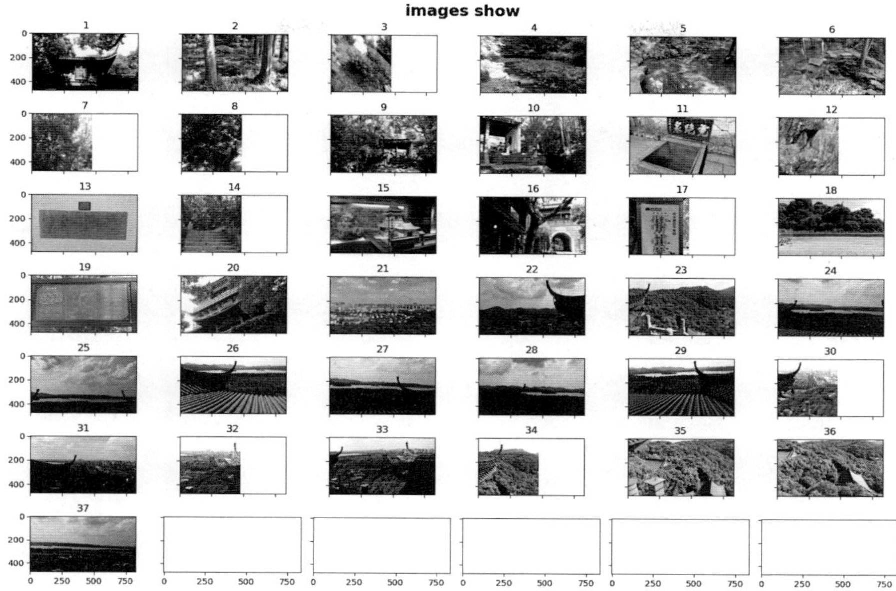

图3.13-3 查看调研拍摄图像

```
    fig, axes = plt.subplots(rows, columns, sharex=True,
                             sharey=True, figsize=figsize)  #布局多个子图，每个子图显示一幅图像
    ax = axes.flatten()  #降至1维，便于循环操作子图
    for i in range(len(imgsFn_lst)):
        img_path = os.path.join(imgs_root, imgsFn_lst[i])  #获取图像的路径
        img_array = Image.open(img_path)  #读取图像为数组，值为RGB格式0 ~ 255
        img_resize = img_array.resize(
            [int(scale * s) for s in img_array.size])  #传入图像的数组，调整图片大小
        ax[i].imshow(img_resize)  #显示图像
        ax[i].set_title(i+1)
    fig.tight_layout()  #自动调整子图参数，使其填充整个图像区域
    fig.suptitle("images show", fontsize=14, fontweight='bold', y=1.02)
    plt.show()
imgs_fn = util_misc.filePath_extraction(
    './data/default_20170720081441', ["jpg"])
imgs_root = list(imgs_fn.keys())[0]
imgsFn_lst = imgs_fn[imgs_root]
columns = 6
scale = 0.2
imgs_layoutShow(imgs_root, imgsFn_lst, columns, scale)
```

上述定义的图像显示函数，存在空白的子图。图像文件路径处理的参数包括了两个：一个是根目录imgs_root；一个是文件名列表imgsFn_lst。下述更新的函数imgs_layoutShow_FPList()则解决了上述两个问题（图3.13-4）。更新的代码块为：

```
invisible_num=rows*columns-len(imgs_fp_list)
if invisible_num>0:
    for i in range(invisible_num):
        ax.flat[-(i+1)].set_visible(False)
```

通过获取空的子图索引，配置不显示该子图达到目的。同时，输入的参数imgs_fp_lis为图片文件路径名，可以通过glob库提取。

```
import glob
def imgs_layoutShow_FPList(imgs_fp_list, columns, scale, figsize=(15, 10)):
    '''
    function - 显示一个文件夹下所有图片，便于查看。
```

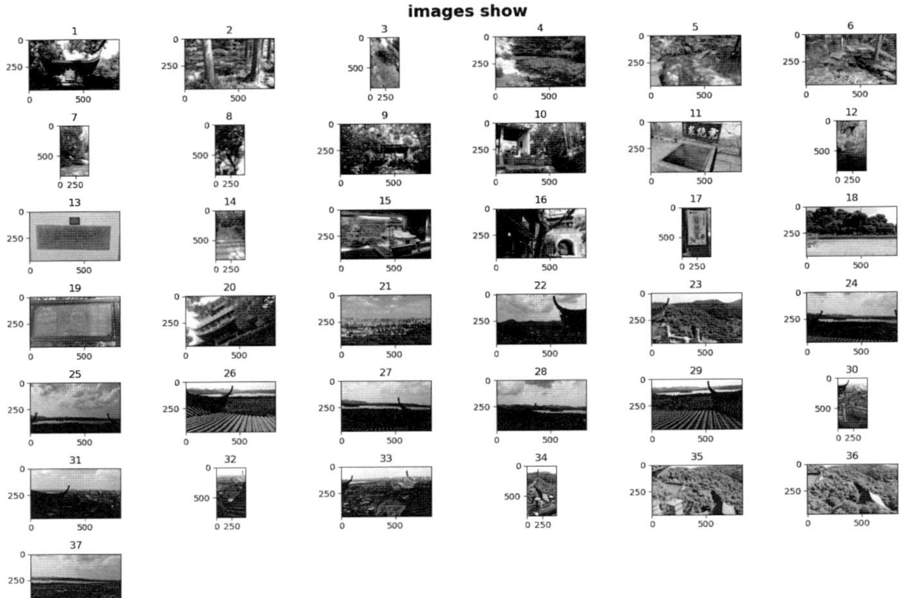

图3.13-4 查看调研拍摄图像（调整代码后）

```
    Params:
        imgs_fp_list - 图像文件路径名列表; list(string)
        columns - 显示列数; int
        scale - 调整图像大小比例因子; float
        figsize - 打印图表大小。The default is (15,10); tuple(int)

    Returns:
        None
    '''
    import math
    import os
    import matplotlib.pyplot as plt
    from PIL import Image
    rows = math.ceil(len(imgs_fp_list)/columns)
    fig, axes = plt.subplots(
        rows, columns, figsize=figsize,)  # 布局多个子图，每个子图显示一幅图像
    ax = axes.flatten()  # 降至 1 维，便于循环操作子图
    for i in range(len(imgs_fp_list)):
        img_path = imgs_fp_list[i]  # 获取图像的路径
        img_array = Image.open(img_path)  # 读取图像为数组，值为 RGB 格式 0 ~ 255
        img_resize = img_array.resize(
            [int(scale * s) for s in img_array.size])  # 传入图像的数组，调整图片大小
        ax[i].imshow(img_resize,)  # 显示图像 aspect=' auto'
        ax[i].set_title(i+1)
    invisible_num = rows*columns-len(imgs_fp_list)
    if invisible_num > 0:
        for i in range(invisible_num):
            ax.flat[-(i+1)].set_visible(False)
    fig.tight_layout()  # 自动调整子图参数，使之填充整个图像区域
    fig.suptitle("images show", fontsize=14, fontweight='bold', y=1.02)
    plt.show()
imgs_dougong_fps = glob.glob('./data/default_20170720081441/*.jpg')
columns = 6
scale = 0.2
imgs_layoutShow_FPList(imgs_dougong_fps, columns, scale)
```

3.13.1.2 Exif（Exchangeable image file format）可交换图像格式

Exif是专门为数码相机相片设定的档案格式，可以记录照片的属性和拍摄信息。根据设置，用手机拍摄的照片通常包含Exif信息。其中，也可能包括GPS位置信息。Exif包括哪些信息内容，

可以通过from PIL.ExifTags import TAGS调入TAGS，打印查看。其中，相对比较关键用于数据分析的一些信息包括拍摄的时间、图像大小、GPS位置数据和记录时间等。

```
import os
def img_exif_info(img_fp, printing=True):
    '''
    function - 提取数码照片的属性信息和拍摄数据，即可交换图像文件格式(Exchangeable image
file format, Exif)

    Params:
        img_fp - 一个图像的文件路径；string
        printing - 是否打印。The default is True；bool

    Returns:
        exif_ - 提取的照片信息结果；dict
    '''
    from PIL import Image
    from PIL.ExifTags import TAGS
    from datetime import datetime
    import time
    img = Image.open(img_fp,)
    try:
        img_exif = img._getexif()
        exif_ = {TAGS[k]: v for k, v in img_exif.items() if k in TAGS}
        # 由 2017:07:20 09:16:58 格式时间，转换为时间戳，用于比较时间先后。
        time_lst = [int(i) for i in re.split(' |:', exif_['DateTimeOriginal'])]
        time_tuple = datetime.timetuple(datetime(
            time_lst[0], time_lst[1], time_lst[2], time_lst[3], time_lst[4],
time_lst[5],))
        time_stamp = time.mktime(time_tuple)
        exif_["timestamp"] = time_stamp
    except ValueError:
        print("exif not found!")
    for tag_id in img_exif:  # 取 Exif 信息（iterating over all EXIF data fields）
   # 获取标签名（get the tag name, instead of human unreadable tag id）
        tag = TAGS.get(tag_id, tag_id)
        data = img_exif.get(tag_id)
        if isinstance(data, bytes):  # 解码 decode bytes
            try:
                data = data.decode()
            except ValueError:
                data = "tag:%s data not found." % tag
        if printing:
            print(f"{tag:30}:{data}")

    # 将度转换为浮点数，Decimal Degrees = Degrees + minutes/60 + seconds/3600
    if 'GPSInfo' in exif_:
        GPSInfo = exif_['GPSInfo']
        geo_coodinate = {
            "GPSLatitude": float(GPSInfo[2][0]+GPSInfo[2][1]/60+GPSInfo[2]
[2]/3600),
            "GPSLongitude": float(GPSInfo[4][0]+GPSInfo[4][1]/60+GPSInfo[4]
[2]/3600),
            "GPSAltitude": GPSInfo[6],
            # 字符形式
            "GPSTimeStamp_str": "%d:%f:%f" % (GPSInfo[7][0], GPSInfo[7][1]/10,
GPSInfo[7][2]/100),
            # 浮点形式
            "GPSTimeStamp": float(GPSInfo[7][0]+GPSInfo[7][1]/10+GPSInfo[7]
[2]/100),
            "GPSImgDirection": GPSInfo[17],
            "GPSDestBearing": GPSInfo[24],
            "GPSDateStamp": GPSInfo[29],
            "GPSHPositioningError": GPSInfo[31],
        }
        if printing:
            print("_"*50)
            print(geo_coodinate)
        return exif_, geo_coodinate
    else:
        return exif_
img_example_1 = os.path.join(imgs_root, imgsFn_lst[0])
img_exif_1 = img_exif_info(img_example_1)
```

```
--------------------------------------------------------------------------------
                           544:0
                           545:0
                           546:0
                           547:0
                           548:1
                           549:
ResolutionUnit                :2
ExifOffset                    :414
ImageDescription              :
Make                          :HTC
Model                         :HTC D830u
Software                      :MediaTek Camera Application

Orientation                   :1
DateTime                      :2017:07:20 09:16:58
YCbCrPositioning              :2
XResolution                   :72.0
YResolution                   :72.0
ExifVersion                   :0220
ComponentsConfiguration       :
FlashPixVersion               :0100
DateTimeOriginal              :2017:07:20 09:16:58
DateTimeDigitized             :2017:07:20 09:16:58
ExposureBiasValue             :0.0
ColorSpace                    :1
MeteringMode                  :2
LightSource                   :255
Flash                         :0
FocalLength                   :3.79
ExifImageWidth                :4160
ExifImageHeight               :2368
ExifInteroperabilityOffset    :1478
SceneCaptureType              :0
Contrast                      :0
SubsecTime                    :42
SubsecTimeOriginal            :42
SubsecTimeDigitized           :42
Saturation                    :0
Sharpness                     :0
ExposureTime                  :0.002756
FNumber                       :2.0
ExposureProgram               :0
ISOSpeedRatings               :100
ExposureMode                  :0
WhiteBalance                  :0
DigitalZoomRatio              :1.0
MakerNote                  :tag:MakerNote data not found.
```

上述图片并未记录有GPS地理位置信息。下述调研的图像则可以查看到GPS信息。GPS信息的记录格式有可能不同，例如下述获取的“GPSInfo”中，2: (41.0, 52.0, 55.38)。但是，也有可能为(19, 1), (31, 1), (5139, 100))格式。注意，自定义的Exif数据提取的函数，仅实现了第一种情况。

```
img_ChicagoDowntown_root=r'./data/imgs_ChicagoDowntown'
img_example_2=os.path.join(img_ChicagoDowntown_root,r'2019-10-11_120110.jpg')
img_exif_2,geo_coodinate=img_exif_info(img_example_2,)
GPSInfo                       :{1: 'N', 2: (41.0, 52.0, 55.38), 3: 'W', 4: (87.0,
37.0, 26.43), 5: b'\x00', 6: 182.35323716873532, 7: (17.0, 1.0, 9.99), 12: 'K',
13: 0.0, 16: 'T', 17: 177.5288773523686, 23: 'T', 24: 177.5288773523686, 29:
'2019:10:11', 31: 10.0}
ResolutionUnit                :2
ExifOffset                    :218
Make                          :Apple
Model                         :iPhone X
Software                      :https://heic.online
Orientation                   :1
DateTime                      :2019:10:11 12:01:11
YCbCrPositioning              :1
XResolution                   :72.0
YResolution                   :72.0
ExifVersion                   :0221
ComponentsConfiguration       :
ShutterSpeedValue             :6.909027361693629
DateTimeOriginal              :2019:10:11 12:01:11
```

```
DateTimeDigitized             :2019:10:11 12:01:11
ApertureValue                 :1.6959938128383605
BrightnessValue               :5.888149338229669
ExposureBiasValue             :0.0
MeteringMode                  :5
Flash                         :24
FocalLength                   :4.0
ColorSpace                    :65535
ExifImageWidth                :4032
FocalLengthIn35mmFilm         :28
SceneCaptureType              :0
ExifImageHeight               :3024
SubsecTimeOriginal            :201
SubsecTimeDigitized           :201
SubjectLocation               :(2015, 1511, 2217, 1330)
SensingMethod                 :2
ExposureTime                  :0.008333333333333333
FNumber                       :1.8
SceneType                     :
ExposureProgram               :2
CustomRendered                :2
ISOSpeedRatings               :25
ExposureMode                  :0
FlashPixVersion               :0100
WhiteBalance                  :0
LensSpecification             :(4.0, 6.0, 1.8, 2.4)
LensMake                      :Apple
LensModel                     :iPhone X back dual camera 4mm f/1.8
MakerNote                  :tag:MakerNote data not found.
__________________________________________________
{'GPSLatitude': 41.88205, 'GPSLongitude': 87.62400833333334, 'GPSAltitude':
182.35323716873532, 'GPSTimeStamp_str': '17:0.100000:0.099900', 'GPSTimeStamp':
17.1999, 'GPSImgDirection': 177.5288773523686, 'GPSDestBearing':
177.5288773523686, 'GPSDateStamp': '2019:10:11', 'GPSHPositioningError': 10.0}
```

调用上述定义的imgs_layoutShow_FPList()函数，排布显示芝加哥市中心调研图像见图3.13–5。

```
imgs_ChicagoDowntown_fps=glob.glob('./data/imgs_ChicagoDowntown/*.jpg')
columns=6;scale=0.2
imgs_layoutShow_FPList(imgs_ChicagoDowntown_fps,columns,scale,figsize=(15,15))
```

在Exif信息提取函数中，根据"DateTimeOriginal:2017:07:20 09:16:58"时间信息，计算时间戳，用于图像按拍摄时间排序，可以绘制图片拍摄的行走路径（图3.13–6）。此次绘制使用Plotly库提供的go方法实现。该方法在调用地图时，不需要提供mapbox地图数据的访问许可（access token）。

```
import pandas as pd
import plotly.graph_objects as go
imgs_ChicagoDowntown_coordi = []
for fn in imgs_ChicagoDowntown_fps:
    img_exif_2, geo_coodinate = img_exif_info(fn, printing=False)
    imgs_ChicagoDowntown_coordi.append(
        (geo_coodinate['GPSLatitude'], -geo_coodinate['GPSLongitude'], geo_
coodinate['GPSAltitude'], img_exif_2['timestamp']))  #注意经度负号

pd.set_option('display.float_format', lambda x: '%.5f' % x)
imgs_ChicagoDowntown_coordi_df = pd.DataFrame(data=imgs_ChicagoDowntown_coordi,
columns=[
                                                    "lat", "lon", "elevation",
'timestamp']).sort_values(by=['timestamp'])  #按图片拍摄时间戳排序

fig = go.Figure(go.Scattermapbox(mode="markers+lines", lat=imgs_ChicagoDowntown_
coordi_df.lat,
                lon=imgs_ChicagoDowntown_coordi_df.lon, marker={'size': 10}))  #
亦可以选择列，通过 size=" " 配置增加显示信息
fig.update_layout(
    margin={'l': 0, 't': 0, 'b': 0, 'r': 0},
    mapbox={
        'style': "stamen-terrain",
        'center': {'lon': -87.62401, 'lat': 41.88205},
        'zoom': 14})
fig.show()
```

图3.13-5 查看芝加哥调研拍摄图像

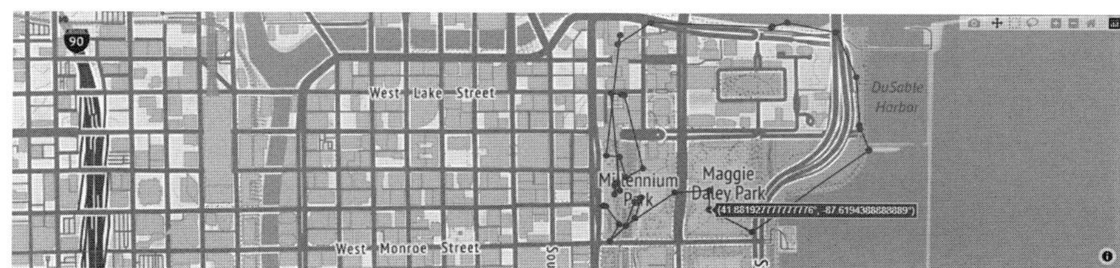

图3.13-6 芝加哥调研行走路径

3.13.1.3 RGB色彩的三维图示

包含色彩信息（RGB）的数据投射到三维空间中，可以通过判断区域色彩在三维空间域中的分布情况来把握Red、Green 和Blue 色彩分量的变化情况（图3.13-7）。为了增加计算速度，同样传入了`scale`参数，可以根据分析目的调整图像大小。使用PIL库的`Image.open()`方法读取图像时，返回的数组颜色值位于［0,255］区间，在用matplotlib打印图像时，通常颜色值位于［0,1］区间，因此需要转换，直接除以值255。

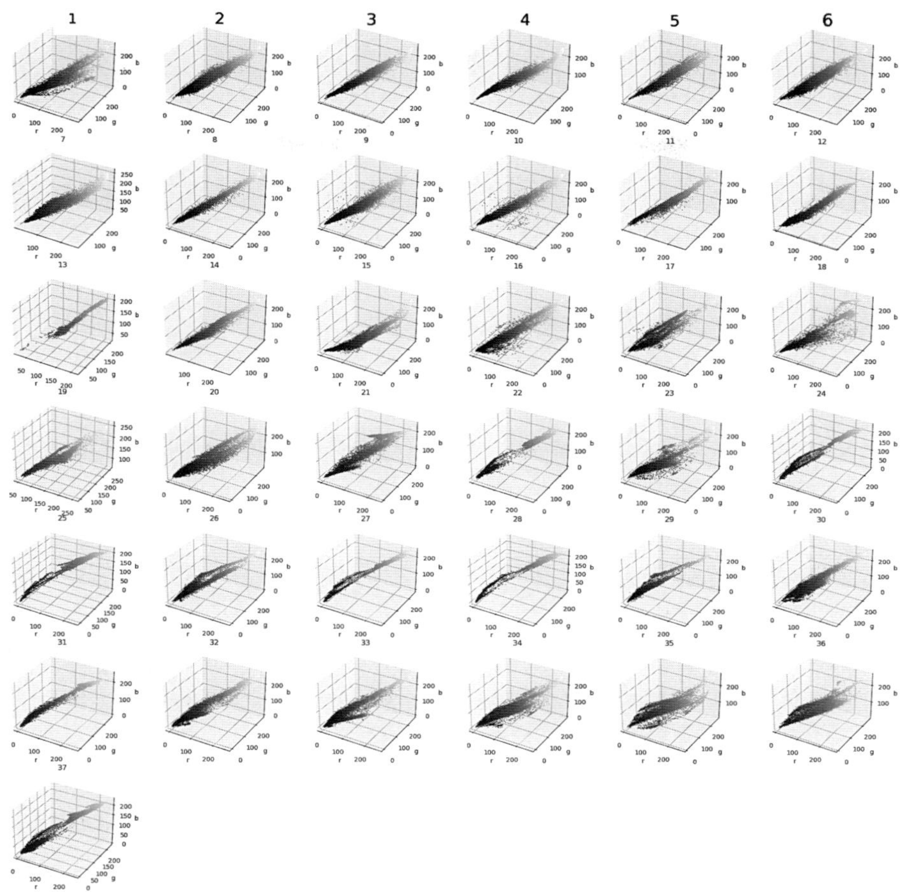

图3.13-7 RGB色彩的三维图示

```
def imgs_colorSpace(imgs_root, imgsFn_lst, columns, scale, figsize=(15, 10)):
    '''
    function - 将像素 RGB 颜色投射到色彩空间中，直观感受图像颜色的分布。

    Params:
        imgs_root - 图像所在根目录，string
        imgsFn_lst - 图像名列表；list(string)
        columns - 列数；int

    Returns:
        None
    '''
    import math
    import os
    import matplotlib.pyplot as plt
    from PIL import Image
    import numpy as np
    from tqdm import tqdm
    rows = math.ceil(len(imgsFn_lst)/columns)
    fig = plt.figure()
    for i in tqdm(range(len(imgsFn_lst))):
        #不断增加子图，并设置投影为 3d 模式，可以显示三维坐标空间
        ax = fig.add_subplot(rows, columns, i+1, projection='3d')
        img = os.path.join(imgs_root, imgsFn_lst[i])
        img_path = os.path.join(imgs_root, imgsFn_lst[i])  #获取图像的路径
        img_array = Image.open(img_path)  #读取图像为数组，值为 RGB 格式 0 ~ 255
        img_resize = img_array.resize(
```

```
            [int(scale * s) for s in img_array.size])  #传入图像的数组，调整图片大小
        img_array = np.asarray(img_resize)
        ax.scatter(img_array[:, :, 0], img_array[:, :, 1], img_array[:, :, 2],
c=(
            img_array/255).reshape(-1, 3), marker='+', s=0.1)  #用RGB的三个分量值
作为颜色的空间坐标，并显示其颜色。设置颜色时，需要将0～255缩放至0～1区间
        ax.set_xlabel('r', labelpad=1)
        ax.set_ylabel('g')
        ax.set_zlabel('b', labelpad=2)
        ax.set_title(i+1)
        fig.set_figheight(figsize[0])
        fig.set_figwidth(figsize[1])
    print("Ready to show...")
    fig.tight_layout()
    plt.show()
imgs_fn = util_misc.filePath_extraction(
    r'./data/default_20170720081441', ["jpg"])
imgs_root = list(imgs_fn.keys())[0]
imgsFn_lst = imgs_fn[imgs_root]
columns = 6
scale = 0.05
imgs_colorSpace(imgs_root, imgsFn_lst, columns, scale, figsize=(20, 20))
---------------------------------------------------------------------------
 100%|██████████████████| 37/37 [00:13<00:00,  2.67it/s]
Ready to show...
```

3.13.2 聚类（Clustering）

监督学习（Supervised learning），是机器学习的一种方法，可以学习训练数据集建立学习模型用于预测新的实例。回归属于监督学习，所用到的数据集被标识分类，通常包括特征值（自变量）和目标值（标签，因变量）。非监督学习（Unsupervised learning），则没有给定事先标记的数据集，自动对数据集进行分类或分群。聚类则属于非监督学习。

聚类是把相似的对象通过静态分类的方法分成不同的组别或者更多的子集（subset），这样让在同一个子集中的成员对象都有相似的一些属性。聚类涉及的算法很多。K-Means是常见算法的一种，通过尝试将样本分离到n个方差相等的组中对数据进行聚类，最小化指标（criterion），例如聚类内平方和指标（within-cluster sum-of-squares criterion），其公式可表达为：$\sum_{i=0}^{n} \min(\|x_i - \mu_j\|^2) : \mu_j \in C$。

图3.13-8中假设了两个簇C_0和C_1，即$k=2$，为分簇数量（簇数）。首先随机放置两个质心（centroid），并标识为old_centroid。通过逐个计算每个点分别到两个质心的距离，比较大小，将点归属于距离最近的质心（代表簇或组分类），例如对于点a，其到质心C_0距离d_0小于到质心C_1的距离，因此点a归属于质心d_0所代表的簇。将所有的点根据距离远近归属于不同的质心所代表的类之后，由所归属簇中点的均值作为新的质心，图中用new_centroid假设标识。分别计算旧质心和新质心的距离，如果所有簇质心的新旧距离值为0，则意味质心没有发生变化，即完成聚类；否则，用新的质心重复上一轮的计算，直至质心新旧距离值为0为止。只要有足够的时间，

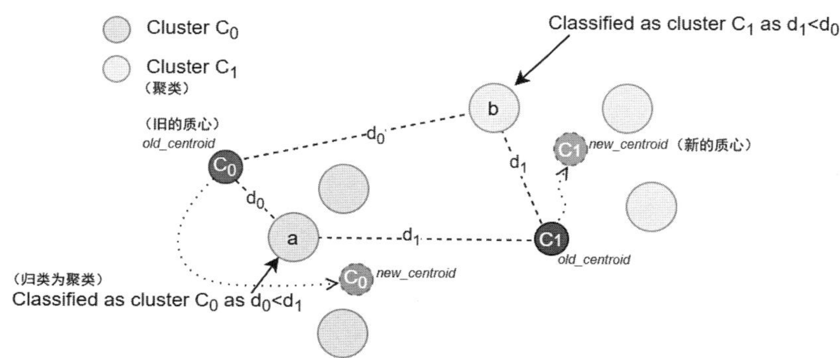

图3.13-8　K-Means聚类算法机制（引自参考文献[1]）

K-Means总是收敛的，但它可能是一个局部最小值，这高度依赖于质心的初始化。因此，计算通常要进行多次，并对质心进行不同的初始化。

下述代码根据上述计算原理自定义聚类函数，同时使用sklearn库的KMeans方法直接实现，比较两者之间的差异（图3.13-9）。自定义的K-Means所计算的结果有多种可能，通过多次运行会获得与sklearn库一致的结果。sklearn库的KMeans方法通过init='k-means++，使得质心彼此之间保持距离，解决了局部最小值的问题，证明比随机初始化可以得到更好的结果。

不配置忽略 warning 警告信息，将提示：KMeans is known to have a memory leak on Windows with MKL, when there are less chunks than available threads. You can avoid it by setting the environment variable OMP_NUM_THREADS=1.

```
import numpy as np
import warnings
warnings.filterwarnings('ignore')
class K_Means:
    '''
    class - 定义 K-Means 算法

    Params:
        X - 待分簇的数据（数组）
        k - 分簇数量
        figsize - Matplotlib 图表大小
    '''
    def __init__(self, X, k, figsize):
        self.X = X
        self.k = k
        self.figsize = figsize
    def euclidean_distance(self, a, b, ax=1):
        '''
        function - 计算两点距离。To calculate the distance between two points

        Params:
        a - 2 维度数组，例如 [[3,4]
                              [5,6]
                              [1,4]]
        b - 2 维度数组
        ax - 计算轴
        '''
        import numpy as np
        return np.linalg.norm(a-b, axis=ax)
    def update(self, ax):
        '''
        function - K-Means 算法
        '''
        from copy import deepcopy
        import numpy as np
```

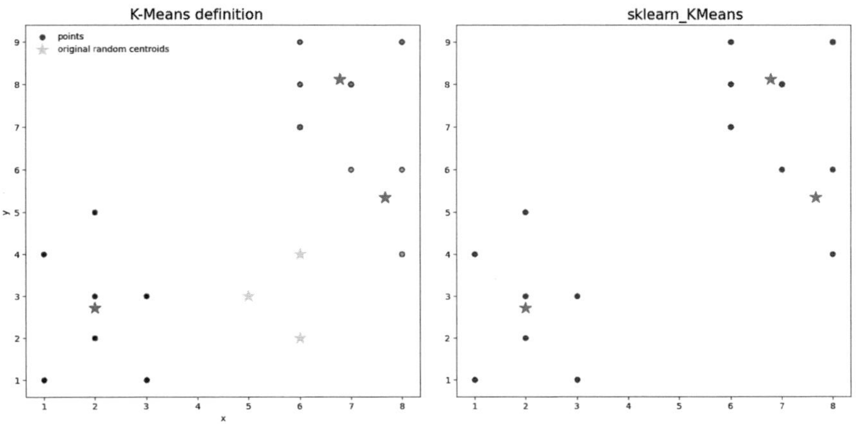

图3.13-9 聚类示例结果，自定义（左）；Sklearn库（右）

```
        # 生成随机质心（ generate k random points (centroids) ）
        Cx = np.random.randint(np.min(X[:, 0]), np.max(X[:, 0]), size=self.k)
        Cy = np.random.randint(np.min(X[:, 1]), np.max(X[:, 1]), size=self.k)
        ax.scatter(Cx, Cy, label="original random centroids",
                   marker='*', c='gainsboro', s=200)

        # 质心数组 （represent the k centroids as a matrix ）
        C = np.array(list(zip(Cx, Cy)), dtype=np.float64)
        # 建立同质心数组形状，值为 0 的数组（create a matrix of 0 with same dimension as C (centroids) ）
        C_prev = np.zeros(C.shape)
        # 存储每个点所属子群（to store the cluster each point belongs to ）
        clusters = np.zeros(len(X))
        # 计算质心与 C_prev 之间的距离（measure the distance between the centroids and C_prev ）
        distance_differences = self.euclidean_distance(C, C_prev)

        # 循环计算，缩小前一步和后一步质心距离的差异 （loop as long as there is still a difference in distance between the previous and current centroids ）
        count = 0
        while distance_differences.any() != 0:
            print("epoch:%d" % count)
            # 将每个值分配到最近的簇（assign each value to its closest cluster ）
            for i in range(len(self.X)):
                distances = self.euclidean_distance(self.X[i], C)
                # 延着一个轴，返回最小值索引（returns the indices of the minimum values along an axis ）
                cluster = np.argmin(distances)
                clusters[i] = cluster

            C_prev = deepcopy(C)  # 存储前一质心（store the prev centroids ）

            # 通过取均值寻找新的质心（find the new centroids by taking the average value ）
            for i in range(k):
                # 取簇 i 中的所有点（take all the points in cluster i ）
                points = [X[j] for j in range(len(X)) if clusters[j] == i]
                if len(points) != 0:
                    C[i] = np.mean(points, axis=0)

            # 计算前一与后一质心的距离（find the distances between the old centroids and the new centroids ）
            distance_differences = self.euclidean_distance(C, C_prev)
            print("distance_differences:", distance_differences)
            count += 1

        # 打印散点图（plot the scatter plot ）
        colors = ['b', 'r', 'y', 'g', 'c', 'm']
        for i in range(k):
            points = np.array([X[j]
                               for j in range(len(X)) if clusters[j] == i])
            if len(points) > 0:
                ax.scatter(points[:, 0], points[:, 1], s=10, c=colors[i])
            else:
                # 这意味着其中一个簇没有点 （this means that one of the clusters has no points ）
                print("Plesae regenerate your centroids again.")
            ax.scatter(C[:, 0], C[:, 1], marker='*', s=200, c='red')
    def sklearn_KMeans(self, ax):
        '''
        function - 使用 Sklearn 库的 KMeans 算法聚类
        '''
        from sklearn.cluster import KMeans
        kmeans = KMeans(n_clusters=self.k)
        kmeans = kmeans.fit(self.X)
        labels = kmeans.predict(self.X)
        centroids = kmeans.cluster_centers_

        c = ['b', 'r', 'y', 'g', 'c', 'm']
        colors = [c[i] for i in labels]
        ax.scatter(centroids[:, 0], centroids[:, 1],
                   marker='*', s=200, c='red')

        print("预测 (7,5) 的簇为 : %d" % kmeans.predict([[7, 5]])[0])

    def execution(self):
        '''
        function - 执行
        '''
        %matplotlib inline
```

```
        import matplotlib.pyplot as plt
        fig, axs = plt.subplots(1, 2, figsize=self.figsize)
        axs[0].scatter(self.X[:, 0], self.X[:, 1], label="points")
        axs[0].set_title(r'K-Means definition', fontsize=15)
        self.update(axs[0])
        axs[1].scatter(self.X[:, 0], self.X[:, 1], label="points")
        axs[1].set_title(r'sklearn_KMeans', fontsize=15)
        self.sklearn_KMeans(axs[1])
        axs[0].set_xlabel('x')
        axs[0].set_ylabel('y')
        axs[0].legend(loc='upper left', frameon=False)
        plt.show()
kmeans_dataset = [(1, 1), (2, 2), (2, 3), (1, 4), (3, 3), (6, 7), (7, 8), (6, 8),
(7, 6),
                  (6, 9), (2, 5), (7, 8), (8, 9), (6, 7), (7, 8), (3, 1), (8, 4),
(8, 6), (8, 9)]
X = np.array(kmeans_dataset)
k = 3  #配置分组的数量（亦随机生成中心的数量）
figsize = (18, 8)
K = K_Means(X, k, figsize)
K.execution()
----------------------------------------------------------------------------
epoch:0
distance_differences: [3.01357473 3.56000156 0.        ]
epoch:1
distance_differences: [0.         0.32363651 2.82842712]
epoch:2
distance_differences: [0.         0.20429277 1.        ]
epoch:3
distance_differences: [0.         0.21227748 0.47140452]
epoch:4
distance_differences: [0. 0. 0.]
预测 (7,5) 的簇为: 2
```

- **聚类算法比较**

sklearn官网聚类部分[①]提供了一组代码，比较多个不同聚类算法，其归结如表3.13-2所示。

① sklearn官网聚类部分，聚类算法比较（https://scikit-learn.org/stable/modules/clustering.html#clustering）。

表3.13-2 不同聚类算法比较

Method name（方法名称）	**Parameters**（参数）	**Scalability**（扩展性）	**Usecase**（用例）	**Geometry (metric used)**（几何）
K-Means	number of clusters	Very large n_samples, medium n_clusters with MiniBatch code	General-purpose, even cluster size, flat geometry, not too many clusters（通用的，均匀的簇大小，平坦的几何，没有太多的簇）	Distances between points（点间距离）
Affinity propagation	damping, sample preference	Not scalable with n_samples	Many clusters, uneven cluster size, non-flat geometry（很多簇，簇大小不均匀，非平坦的几何）	Graph distance（e.g. nearest-neighbor graph）（图的距离，如最近邻图）
Mean-shift	bandwidth	Not scalable with n_samples	Many clusters, uneven cluster size, non-flat geometry（很多簇，簇大小不均匀，非平坦的几何）	Distances between points（点间距离）
Spectral clustering	number of clusters	Medium n_samples, small n_clusters	Few clusters, even cluster size, non-flat geometry（少簇，均匀簇大小，非平坦的几何）	Graph distance（e.g. nearest-neighbor graph）（图的距离，如最近邻图）
Ward hierarchical clustering	number of clusters or distance threshold	Large n_samples and n_clusters	Many clusters, possibly connectivity constraints（许多簇，可能存在连接性约束）	Distances between points（点间距离）

续表

Method name （方法名称）	**Parameters** （参数）	**Scalability** （扩展性）	**Usecase** （用例）	**Geometry (metric used)**（几何）
Agglomerative clustering	number of clusters or distance threshold, linkage type, distance	Large n_samples and n_clusters	Many clusters, possibly connectivity constraints, non Euclidean distances （许多簇，可能存在连接性约束，非欧几里德距离）	Any pairwise distance （任意两两距离）
DBSCAN	neighborhood size	Very large n_samples, medium n_clusters	Non-flat geometry, uneven cluster sizes （非平坦的几何，不均匀的簇大小）	Distances between nearest points （最近点之间的距离）
OPTICS	minimum cluster membership	Very large n_samples, large n_clusters	Non-flat geometry, uneven cluster sizes, variable cluster density （非平坦的几何，不均匀的簇大小，可变簇密度）	Distances between points （点间距离）
Gaussian mixtures	many	Not scalable	Flat geometry, good for density estimation （平坦的几何，适合密度估计）	Mahalanobis distances to centers （到中心的马氏距离）
Birch	branching factor, threshold, optional global clusterer.	Large n_clusters and n_samples	Large dataset, outlier removal, data reduction （大数据集，异常值去除，数据缩减）	Euclidean distance between points （点间的欧氏距离）

官网代码计算结果如图3.13–10所示。

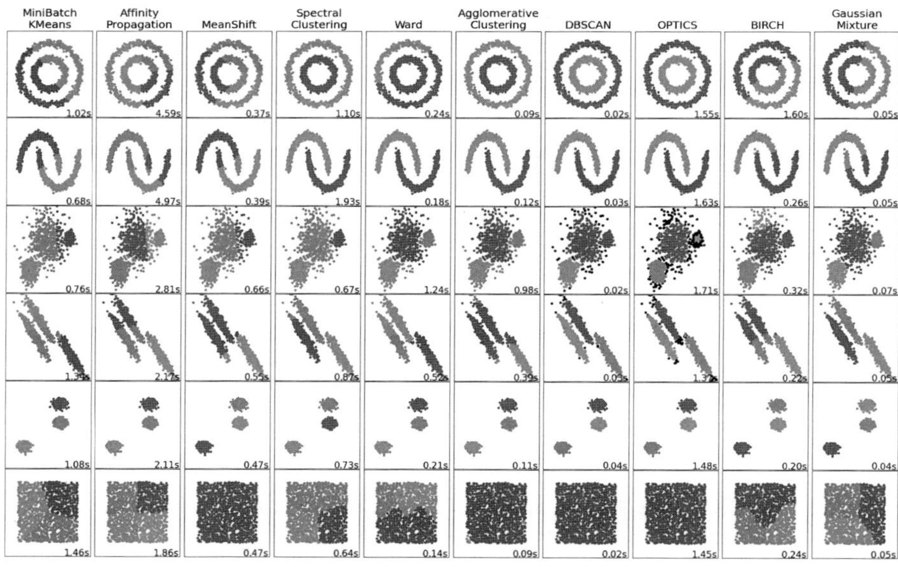

图3.13–10 不同聚类算法结果

3.13.3 聚类图像主题色

城市色彩，也称“城市环境色彩”，泛指城市各个构成要素公共空间部分所呈现出的色彩面貌总和。城市色彩包含大量复杂多变的元素，因此必须科学地调查与分析，才能实现有效引导和规划发展。在城市色彩相关的研究上，主要有：分析城市色彩特征，调查与定量分析，更新与保护机制研究；景观环境色彩构成，以及利用MATLAB计算插值与应用回归算法实现城市色彩主色调意向图的自动填充，得到城市色彩主色调理想色彩地图的研究。部分传统的研究由于受到数据分析技术的限制，对于批量的城市影像提取主题色时偏重手工提取，不仅增加时间成本，也影响分析的精度；同时，在数据分析方法和数据信息表达上亦受到限制。因此，有必要借助机器学习中的聚类等方法，自主聚类主题色，通过Python的matplotlib标准库实现数据可视化增强表达。

首先，在Python程序语言中批量读取图像，因为所拍摄的图像分辨率较高，而色彩分析不需要这样的高精度，因此通过压缩图像，降低图像大小来节约分析的时间。然后，设置色彩主题色聚类的数量为7，获取每幅图像的7个主题色。采用KMeans聚类算法分类色彩提取主题色（图3.13-11），提取所有图像的主题色之后汇总于一个数组中。在数据增强可视化方面，设计了散点形式打印主题色（图3.13-12），直观反映城市色彩印象。通过城市主题色的提取、色彩印象感官的呈现来研究城市色彩，可以针对不同的城市空间、不同的调研时间，分析色彩的变化。

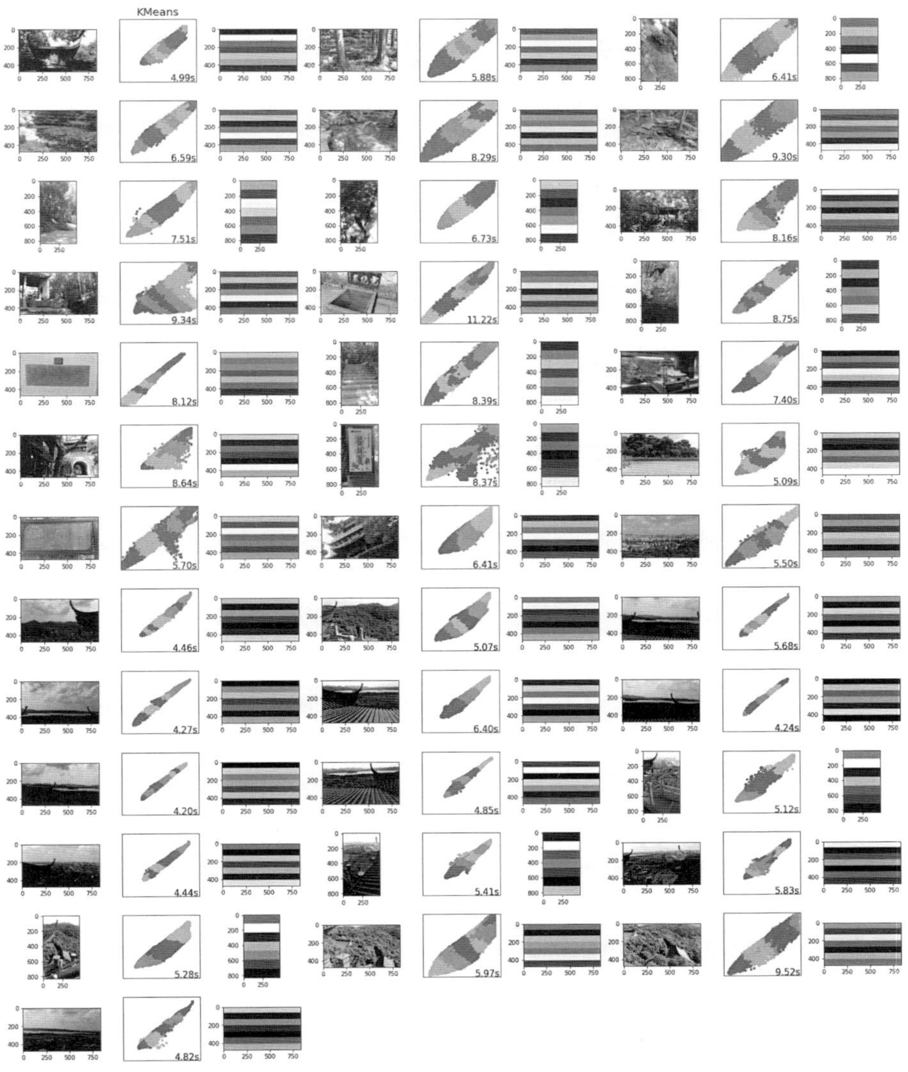

图3.13-11 提取调研图像主题色

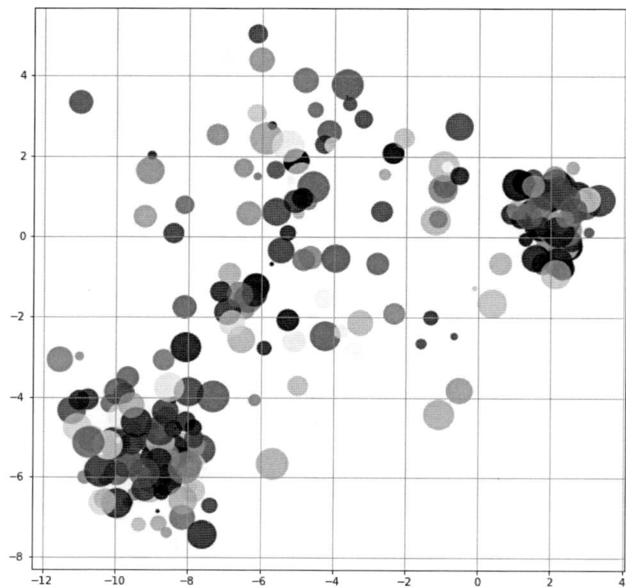

图3.13-12 调研图像主题色的散点图形式表述

聚类部分，参考了上述sklearn提供的*Comparing different clustering algorithms on toy datasets*代码，从下面代码中也可以观察到代码迁移的痕迹。

```
import util_misc
def img_rescale(img_path, scale):
    '''
    function - 读取与压缩图像，返回 2 维度数组

    Params:
        img_path - 待处理图像路径；lstring
        scale - 图像缩放比例因子；float

    Returns:
        img_3d - 返回三维图像数组；ndarray
        img_2d - 返回二维图像数组；ndarray
    '''
    from PIL import Image
    import numpy as np

    img = Image.open(img_path)  # 读取图像为数组，值为 RGB 格式 0-255
    img_resize = img.resize([int(scale * s)
                             for s in img.size])  # 传入图像的数组，调整图片大小
    img_3d = np.array(img_resize)
    h, w, d = img_3d.shape
    img_2d = np.reshape(img_3d, (h*w, d))  # 调整数组形状为 2 维

    return img_3d, img_2d
def img_theme_color(imgs_root, imgsFn_lst, columns, scale,):
    '''
    function - 聚类的方法提取图像主题色，并打印图像、聚类预测类的二维显示和主题色带

    Params:
        imgs_root - 图像所在根目录；string
        imgsFn_lst - 图像名列表；list(string)
        columns - 列数；int

    Returns:
        themes - 图像主题色；array
        pred - 预测的类标；array
    '''
```

```
import os
import time
import warnings
import numpy as np
import matplotlib.pyplot as plt
from tqdm import tqdm
from sklearn.preprocessing import StandardScaler
from sklearn import cluster, datasets, mixture
from itertools import cycle, islice

# 设置聚类参数，本实验中仅使用了 KMeans 算法，其他算法可以自行尝试
kmeans_paras = {'quantile': .3,
                'eps': .3,
                'damping': .9,
                'preference': -200,
                'n_neighbors': 10,
                'n_clusters': 7}
imgsPath_lst = [os.path.join(imgs_root, p) for p in imgsFn_lst]
imgs_rescale = [(img_rescale(img, scale)) for img in imgsPath_lst]
datasets = [((i[1], None), {})
            for i in imgs_rescale]  # 基于 img_2d 的图像数据，用于聚类计算
img_lst = [i[0] for i in imgs_rescale]  # 基于 img_3d 的图像数据，用于图像显示

# 建立 0 占位的数组，用于后面主题数据的追加。n_clusters 为提取主题色的聚类数量，此处为 7，轴 2 为
3，是色彩的 RGB 数值
themes = np.zeros((kmeans_paras['n_clusters'], 3))
# 可以 1 次性提取元组索引值相同的值，img 就是 img_3d，而 pix 是 img_2d
(img_3d, img_2d) = imgs_rescale[0]
img2d_V, img2d_H = img_2d.shape  # 获取 img_2d 数据的形状，用于 pred 预测初始数组的建立
# 建立 0 占位的 pred 预测数组，用于后面预测结果数据的追加，即图像中每一个像素点属于设置的 7 个聚类
中的哪一组，预测给定类标
pred = np.zeros((img2d_V))

# 图表大小的设置，根据图像的数量来设置高度，宽度为 3 组 9 个子图，每组包括图像、预测值散点图和
主题色
plt.figure(figsize=(6*3+3, len(imgsPath_lst)*2))
plt.subplots_adjust(left=.02, right=.98, bottom=.001,
                    top=.96, wspace=.3, hspace=.3)  # 调整图，避免横纵向坐标重叠
subplot_num = 1  # 子图的计数

# 循环 pixData 数据，即待预测的每个图像数据。enumerate() 函数将可迭代对象组成一个索引序列，可以
同时获取索引和值。其中，i_dataset 为索引，从整数 0 开始
for i_dataset, (dataset, algo_params) in tqdm(enumerate(datasets)):
    X, y = dataset  # 用于机器学习的数据一般包括特征值和类标，此次实验为无监督分类的聚类
实验，没有类标，并将其在前文中设置为 None 对象
# 标准化数据仅用于二维图表的散点，可视化预测值，而不用于聚类，聚类数据保持色彩的 0 ~ 255 值范围
    Xstd = StandardScaler().fit_transform(X)
    # 此次实验使用 KMeans 算法，参数为 n_clusters 一项。不同算法计算效率不同，例如
MiniBatchKMeans 和 KMeans 算法计算较快
    km = cluster.KMeans(n_clusters=kmeans_paras['n_clusters'])
    clustering_algorithms = (('KMeans', km),)
    for name, algorithm in clustering_algorithms:
        t0 = time.time()
        # 警告错误，使用 warning 库
        with warnings.catch_warnings():
            warnings.filterwarnings(
                "ignore",
                message="the number of connected components of the " +
                "connectivity matrix is [0-9]{1,2}" +
                " > 1. Completing it to avoid stopping the tree early.",
                category=UserWarning)
            warnings.filterwarnings(
                "ignore",
                message="Graph is not fully connected, spectral embedding" +
                " may not work as expected.",
                category=UserWarning)
            algorithm.fit(X)  # 通过 fit 函数执行聚类算法

        quantize = np.array(algorithm.cluster_centers_,
                            dtype=np.uint8)  # 返回聚类的中心，为主题色
        themes = np.vstack((themes, quantize))  # 将计算获取的每一图像主题色追加到
themes 数组中
        t1 = time.time()  # 计算聚类算法所需时间
        ''' 获取预测值 / 分类类标 '''
```

```
            if hasattr(algorithm, 'labels_'):
                y_pred = algorithm.labels_.astype(int)
            else:
                y_pred = algorithm.predict(X)
            pred = np.hstack((pred, y_pred))  #将计算获取的每一图像聚类预测结果追加到
pred 数组中
            fig_width = (len(clustering_algorithms)+2)*3  #水平向子图数
            plt.subplot(len(datasets), fig_width, subplot_num)
            plt.imshow(img_lst[i_dataset])  #图像显示子图

            plt.subplot(len(datasets), fig_width, subplot_num+1)
            if i_dataset == 0:
                plt.title(name, size=18)
            colors = np.array(list(islice(cycle(['#377eb8', '#ff7f00', '#4daf4a',
'#f781bf', '#a65628',
                                         '#984ea3', '#999999', '#e41a1c', '#dede00']),
int(max(y_pred) + 1))))  #设置预测类标分类颜色
            plt.scatter(Xstd[:, 0], Xstd[:, 1], s=10,
                        color=colors[y_pred])  #预测类标子图
            plt.xlim(-2.5, 2.5)
            plt.ylim(-2.5, 2.5)
            plt.xticks(())
            plt.yticks(())
            plt.text(.99, .01, ('%.2fs' % (t1 - t0)).lstrip('0'), transform=plt.
gca(
            ).transAxes, size=15, horizontalalignment='right')  #子图中显示聚类计算
时间长度、
            #图像主题色子图参数配置
            plt.subplot(len(datasets), fig_width, subplot_num+2)
            t = 1
            pale = np.zeros(img_lst[i_dataset].shape, dtype=np.uint8)
            h, w, _ = pale.shape
            ph = h/len(quantize)
            for y in range(h):
                pale[y, ::] = np.array(quantize[int(y/ph)], dtype=np.uint8)
            plt.imshow(pale)
            t += 1
            subplot_num += 3
    plt.show()
    return themes, pred
imgs_fn = util_misc.filePath_extraction(
    r'./data/default_20170720081441', ["jpg"])
imgs_root = list(imgs_fn.keys())[0]
imgsFn_lst = imgs_fn[imgs_root]
columns = 6
scale = 0.2
themes, pred = img_theme_color(imgs_root, imgsFn_lst, columns, scale,)
-------------------------------------------------------------------------------
37it [06:46, 10.99s/it]
```

聚类所有图像获取主题色后，可以将计算的结果包括主题色，以及预测的类标保存在硬盘中，避免重复计算。

```
def save_as_json(array, save_root, fn):
    '''
    function - 保存文件，将文件存储为 json 数据格式

    Params:
        array - 待保存的数组; array
        save_root - 文件保存的根目录; string
        fn - 保存的文件名; string

    Returns:
        None
    '''
    import time
    import os
    import json
    json_file = open(os.path.join(save_root, r'%s_' %
                     fn+str(time.time()))+'.json', 'w')
    json.dump(array.tolist(), json_file)  #将 numpy 数组转换为列表后存储为 json 数据格式
    json_file.close()
save_root = './data_processed'
save_as_json(themes, save_root, 'themes')
```

```
save_as_json(pred, save_root, 'themes_pred')
--------------------------------------------------------------------------------
```

以随机散点图的形式显示色彩。

```
import json
import numpy as np
def themeColor_impression(theme_color):
    '''
    function - 显示所有图像主题色，获取总体印象

    Params:
        theme_color - 主题色数组；array

    Returns:
        None
    '''
    from sklearn import datasets
    from numpy.random import rand
    import matplotlib.pyplot as plt
    n_samples = theme_color.shape[0]
    random_state = 170  #可为默认，不设置该参数，获得随机图形
    #利用 scikit 的 datasets 数据集构建有差异变化的斑点
    varied = datasets.make_blobs(n_samples=n_samples, cluster_std=[
                                 1.0, 2.5, 0.5], random_state=random_state)
    (x, y) = varied
    fig, ax = plt.subplots(figsize=(10, 10))
    scale = 1000.0*rand(n_samples)  #设置斑点随机大小
    ax.scatter(x[..., 0], x[..., 1], c=theme_color/255, s=scale,
               alpha=0.7, edgecolors='none')  #将主题色赋予斑点
    ax.grid(True)
    plt.show()
with open("./data_processed/themes_1641623409.4860697.json") as f:
    themes = json.load(f)
themeColor_impression(np.array(themes))
```

参考文献（References）:

[1] Wei-Meng Lee (2019). Python Machine Learning. Wiley.
[2] Bonaccorso, G. (2020;2018). Mastering Machine Learning Algorithms Expert techniques for implementing popular machine learning algorithms, fine-tuning your models, and understanding how they work. Packt Publishing.

3.14 图像分类器、识别器与机器学习模型网络实验应用平台部署

该部分的目的是阐述如何将机器学习模型部署到网络平台上，同时包括信息的收集，因此主要包括两大部分内容：一是，应用Flask构建网络应用平台，这包括网络平台的基本架构、问卷调研和将机器学习模型图像识别器嵌入到网络中；二是，构建图像分类器和识别器，这将涉及机器学习相关内容下的分类模型决策树（Decision trees）和随机森林（Random forests），而作为学习的样本特征的提取则使用了视觉词袋的方法来构建图像映射特征用于模型的训练。

3.14.1 用Flask构建实验用网络应用平台

在很多场景中，都需要借助网络完成相关任务，例如展示研究内容、开展问卷调查收集数据、提供服务（例如机器学习或深度学习中已训练好的模型在线预测等）。因为这些任务需要更多的自由性，例如读写和处理数据，也需要根据不同的任务调整网页内容，而使用类似WIX[①]快速网页构建服务提供的方式很难满足数据分析、可视化、模型预测服务及布局调研信息等方面要求，Flask[②]则为空间数据分析研究最好的网络实验构建平台之一。一方面，Flask是应用Python语言编写，是数据分析、大数据分析、机器学习和深度学习广泛使用的语言；同时，Flask是一个轻量级的可定制框架，灵活、轻便、安全，容易上手，能够快速地学习并实现自行搭建网络平台完成相关网络实验部署；Flask也并不限制使用何种数据库，何种模板样式，具有强劲的自由拓展性，实现自由的定制需求。

学习Flask推荐阅读官方文档[③]，以及教材*Flask Web Development: Developing Web Applications with Python*（FWD）。城市空间数据分析网络实验平台的建设，是以上述教材提供的案例为基础（包含社交博客搭建），在此之上扩展不同的实验任务。网络实验的内容及架构代码位于GitHub上caDesign_ExperimentPlatform[④]代码仓库中。

Flask目前已经完全的整合进了PyCharm[⑤]，因此推荐使用PyCharm构建Flask网络实验平台，无须自行搭建Flask的Python环境。同时，对于网页的开发，PyCharm提供了非常友好的书写代码的环境，能够节约研究者搭建的时间。FWD教材提供的案例电子邮件传输协议（SMTP）基于Google Gmail，如果在中国可以使用QQ邮箱提供的服务。FWD教材的写作方式是按照学习的过程递进地讲述，后续代码要基于前述的代码，因此对于初学者跳跃式的阅读并不是很好的选择。虽然递进地讲述符合学习的规律，但是作者并未给出整体的结构框架，丰富的代码类和函数不断更新，变量之间关系的复杂性，容易让读者失去方向，因此给出代码工程的统一建模语言（Unified Modeling Language，UML）很必要。

借助代码逆向工程（Reverse Engineering），使用Visual Paradigm（VP）[⑥]完成部分代码的逆向工程实现。目前，除了VP，还有Pyreverse[⑦]等大量逆向工程工具自动生成UML图表。但是，因为代码的复杂性和结构的丰富性，目前很难找到自动生成全部关联的工具，因此图3.14–1对于FWD教材案例的UML图表绘制，大部分为手工添加。对Flask结构的把握，主要包括：

A-配置+初始化（主程序）；

B-模板和页面；

C-路由与视图函数；

D-Web表单；

E-数据库模型（基于SQLAlchemy）；

F-数据库（SQLite）等7大部分。

Flask的编写也是抓住这几个部分的关系，完成不同功能实现。因为代码书写的关键是不断地调试来验证已有代码是否顺利运行，达到书写的目的。而Flask是网页的开发，因此对于代码的验证，模板页面部分需要查看页面是否显示正常；数据处理部分仍然可以不断地用`print()`方法，打印的结果会显示在运行窗口下。建议代码书写前执行`set FLASK_DEBUG=1`，这样运行窗口下的错误提示，也将在页面中显示。对于简单的Flask开发，因为实现功能简单，因此实现函数通常位于同一文件之下。但是，如果项目工程比较大，众多实现功能如果不加以明确区分，往往造成代码的混乱，因此Flask引入了蓝本（blueprint）的概念。简单来讲，就是将不同功能实现放置

① WIX，创建、设计、管理自定义网站开发平台（https://www.wix.com/）。

② Flask用Python编写的微型网络框架，为微框架（microframework），因为它没有数据库抽象层、表单验证或其他任何预先存在的第三方库提供通用功能的组件。然而，Flask支持扩展，可以增加应用功能，例如对象关系映射器、表单验证、上传处理、各种开放的认证技术和一些常见的框架相关工具都有扩展[（https://pypi.org/project/Flask/）。

③ Flask推荐阅读官方文档（https://flask.palletsprojects.com/en/2.2.x/）。

④ caDesign_ExperimentPlatform，项目实验GitHub仓库地址（https://github.com/richieBao/caDesign_ExperimentPlatform）。

⑤ PyCharm，用于计算机编程的集成开发环境(IDE)，主要用于Python语言开发，提供代码分析、图形化调试器，集成测试器、集成版本控制系统，并支持使用Django进行网页开发（https://www.jetbrains.com/pycharm/）。

⑥ Visual Paradigm（VP），一套设计、分析和管理工具，推动IT项目开发和数字化转型，包括代码逆向工程（https://www.visual-paradigm.com/）。

⑦ Pyreverse，一套用于逆向工程的Python代码的工具，使用一个Python项目的类层次结构表示，可以用来提取任何信息（如生成UML图或单元测试，如pyargo和py2tests）（https://pypi.org/project/pyreverse/）。

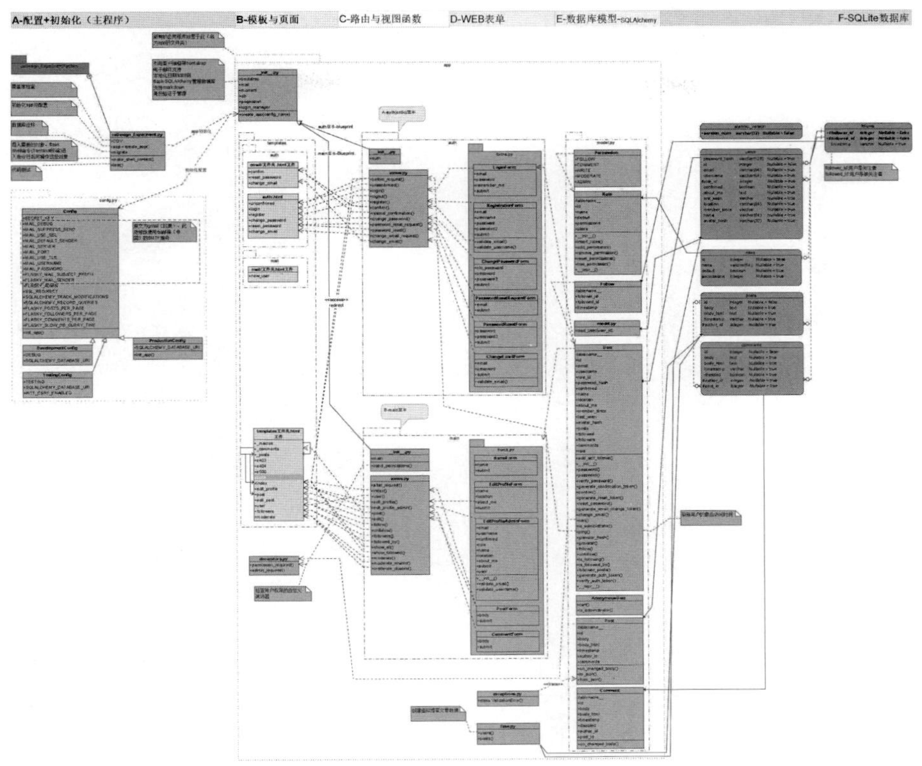

图3.14-1 实验用网络应用平台的UML
（大图获取地址：caDesign_ExperimentPlatform，https://github.com/richieBao/caDesign_ExperimentPlatform）

于不同的文件夹（包）下，并构建子文件夹下代码与主程序代码的关联，可以互相调用方法、函数和属性。网络实验部署是采用蓝图的大型应用工厂，使用一个应用进程得到多个应用实例，易于操作。对于Flask大型应用的把握，需要一开始查看文件夹的结构，这反映了当前应用是如何用Flask架构的。

对Flask的理解需要把握应用包（文件夹）的结构，蓝本实现的方法，配置与初始化的关系，路由与视图函数的关系，视图函数与模板页面的关系，Web表单与视图函数的关系，应用数据库模型（表单）读写数据库（SQLite）和与视图函数的关系，以及显示和隐式的代码关系。对隐式代码关系的理解尤为重要，因为突然冒出来的属性变量往往找不到源头，这因为Flask已经帮助完成了相关的任务，只给了输出。同时，因为涉及页面模板，需要动态地读写视图函数的数据，同时也需要在页面模板内处理数据，Flask应用的Jinjia①模板引擎实现这些功能，其语法也遵循大部分程序语言的结构。

网络实验平台的工程名为“caDesign_ExperimentPlatform”（即根目录），配置（config.py）和主程序（caDesign_Experiment.py），SQLite数据库（data-dev.sqlite）位于根目录下，migrations为数据库迁移文件夹（可以管理工程版本），test文件夹为测试内容，venv是应用PyCharm建立Flask工程时自动创建的Python环境。所有应用位于app文件包下，static文件夹（系统生成）放置图片，CSS等文件，templates文件夹（系统生成）放置HTML的页面模板，main,auth为FWD教材社交博客的功能应用，data文件夹用于放置相关数据，visual_perception文件夹为视觉感知-基于图像空间分类实验的网络实现，如图3.14-2所示。

下述仅列出了主要的模板页面（图3.14-3），类似博客等页面是嵌套在主页等模板页面中。为了便于将其部署于不同模板页面下，通常将此类模板设计成单独的子模板，其模板名为“_post.html”，方便迁移。此外，还有403、404、500等错误页面模板。

① Jinjia，仿照Django模板，一种现代的、对设计师友好的Python模板语言（引擎），其使用范围广，而且有可选择的沙盒式模板（sandboxed template）执行环境，快速安全（https://jinja.palletsprojects.com/en/2.11.x/）。

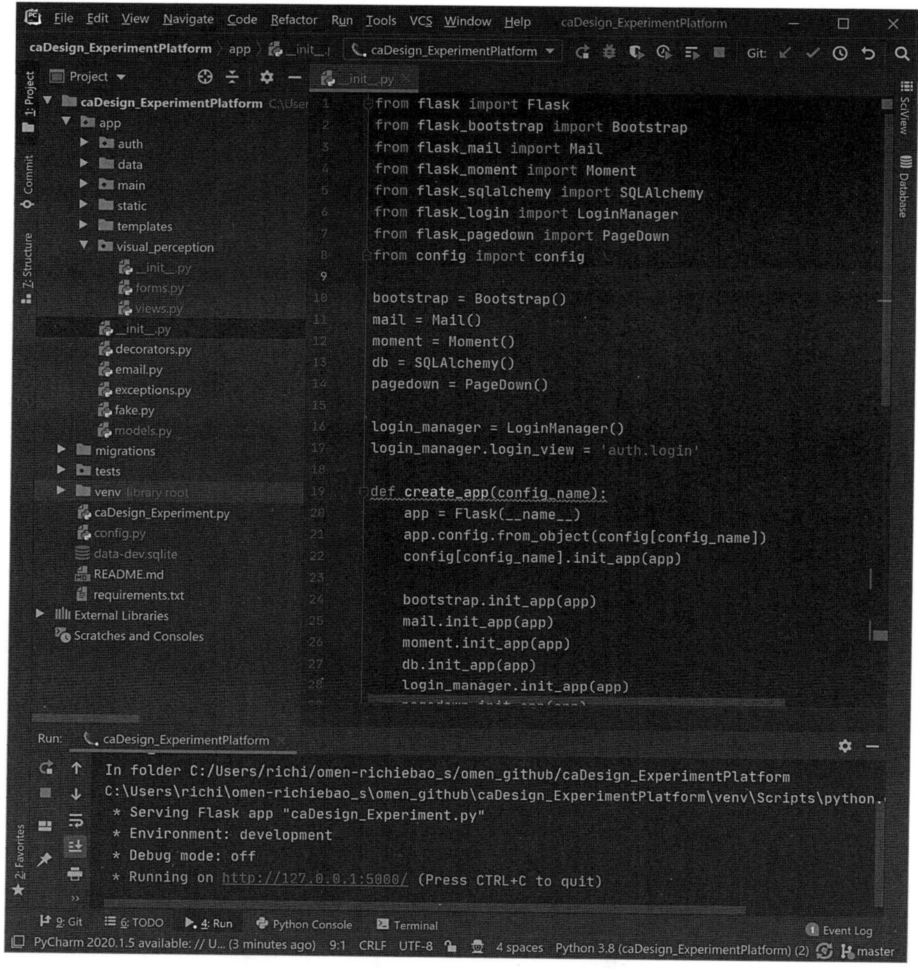

图3.14-2 实验用网络应用平台文件结构

3.14.2 问卷调研

计算机视觉的发展应用领域日益广泛，例如机器人、无人驾驶、文档分析、医疗诊断等智能自主系统，其在规划设计领域的作用也日益凸显，尤其百度、Google的街景图像，以及无人驾驶项目带来的大量序列图像和社交网络的图像，都推动着计算机视觉在规划领域潜在的应用前景。视觉感知部分包括系列实验，例如基于图像的空间分类与城市空间类别分布、图像分割下空间分类识别、视觉评价、绿量研究，以及遥感影像用地类型解译和依据标准的空间生成等内容。

基于图像的空间分类，方法一是，应用Star、SIFT提取特征点（关键点）和描述子，聚类（K-Means）图像特征，进而建立视觉词袋（bag-of-words，BOW）。BOW作为特征向量，输入到图像分类器（例如应用Extremely randomized trees，Extra-Trees/ET）进行训练。训练好的模型作为图像识别器预测新的图像，并应用到更广泛的城市区域内，通过预测的空间分类研究城市类别分布。这个基于图像分类的空间类型可以根据不同的目的分类，例如研究城市空间地面视野的郁密度、空间的开合程度，可以分类有林荫道、窄巷（步行为主）多建筑、窄巷有林木、宽道（1-2条）多建筑、宽道多林木、干道（大于3条，4条居多）多建筑、干道多林木、干道开阔等。方法二是，并不计算图像特征，而是直接应用深度学习的方法训练模型。

如果将多项视觉感知的子项研究综合起来，结合非视觉感知类的分析技术，能够进一步拓展

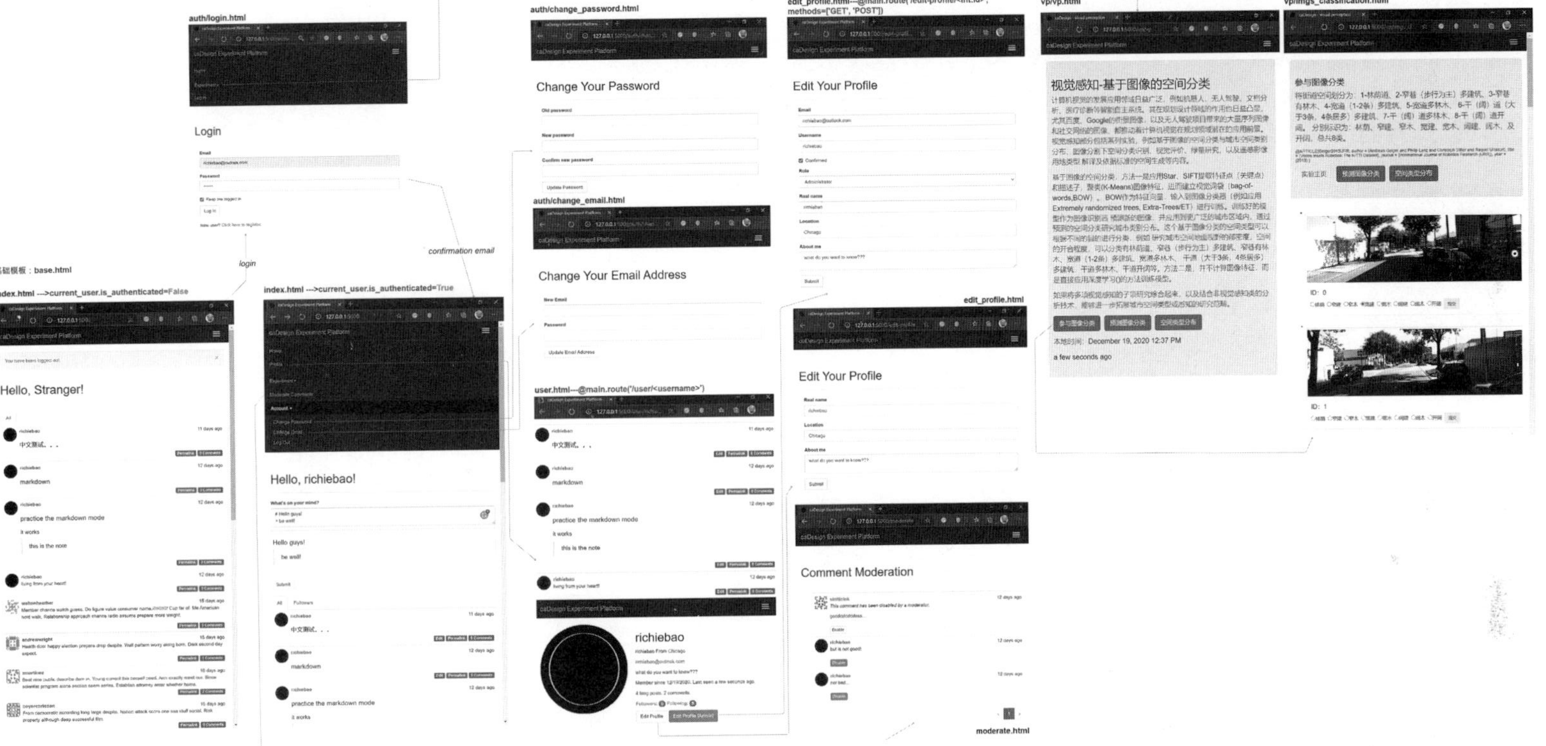

图3.14-3 实验用网络应用平台主要模板页面

（大图获取地址：caDesign_ExperimentPlatform，https://github.com/richieBao/caDesign_ExperimentPlatform）

城市空间类型或感知的研究范畴。视觉感知-基于图像的空间分类：问卷调研部分，是使用KITTI数据集中的图像作为城市空间识别的素材，并以FWD教材案例为网络实验平台的基础，在此基础上扩展实验部分内容。其下代码是迁移了指定路径下返回所有文件夹及其下文件结构的代码，列出了caDesign_ExperimentPlatform网络实验平台下app应用文件夹的文件结构，在FWD教材案例基础上增加了文件夹（蓝本）visual_perception，该阶段包括__init__.py，forms.py和views.py3个文件模块。其中，forms.py中基于wtforms库定义了问卷调研表格。因为单选按钮为纵向排列，并没有使用，而是直接在.html模板中直接自行定义。模板文件夹templates下新增了vp文件夹，该阶段包括vp.html，imgs_classification.html两个文件，分别为“视觉感知-基于图像的空间分类”的说明导航首页，以及“参与图像分类”的问卷调研页。

```
from pathlib import Path
import util_misc
app_root = r'C:\Users\richi\omen_richiebao\omen_github\caDesign_
ExperimentPlatform\app'
paths = util_misc.DisplayablePath.make_tree(Path(app_root))
for path in paths:
    print(path.displayable())
--------------------------------------------------------------------------------
app/
├── __init__.py
├── __pycache__/
│   ├── __init__.cpython-39.pyc
│   ├── decorators.cpython-39.pyc
│   ├── email.cpython-39.pyc
│   ├── exceptions.cpython-39.pyc
│   ├── fake.cpython-39.pyc
│   └── models.cpython-39.pyc
├── auth/
│   ├── __init__.py
│   ├── __pycache__/
│   │   ├── __init__.cpython-39.pyc
│   │   ├── forms.cpython-39.pyc
│   │   └── views.cpython-39.pyc
│   ├── forms.py
│   └── views.py
├── data/
│   └── info_2011_09_26_drive_0009_sync.pkl
├── decorators.py
├── email.py
├── exceptions.py
├── fake.py
├── main/
│   ├── __init__.py
│   ├── __pycache__/
│   │   ├── __init__.cpython-39.pyc
│   │   ├── errors.cpython-39.pyc
│   │   ├── forms.cpython-39.pyc
│   │   └── views.cpython-39.pyc
│   ├── errors.py
│   ├── forms.py
│   └── views.py
├── models.py
├── static/
│   ├── favicon.ico
│   ├── KITTI/
│   │   ├── 2011_09_26_drive_0002_sync/
│   │   │   ├── 0000000006.png
│   │   │   ├── 0000000008.png
│   │   │   ├── 0000000018.png
│   │   │   ├── 0000000020.png
│   │   │   ├── 0000000031.png
│   │   │   ├── 0000000038.png
│   │   │   ├── 0000000045.png
│   │   │   └── 0000000069.png
│   │   ├── 2011_09_26_drive_0005_sync/
│   │   │   ├── 0000000001.png
│   │   │   ├── 0000000006.png
│   └── styles.css
├── templates/
│   ├── 403.html
```

```
│   ├── 404.html
│   ├── 500.html
│   ├── _comments.html
│   ├── _macros.html
│   ├── _posts.html
│   ├── auth/
│   │   ├── change_email.html
│   │   ├── change_password.html
│   │   ├── email/
│   │   │   ├── change_email.html
│   │   │   ├── change_email.txt
│   │   │   ├── confirm.html
│   │   │   ├── confirm.txt
│   │   │   ├── reset_password.html
│   │   │   └── reset_password.txt
│   │   ├── login.html
│   │   ├── register.html
│   │   ├── reset_password.html
│   │   └── unconfirmed.html
│   ├── base.html
│   ├── edit_post.html
│   ├── edit_profile.html
│   ├── followers.html
│   ├── index.html
│   ├── mail/
│   │   ├── new_user.html
│   │   └── new_user.txt
│   ├── moderate.html
│   ├── post.html
│   ├── user.html
│   └── vp/
│       ├── img_prediction.html
│       ├── imgs_classification.html
│       └── vp.html
├── uploads/
│   ├── 0000000020.png
│   ├── 0000000020_1.png
│   ├── 0000000030.png
│   ├── 0000000069.png
│   ├── 0000000069_1.png
│   ├── 01.png
│   ├── 02.png
│   ├── 05.png
│   └── 06.png
└── visual_perception/
    ├── __init__.py
    ├── __pycache__/
    │   ├── __init__.cpython-39.pyc
    │   ├── forms.cpython-39.pyc
    │   ├── ImageTag_extractor.cpython-39.pyc
    │   └── views.cpython-39.pyc
    ├── forms.py
    ├── ImageTag_extractor.py
    ├── views.py
    └── vp_model/
        ├── ERF_clf.pkl
        ├── visual_BOW.pkl
        └── visual_feature.pkl
```

3.14.2.1 配置视觉感知-基于图像的空间分类蓝本

当建立一个新的子项目时，是在app下建立一个独属的文件夹（蓝本/子包），此处视觉感知实验的蓝本文件夹名为visual_perception。并在子包中新建__init__.py文件，调入Blueprint，实例化一个蓝本类对象，并调入views(.py)。把路由与蓝本关联起来，即在全局作用域下可以使用该蓝本下的view.py下视图函数及相关方法。同时，需要在app下的__init__.py文件内对应增加配置。

```
# app/visual_perception/__init__.py
from flask import Blueprint

visual_perception = Blueprint('visual_perception', __name__) #参数：蓝本名称，和蓝本所在的包或模块，默认使用 __name__。
from . import views
```

```python
# app/__init__.py
from .visual_perception import visual_perception as vp_blueprint
app.register_blueprint(vp_blueprint, url_prefix='/vp')
```

3.14.2.2 定义SQLite表结构，写入图像信息

- 定义SQLite表结构

定义两个表结构（模型），class vp_imgs(db.Model)用于存储原始图像的信息，包括图像路径（位于static文件夹下），经纬度和高程信息。class vp_classification(db.Model)表结构，用于存储分类的结果，定义有8个分类，对应c1 ~ c8；同时，也引入图像路径字段imgs_fp，及分类结果classification字段。一开始时，只有vp_imgs表中有数据，是处理好的图像信息，直接写入到该表中。而vp_classification表没有数据，只有在模板页面中点击了单选按钮，选择分类提交后，将该信息写入到该表中。同时，这两个表之间建立了一对一的关系，指定的外键为db.ForeignKey('vp_imgs.index')，并建立了双向关系（back_populates="vp_imgs"，和back_populates="vp_classification"）。这样，可以直接通过一个表的信息读取另一个表的字段信息。例如，在imgs_classification.html模板中，{{ img_info.vp_classification.classification}}的Jinja语句下通过“vp_imgs”表的一个行信息“img_info”，链接到“vp_classification”表中的分类信息“classification”，从而可以在页面中显示当前图像的分类信息。

定义好表结构后，可以在PyCharm的Terminal终端下敲入：flask shell启动shell会话，然后执行from caDesign_Experiment import db调入数据库实例化对象db，“caDesign_Experiment”为对应的app名，再执行db.create_all()，将新建的表结构写入到SQLite数据库（属于库已有的表保持不变）。

```python
# app/models.py
class vp_imgs(db.Model):
    __tablename__ = 'vp_imgs'
    index = db.Column(db.Integer, primary_key=True, autoincrement=True)
    imgs_fp = db.Column(db.Text, unique=True, nullable=False)
    lat = db.Column(db.Float)
    lon = db.Column(db.Float)
    alt = db.Column(db.Float)
    vp_classification = db.relationship(
        'vp_classification', uselist=False, back_populates="vp_imgs")
class vp_classification(db.Model):
    __tablename__ = 'vp_classification'
    id = db.Column(db.Integer, primary_key=True, autoincrement=True)
    imgs_fp = db.Column(db.Text, unique=True, nullable=False)
    c_1 = db.Column(db.Integer)
    c_2 = db.Column(db.Integer)
    c_3 = db.Column(db.Integer)
    c_4 = db.Column(db.Integer)
    c_5 = db.Column(db.Integer)
    c_6 = db.Column(db.Integer)
    c_7 = db.Column(db.Integer)
    c_8 = db.Column(db.Integer)
    classification = db.Column(db.Text)
    timestamp = db.Column(db.DateTime, default=datetime.now)
    index = db.Column(db.Integer, db.ForeignKey('vp_imgs.index'))
    vp_imgs = db.relationship('vp_imgs', back_populates="vp_classification")
```

- 写入图像信息

先定义表结构，并写入数据库后再向表中写入数据，而不是应用Pandas的方法直接默认表结构写入到数据库。这是因为要建立表之间的关系，默认的方式则无法建立表关系。定义函数imgs_compression_cv()实现图像的压缩，因为图像要在网络页面中显示，较大的图像加载速度慢，影响体验。函数KITTI_info_gap()是针对KITTI数据集的操作，因为该数据集是用于无人驾驶场景下计算机视觉算法评测数据集，图像连续，因此通过该函数可以处理KITTI提供的.txt（包含经纬度，高程等信息）文件，保持数据与图像压缩函数给定的参数gap保持一致，即隔一段距离提取一张图像。KITTI_info2sqlite()函数，则是将提取的图像信息写入到数据库表中，因为表已经存在，因此使用method="append"方法。

```
def imgs_compression_cv(imgs_root, imwrite_root, imgsPath_fp, gap=1, png_
compression=9, jpg_quality=100):
    '''
    function - 使用OpenCV的方法压缩保存图像

    Params:
        imgs_root - 待处理的图像文件根目录; string
        imwrite_root - 图像保存根目录; string
        gap - 无人驾驶场景下的图像通常是紧密连续的，可以剔除部分图像避免干扰，默认值为1; int
        png_compression - png格式压缩值，默认为9。值域为 0 - 9(越大表示尺寸越小，压缩时
间越长); int
        jpg_quality - jpg格式压缩值，默认为100。值域为 0 - 100(越高越好); int

    Returns:
        imgs_save_fp - 保存压缩图像文件路径列表; list(string)
    '''
    from pathlib import Path
    import cv2 as cv
    import numpy as np
    from tqdm import tqdm
    import os
    import pandas as pd
    if not os.path.exists(imwrite_root):
        os.makedirs(imwrite_root)
    imgs_root = Path(imgs_root)
    imgs_fp = [p for p in imgs_root.iterdir()][::gap]
    imgs_save_fp = []
    for img_fp in tqdm(imgs_fp):
        img_save_fp = str(Path(imwrite_root).joinpath(img_fp.name))
        img = cv.imread(str(img_fp))
        if img_fp.suffix == '.png':
            cv.imwrite(img_save_fp, img, [
                       int(cv.IMWRITE_PNG_COMPRESSION), png_compression])
            imgs_save_fp.append(img_save_fp)
        elif img_fp.suffix == '.jpg':
            cv.imwrite(img_save_fp, img, [
                       int(cv.IMWRITE_JPEG_QUALITY), jpg_quality])
            imgs_save_fp.append(strimg_save_fp)
        else:
            print("Only .jpg and .png format files are supported.")
    pd.DataFrame(imgs_save_fp, columns=['imgs_fp']).to_pickle(imgsPath_fp)
    return imgs_save_fp
def KITTI_info_gap(KITTI_info_fp, save_fp, gap=1):
    '''
    function - 读取KITTI文件信息，1-包括经纬度，惯性导航系统信息等的.txt文件。只返回经纬度、
海拔信息

    Params:
        KITTI_info_fp - 数据文件路径; string
        save_fp - 文件保存路径; string
        gap - 间隔连续剔除部分图像避免干扰，默认值为1; int

    Returns:
        drive_info_coordi - 返回经纬度和海拔信息; DataFrame

    '''
    import pandas as pd
    from pathlib import Path
    txt_root = Path(KITTI_info_fp)
    txt_fp = [str(p) for p in txt_root.iterdir()][::gap]
    columns = ["lat", "lon", "alt", "roll", "pitch", "yaw", "vn", "ve", "vf", "vl",
"vu", "ax", "ay", "ay", "af", "al", "au",
               "wx", "wy", "wz", "wf", "wl", "wu", "pos_accuracy", "vel_accuracy", "navstat",
"numsats", "posmode", "velmode", "orimode"]
    drive_info = pd.concat(
        [pd.read_csv(item, delimiter=' ', header=None) for item in txt_fp],
axis=0)
    drive_info.columns = columns
    drive_info = drive_info.reset_index()
    drive_info_coordi = drive_info[["lat", "lon", "alt"]]
    drive_info_coordi.to_pickle(save_fp)
    return drive_info_coordi
def KITTI_info2sqlite(imgsPath_fp, info_fp, replace_path, db_fp, table,
method='fail'):
```

```
    '''
    function - 将 KITTI 图像路径与经纬度信息对应起来，并存入 SQLite 数据库

    Params:
        imgsPath_fp - 图像文件路径；string
        info_fp - 图像信息文件路径；string
        replace_path - 替换路径名；string
        db_fp - SQLite 数据库路径；string
        table - 数据库表名；string
        method - 包括 fail，replace，append 等。The default is'fail'；string

    Returns:
        None
    '''
    from pathlib import Path
    import pandas as pd
    from sqlalchemy import create_engine
    imgsPath = pd.read_pickle(imgsPath_fp)
    # flask Jinja 的 url_for 仅支持' /,因此需要替换' \\'
    imgsPath_replace = imgsPath.imgs_fp.apply(lambda row: str(
        Path(replace_path).joinpath(Path(row).name)).replace('\\', '/'))
    info = pd.read_pickle(info_fp)
    imgs_df = pd.concat([imgsPath_replace, info], axis=1)
    engine = create_engine(
        'sqlite:///'+'\\\\'.join(db_fp.split('\\')), echo=True)
    try:
        imgs_df.to_sql('%s' % table, con=engine,
                       index=False, if_exists="%s" % method)
        print("if_exists=%s:------Data has been written to the database!" %
method)
    except:
        print("_"*15, 'the %s table has been existed...' % table)
if __name__ == "__main__":
    imgs_paths = [r'2011_09_26_drive_0009_sync',
                  r'2011_09_29_drive_0071_sync',
                  r'2011_09_28_drive_0001_sync',
                  r'2011_09_29_drive_0026_sync',
                  r'2011_09_28_drive_0002_sync',
                  r'2011_09_26_drive_0117_sync',
                  r'2011_09_26_drive_0113_sync',
                  r'2011_09_26_drive_0106_sync',
                  r'2011_09_26_drive_0104_sync',
                  r'2011_09_26_drive_0096_sync',
                  r'2011_09_26_drive_0095_sync',
                  r'2011_09_26_drive_0093_sync',
                  r'2011_09_26_drive_0084_sync',
                  r'2011_09_26_drive_0060_sync',
                  r'2011_09_26_drive_0059_sync',
                  r'2011_09_26_drive_0057_sync',
                  r'2011_09_26_drive_0056_sync',
                  r'2011_09_26_drive_0051_sync',
                  r'2011_09_26_drive_0048_sync',
                  r'2011_09_26_drive_0018_sync',
                  r'2011_09_26_drive_0017_sync',
                  r'2011_09_26_drive_0014_sync',
                  r'2011_09_26_drive_0013_sync',
                  r'2011_09_26_drive_0011_sync',
                  r'2011_09_26_drive_0005_sync',
                  r'2011_09_26_drive_0002_sync',
                  ]
    g = 10
    i = 0
    for p in imgs_paths:
        print("\n%d-%s--- 处理中 ..." % (-(len(imgs_paths[1:])-i), p))
        imgs_path = p
        # A - 使用 OpenCV 的方法压缩保存图像
        imgs_root = r'D:\dataset\KITTI\%s\image_03\data' % imgs_path
        imwrite_root = r'D:\dataset\KITTI\imgs_compression\%s' % imgs_path
        imgsPath_fp = r'D:\dataset\KITTI\imgs_compression\imgsPath_%s.pkl' %
imgs_path
        imgs_save_fp = imgs_compression_cv(
            imgs_root, imwrite_root, imgsPath_fp, gap=g)

        # B - 读取 KITTI 经纬度信息，可以指定间隔提取距离
```

```
        KITTI_info_fp = r'D:\dataset\KITTI\%s\oxts\data' % imgs_path
        save_fp = r'D:\dataset\KITTI\imgs_compression\info_%s.pkl' % imgs_path
        drive_info = KITTI_info_gap(KITTI_info_fp, save_fp, gap=g)

        #C - 将文件路径信息写入数据库
        imgsPath_fp = r'D:\dataset\KITTI\imgs_compression\imgsPath_%s.pkl' %
imgs_path
        info_fp = r'D:\dataset\KITTI\imgs_compression\info_%s.pkl' % imgs_path
        replace_path = r'KITTI\%s' % imgs_path
        db_fp = r'C:\Users\richi\omen-richiebao_s\omen_github\caDesign_
ExperimentPlatform\data-dev.sqlite'
        KITTI_info2sqlite(imgsPath_fp, info_fp, replace_path,
                          db_fp, table='vp_imgs', method="append")
        i += 1
```

3.14.2.3 定义路由和视图函数

配置完蓝本之后，在app/visual_perception/views.py文件下定义视图函数，指定路由（对应的模板页面）。首先，可以定义一个简单的视图函数，例如用Flask官方那个最简单的代码来测试是否蓝本配置成功：

```
@app.route('/hello')
def hello_world():
    return 'Hello, World!'
```

只是需要重新分配一个路由，例如上述修改为`@app.route('/hello')`，这样可以在`http://127.0.0.1:5000/hello`统一资源定位系统（Uniform Resource Locater，URL），即网页地址下打开。如果返回显示“Hello, World!”，则可以说明蓝本配置无误。目前，包括两个模板页面vp.html、imgs_classification.html，分别对应视图函数`vp()`和`imgs_classification()`。`vp()`视图函数网页地址（路由）指向`"/vp"`，对应模板vp.html下的内容比较简单，只是发布该实验项目的说明，并向模板中传入了`current_time=datetime.datetime.utcnow()`参数，可以在模板页面下显示本地时间。`imgs_classification()`视图函数页面指向`"/imgs_classification"`，对应模板页面imgs_classification.html，因为对图像分类，需要读写数据库并显示图像，表单提交等动作要稍显复杂。

首先，确定分类的标准，将街道空间划分为：

1. 林荫道；
2. 窄巷（步行为主）多建筑；
3. 窄巷有林木；
4. 宽道（1～2条）多建筑；
5. 宽道多林木；
6. 干（阔）道（大于3条，4条居多）多建筑；
7. 干（阔）道多林木；
8. 干（阔）道开阔。

分别标识为：林荫、窄建、窄木、宽建、宽木、阔建、阔木及开阔，总共8类。根据分类信息，可以在模板中定义表单，定义对应的数据库表结构。视图函数则需要思考“问卷调研”的动作：

1. 首先，是读取图像路径信息，传入模板后可以显示图像。
2. 每一图像下对应表单，有8个单选按钮，当选择其中之一后，点击提交按钮，表单信息将返回到视图函数中（POST）。
3. 在视图函数中读取表单信息，将其写入到数据库中（这涉及数据表结构的设计），不同人点击的分类可能不同。当单击一个分类，则对应写入该分类的数据库表字段中，计数方式为累加。确定图像的最终分类是对应哪个分类的累加数最多。
4. 因为对图像给了分类，分类信息也被写入到对应表中，则可以将分类结果显示在页面中。

```
# app/visual_perception/views.py
from . import visual_perception
from flask import render_template, url_for, request, session, current_app,
redirect
import datetime
from .. import db
```

```
from ..models import vp_imgs, vp_classification
import pandas as pd
from .forms import imgs_classi
@visual_perception.route("/vp", methods=['GET', 'POST'])
def vp():
    return render_template('vp/vp.html', current_time=datetime.datetime.utcnow())
@visual_perception.route("/imgs_classification", methods=['GET', 'POST'])
def imgs_classification():
    page = request.args.get('page', 1, type=int)
    query = vp_imgs.query
    pagination = query.order_by(vp_imgs.index).paginate(
        page, per_page=current_app.config['FLASKY_POSTS_PER_PAGE'], error_
out=False)
    imgs_info = pagination.items
    vp_classi = vp_classification.query.all()
    if request.method == 'GET':
        return render_template('vp/imgs_classification.html', imgs_info=imgs_
info, vp_classi=vp_classi, pagination=pagination)
    else:
        img_index = request.form.get('img_index')
        img_fp = request.form.get('img_fp')
        classi = int(request.form.get('classi'))
        img_current = vp_classification.query.filter(
            vp_classification.imgs_fp == img_fp).first()
        classi_dic_value = {1: 0, 2: 0, 3: 0, 4: 0, 5: 0, 6: 0, 7: 0, 8: 0}
        # 中文一定要加 u，即 unicode( str_name )，否则服务器段如果是 py2.7，会提示错误。
        classi_dic_name = {1: u'林荫', 2: u'窄建', 3: u'窄木',
                           4: u'宽建', 5: u'宽木', 6: u'阔建', 7: u'阔木', 8: u'开阔'}
        classi_dic_value.update({classi: 1})
        img_classification = classi_dic_name[classi]
        if not img_current:
            img_classi_info = vp_classification(imgs_fp=img_fp,
                                                c_1=classi_dic_value[1],
                                                c_2=classi_dic_value[2],
                                                c_3=classi_dic_value[3],
                                                c_4=classi_dic_value[4],
                                                c_5=classi_dic_value[5],
                                                c_6=classi_dic_value[6],
                                                c_7=classi_dic_value[7],
                                                c_8=classi_dic_value[8],
                                                classification=img_
classification, index=img_index)
            db.session.add(img_classi_info)
            db.session.commit()
        else:
            query_results = [{1: c_1, 2: c_2, 3: c_3, 4: c_4, 5: c_5, 6: c_6, 7:
c_7, 8: c_8} for c_1, c_2, c_3, c_4, c_5, c_6, c_7, c_8 in db.session.query(
                vp_classification.c_1, vp_classification.c_2, vp_classification.
c_3, vp_classification.c_4, vp_classification.c_5, vp_classification.c_6, vp_
classification.c_7, vp_classification.c_8).filter(vp_classification.imgs_fp ==
img_fp)][0]
            query_results_update = pd.DataFrame.from_dict(query_results,
orient='index').add(
                pd.DataFrame.from_dict(classi_dic_value, orient='index'))
            img_classification_ = classi_dic_name[query_results_update.idxmax()[
                0]]
            query_results_update_dic = query_results_update.squeeze(
                'columns').to_dict()
            query_results_update_dic.update(
                {'classification': img_classification_})
            query_results_update_dic_ = dict(zip(
              ['c_1', 'c_2', 'c_3', 'c_4', 'c_5', 'c_6', 'c_7', 'c_8', 'classification'],
query_results_update_dic.values()))
            vp_classification.query.filter_by(
                imgs_fp=img_fp).update(query_results_update_dic_)
            db.session.commit()
    return render_template('vp/imgs_classification.html', imgs_info=imgs_info,
vp_classi=vp_classi, pagination=pagination)
```

3.14.2.4 定义模板

模板的定义中需要注意对Jinja（2）的使用，Jinja是现代而又设计友好，用于Python的模板语言，起源于Django，结果如图3.14-4所示。

图3.14-4 模板页面

```
<!--templates/vp/vp.html -->
{% extends "base.html" %}
{% import "bootstrap/wtf.html" as wtf %}
{% block title %}caDesign - visual perception{% endblock %}
{% block page_content %}
    <div class="jumbotron">
        <h1>视觉感知 - 基于图像的空间分类</h1>
        <p>
            计算机视觉的发展应用邻域日益广泛，例如机器人、无人驾驶、文档分析、医疗诊断等智能自
主系统。其在规划设计领域的作用也日益凸显，尤其百度、Google 的街景图像，以及无人驾驶项目带来的大量
序列图像和社交网络的图像，都推动着计算机视觉在规划领域潜在的应用前景。
    视觉感知部分包括系列实验，例如基于图像的空间分类与城市空间类别分布、图像分割下空间分类识别、
视觉评价、绿量研究，以及遥感影像用地类型解译及依据标准的空间生成等内容。
        </p>
        <p>
            基于图像的空间分类，方法一是应用 Star、SIFT 提取特征点（关键点）和描述子，聚类（K-Means）
图像特征，进而建立视觉词袋（bag-of-words,BOW）。
    BOW 作为特征向量，输入到图像分类器（例如，应用 Extremely randomized trees, Extra-Trees/ET）进行训练。
训练好的模型作为图像识别器预测新的图像，并应用到更广泛的城市区域内，通过预测的空间分类研究城市类
别分布。这个基于图像分类的空间类型，可以根据不同的目的进行分类。例如，研究城市空间地面视野的郁密
度、空间的开合程度，可以分类有林荫道、窄巷（步行为主）多建筑、窄巷有林木、宽道（1 ~ 2 条）多建筑、
宽道多林木、
    干道（大于 3 条，4 条居多）多建筑、干道多林木、干道开阔等。方法二是，并不计算图像特征，而是
直接应用深度学习 () 的方法训练模型。
        </p>
        <p>
            如果将多项视觉感知的子项研究综合起来，以及结合非视觉感知类的分析技术，能够进一步拓
展城市空间类型或感知的研究范畴。
        </p>
        <p>
            <a class="btn btn-primary btn-lg" href="{{ url_for('visual_
perception.imgs_classification') }}" role="button">参与图像分类</a>
             <a class="btn btn-primary btn-lg" href="{{ url_for('visual_
perception.vp') }}" role="button">预测图像分类</a>
             <a class="btn btn-primary btn-lg" href="{{ url_for('visual_
perception.vp') }}" role="button">空间类型分布</a>
        </p>
        <p>本地时间：{{ moment(current_time).format('LLL') }}</p>
        <p>{{ moment(current_time).fromNow(refresh=True) }}</p>
    </div>
{% endblock %}
--------------------------------------------------------------------------------
<!--templates/vp/imgs_classification.html -->
{% extends "base.html" %}
{% import "bootstrap/wtf.html" as wtf %}
{% import "_macros.html" as macros %}
{% block title %}caDesign - visual perception{% endblock %}
{% block page_content %}
    <div class="jumbotron">
        <h3>参与图像分类</h3>
        <p>
            将街道空间划分为：1- 林荫道、2- 窄巷（步行为主）多建筑、3- 窄巷有林木、4- 宽道（1-2 条）
多建筑、5- 宽道多林木、6- 干（阔）道（大于 3 条，4 条居多）多建筑、7- 干（阔）道多林木、8- 干（阔）道开阔。
```

```
分别标识为：林荫、窄建、窄木、宽建、宽木、阔建、阔木及开阔，总共 8 类。
        </p>
        <h6>@ARTICLE{Geiger2013IJRR, author = {Andreas Geiger and Philip Lenz
and Christoph Stiller and Raquel Urtasun}, title = {Vision meets Robotics: The
KITTI Dataset}, journal = {International Journal of Robotics Research (IJRR)},
year = {2013} }</h6>
        <p>
            <a class="btn btn-secondary btn-lg" href="{{ url_for('visual_
perception.vp') }}" role="button">实验主页 </a>
             <a class="btn btn-primary btn-lg" href="{{ url_for('visual_
perception.vp') }}" role="button">预测图像分类 </a>
             <a class="btn btn-primary btn-lg" href="{{ url_for('visual_
perception.vp') }}" role="button">空间类型分布 </a>
        </p>
    </div>
    <ul class="question-list-group">
        {% for img_info in imgs_info %}
            <li style="float:left">
                <div class="row">
                    <div class=" col-md-10">
                        <div class="thumbnail">
                            <img src="{{ url_for('static',filename=img_info.
imgs_fp[7:]) }}" alt="">
                            <div class="caption">
                                <h4>ID: {{ img_info.index }}
                                    {{ img_info.vp_classification.
classification}}
                                </h4>
                                        <iframe name="formDestination"
class="iframe", style="display:none;"></iframe>
                                        <form action="" method="post"
target="formDestination">
                                        <input type="radio" name="classi"
value="1"/> 林荫  
                                        <input type="radio" name="classi"
value="2"/> 窄建  
                                        <input type="radio" name="classi"
value="3"/> 窄木  
                                        <input type="radio" name="classi"
value="4"/> 宽建  
                                        <input type="radio" name="classi"
value="5"/> 宽木  
                                        <input type="radio" name="classi"
value="6"/> 阔建  
                                        <input type="radio" name="classi"
value="7"/> 阔木  
                                        <input type="radio" name="classi"
value="8"/> 开阔  

                                        <input type="hidden" name="img_index"
value="{{img_info.index }}">
                                        <input type="hidden" name="img_fp"
value="{{ img_info.imgs_fp}}">

                                        <input type="submit" value=" 提交 "
class="btn btn-secondary btn-sm" onClick="this.form.submit(); this.disabled=true;
this.value=' 已提交 '; ">
                                    </form>
                            </div>
                        </div>
                    </div>
                </div>
            </li>
        {% endfor %}
    </ul>
{% if pagination %}
<div class="pagination">
    {{ macros.pagination_widget(pagination, 'visual_perception.imgs_
classification') }}
</div>
{% endif %}
{% endblock %}
```

3.14.3 视觉词袋与构建图像映射特征

计算每一张图像的关键点描述子（一个关键点描述子有128维，一幅图像有多个关键点），把所有图像的关键点描述子集合在一起，就是所有图像特征的集合。因为每一关键点描述子均不同，因此使用聚类的方法指定聚类的数量（32）聚合特征，即聚合所有图像的关键描述子为指定数量构造码本（Bag of Words，BOW），并建立聚类模型。该码本就是所有图像特征（聚类后）的集合，因为每一图像的关键点描述子都可以在该码本中找到对应的特征（32个聚类的1个或多个聚类组合），因此可以用码本编码每一图像（用聚类模型预测）。保持码本的形状（shape，32），将一幅图像的所有关键点描述子对应到码本的编号上，因为聚类的结果，一个编号通常对应有多个关键点，计算所有对应到编号上关键点的频数（有的编码频数也会为0，即没有对应的关键点），这个对应码本编号频数的结果就反映了该图像特征。上述过程可由图3.14–5描述。

码本是将所有可用的码（聚类）放在一起，组成类似字典的表，用序号给不同的码编号（或者是列表的排序）。如果要对一个单词（或一句话）编码，则可以应用码本编号。

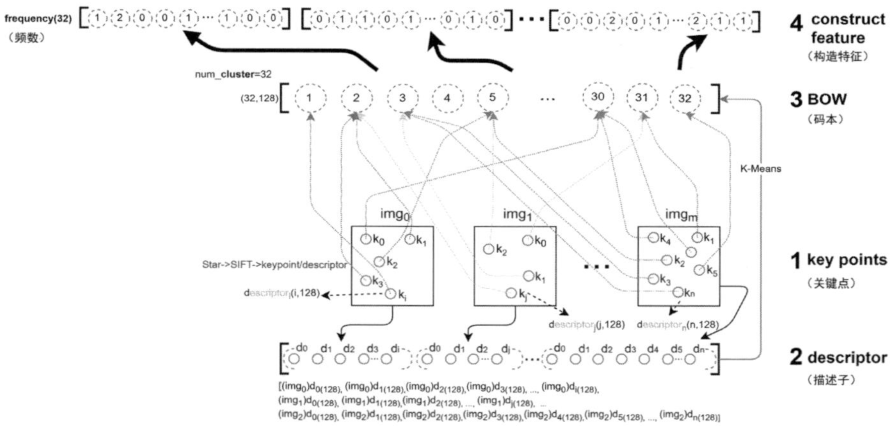

图3.14–5 构建图像映射特征的过程逻辑

读取“问卷调研”的数据库结果（表3.14–1）。因为数据库中存储的图像路径为相对路径，而此时的代码可能位于他处，因此将其转换为绝对路径。

表3.14–1 问卷调研数据库

	id	imgs_fp	c_1	c_2	c_3	c_4	c_5	c_6	c_7	c_8	classification	timestamp	index	absolute_imags_fp
0	1	KITTI/2011_09_26_drive_0009_sync/0000000000.png	0	0	0	0	1	0	0	0	宽木	2020-12-20 13:10:36.725045	1	C:\...\caDesign_ExperimentPlatform\app\static\KITTI/2011_09_26_drive_0009_sync/0000000000.png
1	2	KITTI/2011_09_26_drive_0009_sync/0000000010.png	0	0	0	0	1	0	0	0	宽木	2020-12-20 13:10:42.138661	2	C:\...\caDesign_ExperimentPlatform\app\static\KITTI/2011_09_26_drive_0009_sync/0000000010.png

```
import sqlite3
import pandas as pd
import util_misc
import os
db_fp = r'C:\Users\richi\omen_richiebao\omen_github\caDesign_ExperimentPlatform\
```

```
data-dev.sqlite'
vp_classification = pd.read_sql_table(
    'vp_classification', 'sqlite:///%s' % db_fp)  # pd.read_sql_table 从数据库中读取指定的表
vp_classification['absolute_imags_fp'] = vp_classification.apply(lambda row:
os.path.join(
    r'C:\Users\richi\omen-richiebao_s\omen_github\caDesign_ExperimentPlatform\
app\static', row.imgs_fp), axis=1)
util_misc.print_html(vp_classification)
print("分类频数统计：\n",vp_classification.classification.value_counts())
-----------------------------------------------------------------------------
分类频数统计：
阔木    189
窄建    188
宽木    104
开阔     85
窄木     78
宽建     60
阔建     47
林荫     47
Name: classification, dtype: int64
```

以一张图像的分类结果为键，以图像绝对路径为名建立对应到一张图像的字典。然后，将所有图像的字典放置于一个列表中，用于构造所有图像的码本和计算每一图像的特征映射。

```
def load_training_data(imgs_df, classi_field, classi_list, path_field):
    '''
    function - 按照分类提取图像路径与规律

    Params:
        imgs_df - 由 Pandas 读取的数据库表数据，含分类信息；DataFrame
        classi_field - 分类字段；string
        classi_list - 分类列表；list(string)
        path_field - 图像绝对路径名；string

    Returns:
        训练数据；list(dict)
    '''
    import pandas as pd
    imgs_group = imgs_df.groupby(['classification'])
    training_data = [[{'object_class': classi, 'image_path': row} for row in
imgs_group.get_group(
        classi)[path_field].tolist()] for classi in classi_list]
    def flatten_lst(lst): return [m for n_lst in lst for m in flatten_lst(
        n_lst)] if type(lst) is list else [lst]
    return flatten_lst(training_data)

classi_list = ['林荫', '窄建', '窄木', '宽建', '宽木', '阔建', '阔木', '开阔']
classi_field = 'classification'
path_field = 'absolute_imags_fp'
training_data = load_training_data(
    vp_classification, classi_field, classi_list, path_field)
print(training_data[:5])
-----------------------------------------------------------------------------
[{'object_class': '林荫', 'image_path': 'C:\\Users\\richi\\omen-richiebao_
s\\omen_github\\caDesign_ExperimentPlatform\\app\\static\\KITTI/2011_09_26_
drive_0117_sync/0000000657.png'}, ..., {'object_class': '林荫', 'image_path':
'C:\\Users\\richi\\omen-richiebao_s\\omen_github\\caDesign_ExperimentPlatform\\
app\\static\\KITTI/2011_09_26_drive_0117_sync/0000000196.png'}]
```

如果代码量比较大，逐行地分析代码之间互相调用的关系是费时和费力的事情，因此借助逆向工程。代码逆向工程的工具很多，这里使用了Sourcetrail[①]工具。定义feature_builder_BOW构造视觉码本和提取图像映射特征的类之后，主要调用了两个方法，先用`feature_builder_BOW().get_visual_BOW(training_data,)`构造码本，再用`feature_builder_BOW().get_feature_map(training_data,kmeans)`返回每一图像映射特征。从下述逆向工程分析结果（图3.14-6）能够很清楚地梳理出两次类方法的调用所关联调用的其他类中函数。`.get_visual_BOW`函数内调用有`extract_features()`和`visual_BOW()`两个函数，同时显示了调用的内嵌方法，`print`、`enumerate`、`tqdm`等；`.get_feature_map`则调用有`extract_features()`和`normalize()`两个函数，同时使用了类变量`self.num_clusters`。也可以查看`extract_features`分别被`.get_visual_BOW`和`.get_feature_map`调用。

① Sourcetrail，是一个免费和开源的跨平台源码资源管理器，支持C、C++、Java和Python编程语言，可视化动态生成任何选定类型、函数、变量等的交互式代码映射，并显示与其余代码库的所有依存关系。（https://github.com/CoatiSoftware/Sourcetrail）。

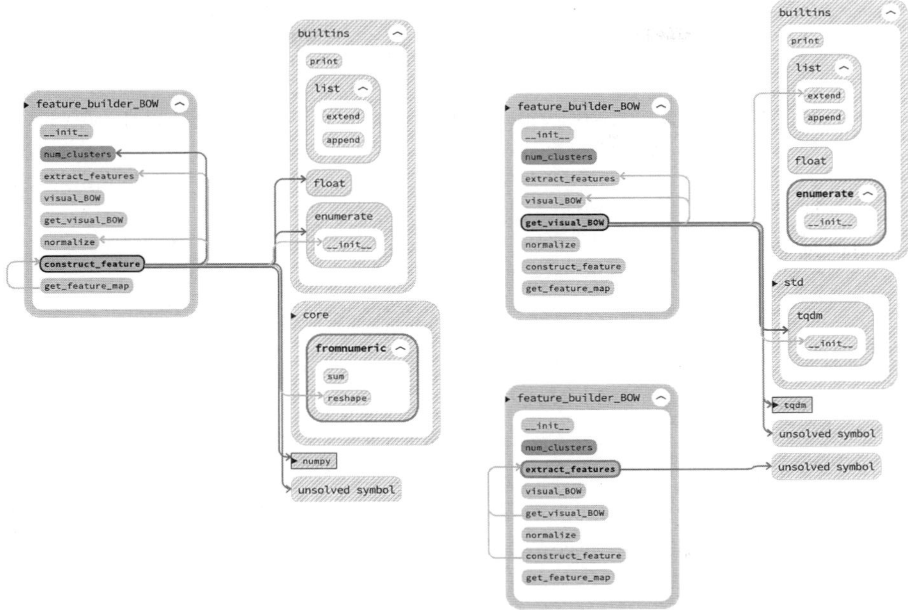

图3.14-6　定义程序调用依赖关系（代码逆向工程）

```
import util_misc
import pickle
class Feature_builder_BOW:
    '''
    class - 根据所有图像关键点描述子聚类建立图像视觉词袋，获取每一图像的特征（码本）映射的频数
统计
    '''
    def __init__(self, num_cluster=32):
        self.num_clusters = num_cluster
    def extract_features(self, img):
        '''
        function - 提取图像特征

        Params:
            img - 读取的图像
        '''
        import cv2 as cv
        star_detector = cv.xfeatures2d.StarDetector_create()
        key_points = star_detector.detect(img)
        img_gray = cv.cvtColor(img, cv.COLOR_BGR2GRAY)
        kp, des = cv.xfeatures2d.SIFT_create().compute(
            img_gray, key_points)  # SIFT 特征提取器提取特征
        return des
    def visual_BOW(self, des_all):
        '''
        function - 聚类所有图像的特征（描述子 /SIFT），建立视觉词袋

        Params:
            des_all - 所有图像的关键点描述子
        '''
        from sklearn.cluster import KMeans
        print("start KMean...")
        kmeans = KMeans(self.num_clusters)
        kmeans = kmeans.fit(des_all)
        # centroids=kmeans.cluster_centers_
        print("end KMean...")
        return kmeans
    def get_visual_BOW(self, training_data):
        '''
        function - 提取图像特征，返回所有图像关键点聚类视觉词袋
```

```
        Params:
            training_data - 训练数据集
        '''
        import cv2 as cv
        from tqdm import tqdm
        des_all = []
        for item in tqdm(training_data):
            classi_judge = item['object_class']
            img = cv.imread(item['image_path'])
            des = self.extract_features(img)
            des_all.extend(des)
        kmeans = self.visual_BOW(des_all)
        return kmeans
    def normalize(self, input_data):
        '''
        fuction - 归一化数据

        Params:
            input_data - 待归一化的数组
        '''
        import numpy as np
        sum_input = np.sum(input_data)
        if sum_input > 0:
            return input_data/sum_input  # 单一数值 / 总体数值之和，最终数值范围 [0,1]
        else:
            return input_data
    def construct_feature(self, img, kmeans):
        '''
        function - 使用聚类的视觉词袋构建图像特征（构造码本）

        Paras:
            img - 读取的单张图像
            kmeans - 已训练的聚类模型
        '''
        import numpy as np
        des = self.extract_features(img)
        labels = kmeans.predict(des.astype(float))  # 对特征执行聚类预测类标
        feature_vector = np.zeros(self.num_clusters)
        for i, item in enumerate(feature_vector):  # 计算特征聚类出现的频数 / 直方图
            feature_vector[labels[i]] += 1
        feature_vector_ = np.reshape(
            feature_vector, ((1, feature_vector.shape[0])))
        return self.normalize(feature_vector_)
    def get_feature_map(self, training_data, kmeans):
        '''
        function - 返回每个图像的特征映射（码本映射）

        Paras:
            training_data - 训练数据集
            kmeans - 已训练的聚类模型
        '''
        import cv2 as cv
        feature_map = []
        for item in training_data:
            temp_dict = {}
            temp_dict['object_class'] = item['object_class']
            img = cv.imread(item['image_path'])
            temp_dict['feature_vector'] = self.construct_feature(img, kmeans)
            if temp_dict['feature_vector'] is not None:
                feature_map.append(temp_dict)
        return feature_map
s_t = util_misc.start_time()

# 修改发生了改变的路径
training_data_ = []
for i in training_data:
    path_update = i['image_path'].replace(
        r"C:\Users\richi\omen-richiebao_s", r"C:\Users\richi\omen_richiebao")
    i.update({'image_path': path_update})
    training_data_.append(i)
kmeans = Feature_builder_BOW().get_visual_BOW(training_data_,)
print("_"*50)

with open('./model/visual_BOW.pkl', 'wb') as f:  # 使用 with 结构避免手动的文件关闭操作
```

```
    pickle.dump(kmeans, f)  #存储 kmeans 聚类模型

feature_map = Feature_builder_BOW().get_feature_map(training_data_, kmeans)
with open('./model/visual_feature.pkl', 'wb') as f:
    pickle.dump(feature_map, f)  #存储图像特征

util_misc.duration(s_t)
start time: 2022-01-08 18:01:30.911273
--------------------------------------------------------------------------------
100%|██████████| 798/798 [00:57<00:00, 13.91it/s]
start KMean...
end KMean...
--------------------------------------------------------------------
end time: 2022-01-08 18:08:38.278630
Total time spend:7.12 minutes
```

3.14.4 决策树（Decision trees）与随机森林（Random forests）

3.14.4.1 决策树

参考*Mastering Machine Learning with scikit-learn*，首先，录入猫狗分类的特征数据集catDog_trainingData_df，所录入的数据为文本类型，将其通过Pandas库提供的pd.get_dummies方法转换为独热编码（One-Hot Encoding），主要应用于数据集的特征列；同时，也应用了sklearn库的preprocessing.LabelEncoder()将其转换为整数编码，主要用于数据集的类标列。

决策树（见下文中定义函数decisionTree_structure()计算的流程图3.14-9）测试特征的内部节点，用盒子表示。节点之间通过边来连接，边表示了测试的可能输出，(根据阈值）将训练实例分到不同的子集中。子节点应用特征值继续测试训练实例的子集，直到满足一个停止标准。分类任务中，决策树中不在分支的节点为叶节点，表示类别（如果是在回归任务中，一个叶节点包含多个实例，这些实例对应的响应变量值可以通过求均值来估计这个叶节点对应的响应变量）；带有分支的节点通常称为分支节点（或子节点）。在决策树构建完成后，对于一个测试实例进行预测，只需要从根节点顺着对应的边到达某个叶节点。

训练决策树的优化算法，使用Ross Quinlan发明的迭代二叉树3代（Iterative Dichotomiser 3，ID3）的算法。

```
import pandas as pd
catDog_trainingData_df = pd.DataFrame({'plays_Fetch': ['Yes', 'No', 'No', 'No', 'No',
'No', 'No', 'No', 'No', 'Yes', 'Yes', 'No', 'Yes', 'Yes'],
                                        'is_grumpy': ['No', 'Yes', 'Yes', 'Yes', 'No',
'Yes', 'Yes', 'No', 'Yes', 'No', 'No', 'No', 'Yes', 'Yes'],
                                         'fvorite_food': ['bacon', 'dog_food', 'cat_
food', 'bacon', 'cat_food', 'bacon', 'cat_food', 'dog_food', 'cat_food', 'dog_
food', 'bacon', 'cat_food', 'cat_food', 'bacon'],
                                          'species': ['dog', 'dog', 'cat', 'cat',
'cat', 'cat', 'cat', 'dog', 'cat', 'dog', 'dog', 'cat', 'cat', 'dog']})
print("原始数据：\n", catDog_trainingData_df)
--------------------------------------------------------------------------------
原始数据：
   plays_Fetch is_grumpy fvorite_food species
0          Yes        No        bacon     dog
1           No       Yes     dog_food     dog
2           No       Yes     cat_food     cat
3           No       Yes        bacon     cat
4           No        No     cat_food     cat
5           No       Yes        bacon     cat
6           No       Yes     cat_food     cat
7           No        No     dog_food     dog
8           No       Yes     cat_food     cat
9          Yes        No     dog_food     dog
10         Yes        No        bacon     dog
11          No        No     cat_food     cat
12         Yes       Yes     cat_food     cat
13         Yes       Yes        bacon     dog
```

数据集的特征值完成独热编码（表3.14-2）后，或增加相应的列，对于仅存在有两个分类的

列（例如“Yes”和“No”），虽然增加了新的一列，但是与单独一列实质上并没有区别（非1即0），因此在决策树中使用两者中的任何一列都是一样的（例如play_No或play_Yes）。但是，分类在3类及其以上者，当用独热编码完成转换后，新增列之间是不能互相替换，例如food_bacon、food_cat food和food_dog food。

表3.14-2 特征值的独热编码

	play_No	play_Yes	grumpy_No	grumpy_Yes	food_bacon	food_cat_food	food_dog_food	species_cat	species_dog
0	0	1	1	0	1	0	0	0	1
1	1	0	0	1	0	0	1	0	1
2	1	0	0	1	0	1	0	1	0
3	1	0	0	1	1	0	0	1	0
4	1	0	1	0	0	1	0	1	0
5	1	0	0	1	1	0	0	1	0
6	1	0	0	1	0	1	0	1	0
7	1	0	1	0	0	0	1	0	1
8	1	0	0	1	0	1	0	1	0
9	0	1	1	0	0	0	1	0	1
10	0	1	1	0	1	0	0	0	1
11	1	0	1	0	0	1	0	1	0
12	0	1	0	1	0	1	0	1	0
13	0	1	0	1	1	0	0	0	1

```
import util_misc
def df_multiColumns_LabelEncoder(df, columns=None):
    '''
    function - 根据指定的（多个）列，将分类转换为整数表示，区间为 [0, 分类数 -1]

    Params:
        df - DataFrame 格式数据；DataFrame
        columns - 指定待转换的列名列表；list(string)

    Returns:
        output - 分类整数编码；DataFrame
    '''
    from sklearn import preprocessing
    output = df.copy()
    if columns is not None:
        for col in columns:
            output[col] = preprocessing.LabelEncoder(
            ).fit_transform(output[col])
    else:
        for column_name, col in output.iteritems():
            output[column_name] = preprocessing.LabelEncoder().fit_transform(col)
    return output
catDog_trainingData_df_encoder = df_multiColumns_LabelEncoder(
    df=catDog_trainingData_df, columns=['plays_Fetch', 'is_grumpy', 'fvorite_
food', 'species'])
print("encoder of each column\n", catDog_trainingData_df_encoder)
catDog_trainingData_dummies = pd.get_dummies(catDog_trainingData_df, prefix=[
                                    'play', 'grumpy', 'food', 'species'])
print("fequency of each column:\n",
      catDog_trainingData_dummies.apply(pd.Series.value_counts))
print('one-hot(dummies):\n')
util_misc.print_html(catDog_trainingData_dummies, row_numbers=14)
--------------------------------------------------------------------------------
encoder of each column
     plays_Fetch  is_grumpy  fvorite_food  species
0            1          0             0        1
1            0          1             2        1
2            0          1             1        0
3            0          1             0        0
4            0          0             1        0
```

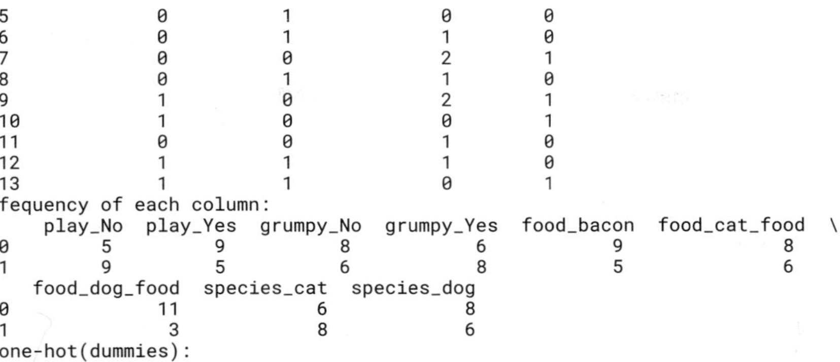

```
5           0           1           0       0
6           0           1           1       0
7           0           0           2       1
8           0           1           1       0
9           1           0           2       1
10          1           0           0       1
11          0           0           1       0
12          1           1           1       0
13          1           1           0       1
fequency of each column:
   play_No  play_Yes  grumpy_No  grumpy_Yes  food_bacon  food_cat_food  \
0        5         9          8           6           9              8
1        9         5          6           8           5              6
   food_dog_food  species_cat  species_dog
0             11            6            8
1              3            8            6
one-hot(dummies):
```

决策树通过检测一个特征序列的值来估计响应变量的值。即能产出只包含猫和只包含狗的子集的检测，要优于一个产出中同时包含猫狗的检测。因为一个子集中的成员同时包含不同的类，无法确定实例的分类。对于这个检测，可以使用熵的衡量方式量化不确定性的程度（单位为比特，bits）。其公式为：$H(X)=-\sum_{i=1}^{n}P(x_i)log_bP(x_i)$。式中，$P(x_i)$ 是输出i的概率，b常见值为2、e和10。由于一个小于1的数值的对数为负数，求和为负数，因此取反。为了方便计算一个对象的熵来查看计算的流程，定义entropy_compomnent()函数实现。

找出对分类动物最有帮助的特征，即找出能把熵降到最低的特征（熵值越大，类别分布越均匀；熵值越小，类别分布越集中，即分布不均匀）。下述代码的过程是按照图3.14-7的执行指向过程计算，在层-A根节点中，根据类标计算猫（8只）狗（6只）熵值为0.985。先利用food_cat food特征列，左子节点不吃猫食（0）对应的狗分类为6只，猫分类为2只，信息熵为0.811；右子节点吃猫食（1）对应的狗分类为0只，猫分类为6只，信息熵为0.0；因为左子节点信息熵0.811>0.5，因此需要对左子节点继续检测。选择grumpy_Yes特征列对应左子节点，即food_cat food中值为0（不吃猫食）的行（实例），包括6只狗和2只猫。这8个实例可以根据grumpy_Yes特征列，即是否脾气暴躁（grumpy）划分为两类：其中，脾气暴躁（1）的为4个实例（右子节点），而脾气不暴躁（0）也为4个实例（左子节点）。将其分布对应到类标列，可以得知脾气暴躁的4个实例中，为猫分类的有2个，为狗分类的亦有2个，信息熵为1；而脾气不暴躁的4个实例中，为猫分类的为0，为狗的分类为4，信息熵为0。就是说明，如果不吃猫食，而脾气不暴躁的实例分类为狗。

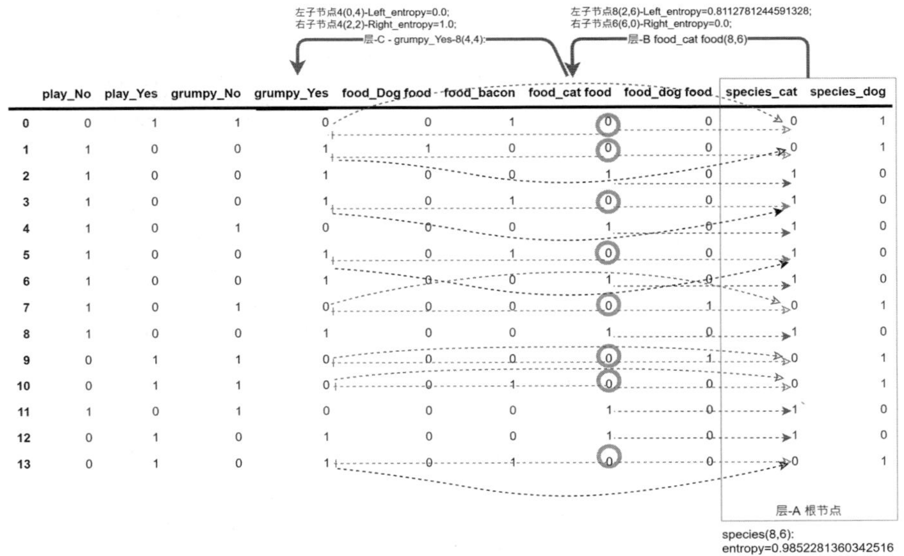

	play_No	play_Yes	grumpy_No	grumpy_Yes	food_Dog food	food_bacon	food_cat food	food_dog food	species_cat	species_dog
0	0	1	1	0	0	1	0	0	0	1
1	1	0	0	1	1	0	0	0	0	1
2	1	0	0	1	0	0	1	0	1	0
3	1	0	0	1	0	1	0	0	1	0
4	1	0	1	0	0	0	1	0	1	0
5	1	0	0	1	0	1	0	0	1	0
6	1	0	0	1	0	0	1	0	1	0
7	1	0	1	0	0	0	0	1	0	1
8	1	0	0	1	0	0	1	0	1	0
9	0	1	1	0	0	0	0	1	0	1
10	0	1	1	0	0	1	0	0	0	1
11	1	0	1	0	0	0	1	0	1	0
12	0	1	0	1	0	0	1	0	1	0
13	0	1	0	1	0	1	0	0	0	1

图3.14-7 决策树过程

上述的决策树，根据两个特征列，food_cat food和grumpy_Yes判断物种分类（类标）。因为层-C存在熵值为1的右子节点，需要继续删选特征，继续检测。例如，想判断第12行实例（样本）的类标，已知food_cat food的值为1（吃猫食），grumpy_Yes的值为1（脾气暴躁）。通过层-B可以判断出，该实例会被分配到右子节点（吃猫食），该层右子节点信息熵为0，因此可以判断，该实例为猫，与实际相符。再例如，对于第10行实例，已知food_cat food的值为0（不吃猫食），grumpy_Yes的值为0（脾气不暴躁），通过层-B可以判断出，该实例会被分配到左子节点（不吃猫食），因为该节点信息熵大于0.5，因此继续用层-C检测，由脾气不暴躁，将其指向左子节点，该节点信息熵为0。因此，可以判定该实例为狗，与实际相符。

- 信息增益（Information gain）

上述的计算直接选择了两个特征列，实际上是需要判断选择哪些特征列用于检测，以减少分类的不确定性。在层-B中，产生了两个子集（子节点），熵分别为0.811和0.0，其平均熵为$(0.811+0.0)/2=0.4055$，而根节点的熵为0.985，最大不确定性的熵为1。衡量熵的减少可以使用信息增益的指标，其公式为：$IG(T,a)=H(T)-\sum_{v\in vals(a)}\frac{|\{x\in T|x_a=v\}|}{|T|}H(\{x\in T|x_a=v\})$。式中，$X_a\in vals(a)$表示实例$x$对应的特征$a$的值，$x\in T|x_a=v$表示特征$a$的值等于$v$的实例数量，$H(\{x\in T|x_a=v\})$是特征$a$的值等于$v$的实例的子集的熵。

自定义信息增益函数IG()，计算结果（表3.14-3）中food_cat_food特征列的值0.463587为所有特征列里最小信息熵，因此用该特征列检测。

表3.14-3　信息增益（IG）

	test	Parent_entropu	first_child_entropy	second_child_entropy	Weighted_average_expression	IG
0	play_No	0.985228	0.721928	0.764205	0.721928*5/14+0.764205*9/14	0.749106
1	play_Yes	0.985228	0.764205	0.721928	0.764205*9/14+0.721928*5/14	0.749106
2	grumpy_No	0.985228	0.811278	0.918296	0.811278*8/14+0.918296*6/14	0.857143
3	grumpy_Yes	0.985228	0.918296	0.811278	0.918296*6/14+0.811278*8/14	0.857143
4	food_bacon	0.985228	0.918296	0.970951	0.918296*9/14+0.970951*5/14	0.937101
5	food_cat_food	0.985228	0.811278	0.000000	0.811278*8/14+0.000000*6/14	0.463587
6	food_dog_food	0.985228	0.845351	0.000000	0.845351*11/14+0.000000*3/14	0.664204

- 基尼不纯度（Gini impurity）

除了通过创建能产生最大信息增益的节点来创建一个决策树，还可以用启发性算法基尼不纯度（Gini impurity）衡量一个集合中类的比例，其公式为：$Gini(t)=1-\sum_{i=1}^{j}P(i|t)^2$。式中，$j$是类的数量，$t$是节点对应的实例子集，$P(i|t)$是从节点的子集中选择一个属于类$i$元素的概率。当集合中所有元素都属于同一类时，选择任一元素属于这个类的概率均为1，因此Gini值为0。和熵一样，当每个被选择的类概率都相等时，Gini达到最大值，其最大值依赖可能类的数量，公式为：$Gini_{max}=1-\frac{1}{n}$。如果分类问题包括两个类，Gini的最大值等于1/2。

sklearn库中DecisionTreeClassifier算法在参数criterion中，给出了上述两种方法{"gini", "entropy"}，可以自行配置。

```python
def entropy_compomnent(numerator, denominator):
    '''
    function - 计算信息熵分量

    Params:
        numerator - 分子;
        denominator - 分母;

    Returns:
        信息熵分量; float
    '''
    import math
    if numerator != 0:
        return -numerator/denominator*math.log2(numerator/denominator)
    elif numerator == 0:
        return 0
```

```
print('层-A 根节点 - 类标 species(8,6):entropy={entropy}'.format(
    entropy=(entropy_compomnent(6, 14)+entropy_compomnent(8, 14))))
print('层-B - food_cat food(8,6):左子节点 8(2,6)-Left_entropy={L_entropy};右子节点 6(6,0)-Right_entropy={R_entropy};'.format(
    L_entropy=(entropy_compomnent(2, 8)+entropy_compomnent(6, 8)), R_entropy=(entropy_compomnent(0, 6)+entropy_compomnent(6, 6))))
print('层-C - grumpy_Yes-8(4,4):左子节点 4(0,4)-Left_entropy={L_entropy};右子节点 4(2,2)-Right_entropy={R_entropy};'.format(
    L_entropy=(entropy_compomnent(0, 4)+entropy_compomnent(4, 4)), R_entropy=(entropy_compomnent(2, 4)+entropy_compomnent(2, 4))))
---------------------------------------------------------------------------------
```

层-A 根节点 - 类标 species(8,6):entropy=0.9852281360342516

层-B - food_cat food(8,6): 左子节点 8(2,6)-Left_entropy=0.8112781244591328; 右子节点 6(6,0)-Right_entropy=0.0;

层-C - grumpy_Yes-8(4,4): 左子节点 4(0,4)-Left_entropy=0.0; 右子节点 4(2,2)-Right_entropy=1.0;

```
def IG(df_dummies):
    '''
    function - 计算信息增量（IG）

    Params:
        df_dummies - DataFrame格式，独热编码的特征值；DataFrame

    Returns:
        cal_info_df - 信息增益（Information gain）；DataFrame
    '''
    import pandas as pd
    weighted_frequency = df_dummies.apply(pd.Series.value_counts)
    weighted_sum = weighted_frequency.sum(axis=0)
    feature_columns = weighted_frequency.columns.tolist()
    Parent_entropy = entropy_compomnent(
        weighted_frequency[feature_columns[-1]][0], 14)+entropy_compomnent(weighted_frequency[feature_columns[-1]][1], 14)
    cal_info = []
    for feature in feature_columns[:-2]:
        v_0_frequency = df_dummies.query('%s==0' % feature).iloc[:, -1].value_counts().reindex(
            df_dummies[feature].unique(), fill_value=0)  #频数可能为0，如果为0则会被舍弃（value_counts），因此需要补回（.reindex）
        v_1_frequency = df_dummies.query(
            '%s==1' % feature).iloc[:, -1].value_counts().reindex(df_dummies[feature].unique(), fill_value=0)
        first_child_entropy = entropy_compomnent(v_0_frequency[0], v_0_frequency.sum(
            axis=0))+entropy_compomnent(v_0_frequency[1], v_0_frequency.sum(axis=0))
        second_child_entropy = entropy_compomnent(v_1_frequency[0], v_1_frequency.sum(
            axis=0))+entropy_compomnent(v_1_frequency[1], v_1_frequency.sum(axis=0))
        cal_dic = {'test': feature,
                   'Parent_entropu': Parent_entropy,
                   'first_child_entropy': first_child_entropy,
                   'second_child_entropy': second_child_entropy,
                   'Weighted_average_expression': '%f*%d/%d+%f*%d/%d' % (first_child_entropy, weighted_frequency[feature][0], weighted_sum.loc[feature], second_child_entropy, weighted_frequency[feature][1], weighted_sum.loc[feature]),
                   'IG': first_child_entropy*(weighted_frequency[feature][0]/weighted_sum.loc[feature])+second_child_entropy*(weighted_frequency[feature][1]/weighted_sum.loc[feature])
                   }
        cal_info.append(cal_dic)
    cal_info_df = pd.DataFrame.from_dict(cal_info)
    return cal_info_df
cal_info_df = IG(df_dummies=catDog_trainingData_dummies,)
util_misc.print_html(cal_info_df, row_numbers=7)
```

使用sklearn库的DecisionTreeClassifier实现决策树及打印决策树流程图表。

- 交叉验证（cross_val_score）

训练和测试数据集如果相同，即在相同的数据上训练和测试，模型（估计器，estimator）只会重复它刚刚看到的样本标签，可能会获得完美的分数。但是，无法正确预测其他的数据，这种情况称为过拟合。因此，在训练机器学习模型时，通常的做法是将数据集分为训练和测试数据集。但是，手动配置超参（数）时，因为参数可以调整，以使估计器达到最优，使得存在测试集过拟合的风险。这样，关于测试集的"知识"学习就会"泄露"到模型中，评估指标就不再报告泛化性能。为了解决这个问题，数据集被切分为训练数据集、验证数据集和测试数据集。然而，将数据集划分为三个集合，大大减少了学习模型的样本数量，以及训练和验证数据集组合随机选择的机会。解决这一问题的方案可以应用交叉验证（cross val score，CV）[①]的方法。

① 交叉验证（cross_val_score），通过交叉验证来评估一个分数（https://scikit-learn.org/stable/modules/generated/sklearn.model_selection.cross_val_score.html）。

在CV方法中，测试数据集仍然用于最终的模型评估，但是不再需要验证数据集。被称为k-fold（k-倍/重/折）的CV中，数据集被分成k个更小的集合，对于每k个folds，k-1个folds用于训练集，1个fold用于测试集。得到的模型在剩余的数据上进行验证（也就是说，它被用作一个测试集来计算性能度量，比如精度）。如图3.14-8所示（颜色表示数据集类型）。

```
X = catDog_trainingData_dummies[catDog_trainingData_dummies.columns[:-2]].to_
numpy()
y = catDog_trainingData_df_encoder[catDog_trainingData_df_encoder.columns[-1]].
to_numpy()
def decisionTree_structure(X, y, criterion='entropy', cv=None, figsize=(6, 6)):
    '''
    function - 使用决策树分类，并打印决策树流程图表。迁移于 Sklearn 的 'Understanding the
decision tree structure', https://scikit-learn.org/stable/auto_examples/tree/
plot_unveil_tree_structure.html#sphx-glr-auto-examples-tree-plot-unveil-tree-
structure-py
    Params:
        X - 数据集 - 特征值（解释变量）; ndarray
        y- 数据集 - 类标 / 标签（响应变量）; ndarray
        criterion - DecisionTreeClassifier 参数，衡量拆分的质量，即衡量哪一项检测最能减
少分类的不确定性; string
        cv - cross_val_score 参数，确定交叉验证分割策略，默认值为 None，即 5-fole(折)的
交叉验证; int

    Returns:
        clf - 返回决策树模型
    '''
    import numpy as np
    from matplotlib import pyplot as plt
    from sklearn.model_selection import train_test_split
    from sklearn.datasets import load_iris
    from sklearn.tree import DecisionTreeClassifier
```

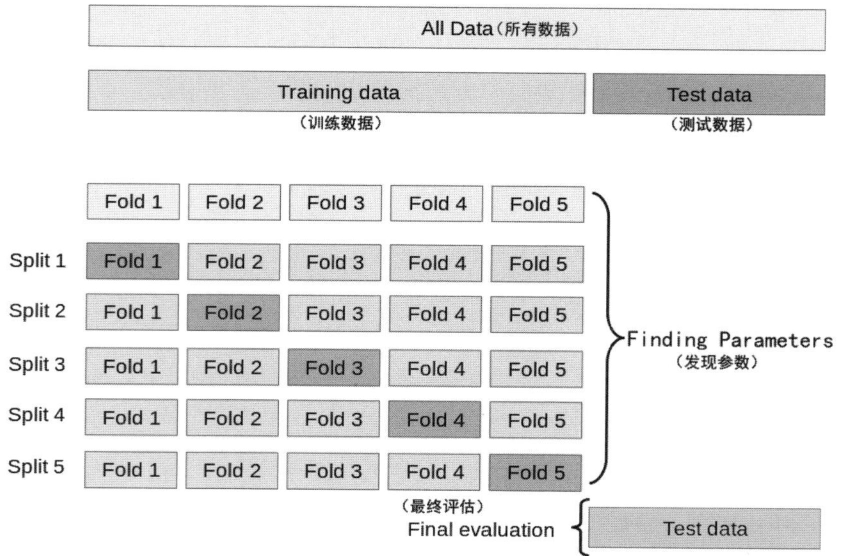

图3.14-8 交叉验证图解（引子sklearn官网——Cross-validation: evaluating estimator performance，评估估计器的性能https://scikit-learn.org/stable/modules/cross_validation.html）

```
    from sklearn import tree
    from sklearn.model_selection import cross_val_score
    X_train, y_train = X, y
    clf = DecisionTreeClassifier(
        criterion=criterion, max_leaf_nodes=3, random_state=0)
    clf.fit(X_train, y_train)
    n_nodes = clf.tree_.node_count
    children_left = clf.tree_.children_left
    children_right = clf.tree_.children_right
    feature = clf.tree_.feature
    threshold = clf.tree_.threshold
    print("n_nodes:{n_nodes},\nchildren_left:{children_left},\nchildren_
right={children_right},\nthreshold={threshold}".format(
        n_nodes=n_nodes, children_left=children_left, children_right=children_
right, threshold=threshold))
    print("_"*50)
    node_depth = np.zeros(shape=n_nodes, dtype=np.int64)
    is_leaves = np.zeros(shape=n_nodes, dtype=bool)
    stack = [(0, 0)]  # 开始于根节点 id(0) 和深度 (0)。

    while len(stack) > 0:
        # `pop` 确保每个节点只访问一次
        node_id, depth = stack.pop()
        node_depth[node_id] = depth

        # 如果一个节点的左子节点和右子节点不相同，就产生一个拆分节点
        is_split_node = children_left[node_id] != children_right[node_id]
        # 如果是拆分节点，就将左、右子节点和深度（depth）附加到 stack 中，从而可以循环遍历它们
        if is_split_node:
            stack.append((children_left[node_id], depth + 1))
            stack.append((children_right[node_id], depth + 1))
        else:
            is_leaves[node_id] = True
    print("The binary tree structure has {n} nodes and has "
          "the following tree structure:\n".format(n=n_nodes))
    for i in range(n_nodes):
        if is_leaves[i]:
            print("{space}node={node} is a leaf node.".format(
                space=node_depth[i] * "\t", node=i))
        else:
            print("{space}node={node} is a split node: "
                  "go to node {left} if X[:, {feature}] <= {threshold} "
                  "else to node {right}.".format(
                      space=node_depth[i] * "\t",
                      node=i,
                      left=children_left[i],
                      feature=feature[i],
                      threshold=threshold[i],
                      right=children_right[i]))
    plt.figure(figsize=figsize)
    tree.plot_tree(clf)
    plt.show()
    CV_scores = cross_val_score(clf, X, y, cv=cv)
    print('cross_val_score:\n', CV_scores)  # 交叉验证每次运行的估计器得分数组
    print("%0.2f accuracy with a standard deviation of %0.2f" %
          (CV_scores.mean(), CV_scores.std()))  # 同时给出了平均得分，和标准差
    return clf
clf = decisionTree_structure(X, y)
----------------------------------------------------------------------------
n_nodes:5,
children_left:[ 1  3 -1 -1 -1],
children_right=[ 2  4 -1 -1 -1],
threshold=[ 0.5  0.5 -2.  -2.  -2. ]
__________________________________________________
The binary tree structure has 5 nodes and has the following tree structure:
node=0 is a split node: go to node 1 if X[:, 5] <= 0.5 else to node 2.
    node=1 is a split node: go to node 3 if X[:, 3] <= 0.5 else to node 4.
    node=2 is a leaf node.
        node=3 is a leaf node.
        node=4 is a leaf node.
----------------------------------------------------------------------------
cross_val_score:
 [0.66666667 0.66666667 0.66666667 1.         0.5       ]
0.70 accuracy with a standard deviation of 0.16
```

3.14.4.2 随机森林（Random forests）

决策树（图3.14–9）属于勤奋学习模型（eager learners），与之相反的是KNN算法这样的惰性学习模型（lazy learners）。决策树学习算法会产生出完美拟合每一个训练实例的巨型复杂的决策树模型，而无法对真实的关系进行泛化，即容易过拟合。解决决策树过拟合的方法可以通过剪枝的方法，移除决策树中过深的节点和叶子，或者从训练数据和特征的子集中创建多棵决策树构成多个模型的集成（集成是指多个估计器的组合），被称为随机森林的决策树集合。创建集成的方法有套袋法（bagging）、推进法（boosting）和堆叠法（stacking）。

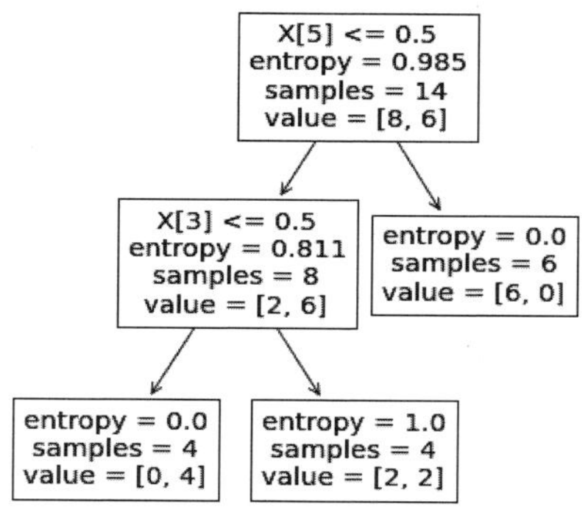

图3.14–9 决策树流程图

- 分类报告（classification_report）

sklearn.metrics.classification_report用于显示分类指标的文本报告，在报告中显示有每个类的精确度（precision）、召回率（recall）、F1分数（f1-score）等信息。精确度和召回率可以从Precision and recall[①]中获取详细的解释，给出的阐释图3.14–10可以一目了然地理解计算方法。实心小圆代表类标狗总计12只，空心小圆代表类标猫总计10只。通过假设的模型预测，得到正确预测为狗的有5只，正确预测为猫的为3只，其他的均为错误预测。因此，对于分类狗而言，$precision=\frac{5}{8}$，$recall=\frac{5}{12}$。f1-score是精确度和召回率的调和平均值，$\frac{2}{F_1}=\frac{1}{P}+\frac{1}{R}\longrightarrow F_1=2\frac{P\times R}{P+R}$。

support字段为每个标签出现的次数。avg行为均值和加权均值（参数sample_weight 配置权重值，默认为None），avg行对应的support为标签总和。

从计算结果可知，使用随机森林算法的各项指标均高于决策树算法的结果。

```
from sklearn.tree import DecisionTreeClassifier
from sklearn.ensemble import RandomForestClassifier
from sklearn.datasets import make_classification
from sklearn.model_selection import train_test_split
from sklearn.metrics import classification_report
X, y = make_classification(n_samples=1000, n_features=100,
                           n_informative=20, n_clusters_per_class=2, random_
state=11)
X_train, X_test, y_train, y_test = train_test_split(X, y, random_state=11)
clf_DTC = DecisionTreeClassifier(random_state=11)
clf_DTC.fit(X_train, y_train)
predictions_DTC = clf_DTC.predict(X_test)
print(classification_report(y_test, predictions_DTC))
--------------------------------------------------------------------------------
              precision    recall  f1-score   support
           0       0.73      0.66      0.69       127
           1       0.68      0.75      0.71       123
    accuracy                           0.70       250
   macro avg       0.71      0.70      0.70       250
weighted avg       0.71      0.70      0.70       250
--------------------------------------------------------------------------------
clf_RFC = RandomForestClassifier(n_estimators=10, random_state=11)
clf_RFC.fit(X_train, y_train)
predictions_RFC = clf_RFC.predict(X_test)
```

① Precision and recall（Wikipedia），https://en.wikipedia.org/wiki/Precision_and_recall）。

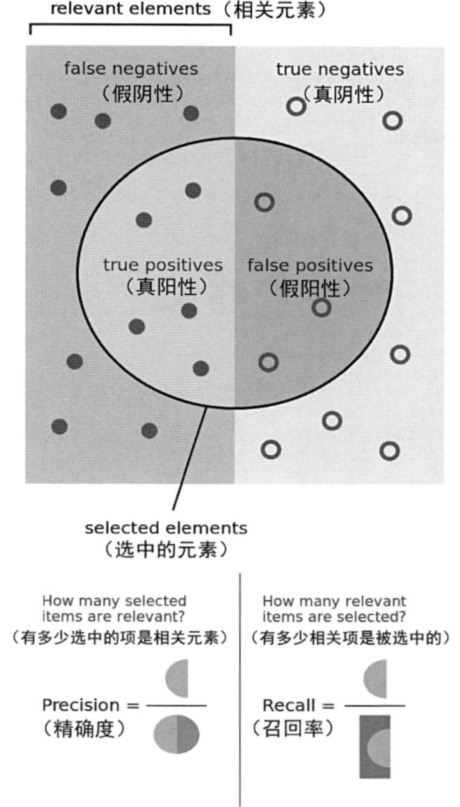

图3.14-10 精确度和召回率图解（引子 Wikipedia——Precision and recall，https://en.wikipedia.org/wiki/Precision_and_recall）

```
print(classification_report(y_test, predictions_RFC))
--------------------------------------------------------------------------------
              precision    recall  f1-score   support
           0       0.74      0.83      0.79       127
           1       0.80      0.70      0.75       123
    accuracy                           0.77       250
   macro avg       0.77      0.77      0.77       250
weighted avg       0.77      0.77      0.77       250
```

3.14.5 图像分类_识别器

3.14.5.1 图像分类器

应用极端随机森林（extremely randomized trees，Extra-Trees）训练分类模型，注意应用了`preprocessing.LabelEncoder()`方法编码。如果要映射回原来的类标，可以执行`.inverse_transform`实现。从对估计器的评测结果来看，f1-score的平均得分为0.62，分项得分中“开阔”“窄建”和“阔建”分类的预测得分大于0.7，相对较好；而“林荫”“窄木”“宽建”和“宽木”都小于0.5，预测精度并不理想。这里的一个主要原因是图像分类的确定，这几个分类中有很多图像并不容易区分之间的差异，也就导致了对图像的分类选择并不专业，最终致使建立的预测模型的预测精度并不是很好，可以尝试从新建立分类标准，提供数据集的类标精度。

```
import numpy as np
```

```
import pickle
from sklearn.model_selection import train_test_split
class ERF_trainer:
    '''
    class - 用极端随机森林训练图像分类器
    '''
    def __init__(self, X, label_words, save_path):
        from sklearn import preprocessing
        from sklearn.ensemble import ExtraTreesClassifier
        import os
        import pickle
        print('Start training...')
        self.le = preprocessing.LabelEncoder()
        # http://scikit-learn.org/stable/modules/generated/sklearn.ensemble.ExtraTreesClassifier.html
        self.clf = ExtraTreesClassifier(
            n_estimators=100, max_depth=16, random_state=0)
        y = self.encode_labels(label_words)
        self.clf.fit(np.asarray(X), y)
        with open(os.path.join(save_path, 'ERF_clf.pkl'), 'wb') as f:  #存储训练好
的图像分类器模型
            pickle.dump(self.clf, f)
        print("end training and saved estimator.")
    def encode_labels(self, label_words):
        '''
        function - 对标签编码及训练分类器
        '''
        self.le.fit(label_words)
        return np.array(self.le.transform(label_words), dtype=np.float64)
    def classify(self, X):
        '''
        function - 对未知数据的预测分类
        '''
        label_nums = self.clf.predict(np.asarray(X))
        label_words = self.le.inverse_transform([int(x) for x in label_nums])
        return label_words
feature_map_fp = './model/visual_feature.pkl'
with open(feature_map_fp, 'rb') as f:
    feature_map = pickle.load(f)  #读取存储的图像特征
label_words = [x['object_class'] for x in feature_map]
dim_size = feature_map[0]['feature_vector'].shape[1]
X = [np.reshape(x['feature_vector'], (dim_size,)) for x in feature_map]
save_path = './model'
X_train, X_test, y_train, y_test = train_test_split(
    X, label_words, test_size=0.3, random_state=42)
erf = ERF_trainer(X_train, y_train, save_path)
--------------------------------------------------------------------------------
Start training...
end training and saved estimator.
--------------------------------------------------------------------------------
from sklearn.metrics import classification_report
print(classification_report(y_test,erf.classify(X_test)))
--------------------------------------------------------------------------------
              precision    recall  f1-score   support
          宽建       0.44      0.21      0.29        19
          宽木       0.55      0.39      0.46        28
          开阔       1.00      0.68      0.81        28
          林荫       0.20      0.17      0.18        12
          窄建       0.74      0.93      0.83        61
          窄木       0.25      0.09      0.13        22
          阔建       0.88      0.64      0.74        11
          阔木       0.54      0.81      0.65        59
    accuracy                           0.62       240
   macro avg       0.57      0.49      0.51       240
weighted avg       0.61      0.62      0.59       240
```

3.14.5.2 **图像识别器**

图像识别器需要调用3个已经保存的文件，第一个是“ERF_clf.pkl”，保存有应用极端随机森林算法训练的图像分类器模型，用于类别预测；第二个是“visual_BOW.pkl”，保存有视觉词袋KMeans聚类模型，用于构建图像特征（位于前述的Feature_builder_BOW类中）的参数输入；第三个是“visual_feature.pkl”，存储的是图像特征，主要是读取类标转换为整型编码；然后，应

用inverse_transform方法，将预测的整型编码转换为原始的类标。

```
import os
import cv2 as cv
class ImageTag_extractor:
    '''
    class - 图像识别器，基于图像分类模型、视觉词袋及图像特征
    '''
    def __init__(self, ERF_clf_fp, visual_BOW_fp, visual_feature_fp):
        from sklearn import preprocessing
        import pickle
        with open(ERF_clf_fp, 'rb') as f:  #读取存储的图像分类器模型
            self.clf = pickle.load(f)

        with open(visual_BOW_fp, 'rb') as f:  #读取存储的聚类模型和聚类中心点
            self.kmeans = pickle.load(f)

        '''对标签编码'''
        with open(visual_feature_fp, 'rb') as f:
            self.feature_map = pickle.load(f)
        self.label_words = [x['object_class'] for x in self.feature_map]
        self.le = preprocessing.LabelEncoder()
        self.le.fit(self.label_words)
    def predict(self, img):
        import util_A
        import numpy as np
        feature_vector = util_A.feature_builder_BOW().construct_feature(
            img, self.kmeans)  #提取图像特征，之前定义的 Feature_builder_BOW() 类，可放置于
util.py 文件中，方便调用
        label_nums = self.clf.predict(np.asarray(feature_vector)) #进行图像识别 / 分类
        image_tag = self.le.inverse_transform(
            [int(x) for x in label_nums])[0]  #获取图像分类标签
        return image_tag
ERF_clf_fp = r'./model/ERF_clf.pkl'
visual_BOW_fp = './model/visual_BOW.pkl'
visual_feature_fp = './model/visual_feature.pkl'
imgs_fp = r'C:\Users\richi\omen_richiebao\omen_github\USDA_CH_final\USDA\
notebook\data\kitti'
imgs_ = [os.path.join(imgs_fp, f) for f in os.listdir(imgs_fp)]
imgs_pred_tag = {fn: ImageTag_extractor(
    ERF_clf_fp, visual_BOW_fp, visual_feature_fp).predict(cv.imread(fn)) for fn
in imgs_}
```

在GoogleEarth的街景（Street view）德国随机城市下，随机的截取了6张尺寸大小不一的图像，应用图像识别器预测分类，其结果如图3.14-11所示。

```
import matplotlib.pyplot as plt
from PIL import Image
import matplotlib
matplotlib.rcParams['font.family'] = ['SimHei']
fig, axes = plt.subplots(2, 3, sharex=True, sharey=True,
                         figsize=(25, 10))  #布局多个子图，每个子图显示一幅图像
ax = axes.flatten()  #降至1维，便于循环操作子图
```

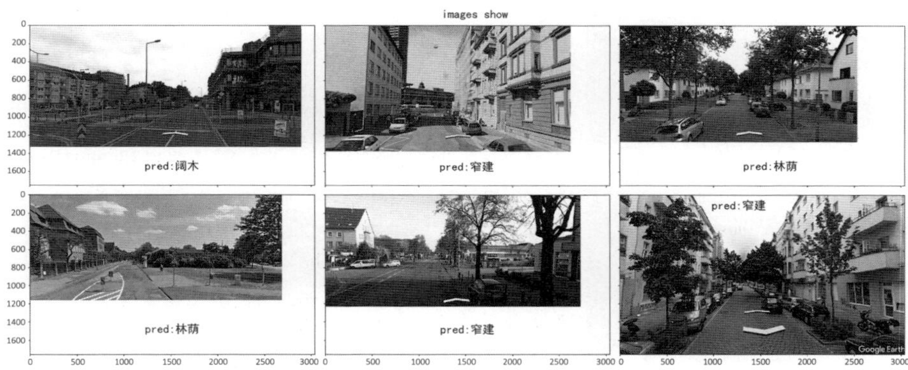

图3.14-11　应用图像识别器预测分类结果

```
i = 0
for f, tag in imgs_pred_tag.items():
    img_array = Image.open(f)
    ax[i].imshow(img_array)  # 显示图像
    ax[i].set_title("pred:{tag}".format(tag=tag))
    i += 1
fig.tight_layout()  # 自动调整子图参数，使其填充整个图像区域
fig.suptitle("images show", fontsize=14, fontweight='bold', y=1.02)
plt.show()
```

3.14.5.3 嵌入图像识别器到网络实验平台

将估计器部署到网络实验平台，需要将图像识别器及其相关的代码和文件（估计器、视觉词袋和图像特征）整合起来。下述将ImageTag_extractor和Feature_builder_BOW类置于同一个文件中（app/visual_perception/ImageTag_extractor.py）；同时，将Feature_builder_BOW类中不需要的功能移除，保持代码简洁，避免干扰，并增加了ImageTag_extractor_execution类，方便外部调用该类，直接执行预测。而"ERF_clf.pkl""visual_BOW.pkl"和"visual_feature.pkl"三个必需文件置于文件夹visual_perception/vp_model中。

预测图像分类的功能是可以在页面端上传一幅图像后，用图像识别器（已经训练好的图像分类器）预测分类，因此增加了一个新的文件夹uploads用于保存上传的图像文件。图像上传的配置主要使用flask_uploads库实现，需要在app/__init__.py文件中增加相应的配置。

嵌入到网络实验平台的图像识别器（图3.14-12）。

```
# app/visual_perception/ImageTag_extractor.py
class feature_builder_BOW:
```

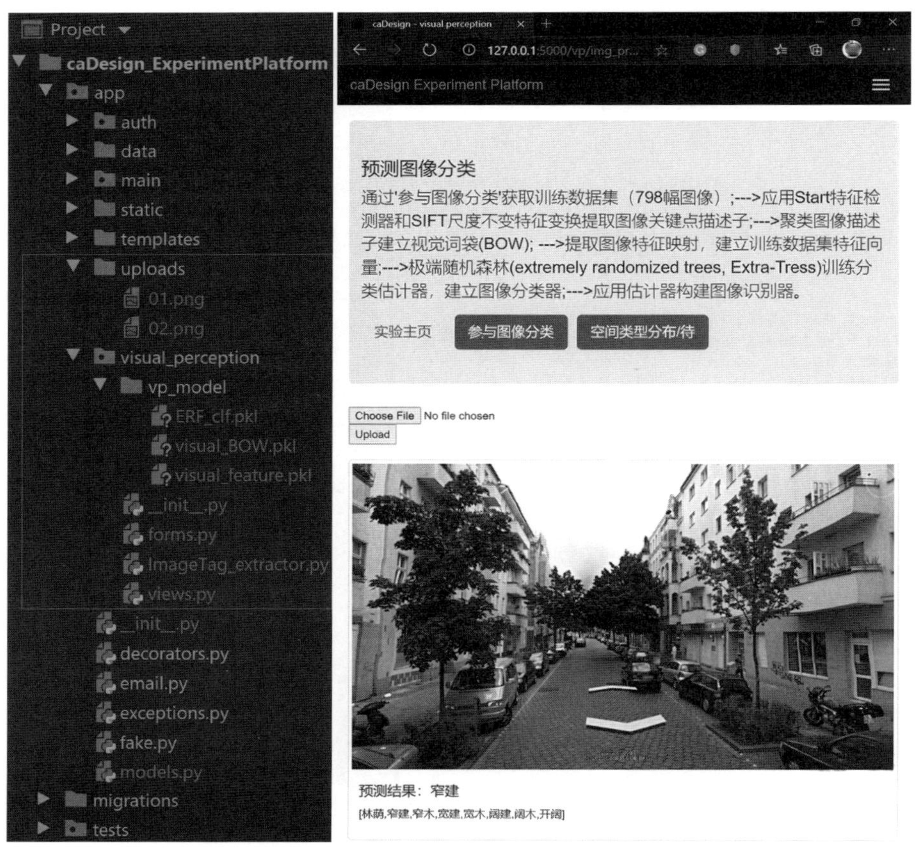

图3.14-12 嵌入到网络实验平台的图像识别器

```
    '''
    class - （仅保留 construct_feature 及关联部分函数）根据所有图像关键点描述子聚类建立图像视觉词袋，获取每一图像的特征（码本）映射的频数统计
    '''
    def __init__(self, num_cluster=32):
        self.num_clusters = num_cluster
    def extract_features(self, img):
        '''
        function - 提取图像特征

        Paras:
        img - 读取的图像
        '''
        import cv2 as cv
        star_detector = cv.xfeatures2d.StarDetector_create()
        key_points = star_detector.detect(img)
        img_gray = cv.cvtColor(img, cv.COLOR_BGR2GRAY)
        kp, des = cv.xfeatures2d.SIFT_create().compute(
            img_gray, key_points)  # SIFT 特征提取器提取特征
        return des
    def normalize(self, input_data):
        '''
        fuction - 归一化数据

        input_data - 待归一化的数组
        '''
        import numpy as np
        sum_input = np.sum(input_data)
        if sum_input > 0:
            return input_data / sum_input  # 单一数值 / 总体数值之和，最终数值范围 [0,1]
        else:
            return input_data
    def construct_feature(self, img, kmeans):
        '''
        function - 使用聚类的视觉词袋构建图像特征（构造码本）

        Paras:
        img - 读取的单张图像
        kmeans - 已训练的聚类模型
        '''
        import numpy as np
        des = self.extract_features(img)
        labels = kmeans.predict(des.astype(np.float))  # 对特征执行聚类预测类标
        feature_vector = np.zeros(self.num_clusters)
        for i, item in enumerate(feature_vector):  # 计算特征聚类出现的频数 / 直方图
            feature_vector[labels[i]] += 1
        feature_vector_ = np.reshape(
            feature_vector, ((1, feature_vector.shape[0])))
        return self.normalize(feature_vector_)
class ImageTag_extractor:
    '''
    class - 图像识别器，基于图像分类模型、视觉词袋及图像特征
    '''
    def __init__(self, ERF_clf_fp, visual_BOW_fp, visual_feature_fp):
        from sklearn import preprocessing
        import pickle
        with open(ERF_clf_fp, 'rb') as f:  # 读取存储的图像分类器模型
            self.clf = pickle.load(f)

        with open(visual_BOW_fp, 'rb') as f:  # 读取存储的聚类模型和聚类中心点
            self.kmeans = pickle.load(f)

        ''' 对标签编码 '''
        with open(visual_feature_fp, 'rb') as f:
            self.feature_map = pickle.load(f)
        self.label_words = [x['object_class'] for x in self.feature_map]
        self.le = preprocessing.LabelEncoder()
        self.le.fit(self.label_words)
    def predict(self, img):
        import numpy as np
        # 提取图像特征，之前定义的 Feature_builder_BOW() 类，可放置于 util.py 文件中，方便调用
        feature_vector = feature_builder_BOW().construct_feature(img, self.
kmeans)
        label_nums = self.clf.predict(np.asarray(feature_vector))  # 进行图像识别 / 分类
```

```
        image_tag = self.le.inverse_transform(
            [int(x) for x in label_nums])[0]  # 获取图像分类标签
        return image_tag
class ImageTag_extractor_execution:
    def __init__(self, img_url):
        self.img_url = img_url
        self.ERF_clf_fp = 'app/visual_perception/vp_model/ERF_clf.pkl'
        self.visual_BOW_fp = 'app/visual_perception/vp_model/visual_BOW.pkl'
        self.visual_feature_fp = 'app/visual_perception/vp_model/visual_feature.pkl'
    def execution(self):
        import cv2 as cv
        print("*"*50)
        print(self.img_url)
        imgs_pred_tag = ImageTag_extractor(
            self.ERF_clf_fp, self.visual_BOW_fp, self.visual_feature_fp).
predict(cv.imread(self.img_url))
        return imgs_pred_tag
```

配置flask_uploads。

```
# app/__init__.py
# ...
from flask_uploads import UploadSet, configure_uploads, IMAGES, patch_request_
class
import os
# ...
photos = UploadSet('photos', IMAGES)
def create_app(config_name):
    # ...
    basedir = os.path.abspath(os.path.dirname(__file__))
    app.config['UPLOADED_PHOTOS_DEST'] = os.path.join(
        basedir, 'uploads')  # you' ll need to create a folder named uploads
    configure_uploads(app, photos)
    patch_request_class(app)
    # ...
```

配置图像文件上传Web表达。

```
#app/visual_perception/forms.py
#...
from .. import photos
#...
class upload_img(FlaskForm):
    photo=FileField(validators=[FileAllowed(photos, 'Image only!'),
FileRequired('File was empty!')])
    submit=SubmitField('Upload')
```

配置路由和视图函数，在视图函数中调用图像识别器。

```
# app/visual_perception/views.py
# ...
from .. import photos
from .ImageTag_extractor import ImageTag_extractor_execution
# ...
@visual_perception.route("/img_prediction", methods=['GET', 'POST'])
def img_prediction():
    import os
    form = upload_img()
    if form.validate_on_submit():
        img_name = photos.save(form.photo.data)
        img_url = photos.url(img_name)
        img_fp = os.path.join(
            current_app.config['UPLOADED_PHOTOS_DEST'], img_name)
        imgs_pred_tag = ImageTag_extractor_execution(img_fp).execution()
        return render_template('vp/img_prediction.html', form=form, img_url=img_
url, imgs_pred_tag=imgs_pred_tag)
    else:
        img_url = None
        return render_template('vp/img_prediction.html', form=form, img_url=img_url)
```

配置“预测图像分类”的模板页面。

```
<!--templates/vp/img_prediction.html -->
{% extends "base.html" %}
{% import "bootstrap/wtf.html" as wtf %}
```

```
{% import "_macros.html" as macros %}
{% block title %}caDesign - visual perception{% endblock %}
{% block page_content %}
    <div class="jumbotron">
        <h3> 预测图像分类 </h3>

        <p>
            通过'参与图像分类'获取训练数据集（798 幅图像）;---> 应用 Start 特征检测器和 SIFT 尺度
不变特征变换提取图像关键点描述子 ;---> 聚类图像描述子建立视觉词袋 (BOW);
    ---> 提取图像特征映射，建立训练数据集特征向量 ;---> 极端随机森林 (extremely randomized trees, Extra-
Tress) 训练分类估计器，建立图像分类器 ;---> 应用估计器构建图像识别器。
        </p>
        <p>
            <a class="btn btn-secondary btn-lg" href="{{ url_for('visual_
perception.vp') }}" role="button"> 实验主页 </a>
             <a class="btn btn-primary btn-lg" href="{{ url_for('visual_
perception.imgs_classification') }}" role="button"> 参与图像分类 </a>
             <a class="btn btn-primary btn-lg" href="{{ url_for('visual_
perception.vp') }}" role="button"> 空间类型分布 / 待 </a>
        </p>
    </div>
    <form method="POST" enctype="multipart/form-data">
         {{ form.hidden_tag() }}
         {{ form.photo }}
         {% for error in form.photo.errors %}
             <span style="color: red;">{{ error }}</span>
         {% endfor %}
         {{ form.submit }}
    </form>
    {% if img_url %}
        <br>
        <div class="thumbnail">
            <img src="{{ img_url }}" >
            <div class="caption">
                <h4> 预测结果：{{ imgs_pred_tag }}</h4>
                <p>[ 林荫 , 窄建 , 窄木 , 宽建 , 宽木 , 阔建 , 阔木 , 开阔 ] </p>
            </div>
        </div>
    {% endif %}
{% endblock %}
```

3.14.5.4 部署

在服务器（Linux系统）下，部署基于Flask构建的网络实验平台。其基本流程是：

1. 建立Python虚拟环境，并根据生成的requirements.txt文件安装库配置环境；
2. 安装nginx、uwsgi、flask；
3. 建立一个简单的App应用，测试程序；
4. 本地项目上传至服务器虚拟环境目录下；
5. 建立uWSGI入口点（entry points）；
6. 配置uWSGI；
7. 建立Upstart Script；
8. 配置Nginx；supervisor进程守护程序。

虽然网络上可以搜索到大量在Linux系统服务器部署Flask的教程，但是因为系统可能会存在不同，说明文字上的不清晰，Flask文件结构的差异，Python环境的变化，尤其库版本的变化，通常在部署时很难一帆风顺。即使给出的配置步骤没有问题，部署的过程也会容易出现不经意的错误，因此需要耐心地跟随步骤操作；同时，最好用markdown工具将步骤所用到的命令记录下来，方便查看。

参考文献（References）：

[1] Miguel Grinberg (2018). Flask Web Development: Developing Web Applications with Python. O'Reilly Media.

[2] Gavin Hackeling (2017). Mastering Machine Learning with scikit-learn: Apply effective learning algorithms to real-world problems using scikit-learn. Packt Publishing.（Cavin Hackeling.张浩然译. scikit-learning机器学习[M]. 北京: 人民邮电出版社, 2019.2.）

3-F 深度学习试验

3.15 从解析解到数值解，从机器学习到深度学习

解析解（analytical solution），又称闭式解，是可以用解析表达式来表达的解（有时，也称为公式解）。在数学上，如果一个方程或者方程组存在的某些解，是由有限次常见运算的组合给出的形式，则称该方程存在解析解。二次方程的根就是一个解析解的典型例子。当解析解不存在时，比如五次及更高次的代数方程，则该方程只能用数值分析的方法求解近似值（有限元的方法、数值逼近、差值的方法，大多数深度学习通过优化算法有限次迭代模型参数来尽可能降低损失函数的值），则是数值解（numerical solution）。例如，大多数偏微分方程，尤其是非线性偏微分方程。

在“简单回归、多元回归”部分，使用解析解的方法（对真实值与预测值之差的平方和，即残差平方和求微分或偏微分）解得回归系数；在“梯度下降法”部分，则应用了数值解的优化算法梯度下降法，通过配置学习率、精度和随机初始化系数，定义模型、定义损失函数（残差平方和为一种损失函数，又称代价函数）、定义梯度下降函数（对损失函数求梯度，即对模型变量微分）。迭代训练模型，直至损失函数计算的结果小于预设值，解得模型权重值。

机器学习（machine learning）是人工智能的一个分支。机器学习理论主要是设计和分析一些让计算机可以自动“学习”的算法。机器学习算法是一类从数据中自动分析获得规律，并利用规律对未知数据进行预测的算法。常用的机器学习库是scikit-learn，包括数据预处理，以及聚类、回归和分类三大方向，其中提供了SGDRegressor随机梯度下降方法。深度学习（deep learning）是机器学习的一个分支，是一种以人工神经网络为结构，对资料数据进行表征学习的算法。深度学习库推荐使用PyTorch，是以Python优先的深度学习框架，其设计符合人类的思维方式（而TensorFlow因为不符合人们的思维习惯，很难应用，但后来结合到Keras框架，确立为Tensorflow高阶API，使得应用TensorFlow的环境得以改善）。

PyTorch的安装可以查看其官网Get Started[①]。根据机器配置情况，包括系统（Linux、Mac和Windows）、安装所用的包管理器（Conda、Pip、LibTorch和Source）、所用语言（Python或者C++/Java）和计算平台（GPU（CUDA）或CPU）等，生成安装代码，例如`pip3 install torch torchvision torchaudio --index-url https://download.pytorch.org/whl/cu118`。

① PyTorch的安装方法（Get Started），（https://pytorch.org/get-started/locally）。

3.15.1 张量（tensor）

最常用的数据格式是使用NumPy库提供的数组（array）形式（也是机器学习库Sklearn数据集，以及数据处理的格式），以及Pandas库提供的DataFrame（常配合地理信息系统中的table/表及数据库使用）、Series数据格式。在深度学习（PyTorch[②]库或TensorFlow[③]）中，引入了张量（tensor），可以用于GPU计算，并有自动求梯度等更多功能。张量和数组类似，0维（一个数，rank=0，rank为维度数）张量，即为标量（Scalar），1维张量（rank=1）即为向量/矢量（Vector），2维张量（rank=2）即为矩阵（Matrix），3维张量（rank=3）即为矩阵数组。深度学习中，张量可以看作一个多维数组（multi-dimentional array）。

② 深度学习库PyTorch官网首页，（https://pytorch.org）。
③ 深度学习库TensorFlow]官网首页，（https://www.tensorflow.or）。

```
import torch
print("_"*50, "A-tensor 创建")
# 01- 未初始化，shape=（5，3）的 tensor
t_a = torch.empty(5, 3)
print("01- 创建未初始化，shape=（5，3）的 tensor:\nt_a={},\nshape={}".format(t_a, t_
a.shape))
# 02- 随机初始化 shape=（5，3）的 tensor
t_b = torch.rand(5, 3)
print("02- 随机初始化 shape=（5，3）的 tensor:\nt_b={}".format(t_b))
# 03- 数据类型为 long，全 0 的 tensor
t_c = torch.zeros(5, 3, dtype=torch.long)
print("03- 数据类型为 long，全 0 的 tensor:\nt_c={}".format(t_c))
# 04- 由给定数据直接创建
t_d = torch.tensor([[3.1415, 0], [9, 2.71828]])
```

```
print("04-由给定数据直接创建 :\nt_d={}".format(t_d))
# 05-tensor.new_ones() 方法重用已有 tensor，保持数据类型，及 torch.device(CPU 或 GPU)
t_e = t_c.new_ones(4, 3)
print("05-tensor.new_ones() 方法重用已有 tensor，保持数据类型，及 torch.device(CPU 或 GPU):\nt_e={}\ndtype={},device={}".format(t_e, t_e.dtype, t_e.device))
# 06-torch.randn_like() 方法重用已有 tensor, 并可重定义数据类型
t_f = torch.randn_like(t_c, dtype=torch.float)
print("#06-torch.randn_like() 方法重用已有 tensor, 并可重定义数据类型 :\nt_f={},\
ndtype={}".format(t_f, t_f.dtype))
# 07-tensor.size() 和 tensor.shape 方法查看 tensor 形状
print("07-tensor.size() 和 tensor.shape 方法查看 tensor 形状 :\nt_f.size()={},t_f.shape={}".format(t_f.size(), t_f.shape))
-------------------------------------------------------------------
A-tensor 创建
01-创建未初始化，shape=（5，3）的 tensor:
t_a=tensor([[0., 0., 0.],
        [0., 0., 0.],
        [0., 0., 0.],
        [0., 0., 0.],
        [0., 0., 0.]]),
shape=torch.Size([5, 3])
02-随机初始化 shape=（5，3）的 tensor:
t_b=tensor([[0.4364, 0.0622, 0.8621],
        [0.2628, 0.5856, 0.4788],
        [0.4085, 0.8417, 0.7704],
        [0.0316, 0.3081, 0.3179],
        [0.4681, 0.3444, 0.0639]])
03-数据类型为 long，全 0 的 tensor:
t_c=tensor([[0, 0, 0],
        [0, 0, 0],
        [0, 0, 0],
        [0, 0, 0],
        [0, 0, 0]])
04-由给定数据直接创建 :
t_d=tensor([[3.1415, 0.0000],
        [9.0000, 2.7183]])
05-tensor.new_ones() 方法重用已有 tensor，保持数据类型，以及 torch.device(CPU 或 GPU):
t_e=tensor([[1, 1, 1],
        [1, 1, 1],
        [1, 1, 1],
        [1, 1, 1]])
dtype=torch.int64,device=cpu
#06-torch.randn_like() 方法重用已有 tensor，并可重定义数据类型 :
t_f=tensor([[ 0.9716,  0.5021, -0.9884],
        [ 0.6704,  0.8553,  1.4779],
        [ 0.3652,  1.5988, -0.7453],
        [-0.3100, -1.4392,  0.4762],
        [-0.0064, -0.3408,  2.0674]]),
dtype=torch.float32
07-tensor.size() 和 tensor.shape 方法查看 tensor 形状 :
t_f.size()=torch.Size([5, 3]),t_f.shape=torch.Size([5, 3])
--------------------------------------------------------------------------------
import numpy as np
print("_"*50, "B-tensor 操作 ")
# 08-加法形式 -1-' +'
print("08-加法形式 -1-'+':\nt_a+t_b={}".format(t_a+t_b))
# 09-加法形式 -2-torch.add()
print("09-加法形式 -2-torch.add():\ntorch.add(t_a,t_b)={}".format(torch.add(t_a,
t_b)))
# 10-加法形式 -3-原地结果替换 inplace/add_(PyTorch 原地操作 inplace 都有后缀 _)
print("10-加法形式 -3-原地结果替换 inplace/add_:\nt_a.add_(t_b)={}, \nt_a={}".
format(t_a.add_(t_b), t_a))
# 11-索引，共享存储地址
t_g = t_a[0, :]
print("11-索引，共享存储地址 :\nt_a[0,:]={}".format(t_g))
t_g += 1
print("t_g+=1 后 ,tg={},t_a[0,:]={}".format(t_g, t_a[0, :]))
# 12-view() 方法改变 tensor 形状（shape），但为同一存储地址
print("12-view() 方法改变 tensor 形状（shape），但为同一存储地址 :\nt_a.shape={},t_a.view(15).shape={},t_a.view(-1,5).shape={}".format(
    t_a.shape, t_a.view(15).shape, t_a.view(-1, 5).shape))
# 13-clone()
print("13-clone():\nid(t_a)==id(t_a.clone())={}".format(id(t_a) == id(t_a.clone())))
```

```
print("_"*50, "C-tensor 广播机制 ")
# 14-广播机制
t_h = torch.arange(1, 3).view(1, 2)
t_i = torch.arange(1, 4).view(3, 1)
print("14-广播机制 :\nt_h={}\nt_i={}\nt_h.shape={},t_i.shape={}\nt_h+t_i={}".
format(t_h,
      t_i, t_h.shape, t_i.shape, t_h+t_i))

print("_"*50, "D-inplace 原地操作、节约内存 ")
# 15-inplace——out 参数
print("15-inplace -- out 参数 :\nid(t_a)==id(torch.add(t_a,t_b,out=t_a))={}".
format(id(t_a)
       == id(torch.add(t_a, t_b, out=t_a))))
# 16-inplace——+=(add_())
t_j = t_a
t_a += 1
print("16-inplace -- +=(add_()):\nid(t_a)==id(t_j)={}".format(id(t_a) == id(t_j)))
# 17-inplace——[:]
t_k = t_a
t_a[:] = t_a+t_b
print("17-inplace -- [:]:\nid(t_a)==id(t_k)={}".format(id(t_a) == id(t_k)))
print("_"*50, "E-tensor<-->array(NumPy)")
# 18-tensor-->array
print("18-tensor-->array:\ntype(t_a)={},type(t_a.numpy())={}".format(type(t_a),
type(t_a.numpy())))
# 19-array-->tenssor
array = np.arange(15).reshape(3, 5)
print("19-array-->tenssor:\ntype(array)={},type(torch.from_numpy(array))={}".
format(
    type(array), type(torch.from_numpy(array))))
```

```
------------------------------------------------------------------
B-tensor 操作
08-加法形式 -1-'+':
t_a+t_b=tensor([[0.4364, 0.0622, 0.8621],
        [0.2628, 0.5856, 0.4788],
        [0.4085, 0.8417, 0.7704],
        [0.0316, 0.3081, 0.3179],
        [0.4681, 0.3444, 0.0639]])
09-加法形式 -2-torch.add():
torch.add(t_a,t_b)=tensor([[0.4364, 0.0622, 0.8621],
        [0.2628, 0.5856, 0.4788],
        [0.4085, 0.8417, 0.7704],
        [0.0316, 0.3081, 0.3179],
        [0.4681, 0.3444, 0.0639]])
10-加法形式 -3-原地结果替换 inplace/add_:
t_a.add_(t_b)=tensor([[0.4364, 0.0622, 0.8621],
        [0.2628, 0.5856, 0.4788],
        [0.4085, 0.8417, 0.7704],
        [0.0316, 0.3081, 0.3179],
        [0.4681, 0.3444, 0.0639]]),
t_a=tensor([[0.4364, 0.0622, 0.8621],
        [0.2628, 0.5856, 0.4788],
        [0.4085, 0.8417, 0.7704],
        [0.0316, 0.3081, 0.3179],
        [0.4681, 0.3444, 0.0639]])
11-索引，共享存储地址 :
t_a[0,:]=tensor([0.4364, 0.0622, 0.8621])
t_g+=1 后 ,tg=tensor([1.4364, 1.0622, 1.8621]),t_a[0,:]=tensor([1.4364, 1.0622,
1.8621])
12-view() 方法改变 tensor 形状（shape），但为同一存储地址 :
t_a.shape=torch.Size([5, 3]),t_a.view(15).shape=torch.Size([15]),t_a.view(-1,5).
shape=torch.Size([3, 5])
13-clone():
id(t_a)==id(t_a.clone())=False
------------------------------------------------------------------
C-tensor 广播机制
14-广播机制 :
t_h=tensor([[1, 2]])
t_i=tensor([[1],
        [2],
        [3]])
t_h.shape=torch.Size([1, 2]),t_i.shape=torch.Size([3, 1])
t_h+t_i=tensor([[2, 3],
        [3, 4],
```

```
        [4, 5]])
------------------------------------------------------------------
D-inplace 原地操作，节约内存
15-inplace -- out 参数：
id(t_a)==id(torch.add(t_a,t_b,out=t_a))=True
16-inplace -- +=(add_()):
id(t_a)==id(t_j)=True
17-inplace -- [:]:
id(t_a)==id(t_k)=True
------------------------------------------------------------------
E-tensor<-->array(NumPy)
18-tensor-->array:
type(t_a)=<class 'torch.Tensor'>,type(t_a.numpy())=<class 'numpy.ndarray'>
19-array-->tenssor:
type(array)=<class 'numpy.ndarray'>,type(torch.from_numpy(array))=<class 'torch.
Tensor'>
------------------------------------------------------------------------------
if torch.cuda.is_available():
    device = torch.device("cuda")
    print("device={}".format(device))
    x = torch.tensor([[3.1415, 0], [9, 2.71828]], device=device)
    print("x={}".format(x))
    y = torch.rand(2, 2)
    print("默认 CPU,y={}".format(y))
    y = y.to(device)
    print(".to(device), y={}".format(y))
    print("GPU 运算 ,x+y={}".format(x+y))
    print(".to('cpu'),(x+y).to('cpu',torch.double)={}".format((x+y).to("cpu",
torch.double)))
------------------------------------------------------------------------------
device=cuda
x=tensor([[3.1415, 0.0000],
        [9.0000, 2.7183]], device='cuda:0')
默认 CPU,y=tensor([[0.3476, 0.3809],
        [0.3052, 0.5888]])
.to(device), y=tensor([[0.3476, 0.3809],
        [0.3052, 0.5888]], device='cuda:0')
GPU 运算 ,x+y=tensor([[3.4891, 0.3809],
        [9.3052, 3.3071]], device='cuda:0')
.to('cpu'),(x+y).to('cpu',torch.double)=tensor([[3.4891, 0.3809],
        [9.3052, 3.3071]], dtype=torch.float64)
```

3.15.2 微积分-链式法则（chain rule）

在“编程微积分”部分，求导的基本公式中给出了复合函数的微分公式为：$\{g(f(x))\}'=g'(f(x))f'(x)$，也表示为：$\frac{\mathrm{d}y}{\mathrm{d}x}=\frac{\mathrm{d}y}{\mathrm{d}u}\cdot\frac{\mathrm{d}u}{\mathrm{d}x}$。链式法则表述为，两个函数组合起来的复合函数。导数等于内层函数代入外层函数的导函数，乘以内层函数的导函数。下述验证代码中，定义了复合函数为$5\times\sin(x^3+5)$，将其分解为外层函数：$5\times\sin(x)$和内层函数：x^3+5。使用SymPy库diff方法求导，分别对各个分解函数求导。求导后，外层导函数带入内层函数，最后求积即为应用链式法则求复合函数的结果（图3.15–1）。其结果与直接对复合函数求导的结果保持一致。

链式法则应用于深度学习神经网络的反向传播计算。

```
import sympy
from sympy import diff, pprint
import matplotlib.pyplot as plt
import numpy as np
x, x_i, x_o = sympy.symbols('x x_i,x_o')
composite_func = 5*sympy.sin(x**3+5)
composite_func_ = sympy.lambdify(x, composite_func, "numpy")
print("复合函数：")
pprint(composite_func)
inner_func = x_i**3+5
inner_func_ = sympy.lambdify(x_i, inner_func, "numpy")
print("内层函数：")
pprint(inner_func)
outer_func = 5*sympy.sin(x_o)
outer_func_ = sympy.lambdify(x_o, outer_func, "numpy")
```

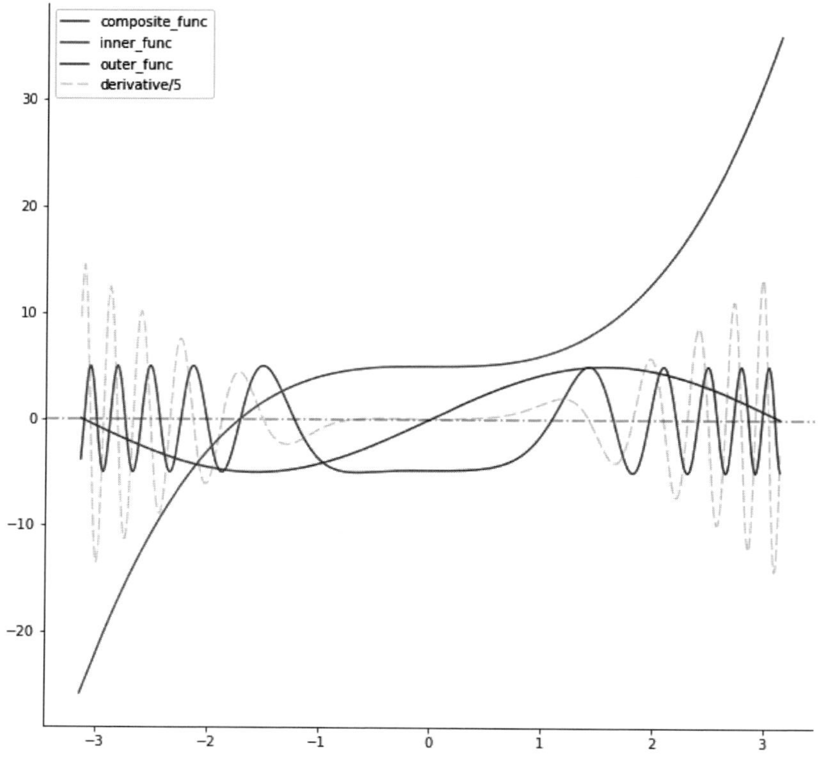

图3.15-1 复合函数及其求导结果

```
print("外层函数：")
pprint(outer_func)

#求导
d_composite = diff(composite_func, x)
d_composite_ = sympy.lambdify(x, d_composite, "numpy")
print("复合函数-求导：")
pprint(d_composite)
d_chain_rule_form_1 = diff(outer_func, x_o).subs(
    x_o, inner_func)*diff(inner_func, x_i)
print("链式法则-求导：")
pprint(d_chain_rule_form_1)
t = np.arange(-np.pi, np.pi, 0.01)
y_composite = composite_func_(t)
y_inner = inner_func_(t)
y_outer = outer_func_(t)
fig, ax = plt.subplots(figsize=(10, 10))
ax.plot(t, y_composite, label="composite_func")
ax.plot(t, y_inner, label="inner_func")
ax.plot(t, y_outer, label="outer_func")
ax.plot(t, d_composite_(t)/10, label="derivative/5",
        dashes=[6, 2], color='gray', alpha=0.3)
ax.axhline(y=0, color='r', linestyle='-.', alpha=0.5)
ax.legend(loc='upper left')
ax.spines['right'].set_visible(False)
ax.spines['top'].set_visible(False)
plt.show()
--------------------------------------------------------------------------------
```

复合函数：

$$5\cdot\sin\left(x^3+5\right)$$

内层函数：

$x_i^3 \ + 5$
外层函数：
$5 \cdot \sin(x_o)$
复合函数 - 求导：
$15 \cdot x^2 \cdot \cos\left(x^3 + 5\right)$
链式法则 - 求导：
$15 \cdot x_i^2 \cdot \cos\left(x_i^3 + 5\right)$

3.15.3 激活函数（activation function）

在多层神经网络中（多层感知机，multilayer perception，MLP），如果不使用激活函数，则每一层节点（神经元）的输入都是上一层输出的线性函数，那么无论神经网络有多少层，输出都是输入的线性组合。即为原始的感知机（单层感知机，perception），其网络的逼近能力有限。因此引入了非线性函数作为激活函数，使神经网络可以逼近任意函数。常用的激活函数有ReLU（rectified linear unit）、sigmoid和tanh函数。

给定元素x，ReLU函数的定义为：$ReLU(x) = \max(x, 0)$，可知ReLU函数只保留正数元素，并将负数元素清零，为两段线性函数。ReLU函数的导数对应在负数时，为0，正数时为1。值为0时，ReLU函数不可导，在此处将其导数配置为0，见图3.15–2（左）。

sigmoid函数可以将元素的值变换到0 ~ 1之间，其公式为：$\text{sigmoid}(x) = \frac{1}{1+\exp(-x)}$。sigmoid函数在早期的神经网络中较为普遍，目前逐渐被更为简单的ReLU函数取代。当输入接近0时，sigmoid函数接近线性变换。sigmoid函数的导数，在输入为0时，导数达到最大值0.25；当输入偏离0时，导数趋近于0，见图3.15–2（中）。

tanh（双曲正切）函数可以将元素的值变换到-1 ~ 1之间。其公式为：$\tanh(x) = \frac{1-\exp(-2x)}{1+\exp(-2x)}$，当输入接近0时，tanh函数接近线性变换。同时，tanh函数在坐标系原点上对称。依据链式法则，tanh函数的导数为：$\tanh(x) = 1 - \tanh^2(x)$。当输入为0时，tanh的导数达到最大值1；当输入偏离0时，tanh函数的导数趋近于0，见图3.15–2（右）。

```
import torch
import matplotlib.pyplot as plt
import torch.nn as nn
fig, axs = plt.subplots(1, 3, figsize=(28, 5))
# A-ReLU
x = torch.arange(-8.0, 8.0, 0.1, requires_grad=True)
y_relu = x.relu()
axs[0].plot(x.detach().numpy(), y_relu.detach().numpy(), label="ReLU")

# ReLU 函数的导数
y_relu.sum().backward()
axs[0].plot(x.detach().numpy(), x.grad.detach().numpy(),
            label="grad of ReLU", linestyle='-.')
# B-sigmoid
y_sigmoid = x.sigmoid()
axs[1].plot(x.detach().numpy(), y_sigmoid.detach().numpy(), label="sigmoid")

# sigmoid 函数的导函数
x.grad.zero_()  # 参数梯度置零
y_sigmoid.sum().backward()
```

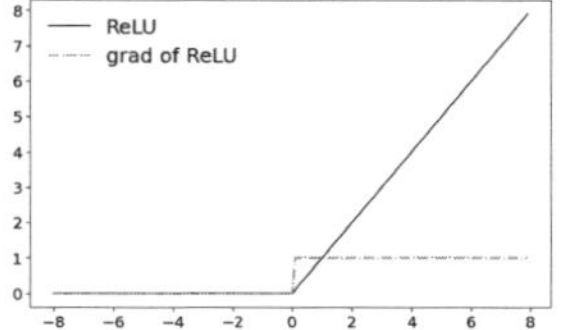

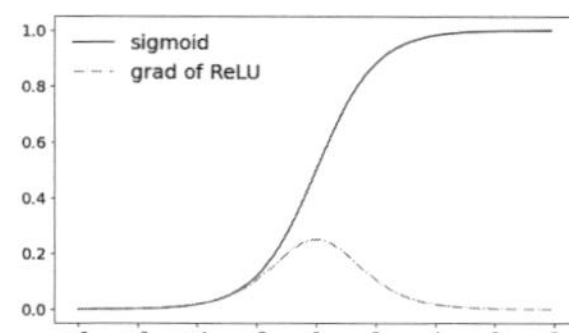

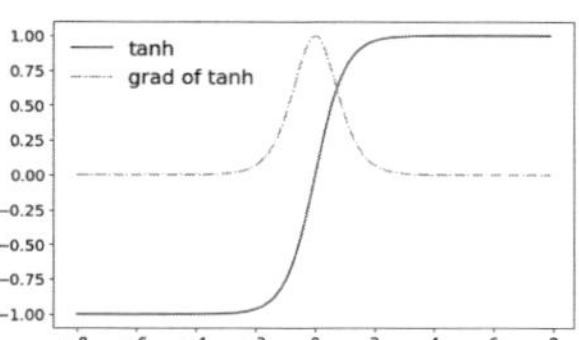

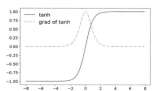

图3.15–2　激活函数，ReLU（左）；sigmoid（中）；tanh（右）

```
axs[1].plot(x.detach().numpy(), x.grad.detach().numpy(),
            label="grad of sigmoid", linestyle='-.')
# C-tanh
y_tanh = x.tanh()
axs[2].plot(x.detach().numpy(), y_tanh.detach().numpy(), label="tanh")

# tanh 函数的导函数
x.grad.zero_()
y_tanh.sum().backward()
axs[2].plot(x.detach().numpy(), x.grad.detach().numpy(),
            label="grad of tanh", linestyle='-.')
axs[0].legend(loc='upper left', frameon=False)
axs[1].legend(loc='upper left', frameon=False)
axs[2].legend(loc='upper left', frameon=False)
plt.show()
```

3.15.4 前向（正向）传播（forward propagation）与后向（反向）传播（back propagation）

构建一个典型的三层神经网络（图3.15–3），包括输入层（input layer）包含两个神经元（neuron）i_1, i_2，和一个偏置（偏差/截距，bias）项b_1；隐含层（hidden layter）包含两个神经元h_1, h_2，和一个偏置b_2；以及输出层（output layer）o_1, o_2。假设数据集仅包含一个特征向量（feature）并含两个值，即输入数据：$i_1 = 0.05, i_2 = 0.10$；对应类标（label），即输出数据：$o_1 = 0.01, o_2 = 0.99$；同时，随机初始化权重值：$w_1 = 0.15, w_2 = 0.20, w_3 = 0.25, w_4 = 0.30, w_5 = 0.40, w_6 = 0.45, w_7 = 0.50, w_8 = 0.55$；偏置值为$b_1 = 0.35, b_2 = 0.60$。

通过神经网络，输入值经过与权重及偏置值的计算，使得结果与类标接近，从而最终确定权重值。

Step-0：初始化

初始化特征值、类标及权重值和偏置

```
i_1,i_2 = 0.05,0.10
w_1,w_2,w_3,w_4,w_5,w_6,w_7,w_8 = 0.15,0.20,0.25,0.30,0.40,0.45,0.50,0.55
b_1,b_2 = 0.35,0.60
o_1,o_2 = 0.01,0.99
```

Step-1：前向传播

- 1. 输入层—>隐含层

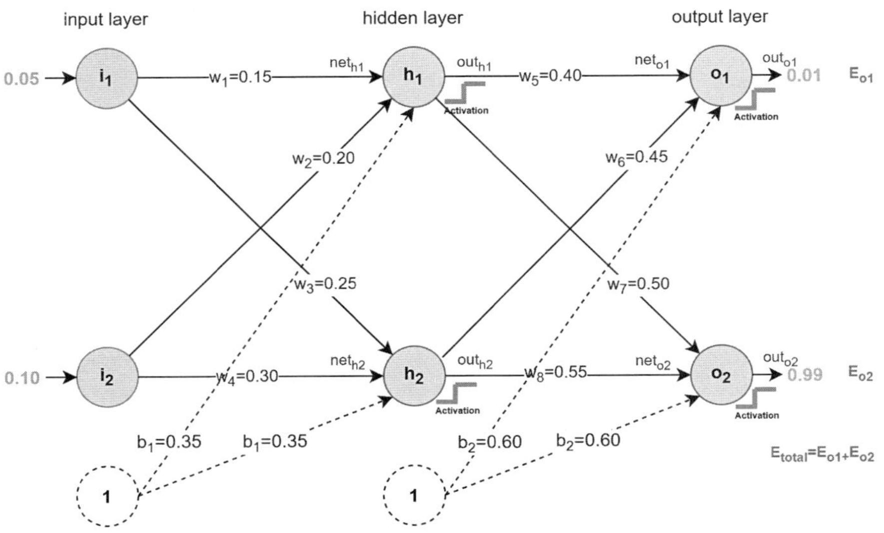

图3.15–3　三层神经网络结构示例

计算神经元h1的输入加权和

```
net_h_1 = w_1*i_1+w_2*i_2+b_1*1
print("net_h_1={}".format(net_h_1))
---------------------------------------------------------------------------
net_h_1=0.3775
```

对h1应用激活函数-sigmoid

```
def sigmoid(x):
    '''
    function - sigmoid 函数
    '''
    import math
    return 1/(1+math.exp(-x))
out_h_1 = sigmoid(net_h_1)
print("out_h_1={}".format(out_h_1))
---------------------------------------------------------------------------
out_h_1=0.5932699921071872
```

同理，计算神经元h2的输入加权和，并应用sigmoid函数

```
net_h_2 = w_3*i_1+w_4*i_2+b_1*1
out_h_2 = sigmoid(net_h_2)
print("out_h_2={}".format(out_h_2))
---------------------------------------------------------------------------
out_h_2=0.596884378259767
```

- 2. 隐含层—>输出层

计算神经元o1、o2的值

```
net_o_1 = w_5*out_h_1+w_6*out_h_2+b_2*1
out_o_1 = sigmoid(net_o_1)
print("out_o_1={}".format(out_o_1))
net_o_2 = w_7*out_h_1+w_8*out_h_2+b_2*1
out_o_2 = sigmoid(net_o_2)
print("out_o_2={}".format(out_o_2))
---------------------------------------------------------------------------
out_o_1=0.7513650695523157
out_o_2=0.7729284653214625
```

随机初始化权重值，逐层计算输入加权和（Summation and Bias）：$\sum_{i=1}^{m}(w_i x_i)+bias$，并应用激活函数（sigmoid），获得输出值out_o_1和out_o_2，与实际值$o_1=0.01, o_2=0.99$相差还很远，现在对误差进行反向传播，更新权重/权值，重新计算输出。

Step-2：反向传播

- 1．计算总误差（残差平方误差，Residual square error）

分别计算各个输出（此时，有2个输出）的误差，再求和。

```
E_o_1 = 1/2*(out_o_1-o_1)**2
E_o_2 = 1/2*(out_o_2-o_2)**2
E_total = E_o_1+E_o_2
print("E_total={}".format(E_total))
---------------------------------------------------------------------------
E_total=0.2983711087600027
```

- 2．隐含层—>输出层的权值更新

以w_5权重值参数为例，计算w_5对整体误差的影响，用整体误差对w_5求偏导，应用微分链式法则，其公式为：$\frac{\partial E_{\text{total}}}{\partial \text{w}_5}=\frac{\partial E_{\text{total}}}{\partial \text{out}_{o1}}\times\frac{\partial \text{out}_{o1}}{\partial \text{net}_{o1}}\times\frac{\partial \text{net}_{o1}}{\partial \text{w}_5}$，分别是对总误差计算公式、激活函数和加权和函数求导。只是各个输入值分别对应w_5计算路径下的各个对应值，可以很容易地从以上神经网络结构图中确定，即函数的输入端值。总误差计算公式导函数对应out_{o1}即out_o_1，激活函数导函数对应net_{o1}即net_o_1，加权和函数导函数对应out_{h1}即out_h_1。

```
from sympy.functions import exp
import sympy
from sympy import diff, pprint

# w_5- 链式法则 -01- 总误差函数（公式）对 out_o_1 的偏导数
out_o_1_, o_1_, out_o_2_, o_2_ = sympy.symbols('out_o_1_ o_1_,out_o_2_ o_2_')
d_E_total_2_out_o_1 = diff(1/2*(out_o_1_-o_1_)**2 +
```

```
                              1/2*(out_o_2_-o_2_)**2, out_o_1_)
d_E_total_2_out_o_1_value = d_E_total_2_out_o_1.subs(
    [(o_1_, o_1), (out_o_1_, out_o_1)])
print("d_E_total_2_out_o_1_value={}".format(d_E_total_2_out_o_1_value))

#w_5- 链式法则 -02- 激活函数对 net_o_1 的偏导数
x_sigmoid = sympy.symbols('x_sigmoid')
d_activation = diff(1/(1+exp(-x_sigmoid)), x_sigmoid)
d_out_o_1_2_net_o_1_value = d_activation.subs([(x_sigmoid, net_o_1)])
print("d_out_o_1_2_activation_value={}".format(d_out_o_1_2_net_o_1_value))

#w_5- 链式法则 -03- 隐藏层 h1，加权和函数（公式）对 out_h_1 的偏导数
w_5_, out_h_1_, w_6_, out_h_2_, b_2_ = sympy.symbols(
    'w_5,out_h_1,w_6,out_h_2,b_2')
d_net_o_1_2_w_5 = diff(w_5_*out_h_1_+w_6_*out_h_2_+b_2_*1, w_5_)
d_net_o_1_2_w_5_value = d_net_o_1_2_w_5.subs([(out_h_1_, out_h_1)])
print("d_net_o_1_2_w_5_value={}".format(d_net_o_1_2_w_5_value))

#w_5- 链式法则 - 各部分相乘
d_E_total_2_w_5 = d_E_total_2_out_o_1_value * \
    d_out_o_1_2_net_o_1_value*d_net_o_1_2_w_5_value
print("d_E_total_2_w_5={}".format(d_E_total_2_w_5))

#w_5- 权重值更新
lr = 0.5  #学习速率
w_5_update = w_5-lr*d_E_total_2_w_5
print("w_5_update={}".format(w_5_update))
--------------------------------------------------------------------------------
d_E_total_2_out_o_1_value=0.741365069552316
d_out_o_1_2_activation_value=0.186815601808960
d_net_o_1_2_w_5_value=0.593269992107187
d_E_total_2_w_5=0.0821670405642308
w_5_update=0.358916479717885
```

所有权值的偏导求法基本相同。为了减少重复的代码，可以定义公用的导函数。只是不同的权值，根据其在神经网络结构下的路径，调整链式法则中的各个偏导函数，并对应替换值。

```
#定义总误差偏导
def partialD_E_total_prediction(true_value_, predicted_value_):
    '''
    function - 定义总误差偏导
    '''
    import sympy
    from sympy import diff
    true_value, predicted_value = sympy.symbols('true_value predicted_value')
    partialD_E_total_prediction = diff(
        1/2*(predicted_value-true_value)**2, predicted_value)
    return partialD_E_total_prediction.subs([(predicted_value, predicted_
value_), (true_value, true_value_)])

#定义激活函数偏导
def partialD_activation(x):
    '''
    function - 定义激活函数偏导
    '''
    import sympy
    from sympy import diff
    from sympy.functions import exp
    x_sigmoid = sympy.symbols('x_sigmoid')
    partialD_activation = diff(1/(1+exp(-x_sigmoid)), x_sigmoid)
    return partialD_activation.subs([(x_sigmoid, x)])

#定义加权和偏导
def partialD_weightedSUM(w_):
    '''
    fucntion - 定义加权和偏导
    '''
    import sympy
    from sympy import diff
    w, x_w = sympy.symbols('w x_w')
    partialD_weightedSUM = diff(w*x_w, w)
    return partialD_weightedSUM.subs([(x_w, w_)])
w_5_update = w_5-lr*partialD_E_total_prediction(
    o_1, out_o_1)*partialD_activation(net_o_1)*partialD_weightedSUM(out_h_1)
```

```
w_6_update = w_6-lr*partialD_E_total_prediction(
    o_1, out_o_1)*partialD_activation(net_o_1)*partialD_weightedSUM(out_h_2)
w_7_update = w_7-lr*partialD_E_total_prediction(
    o_2, out_o_2)*partialD_activation(net_o_2)*partialD_weightedSUM(out_h_1)
w_8_update = w_8-lr*partialD_E_total_prediction(
    o_2, out_o_2)*partialD_activation(net_o_2)*partialD_weightedSUM(out_h_2)
print("w_5_update={}\nw_6_update={}\nw_7_update={}\nw_8_update={}".format(
    w_5_update, w_6_update, w_7_update, w_8_update))
--------------------------------------------------------------------------------
w_5_update=0.358916479717885
w_6_update=0.408666186076233
w_7_update=0.511301270238738
w_8_update=0.561370121107989
```

- 3. 输入层—>隐含层的权值更新

因为隐含层out_{h1}会受到E_{o1}, E_{o2}两个地方传来的误差，因此分别计算并求和，再与其他路径下的导数求积。out_{h2}与之相同。

```
w_1_update = w_1-lr*(partialD_E_total_prediction(o_1, out_o_1)*partialD_
activation(net_o_1)*partialD_weightedSUM(w_5)+partialD_E_total_prediction(
    o_2, out_o_2)*partialD_activation(net_o_2)*partialD_
weightedSUM(w_7))*partialD_activation(net_h_1)*partialD_weightedSUM(i_1)
w_2_update = w_2-lr*(partialD_E_total_prediction(o_1, out_o_1)*partialD_
activation(net_o_1)*partialD_weightedSUM(w_5)+partialD_E_total_prediction(
    o_2, out_o_2)*partialD_activation(net_o_2)*partialD_
weightedSUM(w_7))*partialD_activation(net_h_1)*partialD_weightedSUM(i_2)
w_3_update = w_3-lr*(partialD_E_total_prediction(o_1, out_o_1)*partialD_
activation(net_o_1)*partialD_weightedSUM(w_6)+partialD_E_total_prediction(
    o_2, out_o_2)*partialD_activation(net_o_2)*partialD_
weightedSUM(w_8))*partialD_activation(net_h_2)*partialD_weightedSUM(i_1)
w_4_update = w_4-lr*(partialD_E_total_prediction(o_1, out_o_1)*partialD_
activation(net_o_1)*partialD_weightedSUM(w_6)+partialD_E_total_prediction(
    o_2, out_o_2)*partialD_activation(net_o_2)*partialD_
weightedSUM(w_8))*partialD_activation(net_h_2)*partialD_weightedSUM(i_2)
print("w_1_update={}\nw_2_update={}\nw_3_update={}\nw_4_update={}".format(
    w_1_update, w_2_update, w_3_update, w_4_update))
--------------------------------------------------------------------------------
w_1_update=0.149780716132763
w_2_update=0.199561432265526
w_3_update=0.249751143632370
w_4_update=0.299502287264739
```

至此，误差的反向传播全部计算完，并更新权重值，重复前向传播。为了方便计算，将前向传播定义为一个函数，计算结果显示总误差为0.29102777369359933，较之前次0.2983711087600027有所下降。

```
def forward_func(input_list, weight_list, bias_list, output_list):
    '''
    function - 三层神经网络，各层 2 个神经元的示例，前向传播函数
    '''
    def sigmoid(x):
        '''
        function - sigmoid 函数
        '''
        import math
        return 1/(1+math.exp(-x))
    i_1, i_2 = input_list
    w_1, w_2, w_3, w_4, w_5, w_6, w_7, w_8 = weight_list
    b_1, b_2, b_3, b_4 = bias_list
    o_1, o_2 = output_list
    net_h_1 = w_1*i_1+w_2*i_2+b_1*1
    out_h_1 = sigmoid(net_h_1)
    net_h_2 = w_3*i_1+w_4*i_2+b_2*1
    out_h_2 = sigmoid(net_h_2)
    net_o_1 = w_5*out_h_1+w_6*out_h_2+b_3*1
    out_o_1 = sigmoid(net_o_1)
    net_o_2 = w_7*out_h_1+w_8*out_h_2+b_4*1
    out_o_2 = sigmoid(net_o_2)
    E_total = 1/2*(out_o_1-o_1)**2+1/2*(out_o_2-o_2)**2
    print("out_o_1={}\nout_o_2={}\nE_total={}".format(out_o_1, out_o_2, E_
total))
    return out_o_1, out_o_2, E_total
out_o_1, out_o_2, E_total = forward_func(input_list=[0.05, 0.10], weight_list=[
```

```
                                              w_1_update, w_2_update, w_3_update,
w_4_update, w_5_update, w_6_update, w_7_update, w_8_update], bias_list=[0.35,
0.35, 0.60, 0.60], output_list=[0.01, 0.99])
--------------------------------------------------------------------------------
out_o_1=0.7420881111907824
out_o_2=0.7752849682944595
E_total=0.29102777369359933
```

- 多层感知机-代码整合

将上述的分解过程整合，实现迭代计算求取权重值。代码的结构包括三个类，NEURON类定义神经元前向传播和反向传播的函数；NEURAL_LAYER类是基于NEURON的层级神经元汇总；NEURAL_NETWORK类构建神经网络结构，初始化权值并更新权值。此处的输入值保持不变为［0.05,0.1］，输出值修改为［0.01,0.09］，能够更好地观察收敛的过程。如果仍然保持输出值为［0.01,0.99］，因为偏置保持不变，造成损失函数下降的趋势不明显。

```
from tqdm import tqdm
import numpy as np
class NEURON:
    '''
    class - 每一神经元的输入加权和、激活函数，以及输出计算
    '''
    def __init__(self, bias):
        '''
        function - 初始化权重和偏置
        '''
        self.bias = bias
        self.weights = []
    def activation(self, x_sigmoid):
        '''
        function - 定义 sigmoid 激活函数
        '''
        import math
        return 1/(1+math.exp(-x_sigmoid))
    def net_input(self):
        '''
        function - 每一神经元的输入加权和，inputs，以及 weights 数组形状为 (-1,1)
        '''
        import numpy as np
        y = np.array(self.inputs).reshape(-1, 1) * \
            np.array(self.weights).reshape(-1, 1)
        return y.sum()+self.bias
    def net_output(self, inputs):
        '''
        function - 每一神经元，首先计算输入加权和，然后应用激活函数获得输出
        '''
        self.inputs = inputs
        self.output = self.activation(self.net_input())
        return self.output
    def neuron_error(self, predicted_output):
        '''
        function - 每一神经元的误差（平方差）
        '''
        return 0.5*(predicted_output-self.output)**2
    def pd_activation(self):
        '''
        function - 定义神经元激活函数偏导
        '''
        import sympy
        from sympy import diff
        from sympy.functions import exp
        x_sigmoid = sympy.symbols('x_sigmoid')
        partialD_activation = diff(1/(1+exp(-x_sigmoid)), x_sigmoid)
        # 由 sympy 的 diff 方法获取 sigmoid 导函数
        return partialD_activation.subs([(x_sigmoid, self.output)])
        # return self.output * (1 - self.output) # 推导公式

    def pd_E_prediction(self, predicted_output):
        '''
        function - 定义神经元误差偏导
        '''
        import sympy
        from sympy import diff
```

```
        true_value, predicted_value = sympy.symbols(
            'true_value predicted_value')
        partialD_E_total_prediction = diff(
            0.5*(predicted_value-true_value)**2, true_value)
        return partialD_E_total_prediction.subs([(predicted_value, predicted_
output), (true_value, self.output)])
    def pd_net_output(self, predicted_output):
        '''
        function - 每一神经元误差偏导，由链式法则计算，包括神经元激活函数偏导和神经元总误差
偏导
        '''
        return self.pd_activation()*self.pd_E_prediction(predicted_output)
    def pd_weightedSUM(self, index):
        '''
        fucntion - 定义神经元加权和偏导
        '''
        import sympy
        from sympy import diff
        w, x_w = sympy.symbols('w x_w')
        partialD_weightedSUM = diff(w*x_w, w)
        return partialD_weightedSUM.subs([(x_w, self.inputs[index])])
class NEURAL_LAYER:
    '''
    class - 神经网络每一层的定义
    '''
    def __init__(self, num_neurons, bias):
        '''
        fucntion - 初始化每层偏置及神经元
        '''
        self.bias = bias if bias else random.random()  #同一层各个神经元共享同一个偏置
        self.neurons = [NEURON(self.bias) for i in range(num_neurons)]
    def layer_info(self):
        '''
        function - 查看每层的神经元信息
        '''
        print('neurons:{}'.format(len(self.neurons)))
        for i in range(len(self.neurons)):
            print('neuron-{}'.format(i))
            print('weights:{}'.format([self.neurons[i].weights[j]
                  for j in range(len(self.neurons[i].weights))]))
            print('bias:{}'.format(self.bias))
    def layer_output(self, inputs):
        outputs = [neuron.net_output(inputs) for neuron in self.neurons]
        return outputs
class NEURAL_NETWORK:
    '''
    class - 构建多层感知机（神经网络）
    '''
    def __init__(self, learning_rate, num_inputs, num_hidden, num_outputs,
hiddenLayer_weights=None, hiddenLayer_bias=None, outputLayer_weights=None,
outputLayer_bias=None):
        self.lr = learning_rate
        self.num_inputs = num_inputs
        self.hidden_layer = NEURAL_LAYER(num_hidden, hiddenLayer_bias)
        self.output_layer = NEURAL_LAYER(num_inputs, outputLayer_bias)
        self.weights_init(hiddenLayer_weights,
                          self.hidden_layer, self.num_inputs)
        self.weights_init(outputLayer_weights, self.output_layer,
                          len(self.hidden_layer.neurons))
    def weights_init(self, weights, neu_layer, num_previous):
        '''
        fucntion - 初始化权值
        '''
        weights_num = 0
        for i in range(len(neu_layer.neurons)):
            for j in range(num_previous):
                if not weights:
                    neu_layer.neurons[i].weights.append(random.random())
                else:
                    neu_layer.neurons[i].weights.append(weights[weights_num])
                weights_num += 1
    def neural_info(self):
        '''
        function - 查看神经网络结构
        '''
```

```
            print("_"*50)
            print("inputs number:{}".format(self.num_inputs))
            print("_"*50)
            print("hidden layer:{}".format(self.hidden_layer.layer_info()))
            print("_"*50)
            print("output layer:{}".format(self.output_layer.layer_info()))
            print("_"*50)
        def forward_propagation(self, inputs):
            '''
            fucntion - 前向传播
            '''
            hiddenLayer_outputs = self.hidden_layer.layer_output(inputs)
            return self.output_layer.layer_output(hiddenLayer_outputs)
        def train(self, training_inputs, training_outputs):
            '''
            function - 训练，迭代更新权值
            '''
            self.forward_propagation(training_inputs)

            # 1. 输出层神经元的值
            pd_output_values = [0]*len(self.output_layer.neurons)
            for i in range(len(self.output_layer.neurons)):
                # ∂E/∂zj
                pd_output_values[i] = self.output_layer.neurons[i].pd_net_output(
                    training_outputs[i])
          # print(pd_output_values)

            # 2. 隐含层神经元的值
            hidden_values = [0]*len(self.hidden_layer.neurons)
            for i in range(len(self.hidden_layer.neurons)):
                # dE/dyj = Σ ∂E/∂zj * ∂z/∂yj = Σ ∂E/∂zj * wij
                d_errors = 0
                for j in range(len(self.output_layer.neurons)):
                    d_errors += pd_output_values[j] * \
                        self.output_layer.neurons[j].weights[i]
                # ∂E/∂zj = dE/dyj * ∂zj/∂
                hidden_values[i] = d_errors * \
                    self.hidden_layer.neurons[i].pd_activation()
            # print(hidden_values)

            # 3. 更新输出层权重系数
            for i in range(len(self.output_layer.neurons)):
                for j in range(len(self.output_layer.neurons[i].weights)):
                    # ∂Ej/∂wij = ∂E/∂zj * ∂zj/∂wij
                    weights_update = pd_output_values[i] * \
                        self.output_layer.neurons[i].pd_weightedSUM(j)
                    # Δw = α * ∂Ej/∂wi
                    self.output_layer.neurons[i].weights[j] -= self.lr * \
                        weights_update

            # 4. 更新隐含层的权重系数
            for i in range(len(self.hidden_layer.neurons)):
                for j in range(len(self.hidden_layer.neurons[i].weights)):
                    # ∂Ej /∂wi = ∂E/∂zj * ∂zj /∂wi
                    weights_update = hidden_values[i] * \
                        self.hidden_layer.neurons[i].pd_weightedSUM(j)
                    # Δw = α * ∂Ej /∂wi
                    self.hidden_layer.neurons[i].weights[j] -= self.lr * \
                        weights_update
        def total_error(self, training_sets):
            total_error = 0
            for i in range(len(training_sets)):
                training_inputs, training_outputs = training_sets[i]
                self.forward_propagation(training_inputs)
                for j in range(len(training_outputs)):
                    total_error += self.output_layer.neurons[j].neuron_error(
                        training_outputs[j])
            return total_error
learning_rate = 0.5
nn = NEURAL_NETWORK(learning_rate, 2, 2, 2, hiddenLayer_weights=[
                    0.15, 0.2, 0.25, 0.3], hiddenLayer_bias=0.35, outputLayer_
weights=[0.4, 0.45, 0.5, 0.55], outputLayer_bias=0.6)
nn.neural_info()
for i in tqdm(range(10000)):
```

```
    nn.train([0.05, 0.1], [0.01, 0.09])
    if i % 1000 == 0:
        print(i, round(nn.total_error([[[0.05, 0.1], [0.01, 0.09]]]), 9))
hidenLayer_weights_ = [nn.hidden_layer.neurons[i].weights for i in range(
    len(nn.hidden_layer.neurons))]
output_layer_weights_ = [nn.output_layer.neurons[i].weights for i in range(
    len(nn.output_layer.neurons))]
print("hidenLayer_weights_={}\noutput_layer_weights_={}".format(
    hidenLayer_weights_, output_layer_weights_))
------------------------------------------------------------------------------
inputs number:2
------------------------------------------------
neurons:2
neuron-0
weights:[0.15, 0.2]
bias:0.35
neuron-1
weights:[0.25, 0.3]
bias:0.35
hidden layer:None
------------------------------------------------
neurons:2
neuron-0
weights:[0.4, 0.45]
bias:0.6
neuron-1
weights:[0.5, 0.55]
bias:0.6
output layer:None
------------------------------------------------
  0%|          | 14/10000 [00:00<02:28, 67.03it/s]
0 0.493712428
 10%|█         | 1015/10000 [00:14<01:57, 76.52it/s]
1000 2.4963e-05
 20%|██        | 2016/10000 [00:28<01:48, 73.30it/s]
2000 1.878e-06
 30%|███       | 3011/10000 [00:42<01:39, 70.08it/s]
...
8000 0.0
 90%|█████████ | 9011/10000 [02:01<00:13, 71.49it/s]
9000 0.0
100%|██████████| 10000/10000 [02:14<00:00, 74.62it/s]
hidenLayer_weights_=[[0.374907950905178, 0.649815901810356], [0.469098467292651,
0.738196934585302]]
output_layer_weights_=[[-4.28132748279580, -4.25790301529841],
[-2.41120108789009, -2.37807889000506]]
```

3.15.5 自动求梯度（gradient）

在上述前向和反向传播中，反向传播的梯度计算是应用SymPy的`diff`方法求得导函数计算。因为在深度学习中经常要对函数求梯度，因此PyTorch提供了autograd，可以根据输入和前向传播过程自动构建计算图，并执行反向传播。定义张量时，如果配置参数`requires_grad=True`，将追踪其上的所有操作，因此可以应用链式法则进行梯度传播计算。完成前向传播计算后，可以调用`.backward()`来完成梯度计算。梯度将累积到grad属性中。如果不想被追踪，可以执行`.detach()`，将其从追踪记录中剥离开来；或者，使用`with torch.no_grad()`将不想追踪的操作代码包裹起来（例如，在评估模型时）。

注意：grad在反向传播过程中是累加的（accumulated），一般在反向传播之前，把应用`x.grad.data.zero_()`梯度清零。

梯度，gradient，一种关于多元导数的概况。一元（单变量）函数的导数是标量值函数，而多元函数的梯度是向量值函数。多元可微函数f在点P上的梯度，是以f在P上的偏导数为分量的向量。一元函数的导数表示这个函数图形切线的斜率。如果多元函数在点P上的梯度不是零向量，则它的方向是这个函数在P上最大增长的方向，它的量是在这个方向上的增长率。

```
import torch
# 01- 定义包含梯度追踪的张量
# x.requires_grad_(True) 的方式可以将未指定 requires_grad = False 的张量转换为追踪梯度传播
```

```
x = torch.tensor([3.0, 2.0, 1.0, 4.0], requires_grad=True)
print("01- 定义包含梯度追踪的张量 :")
print("x={}\nx.grad_fn={}\nx.requires_grad={}".format(
    x, x.grad_fn, x.requires_grad))
print("_"*50)

# 02- 追踪运算的梯度
print("02- 追踪运算的梯度 :")
y = (x+1)**3+2
z = y.view(2, 2)
print("z={}\nz.grad_fn={}".format(z, z.grad_fn))
print("x.is_leaf={},y.is_leaf={},z.is_leaf={}".format(
    x.is_leaf, y.is_leaf, z.is_leaf))
print("_"*50)

# 03-z( 张量 ) 关于 x 的梯度
print("03-z( 张量 ) 关于 x 的梯度 :")
v = torch.tensor([[1.0, 0.1], [0.01, 0.001]], dtype=torch.float)
z.backward(v)  # 因为 z 不是标量，在调用 backward 时需要传入一个和 z 同形状的权值向量（权值）进行
加权求和得到一个标量
print("dz/dx={}".format(x.grad))  # x.grad 是和 x 同形的张量
print("_"*50)

# 04- 中断梯度追踪
print("04- 中断梯度追踪 :")
x_1 = torch.tensor(1.0, requires_grad=True)
y_1 = x_1**2
with torch.no_grad():
    y_2 = x_1**3
y_3 = y_1+y_2
print("x.requires_grad={}\ny_1={};y_1.requires_grad={}\ny_2={};y_2.requires_
grad={}\ny_3={};y_3.requires_grad={}\n".format(
    x.requires_grad, y_1, y_1.requires_grad, y_2, y_2.requires_grad, y_3, y_3.
requires_grad))
y_3.backward()
print("dy_3/dx={}".format(x_1.grad))  # y_2 梯度并未被追踪，与 y_2 有关的梯度并不会回传
print("_"*50)

# 05- 不影响反向传播下修改张量值，tensor.data
print("05- 不影响反向传播下修改张量值，tensor.data:")
x_2 = torch.ones(1, requires_grad=True)
print("x_2.data={}\nx_2.data.requires_grad={}".format(x_2, x_2.data.requires_
grad))
y_4 = 2*x_2
x_2.data *= 100  # 只改变了值，不会记录在计算图中（autograd 记录），不影响梯度传播
y_4.backward()
print("x_2={}\nx_2.grad={}".format(x_2, x_2.grad))
```

```
-----------------------------------------------------------------------------------
01- 定义包含梯度追踪的张量 :
x=tensor([3., 2., 1., 4.], requires_grad=True)
x.grad_fn=None
x.requires_grad=True
_____________________________________________________________
02- 追踪运算的梯度 :
z=tensor([[ 66.,  29.],
        [ 10., 127.]], grad_fn=<ViewBackward0>)
z.grad_fn=<ViewBackward0 object at 0x0000024EE6A4B9D0>
x.is_leaf=True,y.is_leaf=False,z.is_leaf=False
_____________________________________________________________
03-z( 张量 ) 关于 x 的梯度 :
dz/dx=tensor([48.0000,  2.7000,  0.1200,  0.0750])
_____________________________________________________________
04- 中断梯度追踪 :
x.requires_grad=True
y_1=1.0;y_1.requires_grad=True
y_2=1.0;y_2.requires_grad=False
y_3=2.0;y_3.requires_grad=True
dy_3/dx=2.0
_____________________________________________________________
05- 不影响反向传播下修改张量值，tensor.data:
x_2.data=tensor([1.], requires_grad=True)
x_2.data.requires_grad=False
x_2=tensor([100.], requires_grad=True)
x_2.grad=tensor([2.])
```

3.15.6 用PyTorch构建多层感知机（多层神经网络）

3.15.6.1 自定义激活函数、模型、损失函数及梯度下降法

PyTorch自动求梯度的方式，让神经网络模型的构建变得轻松，可以将更多的注意力放在模型的构建上，而不是梯度的计算上。上述实现了典型三层神经网络的逐步计算及代码整合，对应输入输出值、权值和偏置、激活函数、损失函数、模型结构及权值更新（优化算法：梯度下降法）保持不变，应用PyTorch分别重新定义。从迭代计算结果来看，因为偏置值也得以更新，损失函数下降得比较快，迅速地收敛。

```
import torch
import numpy as np

# A- 训练数据
X = torch.tensor([0.05, 0.1])
y = torch.tensor([0.01, 0.9])

# B- 定义模型参数
num_inputs, num_outputs, num_hiddens = 2, 2, 2
W1 = torch.tensor([[0.15, 0.2], [0.25, 0.3]])  # hiddenLayer_weights
b1 = torch.tensor([0.35, 0.35])  # hiddenLayer_bias
W2 = torch.tensor([[0.4, 0.45], [0.5, 0.55]])  # outputLayer_weights
b2 = torch.tensor([0.6, 0.6])  # outputLayer_bias
params = [W1, b1, W2, b2]
for param in params:
    param.requires_grad_(requires_grad=True)

# C- 定义激活函数
def sigmoid(X):
    return 1/(1+torch.exp(-X))

# D- 定义模型
def net(X):
    X = X.view((-1, num_inputs))
    H = sigmoid(torch.matmul(X, W1)+b1)
    return torch.matmul(H, W2)+b2

# E- 定义损失函数
def loss(y_hat, y):
    return 0.5*(y_hat-y)**2

# F- 优化算法
def sgd(params, lr):
    for param in params:
        param.data -= lr*param.grad

# G- 训练模型
def train(net, X, y, loss, num_epochs, params=None, lr=None, optimizer=None):
    train_l_sum, train_acc_sum, n = 0.0, 0.0, 0
    for epoch in range(num_epochs):
        y_hat = net(X)
        l = loss(y_hat, y).sum()
        # 梯度清零
        if params is not None and params[0].grad is not None:
            for param in params:
                param.grad.data.zero_()
        l.backward()
        sgd(params, lr)
        print('epoch %d, loss %.9f' % (epoch+1, l))
    return net, params
num_epochs, lr = 5, 0.5
net_, params_ = train(net, X, y, loss, num_epochs, params, lr)
print("应用训练好的模型验证预测值：\nnet_(x)={}".format(net_(X)))
print("参数（权值+偏置）更新结果：\nparams_={}".format(params_))
---------------------------------------------------------------------------
epoch 1, loss 0.677512288
epoch 2, loss 0.013031594
epoch 3, loss 0.000361601
epoch 4, loss 0.000010227
epoch 5, loss 0.000000290
应用训练好的模型验证预测值：
```

```
net_(x)=tensor([[0.0101, 0.9000]], grad_fn=<AddBackward0>)
参数（权值＋偏置）更新结果：
params_=[tensor([[0.1463, 0.1954],
        [0.2427, 0.2907]], requires_grad=True), tensor([0.2770, 0.2573],
requires_grad=True), tensor([[0.0102, 0.3532],
        [0.1094, 0.4529]], requires_grad=True), tensor([-0.0585,  0.4366],
requires_grad=True)]
```

3.15.6.2 直接使用PyTorch提供的函数

PyTorch已经内置了多种激活函数、损失函数、优化算法及各类模型算法，可以直接调用参与计算。

```
import torch
from torch import nn

# A- 训练数据
X = torch.tensor([0.05, 0.1])
y = torch.tensor([0.01, 0.9])

# B- 定义模型参数：权值和偏置此次随机初始化
num_inputs, num_outputs, num_hiddens = 2, 2, 2

# C- 定义模型
net = nn.Sequential(
    nn.Linear(num_inputs, num_hiddens),
    nn.Sigmoid()
)

# D- 初始化参数：可以省略，pytorch 定义模型时，已经初始化
# for params in net.parameters():
            # nn.init.normal_(params,mean=0,std=0.01)

# E- 定义损失函数 - 均方误差损失
loss = nn.MSELoss()

# F- 优化算法
optimizer = torch.optim.SGD(net.parameters(), lr=0.5)

# G- 训练模型
def train(net, X, y, loss, num_epochs, params=None, lr=None, optimizer=None):
    train_l_sum, train_acc_sum, n = 0.0, 0.0, 0
    for epoch in range(num_epochs):
        y_hat = net(X)
        l = loss(y_hat, y).sum()
        # 梯度清零
        if optimizer is not None:
            optimizer.zero_grad()
        elif params is not None and params[0].grad is not None:
            for param in params:
                param.grad.data.zero_()
        # 自动求梯度
        l.backward()
        optimizer.step()
        if epoch % 1000 == 0:
            print('epoch %d, loss %.9f' % (epoch+1, l))
    return net, params
num_epochs = 10000
net, params = train(net, X, y, loss, num_epochs, params, lr, optimizer)
print("应用训练好的模型验证预测值：\nnet_(x)={}".format(net_(X)))
print("参数（权值＋偏置）更新结果：\nparams_={}".format(params_))
--------------------------------------------------------------------------
epoch 1, loss 0.215000793
epoch 1001, loss 0.000375652
epoch 2001, loss 0.000146600
epoch 3001, loss 0.000081315
epoch 4001, loss 0.000052157
epoch 5001, loss 0.000036260
epoch 6001, loss 0.000026540
epoch 7001, loss 0.000020135
epoch 8001, loss 0.000015684
epoch 9001, loss 0.000012468
应用训练好的模型验证预测值：
```

```
net_(x)=tensor([[0.0101, 0.9000]], grad_fn=<AddBackward0>)
参数（权值+偏置）更新结果：
params_=[tensor([[0.1463, 0.1954],
        [0.2427, 0.2907]], requires_grad=True), tensor([0.2770, 0.2573],
requires_grad=True), tensor([[0.0102, 0.3532],
        [0.1094, 0.4529]], requires_grad=True), tensor([-0.0585,  0.4366],
requires_grad=True)]
```

3.15.7 runx.logx-深度学习实验管理（Deep Learning Experiment Management）与模型构建

runx是NVIDIA开源的一款深度学习实验管理工具。可以帮助深度学习研究者自动执行一些常见的任务，例如参数扫描、输出日志记录、保存训练模型等。该库包括3个模块：runx、logx[①]及sumx。其中，最为常用的是logx模块，可以应用`logx.metric()`保存metrics（以字典格式保存的指定评估参数等，例如损失函数值、学习率等）；应用`logx.msg()`保存message（任意指定的信息）；应用`logx.save_model()`保存模型checkpoint，并可以指定一个模型精度评估指标，保存精度最好的模型；以及，初始化时配置`tensorboard=True`，写入tensorboard文件，可以应用tensorboard打开训练时的一些信息图表，例如损失曲线、学习率、输入输出图片（`logx.add_image()`）、卷积核的参数分布等。这些信息能够帮助监督网络的训练过程，为参数优化提供帮助。logx保存文件位于指定的文件夹下，包括best_checkpoint_ep1000.pth，last_checkpoint_ep1000.pth网络模型，`logging.log`日志，`metrics.csv`评估参数，以及events.out.tfevents.1609820463.LAPTOP-GH6EM1TC文件。

① runx.logx，NVIDIA开源的一款深度学习实验管理工具（https://github.com/NVIDIA/runx#introduction--a-simple-example）。

下述代码也涵盖了模型构建的几种方法，包括继承`nn.Module`类构造模型，使用`nn.Sequential`类构造模型（又包括`.add_module`方式和`OrderedDict`方式），使用`nn-ModuleList`类构造模型，以及使用`nn.ModuleDict`类构造模型。其中，优先使用`nn.Sequential`类构造模型方法。如果需要增加模型构造的灵活性，可以使用继承`nn.Module`类构造模型。

生成数据集时，使用了一个二元一次函数，根据随机生成的特征值（特征数为2，即输入值）生成对应的类标（即输出值）。但是，在神经网络模型构建时，并未使用单纯的一个线性模型，而是构建了线性模型->激活函数->线性模型的神经网络结构，用以说明模型构建的方法。如果希望验证一个层线性回归的参数是否与生成数据集类标的方程权值和偏置趋于一致，可以仅保留第一个线性模型，而移除激活函数及第2个线性模型。

将数据集划分了训练数据集和验证数据集，在训练过程中的每一epoch增加了验证数据集根据已训练的网络预测输出值，并应用损失函数`nn.MSELoss()`均方误差的累加和，用于网络评估。

```
import torch.optim as optim
from torch.nn import init
from collections import OrderedDict
import torch
from runx.logx import logx
import torch.utils.data as Data
from torch import nn
import numpy as np
from runx.logx import logx

#初始化 logx
logx.initialize(logdir="./logs/", #指定日志文件保持路径（如果不指定，则自动创建）
                coolname=True,     #是否在 logdir 下生成一个独有的目录
                hparams=None,      #配合 runx 使用，保存超参数
                tensorboard=True, #是否自动写入 tensorboard 文件
                no_timestamp=False,#是否不启用时间戳命名
                global_rank=0,
                eager_flush=True,
                )

#A-生成数据集
num_inputs = 2  #包含的特征数
num_examples = 5000
true_w = [3, -2.3]
true_b = 5.9
features = torch.tensor(np.random.normal(
    0, 1, (num_examples, num_inputs)), dtype=torch.float)
```

```
labels = true_w[0]*features[:, 0]+true_w[1] * \
    features[:, 1]+true_b  #由回归方程：y=w1*x1+-w2*x2+b，计算输出值
labels += torch.tensor(np.random.normal(0, 0.01,
                       size=labels.size()), dtype=torch.float)  #变化输出值
#可以将需要查看的对象以文本的形式保存在 logging.log 文件，此处保存了生成类标的线性方程
logx.msg(r"the expression generating labels:w1*x1+-w2*x2+b")

#B-读取数据（小批量 - 随机读取指定数量 /batch_size 的样本）
batch_size = 100
dataset = Data.TensorDataset(features, labels)  #建立 PyTorch 格式的数据集（组合输入与输出值）
train_size, val_size = [4000, 1000]
train_dataset, val_dataset = torch.utils.data.random_split(
    dataset, [train_size, val_size])
train_data_iter = Data.DataLoader(
    train_dataset, batch_size, shuffle=True)  #随机读取小批量
val_data_iter = Data.DataLoader(val_dataset, batch_size, shuffle=True)

#C-定义模型
#方法 -01- 继承 nn.Module 类构造模型
class MLP(nn.Module):  #继承 nn.Module 父类
    #声明带有模型参数的层
    def __init__(self, n_feature):
        super(MLP, self).__init__()
        #隐藏层 ,torch.nn.Linear(in_features: int, out_features: int, bias: bool = True)，参数指定特征数（输入）及类标数（输出）
        self.hidden = nn.Linear(n_feature, 2)
        self.activation = nn.ReLU()  #激活层
        self.output = nn.Linear(2, 1)  #输出层

    #定义前向传播，根据输入 x 计算返回所需要的模型输出
    def forward(self, x):
        y = self.activation(self.hidden(x))
        return self.output(y)
net = MLP(num_inputs)
print("nn.Module 构造模型：")
logx.msg("net:{}".format(net))  #此处保存了神经网络结构
print("_"*50)

#方法 -02- 使用 nn.Sequential 类构造模型。
net_sequential = nn.Sequential(
    nn.Linear(num_inputs, 2),
    nn.ReLU(),
    nn.Linear(2, 1)
)
print("nn.Sequential 构造模型：", net_sequential)

#nn.Sequential 的 .add_module 方式
net_sequential_ = nn.Sequential()
net_sequential_.add_module('hidden', nn.Linear(num_inputs, 2))
net_sequential_.add_module('activation', nn.ReLU())
net_sequential_.add_module("output", nn.Linear(2, 1))
print("nn.Sequential()-->.add_module 方式：", net_sequential_)

#nn.Sequential 的 OrderedDict 方式
net_sequential_orderDict = nn.Sequential(
    OrderedDict([
        ('hidden', nn.Linear(num_inputs, 2)),
        ('activation', nn.ReLU()),
        ("output", nn.Linear(2, 1))
    ])
)
print("nn.Sequential-->OrderedDict 方式：", net_sequential_orderDict)
print("_"*50)

#方法 -03- 使用 nn-ModuleList 类构造模型。注意 nn.ModuleList 仅仅是一个存储各类模块的列表，这些模块之间没有联系，也没有顺序，没有实现 forward 功能；但是，加入到 nn.ModuleList 中模块的参数会被自动添加到整个网络。
net_moduleList = nn.ModuleList([nn.Linear(num_inputs, 2), nn.ReLU()])
net_moduleList.append(nn.Linear(2, 1))  #可以像列表已有追加层
print("nn.ModuleList 构造模型：", net_moduleList)
print("_"*50)

#方法 -04- 使用 nn.ModuleDict 类构造模型。注意 nn.ModuleDict 与 nn.ModuleList 类似，模块间没有关联，没
```

有实现 forward 功能，但是参数会被自动添加到整个网络中。

```
net_moduleDict = nn.ModuleDict({
    'hidden': nn.Linear(num_inputs, 2),
    'activation': nn.ReLU()
})
net_moduleDict['output'] = nn.Linear(2, 1)  #象字典一样添加模块
print("nn.ModuleDict 构造模型：", net_moduleDict)
print("_"*50)

# D- 查看参数
for param in net.parameters():
    print(param)
print("_"*50)

# E- 初始化模型参数
# 权值初始化为均值为 0，标准差为 0.01 的正态分布。如果是用 net_sequential, 则可以使用 net[0].weight 指定权值；如果层自定义了名称，则可以用 .layerName 来指定
init.normal_(net.hidden.weight, mean=0, std=0.01)
init.constant_(net.hidden.bias, val=0)  # 偏置初始化为 0

print(" 初始化参数后：")
for param in net.parameters():
    print(param)
print("_"*50)

# F- 定义损失函数
loss = nn.MSELoss()

# J- 定义优化算法
optimizer = optim.SGD(net.parameters(), lr=0.03)
print("optimizer:", optimizer)

# 配置不同的学习率 - 方法 -01
optimizer_ = optim.SGD([
    {'params': net.hidden.parameters(), 'lr': 0.03},
    {'params': net.output.parameters(), 'lr': 0.01}
], lr=0.02)  # 如果对某个参数不指定学习率，就使用最外层的默认学习率
print(" 配置不同的学习率 -optimizer_:", optimizer_)

# 配置不同的学习率 - 方法 -02- 新建优化器
for param_group in optimizer_.param_groups:
    param_group['lr'] *= 0.1  # 学习率为之前的 0.1 倍
print(" 新建优化器调整学习率 -optimizer_:", optimizer_)
print("_"*50)

# H- 训练模型
num_epochs = 1000
best_loss = np.inf
for epoch in range(1, num_epochs+1):
    loss_acc_val = 0
    for X, y in train_data_iter:
        output = net(X)
        l = loss(output, y.view(-1, 1))
        optimizer.zero_grad()  # 梯度清零，
        l.backward()
        optimizer.step()

    # 验证数据集，计算预测值与真实值之间的均方误差累加和（用 MSELoss 损失函数）
    with torch.no_grad():
        for X_, y_ in val_data_iter:
            y_pred = net(X_)
            valid_loss = loss(y_pred, y_.view(-1, 1))
            loss_acc_val += valid_loss
    # print( 'epoch %d, loss: %f' % (epoch, l.item()))
    if epoch % 100 == 0:
        # logx.msg 也会打印待保存的结果，因此注释掉上行的 print
        logx.msg('epoch %d, loss: %f' % (epoch, l.item()))
        print(" 验证数据集 - 精度 -loss: {}".format(loss_acc_val))

    metrics = {'loss': l.item(),
               'lr': optimizer.param_groups[-1]['lr']}
    curr_iter = epoch*len(train_data_iter)
    # print( "+" *50)
  # print(metrics,curr_iter)
```

```
    # 对传入字典中的数据进行保存记录，参数 phase 可以选择'train'和'val'；参数 metrics 为传入的字典；参数 epoch 表示全局轮次。若开启了 tensorboard，则自动增加。
    logx.metric(phase="train", metrics=metrics, epoch=curr_iter)
    if loss_acc_val < best_loss:
        best_loss = loss_acc_val
    save_dict = {"state_dict": net.state_dict(),
                 'epoch': epoch+1,
                 'optimizer': optimizer.state_dict()
                 }
    ''' logx.save_model 在 JupyterLab 环境下运行目前会出错，可以在 Spyder 下运行
    logx.save_model(
        save_dict=save_dict,      # checkpoint 字典形式保存，包括 epoch，state_dict 等信息
        metric=best_loss,         # 保存评估指标
        epoch=epoch,              # 当前轮次
        higher_better=False,      # 是否更高更好，例如准确率
        delete_old=True           # 是否删除旧模型
        )
    '''
```

```
--------------------------------------------------------------------------------
the expression generating labels:w1*x1+-w2*x2+b
nn.Module 构造模型：
net:MLP(
  (hidden): Linear(in_features=2, out_features=2, bias=True)
  (activation): ReLU()
  (output): Linear(in_features=2, out_features=1, bias=True)
)
----------------------------------------------------------------------
nn.Sequential 构造模型： Sequential(
  (0): Linear(in_features=2, out_features=2, bias=True)
  (1): ReLU()
  (2): Linear(in_features=2, out_features=1, bias=True)
)
nn.Sequential()-->.add_module 方式： Sequential(
  (hidden): Linear(in_features=2, out_features=2, bias=True)
  (activation): ReLU()
  (output): Linear(in_features=2, out_features=1, bias=True)
)
nn.Sequential-->OrderedDict 方式： Sequential(
  (hidden): Linear(in_features=2, out_features=2, bias=True)
  (activation): ReLU()
  (output): Linear(in_features=2, out_features=1, bias=True)
)
----------------------------------------------------------------------
nn.ModuleList 构造模型： ModuleList(
  (0): Linear(in_features=2, out_features=2, bias=True)
  (1): ReLU()
  (2): Linear(in_features=2, out_features=1, bias=True)
)
----------------------------------------------------------------------
nn.ModuleDict 构造模型： ModuleDict(
  (hidden): Linear(in_features=2, out_features=2, bias=True)
  (activation): ReLU()
  (output): Linear(in_features=2, out_features=1, bias=True)
)
----------------------------------------------------------------------
Parameter containing:
tensor([[0.3069, 0.2901],
        [0.0576, 0.5021]], requires_grad=True)
Parameter containing:
tensor([-0.6169, -0.1993], requires_grad=True)
Parameter containing:
tensor([[ 0.0393, -0.1843]], requires_grad=True)
Parameter containing:
tensor([-0.4607], requires_grad=True)
----------------------------------------------------------------------
初始化参数后：
Parameter containing:
tensor([[-0.0077, -0.0019],
        [-0.0037, -0.0100]], requires_grad=True)
Parameter containing:
tensor([0., 0.], requires_grad=True)
Parameter containing:
tensor([[ 0.0393, -0.1843]], requires_grad=True)
```

```
Parameter containing:
tensor([-0.4607], requires_grad=True)
-------------------------------------------------------------------
optimizer: SGD (
Parameter Group 0
    dampening: 0
    lr: 0.03
    momentum: 0
    nesterov: False
    weight_decay: 0
)
配置不同的学习率-optimizer_: SGD (
Parameter Group 0
    dampening: 0
    lr: 0.03
    momentum: 0
    nesterov: False
    weight_decay: 0
Parameter Group 1
    dampening: 0
    lr: 0.01
    momentum: 0
    nesterov: False
    weight_decay: 0
)
新建优化器调整学习率-optimizer_: SGD (
Parameter Group 0
    dampening: 0
    lr: 0.003
    momentum: 0
    nesterov: False
    weight_decay: 0
Parameter Group 1
    dampening: 0
    lr: 0.001
    momentum: 0
    nesterov: False
    weight_decay: 0
)
-------------------------------------------------------------------
epoch 100, loss: 0.000083
验证数据集-精度-loss：0.001036209985613823
epoch 200, loss: 0.000092
验证数据集-精度-loss：0.0011525173904374242
...
epoch 900, loss: 0.000112
验证数据集-精度-loss：0.000968275882769376
epoch 1000, loss: 0.000092
验证数据集-精度-loss：0.000963839003816247
--------------------------------------------------------------------------------
print("params_hidden:{}\nbias:{}\nparams_output:{}\nbias:{}".format(net.hidden.
weight,net.hidden.bias,net.output.weight,net.output.bias))
--------------------------------------------------------------------------------
params_hidden:Parameter containing:
tensor([[ 0.6233, -0.4065],
        [-0.4955,  0.4709]], requires_grad=True)
bias:Parameter containing:
tensor([2.8779, 2.2247], requires_grad=True)
params_output:Parameter containing:
tensor([[ 2.9630, -2.3273]], requires_grad=True)
bias:Parameter containing:
tensor([2.5503], requires_grad=True)
```

在cmd命令行中输入`tensorboard --logdir=logs`，可以根据提示在http://localhost:6006/页面下打开tensorboard，并显示以下损失曲线和学习率（图3.15–4）。

机器学习（以scikit-learn[①]库为主）各类分析数据规律的算法（聚类、回归和分类），到深度学习（以pytorch人工神经网络（多层感知机））的解读。从典型三层神经网络的逐步构建，代码整合中可以深度地理解前向传播和反向传播，以及梯度计算是如何实现权值（参数）更新，并应用梯度下降法（一种优化算法）实现迭代，降低损失函数（代价函数，例如平方误差），最终确定权值的过程。进一步应用同一神经网络结构，用PyTorch库实现，包括自定义函数实现和直接使用PyTorch的内置算法实现。通过同一神经网络结构的逐步构建、代码整合和PyTorch

① scikit-learn，机器学习库，（https://scikit-learn.org/stable）。

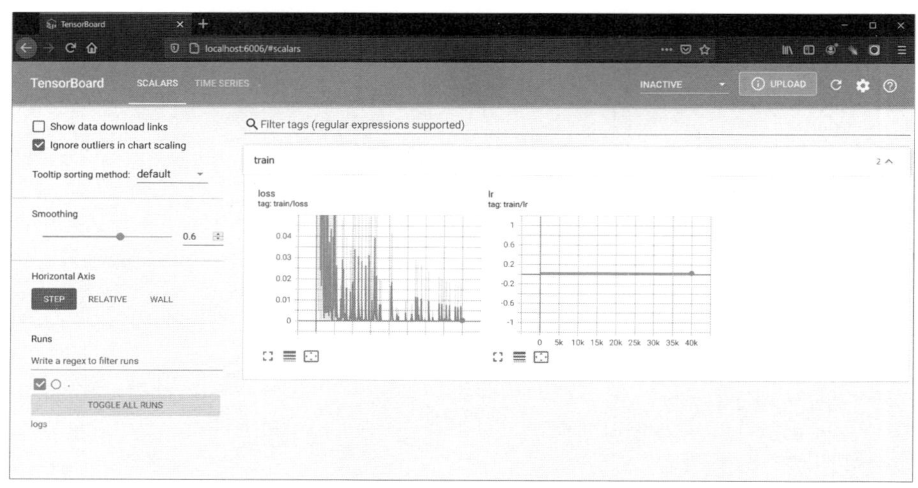

图3.15-4 tensorboard 信息图表

实现，能够清晰地理解深度学习的核心思想，有益于在各个专业领域更准确的构建和应用神经网络。

国内的深度学习框架探索也在发展，例如飞桨 PaddlePaddle[①]，是集深度学习核心框架、工具组件和服务平台为一体的开源深度学习平台。

① 飞桨PaddlePaddle，国内深度学习框架，（https://www.paddlepaddle.org.cn）。

参考文献（References）:

[1] Aston Zhang, & Alexander J. Smola (2021). Dive into Deep Learning. d2l.ai.；中文版-阿斯顿．张，李沐，扎卡里．C. 立顿，亚历山大．J．斯莫拉．动手深度学习[M]．北京：人民邮电出版社，2019-06-01．https://zh.d2l.ai/.
[2] 一文弄懂神经网络中的反向传播法——BackPropagation，https://www.cnblogs.com/charlotte77/p/5629865.html.

3.16 逻辑回归二分类到SoftMax回归多分类

SoftMax回归多分类是逻辑回归二分类的一种推广，逻辑回归通过最大似然估计更新参数（实际上通常使用梯度下降法）。最大似然估计是似然函数联合概率分布对应的最大值，而似然函数描述伯努利分布中概率（参数p）的概率分布，伯努利分布为离散型概率分布。因此，对于SoftMax的理解最好是从伯努利分布开始，逐步地层层剥离。

3.16.1 伯努利分布（Bernouli distribution）

伯努利分布（Bernoulli distribution），又名两点分布或者0-1分布，是一个离散型概率分布（为纪念瑞士科学家Jakob I. Bernoulli而命名）。其实验对象只有两种可能结果，例如从不知比例只有黑色和白色球的罐子里取球（取出后需要放回后，再取球），白色为1，黑色为0。为白的概率记作p（$0\leqslant p\leqslant 1$），为黑的概率则为$q=1-p$。其概率质量函数为：$f_x=p^x(1-p)^{1-x}=\begin{cases}p & if \quad x=1\\ q & if \quad x=0\end{cases}$。期望值为：$E[X]=\sum_{i=0}^{1}x_i f_X(x)=0+p=p$。其方差为：$var[X]=\sum_{i=0}^{1}(x_i-E[X])^2 f_X(x)=(0-p)^2(1-p)+(1-p)^2 p=p(1-p)=pq$。

指定一种结果（例如为白球）的概率$p=0.3$，可以使用scipy.stats库中的`bernoulli`进行相关计算，例如打印概率质量函数曲线（图3.16-1），生成符合概率p的随机数组等。如果定义$p=0.3$，则生成的随机数组包含0和1两个元素，其比例围绕7：3上下浮动。

概率质量函数（probability mass function，pmf），是离散随机变量在各特定取值上的概率。概率质量函数和概率密度函数不同之处在于：概率质量函数是对离散随机变量定义的，本身代表该值的概率；概率密度函数是对连续随机变量定义的，本身不是概率，只是对连续随机变量的概率密度函数在某区间内进行积分后才是概率。

期望值：在概率论和统计学中，一个离散型随机变量的期望值（或数学期望，亦简称期望）是试验中每次可能结果乘以其结果概率的总和。期望值是该变量输出值的加权平均，可能与每一个结果都不相等，并不一定包含于其分布值域，也并不一定等于值域均值。

```
from scipy.stats import bernoulli
import matplotlib.pyplot as plt
import numpy as np
fig, ax = plt.subplots(1, 1)
p = 0.3
mean, var, skew, kurt = bernoulli.stats(p, moments='mvsk')
#打印概率质量函数/分布，pmf
x = np.arange(bernoulli.ppf(0.01, p), bernoulli.ppf(0.99, p))
ax.plot(x, bernoulli.pmf(x, p), 'bo', ms=8, label='bernoulli pmf')
ax.vlines(x, 0, bernoulli.pmf(x, p), colors='b', lw=5, alpha=0.5)
rv = bernoulli(p)
ax.vlines(x, 0, rv.pmf(x), colors='k',
```

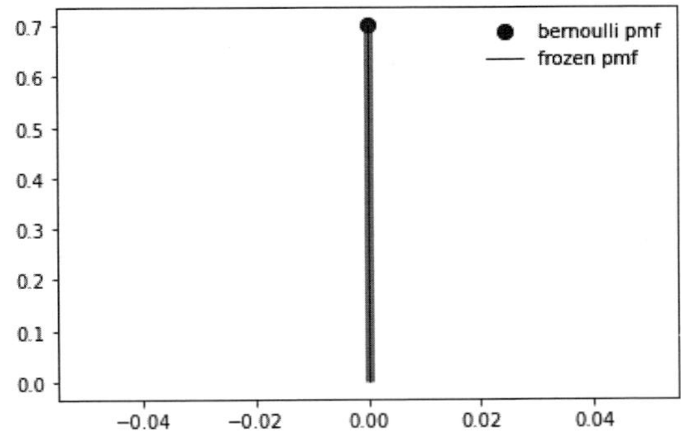

图3.16-1 伯努利分布概率质量函数曲线

```
        linestyles='-', lw=1, label='frozen pmf')
ax.legend(loc='best', frameon=False)
plt.show()

# 查看 cdf 和 ppf 的精度（Check accuracy of cdf and ppf）。cdf(k,p,loc=0),Cumulative distribution function;ppf(q, p, loc=0),Percent point function (inverse of cdf — percentiles).
prob = bernoulli.cdf(x, p)
print(".ppf(.cdf)={}".format(np.allclose(x, bernoulli.ppf(prob, p))))

# 生成符合指定概率 p 的随机数组（Generate random numbers）
r = bernoulli.rvs(p, size=100)
unique_elements, counts_elements = np.unique(r, return_counts=True)
print("生成符合伯努利分布的数组(p={})：\n{}\n包含元素为：{}，对应频数为：{}".format(
    p, r, unique_elements, counts_elements))
.ppf(.cdf)=True
生成符合伯努利分布的数组(p=0.3)：
[0 1 0 0 1 1 0 1 1 0 1 0 0 0 0 1 0 0 0 1 0 0 1 1 0 0 1 0 1 0 1 0 1 0 0 0 0 0 1 0
 0 1 0 0 1 0 1 1 1 0 0 1 0 0 1 0 0 0 1 0 1 0 0 1 0 0 1 0 1 1 0 0 0 1 0 0 0 0 0 1
 0 0 0 0 0 1 1 0 1 0 1 1 0 0 1 0 1 0 1 1]
包含元素为：[0 1]，对应频数为：[61 39]
```

3.16.2 似然函数（Likelihood function）

在伯努利分布中，应用scipy.stats.bernoulli.rvs生成符合$p=0.3$，包含100个要素的随机数组。这个过程可以用实际案例描述为黑罐子里取黑球和白球（两种球的数量比例满足0.3和1–0.3=0.7）的过程（或者抛掷硬币的过程，如果是抛掷硬币，为1，即正面朝上的概率为0.3时，说明硬币分布不均匀，正面部分的质量更高，更容易让背面朝上），取得白球，即值为1时的概率为0.3；而取得黑球，即值为0时的概率为1–0.3=0.7。即，已知参数$p=0.3$，罐子里白球和黑球的比例大约为3∶7（球的总数为任意数），可以根据已知参数求得随机变量的输出结果的概率。例如，随机抽取100次后，出现白球的概率是多少（或黑球的概率是多少）。这个过程就是概率的描述；对于上述过程的近似反过程，如只知道指定数量的随机数组，即已知随机变量输出结果（随机抽取100次后，以数组表述黑白球的抽取结果），此时并不知道参数p，而需要求得参数p取值的概率分布（即估计参数p的可能性，p的取值范围在$0\leqslant p\leqslant 1$），这个过程就是似然的描述。概率函数和似然函数具有一样的模型，区别在于将谁看作变量，将谁看作参数。如果概率函数记作：$P(X|p_x)$，其中已知$p_x=0.3$；则似然函数为：$L(p_x|X)$，其中已知X，为一个包含值1和0（两种情况结果）的随机数组。

概率描述的过程，是应用概率质量函数/分布的公式，代入取得的白球概率$p=0.3$，计算黑球（x=0）的概率为$q=0.3^0\times(1–0.3)^{1-0}=0.7$，可以推断出在随机取球100次之后，所取得白、黑球（事件）的概率为0.3和0.7。而似然描述的过程，是根据随机抽取100次后，用表述黑白球抽取结果的数组中的值（1，和0），即x的值，逐一代入概率质量函数/分布的公式后求积。因为x值为1或为0，因此会出现两种结果，当$x=0$时，结果为$p^0(1-p)^{1-0}=1-p$；当$x=1$时，结果为$p^1(1-p)^{1-1}=p$。如果随机抽取的球里，白球有29个，黑球有71个，则似然函数为$L(p_x|X)=p^{29}(1-p)^{71}$。参数$p$位于0～1之间，因此生成0-1连续$p$参数值(就是抽取黑白球的概率值)，将其代入似然函数，可以绘制抽取黑白球概率值的概率分布（图3.16–2）。

```
import sympy
x_, p_ = sympy.symbols('x p', positive=True)
pmf = p_**x_*(1-p_)**(1-x_)  # 构建概率质量函数 / 分布的公式
print("(离散型)概率质量函数, pmf={}".format(pmf))
L = np.prod([pmf.subs(x_, i) for i in r])  # 推导似然函数 / 分布公式。np.prod() 用于计算所有元素的乘积
print("似然函数,L={}".format(L))

fig, ax = plt.subplots(1, 1)
p_x = np.arange(0, 1, 0.01)  # 生成 0-1 连续的概率值
L_ = sympy.lambdify(p_, L, "numpy")
ax.plot(p_x, L_(p_x), 'b-', ms=8, label='Likelihood function distribution')
ax.legend(loc='best', frameon=False)
plt.show()
--------------------------------------------------------------------------------
(离散型)概率质量函数, pmf=p**x*(1 - p)**(1 - x)
似然函数, L=p**39*(1 - p)**61
```

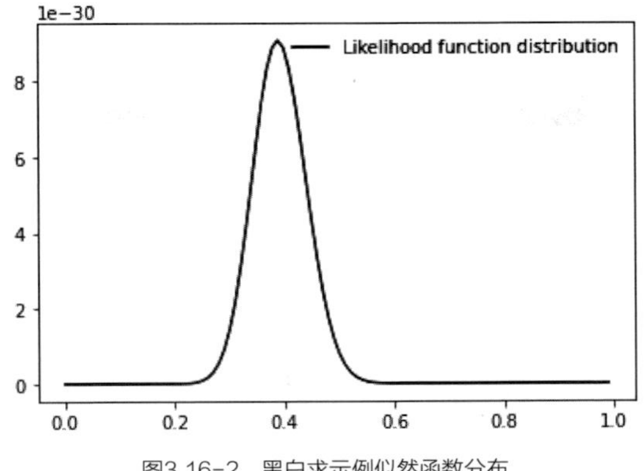

图3.16-2 黑白求示例似然函数分布

3.16.3 最大/极大似然估计（Maximum Likelihood Estimation，MLE）

给定一个概率分布D，已知其概率密度函数（连续分布，Probability Density Function，pdf）或概率质量函数（离散分布，Probability Mass Function，pmf）为f_D，及一个分布参数θ（即上述p值）。则可以从这个分布中抽取n个采样值，$X_1,X_2,...,X_n$，利用f_D计算其似然函数：$L(\theta|x_1,x_2,...,x_n)=f_\theta(x_1,x_2,...,x_n)$。若$D$是离散分布，$f_\theta$即是在参数为$\theta$时观测到这一采样的概率。若其是连续分布，$f_\theta$则为，$X_1,X_2,...,X_n$联合分布的概率密度函数在观测值处的取值。一旦获得，$X_1,X_2,...,X_n$，就可以求得一个关于θ的估计。这正是上述描述的似然函数。似然函数可进一步表示为$L(\theta|X)=P(X|\theta)=\prod_{i=1}^{n}P(x_i|\theta)$，其中$\prod_{i=1}^{n}$表示为元素积。而最大似然估计是寻找关于$\theta$的最可能的值，即在所有$\theta$取值中（0～1），寻找一个值使这个采样的“可能性”最大化，就是在θ的所有可能取值中寻找一个值使得似然函数取到最大值。这个使可能性最大的$\widehat{\theta}$值即称为θ的最大似然估计。最大似然估计是样本的函数。

求最大似然估计就是对似然函数求导，并令其导函数为0时的解，即曲线变化趋于0的位置。其导函数为$-71\times p^{29}\times(1-p)^{70}+29\times p^{28}\times(1-p)^{71}$，计算相对复杂。而最大化一个似然函数同最大化它的自然对数是等价的（图3.16-3），因此对似然函数取对数，其似然函数对数的导函数为$\frac{-71}{1-p}+\frac{29}{p}$，可见计算量得以大幅度减少；而由似然函数积的方式，转换为似然函数对数和的形式，也进一步减少了计算量。

最后，求得的最大似然估计值为29/100，与生成符合指定概率$p=0.3$的随机数组中的p值基本一致。

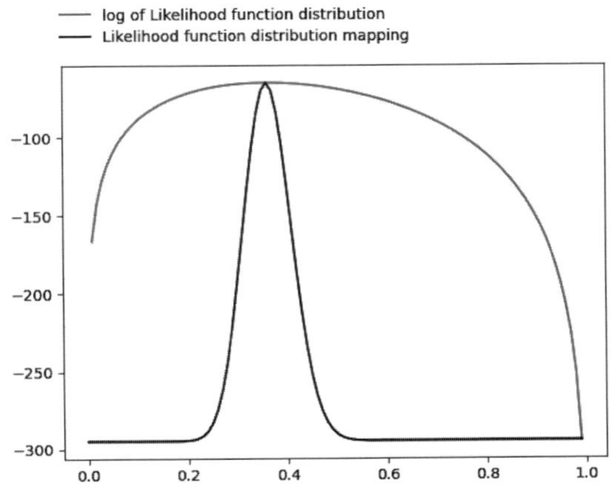

图3.16-3 似然函数及其自然对数

```
print("似然函数的导函数 ={}".format(sympy.diff(L, p_)))
L_max_ = sympy.solve(sympy.diff(L, p_), p_)
print("令似然函数的导数为 0 时，解得最大似然估计 ={}".format(L_max_))

L_log = sympy.expand_log(sympy.log(L))  # sympy.expand_log 来简化数学表达式中的对数项
print("似然函数的对数 ={}".format(L_log))
print("似然函数对数的导函数 ={}".format(sympy.diff(L_log, p_)))
L_log_ = sympy.lambdify(p_, L_log, "numpy")
# sympy.solve 令所有方程等于 0，解方程或方程组。即求曲线最大值的位置
L_max = sympy.solve(sympy.diff(L_log, p_), p_)
print("令似然函数对数的导数为 0 时，解得最大似然估计 ={}".format(L_max))

fig, ax = plt.subplots(1, 1)
L_log_values = L_log_(p_x)
ax.plot(p_x, L_log_values, 'r-', ms=8,
        label='log of Likelihood function distribution')
def rescale_linear(array, new_min, new_max):
    """
    function - 按指定区间缩放 / 映射数组 /Rescale an arrary linearly.

    Paras:
        new_min - 映射区间的最小值
        new_max - 映射区间的最大值
    """
    minimum, maximum = np.min(array), np.max(array)
    m = (new_max - new_min) / (maximum - minimum)
    b = new_min - m * minimum
    return m * array + b
L_log_values = L_log_values[np.isfinite(L_log_values)]
ax.plot(p_x, rescale_linear(L_(p_x), L_log_values.min(), L_log_values.max()),
        'b-', ms=8, label='Likelihood function distribution mapping')
ax.legend(bbox_to_anchor=(0.32, 1), loc=8, borderaxespad=1, frameon=False)
plt.show()
-------------------------------------------------------------------------------
似然函数的导函数 =-61*p**39*(1 - p)**60 + 39*p**38*(1 - p)**61
令似然函数的导数为 0 时，解得最大似然估计 =[39/100, 1]
似然函数的对数 =39*log(p) + log((1 - p)**61)
似然函数对数的导函数 =-61/(1 - p) + 39/p
令似然函数对数的导数为 0 时，解得最大似然估计 =[39/100]

<lambdifygenerated-3>:2: RuntimeWarning: divide by zero encountered in log
  return (39*log(p) + log((1 - p)**61))
```

3.16.4 逻辑回归（Logistic Regression，LR）

逻辑回归模型为：$p(x)=\sigma(t)=\frac{1}{1+\mathrm{e}^{-t}}=\frac{1}{1+\mathrm{e}^{-(\beta_0+\beta_1x_1+\beta_2x_2+\ldots+\beta_2x_\mathrm{n})}}=\frac{1}{1+\mathrm{e}^{-w^\mathrm{T}x}}$，其中$\sigma(t)$就是sigmoid函数（在深度学习中用于激活函数），sigmoid函数可以将线性模型$f(x)=\beta_0+\beta_1x_1+\beta_2x_2+\ldots+\beta_2x_\mathrm{n}=w^\mathrm{T}x$，归一化到［0,1］之间。而因为为二元分类，类标只有两类值1和0，特征值经过加权和（参数全部初始化为1）喂入sigmoid函数后，所得到的值位于［0,1］区间。则该值与类标的差值和就为逻辑回归的损失函数，通过随机梯度下降法更新初始化的参数（并未使用最大似然估计，可能无法解析求解），最终训练所得的参数应该使得逻辑回归模型的预测结果趋近于类标1或0。逻辑回归模型是线性模型和sigmoid函数的组合（图3.16–4），这与解释“反向传播”部分所构建的隐含层和输出层的网络结构是一样的，可以互相印证理解。

随机梯度下降`stocGradAscent1`方法的定义迁移于*Machine learning in action*，其随机梯度下降优化的主要改进包括：每次迭代时，调整更新步长`alpha`值（为学习率`lr`），学习率会越来越小，从而缓解系数的高频波动。同时，为了避免迭代不断减小至0，约束学习率大于一个稍微大点的常数项，对应代码为`alpha = 4/(1.0+j+i)+0.0001`。其中，i和j为迭代次数，随迭代的增加降低`alpha`值。同时，改变样本的优化顺序，即是随机选择样本来更新回归系数。这样，可以减少周期性的波动。对应的代码为`randIndex = int(np.random.uniform(0,len(dataIndex)))`。

当训练求得参数（权重值）后，通过逻辑回归模型预测的结果是位于0到1之间的浮点数，以0.5为界，大于0.5返回类标为1；否则，类标为0作为分类的输出。重新生成分类数据集，用于测试逻辑回归模型精度（图3.16–5）。

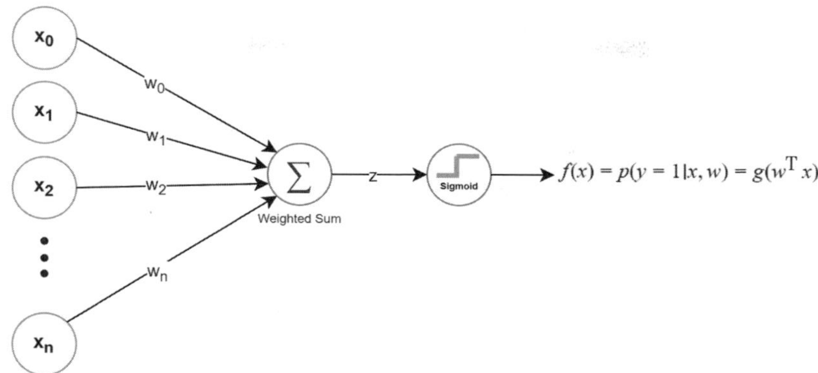

图3.16-4 逻辑回归模型图解

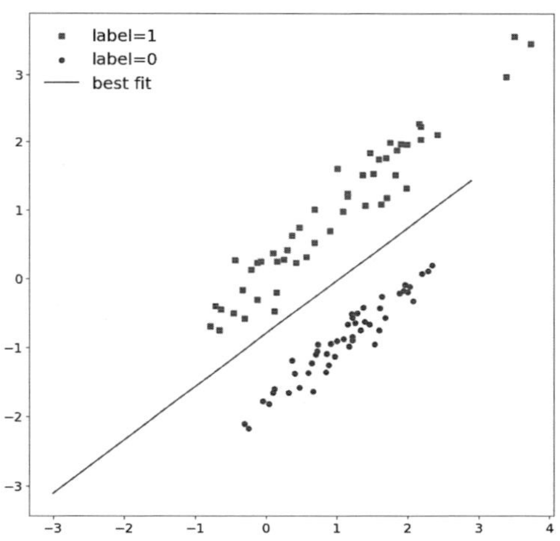

图3.16-5 测试数据集逻辑回归模型预测结果

```
'''
Created on Oct 27, 2010
Logistic Regression Working Module  /https://github.com/pbharrin/
machinelearninginaction/blob/master/Ch05/logRegres.py
@author: Peter
updated on Fri Jan  8 17:42:57 2021 @author: Richie Bao
'''
class LogisticRegression:
    '''
    class - 自定义逻辑回归（二元分类）
    '''
    def __init__(self):
        pass
    def generate_dataset_linear(self, slop, intercept, num=100, multiple=10, 
magnitude=50):
        '''
        function - 根据一元一次函数构建二维离散型分类数据集，类标为 1 和 0
        '''
        import numpy as np
        X_1 = np.random.random(num)-0.5
        X_1 *= multiple
        X_2 = slop*X_1+intercept
        mag_random_val = np.random.random(num)-0.5
        mag_random_val *= magnitude
        mask = mag_random_val >= 0
        y = mask*1
```

```
        X_2 += mag_random_val
        X = np.stack((np.ones(len(X_1)), X_1, X_2), axis=1)
        return X, y
    def make_classification_dataset(self, n_features=2, n_classes=2, n_
samples=100, n_informative=2, n_redundant=0, n_clusters_per_class=1):
        '''
        function - 使用Sklearn提供的make_classification方法，直接构建离散型分类数据集

        Paras:
        参数查看sklearn库make_classification方法
        '''
        from sklearn.datasets import make_classification
        X, y = make_classification(n_samples=n_samples, n_features=n_features,
n_classes=n_classes,
                                   n_redundant=n_redundant, n_informative=n_
informative, n_clusters_per_class=n_clusters_per_class)
        X = np.append(np.ones(len(X)).reshape(-1, 1), X, axis=1)
        return X, y
    def plot(self, X, y, weights, figsize=(10, 10)):
        '''
        function - 绘制逻辑回归结果
        '''
        import matplotlib.pyplot as plt
        fig = plt.figure(figsize=figsize)
        ax = fig.add_subplot(111)
        ax.scatter(X[y == 1][:, 1], X[y == 1][:, 2], s=30,
                   c='red', marker='s', label='label=1')
        ax.scatter(X[y == 0][:, 1], X[y == 0][:, 2],
                   s=30, c='green', label='label=0')
        x = np.arange(-3.0, 3.0, 0.1)
        y = (-weights[0]-weights[1]*x)/weights[2]
        ax.plot(x, y, label='best fit')
        ax.legend(loc='best', frameon=False)
        plt.show()
    def sigmoid(self, x):
        '''
        function - 逻辑回归函数_sigmoid函数
        '''
        return 1.0/(1+np.exp(-x))
    def stocGradAscent1(self, dataMatrix, classLabels, numIter=150):
        '''
        function - 随机梯度下降（优化）
        '''
        from tqdm import tqdm
        m, n = np.shape(dataMatrix)
        weights = np.ones(n)   #初始化为1
        for j in tqdm(range(numIter)):
            dataIndex = list(range(m))
            for i in range(m):
                #apha值随迭代减小
                alpha = 4/(1.0+j+i)+0.0001
                #因为是常数，所以趋于0
                randIndex = int(np.random.uniform(0, len(dataIndex)))
                h = self.sigmoid(sum(dataMatrix[randIndex]*weights))
                error = classLabels[randIndex] - h
                weights = weights + alpha * error * dataMatrix[randIndex]
                del (dataIndex[randIndex])
        return weights
    def classifyVector(self, inX, weights):
        '''
        function - 代入梯度下降法训练的权重值于逻辑回归函数（sigmoid），预测样本特征对应的
类标
        '''
        prob = self.sigmoid(sum(inX*weights))
        if prob > 0.5:
            return 1.0
        else:
            return 0.0
    def test_accuracy(self, X_, y_, weights):
        '''
        function - 测试训练模型
        '''
        from tqdm import tqdm
        m, n = np.shape(X_)
```

```
        matches_num = 0
        for i in tqdm(range(m)):
            pred = self.classifyVector(X_[i, :], weights)
            if pred == np.bool(y_[i]):
                matches_num += 1
        accuracy = float(matches_num)/m
        print("测试数据集精度：{}".format(accuracy))
        return accuracy
LR = LogisticRegression()
X, y = LR.make_classification_dataset(n_features=2, n_classes=2, n_samples=100)
weights = LR.stocGradAscent1(X, y, numIter=150)

# 建立测试数据集，LR.make_classification_dataset 和 LR.generate_dataset_linear 均可以生成测试数据集
X_, y_ = LR.generate_dataset_linear(
    slop=5, intercept=5, num=100, multiple=10, magnitude=50)
accuracy = LR.test_accuracy(X_, y_, weights)
LR.plot(X, y, weights)
--------------------------------------------------------------------------------
100%|██████████| 150/150 [00:00<00:00, 848.40it/s]
  if pred==np.bool(y_[i]):
100%|██████████| 100/100 [00:00<00:00, 132773.16it/s]
测试数据集精度：0.75
```

- sklearn库LogisticRegression方法实现

LogisticRegression方法与上述自定义方法的逻辑回归模型训练的结果保持一致，精度均为0.85。通过不同多次运行（获取不同的训练数据集和测试数据集），各自结果保持一致。混淆矩阵（图3.16-6）分类精度计算结果显示，总共100个样本，真实值为0的类标计55个，误判为类标1的为14个；真实值为1的类标计45个，误判为类标1的为1个。精确度、召回率及调和平均值的结果由classification_report方法计算。

```
from sklearn.linear_model import LogisticRegression
from sklearn.metrics import confusion_matrix, classification_report
import matplotlib.pyplot as plt
logreg = LogisticRegression()
logreg.fit(X, y)
predictions = logreg.predict(X_)
CM = confusion_matrix(y_, predictions, labels=logreg.classes_)
print("confusion_matrix:\n{}".format(CM))
plt.matshow(CM)
plt.title('confusion matrix')
plt.ylabel('True lable')
plt.xlabel('Predicted label')
plt.colorbar()
plt.show()
print(classification_report(y_, predictions))
--------------------------------------------------------------------------------
confusion_matrix:
```

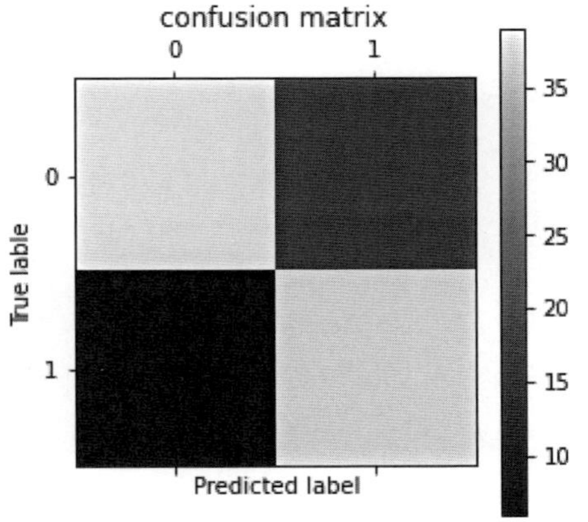

图3.16-6 混淆矩阵

```
[[39 17]
 [ 6 38]]
              precision    recall  f1-score   support
           0       0.87      0.70      0.77        56
           1       0.69      0.86      0.77        44
    accuracy                           0.77       100
   macro avg       0.78      0.78      0.77       100
weighted avg       0.79      0.77      0.77       100
```

3.16.5 SoftMax回归（函数、归一化指数函数）

3.16.5.1 自定义SoftMax回归多分类

- 定义线性函数——linear_weights(self,X)

用于SoftMax函数输入的对象z，是由线性函数计算求取的结果。例如假设有两个特征，每个特征下包含5个样本，$X=\begin{bmatrix}1&2\\4&7\\5&6\\3&6\\12&3\end{bmatrix}$；并假设有3个类标，$y=\begin{bmatrix}0&1&2\end{bmatrix}$。根据类标数量（3）和特征数（2）随机初始化权重值为$W=\begin{bmatrix}7&5&2\\3&4&9\end{bmatrix}$，为2行（特征数）3列（类标数）。将每一个样本（包含两个特征值，例如第一个样本特征值为1，2），分别对应权重矩阵每一列的两个权值（例如第1列，7，3）计算加权和，结果为$1\times7+2\times3=13$，以此类推，对于第一个样本特征值其余对应的权重值计算加权和结果为：13，20。其他样本同上。这个计算的过程就是数组的点积计算，可以由np.dot方法实现。由此根据每个类标对应一个初始化的权值，计算了每个样本对应3个类标的输出值，用于SoftMax函数的输入。

```
X = np.array([[1,2],[4,7],[5,6],[3,6],[12,3]])
print("X=",X)
W = np.array([[7,5,2],[3,4,9]])
print("y=",W)
z = np.dot(X,W)
print("z=",z)
-----------------------------------------------------------------------------
X= [[ 1  2]
 [ 4  7]
 [ 5  6]
 [ 3  6]
 [12  3]]
y= [[7 5 2]
 [3 4 9]]
z= [[13 13 20]
 [49 48 71]
 [53 49 64]
 [39 39 60]
 [93 72 51]]
```

- 定义SoftMax函数——softmax(self,z)

SoftMax回归/函数，或称为归一化指数函数，是逻辑回归/函数（Logistic Regression）的一种推广（图3.16-7）。将一个含任意实数的K维向量z"压缩"到另一个K维实向量$\sigma(z)$中，使得每一个元素的范围都在(0,1)之间，并且所有元素的和为1。该函数的形式通常为：$\sigma(z)_j=\frac{e_j^z}{\sum_{k=1}^{K}e_k^z}\quad for\quad j=1,\ldots,K$。SoftMax函数实际上是有限项离散概率分布的梯度对数归一化，广泛应用于多分类问题方法中。在多项逻辑回归和线性判别分析中，函数的输入是从K个不同的线性函数得到的结果，而样本向量x属于第j个分类的几率为：$P(y=j|x)=\frac{e^{x^{\mathrm{T}}w_j}}{\sum_{k=1}^{K}e^{x^{\mathrm{T}}w_{\mathrm{k}}}}$，这可以被视作$K$个线性函数$x\mapsto x^{\mathrm{T}}w_1,\ldots,x\mapsto x^{\mathrm{T}}w_{\mathrm{k}}$，SoftMax函数的复合。亦可以表述为，对于输入数据$\{(x_1,y_1),(x_2,y_2),\ldots,(x_n,y_n)\}$，有$K$个类别，即$y_i\in\{1,2,\ldots,k\}$，那么SoftMax回归

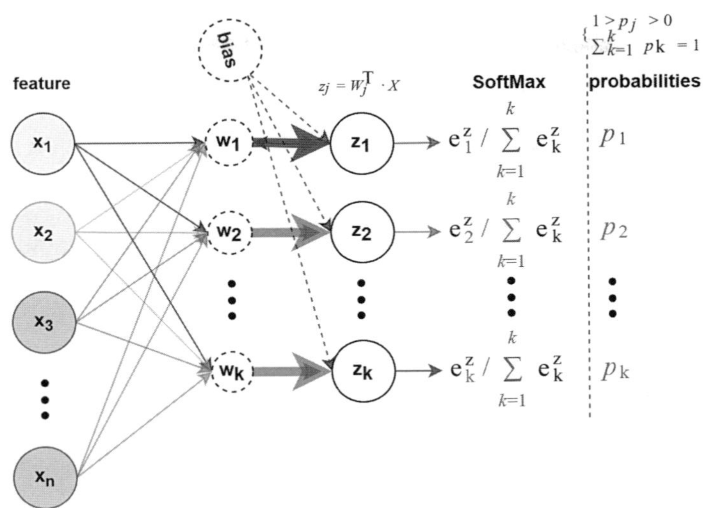

图3.16-7 SoftMax函数多分类图解

主要估算输入数据x_i会属于每一类的概率，即$h_\theta(x_i) = \begin{bmatrix} p(y_i=1 \mid x_i;\theta) \\ p(y_i=2 \mid x_i;\theta) \\ \vdots \\ p(y_i=k \mid x_i;\theta) \end{bmatrix} = \frac{1}{\sum_{k=1}^{K} \mathrm{e}^{\theta_k^{\mathrm{T}} x_i}} \begin{bmatrix} \mathrm{e}^{\theta_1^{\mathrm{T}} x_i} \\ \mathrm{e}^{\theta_2^{\mathrm{T}} x_i} \\ \vdots \\ \mathrm{e}^{\theta_k^{\mathrm{T}} x_i} \end{bmatrix}$，其中$\theta_1,\theta_2,\ldots,\theta_k \in \theta$是模型的参数，乘以$\frac{1}{\sum_{k=1}^{K} \mathrm{e}^{\theta_k^{\mathrm{T}} x_i}}$是为了让概率位于［0,1］并且概率之和为1，SoftMax回归将输入数据x_i归属于类别j的概率为：$p(y_i=j|x_i;\theta) = \frac{\mathrm{e}^{\theta_j^{\mathrm{T}} x_i}}{\sum_{k=1}^{K} \mathrm{e}^{\theta_k^{\mathrm{T}} x_i}}$。

同一问题可以用多种公式表述，如不同的变量符号，或者表述规则，例如上述两种阐述中，模型的权值分别被表述为w和θ，或者以矩阵的方式表述的方程组等。不管公式的书写方式如何，其核心的算法始终保持一致。

● **定义损失函数/交叉熵（cross entropy）损失函数——loss(self,y_one_hot,y_probs)**

SoftMax运算符将输出变换为一个合法的类标预测分布（联合预测概率，预测概率分布），真实标签通过独热编码转换后的形式也是一种类别分布表达（0值为概率为0，1值为概率为1），二者形状保持一致。评估预测概率和真实标签的分布，只需要估计对应正确类别的预测概率。这样衡量两个概率分布差异的测量函数，可以使用交叉熵（cross entropy），公式为：$H(y^{(i)},\hat{y}^{(i)}) = -\sum_{j=1}^{q} y_j^{(i)} \ln \hat{y}_j^{(i)}$。其中，带下标的$y_j^{(i)}$是向量$y^{(i)}$中非0即1的元素（独热编码）；$\hat{y}_j^{(i)}$为联合预测概率$\hat{y}^{(i)}$中对应类标（对应正确类别）的预测概率。

根据交叉熵的定义，SoftMax回归的损失函数（代价函数）计算公式为：$L(\theta) = -\frac{1}{m}[\sum_{i=1}^{m}\sum_{j=1}^{k} 1\{y_i=j\} \ln \frac{\mathrm{e}^{\theta_j^{\mathrm{T}} x_i}}{\sum_{l=1}^{k} \mathrm{e}^{\theta_l^{\mathrm{T}} x_i}}]$，其中$m$为样本数，$k$代表类标数，1{•}是示性函数，1{值为真的表达式}=1，1{值为假的表达式}=0。SoftMax回归的输出值的形状为(m,k)，每一样本（总共m个样本）对应有k（类标数）个概率值，每一类标对应的概率值为p_k（联合预测概率）。p_k概率值位于［0,1］之间，当取对数后，对应值转换到［-inf,0］之间。概率越大的值，即越趋近于1，当取对数后其值越趋近于0。示性函数对应数据集类标列的独热编码(One-Hot Encoding)。例如类标有4类［0,1,2,3］，假设一个样本对应的类标为3，经过独热编码后，表示为［0. 0. 0. 1.］，可理解为对应类标位置的值取1，而没有对应的位置均为0。而该样本经过SoftMax计算后结果为［0.31132239 0.2897376 0.19737688 0.20156313］，即该样本对应各个类标的概率值$p_k, k=0,1,2,3$。对p_k取对数后，结果为［-1.16692628 -1.23877959 -1.62264029 -1.60165265］，概率越大的值，例如0.31132239，其对数结果越趋近于0，对应-1.16692628。将p_k取对数后的值与该样本类标的独热编码相乘，结果为［-0. -0. -0. -1.60165265］，即仅保留该样本对应类标位置概率值的对数，用以表述概率值的误差（如果概率值为1，即100%的正确，其对数为0，即误差为0；如果概率值为0，即100%的错误，其对数为负无穷，即误差无限大）。计算所有样本的误差值后求和平均，即为SoftMax损失函数的结果（图3.16-8），表示预测误差的大小。

示性函数（特征函数，Characteristic function）可以代表不同的概念，最通常且多数统称为指示函数，方程式为：$1_A : X \mapsto \{0,1\}$，其中在集合X中，任意子集A内一点含值 1，于集合$X-A$内一点含值 0。

● **定义梯度下降法，权值更新——gradient(self,X,y_one_hot,y_probs,lr,lambda_)**

利用梯度下降法最小化损失函数，求解θ的梯度，同时因为如果训练数据不多，容易出现过拟合现象，而增加了一个正则项（具体解释见正则化部分），求解梯度公式为（未给推断过程）：$\frac{\partial L(\theta)}{\partial \theta_j}=-\frac{1}{m}[\sum_{i=1}^{m} x_i(1\{y_i=j\}-p(y_i=j|x_i;\theta))]+\lambda\theta_j$，对应的代码为`grad_loss_L=-(1.0/self.m_samples)*np.dot((y_one_hot-y_probs).T,X)+lambda_*self.weights`。

下述实验首先构建离散型分类数据集（图3.16-9）。

```
def make_classification_dataset_(n_features=2, n_classes=2, n_samples=100,
n_informative=2, n_redundant=0, n_clusters_per_class=1, figsize=(10, 10),
plot=True):
    '''
    function - 使用 Sklearn 提供的 make_classification 方法，直接构建离散型分类数据集；并
```

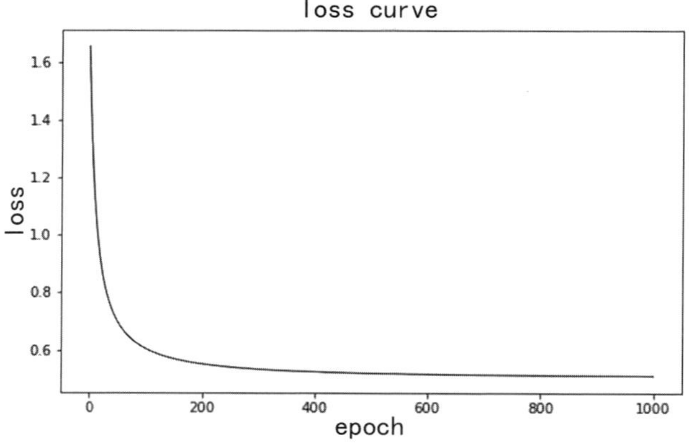

图3.16-8 损失曲线

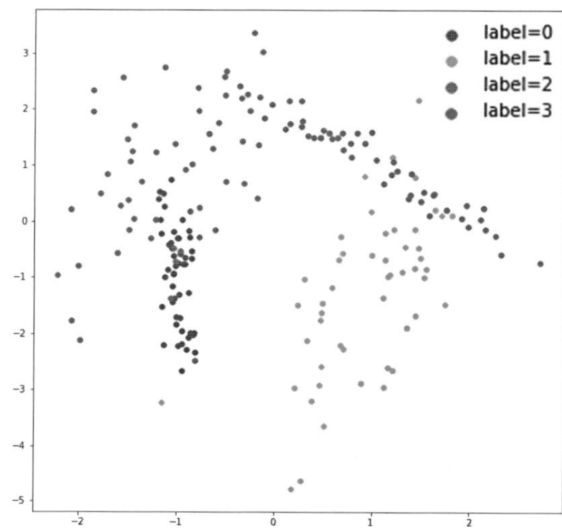

图3.16-9 构建离散型分类数据集结果

打印查看

```
    Paras:
        参数查看 sklearn 库 make_classification 方法
    '''
    from sklearn.datasets import make_classification
    import matplotlib.pyplot as plt
    import numpy as np
    X, y = make_classification(n_samples=n_samples, n_features=n_features, n_
classes=n_classes,
                               n_redundant=n_redundant, n_informative=n_
informative, n_clusters_per_class=n_clusters_per_class)
    print( "类标：" , np.unique(y))

    if plot:
        fig = plt.figure(figsize=figsize)
        ax = fig.add_subplot(111)
        for label in np.unique(y):
            ax.scatter(X[y == label][:, 0], X[y == label][:, 1],
                       s=30, label='label={}'.format(label))
        ax.legend(loc='best', frameon=False)
        plt.show()
    return X, y
X, y = make_classification_dataset_(
    n_features=2, n_redundant=0, n_informative=2, n_clusters_per_class=1, n_
classes=4, n_samples=200)
--------------------------------------------------------------------------------
类标： [0 1 2 3]
```

```
import numpy as np
class softmax_multiClass_classification_UDF:
    '''
    class - 自定义 softmax 回归多分类
    '''
    def __init__(self, m_samples, n_features, n_classes):
        self.m_samples, self.n_features = m_samples, n_features
        self.n_classes = n_classes
        self.weights = np.random.rand(self.n_classes, self.n_features)
        self.loss_overall = list()
    def softmax(self, Z):
        '''
        function - 定义 softmax 函数
        '''
        exp_sum = np.sum(np.exp(Z), axis=1, keepdims=True)
        return np.exp(Z)/exp_sum
    def linear_weights(self, X):
        '''
        function - 定义线性函数
        '''
        return np.dot(X, self.weights.T)
    def loss(self, y_one_hot, y_probs):
        '''
        function - 定义损失函数
        '''
        return -(1.0/self.m_samples)*np.sum(y_one_hot*np.log(y_probs))
    def one_hot(self, y):
        '''
        function - numpy 实现 one-hot-code
        '''
        return np.squeeze(np.eye(self.n_classes)[y.reshape(-1)])
    def gradient(self, X, y_one_hot, y_probs, lr, lambda_):
        '''
        function - 定义梯度下降法，权值更新
        '''
        #求解梯度
        grad_loss_L = -(1.0/self.m_samples) * \
            np.dot((y_one_hot-y_probs).T, X)+lambda_*self.weights
        #更新权值
        grad_loss_L[:, 0] = grad_loss_L[:, 0]-lambda_*self.weights[:, 0]
        self.weights = self.weights-lr*grad_loss_L
    def predict_test(self, X_, y_):
        '''
        function - 预测
        '''
        y_probs_ = self.softmax(self.linear_weights(X_))
```

```
        y_pred = np.argmax(y_probs_, axis=1).reshape((-1, 1))
        accuracy = np.sum(y_pred == y_.reshape((-1, 1)))/len(X_)
        print("accuracy:%.5f" % accuracy)
        return y_pred
    def loss_curve(self, figsize=(8, 5)):
        '''
        function - 更新打印损失曲线
        '''
        import matplotlib.pyplot as plt
        fig = plt.figure(figsize=figsize)
        plt.plot(np.arange(len(self.loss_overall)), self.loss_overall)
        plt.title("loss curve")
        plt.xlabel('epoch')
        plt.ylabel('LOSS')
        plt.show()
    def train(self, X, y, epochs, lr=0.1, lambda_=0.01):
        '''
        function - 训练模型
        '''
        from tqdm import tqdm
        for epoch in tqdm(range(epochs)):
            #执行加权和与softmax函数
            y_probs = self.softmax(self.linear_weights(X))
            #计算损失函数
            y_one_hot = self.one_hot(y.reshape((-1, 1)))
            self.loss_overall.append(self.loss(y_one_hot, y_probs))
            #求解梯度与权值更新
            self.gradient(X, y_one_hot, y_probs, lr, lambda_)
m_samples, n_features, n_classes = X.shape[0], X.shape[1], len(np.unique(y))
sm_classi = softmax_multiClass_classification_UDF(
    m_samples, n_features, n_classes)
epochs = 1000
sm_classi.train(X, y, epochs, lr=0.1, lambda_=0.01,)
sm_classi.loss_curve()
---------------------------------------------------------------------------
100%|██████████| 1000/1000 [00:00<00:00, 13637.49it/s]
X_,y_ = make_classification_dataset_(n_features=2, n_redundant=0, n_
informative=2,n_clusters_per_class=1, n_classes=4,n_samples=100,plot=False)
y_pred = sm_classi.predict_test(X_,y_)
---------------------------------------------------------------------------
类标:  [0 1 2 3]
accuracy:0.87000
```

3.16.5.2 sklearn机器学习库实现SoftMax回归多分类

SoftMax回归多分类sklearn库提供的方法同上述的二分类，应用LogisticRegression实现，只是需要配置multi_class参数（default='auto'，因此也无需配置，会自动识别）；参数solver需要配置为求解多分类的优化项。其测试数据集的计算结果同上述自定义类的精度接近（图3.16-10）。

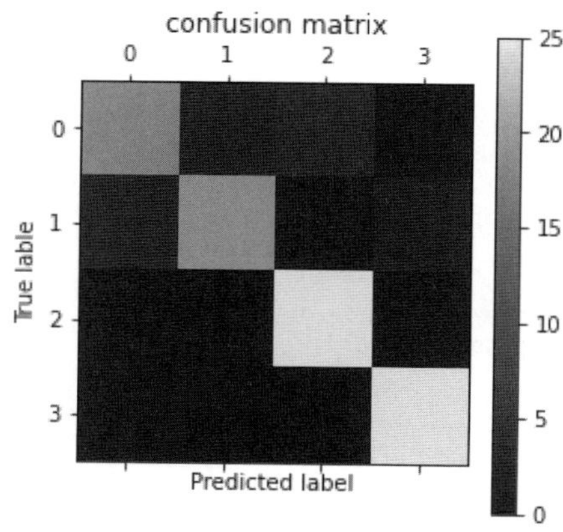

图3.16-10　混淆矩阵

```
from sklearn.linear_model import LogisticRegression
from sklearn.metrics import confusion_matrix, classification_report
import matplotlib.pyplot as plt
logreg_multiClassi = LogisticRegression(
    solver='lbfgs', multi_class="multinomial")
logreg_multiClassi.fit(X, y)
predictions = logreg_multiClassi.predict(X_)
CM = confusion_matrix(y_, predictions, labels=logreg_multiClassi.classes_)
print("confusion_matrix:\n{}".format(CM))
plt.matshow(CM)
plt.title('confusion matrix')
plt.ylabel('True lable')
plt.xlabel('Predicted label')
plt.colorbar()
plt.show()
print(classification_report(y_, predictions))
-----------------------------------------------------------------------------
confusion_matrix:
[[18  3  4  0]
 [ 3 19  1  2]
 [ 0  0 25  0]
 [ 0  0  0 25]]
              precision    recall  f1-score   support
           0       0.86      0.72      0.78        25
           1       0.86      0.76      0.81        25
           2       0.83      1.00      0.91        25
           3       0.93      1.00      0.96        25
    accuracy                           0.87       100
   macro avg       0.87      0.87      0.87       100
weighted avg       0.87      0.87      0.87       100
```

3.16.5.3 PyTorch深度学习库实现SoftMax回归多分类——自定义

应用torch.utils.data库中的TensorDataset可以很方便的建立张量数据集；应用random_split切分数据集为训练、验证和测试数据集；应用DataLoader将数据集转换为可迭代对象，用于模型训练的数据加载。并可以定义在每轮迭代中随机均匀采样多个样本组成一个小批量（指定batch_size参数值），使用小批量计算梯度。也可配置参数num_workers，使用多进程加速数据读取。如果num_workers=0，则为不使用额外的进程来加速读取数据。

PyTorch自定义SoftMax回归的算法同上述阐述，算法保持一致，只是在代码的书写过程中，需要根据PyTorch的语法规则做出调整。尤其PyTorch张量运算自动求梯度的方式，大幅度减轻了代码编写的复杂程度。

首先，构建离散型分类数据集（图3.16-11）。

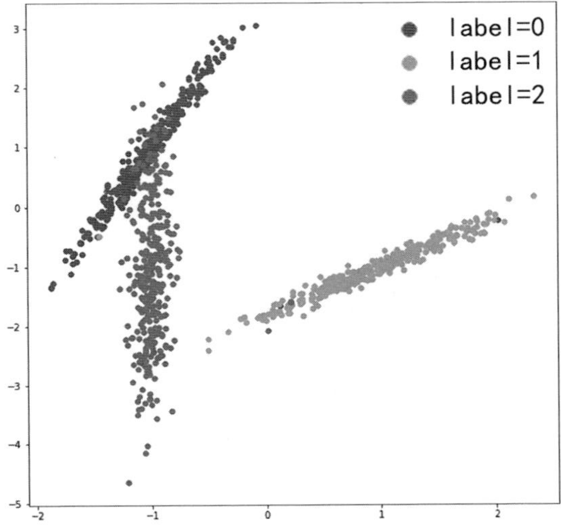

图3.16-11　构建离散型分类数据集结果

```
import torch
import torch.utils.data as data_utils
X, y = make_classification_dataset_(
    n_features=2, n_redundant=0, n_informative=2, n_clusters_per_class=1, n_
classes=3, n_samples=1000)
# 01- 建立数据集
data_iter = data_utils.TensorDataset(
    torch.from_numpy(X).double(), torch.from_numpy(y).long())
# 02- 切分数据集
train_set, val_set = data_utils.random_split(data_iter, [800, 200])
# 03- 配置每批大小，batch size
train_data_loader = data_utils.DataLoader(
    train_set, batch_size=50, shuffle=True, num_workers=2)
val_data_loader = data_utils.DataLoader(
    val_set, batch_size=50, shuffle=True, num_workers=2)
--------------------------------------------------------------------------------
类标： [0 1 2]
import numpy as np
import torchvision
import torch
class softmax_multiClass_classification_UDF_pytorch:
    '''
    class - 自定义 softmax 回归多分类 _PyTorch 版
    '''
    def __init__(self, num_inputs, num_outputs):
        print("num_inputs/feature={},num_outputs/label={}".format(num_inputs,
num_outputs))
        self.num_inputs = num_inputs
        self.num_outputs = num_outputs
        self.W = torch.tensor(np.random.normal(
            0, 0.01, (self.num_inputs, self.num_outputs)), dtype=torch.double,
requires_grad=True)  # 需要对权值求梯度
        self.b = torch.zeros(num_outputs, dtype=torch.float,
                             requires_grad=True)  # 需要对偏置求梯度
        self.params = [self.W, self.b]
    def SoftMax(self, Z):
        '''
        function - 定义 SoftMax 函数
        '''
        Z_exp = Z.exp()
        exp_sum = Z_exp.sum(dim=1, keepdim=True)
        return Z_exp/exp_sum
    def net(self, X):
        '''
        function - 定义模型，含线性模型输入，SoftMax 回归输出
        '''
        return self.SoftMax(torch.mm(X.view((-1, self.num_inputs)), self.W)+self.b)
    def cross_entropy(self, y_pred, y):
        '''
        function - 定义交叉熵损失函数
        '''
        return -torch.log(y_pred.gather(1, y.view(-1, 1)))  # torch.gather，收集输
入的特定维度指定位置的数值。即提取出对应正确类别的预测概率

    def accuracy(self, y_pred, y):
        '''
        function - 定义分类准确率，即正确预测数量与总预测数量之比
        '''
        return (y_pred.argmax(dim=1) == y).float().mean().item()
    def evaluate_accuracy(self, data_iter):
        '''
        funtion - 平均模型 net 在数据集 data_iter 上的准确率
        '''
        accu_sum, n = 0.0, 0
        for X, y in data_iter:
            accu_sum += (self.net(X).argmax(dim=1) == y).float().sum().item()
            n += y.shape[0]
        return accu_sum/n
    def sgd(self, lr):
        '''
        funtion - 梯度下降
        '''
        for param in self.params:
            param.data -= lr*param.grad
```

```
    def train(self, train_iter, epochs, lr, test_iter=None):
        '''
        function - 训练模型
        '''
        from tqdm.auto import tqdm
        for epoch in tqdm(range(epochs)):
            train_l_sum, train_acc_sum, n = 0.0, 0.0, 0
            for X, y in train_iter:
                # 01- 线性模型输入，SoftMax 回归输出
                y_pred = self.net(X)
                # 02- 计算损失函数
                l = self.cross_entropy(y_pred, y).sum()
                # 03- 参数梯度清零
                if self.params is not None and self.params[0].grad is not None:
                    for param in self.params:
                        param.grad.data.zero_()
                # 04- 计算给定张量的梯度和，此处为损失函数的反向传播
                l.backward()
                # 05- 求梯度
                self.sgd(lr)
                # 06- 每批误差和
                train_l_sum += l.item()
                # 07- 每批正确率
                train_acc_sum += (y_pred.argmax(dim=1) == y).sum().item()
                n += y.shape[0]
            if test_iter is not None:
                test_acc = self.evaluate_accuracy(test_iter)
            if epoch % 100 == 0:
                print('epoch %d, loss %.4f, train acc %.3f, test acc %.3f' %
                      (epoch + 1, train_l_sum / n, train_acc_sum / n, test_acc))
num_inputs, num_outputs = X.shape[1], len(np.unique(y))
sm_classi_pytorch = softmax_multiClass_classification_UDF_pytorch(
    num_inputs, num_outputs)
epochs, lr = 1000, 0.1
sm_classi_pytorch.train(train_data_loader, epochs, lr, val_data_loader)
--------------------------------------------------------------------------------
num_inputs/feature=2,num_outputs/label=3
0%|          | 0/1000 [00:00<?, ?it/s]
epoch 1, loss 0.4431, train acc 0.800, test acc 0.850
epoch 101, loss 0.2677, train acc 0.927, test acc 0.945
...
epoch 801, loss 0.2672, train acc 0.936, test acc 0.910
epoch 901, loss 0.2847, train acc 0.944, test acc 0.940
```

3.16.5.4 PyTorch深度学习库实现SoftMax回归多分类——调用已有算法

注意上述分开定义SoftMax回归和交叉熵损失函数可能会造成数值不稳定，而PyTorch提供的torch.nn.CrossEntropyLoss方法，结合了nn.LogSoftmax()和nn.NLLLoss()于单独的一个类中。因此，在定义net网络时，仅含有一个线性模型，而自定义的flattenLayer是用于转换特征输入向量的形状，也置于net网络中。

此时，定义的SoftMax回归多分类方法并未以类的形式出现，而是为单独的函数，可以将flattenLayer、evaluate_accuracy、sgd_v1及train_v1函数置于util_*.py工具文件中，方便日后直接调用。

```
import torch.utils.data as data_utils
import torch
from torch import nn
import numpy as np
from collections import OrderedDict

# 将每批次样本 X 的形状转换为 (batch_size,-1)

class flattenLayer(nn.Module):
    def __init__(self):
        super(flattenLayer, self).__init__()
    def forward(self, x):
        return x.view(x.shape[0], -1)

# 定义模型
net = nn.Sequential(
```

```
    OrderedDict([
        ('flatten', flattenLayer()),
        ('linear', nn.Linear(num_inputs, num_outputs))
    ])
)

#初始化模型的权重参数
nn.init.normal_(net.linear.weight, mean=0, std=0.01)
nn.init.constant_(net.linear.bias, val=0)

# SoftMax, 交叉熵损失函数 CrossEntropyLossntropyLoss()
loss = nn.CrossEntropyLoss()

#优化算法
optimizer = torch.optim.SGD(net.parameters(), lr=0.1)
def evaluate_accuracy(data_iter, net):
    '''
    funtion - 平均模型 net 在数据集 data_iter 上的准确率
    '''
    accu_sum, n = 0.0, 0
    for X, y in data_iter:
        accu_sum += (net(X).argmax(dim=1) == y).float().sum().item()
        n += y.shape[0]
    return accu_sum/n
def sgd_v1(params, lr):
    '''
    funtion - 梯度下降, v1 版
    '''
    for param in params:
        param.data -= lr*param.grad
def train_v1(net, train_iter, test_iter, loss, num_epochs, params=None, lr=None,
optimizer=None, interval_print=100):
    '''
    function - 训练模型, v1 版

    Paras:
        net - 构建的模型结构
        train_iter - 可迭代训练数据集
        test_iter - 可迭代测试数据集
        loss - 损失函数
        num_epochs - 训练迭代次数
        params=None - 初始化模型参数，以列表形式表述，例如 [W,b]
        lr=None, - 学习率
        optimizer=None - 优化函数
        interval_print=100 - 打印反馈信息间隔周期
    '''
    from tqdm.auto import tqdm
    for epoch in tqdm(range(num_epochs)):
        train_l_sum, train_acc_sum, n = 0.0, 0.0, 0
        for X, y in train_iter:
            y_pred = net(X)
            l = loss(y_pred, y).sum()

            #梯度清零
            if optimizer is not None:
                optimizer.zero_grad()
            elif params is not None and params[0].grad is not None:
                for param in params:
                    param.grad.data.zero_()
            l.backward()
            if optimizer is None:
                sgd_v1(params, lr)  #应用自定义的梯度下降法
            else:
                optimizer.step()  #应用 torch.optim.SGD 库的梯度下降法
            train_l_sum += l.item()
            train_acc_sum += (y_pred.argmax(dim=1) == y).sum().item()
            n += y.shape[0]
        if test_iter is not None:
            test_acc = evaluate_accuracy(test_iter, net)
        if epoch % interval_print == 0:
            print('epoch %d, loss %.4f, train acc %.3f, test acc %.3f' %
                  (epoch + 1, train_l_sum / n, train_acc_sum / n, test_acc))
X, y = make_classification_dataset_(n_features=2, n_redundant=0, n_informative=2,
                                    n_clusters_per_class=1, n_classes=3, n_
```

```
samples=1000, plot=False)
# 01- 建立数据集
data_iter = data_utils.TensorDataset(
    torch.from_numpy(X).float(), torch.from_numpy(y).long())
# 02- 切分数据集
train_set, val_set = data_utils.random_split(data_iter, [800, 200])
# 03- 配置每批大小，batch size
train_data_loader = data_utils.DataLoader(
    train_set, batch_size=50, shuffle=True, num_workers=2)
val_data_loader = data_utils.DataLoader(
    val_set, batch_size=50, shuffle=True, num_workers=2)
num_epochs = 500
train_v1(net=net, train_iter=train_data_loader, test_iter=val_data_loader,
         loss=loss, num_epochs=num_epochs, params=None, lr=None,
optimizer=optimizer)
-------------------------------------------------------------------------------
类标： [0 1 2]
0%|          | 0/500 [00:00<?, ?it/s]
epoch 1, loss 0.0166, train acc 0.814, test acc 0.875
epoch 101, loss 0.0051, train acc 0.912, test acc 0.915
epoch 201, loss 0.0050, train acc 0.914, test acc 0.915
epoch 301, loss 0.0050, train acc 0.916, test acc 0.915
epoch 401, loss 0.0050, train acc 0.917, test acc 0.920
```

3.16.5.5 PyTorch_SoftMax回归多分类，用于图像数据集Fashion-MNIST

- **图像数据集Fashion-MNIST**

Fashion-MNIST[①]包括60000个例子的训练集和10000个例子的测试集。每个示例都是一个28 × 28灰度图像（图3.16–12），总共784像素，与来自10个类的标签相关联，可用于替代原始MNIST手写数字数据集[②]，对机器学习算法进行基准测试。图像的像素都有一个与之相关联的像素值，表示该像素的明度和暗度，数字越大表示越暗。像素值为0到255的整数。训练和测试数据集有785列，第一列由标签组成，表示服装的种类。其余的列包含关联图像的像素值。如果要定位一幅图像一个像素的位置，可以应用$x = i \times 28 + j$定位，其中i, j是0到27之间的整数，像素(x)位于一个28 × 28矩阵的第i行和第j列。

① 图像数据集Fashion-MNIST，（https://www.kaggle.com/zalando-research/fashionmnist）。

② MNIST手写数字数据集，（http://yann.lecun.com/exdb/mnist）。

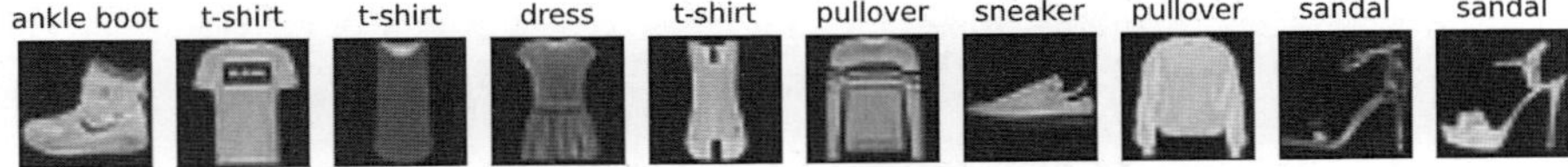

图3.16-12 Fashion-MNIST图像数据集

图像标签（类标）为：

1. T-shirt/top
2. Trouser
3. Pullover
4. Dress
5. Coat
6. Sandal
7. Shirt
8. Sneaker
9. Bag
10. Ankle boot

服务于PyTorch深度学习框架，主要用来构建计算机视觉模型的torchvision[③]包中，`torchvision.datasets`可以用来加载常用的数据集；`torchvision.models`包含常用的模型结果；`torchvision.transforms`用于图片的变换；`torchvision.utils`含有用的工具等。

`.ToTensor()`将所有图像数据shape（h/高,w/宽,c/通道数），像素值位于［0，255］的PIL图片（或者数据类型为`np.unit8`的NumPy数组），转换为tensor张量`shape(c,h,w)`，数据类型为`torch.float32`，像素值位于［0.0,1.0］。

③ torchvision，包含流行的数据集、模型架构和计算机视觉的常见图像变换，https://pytorch.org/vision/stable/index.html）。

```
import torchvision
import torchvision.transforms as transforms
mnist_train = torchvision.datasets.FashionMNIST(
    root='./dataset/', train=True, download=True, transform=transforms.
ToTensor())
mnist_test = torchvision.datasets.FashionMNIST(
    root='./dataset/', train=False, download=True, transform=transforms.
ToTensor())
----------------------------------------------------------------------------
Downloading http://fashion-mnist.s3-website.eu-central-1.amazonaws.com/train-
images-idx3-ubyte.gz
Downloading http://fashion-mnist.s3-website.eu-central-1.amazonaws.com/train-
images-idx3-ubyte.gz to ./dataset/FashionMNIST\raw\train-images-idx3-ubyte.gz
  0%|          | 0/26421880 [00:00<?, ?it/s]
Extracting ./dataset/FashionMNIST\raw\train-images-idx3-ubyte.gz to ./dataset/
FashionMNIST\raw
...
  0%|          | 0/5148 [00:00<?, ?it/s]
Extracting ./dataset/FashionMNIST\raw\t10k-labels-idx1-ubyte.gz to ./dataset/
FashionMNIST\raw
----------------------------------------------------------------------------
import torch.utils.data as data_utils
print("mnist-数据类型：{}\nmnist_train大小={},mnist_test大小={}".
format(type(mnist_train),
      len(mnist_train), len(mnist_test)))
feature, label = mnist_train[0]
def fashionMNIST_label_num2text(labels_int):
    '''
    function - 将Fashion-MNIST数据集，整型类标转换为名称
    '''
    labels_text = ['t-shirt', 'trouser', 'pullover', 'dress',
                   'coat', 'sandal', 'shirt', 'sneaker', 'bag', 'ankle boot']
    return [labels_text[int(i)] for i in labels_int]
print("feature.shape={},lable={}, {}".format(
    feature.shape, label, fashionMNIST_label_num2text([label])))
def fashionMNIST_show(imgs, labels, figsize=(12, 12)):
    '''
    function - 打印显示Fashion-MNIST数据集图像
    '''
    import matplotlib.pyplot as plt
    from IPython import display
    import matplotlib_inline.backend_inline
    matplotlib_inline.backend_inline.set_matplotlib_formats(
        'svg')  #以svg格式打印显示
    _, axs = plt.subplots(1, len(imgs), figsize=figsize)
    for ax, img, label in zip(axs, imgs, labels):
        ax.imshow(img.view((28, 28)).numpy())
        ax.set_title(label)
        ax.axes.get_xaxis().set_visible(False)
        ax.axes.get_yaxis().set_visible(False)
    plt.show()
imgs_len = 10
X = [mnist_train[i][0] for i in range(imgs_len)]
y = [mnist_train[i][1] for i in range(imgs_len)]
fashionMNIST_show(
    imgs=X, labels=fashionMNIST_label_num2text(y), figsize=(12, 12))

#读取小批量

batch_size = 256
num_workers = 4
train_iter = data_utils.DataLoader(
    mnist_train, batch_size=batch_size, shuffle=True, num_workers=num_workers)
test_iter = data_utils.DataLoader(
    mnist_test, batch_size=batch_size, shuffle=False, num_workers=num_workers)
----------------------------------------------------------------------------
mnist-数据类型：<class 'torchvision.datasets.mnist.FashionMNIST'>
mnist_train大小=60000,mnist_test大小=10000
feature.shape=torch.Size([1, 28, 28]),lable=9, ['ankle boot']
```

将上述常用的自定义函数放置于util_A模块中，方便调用，例如flattenLayer类。外部仅定义了net模型，loss损失函数和optimizer优化函数。

```
import util_A
```

```
from torch import nn
from collections import OrderedDict
import torch

#定义模型
num_inputs, num_outputs = 28*28, 10
net = nn.Sequential(
    OrderedDict([
        ('flatten', util_A.flattenLayer()),
        ('linear', nn.Linear(num_inputs, num_outputs))
    ])
)

# SoftMax, 交叉熵损失函数 CrossEntropyLossntropyLoss()
loss = nn.CrossEntropyLoss()

#优化算法
optimizer = torch.optim.SGD(net.parameters(), lr=0.1)
num_epochs = 500
util_A.train_v1(net, train_iter, test_iter, loss, num_epochs,
                params=None, lr=None, optimizer=optimizer, interval_print=100)
--------------------------------------------------------------------------------
  0%|          | 0/500 [00:00<?, ?it/s]
epoch 1, loss 0.0031, train acc 0.746, test acc 0.783
epoch 101, loss 0.0015, train acc 0.869, test acc 0.843
epoch 201, loss 0.0014, train acc 0.871, test acc 0.844
epoch 301, loss 0.0014, train acc 0.874, test acc 0.840
epoch 401, loss 0.0014, train acc 0.875, test acc 0.841
```


参考文献（References）：

[1] Gavin Hackeling (2017). Mastering Machine Learning with scikit-learn: Apply effective learning algorithms to real-world problems using scikit-learn. Packt Publishing. 中文版为：Cavin Hackeling.张浩然译.scikit-learning 机器学习[M].人民邮电出版社.2019.2.

[2] Peter Harrington (2012). Machine Learning in Action. Manning Publications.

[3] Softmax Activation (NOTES ON STATISTICS, PROBABILITY and MATHEMATICS), http://rinterested.github.io/statistics/softmax.html; https://eli.thegreenplace.net/2016/the-softmax-function-and-its-derivative/.

[4] Aston Zhang, & Alexander J. Smola (2021). Dive into Deep Learning. d2l.ai；中文版-阿斯顿．张，李沐，扎卡里．C. 立顿，亚历山大．J. 斯莫拉．动手深度学习[M]．人民邮电出版社，北京，2019-06-01．https://zh.d2l.ai/

3.17 卷积神经网络

3.17.1 卷积神经网络（Convolutional neural network，CNN）——卷积原理与卷积神经网络

在阅读卷积神经网络（CNNs）前，如果已经阅读"卷积""计算机视觉，特征提取和尺度空间"等部分章节，可以更好地理解CNNs。其中，"尺度空间"的概念，通过降采样的不同空间分辨率影像（类似池化，pooling）和不同σ值的高斯核卷积（即卷积计算或称为互相关运算），来提取图像的概貌特征。这与CNNs的多层卷积网络如出一辙。只是CNNs除了卷积层，还可以自由加入其他的数据处理层，提升图像特征捕捉的几率。

一幅图像的意义来自于邻近的一组像素，而不是单个像素自身，因为单个像素并不包含关于整个图像的信息。例如，图3.17-1很好地说明了这两者的差异。一个全连接的神经网络（密集层，dense layer），一层的每个节点都连接到下一层的每个节点，并不能很好地反映节点之间的关系，也具有较大的计算量；但是，卷积网络利用像素之间的空间结构减少了层之间的连接数量，显著提高了训练的速度，减少了模型的参数。更重要的是反映了图像特征像素之间的关系，即图像内容中各个对象是由自身与邻近像素关系决定的空间结构（即特征）所表述。

二维卷积层的卷积计算，就是表征图像的数组shape(c,h,w)（如果为灰度图像则只有一个颜色通道，如果是彩色图像（RGB）则有三个图像通道），每个通道(h,w)二维数组中每一像素，与卷积核（filter/kernal）的加权和计算。这个过程存在一些可变动的因素：一个是步幅（stride），卷积窗口（卷积核）从输入数组（二维图像数组）的最左上方开始，按从左到右，从上往下的顺序，依次在输入数组上滑动，每次滑动的行数和列数称为步幅。步幅越小，代表捕捉不同内容对象的精度越高。图3.17-2卷积核的滑动步幅为2，即行列均跨2步后计算；二是填充（padding），如果不设置填充，并且从左上角滑动，会有一部分卷积核对应空值（没有图像/数据）。同时，要

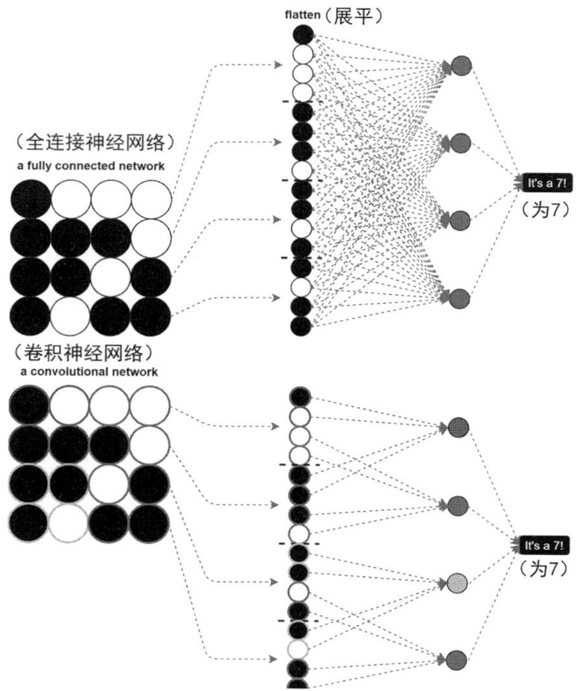

图3.17-1　全连接神经网络与卷积神经网络比较图解（引自参考文献[1]）

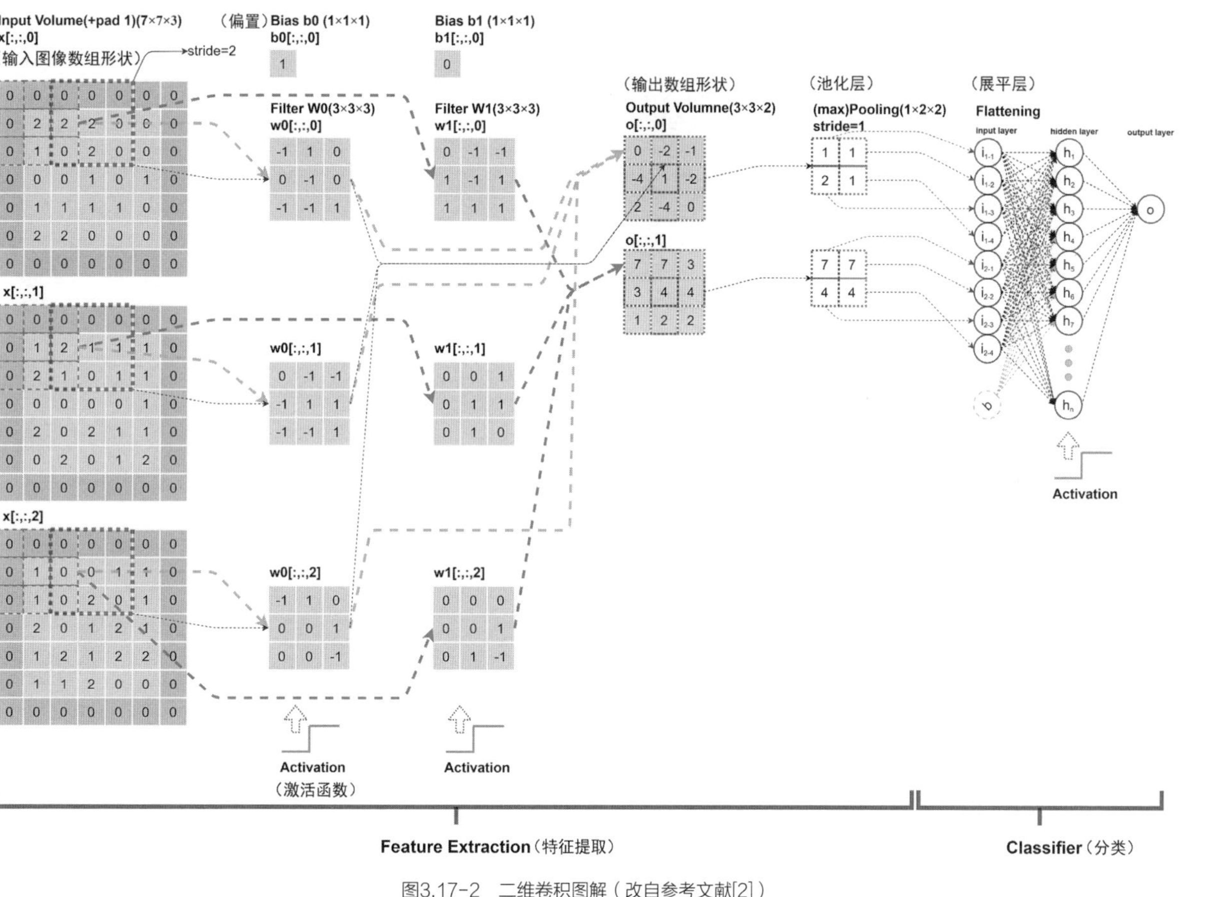

图3.17-2 二维卷积图解（改自参考文献[2]）

保持四周填充的行列数相同，通常使用奇数高宽的卷积核；三是，如果是对图像数据实现卷积，通常包括1个单通道，或3个多通道的情况。对于多通道的计算是可以配置不同的卷积核对应不同的通道，各自通道分别卷积后，计算和（累加）作为结果输出。同时，可以增加并行的新的卷积计算，获取多个输出，例如下图的Filter W0，和Filter W1的（3×3）卷积核，W0和W1各自包含3个卷积核对应3个通道输入，并各自输出。

卷积运算，步幅、填充配置，以及多通道卷积都没有改变图像的空间尺寸（空间分辨率），即尺度空间概念下降采样的表述（可以反映不同对象的尺度大小，或理解为只有在不同的尺度下才可以捕捉到对象的特征）。池化层（pooling）正是降采样在卷积神经网络中的表述，可以降低输入的空间维数，保留输入的深度。在图像卷积过程中，识别出比实际像素信息更多的概念意义，识别保留输入的关键信息，丢弃冗余部分。池化层不仅捕捉尺度空间下的对象特征，同时可以减少训练所需时间，减小模型的参数数量，降低模型复杂度，更好地泛化等。池化层可以用`nn.MaxPool2d`，取最大值；`nn.AvgPool2d`，取均值等方法。

卷积层和池化层都有一个`dilation`（扩张率/膨胀系数）参数，通过`dilation`配置，可以调整卷积核与图像的作用域，即感受野（receptive filed）。图3.17-3中，当3×3的卷积核dilation=1时，没有膨胀效应；当dilation=2时，感受野扩展至7×7；当dilation=24时，感受野扩展至15×15。可以确定，当dilation线性增加时，感受野呈指数增加。

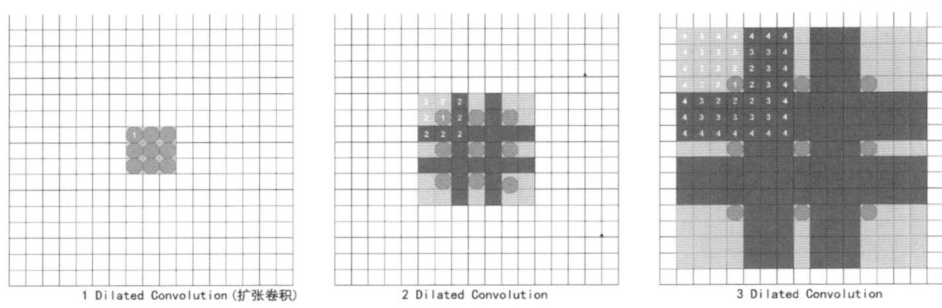

图3.17-3 扩张（率）卷积（引自参考文献[3]）

建立图表中的输入数据`t_input`，张量形状为`(batchsize, nChannels, Height, Width)=(1,3,7,7)`，即只有一幅图像，通道数为3，高度为7，宽度为7。通常查看时，只要确定最后一维内的数据为图像每一行的像素值（由上至下）。PyTorch提供的`nn.Conv2d`卷积方法不需自定义卷积核，其参数为`torch.nn.Conv2d(in_channels: int, out_channels: int, kernel_size: Union[T, Tuple[T, T]], stride: Union[T, Tuple[T, T]] = 1, padding: Union[T, Tuple[T, T]] = 0, dilation: Union[T, Tuple[T, T]] = 1, groups: int = 1, bias: bool = True, padding_mode: str = 'zeros')`。如果需要自定义卷积核，则调用`torch.nn.functional.conv2d`方法，其中`weight`参数即为卷积核，其输入参数为`torch.nn.functional.conv2d(input, weight, bias=None, stride=1, padding=0, dilation=1, groups=1) → Tensor`。

```
import torch
import torch.nn as nn
# batchsize, nChannels, Height, Width
t_input = torch.tensor([[[[0, 0, 0, 0, 0, 0, 0], [0, 2, 2, 2, 0, 0, 0], [0, 1, 0, 2, 0, 0, 0], [0, 0, 0, 1, 0, 1, 0], [0, 1, 1, 1, 1, 0, 0], [0, 2, 2, 0, 0, 0, 0], [0, 0, 0, 0, 0, 0, 0]],
                        [[0, 0, 0, 0, 0, 0, 0], [0, 1, 2, 1, 1, 1, 0], [0, 2, 1, 0, 1, 1, 0], [
                          0, 0, 0, 0, 0, 1, 0], [0, 2, 0, 2, 1, 1, 0], [0, 0, 2, 0, 1, 2, 0], [0, 0, 0, 0, 0, 0, 0]],
                        [[0, 0, 0, 0, 0, 0, 0], [0, 1, 0, 0, 1, 1, 0], [0, 1, 0, 2, 0, 1, 0], [
                          0, 2, 0, 1, 2, 1, 0], [0, 1, 2, 1, 2, 2, 0], [0, 1, 1, 2, 0, 0, 0], [0, 0, 0, 0, 0, 0, 0]],
                        ]], dtype=torch.float)
```

```
print("t_input.shape={}".format(t_input.shape))
conv_2d_3c = nn.Conv2d(in_channels=3, out_channels=2,
                       kernel_size=3, stride=2, padding=0, bias=1)
conv_2d_3c(t_input)
---------------------------------------------------------------------------
t_input.shape=torch.Size([1, 3, 7, 7])
tensor([[[[ 0.0716,  0.4418,  0.2253],
          [ 0.5419,  1.3328,  0.6916],
          [-0.5093, -0.2429,  0.3159]],
         [[-0.5181,  0.4435, -0.2968],
          [-0.5857, -0.9900, -0.4077],
          [-0.4969, -0.1312, -0.1665]]]], grad_fn=<ThnnConv2DBackward0>)
---------------------------------------------------------------
import torch.nn.functional as F
w_0 = torch.tensor([[[[-1, 1, 0],
                      [0, -1, 0],
                      [-1, -1, 1]],
                     [[0, -1, -1],
                      [-1, 1, 1],
                      [-1, -1, 1]],
                     [[-1, 1, 0],
                      [0, 0, 1],
                      [0, 0, -1]]]], dtype=torch.float)
w_1 = torch.tensor([[[[0, -1, -1],
                      [1, -1, 1],
                      [1, 1, 1]],
                     [[0, 0, 1],
                      [0, 1, 1],
                      [0, 1, 0]],
                     [[0, 0, 0],
                      [0, 0, 1],
                      [0, 1, -1]]]], dtype=torch.float)
b_0 = torch.tensor([1])
b_1 = torch.tensor([0])
output_0 = F.conv2d(input=t_input, weight=w_0, stride=2, padding=0, bias=b_0)
output_1 = F.conv2d(input=t_input, weight=w_1, stride=2, padding=0, bias=b_1)
print(output_0)
print(output_1)
---------------------------------------------------------------------------
tensor([[[[ 0., -2., -1.],
          [-4.,  1., -2.],
          [ 2., -4.,  0.]]]])
tensor([[[[7., 7., 3.],
          [3., 4., 4.],
          [1., 2., 2.]]]])
```

最大池化，即设定池化（卷积核）大小，返回覆盖范围内最大值，其参数为torch.nn.MaxPool2d(kernel_size: Union[T, Tuple[T, ...]], stride: Optional[Union[T, Tuple[T, ...]]] = None, padding: Union[T, Tuple[T, ...]] = 0, dilation: Union[T, Tuple[T, ...]] = 1, return_indices: bool = False, ceil_mode: bool = False)。

```
pooling_input = torch.tensor([[[0,-2,-1],[-4,1,-2],[2,-4,0]]],dtype=torch.float)
print("输入数据形状={}".format(pooling_input.shape))

maxPool_2d = nn.MaxPool2d(kernel_size=2,stride=1)
maxPool_2d(pooling_input)
---------------------------------------------------------------------------
输入数据形状=torch.Size([1, 3, 3])
tensor([[[1., 1.],
         [2., 1.]]])
```

3.17.2 卷积_特征提取器到分类器，可视化卷积层/卷积核，及tensorboard

3.17.2.1 卷积层，池化层输出尺寸（形状）计算，及根据输入输出尺寸反推填充pad

构建卷积神经网络，很重要的一步是确定图像经过一次或多次卷积后，输出的尺寸用于分类器（线性模型、全连接层）的输入。在PyTorch的torch.nn.Conv2d[①]类方法说明中，都会给

① torch.nn.Conv2d，（https://pytorch.org/docs/stable/generated/torch.nn.Conv2d.html）。

出卷积输出的尺寸计算公式。但是，手工的计算方式容易出错，并且耗时耗力，因此通常将其编写为代码程序。池化层的输出尺寸计算实际上与卷积的输出相同。但是，为了区分，仍然在conv2d_output_size_calculation类中，增加了pooling_output_shape方法（直接调用conv2d_output_shape）。卷积的方式除了给定h_w图像高宽、kernel_size卷积核（过滤器，filter）、stride步幅，pad填充及dilation膨胀系数，卷积核的初始位置可以分为：以卷积核左上角对位图像左上角第一个像素值开始由左->右，由上->下卷积计算，即conv2d_output_shape函数；以卷积核右下角对位图像左上角第一个像素值开始由左->右，由上->下卷积计算，即convtransp2d_output_shape函数。

同时，也可以根据卷积的输入和输出的尺寸，反推填充的大小，对应卷积核初始位置的不同分别为conv2d_get_padding和convtransp2d_get_padding函数。

对于卷积神经网络net_fashionMNIST的网络结构，如图3.17-4所示：

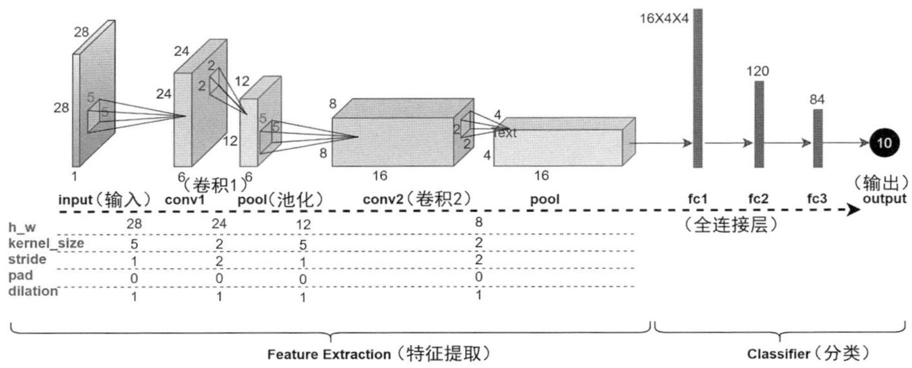

图3.17-4 net_fashionMNIST的卷积神经网络结构

输入图像的大小为28 × 28，经过卷积层conv1->池化层pool->卷积层conv2->池化层pool后，图像尺寸的大小变为4 × 4，该值（卷积层的输出值）由自定义Conv2d_output_size_calculation计算。注意两次池化，其参数值相同，因此卷积神经网络结构定义时，可以仅定义一个池化方法，用于生成不同位置的池化层。计算应用卷积提取特征部分的图像输出大小后，因为conv2 + pool之后，输出的out_channels输出通道数配置为16，因此到分类器部分（线性函数/全连接层/展平层）的输入大小为16 × 4 × 4 = 256。

```
class Conv2d_output_size_calculation:
    '''
    class - PyTorch 卷积层，池化层输出尺寸(shape)计算，及根据输入，输出尺寸(shape)反推
pad 填充大小

    @author:sytelus Shital Shah
    Updated on Tue Jan 12 19:17:22 2021 @author: Richie Bao
    '''
    def num2tuple(self, num):
        '''
        function - 如果num=2，则返回(2,2)；如果num=(2,2)，则返回(2,2).
        '''
        return num if isinstance(num, tuple) else (num, num)
    def conv2d_output_shape(self, h_w, kernel_size=1, stride=1, pad=0,
dilation=1):
        '''
        funciton - 计算 PyTorch 的 nn.Conv2d 卷积方法的输出尺寸。以卷积核左上角对位图像左上
角第一个像素值开始由左 --> 右，由上 --> 下卷积计算
        '''
        import math
        h_w, kernel_size, stride, pad, dilation = self.num2tuple(h_w), \
            self.num2tuple(kernel_size), self.num2tuple(
                stride), self.num2tuple(pad), self.num2tuple(dilation)
        pad = self.num2tuple(pad[0]), self.num2tuple(pad[1])
        h = math.floor((h_w[0] + sum(pad[0]) - dilation[0]
                        * (kernel_size[0]-1) - 1) / stride[0] + 1)
```

```
        w = math.floor((h_w[1] + sum(pad[1]) - dilation[1]
                     * (kernel_size[1]-1) - 1) / stride[1] + 1)
        return h, w
    def convtransp2d_output_shape(self, h_w, kernel_size=1, stride=1, pad=0,
dilation=1, out_pad=0):
        '''
        function - 以卷积核右下角对位图像左上角第一个像素值开始由左 --> 右，由上 --> 下卷积计算
        '''
        h_w, kernel_size, stride, pad, dilation, out_pad = self.num2tuple(h_w), \
            self.num2tuple(kernel_size), self.num2tuple(stride), self.num2tuple(
                pad), self.num2tuple(dilation), self.num2tuple(out_pad)
        pad = self.num2tuple(pad[0]), self.num2tuple(pad[1])
        h = (h_w[0] - 1)*stride[0] - sum(pad[0]) + \
            dilation[0]*(kernel_size[0]-1) + out_pad[0] + 1
        w = (h_w[1] - 1)*stride[1] - sum(pad[1]) + \
            dilation[1]*(kernel_size[1]-1) + out_pad[1] + 1
        return h, w
    def conv2d_get_padding(self, h_w_in, h_w_out, kernel_size=1, stride=1,
dilation=1):
        '''
        function - conv2d_output_shape 方法的逆，求填充 pad
        '''
        import math
        h_w_in, h_w_out, kernel_size, stride, dilation = self.num2tuple(h_w_in),
self.num2tuple(h_w_out), \
            self.num2tuple(kernel_size), self.num2tuple(
                stride), self.num2tuple(dilation)
        p_h = ((h_w_out[0] - 1)*stride[0] - h_w_in[0] +
               dilation[0]*(kernel_size[0]-1) + 1)
        p_w = ((h_w_out[1] - 1)*stride[1] - h_w_in[1] +
               dilation[1]*(kernel_size[1]-1) + 1)

        # ((pad_up, pad_bottom),  (pad_left, pad_right))
        return (math.floor(p_h/2), math.ceil(p_h/2)), (math.floor(p_w/2), math.
ceil(p_w/2))
    def convtransp2d_get_padding(self, h_w_in, h_w_out, kernel_size=1, stride=1,
dilation=1, out_pad=0):
        '''
        function - convtransp2d_output_shape 方法的逆，求填充 pad
        '''
        import math
        h_w_in, h_w_out, kernel_size, stride, dilation, out_pad = self.
num2tuple(h_w_in), self.num2tuple(h_w_out), \
            self.num2tuple(kernel_size), self.num2tuple(
                stride), self.num2tuple(dilation), self.num2tuple(out_pad)
        p_h = -(h_w_out[0] - 1 - out_pad[0] - dilation[0] *
                (kernel_size[0]-1) - (h_w_in[0] - 1)*stride[0]) / 2
        p_w = -(h_w_out[1] - 1 - out_pad[1] - dilation[1] *
                (kernel_size[1]-1) - (h_w_in[1] - 1)*stride[1]) / 2
        return (math.floor(p_h/2), math.ceil(p_h/2)), (math.floor(p_w/2), math.
ceil(p_w/2))
    def pooling_output_shape(self, h_w, kernel_size=1, stride=1, pad=0,
dilation=1):
        '''
        function - pooling 池化层输出尺寸，同 conv2d_output_shape
        '''
        return self.conv2d_output_shape(h_w, kernel_size=kernel_size,
stride=stride, pad=pad, dilation=dilation)
conv2dSize_cal = Conv2d_output_size_calculation()
size_conv1 = conv2dSize_cal.conv2d_output_shape(
    28, kernel_size=5, stride=1, pad=0, dilation=1)
size_pool1 = conv2dSize_cal.pooling_output_shape(
    24, kernel_size=2, stride=2, pad=0, dilation=1)
size_conv2 = conv2dSize_cal.conv2d_output_shape(
    12, kernel_size=5, stride=1, pad=0, dilation=1)
size_pool2 = conv2dSize_cal.pooling_output_shape(
    8, kernel_size=2, stride=2, pad=0, dilation=1)
print("conv1_size={}\nsize_pool1={}\nsize_conv2={}\nsize_pool2={}".format(
    size_conv1, size_pool1, size_conv2, size_pool2))
-----------------------------------------------------------------
conv1_size=(24, 24)
size_pool1=(12, 12)
size_conv2=(8, 8)
size_pool2=(4, 4)
```

在设计卷积层，或者包含很多卷积层时，上述的方法可以进一步改进，定义一个函数可以一次性计算所有的卷积层。输入参数设置时，使用了列表和元组的形式，固定输入参数input、conv和pool，值的参数依次为[kernel_size, stride, pad, dilation]。

```
convs_params = [
    ('input', (28, 28)),
    ('conv', [5, 1, 0, 1]),    # kernel_size, stride, pad, dilation
    ('pool', [2, 2, 0, 1]),
    ('conv', [5, 1, 0, 1]),
    ('pool', [2, 2, 0, 1]),
]
def conv2d_outputSize_A_oneTime(convs_params):
    '''
    fucntion - 一次性计算卷积输出尺寸
    '''
    conv2dSize_cal = Conv2d_output_size_calculation()
    for v in convs_params:
        if v[0] == 'input':
            h_w = v[1]
        elif v[0] == 'conv' or v[0] == 'pool':
            kernel_size, stride, pad, dilation = v[1]
            h_w, kernel_size, stride, pad, dilation = conv2dSize_cal.num2tuple(h_
w), conv2dSize_cal.num2tuple(
                kernel_size), conv2dSize_cal.num2tuple(stride), conv2dSize_cal.
num2tuple(pad), conv2dSize_cal.num2tuple(dilation)
            h_w = conv2dSize_cal.conv2d_output_shape(
                h_w, kernel_size, stride, pad, dilation)
    return h_w
output_h_w = conv2d_outputSize_A_oneTime(convs_params)
print("卷积层输出尺寸 ={}".format(output_h_w))
-----------------------------------------------------------------
卷积层输出尺寸 =(4, 4)
```

3.17.2.2 基于fashionMNIST数据集构建简单的卷积神经网络识别，及tensorboard可视化

此处，构建上图给出的神经网络结构。input->conv1(relu)->pool->conv2(relu)->pool—>fc1(relu)->fc2(relu)->fc3->output。同时，应用tensorboard可以写入并自动的根据写入的内容图示卷积神经网络结构、损失曲线、预测结果、样本信息及自定义的内容等。详细内容可以查看torch.utils.tensorboard[①]。

① torch.utils.tensorboard，（https://pytorch.org/docs/stable/tensorboard.html）。

- 01-下载/读取fashinMNIST数据，以及构建训练、测试可迭代对象（如果已经下载，则直接读取）。

```
def load_fashionMNIST(root, batchsize=4, num_workers=2, resize=None, n_mean=0.5,
n_std=0.5):
    '''
    function - 下载读取 fashionMNIST 数据集，并建立训练、测试可迭代数据集
    '''
    import torchvision
    import torchvision.transforms as transforms

    trans = [transforms.ToTensor(),  # 转换 PIL 图像或 numpy.ndarray 为 tensor 张量
             transforms.Normalize((0.5,), (0.5,))]  # torchvision.transforms.Normalize(mean,
std, inplace=False)，用均值和标准差，标准化张量图像
    if resize:
        trans.append(transforms.Resize(size=resize))
    transform = transforms.Compose(trans)
    mnist_train = torchvision.datasets.FashionMNIST(
        root=root, train=True, download=True, transform=transform)
    mnist_test = torchvision.datasets.FashionMNIST(
        root=root, train=False, download=True, transform=transform)

    # DataLoade- 读取小批量
    import torch.utils.data as data_utils
    batch_size = batchsize
    num_workers = num_workers
    trainloader = data_utils.DataLoader(
        mnist_train, batch_size=batch_size, shuffle=True, num_workers=num_
workers)
    testloader = data_utils.DataLoader(
        mnist_test, batch_size=batch_size, shuffle=False, num_workers=num_
```

```
workers)
    return trainloader, testloader
trainloader, testloader = load_fashionMNIST(root='./dataset/FashionMNIST_norm')
classes = ('T-shirt/top', 'Trouser', 'Pullover', 'Dress', 'Coat',
            'Sandal', 'Shirt', 'Sneaker', 'Bag', 'Ankle Boot')
--------------------------------------------------------------------------------
Downloading http://fashion-mnist.s3-website.eu-central-1.amazonaws.com/train-
images-idx3-ubyte.gz
Downloading http://fashion-mnist.s3-website.eu-central-1.amazonaws.com/train-
images-idx3-ubyte.gz to ./dataset/FashionMNIST_norm\FashionMNIST\raw\train-
images-idx3-ubyte.gz
  0%|          | 0/26421880 [00:00<?, ?it/s]
Extracting ./dataset/FashionMNIST_norm\FashionMNIST\raw\train-images-idx3-ubyte.
gz to ./dataset/FashionMNIST_norm\FashionMNIST\raw
...
--------------------------------------------------------------------------------
```

- 02-定义网络结构，特征提取层+分类器。

```
import torch.optim as optim
import torch.nn as nn
import torch.nn.functional as F
class net_fashionMNIST(nn.Module):
    def __init__(self):
        super(net_fashionMNIST, self).__init__()
        self.conv1 = nn.Conv2d(1, 6, 5)
        self.pool = nn.MaxPool2d(2, 2)
        self.conv2 = nn.Conv2d(6, 16, 5)
        # torch.nn.Linear(in_features: int, out_features: int, bias: bool = True)
        self.fc1 = nn.Linear(16*4*4, 120)
        self.fc2 = nn.Linear(120, 84)
        self.fc3 = nn.Linear(84, 10)
    def forward(self, x):
        x = self.pool(F.relu(self.conv1(x)))
        x = self.pool(F.relu(self.conv2(x)))
        x = x.view(-1, 16*4*4)
        x = F.relu(self.fc1(x))
        x = F.relu(self.fc2(x))
        x = self.fc3(x)
        return x
net_fashionMNIST_ = net_fashionMNIST()
print(net_fashionMNIST_)
```

- 03-定义损失函数核优化算法。

```
criterion = nn.CrossEntropyLoss()  # 定义损失函数
optimizer = optim.SGD(net_fashionMNIST_.parameters(),
                      lr=0.001, momentum=0.9)  # 定义优化算法
--------------------------------------------------------------------------------
net_fashionMNIST(
  (conv1): Conv2d(1, 6, kernel_size=(5, 5), stride=(1, 1))
  (pool): MaxPool2d(kernel_size=2, stride=2, padding=0, dilation=1, ceil_
mode=False)
  (conv2): Conv2d(6, 16, kernel_size=(5, 5), stride=(1, 1))
  (fc1): Linear(in_features=256, out_features=120, bias=True)
  (fc2): Linear(in_features=120, out_features=84, bias=True)
  (fc3): Linear(in_features=84, out_features=10, bias=True)
)
--------------------------------------------------------------------------------
```

- 04-定义显示图像函数（一个batch批次）。

```
from torch.utils.tensorboard import SummaryWriter
import torchvision
def matplotlib_imshow(img, one_channel=False):
    import matplotlib.pyplot as plt
    import numpy as np
    if one_channel:
        img = img.mean(dim=0)
    img = img / 2 + 0.5     # unnormalize
    npimg = img.numpy()
    if one_channel:
        plt.imshow(npimg, cmap="Greys")
    else:
        plt.imshow(np.transpose(npimg, (1, 2, 0)))
```

● 05-调入初始化tensorboard.SummaryWriter（指定数据写入的保存路径），显示图像（图3.17-5），并写入训练图像与模型。

```
writer = SummaryWriter(r'./runs/fashion_mnist_experiment_1')

# 提取图像（数量为 batch_size）（get some random training images）
dataiter = iter(trainloader)
images, labels = dataiter.next()
# 建立图像格网（create grid of images）
img_grid = torchvision.utils.make_grid(images)
# 显示图像（show images）
matplotlib_imshow(img_grid, one_channel=True)
# write to tensorboard
writer.add_image('four_fashion_mnist_images', img_grid)
writer.add_graph(net_fashionMNIST_, images)
writer.close()
```

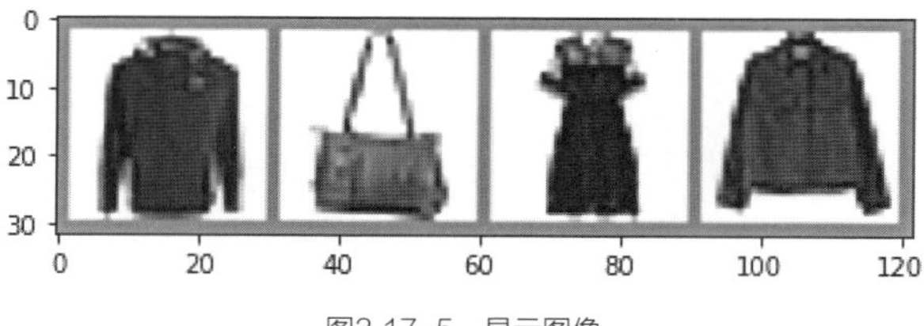

图3.17-5 显示图像

在终端，切换到tensorboard数据保存文件夹runs所在的目录，执行tensorboard --logdir=runs后，通常提示在浏览器中打开http://localhost:6006/地址，可以查看writer.add_image('four_fashion_mnist_images', img_grid)写入的训练图像信息内容（图3.17-6，其语法为：add_image(tag, img_tensor, global_step=None, walltime=None, dataformats='CHW')），如下：

writer.add_graph(net_fashionMNIST_, images)（其语法为：add_graph(model, input_to_model=None, verbose=False)），写入的信息内容下，可以查看网络结构及运算的流程（图3.17-7）。这对模型的构建与调整有所帮助。

下述在训练过程中，riter.add_scalar('training loss',running_loss / 1000, epoch * len(trainloader) + i)（其语法为：add_scalar(tag, scalar_value, global_step=None, walltime=None)），写入损失数据图3.17-8。

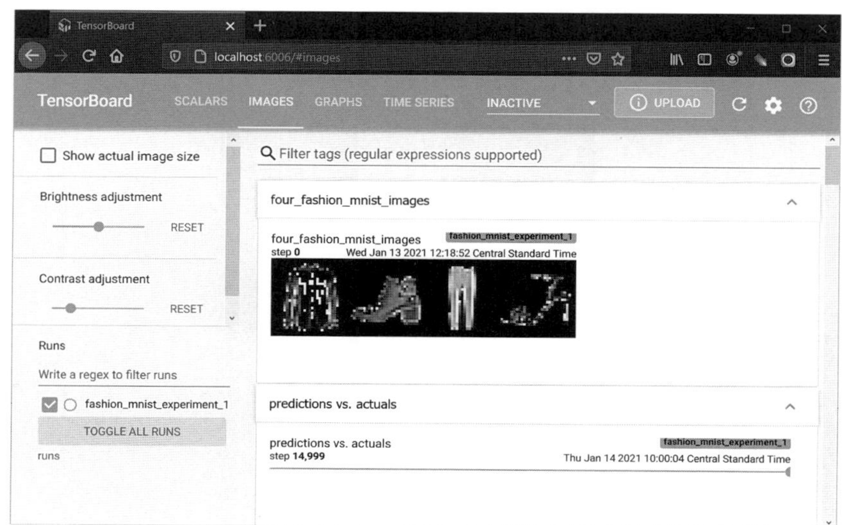

图3.17-6 tensorboard中查看数据

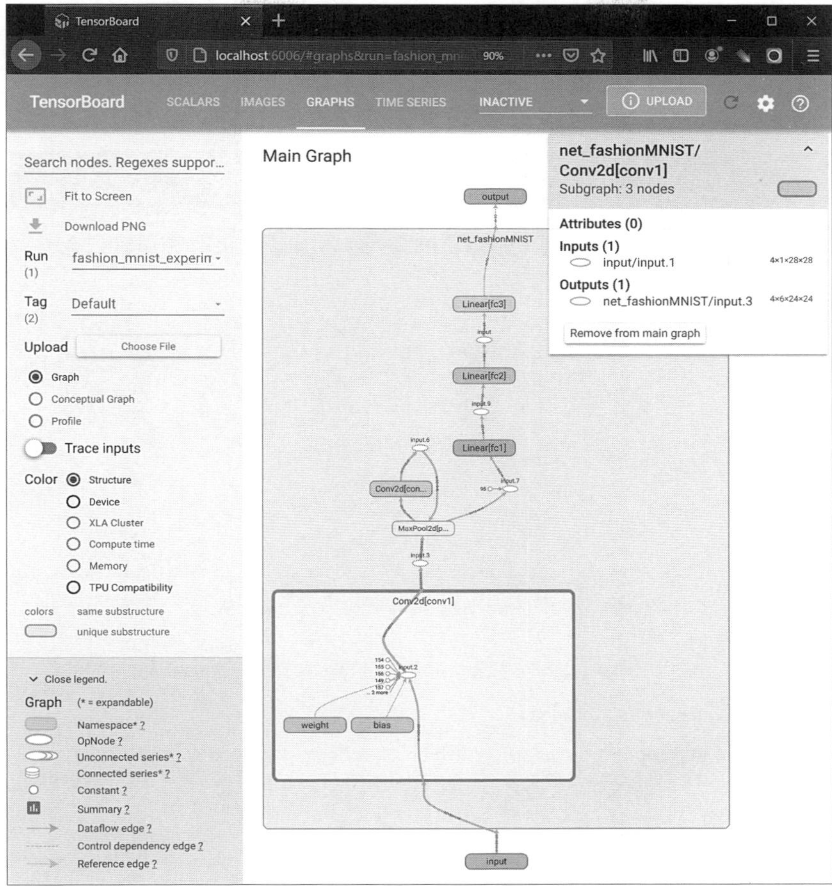

图3.17-7 tensorboard中查看网络结构图

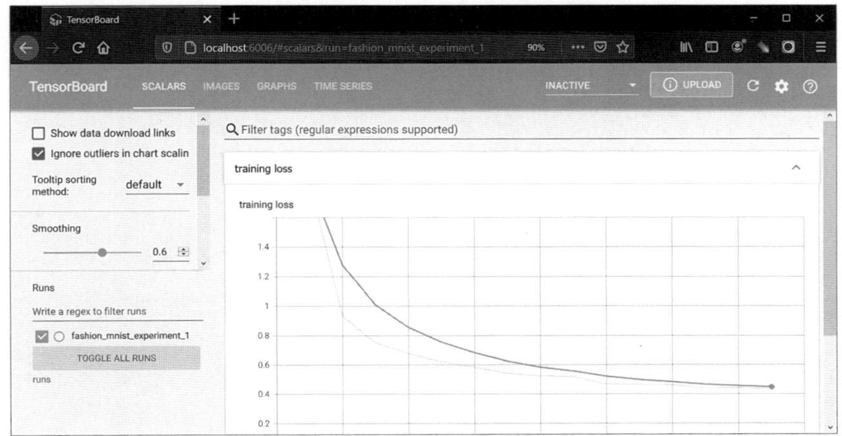

图3.17-8 tensorboard中查看损失曲线

writer.add_figure('predictions vs. actuals',plot_classes_preds(net_fashionMNIST_, inputs, labels),global_step=epoch * len(trainloader) + i)（其语法为：add_figure(tag, figure, global_step=None, close=True, walltime=None)），写入plot_classes_preds返回图表，包括图像、预测值及其概率，如图3.17-9所示。

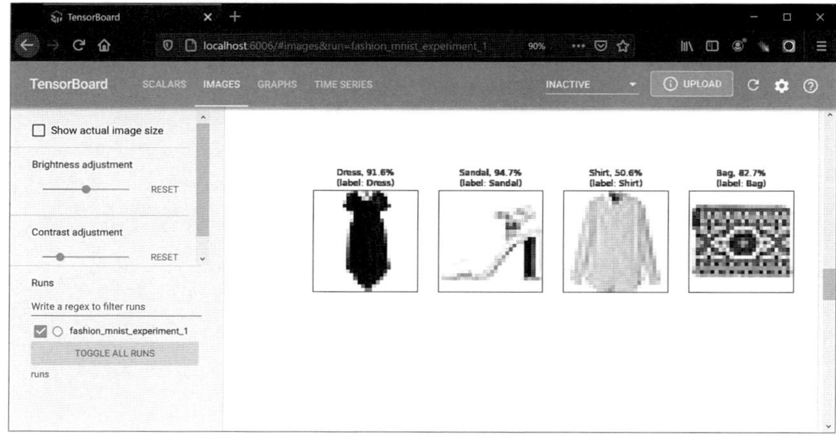

图3.17-9　tensorboard中查看预测值及其概率值

- 06-定义将图像应用训练的网络（模型）预测及其概率待写入tensorboard文件的函数。

```
def images_to_probs(net, images):
    '''
    function - 用训练的网络预测给定的一组图像，并计算相应的概率 （Generates predictions
and corresponding probabilities from a trainednetwork and a list of images）
    '''
    output = net(images)
    #将输出概率转换为预测类
    _, preds_tensor = torch.max(output, 1)
    preds = np.squeeze(preds_tensor.numpy())
    return preds, [F.softmax(el, dim=0)[i].item() for i, el in zip(preds,
output)]
def plot_classes_preds(net, images, labels):
    '''
    function - 用训练的网络预测给定的一组图像，并计算概率后，显示图像、预测值及概率和实际的标签。
    '''
    import matplotlib.pyplot as plt
    preds, probs = images_to_probs(net, images)
    #绘制批处理中的图像，以及预测的和真实的标签
    fig = plt.figure(figsize=(12, 48))
    for idx in np.arange(4):
        ax = fig.add_subplot(1, 4, idx+1, xticks=[], yticks=[])
        matplotlib_imshow(images[idx], one_channel=True)
        ax.set_title("{0}, {1:.1f}%\n(label: {2})".format(
            classes[preds[idx]],
            probs[idx] * 100.0,
            classes[labels[idx]]),
            color=("green" if preds[idx] == labels[idx].item() else "red"))
    return fig
```

- 07-训练模型，同时向tensorboard文件写入相关信息。

```
from tqdm.auto import tqdm
import torch
import numpy as np
running_loss = 0.0
epochs = 1
for epoch in tqdm(range(epochs)): #对数据集进行多次循环
    for i, data in enumerate(trainloader, 0):
```

```
            # 获取输入；数据是一个列表 [inputs, labels]
            inputs, labels = data

            # 将参数梯度归零
            optimizer.zero_grad()

            # 前向（forward）+后向（backward）+优化（optimize）
            outputs = net_fashionMNIST_(inputs)
            loss = criterion(outputs, labels)
            loss.backward()
            optimizer.step()
            running_loss += loss.item()
            if i % 1000 == 999:    # 每 1000 个小批量 ......
                # ... 写入迭代损失
                writer.add_scalar('training loss', running_loss /
                                  1000, epoch * len(trainloader) + i)  # 写入损失数值

                # 记录一个 Matplotlib 图，显示模型对随机小批量的预测
                writer.add_figure('predictions vs. actuals', plot_classes_preds(net_
fashionMNIST_, inputs, labels),
                                  global_step=epoch * len(trainloader) + i)  # 写入自
定义函数 plot_classes_preds 返回的图表，包括图像、预测值及其概率
                loss_temp = running_loss
                running_loss = 0.0
        print("epoch={},running_loss={}".format(epoch, loss_temp))
        loss_temp = 0
print('Finished Training')
--------------------------------------------------------------------------------
  0%|          | 0/1 [00:00<?, ?it/s]
epoch=0,running_loss=365.6714180938143
Finished Training
```

3.17.2.3 可视化卷积层/卷积核

通常，一个卷积层的输出通道（output channel）有6、16、32、64、256等不确定的输出个数及更多的输出个数，即一个卷积层包含输出通道数目的过滤器/卷积核（filter/kernal）；而卷积核通过3 × 3、5 × 5、7 × 7等不确定尺寸（通常为奇数）或者更大尺寸来提取图像的特征。不同的卷积核提取的图像特征不同，或者表述为不同的卷积核关注不同的特征提取，这类似于图像的关键点描述子（位于不同的尺度空间下）。不过因为卷积核的多样性，卷积提取的特征更加丰富多样。通过一个卷积层的多个卷积核（尺度空间的水平向，表述各个像素值与各周边像素值的差异程度），以及多个卷积层（尺度空间的纵深向，表述图像特征所在的（对应的）空间分辨率，例如遥感影像看清建筑轮廓的空间分辨率约为5 ~ 15m，看清行人的空间分辨率约为0.3 ~ 1m。而若要看清人的五官，空间分辨率则约为0.01 ~ 0.05m，这与对象（特征）的尺寸有关）提取了大量的图像特征。将这些图像特征展平（flatten），就构成了该图像的特征集合（feature maps）。

下述定义的conv_retriever类用于取回卷积神经网络中所有卷积层及其权重值（卷积核），取回的卷积层可以直接输入图像数据计算该卷积。例如，上述网络取回的卷积层有两个。函数visualize_convFilter则可以打印显示卷积核。函数visualize_convLayer则能打印显示指定数目的所有卷积图像结果。

```
class Conv_retriever:
    '''
    class - 取回卷积神经网络中所有卷积层及其权重
    '''
    def __init__(self, net):
        self.net = net
        self.model_weights = []  # 在这个列表中保存卷积层的权重
        self.conv_layers = []  # 在这个列表中保存 49 个卷积层
        # 获取所有的模型子节点并转化为列表
        self.model_children = list(net.children())
        self.counter = 0
    def retriever(self):
        import torch.nn as nn
        for i in range(len(self.model_children)):
            if type(self.model_children[i]) == nn.Conv2d:
                self.counter += 1
                self.model_weights.append(self.model_children[i].weight)
                self.conv_layers.append(self.model_children[i])
```

```
            elif type(self.model_children[i]) == nn.Sequential:
                for j in range(len(self.model_children[i])):
                    for child in self.model_children[i][j].children():
                        if type(child) == nn.Conv2d:
                            self.counter += 1
                            self.model_weights.append(child.weight)
                            self.conv_layers.append(child)
conv_retriever_ = Conv_retriever(net_fashionMNIST_)
conv_retriever_.retriever()
# print(conv_retriever_.conv_layers) #conv_retriever_.model_weights
for weight, conv in zip(conv_retriever_.model_weights, conv_retriever_.conv_
layers):
    print(f"CONV: {conv} ====> SHAPE: {weight.shape}")
--------------------------------------------------------------------------------
CONV: Conv2d(1, 6, kernel_size=(5, 5), stride=(1, 1)) ====> SHAPE: torch.
Size([6, 1, 5, 5])
CONV: Conv2d(6, 16, kernel_size=(5, 5), stride=(1, 1)) ====> SHAPE: torch.
Size([16, 6, 5, 5])
```

卷积核显示中，像素越黑，值越小，趋于0；而像素越亮，值越大，趋于255.0（图3.17–10）。因此，越白的像素，在卷积过程中对应位置图像像素的权重越大，对特征影响越重。

```
def visualize_convFilter(conv_layer, model_weight, output_name, figsize=(10,
10)):
    '''
    function - 可视化卷积核 （visualize the conv layer filters）
    '''
    import matplotlib.pyplot as plt
    plt.figure(figsize=figsize)
    kernel_size = conv_layer.kernel_size
    for i, filter in enumerate(model_weight):
        plt.subplot(kernel_size[0]+1, kernel_size[1]+1, i+1)
        plt.imshow(filter[0, :, :].detach(), cmap='gray')
        plt.axis('off')
        plt.savefig(r'./results/%s' % output_name)
    plt.show()
visualize_convFilter(
    conv_retriever_.conv_layers[0], conv_retriever_.model_weights[0], output_
name='fashion_MNIST_filter.png')
```

图3.17–10 卷积核

卷积（层）结果的显示（图3.17–11、图3.17–12），往往可以观察到不同的特征。例如，明显的黑色区域轮廓为商标标识，对象的轮廓也能够通过代表不同颜色的值区分开来等。这为观察不同的卷积核提取了图像哪些特征，为相关研究或者网络调试提供参考。

```
def visualize_convLayer(imgs_batch, conv_layers, model_weights, num_show=6,
figsize=(10, 10)):
    '''
    function - 可视化所有卷积层（卷积结果）
    '''
    import matplotlib.pyplot as plt
    results = [conv_layers[0](imgs_batch)]
    for i in range(1, len(conv_layers)):
        results.append(conv_layers[i](results[-1]))
    outputs = results
    for num_layer in range(len(outputs)):
        plt.figure(figsize=figsize)
        layer_viz = outputs[num_layer][0, :, :, :]
        layer_viz = layer_viz.data
        print("num_layer:{},layer.size={}".format(num_layer, layer_viz.size()))
        print()
        kernel_size = conv_layers[num_layer].kernel_size
        for i, filter in enumerate(layer_viz):
```

 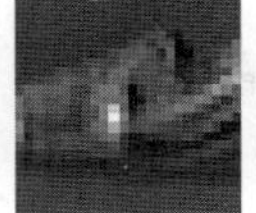  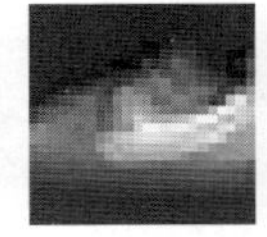  

图3.17-11 卷积结果（0层）

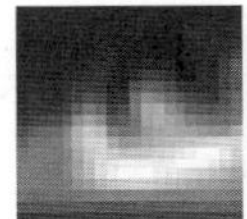  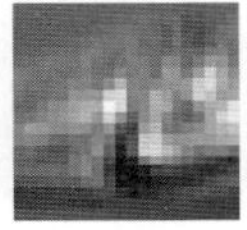 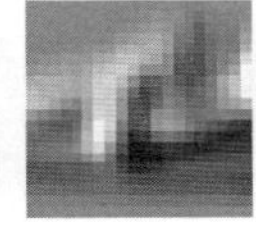  

图3.17-12 卷积结果（1层）

```
            if i == num_show:
                break
            plt.subplot(kernel_size[0]+1, kernel_size[1]+1, i+1)
            plt.imshow(filter, cmap='gray')
            plt.axis('off')
        print(f"Saving layer {num_layer} feature maps...")
        plt.savefig(f'./results/layer_{num_layer}.png')
        plt.show()
        # plt.close() # 如果只保存，不需要显示打印，则可以开启 plt.close()，并注释掉 plt.show()

visualize_convLayer(images, conv_retriever_.conv_layers,
                    conv_retriever_.model_weights, num_show=6)
--------------------------------------------------------------------------------
num_layer:0,layer.size=torch.Size([6, 24, 24])
Saving layer 0 feature maps...
num_layer:1,layer.size=torch.Size([16, 20, 20])
Saving layer 1 feature maps...
```

3.17.3 torchvision.models

torchvision.models库[①]包含有解决不同任务的预定义模型（通过配置参数pretrained=True，可以下载已经训练模型的参数），包括：image classification（图像分类）、pixelwise sementic segmentation（像素语义分割）、object detection（对象检测）、instance segmentation（实例分割）、person keypoint detection（人关键点检测）和video classification（视频分类）等。目前包括的模型如表3.17-1所示。

① torchvision.models，预定义深度学习模型，（https://pytorch.org/vision/0.8/models.html）。

表3.17-1 torchvision.models库模型列表

用途	模型/网络
Classification（分类）	AlexNet，VGG，ResNet，SqueezeNet，DenseNet，Inception v3，GoogLeNet，ShuffleNet v2，MobileNet v2，ResNeXt，Wide ResNet，MNASNet
Semantic Segmentation（语义分割）	FCN ResNet50，ResNet101；DeepLabV3 ResNet50，ResNet101
Object Detection（对象检测），Instance Segmentation and Person Keypoint Detection（实例分割和行人关键点检测）	Faster R-CNN ResNet-50 FPN，Mask R-CNN ResNet-50 FPN
Video classification（视频分类）	ResNet 3D，ResNet Mixed Convolution，ResNet（2+1）D

3.17.3.1 自定义VGG网络

VGG作者研究了在大规模图像识别设置中，卷积深度网络对精度的影响。其贡献是使用非常小的卷积核（3×3）滤波器的架构对深度增加的网络进行全面的评估，表面深度推进到16~19个权重层，可以实现对现有配置的显著改进。同时，该模型可以很好地泛化到其他数据集。VGG的网络结构可以看作不断重复的模块，因此在定义模型时可以先定义规律性的模块，然后给定配置的卷积层数（num_convs），输入通道数（in_channels）和输出通道数（out_channels），调用模块实现完整的架构，避免代码冗长。其中，包括5个卷积模块（block），前2块为单层卷积，后3层为双层卷积。该网络总共8个卷积层和3个全连接层，所以也称为VGG-11。

```
from torch import nn, optim
from torch import nn
import torch
def func_asterisk(*s_position, **d_keywords):
    print("*s_position={}\n**d_keywords={}".format(s_position, d_keywords))
func_asterisk('a', 'b', 'c', 'd', 'e', param_a=1, param_b=2, param_c=3)
 *s_position=('a', 'b', 'c', 'd', 'e')
 **d_keywords={'param_a': 1, 'param_b': 2, 'param_c': 3}

# 将每批次样本 X 的形状转换为 (batch_size,-1)
class FlattenLayer(nn.Module):
    def __init__(self):
        super(FlattenLayer, self).__init__()
    def forward(self, x):
        return x.view(x.shape[0], -1)
device = torch.device('cuda' if torch.cuda.is_available() else 'cpu')
def vgg_block(num_convs, in_channels, out_channels):
    blk = []
    for i in range(num_convs):
        if i == 0:
            blk.append(nn.Conv2d(in_channels, out_channels,
                        kernel_size=3, padding=1))
        else:
            blk.append(nn.Conv2d(out_channels, out_channels,
                        kernel_size=3, padding=1))
        blk.append(nn.ReLU())
    blk.append(nn.MaxPool2d(kernel_size=2, stride=2))
    return nn.Sequential(*blk)
conv_arch = ((1, 1, 64), (1, 64, 128), (2, 128, 256),
             (2, 256, 512), (2, 512, 512))
fc_features = 512*7*7
fc_hidden_units = 4096
def vgg(conv_arch, fc_features, fc_hidden_units=4096):
    net = nn.Sequential()
    for i, (num_convs, in_channels, out_channels) in enumerate(conv_arch):
        net.add_module('vgg_block_'+str(i+1),
                       vgg_block(num_convs, in_channels, out_channels))
    net.add_module('fc', nn.Sequential(FlattenLayer(),
                                 nn.Linear(fc_features, fc_hidden_units),
                                 nn.ReLU(),
                                 nn.Dropout(0.5),
                                 nn.Linear(fc_hidden_units,
                                           fc_hidden_units),
                                 nn.ReLU(),
                                 nn.Dropout(0.5),
                                 nn.Linear(fc_hidden_units, 10)
                                 ))
    return net
VGG_net = vgg(conv_arch, fc_features, fc_hidden_units)
print("VGG 网络结构：\n", VGG_net)
---------------------------------------------------------------------------
VGG 网络结构：
 Sequential(
  (vgg_block_1): Sequential(
    (0): Conv2d(1, 64, kernel_size=(3, 3), stride=(1, 1), padding=(1, 1))
    (1): ReLU()
    (2): MaxPool2d(kernel_size=2, stride=2, padding=0, dilation=1, ceil_
mode=False)
  )
  (vgg_block_2): Sequential(
    (0): Conv2d(64, 128, kernel_size=(3, 3), stride=(1, 1), padding=(1, 1))
    (1): ReLU()
    (2): MaxPool2d(kernel_size=2, stride=2, padding=0, dilation=1, ceil_
mode=False)
  )
  (vgg_block_3): Sequential(
```

```
      (0): Conv2d(128, 256, kernel_size=(3, 3), stride=(1, 1), padding=(1, 1))
      (1): ReLU()
      (2): Conv2d(256, 256, kernel_size=(3, 3), stride=(1, 1), padding=(1, 1))
      (3): ReLU()
      (4): MaxPool2d(kernel_size=2, stride=2, padding=0, dilation=1, ceil_
mode=False)
    )
    (vgg_block_4): Sequential(
      (0): Conv2d(256, 512, kernel_size=(3, 3), stride=(1, 1), padding=(1, 1))
      (1): ReLU()
      (2): Conv2d(512, 512, kernel_size=(3, 3), stride=(1, 1), padding=(1, 1))
      (3): ReLU()
      (4): MaxPool2d(kernel_size=2, stride=2, padding=0, dilation=1, ceil_
mode=False)
    )
    (vgg_block_5): Sequential(
      (0): Conv2d(512, 512, kernel_size=(3, 3), stride=(1, 1), padding=(1, 1))
      (1): ReLU()
      (2): Conv2d(512, 512, kernel_size=(3, 3), stride=(1, 1), padding=(1, 1))
      (3): ReLU()
      (4): MaxPool2d(kernel_size=2, stride=2, padding=0, dilation=1, ceil_
mode=False)
    )
    (fc): Sequential(
      (0): flattenLayer()
      (1): Linear(in_features=25088, out_features=4096, bias=True)
      (2): ReLU()
      (3): Dropout(p=0.5, inplace=False)
      (4): Linear(in_features=4096, out_features=4096, bias=True)
      (5): ReLU()
      (6): Dropout(p=0.5, inplace=False)
      (7): Linear(in_features=4096, out_features=10, bias=True)
    )
)
```

配置完卷积层（特征提取层），可以通过上述定义的conv2d_outputSize_A_oneTime函数，计算全连接层的输入尺寸。计算结果为(7,7)，将其与卷积层最后一层的输出通道数相乘就为全连接层的输入尺寸512 × 7 ×7 = 25088。这与VGG的输入值相同。

```
VGG_convs_params = [
    ('input', (224, 224)),
    ('conv', [3, 1, 1, 1]),   # kernel_size, stride, pad, dilation
    ('pool', [2, 2, 0, 1]),
    ('conv', [3, 1, 1, 1]),
    ('pool', [2, 2, 0, 1]),
    ('conv', [3, 1, 1, 1]),
    ('conv', [3, 1, 1, 1]),
    ('pool', [2, 2, 0, 1]),
    ('conv', [3, 1, 1, 1]),
    ('conv', [3, 1, 1, 1]),
    ('pool', [2, 2, 0, 1]),
    ('conv', [3, 1, 1, 1]),
    ('conv', [3, 1, 1, 1]),
    ('pool', [2, 2, 0, 1])
]
VGG_output_h_w = conv2d_outputSize_A_oneTime(VGG_convs_params)
print("卷积层输出尺寸={}".format(VGG_output_h_w))
---------------------------------------------------------------------------
卷积层输出尺寸=(7, 7)
```

可以先生成随机的样本数据（batchsize，nChannels，Height，Width），逐个循环每一模块，即Sequential对象，获取每一模块计算后数据的形状，从而观察、验证并用于辅助调整模型参数。

```
X = torch.rand(1, 1, 224, 224)
for name, blk in VGG_net.named_children():
    X = blk(X)
    print(name, 'output shape: ', X.shape)
---------------------------------------------------------------------------
vgg_block_1 output shape:  torch.Size([1, 64, 112, 112])
vgg_block_2 output shape:  torch.Size([1, 128, 56, 56])
vgg_block_3 output shape:  torch.Size([1, 256, 28, 28])
```

```
vgg_block_4 output shape:  torch.Size([1, 512, 14, 14])
vgg_block_5 output shape:  torch.Size([1, 512, 7, 7])
fc output shape:  torch.Size([1, 10])
```

VGG的原始输入图像尺寸为224，目前实验的数据为fashionMNIST数据集，一幅图像的尺寸为28 ×28，因此使用torchvision.transforms.Resize的方法调整图像大小。该方法已经包含于自定义load_fashionMNIST中，因此只需要配置输入参数resize=224就可以修改图像尺寸，满足VGG输入数据尺寸的要求。

因为卷积后的图像输出尺寸只与图像的输入尺寸，卷积层、池化层的卷积核大小，填充、步幅等有关，因此可以自由地修改卷积层的输入、输出通道，以及全连接层隐含层的数量。修改时，只需要配置一个缩放比例参数ratio，将所有的相关值除以该值完成新的配置。

```
ratio = 8
small_conv_arch = [(1, 1, 64//ratio), (1, 64//ratio, 128//ratio), (2, 128 //
                                                                   ratio, 256//
ratio), (2, 256//ratio, 512//ratio), (2, 512//ratio, 512//ratio)]
VGG_net_ = vgg(small_conv_arch, fc_features // ratio, fc_hidden_units // ratio)
print("VGG 网络结构 _ 减少通道数：\n", VGG_net_)
-----------------------------------------------------------------------------
VGG 网络结构 _ 减少通道数：
 Sequential(
  (vgg_block_1): Sequential(
    (0): Conv2d(1, 8, kernel_size=(3, 3), stride=(1, 1), padding=(1, 1))
    (1): ReLU()
    (2): MaxPool2d(kernel_size=2, stride=2, padding=0, dilation=1, ceil_
mode=False)
  )
  (vgg_block_2): Sequential(
    (0): Conv2d(8, 16, kernel_size=(3, 3), stride=(1, 1), padding=(1, 1))
    (1): ReLU()
    (2): MaxPool2d(kernel_size=2, stride=2, padding=0, dilation=1, ceil_
mode=False)
  )
  (vgg_block_3): Sequential(
    (0): Conv2d(16, 32, kernel_size=(3, 3), stride=(1, 1), padding=(1, 1))
    (1): ReLU()
    (2): Conv2d(32, 32, kernel_size=(3, 3), stride=(1, 1), padding=(1, 1))
    (3): ReLU()
    (4): MaxPool2d(kernel_size=2, stride=2, padding=0, dilation=1, ceil_
mode=False)
  )
  (vgg_block_4): Sequential(
    (0): Conv2d(32, 64, kernel_size=(3, 3), stride=(1, 1), padding=(1, 1))
    (1): ReLU()
    (2): Conv2d(64, 64, kernel_size=(3, 3), stride=(1, 1), padding=(1, 1))
    (3): ReLU()
    (4): MaxPool2d(kernel_size=2, stride=2, padding=0, dilation=1, ceil_
mode=False)
  )
  (vgg_block_5): Sequential(
    (0): Conv2d(64, 64, kernel_size=(3, 3), stride=(1, 1), padding=(1, 1))
    (1): ReLU()
    (2): Conv2d(64, 64, kernel_size=(3, 3), stride=(1, 1), padding=(1, 1))
    (3): ReLU()
    (4): MaxPool2d(kernel_size=2, stride=2, padding=0, dilation=1, ceil_
mode=False)
  )
  (fc): Sequential(
    (0): flattenLayer()
    (1): Linear(in_features=3136, out_features=512, bias=True)
    (2): ReLU()
    (3): Dropout(p=0.5, inplace=False)
    (4): Linear(in_features=512, out_features=512, bias=True)
    (5): ReLU()
    (6): Dropout(p=0.5, inplace=False)
    (7): Linear(in_features=512, out_features=10, bias=True)
  )
)
-----------------------------------------------------------------------------
batch_size = 64
trainloader, testloader = load_fashionMNIST(
```

```
    root='./dataset/FashionMNIST_norm', batchsize=batch_size, num_workers=2,
resize=224, n_mean=0.5, n_std=0.5)
def dataiter_view(dataiter):
    '''
    function - 查看可迭代数据形状
    '''
    dataiter_ = iter(dataiter)
    images_, labels_ = dataiter_.next()
    print('数据形状：', images_.shape)
    return images_, labels_
images_, labels_ = dataiter_view(testloader)
-----------------------------------------------------------------------------
数据形状： torch.Size([64, 1, 224, 224])
-----------------------------------------------------------------------------
lr, num_epochs = 0.001, 5
optimizer = torch.optim.Adam(VGG_net_.parameters(), lr=lr)
def evaluate_accuracy_V2(data_iter, net, device=None):
    '''
    function - 模型精度计算
    '''
    if device is None and isinstance(net, torch.nn.Module):
        device = list(net.parameters())[0].device  # 如果没指定 device 就使用 net 的 device
    acc_sum, n = 0.0, 0
    with torch.no_grad():
        for X, y in data_iter:
            if isinstance(net, torch.nn.Module):
                net.eval()  # 评估模式，会关闭 dropout
                acc_sum += (net(X.to(device)).argmax(dim=1) ==
                            y.to(device)).float().sum().cpu().item()
                net.train()  # 改回训练模式
            n += y.shape[0]
    return acc_sum/n
def train_v2(net, train_iter, test_iter, optimizer, device, num_epochs):
    '''
    function - 训练模型、v2 版
    '''
    from tqdm.auto import tqdm
    import time
    net = net.to(device)
    print("training on-", device)
    loss = torch.nn.CrossEntropyLoss()
    for epoch in tqdm(range(num_epochs)):
        train_l_sum, train_acc_sum, n, batch_count, start = 0.0, 0.0, 0, 0,
time.time()
        for X, y in train_iter:
            X = X.to(device)
            y = y.to(device)
            y_pred = net(X)
            l = loss(y_pred, y)
            optimizer.zero_grad()
            l.backward()
            optimizer.step()
            train_l_sum += l.cpu().item()
            train_acc_sum += (y_pred.argmax(dim=1) == y).sum().cpu().item()
            n += y.shape[0]
            batch_count += 1
        test_acc = evaluate_accuracy_V2(test_iter, net)
        print('epoch %d, loss %.4f, train acc %.3f, test acc %.3f, time %.1f sec' % (
            epoch + 1, train_l_sum / batch_count, train_acc_sum / n, test_acc,
time.time() - start))
train_v2(net=VGG_net_, train_iter=trainloader, test_iter=testloader,
         optimizer=optimizer, device=device, num_epochs=num_epochs)
-----------------------------------------------------------------------------
training on- cuda
  0%|          | 0/5 [00:00<?, ?it/s]
epoch 1, loss 0.6070, train acc 0.775, test acc 0.874, time 63.3 sec
...
epoch 5, loss 0.2180, train acc 0.922, test acc 0.915, time 107.7 sec
```

3.17.3.2 torchvision.models实现VGG网络

VGG网络包含于torchvision.models库中，因此无须自行配置网络，直接下载使用。

通常，模型也包含预先训练的模型参数，可以配置pretrained=True下载，下载到本地的位置为C:\Users\<your name>\.cache\torch\hub\checkpoints，后缀名为.pth。有时，直接下载的网络并不能直接应用到其他不同的数据集，例如fashionMNIST数据集。该数据集的图像为灰色，即只有一个通道；同时，只有10个标签。因此，需要对应层修改输入、输出大小。在Finetuning Torchvision Models[①]中，包含了模型参数微调的方法。对于VGG而言，修改卷积层，可以通过.features[idx]的方式读取；修改全连接层，可以通过.classifier[idx]的方式修改。需要修改的层为features[0]和classifier[6]这两个层。其他层不需要修改。

注意在训练模型前，需要调整优化函数optimizer=torch.optim.Adam(VGG_model.parameters(),lr=lr)的输入模型参数为当前的网络模型。

① Finetuning Torchvision Models，对torchvision模型进行微调和特征提取，(https://pytorch.org/tutorials/beginner/finetuning_torchvision_models_tutorial.html)。

```
import torchvision
VGG_model = torchvision.models.vgg11(pretrained=False) # 如果为真，返回在 ImageNet 上预先训练的模型
print(VGG_model)
--------------------------------------------------------------------------------
VGG(
  (features): Sequential(
    (0): Conv2d(3, 64, kernel_size=(3, 3), stride=(1, 1), padding=(1, 1))
    (1): ReLU(inplace=True)
    (2): MaxPool2d(kernel_size=2, stride=2, padding=0, dilation=1, ceil_
mode=False)
    (3): Conv2d(64, 128, kernel_size=(3, 3), stride=(1, 1), padding=(1, 1))
    (4): ReLU(inplace=True)
    (5): MaxPool2d(kernel_size=2, stride=2, padding=0, dilation=1, ceil_
mode=False)
    (6): Conv2d(128, 256, kernel_size=(3, 3), stride=(1, 1), padding=(1, 1))
    (7): ReLU(inplace=True)
    (8): Conv2d(256, 256, kernel_size=(3, 3), stride=(1, 1), padding=(1, 1))
    (9): ReLU(inplace=True)
    (10): MaxPool2d(kernel_size=2, stride=2, padding=0, dilation=1, ceil_
mode=False)
    (11): Conv2d(256, 512, kernel_size=(3, 3), stride=(1, 1), padding=(1, 1))
    (12): ReLU(inplace=True)
    (13): Conv2d(512, 512, kernel_size=(3, 3), stride=(1, 1), padding=(1, 1))
    (14): ReLU(inplace=True)
    (15): MaxPool2d(kernel_size=2, stride=2, padding=0, dilation=1, ceil_
mode=False)
    (16): Conv2d(512, 512, kernel_size=(3, 3), stride=(1, 1), padding=(1, 1))
    (17): ReLU(inplace=True)
    (18): Conv2d(512, 512, kernel_size=(3, 3), stride=(1, 1), padding=(1, 1))
    (19): ReLU(inplace=True)
    (20): MaxPool2d(kernel_size=2, stride=2, padding=0, dilation=1, ceil_
mode=False)
  )
  (avgpool): AdaptiveAvgPool2d(output_size=(7, 7))
  (classifier): Sequential(
    (0): Linear(in_features=25088, out_features=4096, bias=True)
    (1): ReLU(inplace=True)
    (2): Dropout(p=0.5, inplace=False)
    (3): Linear(in_features=4096, out_features=4096, bias=True)
    (4): ReLU(inplace=True)
    (5): Dropout(p=0.5, inplace=False)
    (6): Linear(in_features=4096, out_features=1000, bias=True)
  )
)
--------------------------------------------------------------------------------
from torch  import nn
VGG_model.features[0] = nn.Conv2d(1, 64, kernel_size=(3, 3), stride=(1, 1),
padding=(1, 1))
VGG_model.classifier[6] = nn.Linear(in_features=4096, out_features=10, bias=True)
print(VGG_model)
--------------------------------------------------------------------------------
VGG(
  (features): Sequential(
    (0): Conv2d(1, 64, kernel_size=(3, 3), stride=(1, 1), padding=(1, 1))
    (1): ReLU(inplace=True)
    (2): MaxPool2d(kernel_size=2, stride=2, padding=0, dilation=1, ceil_
mode=False)
    (3): Conv2d(64, 128, kernel_size=(3, 3), stride=(1, 1), padding=(1, 1))
    (4): ReLU(inplace=True)
```

```
    (5): MaxPool2d(kernel_size=2, stride=2, padding=0, dilation=1, ceil_
mode=False)
    (6): Conv2d(128, 256, kernel_size=(3, 3), stride=(1, 1), padding=(1, 1))
    (7): ReLU(inplace=True)
    (8): Conv2d(256, 256, kernel_size=(3, 3), stride=(1, 1), padding=(1, 1))
    (9): ReLU(inplace=True)
    (10): MaxPool2d(kernel_size=2, stride=2, padding=0, dilation=1, ceil_
mode=False)
    (11): Conv2d(256, 512, kernel_size=(3, 3), stride=(1, 1), padding=(1, 1))
    (12): ReLU(inplace=True)
    (13): Conv2d(512, 512, kernel_size=(3, 3), stride=(1, 1), padding=(1, 1))
    (14): ReLU(inplace=True)
    (15): MaxPool2d(kernel_size=2, stride=2, padding=0, dilation=1, ceil_
mode=False)
    (16): Conv2d(512, 512, kernel_size=(3, 3), stride=(1, 1), padding=(1, 1))
    (17): ReLU(inplace=True)
    (18): Conv2d(512, 512, kernel_size=(3, 3), stride=(1, 1), padding=(1, 1))
    (19): ReLU(inplace=True)
    (20): MaxPool2d(kernel_size=2, stride=2, padding=0, dilation=1, ceil_
mode=False)
  )
  (avgpool): AdaptiveAvgPool2d(output_size=(7, 7))
  (classifier): Sequential(
    (0): Linear(in_features=25088, out_features=4096, bias=True)
    (1): ReLU(inplace=True)
    (2): Dropout(p=0.5, inplace=False)
    (3): Linear(in_features=4096, out_features=4096, bias=True)
    (4): ReLU(inplace=True)
    (5): Dropout(p=0.5, inplace=False)
    (6): Linear(in_features=4096, out_features=10, bias=True)
  )
)
--------------------------------------------------------------------------------
import torch
device = torch.device('cuda' if torch.cuda.is_available() else 'cpu')
lr, num_epochs = 0.001, 5
optimizer = torch.optim.Adam(VGG_model.parameters(), lr=lr)
train_v2(net=VGG_model, train_iter=trainloader, test_iter=testloader,
         optimizer=optimizer, device=device, num_epochs=num_epochs)
------------------------------------------------------------
training on- cuda
  0%|          | 0/5 [00:00<?, ?it/s]
epoch 1, loss 0.6091, train acc 0.814, test acc 0.876, time 944.2 sec
epoch 2, loss 0.2923, train acc 0.893, test acc 0.895, time 1089.5 sec
epoch 3, loss 0.2509, train acc 0.910, test acc 0.908, time 1063.9 sec
epoch 4, loss 0.2229, train acc 0.919, test acc 0.913, time 1047.5 sec
epoch 5, loss 0.2006, train acc 0.926, test acc 0.919, time 1089.6 sec
```

参考文献（References）:

[1] Jibin Mathew (2020). PyTorch Artificial Intelligence Fundamentals. Packt.

[2] Convolutional Neural Networks, https://cs231n.github.io/convolutional-networks.

[3] Understanding 2D Dilated Convolution Operation with Examples in Numpy and Tensorflow with Interactive Code, https://towardsdatascience.com/understanding-2d-dilated-convolution-operation-with-examples-in-numpy-and-tensorflow-with-d376b3972b25.

[4] [Visualizing Models, Data, and Training with TensorBoard, <https://pytorch.org/tutorials/intermediate/tensorboard_tutorial.html?highlight=fashion%20mnist).

[5] [Visualizing Filters and Feature Maps in Convolutional Neural Networks using PyTorch, https://debuggercafe.com/visualizing-filters-and-feature-maps-in-convolutional-neural-networks-using-pytorch/.

[6] Simonyan, K., & Zisserman, A. (2015). Very Deep Convolutional Networks for Large-Scale Image Recognition. arXiv [Cs.CV]. Retrieved from http://arxiv.org/abs/1409.1556.

3.18 对象检测、实例分割与人流量估算和对象统计

在影像和视频处理方面，深度学习主要涉及图像分类（Image classification）、对象检测（Object detection）、迁移学习（Transfer learnig）、对抗生成（Adversarial generation：DCGAN/deep convolutional generative adversarial network）等。在城市空间分析研究中，每一个方向都能为其提供新的分析技术方法。对象检测是计算机视觉下，分析图像或影像，将其中的对象（例如车辆、行人、动物、桌椅、植被等，通常包括室外环境、室内环境，甚至某一对象的具体再分，例如人脸识别等）标记出来。通过图像对城市环境内对象的识别，可以统计对象的空间分布情况。同时，因为无人驾驶项目大量影像数据的获取，不仅可以对城市不连续的影像分析，亦可以通过连续影像的分析进一步获取目标对象的空间分布变化情况。在下文的分析中，对于对象的分析锁定在两个方向，一个是仅识别行人，实现人流量估计；再者是尽量多的识别对象，分析城市空间下各个对象的分布变化情况，以及对象之间的联系。

目前，PyTorch已经集合了大量分析算法模型，而且提供预训练模型的参数，一定程度上可以跨过训练阶段，直接用于预测。PyTorch也提供了大量参考的代码（还有大量的著作教程）。这都为研究者更快速和轻松地应用已有模型研究提供了便利，从而能够快速地将其应用到各自的研究领域当中，而不是计算机视觉、深度学习算法本身研究的专业。

3.18.1 对象/目标检测（行人）与人流量估算

PyTorch给出的教程（例如TorchVision Object Detection Finetuning Tutorial（对象检测微调教程））通常也提供了Googel Colab版本[①]，可以直接在线运行代码，解决当前没有配置GPU或GPU算力较低的情况，使得深度学习可以快速地反馈运行结果。

微调Penn-Fudan Database [for Pedestrian Detection and Segmentation（用于行人检测和分割）][②]数据库中预训练的Mask R-CNN，用于行人检测和分割。该数据库包含170张图像和345个行人实例。在人流量估算研究中，直接应用Mask R-CNN训练的模型。对该方法的解释直接参看官方内容，仅是保留关键信息的解释。重点在于，通过对行人的目标检测，以KITTI数据集[③]为例，计算行人的空间分布核密度，估算人流量。

3.18.1.1 对象检测（行人）

用!符号直接运行conda（作为shell命令）。有些终端命令，如果无法在JupyterLab下直接实现，则打开Anaconda的终端（terminal）执行。

```
! pip install cython
--------------------------------------------------------------------------------
Requirement already satisfied: cython in c:\users\richi\anaconda3\envs\usda_
database\lib\site-packages (0.29.26)
```

可以直接在浏览器下输入https://www.cis.upenn.edu/~jshi/ped_html/PennFudanPed.zip下载，也可以在终端执行`wget https://www.cis.upenn.edu/~jshi/ped_html/PennFudanPed.zip`，并且解压缩`unzip PennFudanPed.zip`获取PennFudan数据集[④]。其数据库文件夹结构如下：

```
PennFudanPed/
  PedMasks/
    FudanPed00001_mask.png
    FudanPed00002_mask.png
    FudanPed00003_mask.png
    FudanPed00004_mask.png
    ...
  PNGImages/
    FudanPed00001.png
    FudanPed00002.png
    FudanPed00003.png
    FudanPed00004.png
```

① TorchVision Object Detection Finetuning Tutorial的Googel Colab版本，（https://colab.research.google.com/github/pytorch/tutorials/blob/gh-pages/_downloads/torchvision_finetuning_instance_segmentation.ipynb）。

② Penn-Fudan Database（for Pedestrian Detection and Segmentation，（https://www.cis.upenn.edu/~jshi/ped_html）。

③ KITTI数据集，由德国卡尔斯鲁厄理工学院（Karlsruhe Institute of Technology（KIT））和丰田美国技术研究院（Toyota Technological Institute at Chicago（TTI-C））联合创办，是目前国际上最大的自动驾驶场景下的计算机视觉算法评测数据集。该数据集用于评测立体图像（stereo）、光流（optical flow）、视觉测距（visual odometry）、3D物体检测（object detection）和3D跟踪（tracking）等计算机视觉技术在车载环境下的性能。KITTI包含市区、乡村和高速公路等场景采集的真实图像数据，每张图像中最多达15辆车和30个行人，还有各种程度的遮挡与截断，（https://www.cvlibs.net/datasets/kitti）。

④ PennFudan数据集，用于行人检测的图像数据库。取自校园和城市街道等场景，每张图片中至少有一个行人（https://www.cis.upenn.edu/~jshi/ped_html）。

直接打开一张图像（图3.18–1）和其对应的对象分割掩码（图3.18–2）。

图3.18-1 PennFudan数据集图像示例

```
from PIL import Image
import os
PennFudanPed_fp = './dataset/PennFudanPed'
Image.open(os.path.join(PennFudanPed_fp,'PNGImages/PennPed00019.png'))
mask = Image.open(os.path.join(PennFudanPed_fp, 'PedMasks/PennPed00019_mask.
png')
                  ).convert('P')  # Open mask with Pillow, and convert to mode 'P'
mask.putpalette([
    0, 0, 0,        # black background
    255, 0, 0,      # index 1 is red
    255, 255, 0,    # index 2 is yellow
    255, 153, 0,    # index 3 is orange
])
mask
```

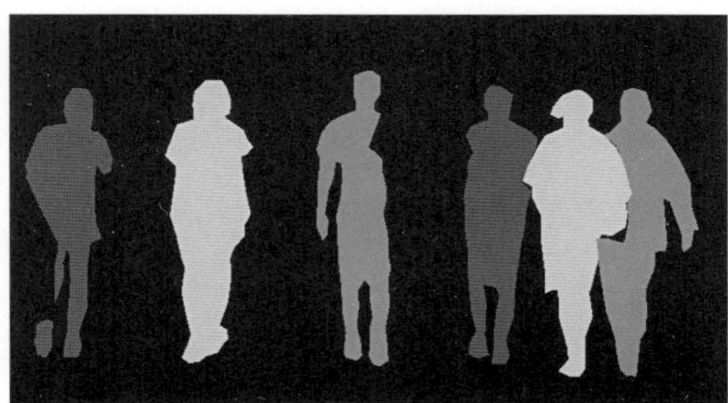

图3.18-2 PennFudan数据集对象分割掩码

定义继承父类torch.utils.data.Dataset，建立数据集（tensor数据类型）作为torch.utils.data.DataLoader输入生成可迭代对象，用于训练模型数据加载的类。

```
import os
import numpy as np
import torch
import torch.utils.data
from PIL import Image
class PennFudanDataset(torch.utils.data.Dataset):
    def __init__(self, root, transforms=None):
        self.root = root
        self.transforms = transforms
        # 加载所有图像文件，并且对他们进行排序，确保图像和掩码文件对应
        self.imgs = list(sorted(os.listdir(os.path.join(root, "PNGImages"))))
        self.masks = list(sorted(os.listdir(os.path.join(root, "PedMasks"))))
    def __getitem__(self, idx):
        # 加载图像和掩码文件
        img_path = os.path.join(self.root, "PNGImages", self.imgs[idx])
        mask_path = os.path.join(self.root, "PedMasks", self.masks[idx])
```

```
        img = Image.open(img_path).convert("RGB")
        # 因为每种颜色对应不同的实例，因此没有将掩码转换为 RGB,
        # 值 0 为背景色
        mask = Image.open(mask_path)
        mask = np.array(mask)
        # 编码实例为不同的颜色
        obj_ids = np.unique(mask)
        # 移除背景
        obj_ids = obj_ids[1:]

        # 将颜色编码的掩码转为一组二进制掩码
        masks = mask == obj_ids[:, None, None]

        # 获取每个掩码的边界框坐标
        num_objs = len(obj_ids)
        boxes = []
        for i in range(num_objs):
            pos = np.where(masks[i])
            xmin = np.min(pos[1])
            xmax = np.max(pos[1])
            ymin = np.min(pos[0])
            ymax = np.max(pos[0])
            boxes.append([xmin, ymin, xmax, ymax])
        boxes = torch.as_tensor(boxes, dtype=torch.float32)
        # 仅有一个类
        labels = torch.ones((num_objs,), dtype=torch.int64)
        masks = torch.as_tensor(masks, dtype=torch.uint8)
        image_id = torch.tensor([idx])
        area = (boxes[:, 3] - boxes[:, 1]) * (boxes[:, 2] - boxes[:, 0])
        # 假设所有实例都不是重叠
        iscrowd = torch.zeros((num_objs,), dtype=torch.int64)
        target = {}
        target["boxes"] = boxes
        target["labels"] = labels
        target["masks"] = masks
        target["image_id"] = image_id
        target["area"] = area
        target["iscrowd"] = iscrowd
        if self.transforms is not None:
            img, target = self.transforms(img, target)
        return img, target
    def __len__(self):
        return len(self.imgs)
```

返回定义的数据集包含一个PIL.Image和一个字典，包含boxes锚框、labels类标、masks分割掩码、image_id图像索引及area锚框面积和iscrowd。

```
dataset = PennFudanDataset(PennFudanPed_fp)
dataset[0]
------------------------------------------------------------------
(<PIL.Image.Image image mode=RGB size=559x536 at 0x1D3DBC2BD00>,
 {'boxes': tensor([[159., 181., 301., 430.],
         [419., 170., 534., 485.]]),
  'labels': tensor([1, 1]),
  'masks': tensor([[[0, 0, 0,  ..., 0, 0, 0],
          [0, 0, 0,  ..., 0, 0, 0],
          [0, 0, 0,  ..., 0, 0, 0],
          ...,
          [0, 0, 0,  ..., 0, 0, 0],
          [0, 0, 0,  ..., 0, 0, 0],
          [0, 0, 0,  ..., 0, 0, 0]],
          [[0, 0, 0,  ..., 0, 0, 0],
          [0, 0, 0,  ..., 0, 0, 0],
          [0, 0, 0,  ..., 0, 0, 0],
          ...,
          [0, 0, 0,  ..., 0, 0, 0],
          [0, 0, 0,  ..., 0, 0, 0],
          [0, 0, 0,  ..., 0, 0, 0]]], dtype=torch.uint8),
  'image_id': tensor([0]),
  'area': tensor([35358., 36225.]),
  'iscrowd': tensor([0, 0])})
```

torchvision.models.detection提供了fasterrcnn_resnet50_fpn、maskrcnn_resnet50_fpn

（含对象分割/对象掩码（mask））对象检测模型，包含基于COCO数据集[①]预先训练的参数。针对特定类别对其微调（finetune）。

COCO数据集，是一个大规模对象检测（object detection）、分割（segmentation）和标注（captioning）数据集，其特征有：

1. 对象分割（Object segmentation）；
2. 上下文识别（Recognition in context）；
3. 超像素分割（Superpixel stuff segmentation）；
4. 330K张图片（>200K个标签）（330K images（>200K labeled））；
5. 150万个对象实例（1.5 million object instances）；
6. 80个对象类（80 object categories）；
7. 91个分割类（91 stuff categories）[②]；
8. 每张图像5个标注（5 captions per image）；
9. 25万人带有关键点（250,000 people with keypoints）。

可以从COCO官网下载数据查看其对象的分类，也可以从http://images.cocodataset.org/annotations/annotations_trainval2014.zip处下载2014年，以及从http://images.cocodataset.org/annotations/annotations_trainval2017.zip下载2017年数据集。解压后打开instances_val2017.json，打印查看对象分类。可以看到其中分类person，其ID为1。

```
import json
annotations_trainval2017_fp = r'./data/instances_val2017.json'
with open(annotations_trainval2017_fp, 'r') as f:
    annotations = json.loads(f.read())
    object_categories = json.dumps(annotations['categories'])
print('COCO 对象分类：\n', object_categories)
---------------------------------------------------------------------------
COCO 对象分类：
 [{"supercategory": "person", "id": 1, "name": "person"}, {"supercategory": "vehicle", "id": 2, "name": "bicycle"}, {"supercategory": "vehicle", "id": 3, "name": "car"}, {"supercategory": "vehicle", "id": 4, "name": "motorcycle"}, {"supercategory": "vehicle", "id": 5, "name": "airplane"}, {"supercategory": "vehicle", "id": 6, "name": "bus"}, {"supercategory": "vehicle", "id": 7, "name": "train"}, {"supercategory": "vehicle", "id": 8, "name": "truck"},...{"supercategory": "indoor", "id": 89, "name": "hair drier"}, {"supercategory": "indoor", "id": 90, "name": "toothbrush"}]
```

精调模型（finetuning）。

```
import torchvision
from torchvision.models.detection.faster_rcnn import FastRCNNPredictor
from torchvision.models.detection.mask_rcnn import MaskRCNNPredictor
def get_instance_segmentation_model(num_classes):
    # 01- 加载在 COCO 上预训练的实例分割模型
    model = torchvision.models.detection.maskrcnn_resnet50_fpn(pretrained=True)

    # 02- 获取分类器的输入特征数
    in_features = model.roi_heads.box_predictor.cls_score.in_features
    # 03- 用新的模型头（model head）替换预训练的模型头
    # 因为仅检测行人分类，包含背景总共两类，用其替换原输入特征数 in_features(1024)
    model.roi_heads.box_predictor = FastRCNNPredictor(in_features, num_classes)

    # 04- 获得掩码分类器的输入特征数
    in_features_mask = model.roi_heads.mask_predictor.conv5_mask.in_channels
    hidden_layer = 256
    # 05- 替换掩码预测器
    model.roi_heads.mask_predictor = MaskRCNNPredictor(
        in_features_mask, hidden_layer, num_classes)
    return model
```

可以下载https://github.com/pytorch/vision.git代码，在references\detection下包含有大量帮助函数，可以简化对象/目标检测训练和评估。将对应的utils.py，transforms.py，coco_eval.py，engine.py，coco_utils.py这5个文件，直接复制到该文件（.ipynb）所在的目录下，执行调入。定义的get_transform函数，可以实现对图像指定方式的变换操作，即图像增广（image augmentation）。PyTorh的transforms[③]提供了大量图像增广的方法。这里，使用了RandomHorizontalFlip实现图像的随机水平向翻转，并且采用ToTensor()方法将图像转换为tensor数据类型。

① COCO数据集，是一个大规模对象检测（object detection）、分割（segmentation）和标注（captioning）数据集（https://cocodataset.org/#home）。

② COCO数据集91个分割类（https://github.com/nightrome/cocostuff）。

③ torchvision.transforms，（https://pytorch.org/vision/0.9/transforms.html）。

```
# 安装依赖库（pip install pycocotools-windows）
from engine import train_one_epoch, evaluate
import utils
import transforms as T
def get_transform(train):
    transforms = []
    # 将图像 (PIL 图像 ) 转换为 PyTorch 张量
    transforms.append(T.ToTensor())
    if train:
        # 在训练过程中，随机翻转训练图像和分割淹没（ground-truth）进行数据增强
        transforms.append(T.RandomHorizontalFlip(0.5))
    return T.Compose(transforms)
```

将建立的数据集通过torch.utils.data.DataLoader方法加载为可迭代对象，用于训练模型的输入数据。

```
dataset = PennFudanDataset(PennFudanPed_fp, get_transform(train=True))
dataset_test = PennFudanDataset(PennFudanPed_fp, get_transform(train=False))
torch.manual_seed(1)
indices = torch.randperm(len(dataset)).tolist()
dataset = torch.utils.data.Subset(dataset, indices[:-50])
dataset_test = torch.utils.data.Subset(dataset_test, indices[-50:])
data_loader = torch.utils.data.DataLoader(
    dataset, batch_size=2, shuffle=True, num_workers=0, collate_fn=utils.
collate_fn)
data_loader_test = torch.utils.data.DataLoader(
    dataset_test, batch_size=1, shuffle=False, num_workers=0, collate_fn=utils.
collate_fn)
```

定义训练模型、优化函数、学习率。指定GPU或者CPU，训练模型。

```
from tqdm.auto import tqdm
device = torch.device('cuda') if torch.cuda.is_available(
) else torch.device('cpu')
num_classes = 2
model = get_instance_segmentation_model(num_classes)
model.to(device)
params = [p for p in model.parameters() if p.requires_grad]  # 提取含梯度的参数
optimizer = torch.optim.SGD(
    params, lr=0.005, momentum=0.9, weight_decay=0.0005)  # 对含梯度的参数执行梯度下
降法 -SGD
# 根据 epoch 训练次数调整学习率的方法。PyTorch 也提供了 torch.optim.lr_scheduler.ReduceLROnPlateau，基于
训练中某些测量值使学习率下降的方法。
lr_scheduler = torch.optim.lr_scheduler.StepLR(
    optimizer, step_size=3, gamma=0.1)
num_epochs = 10
for epoch in tqdm(range(num_epochs)):
    # 训练一个 epoch（全部样本），每 10 次迭代（Iteration）打印一次
    train_one_epoch(model, optimizer, data_loader,
                    device, epoch, print_freq=999)
    # 更新学习率
    lr_scheduler.step()
    # 在测试数据集上进行评估
    if epoch == 5 or epoch == 9:
        evaluate(model, data_loader_test, device=device)
        # 仅保存模型的状态字典 (state_dict)，state_dict 由 PyTorch 自动生成，包含各层可训练参数（通常为
卷积层、线性层），例如权值、偏置等。
        torch.save(model.state_dict(
        ), './model/mask_R_CNN_person/mask_R_CNN_person_stateDict_{}.pth'.
format(epoch))
torch.save(model, './model/mask_R_CNN_person/mask_R_CNN_person_final.pth')  # 保存
整个模型
...
Epoch: [9]  [59/60]  eta: 0:00:01  lr: 0.000005  loss: 0.1484 (0.1511)  loss_
classifier: 0.0188 (0.0194)  loss_box_reg: 0.0263 (0.0252)  loss_mask: 0.0994
(0.1037)  loss_objectness: 0.0003 (0.0007)  loss_rpn_box_reg: 0.0018 (0.0020)
time: 1.5774  data: 0.0319  max mem: 3598
Epoch: [9] Total time: 0:01:36 (1.6148 s / it)
creating index...
index created!
Test:  [ 0/50]  eta: 0:00:07  model_time: 0.1326 (0.1326)  evaluator_time: 0.0030
(0.0030)  time: 0.1476  data: 0.0110  max mem: 3598
Test:  [49/50]  eta: 0:00:00  model_time: 0.1865 (0.1704)  evaluator_time: 0.0040
(0.0056)  time: 0.2054  data: 0.0176  max mem: 3598
```

```
Test: Total time: 0:00:09 (0.1954 s / it)
Averaged stats: model_time: 0.1865 (0.1704)  evaluator_time: 0.0040 (0.0056)
Accumulating evaluation results...
DONE (t=0.01s).
Accumulating evaluation results...
DONE (t=0.01s).
IoU metric: bbox
...
  Average Precision  (AP) @[ IoU=0.50:0.95 | area= large | maxDets=100 ] = 0.772
...
 Average Recall     (AR) @[ IoU=0.50:0.95 | area= large | maxDets=100 ] = 0.813
--------------------------------------------------------------------------------
model_ = get_instance_segmentation_model(num_classes)
model_.load_state_dict(torch.load('./model/mask_R_CNN_person/mask_R_CNN_person_
stateDict_9.pth'))
--------------------------------------------------------------------------------
<All keys matched successfully>
```

Intersection over Union(IoU)交并比，用于评价对象检测、图像分割的精度，计算公式如图3.18–3所示。由于模型的参数变化，例如图像金字塔尺度（image pyramid scale）、滑动窗口大小（sliding window size，即卷积核）、特征提取方法（feature extraction method）等，预测的边界（或锚框）与实际情况（ground-truth）完全匹配是不现实的。而IoU评估指标能很好地表述预测的精度。

除了IoU评价指标，同时提取一幅图像3.18–4测试，测试结果显示能够很好地提取图像场景中的行人对象（图3.18–5）。

注意将模型和数据均转化到指定的同一device（GPU或CPU）下，否则提示数据类型错误。

```
img, _ = dataset_test[34]
model_.to(device)
model_.eval()
with torch.no_grad():
    prediction = model_([img.to(device)])
Image.fromarray(img.mul(255).permute(1, 2, 0).byte().numpy())
```

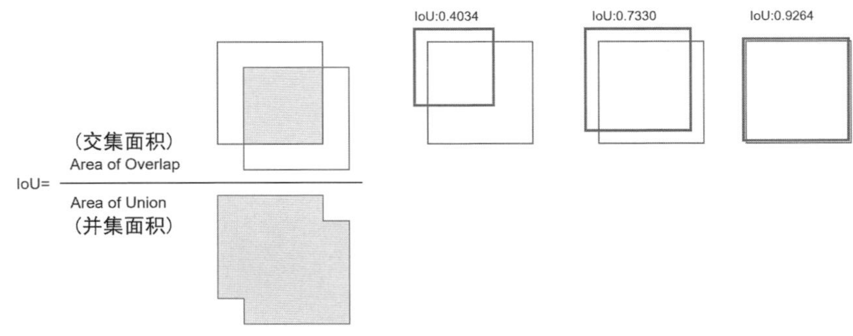

图3.18–3 交并比（IoU）图解

图3.18–4 PennFudan数据集一幅图像

图3.18-5 行人掩码提取结果

行人掩码包含在返回预测值的masks键下，其数据形状为torch.Size([12,1,378,745])。其中，12为预测的行人数量，这与实际的基本吻合（行人中往往存在互相遮掩的情况），可以用于人流量的统计。

```
import torchvision.transforms as transforms
print('掩码形状：',prediction[0]['masks'].shape)
transforms.ToPILImage()(torch.sum(prediction[0]['masks'],dim=0)).convert("RGB")
# Image.fromarray(prediction[0]['masks'][0, 0].mul(255).byte().cpu().numpy())
-------------------------------------------------------------------------------
掩码形状： torch.Size([10, 1, 378, 745])
```

3.18.1.2 **人流量估算**

直接将上述训练后的模型用于无人驾驶场景KITTI数据集，先随机提取一幅含有行人的图3.18-6，用该模型预测，查看预测结果（图3.18-7）。如果基本吻合，则说明该模型可以用于进一步的人流量分析。从观察结果来看，基本能够提取出场景内的行人。

```
img_kitti_fp = r'./data/0000000181.png'
img_kitti = Image.open(img_kitti_fp)
img_kitti
import torchvision.transforms as transforms
device = torch.device(
    'cuda') if torch.cuda.is_available() else torch.device('cpu')
print('device:{}'.format(device))

trans_2tensor = transforms.Compose([transforms.ToTensor(),])  #将图像转化为张量（tensor）
img_kitti_tensor = trans_2tensor(img_kitti)
model_.to(device)
model_.eval()
```

图3.18-6 KITTI数据集影像

图3.18-7 KITTI数据集一幅影像行人掩码提取结果

```
with torch.no_grad():
    img_kitti_pred = model_([img_kitti_tensor.to(device)])
print('估计行人数量={}'.format(img_kitti_pred[0]['masks'].shape[0]))
transforms.ToPILImage()(
    torch.sum(img_kitti_pred[0]['masks'], dim=0)).convert("RGB")
--------------------------------------------------------------------------------
device:cuda
估计行人数量=16
```

● A-提取KITTI数据集图像的位置坐标（表3.18-1）。

```
import util_misc
import util_A
KITTI_info_fp = r'G:\data\2011_09_29_drive_0071_sync\oxts\data'
timestamps_fp = r'G:\data\2011_09_29_drive_0071_sync\image_03\timestamps.txt'
drive_29_0071_info = util_A.KITTI_info(KITTI_info_fp, timestamps_fp)
drive_29_0071_info_coordi = drive_29_0071_info[['lat', 'lon', 'timestamps_']]
util_misc.print_html(drive_29_0071_info_coordi)
```

表3.18-1 KITTI数据集图像位置坐标

	lat	lon	timestamps_
0	49.008650	8.398092	2011-09-29 13:54:59.990872576
1	49.008777	8.397611	2011-09-29 13:55:00.094612992
2	49.009162	8.396541	2011-09-29 13:55:00.198486528
3	49.008962	8.397075	2011-09-29 13:55:00.302340864
4	49.009505	8.395251	2011-09-29 13:55:00.406079232

● B-提取KITTI数据集图像（2011_09_29_drive_0071_sync数据子集），并转换为tensor。

```
import os
from PIL import Image
import torchvision.transforms as transforms
from tqdm.auto import tqdm
drive_29_0071_img_fp = util_misc.filePath_extraction(
    r'G:\data\2011_09_29_drive_0071_sync\image_03\data', ['png'])
drive_29_0071_img_fp_list = util_misc.flatten_lst([[os.path.join(
    k, f) for f in drive_29_0071_img_fp[k]] for k, v in drive_29_0071_img_
fp.items()])
print("2011_09_29_drive_0071_sync-数据子集图像数据：", len(drive_29_0071_img_fp_
list))

trans_2tensor = transforms.Compose([transforms.ToTensor(),])  #将图像转化为张量
（tensor）
drive_29_0071_img_tensor = [trans_2tensor(
    Image.open(i)) for i in tqdm(drive_29_0071_img_fp_list)]
--------------------------------------------------------------------------------
2011_09_29_drive_0071_sync-数据子集图像数据：1059
  0%|          | 0/1059 [00:00<?, ?it/s]
```

● C-加载已训练的模型，并预测数据集图像

```
import torch
device = torch.device(
    'cuda') if torch.cuda.is_available() else torch.device('cpu')
model_entire = torch.load(
    './model/mask_R_CNN_person/mask_R_CNN_person_final.pth')
model_entire.eval()
with torch.no_grad():
    drive_29_0071_img_pred = [model_entire(
        [i.to(device)])[0]['masks'].shape[0] for i in tqdm(drive_29_0071_img_
tensor)]
--------------------------------------------------------------------------------
  0%|          | 0/1059 [00:00<?, ?it/s]
```

● D-显示人流分布密度（图3.18-8）。

```
import plotly.express as px
```

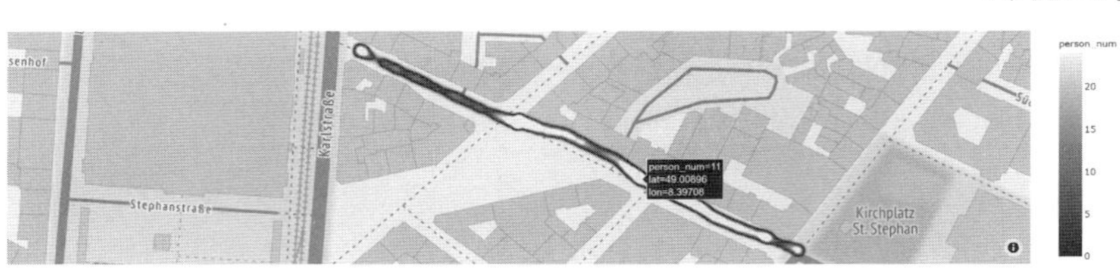

图3.18-8 人流分布密度

```
drive_29_0071_info_coordi['person_num'] = drive_29_0071_img_pred
fig = px.density_mapbox(drive_29_0071_info_coordi, lat='lat', lon='lon',
z='person_num', radius=10,
                        center=dict(lat=49.008645, lon=8.398104), zoom=18,
                        mapbox_style="stamen-terrain")
fig.show()
```

3.18.2 对象实例分割与对象统计

PyTorch图像/语义分割模型（semantic segmentation），包括Faster R-CNN ResNet-50 FPN，Mask R-CNN ResNet-50 FPN，其预先训练的模型采用的数据集为COCO train2017的子集Pascal VOC[①]，包括有20个分类。['__background__', 'aeroplane', 'bicycle', 'bird', 'boat', 'bottle', 'bus', 'car', 'cat', 'chair', 'cow', 'diningtable', 'dog', 'horse', 'motorbike', 'person', 'pottedplant', 'sheep', 'sofa', 'train', 'tvmonitor']等。使用预先训练的模型，输入的图像期望已训练时相同的方式归一化处理。图像映射到[0,1]区间，使用mean =[0.485, 0.456, 0.406]和std =[0.229, 0.224, 0.225]进行归一化。图像变换的处理使用torchvision.transforms实现。

① The PASCAL Visual Object Classes Homepage，(http://host.robots.ox.ac.uk/pascal/VOC)。

通过图像分割可以获取一些城市对象（Pascal VOC有20个分类），对于KITTI数据集而言，就可以获取每一位置的城市对象内容。这样，就可以对城市空间的内容加以统计，并通过建立关联结构分析对象之间的关系（即在多处场景中，同时出现某些对象的可能性大小）。

#关于 TORCH.SQUEEZE[②]

② 关于TORCH.SQUEEZE，(https://pytorch.org/docs/stable/generated/torch.squeeze.html)。

可以根据指定轴，移除轴尺寸为1的轴，与torch.unsqueeze互逆。

```
x = torch.zeros(2, 1, 2, 1, 2)
print('x.size:{}\ndim=default:{}\ndim=0:{}\ndim=1:{}'.format(x.size(), torch.
squeeze(
    x).size(), torch.squeeze(x, dim=0).size(), torch.squeeze(x, dim=1).size()))
--------------------------------------------------------------------------------
x.size:torch.Size([2, 1, 2, 1, 2])
dim=default:torch.Size([2, 2, 2])
dim=0:torch.Size([2, 1, 2, 1, 2])
dim=1:torch.Size([2, 2, 1, 2])
```

3.18.2.1 FCN-RESNET101对象实例分割

直接读取PyTorch预训练的模型，用于新场景的应用。

```
import torch
x = torch.tensor([1, 2, 3, 4])
print('dim=0',torch.unsqueeze(x, 0))
print('dim=1',torch.unsqueeze(x, 1))
--------------------------------------------------------------------------------
dim=0 tensor([[1, 2, 3, 4]])
dim=1 tensor([[1],
        [2],
        [3],
        [4]])
--------------------------------------------------------------------------------
```

- A-加载模型。

```
import torch
```

```
import os
import util_misc
from torchvision import models
fcn = models.segmentation.fcn_resnet101(pretrained=True).eval()
```

- B-加载数据集。

```
drive_29_0071_img_fp = util_misc.filePath_extraction(
    r'G:\data\2011_09_29_drive_0071_sync\image_03\data', ['png'])
drive_29_0071_img_fp_list = util_misc.flatten_lst([[os.path.join(
    k, f) for f in drive_29_0071_img_fp[k]] for k, v in drive_29_0071_img_
fp.items()])
# 指定运算设备 ,GPU 或 CPU
device = torch.device(
    'cuda') if torch.cuda.is_available() else torch.device('cpu')
------------------------------------------------------------------------------
Downloading: "https://download.pytorch.org/models/fcn_resnet101_coco-7ecb50ca.
pth" to C:\Users\richi/.cache\torch\hub\checkpoints\fcn_resnet101_coco-7ecb50ca.
pth
  0%|          | 0.00/208M [00:00<?, ?B/s]
```

- C-映射图像分割的类标（类别）。

```
def decode_segmap_FCN_RESNET101(image, nc=21):
    '''
    function - fcn_resnet101 模型，图像分割的类别给予颜色标识
    '''
    import numpy as np
    label_colors = np.array([(0, 0, 0),  # 0=background
                             # 1=aeroplane, 2=bicycle, 3=bird, 4=boat, 5=bottle
                             (128, 0, 0), (0, 128, 0), (128, 128,
                                                        0), (0, 0, 128), (128, 0,
128),
                             # 6=bus, 7=car, 8=cat, 9=chair, 10=cow
                             (0, 128, 128), (128, 128, 128), (64,
                                                              0, 0), (192, 0, 0),
(64, 128, 0),
                             # 11=dining table, 12=dog, 13=horse, 14=motorbike, 15=person
                             (192, 128, 0), (64, 0, 128), (192, 0,
                                                           128), (64, 128, 128),
(192, 128, 128),
                             # 16=potted plant, 17=sheep, 18=sofa, 19=train, 20=tv/monitor
                            (0, 64, 0), (128, 64, 0), (0, 192, 0), (128, 192, 0),
(0, 64, 128)])
    r = np.zeros_like(image).astype(np.uint8)
    g = np.zeros_like(image).astype(np.uint8)
    b = np.zeros_like(image).astype(np.uint8)
    for l in range(0, nc):
        idx = image == l
        r[idx] = label_colors[l, 0]
        g[idx] = label_colors[l, 1]
        b[idx] = label_colors[l, 2]
    rgb = np.stack([r, g, b], axis=2)
    return rgb
```

- D-预测图像分割结果，并打印。

```
def segmentation_FCN_RESNET101_plot(net, path, show_orig=True, dev='cuda', img_
resize=640, figsize=(20, 20)):
    '''
    function - 应用 torchvision.models.segmentation.fcn_resnet101 预测图像，并打印显
示分割预测结果
    '''
    import matplotlib.pyplot as plt
    from PIL import Image
    import torch
    import torchvision.transforms as T
    img = Image.open(path)
    plt.figure(figsize=figsize)
    if show_orig:
        plt.imshow(img)
        plt.axis('off')
        plt.show()
    # 通过图像增强以获得更好的推断结果
```

```
    trf = T.Compose([
        T.Resize(img_resize),
        # T.CenterCrop(224),
        T.ToTensor(),
        T.Normalize(mean=[0.485, 0.456, 0.406], std=[0.229, 0.224, 0.225])])
    inp = trf(img).unsqueeze(0).to(dev)
    out = net.to(dev)(inp)['out']
    om = torch.argmax(out.squeeze(), dim=0).detach().cpu().numpy()
    rgb = decode_segmap_FCN_RESNET101(om)
    plt.figure(figsize=figsize)
    plt.imshow(rgb)
    plt.axis('off')
    plt.show()
```

- E-提取一幅图像如图3.18–9，预测结果为图3.18–10

```
# 可以通过调整 img_resize 参数，即调整图像大小来减少 GPU 使用量，避免 GPU 溢出
segmentation_FCN_RESNET101_plot(
    fcn, drive_29_0071_img_fp_list[550], dev=device, img_resize=480)
```

图3.18-9　KITTI数据集影像

图3.18-10　KITTI数据集影像对象实例分割结果

- F-计算KITTI-drive_29_0071_img子集的所有图像分割，返回结果并动态显示（图3.18–11）。

```
from tqdm.auto import tqdm
from IPython.display import HTML
import matplotlib.pyplot as plt
def segmentation_FCN_RESNET101_animation(net, paths, save_path='./animation.
mp4', dev='cuda', img_resize=640, interval=150, figsize=(20, 20)):
    '''
    function - 应用 torchvision.models.segmentation.fcn_resnet101 预测图像，并打印显
示分割预测结果的动画
    '''
    import matplotlib.pyplot as plt
    from PIL import Image
    import torch
    import torchvision.transforms as T
    import matplotlib.animation as animation
    from tqdm.auto import tqdm
    plt.figure(figsize=figsize)
    imgs = []
    fig = plt.figure(figsize=figsize)
    for path in tqdm(paths):
        img = Image.open(path)
        trf = T.Compose([T.Resize(img_resize), T.ToTensor(), T.Normalize(
            mean=[0.485, 0.456, 0.406], std=[0.229, 0.224, 0.225])])
        inp = trf(img).unsqueeze(0).to(dev)
        out = net.to(dev)(inp)['out']
```

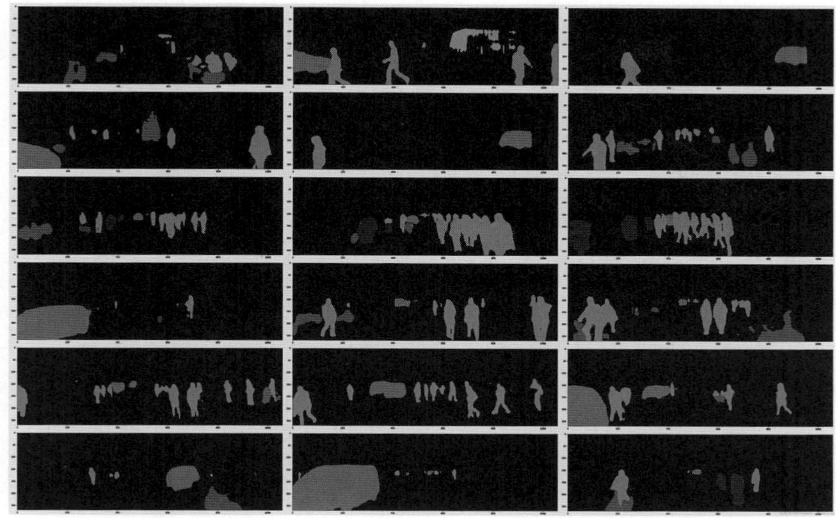

图3.18-11 KITTI子集的所有图像分割部分示例（源自mp4动画）

```
        sementic_seg = torch.argmax(
            out.squeeze(), dim=0).detach().cpu().numpy()
        rgb = decode_segmap_FCN_RESNET101(sementic_seg)
        imgs.append([plt.imshow(rgb, animated=True,)])
    anima = animation.ArtistAnimation(
        fig, imgs, interval=interval, blit=True, repeat_delay=1000)
    anima.save(save_path)
    print(".mp4 saved.")
    return anima, imgs
save_path = r'./results/segmentation_FCN_RESNET101_animation.mp4'
drive_29_0071_img_fp_list.sort()
anima, _ = segmentation_FCN_RESNET101_animation(
    fcn, drive_29_0071_img_fp_list, save_path, dev='cuda', img_resize=320,
figsize=(20, 8))
HTML(anima.to_html5_video())
------------------------------------------------------------------------------
  0%|          | 0/1059 [00:00<?, ?it/s]
.mp4 saved.
```

3.18.2.2 对象统计与关联网络结构

对象统计是确定每一位置空间下存在有哪些物，因为使用的训练模型包括20个分类，因此只能识别这些已训练的对象，而树木、建筑等不能识别；再者，该模型只返回对象掩码（不同对象、不同索引，但是同一对象不可再分，例如对人数的统计不能实现）。但是，对每一位置空间已有对象的识别可以初步判断该空间的特征；并通过统计所有位置下，与每一对象同时存在的其他对象的频数，可以应用NetworkX[①]库构建网络结构，直观地观察对象之间的空间存在关系。并统计每一位置出现对象种类的数量，以热力图的形式可视化，可以初步判断不同地段的对象混杂程度，通常对象种类越丰富的区域，空间表现出的活力越高。

首先调整了预测函数，使返回值为对象实例分割索引及对应的分类名称。

```
def segmentation_FCN_RESNET101(net, path, show_orig=True, dev='cuda', img_
resize=640):
    '''
    function - 应用 torchvision.models.segmentation.fcn_resnet101 预测图像，返回预测
结果
    '''
    from PIL import Image
    import torch
    import torchvision.transforms as T
    import numpy as np
    seg_FCN_RESNET101_classi_mapping = {0: 'background', 1: 'aeroplane', 2: 'bicycle',
```

① NetworkX，Python编程语言软件包，可用于创建、操作和学习复杂网络（图）的结构、动态和功能等（https://networkx.org）。

```
3: 'bird', 4: 'boat', 5: 'bottle', 6: 'bus', 7: 'car', 8: 'cat', 9: 'chair', 10: 'cow',
11: 'dining table', 12: 'dog',
                                 13: 'horse', 14: 'motorbike', 15: 'person',
16: 'potted plant', 17: 'sheep', 18: 'sofa', 19: 'train', 20: 'tv/monitor'}
    img = Image.open(path)
    trf = T.Compose([
        T.Resize(img_resize),
        # T.CenterCrop(224),
        T.ToTensor(),
        T.Normalize(mean=[0.485, 0.456, 0.406], std=[0.229, 0.224, 0.225])])
    inp = trf(img).unsqueeze(0).to(dev)
    out = net.to(dev)(inp)['out']
    sementic_seg = torch.argmax(out.squeeze(), dim=0).detach().cpu().numpy()
    sementic_seg_classi = [seg_FCN_RESNET101_classi_mapping[i]
                           for i in np.unique(sementic_seg)]
    return (np.unique(sementic_seg).tolist(), sementic_seg_classi)
sementic_seg = segmentation_FCN_RESNET101(
    fcn, drive_29_0071_img_fp_list[200], dev=device, img_resize=280)
print('预测的图像包含的对象标签：{}'.format(sementic_seg))
---------------------------------------------------------------------------
预测的图像包含的对象标签：([0, 2, 7, 9, 11, 15], ['background', 'bicycle', 'car',
'chair', 'dining table', 'person'])
```

计算所有图像，返回预测结果。

```
from tqdm.auto import tqdm
drive_29_0071_img_seg_pred = [segmentation_FCN_RESNET101(fcn, img,dev=device,img_
resize=280) for img in tqdm(drive_29_0071_img_fp_list)]
---------------------------------------------------------------------------
  0%|          | 0/1059 [00:00<?, ?it/s]
```

通过查看预测结果，可以发现场景中存在狗和火车。而一些未出现的类，例如sheep、hourse等应该是狗在不同影像位置下识别的错误。其他的分类错误，可能与场景中出现的海报等图画有关。观察最终的网络结构（图3.18-12），因为将每一个对象与其他对象在不同场景中共存的情况进行统计，即计算共存对象的频数。将该频数或者其倍数作为网络边的权重值，并通过粗细显示。因此，可以观察到，线越细的对象在整个1059张图像所代表的位置空间下，出现

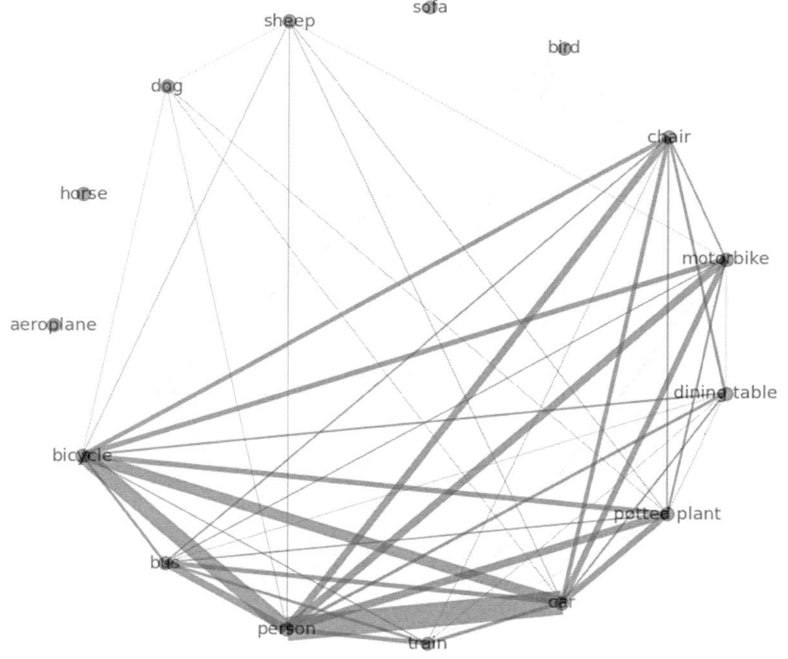

图3.18-12 对象实例分割结果关联网络结构

的位置较少；而线越粗的，则出现位置相对较多。经常同时出现的对象为'person'、'car'和'bicycle'，次之的有'chair'、'motorbike'和'potted plant'等。有些信息的出现是合乎常理，例如场景中骑车的人，因此这些分析结果似乎价值偏弱；但是，'chair'和'potted plant'的出现，可以判定该条街道室外活动的主要内容，餐饮、休闲等。

```
def count_list_frequency(lst):
    '''
    function - 计算列表的频数
    '''
    freq = {}
    for i in lst:
        if (i in freq):
            freq[i] += 1
        else:
            freq[i] = 1
    return freq
def objects_network_PascalVOC(seg_idxs, figsize=(15, 15), layout='spring_
layout', w_ratio=0.5):
    '''
    function - 根据连续的图像分割数据，计算各个对象（真实世界存在的物）与其他对象对应的数量，
构建网络结构，分析相互关系
    '''
    import numpy as np
    import networkx as nx
    import matplotlib.pyplot as plt
    seg_FCN_RESNET101_classi_mapping = {0: 'background', 1: 'aeroplane', 2: 'bicycle',
3: 'bird', 4: 'boat', 5: 'bottle', 6: 'bus', 7: 'car', 8: 'cat', 9: 'chair', 10: 'cow',
11: 'dining table', 12: 'dog',
                                        13: 'horse', 14: 'motorbike', 15: 'person',
16: 'potted plant', 17: 'sheep', 18: 'sofa', 19: 'train', 20: 'tv/monitor'}
    def flatten_lst(lst): return [m for n_lst in lst for m in flatten_lst(
        n_lst)] if type(lst) is list else [lst]
    unique_idxs = np.unique(flatten_lst(seg_idxs)).tolist()
    unique_idxs_ = list(filter(lambda x: x != 0, unique_idxs))
    print('存在的对象有：', [(i, seg_FCN_RESNET101_classi_mapping[i])
           for i in unique_idxs])

    # 01- 收集每一对象所有存在时刻包含的其他对象
    object_associations = {}
    for obj in unique_idxs_:
        obj_associations_list = flatten_lst([i for i in seg_idxs if obj in i])
        obj_associations_list_ = list(
            filter(lambda x: x != 0, obj_associations_list))
        object_associations[obj] = obj_associations_list_
    # print(object_associations)

    # 02- 计算每一对象，包含其他对象的频数
    object_associations_frequency = {}
    for k, v in object_associations.items():
        v_ = list(filter(lambda x: x != k, v))
        object_associations_frequency[k] = count_list_frequency(v_)
    # print(object_associations_frequency)

    # 03- 构建网络，以频数或其倍数为权重
    fig, ax = plt.subplots(figsize=figsize)
    G = nx.Graph()
    layout_dic = {
        'spring_layout': nx.spring_layout,
        'random_layout': nx.random_layout,
        'circular_layout': nx.circular_layout,
        'kamada_kawai_layout': nx.kamada_kawai_layout,
        'shell_layout': nx.shell_layout,
        'spiral_layout': nx.spiral_layout,
    }
    for k, v in object_associations_frequency.items():
        for obj, w in v.items():
            G.add_edge(
                seg_FCN_RESNET101_classi_mapping[k], seg_FCN_RESNET101_classi_
mapping[obj], weight=w*w_ratio)
    pos = layout_dic[layout](G)
    weights = nx.get_edge_attributes(G, 'weight').values()
    nx.draw(G, pos, with_labels=True, font_size=18, alpha=0.4,
            edge_color="r", node_size=200, width=list(weights))
```

```
seg_idxs = [i[0] for i in drive_29_0071_img_seg_pred]
objects_network_PascalVOC(seg_idxs, layout='shell_layout', w_ratio=0.03)
----------------------------------------------------------------
存在的对象有: [(0, 'background'), (1, 'aeroplane'), (2, 'bicycle'), (3, 'bird'), (6, 'bus'), (7, 'car'), (9, 'chair'), (11, 'dining table'), (12, 'dog'), (13, 'horse'), (14, 'motorbike'), (15, 'person'), (16, 'potted plant'), (17, 'sheep'), (18, 'sofa'), (19, 'train')]
```

空间对象种类的丰富程度（图3.18-13），表3.18-2统计了从图像中提取对象种类的数量。

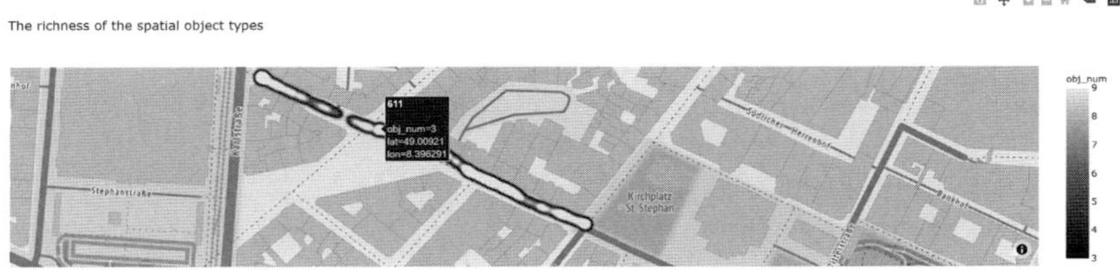

图3.18-13 空间对象种类的丰富程度

表3.18-2 从图像中提取对象种类的数量

	lat	lon	timestamps_	obj_num	idx
0	49.008650	8.398092	2011-09-29 13:54:59.990872576	3	0
1	49.008777	8.397611	2011-09-29 13:55:00.094612992	3	1
2	49.009162	8.396541	2011-09-29 13:55:00.198486528	3	2
3	49.008962	8.397075	2011-09-29 13:55:00.302340864	3	3
4	49.009505	8.395251	2011-09-29 13:55:00.406079232	4	4
...	...	...	...	...	...
1054	49.009215	8.396286	2011-09-29 13:56:49.458599424	4	1054
1055	49.009353	8.395764	2011-09-29 13:56:49.562463744	4	1055
1056	49.008706	8.397888	2011-09-29 13:56:49.666327808	3	1056
1057	49.009215	8.396288	2011-09-29 13:56:49.770316544	3	1057
1058	49.009079	8.396812	2011-09-29 13:56:49.874179584	3	1058

```
import util_A
KITTI_info_fp = r'G:\data\2011_09_29_drive_0071_sync\oxts\data'
timestamps_fp = r'G:\data\2011_09_29_drive_0071_sync\image_03\timestamps.txt'
drive_29_0071_info = util_A.KITTI_info(KITTI_info_fp, timestamps_fp)
drive_29_0071_info_coordi = drive_29_0071_info[['lat', 'lon', 'timestamps_']]
obj_num = [len(i) for i in seg_idxs]
drive_29_0071_info_coordi['obj_num'] = obj_num
drive_29_0071_info_coordi['idx'] = drive_29_0071_info_coordi.index
drive_29_0071_info_coordi
1059 rows × 5 columns
import plotly.express as px
fig = px.density_mapbox(drive_29_0071_info_coordi, lat='lat', lon='lon', z='obj_num', radius=10,
                        center=dict(lat=49.008645, lon=8.398104), zoom=18,
                        mapbox_style="stamen-terrain",
                        title='The richness of the spatial object types',
                        # hover_data=['idx'],
                        hover_name='idx'
                        )
fig.show()
```

参考文献（References）:

[1] TorchVision Object Detection Finetuning Tutorial, https://pytorch.org/tutorials/intermediate/torchvision_tutorial.html.
[2] He, K., Gkioxari, G., Dollár, P., & Girshick, R. (2018). Mask R-CNN. arXiv [Cs.CV]. Retrieved from http://arxiv.org/abs/1703.06870.
[3] Intersection over Union (IoU) for object detection, https://pyimagesearch.com/2016/11/07/intersection-over-union-iou-for-object-detection/.
[4] Ren, S., He, K., Girshick, R., & Sun, J. (2016). Faster R-CNN: Towards Real-Time Object Detection with Region Proposal Networks. arXiv [Cs.CV]. Retrieved from http://arxiv.org/abs/1506.01497.
[5] Fully-Convolutional Network model with ResNet-50 and ResNet-101 backbones, https://pytorch.org/hub/pytorch_vision_fcn_resnet101.
[6] PyTorch for Beginners: Semantic Segmentation using torchvision, https://learnopencv.com/pytorch-for-beginners-semantic-segmentation-using-torchvision/.

3.19 Cityscapes数据集、图像分割与城市空间对象统计

3.19.1 Cityscapes数据集

① Cityscapes数据集，（https://www.cityscapes-dataset.com）。
② Cityscapes Dataset Overview, https://www.cityscapes-dataset.com/dataset-overview/#features.

Cityscapes数据集[①]集中于城市街道场景的语义解释（semantic understanding），这非常适用于对城市空间内容的分析。这个数据集的主要特点有[②]：

1. 多边形标注（Polygonal annotations）
- 密集的语义分割（Dense semantic segmentation）
- 车辆和行人的实例（对象）分割（Instance segmentation for vehicle and people）

2. 复杂性（Complexity）
- 30个分类（30 classes）
- 分类定义如表3.19–1（Class Definitions for a list of all classes）：

表3.19–1　Cityscapes数据集分类

Group（组）	Classes（类）
flat（平面）	road · sidewalk · parking+ · rail track+
human（人）	person · rider
vehicle（车辆）	car · truck · bus · on rails · motorcycle · bicycle · caravan+ · trailer+
construction（建筑）	building · wall · fence · guard rail+ · bridge+ · tunnel+
object（构筑）	pole · pole group+ · traffic sign · traffic light
nature（自然）	vegetation · terrain
sky（天空）	sky
void（无）	ground+ · dynamic+ · static+

* 包含单个实例对象注释。但是，如果多个对象（实例）之间界限不清楚，整个组（crowd/group），包含多个对象被标记在一起，并标注为组，例如车辆组。

+ 该标签不包括在任何评估中，并视为无效（这是关于模型评估的说明）。

3. 多样性（Diversity）
- 包含有50座城市（50 cities）
- 春夏秋等多季节变化（Several months (spring，summer，fall)）
- 白天（Daytime）
- 好/中等天气状况（Good/medium weather conditions）
- 手动选择的帧（Manually selected frames）
- 大量的动态对象（Large number of dynamic objects）
- 多样的场景布局（Varying scene layout）
- 多样的背景（Varying background）

4. 量（Volume）
- 5000张精标注（5000 annotated images with fine annotations）
- 20000张粗标注（20000 annotated images with coarse annotations）

5. 元数据（metadata）
- 前后视频帧。每一幅注释图像都是30帧视频片段（1.8s）中的第20幅图像（Preceding and trailing video frames. Each annotated image is the 20th image from a 30 frame video snippets (1.8s)）
- 对应右侧立体视图（Corresponding right stereo views）
- 坐标（GPS coordinates GPS）
- 汽车里程测量的运动数据（Ego-motion data from vehicle odometry）
- 来自车辆传感器的外部温度（Outside temperature from vehicle sensor）

6. 其他研究者的扩展（Extensions by other researchers）
- 行人锚框标注（Bounding box annotations of people）

- 雨雾图像增广（Images augmented with fog and rain）

7. 基准测试套件和评估服务器（Benchmark suite and evaluation server）

- 像素级语义标签（Pixel-level semantic labeling）
- 实例（对象）级语义标签（Instance-level semantic labeling）
- 展示全景的语义标签（Panoptic semantic labeling）

图像分割训练的数据集在Cityscapes下载页[①]注册后获取。主要包括两个文件：一个是11GB大小的leftImg8bit_trainvaltest.zip，为训练、验证和测试的图像；二是241MB大小的gtFine_trainvaltest.zip，为训练和验证图像数据集对应的精细标注。

读取Cityscapes数据集，理解该数据集的文件结构和数据结构。

下述代码的前半部分迁移于 *Hierarchical Multi-Scale Attention for Semantic Segmentation*，<https://github.com/NVIDIA/semantic-segmentation>。

3.19.1.1 参数管理

应用（App）涉及的参数很多，为了便于管理，使用argparse[②]命令行参数解析包，可以将参数和代码分离开来，方便读取命令行参数，尤其适合于参数的频繁修改。同时，在程序执行过程中，为避免参数变化导致难以调试或难以理解代码。在参数配置完之后，需要将mutable参数转变为immutable参数，迁移代码类`class AttrDict(dict)`实现。`AttrDict`类可以定义类的属性，并通过该类的`immutable`方法，实现类属性的批量类型转换（mutable到immutable，或反之）。

① Cityscape数据下载注册页（https://www.cityscapes-dataset.com/login）。

② argparse，用来解析命令行参数的Python库，使编写用户友好的命令行界面变得容易。可以从所定义的参数中解析这些参数，并会自动生成帮助和用法信息；当提供的参数值无效时，引发错误提示（https://docs.python.org/3/library/argparse.html）。

```
# semantic-segmentation-main\train.py
import argparse
# Argument Parser
parser = argparse.ArgumentParser(description='Semantic Segmentation')
# ...
parser.add_argument('--start_epoch', type=int, default=0)
parser.add_argument('--max_epoch', type=int, default=180)
parser.add_argument('--global_rank', default=0, type=int,
                    help='parameter used by apex library')
parser.add_argument('--test_mode', action='store_true', default=False, help=(
    'Minimum testing to verify nothing failed, ''Runs code for 1 epoch of train
and val'))
parser.add_argument('--init_decoder', default=False, action='store_true',
                    help='initialize decoder with kaiming normal')
parser.add_argument('--syncbn', action='store_true',
                    default=False, help='Use Synchronized BN')
parser.add_argument('--extra_scales', type=str, default='0.5,2.0')
parser.add_argument('--set_cityscapes_root', type=str,
                    default=None, help='override cityscapes default root dir')
parser.add_argument('--dataset', type=str, default='cityscapes',
                    help='cityscapes, mapillary, camvid, kitti')
parser.add_argument('--result_dir', type=str,
                    default='./logs_sementic', help='where to write log output')
parser.add_argument('--cv', type=int, default=0, help=(
    'Cross-validation split id to use. Default # of splits set' ' to 3 in
config'))
parser.add_argument('--crop_size', type=str, default='448',
                    help='dynamically scale training images down to this size')
# 896
parser.add_argument('--scale_min', type=float, default=0.5,
                    help='dynamically scale training images down to this size')
parser.add_argument('--scale_max', type=float, default=2.0,
                    help='dynamically scale training images up to this size')
parser.add_argument('--full_crop_training', action='store_true',
                    default=False, help='Full Crop Training')
parser.add_argument('--pre_size', type=int, default=None,
                    help=('resize long edge of images to this before''
augmentation'))
parser.add_argument('--color_aug', type=float, default=0.25,
                    help='level of color augmentation')
parser.add_argument('--jointwtborder', action='store_true',
                    default=False, help='Enable boundary label relaxation')
parser.add_argument('--dump_augmentation_images', action='store_true',
                    default=False, help='Dump Augmentated Images for sanity
check')
```

```
parser.add_argument('--rmi_loss', action='store_true',
                    default=False, help='use RMI loss')
parser.add_argument('--img_wt_loss', action='store_true',
                    default=False, help='per-image class-weighted loss')
parser.add_argument('--arch', type=str, default='deepv3.DeepV3Plus',
                    help='Network architecture. We have DeepSRNX50V3PlusD
(backbone: ResNeXt50)and deepWV3Plus (backbone: WideResNet38).')
parser.add_argument('--trunk', type=str, default='resnet101',
                    help='trunk model, can be: resnet101 (default), resnet50')
parser.add_argument('--apex', action='store_true', default=False,
                    help='Use Nvidia Apex Distributed Data Parallel')
parser.add_argument('--optimizer', type=str, default='sgd', help='optimizer')
parser.add_argument('--lr', type=float, default=0.002)
parser.add_argument('--weight_decay', type=float, default=1e-4)
parser.add_argument('--momentum', type=float, default=0.9)
parser.add_argument('--lr_schedule', type=str, default='poly',
                    help='name of lr schedule: poly')
parser.add_argument('--poly_exp', type=float, default=1.0,
                    help='polynomial LR exponent')
# ...
# JupyterLab 需要在 args = parser.parse_args() 中传入空的 []，否则引发异常
args = parser.parse_args([])
print("--syncbn=%f" % args.syncbn)
args.world_size = 1
print("增加新的参数-world_size=%d" % args.world_size)
print("args参数解析:", args)
```

```
--syncbn=0.000000
增加新的参数-world_size=1
args参数解析: Namespace(apex=False, arch='deepv3.DeepV3Plus', color_
aug=0.25, crop_size='448', cv=0, dataset='cityscapes', dump_augmentation_
images=False, extra_scales='0.5,2.0', full_crop_training=False, global_rank=0,
img_wt_loss=False, init_decoder=False, jointwtborder=False, lr=0.002, lr_
schedule='poly', max_epoch=180, momentum=0.9, optimizer='sgd', poly_exp=1.0, pre_
size=None, result_dir='./logs_sementic', rmi_loss=False, scale_max=2.0, scale_
min=0.5, set_cityscapes_root=None, start_epoch=0, syncbn=False, test_mode=False,
trunk='resnet101', weight_decay=0.0001, world_size=1)
```

```
# semantic-segmentation-main\utils\attr_dict.py
class AttrDict(dict):
    IMMUTABLE = '__immutable__'
    def __init__(self, *args, **kwargs):
        super(AttrDict, self).__init__(*args, **kwargs)
        self.__dict__[AttrDict.IMMUTABLE] = False
    def __getattr__(self, name):
        if name in self.__dict__:
            return self.__dict__[name]
        elif name in self:
            return self[name]
        else:
            raise AttributeError(name)
    def __setattr__(self, name, value):
        if not self.__dict__[AttrDict.IMMUTABLE]:
            if name in self.__dict__:
                self.__dict__[name] = value
            else:
                self[name] = value
        else:
            raise AttributeError(
                'Attempted to set "{}" to "{}", but AttrDict is immutable'.
                format(name, value)
            )
    def immutable(self, is_immutable):
        """将不可变性（immutability）设置为is_immutable，并递归地将该设置应用于所有嵌套
的AttrDicts.
        """
        self.__dict__[AttrDict.IMMUTABLE] = is_immutable
        #递归的设置不可变状态
        for v in self.__dict__.values():
            if isinstance(v, AttrDict):
                v.immutable(is_immutable)
        for v in self.values():
            if isinstance(v, AttrDict):
```

```
            v.immutable(is_immutable)
    def is_immutable(self):
        return self.__dict__[AttrDict.IMMUTABLE]
```

实例化类AttrDict，使用类的属性存储参数，并配置参数immutable与mutable互相转换的方法。

```
# semantic-segmentation-main\config.py
import os

__C = AttrDict()  #非嵌套字典
cfg = __C
print("cfg=", cfg)
#...
__C.GLOBAL_RANK = 0
__C.EPOCH = 0

#保存一些大文件位置的绝对路径.
__C.ASSETS_PATH = r'G:\data\Cityscapes_assets'

print("参数--非嵌套字典:", cfg)

#可选项的属性参数配置
__C.OPTIONS = AttrDict()  #嵌套字典
__C.OPTIONS.TEST_MODE = False
__C.OPTIONS.INIT_DECODER = False

#模型的属性参数配置
__C.MODEL = AttrDict()
__C.MODEL.EXTRA_SCALES = '0.5,1.5'
__C.MODEL.BNFUNC = None
WEIGHTS_PATH = os.path.join(__C.ASSETS_PATH, 'seg_weights')
__C.MODEL.WRN38_CHECKPOINT = os.path.join(
    WEIGHTS_PATH, 'wider_resnet38.pth.tar')

#数据集的属性参数配置
__C.DATASET = AttrDict()
__C.DATASET.CITYSCAPES_DIR = os.path.join(__C.ASSETS_PATH, 'data/Cityscapes')
__C.DATASET.IGNORE_LABEL = 255
__C.DATASET.NUM_CLASSES = 0
__C.DATASET.CV = 1  # cv_split - 0,1,2,3
__C.DATASET.CUSTOM_COARSE_PROB = None
__C.DATASET.CLASS_UNIFORM_PCT = 0.5
__C.DATASET.NAME = ''
__C.DATASET.CLASS_UNIFORM_TILE = 1024
__C.DATASET.CENTROID_ROOT = os.path.join(__C.ASSETS_PATH, 'uniform_centroids')
__C.DATASET.CITYSCAPES_CUSTOMCOARSE = os.path.join(
    __C.ASSETS_PATH, 'data/Cityscapes/autolabelled')
__C.DATASET.MEAN = [0.485, 0.456, 0.406]
__C.DATASET.STD = [0.229, 0.224, 0.225]
__C.DATASET.DUMP_IMAGES = False

#随机裁剪变换增强
__C.DATASET.TRANSLATE_AUG_FIX = False
#...
print("参数--含嵌套字典:", cfg)
--------------------------------------------------------------------------------
cfg= {}
参数--非嵌套字典: {'GLOBAL_RANK': 0, 'EPOCH': 0, 'ASSETS_PATH': 'G:\\data\\Cityscapes_assets'}
参数--含嵌套字典: {'GLOBAL_RANK': 0, 'EPOCH': 0, 'ASSETS_PATH': 'G:\\data\\Cityscapes_assets', 'OPTIONS': {'TEST_MODE': False, 'INIT_DECODER': False}, 'MODEL': {'EXTRA_SCALES': '0.5,1.5', 'BNFUNC': None, 'WRN38_CHECKPOINT': 'G:\\data\\Cityscapes_assets\\seg_weights\\wider_resnet38.pth.tar'}, 'DATASET': {'CITYSCAPES_DIR': 'G:\\data\\Cityscapes_assets\\data/Cityscapes', 'IGNORE_LABEL': 255, 'NUM_CLASSES': 0, 'CV': 1, 'CUSTOM_COARSE_PROB': None, 'CLASS_UNIFORM_PCT': 0.5, 'NAME': '', 'CLASS_UNIFORM_TILE': 1024, 'CENTROID_ROOT': 'G:\\data\\Cityscapes_assets\\uniform_centroids', 'CITYSCAPES_CUSTOMCOARSE': 'G:\\data\\Cityscapes_assets\\data/Cityscapes/autolabelled', 'MEAN': [0.485, 0.456, 0.406], 'STD': [0.229, 0.224, 0.225], 'DUMP_IMAGES': False, 'TRANSLATE_AUG_FIX': False}}
```

assert_and_infer_cfg函数可以将argparse定义的参数args，有选择性地存储到类AttrDict的实例cfg(__C)中，并配置为immutable类型。

通过AttrDict的`immutable`方法可以转换参数的mutable和immutable的类型，从而修改参数。

```
cfg.immutable(False)
cfg.MODEL.EXTRA_SCALES = '0.9,1.9'
print("mutable -- updated_cfg.MODEL.EXTRA_SCALES=",cfg.MODEL.EXTRA_SCALES)
cfg.immutable(True)
--------------------------------------------------------------------------------
mutable -- updated_cfg.MODEL.EXTRA_SCALES= 0.9,1.9
```

3.19.1.2　cityscapes数据读取

- cityscapes的标签数据处理

Cityscapes数据集的处理工具，可以从cityscapesScripts[①]中查找。下述代码`namedtuple`对象`labels`、创建了不同类别之间（列之间）的映射关系，并通过`ainId2label`、`label2trainid`、`trainId2name`、`trainId2color`和`category2labels`，定义了不同列转换的映射关系，方便后续cityscapes标签的变换。

① cityscapesScripts，该库包含用于检查、准备和评估Cityscapes数据集的模块（https://github.com/mcordts/cityscapesScripts）。

```
# semantic-segmentation-main\datasets\cityscapes_labels.py
from collections import namedtuple

# 一个标签和所有元信息（meta information）
Label = namedtuple( 'Label' , [

    'name'          , # 标签的标识符，例如'car' 'person' 等
                      # 使用'name' 命名唯一的类
    'id'            , # 与标签关联的整型 ID
                      # 被用于表示真实图像的标签
                      # 为 -1 值的 ID 意为这个标签没有 ID（被忽略），例如车牌（涉及公共安全），在
创建真实图像分类时就会标识为 -1
                      # 不要修改这些 IDs，因为这些 IDs 是真实服务器所期望的的值
    'trainId'       , # 这列 IDs 可以随意修改，以满足不同的训练目的。在创建自己的真实图像分类时，
可以在 cityscapesScripts GitHub 仓库中的 preparation 文件夹下寻找创建工具。但是，在验证模型，以及向评估
服务器提交结果时，还是需要使用上述同一的 ID
                      # 对于'trainId'，多个标签可能具有相同的 ID。然后这些标签映射到真实图像中
同一类。例如，对于某些方法，将所有 void 类型的类映射到训练中的同一个 ID 可能是有意义的，值为 255
    'category'      , # 此标签所属类别的名称
    'categoryId'    , # 这个类别的 ID，用于在类别水平上创建真实图像分类
    'hasInstances', # 这个标签用于区分是否有单个实例（对象）
    'ignoreInEval', # 在评估中，像素有作为真实类标的分类被忽略，或者未被忽略
    'color'         , # 类标对应的颜色
    ] )
labels = [
    #       name                     id    trainId   category            catId
hasInstances   ignoreInEval   color
    Label(  'unlabeled'            ,  0 ,      255 , 'void'            , 0       ,
False        , True          , (  0,  0,  0) ),
    Label(  'ego_vehicle'          ,  1 ,      255 , 'void'            , 0       ,
False        , True          , (  0,  0,  0) ),
    Label(  'rectification_border' ,  2 ,      255 , 'void'            , 0       ,
False        , True          , (  0,  0,  0) ),
    Label(  'out_of_roi'           ,  3 ,      255 , 'void'            , 0       ,
False        , True          , (  0,  0,  0) ),
    Label(  'static'               ,  4 ,      255 , 'void'            , 0       ,
False        , True          , (  0,  0,  0) ),
    Label(  'dynamic'              ,  5 ,      255 , 'void'            , 0       ,
False        , True          , (111, 74,  0) ),
    Label(  'ground'               ,  6 ,      255 , 'void'            , 0       ,
False        , True          , ( 81,  0, 81) ),
    Label(  'road'                 ,  7 ,        0 , 'flat'            , 1       ,
False        , False         , (128, 64,128) ),
    Label(  'sidewalk'             ,  8 ,        1 , 'flat'            , 1       ,
False        , False         , (244, 35,232) ),
    Label(  'parking'              ,  9 ,      255 , 'flat'            , 1       ,
False        , True          , (250,170,160) ),
    Label(  'rail_track'           , 10 ,      255 , 'flat'            , 1       ,
False        , True          , (230,150,140) ),
    Label(  'building'             , 11 ,        2 , 'construction'    , 2       ,
False        , False         , ( 70, 70, 70) ),
    Label(  'wall'                 , 12 ,        3 , 'construction'    , 2       ,
False        , False         , (102,102,156) ),
    Label(  'fence'                , 13 ,        4 , 'construction'    , 2       ,
```

```
False          , False          , (190,153,153) ),
    Label(  'guard_rail'            , 14 ,      255 , 'construction'    , 2        ,
False          , True           , (180,165,180) ),
    Label(  'bridge'                , 15 ,      255 , 'construction'    , 2        ,
False          , True           , (150,100,100) ),
    Label(  'tunnel'                , 16 ,      255 , 'construction'    , 2        ,
False          , True           , (150,120, 90) ),
    Label(  'pole'                  , 17 ,        5 , 'object'          , 3        ,
False          , False          , (153,153,153) ),
    Label(  'polegroup'             , 18 ,      255 , 'object'          , 3        ,
False          , True           , (153,153,153) ),
    Label(  'traffic_light'         , 19 ,        6 , 'object'          , 3        ,
False          , False          , (250,170, 30) ),
    Label(  'traffic_sign'          , 20 ,        7 , 'object'          , 3        ,
False          , False          , (220,220,  0) ),
    Label(  'vegetation'            , 21 ,        8 , 'nature'          , 4        ,
False          , False          , (107,142, 35) ),
    Label(  'terrain'               , 22 ,        9 , 'nature'          , 4        ,
False          , False          , (152,251,152) ),
    Label(  'sky'                   , 23 ,       10 , 'sky'             , 5        ,
False          , False          , ( 70,130,180) ),
    Label(  'person'                , 24 ,       11 , 'human'           , 6        ,
True           , False          , (220, 20, 60) ),
    Label(  'rider'                 , 25 ,       12 , 'human'           , 6        ,
True           , False          , (255,  0,  0) ),
    Label(  'car'                   , 26 ,       13 , 'vehicle'         , 7        ,
True           , False          , (  0,  0,142) ),
    Label(  'truck'                 , 27 ,       14 , 'vehicle'         , 7        ,
True           , False          , (  0,  0, 70) ),
    Label(  'bus'                   , 28 ,       15 , 'vehicle'         , 7        ,
True           , False          , (  0, 60,100) ),
    Label(  'caravan'               , 29 ,      255 , 'vehicle'         , 7        ,
True           , True           , (  0,  0, 90) ),
    Label(  'trailer'               , 30 ,      255 , 'vehicle'         , 7        ,
True           , True           , (  0,  0,110) ),
    Label(  'train'                 , 31 ,       16 , 'vehicle'         , 7        ,
True           , False          , (  0, 80,100) ),
    Label(  'motorcycle'            , 32 ,       17 , 'vehicle'         , 7        ,
True           , False          , (  0,  0,230) ),
    Label(  'bicycle'               , 33 ,       18 , 'vehicle'         , 7        ,
True           , False          , (119, 11, 32) ),
    Label(  'license_plate'         , -1 ,       -1 , 'vehicle'         , 7        ,
False          , True           , (  0,  0,142) ),
]

# 名称到标签对象
name2label      = { label.name    : label for label in labels           }
# id 到标签对象
id2label        = { label.id      : label for label in labels           }
# trainId 到标签对象
trainId2label   = { label.trainId : label for label in reversed(labels) }
# 标签到 trainid
label2trainid   = { label.id      : label.trainId for label in labels   }
# trainId 到标签对象
trainId2name   = { label.trainId : label.name for label in labels    }
trainId2color  = { label.trainId : label.color for label in labels      }
# 分类到标签对象列表
category2labels = {}
for label in labels:
    category = label.category
    if category in category2labels:
        category2labels[category].append(label)
    else:
        category2labels[category] = [label]
        def assureSingleInstanceName( name ):
    # 如果名称已知，便不是一个组
    if name in name2label:
        return name
    # 测试是否名称代表一个组
    if not name.endswith("group"):
        return None
    # 移除组
    name = name[:-len("group")]
    # 测试新名称是否存在
```

```
    if not name in name2label:
        return None
    #测试新名称是否表示实际具有实例的标签
    if not name2label[name].hasInstances:
        return None
    #返回
    return name
    print(assureSingleInstanceName('ego_vehicle' ))
--------------------------------------------------------------------------------
ego_vehicle
--------------------------------------------------------------------------------
# semantic-segmentation-main\datasets\cityscapes.py
root = cfg.DATASET.CITYSCAPES_DIR
id_to_trainid = label2trainid
print("id_to_trainid:",id_to_trainid)
print("_"*50)
trainid_to_name = trainId2name
print("trainid_to_name:",trainid_to_name)
--------------------------------------------------------------------------------
id_to_trainid: {0: 255, 1: 255, 2: 255, 3: 255, 4: 255, 5: 255, 6: 255, 7: 0, 8:
1, 9: 255, 10: 255, 11: 2, 12: 3, 13: 4, 14: 255, 15: 255, 16: 255, 17: 5, 18:
255, 19: 6, 20: 7, 21: 8, 22: 9, 23: 10, 24: 11, 25: 12, 26: 13, 27: 14, 28: 15,
29: 255, 30: 255, 31: 16, 32: 17, 33: 18, -1: -1}
__________________________________________________
trainid_to_name: {255: 'trailer', 0: 'road', 1: 'sidewalk', 2: 'building', 3: 'wall',
4: 'fence', 5: 'pole', 6: 'traffic_light', 7: 'traffic_sign', 8: 'vegetation', 9:
'terrain', 10: 'sky', 11: 'person', 12: 'rider', 13: 'car', 14: 'truck', 15: 'bus',
16: 'train', 17: 'motorcycle', 18: 'bicycle', -1: 'license_plate'}
--------------------------------------------------------------------------------
# semantic-segmentation-main\datasets\cityscapes.py
def fill_colormap():
    palette = [128, 64, 128,
               244, 35, 232,
               70, 70, 70,
               102, 102, 156,
               190, 153, 153,
               153, 153, 153,
               250, 170, 30,
               220, 220, 0,
               107, 142, 35,
               152, 251, 152,
               70, 130, 180,
               220, 20, 60,
               255, 0, 0,
               0, 0, 142,
               0, 0, 70,
               0, 60, 100,
               0, 80, 100,
               0, 0, 230,
               119, 11, 32]
    zero_pad = 256 * 3 - len(palette)
    for i in range(zero_pad):
        palette.append(0)
    return palette
color_mapping = fill_colormap()
```

查看Cityscapes数据集图像，如图3.19-1所示。

图3.19-1　查看Cityscapes数据集图像

```
# semantic-segmentation-main\datasets\cityscapes.py
import os
import os.path as path
img_root = path.join(root, 'leftImg8bit_trainvaltest/leftImg8bit')
mask_root = path.join(root, 'gtFine_trainvaltest/gtFine')
import util_misc
imgs_fn = util_misc.filePath_extraction(img_root,["png"])
imgs_root = list(imgs_fn.keys())[0]
imgsFn_lst = imgs_fn[imgs_root]
imgsFn_lst.sort()
imgsFn_lst_ = imgsFn_lst[:2]
columns = 2
scale = 1
util_misc.imgs_layoutShow(imgs_root,imgsFn_lst_,columns,scale,figsize=(10,3))
print(imgsFn_lst_)
['berlin_000000_000019_leftImg8bit.png', 'berlin_000001_000019_leftImg8bit.png']
```

查看Cityscapes数据集图像的语义分割，如图3.19-2所示。

```
mask_fn = util_misc.filePath_extraction(mask_root, ["png"])
mask_root_list = list(mask_fn.keys())
mask_root_list.sort()
mask_root_ = mask_root_list[15]
maskFn_lst = mask_fn[mask_root_]
# maskFn_lst.sort()
maskFn_lst_ = maskFn_lst[:6]
columns = 6
scale = 1
util_misc.imgs_layoutShow(mask_root_, maskFn_lst_,
                          columns, scale, figsize=(30, 3))
print(maskFn_lst_)
['jena_000000_000019_gtFine_color.png', 'jena_000000_000019_gtFine_instanceIds.
png', 'jena_000000_000019_gtFine_labelIds.png', 'jena_000001_000019_gtFine_
color.png', 'jena_000001_000019_gtFine_instanceIds.png', 'jena_000001_000019_
gtFine_labelIds.png']
```

图3.19-2 查看Cityscapes数据集图像的语义分割

3.19.1.3 torchvision.datasets.Cityscapes方法读取cityscapes数据

输入参数root为包含有leftImg8bit_trainextra.zip、leftImg8bit_trainvaltest.zip、gtCoarse.zip和gtFine_trainvaltest.zip所在位置的文件夹。ityscapes数据读取结果如图3.19-3所示。

```
import matplotlib.pyplot as plt
from torchvision.datasets import Cityscapes
dataset = Cityscapes(r'G:\data\cityscapes', split='train',
                     mode='fine', target_type='semantic')
img, smnt = dataset[0]
plt.rcParams["figure.figsize"] = (20, 20)
plt.subplot(121)
plt.imshow(img)
plt.subplot(122)
plt.imshow(smnt)
plt.show()
```

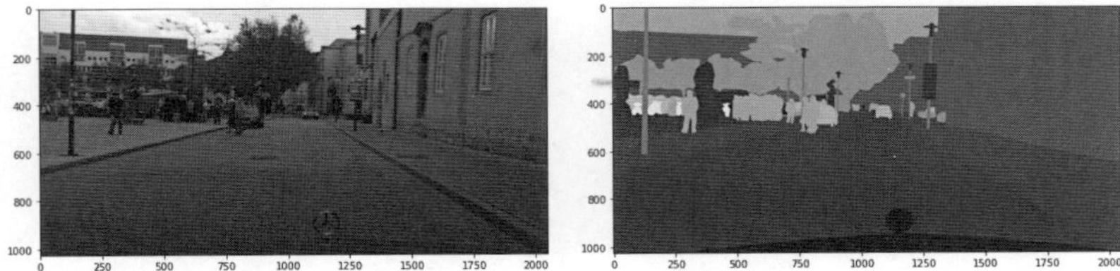

图3.19-3 用torchvision库提供的方法查看cityscapes数据集图像及其语义分割

3.19.2 开放神经网络交换——ONNX与Netron网络可视化工具

3.19.2.1 开放神经网络交换——ONNX

开放神经网络交换——ONNX（Open Neural Network，Exchage）[①]是针对深度学习所设计的开放式文件格式，用于存储训练好的模型。ONNX相信人工智能社区需要更强的互操作性，不会被局限于一种人工智能框架，使得不同类型的人工智能框架（例如PyTorch、MXNet)可以采用相同格式存储模型数据并交互，共享模型。例如，使用PyTorch的人工智能框架（深度学习），并利用`torchvision.models`[②]（预训练）模型库，实现城市空间下行人的对象检测来估算人流量变化；应用对象分割（Instance Segmentation），统计城市空间对象内容，建立关联网络结构分析对象之间的关系。因此，预训练好的模型可以帮助其他领域的研究者迅速应用已有的研究模型，避免重新构建模型网络，以及不菲的训练时间成本（即使有多GPU加速，有些海量数据集或高度复杂的网络，训练时间长度也相当惊人）。非计算机科学领域的研究者，往往也不具有理想的硬件条件，因此预训练好的模型共享，显得尤为重要。这也是将人工智能从研究带到现实的有效途径。

ONNX Model Zoo[③]汇集了已有的大量模型，包括的种类有：

1. 视觉类（Vision）：图像分类（Image Classification），对象检测和图像分割（Object Detection & Image Segmentation），人体，人脸和姿势分析（Body、Face & Gesture Analysis），图像处理（Image Manipulation）；
2. 语言类（Language）：机器理解（Machine Comprehension），机器翻译（Machine Translation），语言模型（Language Modelling）；
3. 其他类：视觉问答&对话（Visual Question Answering & Dialog），语音和音频处理（Speech & Audio Processing）与其他有趣的模型。

① 开放神经网络交换——ONNX，帮助创建和部署深度学习模型（https://onnx.ai/supported-tools.html）。

② torchvision.models，预定义深度学习模型（https://pytorch.org/vision/0.8/models.html）。

③ ONNX Model Zoo，为预训练的深度学习模型库，每个模型都有基于Python编写训练和运行交互解释的Jupyter格式文件（https://github.com/onnx/models）。

3.19.2.2 Netron[④]网络可视化工具

tensorboard可以在深度学习模型建立过程中，帮助分析网络结构，显示训练图像和预测精度及损失曲线等。但是，如果想更加方便地查看模型网络结构，Netron可以直接读取开放神经网络交换格式(.onnx)。`torch.onnx.export`方法可以将PyTorch模型导出为.onnx交换格式，默认`export_params=True`，保存预训练模型的参数。直接用Netron工具打开，可以查看到模型的网络结构（图3.19-4）。

④ Netron，网络可视化工具（https://github.com/lutzroeder/Netron）。

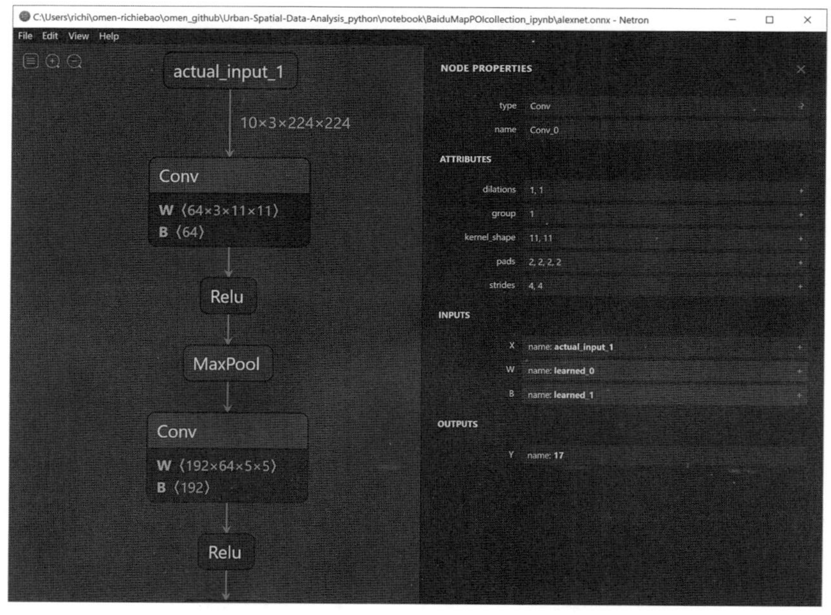

图3.19-4 用 Netron工具查看网络结构

3.19.3 DUC图像分割

目前，torchvision.models图像分割部分的预训练模型主要是针对COCO数据集子集Pascal VOC，包括有20个分类。并且，分类中包括了部分室内物品，这不能够满足室外城市街道环境的语义分割。在ONNX Model Zoo汇集的大量模型中，语义分割部分可获取的模型目前只有DUC（Dense Upsampling Convolution）①，具体实现的方法在Inference demo for DUC models② notebook中给出了阐述，使用的深度学习框架是MXNet③。MXNet在PyTorch和TensorFlow双重夹击下发展缓慢，其架构从安装到应用并不友好，因此不建议使用。但是，因为该DUC模型提供了已经训练好的模型，因此下述仍然使用MXNet框架。具体安装及相关依赖库在其notebook文件中均有说明。

DUC所使用的数据集为针对建筑外街道环境的Cityscapes数据集。

当读者察看该部分时，最好先在`torchvision.models`中查看是否有类似针对Cityscapes数据集的图像分割预训练模型，或者在ONNX Model Zoo中是否有新的更新，以及任何深度学习模型库中查找比较选择，使用最新和预测高分的预训练模型，进行图像的语义分割。

- A-调入依赖库。

```
import mxnet as mx
import cv2 as cv
import numpy as np
import os
from PIL import Image
import math
from collections import namedtuple
from mxnet.contrib.onnx import import_model
import cityscapes_labels
```

- B-图像预处理，其颜色通道RGB均减去RGB的均值，并转换为MXNet的ndarray数据格式。

```
def preprocess(im):
    # 转化为 float32
    test_img = im.astype(np.float32)
    # 用小边界外推图像，以获得 DUC 层后的精确重构图像
    test_shape = [im.shape[0], im.shape[1]]
    cell_shapes = [math.ceil(l / 8)*8 for l in test_shape]
    test_img = cv.copyMakeBorder(test_img, 0, max(0, int(cell_shapes[0]) -
im.shape[0]), 0, max(
        0, int(cell_shapes[1]) - im.shape[1]), cv.BORDER_CONSTANT, value=rgb_
mean)
    test_img = np.transpose(test_img, (2, 0, 1))
    # 减去 RBG 均值
    for i in range(3):
        test_img[i] -= rgb_mean[i]
    test_img = np.expand_dims(test_img, axis=0)
    # 转化为 ndarray
    test_img = mx.ndarray.array(test_img)
    return test_img
```

- `C-get_palette()`：返回用于生成输出分割图的预定义调色板；`colorize()`：使用由模型生成的输出类标和`get_palette()`建立的分割图调色板构建分割映射；`predict()`：向模型传入预处理图像，执行前向传播，并将预测输出数据重新调整为输入图像的形状，使用`colorize()`生成彩色分割的分类掩码图。

```
def get_palette():
    # 从文件中获取训练 id 到颜色的映射
    trainId2colors = {
        label.trainId: label.color for label in cityscapes_labels.labels}
    # 准备和返回调色板
    palette = [0] * 256 * 3
    for trainId in trainId2colors:
        colors = trainId2colors[trainId]
        if trainId == 255:
            colors = (0, 0, 0)
        for i in range(3):
            palette[trainId * 3 + i] = colors[i]
    return palette
def colorize(labels):
    # 用输出标签和调色板生成彩色图像
```

① DUC，基于CNN的语义分割模型（https://github.com/onnx/models/tree/master/vision/object_detection_segmentation/duc）。

② Inference demo for DUC models（https://github.com/onnx/models/blob/master/vision/object_detection_segmentation/duc/dependencies/duc-inference.ipynb）。

③ MXNet，开源深度学习框架（https://mxnet.apache.org/）。

```
        result_img = Image.fromarray(labels).convert('P')
        result_img.putpalette(get_palette())
        return np.array(result_img.convert('RGB'))
def predict(imgs):
    # 获取输入和输出维度
    result_height, result_width = result_shape
    _, _, img_height, img_width = imgs.shape
    #设置降采样率
    ds_rate = 8
    #设置单元宽度
    cell_width = 2
    # 输出标签类的数量
    label_num = 19

    #执行前向传递
    batch = namedtuple('Batch', ['data'])
    mod.forward(batch([imgs]), is_train=False)
    labels = mod.get_outputs()[0].asnumpy().squeeze()

    #重新组织输出
    test_width = int((int(img_width) / ds_rate) * ds_rate)
    test_height = int((int(img_height) / ds_rate) * ds_rate)
    feat_width = int(test_width / ds_rate)
    feat_height = int(test_height / ds_rate)
    labels = labels.reshape((label_num, 4, 4, feat_height, feat_width))
    labels = np.transpose(labels, (0, 3, 1, 4, 2))
    labels = labels.reshape(
        (label_num, int(test_height / cell_width), int(test_width / cell_
width)))
    labels = labels[:, :int(img_height / cell_width),
                    :int(img_width / cell_width)]
    labels = np.transpose(labels, [1, 2, 0])
    labels = cv.resize(labels, (result_width, result_height),
                       interpolation=cv.INTER_LINEAR)
    labels = np.transpose(labels, [2, 0, 1])

    # 获取 softmax 输出
    softmax = labels

    # 获取分类标签
    results = np.argmax(labels, axis=0).astype(np.uint8)
    raw_labels = results

    # 计算置信度得分
    confidence = float(np.max(softmax, axis=0).mean())

    # 生成分割图像
    result_img = Image.fromarray(
        colorize(raw_labels)).resize(result_shape[::-1])

    # 生成混合图像
    blended_img = Image.fromarray(cv.addWeighted(
        im[:, :, ::-1], 0.5, np.array(result_img), 0.5, 0))
    return confidence, result_img, blended_img, raw_labels
```

- D-加载预训练的模型。导入ONNX预训练模型到MxNet中，使用符号文件（symbol file）定义模型，使用参数文件（params file）绑定参数。

```
def get_model(ctx, model_path):
    # 导入 ONNX 模型到 MXNet
    sym, arg, aux = import_model(model_path)
    # 定义网络模块
    mod = mx.mod.Module(symbol=sym, data_names=[
                        'data'], context=ctx, label_names=None)
    # 为网络模型绑定参数
    mod.bind(for_training=False, data_shapes=[
             ('data', (1, 3, im.shape[0], im.shape[1]))], label_shapes=mod._
label_shapes)
    mod.set_params(arg_params=arg, aux_params=aux,
                   allow_missing=True, allow_extra=True)
    return mod
```

- E-给出一张KITTI[①]数据中的影像并显示（图3.19-5）。

① KITTI数据集（https://www.cvlibs.net/datasets/kitti）。

图3.19-5 从KITTI数据集提取一张影像

```
im = cv.imread('./data/0000000389.png')[:, :, ::-1]
# 设置输出形状 ( 与输入形状相同 )
result_shape = [im.shape[0],im.shape[1]]
# 设置输入图像的 RGB 均值 ( 用于均值减法 )
rgb_mean = cv.mean(im)
# 显示输入图像
Image.fromarray(im)
```

● F-可以手动从*https://s3.amazonaws.com/onnx-model-zoo/duc/ResNet101_DUC_HDC.onnx*下载预训练模型。并设置使用GPU，还是CPU。

```
# 下载 ONNX 模型
# mx.test_utils.download('https://s3.amazonaws.com/onnx-model-zoo/duc/ResNet101_DUC_HDC.onnx')
# 配置 cpu 或 gpu
if len(mx.test_utils.list_gpus()) == 0:
    ctx = mx.cpu()
else:
    ctx = mx.gpu(0)
# 加载 ONNX 模型
mod = get_model(ctx, r'./model/ResNet101_DUC_HDC.onnx')
print("The model is loaded...")
--------------------------------------------------------------------------------
The model is loaded...
```

● G-处理输入图像，并执行预测，查看预测结果（图3.19–6）。

```
pre = preprocess(im)
conf,result_img,blended_img,raw = predict(pre)
result_img
```

图3.19-6 图像语义分割结果

● H-混合输出。分割图与真实图叠合，方便观察预测结果（图3.19–7）。

```
blended_img
```

● I-查看精度（confidence score）。为SoftMax回归分类模型输出的联合概率分布最大值。数值位于[0,1]，数值越大，像素属于某一分类可能性越大。

```
print('Confidence = %f' %(conf))
--------------------------------------------------------------------------------
Confidence = 0.929088
```

图3.19-7 图像及其语义分割结果的混合输出

3.19.4 城市空间要素组成、时空量度、绿视率和均衡度

cityscapes数据集，标签/分类包括主要的城市街道场景内容，这为城市空间的分析提供了基础的数据支持，例如对于固定行进流线，视野方向和宽度下，通过标签vegetation可以计算绿视率（Green view index，GVI），当绿视率达到一定水平，会让行人在街道空间中觉得舒适；通过sky可以获知视野下所见天空的比例，这与天空视域因子（Sky View Factor，SVF）可以比较研究；对于其他项，例如car、truck和bus 可以初步判断某一时刻街道的交通情况；person和rider则可以初步判断行人情况。根据待分析的内容，可以有意识地选择对应的要素进行分析，也可以综合考虑所有因素，计算每一位置的信息熵和均衡度，比较不同位置的混杂程度。通常，混杂比较高的位置可能感觉会比较热闹，而低的区域则相对简单和冷清。

因为将DUC预训练模型用于KITTI数据集，无人驾驶项目拍摄的连续图像，是固定行进流线、视野方向和宽度的。这可以保证图像具有统一的属性，避免因为拍摄上下角度变化的问题，使得图像之间不具有比较性。预测计算时间较长，为了避免数据丢失，将DUC预测的结果，conf 精度/概率、result_img 语义分割的掩码图像（颜色区分）、blended_img 叠合掩码和实际的图像、可方便查看分割与实际之间的差异、raw trainId数字索引等分别保存为图像格式及存储在列表下，用pickle保存。语义分割的结果如图3.19-8所示。

```
import glob
drive_29_0071_img_fp_list = glob.glob(
    r"G:/data/2011_09_29_drive_0071_sync/image_03/data/*.png")
drive_29_0071_img_fp_list.sort()
def sementicSeg_DUC_pred(DUC_output_root, img_fps, preprocess, predict):
    '''
    function - DUC 图像分割，及预测图像保存
    '''
    from tqdm.auto import tqdm
```

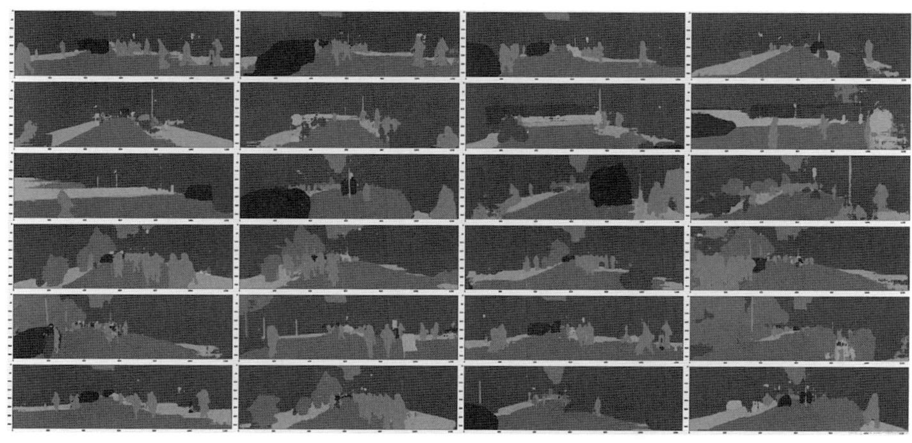

图3.19-8 KITTI数据集序列影像的语义分割结果（.mp4格式导出）

```
    import os
    import pickle
    import cv2 as cv
    from tqdm.auto import tqdm
    DUC_output_root = DUC_output_root
    conf_list = []
    raw_list = []
    for i, img in enumerate(tqdm(img_fps)):
        im = cv.imread(img)[:, :, ::-1]
        # 设置输出形状 ( 与输入形状相同 )
        result_shape = [im.shape[0], im.shape[1]]
        # 设置输入图像的 RGB 均值 ( 用于均值减法 )
        rgb_mean = cv.mean(im)
        pre = preprocess(im)
        conf, result_img, blended_img, raw = predict(pre)
        conf_list.append(conf)
        raw_list.append(raw)
        if not os.path.exists(os.path.join(DUC_output_root, "result_img")):
            os.makedirs(os.path.join(DUC_output_root, "result_img"))
        if not os.path.exists(os.path.join(DUC_output_root, "blended_img")):
            os.makedirs(os.path.join(DUC_output_root, "blended_img"))
        result_img.save(os.path.join(DUC_output_root,
                        "result_img/result_img_{}.png".format(i)))
        blended_img.save(os.path.join(
            DUC_output_root, "blended_img/blended_img_{}.png".format(i)))
    with open(os.path.join(DUC_output_root, 'KITTI_DUC_confi.pkl'), 'wb') as f1:
        pickle.dump(conf_list, f1)
    with open(os.path.join(DUC_output_root, 'KITTI_DUC_raw.pkl'), 'wb') as f2:
        pickle.dump(raw_list, f2)
DUC_output_root = r'G:\data\data_processed\KITTI_DUC'
sementicSeg_DUC_pred(DUC_output_root, img_fps=drive_29_0071_img_fp_list,
                     preprocess=preprocess, predict=predict)
--------------------------------------------------------------------------------
  0%|          | 0/1059 [00:00<?, ?it/s]
--------------------------------------------------------------------------------
import matplotlib.pyplot as plt
import matplotlib.animation as animation
import matplotlib.image as mpimg
from IPython.display import HTML
import os
import glob
DUC_output_root = r'G:\data\data_processed\KITTI_DUC'
result_img_fp_list = glob.glob(DUC_output_root+r'/result_img/*.png')
fig = plt.figure(figsize=(20, 10))
ims = [[plt.imshow(mpimg.imread(f), animated=True)]
       for f in result_img_fp_list[:]]
print("finished reading the imgs.")
# conda install -c conda-forge ffmpeg
ani = animation.ArtistAnimation(
    fig, ims, interval=50, blit=True, repeat_delay=1000)
ani.save(os.path.join(DUC_output_root, r'DUC_result_imgs.mp4'))
print(".mp4 saved.")
HTML(ani.to_html5_video())
--------------------------------------------------------------------------------
finished reading the imgs.
.mp4 saved.
```

对每一位置计算所有对象的频数，可以得知各对象在图像所代表的视野下占的比例。

```
def DUC_pred_frequency_moment(KITTI_DUC_raw_fp, KITTI_DUC_confi_fp):
    '''
    function - 读取 DUC 语义分割结果保存的 'KITTI_DUC_confi.pkl' 和 'KITTI_DUC_raw.pkl'
文件，用于位置图像对象 / 语义分割类别频数统计
    '''
    import pickle
    import numpy as np
    import pandas as pd
    from tqdm.auto import tqdm
    with open(KITTI_DUC_raw_fp, 'rb') as f:
        KITTI_DUC_raw = pickle.load(f)
    with open(KITTI_DUC_confi_fp, 'rb') as f:
        KITTI_DUC_confi = pickle.load(f)
    unique_id_all = np.unique(np.stack(KITTI_DUC_raw))
    print(" 所有出现的 id:{}".format(unique_id_all))  # 对应 trainId
```

表3.19-2 KITTI数据集的经纬度位置信息和时间戳及语义分割对象的频数

	road	sidewalk	building	wall	…	truck	bus	train	motorcycle	bicycle	confidence	trainID	lat	lon	timestamps_
0	84650	10430	319765	1696	…	0	0	0	0	0	0.957152	0	49.008650	8.398092	2011-09-29 13:54:59.990872576
1	86454	9873	322688	2318	…	0	0	0	4	652	0.957426	1	49.008777	8.397611	2011-09-29 13:55:00.094612992
2	87182	12023	325361	2264	…	0	0	0	8	451	0.957314	2	49.009162	8.396541	2011-09-29 13:55:00.198486528
…	…	…	…	…	…	…	…	…	…	…	…	…	…	…	…
1056	122516	14267	241194	14602	…	0	313	0	0	0	0.941350	1056	49.008706	8.397888	2011-09-29 13:56:49.666327808
1057	122223	14577	241842	14740	…	0	714	0	0	0	0.941474	1057	49.009215	8.396288	2011-09-29 13:56:49.770316544
1058	121671	14120	241772	14341	…	0	561	0	0	0	0.939658	1058	49.009079	8.396812	2011-09-29 13:56:49.874179584

```
    id_info_df = pd.DataFrame(columns=unique_id_all)
    for seg in tqdm(KITTI_DUC_raw):
        unique_id, counts_id = np.unique(seg, return_
counts=True)
        id_fre_dic = dict(list(zip(unique_id, counts_id)))
        for i in unique_id_all:
            if i not in unique_id.tolist():
                id_fre_dic.setdefault(i, 0)
        id_info_df = id_info_df.append(id_fre_dic, ignore_
index=True)
    id_info_df["confidence"] = KITTI_DUC_confi
    return id_info_df, unique_id_all
KITTI_DUC_raw_fp = r'G:\data\data_processed\KITTI_DUC\
KITTI_DUC_raw.pkl'
KITTI_DUC_confi_fp = r'G:\data\data_processed\KITTI_DUC\
KITTI_DUC_confi.pkl'
id_info_df, unique_id_all = DUC_pred_frequency_moment(
    KITTI_DUC_raw_fp, KITTI_DUC_confi_fp)
------------------------------------------------------
所有出现的id:[ 0  1  2  3  4  5  6  7  8  9 10 11 12 13 14
15 16 17 18]
  0%|          | 0/1059 [00:00<?, ?it/s]
```

读取KITTI数据集的经纬度位置信息和时间戳，与频数的DataFrame格式数据合并在一个DataFrame之下，见表3.19-2，方便后续数据处理。

```
import util_A
KITTI_info_fp = r'G:\data\2011_09_29_drive_0071_sync\oxts\
data'
timestamps_fp = r'G:\data\2011_09_29_drive_0071_sync\
image_03\timestamps.txt'
drive_29_0071_info = util_A.KITTI_info(KITTI_info_fp,
timestamps_fp)
drive_29_0071_info_coordi = drive_29_0071_info[['lat',
'lon', 'timestamps_']]

# 建立 trainID 到类别的映射字典，trainId2label 定义在 cityscapes 的标签数据处
理部分
trainID_label_mapping = {id_: trainId2label[id_].name for
id_ in unique_id_all}
id_info_df = id_info_df.rename(columns=trainID_label_
mapping)
id_info_df['trainID'] = id_info_df.index
id_info_df = id_info_df.join(drive_29_0071_info_coordi)
id_info_df
------------------------------------------------------
1059 rows × 24 columns
```

打印所有位置时刻（图3.19-9），每一图像包含对象的面积频数，即各对象像素占整体图像像素的比例。如果要查看单个对象，可以点击图例对应项。因为2011_09_29_drive_0071_sync部分数据位于城市的街巷内，因此可以明显的观察到building对象所占的数量较大，次之则为road，其他对象相对较小。vegetation在部分区域有较高的比例。

```
import plotly.express as px
labels = list(trainID_label_mapping.values())
fig = px.line(id_info_df, x="trainID", y=labels,
              hover_data=['confidence', 'lat', 'lon',
'trainID'],
              title='id_info_df'
              )
fig.show()
```

可以打印感兴趣的对象，观察在实际空间地理位置上的分布情况。例如，sky（图3.19-10）和vegetation（图3.19-11）的分布，通过量化的方式能够明确变化方式的具体位置。

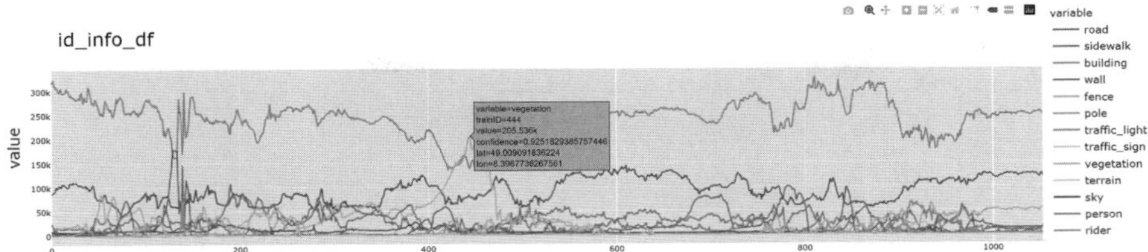

图3.19-9 所有时刻语义分割对象面积（像素数）比例

图3.19-10 天空视域因子（基于图像语义分割）

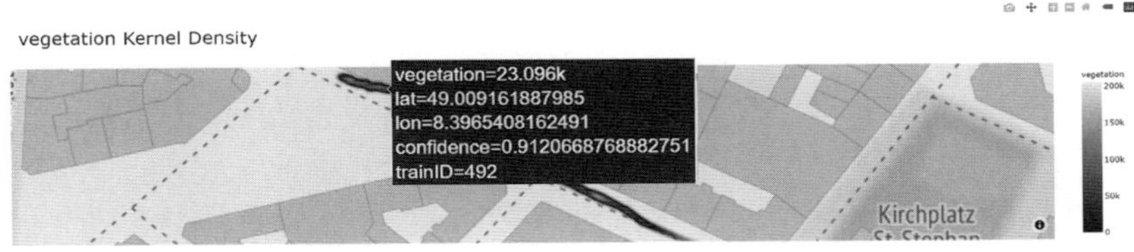

图3.19-11 绿视率（基于图像语义分割）

```
fig = px.density_mapbox(id_info_df, lat='lat', lon='lon', z='sky', radius=10,
                        center=dict(lat=49.008645, lon=8.398104), zoom=18,
                        mapbox_style="stamen-terrain",
                        hover_data=['confidence', 'lat', 'lon', 'trainID'],
                        title=r'sky Kernel Density')
fig.show()
import plotly.express as px
fig = px.density_mapbox(id_info_df, lat='lat', lon='lon', z='vegetation',
radius=10,
                        center=dict(lat=49.008645, lon=8.398104), zoom=18,
                        mapbox_style="stamen-terrain",
                        hover_data=['confidence', 'lat', 'lon', 'trainID'],
                        title=r'vegetation Kernel Density')
fig.show()
```

虽然对象的像素数量可以比较不同对象所占的比例，但是计算各自对象所占的百分比，则更容易得知对象在整个图像视野中比例的变化情况（表3.19–3）。

```
labels = list(trainID_label_mapping.values())
sum_syntax = 'pixels='+''.join("%s+" % ''.join(map(str, x))
                               for x in labels)[:-1]
id_info_df_int = id_info_df[labels].astype(int)
id_info_df_int = id_info_df_int.eval(sum_syntax)
id_info_df_int['vegetation_percent'] = id_info_df_int.apply(
    lambda row: row.vegetation/row.pixels*100, axis=1)
id_info_df_int['sky_percent'] = id_info_df_int.apply(
    lambda row: row.sky/row.pixels*100, axis=1)
id_info_df_int
--------------------------------------------------------------------------
1059 rows × 22 columns
```

打印植被百分比的分布，可以得知该街道植被在前半部分开始逐渐增加，在中心广场部分则有相对较多的树木栽植，但是到了后半段，则迅速减少。同时，也加入了天空的百分比（图3.19–12）。

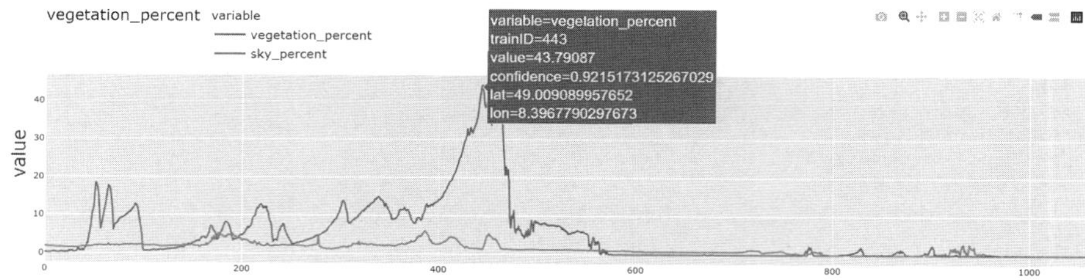

图3.19-12 绿视率和天空视域因子（百分比）

表3.19-3 语义分割对象百分比

	road	sidewalk	building	wall	…	truck	bus	train	motorcycle	bicycle	pixels	vegetation_percent	sky_percent
0	84650	10430	319765	1696	…	0	0	0	0	0	463012	0.559165	1.908590
1	86454	9873	322688	2318	…	0	0	0	4	652	463012	0.365649	1.878137
2	87182	12023	325361	2264	…	0	0	0	8	451	463012	0.269971	1.838829
…	…	…	…	…	…	…	…	…	…	…	…	…	…
1056	122516	14267	241194	14602	…	0	313	0	0	0	463012	0.000000	0.031965
1057	122223	14577	241842	14740	…	0	714	0	0	0	463012	0.000000	0.029589
1058	121671	14120	241772	14341	…	0	561	0	0	0	463012	0.000000	0.021814

```
import plotly.express as px
labels=list(trainID_label_mapping.values())
extracted_labels=['vegetation_percent','sky_
percent']
hover_data=['confidence','lat','lon','trainID']
id_info_df_int[hover_data]=id_info_df[hover_data]
fig = px.line(id_info_df_int, x="trainID",
y=extracted_labels,
              hover_data=hover_data,
              title='vegetation_percent'
             )
fig.show()
```

将“KITTI动态街景视觉感知”部分的代码加入到util_A模块中，此处调用计算视觉感知消失的距离（表3.19-4），获得开始点和消失点的索引，计算每一位置下与感知距离消失位置间均衡度的变化。这一信息的比较可以粗略地得知当前位置到视觉感知消失位置城市空间场景混杂程度的变化。当差值绝对值较大时，说明场景在混杂丰富和简单冷清间互相变换，即视觉消失的位置场景与当前场景有很大不同；如果差值绝对值较小，则说明城市空间对象的混合程度基本保持不变，即视觉消失的位置场景与当前场景类似。

对于需要花时间计算的内容，通常都将其保存到本地磁盘中，需要时直接调用，避免重复计算。

```
import util_A
import glob
drive_29_0071_img_fp_list = glob.glob(
    "G:/data/2011_09_29_drive_0071_sync/image_03/
data/*.png")
drive_29_0071_img_fp_list.sort()
#pip install opencv-python and pip install opencv-contrib-python
dsv_vp = util_A.dynamicStreetView_visualPerception(
    drive_29_0071_img_fp_list[:])
matches_num = dsv_vp.sequence_statistics()
---------------------------------------------------
计算关键点和描述子...
100%|██████████| 1059/1059 [01:16<00:00,
13.87it/s]
计算序列图像匹配数...
100%|██████████| 1058/1058 [31:31<00:00,
1.79s/it]
---------------------------------------------------
import util_A
import pandas as pd
KITTI_info_fp = r'G:\data\2011_09_29_drive_0071_
sync\oxts\data'
timestamps_fp = r'G:\data\2011_09_29_drive_0071_
sync\image_03\timestamps.txt'
drive_29_0071_info = util_A.KITTI_info(KITTI_info_
fp, timestamps_fp)
drive_29_0071_info_coordi = drive_29_0071_
info[['lat', 'lon', 'timestamps_']]
coordi_df = drive_29_0071_info_coordi
```

```
vanishing_gpd = vanishing_position_length(
    matches_num, coordi_df, epsg="EPSG:3857", threshold=0)
vanishing_gpd.to_pickle('./results/drive_29_0071_vanishing_gpd.pkl')
vanishing_gpd_ = pd.read_pickle('./results/drive_29_0071_vanishing_gpd.pkl')
vanishing_gpd_
------------------------------------------------------------------------------
1058 rows × 4 columns
```

表3.19-4 视觉感知消失的距离

	start_idx	end_idx	geometry	length
0	0	91	LINESTRING (934871.288 6276329.185, 934817.770...	11310.685843
1	1	81	LINESTRING (934817.770 6276350.782, 934698.648...	9900.262438
2	2	81	LINESTRING (934698.648 6276416.118, 934758.123...	9764.399529
3	3	81	LINESTRING (934758.123 6276382.241, 934555.096...	9695.952996
4	4	82	LINESTRING (934555.096 6276474.385, 934582.633...	9474.584042
...	...	...	...	...
1053	1053	1053	GEOMETRYCOLLECTION EMPTY	0.000000
1054	1054	1054	GEOMETRYCOLLECTION EMPTY	0.000000
1055	1055	1055	GEOMETRYCOLLECTION EMPTY	0.000000
1056	1056	1056	GEOMETRYCOLLECTION EMPTY	0.000000
1057	1057	1057	GEOMETRYCOLLECTION EMPTY	0.000000

合并语义分割对象百分比和视觉感知消失距离计算结果（表3.19-5）。

```
id_info_df = id_info_df_int.join(vanishing_gpd_)
id_info_df
------------------------------------------------------------------------------
1059 rows × 30 columns
------------------------------------------------------------------------------
id_info_df.to_pickle('./results/DUC_info_drive_29_0071_vanishing_gpd.pkl')
id_info_df_=pd.read_pickle('./results/DUC_info_drive_29_0071_vanishing_gpd.pkl')
```

计算信息熵和均衡度。计算的内容是图像中各个对象的频数或百分比（表3.19-6）。

```
def entroy_df_row(row, labels, id_info_df_):
    '''
    function - 计算 DataFrame 每行的信息熵，用于 df.apply(lambda)
    '''
    import numpy as np
    import math
    labels_percent = row[labels].to_numpy(
    )*1.000/id_info_df_.iloc[[0]][["pixels"]].to_numpy()
    labels_percent = labels_percent[labels_percent != 0]
    entropy = -np.sum(labels_percent*np.log(labels_percent.astype(np.float)))
    max_entropy = math.log(len(labels))
    frank_e = entropy/max_entropy
    return frank_e
id_info_df_['equilibrium'] = id_info_df_.apply(
    lambda row: entroy_df_row(row, labels, id_info_df_), axis=1)
id_info_df_
---------------------------------------------
1059 rows × 31 columns
```

打印均衡度的曲线分布（图3.19-13）。

```
import plotly.express as px
labels = list(trainID_label_mapping.values())
fig = px.line(id_info_df_, x="trainID", y='equilibrium',
              hover_data=['confidence', 'lat', 'lon', 'trainID'],
              title='equilibrium'
              )
fig.show()
```

打印均衡度的空间核密度（图3.19-14）。

表3.19-5　语义分割对象百分比和视觉感知消失距离

	road	...	vegetation_percent	sky_percent	confidence	lat	lon	trainID	start_idx	end_idx	geometry	length
0	84650	...	0.559165	1.908590	0.957152	49.008650	8.398092	0	0.0	91.0	LINESTRING (934871.288 6276329.185, 934817.770...	11310.685843
1	86454	...	0.365649	1.878137	0.957426	49.008777	8.397611	1	1.0	81.0	LINESTRING (934817.770 6276350.782, 934698.648...	9900.262438
2	87182	...	0.269971	1.838829	0.957314	49.009162	8.396541	2	2.0	81.0	LINESTRING (934698.648 6276416.118, 934758.123...	9764.399529
...	...	...	...	...	...	...	...	...	...	...	...	...
1056	122516	...	0.000000	0.031965	0.941350	49.008706	8.397888	1056	1056.0	1056.0	GEOMETRYCOLLECTION EMPTY	0.000000
1057	122223	...	0.000000	0.029589	0.941474	49.009215	8.396288	1057	1057.0	1057.0	GEOMETRYCOLLECTION EMPTY	0.000000
1058	121671	...	0.000000	0.021814	0.939658	49.009079	8.396812	1058	NaN	NaN	None	NaN

表3.19-6　对象频数的均衡度

	road	...	sky_percent	confidence	lat	lon	trainID	start_idx	end_idx	geometry	length	equilibrium
0	84650	...	1.908590	0.957152	49.008650	8.398092	0	0.0	91.0	LINESTRING (934871.288 6276329.185, 934817.770...	11310.685843	0.349989
1	86454	...	1.878137	0.957426	49.008777	8.397611	1	1.0	81.0	LINESTRING (934817.770 6276350.782, 934698.648...	9900.262438	0.345377
...	...	...	...	...	...	...	...	...	...	...	...	...
1057	122223	...	0.029589	0.941474	49.009215	8.396288	1057	1057.0	1057.0	GEOMETRYCOLLECTION EMPTY	0.000000	0.450238
1058	121671	...	0.021814	0.939658	49.009079	8.396812	1058	NaN	NaN	None	NaN	0.450034

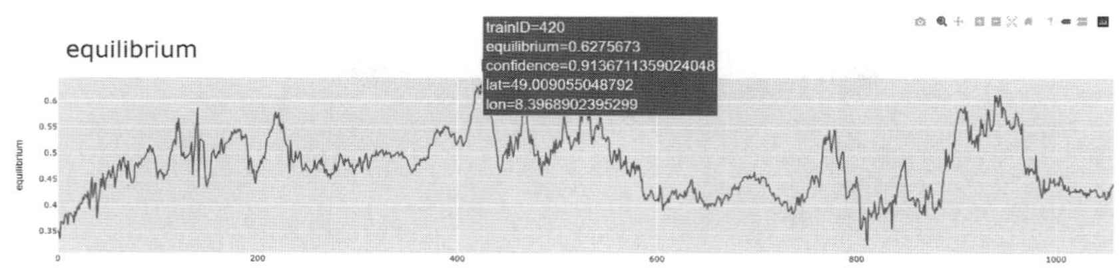

图3.19-13 均衡度曲线

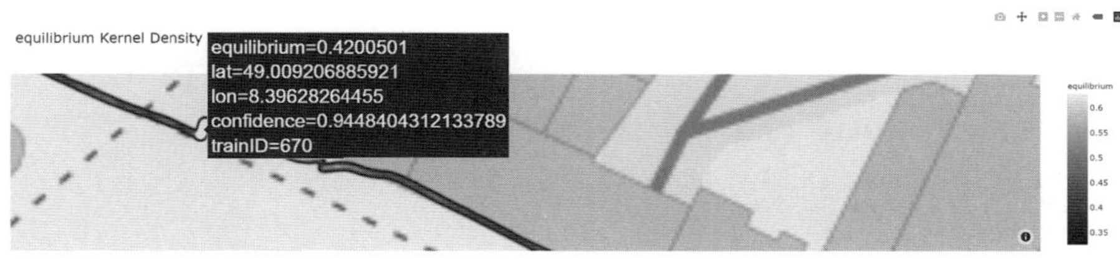

图3.19-14 均衡度的空间核密度

```
import plotly.express as px
fig = px.density_mapbox(id_info_df_, lat='lat', lon='lon', z='equilibrium',
radius=10,
                        center=dict(lat=49.008645, lon=8.398104), zoom=18,
                        mapbox_style="stamen-terrain",
                        hover_data=['confidence', 'lat', 'lon', 'trainID'],
                        title=r'equilibrium Kernel Density')
fig.show()
```

计算视觉感知消失距离对应图像之间的均衡度变化（图3.19-15）。

```
import plotly.express as px
id_info_df_.dropna(inplace=True)
id_info_df_["VP_equilibrium"] = id_info_df_.apply(lambda row: id_info_df_.iloc[[
                                                   int(row.end_idx)]].
equilibrium.values[0]-row.equilibrium, axis=1)
fig = px.density_mapbox(id_info_df_, lat='lat', lon='lon', z='VP_equilibrium',
radius=10,
                        center=dict(lat=49.008645, lon=8.398104), zoom=18,
                        mapbox_style="stamen-terrain",
                        hover_data=['confidence', 'lat', 'lon', 'trainID'],
                        title=r'VP_equilibrium Kernel Density')
fig.show()
```

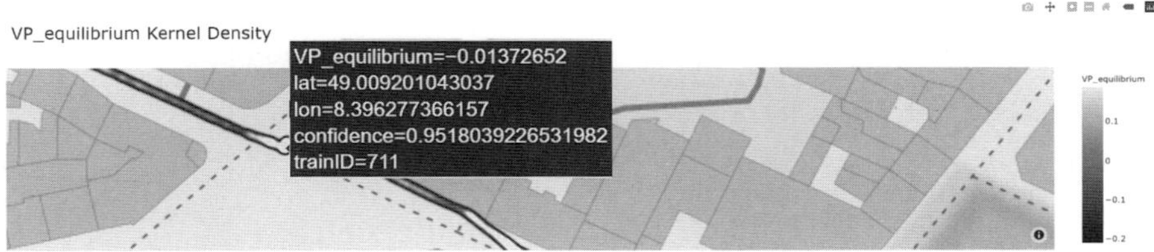

图3.19-15 对应视觉感知消失距离的均衡度变化

附：知识点

知识点-01：`super().__init__()`——继承父类的`init`方法

通过下述实例理解继承父类的方法，对于子类Child_robin虽然继承了父类，可以调用父类的方法（函数），但是因为子类自身的__init__初始化，覆盖了父类的属性，因此无法调用父类属性。对于子类Child_sparrow，增加了super().__init__()方法，从而可以调用父类属性。

```
class Parent:
    def __init__(self, name="bird"):
        self.name = name
class Child_robin(Parent):
    def __init__(self, species="robin"):
        self.species = species
class Child_sparrow(Parent):
    def __init__(self, species="sparrow"):
        self.species = species
        super(Child_sparrow, self).__init__()
p = Parent()
print("获取Parent类属性：name=%s" % p.name)
print("_"*50)
c_r = Child_robin()
print("获取子类Child_robin的属性：species=%s" % c_r.species)
try:
    print("获取父类Parent属性：name=%s" % c_r.name)
except AttributeError as error:
    print(error)
print("_"*50)
c_s = Child_sparrow()
print("获取子类Child_sparrow的属性：species=%s" % c_s.species)
print("获取父类Parent属性：name=%s" % c_s.name)
--------------------------------------------------------------------------------
获取Parent类属性：name=bird
__________________________________________________
获取子类Child_robin的属性：species=robin
'Child_robin' object has no attribute 'name'
__________________________________________________
获取子类Child_sparrow的属性：species=sparrow
获取父类Parent属性：name=bird
```

知识点-02：__getattr__，__setattr__和__delattr__

类Class_A的实例（instance）C_A，通过C_A.attri_a访问实例属性attri_a（对象变量），并返回属性对应值；通过实例的C_A.__dict__可以查看所有实例的属性（即实例的属性存储在__dict__中）。如果预提取实例中不存在的属性，则会调用__getattri__。如果类的变量（属性）定义在初始化函数外部，例如attri_c（类变量），则实例的C_A.__dict__并不包含该属性，但是在类自身的Class_A.__dict__对象中包含该属性。

在实例初始化、重新赋值及增加新的属性时，均会自动调用__setattr__方法，并用self.__dict__[name]=value语句，把属性键值对保存在__dict__对象中。其中，__setattr__(self,name,value)的参数name和value为固定参数，代表属性的键值对。

如果要删除实例的属性键值对，则可以执行del C_A.attri_d，调用__delattr__，用del self.__dict__[name]方法删除属性键值对。

```
class Class_A:
    attri_c = "attri_C"
    def __init__(self, attri_a, attri_b):
        self.attri_a = attri_a
        self.attri_b = attri_b
    def __getattr__(self, attri):
        return ('invoke __getattr__', attri)
    def __setattr__(self, name, value):
        print("invoke __setattr__",)
        self.__dict__[name] = value
    def __delattr__(self, name):
        print("invoke __delattr__",)
        print("deleting `{}`".format(str(name)))
        try:
            del self.__dict__[name]
            print("`{}` deleted".format(str(name)))
        except KeyError as k:
            return None
C_A = Class_A("attri_A", "attri_B")
print("实例C_A包含的属性及其值：", C_A.__dict__)
print("实例C_A已有属性attri_a，则直接返回该属性对应值，不会调用__getattr__，attri_a=",
```

```
C_A.attri_a)
print("实例C_A没有属性attri_none，则调用__getattr__：", C_A.attri_none)
print(Class_A.__dict__)
print("用类的实例C_A提取属性attri_c=%s" % C_A.attri_c,
      ";" "用类自身Class_A直接提取属性attri_c=%s" % Class_A.attri_c)
print("_"*50)
C_A.attri_b = "attri_B_assignment"
print("对已有属性重新赋值：", C_A.__dict__)
C_A.attri_d = "attri_D"
print("增加新的属性，并赋值：", C_A.__dict__)
-----------------------------------------------------------------------------
invoke __setattr__
invoke __setattr__
实例C_A包含的属性及其值： {'attri_a': 'attri_A', 'attri_b': 'attri_B'}
实例C_A已有属性attri_a，则直接返回该属性对应值，不会调用__getattr__，attri_a= attri_A
实例C_A没有属性attri_none，则调用__getattr__： ('invoke __getattr__', 'attri_none')
{'__module__': '__main__', 'attri_c': 'attri_C', '__init__': <function Class_
A.__init__ at 0x0000020817835820>, '__getattr__': <function Class_A.__
getattr__ at 0x0000020817835790>, '__setattr__': <function Class_A.__setattr__
at 0x0000020817835700>, '__delattr__': <function Class_A.__delattr__ at
0x0000020817835670>, '__dict__': <attribute '__dict__' of 'Class_A' objects>, '__
weakref__': <attribute '__weakref__' of 'Class_A' objects>, '__doc__': None}
用类的实例C_A提取属性attri_c=attri_C ；用类自身Class_A直接提取属性attri_c=attri_C
-------------------------------------------------------------------
invoke __setattr__
对已有属性重新赋值： {'attri_a': 'attri_A', 'attri_b': 'attri_B_assignment'}
invoke __setattr__
增加新的属性,并赋值： {'attri_a': 'attri_A', 'attri_b': 'attri_B_assignment', 'attri_
d': 'attri_D'}
-----------------------------------------------------------------------------
print("_"*50)
del C_A.attri_d
print("删除属性attri_d：",C_A.__dict__)
-----------------------------------------------------------------------------
invoke __delattr__
deleting `attri_d`
`attri_d` deleted
删除属性attri_d： {'attri_a': 'attri_A', 'attri_b': 'attri_B_assignment'}
```

知识点-03：mutable（可变）与immutable（不可变）

Python的数据类型分为mutable（可变）与immutable（不可变），mutable就是创建后可以修改，而immutable是创建后不可修改。

对于mutable，下述代码定义了变量a，并将变量b指向了变量a，因此a和b指向同一对象；但是当变量b执行运算后，则变量b指向新的对象（地址）。同样，定义列表lst_a，并将列表lst_b指向列表lst_a，则lst_a和lst_b指向同一对象。即使二者分别追加新的值，仍然指向同一对象。但是，重新定义变量lst_b为新的列表，则lst_b指向新的对象。

```
a = 0
b = a
print("a,b 是否指向同一个对象：id_a={};id_b={}".format(id(a),id(b)),id(a)==id(b))
b += 1
print("b执行运算后，a,b 是否指向同一个对象：",id(a)==id(b))
-----------------------------------------------------------------------------
a,b 是否指向同一个对象：id_a=140716884631264;id_b=140716884631264 True
b执行运算后，a,b 是否指向同一个对象： False
-----------------------------------------------------------------------------
lst_a = [0]
lst_b = lst_a
print("列表lst_a和lst_b是否指向同一个对象：", id(lst_a) == id(lst_b))
lst_b.append(99)
print("lit_b追加值后，列表lst_a和lst_b是否指向同一个对象：", id(lst_a) == id(lst_b))
lst_a.append(79)
print("lit_a追加值后，列表lst_a和lst_b是否指向同一个对象：", id(lst_a) == id(lst_b))
lst_b = [0]
print("lit_b定义新的列表，列表lst_a和lst_b是否指向同一个对象：", id(lst_a) == id(lst_b))
-----------------------------------------------------------------------------
列表lst_a和lst_b是否指向同一个对象： True
lit_b追加值后，列表lst_a和lst_b是否指向同一个对象： True
lit_a追加值后，列表lst_a和lst_b是否指向同一个对象： True
lit_b定义新的列表，列表lst_a和lst_b是否指向同一个对象： False
```

对于immutable，因为自定义的python类型一般都是mutable，如果实现immutable数据类型，通常需要重写对象（object）的__setattr__和__delattr__方法。例如，下述重新定义了__setattr__，并不会将待增加或修改的属性写入__dict__中，而是直接引起TypeError异常。为保证不能删除类实例对象，令__delattr__ = __setattr__。因此，待类immutable实例化为cls对象，修改删除和增加属性值都会引发异常。

```
class immutable:
    def __setattr__(self, *args):
        print("invoke __setattr__")
        raise TypeError("cannot modify the value of immutable instance")
    __delattr__ = __setattr__
    def __init__(self, name, value):
        super(immutable, self).__setattr__(name, value)
cls = immutable("attri_e", "attri_E")
print("实例初始化属性值，并读取 attri_e=%s" % cls.attri_e)
cls.attri_e = "attri_new"
-----------------------------------------------------------------------------
实例初始化属性值，并读取 attri_e=attri_E
invoke __setattr__
TypeError                                 Traceback (most recent call last)
~\AppData\Local\Temp/ipykernel_21220/902156601.py in <module>
      8 cls=immutable("attri_e","attri_E")
      9 print("实例初始化属性值，并读取 attri_e=%s"%cls.attri_e)
---> 10 cls.attri_e="attri_new"
~\AppData\Local\Temp/ipykernel_21220/902156601.py in __setattr__(self, *args)
      2     def __setattr__(self, *args):
      3         print("invoke __setattr__")
----> 4         raise TypeError("cannot modify the value of immutable instance")
      5     __delattr__ = __setattr__
      6     def __init__(self,name,value):
TypeError: cannot modify the value of immutable instance
-----------------------------------------------------------------------------
del cls.attri_e
-----------------------------------------------------------------------------
invoke __setattr__
TypeError                                 Traceback (most recent call last)
~\AppData\Local\Temp/ipykernel_21220/3598793624.py in <module>
----> 1 del cls.attri_e
~\AppData\Local\Temp/ipykernel_21220/902156601.py in __setattr__(self, *args)
      2     def __setattr__(self, *args):
      3         print("invoke __setattr__")
----> 4         raise TypeError("cannot modify the value of immutable instance")
      5     __delattr__ = __setattr__
      6     def __init__(self,name,value):
TypeError: cannot modify the value of immutable instance
-----------------------------------------------------------------------------
cls.attri_f = "attri_F"
-----------------------------------------------------------------------------
invoke __setattr__
TypeError                                 Traceback (most recent call last)
~\AppData\Local\Temp/ipykernel_21220/3100895739.py in <module>
----> 1 cls.attri_f="attri_F"
~\AppData\Local\Temp/ipykernel_21220/902156601.py in __setattr__(self, *args)
      2     def __setattr__(self, *args):
      3         print("invoke __setattr__")
----> 4         raise TypeError("cannot modify the value of immutable instance")
      5     __delattr__ = __setattr__
      6     def __init__(self,name,value):
TypeError: cannot modify the value of immutable instance
```

知识点-04：_variable，__variable和__variable__

Python中，成员函数和变量都是公开的public，在Python中没有public和private方法修饰成员函数或变量。虽然没有支持私有化（priviate），但是可以应用下画线的方法限制成员函数和成员变量的访问权限（尽力避免定义以下画线开头的变量）。_variable单下画线开始的成员变量叫作包含变量，只有类的实例和子类的实例能访问这些变量，并需要通过类的接口访问，不能用from module imort *的方法导入。__variable双下画线开始的成员变量为私有成员，只有类对象自己能访问，子类对象不能访问。__variable__前后双下画线，为Python特殊方法专用的标识，例如__init__()类的构造函数。

```
class private:
    def __init__(self):
        self.attri = 'attri public'
        self._attri = 'attri_singleUnderscore'
        self.__attri = 'attri__doubleUnderscore'
    def func(self):
        return self.attri+' func'
    def _func(self):
        return self._attri+' _func'
    def __func(self):
        return self.__attri+' __func'
    def invoke__func(self):
        return self.__func()
class private_Child(private):
    def __init__(self):
        self.attri_Child = 'attri child'
        super(private_Child, self).__init__()
p = private()
print("类属性--公有成员：attri=%s" % p.attri)
print("类属性--包含变量（单下画线）:_attri=%s" % p._attri)
try:
    print("类属性--私有变量（双下画线）:__attri=%s" % p.__attri)
except AttributeError as error:
    print(error)
print("_"*50)
print("类方法--公有成员：func=", p.func())
print("类方法--单下画线：_func=", p._func())
try:
    print("类方法--双下画线：__func=", p.__func())
except:
    print("没有类方法：__func")
print("_"*50)
p_child = private_Child()
print("子类调用父类单下画线方法：", p_child._func())
try:
    print("子类调用父类双下画线方法：", p_child.__func())
except:
    print("子类没有父类方法：__func()")
------------------------------------------------------------------------------
类属性--公有成员：attri=attri public
类属性--包含变量（单下画线）:_attri=attri_singleUnderscore
'private' object has no attribute '__attri'
__________________________________________________________________
类方法--公有成员：func= attri public func
类方法--单下画线：_func= attri_singleUnderscore _func
没有类方法：__func
__________________________________________________________________
子类调用父类单下画线方法： attri_singleUnderscore _func
子类没有父类方法：__func()
```

知识点-05：collections.namedtuple

collections.namedtuple(typename, field_names, *, verbose=False, rename=False, module=None)[①]，其中参数typename为创建的一个元组子类类名，用于实例化各种元组对象；field_names类似于字典的键（key），通过键提取对应的值（value）；rename默认为False，如果为True，则不能包含有“非Python标识符，Python中的关键字以及重复的name”；如果有，则会默认重命名。

① collections，专门的数据类型，为常规的dict、list、set和tuple数据结构提供了可替代方案（https://docs.python.org/3/library/collections.html）。

```
from collections import namedtuple
# 01-实例化 nametuple 对象
Point = namedtuple('Point', ['x', 'y'])
# 02-使用关键字参数或位置参数初始化 nametuple
p = Point(11, y=22)
print("02-使用关键字参数或位置参数初始化 nametuple:p={}".format(p))
# 03-使用键提取元组元素
print("03-使用键提取元组元素 :p[0]={},p[1]={}".format(p[0], p[1]))
# 04-拆包
x, y = p
print("04-拆包：x,y=p-->x={},y={}".format(x, y))
# 05-instance.key 的方式提取值
print("05-instance.key 的方式提取值 :p.x={},p.y={}".format(p.x, p.y))
# 06-用已有序列或可迭代对象实例化一个 nametuple
lst = [99, 77]
```

```
print("06-用已有序列或可迭代对象实例化一个nametuple:Point._make(lst)={}".format(Point._make(lst)))
# 07-将nametuple对象转换为有序字典OrderDict
print("07-将nametuple对象转换为有序字典OrderDict:p._asdict()={}".format(p._asdict()))
# 08-有序字典转换为nametuple对象
dic = {'x': 11, 'y': 22}
print("08-有序字典转换为nametuple对象 :Point(**dic)={}".format(Point(**dic)))
# 09-替换值
print("09-替换值 :p._replace(x=33) -- >{}".format(p._replace(x=33)))
# 10-获取所有nametuple对象字段名
print("10-获取所有nametuple对象字段名 :p._fields={}".format(p._fields))
--------------------------------------------------------------------------------
02-使用关键字参数或位置参数初始化nametuple:p=Point(x=11, y=22)
03-使用键提取元组元素 :p[0]=11,p[1]=22
04-拆包：x,y=p -- >x=11,y=22
05-instance.key的方式提取值 :p.x=11,p.y=22
06-用已有序列或可迭代对象实例化一个nametuple:Point._make(lst)=Point(x=99, y=77)
07-将nametuple对象转换为有序字典OrderDict:p._asdict()={'x': 11, 'y': 22}
08-有序字典转换为nametuple对象 :Point(**dic)=Point(x=11, y=22)
09-替换值 :p._replace(x=33) -- >Point(x=33, y=22)
10-获取所有nametuple对象字段名 :p._fields=('x', 'y')
```

知识点-06：importlib与getattr

Python标准库importlib[1]，可以导入自定义的对象（.py文件/模块），并支持传入字符串导入模块。首先定义了importlib_func_A.py文件，将其置于datasets文件夹（包）下，使用importlib库，读取模块，并应用模块的基本操作，读取模块中的变量值、类的属性和方法。同时可以应用importlib.util.find_spec查看是否存在模块等。读取模块的属性使用getattr方法。

① importlib（https://docs.python.org/3/library/importlib.html）。

```
# datasets/importlib_func_A.py
# -*- coding: utf-8 -*-
attri_f = "attri_F"
def func_A():
    print("importlib_func_A/func_A")
class cls_A:
    attri_g = "attri_G"
    def func_B():
        print("importlib_func_A/cls_A/func_B")
if __name__ == "__main__":
    func_A()
--------------------------------------------------------------------------------
importlib_func_A/func_A
--------------------------------------------------------------------------------
args.impoftlib_module = "importlib_func_A"
def dynamic_import(package, module):
    import importlib
    '''
    function - 应用importlib调入自定义模块
    '''
    return importlib.import_module('{}.{}'.format(package, module))
module = dynamic_import("dataset", args.impoftlib_module)
print("imported module:", module)
print("调入模块中的变量值：", getattr(module, "attri_f"))
cls_A = getattr(module, 'cls_A')
print("调入模块中的类：", cls_A)
print("调入模块中类的属性：", cls_A.attri_g)
print("_"*50)
print("调入模块中类的方法：")
cls_A.func_B()
--------------------------------------------------------------------------------
imported module: <module 'dataset.importlib_func_A' from 'C:\\Users\\richi\\omen_richiebao\\omen_github\\USDA_CH_final\\USDA\\notebook\\dataset\\importlib_func_A.py'>
调入模块中的变量值： attri_F
调入模块中的类： <class 'dataset.importlib_func_A.cls_A'>
调入模块中类的属性： attri_G
__________________________________________________
调入模块中类的方法：
importlib_func_A/cls_A/func_B
--------------------------------------------------------------------------------
def check_module(package, module):
    '''
```

```
    function - 应用 importlib 查看是否存在模块
    '''
    import importlib
    module_spec = importlib.util.find_spec('{}.{}'.format(package, module))
    if module_spec is None:
        print("Module:{} not found.".format('{}.{}'.format(package, module)))
        return None
    else:
        print("Module:{} can be imported.".format(
            '{}.{}'.format(package, module)))
    return module_spec
check_module('dataset', args.impoftlib_module)
--------------------------------------------------------------------------------
Module:dataset.importlib_func_A can be imported.
ModuleSpec(name='dataset.importlib_func_A', loader=<_frozen_importlib_external.
SourceFileLoader object at 0x0000020817914CA0>, origin='C:\\Users\\richi\\omen_
richiebao\\omen_github\\USDA_CH_final\\USDA\\notebook\\dataset\\importlib_func_
A.py')
```

参考文献（References）:

[1] Cityscapes Dataset Overview, https://www.cityscapes-dataset.com/dataset-overview/.

[2] Tao, A., Sapra, K., & Catanzaro, B. (2020). Hierarchical Multi-Scale Attention for Semantic Segmentation. arXiv [Cs.CV]. Retrieved from http://arxiv.org/abs/2005.10821.

3.20 高分辨率遥感影像解译

3.20.1 无监督土地分类（聚类方法）

遥感影像的各个波段记录了地物的相关信息，那么以波段的数据作为机器学习的训练数据集，喂入相关模型，可以对应解决相关问题。其中之一是使用聚类的方法初步实现无监督土地分类（K-Menas算法）。sentinel-2影像有多个波段，可以尝试使用单个波段，或者多个波段作为特征向量，对比波段的合成显示，估计不同输入数据聚类结果的效果。

sentinel-2影像的信息均记录于MTD_MSIL2A.xml中（可查看sentinel-2部分内容），因此可以从该文件获取各个波段的路径。该文件给出的路径为相对于影像文件夹的相对路径。

```
def Sentinel2_bandFNs(MTD_MSIL2A_fn):
    '''
    funciton - 获取 sentinel-2 波段文件路径，和打印主要信息
    Paras:
        MTD_MSIL2A_fn - MTD_MSIL2A 文件路径
    Returns:
        band_fns_list - 波段相对路径列表
        band_fns_dict - 波段路径为值，反应波段信息的字段为键的字典
    '''
    import xml.etree.ElementTree as ET
    Sentinel2_tree = ET.parse(MTD_MSIL2A_fn)
    Sentinel2_root = Sentinel2_tree.getroot()
    print("GENERATION_TIME:{}\nPRODUCT_TYPE:{}\nPROCESSING_LEVEL:{}".
format(Sentinel2_root[0][0].find('GENERATION_TIME').text,

Sentinel2_root[0][0].find(
'PRODUCT_TYPE').text,
Sentinel2_root[0][0].find(
'PROCESSING_LEVEL').text
))

    print("MTD_MSIL2A.xml 文件父结构 :")
    for child in Sentinel2_root:
        print(child.tag, "-", child.attrib)
    print("_"*50)
    # [elem.text for elem in Sentinel2_root[0][0][11][0][0].iter()]
    band_fns_list = [elem.text for elem in Sentinel2_root.iter('IMAGE_FILE')]
    band_fns_dict = {f.split('_')[-2]+'_'+f.split('_')
                    [-1]: f+'.jp2' for f in band_fns_list}
    print('获取 sentinel-2 波段文件路径 :\n', band_fns_dict)

    return band_fns_list, band_fns_dict
MTD_MSIL2A_fn = r'G:\data\S2B_MSIL2A_20200709T163839_N0214_R126_
T16TDM_20200709T211044.SAFE\MTD_MSIL2A.xml'
band_fns_list, band_fns_dict = Sentinel2_bandFNs(MTD_MSIL2A_fn)
----------------------------------------------------------------------------
GENERATION_TIME:2020-07-09T21:10:44.000000Z
PRODUCT_TYPE:S2MSI2A
PROCESSING_LEVEL:Level-2A
MTD_MSIL2A.xml 文件父结构 :
{https://psd-14.sentinel2.eo.esa.int/PSD/User_Product_Level-2A.xsd}General_Info
- {}
...
{https://psd-14.sentinel2.eo.esa.int/PSD/User_Product_Level-2A.xsd}Quality_
Indicators_Info - {}
__________________________________________________
获取 sentinel-2 波段文件路径 :
 {'B02_10m': 'GRANULE/L2A_T16TDM_A017455_20200709T164859/IMG_DATA/R10m/
T16TDM_20200709T163839_B02_10m.jp2', 'B03_10m': 'GRANULE/L2A_T16TDM_
A017455_20200709T164859/IMG_DATA/R10m/T16TDM_20200709T163839_B03_10m.jp2',
...
 'WVP_60m': 'GRANULE/L2A_T16TDM_A017455_20200709T164859/IMG_DATA/R60m/
T16TDM_20200709T163839_WVP_60m.jp2', 'SCL_60m': 'GRANULE/L2A_T16TDM_
A017455_20200709T164859/IMG_DATA/R60m/T16TDM_20200709T163839_SCL_60m.jp2'}
```

根据返回字典的键，可以提取对应的波段路径名。EarthPy库的`stack`方法可以融合多个波

段，同时会返回波段的元数据，包括：driver驱动，dtype数据类型，nodata空值，width影像宽度，height影像高度，count波段数量，crs坐标系统（投影），transform变换，blockxsizex向单元数量（每个单元的精度为10m，即一个像素代表10m的实际地理空间，小于10m的地物则无法分辨），blockysizey向单元数量。查看影像波段如图3.20-1所示。

```
import os
import matplotlib.pyplot as plt
import earthpy.spatial as es
import earthpy.plot as ep
import geopandas as gpd
imgs_root = r"G:\data\S2B_MSIL2A_20200709T163839_N0214_R126_
T16TDM_20200709T211044.SAFE"
bands_selection = ["B02_10m", "B03_10m", "B04_10m", "B08_10m"]
stack_bands = [os.path.join(imgs_root, band_fns_dict[b])
               for b in bands_selection]
array_stack, meta_data = es.stack(stack_bands)
print("meta_data:\n", meta_data)
ep.plot_bands(array_stack, title=bands_selection,
              cols=array_stack.shape[0], cbar=True, figsize=(10, 10))
plt.show()
--------------------------------------------------------------------------------
meta_data:
 {'driver': 'JP2OpenJPEG', 'dtype': 'uint16', 'nodata': None, 'width': 10980, 'height':
10980, 'count': 4, 'crs': CRS.from_epsg(32616), 'transform': Affine(10.0, 0.0,
399960.0,
       0.0, -10.0, 4700040.0), 'blockxsize': 1024, 'blockysize': 1024, 'tiled':
True}
```

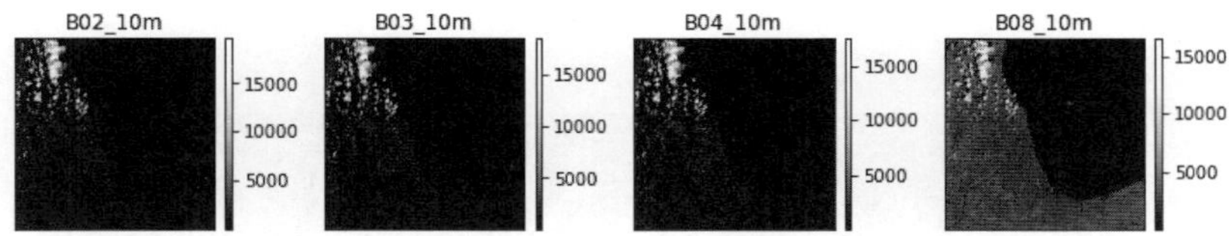

图3.20-1 打印sentinel-2影像波段

在QGIS中读取一个波段，或多个波段的组合显示，绘制裁切边界（设置坐标为WGS84，不配置投影，读取后根据影像的投影再进行定义），用于影像的裁切。裁切文件保存于指定的文件夹下。

```
crop_output_dir = r'G:\data\data_processed\sentinel-2_crop'
imgs_crs = meta_data['crs']
shape_polygon_fp = './data/sentinel2Chicago_boundary/sentinel2Chicago_boundary.
shp'
crop_bound = gpd.read_file(shape_polygon_fp)
crop_bound_proj = crop_bound.to_crs(imgs_crs)
crop_imgs = es.crop_all([os.path.join(imgs_root, f+'.jp2') for f in band_fns_
list],
                        crop_output_dir, crop_bound_proj, overwrite=True)  # 对所
有波段 band 执行裁切
print("finished cropping...")
--------------------------------------------------------------------------------
finished cropping...
```

显示裁切后的影像（图3.20–2）。

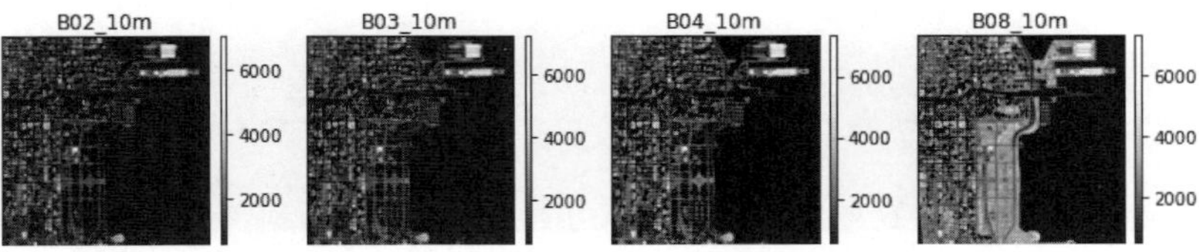

图3.20-2 打印sentinel-2裁切后的影像波段

```
import glob
croppedImgs_fns = glob.glob(crop_output_dir+"/*.jp2")
croppedBands_fnsDict = {
    f.split('_')[-3]+'_'+f.split('_')[-2]: f for f in croppedImgs_fns}
bands_selection_ = ["B02_10m", "B03_10m",
                    "B04_10m", "B08_10m"]  #" AOT_10m" ," WVP_10m"
cropped_stack_bands = [croppedBands_fnsDict[b] for b in bands_selection_]
cropped_array_stack, _ = es.stack(cropped_stack_bands)
ep.plot_bands(cropped_array_stack, title=bands_selection_,
              cols=cropped_array_stack.shape[0], cbar=True, figsize=(10, 10))
plt.show()
```

可以尝试调整不同的聚类数量n_cluster参数，分类越多划分的地物类别也就越细。基于聚类无监督分类的结果并没有明确分类的名称，需要结合已经聚类的结果，根据实际地物情况判别。注意喂入模型数据的形状为（样本数、特征数）。如果理解为矩阵，则每一列为一个特征向量，每一行为一个样本的多个特征值。通常输入的特征数越多，波段数越多，分类的精度可能越好。聚类结果如图3.20-3所示。

```
from sklearn import cluster
import matplotlib.pyplot as plt
X = cropped_array_stack.reshape(
    cropped_array_stack.shape[0], -1).transpose(1, 0)
print(X)
k_means = cluster.KMeans(n_clusters=8)
k_means.fit(X)
X_cluster = k_means.labels_
X_cluster = X_cluster.reshape(cropped_array_stack[0, :, :].shape)
plt.figure(figsize=(8, 8))
plt.imshow(X_cluster, cmap="hsv")
plt.show()
```

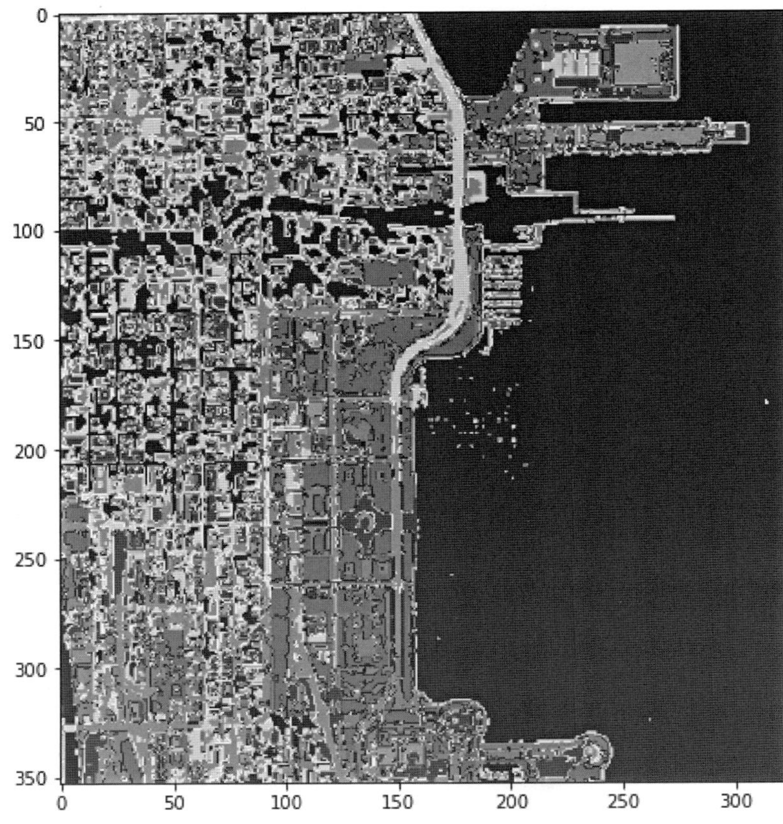

图3.20-3　sentinel-2影像波段聚类结果

```
------------------------------------------------------------------------------
[[2972 3248 3344 3550]
 [2574 2548 2602 2768]
 [1406 1406 1824 1920]
 ...
 [ 825  749  519  484]
 [ 811  738  516  498]
 [ 776  760  538  488]]
```

3.20.2 VGG16卷积神经网络

VGGNet研究了在大规模图像识别环境下，卷积网络深度对识别精度的影响。主要贡献是使用非常小的（3 × 3）卷积滤波器（卷积核）和（2 × 2）的最大池化层反复堆叠，在深度不断增加的网络下的表现评估。当将网络深度推进到16～19个全支持层时（图表的第C、D列为16层，第E列为19层），可以发现识别精度得以显著提升。该项研究最初用于ImageNet[①]数据集，并同时能够很好地泛化到其他数据集。

对VGGNet网络的理解同样可以对应到图像特征提取——尺度不变特征转换下尺度空间（scale space）的概念上。因为不同的地物尺寸不同，因此对于同一地理范围下的影像，分辨率越高，例如0.3～0.5m，则可以识别出行人轮廓。但是，10m的高空分辨率则无法识别。对于通常大于10m的对象，例如建筑、绿地则可以识别。这个变换的分辨率就是尺度空间的纵向深度，由降采样实现。对应到VGG网络上，就是网络深度的不断增加，是由maxpool最大池化层实现。因为不同地物的尺寸多样，但是通常可以形成一个连续的尺寸变化，例如从室外摆放的餐具，过往或静坐的行人，到车辆、建筑，再到农田和成片的林地。因此，为了检测到每一地物对应的尺度空间，采用2 × 2的最大池化能够很好地捕捉到不同的地物。即低分辨率的图像可以忽略掉较小的对象，而专注于该尺度及之上的对象，以此类推。在尺度空间中还有一个水平向，使用不同的卷积核检测同一尺度即深度下地物即图像的特征。不同的卷积核会识别出不同的特征内容，例如对象间的边界形状、颜色的差异变化，以及很多一般常识很难判定但却可以区分对象的特征。因此，在每一深度进行卷积操作时，通常要使用多个不同的卷积核，并随机初始化卷积核数值，以捕捉到对象的特征。这对应深度网络结构中的输出通道数。VGGNet在深度增加过程中，所使用的卷积核大小不变，均为3 × 3。因为深度的逐层增加，不同尺度的地物会被捕捉到，同一大小的卷积核可以检测到不同地物的特征。同时，使用的卷积数量在增加，以适应深度增加，尺度增大，即图像越加模糊时的特征提取。图像的特征并不仅表现在一次卷积的结果上，例如，如果应用一次卷积提取了对象的轮廓边界，那么仍然可以再应用卷积提取对象轮廓边界的特征，以此类推。这可以用于解释每一层深度/尺度下使用多层卷积的原因。

① ImageNet数据集，是按WordNet层次结构（目前只有名词）组织的图像数据库。其中，层次结构的每个节点由成百上千的图像描述。该项目在推进计算机视觉和深度学习研究方面发挥了重要作用。这些数据免费提供给研究人员用于非商业用途（https://image-net.org/）。

表3.20-1 卷积网络配置

ConvNet Configuration（卷积网络配置）					
A	A-LRN	B	C	D	E
11 weight layers	11 weight layers	13 weight layers	16 weight layers	16 weight layers	19 weight layers
input(224 × 224 RGB image)					
conv3-64	conv3-64 LRN	conv3-64 conv3-64	conv3-64 conv3-64	conv3-64 conv3-64	conv3-64 conv3-64
maxpool					
conv3-128	conv3-128	conv3-128 conv3-128	conv3-128 conv3-128	conv3-128 conv3-128	conv3-128 conv3-128
maxpool					
conv3-256 conv3-256	conv3-256 conv3-256	conv3-256 conv3-256	conv3-256 conv3-256 conv1-256	conv3-256 conv3-256 conv3-256	conv3-256 conv3-256 conv3-256 conv3-256
maxpool					
conv3-512 conv3-512	conv3-512 conv3-512	conv3-512 conv3-512	conv3-512 conv3-512 conv1-512	conv3-512 conv3-512 conv3-512	conv3-512 conv3-512 conv3-512 conv3-512

续表

ConvNet Configuration（卷积网络配置）					
A	A-LRN	B	C	D	E
conv3-512 conv3-512	conv3-512 conv3-512	conv3-512 conv3-512	conv3-512 conv3-512 conv1-512	conv3-512 conv3-512 conv3-512	conv3-512 conv3-512 conv3-512 conv3-512
maxpool					
FC-4096					
FC-4096					
FC- 1000					
soft-max					

将表3.20–1的第D列，即VGG16，通过方块序列图3.20–4的形式可以更好地表述、观察层级间的变化。

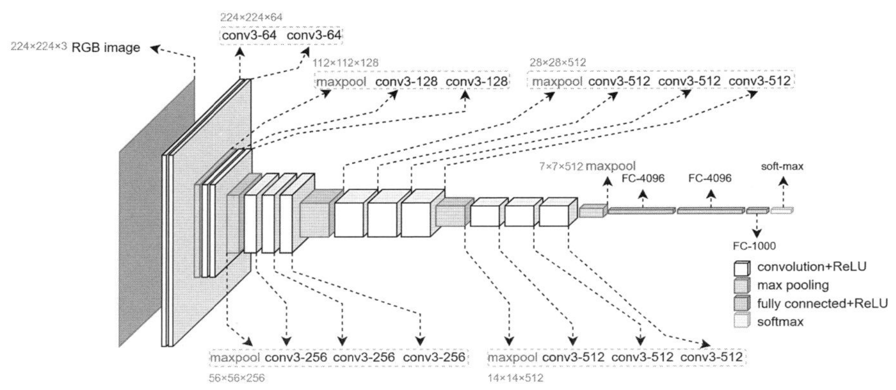

图3.20–4 VGG16卷积神经网络方块序列图解

ImageNet数据集

ImageNet数据集于2007年开始建设，已有超过1500万张图像，2万多个类别，是一个庞大的数据集。它是根据WordNet[①]层次结构（目前只有名词nouns）组织的图像数据库。其中，层次结构的每一个节点都由成百上千张图像描述。ImageNet数据集1000个类别文件可以从imagenet_classes.txt[②]处下载，其分类涉及动植物和各类人造物。

VGGNet预训练模型已经置于torchvision.models[③]模型库中，通过下载该模型，来尝试识别对象（参考VGG-NETS[④]）。

● 01-下载预训练的VGG16模型

```
import torch
model = torch.hub.load('pytorch/vision:v0.6.0', 'vgg16', pretrained=True)
model.eval()
-----------------------------------------------------------------------------------
Downloading: "https://github.com/pytorch/vision/archive/v0.6.0.zip" to C:\Users\richi/.cache\torch\hub\v0.6.0.zip
Downloading: "https://download.pytorch.org/models/vgg16-397923af.pth" to C:\Users\richi/.cache\torch\hub\checkpoints\vgg16-397923af.pth
  0%|          | 0.00/528M [00:00<?, ?B/s]
VGG(
  (features): Sequential(
    (0): Conv2d(3, 64, kernel_size=(3, 3), stride=(1, 1), padding=(1, 1))
    (1): ReLU(inplace=True)
    (2): Conv2d(64, 64, kernel_size=(3, 3), stride=(1, 1), padding=(1, 1))
    (3): ReLU(inplace=True)
    (4): MaxPool2d(kernel_size=2, stride=2, padding=0, dilation=1, ceil_
```

① WordNet，是一个大型的英语词汇数据库。名词、动词、形容词和副词被分组到认知同义词集（同义词集），每个都表达一个不同的概念。同义词集通过概念——语义和词汇关系相互关联。WordNet可以免费公开下载。WordNet的结构使其成为计算语言学和自然语言处理的有用工具（https://wordnet.princeton.edu/）。

② ImageNet数据集1000个类别文件——imagenet_classes.txt，（https://raw.githubusercontent.com/pytorch/hub/master/imagenet_classes.txt）。

③ torchvision.models，预定义深度学习模型，（https://pytorch.org/vision/0.8/models.html）。

④ VGG-NETS，2014年Imagenet ILSVRC挑战赛获奖的ConvNets（https://pytorch.org/hub/pytorch_vision_vgg/）。

```
mode=False)
    (5): Conv2d(64, 128, kernel_size=(3, 3), stride=(1, 1), padding=(1, 1))
    (6): ReLU(inplace=True)
    (7): Conv2d(128, 128, kernel_size=(3, 3), stride=(1, 1), padding=(1, 1))
    (8): ReLU(inplace=True)
    (9): MaxPool2d(kernel_size=2, stride=2, padding=0, dilation=1, ceil_
mode=False)
    (10): Conv2d(128, 256, kernel_size=(3, 3), stride=(1, 1), padding=(1, 1))
    (11): ReLU(inplace=True)
    (12): Conv2d(256, 256, kernel_size=(3, 3), stride=(1, 1), padding=(1, 1))
    (13): ReLU(inplace=True)
    (14): Conv2d(256, 256, kernel_size=(3, 3), stride=(1, 1), padding=(1, 1))
    (15): ReLU(inplace=True)
    (16): MaxPool2d(kernel_size=2, stride=2, padding=0, dilation=1, ceil_
mode=False)
    (17): Conv2d(256, 512, kernel_size=(3, 3), stride=(1, 1), padding=(1, 1))
    (18): ReLU(inplace=True)
    (19): Conv2d(512, 512, kernel_size=(3, 3), stride=(1, 1), padding=(1, 1))
    (20): ReLU(inplace=True)
    (21): Conv2d(512, 512, kernel_size=(3, 3), stride=(1, 1), padding=(1, 1))
    (22): ReLU(inplace=True)
    (23): MaxPool2d(kernel_size=2, stride=2, padding=0, dilation=1, ceil_
mode=False)
    (24): Conv2d(512, 512, kernel_size=(3, 3), stride=(1, 1), padding=(1, 1))
    (25): ReLU(inplace=True)
    (26): Conv2d(512, 512, kernel_size=(3, 3), stride=(1, 1), padding=(1, 1))
    (27): ReLU(inplace=True)
    (28): Conv2d(512, 512, kernel_size=(3, 3), stride=(1, 1), padding=(1, 1))
    (29): ReLU(inplace=True)
    (30): MaxPool2d(kernel_size=2, stride=2, padding=0, dilation=1, ceil_
mode=False)
  )
  (avgpool): AdaptiveAvgPool2d(output_size=(7, 7))
  (classifier): Sequential(
    (0): Linear(in_features=25088, out_features=4096, bias=True)
    (1): ReLU(inplace=True)
    (2): Dropout(p=0.5, inplace=False)
    (3): Linear(in_features=4096, out_features=4096, bias=True)
    (4): ReLU(inplace=True)
    (5): Dropout(p=0.5, inplace=False)
    (6): Linear(in_features=4096, out_features=1000, bias=True)
  )
)
```

- 02-读取一幅图像（图3.20-5）。执行调整图像大小Resize、裁切CenterCrop、转换为张量ToTensor和标准化Normalize等操作，使其满足网络结构的数据输入需求。

```
from PIL import Image
from torchvision import transforms
import matplotlib.pyplot as plt
import numpy as np
cat_01 = r'./data/stuff_01.jpg' # cat_01.png;stuff_01.jpg
cat_img = Image.open(cat_01).convert('RGB')
```

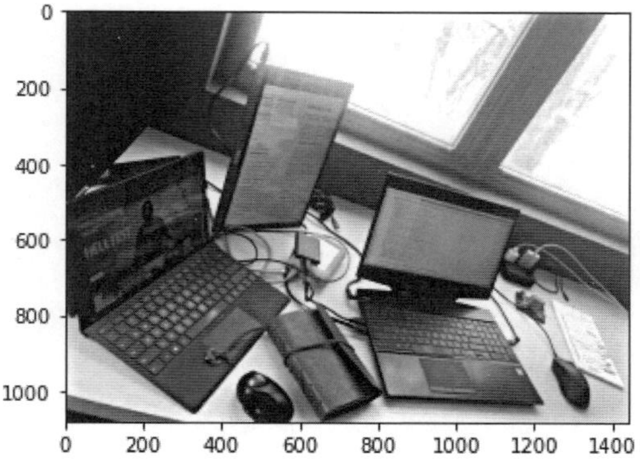

图3.20-5 实验图像

```
plt.imshow(cat_img)
plt.show()
input_image = Image.open(cat_01)
preprocess = transforms.Compose([
    transforms.Resize(256),
    transforms.CenterCrop(224),
    transforms.ToTensor(),
    transforms.Normalize(mean=[0.485, 0.456, 0.406],
                         std=[0.229, 0.224, 0.225]),
])
input_tensor = preprocess(input_image)
# 按模型输入条件建立一个小批次（batch）样本
input_batch = input_tensor.unsqueeze(0)
print("VGG16 输入数据的形状（batchsize, nChannels, Height, Width）：", input_batch.shape)
-----------------------------------------------------------------------------
VGG16 输入数据的形状（batchsize, nChannels, Height, Width）： torch.Size([1, 3, 224, 224])
```

- 03 - 图像中的内容预测概率

```
# 输入数据和模型传入 GPU 以提供计算效率（move the input and model to GPU for speed if available）
if torch.cuda.is_available():
    input_batch = input_batch.to('cuda')
    model.to('cuda')
with torch.no_grad():
    output = model(input_batch)
# 全连接最后一层的线性输出通道数为 1000，对应 ImageNet 数据集的 1000 个分类 （Tensor of shape 1000, with confidence scores over Imagenet's 1000 classes）
# 输出为非标准化得分。为了得到概率，可执行 softmax。
probabilities = torch.nn.functional.softmax(output[0], dim=0)
print(" 预测的 1000 个分类联合概率分布数组的形状：", probabilities.shape)
-----------------------------------------------------------------------------
预测的 1000 个分类联合概率分布数组的形状： torch.Size([1000])
```

- 04 - 打印预测概率分布中最大的前几个对象，可以观察到预测的对象，desktop computer（及monitor、laptop、screen、computer keyboard）、notebook、desk都出现在该图像中。而printer和modem则没有，但是modem和插座的形状比较近似。

```
# 读取 ImageNet 数据集 1000 个类别文件，可以从 [imagenet_classes.txt](https://raw.githubusercontent.com/pytorch/hub/master/imagenet_classes.txt) 处下载
with open("./data/imagenet_classes.txt", "r") as f:
    categories = [s.strip() for s in f.readlines()]
# 显示所预测图像，和前几个最大概率对应的分类名（ Show top categories per image）
top5_prob, top5_catid = torch.topk(probabilities, 10)
for i in range(top5_prob.size(0)):
    print(categories[top5_catid[i]], top5_prob[i].item())
-----------------------------------------------------------------------------
desktop computer 0.17022140324115753
notebook 0.15083454549312592
desk 0.14025698602199554
monitor 0.1317901611328125
laptop 0.10974671691656113
mouse 0.10374213010072708
screen 0.07918370515108109
computer keyboard 0.033995289355516434
printer 0.021042412146925926
modem 0.008885478600859642
```

3.20.3 SegNet遥感影像语义分割/解译

SegNet于2016年提出，核心的概念是将网络划分为encoder编码器网络，decoder解码器网络和一个像素级的分类层SoftMax。编码器网络结构与VGG16的13个特征提取卷积层结构相同。而解码器网络的结构与编码器网络刚好相逆，可以理解为反卷积的过程。每个编码器层都对应一个解码器层，将编码结果的低分辨率特征重新映射到输入时的分辨率，以便进行像素级分类，为每个像素生成类概率，输出不同分类的最大值，得到图像分割图。编码过程是池化层（nn.MaxPool2d(2, return_indices=True)）下采样的过程，而解码过程是提取的特征值上采样(nn.MaxUnpool2d(2))的过程，如图3.20–6所示。

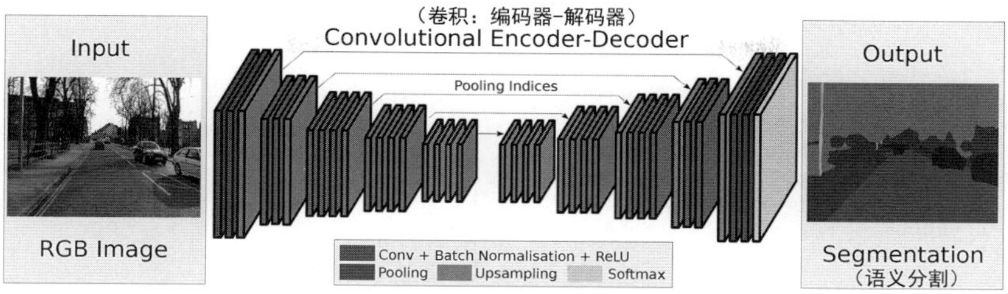

图3.20-6　SegNet图解（引自参考文献[2]）

下采样（pooling）就是池化层的作用，增加网络的深度。对于最大池化层，是提取区域内最大值作为输出，那么就可以得到最大值所在位置的索引。因此，在上采样（upsampling）的过程中（图3.20-7），对于2 ×2池化下采样结果执行上采样时，已经丢失3个权重值。在将特征图放大2倍后，原来特征图的数据会根据下采样时获取的位置索引归位放入。对于池化最大值位置索引，PyTorch的`nn.MaxPool2d()`下`return_indices=True`参数配置可以返回索引值。

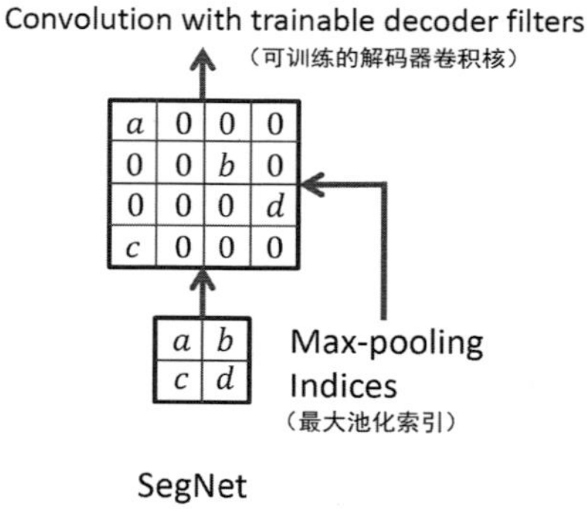

图3.20-7　上采样图解（引自参考文献[2]）

- `nn.MaxPool2d(2, return_indices=True)`

下述代码片段展示了编码器最大池化，以及解码器应用索引值上采样的过程。

```
import torch.nn as nn
pool = nn.MaxPool2d(2, stride=2, return_indices=True)
unpool = nn.MaxUnpool2d(2, stride=2)
input = torch.tensor([[[[0.,  1,  2,  3],
                        [4,  5,  6,  7],
                        [8,  9, 10, 11],
                        [12, 13, 14, 15]]]])
output, indices = pool(input)
print("最大池化索引：\n", indices)
print("最大池化结果：\n", output)

upsampling = unpool(output, indices)
print("根据池化索引上采样结果：\n", upsampling)
--------------------------------------------------------------------------------
最大池化索引：
 tensor([[[[ 5,  7],
          [13, 15]]]])
最大池化结果：
 tensor([[[[ 5.,  7.],
          [13., 15.]]]])
根据池化索引上采样结果：
 tensor([[[[ 0.,  0.,  0.,  0.],
          [ 0.,  5.,  0.,  7.],
          [ 0.,  0.,  0.,  0.],
          [ 0., 13.,  0., 15.]]]])
```

- 01-调入所用的库

```
# 调入库
import numpy as np
from skimage import io
from glob import glob
from tqdm import tqdm_notebook as tqdm
from sklearn.metrics import confusion_matrix
import random
import itertools
import os

# Matplotlib 相关
import matplotlib.pyplot as plt
%matplotlib inline

# Torch 相关
import torch
import torch.nn as nn
import torch.nn.functional as F
import torch.utils.data as data
import torch.optim as optim
import torch.optim.lr_scheduler
import torch.nn.init
from torch.autograd import Variable
```

- 02 - ISPRS数据集及数据查看、预处理和小批量可迭代数据加载

ISPRS[①]是遥感图像数据集。对于遥感图像数据集，因为大量影像的开源和图像解译工具的存在（例如e-Cognition），可以用传统的解译工具建立影像的分割标签，从而建立数据集。因此，目前有大量的遥感影像数据集可以使用，避免自行从新建立。下载的ISPRS数据集包括三个地方，分别为Potsdam、Toronto和Vaihingen。每个对应区域的所有数据放置于以地名命名的文件夹下，包含.tif格式（GTiff驱动），投影为CRS->EPSG:32633 - WGS 84 / UTM zone ?N - Projected的原始影像，及影像标签，标签类别包括roads、buildings、low veg.、trees、cars和clutter等，可以分辨出主要的地物内容。如果该数据集的标签不能满足解译后使用上的需求，可以用其他满足要求的影像数据集替换，或者用传统工具自行解译部分影像用作训练数据集。数据有blue、green、red和NIR四个波段，不过波段已经合成为RGB、IRRG和RGBIR等形式，通常放置于各自单独的文件夹下。下述训练的数据使用的为IRRG合成的波段，即NIR+red+green。Vaihingen区域数据是由德国摄影测量和遥感协会（German Association of Photogrammetry and Remote Sensing，DGPF）用于测试数字航拍数据的子集。图像为8cm地面分辨率。

① ISPRS数据集，遥感影像数据集，（https://www.isprs.org/data）。

建立数据存放的字符串格式化模式，在后续调用`class ISPRS_dataset(torch.utils.data.Dataset)`时使用。将训练集的数据对应到DATA_FOLDE文件夹下，训练集的标签对应到LABEL_FOLDER文件夹下，测试集的数据对应到ERODED_FOLDER文件夹下。

```
# 参数配置
WINDOW_SIZE = (256, 256)  # 窗口大小（Patch size）
STRIDE = 32  # 步幅
IN_CHANNELS = 3  # 输入通道数 (e.g. RGB)
FOLDER = r"G:/data/ISPRS/"  # 数据文件路径
BATCH_SIZE = 10  # 小批量样本数量

LABELS = ["roads", "buildings", "low veg.",
          "trees", "cars", "clutter"]  # 标签名
N_CLASSES = len(LABELS)  # 分类数量
WEIGHTS = torch.ones(N_CLASSES)  # 类平衡的权重
CACHE = True  # 将数据存储在内存中

DATASET = 'Vaihingen'
if DATASET == 'Potsdam':
    MAIN_FOLDER = FOLDER + 'Potsdam/'
    # 如果为 IRRG 数据，则取消对下一行的注释
    # DATA_FOLDER = MAIN_FOLDER + '3_Ortho_IRRG/top_potsdam_{}_IRRG.tif'
    # 为 RGB 数据
    DATA_FOLDER = MAIN_FOLDER + '2_Ortho_RGB/top_potsdam_{}_RGB.tif'
    LABEL_FOLDER = MAIN_FOLDER + '5_Labels_for_participants/top_potsdam_{}_
label.tif'
    ERODED_FOLDER = MAIN_FOLDER + \
        '5_Labels_for_participants_no_Boundary/top_potsdam_{}_label_noBoundary.tif'
```

```
elif DATASET == 'Vaihingen':
    MAIN_FOLDER = FOLDER + 'Vaihingen/'
    DATA_FOLDER = MAIN_FOLDER + 'top/top_mosaic_09cm_area{}.tif'
    LABEL_FOLDER = MAIN_FOLDER + 'gts_for_participants/top_mosaic_09cm_area{}.
tif'
    ERODED_FOLDER = MAIN_FOLDER + \
        'gts_eroded_for_participants/top_mosaic_09cm_area{}_noBoundary.tif'
```

数据查看，包括影像和对应标签（图3.20-8）。定义的函数convert_to_color(arr_2d, palette=palette)和convert_from_color(arr_3d, palette=invert_palette)给定数值和对应RGB颜色值映射字典，实现数值和颜色之间的互相转换。

```
# ISPRS 调色板 （color palette）
# 定义标准的 ISPRS 调色板
palette = {0: (255, 255, 255), # Impervious surfaces (white)
           1: (0, 0, 255),     # Buildings (blue)
           2: (0, 255, 255),   # Low vegetation (cyan)
           3: (0, 255, 0),     # Trees (green)
           4: (255, 255, 0),   # Cars (yellow)
           5: (255, 0, 0),     # Clutter (red)
           6: (0, 0, 0)}       # Undefined (black)
invert_palette = {v: k for k, v in palette.items()}
def convert_to_color(arr_2d, palette=palette):
    """ 数值标签转换为 RGB 颜色标签 （Numeric labels to RGB-color encoding ）"""
    arr_3d = np.zeros((arr_2d.shape[0], arr_2d.shape[1], 3), dtype=np.uint8)
    for c, i in palette.items():
        m = arr_2d == c
        arr_3d[m] = i
    return arr_3d
def convert_from_color(arr_3d, palette=invert_palette):
    """RGB 颜色标签转换为数值标签（灰度图）（RGB-color encoding to grayscale labels ）"""
    arr_2d = np.zeros((arr_3d.shape[0], arr_3d.shape[1]), dtype=np.uint8)
    for c, i in palette.items():
        m = np.all(arr_3d == np.array(c).reshape(1, 1, 3), axis=2)
        arr_2d[m] = i
    return arr_2d

# 从数据集中加载并显示一个瓦片
img = io.imread(r'G:\data\ISPRS\Vaihingen\top/top_mosaic_09cm_area11.tif')
fig = plt.figure()
fig.add_subplot(121)
plt.imshow(img)

# 加载分割标签（ground truth）
gt = io.imread(
    r'G:\data\ISPRS\Vaihingen\gts_for_participants/top_mosaic_09cm_area11.tif')
```

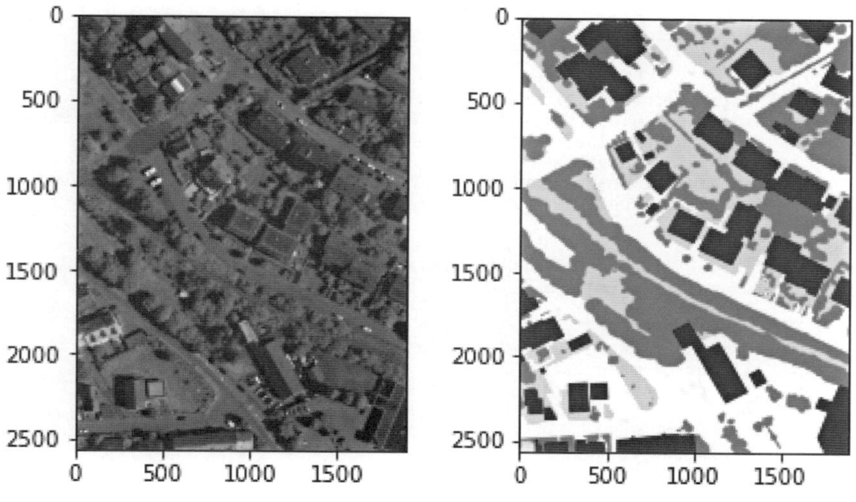

图3.20-8 ISPRS遥感图像数据集影像和标签（语义分割）查看

```
fig.add_subplot(122)
plt.imshow(gt)
plt.show()

# 将分割标签转化为一个数组
array_gt = convert_from_color(gt)
print("Ground truth in numerical format has shape ({},{}) : \n".format(
    *array_gt.shape[:2]), array_gt)
Ground truth in numerical format has shape (2566,1893) :
 [[3 3 3 ... 3 3 3]
 [3 3 3 ... 3 3 3]
 [3 3 3 ... 3 3 3]
 ...
 [2 2 2 ... 1 1 1]
 [2 2 2 ... 1 1 1]
 [2 2 2 ... 1 1 1]]
```

定义小批量可迭代数据加载类，同时执行图像增广（image augmentation），由定义的data_augmentation函数执行随机的翻转和镜像；并标准化数据到［0,1］。同时，标识data_augmentation函数有@classmethod装饰器，即标记该方法为类方法的装饰器。除了由实例对象调用外，可以直接由该类调用。如果作为父类，其子类也可以直接调用父类的类方法。

```
class C:
    @classmethod
    def f(cls, arg_str):
        print(cls, arg_str)
class C_child(C):
    pass
print(C.f("类对象调用类方法 ..."))
c = C()
print(c.f("类实例调用类方法 ..."))
print(C_child.f("子类调用父类的类方法 ..."))
--------------------------------------------------------------------------------
<class '__main__.C'> 类对象调用类方法 ...
None
<class '__main__.C'> 类实例调用类方法 ...
None
<class '__main__.C_child'> 子类调用父类的类方法 ...
None
--------------------------------------------------------------------------------
def get_random_pos(img, window_shape):
    """ 给定窗口大小，随机提取部分图像 （Extract of 2D random patch of shape window_
shape in the image ）"""
    w, h = window_shape
    W, H = img.shape[-2:]
    x1 = random.randint(0, W - w - 1)
    x2 = x1 + w
    y1 = random.randint(0, H - h - 1)
    y2 = y1 + h
    return x1, x2, y1, y2

# ISPRS 数据集
class ISPRS_dataset(torch.utils.data.Dataset):
    def __init__(self, ids, data_files=DATA_FOLDER, label_files=LABEL_FOLDER,
                 cache=False, augmentation=True):
        super(ISPRS_dataset, self).__init__()
        self.augmentation = augmentation
        self.cache = cache

        # 文件列表
        self.data_files = [DATA_FOLDER.format(id) for id in ids]
        self.label_files = [LABEL_FOLDER.format(id) for id in ids]

        # 完整性检查：如果某些文件不存在则引发异常
        for f in self.data_files + self.label_files:
            if not os.path.isfile(f):
                raise KeyError('{} is not a file !'.format(f))

        # 初始化缓存字典
        self.data_cache_ = {}
        self.label_cache_ = {}
    def __len__(self):
        # 默认 epoch 大小为 10 000 个样本
```

```
        return 10000
    @classmethod
    def data_augmentation(cls, *arrays, flip=True, mirror=True):
        will_flip, will_mirror = False, False
        if flip and random.random() < 0.5:
            will_flip = True
        if mirror and random.random() < 0.5:
            will_mirror = True
        results = []
        for array in arrays:
            if will_flip:
                if len(array.shape) == 2:
                    array = array[::-1, :]
                else:
                    array = array[:, ::-1, :]
            if will_mirror:
                if len(array.shape) == 2:
                    array = array[:, ::-1]
                else:
                    array = array[:, :, ::-1]
            results.append(np.copy(array))
        return tuple(results)
    def __getitem__(self, i):
        # 随机选一张图片
        random_idx = random.randint(0, len(self.data_files) - 1)

        # 如果瓦片尚未加载，则放入缓存中
        if random_idx in self.data_cache_.keys():
            data = self.data_cache_[random_idx]
        else:
            # 数据归一化为 [0, 1]
            data = 1/255 * \
                np.asarray(io.imread(self.data_files[random_idx]).transpose(
                    (2, 0, 1)), dtype='float32')
            if self.cache:
                self.data_cache_[random_idx] = data
        if random_idx in self.label_cache_.keys():
            label = self.label_cache_[random_idx]
        else:
            # 标签从 RGB 转换为对应的数值
            label = np.asarray(convert_from_color(
                io.imread(self.label_files[random_idx])), dtype='int64')
            if self.cache:
                self.label_cache_[random_idx] = label

        # 获得一个随机窗口
        x1, x2, y1, y2 = get_random_pos(data, WINDOW_SIZE)
        data_p = data[:, x1:x2, y1:y2]
        label_p = label[x1:x2, y1:y2]

        # 数据增强
        data_p, label_p = self.data_augmentation(data_p, label_p)

        # 返回 torch.Tensor 张量
        return (torch.from_numpy(data_p),
                torch.from_numpy(label_p))
```

加载数据，并切分数据集为训练和测试数据集。

```
# 加载数据集
if DATASET == 'Potsdam':
    all_files = sorted(glob(LABEL_FOLDER.replace('{}', '*')))
    all_ids = ["".join(f.split('')[5:7]) for f in all_files]
elif DATASET == 'Vaihingen':
    all_files = sorted(glob(LABEL_FOLDER.replace('{}', '*')))
    all_ids = [f.split('area')[-1].split('.')[0] for f in all_files]
# 用于切分训练 / 测试数据集的随机瓦片数
train_ids = random.sample(all_ids, 2 * len(all_ids) // 3 + 1)
test_ids = list(set(all_ids) - set(train_ids))
print("Tiles for training : ", train_ids)
print("Tiles for testing : ", test_ids)
train_set = ISPRS_dataset(train_ids, cache=CACHE)
train_loader = torch.utils.data.DataLoader(train_set, batch_size=BATCH_SIZE)
---------------------------------------------------------------------------
```

```
Tiles for training :  ['17', '28', '32', '34', '26', '15', '7', '21', '30', '23', '3']
Tiles for testing :  ['13', '37', '5', '1', '11']
```

- 03 - 定义网络

```
class SegNet(nn.Module):
    # SegNet 网络
    @staticmethod
    def weight_init(m):
        if isinstance(m, nn.Linear):
            torch.nn.init.kaiming_normal(m.weight.data)
    def __init__(self, in_channels=IN_CHANNELS, out_channels=N_CLASSES):
        super(SegNet, self).__init__()
        self.pool = nn.MaxPool2d(2, return_indices=True)
        self.unpool = nn.MaxUnpool2d(2)
        self.conv1_1 = nn.Conv2d(in_channels, 64, 3, padding=1)
        self.conv1_1_bn = nn.BatchNorm2d(64)
        self.conv1_2 = nn.Conv2d(64, 64, 3, padding=1)
        self.conv1_2_bn = nn.BatchNorm2d(64)
        self.conv2_1 = nn.Conv2d(64, 128, 3, padding=1)
        self.conv2_1_bn = nn.BatchNorm2d(128)
        self.conv2_2 = nn.Conv2d(128, 128, 3, padding=1)
        self.conv2_2_bn = nn.BatchNorm2d(128)
        self.conv3_1 = nn.Conv2d(128, 256, 3, padding=1)
        self.conv3_1_bn = nn.BatchNorm2d(256)
        self.conv3_2 = nn.Conv2d(256, 256, 3, padding=1)
        self.conv3_2_bn = nn.BatchNorm2d(256)
        self.conv3_3 = nn.Conv2d(256, 256, 3, padding=1)
        self.conv3_3_bn = nn.BatchNorm2d(256)
        self.conv4_1 = nn.Conv2d(256, 512, 3, padding=1)
        self.conv4_1_bn = nn.BatchNorm2d(512)
        self.conv4_2 = nn.Conv2d(512, 512, 3, padding=1)
        self.conv4_2_bn = nn.BatchNorm2d(512)
        self.conv4_3 = nn.Conv2d(512, 512, 3, padding=1)
        self.conv4_3_bn = nn.BatchNorm2d(512)
        self.conv5_1 = nn.Conv2d(512, 512, 3, padding=1)
        self.conv5_1_bn = nn.BatchNorm2d(512)
        self.conv5_2 = nn.Conv2d(512, 512, 3, padding=1)
        self.conv5_2_bn = nn.BatchNorm2d(512)
        self.conv5_3 = nn.Conv2d(512, 512, 3, padding=1)
        self.conv5_3_bn = nn.BatchNorm2d(512)
        # -----------------------------------------------
        self.conv5_3_D = nn.Conv2d(512, 512, 3, padding=1)
        self.conv5_3_D_bn = nn.BatchNorm2d(512)
        self.conv5_2_D = nn.Conv2d(512, 512, 3, padding=1)
        self.conv5_2_D_bn = nn.BatchNorm2d(512)
        self.conv5_1_D = nn.Conv2d(512, 512, 3, padding=1)
        self.conv5_1_D_bn = nn.BatchNorm2d(512)
        self.conv4_3_D = nn.Conv2d(512, 512, 3, padding=1)
        self.conv4_3_D_bn = nn.BatchNorm2d(512)
        self.conv4_2_D = nn.Conv2d(512, 512, 3, padding=1)
        self.conv4_2_D_bn = nn.BatchNorm2d(512)
        self.conv4_1_D = nn.Conv2d(512, 256, 3, padding=1)
        self.conv4_1_D_bn = nn.BatchNorm2d(256)
        self.conv3_3_D = nn.Conv2d(256, 256, 3, padding=1)
        self.conv3_3_D_bn = nn.BatchNorm2d(256)
        self.conv3_2_D = nn.Conv2d(256, 256, 3, padding=1)
        self.conv3_2_D_bn = nn.BatchNorm2d(256)
        self.conv3_1_D = nn.Conv2d(256, 128, 3, padding=1)
        self.conv3_1_D_bn = nn.BatchNorm2d(128)
        self.conv2_2_D = nn.Conv2d(128, 128, 3, padding=1)
        self.conv2_2_D_bn = nn.BatchNorm2d(128)
        self.conv2_1_D = nn.Conv2d(128, 64, 3, padding=1)
        self.conv2_1_D_bn = nn.BatchNorm2d(64)
        self.conv1_2_D = nn.Conv2d(64, 64, 3, padding=1)
        self.conv1_2_D_bn = nn.BatchNorm2d(64)
        self.conv1_1_D = nn.Conv2d(64, out_channels, 3, padding=1)
        self.apply(self.weight_init)
    def forward(self, x):
        # Encoder block 1
        x = self.conv1_1_bn(F.relu(self.conv1_1(x)))
        x = self.conv1_2_bn(F.relu(self.conv1_2(x)))
        x, mask1 = self.pool(x)
        # Encoder block 2
```

```
        x = self.conv2_1_bn(F.relu(self.conv2_1(x)))
        x = self.conv2_2_bn(F.relu(self.conv2_2(x)))
        x, mask2 = self.pool(x)
        # Encoder block 3
        x = self.conv3_1_bn(F.relu(self.conv3_1(x)))
        x = self.conv3_2_bn(F.relu(self.conv3_2(x)))
        x = self.conv3_3_bn(F.relu(self.conv3_3(x)))
        x, mask3 = self.pool(x)
        # Encoder block 4
        x = self.conv4_1_bn(F.relu(self.conv4_1(x)))
        x = self.conv4_2_bn(F.relu(self.conv4_2(x)))
        x = self.conv4_3_bn(F.relu(self.conv4_3(x)))
        x, mask4 = self.pool(x)
        # Encoder block 5
        x = self.conv5_1_bn(F.relu(self.conv5_1(x)))
        x = self.conv5_2_bn(F.relu(self.conv5_2(x)))
        x = self.conv5_3_bn(F.relu(self.conv5_3(x)))
        x, mask5 = self.pool(x)
        # ----------------------------------------------
        # Decoder block 5
        x = self.unpool(x, mask5)
        x = self.conv5_3_D_bn(F.relu(self.conv5_3_D(x)))
        x = self.conv5_2_D_bn(F.relu(self.conv5_2_D(x)))
        x = self.conv5_1_D_bn(F.relu(self.conv5_1_D(x)))
        # Decoder block 4
        x = self.unpool(x, mask4)
        x = self.conv4_3_D_bn(F.relu(self.conv4_3_D(x)))
        x = self.conv4_2_D_bn(F.relu(self.conv4_2_D(x)))
        x = self.conv4_1_D_bn(F.relu(self.conv4_1_D(x)))
        # Decoder block 3
        x = self.unpool(x, mask3)
        x = self.conv3_3_D_bn(F.relu(self.conv3_3_D(x)))
        x = self.conv3_2_D_bn(F.relu(self.conv3_2_D(x)))
        x = self.conv3_1_D_bn(F.relu(self.conv3_1_D(x)))
        # Decoder block 2
        x = self.unpool(x, mask2)
        x = self.conv2_2_D_bn(F.relu(self.conv2_2_D(x)))
        x = self.conv2_1_D_bn(F.relu(self.conv2_1_D(x)))
        # Decoder block 1
        x = self.unpool(x, mask1)
        x = self.conv1_2_D_bn(F.relu(self.conv1_2_D(x)))
        x = F.log_softmax(self.conv1_1_D(x))
        return x
```

从地址 https://download.pytorch.org/models/vgg16_bn-6c64b313.pth 下载VGG16网络模型预训练参数。因为下载的预训练参数对应的层名与上述模型定义的不同，因此需要一一对位，将权值映射到新的层名上来。然后，应用`net.state_dict().update(mapped_weights)`方法更新权值。

```
# 实例化网络模型
import os
net = SegNet()
try:
    from urllib.request import URLopener
except ImportError:
    from urllib import URLopener

# 从 PyTorch 下载 VGG-16 权重
vgg_url = 'https://download.pytorch.org/models/vgg16_bn-6c64b313.pth'
if not os.path.isfile(r'G:\data\model\vgg16_bn-6c64b313.pth'):
    weights = URLopener().retrieve(vgg_url, r'G:\data\model\vgg16_bn-6c64b313.
pth')
vgg16_weights = torch.load(r'G:\data\model\vgg16_bn-6c64b313.pth')
mapped_weights = {}
for k_vgg, k_segnet in zip(vgg16_weights.keys(), net.state_dict().keys()):
    if "features" in k_vgg:
        mapped_weights[k_segnet] = vgg16_weights[k_vgg]
        print("Mapping {} to {}".format(k_vgg, k_segnet))
try:
    net.state_dict().update(mapped_weights)
    print("_"*50)
    print("Loaded VGG-16 weights in SegNet !")
except:
```

```
    print("Ignore missing keys")
    pass
--------------------------------------------------------------------------------
Mapping features.0.weight to conv1_1.weight
Mapping features.0.bias to conv1_1.bias
...
Mapping features.41.weight to conv5_1_bn.running_mean
Mapping features.41.bias to conv5_1_bn.running_var
Mapping features.41.running_mean to conv5_1_bn.num_batches_tracked
Mapping features.41.running_var to conv5_2.weight
--------------------------------------------------------------------
Loaded VGG-16 weights in SegNet !
--------------------------------------------------------------------------------
if torch.cuda.is_available():
    net.to('cuda')
```

● 04 - 定义训练模型相关函数、损失函数、预测精度及相关度量值（全局精度、F1分数和kappa系数）等内容。

```
def CrossEntropy2d(input, target, weight=None, size_average=True):
    """ 定义损失函数-- 2D 版交叉熵损失 （2D version of the cross entropy loss ）"""
    dim = input.dim()
    if dim == 2:
        return F.cross_entropy(input, target, weight, size_average)
    elif dim == 4:
        output = input.view(input.size(0), input.size(1), -1)
        output = torch.transpose(output, 1, 2).contiguous()
        output = output.view(-1, output.size(2))
        target = target.view(-1)
        return F.cross_entropy(output, target, weight, size_average)
    else:
        raise ValueError('Expected 2 or 4 dimensions (got {})'.format(dim))
def accuracy(input, target):
    ''' 定义预测精度 '''
    return 100 * float(np.count_nonzero(input == target)) / target.size
def sliding_window(top, step=10, window_size=(20, 20)):
    """ 给定步幅、窗口形状、滑动过整幅图像，迭代计算窗口所在图像 x,y 位置值，返回每一切分图像
(patch) 的 x,y 坐标值和高宽大小，即 yield 返回值。参数 step 可以控制切分窗口叠合的程度 （Slide
a window_shape window across the image with a stride of step ）"""
    for x in range(0, top.shape[0], step):
        if x + window_size[0] > top.shape[0]:
            x = top.shape[0] - window_size[0]
        for y in range(0, top.shape[1], step):
            if y + window_size[1] > top.shape[1]:
                y = top.shape[1] - window_size[1]
            yield x, y, window_size[0], window_size[1]
def count_sliding_window(top, step=10, window_size=(20, 20)):
    """ 计算图像滑动给定窗口大小的数量 （Count the number of windows in an image ）"""
    c = 0
    for x in range(0, top.shape[0], step):
        if x + window_size[0] > top.shape[0]:
            x = top.shape[0] - window_size[0]
        for y in range(0, top.shape[1], step):
            if y + window_size[1] > top.shape[1]:
                y = top.shape[1] - window_size[1]
            c += 1
    return c
def grouper(n, iterable):
    """ n 个元素为一批次迭代 """
    it = iter(iterable)
    while True:
        chunk = tuple(itertools.islice(it, n))
        if not chunk:
            return
        yield chunk
def metrics(predictions, gts, label_values=LABELS):
    ''' 预测值度量 '''
    cm = confusion_matrix(
        gts,
        predictions,
        labels=range(len(label_values)))
    print("Confusion matrix :")
    print(cm)
    print("---")
```

```
    # 全局精度 (Compute global accuracy)
    total = sum(sum(cm))
    accuracy = sum([cm[x][x] for x in range(len(cm))])
    accuracy *= 100 / float(total)
    print("{} pixels processed".format(total))
    print("Total accuracy : {}%".format(accuracy))
    print("---")

    # F1 分数 (Compute F1 score)
    F1Score = np.zeros(len(label_values))
    for i in range(len(label_values)):
        try:
            F1Score[i] = 2. * cm[i, i] / (np.sum(cm[i, :]) + np.sum(cm[:, i]))
        except:
            # 如果测试集 i 类中没有元素，则忽略异常
            pass
    print("F1Score :")
    for l_id, score in enumerate(F1Score):
        print("{}: {}".format(label_values[l_id], score))
    print("---")

    # 计算 kappa 系数 (Compute kappa coefficient)
    total = np.sum(cm)
    pa = np.trace(cm) / float(total)
    pe = np.sum(np.sum(cm, axis=0) * np.sum(cm, axis=1)) / float(total*total)
    kappa = (pa - pe) / (1 - pe)
    print("Kappa: " + str(kappa))
    return accuracy
```

使用标准的随机梯度下降算法优化网络的权值。如果调入了预先训练的VGG16模型参数，则可调整学习率。即encoder编码部分（VGG16卷积、特征提取部分）的训练速度为decoder解码器的一半。

```
base_lr = 0.01
params_dict = dict(net.named_parameters())
params = []
for key, value in params_dict.items():
    if '_D' in key:
        # 以适宜的学习率训练解码器权重
        params += [{'params': [value], 'lr': base_lr}]
    else:
        # 训练编码器权值的学习率为 lr / 2 (用 VGG-16 权重值初始化)
        params += [{'params': [value], 'lr': base_lr / 2}]
optimizer = optim.SGD(net.parameters(), lr=base_lr,
                      momentum=0.9, weight_decay=0.0005)
# 定义调度器
scheduler = optim.lr_scheduler.MultiStepLR(optimizer, [25, 35, 45], gamma=0.1)
```

- 05 - 定义测试函数，显示RGB影像，以及对应的真实值和预测值图像。计算metrics函数定义的相关预测度量值。

```
def test(net, test_ids, all=False, stride=WINDOW_SIZE[0], batch_size=BATCH_SIZE,
window_size=WINDOW_SIZE):
    # 在测试数据集上使用网络模型
    test_images = (1 / 255 * np.asarray(io.imread(DATA_FOLDER.format(id)),
                   dtype='float32') for id in test_ids)
    test_labels = (np.asarray(io.imread(LABEL_FOLDER.format(id)),
                   dtype='uint8') for id in test_ids)
    eroded_labels = (convert_from_color(
        io.imread(ERODED_FOLDER.format(id))) for id in test_ids)
    all_preds = []
    all_gts = []

    # 将网络模型切换为推理模式 / 预测模式
    net.eval()
    for img, gt, gt_e in tqdm(zip(test_images, test_labels, eroded_labels),
total=len(test_ids), leave=False):
        pred = np.zeros(img.shape[:2] + (N_CLASSES,))
        total = count_sliding_window(
            img, step=stride, window_size=window_size) // batch_size
        for i, coords in enumerate(tqdm(grouper(batch_size, sliding_window(img,
step=stride, window_size=window_size)), total=total, leave=False)):
```

```
            # 打印进度
            if i > 0 and total > 10 and i % int(10 * total / 100) == 0:
                _pred = np.argmax(pred, axis=-1)
                fig = plt.figure()
                fig.add_subplot(1, 3, 1)
                plt.imshow(np.asarray(255 * img, dtype='uint8'))
                fig.add_subplot(1, 3, 2)
                plt.imshow(convert_to_color(_pred))
                fig.add_subplot(1, 3, 3)
                plt.imshow(gt)
                clear_output()
                plt.show()

            # 构建张量
            image_patches = [
                np.copy(img[x:x+w, y:y+h]).transpose((2, 0, 1)) for x, y, w, h
in coords]
            image_patches = np.asarray(image_patches)
            image_patches = Variable(torch.from_numpy(
                image_patches).cuda(), volatile=True)

            # 进行推理 / 推断 / 预测
            outs = net(image_patches)
            outs = outs.data.cpu().numpy()

            # 预测输出
            for out, (x, y, w, h) in zip(outs, coords):
                out = out.transpose((1, 2, 0))
                pred[x:x+w, y:y+h] += out
            del (outs)
        pred = np.argmax(pred, axis=-1)

        # 显示结果
        clear_output()
        fig = plt.figure()
        fig.add_subplot(1, 3, 1)
        plt.imshow(np.asarray(255 * img, dtype='uint8'))
        fig.add_subplot(1, 3, 2)
        plt.imshow(convert_to_color(pred))
        fig.add_subplot(1, 3, 3)
        plt.imshow(gt)
        plt.show()
        all_preds.append(pred)
        all_gts.append(gt_e)
        clear_output()
        # 精度指标计算
        metrics(pred.ravel(), gt_e.ravel())
        accuracy = metrics(np.concatenate([p.ravel() for p in all_preds]),
np.concatenate(
            [p.ravel() for p in all_gts]).ravel())
    if all:
        return accuracy, all_preds, all_gts
    else:
        return accuracy
```

● 06 - 定义训练函数。输出损失曲线，显示RGB影像，以及对应的真实值和预测值图像（图3.20–9）。打印损失值和精度值，观察模型训练情况。同时，指定文件夹，保存模型参数。

```
from IPython.display import clear_output
def train(net, optimizer, epochs, scheduler=None, weights=WEIGHTS, save_epoch=5):
    losses = np.zeros(1000000)
    mean_losses = np.zeros(100000000)
    weights = weights.cuda()
    criterion = nn.NLLLoss2d(weight=weights)
    iter_ = 0
    for e in range(1, epochs + 1):
        if scheduler is not None:
            scheduler.step()
        net.train()
        for batch_idx, (data, target) in enumerate(train_loader):
            data, target = Variable(data.cuda()), Variable(target.cuda())
            optimizer.zero_grad()
            output = net(data)
            loss = CrossEntropy2d(output, target, weight=weights)
```

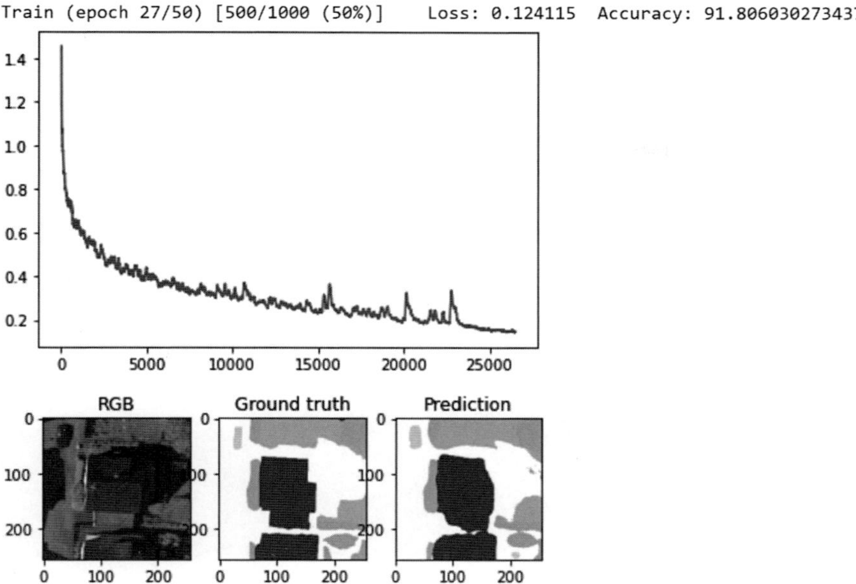

图3.20-9 损失曲线、图像、语义分割的真实值和预测值

```
            loss.backward()
            optimizer.step()
            losses[iter_] = loss.data.item()  # losses[iter_] = loss.data[0]
            mean_losses[iter_] = np.mean(losses[max(0, iter_-100):iter_])
            if iter_ % 100 == 0:
                clear_output()
                rgb = np.asarray(
                    255 * np.transpose(data.data.cpu().numpy()[0], (1, 2, 0)),
dtype='uint8')
                pred = np.argmax(output.data.cpu().numpy()[0], axis=0)
                gt = target.data.cpu().numpy()[0]
                print('Train (epoch {}/{}) [{}/{} ({:.0f}%)]\tLoss: {:.6f}\
tAccuracy: {}'.format(
                    e, epochs, batch_idx, len(train_loader),
                    100. * batch_idx / len(train_loader), loss.data.item(),
accuracy(pred, gt)))  # 100. * batch_idx / len(train_loader), loss.data[0], accuracy(pred, gt)))
                plt.plot(mean_losses[:iter_]) and plt.show()
                fig = plt.figure()
                fig.add_subplot(131)
                plt.imshow(rgb)
                plt.title('RGB')
                fig.add_subplot(132)
                plt.imshow(convert_to_color(gt))
                plt.title('Ground truth')
                fig.add_subplot(133)
                plt.title('Prediction')
                plt.imshow(convert_to_color(pred))
                plt.show()
            iter_ += 1
            del (data, target, loss)
        if e % save_epoch == 0:
            # 以尽可能大的步幅进行验证，以获得更快的计算速度
            acc = test(net, test_ids, all=False, stride=min(WINDOW_SIZE))
            torch.save(net.state_dict(),
                       './model/segnet256_epoch{}_{}'.format(e, acc))
    torch.save(net.state_dict(), './model/segnet_final')
```

● 07 - 训练模型

```
train(net, optimizer, 10, scheduler)
```

```
[[1353141   32015   65375   19629    4922    1733]
 [  34820  796320    9073    3273     670       3]
 [  45405   13111  541334   98071      62       0]
 [   9127    2476   85643 1235041      45       0]
 [   9981     742     776     581   32381     214]
 [      0       0       0       0       0       0]]
---
4395964 pixels processed
Total accuracy : 90.0420704082199%
---
F1Score :
roads: 0.9238699220186195
buildings: 0.9430473175696921
low veg.: 0.7732326608502883
trees: 0.9186125171862234
cars: 0.7825750709926893
clutter: 0.0
---
Kappa: 0.8641704967438216
Confusion matrix :
[[5943500  251376  192249   59281   23200    3986]
 [ 257402 6629996   54712    9507    3639      72]
 [ 370167  111835 4206096  564403     529    3257]
 [  59915   12051  533183 4003598     151       0]
 [  45805    6880    1743     828  110866     883]
 [      0       0       0       0       0       0]]
---
23461110 pixels processed
Total accuracy : 89.05825853934446%
---
F1Score :
roads: 0.9039281827651989
buildings: 0.9493484358580146
low veg.: 0.8211607074003321
trees: 0.8659690705092675
cars: 0.7260617570974819
clutter: 0.0
---
Kappa: 0.8533983408961401
```

● 08 - 加载保存的SegNet模型参数，应用测试数据集测试模型。通过配置stride参数，设置图像被切分为多个小块之间的重叠程度。重叠程度由stride参数和WINDOW_SIZE=(256, 256)参数（即patch大小）确定。

在应用所训练的模型解译新的图像时，新图像的形式应该与训练数据的形式保持一致或近似，这样才能够保证正确的预测结果。例如，图像的高空分辨率及波段合成信息等，能够基本相同。

```
net.load_state_dict(torch.load('./model/segnet_final'))
if torch.cuda.is_available():
    net.to('cuda')
_, all_preds, all_gts = test(net, test_ids, all=True, stride=32)
--------------------------------------------------------------------------------
Confusion matrix :
[[1370081   25031   57259   19254    4022    1168]
 [  29015  804242    7865    2584     453       0]
 [  47582   10867  542282   97228      24       0]
 [   7684    2311   72972 1249350      15       0]
 [  10697     260     726     528   32321     143]
 [      0       0       0       0       0       0]]
---
4395964 pixels processed
Total accuracy : 90.95333810740944%
---
F1Score :
roads: 0.9314341810696175
buildings: 0.9535316888675476
low veg.: 0.7864362436887593
trees: 0.9250072928497495
cars: 0.7930560667402773
clutter: 0.0
---
Kappa: 0.8764381914078332
Confusion matrix :
```

```
[[6007256  209282  180520   56386   18409    1739]
 [ 201880 6694014   48209    8352    2772     101]
 [ 366170   98613 4254461  535807     385     851]
 [  53371   10865  501938 4042672      52       0]
 [  49247    4373    1509     899  110263     714]
 [      0       0       0       0       0       0]]
---
23461110 pixels processed
Total accuracy : 89.97300639228067%
---
F1Score :
roads: 0.9135457843795346
buildings: 0.9581715479898872
low veg.: 0.8307122067878274
trees: 0.8738065240147697
cars: 0.737826462263204
clutter: 0.0
---
Kappa: 0.8656335585255769
```

- 09 - 显示与保存预测的图像分割/解译（图3.20-10）。

```
import matplotlib.pyplot as plt
from tqdm import tqdm
plt.figure(figsize=(20, 5))
i = 0
for p, id_ in tqdm(zip(all_preds, test_ids), total=len(all_preds), leave=False):
    img = convert_to_color(p)
    plt.subplot(1, len(all_preds), i+1)
    plt.imshow(img)
    plt.axis('off')
    io.imsave('./results/segment_pred/inference_tile_{}.png'.format(id_), img)
    i += 1
plt.show()
```

图3.20-10 显示预测图像语义分割结果

参考文献（References）：

[1] Simonyan, K., & Zisserman, A. (2015). Very Deep Convolutional Networks for Large-Scale Image Recognition. arXiv [Cs. CV]. Retrieved from http://arxiv.org/abs/1409.1556.
[2] Badrinarayanan, V., Kendall, A., & Cipolla, R. (2016). SegNet: A Deep Convolutional Encoder-Decoder Architecture for Image Segmentation. arXiv [Cs.CV]. Retrieved from http://arxiv.org/abs/1511.00561.
[3] Deep learning for Earth Observation, https://github.com/nshaud/DeepNetsForEO.

3.21 NAIP航拍影像与分割模型库及Colaboratory和Planetary Computer Hub

3.21.1 NAIP航拍影像和构建图像数据集

（美）国家农业图像项目（National Agriculture Imagery Program，NAIP），由美国农业部农业服务局（the U.S. Department of Agriculture's Farm Service Agency，FSA）通过位于犹他州（Utah）盐湖城（Salt Lack City）航空摄影驻地办公室（Aerial Photography Field Office，APFO）管理。NAIP在农业种植季或植被生长季（leaf on），采集地面采样距离为1m分辨率的航空影像（结合具有地理参照的图像特征进行正射校正）。每个单独的图像瓦片（tile）都为3.75-分钟经度和3.75-分钟纬度的四分四边形，加上每边300m的缓冲距离。NAIP图像包含红、绿和蓝波段，且通常含有近红外波段。美国地质调查局（U.S. Geological Survey，USGS）地球资源观测与科学（Earth Resources Observation and Science，EROS）中心发布格式为GeoTIFF和JPEG2000（内嵌地理坐标信息的压缩文件）的NAIP产品。10：1的有损压缩以减小图像大小，但图像质量也会有所下降。APFO要求图像符合制图标准，所有交付的图像均会用自动化视觉检验图像的质量，确保其准确性且符合规范[①]。数据至少每3年更新一次。

NAIP航拍影像获取图途径可以从USGS的EarthExplorer[②]中检索下载；或者，从微软（Microsoft）的行星计算机（Planetary Computer）[③]上检索下载。Planetary Computer借助云的力量支持全球环境可持续发展决策，集合具有应用接口（APIs）千万亿字节的全球环境数据目录，弹性的科学环境允许用户回答有关数据的全球问题，并将其提交到相关保护利益者手中。例如，官网应用中列举了全球规模的环境监测案例，使数据驱动决策成为可能。这包括全球土地利用和土地覆盖分类（Land use and Land cover，LULC），森林砍伐风险分析，生态系统监测（森林变化、栖息地连通性、土地使用等），保护规划，森林碳（排放）风险评估，基于AI的土地覆盖物分类等。

3.21.1.1 NAIP航拍影像数据下载

此次实验所用数据为特拉华州（Delaware）的全部NAIP数据，约147个单独影像文件。而EarthExplorer不支持批量下载，因此使用TorchGeo[④]库提供的`download_url`方法（基本类似torchvision[⑤]库提供的`download_url`方法）从Planetary Computer云平台下载。首先，在EarthExplorer地图中绘制检索区域（Polygon方式），选择NAIP数据集（Data Sets标签），检索结果的文件名（Entity ID）和区域边界可以导出为Shapefile、CSV、Comma（，）Delimited和IKMZ等文件格式，然后Python（GeoPandas库）读取，根据影像获取年份，移除重叠影像，提取获得最终要下载的文件名。

参数管理使用`AttrDict()`方法（具体查看Cityscapes数据集——参数管理一节）。data子属性存储数据文件路径。文件名Shapefile数据存储于`naip_de_entityID`变量路径名文件夹下，NAIP影像数据存储于`Chesapeake_root`变量路径名文件夹下，对应的土地覆盖数据存储于`Chesapeake_LC`变量路径名文件夹下。

植被生长期（leaf-on），当树木或者灌木等生长有枝叶时采集航拍影像可以获得植被物种特有的光谱反射率（spectral reflection），可用于区分植被类型；也可以检测农作物的长势和健康状况等。植被落叶期（leaf-off），当树木或者灌木等落叶，枝头很少或者没有叶子时采集航拍影像可以更清楚地看到地面特征，有助于绘制建筑物和道路等对象。

```
from util_misc import AttrDict
import os
__C = AttrDict()
args = __C
__C.data = AttrDict()
__C.data.Chesapeake_root = r'E:\data\Delaware'
__C.data.Chesapeake_LC = os.path.join(args.data.Chesapeake_root, 'LC')
__C.data.Chesapeake_imagery = os.path.join(
    args.data.Chesapeake_root, 'imagery')
__C.data.naip_de_entityID = './data/naip_Delaware/naip_63b945b372d740b7.shp'
```

① USGS EROS Archive - Aerial Photography - National Agriculture Imagery Program (NAIP)，(https://www.usgs.gov/centers/eros/science/usgs-eros-archive-aerial-photography-national-agriculture-imagery-program-naip)。w

② EarthExplorer，(https://earthexplorer.usgs.gov/)。

③ 行星计算机（Planetary Computer），(https://planetary-computer.microsoft.com/)。

④ TorchGeo，(https://torchgeo.readthedocs.io/en/stable/user/installation.html)。

⑤ torchvision，(https://pytorch.org/vision/stable/index.htm)。

读取查看从EarthExplorer检索获取的NAIP数据文件名字段，包括NAIP Entit(y)和几何对象geometry字段。

```
import geopandas as gpd
naip_de_entityID_gdf = gpd.read_file(args.data.naip_de_entityID)
naip_de_entityID_gdf.sort_alues(by='Acquisitio',ascending=False,inplace=True)
naip_de_entityID_gdf.head(1).squeeze()
---------------------------------------------------------------------------------
NAIP Entit                    M_3907562_SW_18_060_20211019
State                                                   DE
Agency                                                USDA
Vendor                                       USDA-FSA-APFO
Map Projec                                             UTM
Projection                                             18N
Datum                                                NAD83
Resolution                                0.600000000000000
Units                                                METER
Number of                                                4
Sensor Typ                                            CNIR
Project Na          202112_DELAWARE_NAIP_0X6000M_UTM_CNIR
Acquisitio                                      2021-10-19
Center Lat                               39&deg;01'52.48"N
Center Lon                               75&deg;20'37.51"W
NW Corner                                39&deg;03'51.42"N
NW Corne_1                               75&deg;22'39.55"W
NE Corner                                39&deg;03'52.13"N
NE Corne_1                               75&deg;18'36.62"W
SE Corner                                38&deg;59'53.50"N
SE Corne_1                               75&deg;18'35.58"W
SW Corner                                38&deg;59'52.79"N
SW Corne_1                               75&deg;22'38.28"W
Center L_1                                      39.0312443
Center L_2                                     -75.3437527
NW Corne_2                                      39.0642833
NW Corne_3                                     -75.3776527
NE Corne_2                                      39.0644805
NE Corne_3                     \               -75.3101722
SE Corne_2                                      38.9981944
SE Corne_3                                     -75.3098833
SW Corne_2                                      38.9979971
SW Corne_3                                     -75.3772999
geometry     POLYGON ((-75.3772999 38.9979971, -75.3776527
Name: 1343, dtype: object
```

打印NAIP影像瓦片分布（图3.21-1）可以发现，很多瓦片重叠，一方面是不同年份的影像，另一方面是同一年份不同时间段可能也有重叠，因此需要移除重叠的瓦片。定义IoU_2Polygons()函数计算交并比（Intersection over Union，IoU）；定义drop_overlapping_polygons()函数，根据交并比移除过度重叠的瓦片。移除的方式是保留第一个出现的瓦片，而移除后续与之重叠的瓦片。

```
naip_de_entityID_gdf.boundary.
plot(figsize=(10,10))

def IoU_2Polygons(polygon1, polygon2):
    '''
    计算两个 Poygon（Shapely）对象的交并比

    Parameters
    ----------
    polygon1 : POLYGON（Shapely）
        多边形对象 1.
    polygon2 : POLYGON（Shapely）
        多边形对象 2.

    Returns
    -------
    iou : float
        交并比 Intersection over Union, IoU.

    '''
    from shapely.geometry import Polygon
```

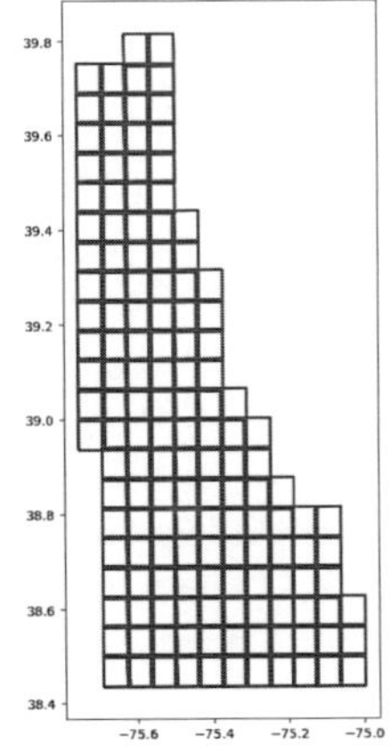

图3.21-1 重叠的NAIP影像瓦片

```
    intersect_area = polygon1.intersection(polygon2).area
    union_area = polygon1.union(polygon2).area
    iou = intersect_area/union_area
    return iou
def drop_overlapping_polygons(gdf_, iou=0.5):
    from tqdm import tqdm
    '''
    移除 GeoDataFrame 格式文件 Polygon 对象重叠的行。保留第一个出现的对象，而移除后面与之重叠的对象

    Parameters
    ----------
    gdf_ : GeoDataFrame
        为 Polygon 对象的地理信息数据 .
    iou : float, optional
        交并比（Intersection over Union, IoU）. The default is 0.5.

    Returns
    -------
    gdf_non_overlapping : GeoDataFrame
        移除重叠的 Polygon 后的 GeoDataFrame 格式数据 .

    '''
    gdf = gdf_.copy(deep=True)
    polygons_dict = gdf['geometry'].to_dict()
    tabu_idx = []
    for idx, row in tqdm(gdf.iterrows(), total=gdf.shape[0]):
        tabu_idx.append(idx)
        polygons_except4one_dict = {
            key: value for key, value in polygons_dict.items() if key not in 
tabu_idx}
        pg_gdf = row.geometry
        for k, pg_dict in polygons_except4one_dict.items():
            iou_2pgs = IoU_2Polygons(pg_dict, pg_gdf)
            if iou_2pgs > iou:
                polygons_dict.pop(k)
    gdf_non_overlapping = gdf.loc[list(polygons_dict.keys())]
    return gdf_non_overlapping
```

提取2018年的数据行，并执行移除重叠多边形的操作。

```
naip_de_entityIDd_2018 = [
    i for i in naip_de_entityID_gdf['Acquisitio'].unique() if i.split('-')[0] == 
'2018']
naip_de_entityID_2018_gdf = naip_de_entityID_gdf.loc[naip_de_entityID_
gdf['Acquisitio'].isin(
    naip_de_entityIDd_2018)]
naip_de_entityID_2018_nonOverlapping_gdf = drop_overlapping_polygons(
    naip_de_entityID_2018_gdf)
---------------------------------------------------------------------------
  return lib.intersection(a, b, **kwargs)
100%|██████████| 162/162 [00:00<00:00, 
199.94it/s]
```

打印处理后的NAIP影像瓦片分布（图3.21-2），检测瓦片的重叠情况，确定已经移除了重叠的瓦片，并且基本能够覆盖整个特拉华州。

```
naip_de_entityID_2018_nonOverlapping_gdf.
boundary.plot(figsize=(10,10))
```

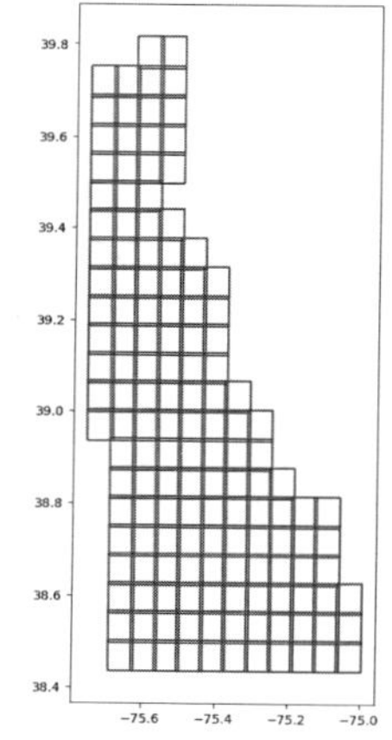

图3.21-2 处理后的NAIP影像瓦片

搜索位于Planetary Computer上的NAIP数据集，NAIP: National Agriculture Imagery Program[①]条目提供了NAIP的信息，并在Example Notebook标签下给出了获取数据的Python代码。这里参考TorchGeo库提供的方法，通过文件名（Entity ID）构建文件下载地址，因为2018年的NAIP数据位于38075和39075两个目录下，因此分别构建下载地址下载对应的影像文件，核心使用download_url方法，从而无须自行构建检索下载工具。

① NAIP: National Agriculture Imagery Program（https://planetarycomputer.microsoft.com/dataset/naip）。

```
# from torchvision.datasets.utils import download_url
from torchgeo.datasets.utils import download_url
naip_38075_url = (
    "https://naipeuwest.blob.core.windows.net/naip/v002/de/2018/
de_060cm_2018/38075/")
naip_39075_url = (
    "https://naipeuwest.blob.core.windows.net/naip/v002/de/2018/
de_060cm_2018/39075/")
tiles = [i.lower()+'.tif' for i in naip_de_entityID_2018_nonOverlapping_gdf['NAIP
Entit'].tolist()]
tiles_38075 = [i for i in tiles if '38075' in i.split('_')[1]]
tiles_39075 = [i for i in tiles if '39075' in i.split('_')[1]]
def naip_download_rul(url, tile, root): return download_url(url+tile, root)
downloaded_tiles = []
failed_tiles = []
for tile in tiles_38075:
    try:
        naip_download_rul(naip_38075_url, tile, args.data.Chesapeake_imagery)
        downloaded_tiles.append(tile)
    except:
        failed_tiles.append(tile)
for tile in tiles_39075:
    try:
        naip_download_rul(naip_39075_url, tile, args.data.Chesapeake_imagery)
        downloaded_tiles.append(tile)
    except:
        failed_tiles.append(tile)
-----------------------------------------------------------------------------
Downloading https://naipeuwest.blob.core.windows.net/naip/v002/de/2018/
de_060cm_2018/38075/m_3807536_ne_18_060_20181104.tif to E:_3807536_
ne_18_060_20181104.tif 100%|████████████████| 508699496/508699496 [14:35<00:00,
580974.44it/s] ...Downloading https://naipeuwest.blob.core.windows.net/naip/
v002/de/2018/de_060cm_2018/38075/m_3807527_ne_18_060_20181104.tif to E:_3807527_
ne_18_060_20181104.tif 100%|████████████████| 530246098/530246098 [14:10<00:00,
623712.25it/s]
```

3.21.1.2 构建数据集并查看样本

几十年来，地球观测卫星、飞机及无人机平台一直在搜集地球表面图像，使得图像具有时空连续性，从而利于人们用遥感影像解决当今人类面临的诸多挑战，例如适应气候变化、自然灾害监测、水资源管理和全球人口日益增长下的粮食安全问题等。计算机视觉领域则包括土地覆盖分类（语义分割）、森林砍伐和洪水监测（变化检测）、冰川流（像素跟踪）、飓风跟踪和强度估计（回归）及建筑和道路检测（物体检测、实例分割）等应用。PyTorch Ecosystem中的TorchGeo库试图同时处理深度学习模型和地理空间数据这两个截然不同领域的专业知识，为处理地理空间数据的PyTorch深度学习库，使得用深度学习处理地理空间数据工作变得简单。

TorchGeo库提供了常用数据的下载工具，目前集成的数据集[①]摘录如表3.21–1所示。

① Geospatial Datasets（TorchGeo）（https://torchgeo.readthedocs.io/en/stable/api/datasets.html）。

表3.21–1 TorchGeo库集成的数据集

Dataset（数据集）	**Type**（类型）	**Source**（源）	**Size（px）**（大小）	**Resolution（m）**（分辨率）
Aboveground Woody Biomass	Masks	Landsat, LiDAR	40000 × 40000	30
Aster Global DEM	Masks	Aster	3601 × 3601	30
Canadian Building Footprints	Geometries	Bing Imagery		
Chesapeake Land Cover	Imagery, Masks	NAIP		1
Global Mangrove Distribution	Masks	Remote Sensing, In Situ Measurements		3
Cropland Data Layer	Masks	Aerial		30
EDDMapS	Points	Citizen Scientists		
EnviroAtlas	Imagery, Masks	NAIP, NLCD, OpenStreetMap		1
Esri2020	Masks	Sentinel-2		10

续表

Dataset（数据集）	Type（类型）	Source（源）	Size（px）（大小）	Resolution（m）（分辨率）
EU-DEM	Masks	Aster, SRTM, Russian Topomaps		25
GBIF	Points	Citizen Scientists		
GlobBiomass	Masks	Landsat	45000 × 45000	100
iNaturalist	Points	Citizen Scientists		
Landsat	Imagery	Landsat	8900 × 8900	30
NAIP	Imagery	Aerial	6100 × 7600	1
Open Buildings	Geometries	Maxar, CNES/Airbus		
Sentinel	Imagery	Sentinel	10000 × 10000	10

TorchGeo库参考PyTorch Ecosystem下的torchvision、PyTorch Lightning等库，同时提供了用于深度学习训练的数据集（Datasets）构建方法、样本采样（Sampler）、数据加载（DataLoader）及模型训练（Training）方式，大幅度简化了直接使用PyTorch构建深度学习模型从数据集（训练、测试、验证等）建立、模型构建（激活函数、模型和损失函数等）到训练的整个流程。NAIP方法可以直接构建NAIP航拍影像的数据集，ChesapeakeDE方法则可以下载来自于切萨皮克保护协会（Chesapeake Conservancy Center，CIC）[①]提供的对应NAIP数据土地覆盖类型数据集。CIC成立于2013年，旨在利用尖端技术赋予保护和修复环境以数据驱动力。通过`dataset=naip & chesapeake`方式可以直接对位叠加NAIP的航拍影像和对应的土地覆盖类型数据为单个的数据集，而且可以用`RandomGeoSampler`等方法采样数据，返回给定大小的影像`image`及对应的分类`mask`。该土地覆盖类型包括13类：

1. Water（水体）：所有开放水域，包括池塘、河流、湖泊（含不依附码头的船只）等，及小型人为的农场池塘和雨水滞留设施等。MMU=25mm²。
2. Emergent Wetlands（湿地）：位于海洋或河口的低植被区域，视觉上可以观察到包含植被的饱和地面，并且位于主要水道（河流、海洋等）。对于弗吉尼亚州（Virginia）的潮汐带，该类湿地通常具有低矮的植被、木本植被和土地贫瘠（裸地）的特征。MMU=225mm²。
3. Tree Canopy（树冠/林冠）：自然演替或人工种植的落叶和常绿木本植被，高度通常超过3m。单株，离散的林木团和紧密连接的单株都计算在内。MMU=9m²。
4. Scrub/Shrub（灌木丛）：落叶和/或常绿木本植被的异质区域，贯穿覆盖有低矮植被和草地，高低变化斑驳分布的灌木丛和幼树为特征，包括离散的灌木丛和紧密联系的单株植被。和由于环境条件发育不良矮小的灌木、幼树等，它们混长于植被较低的异质景观中。MMU=225m²。
5. Low Vegetation（低矮植被）：高度小于约3m的植被，包括草坪、耕地、有或无防水布覆盖的苗圃种植区，近期砍伐的森林管理区，和自然地表覆盖物等。MMU=9m²。
6. Barren（裸地）：由天然土质构成的无植被区域，包括海滩、泥滩和建筑工地的裸露地面等。MMU=25m²。
7. Impervious Structures（不透水结构）：由不透水材料制成高度大于约2m的人造构筑，包括房屋、商城和电力塔等。MMU=9m²。
8. Other Impervious（其他不透水类）：水不能渗透、小于约2m的人造构筑。MMU=9m²。
9. Impervious Road（不透水路面）：为运输使用和需要维护的不透水表面。MMU=9m²。
10. Tree Canopy over Impervious Structure（覆盖于不透水结构上的树冠）：与不透水表面重叠的森林或树木覆盖物，使得不透水结构部分可见或完全不可见。注：不透水表面和树冠层是独立绘制的，通过重叠区域识别。MMU=9m²。
11. Tree Canopy over Other Impervious（覆盖于其他不透水结构上的树冠）：与不透水表面重叠的森林或树木覆盖物，使得不透水结构部分可见或完全不可见。MMU=9m²。
12. Tree Canopy over Impervious Roads（覆盖于不透水路面上的树冠）：与不透水路面重叠的森林或树木覆盖物，使得不透水结构部分可见或完全不可见。MMU=9m²。

① Chesapeake Conservancy（https://www.chesapeakeconservancy.org/conservation-innovation-center/high-resolution-data/）。

13. 254.Aberdeen Proving Ground（阿伯丁实验场）：该类区域无源图像或辅助数据可用，该类存在于马里兰郡的哈福德（Harford，County Maryland）。

PyTorch 的 Ecosystem（生态）由来自学术界、工业界、应用程序开发人员和 ML（Machine Learning）工程师开发的项目、工具和库组成，其目的是支持、加速和帮助对 PyTorch 的探索。

```
from torch.utils.data import DataLoader
from torchgeo.samplers import RandomGeoSampler
from torchgeo.datasets import NAIP, ChesapeakeDE, stack_samples
naip = NAIP(args.data.Chesapeake_imagery)
chesapeake = ChesapeakeDE(args.data.Chesapeake_LC,
                          crs=naip.crs, res=naip.res, download=False)
dataset = naip & chesapeake
sampler = RandomGeoSampler(dataset, size=256, length=10)
dataloader = DataLoader(dataset, sampler=sampler, collate_fn=stack_samples)
for batch in dataloader:
    image = batch["image"]
    target = batch["mask"]
    break
print(image, '\n', target)
--------------------------------------------------------------------------------
tensor([[[[102,  94,  91,  ...,  87,  93,  91],
          [ 96,  89,  92,  ...,  98,  92,  99],
          [ 96,  85,  88,  ...,  96,  96,  99],
          ...,
          [163, 163, 155,  ...,  60,  30,  30],
          [158, 158, 156,  ...,  66,  39,  39],
          [159, 152, 152,  ...,  57,  41,  35]]]], dtype=torch.uint8)
 tensor([[[[5, 5, 5,  ..., 5, 5, 5],
          [5, 5, 5,  ..., 5, 5, 5],
          [5, 5, 5,  ..., 5, 5, 5],
          ...,
          [3, 3, 3,  ..., 8, 8, 8],
          [3, 3, 3,  ..., 8, 8, 8],
          [3, 3, 3,  ..., 8, 8, 8]]]], dtype=torch.uint8)
```

unbind_samples方法与stack_sample方法相逆，将小批量（mini-batch）样本转换成样本列表。

```
from torchgeo.datasets import unbind_samples
sample = unbind_samples(batch)[0]
sample
--------------------------------------------------------------------------------
{'image': tensor([[[107, 118, 149,  ..., 103, 132, 138],
          [174, 159, 187,  ..., 130, 134, 136],
          [190, 202, 207,  ..., 103, 100,  97],
          ...,
          [ 17,  16,  18,  ...,  52,  52,  57],
          [ 17,  18,  17,  ..., 131, 115,  97],
          [ 17,  16,  15,  ..., 191, 180, 174]]], dtype=torch.uint8),
 'crs': CRS.from_epsg(26918),
 'bbox': BoundingBox(minx=494628.6, maxx=494884.6, miny=4264922.399999999,
maxy=4265178.399999999, mint=1534262399.999999, maxt=1534348799.999999),
 'mask': tensor([[[ 3,  3,  3,  ..., 10, 10, 10],
          [ 3,  3,  3,  ..., 10, 10, 10],
          [11, 11,  3,  ..., 10, 10, 10],
          ...,
          [ 8,  8,  8,  ..., 10, 10, 10],
          [ 8,  8,  8,  ..., 10, 10, 10],
          [ 8,  8,  8,  ..., 10, 10, 10]]], dtype=torch.uint8)}
```

分别打印样本的影像（图3.21–3）和对应的分类（标签，图3.21–4）。

```
import matplotlib.pyplot as plt
import torchvision.transforms as T
print(sample['image'].shape)
plt.imshow(T.ToPILImage()(sample['image']));
--------------------------------------------------------------------------------
torch.Size([4, 427, 427])
```

配置土地覆盖类型的颜色字典LC_color_dict，并转换为matplotlib库的cmap颜色图实例，用于地图打印分类颜色。

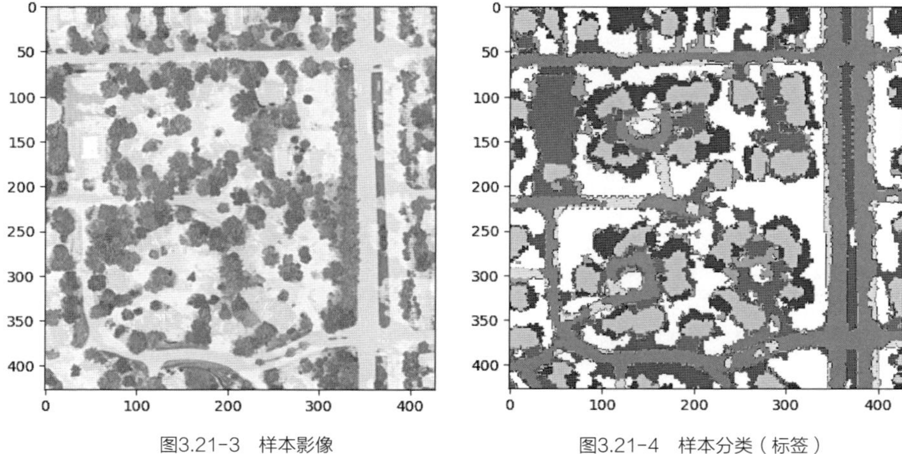

图3.21-3 样本影像　　图3.21-4 样本分类（标签）

```
import matplotlib
import numpy as np
LC_color_dict = {
    0: (0, 0, 0, 0),
    1: (0, 197, 255, 255),
    2: (0, 168, 132, 255),
    3: (38, 115, 0, 255),
    4: (76, 230, 0, 255),
    5: (163, 255, 115, 255),
    6: (255, 170, 0, 255),
    7: (255, 0, 0, 255),
    8: (156, 156, 156, 255),
    9: (0, 0, 0, 255),
    10: (115, 115, 0, 255),
    11: (230, 230, 0, 255),
    12: (255, 255, 115, 255),
    13: (197, 0, 255, 255),
}
cmap_LC, norm = matplotlib.colors.from_levels_and_colors(list(LC_color_dict.
keys(
)), [[v/255 for v in i] for i in LC_color_dict.values()], extend='max')
print(target.shape)
plt.imshow(np.squeeze(target), cmap=cmap_LC)
--------------------------------------------------------------------------
torch.Size([1, 1, 427, 427])
```

可以用naip.plot()方法直接打印unbind_samples后的小批量样本，如图3.21-5所示。

```
sampler_naip = RandomGeoSampler(naip, size=4096, length=3)
dataloader_naip = DataLoader(
    naip, sampler=sampler_naip, collate_fn=stack_samples)
i = 0
for batch in dataloader_naip:
```

图3.21-5 naip.plot()方法直接打印样本图像

```
    sample = unbind_samples(batch)[0]
    naip.plot(sample['image'])
    if i == 3:
        break
    i += 1
```

3.21.1.3 图像数据增强变换（data augmentation transforms）

TorchVison和Kornia[①]均为计算机视觉库，且都提供了图像增强变换的诸多方法。TorchGeo的transforms模块则针对地理空间数据多个栅格波段追加了如下指数：

① Kornia，（https://github.com/kornia/kornia）。

Normalized Burn Ratio（归一化燃烧比值指数）：$NBR = \frac{NIR-SWIR}{NIR+SWIR}$；

Normalized Difference Built-up Index（归一化建筑指数）：$NDBI = \frac{SWIR-NIR}{SWIR+NIR}$；

Normalized Difference Snow Index（归一化雪指数）：$NDSI = \frac{G-SWIR}{G+SWIR}$；

Normalized Difference Vegetation Index（归一化植被指数）：$NDVI = \frac{NIR-R}{NIR+R}$；

Normalized Difference Water Index（归一化水体指数）：$NDWI = \frac{G-NIR}{G+NIR}$；

Standardized Water-Level Index（标准化水位指数）：$SWI = \frac{VRE1-SWIR2}{VRE1+SWIR2}$；

Green Normalized Difference Vegetation Index（绿波段归一化植被指数）：$GNDVI = \frac{NIR-G}{NIR+G}$；

Blue Normalized Difference Vegetation Index（蓝波段归一化植被指数）：$BNDVI = \frac{NIR-B}{NIR+B}$；

Normalized Difference Red Edge Vegetation Index（归一化差分红边植被指数）：$textNDRE = \frac{NIR-VRE1}{NIR+VRE1}$；

Green-Red Normalized Difference Vegetation Index（绿红归一化植被指）数：$GRNDVI = \frac{NIR-(G+R)}{NIR+(G+R)}$；

Green-Blue Normalized Difference Vegetation Index（绿蓝归一化植被指数）：$GBNDVI = \frac{NIR-(G+B)}{NIR+(G+B)}$；

Red-Blue Normalized Difference Vegetation Index（红蓝归一化植被指数）：$RBNDVI = \frac{NIR-(R+B)}{NIR+(R+B)}$。

下述代码调用了kornia库提供的图像增强变换方法，并追加了NDVI和NDWI两个指数变换图像。

```
import torch.nn as nn
import kornia.augmentation as K
from torchgeo.transforms import AugmentationSequential, indices
dataloader = DataLoader(dataset, sampler=sampler, collate_fn=stack_samples)
dataloader = iter(dataloader)
batch = next(dataloader)
x, y = batch["image"], batch["mask"]
augmentations = AugmentationSequential(
    K.RandomHorizontalFlip(p=0.5),
    K.RandomVerticalFlip(p=0.5),
    # K.RandomAffine(degrees=(0, 90), p=0.25),
    # K.RandomGaussianBlur(kernel_size=(3, 3), sigma=(0.1, 2.0), p=0.25),
    # K.RandomResizedCrop(size=(512, 512), scale=(0.8, 1.0), p=0.25),
    data_keys=["image", 'mask'],
)
transforms = nn.Sequential(
    indices.AppendNDVI(index_nir=3, index_red=0),
    indices.AppendNDWI(index_green=1, index_nir=3),
    augmentations,
)
batch = next(dataloader)
print(batch["image"].shape)
batch = transforms(batch)
print(batch["image"].shape)
---------------------------------------------------------------------------
torch.Size([1, 4, 427, 427])
torch.Size([1, 6, 427, 427])
```

nvidia-smi（NVSMI）为NVIDIA显卡（例如Tesla、Quadro、GRID和GeForce等）设备提供监控和管理功能。下述打印的信息显示了显卡驱动（driver）版本（表3.21-2）、CUDA（Compute Unified Device Architecture））版本等相关信息。

```
!nvidia-smi
```

将图像增强变换nn.Sequential对象（按照构造函数中传递的对象顺序添加到模块容器中）通过`sequential.to(device[cuda])`转换到GPU中进行计算。打印原始图像和增强变换后的图像（图3.21-6）。

表3.21-2 显卡驱动版本信息

```
Wed Jan 11 17:54:04 2023
+-----------------------------------------------------------------------------+
| NVIDIA-SMI 512.78       Driver Version: 512.78       CUDA Version: 11.6     |
|-------------------------------+----------------------+----------------------+
| GPU  Name            TCC/WDDM | Bus-Id        Disp.A | Volatile Uncorr. ECC |
| Fan  Temp  Perf  Pwr:Usage/Cap|         Memory-Usage | GPU-Util  Compute M. |
|                               |                      |               MIG M. |
|===============================+======================+======================|
|   0  NVIDIA GeForce ... WDDM  | 00000000:01:00.0  On |                  N/A |
| N/A   64C    P0    37W /  N/A |   3320MiB /  8192MiB |      1%      Default |
|                               |                      |                  N/A |
+-------------------------------+----------------------+----------------------+

+-----------------------------------------------------------------------------+
| Processes:                                                                  |
|  GPU   GI   CI        PID   Type   Process name                  GPU Memory |
|        ID   ID                                                   Usage      |
|=============================================================================|
|    0   N/A  N/A      1144    C+G   ...ekyb3d8bbwe\HxOutlook.exe    N/A      |
|    0   N/A  N/A      5804    C+G   ...n1h2txyewy\SearchHost.exe    N/A      |
|    0   N/A  N/A      6956    C+G   ...d\runtime\WeChatAppEx.exe    N/A      |
|    0   N/A  N/A      8908    C+G   ...zilla Firefox\firefox.exe    N/A      |
|    0   N/A  N/A      9428    C+G   ...zilla Firefox\firefox.exe    N/A      |
|    0   N/A  N/A     12076    C+G   C:\Windows\explorer.exe         N/A      |
|    0   N/A  N/A     12320    C+G   ...462.76\msedgewebview2.exe    N/A      |
|    0   N/A  N/A     13148    C+G   ...artMenuExperienceHost.exe    N/A      |
|    0   N/A  N/A     13856    C+G   ...lPanel\SystemSettings.exe    N/A      |
|    0   N/A  N/A     14928    C+G   ...e\Current\LogiOverlay.exe    N/A      |
|    0   N/A  N/A     15388    C+G   ...ystemEventUtilityHost.exe    N/A      |
|    0   N/A  N/A     15912    C+G   ...2txyewy\TextInputHost.exe    N/A      |
|    0   N/A  N/A     16884    C+G   ...e\PhoneExperienceHost.exe    N/A      |
|    0   N/A  N/A     17076    C+G   ...y\ShellExperienceHost.exe    N/A      |
|    0   N/A  N/A     17812    C+G   ...ge\Application\msedge.exe    N/A      |
|    0   N/A  N/A     18776    C+G   ...ram Files\LGHUB\lghub.exe    N/A      |
|    0   N/A  N/A     19372    C+G   ...ray\lghub_system_tray.exe    N/A      |
|    0   N/A  N/A     20280    C+G   ...arkupHero\Markup Hero.exe    N/A      |
|    0   N/A  N/A     21124    C+G   ...cw5n1h2txyewy\LockApp.exe    N/A      |
|    0   N/A  N/A     22632    C+G   ...me\Application\chrome.exe    N/A      |
|    0   N/A  N/A     27516    C+G   ...tracted\WechatBrowser.exe    N/A      |
|    0   N/A  N/A     27820    C+G   ...wekyb3d8bbwe\Video.UI.exe    N/A      |
|    0   N/A  N/A     31488    C+G   ...root\Office16\WINWORD.EXE    N/A      |
|    0   N/A  N/A     34900    C+G   ...icrosoft VS Code\Code.exe    N/A      |
+-----------------------------------------------------------------------------+
```

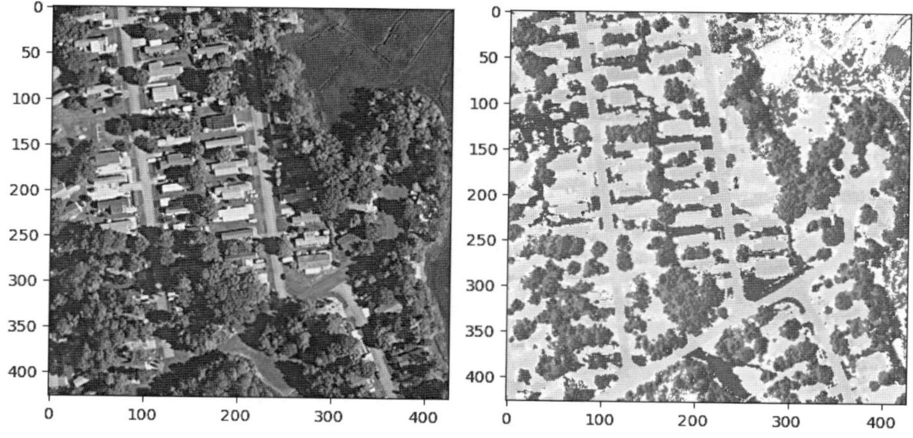

图3.21-6 原始图像（左）与增强变换后的图像（右）

```
import torch
from copy import deepcopy
device = "cuda" if torch.cuda.is_available() else "cpu"
print(device)
transforms_gpu = deepcopy(transforms).to(device)
------------------------------------------------------------------------------
cuda
------------------------------------------------------------------------------
import copy
sampler_naip = RandomGeoSampler(naip, size=4096, length=1)
dataloader = DataLoader(dataset, sampler=sampler, collate_fn=stack_samples)
fig, axs = plt.subplots(2, figsize=(10, 10))
for n in dataloader:
    batch = copy.deepcopy(n)
    print(batch['image'].shape)
    axs[0].imshow(T.ToPILImage()(batch['image'][0][:3]))
    batch = transforms_gpu(batch)
    print(batch['image'].shape)
    axs[1].imshow(T.ToPILImage()(batch['image'][0][2:6]))
    break
------------------------------------------------------------------------------
torch.Size([1, 4, 427, 427])
torch.Size([1, 6, 427, 427])
```

打印真实分类（图3.21-7）。

```
plt.imshow(T.ToPILImage()(batch['mask'][0]));
```

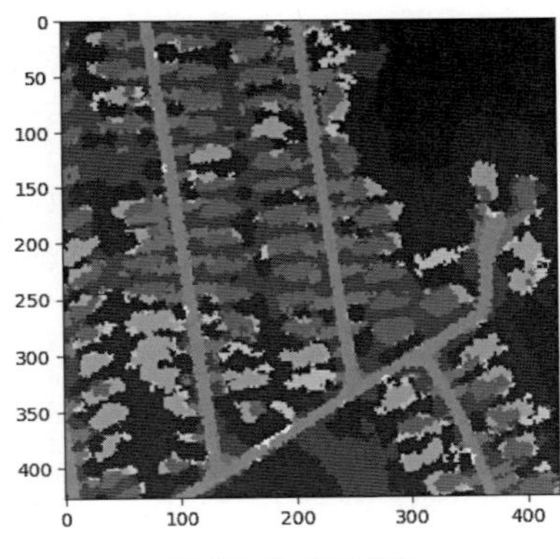

图3.21-7　真实分类

3.21.2　分割模型库及Colaboratory和Planetary Computer Hub

3.21.2.1　分割模型（Segmentation Models）

深度学习的不断发展更新，新的网络不断涌现，大量被证实可用的深度学习网络广泛传播，因此集成已有成熟的网络（通常包含预训练参数权重）库为网络的实际应用带来便利，例如前文涉及的ONNX Model Zoo①、torchvision.models②等，以及PyTorch Image Models（TIMM）③、Transformers④和Segmentation Models⑤等，且不仅上述所限。以图像语义分割为主的Segmentation Models为例，可以仅用几行代码调用预训练的模型，涉及的分割网络包括Unet、Unet++、MAnet、Linknet、FPN、PSPNet、PAN、DeepLabV3和DeepLabV3+等；而且，包括高达124个可用的编码器（encoders），所有编码器都有预训练的参数权重，以实现更快、更好的模型训练收敛，例如ResNet（imagenet / ssl / swsl）、ResNeXt（imagenet / instagram / ssl / swsl）、ResNeSt（imagenet）、Res2Ne(X)t（imagenet）、RegNet(x/y)（imagenet）、GERNet（imagenet）、SE-Net（imagenet）、SK-ResNe(X)t（imagenet）、DenseNet（imagenet）、Inception（imagenet / imagenet+background）、EfficientNet（imagenet / advprop / noisy-student）、MobileNet（imagenet）、DPN（imagenet+5k）、VGG（imagenet）、Mix Vision Transformer（imagenet）和MobileOne（imagenet）等。其中，预训

① ONNX Model Zoo（https://github.com/onnx/models）。
② torchvision.models（https://pytorch.org/vision/stable/models.html）。
③ PyTorch Image Models（https://github.com/rwightman/pytorch-image-models）。
④ Transformers（https://github.com/huggingface/transformers）。
⑤ Segmentation Models（https://smp.readthedocs.io/en/stable/index.html）。

练模型最常用的图像数据库为ImageNet[1]（仅包含名词），为按照WordNet[2]层次结构组织的图像数据库，其中层次结构的每个节点由成百上千的图像描述。该项目在推进计算机视觉和深度学习研究方面发挥了重要作用（ImageNet图像数据库开源用于非商业用途的研究）。

下述示例调用网络构建模型和预测（假值）的代码，并打印了该Unet网络模型。其中，编码器使用了resnet34，预训练参数权重来自于图像数据库imagenet。预测的结果为mask掩码。

① ImageNet（https://www.image-net.org/index.php）。

② WordNet（Wikipedia），为200多种语言中单词之间语义关系的词汇数据库（https://en.wikipedia.org/wiki/WordNet）。

```
import segmentation_models_pytorch as smp
import torch
model = smp.Unet(
    encoder_name="resnet34",        # 选择编码器，例如 :mobilenet_v2 or efficientnet-b7
    # 使用 imagenet 预训练权重进行编码器初始化
    encoder_weights="imagenet",
    # 模型输入通道 (1 个为灰度图像，3 个为 RGB 等 )
    in_channels=3,
    # 模型输出通道 ( 数据集中的类数量 )
    classes=13,
)
mask = model(torch.ones([1, 3, 64, 64]))
print(mask[:, 0, :])
print('-'*50)
print(model)
--------------------------------------------------------------------------------
tensor([[[-0.0060, -0.1177, -0.1687,  ...,  0.3846,  0.0027,  0.0376],
         [ 0.4949, -0.3818,  0.0144,  ...,  0.3487,  0.0475, -0.0637],
         [ 0.1829, -0.9314, -0.4307,  ...,  1.1400,  0.3028, -0.0593],
         ...,
         [ 0.0888, -0.0988, -0.2369,  ..., -0.4098, -0.2458, -0.5457],
         [-0.6411, -0.4427, -0.9519,  ..., -0.9773, -0.4209, -0.2852],
         [-0.1010, -0.4232, -0.8313,  ..., -0.2045,  0.1289,  0.0031]]],
       grad_fn=<SliceBackward0>)
--------------------------------------------------------------------------------
Unet(
  (encoder): ResNetEncoder(
    (conv1): Conv2d(3, 64, kernel_size=(7, 7), stride=(2, 2), padding=(3, 3), 
bias=False)
    (bn1): BatchNorm2d(64, eps=1e-05, momentum=0.1, affine=True, track_running_
stats=True)
    (relu): ReLU(inplace=True)
    (maxpool): MaxPool2d(kernel_size=3, stride=2, padding=1, dilation=1, ceil_
mode=False)
    (layer1): Sequential(
      (0): BasicBlock(
        (conv1): Conv2d(64, 64, kernel_size=(3, 3), stride=(1, 1), padding=(1, 
1), bias=False)
        (bn1): BatchNorm2d(64, eps=1e-05, momentum=0.1, affine=True, track_
running_stats=True)
        (relu): ReLU(inplace=True)
        (conv2): Conv2d(64, 64, kernel_size=(3, 3), stride=(1, 1), padding=(1, 
1), bias=False)
        (bn2): BatchNorm2d(64, eps=1e-05, momentum=0.1, affine=True, track_
running_stats=True)
      )
      (1): BasicBlock(
        (conv1): Conv2d(64, 64, kernel_size=(3, 3), stride=(1, 1), padding=(1, 
1), bias=False)
        (bn1): BatchNorm2d(64, eps=1e-05, momentum=0.1, affine=True, track_
running_stats=True)
        (relu): ReLU(inplace=True)
        (conv2): Conv2d(64, 64, kernel_size=(3, 3), stride=(1, 1), padding=(1, 
1), bias=False)
        (bn2): BatchNorm2d(64, eps=1e-05, momentum=0.1, affine=True, track_
running_stats=True)
      )
      (2): BasicBlock(
        (conv1): Conv2d(64, 64, kernel_size=(3, 3), stride=(1, 1), padding=(1, 
1), bias=False)
        (bn1): BatchNorm2d(64, eps=1e-05, momentum=0.1, affine=True, track_
running_stats=True)
        (relu): ReLU(inplace=True)
        (conv2): Conv2d(64, 64, kernel_size=(3, 3), stride=(1, 1), padding=(1, 
1), bias=False)
        (bn2): BatchNorm2d(64, eps=1e-05, momentum=0.1, affine=True, track_
```

```
running_stats=True)
      )
    )
    (layer2): Sequential(
      (0): BasicBlock(
        (conv1): Conv2d(64, 128, kernel_size=(3, 3), stride=(2, 2), padding=(1, 1), bias=False)
        (bn1): BatchNorm2d(128, eps=1e-05, momentum=0.1, affine=True, track_running_stats=True)
        (relu): ReLU(inplace=True)
        (conv2): Conv2d(128, 128, kernel_size=(3, 3), stride=(1, 1), padding=(1, 1), bias=False)
        (bn2): BatchNorm2d(128, eps=1e-05, momentum=0.1, affine=True, track_running_stats=True)
        (downsample): Sequential(
          (0): Conv2d(64, 128, kernel_size=(1, 1), stride=(2, 2), bias=False)
          (1): BatchNorm2d(128, eps=1e-05, momentum=0.1, affine=True, track_running_stats=True)
        )
      )
      (1): BasicBlock(
        (conv1): Conv2d(128, 128, kernel_size=(3, 3), stride=(1, 1), padding=(1, 1), bias=False)
        (bn1): BatchNorm2d(128, eps=1e-05, momentum=0.1, affine=True, track_running_stats=True)
        (relu): ReLU(inplace=True)
        (conv2): Conv2d(128, 128, kernel_size=(3, 3), stride=(1, 1), padding=(1, 1), bias=False)
        (bn2): BatchNorm2d(128, eps=1e-05, momentum=0.1, affine=True, track_running_stats=True)
      )
      (2): BasicBlock(
        (conv1): Conv2d(128, 128, kernel_size=(3, 3), stride=(1, 1), padding=(1, 1), bias=False)
        (bn1): BatchNorm2d(128, eps=1e-05, momentum=0.1, affine=True, track_running_stats=True)
        (relu): ReLU(inplace=True)
        (conv2): Conv2d(128, 128, kernel_size=(3, 3), stride=(1, 1), padding=(1, 1), bias=False)
        (bn2): BatchNorm2d(128, eps=1e-05, momentum=0.1, affine=True, track_running_stats=True)
      )
      (3): BasicBlock(
        (conv1): Conv2d(128, 128, kernel_size=(3, 3), stride=(1, 1), padding=(1, 1), bias=False)
        (bn1): BatchNorm2d(128, eps=1e-05, momentum=0.1, affine=True, track_running_stats=True)
        (relu): ReLU(inplace=True)
        (conv2): Conv2d(128, 128, kernel_size=(3, 3), stride=(1, 1), padding=(1, 1), bias=False)
        (bn2): BatchNorm2d(128, eps=1e-05, momentum=0.1, affine=True, track_running_stats=True)
      )
    )
    (layer3): Sequential(
      (0): BasicBlock(
        (conv1): Conv2d(128, 256, kernel_size=(3, 3), stride=(2, 2), padding=(1, 1), bias=False)
        (bn1): BatchNorm2d(256, eps=1e-05, momentum=0.1, affine=True, track_running_stats=True)
        (relu): ReLU(inplace=True)
        (conv2): Conv2d(256, 256, kernel_size=(3, 3), stride=(1, 1), padding=(1, 1), bias=False)
        (bn2): BatchNorm2d(256, eps=1e-05, momentum=0.1, affine=True, track_running_stats=True)
        (downsample): Sequential(
          (0): Conv2d(128, 256, kernel_size=(1, 1), stride=(2, 2), bias=False)
          (1): BatchNorm2d(256, eps=1e-05, momentum=0.1, affine=True, track_running_stats=True)
        )
      )
      (1): BasicBlock(
        (conv1): Conv2d(256, 256, kernel_size=(3, 3), stride=(1, 1), padding=(1, 1), bias=False)
```

```
        (bn1): BatchNorm2d(256, eps=1e-05, momentum=0.1, affine=True, track_
running_stats=True)
        (relu): ReLU(inplace=True)
        (conv2): Conv2d(256, 256, kernel_size=(3, 3), stride=(1, 1), padding=(1,
1), bias=False)
        (bn2): BatchNorm2d(256, eps=1e-05, momentum=0.1, affine=True, track_
running_stats=True)
      )
      (2): BasicBlock(
        (conv1): Conv2d(256, 256, kernel_size=(3, 3), stride=(1, 1), padding=(1,
1), bias=False)
        (bn1): BatchNorm2d(256, eps=1e-05, momentum=0.1, affine=True, track_
running_stats=True)
        (relu): ReLU(inplace=True)
        (conv2): Conv2d(256, 256, kernel_size=(3, 3), stride=(1, 1), padding=(1,
1), bias=False)
        (bn2): BatchNorm2d(256, eps=1e-05, momentum=0.1, affine=True, track_
running_stats=True)
      )
      (3): BasicBlock(
        (conv1): Conv2d(256, 256, kernel_size=(3, 3), stride=(1, 1), padding=(1,
1), bias=False)
        (bn1): BatchNorm2d(256, eps=1e-05, momentum=0.1, affine=True, track_
running_stats=True)
        (relu): ReLU(inplace=True)
        (conv2): Conv2d(256, 256, kernel_size=(3, 3), stride=(1, 1), padding=(1,
1), bias=False)
        (bn2): BatchNorm2d(256, eps=1e-05, momentum=0.1, affine=True, track_
running_stats=True)
      )
      (4): BasicBlock(
        (conv1): Conv2d(256, 256, kernel_size=(3, 3), stride=(1, 1), padding=(1,
1), bias=False)
        (bn1): BatchNorm2d(256, eps=1e-05, momentum=0.1, affine=True, track_
running_stats=True)
        (relu): ReLU(inplace=True)
        (conv2): Conv2d(256, 256, kernel_size=(3, 3), stride=(1, 1), padding=(1,
1), bias=False)
        (bn2): BatchNorm2d(256, eps=1e-05, momentum=0.1, affine=True, track_
running_stats=True)
      )
      (5): BasicBlock(
        (conv1): Conv2d(256, 256, kernel_size=(3, 3), stride=(1, 1), padding=(1,
1), bias=False)
        (bn1): BatchNorm2d(256, eps=1e-05, momentum=0.1, affine=True, track_
running_stats=True)
        (relu): ReLU(inplace=True)
        (conv2): Conv2d(256, 256, kernel_size=(3, 3), stride=(1, 1), padding=(1,
1), bias=False)
        (bn2): BatchNorm2d(256, eps=1e-05, momentum=0.1, affine=True, track_
running_stats=True)
      )
    )
    (layer4): Sequential(
      (0): BasicBlock(
        (conv1): Conv2d(256, 512, kernel_size=(3, 3), stride=(2, 2), padding=(1,
1), bias=False)
        (bn1): BatchNorm2d(512, eps=1e-05, momentum=0.1, affine=True, track_
running_stats=True)
        (relu): ReLU(inplace=True)
        (conv2): Conv2d(512, 512, kernel_size=(3, 3), stride=(1, 1), padding=(1,
1), bias=False)
        (bn2): BatchNorm2d(512, eps=1e-05, momentum=0.1, affine=True, track_
running_stats=True)
        (downsample): Sequential(
          (0): Conv2d(256, 512, kernel_size=(1, 1), stride=(2, 2), bias=False)
          (1): BatchNorm2d(512, eps=1e-05, momentum=0.1, affine=True, track_
running_stats=True)
        )
      )
      (1): BasicBlock(
        (conv1): Conv2d(512, 512, kernel_size=(3, 3), stride=(1, 1), padding=(1,
1), bias=False)
        (bn1): BatchNorm2d(512, eps=1e-05, momentum=0.1, affine=True, track_
```

```
running_stats=True)
        (relu): ReLU(inplace=True)
        (conv2): Conv2d(512, 512, kernel_size=(3, 3), stride=(1, 1), padding=(1,
1), bias=False)
        (bn2): BatchNorm2d(512, eps=1e-05, momentum=0.1, affine=True, track_
running_stats=True)
      )
      (2): BasicBlock(
        (conv1): Conv2d(512, 512, kernel_size=(3, 3), stride=(1, 1), padding=(1,
1), bias=False)
        (bn1): BatchNorm2d(512, eps=1e-05, momentum=0.1, affine=True, track_
running_stats=True)
        (relu): ReLU(inplace=True)
        (conv2): Conv2d(512, 512, kernel_size=(3, 3), stride=(1, 1), padding=(1,
1), bias=False)
        (bn2): BatchNorm2d(512, eps=1e-05, momentum=0.1, affine=True, track_
running_stats=True)
      )
    )
  )
  (decoder): UnetDecoder(
    (center): Identity()
    (blocks): ModuleList(
      (0): DecoderBlock(
        (conv1): Conv2dReLU(
          (0): Conv2d(768, 256, kernel_size=(3, 3), stride=(1, 1), padding=(1,
1), bias=False)
          (1): BatchNorm2d(256, eps=1e-05, momentum=0.1, affine=True, track_
running_stats=True)
          (2): ReLU(inplace=True)
        )
        (attention1): Attention(
          (attention): Identity()
        )
        (conv2): Conv2dReLU(
          (0): Conv2d(256, 256, kernel_size=(3, 3), stride=(1, 1), padding=(1,
1), bias=False)
          (1): BatchNorm2d(256, eps=1e-05, momentum=0.1, affine=True, track_
running_stats=True)
          (2): ReLU(inplace=True)
        )
        (attention2): Attention(
          (attention): Identity()
        )
      )
      (1): DecoderBlock(
        (conv1): Conv2dReLU(
          (0): Conv2d(384, 128, kernel_size=(3, 3), stride=(1, 1), padding=(1,
1), bias=False)
          (1): BatchNorm2d(128, eps=1e-05, momentum=0.1, affine=True, track_
running_stats=True)
          (2): ReLU(inplace=True)
        )
        (attention1): Attention(
          (attention): Identity()
        )
        (conv2): Conv2dReLU(
          (0): Conv2d(128, 128, kernel_size=(3, 3), stride=(1, 1), padding=(1,
1), bias=False)
          (1): BatchNorm2d(128, eps=1e-05, momentum=0.1, affine=True, track_
running_stats=True)
          (2): ReLU(inplace=True)
        )
        (attention2): Attention(
          (attention): Identity()
        )
      )
      (2): DecoderBlock(
        (conv1): Conv2dReLU(
          (0): Conv2d(192, 64, kernel_size=(3, 3), stride=(1, 1), padding=(1, 1),
bias=False)
          (1): BatchNorm2d(64, eps=1e-05, momentum=0.1, affine=True, track_
running_stats=True)
          (2): ReLU(inplace=True)
```

```
        )
        (attention1): Attention(
          (attention): Identity()
        )
        (conv2): Conv2dReLU(
          (0): Conv2d(64, 64, kernel_size=(3, 3), stride=(1, 1), padding=(1, 1),
bias=False)
          (1): BatchNorm2d(64, eps=1e-05, momentum=0.1, affine=True, track_
running_stats=True)
          (2): ReLU(inplace=True)
        )
        (attention2): Attention(
          (attention): Identity()
        )
      )
      (3): DecoderBlock(
        (conv1): Conv2dReLU(
          (0): Conv2d(128, 32, kernel_size=(3, 3), stride=(1, 1), padding=(1, 1),
bias=False)
          (1): BatchNorm2d(32, eps=1e-05, momentum=0.1, affine=True, track_
running_stats=True)
          (2): ReLU(inplace=True)
        )
        (attention1): Attention(
          (attention): Identity()
        )
        (conv2): Conv2dReLU(
          (0): Conv2d(32, 32, kernel_size=(3, 3), stride=(1, 1), padding=(1, 1),
bias=False)
          (1): BatchNorm2d(32, eps=1e-05, momentum=0.1, affine=True, track_
running_stats=True)
          (2): ReLU(inplace=True)
        )
        (attention2): Attention(
          (attention): Identity()
        )
      )
      (4): DecoderBlock(
        (conv1): Conv2dReLU(
          (0): Conv2d(32, 16, kernel_size=(3, 3), stride=(1, 1), padding=(1, 1),
bias=False)
          (1): BatchNorm2d(16, eps=1e-05, momentum=0.1, affine=True, track_
running_stats=True)
          (2): ReLU(inplace=True)
        )
        (attention1): Attention(
          (attention): Identity()
        )
        (conv2): Conv2dReLU(
          (0): Conv2d(16, 16, kernel_size=(3, 3), stride=(1, 1), padding=(1, 1),
bias=False)
          (1): BatchNorm2d(16, eps=1e-05, momentum=0.1, affine=True, track_
running_stats=True)
          (2): ReLU(inplace=True)
        )
        (attention2): Attention(
          (attention): Identity()
        )
      )
    )
  )
  (segmentation_head): SegmentationHead(
    (0): Conv2d(16, 13, kernel_size=(3, 3), stride=(1, 1), padding=(1, 1))
    (1): Identity()
    (2): Activation(
      (activation): Identity()
    )
  )
)
```

所有编码器都有预训练的权重。以与权重训练相同的方式准备数据，可能使得训练更快地收敛，且获得更好的预测准确度（各类测量指标，例如precision（查准率）、recall（召回率）、F1-Score（（F1-分数）和Intersection over Uion（IOU）等）。但是，如果训练整个模型，而不仅是解码器，则没有这个必要。

```
from segmentation_models_pytorch.encoders import get_preprocessing_fn
preprocess_input = get_preprocessing_fn('resnet18', pretrained='imagenet')
print(preprocess_input)
--------------------------------------------------------------------------------
functools.partial(<function preprocess_input at 0x000001B11D4C4AF0>, input_
space='RGB', input_range=[0, 1], mean=[0.485, 0.456, 0.406], std=[0.229, 0.224,
0.225])
```

输入通道in_channels数量根据训练图像数据确定，如果预训练权重来自于ImageNet，第一个卷积层的权重将被重复使用。对于单通道（in_channels=1）的情况，为第一个卷积层的权重之和；否则，通道权重类似为new_weight[:, i] = pretrained_weight[:, i % 3]，然后用new_weight * 3 / new_in_channels缩放。

```
model = smp.FPN('resnet34', in_channels=1)
mask = model(torch.ones([1, 1, 64, 64]))
print(mask,mask.shape)
--------------------------------------------------------------------------------
tensor([[[[0.6180, 0.5185,  ..., 0.6451, 0.4130, 0.1810],
          [0.6129, 0.5560,  ..., 1.0183, 0.7343, 0.4504],
          [0.6078, 0.5934,  ..., 1.3915, 1.0556, 0.7197],
          ...,
          [1.6890, 1.6174,  ..., 2.0490, 1.9673, 1.8856],
          [1.8117, 1.7673,  ..., 2.1160, 2.0245, 1.9329],
          [1.9345, 1.9173,  ..., 2.1830, 2.0816, 1.9802]]]],
       grad_fn=<UpsampleBilinear2DBackward1>) torch.Size([1, 1, 64, 64])
```

所有模型都支持aux_params辅助分类输出参数，默认值为None，不创建分类辅助输出；如果配置了该参数，模型不仅产生掩码（mask），并且产生分类标签（联合预测概率）。分类头（Classification head） 由GlobalPooling->Dropout(optional)->Linear->Activation(optional)等层组成。

```
aux_params = dict(
    pooling='avg',              # 'avg' 和 'max' 中一个
    dropout=0.5,                # 丢弃比率，默认为无
    activation='sigmoid',       # 激活函数，默认为无
    classes=3,                  # 定义输出标签的数量
)
model = smp.Unet('resnet34', classes=3, aux_params=aux_params)
mask, label = model(torch.ones([1, 3, 64, 64]))
print(mask.shape)
print(label)
--------------------------------------------------------------------------------
torch.Size([1, 3, 64, 64])
tensor([[0.0278, 0.3745, 0.2524]], grad_fn=<SigmoidBackward0>)
```

encoder_depth深度参数可以配置下采样大小，较小的值可以减轻模型的计算量。

```
model=smp.DeepLabV3Plus('resnet34', encoder_depth=3)
```

3.21.2.2 Colaboratory和Planetary Computer Hub

深度学习神经网络的参数量通常在百万级之上（例如，包括卷积层参数、全连接层参数、输出层参数及训练和测试的参数等），因此对硬件通常有必要的需求；而神经网络是高度并行的，那么适合于并行计算的GPU（Graphics processing unit）图形处理器成为深度学习的重要依赖。在CPU上需要花费几个小时完成的训练任务，在GPU上只需要10几分钟；那么，对于GPU需要几天甚至几周完成的训练任务，很难想象用CPU计算完成。目前，个人电脑根据需要通常配置有GPU，但是其显存大小、位宽、带宽和计算能力影响到GPU的效能，高的配置可以减小深度学习网络训练的时间，尤其无法避免反复调整参数的情况下，高效率的GPU配置将大幅度减少等待时间。高配置的GPU往往价格偏高，对于个人使用可能会受到限制；如果科研团队配置有高配的GPU，但是数量可能有限。如果有经常训练模型的需求，团队公用的GPU也限制了使用的自由，因此Colaborator、CoLab[①]（由Google推出）和Planetary Computer Hub、PcHub[②]（由Microsoft推出）等云端提供的算力则解决了这一问题；同时，以社区方式的平台建立，也大大增加了深度学习相关研究的推进和开源分享。CoLab和PcHub均集成了Jupyter笔记服务，并且预先安装了常用

① Colaborator，CoLab，（https://colab.research.google.com/）。

② Planetary Computer Hub，PcHub，（https://pccompute.westeurope.cloudapp.azure.com/compute/hub）。

的Python库，这包括PyTorch和TenserFlow等深度学习库。

CoLab通常配合Google Drive使用，可以实现数据集和训练模型的无缝连接，避免数据丢失（重新下载或上传），尤其是训练好的模型丢失造成的损失。CoLab的另一优势在于可以访问计算机系统层，这对于需要修改Python库代码的用户而言是必要的。相比CoLab，Hub无法访问系统层，但Hub集成了很多全球环境数据目录，例如naip、sentinel、modis、landsat、us-census、noaa、eclipse等几十种，可以方便、快捷地调用数据用于相关分析。

以NAIP航拍影像语义分割为例，使用CoLab书写代码、训练模型，配合Google Drive存储数据和训练结果。分割模型调用Segmentation Models库提供的Unet网络，使用TorchGeo和Pytorcch Lighting库处理。因为TorchGeo库目前处于开发上升期，因此有些不方便应用的方法，则根据本次实验进行调整。Python库无以计数，不同的库处于不同的状态，例如Pytorch、Pandas等稳定型，有大量使用者和大量开发者维护；有些库则处于开发上升期，代码变更可能会比较频繁；有些方法在前一版本存在，在后一版本可能就会取消，又或者变更到其他模块，或者修改了名称，以及调整了参数和返回值等；还有些库已经停止了维护及更新，但有些功能在特定的分析中却非常有用。因此，面对不同库的状态，进行数据分析时，往往也会对调用的库进行修改。修改的方式主要包括两种：一种是，直接在安装环境下修改库代码；另一种是迁移库代码，结合自身分析调整代码等。

下述代码为 CoLab 中执行代码

● 库的安装

因为CoLab没有预先安装TorchGeo库，因此需要安装；当前版本的Matplotlib库的`plt.imshow`方法提示错误，因此更新到3.6.2版本。

```
%pip install torchgeo tensorboard pytorch_lightning
```

如果要修改TorchGeo库的部分代码，满足分析需要，可以通过`!pip show torchgeo`获取该库位置（本次试验未修改安装库的代码，但是通过代码迁移修改的方式实现）。

```
!pip show torchgeo
----------------------------------------------------------------------------------
Name: torchgeo
Version: 0.4.1
Summary: TorchGeo: datasets, samplers, transforms, and pre-trained models for
geospatial data
Home-page: https://github.com/microsoft/torchgeo
Author: Adam J. Stewart
Author-email: ajstewart426@gmail.com
License:
Location: /usr/local/lib/python3.10/dist-packages
Requires: einops, fiona, kornia, lightning, matplotlib, numpy, pillow, pyproj,
rasterio, rtree, scikit-learn, segmentation-models-pytorch, shapely, timm,
torch, torchmetrics, torchvision
Required-by:
----------------------------------------------------------------------------------
%matplotlib inline
%load_ext tensorboard
import os
from torchvision.transforms import Compose
from torchgeo.datasets import RasterDataset,stack_samples,unbind_samples
from torchgeo.datasets.utils import download_url
from torchgeo.transforms import indices,AugmentationSequential
import kornia.augmentation as K
from torchgeo.samplers import RandomGeoSampler
from torch.utils.data import DataLoader
import matplotlib
import numpy as np
import matplotlib.pyplot as plt
from torchgeo.datamodules import NAIPChesapeakeDataModule
import torch
from lightning.pytorch import Trainer
from lightning.pytorch.callbacks import EarlyStopping, ModelCheckpoint
from lightning.pytorch.loggers import TensorBoardLogger
from torchgeo.trainers import SemanticSegmentationTask
from os import path
import torch.nn as nn
import torchvision.transforms as T
```

```
from pytorch_lightning.loggers import CSVLogger
import torchgeo
from lightning.pytorch import LightningDataModule
from torchgeo.datasets import BoundingBox
import csv
from torchgeo.samplers.batch import RandomBatchGeoSampler
from torchgeo.samplers.single import GridGeoSampler
-----------------------------------------------------------------
The tensorboard extension is already loaded. To reload it, use:
  %reload_ext tensorboard
```

- 连接Google Drive

为避免数据丢失，方便数据调用，连接Google Drive到CoLab，文件夹会加载到CoLab本地显示，用法同本地文件。

```
from google.colab import drive
drive.mount('/content/gdrive')
-----------------------------------------------------------------------------
Mounted at /content/gdrive
```

- 配置数据存储路径

数据文件、训练过程和结果文件均存储于Google Drive中，文件夹可以在Googel Drive中建立，也可以直接在CoLab中建立。文件路径可以在对应文件上右键Copy path直接获取。

```
root = '/content/gdrive/MyDrive/dataset/naipNdelaware4seg'
imagery_data = os.path.join(root, "imagery") # 存储 NAIP 航拍影像数据
LC_data = os.path.join(root, "LC") # 存储 NAIP 土地覆盖分类数据
data_dir = os.path.join(root, "training") # 存储模型训练过程和结果文件
```

- NAIP图像数据下载

训练时，未使用全部Delaware（特拉华州（位于美国东部））区域影像，仅选取了4个瓦片。将瓦片编号录入文本文件naipEntityID_selection.txt下，传至Google Drive用于读取并下载对应图像数据至Google Drive。下载时直接调用torchvision库的`download_url`方法。

图3.21-8为Delaware区域13类土地覆盖类型，以及所用的NAIP 7个影像瓦片。

```
naipEntityID_selection_fn = os.path.join(root,'naipEntityID_selection.txt')
with open(naipEntityID_selection_fn,'r') as f:
    naipEntityID_selection = f.readlines()
naipEntityID_selection = [line.rstrip() for line in naipEntityID_selection]
naipEntityID_selection
-----------------------------------------------------------------------------
['m_3807505_se_18_060_20180827.tif',
 'm_3807505_sw_18_060_20180815.tif',
 'm_3807504_se_18_060_20180815.tif',
 'm_3807504_sw_18_060_20180815.tif',
 'm_3807532_nw_18_060_20180815.tif',
 'm_3807531_nw_18_060_20180815.tif',
 'm_3807531_ne_18_060_20180815.tif']
-----------------------------------------------------------------------------
naip_38075_url = (
    "https://naipeuwest.blob.core.windows.net/naip/v002/de/2018/
de_060cm_2018/38075/")
def naip_download_rul(url, tile, root): return download_url(url+tile, root)
for tile in naipEntityID_selection:
    try:
        naip_download_rul(naip_38075_url, tile, imagery_data)
    except:
        print(f'Can not access to:{tile}')
-----------------------------------------------------------------------------
Using downloaded and verified file: /content/gdrive/MyDrive/dataset/
naipNdelaware4seg/imagery/m_3807505_se_18_060_20180827.tif
...
Downloading https://naipeuwest.blob.core.windows.net/naip/v002/de/2018/
de_060cm_2018/38075/m_3807532_nw_18_060_20180815.tif to /content/gdrive/MyDrive/
dataset/naipNdelaware4seg/imagery/m_3807532_nw_18_060_20180815.tif
100%|██████████| 489103340/489103340 [00:42<00:00, 11388534.38it/s]
...
```

- 对应NAIP的土地覆盖类型数据下载

TorchGeo库下载土地覆盖类型数据时，直接将其转换为训练数据集，这里单独下载为初始的

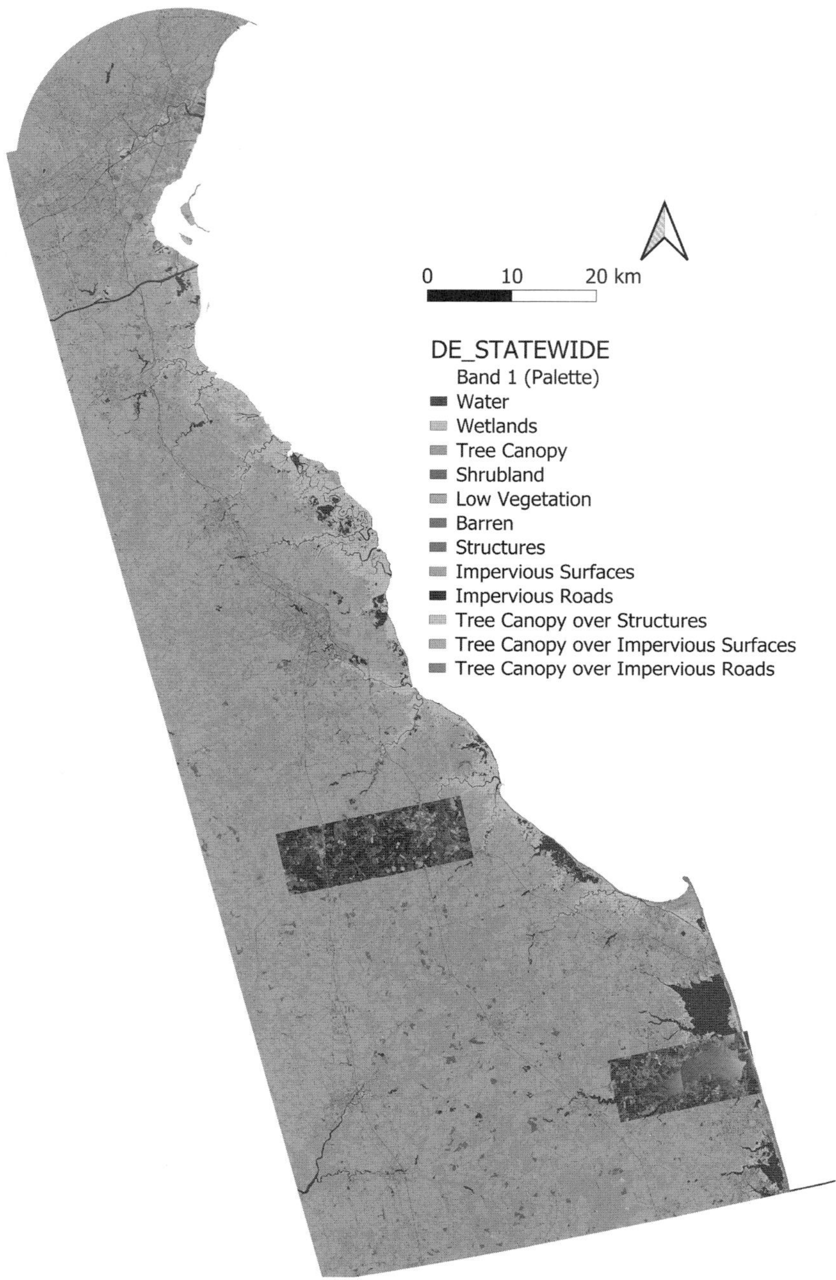

图3.21-8 土地覆盖及NAIP影像瓦片

ZIP压缩文件，解压后再单独建立训练数据集。

```
chesapeakebay_landcover_url = "https://cicwebresources.blob.core.windows.net/
chesapeakebaylandcover"
base_folder = "DE"
zipfile = "_DE_STATEWIDE.zip"
chesapeakebay_landcover_url += f"/{base_folder}/{zipfile}"
download_url(chesapeakebay_landcover_url, LC_data, filename=zipfile)
```

```
--------------------------------------------------------------------------------
Using downloaded and verified file: /content/gdrive/MyDrive/dataset/
naipNdelaware4seg/LC/_DE_STATEWIDE.zip
```

- 迁移解压缩代码

因为下载的土地覆盖类型数据为ZIP格式压缩文件，因此迁移TorchGeo库已有解压缩代码，位于torchgeo.datasets.utils模块。

```
import tarfile
from typing import (
    Any,
    Dict,
    Iterable,
    Iterator,
    List,
    Optional,
    Sequence,
    Tuple,
    Union,
    cast,
    overload,
)
class _rarfile:
    class RarFile:
        def __init__(self, *args: Any, **kwargs: Any) -> None:
            self.args = args
            self.kwargs = kwargs
        def __enter__(self) -> Any:
            try:
                import rarfile
            except ImportError:
                raise ImportError(
                    "rarfile is not installed and is required to extract this
dataset"
                )
            return rarfile.RarFile(*self.args, **self.kwargs)
        def __exit__(self, exc_type: None, exc_value: None, traceback: None) ->
None:
            pass
class _zipfile:
    class ZipFile:
        def __init__(self, *args: Any, **kwargs: Any) -> None:
            self.args = args
            self.kwargs = kwargs
        def __enter__(self) -> Any:
            try:
                # 支持普通 zip 文件，专有的 deflate64 压缩算法
                import zipfile_deflate64 as zipfile
            except ImportError:
                # 只支持常规的 zip 文件
                import zipfile
            return zipfile.ZipFile(*self.args, **self.kwargs)
        def __exit__(self, exc_type: None, exc_value: None, traceback: None) ->
None:
            pass
def extract_archive(src: str, dst: Optional[str] = None) -> None:
    """解压缩 .

    Args:
        src: 要提取的压缩文件
        dst: 要提取到的目录（默认为 src 的目录名）

    Raises:
        RuntimeError: 如果 SRC 文件有未知的存档 / 压缩计划
    """
    if dst is None:
        dst = os.path.dirname(src)
    suffix_and_extractor: List[Tuple[Union[str, Tuple[str, ...]], Any]] = [
        (".rar", _rarfile.RarFile),
        (
            (".tar", ".tar.gz", ".tar.bz2", ".tar.xz",
             ".tgz", ".tbz2", ".tbz", ".txz"),
            tarfile.open,
```

```
        ),
        (".zip", _zipfile.ZipFile),
    ]
    for suffix, extractor in suffix_and_extractor:
        if src.endswith(suffix):
            with extractor(src, "r") as f:
                f.extractall(dst)
            return
    suffix_and_decompressor: List[Tuple[str, Any]] = [
        (".bz2", bz2.open),
        (".gz", gzip.open),
        (".xz", lzma.open),
    ]
    for suffix, decompressor in suffix_and_decompressor:
        if src.endswith(suffix):
            dst = os.path.join(dst, os.path.basename(src).replace(suffix, ""))
            with decompressor(src, "rb") as sf, open(dst, "wb") as df:
                df.write(sf.read())
            return
    raise RuntimeError("src file has unknown archival/compression scheme")
--------------------------------------------------------------------------------
extract_archive(os.path.join(LC_data, zipfile))
```

- 构建包含图像增强变换的训练数据集

TorchGeo库设计时对图像增强变换的操作有两个途径：一个是构建训练数据集时，且融入到了具体的数据类型下，例如torchgeo.datasets.NAIP(root, crs=None, res=None, transforms=None, cache=True)集成了NAIP图像下载、图像增强变换和构建训练数据集等多个功能；另一个是用DataLoader加载小批量训练样本时，直接对小批量样本执行图像增强变换操作。在后续训练时，调用pytorch_lightning库提供的Trainer类，其fit(model, train_dataloaders=None, val_dataloaders=None, datamodule=None, ckpt_path=None)方法提供datamodule，为LightningDataModule类的一个实例对象。TorchGeop构建的NAIPChesapeakeDataModule(pl.LightningDataModule)类继承了LightningDataModule类，NAIPChesapeakeDataModule构建时融合了土地覆盖类型数据的下载，并构建训练数据集和NAIP训练数据集的构建，以及一些图像增强变换操作。这些融合的方式虽然大幅度增加了数据处理的效率，但混杂也带来了很多数据处理分步的不清晰和不便利，因此下述迁移更新了NAIPChesapeakeDataModule类，定义为为geo_datamodule(pl.LightningDataModule)。新定义的geo_datamodule类，拆离了训练数据集构建，从而把图像增强变换仅作用于训练数据集构建过程中，而不再混杂于datamodule的构建过程。这样，也方便观察图像增强变换后的数据结果，以及用于图像预测时输入数据的变换。

土地覆盖类型中，树冠类有很多子类很难通过航拍影像识别，如果仍然保留进行训练将会影响训练收敛和预测精度，因此定义labes_merge()函数作为图像增强变换输入条件，合并部分分类。下述定义的变量label_merge_mapping合并了树冠类的各个子类，及灌丛和低矮植被，和不透水人造对象。同时，调整了标签数值，排序为1～7，缩减为7个土地覆盖类；另外，如果读取区域为空值，对应值则为 0。因此，分类数量由13类缩减为8类（含空值类），以便提高训练收敛的速度和预测精度。

```
# 迁移更新于 TorchGeo 库：torchgeo.datamodules.naip 模块

def chesapeake_transform(sample: Dict[str, Any]) -> Dict[str, Any]:
    """ 转换来自 Chesapeake 数据集的单个样本 .

    Args:
        sample: Chesapeake 掩码字典

    Returns:
        预处理的 Chesapeake 数据
    """
    sample["mask"] = sample["mask"].long()[0]
    return sample
def remove_bbox(sample: Dict[str, Any]) -> Dict[str, Any]:
    """ 从样本中移除边界框属性 .

    Args:
        sample: 带有地理元数据的字典
```

```
    Returns
        没有 bbox 属性的样本
    """
    del sample["bbox"]
    return sample

def labes_merge(sample: Dict[str, Any]) -> Dict[str, Any]:
    """ 合并土地覆盖分类 .

    Args:
        sample: 带有地理元数据的字典

    Returns
        没有 bbox 属性的样本
    """
    label_merge_mapping = {
        0: 0,    # Null
        1: 1,    # Water ->Water
        2: 2,    # Emergent Wetlands -> Emergent Wetlands
        3: 3,    # Tree Canopy -> Tree Canopy
        4: 4,    # Scrub/Shrub -> Scrub/Shrub
        5: 4,    # Low Vegetation -> Scrub/Shrub
        6: 5,    # Barren -> Barren(6->5)
        7: 6,    # Impervious Structures -> Impervious Structures(7->6)
        8: 6,    # Other Impervious -> Impervious Structures(7->6)
        9: 6,    # Impervious Road -> Impervious Road(9->7)
        10: 3,    # Tree Canopy over Impervious Structure -> Tree Canopy
        11: 3,    # Tree Canopy over Other Impervious -> Tree Canopy
        12: 3,    # Tree Canopy over Impervious Roads -> Tree Canopy
        254: 7,    # Aberdeen Proving Ground -> Aberdeen Proving Ground(254->7)
    }
    old = sample["mask"]
    indexer = np.array([label_merge_mapping.get(i, -1)
                        for i in range(old.min(), old.max() + 1)])
    new = torch.from_numpy(indexer[(old - old.min())])
    sample["mask"] = new
    return sample
```

TorchGeo定义RasterDataset类时，输入的参数部分为全局变量（位于类下而于函数外），因此定义类的方式对变量进行赋值，例如对土地覆盖类型构建TorchGeo的RasterDataset栅格数据集，建立类delaware_lc_rd(RasterDataset)，继承RasterDataset类，并对父类的filename_glob变量进行更新。同时，增加迁移的上述两个变换chesapeake_transform和remove_bbox。

```
chesapeak_transforms = Compose([chesapeake_transform,remove_bbox,labes_merge])
class delaware_lc_rd(RasterDataset):
    filename_glob = "DE_STATEWIDE.tif"
    is_image = False
chesapeake = delaware_lc_rd(LC_data,transforms=chesapeak_transforms)
print(f"crs:{chesapeake.crs}\nres:{chesapeake.res}")
--------------------------------------------------------------------------------
crs:ESRI:102039
res:1.0
```

同样的方式操作NAIP图像。在图像增强变换时，增加了NDVI一个指数层（同时也备用给出了NDWI）。因为TorchGeo的transforms模块中计算指数时，将其结果追加到dim=0的维度上，从而构建数据加载器时，本应输入维度为4维，变成了5维，如(1,1,5,256,256)，而应为(1,5,256,256)。因此，直接迁移TorchGeo库对应代码，修改AppendNormalizedDifferenceIndex类的返回值return input为return input[0]，并直接调用。

```
    # 图像增强变换的内容和组合结构需要根据训练效果不断调整实验，使得训练收敛且提高预测精度。

from typing import Optional
import torch
from kornia.augmentation import IntensityAugmentationBase2D
from torch import Tensor
_EPSILON = 1e-10
class AppendNormalizedDifferenceIndex( IntensityAugmentationBase2D ):
    r""" 追加归一化差分指数到图像张量 .
```

```
    指数计算方法如下：

        NDI = (A−B)/(A+B)

    .. 版本 :: 0.2
    """
    def __init__(self, index_a: int, index_b: int) -> None:
        """ 初始化一个新的变换实例 .

        Args:
            index_a: 引用波段通道索引
            index_b: 差分波段通道索引
        """
        super().__init__(p=1)
        self.flags = {"index_a": index_a, "index_b": index_b}
    def apply_transform(
        self,
        input: Tensor,
        params: dict[str, Tensor],
        flags: dict[str, int],
        transform: Optional[Tensor] = None,
    ) -> Tensor:
        """ 应用变换 .

        Args:
            input: 输入张量
            params: 生成的参数
            flags: 静态参数
            transform: 地理变换张量

        Returns:
            增广输入
        """
        band_a = input[..., flags["index_a"], :, :]
        band_b = input[..., flags["index_b"], :, :]
        ndi = (band_a - band_b) / (band_a + band_b + _EPSILON)
        ndi = torch.unsqueeze(ndi, -3)
        input = torch.cat((input, ndi), dim=-3)
        return input[0]

class AppendNDVI(AppendNormalizedDifferenceIndex):
    r""" 追加归一化植被指数 （Normalized Difference Vegetation Index, NDVI）.

    指数计算方法如下：

        NDVI = (NIR−R)/(NIR+R)

    *  参考: https://doi.org/10.1016/0034-4257(79)90013-0
    """

    def __init__(self, index_nir: int, index_red: int) -> None:
        """ 初始化一个新的变换实例 .

        Args:
            index_nir: 图像中近红外 (NIR) 波段索引
            index_red: 图像中红色波段索引
        """
        super().__init__(index_a=index_nir, index_b=index_red)

class AppendNDWI(AppendNormalizedDifferenceIndex):
    r""" 追加归一化水体指数 （Normalized Difference Water Index , NDWI）.

    指数计算方法如下：

        NDWI = (G−NIR)/(G+NIR)

    * 参考:  https://doi.org/10.1080/01431169608948714
    """
    def __init__(self, index_green: int, index_nir: int) -> None:
        """ 初始化一个新的变换实例 .

        Args:
            index_green: 图像中绿色波段的索引
            index_nir: 图像中近红外 (NIR) 波段的索引
```

```
        """
        super().__init__(index_a=index_green, index_b=index_nir)
```

定义继承父类RasterDataset的naip_rd类实现NAIP影像数据集定义。图像增强变换除了应用NDVI等指数试验外，增加了自定义的naip_preprocess()变换，将图像4个波段值归一化。

```
def naip_preprocess(sample: Dict[str, Any]) -> Dict[str, Any]:
    """ 变换 NAIP 数据集的一个样本 .

    Args:
        sample: NAIP 影像字典

    Returns:
        预处理的 NAIP 数据
    """
    sample["image"] = sample["image"].float()
    sample["image"] /= 255.0
    return sample
class naip_rd(RasterDataset):
    filename_glob = "m_*.*"
    filename_regex = r"""
        ^m
        _(?P<quadrangle>\d+)
        _(?P<quarter_quad>[a-z]+)
        _(?P<utm_zone>\d+)
        _(?P<resolution>\d+)
        _(?P<date>\d+)
        (?:_(?P<processing_date>\d+))?
        \..*$
    """
    is_image = True
    all_bands = ["R", "G", "B", "NIR"]
    rgb_bands = ["R", "G", "B"]
ndvi = AppendNDVI(index_nir=3, index_red=0)
ndwi = AppendNDWI(index_green=1, index_nir=3)
naip_transforms = Compose([
    naip_preprocess,
    remove_bbox,
    AugmentationSequential(ndvi, data_keys=["image"])
])
naip = naip_rd(imagery_data, chesapeake.crs,
               chesapeake.res, transforms=naip_transforms)
print(f"crs:{naip.crs}\nres:{naip.res}")
--------------------------------------------------------------------------------
crs:ESRI:102039
res:1.0
```

构建包含图像和类标的训练数据集。TorchGeo库重写了__and__、__or__等方法，可以直接对地理空间数据集执行交集和并集的操作。

```
dataset=chesapeake & naip
```

- 查看图像增强变换后的数据集样本

因为增加了NDVI一个指数层，因此数组的维度为［1, 5, 256, 256］，即包括R、G、B、Nir及NDVI这5个波段（图3.21–9）。

```
sampler = RandomGeoSampler(dataset,size=256, length=10)
dataloader = DataLoader(dataset, sampler=sampler, collate_fn=stack_samples)
for batch in dataloader:
    sample = unbind_samples(batch)[0]
    print(sample.keys())
    print(sample['image'].shape)
    break
--------------------------------------------------------------------------------
dict_keys(['crs', 'mask', 'image'])
torch.Size([5, 256, 256])
```

```
    # 注意下述所用生成的数据集含有 NDWI 指数，即含 6 个通道。

sampler = RandomGeoSampler(dataset,size=256, length=10)
dataloader = DataLoader(dataset, sampler=sampler, collate_fn=stack_samples)
```

图3.21-9　图像增强变换后的数据集样本，RGB合成影像（左）；Nir（中）；NDVI（右）

```
for batch in dataloader:
    sample = unbind_samples(batch)[0]
    print(sample.keys())
    print(sample['image'].shape)
    fig, axes = plt.subplots(1, 3,figsize=(20,10))
    axes[0].imshow(T.ToPILImage()(sample['image'][:3]))
    axes[1].imshow(T.ToPILImage()(sample['image'][4]))
    axes[2].imshow(T.ToPILImage()(sample['image'][5]))
    plt.show()
    break
--------------------------------------------------------------------------------
dict_keys(['crs', 'mask', 'image'])
torch.Size([6, 256, 256])
np.unique(sample['mask'])
--------------------------------------------------------------------------------
array([3, 4, 5, 6, 7])
```

查看对应示例样本图像的分类数据（图3.21-10）。

```
LC_color_dict={
    # 0: (0, 0, 0, 0),
    1: (30, 136, 229,255),
    2: (230, 238, 156, 255),
    3: (46, 125, 50,255),
    4: (205, 220, 57, 255),
    5: (176, 190, 197, 255),
    6: (66, 66, 66, 255),
    7: (189, 189, 189, 255),
    }
LC_color_dict_selection = {k:LC_color_dict[k] for k,v in LC_color_dict.items()}
cmap_LC, norm = matplotlib.colors.from_levels_and_colors(list(LC_color_dict_selection.keys()),[[v/255 for v in i] for i in LC_color_dict_selection.values()],extend='max')
plt.imshow(np.squeeze(sample['mask']),cmap=cmap_LC);
```

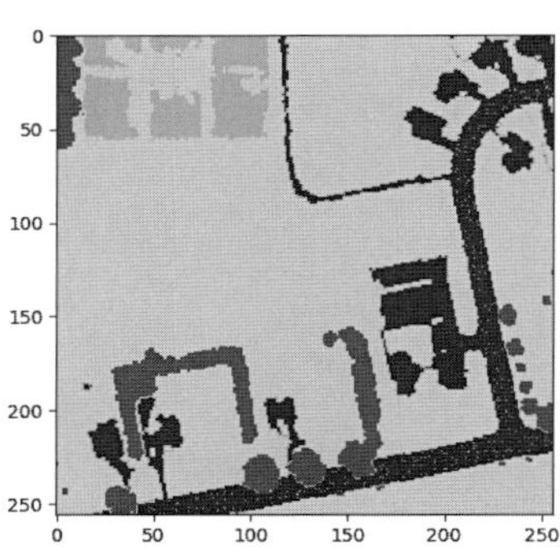

图3.21-10　对应示例样本图像的分类数据

● 更新datamodule类

新定义的geo_datamodule类，移除了原NAIPChesapeakeDataModule类中的数据集前期处理部分内容（包括数据下载、图像增强变换等），而集中于训练数据切分为训练数据集（train_dataloader）、测试数据集（test_dataloader）和验证数据集（val_dataloader）的Dataloder对象的构建上。

```
class Geo_datamodule(LightningDataModule):
    """TorchGeo 数据集的 LightningDataModule 实现 .

    将数据集切分为 train（训练）/val（验证）/test（测试）.
    """
    length = 1000
    stride = 128
    def __init__(
        self,
        ds_image: torchgeo.datasets,
```

```
        ds_label: torchgeo.datasets,
        batch_size: int = 64,
        num_workers: int = 0,
        patch_size: int = 256,
        **kwargs: Any,
    ) -> None:
        """ 为基于 dataloader 的 TorchGeo 数据集初始化 LightningDataModule.

        Args:
            ds_image: TorchGeo RasterDataset 格式的图像数据集
            ds_label: TorchGeo RasterDataset 格式的目标（标签）数据集
            batch_size: 要在所有已创建的 dataloader 中使用的批次大小
            num_workers: 要在所有已创建的 dataloader 中使用的线程数
            patch_size: 一个样本图像大小
        """
        super().__init__()
        self.ds_image = ds_image
        self.ds_label = ds_label
        self.batch_size = batch_size
        self.num_workers = num_workers
        self.patch_size = patch_size
    def setup(self, stage: Optional[str] = None) -> None:
        """ 初始化主 Dataset 对象 .

        每次运行每个 GPU 调用一次这个方法 .

        Args:
            stage: 状态配置
        """
        self.dataset = self.ds_label & self.ds_image
        roi = self.dataset.bounds
        midx = roi.minx + (roi.maxx - roi.minx) / 2
        midy = roi.miny + (roi.maxy - roi.miny) / 2
        train_roi = BoundingBox(roi.minx, midx, roi.miny,
                                roi.maxy, roi.mint, roi.maxt)
        val_roi = BoundingBox(midx, roi.maxx, roi.miny,
                              midy, roi.mint, roi.maxt)
        test_roi = BoundingBox(roi.minx, roi.maxx, midy,
                               roi.maxy, roi.mint, roi.maxt)
        self.train_sampler = RandomBatchGeoSampler(
            self.ds_image, self.patch_size, self.batch_size, self.length, train_roi
        )
        self.val_sampler = GridGeoSampler(
            self.ds_image, self.patch_size, self.stride, val_roi)
        self.test_sampler = GridGeoSampler(
            self.ds_image, self.patch_size, self.stride, test_roi)
    def train_dataloader(self) -> DataLoader[Any]:
        """ 返回一个用于训练的 DataLoader.

        Returns:
            训练数据集加载器
        """
        return DataLoader(
            self.dataset,
            batch_sampler=self.train_sampler,
            num_workers=self.num_workers,
            collate_fn=stack_samples,
        )

    def val_dataloader(self) -> DataLoader[Any]:
        """ 返回一个用于验证的 DataLoader.

        Returns:
            验证数据集加载器
        """
        return DataLoader(
            self.dataset,
            batch_size=self.batch_size,
            sampler=self.val_sampler,
            num_workers=self.num_workers,
            collate_fn=stack_samples,
        )
    def test_dataloader(self) -> DataLoader[Any]:
        """ 返回一个用于测试的 DataLoader.
```

```
        Returns:
            测试数据集加载器
        """
        return DataLoader(
            self.dataset,
            batch_size=self.batch_size,
            sampler=self.test_sampler,
            num_workers=self.num_workers,
            collate_fn=stack_samples,
        )
    def plot(
        self,
        sample: dict[str, Any],
        show_titles: bool = True,
        suptitle: Optional[str] = None,
    ) -> plt.Figure:
        """ 从数据集中绘制一个样本 .

        Args:
            sample: 返回一个样本，由 :meth:`RasterDataset.__getitem__`
            show_titles: 指示是否在每个面板上方显示标题的标志
            suptitle: 用作副标题的可选字符串

        Returns:
            渲染样本图像的 matplotlib 图表

        """
        image = sample["image"][0:3, :, :].permute(1, 2, 0)
        fig, ax = plt.subplots(nrows=1, ncols=1, figsize=(4, 4))
        ax.imshow(image)
        ax.axis("off")
        if show_titles:
            ax.set_title("Image")
        if suptitle is not None:
            plt.suptitle(suptitle)
        return fig
datamodule = Geo_datamodule(
    ds_image=naip,
    ds_label=chesapeake,
    batch_size=64,
    patch_size=256
)
```

- 构建深度学习网络模型

TorchGeo的深度学习网络模型构建SemanticSegmentationTask类继承了pytorch_lightning的LightningModule类，并调用了segmentation_models_pytorch库的unet模型。配置了aux_params参数，返回掩码和分类标签。因为增加了NDVI一个指数层，因此总共5个通道，配置in_channels=5。调整后的土地覆盖类型共计8类，因此配置num_classes=8。

```
aux_params = dict(
    pooling='avg',              # 'avg' 和 'max' 中一个
    dropout=0.5,                # 丢弃比率，默认为无
    activation='sigmoid',       # 激活函数，默认为无
    )
task = SemanticSegmentationTask(
    model='unet',
    backbone='resnet34',
    weights='imagenet',
    pretrained=True,
    in_channels=5,
    num_classes=8,
    ignore_index=0,
    loss='ce', # 'jaccard'
    learning_rate=0.1,
    learning_rate_schedule_patience=5,
    aux_params=aux_params,
    )
```

- 配置训练参数

PyTorch Lightning库提供的Trainer方法大幅度简化了训练参数的配置。如果不希望部分参

数自动化处理，也可以自行配置。下述配置了ModelCheckpoint，通过配置moniter参数监测给定参数的值，定期保存模型。在LightningModule中，使用log()或log_dict()记录的每个指标都可以为监控的对象；配置EarlyStopping，监控给定的指标，当该指标停止改进时则停止训练，这非常有利于了解到模型训练的程度，避免无效的训练，利于反复调整参数实验；同时，配置了训练文件、过程网络模型等保存的本地位置（本次存储到Google Drive中）default_root_dir；用min_epochs和max_epochs配置最小和最大的迭代次数；fast_dev_run为一种单元测试；配置accelerator="gpu"，支持传递不同的加速器类型，例如“cpu”、“gpu”、“tpu”、“ipu”、“hpu”、“mps”、“auto”，以及自定义的加速器实例等。Trainer方法提供的参数高达50多个，可以从官网查看应用。

在配置日志时，给出了两种途径：一种是使用CSVLogger方法，以YAML或CSV格式存储测量指标，用于后续读取分析；另一种是使用TensorBoardLogger方法，后续可以直接用于训练报告查看。

```
accelerator = “gpu” if torch.cuda.is_available() else “cpu”
checkpoint_callback = ModelCheckpoint(
    monitor=”val_loss”, dirpath=data_dir, save_top_k=1, save_last=True)  #
,save_on_train_epoch_end=True
early_stopping_callback = EarlyStopping(
    monitor=”val_loss”, min_delta=0.00, patience=10)
# logger=CSVLogger(save_dir=data_dir, name=” segmentation_unet” )
logger = TensorBoardLogger(
    save_dir=data_dir, name=”tensorboard_logs”, version=1,)
in_tests = “PYTEST_CURRENT_TEST” in os.environ
trainer = Trainer(
    callbacks=[checkpoint_callback, early_stopping_callback],
    logger=logger,
    default_root_dir=data_dir,
    min_epochs=1,
    max_epochs=300,
    fast_dev_run=in_tests,
    accelerator=accelerator,
    # limit_val_batches=500,
)
--------------------------------------------------------------------------------
INFO:pytorch_lightning.utilities.rank_zero:GPU available: True (cuda), used:
True
INFO:pytorch_lightning.utilities.rank_zero:TPU available: False, using: 0 TPU
cores
INFO:pytorch_lightning.utilities.rank_zero:IPU available: False, using: 0 IPUs
INFO:pytorch_lightning.utilities.rank_zero:HPU available: False, using: 0 HPUs
```

- 网络模型训练

fit方法传入的参数ckpt_path，只对已经训练且保存的模型权重值有效，因此需要检查该文件是否已经存在。利用CoLab提供的GPU算力，模型达到给定指标不再有效提升时所需要的时间约为1小时42分钟（CoLab Pro版）。

```
ckpt_path = os.path.join(data_dir,'last.ckpt')
if path.exists(ckpt_path):
    print('ckpt file exists.')
    trainer.fit(model=task,datamodule=datamodule,ckpt_path=ckpt_path)
else:
    trainer.fit(model=task,datamodule=datamodule)
--------------------------------------------------------------------------------
INFO: LOCAL_RANK: 0 - CUDA_VISIBLE_DEVICES: [0]
INFO:lightning.pytorch.accelerators.cuda:LOCAL_RANK: 0 - CUDA_VISIBLE_DEVICES:
[0]
INFO:
  | Name            | Type              | Params
-------------------------------------------------------
0 | model           | Unet              | 24.4 M
1 | loss            | CrossEntropyLoss  | 0
2 | train_metrics   | MetricCollection  | 0
3 | val_metrics     | MetricCollection  | 0
4 | test_metrics    | MetricCollection  | 0
-------------------------------------------------------
24.4 M    Trainable params
0         Non-trainable params
```

```
24.4 M    Total params
97.775    Total estimated model params size (MB)
INFO:lightning.pytorch.callbacks.model_summary:
  | Name          | Type             | Params
------------------------------------------------
0 | model         | Unet             | 24.4 M
1 | loss          | CrossEntropyLoss | 0
2 | train_metrics | MetricCollection | 0
3 | val_metrics   | MetricCollection | 0
4 | test_metrics  | MetricCollection | 0
------------------------------------------------
24.4 M    Trainable params
0         Non-trainable params
24.4 M    Total params
97.775    Total estimated model params size (MB)
Sanity Checking: 0it [00:00, ?it/s]
```

- 从日志（Log）中提取训练测量指标

1. 读取CSVLogger方法的日志

测量指标已经存入Google Drive文件夹下，读取metrics.csv文件数据，并打印Train loss和Validation Accuracy指标（图3.21-11）。可以初步判断深度学习网络模型有效收敛，且验证数据集上的预测精度能够达到约0.8以上。

```
if not in_tests:
    train_steps = []
    train_rmse = []
    val_steps = []
    val_rmse = []
    with open(
        os.path.join(data_dir, "segmentation_unet",
                     "version_0", "metrics.csv"), "r"
    ) as f:
        csv_reader = csv.DictReader(f, delimiter=",")
        for i, row in enumerate(csv_reader):
            try:
                train_rmse.append(float(row["train_loss"]))
                train_steps.append(i)
            except ValueError:  # 在训练 RMSE 为空处忽略行
                pass
            try:
                val_rmse.append(float(row["val_Accuracy"]))
                val_steps.append(i)
            except ValueError:  # 在验证 RMSE 为空处忽略行
                pass
if not in_tests:
    plt.figure()
    plt.plot(train_steps, train_rmse, label="Train loss")
    plt.plot(val_steps, val_rmse, label="Validation Accuracy")
    plt.legend(fontsize=15)
    plt.xlabel("Batches", fontsize=15)
    plt.ylabel("Loss", fontsize=15)
```

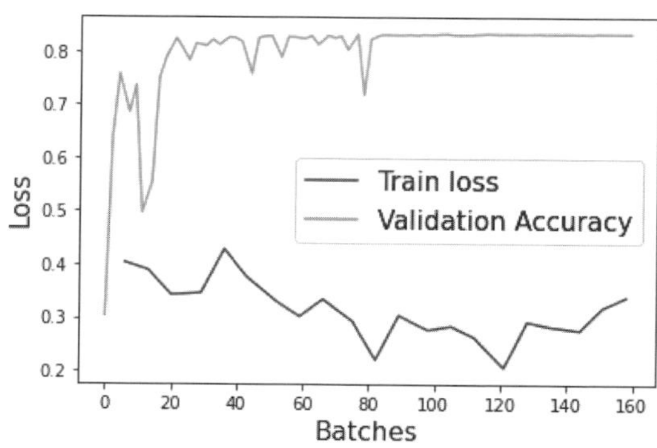

图3.21-11　损失曲线和验证精度

```
plt.show()
plt.close()
```

2. 读取TensorBoardLogger方法的日志（图3.21-12）

```
%tensorboard --logdir "$data_dir"
```

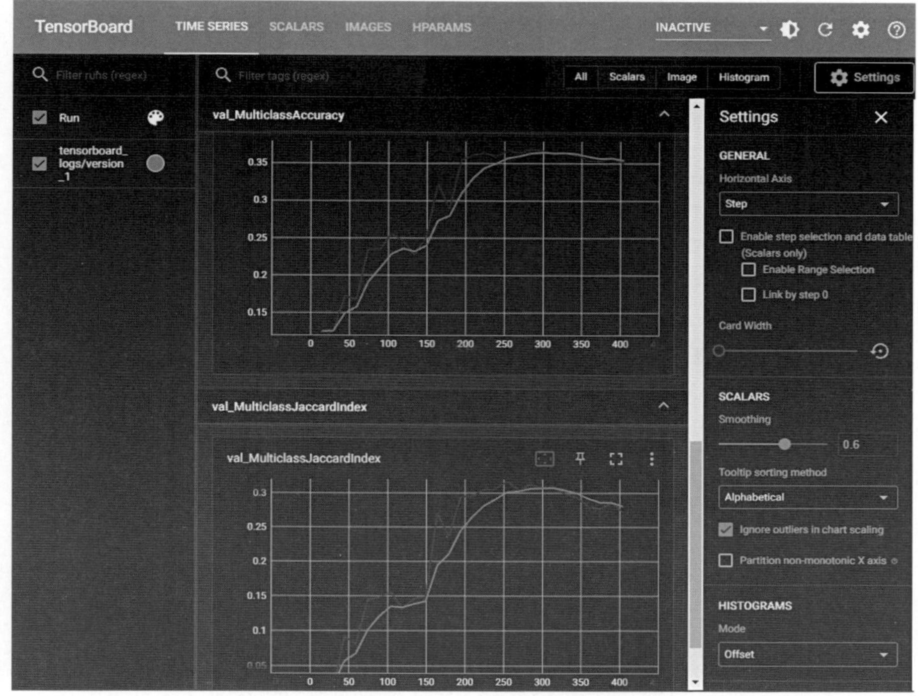

图3.21-12 TensorBoard日志

- TorchMetrics 指标度量库

TorchMetrics①集合了100+个指标实现，且提供了一个易于使用的API（Application Programming Interface）用于自定义指标。该API提供了可以增加再现性（reproducibility）的一个标准化接口；并兼容分布式训练；可以在批次（batch）间自动积累；在多个设备间自动同步；并且经过了严格的测试。

① TorchMetrics (https://torchmetrics.readthedocs.io/en/stable/)。

上述影像分割实验中，TorchGeo库在SemanticSegmentationTask类中引入了TorchMetrics提供的MulticlassAccuracy和MulticlassJaccardIndex两个指标。

1）MulticlassAccuracy为多分类精度计算，公式为$\text{Accuracy} = \frac{1}{N}\sum_i^N 1(y_i = \hat{y}_i)$。式中，$y$为真实标签（目标值），$\hat{y}$为预测值。如图3.21-12中val_MulticlassAccuracy标签部分所示；

2）MulticlassJaccardIndex为多分类Jaccard指数，即交并集，也称为Jaccard相似系数，公式为$J(A,B) = \frac{|A\cap B|}{|A\cup B|}$。如有，

真实值\预测值	A	B	C
A	AA	AB	Ac
B	BA	BB	BC
C	CA	CB	CC

则多分类Jaccard指数计算可表述为$\frac{AA+BB+CC}{AA+AB+AC+BA+BB+BC+CA+CB+CC}$；Jaccard 平均得分计算可表述为$\frac{1}{3}\left(\frac{AA}{AA+AB+AC+BA+CA}+\frac{BB}{AB+BA+BB+BC+CB}+\frac{CC}{AC+BC+CA+CB+CC}\right)$。

- 测试数据集测试

用测试数据集测试，测试结果显示测试精度（test_Accuracy）为0.820。

```
trainer.test(model=task, datamodule=datamodule)
--------------------------------------------------------------------------------
[{'test_loss': 0.45782798528671265,
  'test_MulticlassAccuracy': 0.362870454788208,
  'test_MulticlassJaccardIndex': 0.35294270515441895}]
```

- 用训练的模型预测

这里并未使用用于训练下载的4个NAIP图像瓦片，而是下载了新的16个图像瓦片，新建立数据集，随机采样（图3.21-13）用已经训练好的模型进行预测。从预测结果来看，所提取的样本包括1、3、4、6三个分类，对应水体、林冠、灌木丛和不透水区域，目视分类图结果（图3.21-14），能够较好地对原始的NAIP图像进行语义分割（解译）。

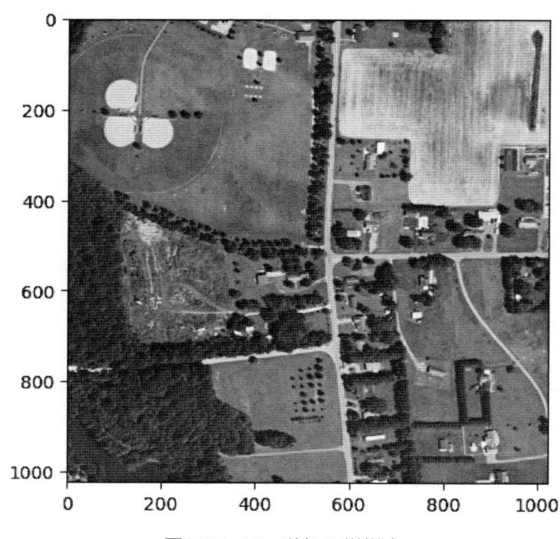

图3.21-13　随机采样样本

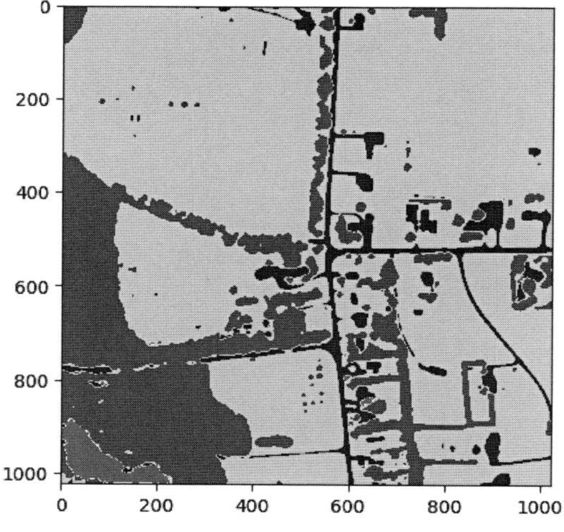

图3.21-14　土地覆盖（分类\标签）预测结果

```
from torchgeo.samplers import GridGeoSampler, RandomGeoSampler
from torchgeo.datasets import NAIP, stack_samples
from torch.utils.data import DataLoader
import torch
import matplotlib.pyplot as plt
import torchvision.transforms as T
import matplotlib
import numpy as np
X_pre = NAIP(r'E:\data\Delaware\imagery_test', transforms=naip_transforms)
X_sample = RandomGeoSampler(X_pre, size=1024, length=100)
X_dataloader = DataLoader(X_pre, sampler=X_sample, collate_fn=stack_samples)
X_dataloader_ = iter(X_dataloader)
X_batch = next(X_dataloader_)
X = X_batch["image"]  # .float()
plt.imshow(T.ToPILImage()(X[0][:3]))
unet_model = task.load_from_checkpoint(os.path.join(data_dir,'last.ckpt'))
unet_model.freeze()
y_probs = unet_model(X)
print(y_probs.shape)
--------------------------------------------------------------------------------
torch.Size([1, 8, 1024, 1024])
```

将联合预测概率转换为对应的土地覆盖类型标签（即概率最大对应的列）。

```
import numpy as np
y_pred = np.argmax(y_probs,axis=1)
print(y_pred.shape,'\n',y_pred,'\n',y_pred.unique())
--------------------------------------------------------------------------------
torch.Size([1, 1024, 1024])
 tensor([[[4, 4, 4,  ..., 4, 4, 4],
         [4, 4, 4,  ..., 4, 4, 4],
         [4, 4, 3,  ..., 4, 4, 4],
         ...,
         [4, 4, 4,  ..., 4, 4, 4],
         [4, 4, 4,  ..., 4, 4, 4],
         [4, 4, 4,  ..., 4, 4, 4]]])
 tensor([1, 3, 4, 6])
--------------------------------------------------------------------------------
LC_color_dict_selection = {k:LC_color_dict[k] for k,v in LC_color_dict.items()}
cmap_LC, norm = matplotlib.colors.from_levels_and_colors(list(LC_color_dict_
selection.keys()),[[v/255 for v in i] for i in LC_color_dict_selection.
values()],extend='max')
plt.imshow(np.squeeze(y_pred),cmap=cmap_LC);
```

参考文献（References）：

[1] Geospatial deep learning with TorchGeo，https://pytorch.org/blog/geospatial-deep-learning-with-torchgeo/.

[2] Conservation Innovation Center (CIC). LC Class Descriptions, https://www.chesapeakeconservancy.org/conservation-innovation-center/high-resolution-data/lulc-data-project-2022/.

3-G 点云数据处理与内存管理

3.22 点云数据处理

点云（point cloud）是使用三维扫描仪获取的资料，当然设计的三维模型也可以转换为点云数据。其中，三维对象以点的形式记录，每个点为一个三维坐标，同时可能包含颜色信息（RGB），或物体反射面的强度（intensity）。强度信息是激光扫描仪接收装置采集到的回波强度，与目标的表面材质、粗糙度、入射角方向以及仪器的发射能量，激光波长有关。点云数据格式比较丰富，常用的包括.xyz(.xyzn，.xyzrgb)、.las、.ply、.pcd、.pts 等，也包括一些关联格式的存储类型，例如基于numpy存储的array数组.numpy(.npu)，基于matlab格式存储的.matlab数组格式，当然也有基于文本存储的.txt文件。注意虽然有些存储类型后缀名不同。但实际上，数据格式可能相同。在地理空间数据中，常使用.las格式的数据。LAS（LASer）格式是由美国摄影测量和遥感协会（American Society for Photogrammetry and Remote Sensing，ASPRS）制定的激光雷达点云数据的交换和归档文件格式，被认为是激光雷达数据的行业标准。LAS格式点云数据包括多个版本，最近的为LAS 1.4（2011年11月14日），不同的版本点云数据包括的信息也许不同，需要注意这点。LAS通常包括由整数值标识的分类信息（LAS1.1及之后的版本），其1.1-1.4 LAS类别代码如表3.22-1所示。

表3.22-1 LAS类别代码

分类值/classification value	类别
0	不被用于分类（Never classified）
1	未被定义（unassigned）
2	地面（ground）
3	低矮树木（low vegetation）
4	中等树木（medium vegetation）
5	高的树木（high vegetation）
6	建筑（building）
7	低的点（low point）
8	保留（reserved）
9	水体（water）
10	铁路（rail）
11	道路表面（road surface）
12	保留（reserved）
13	金属丝防护（屏蔽）[wire-guard(shield)]
14	导线（相）[wire-conductor(phase)]
15	输电杆塔（transmission tower）
16	电线连接器（绝缘子）[wire-structure connector(insulator)]
17	桥面（bridge deck）
18	高噪声（high noise）
19-63	保留（reserved）
64-255	用户定义（user definable）

处理点云数据的Python库比较多，常用的包括PDAL①、PCL②、open3D③等。其中，PDAL可以处理.las格式数据，读取后也可以存储为其他格式数据，使用其他库的功能处理也未尝不可。

此次实验数据为伊利诺伊州草原地质调查研究所（Illinois state geological survey - prairie research institute）发布的伊利诺伊州.las格式的激光雷达数据④。研究的目标区域为芝加哥城及其周边。因为分辨率为1m，研究区域部分数据量高达1.4T。其中，每一单元（tile）基本为2501 × 2501，大约1G，最小的也有几百M。对于普通的计算机配置，处理大数据，通常要判断内存所能支持的容量（很多程序在处理数据时，可以不将其全部读入内存，例如h5py格式数据的批量写入

① PDAL - Point Data Abstraction Library，是一个用于转换和操作点云数据的C++库，类似于处理栅格和矢量数据的GDAL库（https://pdal.io）。
② PCL（Point Cloud Library），是一个独立的、大规模的开放式项目，用于2D/3D图像和点云处理（https://pointclouds.org）。
③ open3D，是一个支持快速开发处理3D数据的开源库。Open3D前端公开了一组在C++和Python中精心挑选的数据结构和算法。后端使用经过高度优化的多线程（http://www.open3d.org/docs/release/introduction.html）。
④ 伊利诺伊州.las格式的激光雷达数据（https://www.arcgis.com/apps/webappviewer/index.html?id=44eb65c92c944f3e8b231eb1e2814f4d）。

和批量读取；rasterio库提供有windows功能，可以分区读取单独的栅格文件）；以及CPU的计算速度，分批的处理可以避免因为处理中断造成全部数据丢失。在阐述点云数据处理时，并不处理芝加哥城所有区域数据，仅以IIT（Illinois Institute of Technology）校园为核心，一定范围内的数据处理为例。下载的点云数据包括的单元编号（文件）有（总共73个单元文件，计26.2GB）：

```
LAS_16758900.las   LAS_17008900.las   LAS_17258900.las   LAS_17508900.las
LAS_17758900.las   LAS_18008900.las   LAS_18258900.las   LAS_16758875.las
LAS_17008875.las   LAS_17258875.las   LAS_17508875.las   LAS_17758875.las
LAS_18008875.las   LAS_18258875.las   LAS_16758850.las   LAS_17008850.las
LAS_17258850.las   LAS_17508850.las   LAS_17758850.las   LAS_18008850.las
LAS_18258850.las   LAS_16758825.las   LAS_17008825.las   LAS_17258825.las
LAS_17508825.las   LAS_17758825.las   LAS_18008825.las   LAS_18258825.las
LAS_16758800.las   LAS_17008800.las   LAS_17258800.las   LAS_17508800.las
LAS_17758800.las   LAS_18008800.las   LAS_18258800.las   LAS_18508800.las
LAS_16758775.las   LAS_17008775.las   LAS_17258775.las   LAS_17508775.las
LAS_17758775.las   LAS_18008775.las   LAS_18258775.las   LAS_18508775.las
LAS_16758750.las   LAS_17008750.las   LAS_17258750.las   LAS_17508750.las
LAS_17758750.las   LAS_18008750.las   LAS_18258750.las   LAS_18508750.las
LAS_18758750.las   LAS_16758725.las   LAS_17008725.las   LAS_17258725.las
LAS_17508725.las   LAS_17758725.las   LAS_18008725.las   LAS_18258725.las
LAS_18508725.las   LAS_18758725.las   LAS_19008725.las   LAS_16758700.las
LAS_17008700.las   LAS_17258700.las   LAS_17508700.las   LAS_17758700.las
LAS_18008700.las   LAS_18258700.las   LAS_18508700.las   LAS_18758700.las
LAS_19008700.las
```

3.22.1 点云数据处理

3.22.1.1 查看点云数据信息

- **PDAL的主要参数配置**

1. Dimensions①，维度，该参数给出了可能存储的不同信息，可以基于维度配置type类型，例如维度配置为"dimension"： "X"，可以配置"type"： "filters.sort"，即依据给出的维度，排序返回的点云。常用的包括'Classification'（分类数据）；'Density'（点密度估计）；'GpsTime'（获取该点的GPS时间）；'Intensity'（物体反射面的强度）和X、Y、Z（坐标）等。下述代码pipeline.arrays返回的列表数组中，包含有dtype=[('X', '<f8'), ('Y', '<f8'), ('Z', '<f8'), ('Intensity', '<u2'), ('ReturnNumber', 'u1'), ('NumberOfReturns', 'u1'), ('ScanDirectionFlag', 'u1'), ('EdgeOfFlightLine', 'u1'), ('Classification', 'u1'), ('ScanAngleRank', '<f4'), ('UserData', 'u1'), ('PointSourceId', '<u2'), ('GpsTime', '<f8'), ('ScanChannel', 'u1'), ('ClassFlags', 'u1')])]，可以明确LAS点云包括哪些维度。

2. Filters②，过滤器，给定操作数据的方式，可以删除、修改、重组数据流。有些过滤器需要在对应的维度上实现，例如在XYZ坐标上实现重投影等。常用的过滤器有：

create部分：

1. filters.approximatecoplanar，基于k近邻估计点平面性；
2. filters.cluster，利用欧氏距离度量提取和标记聚类；
3. filters.dbscan，基于密度的空间聚类；
4. filters.covariancefeatures，基于一个点邻域的协方差计算局部特征；
5. filters.eigenvalues，基于k最近邻计算点特征值；
6. filters.nndistance，根据最近邻计算距离指数；
7. filters.radialdensity，给定距离内的点的密度。

Order部分：

1. filters.mortonorder，使用Morton排序XY数据；
2. filters.randomize，随机化视图中的点；
3. filters.sort，基于给定的维度排序数据。

Move部分：

1. filters.reprojection，使用GDAL将数据从一个坐标系重新投影到另一个坐标系；
2. filters.transformation，使用4×4变换矩阵变换每个点。

① PDAL的主要参数配置——Dimensions，（https://pdal.io/en/2.6.0/dimensions.html）。

② PDAL的主要参数配置——Filters（https://pdal.io/en/2.6.0/stages/filters.html）。

Cull部分：

1. filters.crop，根据边界框或一个多边形，过滤点；
2. filters.iqr，剔除给定维度上四分位范围外的点；
3. filters.locate，给定维度，通过min/max返回一个点；
4. filters.sample，执行泊松采样并只返回输入点的一个子集；
5. filters.voxelcenternearestneighbor，返回每个体素内最靠近体素中心的点；
6. filters.voxelcentroidnearestneighbor，返回每个体素内最接近体素质心的点。

Join部分：

filters.merge，将来自两个不同读取器的数据合并到一个流中。

Mesh部分：

1. filters.delaunay，使用Delaunay三角化创建mesh;
2. filters.gridprojection，使用网格投影方法创建mesh;
3. filters.poisson，使用泊松曲面重建算法创建mesh。

Languages部分：

filters.python，在pipeline中嵌入Python代码。

Metadata部分：

filters.stats，计算每个维度的统计信息（均值、最小值、最大值等）。

1. type-readers[①]-writers[②]，读写类型，例如通过"type":"writers.gdal"，可以使用"gdaldriver":"GTiff"驱动，使用差值算法从点云创建栅格数据。常用保存的数据类型有，writers.gdal、writers.las、writers.ogr、writers.pgpointcloud、writers.ply、writers.sqlite和writers.text等。

2. type，通常配合Filters过滤器使用。例如，如果配置"type":"filters.crop"，则可以设置"bounds":"([0,100],[0,100])"边界框进行裁切。

3. output_type，是给出数据计算的方式，例如mean、min、max、idx、count、stdev、all和idw等。

4. resolution，指定输出栅格的精度，例如1、10等。

5. filename，指定保存文件的名称。

6. data_type[③]，保存的数据类型，例如int8、int16、unint8、float和double等。

7. limits，数据限制，例如配置过滤器为"type":"filters.range"，则"limits":"Z[0:],Classification[6:6]"，仅提取标识为6，即建筑分类的点和建筑的Z值。

① PDAL的主要参数配置——readers，（https://pdal.io/en/2.6.0/stages/readers.html）。

② PDAL的主要参数配置——writers，（https://pdal.io/en/2.6.0/stages/writers.html）。

③ PDAL的主要参数配置——data_type，（https://pdal.io/en/2.6.0/types.html）。

PDAL是命令行工具，在Anaconda中打开对应环境的终端，输入下述命令，会获得一个点的信息。命令行操作模式可以避免大批量数据读入内存，造成溢出。只是不方便查看数据，因此采用何种方式，可以依据具体情况确定。

```
pdal info G:\data\IIT_lidarPtClouds\rawPtClouds\LAS_16758900.las -p 0{
"file_size": 549264685,
"filename": "G:\\data\\IIT_lidarPtClouds\\rawPtClouds\\LAS_16758900.las",
"now": "2022-01-13T16:44:02+0800",
"pdal_version": "2.3.0 (git-version: Release)",
"points":
{
    "point":
    {
    "ClassFlags": 0,
    "Classification": 3,
    "EdgeOfFlightLine": 0,
    "GpsTime": 179803760,
    "Intensity": 1395,
    "NumberOfReturns": 1,
    "PointId": 0,
    "PointSourceId": 0,
    "ReturnNumber": 1,
    "ScanAngleRank": 15,
    "ScanChannel": 0,
    "ScanDirectionFlag": 0,
    "UserData": 0,
    "X": 1167506.44,
    "Y": 1892449.7,
```

```
    "Z": 594.62
    }
},
"reader": "readers.las"
}
```

- pipeline

通常点云数据处理过程中，包括读取、处理、写入等操作。为了方便处理流程，PDAL引入pipeline概念，可以将多个操作堆叠在一个由JSON数据格式定义的数据流中。这对于复杂的处理流程而言，具有更大的优势。同时，PDAL也提供了Python模式。可以在Python中调入PDAL库，以及定义pipeline操作流程。例如，如下官网提供的一个简单案例，包括读取LAS文件（`"%s"%separate_las`），配置维度为点云x坐标（`"dimension": "X"`），并依据x坐标排序返回的数组（`"type": "filters.sort"`）等操作。执行pipeline（`pipeline.execute()`）之后，pipeline对象返回点云具有维度的值，其`dtypes`项返回了点云具有的维度，对应返回的数组信息。这一个单元包含点的数量为count=18721702个点。

metadata元数据，可以打印查看，包括有坐标投影信息。

```
import os
import pdal
import util_misc
dirpath = r"G:\data\IIT_lidarPtClouds\rawPtClouds"
fileType = ["las"]
las_paths = util_misc.filePath_extraction(dirpath, fileType)
s_t = util_misc.start_time()
separate_las = os.path.join(list(las_paths.keys())[0], list(las_paths.values())[
                              0][32]).replace("\\", "/")  #注意文件名路径中"\"和"/"，
不同库支持的类型可能有所不同，需自行调整

json = """
[
    "%s",
    {
        "type": "filters.sort",
        "dimension": "X"
    }
]
""" % separate_las
pipeline = pdal.Pipeline(json)
count = pipeline.execute()
print("pts count:", count)
arrays = pipeline.arrays
print("arrays:", arrays)
metadata = pipeline.metadata
log = pipeline.log
print("complete .las reading ")
util_misc.duration(s_t)
--------------------------------------------------------------------------------
start time: 2022-01-13 16:51:29.329062
pts count: 16677942
arrays: [array([(1175000., 1884958.18, 634.81,  9832, 1, 1, 0, 0, 5, 15., 0, 0,
1.78265216e+08, 0, 0),
       ...,
       (1177500., 1882501.12, 596.34, 44193, 1, 1, 0, 0, 2, 15., 0, 0,
1.78239520e+08, 0, 0)],
      dtype=[('X', '<f8'), ('Y', '<f8'), ('Z', '<f8'), ('Intensity', '<u2'),
('ReturnNumber', 'u1'), ('NumberOfReturns', 'u1'), ('ScanDirectionFlag', 'u1'),
('EdgeOfFlightLine', 'u1'), ('Classification', 'u1'), ('ScanAngleRank', '<f4'),
('UserData', 'u1'), ('PointSourceId', '<u2'), ('GpsTime', '<f8'), ('ScanChannel',
'u1'), ('ClassFlags', 'u1')])]
complete .las reading
end time: 2022-01-13 16:51:57.974012
Total time spend:0.47 minutes
```

PDAL处理后，可以读取`pipeline`对象的属性，因为一个点包含多个信息。为方便查看，可以将点云数组转换为DataFrame格式数据（表3.22-2）。

```
import pandas as pd
pts_df = pd.DataFrame(arrays[0])
util_misc.print_html(pts_df)
```

表3.22-2 点云信息

	X	Y	Z	Intensity	Return Number	NumberOf Returns	Scan DirectionFlag	EdgeOf FlightLine	Classification	ScanAngle Rank	UserData	Point SourceId	GpsTime	Scan Channel	ClassFlags
0	1175000	1884958.18	634.81	9832	1	1	0	0	5	15	0	0	178265216	0	0
1	1175000	1884941.11	644.75	18604	1	1	0	0	5	15	0	0	178265216	0	0
2	1175000	1884931.3	641.59	4836	1	1	0	0	5	15	0	0	178265232	0	0
3	1175000	1884936.81	644.47	15047	1	1	0	0	5	15	0	0	179806048	0	0
4	1175000	1884948.02	635.57	2700	1	1	0	0	5	15	0	0	179805664	0	0

除了直接查看数据内容，可以借助open3d库打印三维点云((图3.22-1)，互动观察。但是因为最初使用PDAL读取LAS点云(open3d目前不支持读取LAS格式点云数据)，需要将读取的点云数据转换为open3d支持的格式。显示的色彩代表点云的高度信息。

```
import open3d as o3d
o3d_pts = o3d.geometry.PointCloud()
o3d_pts.points = o3d.utility.Vector3dVector(pts_
df[['X','Y','Z']].to_numpy())
o3d.visualization.draw_geometries([o3d_pts])
-----------------------------------------------------------
Jupyter environment detected. Enabling Open3D
WebVisualizer.
[Open3D INFO] WebRTC GUI backend enabled.
[Open3D INFO] WebRTCWindowSystem: HTTP handshake
server disabled.
```

3.22.1.2 建立数字表面模型与分类栅格

三维点云数据是三维格式数据，可以提取信息将其转换为对应的二维栅格数据，方便数据分析。点云数据转二维栅格数据最为常用的包括生成分类栅格数据，即地表覆盖类型；二是，提取地物高度，例如提取建筑物高度信息，植被高度信息等；三是，生成DEM（Digital Elevation Model）、DTM（Digital Terrain Model）等。

- DEM——数字高程模型，为去除自然和建筑对象的裸地表面高程。
- DTM——数字地形（或地面）模型，在DEM基础上增加自然地形的矢量特征，如河流和山脊。DEM和DTM很多时候并未区分明确，具体由数据所包含的内容来确定。
- DSM——数字表面模型（Digital Surface Model，DSM），同时捕捉地面、自然（例如树木）及人造物（例如建筑）特征。

将对点云数据所要处理的内容定义在一个函数中，每一处理内容为一个pipeline，由json格式定义。下述代码定义了三项内容：一是，提取地物覆盖分类信息，用于建立分类栅格；二是，提取高程信息，用于建立DSM；三是，仅提取ground地表类型的高程，可以建立DEM。为了方便函数内日后不断增加新的提取内容，定义输入参数json_combo管理和判断所要计算的pipleline，灵活处理函数。当增加新pipeline时，避免较大的改动。

对于文件很大的地理空间信息数据，通常在处理过程中，完成一个主要的数据处理后就将其置于本地磁盘中，使用时再读取。因此，处理完一个点云单元（tile）之后，即刻将其保存到磁盘中，并不驻留于内存里，避免内存溢出。

```
import os
def las_info_extraction(las_fp, json_combo):
    '''
    function - 转换单个 .las 点云数据为分类栅格数据，和 DSM 栅
格数据等

    Paras:
        las_fp - .las 格式文件路径
        save_path - 保存路径列表，分类 DSM 存储与不同路径下
    '''
    import pdal
    pipeline_dict = {}
    if 'json_classification' in json_combo.keys():
        # pipeline- 用于建立分类栅格
        json_classification = """
```

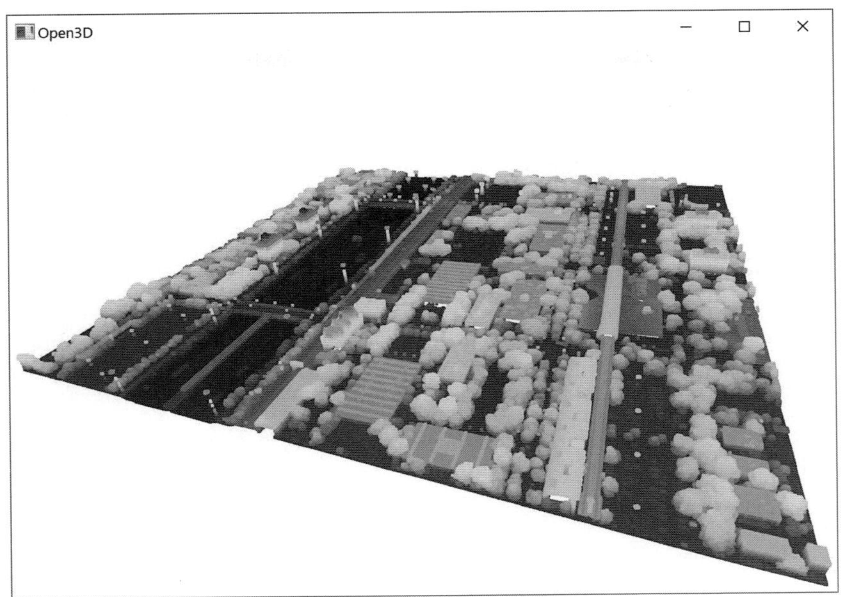

图3.22-1 显示三维点云数据

```
        {
        "pipeline": [
                        "%s",
                        {
                        "filename":"%s",
                        "type":"writers.gdal",
                        "dimension":"Classification",
                        "data_type":"uint16_t",
                        "output_type":"mean",
                        "resolution": 1
                        }
        ]
        }""" % (las_fp, json_combo['json_classification'])
    pipeline_dict['json_classification'] = json_classification
if "json_DSM" in json_combo.keys():
    # pipeline- 用于建立 DSM 栅格数据
    json_DSM = """
        {
        "pipeline": [
                        "%s",
                        {
                        "filename":"%s",
                        "gdaldriver":"GTiff",
                        "type":"writers.gdal",
                        "output_type":"mean",
                        "resolution": 1
                        }
        ]
        }""" % (las_fp, json_combo['json_DSM'])
    pipeline_dict['json_DSM'] = json_DSM
if 'json_ground' in json_combo.keys():
    # pipelin- 用于提取 ground 地表
    json_ground = """
        {
        "pipeline": [
                        "%s",
                        {
                        "type":"filters.range",
                        "limits":"Classification[2:2]"
                        },
                        {
                        "filename":"%s",
```

```
                                "gdaldriver":"GTiff",
                                "type":"writers.gdal",
                                "output_type":"mean",
                                "resolution": 1
                                }
                ]
                }""" % (las_fp, json_combo['json_ground'])
            pipeline_dict['json_ground'] = json_ground
        for k, json in pipeline_dict.items():
            pipeline = pdal.Pipeline(json)
            try:
                pipeline.execute()
            except:
                print("\n An exception occurred,the file name:%s" % las_fp)
        print("finished conversion...")
dirpath = r"G:\data\IIT_lidarPtClouds\rawPtClouds"
las_fp = os.path.join(dirpath, 'LAS_17508825.las').replace("\\", "/")
workspace = r'G:\data\IIT_lidarPtClouds'
json_combo = {"json_classification": os.path.join(workspace, 'classification_
DSM\LAS_17508825_classification.tif').replace("\\", "/"),
              "json_DSM": os.path.join(workspace, 'classification_DSM\
LAS_17508825_DSM.tif').replace("\\", "/")}  #配置输入参数
las_info_extraction(las_fp, json_combo)
---------------------------------------------------------------------------------
finished conversion...
```

读取保存的DSM栅格文件，用Earthpy库打印查看（图3.22-2），因为数据中可能存在异常值，造成显示上的灰度，因此可以用分位数（np.quantile）的方法配置vmin和vmax参数。

```
import rasterio as rio
import os
import numpy as np
import earthpy.plot as ep
workspace = r'G:\data\IIT_lidarPtClouds'
with rio.open(os.path.join(workspace,'classification_DSM\LAS_17508825_DSM.tif'))
as DSM_src:
    DSM_array=DSM_src.read(1)
titles = ["LAS_17508825_DTM"]
ep.plot_bands(DSM_array, cmap="turbo", cols=1, title=titles, vmin=np.
quantile(DSM_array,0.1), vmax=np.quantile(DSM_array,0.9))
```

图3.22-2　显示DSM栅格数据

同样，读取保存的分类栅格数据，但是需要自行定义打印显示函数，根据整数指示的类别打印（图3.22-3）。其中，类别由LAS格式给定的分类标识确定，颜色可以根据显示所要达到的效果自行定义。并增加了图例，方便查看颜色所对应的分类。

```
import os
def las_classification_plotWithLegend(las_fp):
    '''
```

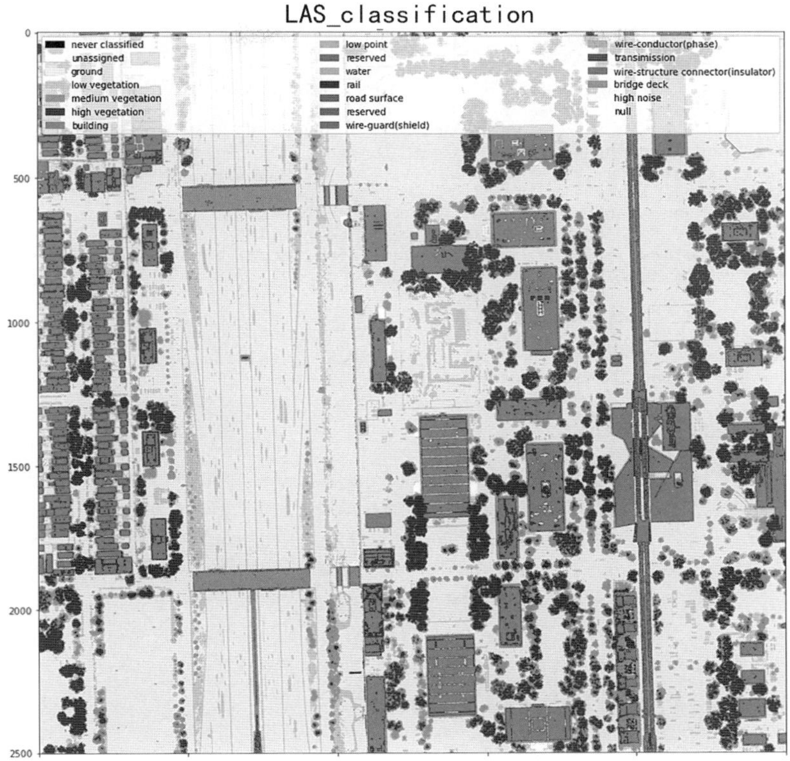

图3.22-3 分类栅格数据

```
    function - 显示由 .las 文件生成的分类栅格文件，并显示图例

    Paras:
        las_fp - 分类文件路径
    '''
    import rasterio as rio
    import pandas as pd
    import numpy as np
    import matplotlib.pyplot as plt
    from matplotlib.colors import ListedColormap
    from matplotlib import colors
    from matplotlib.patches import Rectangle
    with rio.open(las_fp) as classi_src:
        classi_array = classi_src.read(1)
    las_classi_colorName = {0: 'black', 1: 'white', 2: 'beige', 3: 'palegreen', 4:
'lime', 5: 'green', 6: 'tomato', 7: 'silver', 8: 'grey', 9: 'lightskyblue',
                            10: 'purple', 11: 'slategray', 12: 'grey', 13: 'cadetblue',
14: 'lightsteelblue', 15: 'brown', 16: 'indianred', 17: 'darkkhaki', 18: 'azure',
9999: 'white'}
    las_classi_colorRGB = pd.DataFrame({key: colors.hex2color(
        colors.cnames[las_classi_colorName[key]]) for key in las_classi_
colorName.keys()})
    classi_array_color = [pd.DataFrame(classi_array).replace(
        las_classi_colorRGB.iloc[idx]).to_numpy() for idx in las_classi_colorRGB.
index]
    classi_array_color_ = np.concatenate(
        [np.expand_dims(i, axis=-1) for i in classi_array_color], axis=-1)
    fig, ax = plt.subplots(figsize=(12, 12))
    im = ax.imshow(classi_array_color_, )
    ax.set_title(
        "LAS_classification",
        fontsize=14,
    )
```

```
    #增加图例
    color_legend = pd.DataFrame(
        las_classi_colorName.items(), columns=["id", "color"])
    las_classi_name = {0: 'never classified', 1: 'unassigned', 2: 'ground', 3: 'low
vegetation', 4: 'medium vegetation', 5: 'high vegetation', 6: 'building', 7: 'low
point', 8: 'reserved', 9: 'water', 10: 'rail',
                       11: 'road surface', 12: 'reserved', 13: 'wire-guard(shield)',
14: 'wire-conductor(phase)', 15: 'transimission', 16: 'wire-structure
connector(insulator)', 17: 'bridge deck', 18: 'high noise', 9999: 'null'}
    color_legend['label'] = las_classi_name.values()
    classi_lengend = [Rectangle((0, 0), 1, 1, color=c)
                      for c in color_legend['color']]
    ax.legend(classi_lengend, color_legend.label, mode='expand', ncol=3)
    plt.tight_layout()
    plt.show()
workspace = r'G:\data\IIT_lidarPtClouds'
las_fp = os.path.join(
    workspace, 'classification_DSM\LAS_17508825_classification.tif')
las_classification_plotWithLegend(las_fp)
```

3.22.1.3 批量处理LAS点云单元

通过一个点云单元的代码调试，完成对一个单元的点云数据处理。为了能够批量处理所有点云单元，建立批量处理点云数据的函数。该函数直接调用上述单个点云单元处理函数，仅梳理所有点云单元文件的读取和保存路径。为了查看计算进度，使用tqdm库可以将循环计算过程以进度条的方式显示，明确完成所有数据计算大概所要花费的时间。同时，以调用自定义的start_time()和duration(s_t)方法，计算具体的时长。

```
import util_misc
def las_info_extraction_combo(las_dirPath, json_combo_):
    '''
    function - 批量转换 .las 点云数据为 DSM 和分类栅格

    Paras:
        las_dirPath - LAS 文件路径
        save_path - 保存路径
    '''
    import util_misc
    import util_A
    import os
    import re
    from tqdm import tqdm
    file_type = ['las']
    las_fn = util_misc.filePath_extraction(las_dirPath, file_type)
    '''展平列表函数'''
    def flatten_lst(lst): return [m for n_lst in lst for m in flatten_lst(
        n_lst)] if type(lst) is list else [lst]
    las_fn_list = flatten_lst([[os.path.join(k, las_fn[k][i])
                                for i in range(len(las_fn[k]))] for k in las_
fn.keys()])
    pattern = re.compile(r'[_](.*?)[.]', re.S)
    for i in tqdm(las_fn_list):
        fn_num = re.findall(pattern, i.split("\\")[-1])[0]  #提取文件名字符串中的数字
        #注意文件名路径中" \" 和" /" ，不同库支持的类型可能有所不同，需自行调整
        json_combo = {key: os.path.join(json_combo_[key], "%s_%s.tif" % (os.
path.split(
            json_combo_[key])[-1], fn_num)).replace("\\", "/") for key in json_
combo_.keys()}
        util_A.las_info_extraction(i.replace("\\", "/"), json_combo)
dirpath = r"G:\data\IIT_lidarPtClouds\rawPtClouds"
json_combo_ = {"json_classification": r'G:\data\IIT_lidarPtClouds\
classification'.replace(
    "\\", "/"), "json_DSM": r'G:\data\IIT_lidarPtClouds\DSM'.replace("\\", "/")}
#配置输入参数

s_t = util_misc.start_time()
las_info_extraction_combo(dirpath, json_combo_)
util_misc.duration(s_t)
------------------------------------------------------------------------
start time: 2022-01-13 19:07:54.440621
100%|██████████| 73/73 [19:20<00:00, 15.89s/it]
```

```
end time: 2022-01-13 19:27:14.586610
Total time spend:19.33 minutes
```

- **合并栅格数据**

以点云单元形式批量处理完所有的点云数据，生成同点云单元数量的多个DSM文件和多个分类文件后，需要将其合并成一个完整的栅格文件。合并的方法主要使用rasterio库提供的merge方法。同时，需要注意，要配置压缩及保存类型；否则，合并后的栅格文件可能非常大，例如本次合并所有的栅格后，文件大小约为4.5GB。但是，配置"compress":'lzw'和"dtype":get_minimum_int_dtype(mosaic)后，文件大小仅为201MB，大幅度压缩了文件，有利于节约硬盘空间及存读速度。

在配置文件保存类型时，迁移了rasterio库给出的函数get_minimum_int_dtype(values)，自行依据数组数值确定所要保存的文件类型，从而避免了自行定义。

```
import util_misc
def raster_mosaic(dir_path, out_fp,):
    '''
    function - 合并多个栅格为一个

    Paras:
        dir_path - 栅格根目录
        out-fp - 保存路径

    return:
        out_trans - 返回变换信息
    '''
    import rasterio
    import glob
    import os
    from rasterio.merge import merge

    # 迁移 rasterio 提供的定义数组最小数据类型的函数
    def get_minimum_int_dtype(values):
        """
        使用范围检查来确定表示值所需的最小整数数据类型 .

        :param values: numpy 数组（array）
        :return: 命名数据类型，用于创建 numpy dtype（数据类型）
        """
        min_value = values.min()
        max_value = values.max()
        if min_value >= 0:
            if max_value <= 255:
                return rasterio.uint8
            elif max_value <= 65535:
                return rasterio.uint16
            elif max_value <= 4294967295:
                return rasterio.uint32
        elif min_value >= -32768 and max_value <= 32767:
            return rasterio.int16
        elif min_value >= -2147483648 and max_value <= 2147483647:
            return rasterio.int32

    search_criteria = "*.tif"  # 搜寻所要合并的栅格 .tif 文件
    fp_pattern = os.path.join(dir_path, search_criteria)
    fps = glob.glob(fp_pattern)  # 使用 glob 库搜索指定模式的文件
    src_files_to_mosaic = []
    for fp in fps:
        src = rasterio.open(fp)
        src_files_to_mosaic.append(src)
    mosaic, out_trans = merge(src_files_to_mosaic)  # merge 函数返回一个栅格数组，以及
转换信息

    # 获得元数据
    out_meta = src.meta.copy()
    # 更新元数据
    data_type = get_minimum_int_dtype(mosaic)
    out_meta.update({"driver": "GTiff",
                     "height": mosaic.shape[1],
                     "width": mosaic.shape[2],
                     "transform": out_trans,
```

```
                        #通过压缩和配置存储类型，减小存储文件大小
                        "compress": 'lzw',
                        "dtype": get_minimum_int_dtype(mosaic),
                        }
                       )
    with rasterio.open(out_fp, "w", **out_meta) as dest:
        dest.write(mosaic.astype(data_type))
    return out_trans
DSM_dir_path = r'G:\data\IIT_lidarPtClouds\DSM'
DSM_out_fp = r'G:\data\IIT_lidarPtClouds\mosaic\DSM_mosaic.tif'
s_t = util_misc.start_time()
out_trans = raster_mosaic(DSM_dir_path, DSM_out_fp)
util_misc.duration(s_t)
--------------------------------------------------------------------------------
start time: 2022-01-13 19:34:23.680848
end time: 2022-01-13 19:35:13.451225
Total time spend:0.82 minutes
```

依据上述同样的方法，读取、打印和查看合并后的DSM栅格（图3.22-4、图3.22-5）。

```
import rasterio as rio
import earthpy.plot as ep
import numpy as np
DSM_fp = r'G:\data\IIT_lidarPtClouds\mosaic\DSM_mosaic.tif'
with rio.open(DSM_fp) as DSM_src:
    mosaic_DSM_array = DSM_src.read(1)
titles = ["mosaic_DTM"]
ep.plot_bands(mosaic_DSM_array, cmap="turbo", cols=1, title=titles, vmin=np.
quantile(mosaic_DSM_array,0.25), vmax=np.quantile(mosaic_DSM_array,0.95))
```

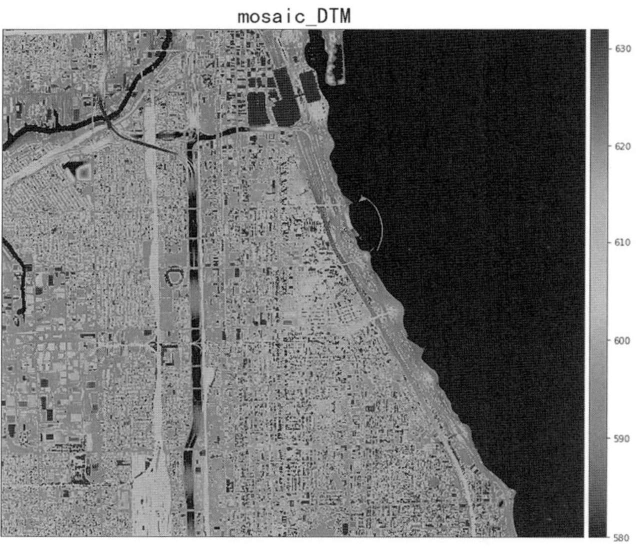

图3.22-4　合并的DSM栅格数据

```
import util_misc,util_A
classi_dir_path = r'G:\data\IIT_lidarPtClouds\classification'
classi_out_fp = r'G:\data\IIT_lidarPtClouds\mosaic\classification_mosaic.tif'
s_t = util_misc.start_time()
out_trans = util_A.raster_mosaic(classi_dir_path,classi_out_fp)
util_misc.duration(s_t)
--------------------------------------------------------------------------------
start time: 2022-01-13 19:37:32.662113
end time: 2022-01-13 19:37:49.891427
Total time spend:0.28 minutes
--------------------------------------------------------------------------------
import util_A
from skimage.transform import rescale
```

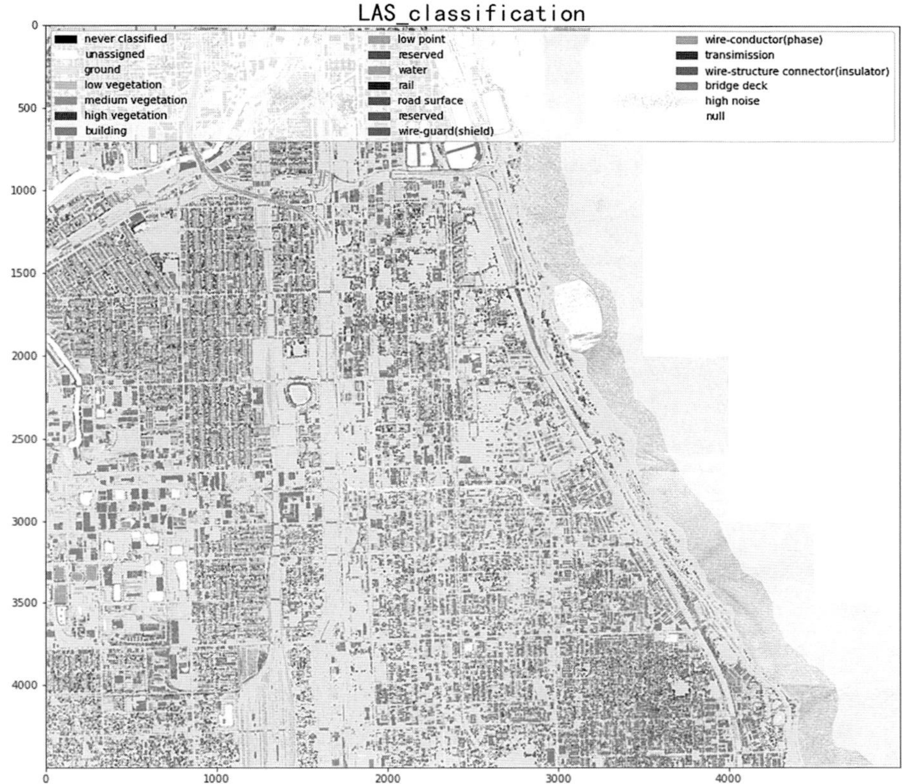

图3.22-5 合并分类栅格数据

```
import rasterio as rio
import earthpy.plot as ep
import numpy as np
classi_fp = r'G:\data\IIT_lidarPtClouds\mosaic\classification_mosaic.tif'
with rio.open(classi_fp) as classi_src:
    mosaic_classi_array = classi_src.read(1)
mosaic_classi_array_rescaled = rescale(
    mosaic_classi_array, 0.2, anti_aliasing=False, preserve_range=True)
print("original shape:", mosaic_classi_array.shape)
print("rescaled shape:", mosaic_classi_array_rescaled.shape)
util_A.las_classification_plotWithLegend_(mosaic_classi_array_rescaled)
--------------------------------------------------------------------------------
original shape: (22501, 25001)
rescaled shape: (4500, 5000)
Clipping input data to the valid range for imshow with RGB data ([0..1] for
floats or [0..255] for integers).
```

3.22.2 建筑高度提取

建筑高度提取的流程为：

1. 将DSM栅格重投影为该区域（芝加哥）Landsat所定义的坐标投影系统，统一投影坐标系；
2. 从Chicago Data Portal、CDP[①]获取Building Footprints (current)文件的SHP格式Polygon建筑分布；
3. 依据DSM栅格的范围（extent）裁切SHP格式的建筑分布矢量数据，并定义投影同重投影后的DSM栅格文件；
4. PSAL提取地面（ground）信息，并存储为栅格；

① Chicago Data Portal，CDP，芝加哥市的开放数据门户，可以找到城市数据，创建关于城市的地图和图表用于相关分析研究（https://data.cityofchicago.org）。

5. 插值所有单个的ground栅格并合并，重投影同DSM投影后保存；
6. 根据分类栅格数据，从DSM中提取建筑区域的高程数据；
7. 用裁切后的建筑矢量数据，使用rasterstats库提供的zonal_stats方法，分别提取DSM和ground栅格数据高程信息，统计方式为median（中位数）；
8. 用区域统计提取的DSM-ground，即为建筑高度数据；
9. 将建筑高度数据写入GeoDataFrame，并保存为.shp文件，备日后分析使用。

3.22.2.1 定义获取栅格投影函数和栅格重投影函数

投影和重投影的方法在Landsat遥感影像处理部分使用过，可以结合查看。

```
def get_crs_raster(raster_fp):
    '''
    function - 获取给定栅格的投影坐标 -crs.

    Paras:
        raster_fp - 给定栅格文件的路径
    '''
    import rasterio as rio
    with rio.open(raster_fp) as raster_crs:
        raster_profile = raster_crs.profile
        return raster_profile['crs']

# 使用的为 Landsat 部分处理的遥感影像，可以自行下载对应区域的 Landsat，作为参数输入，获取其投影
ref_raster = r'G:\data\Landsat\data_processed\DE_Chicago.tif'
dst_crs = get_crs_raster(ref_raster)
print("dst_crs:", dst_crs)
----------------------------------------------------------------------------
dst_crs: EPSG:32616
----------------------------------------------------------------------------
def raster_reprojection(raster_fp, dst_crs, save_path):
    '''
    function - 转换栅格投影

    Paras:
        raster_fp - 待转换投影的栅格
        dst_crs - 目标投影
        save_path - 保存路径
    '''
    from rasterio.warp import calculate_default_transform, reproject, Resampling
    import rasterio as rio
    with rio.open(raster_fp) as src:
        transform, width, height = calculate_default_transform(
            src.crs, dst_crs, src.width, src.height, *src.bounds)
        kwargs = src.meta.copy()
        kwargs.update({
            'crs': dst_crs,
            'transform': transform,
            'width': width,
            'height': height
        })
        with rio.open(save_path, 'w', **kwargs) as dst:
            for i in range(1, src.count + 1):
                reproject(
                    source=rio.band(src, i),
                    destination=rio.band(dst, i),
                    src_transform=src.transform,
                    src_crs=src.crs,
                    dst_transform=transform,
                    dst_crs=dst_crs,
                    resampling=Resampling.nearest)
    print("finished reprojecting...")
DTM_fp = r'G:\data\IIT_lidarPtClouds\mosaic\DSM_mosaic.tif'
DTM_reprojection_fp = r'G:\data\IIT_lidarPtClouds\mosaic\DSM_mosaic_reprojection.
tif'
dst_crs = dst_crs
raster_reprojection(DTM_fp, dst_crs, DTM_reprojection_fp)
----------------------------------------------------------------------------
finished reprojecting...
```

3.22.2.2 按照给定的栅格，获取栅格的范围来裁切SHP格式文件

在使用gpd.clip(vector_projection_,poly_gdf)方法裁切矢量数据时，需要清理数据，包括vector.dropna(subset=["geometry"], inplace=True)清理空值；及polygon_bool=vector_projection.geometry.apply(lambda row:True if type(row)==type_Polygon and row.is_valid else False)清理无效的Polygon对象，和不为shapely.geometry.polygon.Polygon格式的数据。只有清理完数据后才能够执行裁切，否则会提示错误。

```
def clip_shp_withRasterExtent(vector_shp_fp, reference_raster_fp, save_path):
    '''
    function - 根据给定栅格的范围，裁切 .shp 格式数据，并定义投影同给定栅格

    Paras:
        vector_shp_fp - 待裁切的 vector 文件路劲
        reference_raster_fp - 参考栅格，extent 及投影
        save_path - 保存路径

    return:
        poly_gdf - 返回裁切边界
    '''
    import rasterio as rio
    from rasterio.plot import plotting_extent
    import geopandas as gpd
    import pandas as pd
    from shapely.geometry import Polygon
    import shapely
    vector = gpd.read_file(vector_shp_fp)
    print("before dropna:", vector.shape)
    vector.dropna(subset=["geometry"], inplace=True)
    print("after dropna:", vector.shape)
    with rio.open(reference_raster_fp) as src:
        raster_extent = plotting_extent(src)
        print("extent:", raster_extent)
        raster_profile = src.profile
        crs = raster_profile['crs']
        print("crs:", crs)
        extent = raster_extent
        polygon = Polygon([(extent[0], extent[2]), (extent[0], extent[3]),
                           (extent[1], extent[3]), (extent[1], extent[2]),
(extent[0], extent[2])])
        # poly_gdf = gpd.GeoDataFrame([1],geometry=[polygon],crs=crs)
        poly_gdf = gpd.GeoDataFrame(
            {'name': [1], 'geometry': [polygon]}, crs=crs)
        vector_projection = vector.to_crs(crs)

    # 移除非 Polygon 类型的行，和无效的 Polygon( 用 .is_valid 验证 )，否则无法执行 .clip
    type_Polygon = shapely.geometry.polygon.Polygon
    polygon_bool = vector_projection.geometry.apply(
        lambda row: True if type(row) == type_Polygon and row.is_valid else
False)
    vector_projection_ = vector_projection[polygon_bool]
    vector_clip = gpd.clip(vector_projection_, poly_gdf)
    vector_clip.to_file(save_path)
    print("finished clipping and projection...")
    return poly_gdf
DTM_reprojection_fp = r'G:\data\IIT_lidarPtClouds\mosaic\DSM_mosaic_reprojection.
tif'
vector_shp_fp = r'G:\data\Building Footprints\Building Footprints.shp'
save_path = r'G:\data\building_footprints_clip_projection\building_footprints_
clip_projection.shp'
poly_gdf = clip_shp_withRasterExtent(
    vector_shp_fp, DTM_reprojection_fp, save_path)
-------------------------------------------------------------------------------
before dropna: (820606, 50)
after dropna: (820600, 50)
extent: (445062.87208577903, 452785.6657144039, 4627534.041486925,
4634507.0108712595)
crs: EPSG:32616
finished clipping and projection...
```

- **查看处理后的SHP格式建筑矢量数据**

除了直接查看建筑轮廓数据（图3.22-6），可以叠加打印DSM的栅格数据和建筑矢量数据（图3.22-7），确定二者在地理空间坐标保持一致的条件下相互吻合。说明数据处理正确，否则需要返回查看之前的代码，确定出错的位置，调整代码重新计算。

图3.22-6　芝加哥建筑轮廓数据

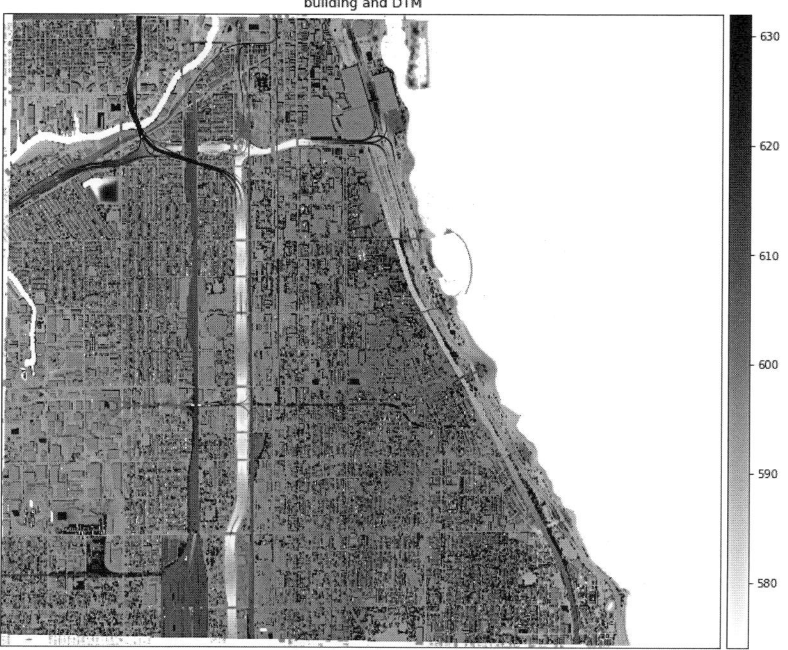

图3.22-7　叠加打印芝加哥建筑轮廓数据和DSM数据

```
import matplotlib.pyplot as plt
import geopandas as gpd
vector = gpd.read_file(vector_shp_fp)
vector.plot(figsize=(12,12))
import matplotlib.pyplot as plt
import earthpy.plot as ep
import rasterio as rio
from rasterio.plot import plotting_extent
import numpy as np
import geopandas as gpd
fig, ax = plt.subplots(figsize=(12, 12))
DTM_reprojection_fp = r'G:\data\IIT_lidarPtClouds\mosaic\DSM_mosaic_reprojection.
tif'
with rio.open(DTM_reprojection_fp) as DTM_src:
    mosaic_DTM_array = DTM_src.read(1)
    plot_extent = plotting_extent(DTM_src)
titles = ["building and DTM"]
ep.plot_bands(mosaic_DTM_array, cmap="binary", cols=1, title=titles, vmin=np.
quantile(
    mosaic_DTM_array, 0.25), vmax=np.quantile(mosaic_DTM_array, 0.95), ax=ax,
extent=plot_extent)
building_clipped_fp = r'G:\data\building_footprints_clip_projection\building_
footprints_clip_projection.shp'
vector = gpd.read_file(building_clipped_fp)
vector.plot(ax=ax, color='tomato')
plt.show()
```

3.22.2.3 根据分类栅格数据，从DSM中提取建筑区域的高程数据

因为建筑矢量数据每个建筑polygon并不一定仅包括分类为建筑的DSM栅格，可能包括其他分类数据，因此需要DSM仅保留建筑高程信息，避免计算误差。配合使用`np.where()`实现。

```
import util_A
classi_fp = r'G:\data\IIT_lidarPtClouds\mosaic\classification_mosaic.tif'
classi_reprojection_fp = r'G:\data\IIT_lidarPtClouds\mosaic\classi_mosaic_
reprojection.tif'
dst_crs = dst_crs
util_A.raster_reprojection(classi_fp,dst_crs,classi_reprojection_fp)
--------------------------------------------------------------------------------
finished reprojecting...
--------------------------------------------------------------------------------
import util_misc
s_t = util_misc.start_time()
classi_reprojection_fp = r'G:\data\IIT_lidarPtClouds\mosaic\classi_mosaic_
reprojection.tif'
with rio.open(classi_reprojection_fp) as classi_src:
    classi_reprojection = classi_src.read(1)
    out_meta = classi_src.meta.copy()
building_DSM = np.where(classi_reprojection == 6,
                        mosaic_DTM_array, np.nan)  #仅保留建筑高程信息
building_DSM_fp = r'G:\data\IIT_lidarPtClouds\mosaic\building_DSM.tif'
with rio.open(building_DSM_fp, "w", **out_meta) as dest:
    dest.write(building_DSM.astype(rio.uint16), 1)
util_misc.duration(s_t)
--------------------------------------------------------------------------------
start time: 2022-01-13 20:22:29.961764
end time: 2022-01-13 20:22:38.113300
Total time spend:0.13 minutes
```

- **提取ground并插值，合并及重投影，查看数据**

--提取

```
import util_A
las_dirPath = r"G:\data\IIT_lidarPtClouds\rawPtClouds"
json_combo_ = {"json_ground":r'G:\data\IIT_lidarPtClouds\ground'}
util_A.las_info_extraction_combo(las_dirPath,json_combo_)
----------------------------------------------------------------------------
100%|██████████| 73/73 [06:59<00:00,  5.74s/it]
```

--插值

插值使用了rasterio库提供的`fillnodata`方法。该方法是对每个像素，在四个方向上以圆锥形搜索值，根据反向距离加权计算插值。一旦完成所有插值，可以使用插值像素上的3 × 3平均

过滤器迭代，平滑数据。这种算法通常适宜于连续变化的栅格，例如DEM及填补小的空洞。

```
import util_misc
def rasters_interpolation(raster_path, save_path, max_search_distance=400,
smoothing_iteration=0):
    '''
    function - 使用 rasterio.fill 的插值方法，补全缺失的数据

    Paras:
        raster_path - 栅格根目录
        save_path - 保持的目录
    '''
    import rasterio
    import os
    from rasterio.fill import fillnodata
    import glob
    from tqdm import tqdm

    search_criteria = "*.tif"  #搜寻所要合并的栅格 .tif 文件
    fp_pattern = os.path.join(raster_path, search_criteria)
    fps = glob.glob(fp_pattern)  #使用 glob 库搜索指定模式的文件

    for fp in tqdm(fps):
        with rasterio.open(fp, 'r') as src:
            data = src.read(1, masked=True)
            msk = src.read_masks(1)
            #配置 max_search_distance 参数，或者多次执行插值，补全较大数据缺失区域
            fill_raster = fillnodata(
                data, msk, max_search_distance=max_search_distance, smoothing_
iterations=0)
            out_meta = src.meta.copy()
            with rasterio.open(os.path.join(save_path, "interplate_%s" % os.path.
basename(fp)), "w", **out_meta) as dest:
                dest.write(fill_raster, 1)
raster_path = r'G:\data\IIT_lidarPtClouds\ground'
save_path = r'G:\data\IIT_lidarPtClouds\ground_interpolation'
s_t = util_misc.start_time()
rasters_interpolation(raster_path, save_path,
                      max_search_distance=400, smoothing_iteration=0)
util_misc.duration(s_t)
--------------------------------------------------------------------------------
start time: 2022-01-13 20:32:00.967135
100%|████████████████| 73/73 [04:56<00:00,  4.06s/it]
end time: 2022-01-13 20:36:58.039148
Total time spend:4.95 minutes
```

--合并

```
import util_misc,util_A
ground_dir_path = r'G:\data\IIT_lidarPtClouds\ground_interpolation'
ground_out_fp = r'G:\data\IIT_lidarPtClouds\mosaic\ground_mosaic.tif'
s_t = util_misc.start_time()
out_trans = util_A.raster_mosaic(ground_dir_path,ground_out_fp)
util_misc.duration(s_t)
--------------------------------------------------------------------------------
start time: 2022-01-13 20:38:01.332643
end time: 2022-01-13 20:38:22.115885
Total time spend:0.33 minutes
```

--重投影

```
import util_A
ref_raster = r'G:\data\Landsat\data_processed\DE_Chicago.tif'  #使用的为 Landsat 部分
处理的遥感影像，可以自行下载对应区域的 Landsat，作为参数输入，获取其投影
dst_crs = util_A.get_crs_raster(ref_raster)
print("dst_crs:",dst_crs)
ground_fp = r'G:\data\IIT_lidarPtClouds\mosaic\ground_mosaic.tif'
ground_reprojection_fp = r'G:\data\IIT_lidarPtClouds\mosaic\ground_mosaic_
reprojection.tif'
util_A.raster_reprojection(ground_fp,dst_crs,ground_reprojection_fp)
--------------------------------------------------------------------------------
dst_crs: EPSG:32616
finished reprojecting...
```

--查看数据

为了方便栅格数据的打印查看（图3.22-8），将其定义为一个函数，方便调用。

```
def raster_show(raster_fp, title='raster', vmin_vmax=[0.25, 0.95], cmap="turbo"):
    '''
    function - 使用earthpy库显示遥感影像（一个波段）

    Paras:
        raster_fp - 输入栅格路径
        vmin_vmax - 调整显示区间
    '''
    import rasterio as rio
    import earthpy.plot as ep
    import numpy as np
    with rio.open(raster_fp) as src:
        array = src.read(1)
    titles = [title]
    ep.plot_bands(array, cmap=cmap, cols=1, title=titles, vmin=np.quantile(
        array, vmin_vmax[0]), vmax=np.quantile(array, vmin_vmax[1]))
raster_fp = r'G:\data\IIT_lidarPtClouds\mosaic\ground_mosaic_reprojection.tif'
raster_show(raster_fp)
```

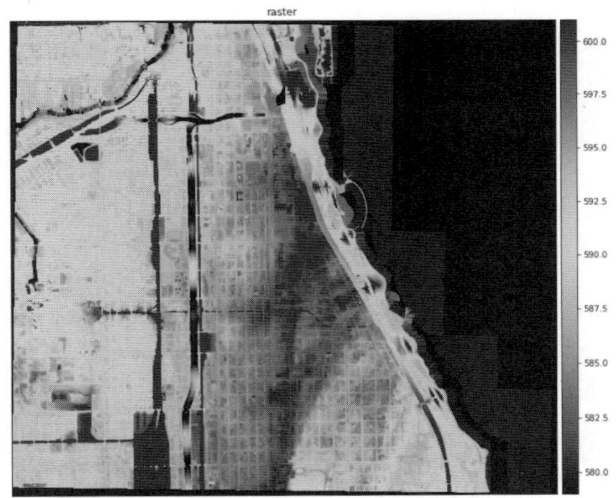

图3.22-8　高程数据

3.22.2.4　区域统计，计算建筑高度

使用rasterstats库的zonal_stats方法，提取DSM和ground栅格高程数据。

```
from rasterstats import zonal_stats
import util_misc
building_clipped_fp = r'G:\data\building_footprints_clip_projection\building_
footprints_clip_projection.shp'
s_t = util_misc.start_time()
building_DTM_fp = r'G:\data\IIT_lidarPtClouds\mosaic\building_DSM.tif'
stats_DTM = zonal_stats(building_clipped_fp, building_DTM_fp,stats="median")
ground_mosaic_fp = r'G:\data\IIT_lidarPtClouds\mosaic\ground_mosaic_reprojection.
tif'
stats_ground = zonal_stats(building_clipped_fp, ground_mosaic_fp,stats="median")
util_misc.duration(s_t)
--------------------------------------------------------------------------------
start time: 2022-01-13 20:45:20.199333
end time: 2022-01-13 20:53:11.112515
Total time spend:7.83 minutes
```

建筑高度=DSM提取的高程-ground提取的高程。为方便计算将其转换为Pandas的DataFrame数据格式，应用.apply及lambda函数进行计算。并将计算结果增加到建筑矢量数据的GeoDataFrame中，另存为SHP格式数据。

```
import numpy as np
import pandas as pd
import geopandas as gpd
building_height_df = pd.DataFrame({'dtm': [k['median'] for k in stats_DTM], 'ground': [
                                    k['median'] for k in stats_ground]})
building_height_df['height'] = building_height_df.apply(
    lambda row: row.dtm-row.ground if row.dtm > row.ground else -9999, axis=1)
print(building_height_df[:10])
building_clipped_fp = r'G:\data\building_footprints_clip_projection\building_
footprints_clip_projection.shp'
vector = gpd.read_file(building_clipped_fp)
vector['height'] = building_height_df['height']
vector.to_file(
    r'G:\data\building_footprints_height\building_footprints_height.shp')
print("finished computation and save...")
-----------------------------------------------------------------------------
     dtm  ground   height
0  625.0   593.0     32.0
1  625.0   591.0     34.0
2  626.0   595.0     31.0
3  625.0   593.0     32.0
4  625.0   591.0     34.0
5  625.0   591.0     34.0
6  625.0   591.0     34.0
7    0.0   594.0  -9999.0
8  630.0   592.0     38.0
9  626.0   592.0     34.0
finished computation and save...
```

打开，与打印查看计算结果（图3.22–9）。

```
import geopandas as gpd
import matplotlib.pyplot as plt
from mpl_toolkits.axes_grid1 import make_axes_locatable
```

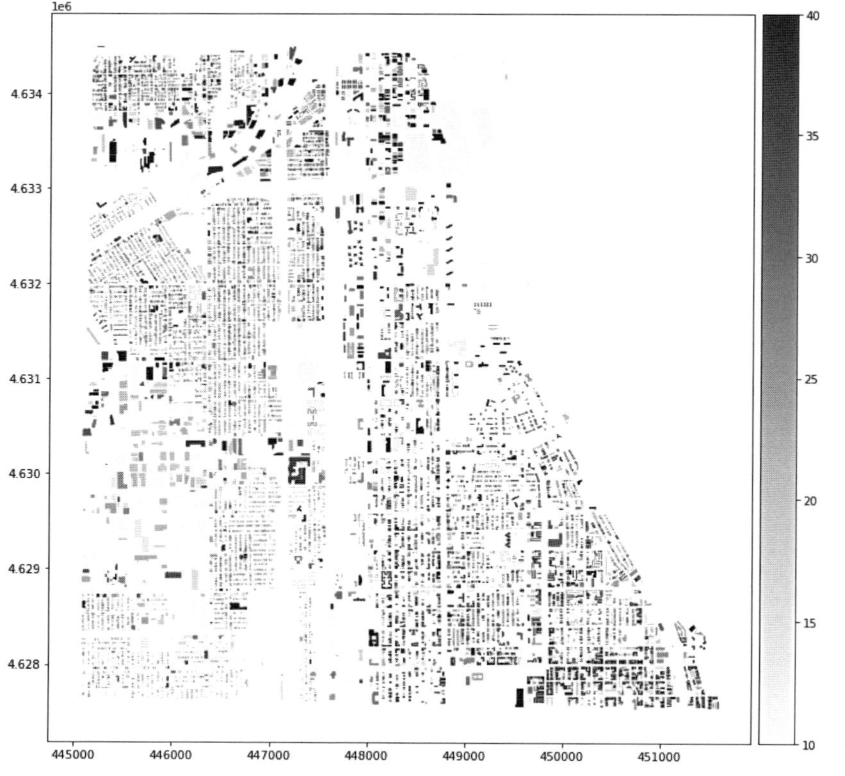

图3.22–9　建筑高度

```
import numpy as np
building_footprints_height_fp = r'G:\data\building_footprints_height\building_
footprints_height.shp'
building_footprints_height = gpd.read_file(building_footprints_height_fp)
fig, ax = plt.subplots(figsize=(12, 12))
divider = make_axes_locatable(ax)
cax_1 = divider.append_axes("right", size="5%", pad=0.1)
building_footprints_height.plot(column='height', ax=ax, cax=cax_1, legend=True,
cmap='OrRd', vmin=np.quantile(
    building_footprints_height.height, 0.25), vmax=np.quantile(building_
footprints_height.height, 0.95))  # 'OrRd' ,' PuOr'
```

参考文献（References）:

[1] PDAL: Point cloud Data Abstraction Library, https://pdal. io/en/2.6.0/.

3.23 天空视域因子计算与内存管理

3.23.1 天空视域因子（Sky View Factor，SVF）计算方法

天空视域因子用于描述三维空间形态的数值，是空间中某一点上可见天空与以该点为中心整个半球之间的比率。通常，值趋近于1，表明视域开阔；趋近于0，则封闭。天空视域因子广泛应用于城市热岛效应，城市能量平衡，和城市小气候等相关的研究中。虽然QGIS等平台也提供有SVF计算工具，例如`QGIS:SAGA:Sky View Factor`，但是计算大尺度，高分辨率DSM栅格数据的SVF通常会溢出，且计算缺乏灵活性。因此，有必要直接在Python下定义SVF计算方法。利用DSM获取城市下垫面SVF的计算公式为：$SVF=1-\frac{\sum_{i=1}^{n}\sin\gamma_i}{n}$。其中，$\gamma_i$为不同障碍物的高度角，$n$为搜索方向的数量。已知栅格单元的高度（DSM值），以及障碍物的高度和观察点到障碍物的距离，可以通过三角函数计算γ_i值。

为最终实现基于DSM计算SVF的代码编写，首先通过自定义一个小数据量的数组（代表DSM栅格），根据上述计算过程编写代码，快速的实现和检验方法的可行性后，再将其迁移到最终定义的SVF计算的类中。SVF计算的基本参数包括观察点、基于观察点向四周环顾一周的视线数量及其距离半径（视距），每根视线等分的数量。程序中比较关键的代码行是计算不同点或位置的坐标值，并将其转换为基于栅格（数组）的相对坐标值（整数型）。前者直接使用三角函数方法计算，后者借助`scipy.ndimage.map_coordinates()`方法实现。

- A-配置基本参数

```
import numpy as np

z_value = np.round(np.random.rand(10,10)*0.5,2) #生成栅格（矩阵、数组）的随机高程值
print("random z value:\n{}".format(z_value))
observation_spot = [4,4] #定义观察点
division_num = 5 #视线等分的数量
sight_distance = 3 #视线距离（视距）
sight_line_num = 16 #环顾一周视线数
--------------------------------------------------------------------------------
random z value:
[[0.06 0.37 0.26 0.06 0.48 0.42 0.11 0.35 0.08 0.31]
 [0.03 0.34 0.23 0.3  0.42 0.49 0.35 0.38 0.44 0.04]
 [0.25 0.23 0.39 0.26 0.2  0.03 0.3  0.43 0.15 0.43]
 [0.16 0.13 0.49 0.06 0.1  0.4  0.44 0.29 0.28 0.01]
 [0.12 0.28 0.42 0.31 0.08 0.01 0.03 0.15 0.46 0.48]
 [0.32 0.27 0.04 0.31 0.5  0.17 0.34 0.19 0.03 0.02]
 [0.17 0.46 0.05 0.19 0.   0.03 0.42 0.06 0.06 0.07]
 [0.26 0.21 0.22 0.21 0.39 0.14 0.1  0.32 0.45 0.12]
 [0.35 0.04 0.34 0.04 0.07 0.03 0.09 0.28 0.17 0.13]
 [0.03 0.07 0.16 0.5  0.41 0.15 0.36 0.48 0.05 0.16]]
--------------------------------------------------------------------------------
```

- B-等分视域，获取点坐标（图3.23–1）

```
def circle_division(observation_spot, sight_distance, sight_line_num):
    '''
    function - 给定观测位置点，视距和视线数量，等分圆，返回等分坐标点列表

    Paras:
        observation_spot - 观察点
        sight_distance - 视线距离（视距）
        sight_line_num - 环顾一周视线数
    '''
    import math
    import matplotlib.pyplot as plt
    angle_s = 360/sight_line_num
    angle_list = [i*angle_s for i in range(sight_line_num)]
    print("angle_list={}".format(angle_list))
    coordi_list = []
    for angle in angle_list:
        opposite = math.sin(math.radians(angle))*sight_distance
        adjacent = math.cos(math.radians(angle))*sight_distance
```

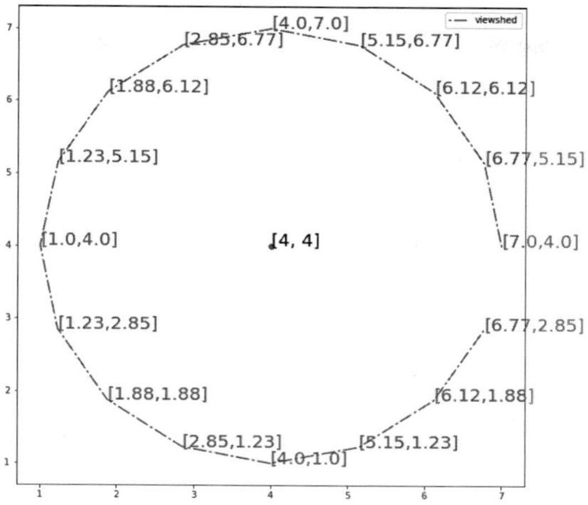

图3.23-1 等分坐标点

```
        coordi_list.append(
            (adjacent+observation_spot[0], opposite+observation_spot[1]))
    fig, ax = plt.subplots(figsize=(10, 10))
    x = [i[0] for i in coordi_list]
    y = [i[1] for i in coordi_list]
    # 使用 set_dashes() 修改现有直线为虚线
    line1, = ax.plot(x, y, label='viewshed')
    # 2pt line（直线）, 2pt break（打断）, 10pt line, 2pt break
    line1.set_dashes([2, 2, 10, 2])
    ax.legend()
    for i in range(len(x)):
        ax.text(x[i], y[i], "[%s,%s]" %
                (round(x[i], 2), round(y[i], 2)), fontsize=20, color="r")
    ax.text(observation_spot[0], observation_spot[1],
            observation_spot, fontsize=20)
    ax.plot(observation_spot[0], observation_spot[1], "ro")
    plt.show()
    return coordi_list
coordi_list = circle_division(observation_spot, sight_distance, sight_line_num)
--------------------------------------------------------------------------------
angle_list=[0.0, 22.5, 45.0, 67.5, 90.0, 112.5, 135.0, 157.5, 180.0, 202.5,
225.0, 247.5, 270.0, 292.5, 315.0, 337.5]
```

- C-根据视线提取栅格（数组）对应位置的高程（数组对应位置值），即批量提取截面高程数据（图3.23–2）

```
def line_profile(z_value, observation_spot, end_point, division_num):
    '''
    function - 获取与视线相交单元栅格（数组对应位置）的栅格值（数组值）

    Paras:
        z_value - 高程数组
        observation_spot - 观察点
        end_point - 视线末尾点
        division_num - 视线等分的数量
    '''
    import scipy.ndimage
    z = z_value
    x0, y0 = observation_spot
    x1, y1 = end_point
    num = division_num+1
    x, y = np.linspace(x0, x1, num), np.linspace(y0, y1, num)

    # 提取沿线的值，使用三次插值
    # 通过插值将输入数组映射到新的坐标 .
    zi = scipy.ndimage.map_coordinates(
        z, np.vstack((x, y)), cval=0, mode="nearest", order=0)
    x_around = np.around(x).astype(int)
```

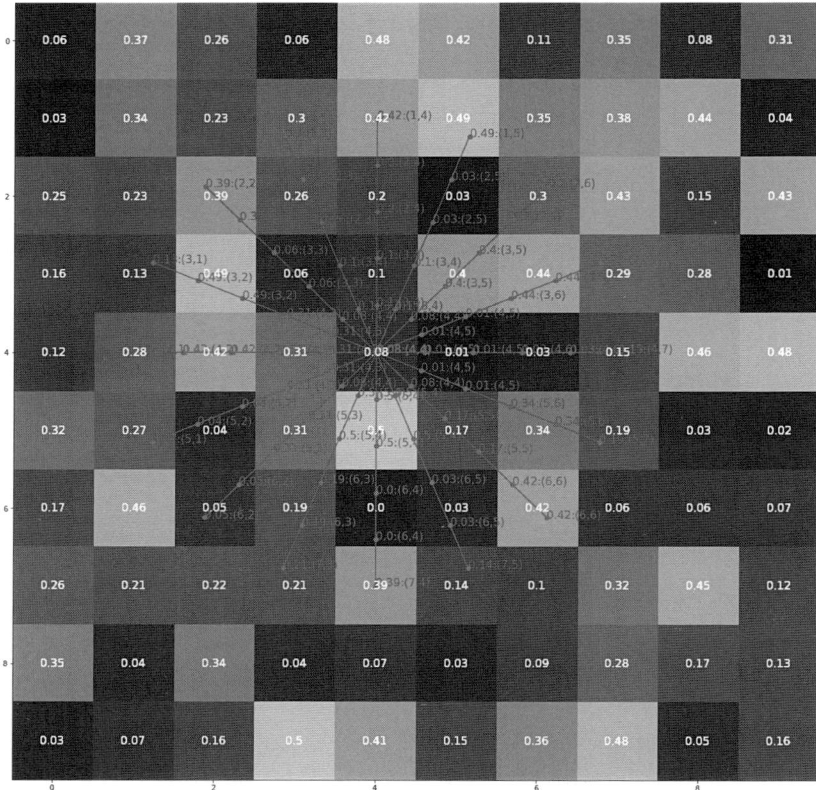

图3.23-2　截面高程

```
    y_around = np.around(y).astype(int)
    return zi, (x, y)
def combo_profile(z_value, observation_spot, coordi_list, division_num):
    '''
    function - 批量提取视线与高程数组（栅格）相交位置高程（数组对应位置值）

    Paras:
        z_value - 高程数组
        observation_spot - 观察点
        coordi_list - 坐标值列表（视线末端点）
        division_num - 视线等分的数量
    '''
    import matplotlib.pyplot as plt
    z_list = []
    sub_coordi_list = []
    for i in coordi_list:
        zi, sub_coordi = line_profile(
            z_value, observation_spot, i, division_num)
        z_list.append(zi)
        sub_coordi_list.append(sub_coordi)
    fig, axes = plt.subplots(nrows=1, figsize=z_value.shape*np.array([2]))
    axes.imshow(z_value)
    for n in range(len(sub_coordi_list)):
        x, y = sub_coordi_list[n]
        axes.plot(y, x, 'ro-')
        axes.axis('image')
        for i in range(z_value.shape[0]):
            for j in range(z_value.shape[1]):
                axes.text(j, i, z_value[i, j], ha="center",
                          va="center", color="w", size=15)
        for i in range(len(x)):
            axes.text(y[i], x[i], "%s:(%s,%s)" % (
```

```
                round(z_list[n][i], 2), round(x[i]), round(y[i])), fontsize=15,
color="r")
    plt.show()
    return z_list
z_list = combo_profile(z_value, observation_spot, coordi_list, division_num)
```

- D-计算SVF

```
def SVF(sight_distance, division_num, z_list):
    '''
    function - 计算天空视域因子（Sky View Factor, SVF）

    Paras:
        sight_distance - 视线距离（视距）
        division_num - 视线等分的数量
        z_list - 截面高程数组
    '''
    import math
    segment = sight_distance/division_num
    distance_list = [i*segment for i in range(division_num+1)]
    # print(distance_list)
    distance_list = distance_list[1:]
    sin_value_list = []
    for i in z_list:
        sin_maximum = 0
        for j in range(len(distance_list)):
            sin_temp = (
                i[j+1]-i[0])/math.sqrt(math.pow(distance_list[j], 2)+math.
pow(i[j+1]-i[0], 2))
            if sin_temp > sin_maximum:
                sin_maximum = sin_temp
            else:
                pass
        sin_value_list.append(sin_maximum)
    SVF_value = 1-sum(sin_value_list)/len(z_list)
    print("SVF_value=%s" % SVF_value)
    return SVF_value
SVF_value = SVF(sight_distance, division_num, z_list)
--------------------------------------------------------------------------------
SVF_value=0.7363503424243715
```

3.23.2 基于DSM计算SVF

上述SVF计算过程中使用了for循环，这将会大幅度增加计算时间，例如对于配置为：16G内存，可用约13G；Intel Core i7-8650U CPU @1.90GHz; 含大容量外置硬盘用于数据存储的条件下，（392，380）即148960个值计算时长约为1min；（4428，4460）即19748880个值时为2hs。当计算数据量（栅格单元数）为3119035612时(cell size=3 × 3)，for循环很难实现计算，因此必须转换为numpy数组形式逐次批量计算，避免逐个循环计算。其中，核心的SVF计算方法（公式）并没有改变。最终，区域DSM计算结果如图3.23–3所示。

在计算过程中，使用rasterio库Window方法切分栅格为多个子栅格逐个读取计算（如图3.23–4），减少单个栅格的SVF计算量，并逐一保存。当计算完所有子栅格后，将其合并为单独一个或者2～3个栅格（当一个栅格存储文件过大时）。使用切分为子栅格计算SVF，存在一个问题，子栅格边缘位置的单元因为缺失邻近栅格，造成计算不准确。这时，可以衡量计算机的硬件条件，尽量让子栅格尽量大，减少误差的位置数量。或者配置Window时叠错处理，计算完后移除边缘。

```
class SVF_DSM:
    '''
    class - 由 DSM 栅格计算 SVF（适用于高分辨率大尺度栅格运算）
    '''
    def __init__(self, dsm_fp, save_root, sight_distance, sight_line_num,
division_num):
        self.dsm_fp = dsm_fp
        self.save_root = save_root
        self.sight_distance = sight_distance
        self.sight_line_num = sight_line_num
```

图3.23-3 SVF

```
        self.division_num = division_num
    def raster_properties(self):
        '''
        function - 读取栅格，查看属性值，返回需要的属性（栅格总宽高）
        '''
        import rasterio as rio
        raster = rio.open(self.dsm_fp)
        print("type:", type(raster))
        print("transform:", raster.transform)
        print("[width,height]:", raster.width, raster.height)
        print("number of bands:", raster.count)
        print("bounds:", raster.bounds)
        print("driver:", raster.driver)
        print("no data values:", raster.nodatavals)
        print("_"*50)
        return raster.width, raster.height
    def divide_chunks(self, l, n):
        '''
        function - 递归分组列表数据
        '''
```

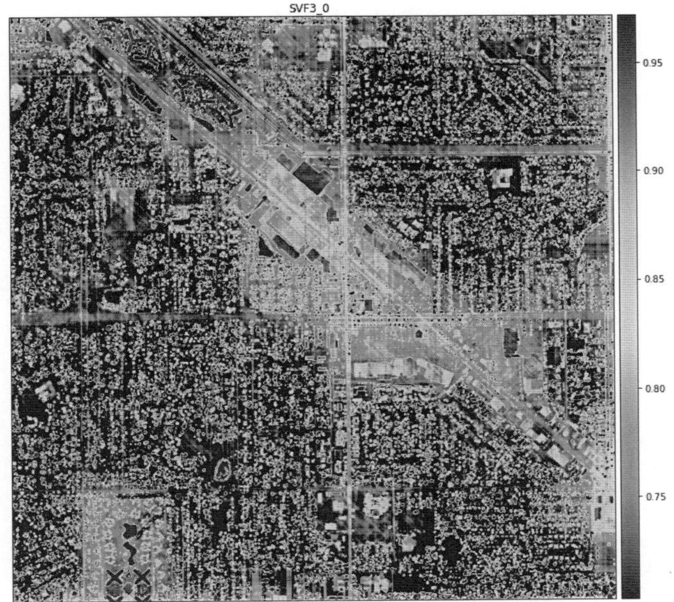

图3.23-4 子区域SVF计算

```
        for i in range(0, len(l), n):  # looping till length l
            yield l[i:i + n]
    def rasterio_windows(self, total_width, total_height, sub_width, sub_
height):
        '''
        function - 建立用于 rasterio 库分批读取一个较大 raster 数据的 windows 列表（尤其要
处理较大单独的 raster 数据时，避免内存溢出）

        Paras:
            total_width - 栅格总宽
            total_height - 栅格总高
            sub_width - 切分的子栅格宽
            sub_height - 切分的子栅格高
        '''
        from rasterio.windows import Window
        w_n = list(self.divide_chunks(list(range(total_width)), sub_width))
        h_n = list(self.divide_chunks(list(range(total_height)), sub_height))
        wins = [Window(w[0], h[0], len(w), len(h)) for h in h_n for w in w_n]
        print("raster windows amount:", len(wins))
        return wins
    def array_coordi(self, raster_array):
        '''
        function - 计算栅格单元（数组）的相对坐标
        '''
        import numpy as np
        relative_cell_coords = np.indices(raster_array.shape)
        relative_cell_coords2D = np.stack(
            relative_cell_coords, axis=2).reshape(-1, 2)
        del relative_cell_coords
        # print(relative_cell_coords2D)
        return relative_cell_coords2D
    def circle_division(self, observation_spot):
        '''
        function - 给定观测位置点，视距和视线数量，等分圆，返回等分坐标点数组
        '''
        import numpy as np
        angle_s = 360/self.sight_line_num
        angle_array = np.array(
            [i*angle_s for i in range(self.sight_line_num)], dtype=np.float32)
        opposite = np.sin(np.radians(angle_array),
                          dtype=np.float16)*self.sight_distance
```

```
            opposite = opposite.astype(np.float32)
            yCoordi = np.add(
                opposite, observation_spot[:, 1].reshape(-1, 1), dtype=np.float32)
            del opposite
            adjacent = np.cos(np.radians(angle_array),
                              dtype=np.float16)*self.sight_distance
            xCoordi = np.add(
                adjacent, observation_spot[:, 0].reshape(-1, 1), dtype=np.float32)
            del adjacent, angle_array
            coordi_array = np.stack((xCoordi, yCoordi), axis=-1)
            return coordi_array
        def line_profile(self, z_value, observation_spot, end_point):
            '''
            function - 获取与视线相交单元栅格（数组对应位置）的栅格值（数组值），即数组延直线截面
提取单元值

            Paras:
                z_value - DSM 栅格（含高程信息）
                observation_spot - 观察点数组
                end_point -  视线末尾点数组
            '''
            import numpy as np
            import scipy.ndimage
            num = self.division_num+1
            x0 = observation_spot[:, 0].reshape(-1, 1)
            x1 = end_point[:, :, 0]
            # 可以不用修改数组类型。出于内存优化考虑，会加快后续 np.stack 的计算速度
            x = np.linspace(x0, x1, num, dtype=int)
            del x0, x1
            y0 = observation_spot[:, 1].reshape(-1, 1)
            y1 = end_point[:, :, 1]
            y = np.linspace(y0, y1, num, dtype=int)
            del y0, y1
            xStack = np.stack(x, axis=-1)
            yStack = np.stack(y, axis=-1)
            del x, y
            zi = scipy.ndimage.map_coordinates(
                z_value, [xStack, yStack], cval=0, mode="nearest", order=0)  # 根据数
组索引值，提取实际值
            del xStack, yStack
            return zi
        def SVF(self, z_list):
            '''
            function - 计算天空视域因子（Sky View Factor, SVF）
            '''
            import numpy as np
            segment = self.sight_distance/self.sight_line_num
            distance_list = np.array(
                [i*segment for i in range(self.sight_line_num+1)], dtype=np.float32)
            distance_list = distance_list[1:]
            distance_list = np.expand_dims(distance_list, axis=1)
            z_list_sub = z_list[:, :, 1:]
            z_list_origin = z_list[:, :, 0]
            sin_values = np.true_divide(np.subtract(z_list_sub, np.expand_dims(z_
list_origin, axis=2)), np.sqrt(np.add(
                np.power(distance_list, 2), np.power(np.subtract(z_list_sub,
np.expand_dims(z_list_origin, axis=2)), 2))))
            sin_max_value = np.amax(sin_values, axis=2)
            del sin_values
            SVF_value = 1 - \
                np.true_divide(np.sum(sin_max_value, axis=1),
                               sin_max_value.shape[-1])
            return SVF_value.astype(np.float32)
        def svf_wins(self, wins_list):
            '''
            function - 计算 SVF 的主程序，并保存 SVF 子栅格文件
            '''
            from tqdm import tqdm
            import rasterio as rio
            import datetime
            import gc
            import os
            from rasterio.windows import Window
            import warnings
```

```
        # suppress warnings
        warnings.filterwarnings('ignore')
        i = 0
        for win in tqdm(wins_list):
            with rio.open(self.dsm_fp, "r+") as src:
                src.nodata = -1
                w = src.read(1, window=win)
                profile = src.profile
                win_transform = src.window_transform(win)

            ''' 计算部分 '''
            # print(w.shape)
            relative_cell_coords2D = self.array_coordi(w)

            # B- 等分视域，获取点坐标
            a_T = datetime.datetime.now()
            coordi_array = self.circle_division(relative_cell_coords2D)
            b_T = datetime.datetime.now()
            print("circle_division-time span:{}".format(b_T-a_T))
            gc.collect()

            # C- 根据视线提取栅格（数组）对应位置的高程（数组对应位置值），即批量提取截面高程数据
            c_T = datetime.datetime.now()
            zi = self.line_profile(w, relative_cell_coords2D, coordi_array)
            d_T = datetime.datetime.now()
            print("lineProfile-time span:{}".format(d_T-c_T))
            gc.collect()
            del coordi_array

            # D- 计算 SVF
            e_T = datetime.datetime.now()
            SVF_value = self.SVF(zi)
            f_T = datetime.datetime.now()
            print("SVF-time span:{}".format(d_T-c_T))
            gc.collect()
            del zi
            profile.update(
                width=win.width,
                height=win.height,
                count=1,
                transform=win_transform,
                compress='lzw',
                dtype=rio.float32
            )
            with rio.open(os.path.join(self.save_root, "SVF3_%d.tif" % i), 'w',
**profile) as dst:
                dst.write(SVF_value.reshape(w.shape), window=Window(
                    0, 0, win.width, win.height), indexes=1)
            del SVF_value
            g_T = datetime.datetime.now()
            print("total-time span:{}".format(g_T-a_T))
            i += 1
            break
dsm_fp = r"G:\data\DSM_pixel_3.tif"
save_root = r"G:\data\data_processed\SVF\SFV3_A"
# A-SVF 配置基本参数
raster_resolution = 3  # 计算栅格的分辨率
sight_distance = 100*raster_resolution  # 扫描半径
sight_line_num = 8  # 扫描截面数量 36
division_num = 30  # 每条扫描线的等分数量 50

SVF = SVF_DSM(dsm_fp, save_root, sight_distance, sight_line_num, division_num)
total_width, total_height = SVF.raster_properties()
print("total_width={},total_height={}".format(total_width, total_height))
sub_width, sub_height = 3000, 3000
wins_list = SVF.rasterio_windows(
    total_width, total_height, sub_width, sub_height)
print("windows example:{}".format(wins_list[:5]))
SVF.svf_wins(wins_list)
---------------------------------------------------------------------------------
type: <class 'rasterio.io.DatasetReader'>
transform: | 3.00, 0.00, 1075000.00|
| 0.00,-3.00, 1987501.00|
| 0.00, 0.00, 1.00|
```

```
[width,height]: 46667 60834
number of bands: 1
bounds: BoundingBox(left=1075000.0, bottom=1805000.0, right=1215001.0,
top=1987501.0)
driver: GTiff
no data values: (-1.0,)
--------------------------------------------------------------------
total_width=46667,total_height=60834
raster windows amount: 336
windows example:[Window(col_off=0, row_off=0, width=3000, height=3000),
Window(col_off=3000, row_off=0, width=3000, height=3000), Window(col_off=6000,
row_off=0, width=3000, height=3000), Window(col_off=9000, row_off=0, width=3000,
height=3000), Window(col_off=12000, row_off=0, width=3000, height=3000)]
  0%|          | 0/336 [00:00<?, ?it/s]
circle_division-time span:0:00:00.658197
lineProfile-time span:0:03:49.709210
SVF-time span:0:03:49.709210
  0%|          | 0/336 [04:54<?, ?it/s]
total-time span:0:04:52.874828
--------------------------------------------------------------------------------
import util_A
SVF3_0_fp = r"G:\data\data_processed\SVF\SFV3_A\SVF3_0.tif"
util_A.raster_show(SVF3_0_fp,title='SVF3_0')
```

3.23.3 内存管理

常规电脑的硬件配置通常有个限度。因此，在处理大数据时，可能因为数据量、计算量造成内存溢出。例如，np.arange(6000000000,dtype=np.float64)数组，预计占用45.7GB。如果内存小于该值，将溢出。6000000000数据量，相当于77459.6m × 77459.6m的城市区域。如果在这个过程中，存在其他计算，则将大幅度增加所用内存，因此如何管理内存与释放内存，在大数据处理过程中显得尤为重要。

- 对较大数据进行数据分析的几点建议：

1. 避免使用for循环，尽量使用array（numpy）直接数组间计算或者使用DataFrame（Pandas），可以大幅度增加计算速度。但是，会占用较大内存，须使用内存管理/减压工具处理。
2. 数据分批处理，并保存于硬盘中。逐一处理完后，读取所有文件进行后续处理。须平衡分批与一次性数组计算量，每次数组大计算速度快，但占用内存多；如果增加批次，则会降低单次数组量，但会增加计算时间。
3. 如果不必要，不须保存中间过渡的大数组。通常应用后，del释放内存，仅保留和存储必要的计算结果或中间结果。大数组尽量使用h5py保存于硬盘中。并尽量避免使用np.save工具，该工具保存的数据占据磁盘空间较大。
4. 大数据处理过程中，尽量避免使用matplotlib查看数据。只有必要分析时，再单独处理。
5. 为了减缓内存，数据及过程数据存储于大的硬盘中。根据自身数据大小，可以准备高容量的外置硬盘使用。另外，固态硬盘具有更快的读写速度。
6. 使用虚拟内存，可以有效缓解内存压力。
7. 有必要使用GPU计算，尤其训练深度学习模型。
8. 优化算法。

3.23.3.1 psutil

psutil（python system and process utilities）[①]用于在Python中检索有关正在运行的进程和系统利用率(CPU、内存、磁盘、网络、传感器)的信息。主要用于系统监控、分析、限制进程资源和管理运行的进程（更详细的内容须查看psutil文档）。

① psutil（https://psutil.readthedocs.io/en/latest）。

```
import psutil
# 当配置 logical=True 时，返回 logical CPUs 的数量。同 os.cpu_count()
print("cpu_count:{}".format(psutil.cpu_count(logical=True)))
mem = psutil.virtual_memory()
print("virtual_memory:{}".format(mem))  # 以元组的形式返回系统内存使用的统计信息
THRESHOLD = 100 * 1024 * 1024  # 100MB
```

```
if mem.available <= THRESHOLD:
    print("warning")

# 返回挂载的磁盘分区（mounted partitions）有关信息
print("disk_partitions:{}".format(psutil.disk_partitions()))
# 返回包含给定路径的分区磁盘使用统计信息，包含以字节表示的总空间、已用空间和空闲空间，以及使用百分比
print("disk_usage:{}".format(psutil.disk_usage('/')))
--------------------------------------------------------------------------------
cpu_count:16
virtual_memory:svmem(total=51410481152, available=38877888512, percent=24.4,
used=12532592640, free=38877888512)
disk_partitions:[sdiskpart(device='C:\\', mountpoint='C:\\', fstype='NTFS',
opts='rw,fixed'), sdiskpart(device='E:\\', mountpoint='E:\\', fstype='exFAT',
opts='rw,fixed'), sdiskpart(device='F:\\', mountpoint='F:\\', fstype='NTFS',
opts='rw,fixed'), sdiskpart(device='G:\\', mountpoint='G:\\', fstype='exFAT',
opts='rw,fixed')]
disk_usage:sdiskusage(total=511101923328, used=473427255296, free=37674668032,
percent=92.6)
```

3.23.3.2 h5py

HDF5（python,h5py）[①]是HDF5二进制数据格式的Python接口。HDF5可以存储大量数值数据（numerical data），并可以轻松地用NumPy操作这些数据。例如，可以将存储在磁盘上的tb级数据集切片；成千上万的数据集可以存储在一个文件中，根据需要进行分类和标记（更详细的内容需查看h5py文档）。

① HDF5 for Python，h5py包是HDF5二进制数据格式的Python接口（https://www.h5py.org）。

```
import h5py
import os
import numpy as np

# 建立 hdf5 文件路径
data_save_root = r"G:\data\data_processed\h5py"
hdf5_fp = os.path.join(data_save_root, "h5py_experi.hdf5")
if os.path.exists(hdf5_fp):
    os.remove(hdf5_fp)
else:
    print("Can note delete the file as it does not exists. to built a new one!")

# 生成随机的实验数据
data_a = np.random.rand(10000, 10000)
data_b = np.random.rand(10000, 10000)
data_c = np.random.rand(10000, 10000)
data_d = np.random.rand(10000, 10000)
data_e = np.random.randint(100, size=(100, 100))

# 用 f=h5py.File() 和 f.close()，打开写入数据文件，并关闭文件
# create a file by setting the mode to w when the File object is initialized. Some other modes are a (for read/write/create access), and r+ (for read/write access).
f = h5py.File(hdf5_fp, "w")
dset_a = f.create_dataset("dataset_a", data=data_a)
dset_b = f.create_dataset("dataset_b", data=data_b)
print("dset_a name={}; dset_b name={}".format(dset_a.name, dset_b.name))
print("f name={}".format(f.name))  # 可以看作默认组（group）
f.close()

# 可以用 with h5py.File() as f: 的方法，读写数据
with h5py.File(hdf5_fp, "a") as f:
    # 建立子组 / 群（subgroup），子组中可以写入若干数据集 (dataset)
    grp = f.create_group("subgroup")
    print("grp(group) name={}".format(grp.name))
    dset2 = grp.create_dataset(
        "dataset_c", (10000, 10000), dtype='f', data=data_c)
    print("dset2 name={}".format(dset2.name))
    grp.create_dataset("dataset_d", data=data_d)
    print("subgroup keys={}".format(list(f["subgroup"].keys())))
with h5py.File(hdf5_fp, "a") as f:
    # 直接建立子组和子组下的数据集
    dset3 = f.create_dataset('subgroup2/dataset_three',
                             (100, 100), dtype='i', data=data_e)
    dset3.attrs["attri_a"] = "attri_A"  # 可以配置属性字段
    dset3.attrs["attri_b"] = "attri_B"
```

```
    print("dset3 name={}".format(dset3.name))
    print("dset3.attrs[\"attri_a\"]={}".format(dset3.attrs["attri_a"]))
    print("dset3.attrs list={}".format(list(dset3.attrs)))
--------------------------------------------------------------------------------
dset_a name=/dataset_a; dset_b name=/dataset_b
f name=/
grp(group) name=/subgroup
dset2 name=/subgroup/dataset_c
subgroup keys=['dataset_c', 'dataset_d']
dset3 name=/subgroup2/dataset_three
dset3.attrs["attri_a"]=attri_A
dset3.attrs list=['attri_a', 'attri_b']
--------------------------------------------------------------------------------
def visit_func(name, node):
    '''
    function - 打印 HDF5 文件数据集和子组信息
    '''
    print('Full object pathname is:', node.name)
    if isinstance(node, h5py.Group):
        print('Object:', name, 'is a Group\n')
    elif isinstance(node, h5py.Dataset):
        print('Object:', name, 'is a Dataset\n')
    else:
        print('Object:', name, 'is an unknown type\n')
def get_dataset_keys(f):
    '''
    function - 返回 HDF5 文件数据集路径（包括子组内数据集）
    '''
    keys = []
    f.visit(lambda key: keys.append(key) if isinstance(
        f[key], h5py.Dataset) else None)
    return keys

# 查看 HDF5 文件，并读取数据集
with h5py.File(hdf5_fp, "r+") as f:
    print("f_keys={}".format(list(f.keys())))
    print("_"*50)
    datasets_description = f.visititems(visit_func)
    dataset_keys = get_dataset_keys(f)
    print("dataset keys={}".format(dataset_keys))
    dataset_three = f.get(dataset_keys[-1])[:]
print("_"*50)
print("'subgroup2/dataset_three' array:\n{}".format(dataset_three))
--------------------------------------------------------------------------------
f_keys=['dataset_a', 'dataset_b', 'subgroup', 'subgroup2']
__________________________________________________
Full object pathname is: /dataset_a
Object: dataset_a is a Dataset
Full object pathname is: /dataset_b
Object: dataset_b is a Dataset
Full object pathname is: /subgroup
Object: subgroup is a Group
Full object pathname is: /subgroup/dataset_c
Object: subgroup/dataset_c is a Dataset
Full object pathname is: /subgroup/dataset_d
Object: subgroup/dataset_d is a Dataset
Full object pathname is: /subgroup2
Object: subgroup2 is a Group
Full object pathname is: /subgroup2/dataset_three
Object: subgroup2/dataset_three is a Dataset
dataset keys=['dataset_a', 'dataset_b', 'subgroup/dataset_c', 'subgroup/dataset_
d', 'subgroup2/dataset_three']
__________________________________________________
'subgroup2/dataset_three' array:
[[98 15 44 ... 51  9 14]
 [89 98 29 ... 29 91 57]
 [45 54 54 ... 92 52 45]
 ...
 [12 92  2 ... 45 86 61]
 [ 9 36 42 ... 57 90 75]
 [33 62 23 ... 31 14 98]]
```

3.23.3.3 PyTables

PyTables[①]是一个用于管理分层数据集的库，旨在高效、轻松地处理极其大量的数据。PyTables是在HDF5库的基础上构建的，使用Python语言和NumPy库，可以交互式浏览、处理和搜索大量数据。该库优化内存和磁盘资源，使得数据占用的空间得以优化（更详细的内容需查看PyTables文档）。

① PyTables（https://pypi.org/project/tables）。

#对于 DataFrame(pandas) 可以使用 DataFrame.to_hdf() 和 pandas.read_hdf() 来读写 HDF5 文件。pandas 是使用 PyTables 完成读写 HDF5 文件。

```
import tables as tb
import numpy as np
# Define a user record to characterize some kind of particles
class Particle(tb.IsDescription):
    name = tb.StringCol(16)         # 16-character String
    idnumber = tb.Int64Col()        # Signed 64-bit integer
    ADCcount = tb.UInt16Col()       # Unsigned short integer
    TDCcount = tb.UInt8Col()        # unsigned byte
    grid_i = tb.Int32Col()          # integer
    grid_j = tb.Int32Col()          # integer
    pressure = tb.Float32Col()      # float  (single-precision)
    energy = tb.Float64Col()        # double (double-precision)
print('-**-**-**-**-**-**- file creation  -**-**-**-**-**-**-**-')
# HDF5 文件名的名称
filename = "G:\data\data_processed\h5py\pytables_experi.h5"
# 以 "w" 可写模式打开文件
h5file = tb.open_file(filename, mode="w", title="pytables_experi")
print("Creating file:", filename)
print()
print('-**-**-**-**-**-**- group and table creation  -**-**-**-**-**-**-**-')
# 在 "/" (根目录) 下创建一个新组
group = h5file.create_group("/", 'detector', 'Detector information')
print("Group '/detector' created")

# 创建一个表
table = h5file.create_table(group, 'readout', Particle, "Readout example")
print("Table '/detector/readout' created")

# 打印文件
print(h5file)
print(repr(h5file))

# 获取表中记录对象的快捷方式
particle = table.row

# 填充表格
for i in range(10):
    particle['name'] = 'Particle: %6d' % (i)
    particle['TDCcount'] = i % 256
    particle['ADCcount'] = (i * 256) % (1 << 16)
    particle['grid_i'] = i
    particle['grid_j'] = 10 - i
    particle['pressure'] = float(i * i)
    particle['energy'] = float(particle['pressure'] ** 4)
    particle['idnumber'] = i * (2 ** 34)
    particle.append()

# 刷新表的缓冲区
table.flush()
print()
print('-**-**-**-**-**-**-**- table data reading & selection  -**-**-**-**-**-')
# 从表中读取实际数据，收集 TDCcount 字段大于 3 且压力小于 50 的条目的压力值
xs = [x for x in table.iterrows() if x['TDCcount'] >
      3 and 20 <= x['pressure'] < 50]
pressure = [x['pressure'] for x in xs]
print("Last record read:")
print(repr(xs[-1]))
print("Field pressure elements satisfying the cuts:")
print(repr(pressure))

# 根据截断读取对应名称
names = [x['name'] for x in table.where(
```

```
    """(TDCcount > 3) & (20 <= pressure) & (pressure < 50)""")]
print("Field names elements satisfying the cuts:")
print(repr(names))
print()
print('-**-**-**-**-**-**- array object creation  -**-**-**-**-**-**-**-')
print("Creating a new group called '/columns' to hold new arrays")
gcolumns = h5file.create_group(h5file.root, "columns", "Pressure and Name")
print("Creating an array called 'pressure' under '/columns' group")
h5file.create_array(gcolumns, 'pressure', np.array(
    pressure), "Pressure column selection")
print(repr(h5file.root.columns.pressure))
print("Creating another array called 'name' under '/columns' group")
h5file.create_array(gcolumns, 'name', names, "Name column selection")
print(repr(h5file.root.columns.name))
print("HDF5 file:")
print(h5file)

#关闭文件
h5file.close()
print("File '" + filename + "' created")
--------------------------------------------------------------------------------
-**-**-**-**-**-**- file creation  -**-**-**-**-**-**-**-
Creating file: G:\data\data_processed\h5py\pytables_experi.h5
-**-**-**-**-**- group and table creation  -**-**-**-**-**-**-**-
Group '/detector' created
Table '/detector/readout' created
G:\data\data_processed\h5py\pytables_experi.h5 (File) 'pytables_experi'
Last modif.: 'Sat Jan 15 14:34:30 2022'
Object Tree:
/ (RootGroup) 'pytables_experi'
/detector (Group) 'Detector information'
/detector/readout (Table(0,)) 'Readout example'
File(filename=G:\data\data_processed\h5py\pytables_experi.h5, title='pytables_
experi', mode='w', root_uep='/', filters=Filters(complevel=0, shuffle=False,
bitshuffle=False, fletcher32=False, least_significant_digit=None))
/ (RootGroup) 'pytables_experi'
/detector (Group) 'Detector information'
/detector/readout (Table(0,)) 'Readout example'
  description := {
  "ADCcount": UInt16Col(shape=(), dflt=0, pos=0),
  "TDCcount": UInt8Col(shape=(), dflt=0, pos=1),
  "energy": Float64Col(shape=(), dflt=0.0, pos=2),
  "grid_i": Int32Col(shape=(), dflt=0, pos=3),
  "grid_j": Int32Col(shape=(), dflt=0, pos=4),
  "idnumber": Int64Col(shape=(), dflt=0, pos=5),
  "name": StringCol(itemsize=16, shape=(), dflt=b'', pos=6),
  "pressure": Float32Col(shape=(), dflt=0.0, pos=7)}
  byteorder := 'little'
  chunkshape := (1394,)
-**-**-**-**-**-**- table data reading & selection  -**-**-**-**-**-
Last record read:
/detector/readout.row (Row), pointing to row #9
Field pressure elements satisfying the cuts:
[81.0, 81.0, 81.0]
Field names elements satisfying the cuts:
[b'Particle:      5', b'Particle:      6', b'Particle:      7']
-**-**-**-**-**-**- array object creation  -**-**-**-**-**-**-**-
Creating a new group called '/columns' to hold new arrays
Creating an array called 'pressure' under '/columns' group
/columns/pressure (Array(3,)) 'Pressure column selection'
  atom := Float64Atom(shape=(), dflt=0.0)
  maindim := 0
  flavor := 'numpy'
  byteorder := 'little'
  chunkshape := None
Creating another array called 'name' under '/columns' group
/columns/name (Array(3,)) 'Name column selection'
  atom := StringAtom(itemsize=16, shape=(), dflt=b'')
  maindim := 0
  flavor := 'python'
  byteorder := 'irrelevant'
  chunkshape := None
HDF5 file:
G:\data\data_processed\h5py\pytables_experi.h5 (File) 'pytables_experi'
```

```
Last modif.: 'Sat Jan 15 14:34:30 2022'
Object Tree:
/ (RootGroup) 'pytables_experi'
/columns (Group) 'Pressure and Name'
/columns/name (Array(3,)) 'Name column selection'
/columns/pressure (Array(3,)) 'Pressure column selection'
/detector (Group) 'Detector information'
/detector/readout (Table(10,)) 'Readout example'
File 'G:\data\data_processed\h5py\pytables_experi.h5' created
```

3.23.3.4 memory_profiler

memory_profiler [①]用于监控进程的内存消耗，以及逐行分析Python程序的内存消耗（更详细的内容需查看memory_profiler文档）。

① memory_profiler（https://pypi.org/project/memory-profiler）。

```
@profile
def my_func():
    a = [1] * (10 ** 6)
    b = [2] * (2 * 10 ** 7)
    del b
    return a
if __name__ == '__main__':
    my_func()
```

将上述官网提供的示例程序，单独保存为.py文件后，在终端执行`mprof run <executable>`及`mprof plot`，可以获取如图3.23-5所示。

执行`python -m memory_profiler <executable>`，可以逐行观察计算所占内存量，用于观察代码优化结果。计算结果如下：

```
Line #  Mem usage   Increment  Occurences   Line Contents
=========================================================
     1   40.832 MiB   40.832 MiB           1   @profile
     2                                     def my_func():
     3   48.465 MiB    7.633 MiB 1     a = [1] * (10 ** 6)
     4  201.055 MiB  152.590 MiB 1 b = [2]* (2 * 10 ** 7)
     5   48.465 MiB -152.590 MiB           1       del b
     6   48.465 MiB    0.000 MiB           1     return a
```

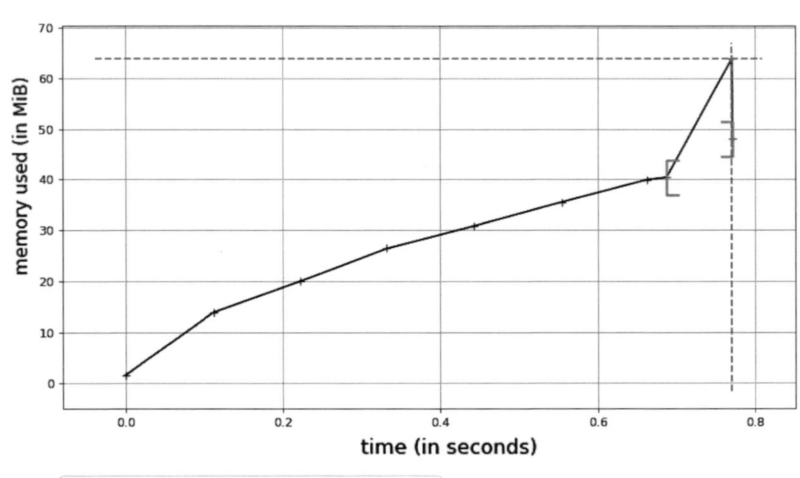

图3.23-5 进程监控

参考文献（References）:

[1] 段欣，胡德勇，曹诗颂，于琛，张亚妮．城区复杂下垫面天空视域因子参数化方法——以北京鸟巢周边地区为例[J]. 国土资源遥感，2019，31（03）：29-35.

[2] Böhner, J., & Antonić, O. (2009). Chapter 8 Land-Surface Parameters Specific to Topo-Climatology. Geomorphometry - Concepts, Software, Applications, 195–226. doi:10.1016/s0166-2481(08)00008-1

[3] Häntzschel, J., Goldberg, V., & Bernhofer, C. (2005). GIS-based regionalisation of radiation, temperature and coupling measures in complex terrain for low mountain ranges. Meteorological Applications, 12(1), 33–42. doi:10.1017/s1350482705001489

[4] PyTables, tutorial1-1.py，https://github.com/PyTables/PyTables/blob/master/examples/tutorial1-1.py.

附：Python基础核心

学写代码的方式

附：Python基础核心

学写代码的方式

对于从未接触过Python，或者对Python代码只有些了解的初学者，究竟如何进入到Python代码的世界里，并能够应用于自身的专业解决实际的问题？

● 调查目前为止所有的学习方式，主要包括：

1. 教材类。对Python语言系统的阐述，从数据类型结构至基本语句，以及函数和类，到标准库和扩展库，最后为高级的专项，例如机器学习、深度学习等。此类方法多出现在系统性阐述的教材（图书）中。该类方法通常适用于具有较强自修能力的学习者，可以从头阅读教材，逐行敲入代码跟随练习。
2. 在线视频课程。该类视频课程教学者通常会选择一本教材，或者自行编写教学大纲、教案进行讲解。此类的教学途径通常适合于自修能力相对较弱的学习者，可以从教学者的讲解中直接获取信息，减少阅读教材信息再提取加工的过程。
3. 在线代码交互练习。可以跟随章节进度，对照案例或者提示在线敲写代码，并实时反馈结果。这种学习的方式，因为互动信息反馈，具有游戏性，初学者往往倾向于优先选择此类途径。
4. 表格形式的卡片。此类方式可以理解为对系统阐述内容的教材类重要内容的提取，转换为表格卡片的形式。这么处理的好处是去除一些目前不必要而日后需要再了解的内容，将更多精力用于重点核心内容上。
5. 练习形式。类似表格卡片形式，将传统系统阐述的内容分解为几十，甚至更多的练习小节，结合文字阐释，强调练习的重要性。
6. 自由组合上述的教学途径。通过使用不同教学手段，可以更综合地激发学习者的学习兴趣，提高学习效率和学习效果。

● 调查目前为止Python教学的主要示例内容，主要包括：金融类、网页类、计算机（软件开发类和嵌入式系统）、数据分析类（含金融类，地理信息和大数据等）、算法类、游戏类，艺术类等。

根据当前Python教学途径及示例内容，对于城乡规划、风景园林、生态规划（林业、水利，湿地、草原、海洋）及建筑等归属自然资源部的相关专业，在开展Python编程语言学习、解决本专业的问题时，示例的内容最好能够更贴合专业内容，而学习方式上则可以任意组合。

1 示例数据以专业相关为主

如果能尽量以本专业（规划设计）相关的数据作为Python学习的数据内容，以处理本专业问题为导向学习代码的基本核心，这将能够更好地引导、学习Python基础知识后，不知道如何应用Python处理专业问题的弊病。同时，Python基础核心部分内容是为没有或者基础较弱的读者提供阅读《城市空间数据分析方法》的先导知识补充。

在示例数据选择上，优先考虑《城市空间数据分析方法》基础实验和专项研究部分的内容。如果数据无法满足要求，或者不能很好地解释待说明的知识点，将会调整选择的数据，仍然尽量保持为专业相关数据。

2 结合表格卡片形式的知识点内容

基础核心除了作为先导知识补充之外，仍需要解决两个问题：其一，可以作为知识点查询工具，当遗忘或寻找解决问题的方法时，可以从这些卡片中搜索定位；其二，知识点内容可以不断拓展更新，这不仅包括基础知识点，还会增加标准库及扩展库部分、任何有价值的算法或解决问题的逻辑。这将有助于不断积累实践中会用到的各类知识内容，也避免同类问题重复解决。

表格卡片的形式能让读者可以根据卡片表格形式，增加自己关注的知识点，不断积累。通常，一组内容组成一张卡片，这种将知识切分成微内容（碎片）的方式，能够让初学者每完成一张卡片，就会产生一定的成就感，有意愿不断地学习每张卡片，并可以根据自己的时间弹性调整学习时段；根据自己掌握的水平，挑选适合的卡片；根据要解决的问题，搜索查询卡片内容。

卡片完成的数量，可以作为衡量代码水平（完成度）的因素之一。每完成部分卡片，就该部分内容可以增加小节测试题，作为衡量代码水平的又一因素。综合卡片数量和测验，制定可以衡量学习者编码水平的分数级别。这有助于帮助初学者粗略地了解自己的代码水平，激励初学者不断学习，提高代码的书写水平。

3 搜索与英文

在代码世界中，通常用google搜索引擎，能够较为方便、准确地定位到待搜索的答案，从而快速解决问题，节约时间。同时，Python编程语言为英语书写，因此在写作该部分内容时，对于主要的词汇会同时给出英文，方便英文搜索时使用，并接轨英文这一国际通用语言。

4 完成度自测

等级	ID	卡片名称	得分（是否完成）	备注
Ⅰ 级	1	PCS_1 善用print()，基础运算，变量及赋值	10（ ）	
	2	PCS_2 数据结构-list_tuple_dict_set	10（ ）	
	3	PCS_3 数据结构-string	10（ ）	
	4	PCS_4 基本语句-if_for_while_comprehension	10（ ）	
	5	PCS_5 函数-def_scope_args	10（ ）	
	6	PCS_6 函数-recursion_lambda_generator	10（ ）	
	7	PCS_7 模块与包及发布-module_package_PypI	10（ ）	
	8	PCS_8 (OOP)类Classes-定义，继承，init()构造方法，私有变量/方法	10（ ）	
	9	PCS-9 (OOP)类Classes-Decorators(装饰器)_Slot	10（ ）	
	10	PCS-10 异常-Errors and Exceptions	10（ ）	
总计			100（ ）	
[Python Cheat Sheet，PCS]				

注：Python 基础核心 以表格卡片的形式写作，主要受到 *Coffee Break Python* 一书及配套在线交互学习 finxter-puzzle training① 的影响。该作者在书中阐述了基于谜题学习方法（Puzzle-based Learning）对学习者的有效性，从克服知识鸿沟（Overcome the Knowledge Gap）、灵光闪现（Embrace the Eureka Moment）、分而治之（Divide and Conquer），从即时反馈中改进（Improve From Immediate Feedback）、衡量你的技能（Measure Your Skills）、个性化学习（Individualized Learning）、小即是美（Small is Beautiful）、主动学习胜过被动学习（Active Beats Passive Learning），以及让代码成为一等公民（Make Code a First-class Citizen）等多个方法，阐述了其价值。就 Python 基础核心而言，采纳了表格卡片的形式，并借鉴了评级方式。卡片的内容则就规划设计专业和作为《城市空间数据分析方法》的先导知识补充及其合理性做出了调整。

① finxter-puzzle training（https://app.finxter.com/learn/computer/science/）。

参考文献（References）：

[1] Mayer, C. Coffee break python: 50 workouts to kickstart your rapid code understanding in python[M]. September 16, 2018.
[2] Shaw, Z. A. Learn python3 the hard way: A very simple introduction to the terrifyingly beautiful world of computers and code[M]. Addison-Wesley Professional,June 26, 2017.
[3] Shaw, Z. A. Learn more python the hard way: The next step for new python programmers[M]. Addison-Wesley Professional, September 1, 2017.
[4] Publishing, A. Python programming for beginners: The ultimate guide for beginners to learn python programming[M].October 26, 2022.
[5] Rao, B. N. Learning python[M]. CyberPlus Infotech Pvt. Ltd, February 14, 2021.
[6] Snowden, J. Python for beginners A practical guide for the people who want to learn python the right and simple way[M]. Independently published, December 2, 2020.
[7] 包瑞清．学习PYTHON——做个有编程能力的设计师．南京：江苏凤凰科学技术出版社．2015.

PCS-1. 善用print( )，基础运算，变量及赋值

PCS-1（　）：1.1 不知道运行结果，就很难写代码——善用print( )，第一个Python代码；1.2 增加注释的必要性；1.3 基本的数据类型（Basic Data Types）及运算（Operations）；1.4 变量及赋值（Variables and assignment）

知识点：

1.1 不知道运行结果，就很难写代码——善用print( )；第一个Python代码

1

描述：

print()是Python语言中使用最为频繁的语句，在代码编写、调试过程中，要不断地用该语句来查看数据的值、数据的变化、数据的结构、变量所代表的内容、监测程序进程和显示结果等。通过print()实时查看数据反馈，才可知道代码编写的目的是否达到并做出反馈。善用print()是用Python写代码的基础。

print('Hello World!') 代码（打印显示一行字符串）基本成为所有类型编程语言第一行代码的标配，标志着正式开启代码学习的篇章。

代码段：

```
print('Hello World!')
```

运算结果：

```
Hello World!
```

2

描述：

实际应用print()时，通常是打印变量来查看变量值，在代码调试时可以直接打印变量，例如print(v_sum)。下述代码则增加了解释的字符串配合打印变量，使用的是字符串格式化方法中的f"{}"方式。

代码段：

```
v_1 = 10
v_2 = 7
v_sum = v_1+v_2
print(f"The result of the calculation is {v_sum}")
```

运算结果：

```
The result of the calculation is 17.
```

3

描述：

用print()查看当前Python版本。这里调入了Python的一个标准库①platform②的python_version方法查看Python版本。调入库的方法，使用import。如果仅调入库中的一个方法，则可以使用from "library name" import "method"。

代码段：

```
from platform import python_version

print(python_version())
```

运算结果：

```
3.8.13
```

4

描述：

print("_"*50)，这里对字符"_"乘以了一个数字，则复制该字符多少个；对于字符串可以使用双引号，也可以使用单引号。但是，希望内部字符包括单引号时，则外部使用双引号，而

① 标准库，The Python Standard Library，Python标准库非常庞大，提供的组件涉及范围十分广泛，包含了多个内置模块（以C编写），可以用来实现系统级功能，例如文件I/O；也有大量Python编写的模块，提供了日常编程中许多问题的标准解决方案；其中，有些模块经过专门设计，通过将特定平台功能抽象化为平台中立的API来鼓励和加强Python程序的可移植性。Windows版本的Python安装程序通常包含整个标准库，往往还包含许多额外组件。对于类Unix操作系统，Python通常会分成一系列的软件包，因此可能需要使用操作系统所提供的包管理工具来获取部分或全部可选组件（https://docs.python.org/3/library/index.html）。

② platform，Access to underlying platform's identifying data，访问底层平台的识别数据（https://docs.python.org/3/library/platform.html?highlight=platform #module-platform）。

内部使用单引号。如果内容包括双引号时，则需要借助转义字符（escape character）\实现转义，即将Python的特殊字符，例如表征字符串的双引号转换为普通字符串使用。当然，也可以配合使用三引号。如果语句位于同一行，直接可以用;号分割。但是，通常不会这么做，因为这使得代码的可读性变弱；右斜杠（backslash，\）可以将长文本切为多段输入，输出字符串不断行。

代码段：

```
print("Hello Python!")
print("_"*50)
print("编程让设计更具'创造力! '")
print("Everybody should learn how to code a computer, because it teaches you how to think, and allows designers more creative!")
print("成为工具的\"建构者! \"")
print("""You must "type" each of these excercises in, mannually. \
If you copy and paste, you might as well as not even do them.""")
```

运算结果：

```
Hello Python!
__________________________________________________________________
编程让设计更具'创造力! '
Everybody should learn how to code a computer, because it teaches you how to
think, and allows designers more creative!
成为工具的"建构者! "
You must "type" each of these excercises in, mannually. If you copy and paste,
you might as well as not even do them.
```

知识点：

1.2 增加注释的必要性

5

描述：

注释包括单行注释，使用井号（hash，#）开头；多行注释，使用''' comments '''或""" comments """。注释并不会被执行，解释器将忽略注释的所有内容。注释的目的：

1. 为作者的注解，方便日后查看已经写过的代码含义，避免重新解读（尤其对于复杂或不易理解的逻辑和算法）。
2. 方便交流。他人阅读该代码时，可以快速地知道代码书写的目的或逻辑。
3. 传递代码书写作者、日期、版权等辅助信息。
4. 书写函数时，以注释的方式说明函数的功用、输入参数和返回变量的数据类型及说明等。

注：用于函数说明时，如果是使用Spyder交互式解释器编写代码，函数名行后回车，会提示是否书写函数说明，并自动配置下述格式，作者仅需要输入必要信息。

代码段：

```
# 1- 作者备忘注释及说明方便交流
data_path = './data'  # 配置数据存储位置

# 2- 辅助信息
"""
Created on Tue Feb 15 09:58:38 2022

@author: Richie Bao
"""

# 3- 用于函数说明
def cfg_load_yaml(ymlf_fp):
    '''
    读取 yaml 格式的配置文件

    Parameters
    ----------
    ymlf_fp : string
        配置文件路径

    Returns
    -------
    cfg : yaml-dict
```

```
    读取到 python 中的配置信息
    '''
    import yaml
    with open(ymlf_fp, 'r') as ymlfile:
        cfg = yaml.safe_load(ymlfile)
    return cfg
```

知识点:

1.3 基本的数据类型(Basic Data Types)及运算(Operations)

●●6

描述:

代码处理的对象就是数据,基本的数据类型包括整数(Integer,int)、实数(浮点型)(Real numbers,float)、复数(Complex numbers,complex)、字符(String,str)和布尔(Boolean,bool)。各种数据类型,都可以通过Python内置函数(方法)type查看数据类型。

注:内置函数为可以直接调用的函数,直接使用而无需导入库(模块)。

代码段:

```
print(type(7))
print(type(3.1415926))
print(type(3+6j))
print(type('Small is Beautiful'))
print(type(True), type(False))
```

运算结果:

```
<class 'int'>
<class 'float'>
<class 'complex'>
<class 'str'>
<class 'bool'> <class 'bool'>
```

——————————————————————————————————7

● 变换数据类型

描述:

int(value,base),其中base基数默认为10。float(value)只有一个输入参数。可已用内置函数转二进制、十进制和十六进制,其计算结果类型表述中0b代表二进制,0o代表八进制,ox代表十六进制。

代码段:

```
print(int(3.1415926))
print(int(2.7182818))  # 直接使用 int() 会自动向下取整
print(int("255", 10))  # 字符串转整数。如果字符串内容为浮点数,则会提示错误

print(bin(12))  # 转二进制(binary)
print(oct(12))  # 转十进制(octal)
print(hex(12))  # 转十六进制(hexadecimal)

print("_"*50)
print(float(64))
print(float("1.618034"))
print("_"*50)
print(complex(10))
print(complex("10+3j"))
print("_"*50)
print(bool(0))
print(bool(1))
print(bool())
print(bool(""))
print(bool("values"))
print("_"*50)
print(str(3.1415926), ":", type(str(3.1415926)))
```

运算结果:

```
3
2
```

```
255
0b1100
0o14
0xc
--------------------------------------------------------------------------------
64.0
1.618034
--------------------------------------------------------------------------------
(10+0j)
(10+3j)
--------------------------------------------------------------------------------
False
True
False
False
True
--------------------------------------------------------------------------------
3.1415926 : <class 'str'>
```

8

- 运算类型（Types of Operators）

描述：

6+7=13中，数值（numerical values）6和7为操作数（operands）；+为运算符/操作符（operators）。

1．算数运算符（表pcs.1-1）

表pcs.1-1　算数运算符

运算（**Syntax**）	说明（**Description**）
a+b	a加b (Addition)
a-b	a减b (Subtraction)
a*b	a乘以b (Multiplication)
a/b	a除以b (Division)
a//b	a除以b后向下取整 (Floor Divisiont)
a**b	a的b次方 (Exponential/Power)
a%b	模运算（Modulus）取模运算是计算两个数相除之后的余数

代码段：

```
print(15//7)
print(15 % 7)
```

运算结果：

```
2
1
```

9

描述：

2．比较运算符（Comparison/Relational Operators）

比较运算结果为布尔值（True 或False），如表pcs.1-2所示。

表pcs.1-2　比较运算符

运算（**Syntax**）	说明（**Description**）
a>b、a>=b	如果a大于（或大于等于）b，则结果为True (Greater than, Greater than or equal to)
a<b、a<=b	如果a小于等于（或小于）b，则结果为True (Lesser than, Lesser than or equal to)
a==b	如果a等于b，则结果为True (Equal to)
a!=b	如果a不等于b，则结果为True (Not equals to)

代码段：

```
print(6 != 7)
```

```
print(6 == 7)
print(“_”*50)
print(“six” != “seven”)
print(“six” == “six”)
print(“_”*50)
print(2.718 == 2.718000)
```

运算结果：

```
True
False
__________________________________________________________________________________
True
True
__________________________________________________________________________________
True
```

10

描述：

3. 赋值运算符（Assignment Operators）

赋值运算符相当于将等号右边的值按运算符计算到等号左边值，此时a为变量，而不是具体的值，计算后的值再赋值给变量a，如表pcs.1-3所示。

表pcs.1-3 赋值运算符

运算（Syntax）	等价于（Syntax Equivalence）
a+=b	a=a+b
a-=b	a=a-b
a*=b	a=a*b
a/=b	a=a/b)
a//=b	a=a//b
a**=b	a=a**b
a%=b	a=a%b

代码段：

```
i = 0
i += 1
print(i)
i += 1
print(i)
```

运算结果：

```
1
2
```

11

描述：

4. 逻辑运算符（Logical Operators），如表pcs.1-4所示

表pcs.1-4 逻辑运算符

运算（Syntax）	说明（Description）
a and b	都为True时，返回True
a or b	至少一个为True时，返回True
not a	为True时返回False，为False时返回True

代码段：

```
print(True and True)
```

```
print(True and False)
print(True or True)
print(True or False)
print(False and False)
print(not True)
print(not False)
```

运算结果：

```
True
False
True
True
False
False
True
```

12

描述：

5. 按位运算符（Bitwise Operators）

按位运算符通常用于嵌入式系统，多个输入输出端口（高低电平）表示的命令操作中，在数据分析领域使用暂不常见。但&和|可以替代and和or逻辑运算符使用，如表pcs.1-5所示。

表pcs.1-5 按位运算符

运算（Syntax）	说明（Description）
a & b	如果a和b均为True，则结果为True。对于整数（二进制），执行按位与操作。(Bitwise AND)
a | b	如果a和b任意一个为True，返回True。对于整数（二进制），执行按位或操作。(Bitwise OR)
a^b	为True时返回False，为False时返回True。对于布尔值，如果a或b为True（但不都为True），则结果为True。对于整数（二进制），执行按位异或操作。(Bitwise XOR)
~a	对于整数（二进制），执行按位取反操作。(Bitwiese NOT)
a<<b	对于整数（二进制），对a执行按位左移b个位操作。(Bitwise left shift)
a>>b	对于整数（二进制），对a执行按位右移b个位操作。(Bitwise right shift)

代码段：

```
print(True & True)
print(True & False)
print(True | True)
print(True | False)
print(False | False)
print("_"*50)
print(bin(7))
print(bin(0b0111))
print(0b0111)  #会自动转换为十进制
print("_"*50)
print(bin(~0b0111))
print("_"*50)
print(bin(0b0111 << 1))
print(bin(0b0111 >> 1))
```

运算结果：

```
True
False
True
True
False
__________________________________________________
0b111
0b111
7
__________________________________________________
```

```
-0b1000
--------------------------------------------------------------------------------
0b1110
0b11
```

13

描述：

6. 成员运算符（Membership Operators）

用于判断一个对象是否在Python序列中（例如，string/字符串、list/列表、tuple/元组和array/数组等），如表pcs.1-6所示。

表pcs.1-6　成员运算符

运算（Syntax）	说明（Description）
a in b	如果a在序列b中，则为True
a not in b	如果a不在序列b中，则为True

代码段：

```
lst = [1, 3, 6, 7, 9]
string = "python supports two membership operators, in and not in."
print(2 in lst)
print(3 in lst)
print("in" in string)
print("is not" not in string)
```

运算结果：

```
False
True
True
True
```

14

描述：

7. 同一运算符（Identity Operators）

用于判断两个对象（例如变量）是否使用同一位置索引内存。可用内置函数id()查看对象唯一标识，即获取对象的内存地址，如表pcs.1-7所示。

表pcs.1-7　同一运算符

运算（Syntax）	说明（Description）
a is b	如果变量a和b指向同一个Python对象，则结果为True
a is not b	如果变量a和b指向不同的Python对象，则结果为True

代码段：

```
a = 7
b = a
c = 7
d = 9
print(b is a)
print(c is a)
print(d is a)
print(id(a), id(b), id(c), id(d))
```

运算结果：

```
True
True
False
140730917330880 140730917330880 140730917330880 140730917330944
```

15

描述：

● 运算符优先级（Precedence and Associativity Rule of Operators）

包括多个运算符时，优先顺序如下表。L2R表示Left to right（从左到右）；R2L表示Right to left（从右到左），如表pcs.1-8所示。

表pcs.1-8　运算符优先级

运算符（**Operator**）	说明（**Description**）	结合性（**Associativity**）
()	圆括号（Parentheses）	L2R
**	幂（Exponential/Power）	R2L
+x,-x,~x	一元加（Unary Addition），一元减（Unary Subtraction），按位取反（Bitwise NOT）	L2R
*,/,//,%	乘（Multiplication），除（Division），向下取整除（Floor Division），取模运算（Modulus）	L2R
+,-	算数加（Arithmetic addition），算数减（Arithmetic subtraction）	L2R
<<,>>	按位左移（Bitwise shift left），按位右移（Bitwise shift right）	L2R
&	按位与（Bitwise AND）	L2R
^	按位或（Bitwise OR）	L2R
|	按位异或（Bitwise XOR）	L2R
==,!=,>,>=,<,<=	比较运算符（Relational operators）	L2R
=,+=,-=,*=, /=,//=,**=,%=	赋值运算符（Assignment and Augmented assignment operators））	R2L
in, not in, is, is not	成员，同一运算符（Membership, Identity operators）	L2R
Not	逻辑非（Logical NOT）	L2R
And	逻辑与（Logical AND）	L2R
Or	逻辑或（Logical OR）	L2R

代码段：

```
x = 7
print((x**2-2*x-3)/2)
```

运算结果：

```
16.0
```

知识点：

1.4 变量及赋值（Variables and assignment）

16

描述：

代码读起来应该像流畅的英语散文，而不是加密的密码。好的易读的变量名的定义正是让代码变的流畅的基础。变量名不能以数字和特殊字符为开头，也不可以内置的函数名定义，也不存在空格，如果由几个单词或数字组成变量名，通常由下划线连接，或者每一新单词首字母大写（通常尽量保持一种风格）。变量名定义不符合规范时，解释器会提示错误。

代码段：

```
func = 2*y+1  # 当程序逐行从上至下运行时，注意变量定义的顺序
y = 5
```

```
print(func)
```

运算结果：

```
11
```

17

描述：

如果想把变量值作为字符串的一部分打印出来，可以使用字符串格式化方法，例如%形式，或者'{}'.format(variable)方式及f"{}"方法。

代码段：

```
x = 5.0
monadic_equation = 2*x+1
print("monadic_equation=", monadic_equation)
print("monadic_equation=%.2f" % monadic_equation)  #%字符串格式化方法
print("monadic_equation={:.2f}".format(monadic_equation))  # format()字符串格式化方法
print(f"monadic_equation={monadic_equation:.2f}")  # f" {}" 字符串格式化方法
```

运算结果：

```
monadic_equation= 11.0
monadic_equation=11.00
monadic_equation=11.00
monadic_equation=11.00
```

18

代码段：

```
city_name = "Xi'an"
coordinate_longitude = 108.942292
coordiante_latitude = 34.261013
print("The longitude of the Xi'an coordinate is {lon:.2f}, and the latitude is {lat}.".format(
    lon=coordinate_longitude, lat=coordiante_latitude))
```

运算结果：

```
The longitude of the Xi'an coordinate is 108.94, and the latitude is 34.261013.
```

19

代码段：

```
#序列解包（unpacking）。尝试，x,y,*z=0,1,2,3,4,5,6; x,y,*z=0,1; (x,y),(a,b)=(0,1),(2,3)
x, y, b = 2, 5, 7
func_2 = 2*x+3*y+b
print("func_2={}".format(x, y, b, func_2))
```

运算结果：

```
func_2=2
```

20

代码段：

```
landuseName = 'General_Industrial'
landuseID = 3
GIndustrial_area = 5700
GIndustrial_greenArea = 3214
GIndustrial_GSR = GIndustrial_greenArea / \
    GIndustrial_area*100  # green space ratio（GSR）
print("GIndustrial_GSR={:.3f}%".format(GIndustrial_GSR))
```

运算结果：

```
GIndustrial_GSR=56.386%
```

PCS-2. 数据结构（Data Structure）-list，tuple，dict，set

PCS-2（ ）：2.1 列表（List）、元组（Tuple）和字典（Dictionary）及集合（Set）；2.2 列表（List）；2.3 元组（Tuple）；2.4 字典（Dictionary）；2.5 集合（Set）

知识点：

2.1 列表（List）、元组（Tuple）和字典（Dictionary）及集合（Set）

●●1

描述：

对于一组序列（sequence）数据（有序的），包含有一个或者多个值时，如果要赋予一个变量，则需要使用列表进行存储。同时，可以对数据进行组织管理，即为可变的（mutable），这包括使用索引（index）、分片（slicing）及提供的内置（built-in）函数或方法等途径，实现数据提取、插入、检索和列表间的运算等。

元组类似于列表，只是元组不可修改数据，但可以提取项值；以及用内置函数查看数据属性，例如数据长度、最大最小值等。

如果有一组空气污染物浓度的数据，希望可以通过检索污染物名称（string类型）提取数值，而并不要求浓度数据为序列型。则需要使用字典类型存储数据，并可赋予一个变量。对字典实现数据的组织管理，通过对键值操作、内置的函数和方法实现。

集合为一组无序不重复元素集，可以进行交集、差集和并集等操作。在某些算法中，可以简化运算复杂程度，例如在“蚁群算法（Ant Colony Optimization，ACO）”中或旅行商问题，建立禁忌城市集合`TabuCitySet`。通过差集，每次迭代逐次减去已经走过的城市，那么后续计算只需考虑差集后的集合，至集合为空完成计算。

列表、元组和字典存储的值可以是字符串、数值，也可以是列表、元组或字典自身，或者其他数据类型和结构。对于包含列表的列表为嵌套列表（nested list）。集合通常由列表或字符串转换而来，但是不可包含嵌套列表；否则，会提示错误`TypeError: unhashable type: 'list'`。

究竟选择哪类数据结构用于组织数据分析，需要根据具体分析的内容、途径，优先考虑便于操作的类型或者组合。

注：当实际工作研究时，对于城市空间数据或用sklearn（scikit-learn）机器学习，以及PyTorch深度学习时，扩展库NumPy和Pandas（及GeoPandas）提供的数组（array）和表（DataFrame，Series）数据结构无法避免[①]。

① sklearn（scikit-learn）机器学习，（https://scikit-learn.org/stable）；PyTorch，（https://pytorch.org）；NumPy，（https://numpy.org）；pandas，（<(https://pandas.pydata.org>）；GeoPandas，（https://geopandas.org/en/stable）。

知识点：

2.2 列表（List）

●●2

描述：

语法为`[v0,v1,v2,...,vn]`。默认索引关系为从0开始，递增1计算，也可以逆序（Negative Indexing），如图pcs.2-1所示。

```
0(-10) 1(-9) 2(-8) 3(-7) 4(-6) 5(-5) 6(-4) 7(-3) 8(-2) 9(-1)
['d', 'e', 'f', 'g', 'h', 'i', 'j', 'k', 'l', 'm']
```

图pcs.2-1 列表索引图解

- 建立列表

一种是按照语法直接书写并赋予变量，或用`list()`实现。用内置函数的方法时，如果是字符串，会自动切分为独个字母；如果是与`range()`配合，会产生一个给定始末和步幅值的序列。

代码段：

```
letters_1 = [‘d’, ‘e’, ‘f’, ‘g’, ‘h’, ‘i’, ‘j’, ‘k’, ‘l’, ‘m’]
print(letters_1)
letters_2 = list(‘defghijklm’)
print(letters_2)

# 内置函数，语法 range(start, stop, step)，给定始末和步幅值，构建序列
sequence_range = range(5, 20, 3)
print(sequence_range)
sequence_lst = list(range(5, 20, 3))
print(sequence_lst)
```

运算结果：

```
[‘d’, ‘e’, ‘f’, ‘g’, ‘h’, ‘i’, ‘j’, ‘k’, ‘l’, ‘m’]
[‘d’, ‘e’, ‘f’, ‘g’, ‘h’, ‘i’, ‘j’, ‘k’, ‘l’, ‘m’]
range(5, 20, 3)
[5, 8, 11, 14, 17]
```

3

描述：

enumerate(iterable, start=0)方法，第一个参数为iterable可迭代对象（列表或元组）；第二个参数为计数起始值，默认为0。返回的枚举对象为一个元组列表，每个元组第一个对象为计数值，第二个对象为对应计数的可迭代对象数值。enumerate()函数通常用于循环语句中，循环可迭代对象同时，返回索引值对象，用于计数或者作为其他对应数据提取值时的索引对象。返回的枚举对象<enumerate object at 0x000001C6EDEC1500>可以用list()方法提取具体值。

代码段：

```
idx_value = enumerate(letters_1, start=0)
print(idx_value)
print(list(idx_value))
```

运算结果：

```
<enumerate object at 0x000001C6EDEE3B00>
[(0, ‘d’), (1, ‘e’), (2, ‘f’), (3, ‘g’), (4, ‘h’), (5, ‘i’), (6, ‘j’), (7, ‘k’), (8,
‘l’), (9, ‘m’)]
```

4

描述：

提取值-用索引或者分片（slicing）的方式

给定索引可以直接提取对应索引的值。而slicing的方式需要给始末索引值及步幅值，语法为list[a:b:c]，a为开始值，b为结束值，c为步幅值。三个参数可以配置负值逆序。

下述案例中，首先用了一个实际研究项目的数据来说明用列表存储多组数据的方式。可以将各组数据单独存储于单个列表中；亦可以用嵌套列表的形式将其存储于一个列表中。不过，对于多组数据的存储，通常使用dict、array（NumPy）或DataFrame（Pandas）的形式处理更加便捷。

注：在解释数据结构具体操作方式时，使用简单的数据形式，较之具体研究项目的数据更加简单，易于专注于操作本身。但是，同样给出一个具体的案例，来说明在具体数据分析时数据是如何存储的。

多组数据单独存储各自列表的方式。

代码段：

```
# 这组数据来自城市环境传感器（AoT, array of things），提取了部分数据，包括字段有：地址（string）、ID 编号 (string)、经纬度 (float) 共 4 类值，以单独列表形式存储。
coordi_lon = [-87.628, -87.616, -87.631, -87.59, -
              87.711, -87.628, -87.586, -87.713, -87.676, -87.624]
coordi_lat = [41.878, 41.858, 41.926, 41.81,
              41.866, 41.883, 41.781, 41.751, 41.852, 41.736]
node_id = [‘ba46’, ‘ba3b’, ‘f02f’, ‘ba8f’, ‘ba16’,
           ‘7e5d’, ‘ba8b’, ‘ba13’, ‘ba18’, ‘bc10’]
address = [‘State St & Jackson Blvd Chicago IL’, ‘18th St & Lake Shore Dr
Chicago IL’, ‘Lake Shore Drive & Fullerton Ave Chicago IL’, ‘Cornell & 47th St
Chicago IL’, ‘Homan Ave & Roosevelt Rd Chicago IL’,
           ‘State St & Washington St Chicago IL’, ‘Stony Island Ave & 63rd St
Chicago IL’, ‘7801 S Lawndale Ave Chicago IL’, ‘Damen Ave & Cermak Chicago IL’,
```

```
'State St & 87th Chicago IL']
```

5

描述：

'{0},{1}'.format(v0,v1)该种字符串格式化的方式，可以给定索引值。这样，就可以不受位置影响，比较方便地增加{}待格式内容。

如果打印字符串语句比较长，编辑起来不是很方便，则可以用\将其分成多段处理。打印字符串时，如果需要换行打印显示，则加入\n换行符处理。

代码段：

```
# 提取索引值为 2 对应的数值
idx_AoT = 2
idx_lon = coordi_lon[idx_AoT]
idx_lat = coordi_lat[idx_AoT]
idx_nodeID = node_id[idx_AoT]
idx_address = address[idx_AoT]
print('values with the index 2: \
      \nid={0};\nlon={2};\nlat={3};\naddress={1}'.format(idx_nodeID, idx_
address, idx_lon, idx_lat))

# 提取分片数据，slice(2,5) 为即为 [2,3,4]
slice_AoT = slice(2, 5)  # 默认步幅为 1
print("_"*50)
print(idxes_AoT)
slice_lon = coordi_lon[slice_AoT]
slice_lat = coordi_lat[slice_AoT]
slice_nodeID = node_id[slice_AoT]
slice_address = address[slice_AoT]
print('values with the index {4}: \
      \nid={0};\nlon={2};\nlat={3};\naddress={1}'.format(slice_nodeID, slice_
address, slice_lon, slice_lat, list(range(2, 5))))
```

运算结果：

```
values with the index 2:
id=f02f;
lon=-87.631;
lat=41.926;
address=Lake Shore Drive & Fullerton Ave Chicago IL
__________________________________________________
slice(2, 5, None)
values with the index [2, 3, 4]:
id=['f02f', 'ba8f', 'ba16'];
lon=[-87.631, -87.59, -87.711];
lat=[41.926, 41.81, 41.866];
address=['Lake Shore Drive & Fullerton Ave Chicago IL', 'Cornell & 47th St
Chicago IL', 'Homan Ave & Roosevelt Rd Chicago IL']
```

6

描述：

以嵌套列表的形式存储多组数据。很多时候，为了方便获知变量的数据结构，而不用打印print()或者用type()的方式查看数据，则可以在起变量名时，直接增加一个可以表明数据结构的单词或缩写，例如list->lst，dict->dict，DataFrame->df，GeoDataFrame->gdf，array->array等。

代码段：

```
AoT_info_lst = [['001e0610ba46', -87.627678, 41.87837699999999, 'State St &
Jackson Blvd Chicago IL'],
                ['001e0610ba3b', -87.616055, 41.85813599999999,
                    '18th St & Lake Shore Dr Chicago IL'],
                ['001e0610f02f', -87.6307578, 41.926261399999994,
                    'Lake Shore Drive & Fullerton Ave Chicago IL'],
                ['001e0610ba8f', -87.590228, 41.81034199999999,
                    'Cornell & 47th St Chicago IL'],
                ['001e0610ba16', -87.710543, 41.866349,
                    'Homan Ave & Roosevelt Rd Chicago IL'],
                ['001e06107e5d', -87.6277685, 41.88320529999999,
                 'State St & Washington St Chicago IL'],
                ['001e0610ba8b', -87.58645600000001, 41.7806,
```

```
                'Stony Island Ave & 63rd St Chicago IL'],
               ['001e0610ba13', -87.71299, 41.75123799999999,
                   '7801 S Lawndale Ave Chicago IL'],
               ['001e0610ba18', -87.675825, 41.85217899999999,
                   'Damen Ave & Cermak Chicago IL'],
               ['001e0610bc10', -87.62417900000001, 41.73631399999999, 'State
St & 87th Chicago IL']]
print(AoT_info)
```

运算结果:

```
[['001e0610ba46', -87.627678, 41.87837699999999, 'State St & Jackson Blvd Chicago
IL'], ['001e0610ba3b', -87.616055, 41.85813599999999, '18th St & Lake Shore Dr
Chicago IL'], ['001e0610f02f', -87.6307578, 41.926261399999994, 'Lake Shore Drive
& Fullerton Ave Chicago IL'], ['001e0610ba8f', -87.590228, 41.81034199999999,
'Cornell & 47th St Chicago IL'], ['001e0610ba16', -87.710543, 41.866349, 'Homan
Ave & Roosevelt Rd Chicago IL'], ['001e06107e5d', -87.6277685, 41.88320529999999,
'State St & Washington St Chicago IL'], ['001e0610ba8b', -87.58645600000001,
41.7806, 'Stony Island Ave & 63rd St Chicago IL'], ['001e0610ba13', -87.71299,
41.75123799999999, '7801 S Lawndale Ave Chicago IL'], ['001e0610ba18',
-87.675825, 41.85217899999999, 'Damen Ave & Cermak Chicago IL'], ['001e0610bc10',
-87.62417900000001, 41.73631399999999, 'State St & 87th Chicago IL']]
```

7

描述:

含有括号、大括号的语句，其中的内容可以按逗号分段对其书写，使得代码易读。

代码段:

```
print('values with the index 2: \
      \nid={0};\nlon={2};\nlat={3};\naddress={1}'.format(AoT_info_lst[2][0],
                                                         AoT_info_lst[2][1],
                                                         AoT_info_lst[2][2],
                                                         AoT_info_lst[2][3]))
```

运算结果:

```
values with the index 2:
id=001e0610f02f;
lon=41.926261399999994;
lat=Lake Shore Drive & Fullerton Ave Chicago IL;
address=-87.6307578
```

8

描述:

letters_1直接敲入的字母，可以用list(map(chr,range(100,110))) 语句提取。其中，chr(i, /)，是给定索引值，返回Unicode[1]（统一码、万国码、单一码）字符串。map(func, *iterables) 函数，输入的第一个参数func为一个函数对象，第二个参数*iterables为可迭代对象（列表、元组等），以迭代器方式逐个返回给定迭代对象的函数计算值。因为返回的是迭代器，本例返回值为 <map object at 0x000001C6EDDB96D0>，可以通过list(iterate object)方式读取。这个函数可以简单地理解为，一次性地给定函数多个参数值，返回所有函数计算值。在案例中，又给出了两个小案例，以方便理解。

注：所有函数、关键字等，都可以通过 help() 函数查看说明文档。

① Unicode，The Unicode Standard，是一个信息技术标准，用于对世界上大多数书写系统所表达的文本进行统一编码、表示和处理。该系统由Unicode Consortium 维护，截至目前的版本（15.0）共定义了149186个字符，涵盖了161种现代和历史文字，以及符号、表情符号（包括颜色）和非视觉控制和格式化代码（https://home.unicode.org/）。

代码段:

```
print(map(chr, range(100, 110)))  #返回的为迭代对象
lst = list(map(chr, range(100, 110)))
print(lst)
print("_"*50)
print(chr(100))  #打印索引值为 100 的 unicode 对象

print("_"*50)
lst4len = ['python', 'C++', 'C', 'hello guys!']
string_len = list(map(len, lst4len))
print(string_len)
lst4max = [[1, 2], [5, 77], [33, 65]]
max_v = list(map(max, lst4max))
print(max_v)
```

运算结果：

```
<map object at 0x000001C6EDEDFDF0>
[‘d’, ‘e’, ‘f’, ‘g’, ‘h’, ‘i’, ‘j’, ‘k’, ‘l’, ‘m’]
------------------------------------------------------------------
d
------------------------------------------------------------------
[6, 3, 1, 11]
[2, 77, 65]
```

9

代码段：

```
help(map)
```

运算结果：

```
Help on class map in module builtins:
class map(object)
 |  map(func, *iterables) --> map object
 |
 |  Make an iterator that computes the function using arguments from
 |  each of the iterables.  Stops when the shortest iterable is exhausted.
 |
 |  Methods defined here:
 |
 |  __getattribute__(self, name, /)
 |      Return getattr(self, name).
 |
 |  __iter__(self, /)
 |      Implement iter(self).
 |
 |  __next__(self, /)
 |      Implement next(self).
 |
 |  __reduce__(...)
 |      Return state information for pickling.
 |
 |  ----------------------------------------------------------------------
 |  Static methods defined here:
 |
 |  __new__(*args, **kwargs) from builtins.type
 |      Create and return a new object.  See help(type) for accurate signature.
```

10

描述：

分片练习

代码段：

```
print(lst[3:6:1])
print(lst[3:6])
print(lst[3:6:2])
print(lst[-3:-1])
print(lst[-3:])
print(lst[-1])
print(lst[:3])
print(lst[:])
```

运算结果：

```
[‘g’, ‘h’, ‘i’]
[‘g’, ‘h’, ‘i’]
[‘g’, ‘i’]
[‘k’, ‘l’]
[‘k’, ‘l’, ‘m’]
m
[‘d’, ‘e’, ‘f’]
[‘d’, ‘e’, ‘f’, ‘g’, ‘h’, ‘i’, ‘j’, ‘k’, ‘l’, ‘m’]
```

11

描述：

元素赋值+分片赋值+删除元素+列表相加+列表的乘法+成员运算符

代码段：

```
lst[5] = 99  #元素赋值
print(lst)
lst_none = lst+[None]*6  #列表相加，以及列表的乘法
print(lst_none)
lst_none[13] = 2015
print(lst_none)
lst_none[-6:-3] = list(range(100, 104, 2))  #分片赋值
print(lst_none)
lst_none[1:1] = [0, 0, 0, 12]
print(lst_none)
del lst_none[-2:]  #删除元素
print(lst_none)

print([1000, 1200, 1500] in lst_none)  #成员运算符
print([1000, 1200, 1500] not in lst_none)  #成员运算符
```

运算结果：

```
[‘*’, ‘)’, 99, ‘h’, ‘g’, 99, [1000, 1200, 1500], ‘e’, ‘d’]
[‘*’, ‘)’, 99, ‘h’, ‘g’, 99, [1000, 1200, 1500], ‘e’, ‘d’, None, None, None,
None, None, None]
[‘*’, ‘)’, 99, ‘h’, ‘g’, 99, [1000, 1200, 1500], ‘e’, ‘d’, None, None, None,
None, 2015, None]
[‘*’, ‘)’, 99, ‘h’, ‘g’, 99, [1000, 1200, 1500], ‘e’, ‘d’, 100, 102, None, 2015,
None]
[‘*’, 0, 0, 0, 12, ‘)’, 99, ‘h’, ‘g’, 99, [1000, 1200, 1500], ‘e’, ‘d’, 100,
102, None, 2015, None]
[‘*’, 0, 0, 0, 12, ‘)’, 99, ‘h’, ‘g’, 99, [1000, 1200, 1500], ‘e’, ‘d’, 100,
102, None]
True
False
```

12

描述：

● 内置函数

可以用于列表的内置函数，除了构建列表的list()函数外，可以用len()函数计算列表长度，max()函数返回列表最大值，min()函数返回列表最小值等。表述为函数时，通常是以func(auguments)语法操作，函数可以理解为将要执行的动作，输入参数为被执行的对象。

练习时，如果手工随机输入一组数据，即使数值比较简单，但是还是会消磨掉一些对技能增长无用的时间，因此练习代码需要随机数值就可以，那就用代码自动生成一组数据。下述调入random库，使用该库提供的sample(population, k)方法，随机生成一组值。其中，第一个参数population为序列（列表、元组等）或集合（set）的一组数；第二个参数为随机提取的数量。返回值为在给定的一组数中随机抽取k个，组成一个新的列表。每运行一次抽取的随机数都会发生改变，如果希望每次运行结果保持一致，则需要用seed(a=None, version=2)固定一个随机种子。关于参数的解释，可以用help(random.seed)方法查看。

代码段：

```
import random
letters_lst = list(map(chr, range(60, 70)))
print(lst)
random.seed(10)
numericalVals_lst = random.sample(range(1, 200), 10)
print(numericalVals_lst)
print(help(random.sample))
```

运算结果：

```
[‘<’, ‘=’, ‘>’, ‘?’, ‘@’, ‘A’, ‘B’, ‘C’, ‘D’, ‘E’]
[147, 9, 110, 124, 148, 4, 53, 119, 126, 72]
Help on method sample in module random:
sample(population, k) method of random.Random instance
    Chooses k unique random elements from a population sequence or set.
    Returns a new list containing elements from the population while
    leaving the original population unchanged.  The resulting list is
    in selection order so that all sub-slices will also be valid random
    samples.  This allows raffle winners (the sample) to be partitioned
```

```
    into grand prize and second place winners (the subslices).
    Members of the population need not be hashable or unique.  If the
    population contains repeats, then each occurrence is a possible
    selection in the sample.
    To choose a sample in a range of integers, use range as an argument.
    This is especially fast and space efficient for sampling from a
    large population:   sample(range(10000000), 60)
None
```

13

代码段：

```
print(len(letters_lst), len(numericalVals_lst))
print(max(letters_lst), max(numericalVals_lst))
print(min(letters_lst), min(numericalVals_lst))
```

运算结果：

```
10 10
E 148
< 4
```

14

描述：

● 内置方法

列表的方法使用的语法为lst.method()，包括：

append()，在列表末尾追加新的对象；

extend()，在列表的末尾一次性追加另一个序列中的多个值；

insert()，将对象插入到列表中；

pop()，根据指定的索引值移除列表中的项值，并返回该项值，在默认无参数的条件下移除最后一个；

remove()，移除项值，输入参数为指定的项值而不是索引值；

count()，统计某个元素在列表中出现的次数；

index()，从列表中找到某一个值第一个匹配项的索引位置；

reverse()，翻转列表；

clear()，清空列表；

sort()，是按一定的顺序重新排序；

copy()，复制列表；

查看帮助文件时，使用help(list.method)语法。

代码段：

```
lst = list(map(chr, range(100, 105)))
print(lst)
lst.append(99)
print(lst)
lst.append(list(range(50, 80, 5)))
print(lst)
lst_b = [‘*’, ‘)’, ‘*’]
print(lst)
lst.extend(lst_b)
print(lst)
lst.count(‘*’)
print(lst)
lst.index(‘e’)
print(lst)
lst.insert(2, [1000, 1200, 1500])
print(lst)
lst.pop(7)
print(lst)
lst.remove(‘*’)
print(lst)
lst.reverse()
print(lst)
lst.sort()  # 将提示错误，TypeError: ‘<’ not supported between instances of ‘int’ and ‘str’
```

运算结果：

```
['d', 'e', 'f', 'g', 'h']
['d', 'e', 'f', 'g', 'h', 99]
['d', 'e', 'f', 'g', 'h', 99, [50, 55, 60, 65, 70, 75]]
['d', 'e', 'f', 'g', 'h', 99, [50, 55, 60, 65, 70, 75]]
['d', 'e', 'f', 'g', 'h', 99, [50, 55, 60, 65, 70, 75], '*', ')', '*']
['d', 'e', 'f', 'g', 'h', 99, [50, 55, 60, 65, 70, 75], '*', ')', '*']
['d', 'e', 'f', 'g', 'h', 99, [50, 55, 60, 65, 70, 75], '*', ')', '*']
['d', 'e', [1000, 1200, 1500], 'f', 'g', 'h', 99, [50, 55, 60, 65, 70, 75], '*',
')', '*']
['d', 'e', [1000, 1200, 1500], 'f', 'g', 'h', 99, '*', ')', '*']
['d', 'e', [1000, 1200, 1500], 'f', 'g', 'h', 99, ')', '*']
['*', ')', 99, 'h', 'g', 'f', [1000, 1200, 1500], 'e', 'd']
---------------------------------------------------------------------------
TypeError                                 Traceback (most recent call last)
Input In [141], in <cell line: 23>()
     21 lst.reverse()
     22 print(lst)
---> 23 lst.sort()
TypeError: '<' not supported between instances of 'int' and 'str'
```

15

代码段：

```
print(lst)
r_lst = random.sample(range(10), 6)
print(r_lst)
r_lst.sort()
print(r_lst)
lst_copy = r_lst.copy()
print(lst_copy)
lst_copy.clear()
print(lst_copy)
```

运算结果：

```
['*', ')', 99, 'h', 'g', 'f', [1000, 1200, 1500], 'e', 'd']
[7, 2, 4, 5, 8, 1]
[1, 2, 4, 5, 7, 8]
[1, 2, 4, 5, 7, 8]
[]
```

16

代码段：

```
help(list.extend)
```

运算结果：

```
Help on method_descriptor:
extend(self, iterable, /)
    Extend list by appending elements from the iterable.
```

知识点：

2.3 元组（Tuple）

17

描述：

元组的语法为(v0,v1,v2,...,vn)，不同于列表的中括号，为小括号表示，中间逗号隔开。元组不能够修改数据是其主要特性。元组的建立方法一种是2,5,6,在成员对象末尾直接加一个逗号；或则使用tuple(iterable=(), /)函数，参数为可迭代对象；亦可直接敲入元组语法(2,5,6)。元组含有两个方法，一个是count()，用于统计给定项值的数量；另一个是index()，用于返回给定项值的第一个出现项值的索引值。内嵌函数除构建函数外，与列表相同。

注：对于代码，所有语法中所涉及符号，均为英文格式。

代码段：

```
tup = 2, 5, 6,
print(tup)
print(type(tup))
print(type((2, 5, 6)))
print(3*(20*3))
print(3*(20*3,))
tup_0 = tuple([5, 8, 9])
print(tup_0)
tup_1 = tuple((5, 8, 9))
print(tup_1)
tup_0N1 = tup_0+tup_1
print(tup_0N1)
print(tup_0N1.count(5))
print(tup_0N1.index(8))

#尝试修改项值时，将提示错误，TypeError: 'tuple' object does not support item assignment
tup_0N1[2] = 9999
```

运算结果：

```
(2, 5, 6)
<class 'tuple'>
<class 'tuple'>
180
(60, 60, 60)
(5, 8, 9)
(5, 8, 9)
(5, 8, 9, 5, 8, 9)
2
1
---------------------------------------------------------------------------
TypeError                                 Traceback (most recent call last)
Input In [166], in <cell line: 20>()
     17 print(tup_0N1.count(5))
     18 print(tup_0N1.index(8))
---> 20 tup_0N1[2]=9999
TypeError: 'tuple' object does not support item assignment
```

知识点：

2.4 字典（Dictionary）

18

描述：

字典语法为{k0:v0,k1:v1,k2:v2,...,kn:vn}。其中，k为键，v为值，键值对之间：冒号分割，大括号括起。键的类型可以为数值型或者字符串类型。字典是无序的，键值对的位置可能会发生变动。

字典键的存在，相当于数据库中用于检索的字段名，方便数据组织、查询、管理，避免像列表在存储数据时，需要根据索引值定位项值，而索引值是不具有自定义含义的默认整数。解释列表应用数据的途径时，使用了实际研究项目的数据，将其转换为三种以字典形式存储数据的方式。第一种，AoT_info_dict_1存储方式，以'id','lon','lat','address'为键，可以快速地提取对应的列表形式的值；第二种，AoT_info_dict_2存储方式，以node_id列表的'ID'值为键，以对应'ID'值的经纬度和地址组成的元组为值，方便通过'ID'值查询对应信息；第三种，AoT_info_dict_3存储方式，是嵌套字典，第一层仍旧以'ID'值为键，嵌套的内层字典则以'lon','lat'和'address'为键。这样处理的好处是，可以根据具有含义的键方便搜索具体细分的值，而不用根据列表或者元组无意义的索引值进行定位。

通过4个列表构建上述三类字典存储形式时，采用了不同的方式，这包括dict()函数构建字典；同时，配合使用了zip(*iterables)函数。zip()函数是将多个序列对象，按照索引值一一对应组合，如下述帮助文件提供的案例。对于嵌套字典的构建方式，是列表推导式（comprehensions）的方法，这将在基本语句部分深入解释。

对于复杂的代码语句，如果不能直观地看出具体的逻辑，可以将其拆分为部分，逐部分打印查看数据及数据的变化，从而理解代码所要解决的逻辑。例如，案例中拆分AoT_info_dict_2

构建代码。

代码段：

```
print(list(zip('abcdefg', range(3), range(4)))) # zip() 帮助文件提供的案例
```

运算结果：

```
[('a', 0, 0), ('b', 1, 1), ('c', 2, 2)]
```

19

代码段：

```
coordi_lon = [-87.628, -87.616, -87.631, -87.59,
              -87.711, -87.628, -87.586, -87.713, -87.676, -87.624]
coordi_lat = [41.878, 41.858, 41.926, 41.81,
              41.866, 41.883, 41.781, 41.751, 41.852, 41.736]
node_id = ['ba46', 'ba3b', 'f02f', 'ba8f', 'ba16',
           '7e5d', 'ba8b', 'ba13', 'ba18', 'bc10']
address = ['State St & Jackson Blvd Chicago IL', '18th St & Lake Shore Dr
Chicago IL', 'Lake Shore Drive & Fullerton Ave Chicago IL', 'Cornell & 47th St
Chicago IL', 'Homan Ave & Roosevelt Rd Chicago IL',
           'State St & Washington St Chicago IL', 'Stony Island Ave & 63rd St
Chicago IL', '7801 S Lawndale Ave Chicago IL', 'Damen Ave & Cermak Chicago IL',
'State St & 87th Chicago IL']
AoT_info_dict_1 = dict(zip(['id', 'lon', 'lat', 'address'], [
                      node_id, coordi_lon, coordi_lat, address]))
print(AoT_info_dict_1)
AoT_info_dict_2 = dict(zip(node_id, zip(coordi_lon, coordi_lat, address)))
print("_"*50)
print(AoT_info_dict_2)
AoT_info_dict_3 = {id: {'lon': lon, 'lat': lat, 'address': addr}
                   for id, lon, lat, addr in zip(node_id, coordi_lon, coordi_
lat, address)}
print("_"*50)
print(AoT_info_dict_3)
```

运算结果：

```
{'id': ['ba46', 'ba3b', 'f02f', 'ba8f', 'ba16', '7e5d', 'ba8b', 'ba13', 'ba18',
'bc10'], 'lon': [-87.628, -87.616, -87.631, -87.59, -87.711, -87.628, -87.586,
-87.713, -87.676, -87.624], 'lat': [41.878, 41.858, 41.926, 41.81, 41.866,
41.883, 41.781, 41.751, 41.852, 41.736], 'address': ['State St & Jackson Blvd
Chicago IL', '18th St & Lake Shore Dr Chicago IL', 'Lake Shore Drive & Fullerton
Ave Chicago IL', 'Cornell & 47th St Chicago IL', 'Homan Ave & Roosevelt Rd
Chicago IL', 'State St & Washington St Chicago IL', 'Stony Island Ave & 63rd St
Chicago IL', '7801 S Lawndale Ave Chicago IL', 'Damen Ave & Cermak Chicago IL',
'State St & 87th Chicago IL']}
__________________________________________________
{'ba46': (-87.628, 41.878, 'State St & Jackson Blvd Chicago IL'), 'ba3b':
(-87.616, 41.858, '18th St & Lake Shore Dr Chicago IL'), 'f02f': (-87.631,
41.926, 'Lake Shore Drive & Fullerton Ave Chicago IL'), 'ba8f': (-87.59, 41.81,
'Cornell & 47th St Chicago IL'), 'ba16': (-87.711, 41.866, 'Homan Ave & Roosevelt
Rd Chicago IL'), '7e5d': (-87.628, 41.883, 'State St & Washington St Chicago
IL'), 'ba8b': (-87.586, 41.781, 'Stony Island Ave & 63rd St Chicago IL'), 'ba13':
(-87.713, 41.751, '7801 S Lawndale Ave Chicago IL'), 'ba18': (-87.676, 41.852,
'Damen Ave & Cermak Chicago IL'), 'bc10': (-87.624, 41.736, 'State St & 87th
Chicago IL')}

__________________________________________________
{'ba46': {'lon': -87.628, 'lat': 41.878, 'address': 'State St & Jackson Blvd
Chicago IL'}, 'ba3b': {'lon': -87.616, 'lat': 41.858, 'address': '18th St &
Lake Shore Dr Chicago IL'}, 'f02f': {'lon': -87.631, 'lat': 41.926, 'address':
'Lake Shore Drive & Fullerton Ave Chicago IL'}, 'ba8f': {'lon': -87.59, 'lat':
41.81, 'address': 'Cornell & 47th St Chicago IL'}, 'ba16': {'lon': -87.711,
'lat': 41.866, 'address': 'Homan Ave & Roosevelt Rd Chicago IL'}, '7e5d': {'lon':
-87.628, 'lat': 41.883, 'address': 'State St & Washington St Chicago IL'}, 'ba8b':
{'lon': -87.586, 'lat': 41.781, 'address': 'Stony Island Ave & 63rd St Chicago
IL'}, 'ba13': {'lon': -87.713, 'lat': 41.751, 'address': '7801 S Lawndale Ave
Chicago IL'}, 'ba18': {'lon': -87.676, 'lat': 41.852, 'address': 'Damen Ave &
Cermak Chicago IL'}, 'bc10': {'lon': -87.624, 'lat': 41.736, 'address': 'State
St & 87th Chicago IL'}}
```

20

代码段：

```
print(‘node_id={}’.format(AoT_info_dict_1[‘id’]))
ba8f_info_from2 = AoT_info_dict_2[‘ba8f’]
print(“_”*50)
print(ba8f_info_from2)
print(‘ba8f_lat={},lon={}’.format(ba8f_info_from2[1], ba8f_info_from2[0]))
ba8f_info_from3 = AoT_info_dict_3[‘ba8f’]
print(“_”*50)
print(ba8f_info_from3)
print(‘ba8f_lat={},lon={}’.format(
    ba8f_info_from3[‘lat’], ba8f_info_from3[‘lon’]))
```

运算结果：

```
node_id=[‘ba46’, ‘ba3b’, ‘f02f’, ‘ba8f’, ‘ba16’, ‘7e5d’, ‘ba8b’, ‘ba13’, ‘ba18’, ‘bc10’]
__________________________________________________
(-87.59, 41.81, ‘Cornell & 47th St Chicago IL’)
ba8f_lat=41.81,lon=-87.59
__________________________________________________
{‘lon’: -87.59, ‘lat’: 41.81, ‘address’: ‘Cornell & 47th St Chicago IL’}
ba8f_lat=41.81,lon=-87.59
```

21

描述：

拆分代码，易于查看内在数据组织方式。这里需要注意，对于enumerate()、zip()、map()等函数，返回值为迭代器（iterator）。迭代器的特点是只能前进，而不能后退。当遍历所有值后，引发StopIteration异常，变量值为空。具体解释查看迭代器部分。这里，为了避免list(iterator)读取迭代器值后为空，使用了itertools库提供的tee()方法，将迭代器复制为独立的两份，一份用于打印查看数据；另一份用于后续的输入参数。

代码段：

```
from itertools import tee
AoT_info_dict_2 = dict(zip(node_id, zip(coordi_lon, coordi_lat, address)))
print(AoT_info_dict_2)
coordi_address = zip(coordi_lon, coordi_lat, address)
coordi_address_first, coordi_address_second = tee(coordi_address)
print(“_”*50)
print(list(coordi_address_first))
id_coordi_address = zip(node_id, coordi_address_second)
print(“_”*50)
id_coordi_address_first, id_coordi_address_second = tee(id_coordi_address)
print(list(id_coordi_address_first))
AoT_info_dict_2_split = dict(id_coordi_address_second)
print(“_”*50)
print(AoT_info_dict_2_split)
```

运算结果：

```
{‘ba46’: (-87.628, 41.878, ‘State St & Jackson Blvd Chicago IL’), ‘ba3b’: (-87.616,
41.858, ‘18th St & Lake Shore Dr Chicago IL’), ‘f02f’: (-87.631, 41.926, ‘Lake
Shore Drive & Fullerton Ave Chicago IL’), ‘ba8f’: (-87.59, 41.81, ‘Cornell & 47th
St Chicago IL’), ‘ba16’: (-87.711, 41.866, ‘Homan Ave & Roosevelt Rd Chicago
IL’), ‘7e5d’: (-87.628, 41.883, ‘State St & Washington St Chicago IL’), ‘ba8b’:
(-87.586, 41.781, ‘Stony Island Ave & 63rd St Chicago IL’), ‘ba13’: (-87.713,
41.751, ‘7801 S Lawndale Ave Chicago IL’), ‘ba18’: (-87.676, 41.852, ‘Damen Ave
& Cermak Chicago IL’), ‘bc10’: (-87.624, 41.736, ‘State St & 87th Chicago IL’)}
__________________________________________________
[(-87.628, 41.878, ‘State St & Jackson Blvd Chicago IL’), (-87.616, 41.858, ‘18th
St & Lake Shore Dr Chicago IL’), (-87.631, 41.926, ‘Lake Shore Drive & Fullerton
Ave Chicago IL’), (-87.59, 41.81, ‘Cornell & 47th St Chicago IL’), (-87.711,
41.866, ‘Homan Ave & Roosevelt Rd Chicago IL’), (-87.628, 41.883, ‘State St &
Washington St Chicago IL’), (-87.586, 41.781, ‘Stony Island Ave & 63rd St Chicago
IL’), (-87.713, 41.751, ‘7801 S Lawndale Ave Chicago IL’), (-87.676, 41.852, ‘Damen
Ave & Cermak Chicago IL’), (-87.624, 41.736, ‘State St & 87th Chicago IL’)]
__________________________________________________
[(‘ba46’, (-87.628, 41.878, ‘State St & Jackson Blvd Chicago IL’)), (‘ba3b’,
(-87.616, 41.858, ‘18th St & Lake Shore Dr Chicago IL’)), (‘f02f’, (-87.631,
41.926, ‘Lake Shore Drive & Fullerton Ave Chicago IL’)), (‘ba8f’, (-87.59,
41.81, ‘Cornell & 47th St Chicago IL’)), (‘ba16’, (-87.711, 41.866, ‘Homan Ave
& Roosevelt Rd Chicago IL’)), (‘7e5d’, (-87.628, 41.883, ‘State St & Washington
```

```
St Chicago IL')), ('ba8b', (-87.586, 41.781, 'Stony Island Ave & 63rd St Chicago IL')), ('ba13', (-87.713, 41.751, '7801 S Lawndale Ave Chicago IL')), ('ba18', (-87.676, 41.852, 'Damen Ave & Cermak Chicago IL')), ('bc10', (-87.624, 41.736, 'State St & 87th Chicago IL'))]
--------------------------------------------------------------------------------
{'ba46': (-87.628, 41.878, 'State St & Jackson Blvd Chicago IL'), 'ba3b': (-87.616, 41.858, '18th St & Lake Shore Dr Chicago IL'), 'f02f': (-87.631, 41.926, 'Lake Shore Drive & Fullerton Ave Chicago IL'), 'ba8f': (-87.59, 41.81, 'Cornell & 47th St Chicago IL'), 'ba16': (-87.711, 41.866, 'Homan Ave & Roosevelt Rd Chicago IL'), '7e5d': (-87.628, 41.883, 'State St & Washington St Chicago IL'), 'ba8b': (-87.586, 41.781, 'Stony Island Ave & 63rd St Chicago IL'), 'ba13': (-87.713, 41.751, '7801 S Lawndale Ave Chicago IL'), 'ba18': (-87.676, 41.852, 'Damen Ave & Cermak Chicago IL'), 'bc10': (-87.624, 41.736, 'State St & 87th Chicago IL')}
```

22

描述：

- 字典的基本操作

构建字典可以直接使用字典语法构建、dict()方式键值对构建外，还可以构建空的字典{}后更新字典，或者dict()方式下以赋值的方式给出键值对的形式。

random.sample()是从一个序列中随机抽取数值。random.random()方法是随机生成位于区间［0, 1）内的一个随机数；random.uniform(a, b)方法则是随机生成给定区间内的一个随机数，为浮点型，是否包含区间边界b，依赖于四舍五入的结果。

代码段：

```python
items_lst = [(0, [0, 1, 2, 3, 4]), ('key', [[100, 157, 150]]), (2, 'python')]
items_dict = dict(items_lst)  #按照键值对的方式建立字典
print(items_dict)
print(items_dict[0])  #根据键提取值
print(items_dict['key'])

items_dict[2] = 'C'  #指定键，替换新值
print(items_dict)
random.seed(10)
items_dict[3] = (random.random(), random.uniform(200, 300))
print(items_dict)

del items_dict[0]  #删除键值对，该类方法尽量避免使用
print(items_dict)

print('key' in items_dict)  #成员运算符
print('key' not in items_dict)  #成员运算符

print("_"*50)
CO_dict_1 = {}  #建立空字典后，更新键值对
CO_dict_1['name'] = 'CO'
CO_dict_1['concentration'] = 2.1
CO_dict_1['node_id'] = 'ba46'
print(CO_dict_1)

CO_dict_2 = dict(name='CO', concentration=2.1, node_id='ba46')  #以变量赋值的方式构建字典
print(CO_dict_2)

CO_dict_2['concentration'] += 1  #用赋值运算符自加的方式更新对应键的值
print(CO_dict_2)
```

运算结果：

```
{0: [0, 1, 2, 3, 4], 'key': [[100, 157, 150]], 2: 'python'}
[0, 1, 2, 3, 4]
[[100, 157, 150]]
{0: [0, 1, 2, 3, 4], 'key': [[100, 157, 150]], 2: 'C'}
{0: [0, 1, 2, 3, 4], 'key': [[100, 157, 150]], 2: 'C', 3: (0.5714025946899135, 242.88890546751145)}
{'key': [[100, 157, 150]], 2: 'C', 3: (0.5714025946899135, 242.88890546751145)}
True
False
--------------------------------------------------------------------------------
{'name': 'CO', 'concentration': 2.1, 'node_id': 'ba46'}
{'name': 'CO', 'concentration': 2.1, 'node_id': 'ba46'}
{'name': 'CO', 'concentration': 3.1, 'node_id': 'ba46'}
```

23

描述：

- 内置函数

len()函数，返回键值对的数量；

sorted()函数，返回排序后的键列表；

any()函数，如果字典中存在任何对象则返回True，如果为空字典{}则返回False;

all()函数，为如果字典中的键非零或为空字典，则返回True；如果存在键为0，则返回False。

代码段：

```
n = 6
dict_A = dict(zip(random.sample(range(0, 10, 1), n),
              random.sample(range(100, 110, 1), n)))
print(dict_A)
print(len(dict_A))
print(sorted(dict_A))
print(any(dict_A))
print(any({}))
print("_"*50)
print(all(dict_A))
print(all({}))
dict_B = {5: 0, 1: 5}  # 键非 0，而值有 0 时
print(all(dict_B))
dict_C = {0: 7, 1: 5}  # 键有 0，而值非 0
print(all(dict_C))
```

运算结果：

```
{9: 102, 0: 100, 3: 107, 7: 109, 6: 108, 2: 101}
6
[0, 2, 3, 6, 7, 9]
True
False
__________________________________________________
False
True
True
False
```

24

描述：

- 内置方法

D.clear()，清除字典中所有的键/值项，但是这个方法属于原地操作，并不返回值。

D.copy()，可以复制字典，但是属于浅复制，当复制的字典已有键/值发生改变时，被复制的字典也会随之发生改变。如果增加新的键值对，则不会发生变化。

D.get()，可以根据指定的键返回值，但是如果指定的键在被访问的字典中没有，则不返回任何值。

D.items()，将所有的字典项即键/值对以列表方式返回。

D.keys()，返回字典的键对象列表。

D.values()，返回字典的值对象列表。

D.pop()，根据指定的键返回值，并同时移除字典中对应的键值对。

D.popitem()，随机弹出键值对，并同时移除字典中对应的键值对。

D.setdefault()，更新的键值对，其键不在原有字典键列表中，则增加该键值对，否则不增加。

D.update()，用一个字典更新另一字典，键值相同的项将被替换。

记忆字典方法时，最好是按照类似项分组记忆，例如，D.keys()+D.values()+D.items()，D.setdefaulte()+D.update()，D.pop()+D.popitems()等。

关于浅复制（shallow copy）和深复制（deep copy）及赋值变量

在数据分析时，以赋值的方式将一个变量的值传递给一个新的变量，通常用在类定义时值的初始化。一般情况下，通常一组值对应一个变量，避免变量混淆。如果要变化原始数据，通

常直接操作原始数据后赋予给一个新的变量，原始数据不会受到影响，因此不会用D.copy()、copy.copy()或copy.deepcopy()复制的方法。先复制后操作，这是完全不必要的。但是，在有些情况下，例如DataFrame数据结构，即使对原始数据操作赋予给新变量后，原始数据也会发生变化，而不希望改变原始数据时，则需要进行复制。复制时，可以用字典的内置方法D.copy()实现，也可以用copy中的方法库实现。复制包括浅复制和深复制，从下述代码示例中可以观察到，浅复制会在已有键值对变动下影响已复制对象，反之亦然；深复制则可以理解为复制对象和被复制对象是两个完全不相关的变量，即使值相同。

代码段：

```
import copy
lst_A = list(range(6, 20, 3))
lst_B = list(range(100, 150, 15))
d = {0: lst_A, 1: lst_B}
print(d)
d_copy = d
print(d_copy)
return_v = d.clear()
print(d, d_copy, return_v)  # 直接赋值时，原变量发生改变，被赋值的变量也会发生变化

d[5] = list(range(1, 9, 2))
print("_"*50)
print(d)
d_copy = d.copy()
print(d_copy)
d[8] = [5, 7]
print(d, '---', d_copy)  # 用 copy() 复制的方法，增加新键值对时，赋值的变量不会增加新键值对
d[5].append(9999)
print(d, '---', d_copy)  # 用 copy() 复制的方法，已有键值对发生改变时，赋值的变量已有的键值对
发生变化
d_copy[5].pop()
print(d, '---', d_copy)  # 复制对象已有键值对发生变化时，被复制对象也发生变化。
d_copy['new_key'] = 'new_value'
print(d, '---', d_copy)  # 复制对象增加新键值对后，被复制对象不会发生变化。

print('_'*50)
d_deepCopy = copy.deepcopy(d)
print(d, '---', d_deepCopy)
d[5][0] *= 100
print(d, '---', d_deepCopy)  # 使用 deepcopy() 方法，被复制对象不会受到原始字典变动的影响
```

运算结果：

```
{0: [6, 9, 12, 15, 18], 1: [100, 115, 130, 145]}
{0: [6, 9, 12, 15, 18], 1: [100, 115, 130, 145]}
{} {} None
--------------------------------------------------
{5: [1, 3, 5, 7]}
{5: [1, 3, 5, 7]}
{5: [1, 3, 5, 7], 8: [5, 7]} --- {5: [1, 3, 5, 7]}
{5: [1, 3, 5, 7, 9999], 8: [5, 7]} --- {5: [1, 3, 5, 7, 9999]}
{5: [1, 3, 5, 7], 8: [5, 7]} --- {5: [1, 3, 5, 7]}
{5: [1, 3, 5, 7], 8: [5, 7]} --- {5: [1, 3, 5, 7], 'new_key': 'new_value'}
--------------------------------------------------
{5: [1, 3, 5, 7], 8: [5, 7]} --- {5: [1, 3, 5, 7], 8: [5, 7]}
{5: [100, 3, 5, 7], 8: [5, 7]} --- {5: [1, 3, 5, 7], 8: [5, 7]}
```

25

代码段：

```
random.seed(10)
D = dict(zip(random.sample(range(0, 10, 1), 7),
         random.sample(range(100, 110, 1), 7)))
print(D)
print(D.get(9))
print(D.items())
print(D.keys())
print(D.values())
print(D.pop(0))
print(D)
print(D.popitem())
print(D)
```

运算结果：

```
{9: 107, 0: 109, 6: 104, 3: 105, 4: 101, 8: 100, 1: 103}
107
dict_items([(9, 107), (0, 109), (6, 104), (3, 105), (4, 101), (8, 100), (1, 103)])
dict_keys([9, 0, 6, 3, 4, 8, 1])
dict_values([107, 109, 104, 105, 101, 100, 103])
109
{9: 107, 6: 104, 3: 105, 4: 101, 8: 100, 1: 103}
(1, 103)
{9: 107, 6: 104, 3: 105, 4: 101, 8: 100}
```

26

描述：

● 更新字典的方法——D.setdefault()和D.update()

定义函数时，有时会预先以字典的形式配置一些参数值传入参数，作为默认值。但是，会根据具体情况调整某些配置参数，通常会使用D.update()方法更新字典，即以键为参数名，以键对应的值为参数值。如果需要增加新的参数，则也可以使用D.setdefault()方法，可以避免对原有参数无意中的修改。

代码段：

```
random.seed(10)
print("_"*50)
D = dict(zip(random.sample(range(0, 10, 1), 7),
         random.sample(range(100, 110, 1), 7)))
print(D)
return_v = D.setdefault(22, [55, 66])
print(return_v)
print(D)
D.setdefault(22, 'update value')
print(D)
print("_"*50)
D.update({22: 'update value'})
print(D)
```

运算结果：

```
__________________________________________________
{9: 107, 0: 109, 6: 104, 3: 105, 4: 101, 8: 100, 1: 103} [55, 66] {9: 107, 0:
109, 6: 104, 3: 105, 4: 101, 8: 100, 1: 103, 22: [55, 66]} {9: 107, 0: 109, 6:
104, 3: 105, 4: 101, 8: 100, 1: 103, 22: [55, 66]}
__________________________________________________
{9: 107, 0: 109, 6: 104, 3: 105, 4: 101, 8: 100, 1: 103, 22: 'update value'}
```

27

描述：

这是一组实际研究的数据，在检索百度地图的兴趣点（point of interest，POI）数据时，需要指定query、page_size、scope等参数值。对于POI数据划分，有很多类，为poi_classificationName列表值。在每次检索数据时，query对象即为分类对象，只可以指定一个类。因此，为了避免重复工作，用字典更新的方法更新query参数。

注：这里用到了后续才会阐述的 for 循环语句。

代码段：

```
query_dic = {
    'query': '旅游景点',
    'page_size': '20',
    'scope': 2,
}

poi_classificationName = ["美食", "酒店", "购物", "生活服务", "丽人", "旅游景点",
"休闲娱乐"]
for i in poi_classificationName:
    query_dic.update({'query': i})
    print(query_dic)
```

运算结果：

```
{'query': '美食', 'page_size': '20', 'scope': 2}
```

```
{'query': '酒店', 'page_size': '20', 'scope': 2}
{'query': '购物', 'page_size': '20', 'scope': 2}
{'query': '生活服务', 'page_size': '20', 'scope': 2}
{'query': '丽人', 'page_size': '20', 'scope': 2}
{'query': '旅游景点', 'page_size': '20', 'scope': 2}
{'query': '休闲娱乐', 'page_size': '20', 'scope': 2}
```

知识点:

2.5 集合(Set)

28

描述:

集合最大的特点是集合中不含重复的元素，如果通过序列数据构建集合时存在重复元素，则构建的集合会只保留一个。集合不具有像列表索引值的属性，不能用索引值提取项值。集合元素的唯一性及集合间的运算，可以非常方便地处理一些元素具有唯一性变化的数据。例如，下面用了旅行商问题的数据，构建一个城市集合cities_set。这个数据用列表也可以，示例是说明直接构建集合的方式。tabu_cities_set集合包括所有城市，current_cities_set建立了一个空集合。当循环cities_set元素时，current_cities_set会通过add()方法增加每次循环的城市名，而tabu_cities_set集合则会移除每次循环的城市名。注意，这里-自减运算中，减去的是current_cities_set集合，也可以减去set([city])，即每次循环的一个城市；也可以用discard()方法移除。

代码段:

```
cities_set = {'Beijing', 'Chicago', 'Xian', 'San Francisco', 'San Diego'}
tabu_cities_set = set(['Beijing', 'Chicago', 'Xian',
                       'San Francisco', 'San Diego'])
current_cities_set = set()
print(cities_set)
print(tabu_cities_set)
print(current_cities_lst)
for i, city in enumerate(cities_set):
    current_cities_set.update([city])
    tabu_cities_set -= set(current_cities_set)
    print("_"*50, i)
    print('current_cities:{};\ntabu_citeis:{}'.format(
        current_cities_set, tabu_cities_set))
```

运算结果:

```
{'Beijing', 'Chicago', 'San Francisco', 'San Diego', 'Xian'}
{'Beijing', 'Chicago', 'San Francisco', 'San Diego', 'Xian'}
['Beijing', 'Chicago', 'San Francisco', 'San Diego', 'Xian']
__________________________________________________ 0
current_cities:{'Beijing'};
tabu_citeis:{'Chicago', 'San Francisco', 'San Diego', 'Xian'}
__________________________________________________ 1
current_cities:{'Beijing', 'Chicago'};
tabu_citeis:{'San Francisco', 'San Diego', 'Xian'}
__________________________________________________ 2
current_cities:{'Beijing', 'Chicago', 'San Francisco'};
tabu_citeis:{'San Diego', 'Xian'}
__________________________________________________ 3
current_cities:{'Beijing', 'Chicago', 'San Diego', 'San Francisco'};
tabu_citeis:{'Xian'}
__________________________________________________ 4
current_cities:{'Beijing', 'Chicago', 'San Francisco', 'San Diego', 'Xian'};
tabu_citeis:set()
```

29

描述:

- 集合的方法

S.add()，向集合中增加元素；

S.clear()，移除所有元素，成空集合；

S.copy()，复制集合；

S.pop()，返回第一个元素，同时原集合移除该元素；如果为空集合，则引发KeyError异常；

S.remove()，移除一个集合元素，如果输入参数值不在集合中，则引发KeyError异常；

S.discard()，移除一个集合元素，如果输入参数值不在集合中，则无变化，也并不提示错误；可以比较S.remove()方法；

S.update()，更新集合，输入参数为序列或集合；如果存在相同元素，则保持不变。

代码段：

```
# 不能包含嵌套列表，否则提示错误，TypeError: unhashable type: 'list'
lst4set = [1, 1, 1, 2, 2, 6, 33, 8, 9, 55, 0, 'string', (33, 55, 99)]
print(lst4set)
set_1 = set(lst4set)
print(set_1)

empty_set = set()  # empty_dict={} 是空字典建立的方法
print(empty_set)
print('_'*50)
set_1.add('new value')
print(set_1)
set_1.update([1, 2, 'update'])
print(set_1)  # 更新时，重复的元素时保持不变的，新元素被增加到集合中
set_1.update()
set_1.update([1, 2, 'lst_v'], {1, 32, 56, 78})
print(set_1)  # 更新时，可以传入列表，也可以传入集合
set_1.remove((33, 55, 99))
print(set_1)
set_1.remove('new value')
print(set_1)
set_1.discard(1)
print(set_1)
# 如果使用 set_1.remove(33333)，则提示错误为 KeyError: 33333。discard 方法则忽略
set_1.discard(33333)
print(set_1)
pop_v = set_1.pop()  # 返回第一个元素，并从集合中移除
print(pop_v)
print(set_1)
set_1.clear()
print(set_1)
```

运算结果：

```
[1, 1, 1, 2, 2, 6, 33, 8, 9, 55, 0, 'string', (33, 55, 99)]
{0, 1, 2, 33, 6, 8, 9, 'string', 55, (33, 55, 99)}
set()
__________________________________________________________________________
{0, 1, 2, 33, 6, 'new value', 8, 9, 'string', 55, (33, 55, 99)}
{0, 1, 2, 33, 'update', 6, 'new value', 8, 9, 'string', 55, (33, 55, 99)}
{0, 1, 2, 33, 'lst_v', 'update', 6, 'new value', 8, 9, 32, 78, 'string', 55, 56,
(33, 55, 99)}
{0, 1, 2, 33, 'lst_v', 'update', 6, 'new value', 8, 9, 32, 78, 'string', 55, 56}
{0, 1, 2, 33, 'lst_v', 'update', 6, 8, 9, 32, 78, 'string', 55, 56}
{0, 2, 33, 'lst_v', 'update', 6, 8, 9, 32, 78, 'string', 55, 56}
{0, 2, 33, 'lst_v', 'update', 6, 8, 9, 32, 78, 'string', 55, 56}
0
{2, 33, 'lst_v', 'update', 6, 8, 9, 32, 78, 'string', 55, 56}
set()
```

30

描述：

S1.uion(S2)，两个集合的并集运算，同符号|；

S1.intersection(S2)，两个集合的交集运算，同符号&；

S1.difference(S2)，两个（多个）集合的差集运算，同符号-；

S1.symmetric_difference(S2)，返回两个集合中不重复的元素为一个新集合，同符号^；

S1.intersection_update(S2)，同S1.intersection(S2)，只是对S1本地更新；

S1.differnce_update(S2)，同S1.difference(S2)，只是对S1本地更新；

S1.symmetric_difference_update(S2)，同S1.symmetric_difference(S2)，只是对S1本地更新；

S.isdisjoint()，判断两个集合是否包括相同的元素，如果包含返回False，不包含返回True；

S.issubset()，判断一个数据集是否是另一个的子集，同符号<=；

S.issuperset()，判断一个数据集是否是另一个的超集（父集），同符号>=。

代码段：

```
S_1 = set([1, 2, 3, 4, 5])
S_2 = set([4, 5, 6, 7, 8])
S_1_copy1, S_1_copy2, S_1_copy3 = [S_1.copy() for i in range(3)]
S_2_copy1, S_2_copy2, S_2_copy3 = [S_2.copy() for i in range(3)]
print(S_1)
print(S_2)
print(“_”*50)

#并集
print(S_1 | S_2)
print(S_1.union(S_2))
print(‘_’*50)

#交集
print(S_1 & S_2)
print(S_1.intersection(S_2))
print(‘_’*50)

#差集
print(S_1 - S_2)
print(S_1.difference(S_2))
print(S_2-S_1)
print(S_2.difference(S_1))
print(‘_’*50)

#symmetric 对称差集
print(S_1 ^ S_2)
print(S_1.symmetric_difference(S_2))
print(‘_’*50)
print(“#”*50)
S_1_copy1.intersection_update(S_2_copy1)
print(S_1_copy1)
S_1_copy2.difference_update(S_2_copy2)
print(S_1_copy2)
S_1_copy3.symmetric_difference_update(S_2_copy3)
print(S_1_copy3)
print(“_”*50)
print(S_1.isdisjoint(S_2))
print(set([1, 2]).issubset(S_1))
print(S_1.issuperset(set([1, 2])))
```

31

描述：

- frozenset()——不可变（immutable）集合

列表为可变（mutable）序列对象，元组为不可变（immutable）序列对象。对于集合而言，可以用frozenset( )构建不可变集合，支持FS.copy()、FS1.difference(FS2)、FS1.intersection(FS2)、FS1.isdisjoint(FS2)、FS1.issubset(FS2)、FS1.issuperset(FS2)、FS1.symmetric_difference(FS2)、FS1.union(FS2)等方法，但是不支持remove()、discard()等对集合元素产生变动的方法。

代码段：

```
FS_1 = frozenset([1, 2, 3, 4, 5])
FS_2 = frozenset([4, 5, 6, 7, 8])
print(FS_1.isdisjoint(FS_2))
print(FS_1.difference(FS_2))
print(FS_1 | FS_2)
```

运算结果：

```
False
frozenset({1, 2, 3})
frozenset({1, 2, 3, 4, 5, 6, 7, 8})
```

PCS-3. 数据结构（Data Structure）-string

PCS-3（　）：3.1 字符串（String）与文本文件；3.2 文件的打开与读写——用open( )_python内置函数（built-in functions）；3.3 常用字符串操作方法；3.4 字符串格式化；3.5 正则表达式（regular expression，re）

知识点：

3.1 字符串（String）与文本文件

1

描述：

数据处理时，必然会遇到包含字符串的文件，或者数值以文本方式存储，读取后再转换为数值型。不管是从文本中提取数据，还是图表中的文字表述，及对文本内容的分析，都需要知道如何处理字符串。字符串处理的方法途径异常繁多，各类模式匹配符号组合表述技巧性较强。除了常规用到的处理方式外，不经常用到或从未用过的方式则很难记住，因此字符串处理部分以查阅为主，当遇到要处理的字符串时，可以根据要处理要求从PCS-3或相关文件，尤其搜索引擎中找到处理方法的答案。当然，对于经常用到的字符串处理方法，需要有意识的练习记忆，避免每次重复搜索。

注：互联网是对字符串处理最为频繁的领域，城市空间数据分析和数字化设计领域相对较少。

字符串处理除了Python内置方法外，主要使用正则表达式（Regular expression operations，re）[①]。

很多时，用文本文件存储数据，因文本存储的格式不同会表述为不同的文件格式，例如没有格式限制的TXT文件（通常按行记录数据，逗号、分号或空格分隔字段），后缀名为.txt；逗号分隔的CSV（comma-separated values）[②]文件格式（每行为一组数据，逗号隔离字段），后缀名为.csv；JSON（JavaScript Object Notation）[③]，开放的标准文件格式和数据交换格式，以属性-值对和数组的形式记录，后缀名为.json。存储数据的方式很多，也可以自定义存储格式和后缀名，不过以常规标准的格式存储数据方便数据交换，因为常用的格式通常已有大量写好的读写代码，例如`pandas.read_csv()`、`pandas.DataFrame.to_csv`等[④]；又或者CSV库[⑤]，通过`import csv`调入库读写方法等，这都极大方便地增加了书写代码的效率，同时也尽量避免了读写错误。

注：快速查看文本文件（也常用于查看代码）推荐使用 notepad++[⑥] 工具。

① re, Regular expression operations（正则表达式），用于匹配和查找字符串（https://docs.python.org/3/library/re.html）。

② CSV, comma-separated values。是一个以逗号分隔的文本文件（https://en.wikipedia.org/wiki/Comma-separated_values）。

③ JSON, JavaScript Object Notation，是一种开放的标准文件格式和数据交换格式，使用人类可读的文本来存储和传输由属性-值对和数组（或其他可序列化的值）组成的数据对象。它是一种常见的数据格式，在电子数据交换中具有多种用途，包括网络应用程序与服务器的数据交换（https://en.wikipedia.org/wiki/JSON）。

④ pandas.read_csv( ),（https://pandas.pydata.org/docs/reference/api/pandas.read_csv.html）；pandas.DataFrame.to_csv,（https://pandas.pydata.org/docs/reference/api/pandas.DataFrame.to_csv.html）。

⑤ CSV库，实现了读取和写入CSV格式的表格数据的类（https://docs.python.org/3/library/csv.html）。

⑥ notepad++，是一个自由的源代码编辑器和记事本的替代品，支持多种语言（https://notepad-plus-plus.org/）。

知识点：

3.2 文件的打开与读写——用open( )_python内置函数（built-in functions）

2

描述：

`open(file, mode='r', buffering=-1, encoding=None, errors=None, newline=None, closefd=True, opener=None)`，在`open()`函数的参数配置上通常只使用3个参数——`file`、`mode`和`encoding`。其中，`file`为文件存储位置路径；`mode`为打开文件的模式，包括基本模式和组合模式。基本模式的字符含义如表pcs.3-1所示。

表pcs.3-1　open( )参数基本模式的字符含义

字符（character）	含义（meaning）
'r'	以只读方式打开文件，读取文件内容的指针位于文件的开始。为默认模式
'w'	以只写模式打开文件，如果文件存在，则会清空文件中已有内容；如果文件不存在，则创建新文件
'x'	建立一个新文件，并以写模式打开。如果文件存在，则报错
'a'	写模式，如果文件中存在内容，则在其后追加新内容

续表

字符（character）	含义（meaning）
'b'	二进制模式，需要' r'、' w'、' a' 等字符模式配合使用。为以二进制格式，读写文件，通常用于非文本文件，例如影音图像，或以二进制存储的数据等
't'	文本模式，为默认。如果要以二进制读写，加符号' b'
'+'	打开一个文件进行更新，可读、可写

在基本模式之上，可以组合为'rb'、'r+'、'rb+'、'wb'、'w+'、'wb+'、'ab'、'a+'、'ab+'等多种组合模式，组合后含义为单独字符模式含义的组合。

如果不配置encoding参数，文件打开时报错（经常出现在含有中文字符的文件中），则需要指定该参数值，为打开该文件所使用的编码格式。编码格式通常配置为utf-8（有时也写为utf8）。对于utf-8无法识别含有中文的文件，通常可以尝试配置为GBK、Chinese Internal Code Specification（汉字内码扩展规范）。

f=open()打开文件的返回值是一个文件对象句柄（help(open)给出的解释是 Open file and return a stream），并将其赋给自定义变量f，通过句柄（该变量）对文件进行操作。此时，f包含操作文件内容的多个属性和方法：

- f属性

f.name，返回文件名称；

f.mode，返回文件打开时的文件打开模式；

f.encoding，返回文件打开时使用的编码格式；

f.closed，判断文件是否已经关闭。

这里，例举了实际研究项目的数据为从百度地图应用中检索下载的POI。其中，第一行为1,0101000020897F000008D599CB26E312418530CECF16EB4C41,美香源,34.23709808337344,108.93100212046282,美食;中餐厅,0,,309449.6988290106,3790381.623479905，按逗号分割，各个字段名为序号，ID，名称，维度，经度，一级行业分类；二级行业分类，评分，均价，地理坐标投影后y值，地理坐标投影后X值（这是经过处理了的数据，并非下载的原数据）

代码段：

```
xian_poi_fn = './data/xian_poi.csv'   # 存储有西安 POI, point of interesting 兴趣点数据
xian_poi_f = open(xian_poi_fn, 'r', encoding="utf-8")  # 以只读方式打开文件
print(xian_poi_f)
print('_'*50)
print(xian_poi_f.name)
print(xian_poi_f.mode)
print(xian_poi_f.encoding)
print(xian_poi_f.closed)
```

运算结果：

```
<_io.TextIOWrapper name='./data/xian_poi.csv' mode='r' encoding='utf-8'>
--------------------------------------------------------------------------
./data/xian_poi.csv
r
utf-8
False
```

3

描述：

- f 读取方法

f.read(size=-1, /)，参数size未指定或为负值时返回整个文件，否则到指定字符长度位置或到文件末尾（EOF，end of file）；

f.readline(size=-1, /)，读取一行，即读到换行符或者EOF。如果给定size，则按长度读取；

f.readlines()，返回所有行的一个列表；

f.tell()，返回当前读取文件的位置；

f.clost()，关闭文件。

注：类定义时，如果以f.attribute方式返回值，则为属性，例如f.name；如果以f.function()方式，即调用类的方法，例如f.read()。

代码段：

```
print(xian_poi_f.read(57))
print("_"*10, xian_poi_f.tell())
print(xian_poi_f.readline())  #从上一语句 57 处继续读，读到该行结束
print("_"*10, xian_poi_f.tell())
print(xian_poi_f.readline())  #继续读写一行文本内容
print('_'*50)
print(xian_poi_f.readlines()[:5])  #这里只打印了返回列表的前 5 行
print("_"*10, xian_poi_f.tell())
print('_'*50)
xian_poi_f.close()
print(xian_poi_f.closed)
```

运算结果：

```
1,0101000020897F000008D599CB26E312418530CECF16EB4C41,美香源,
__________ 63
34.23709808337344,108.93100212046282,美食;中餐厅,0,,309449.6988290106,3790381.623
479905
__________ 156
2,0101000020897F0000B038F4CF9BE31241389AD606BFEC4C41,雷记澄城水盆羊肉(红樱路店),34.
244750060429915,108.93113243785623,美食;中餐厅,3.9,20.0,309478.95308006834,379123
0.053424146
__________________________________________________________________________
['3,0101000020897F00005C24B17F97E3124160F86A81E5EC4C41,段府农家菠菜面(红缨路店),34.
245443456204875,108.93110375582032,美食;中餐厅,4,,309477.8746991807,3791307.0110
76972\n', '4,0101000020897F000040F7FD9433E412412CD771E043EC4C41,平价餐厅(友谊西路
店),34.24253719175486,108.93159854262964,美食;中餐厅,4.2,26.0,309516.8955000527,3
790983.753474137\n', '5,0101000020897F0000EFFE229C56E61241ABEE21C2D7EC4C41,山妹川
菜,34.24522784875664,108.93301750888804,美食;中餐厅,3.5,22.0,309653.6524772485,3
791279.5166605315\n', '6,0101000020897F0000522963D2BFE312419E6ED4B833EC4C41,老三
澄合羊肉水盆,34.24224069599151,108.93129159808142,美食;中餐厅,3,,309487.9554563957
5,3790951.443982913\n', '7,0101000020897F00008C6ACABF13E412416B0C7D7B09ED4C41,湘
村菜馆(红缨路店),34.24609764121232,108.9314250012306,美食;中餐厅,4.2,20.0,309508.9
37295594,3791378.9647536776\n']
__________ 2470810
__________________________________________________________________________
True
```

4

描述：

- f写入方法

f.write()，将字符串写入到文本，如果是数值等数据，需要将其转换为字符串后再写入；

f.writelines()，将字符串列表逐行写入到文件；

f.flush()，将数据刷至硬盘；通常在f.close()文件关闭时，会自动一次性刷至硬盘，除非特殊需求，否则不用执行f.flush()；

seek(cookie, whence=0, /)，更改当前读写位置，为字节偏移量（byte offset）；whence为0时（默认值），代表从文件开始定位算起；为1时，以当前位置定位算起；为2时，以文件末尾定位算起。

在下述的示例中，poi_1PieceOFdata变量只存储了一行数据；而poi_2PiecesOFdata变量存储了两行数据，行之间用换行符\n完成换行动作。poi_piecesOFdata.flush()会将先写入的一行数据刷至硬盘文件中，因为使用了w+模式，因此可以用poi_piecesOFdata.flush()方法定位到文本开始，再用poi_piecesOFdata.read()方法查看，否则返回内容为空。也可以用外部notepad++等工具，打开查看内容。而后，将包含两行数据的poi_2PiecesOFdata变量写入，并调用poi_piecesOFdata.close()方法，将后续写入的数据刷至硬盘文件中。

代码段：

```
poi_1PieceOFdata = '2,0101000020897F0000B038F4CF9BE31241389AD606BFEC4C41,雷记澄城
水盆羊肉(红樱路店),34.244750060429915,108.93113243785623,美食;中餐厅,3.9,20.0,3094
78.95308006834,3791230.05342414'
```

```
poi_2PiecesOFdata = '\n3,0101000020897F00005C24B17F97E3124160F86A81E5EC
4C41,段府农家菠菜面(红缨路店),34.245443456204875,108.93110375582032,美食;中餐
厅,4,,309477.8746991807,3791307.011076972,\n4,0101000020897F000040F7FD9433E41241
2CD771E043EC4C41,平价餐厅(友谊西路店),34.24253719175486,108.93159854262964,美食;
中餐厅,4.2,26.0,309516.8955000527,3790983.753474137'
poi_piecesOFdata_fn = './data/poi_piecesOFdata.csv'
poi_piecesOFdata = open(poi_piecesOFdata_fn, 'w+', encoding='utf-8')
poi_piecesOFdata.write(poi_1PieceOFdata)
poi_piecesOFdata.flush()
```

5

描述：

注意，写入文本内容后，读写位置位于文件末尾，不通过f.seek()指定开始位置，读取的内容会为空。

代码段：

```
poi_piecesOFdata.seek(0)
print(poi_piecesOFdata.read())
```

运算结果：

```
2,0101000020897F0000B038F4CF9BE31241389AD606BFEC4C41,雷记澄城水盆羊肉(红樱路店),34.
244750060429915,108.93113243785623,美食;中餐厅,3.9,20.0,309478.95308006834,379123
0.05342414
```

6

描述：

不管读或者写，当完成读写动作后，需要调用f.close()关闭文件。

代码段：

```
poi_piecesOFdata.write(poi_2PiecesOFdata)

poi_listOFdata = ['5,0101000020897F0000EFFE229C56E61241ABEE21C2D7EC4C41,山妹川菜,
34.24522784875664,108.93301750888804,美食;中餐厅,3.5,22.0,309653.6524772485,
3791279.5166605315\n','6,0101000020897F0000522963D2BFE312419E6ED4B833EC4C41,老三
澄合羊肉水盆,34.24224069599151,108.93129159808142,美食;中餐厅,3,,309487.95545639575,
3790951.443982913\n','7,0101000020897F00008C6ACABF13E412416B0C7D7B09ED4C41,湘村菜
馆(红缨路店),34.24609764121232,108.9314250012306,美食;中餐厅,4.2,20.0,309508.9372
95594,3791378.9647536776\n']
poi_piecesOFdata.write('\n')  #因为写入两行时，末尾为写入'\n'换行符。因此单独写入，避免
后续写入内容未起新行
poi_piecesOFdata.writelines(poi_listOFdata)
poi_piecesOFdata.close()
```

7

描述：

用with open(fn, mode) as f:上下文管理的方式打开文件，则不需要调用f.close()的方式关闭文件，也可以避免文件读写时可能产生的IOError。

这里将读取的CSV格式数据转换为字典格式，格式样式为{ID:{'name':name,'coordi':{'lat':lat,'lon':lon}}}。其中，有3层嵌套，同时将字符串格式的经纬度使用float()方法转换为浮点型。具体方法是应用字符串处理中的S.split()将字符串切分为字段列表后循环提取需要的数据内容。注意，这里提前应用了非常好用的匿名函数（lambda）及列表推导式（comprehensions）。

注：同样，可以将 poi_info_dict={S_split(S)[0]:{'name':S_split(S)[2],'coordi':{'lat':float(S_split(S)[3]),'lon':float(S_split(S)[4])}} for S in poi_lst}这个语句用 for 循环的方式拆分处理。

代码段：

```
poi_piecesOFdata_fn = './data/poi_piecesOFdata.csv'
with open(poi_piecesOFdata_fn, 'r', encoding='utf-8') as f:
    poi_lst = f.readlines()
print(poi_lst)
def S_split(S): return S.split(",")
poi_info_dict = {S_split(S)[0]: {'name': S_split(S)[2], 'coordi': {
    'lat': float(S_split(S)[3]), 'lon': float(S_split(S)[4])}} for S in poi_lst}
print("_"*50)
```

```
print(poi_info_dict)
```

运算结果：

```
['2,0101000020897F0000B038F4CF9BE31241389AD606BFEC4C41,雷记澄城水盆羊肉（红樱路店）,34.244750060429915,108.93113243785623,美食；中餐厅,3.9,20.0,309478.95308006834,3791230.05342414\n', '3,0101000020897F00005C24B17F97E3124160F86A81E5EC4C41,段府农家菠菜面（红缨路店）,34.245443456204875,108.93110375582032,美食；中餐厅,4,,309477.8746991807,3791307.011076972,\n', '4,0101000020897F000040F7FD9433E412412CD771E043EC4C41,平价餐厅（友谊西路店）,34.24253719175486,108.93159854262964,美食；中餐厅,4.2,26.0,309516.8955000527,3790983.753474137\n', '5,0101000020897F0000EFFE229C56E61241ABEE21C2D7EC4C41,山妹川菜,34.24522784875664,108.93301750888804,美食；中餐厅,3.5,22.0,309653.6524772485,3791279.5166605315\n', '6,0101000020897F0000522963D2BFE312419E6ED4B833EC4C41,老三澄合羊肉水盆,34.24224069599151,108.93129159808142,美食；中餐厅,3,,309487.95545639575,3790951.443982913\n', '7,0101000020897F00008C6ACABF13E412416B0C7D7B09ED4C41,湘村菜馆（红缨路店）,34.24609764121232,108.9314250012306,美食；中餐厅,4.2,20.0,309508.937295594,3791378.9647536776\n']
--------------------------------------------------------------------------------
{'2': {'name': '雷记澄城水盆羊肉（红樱路店）', 'coordi': {'lat': 34.244750060429915, 'lon': 108.93113243785623}}, '3': {'name': '段府农家菠菜面（红缨路店）', 'coordi': {'lat': 34.245443456204875, 'lon': 108.93110375582032}}, '4': {'name': '平价餐厅（友谊西路店）', 'coordi': {'lat': 34.24253719175486, 'lon': 108.93159854262964}}, '5': {'name': '山妹川菜', 'coordi': {'lat': 34.24522784875664, 'lon': 108.93301750888804}}, '6': {'name': '老三澄合羊肉水盆', 'coordi': {'lat': 34.24224069599151, 'lon': 108.93129159808142}}, '7': {'name': '湘村菜馆(红缨路店)', 'coordi': {'lat': 34.24609764121232, 'lon': 108.9314250012306}}}
```

知识点：

3.3 常用字符串操作方法

8

描述：

表pcs.3-2中融合了字符串常用操作的方法，这包括字符串的运算、函数和方法。

表pcs.3-2 字符串常用操作方法

操作	解释
S=''	建立空字符
"''"	双引号与单引号嵌套使用
bool('')	可以用于检查是否为空字符
\t \n	转义字符（escape, string backslash characters）中常用到的字符，制表符（Horioznţal tab）和换行符（Newline/linefeed）
S1+S2	合并字符串
S*n	复制字符串
S[idx]	按索引（字符位置）提取字符
S[start:end]	分片方式提取字符串
len(S)	计算字符串长度
r'string'	原始字符串（无转义）
S.split(sep=None, maxsplit=-1)	按分隔符（delimiter）切分字符串为字段列表
'%s'%String	%形式格式字符串
'{}'.format(value)	format()方法格式字符串
S.find(sub[, start[, end]])	寻找给定字段的开始索引值
S.strip()	移除字符串中前后的空白（空格）
S.lstrip()	移除字符串中左端的空白
S.rstrip()	移除字符串中右端的空白
S.isdigit()	判断字符串是否为整数字符串
S.lower()	将字符串小写
S.upper()	将字符串大写

续表

操作	解释
S.endswith(suffix[, start[, end]])	判断字符串末尾字符，返回布尔值
S.encode(encoding = 'utf-8', errors='strict')	字符串编码
S.decode()	字符串解码
str in S	成员运算符，给定字符或字段是否在字符串中，返回布尔值
str not in S	成员运算符，给定字符或字段是否不在字符串中，返回布尔值
map(ord,S)	返回给定字符在Unicode中的码值
[s for s in S]	用列表推导式循环拆解字符串为单个字符
's'.join(iterable, /)	给定分隔符，合并字段列表为一个字符串

字符串的方法还有很多，罗列如表pcs.3-3所示，方便查询，或查看Python String Methods①等在线文件。

① Python String Methods，Python有一组内置的方法，可以在字符串上使用（https://www.w3schools.com/python/python_ref_string.asp）。

表pcs.3-3 字符串方法

1	2	3
S.capitalize()	S.ljust(width [, fill])	S.casefold()
S.count(sub [, start [, end]])	S.maketrans(x[, y[, z]])	S.encode([encoding [,errors]])
S.expandtabs([tabsize])	S.rfind(sub [,start [,end]])	S.find(sub [, start [, end]])
S.index(sub [, start [, end]])	S.rpartition(sep)	S.isalnum()
S.isdecimal()	S.split([sep [,maxsplit]])	S.isdigit()
S.islower()	S.strip([chars])	S.isnumeric()
S.isspace()	S.translate(map)	S.istitle()
S.join(iterable)		
4	**5**	**6**
S.lower()	S.center(width [, fill])	S.lstrip([chars])
S.partition(sep)	S.endswith(suffix [, start [, end]])	S.replace(old, new [, count])
S.rindex(sub [, start [, end]])	S.format(fmtstr, *args, **kwargs)	S.rjust(width [, fill])
S.rsplit([sep[, maxsplit]])	S.isalpha()	S.rstrip([chars])
S.splitlines([keepends])	S.isidentifier()	S.startswith(prefix [, start [, end]])
S.swapcase()	S.isprintable()	S.title()
S.upper()	S.isupper()	S.zfill(width)

代码段：

```
S = ''
print(bool(S))
print(S)
print("_"*50)
S = "coordi:'34.244750060429915,108.93113243785623'"
print(S)
print(bool(S))
print("_"*50)
S = 'ID:2,\tname:restaurant\tscore:5\nID:3,\tname:hotel\tscore:3'
print(S)
```

```
S = """___triple-quoted block strings___"""
print(S)
print("_"*50)
S = '\ID\name'
print(S)
print("_"*25)
S = r'\ID\name'
print(S)
print("_"*50)
S1 = 'category:'
S2 = 'restaurant'
print(S1+S2)
S = 'name,'*3
print(S)
S_split_lst = S.split(",")
print(S_split_lst)
print("_"*50)
S_poi = '2,雷记澄城水盆羊肉(红樱路店),34.244750060429915,108.93113243785623,美食;
中餐厅,3.9,20.0,309478.95308006834,3791230.05342414\n'
print(S_poi[2])
print(S_poi[2:10])
print('string length={}'.format(len(S_poi)))
print('name=%s' % S_poi[2:10])
lat_start_position = S_poi.find('34.244750060429915')
lat_end_position = S_poi.find('108.93113243785623')-1
print(lat_start_position)
print(S_poi[lat_start_position:lat_end_position])
print("_"*50)
S_rstrip = "   34.244   ".strip()
print("{1}={0};".format(S_rstrip, 'lat'))
S_rstrip = "   34.244   ".lstrip()
print("{1}={0};".format(S_rstrip, 'lat'))
S_rstrip = "   34.244   ".rstrip()
print("{1}={0};".format(S_rstrip, 'lat'))
print("_"*50)
print('name:108.931'.replace('name', 'lon'))
print('108.931'.isdigit())
print('108'.isdigit())
print("_"*50)
print('code'.upper())
print('CODE'.lower())
S_poi_lst = S_poi.split(",")
print(S_poi_lst)
print('_'.join(S_poi_lst))

S = '美食;中餐厅'
encode_S = S.encode('GBK')
print(encode_S)
decode_S = encode_S.decode('GBK')
print(decode_S)
ord_s = map(ord, ['S', 'a'])
print(list(ord_s))
print("_"*50)
print('p' in 'python')
print('j' in 'python')
print('j' not in 'python')
print([s for s in 'python'])
print('python'.endswith('on'))
```

运算结果：

```
False
__________________________________________________________________________________
coordi:'34.244750060429915,108.93113243785623'
True
__________________________________________________________________________________
ID:2,    name:restaurant   score:5
ID:3,    name:hotel        score:3
___triple-quoted block strings___
__________________________________________________________________________________
\ID
ame
__________________________________________________________________________________
\ID\name
```

```
--------------------------------------------------------------------------------
category:restaurant
name,name,name,
['name', 'name', 'name', '']
--------------------------------------------------------------------------------
雷
雷记澄城水盆羊肉
string length=107
name=雷记澄城水盆羊肉
17
34.244750060429915
--------------------------------------------------------------------------------
lat=34.244;
lat=34.244    ;
lat=   34.244;
--------------------------------------------------------------------------------
lon:108.931
False
True
--------------------------------------------------------------------------------
CODE
code
['2', '雷记澄城水盆羊肉(红樱路店)', '34.244750060429915', '108.93113243785623', '美食;中餐厅', '3.9', '20.0', '309478.95308006834', '3791230.05342414\n']
2_雷记澄城水盆羊肉(红樱路店)_34.244750060429915_108.93113243785623_美食;中餐厅_3.9_20.0_309478.95308006834_3791230.05342414
b'\xc3\xc0\xca\xb3;\xd6\xd0\xb2\xcd\xcc\xfc'
美食;中餐厅
[83, 97]
--------------------------------------------------------------------------------
True
False
True
['p', 'y', 't', 'h', 'o', 'n']
True
```

9

代码段：

```
help(ord)
```

运算结果：

```
Help on built-in function ord in module builtins:
ord(c, /)
    Return the Unicode code point for a one-character string.
```

10

描述：

- \转义字符（String backslash characters）

转义字符\n可以转义很多字符，例如\n表示换行，\t表示制表符等。字符\本身也需要转义，用\\表示。如果字符串里有很多字符需要转义，则直接使用无转义的原始字符串r""达到目的。这在表述文件路径时经常使用，例如r'.\data\xian_poi.csv'（也可使用做斜杠'./data/poi_piecesOFdata.csv'，则不用原始字符串）。而如果字符串中有很多换行，为了避免每次末尾敲入\n，可以使用"""line1,line2,...,lineN"""表述，如表pcs.3-4所示。

表pcs.3-4 \转义字符

转义字符（escape character）	意义
\a	响铃 Bell
\b	退格，将当前位置移到前一列 Backspace
\f	换页，将当前位置移到下页开头 Formfeed
\n	换行，将当前位置移到下一行开头 Newline（linefeed）
\r	回车，将当前位置移到本行开头 Carriage return
\t	水平制表符 Horizontal tap
\v	垂直制表符 Vertical tap

续表

转义字符（escape character）	意义
\\	代表一个反斜线字符\ Backslash
\'	代表一个单引号 Single quote
\"	代表一个双引号 Double quote
\0	空字符 Null:binary 0 character（doesn't end string）
\xhh	十六进制所代表的任意字符 Character with hex value hh（exactly 2 digits）
\newline	忽略（续行）Ignored（continuation line）

代码段：

```
S = """
line1,
line2,
line2
"""
print(S)
print("_"*50)
```

打印转义字符对应的 Unicode 码值

```
print(
    list(map(ord, ['\a', '\b', '\f', '\n', '\r', '\t', '\v', '\\', '\'', '\"',])))
```

运算结果：

```
line1,
line2,
line2
__________________________________________________
[7, 8, 12, 10, 13, 9, 11, 92, 39, 34]
```

知识点：

3.4 字符串格式化

●●11

描述：

字符串格式化在数据分析领域可以用于以文本方式存储格式化后的数据，方便后续数据读取分析；更经常用于图表中的文字表达，这也包括动态交互内容；也用于代码调试过程中print()打印字符，标识打印变量名，格式化数值，方便查看，或者用于交流。

● % 的方式

'string'%value/(values)/{Ks:Vs}的格式化语句语法为%[(keyname)][flags][width][.precision]typecode，如果格式化右侧提供的数据结构为字典形式，则keyname为字典键名索引；如果提供的为列表，则按顺序索引；也可以为单个值。flags标记包括，-：在指定字符宽度时，当字符位数小于宽度则字符左对齐，末尾空格；+：在数值前添加整数或负数符号；0：在指定字符宽度时，当字符位数小于宽度则在字符前用0填充；如果为空格，则在前添加空格符号位。width为字符宽度。.precision为数值精度（保留小数点位数）。typecode为转换类型代码（conversion type codes），如表pcs.3-5所示。

表pcs.3-5 % 字符串格式化转换类型代码

代码（code）	含义
s	字符串，或将非字符类型对象用str()转换为字符串
r	同s，不过用repr()函数转换非字符型对象为字符串
c	参数为单个字符或者字符的Unicode码时，将Unicode码转换为对应的字符
d	参数为数值时，转换为带有符号的十进制整数
i	同d转换数值为整数

续表

代码（code）	含义
u	同d转换数值为整数
o	参数为数值时，转换为带有符号的八进制整数
x	参数为数值时，转换为带有符号的十六进制整数，字母小写
X	参数为数值时，转换为带有符号的十六进制整数，字母大写
e	将数值转换为科学计数法格式，字母小写
E	将数值转换为科学计数法格式，字母大写
f	将数值转换为十进制浮点数
F	同f，将数值转换为十进制浮点数
g	浮点格式。如果指数小于-1或不小于精度（默认为6）使用指数格式，否则使用十进制格式
G	同g
%	%%即为字符%

注：如果是使用 Python 官网提供的 IDLE Shell①，下述示例中的 `from scipy.stats import norm`，需要安装 SciPy② 库，对于 windows 系统，在命令提示符（Command Prompt）下敲入 `py -3 -m pip install scipy` 进行安装。另外，IDLE Shell 可能无法输入中文。推荐使用 anaconda③ 这一专门用于数据分析、科学计算的（数据科学，data science）解释器。

① IDLE Shell，在Python官网下载解释器（https://www.python.org/downloads/）。

② SciPy，Python中科学计算的基本算法库（https://scipy.org/）。

③ Anaconda，包含了Python的环境管理、代码编辑器、包管理等软件平台（https://www.anaconda.com/）。

数据分析时，会涉及很多计算结果显示查看，尤其用于交流的代码。下述是正态分布（normal distribution/Gaussian distribution）的概率计算，调用了SciPy库的`norm.sf(x,loc,scale)`，`norm.cdf()`和`norm.ppf()`的方法，计算给定值(x)，给定正态分布均值（loc）和标准差（scale），求取大于等于（`sf`）或小于等于（`cdf`）给定值的概率；反之，求取满足概率的值（`ppf`）。

代码段：

```
from scipy.stats import norm

print("用 .sf 计算值大于或等于 0.7 待概率为: %s", norm.sf(0.7, 0, 1))
print("用 .cdf 计算值小于或等于 0.7 的概率为: %f" % norm.cdf(0.7, 0, 1))
print("可以观察到 .cdf (<=0.7) 概率结果 +.sf(>=0.7) 概率结果为: %.3f" %
      (norm.cdf(113, 0, 1)+norm.sf(113, 0, 1)))
print("用 .ppf 找到给定概率值为 0.758036(约 75.80%%) 的数值为: %e" % norm.ppf(0.758036, 0,
1))
```

运算结果：

```
用 .sf 计算值大于或等于 0.7 待概率为: %s 0.24196365222307303
用 .cdf 计算值小于或等于 0.7 的概率为: 0.758036
可以观察到 .cdf (<=0.7) 概率结果 +.sf(>=0.7) 概率结果为: 1.000
用 .ppf 找到给定概率值为 0.758036(约 75.80%) 的数值为: 6.999989e-01
```

12

描述：

数据分析必不可少的表达方式是图表，python可以调用的各类图表扩展库不少，其中最为基础和常用的是Matplotlib④。对于此类库通常不必记忆，一般是在需要图表表述数据分析过程、结果，传达研究发现时，查看各类图表库的示例，或者网络分享的示例，直接复用该代码，加以调整，替换数据，进一步调整表达风格，例如颜色、字体、线型、图样等，完成对自身数据分析的表达。下述表述正态分布的图表表达就是复用Matplotlib曲线示例部分代码，替换数据，调整形式而成。对于Matplotlib中常用的语句和参数，如果经常用到则会被记住；不常用的，只要搜索找到可复用的代码即可。

④ Matplotlib，Python图表库，课创建静态、动画和交互式数据可视化（https://matplotlib.org/）。

图pcs.3-1除了表达均值为0、标准差为1的正态分布曲线，同时增加了数值0.7的位置表述垂直虚线，并增加了注释。图表文字的代码则是使用了%的字符串格式化方式，如图例部分增加了均值和标准差的显示，注释上增加了小于等于0.7的概率值说明。

注：图表会在后续的各类数据分析中必不可少的加以应用，以便直观地表述各类数据分析，佐证研究成果。不同的分析内容和表述目的会比较选择适合的图表表述方式。

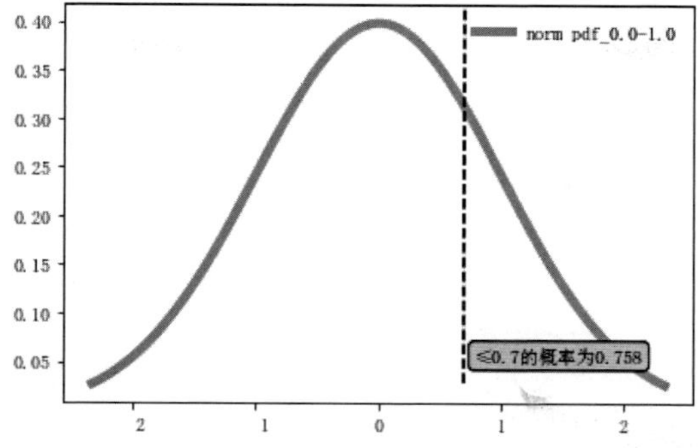

图pcs.3-1 正态分布曲线和概率

代码段：

```
import numpy as np
import matplotlib
import matplotlib.pyplot as plt

matplotlib.rcParams['font.family'] = ['SimSun']  #解决中文乱字符

fig, ax = plt.subplots(1, 1)
mean, var, skew, kurt = norm.stats(moments='mvsk')
print('mean=%s, var=%s, skew=%s, kurt=%s\n' %
      (mean, var, skew, kurt))  #验证符合标准正态分布的相关统计量
# norm.ppf 百分比点函数 - Percent point function (inverse of cdf — percentiles)
x = np.linspace(norm.ppf(0.01), norm.ppf(0.99), 100)
ax.plot(x, norm.pdf(x), 'r-', lw=5, alpha=0.6, label='norm pdf_%s-%s' %
        (mean, var))  # norm.pdf 为概率密度函数
ax.legend(loc='best', frameon=False)
ax.axvline(x=0.7, ymin=0.05, color='k', linestyle='--')
bbox = dict(boxstyle="round", fc="0.8")
ax.annotate("≤%s 的概率为 %.3f" %
            (0.7, norm.cdf(0.7, 0, 1)), (0.78, 0.05), bbox=bbox)
plt.show()
```

运算结果：

```
mean=0.0, var=1.0, skew=0.0, kurt=0.0
```

13

描述：

用字符串格式化的方式组织数据，并写入文件。这里第一行写入的为字段名，其他每一行为一组数据，对应字段名使用制表符\t格式化数据，并在每一行末增加\n换行符。因为这里用制表符分割字符串，并没有使用逗号等分隔符，因此格式化字符串连在一起，阅读起来需要仔细分析字符、转义字符和格式化字符，以及各类标示符。

代码段：

```
poi_info_dict = {'2': {'name': '雷记澄城水盆羊肉(红樱路店)', 'coordi': {'lat':
34.244750060429915, 'lon': 108.93113243785623}}, '3': {'name': '段府农家菠菜面
(红缨路店)', 'coordi': {'lat': 34.245443456204875, 'lon': 108.93110375582032}},
'4': {'name': '平价餐厅(友谊西路店)', 'coordi': {'lat': 34.24253719175486, 'lon':
108.93159854262964}}, '5': {
    'name': '山妹川菜', 'coordi': {'lat': 34.24522784875664, 'lon':
108.93301750888804}}, '6': {'name': '老三澄合羊肉水盆', 'coordi': {'lat':
34.24224069599151, 'lon': 108.93129159808142}}, '7': {'name': '湘村菜馆(红缨路店)',
'coordi': {'lat': 34.24609764121232, 'lon': 108.9314250012306}}}
poi_info_lst = ['%s\t%s\t%.5f\t%.5f\t\n' % (
    k, v['name'], v['coordi']['lat'], v['coordi']['lon']) for k, v in poi_info_
dict.items()]
poi_info_lst_fn = './data/poi_info_dict.txt'
with open(poi_info_lst_fn, 'w', encoding='utf8') as f:
    f.write('%s\t%s\t%s\t%s\t\n' % ('ID', 'name', 'lat', 'lon'))
    f.writelines(poi_info_lst)
with open(poi_info_lst_fn, 'r', encoding='utf8') as f:
```

```
print(f.read())
```

运算结果：

```
ID      name     lat      lon
2       雷记澄城水盆羊肉（红樱路店）  34.24475    108.93113
3       段府农家菠菜面（红缨路店）    34.24544    108.93110
4       平价餐厅（友谊西路店）34.24254        108.93160
5       山妹川菜  34.24523          108.93302
6       老三澄合羊肉水盆        34.24224        108.93129
7       湘村菜馆（红缨路店）    34.24610        108.93143
```

14

代码段：

```
import datetime
today = datetime.datetime.now()
print('%s' % today)
print('%r' % today)
print(ord('a'))
print('%c' % 97)
print('%c' % 'a')
print('%d' % 99.35)
print('%i' % 99.35)
print('%u' % 99.35)
print('%o' % 109)
print('%x' % 109)
print('%X' % 109)
print('%e' % (math.pi*10**6))
print('%E' % (math.pi*10**6))
print('%f' % math.pi)
print('%F' % math.pi)
print('%f' % 0x6D)
print('%f' % 0o155)
print('%g' % (3.30*10**10))
print('%g' % (3.30*10**5))
print('%G' % (3.30*10**5))
print('%.3f%%' % (3.0/11.0*100))
```

运算结果：

```
2022-07-22 17:21:38.978794
datetime.datetime(2022, 7, 22, 17, 21, 38, 978794)
97
a
a
99
99
99
155
6d
6D
3.141593e+06
3.141593E+06
3.141593
3.141593
109.000000
109.000000
3.3e+10
330000
330000
27.273%
```

15

代码段：

```
import math

print('name:%s,category:%s,score:%s' % ('湘村菜馆', '美食_中餐厅', 4))
info_dict = {'name': '湘村菜馆', 'category': '美食_中餐厅', 'score': 4}
print('name:%(name)s,category:%(category)s,score:%(score)s' % info_dict)
print('_'*50)
print('%+-10.3f:)' % -math.pi)
print('%+-10.3f:)' % math.pi)
print('%+-10.*f:)' % (3, math.pi))
```

```
print('%010.3f:)' % math.pi)
```

运算结果：

```
name:湘村菜馆,category:美食_中餐厅,score:4
name:湘村菜馆,category:美食_中餐厅,score:4
--------------------------------------------------------------------------
-3.142    :)
+3.142    :)
+3.142    :)
000003.142:)
```

16

描述：

- format()的方式

format()支持位置索引和关键字，而且可以自由搭配进行格式化，从而形成多种格式化方式。对format()格式化的字符串配置宽度和数值精度，一般语法为{idx/keyname:witdh/.precision}，中间由:分割，右侧配置相关参数。

代码段：

```
import sys
template = 'name:{0},category:{1},score:{2}'
print(template.format('湘村菜馆', '美食_中餐厅', 4))

template = 'name:{},category:{},score:{}'
print(template.format('湘村菜馆', '美食_中餐厅', 4))

template = 'name:{name},category:{category},score:{score}'
print(template.format(name='湘村菜馆', category='美食_中餐厅', score=4))

info_dict = {'name': '湘村菜馆', 'category': '美食_中餐厅', 'score': 4}
template = 'name:{0[name]},category:{0[category]},score:{0[score]}'
print(template.format(info_dict))

template = 'name:%(name)s,category:%(category)s,score:%(score)s'
print(template % dict(name='湘村菜馆', category='美食_中餐厅', score=4))

template = 'name:{0},category:{category},score:{score}'
print(template.format('湘村菜馆', category='美食_中餐厅', score=4))

print('My {1[name]} runs {0.platform}.'.format(sys, {'name': 'Omen'}))

info_lst = ['湘村菜馆', '美食_中餐厅']
print('name:{0[0]},category:{0[1]},score:{1}'.format(info_lst, 4))
```

运算结果：

```
name:湘村菜馆,category:美食_中餐厅,score:4
name:湘村菜馆,category:美食_中餐厅,score:4
name:湘村菜馆,category:美食_中餐厅,score:4
name:湘村菜馆,category:美食_中餐厅,score:4
name:湘村菜馆,category:美食_中餐厅,score:4
name:湘村菜馆,category:美食_中餐厅,score:4
My Omen runs win32.
name:湘村菜馆,category:美食_中餐厅,score:4
```

17

代码段：

```
print('{0:10}={1:5}'.format('pi', math.pi))
print('{0:>10}={1:5}'.format('pi', math.pi))
print('{0:<10}={1:5}'.format('pi', math.pi))
print('{0}={1:.3f}'.format('pi', math.pi))
```

运算结果：

```
pi        =3.141592653589793
        pi=3.141592653589793
pi        =3.141592653589793
pi=3.142
```

18

描述：

- f"{}"（f-string）的方式

f-string的{}中为变量名，直接进行格式化。也可以为Python的表达式或者函数（包括lambda匿名函数）及方法等。

代码段：

```
number = 2
where = "sea"
print(f"{number} of us are gone to {where}.")
print("-"*50)
print(f"{17-2}")  # {} 之内可以放置任何有效的 Python 表达式

print("-"*50)
def to_uppercase(words):
    return words.upper()

print(f"{number} of us are gone to {to_uppercase(where)}.")  # {} 之内可以调用函数
print(f"{number} of us are gone to {where.upper()}.")  # {} 之内可以直接调用方法
# {} 之内可以使用 lambda 匿名函数
print(f"{(lambda a,b:a+b)(1,1)} of us are gone to {where}.")
```

运算结果：

```
2 of us are gone to sea.
--------------------------------------------------
15
--------------------------------------------------
2 of us are gone to SEA.
2 of us are gone to SEA.
2 of us are gone to sea.
```

19

描述：

f-string自定义格式：对齐、宽度、符号、补零、精度和进位制等与%和format()方式基本相同，其基本格式为{content:format}，配置宽度和精度时为{content:width.precision}。

代码段：

```
pi = math.pi
print(f"The number pi approximately equal to {pi:.3f}.")  # 配置数值精度
print(f"approximately equal to {pi:12.3f}")  # 同时配置宽度和精度
print(f"approximately equal to {pi:012.3f}")  # 以 0 填充宽度
print(f"approximately equal to {pi:.3e}")  # 用科学计数法
print(f"approximately equal to {pi:.3%}")  # 百分比形式
print(f"approximately equal to {pi:.3g}")  # 为有效位数（小数点前位数 + 小数点后位数）

s = "world"
print(f"Hello {s:10s}!")  # 字符串形式
```

运算结果：

```
The number pi approximately equal to 3.142.
approximately equal to        3.142
approximately equal to 00000003.142
approximately equal to 3.142e+00
approximately equal to 314.159%
approximately equal to 3.14
Hello world     !
```

20

描述：

格式化方法可以使用时间格式化方式。

代码段：

```
dt = datetime.datetime.today()
print(dt)
print(f'The time is {dt:%Y-%m-%d (%a) %H:%M:%S}')
```

运算结果：

```
2022-11-02 08:31:21.267394
```

```
The time is 2022-11-02 (Wed) 08:31:21
```

知识点：

3.5 正则表达式（regular expression，re）

●●●21

描述：

字符串处理常用到标准库模块中的re，regular expression（正则表达式），re非常强大，可以处理更复杂的字符串，本质是可以匹配文本片段的模式。最简单的re是普通字符串，即大多数字母和字符一般都会和自身匹配，例如'python'可以匹配字符串'python'。

● 字符匹配-模式语法

re可以使用特殊字符的方式匹配一个或者多于一个的字符串，例如使用点号.，可以匹配除了换行符之外的任何字符；但是，只匹配一个字母，多于一个或者零个都不会匹配。点号特殊字符只匹配一个字符。如果希望匹配多个可以使用*星号，匹配前面表达式的0个或者多个副本，并匹配尽可能多的副本；而+加号则匹配至少1个或者多个副本；？问号也是匹配0个或者多个副本。如果想确定具体匹配的数量区间，可以使用{m,n}的方式，即匹配前面表达式的第m～n各副本，如果省略了m，则默认值为0；如果省略了n，则默认设置为无穷大。

在使用*、+、?、{m,n}时，如果模式为r'Hello Py*thon!'，则*星号只对星号之前的一个字符y进行匹配。如果希望同时对P也进行匹配，则需要使用[]中括号字符集把Py括起来即[Py]，完整的模式为r'Hell [Py]*thon!'。还可以应用于更加广泛的范围，例如[a-z]能够匹配a到z的任意一个字符，甚至[a-zA-Z0-9]的使用方式可以匹配任意大小写字母和数字。同时，可以配合使用^字符放置于字符集的开头反转字符集，例如[^abc]则是匹配除了a、b、c之外的字符。

在建立re表达式时，希望能够有选择性地匹配几种不同的情况，例如即匹配字符'python'又匹配'grasshopper'，为同时匹配'python'和'grasshopper'，那么就需要使用|管道符号，re表达式可以写为'python|grasshopper'。如果仅是对部分模式使用管道符号即选择符，可以用圆括号括起需要的部分，例如'p(ython|erl)'。

在匹配字符串时，有时仅需要在开头或者结尾处匹配。这时，可以使用脱字符^标记开始，使用美元符号$标记结尾。

主要使用的re特殊字符如表pcs.3-6。

表pcs.3-6 re特殊字符列表

字符	描述
.	匹配除换行符外任何字符串
^	匹配字符串的开始标志
$	匹配字符串的结束标志
*	匹配前面表达式的0个或多个副本，匹配尽可能多的副本。例如ab*会匹配a，ab，或者abb，abbb等尽可能多（任何数量）的跟随a后b的副本
+	匹配前面表达式的1个或多个副本，匹配尽可能多的副本。例如ab*会匹配除了a外的ab，或者abb，abbb等尽可能多（任何数量）的跟随a后b的副本
?	匹配前面表达式的0个或多个副本，例如ab?将匹配a或者ab
*?	匹配前面表达式的0个或多个副本，匹配尽可能少的副本
+?	匹配前面表达式的1个或多个副本，匹配尽可能少的副本
??	匹配前面表达式的0个或1个副本，匹配尽可能少的副本
{m}	准确匹配前面表达式的m个副本。例如a{6}会精确匹配6个a，而不是5个或其他
{m,n}	匹配前面表达式的第m到n个副本，匹配尽可能多的副本。如果省略了m，则默认为0；如果省略了n，默认为无穷大。例如a{3,5}会匹配3-5个a字符。而a{4,}b会匹配aaaab，甚至无以计数前置b的a字符，但不会匹配aaab，因为a 的数量少于了4
{m,n}?	匹配前面表达式的第m到n个副本，匹配尽可能少的副本。例如a{3,5}会匹配5个a字符；但是，a{3,5}?，只会匹配3个a字符

续表

字符	描述
[...]	匹配一组字符，如'[abcdef]',或'[a-zA-Z]'。特殊字符，例如*在字符集中将失去特殊字符意义，例如[(+*)]会匹配(，+，*或)
[^...]	匹配集合中未包含的字符，例如[^5]将匹配除了5之外的所有字符
A|B	匹配A或B
(...)	匹配圆括号中的re表达式（圆括号中的内容为一个分组），并保存匹配的子字符串。在匹配时，分组中的内容可以使用所获取的MatchObject对象的group()方法获取
(?aiLmsux)	扩展标记法，以?符号开头，其后第一个字符决定采用什么样的语法。其中，a只匹配ASCII字符re.A(re.ASCII)；i忽略大小写re.I(re.IGNORECASE)；L 由当前语言区域决定re.L (locale dependent)；m多行模式re.M(re.MULTILINE)；s为.匹配全部字符re.S(re.DOTALL)；u Unicode匹配，Python3默认开启这个模式re.U；x冗长模式re.X(re.VERBOSE)
(?:...)	常规括号的非捕获版本（A non-capturing version of regular parentheses.），匹配括号内的任何re表达式，但是分组所匹配的子字符串不能再执行匹配后获取或是在之后的模式种被引用
(?P<name>...)	类似于常规括号，但组匹配的子字符串可通过符号组名成访问。组名必须是有效的Python标示符，并且每个组名只能在re表达式中定义一次
(?P=name)	对命名组的反向引用。它匹配与早先名为name的组匹配到的任何文本（字符串）
(?#...)	注释信息，里面的内容会被忽略
(?=...)	只有在括号中的模式匹配时，才匹配前面的表达式，为a lookahead assertion，前视断言
(?!...)	只有在括号中的模式不匹配时，才匹配前面的表达式，为a negative lookahead assertion，前视取反
(?<=...)	如果括号后面的表达式前面的值与括号中的模式匹配，则匹配该表达式，为a positive lookbehand assersion
(?<!...)	如果括号后面的表达式前面的值与括号中的模式不匹配，则匹配该表达式，为a negative lookbehand assersion
(?(id/name)yes-pattern|no-pattern)	检查id或name标识的re表达式组是否存在。如果存在，则匹配re表达式的yes-pattern；否则，匹配可选的表达式no-pattern

一些用\开始的特殊字符所表示的预定义字符集通常是很有用的，例如数字集、字母集或其他非空字符集，如表pcs.3-7所示。

表pcs.3-7　\开始的特殊字符所表示的预定义字符集

字符	描述
\number	匹配相同组编号的组内容。组编号范围为1～99，从左侧开始
\A	仅匹配字符串的开始标志
\b	匹配空字符串，但只匹配单词的开头和结尾。例如r'\bfoo\b'匹配'foo'，'foo.'，'bar foo baz'，但是不会匹配'foobar'，或者'foo3'
\B	匹配空字符串，但仅当它不在单词的开头或结尾时。例如r'py\B'匹配'python'，'py3'，'py2'，但是不会匹配'py'，'py.'或者'py!'
\d	匹配任何Unicode中的十进制数，等同于r'[0-9]'
\D	匹配任何非十进制数的字符，等同于r'[^0-9]'
\s	匹配Unicode空白字符，包括['\t','\n','\r','\f','\v']及许多其他字符
\S	匹配任何非空格字符
\w	匹配Unicode单词字符，这包括可以成为任何语言中单词部分的大多数字符，及数字和下画线。如果使用ASCII标志，仅匹配'[a-zA-Z0-9]'
\W	匹配\w 中定义集合中不包含的字符
\z	仅匹配字符串的末尾
\\	匹配反斜杠本身

代码段：

```
import re

kml_description = '<description>线路开始时间：2017-07-20 08:14:41，结束时间：2017-07-20 20:53:03，线路长度：197801。由 GPS 工具箱导出。</description>'
pattern = 'description'
#使用 re.findall() 方法以列表形式返回给定模式的所有匹配项
print(re.findall(pattern, kml_description))
# .
pattern = '.description'
print(re.findall(pattern, kml_description))
# ? +
pattern = r'w?cadesign\.cn, w+\.cadesign\.cn'  #用转义字符使用点号，而不是用作特殊字符
text = 'cadesign.cn, www.cadesign.cn'
#？号可以匹配 0 个或者多个字符，因此即使不存在字符'w'，也会匹配'cadesign.cn；+号需要匹配至少一个，并尽可能多地匹配，因此可以提取出'www.cadesign.cn'
print(re.findall(pattern, text))
# {m}
pattern = r'w{2}\.cadesign\.cn'
print(re.findall(pattern, text))
# [...]
pattern = '[Py]*thon!'
textA = 'Hello Python!'
textB = 'Hello Pthon!'
textC = 'Hello ython!'
textD = 'Hello thon!'
print(re.findall(pattern, textA))
print(re.findall(pattern, textB))
print(re.findall(pattern, textC))
print(re.findall(pattern, textD))
# A|B
pattern = '<description>|</description>'
print(re.findall(pattern, kml_description))
```

运算结果：

```
['description', 'description']
['<description', '/description']
['cadesign.cn, www.cadesign.cn']
['ww.cadesign.cn']
['Python!']
['Pthon!']
['ython!']
['thon!']
['<description>', '</description>']
```

22

代码段：

```
# (?aiLmsux)
print(re.findall('(?i)ab', 'Ab'))  #i- 为忽略大小写
print(re.findall('(?si)ab.', 'Ab\n'))  #s 为 . 匹配了全部字符，包括换行符，i 忽略大小写。连用了 s 和 i

print(re.findall('^a.', 'ab\nac'))
print(re.findall('(?m)^a.', 'ab\nac'))  #m 为多行模式

print(re.findall('(?x)\d+\.\d*', '3.1415926nan'))  #x 为冗长模式
```

运算结果：

```
['Ab']
['Ab\n']
['ab']
['ab', 'ac']
['3.1415926']
```

23

代码段：

```
# (?:...)
print(re.findall('(abc){2}', 'abcabc'))  #常规捕获版本，捕获到 () 分组内的匹配字符
print(re.findall('(?:abc)', 'abcabc'))  #非捕获版本，将 () 分组视为一个整体
```

```
print(re.findall(‘(a(bc))cbs’, ‘abccbs’))
print(re.findall(‘(a(?:bc))cbs’, ‘abccbs’))  #嵌套捕获模式

print(re.findall(‘(abc)|cbs’, ‘abccbs’))
print(re.findall(‘(abc)|cbs’, ‘cbs’))
print(re.findall(‘(?:abc)|cbs’, ‘cbs’))
```

运算结果：

```
[‘abc’]
[‘abc’, ‘abc’]
[(‘abc’, ‘bc’)]
[‘abc’]
[‘abc’, ‘’]
[‘’]
[‘cbs’]
```

24

代码段：

```
# (?P<name>...) 与 (?P=name)，和 (?#...)
print(re.findall(‘(?P<name>abc)\\1’, ‘abcabc’))
print(re.findall(‘(?P<gname>abc)\d+(?P=gname)(?# 后面的 gname 匹配到前面的匹配到的字符
abc)’, ‘abc996abc’))
```

运算结果：

```
[‘abc’]
[‘abc’]
```

25

代码段：

```
# (?=...) 与 (?!...)
print(re.findall(‘Isaac (?=Asimov)’, ‘Isaac Asimov’))
print(re.findall(‘Isaac (?=Asimov)’, ‘Isaac Asi’))
print(re.findall(‘Isaac (?!Asimov)’, ‘Isaac Asi’))
```

运算结果：

```
[‘Isaac ’]
[]
[‘Isaac ’]
```

26

代码段：

```
# (?<=...) 与 (?<!...)
m = re.search(‘(?<=abc)def’, ‘abcdef’)
print(m.group(0))
m = re.search(r’(?<=-)\w+’, ‘spam-egg’)
print(m.group(0))
```

运算结果：

```
def
egg
```

27

描述：

正则表达式的模式需要配合正则表达式的方法使用，主要方法的解释如下：

`re.findall(pattern, string)`，以列表形式返回给定模式的所有匹配项。

`re.search(pattern,string)`，会在给定字符串中寻找第一个匹配给定正则表达式的子字符串，返回匹配对象（Match Object）。如果字符串中没有位置与模式匹配，则返回None。

`re.match(pattern,string)`，会在给定字符串的开头匹配正则表达式，返回匹配对象。如果字符串中没有位置与模式匹配，则返回None。

`re.fullmatch(pattern,string)`，如果整个字符串与正则表达式模式匹配，则返回相应的匹配对象。如果字符串与模式不匹配，则返回None。

`re.split(pattern,string[,maxsplit=0])`，按出现的模式拆分字符串。如果在模式中使用捕获括号，则模式中所有组的文本也会作为结果列表的一部分返回。如果maxsplit不为

零，则最多发生maxsplit个拆分，并将字符串的其余部分作为列表的最后一个元素返回。

re.sub(pattern,repl,string)，使用给定的替换内容repl将匹配模式pattern的子字符串替换掉。如果未找到该模式，则字符串原样返回。repl可以是字符串或函数。

re.subn(pattern, repl, string, count=0,flags=0)，同re.sub()，只是返回一个元组为(new_string, number_of_subs_made)。

re.escape(string)，可以对字符串中所有可能被解释为正则运算符的字符进行转义，避免输入较多的反斜杠。

re.compile(pattern)，可以将以字符串书写的正则表达式转换为模式对象，例如转换为模式对象后可以直接使用pattern.search(string)的方法，这与re.search(pattern,string)方式一样。因为使用re模块的方法时，不管是re.search()还是re.math()都会在内部将字符串表示的正则表达式转换为正则表达式模式对象，因此re.compile()的方法可以避免每次使用模式时都得重新转化的过程。

re.purge()，清除正则表达式缓存。

代码段：

```
print(re.findall(r'\bf[a-z]*', 'which foot or hand fell fastest'))
print(re.findall(r'(\w+)=(\d+)', 'set width=20 and height=10'))
pattern = '[a-z]+'
text = '<coordinates>120.130095,30.21169,20.5</coordinates>'
print(re.findall(pattern, text))
pattern = r'(?x)\d+\.\d*'
print(re.findall(pattern, text))
pattern = 'description'
text = '<description>GPS 工具箱导出数据 </description>'
print(re.search(pattern, text))
print(re.search(pattern, text).group())
print(re.match(pattern, text))
if re.search(pattern, text):
    print('found a match')
else:
    print('no match')
pattern = 'g...s'
text = 'geeks'
print(re.fullmatch(pattern, text))
```

运算结果：

```
['foot', 'fell', 'fastest']
[('width', '20'), ('height', '10')]
['coordinates', 'coordinates']
['120.130095', '30.21169', '20.5']
<re.Match object; span=(1, 12), match='description'>
description
None
found a match
<re.Match object; span=(0, 5), match='geeks'>
```

28

描述：

re.search()和re.match()返回的MatchObject实例对象包含关于分组内容的信息和匹配值的位置数据。组就是放置在圆括号内的子模式。可以通过m.group()返回组，m.start()获取组的开始索引值，m.end()则获取结束位置索引值，m.span()返回区间值。

代码段：

```
# 对字符串进行模式匹配，返回 MatchObject 对象
m = re.match(r'www\.(.*)\..{3}', 'www.python.org')
print(m.group(1))
print(m.start(1))
print(m.end(1))
print(m.span(1))
```

运算结果：

```
python
4
10
```

```
(4, 10)
```

29

描述：

下述应用了手机APP记录调研路径，存储为KML数据格式，在Python中读取，提取需要数据内容的简化示例。对于KML格式的数据，其后缀名通常为.kml，与KMZ一样是Google Earth所使用的点、线和面的地标文件格式。记录的数据均由类似<coordinates>...</coordinates>、<description>...</description>、<name>...</name>等<identifier>text</identifier>模式构成，这有助于数据的提取。

代码段：

```
pattern_coordi = re.compile('<coordinates>(.*?)</coordinates>')
pattern_description = re.compile('<description>(.*?)</description>')
coordi_text = '<coordinates>120.132007,30.300508,9.7</coordinates>'
description_text = '<description>线路开始时间：2017-07-20 08:14:41，结束时间：2017-07-20 20:53:03，线路长度：197801。由 GPS 工具箱导出。</description>'

print(pattern_coordi.findall(coordi_text))
print(pattern_description.findall(description_text))
```

运算结果：

```
['120.132007,30.300508,9.7']
['线路开始时间：2017-07-20 08:14:41，结束时间：2017-07-20 20:53:03，线路长度：197801。由 GPS 工具箱导出。']
```

30

代码段：

```
print(re.split(r'\W+', 'Words, words, words.'))
print(re.split(r'(\W+)', 'Words, words, words.'))  #匹配分组文本作为列表一部分返回
print(re.split(r'\W+', 'Words, words, words.', 1))  #只拆分了一次，余下部分作为列表一部分返回
print(re.split('[a-f]+', '0a3B9', flags=re.IGNORECASE))
print(re.split('[a-f]+', '0a3B9'))
```

运算结果：

```
['Words', 'words', 'words', '']
['Words', ', ', 'words', ', ', 'words', '.', '']
['Words', 'words, words.']
['0', '3', '9']
['0', '3B9']
```

31

描述：

如果分隔符中有捕获组并且它在字符串的开头匹配，则结果将以空字符串开头。这同样适用于字符串的结尾。

模式的空匹配仅在与先前的空匹配不相邻时，才拆分字符串。

代码段：

```
print(re.split(r'(\W+)', '...words, words...'))
print("_"*50)
print(re.split(r'\b', 'Words, words, words.'))
print(re.split(r'\W+', '...words...'))
print(re.split(r'\W*', '...words...'))  #*匹配0个或多个，+匹配1个或多个
print(re.split(r'(\W*)', '...words...'))
```

运算结果：

```
['', '...', 'words', ', ', 'words', '...', '']
__________________________________________________
['', 'Words', ', ', 'words', ', ', 'words', '.']
['', 'words', '']
['', '', 'w', 'o', 'r', 'd', 's', '', '']
['', '...', '', '', 'w', '', 'o', '', 'r', '', 'd', '', 's', '...', '', '', '']
```

32

描述：

如果repl是一个函数，则每次出现不重叠的模式时都会调用它。该函数采用单个匹配对象参数，并返回替换字符串。

代码段：

```
print(re.sub(r'def\s+([a-zA-Z_][a-zA-Z_0-9]*)\s*\(\s*\):',
             r'static PyObject*\npy_\1(void)\n{',
             'def myfunc():'))

#这里用到了一个自定义函数
def dashrepl(matchobj):
    if matchobj.group(0) == '-':
        return ' '
    else:
        return '-'
print(re.sub('-{1,2}', dashrepl, 'pro----gram-files'))
print(re.sub(r'\sAND\s', ' & ', 'Baked Beans And Spam', flags=re.IGNORECASE))
```

运算结果：

```
static PyObject*
py_myfunc(void)
{
pro--gram files
Baked Beans & Spam
```

33

描述：

转义模式中的特殊字符。例如，匹配文本中可能包含正则表达式元字符的任意文字字符串。

代码段：

```
import string
print(re.escape('https://www.python.org'))
legal_chars = string.ascii_lowercase + string.digits + "!#$%&'*+-.^_`|~:"
print('[%s]+' % re.escape(legal_chars))
operators = ['+', '-', '*', '/', '**']
print('|'.join(map(re.escape, sorted(operators, reverse=True))))
```

运算结果：

```
https://www\.python\.org
[abcdefghijklmnopqrstuvwxyz0123456789!\#\$%\&' \*\+\-\.\^_`\|\~:]+
/|\-|\+|\*\*|\*
```

34

代码段：

```
text = '<coordinates>120.130095,30.21169,20.5</coordinates>'
pattern = re.compile(r'(?x)\d+\.\d*')
pattern.findall(text)
```

运算结果：

```
['120.130095', '30.21169', '20.5']
```

PCS-4. 基本语句_条件语句（if、elif、else），循环语句（for、while）和列表推导式（comprehension）

PCS-4（　）：4.1 缩进（indentation）和代码（语句）块（code block）；4.2 条件语句；4.3 循环语句_for loops 模式；4.4 循环语句_while 模式；4.5 列表推导式（comprehension）

知识点：

4.1 缩进（indentation）和代码（语句）块（code block）

1

描述：

代码语句的书写上，Python与C、C++等语言最大的不同是Python强制缩进。这样的好处是任何人书写的Python代码都具有统一的“格式”，无须另行规定基本的代码书写规范，方便代码的传播和复用；同时，缩进的方式使得代码易读，不用特意去寻找语句块结束的标志，通过段落就可以轻易判断。

可以把一行代码看作一个基本单元，即为一个可执行的语句。语句通常是从上至下逐行执行。当定义函数和类之后，调用函数或方法则可以跳转执行语句，但跳转后仍是从上至下执行语句，结束调用方法后从之前调用的语句部分从新向下执行；或者，遇到条件语句，需要根据条件判断将要执行哪部分语句；或者，遇到循环语句和递归算法，将从循环位置反复执行同一语句或语句块，直至遇到结束条件；代码也可以不在同一个文件中，通过调用其他文件中的代码，语句的执行顺序也会跳转。

代码块是一段可以作为一个单元执行的一行或多行语句（程序文本），例如一个条件或循环的代码段、一个函数、一个类等；一个代码块也可以包含其他代码块，或调用执行其他代码块，因此对于代码块的定义相对比较宽泛，可以根据是否执行了一个任务来确定，无关任务的大小。

知识点：

4.2 条件语句

2

描述：

条件语句的基本语法如下：

```
if test1:
    statement1
elif test2:
    statement2
elif test3:
    statement3
...
else:
    statements
```

同一条件代码块，if通常只用一次，elif可以执行0次或多次，else为不满足上述所有条件后，执行的语句，也可以不调用，但最好通过else表明其他情况如何处理。

下面应用了《漫画统计学》[1]“美味拉面畅销前50”上刊载的拉面馆的拉面价格数据，并将其存储在了ramen_price_lst列表中。这里给了一个input()内置函数来在外部交互输入指令，这里的指令就是条件语句中的test部分，如果输入指令满足if或elif后的要求，则对应执行该语句缩进后的代码。如果都不满足则执行else后的语句，提示"Please enter the correct command:("。

注：数据分析时很少用到 input() 函数来外部输入参数值，而通常使用交互图表，一般选择 tkinter GUI（Graphical User Inteface）① 工具包、Plotly② 自定义控件、Pygame 游戏编程模块 ③、gradio④ 及 Web 界面演示机器学习模型等既有成熟、完善的库来处理。

① tkinter，是Tcl/Tk GUI工具包的标准Python接口（https://docs.python.org/3/library/tkinter.html）。
② Plotly，是一个通过数据应用实现数据驱动决策的实践者，构建有支持多种编程语言的图表库（https://plotly.com/python/#controls）。
③ Pygame，是一套跨平台的Python模块，用于编写视频游戏。它包括计算机图形和声音库，旨在与Python编程语言一起使用（https://www.pygame.org）。
④ gradio，是演示机器学习模型的最快方式，提供一个友好的网络界面，任何人都可以在任何地方使用该模型（https://gradio.app/）。

代码段：

```
import numpy as np
ramen_price_lst = [700, 850, 600, 650, 980, 750, 500, 890, 880, 700, 890, 720,
680, 650, 790, 670, 680, 900, 880, 720, 850, 700, 780, 850, 750,
                    80, 590, 650, 580, 750, 800, 550, 750, 700, 600, 800, 800,
880, 790, 790, 780, 600, 690, 680, 650, 890, 930, 650, 777, 700]
command = input("Enter your command('mean,std,max,min,median'):")
if command == 'mean':
    print(np.mean(ramen_price_lst))
elif command == 'std':
    print(np.std(ramen_price_lst))
elif command == 'max':
    print(np.max(ramen_price_lst))
elif command == 'min':
    print(np.min(ramen_price_lst))
elif command == 'median':
    print(np.median(ramen_price_lst))
else:
    print("Please enter the correct command:(")
```

运算结果：

```
Enter your command('mean,std,max,min,median'):
```

3

描述：

- 嵌套条件语句（Nested if statements）

一个条件下可以再嵌套多个条件，例如下述代码外层条件语句是判断变量price_x值是否属于列表ramen_price_lst，如果属于则打印该价格，并执行嵌套条件语句块，判断该值是否大于或者小于等于平均价格；回到外层条件，如果price_x值不属于列表ramen_price_lst，则寻找最近值，这里使用了一个lambda匿名函数计算绝对值的功能，并将其作为min(iterable, *[, default=obj, key=func])函数的key参数值，即比较的是匿名函数所定义返回值（差值的绝对值）的最小值，并返回对应绝对值最小的价格列表中的值。打印该值，同时执行嵌套条件，与if下嵌套条件一样来判断大于或者小于等于价格均值。

代码段：

```
ramen_price_lst = [700, 850, 600, 650, 980, 750, 500, 890, 880, 700, 890, 720,
680, 650, 790, 670, 680, 900, 880, 720, 850, 700, 780, 850, 750,
                    80, 590, 650, 580, 750, 800, 550, 750, 700, 600, 800, 800,
880, 790, 790, 780, 600, 690, 680, 650, 890, 930, 650, 777, 700]
price_x = 200.68
def abs_difference_func(value): return abs(value-price_x)
if price_x in ramen_price_lst:
    print(price_x)
    if price_x > np.mean(ramen_price_lst):
        print('The price is higher than the average price.')
    else:
        print('The price is lower than the average price.')
else:
    print('%.3f is no in ramen_price_lst.' % price_x)
    closest_value = min(ramen_price_lst, key=abs_difference_func)
    print('the nearest value to %s is %s.' % (price_x, closest_value))
    # 定义了与 if 中同样的功能代码块，不过将 price_X 替换为 closest_value
    price_mean = np.mean(ramen_price_lst)
    if closest_value > price_mean:
        print('The price is higher than the average price.')
    else:
        print('The price is lower than the average price %s.' % price_mean)
```

运算结果：

```
200.680 is no in ramen_price_lst.
the nearest value to 200.68 is 80.
The price is lower than the average price 729.34.
```

4

描述：

+尝试下定义函数的优势（下一PCS预热）

如果要重复比较不同值和不同列表值的关系，返回列表最近值，那么上述的代码使用起来不方便，还会很烦琐，也很难分享，不易被其他程序调用（代码复用），因此需要将这一功能代码块定义为函数形式。从下述转换为函数后的代码，可以观察到几个需要注意的点：

1. 关于变量名和函数名的命名，可以发现下述的变量名并没有延续上一代码段定义的各类名称，这包括变量名、函数名及参数名。因为该函数代码的主要功能是比较一个值和一个列表中的值的关系，给的数据不一定是拉面价格，因此函数中各个名称的定义应该尽量通用化，主要表述和反应定义函数所要解决的内容或问题。
2. 对于重复的代码段或变量，通常不会重复书写，例如上述代码中内层的两个判断与均值大小的条件语句块重复书写，因此将其定义为单独的匿名函数`comparisonOF2values`方便调用。也可以看到列表均值的计算`np.mean(ramen_price_lst)`被书写了两次，可以将该计算语句放置于条件代码块之外赋值给单独变量名。之后，只需要用该变量就可，避免重复较长语句的书写。
3. 函数内的打印语句文字，同变量名的定义一样应通用化。

代码段：

```
def value2values_comparison(x, lst):
    import numpy as np
    lst_mean = np.mean(lst)
    abs_difference_func = lambda value:abs(value-x)
    comparisonOF2values = lambda v1,v2:print(‘x is higher than the average %s of the list.’%lst_mean) if v1>v2 else print(‘x is lower than the average %s of the list.’%lst_mean)
    if x in lst:
        print(“%s in the given list.” % x)
        comparisonOF2values(x, lst_mean)
        return x
    else:
        print(‘%.3f is not in the list.’ % x)
        closest_value = min(lst, key=abs_difference_func)
        print(‘the nearest value to %s is %s.’ % (x, closest_value))
        print(“_”*50)
        comparisonOF2values(closest_value, lst_mean)
        return closest_value
ramen_price_lst = [700, 850, 600, 650, 980, 750, 500, 890, 880, 700, 890, 720, 680, 650, 790, 670, 680, 900, 880, 720, 850, 700, 780, 850, 750,
                   80, 590, 650, 580, 750, 800, 550, 750, 700, 600, 800, 800, 880, 790, 790, 780, 600, 690, 680, 650, 890, 930, 650, 777, 700]
price_x = 200.68
price_x_closestValue = value2values_comparison(price_x, ramen_price_lst)
print(price_x_closestValue)
```

运算结果：

```
200.680 is not in the list.
the nearest value to 200.68 is 80.
__________________________________________________
x is lower than the average 729.34 of the list.
80
```

5

代码段：

```
price_x_closestValue = value2values_comparison(890, ramen_price_lst)
print(price_x_closestValue)
```

运算结果：

```
890 in the given list.
x is higher than the average 729.34 of the list.
890
```

6

代码段：

```
price_x_closestValue = value2values_comparison(
    78, [3, 4, 5, 733, 66, 22, 99, 88, 11])
print(price_x_closestValue)
```

运算结果：

```
78.000 is not in the list.
the nearest value to 78 is 88.
--------------------------------------------------------------------------------
x is lower than the average 729.34 of the list.
88
```

7

描述：

● 三元表达式（Ternary Expression）

形如`variable=v1 if test else v2`的语句即为三元表达式，该语句等同于：

```
if test:
    variable=v1
else:
    variable=v2
```

三元表达式通常用于较简单的条件语句，因为用一行表述较之多行书写更为便捷；但是对于较长、较复杂的条件语句，则建议按常规缩进书写。

代码段：

```
v1 = 33.5
v2 = 78.3
max_v1Nv2 = v1 if v1 > v2 else v2
print(max_v1Nv2)
```

运算结果：

```
78.3
```

8

描述：

● 用;连接简单的语句为一行

如果语句非常的简单，则可以使用;将其连接置于一行。下述示例还包括了一个简单的三元表达。

代码段：

```
x = 3.5
y = 7.8
print(x if x > y else y)
```

运算结果：

```
7.8
```

9

描述：

● 条件语句与比较运算符、逻辑运算符和成员运算符

条件语句通常会用逻辑运算符连接多个比较运算符或其他条件，实现条件判断的目的。

代码段：

```
a, b, c = 23, 57, 68
lst = [23, 77, 96]
if a < b and c > b:
    print('a is less than c.')
if a < b or b > c:
    print('b is not sure greater than c.')
if a not in lst:
    print('a not in list')
else:
    print('a in list')
if a in lst:
    print('a in list')
else:
    print('a not in list')
```

运算结果：

```
a is less than c.
b is not sure greater than c.
```

```
a in list
a in list
```

10

描述：

很多变量可以直接用于条件之后，简化条件书写，例如下述是否为空列表的判断，一个直接使用变量，一个则计算列表的长度来判断是否为0，从而证实是否为空列表。

代码段：

```
if 1:
    print('return true.')
if 0:
    print('This statement will not be executed.')
else:
    print('It is 0.')
if True:
    print('This statement is executed!')
if '':
    print('Empty string...')
else:
    print('This test is an empty string.')
empty_lst = []
if not empty_lst:
    print('This is an empty list!!!')
if len(empty_lst) == 0:
    print('This is an empty list!!!')
lst = [3, 4, 5]
if lst:
    print('This is not an empty list!!!')
```

运算结果：

```
return true.
It is 0.
This statement is executed!
This test is an empty string.
This is an empty list!!!
This is an empty list!!!
This is not an empty list!!!
```

知识点：

4.3 循环语句_for loops 模式

11

描述：

for 循环的基本语法为：

```
for target in object:
    statements
else: # 可选部分
    statements # 如果 for 循环没有被中断（break）
```

object 为序列或者任何可迭代的对象，例如strings、lists、tuples、dict和其他内置可迭代（iterable）对象，如zip()、map()等返回的可迭代对象。

● 循环列表与enumerate()

在解释循环语句时，使用了城市景观数据集（The Cityscapes Dataset）[①]的标签数据（Cityscapes数据集集中于城市街道场景的语义解释（semantic understanding），如图像语义分割、对象检测等深度学习模型的训练，这非常适用于对城市空间内容的分析）。为了方便数据的处理，将标签数据存储为namedtuple数据格式（结构），namedtuple由Python内置库collections[②]提供，一般翻译为具名元组。Python内置数据结构tuple（元组），不能像表格抬头（例如Pandas的DataFrame数据结构）一样为数据指定字段名（列名），因此不能够很好地管理数据，这包括对于数据的存储更新和提取，因此collections.namedtuple类型的数据结构就解决了这个问题。namedtuple(typename, field_names, *, rename=False, defaults=None, module=None)，定义namedtuple的输入参数中typename为元组的名称，filed_names

① The Cityscapes Dataset，大规模数据集，包含了50个不同城市的街道场景中记录的各种立体（stereo）视频序列，除了20000个弱注释帧的大集合外，还有5000帧的高质量像素级注释（https://www.cityscapes-dataset.com/）。

② collections，这个模块实现了专门的容器数据类型，提供了Python的通用内置容器（数据结构），dict、list、set和tuple的替代格式（https://docs.python.org/3/library/collections.html）。

为元组中元素的名称，rename为如果元素名称含有Python的关键字，则必须配置该参数为rename=True。使用namedtuple首先定义一个namedtuple对象，例如示例中的Label对象，然后应用该对象定义不同的namedtuple变量存储数据，例如label_building和label_caravan。可以通过类属性值（object.attribute）的途径读取字段值，以及更新字段值。

注：collections 库提供有专门的容器数据类型（container datatype），即数据结构，为 dict、 list、 set 和 tuple 提供了可替代数据存储管理方式。

代码段：

```
from collections import namedtuple
Label = namedtuple(‘label’, [‘name’, ‘id’, ‘trainID’, ‘category’,
                   ‘categoryID’, ‘hasInstances’, ‘igoreInEval’, ‘color’])
print(Label)
print(“_”*50)
label_building = Label(‘building’, 11, 2, ‘construction’,
                       2, False, False, (70, 70, 70))
print(label)
print(label_building._fields)
print(“_”*50)
print(label_building.name)
print(label_building.id)
print(label_building.category)
print(label_building.color)
print(“_”*50)
caravan_lst = [‘caravan’, 29, 255, ‘vehicle’, 7, True, True, (0,  0, 90)]
label_caravan = Label._make(caravan_lst)
print(label_caravan.name)
print(label_caravan.id)
print(label_caravan.category)
print(label_caravan.color)
label_caravan = label_caravan._replace(
    category=’schooner’, color=(30, 30, 60))  # 替换属性值
print(label_caravan.category)
print(label_caravan.color)
print(“_”*50)
caravan_dict = label_caravan._asdict()  # 将 nametuple 转换为 dict
print(caravan_dict)
```

运算结果：

```
<class ‘__main__.label’>
__________________________________________________________________________________
label(name=’building’, id=11, trainID=2, category=’construction’, categoryID=2,
hasInstances=False, igoreInEval=False, color=(70, 70, 70))
(‘name’, ‘id’, ‘trainID’, ‘category’, ‘categoryID’, ‘hasInstances’, ‘igoreInEval’,
‘color’)
__________________________________________________________________________________
building
11
construction
(70, 70, 70)
__________________________________________________________________________________
caravan
29
vehicle
(0, 0, 90)
schooner
(30, 30, 60)
__________________________________________________________________________________
{‘name’: ‘caravan’, ‘id’: 29, ‘trainID’: 255, ‘category’: ‘schooner’,
‘categoryID’: 7, ‘hasInstances’: True, ‘igoreInEval’: True, ‘color’: (30, 30, 60)}
```

12

描述：

cityscapes的标签数据以namedtuple列表形式存储，列表中的每一个值就为一个namedtuple对象，具有相同的字段名称。通过namedtuple读取值的方法，并配合列表推导式很容易提取各个字段名为单独的列表，或两个到多个字段名提取为字典的模式，建立不同字段之间的映射。为了清晰地观察数据，在输入数据时，有意识地将其各列对齐，每一列就为一个具有名称的元素，例如name字段列对齐，方便观察名称。注意，这里修改了color字段的值，使用了ANSI Escape Sequences/Codes，ANSI code[①]，可翻译为ANSI转义序列/代码，ANSI code用于控制光标位置、颜

① ANSI Escape Sequences/Codes，ANSI code，是一种带内信号（in-band signaling）标准，用于控制视频文本终端和终端仿真器上的光标位置、颜色、字体风格和其他选项。某些字节序列，大多数以ASCII转义字符和括号字符开始，被嵌入文本中。终端将这些序列解释为命令，而不是逐字显示的文本（https://en.wikipedia.org/wiki/ANSI_escape_code）。

色和字体样式，也包括视频文本终端或终端仿真器。某些字节序列（大多数以ASCII转义字符和括号字符开头）被嵌入到文本中。终端将这些序列解释为命令，而不是逐字显示的文本。

在Python解释器中显示字体的颜色，包括16色模式（8个字体颜色和8个背景颜色）和256色模式。16色字符串格式化的模式示例为`print('\033[2;31;43m CHEESY \033[0;0m')`，或`print('\x1b[2;31;43m CHEESY \x1b[0;0m')`，其中`\033[0;0m')`是重置终端打印颜色为默认，防止继续打印设置的颜色，各字符含义如图pcs.4-1所示。

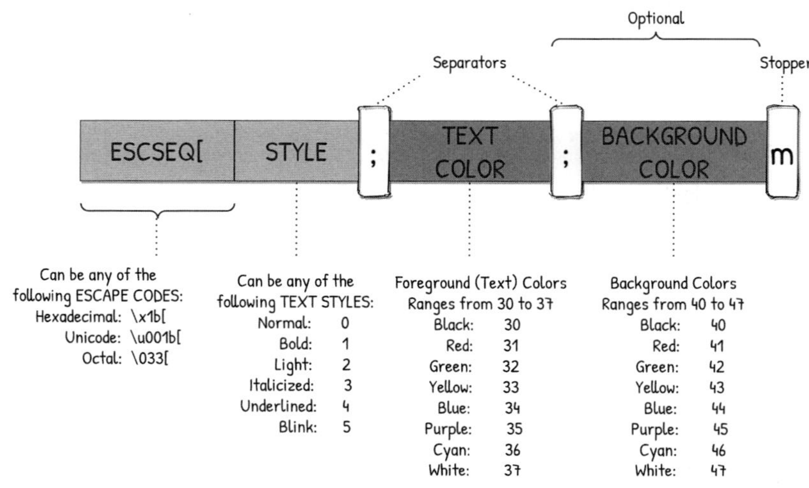

图pcs.4-1 Python解释器中显示字体颜色字符含义（16色模式）

256色打印方式例如`print("\033[48;5;236m\033[38;5;231mStack \033[38;5;208mAbuse\033[0;0m")`，各字符含义如图pcs.4-2。

注：参考 How to Print Colored Text in Python[①]；American National Standards Institute , ANSI

① How to Print Colored Text in Python（https://stackabuse.com/how-to-print-colored-text-in-python/）。

代码段：

```
Label = namedtuple('label', ['name', 'id', 'trainId', 'category',
                   'catId', 'hasInstances', 'igoreInEval', 'color'])
labels = [
    #       name                     id    trainId   category            catId
hasInstances   ignoreInEval   color
```

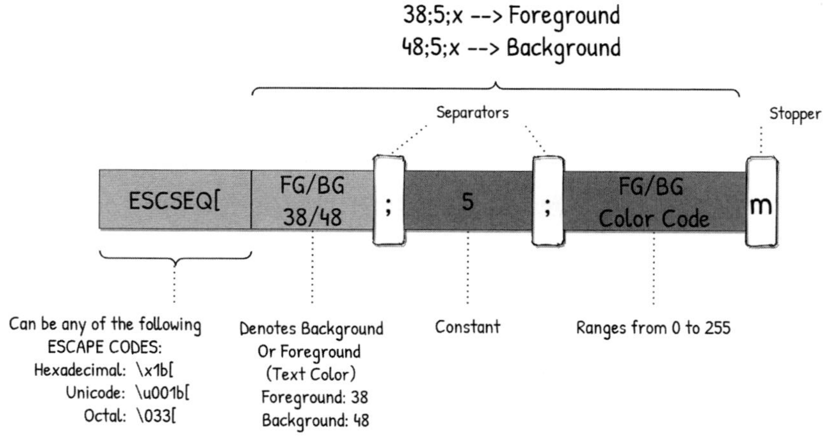

图pcs.4-2 Python解释器中显示字体颜色字符含义（256色模式）

```
    Label(‘unlabeled’,  0,      255, ‘void’, 0, False, True, (0, 30,  47)),
    Label(‘ego_vehicle’,  1,      255, ‘void’, 0, False, True, (0, 31,  46)),
    Label(‘rectification_border’,  2,      255,
        ‘void’, 0, False, True, (0, 32,  45)),
    Label(‘out_of_roi’,  3,      255, ‘void’, 0, False, True, (0, 33,  44)),
    Label(‘static’,  4,      255, ‘void’, 0, False, True, (0, 34,  43)),
    Label(‘dynamic’,  5,      255, ‘void’, 0, False, True, (1, 35,  42)),
    Label(‘ground’,  6,      255, ‘void’, 0, False, True, (1, 36,  41)),
    Label(‘road’,  7,      0, ‘flat’, 1, False, False, (1, 37,  40)),
    Label(‘sidewalk’,  8,      1, ‘flat’, 1, False, False, (1, 30,  41)),
    Label(‘parking’,  9,      255, ‘flat’, 1, False, True, (1, 31,  42)),
    Label(‘rail_track’, 10,      255, ‘flat’, 1, False, True, (2, 32,  43)),
    Label(‘building’, 11,      2, ‘construction’,
        2, False, False, (2, 33,  44)),
    Label(‘wall’, 12,      3, ‘construction’, 2, False, False, (2, 34,  45)),
    Label(‘fence’, 13,      4, ‘construction’,
        2, False, False, (2, 35,  46)),
    Label(‘guard_rail’, 14,      255, ‘construction’,
        2, False, True, (2, 36,  47)),
    Label(‘bridge’, 15,      255, ‘construction’, 2, False, True, (3, 37,  40)),
    Label(‘tunnel’, 16,      255, ‘construction’, 2, False, True, (3, 30,  42)),
    Label(‘pole’, 17,      5, ‘object’, 3, False, False, (3, 31,  43)),
    Label(‘polegroup’, 18,      255, ‘object’, 3, False, True, (3, 32,  44)),
    Label(‘traffic_light’, 19,      6, ‘object’,
        3, False, False, (3, 33,  45)),
    Label(‘traffic_sign’, 20,      7, ‘object’,
        3, False, False, (4, 34,  46)),
    Label(‘vegetation’, 21,      8, ‘nature’, 4, False, False, (4, 35,  47)),
    Label(‘terrain’, 22,      9, ‘nature’, 4, False, False, (4, 36,  40)),
    Label(‘sky’, 23,      10, ‘sky’, 5, False, False, (4, 37,  41)),
    Label(‘person’, 24,      11, ‘human’, 6, True, False, (4, 30,  43)),
    Label(‘rider’, 25,      12, ‘human’, 6, True, False, (5, 31,  44)),
    Label(‘car’, 26,      13, ‘vehicle’, 7, True, False, (5, 32,  45)),
    Label(‘truck’, 27,      14, ‘vehicle’, 7, True, False, (5, 33,  46)),
    Label(‘bus’, 28,      15, ‘vehicle’, 7, True, False, (5, 34,  47)),
    Label(‘caravan’, 29,      255, ‘vehicle’, 7, True, True, (5, 35,  42)),
    Label(‘trailer’, 30,      255, ‘vehicle’, 7, True, True, (5, 36,  41)),
    Label(‘train’, 31,      16, ‘vehicle’, 7, True, False, (0, 37,  40)),
    Label(‘motorcycle’, 32,      17, ‘vehicle’, 7, True, False, (1, 30,  44)),
    Label(‘bicycle’, 33,      18, ‘vehicle’, 7, True, False, (2, 32,  45)),
    Label(‘license_plate’, -1,      -1,
        ‘vehicle’, 7, False, True, (3, 33,  47)),
]
print(labels[:3])
```

运算结果：

```
[label(name='unlabeled', id=0, trainId=255, category='void', catId=0,
hasInstances=False, igoreInEval=True, color=(0, 30, 47)), label(name='ego_
vehicle', id=1, trainId=255, category='void', catId=0, hasInstances=False,
igoreInEval=True, color=(0, 31, 46)), label(name='rectification_border', id=2,
trainId=255, category='void', catId=0, hasInstances=False, igoreInEval=True,
color=(0, 32, 45))]
```

13

代码段：

```
print(‘\x1b[2;31;43m CHEESY \x1b[0;0m’)
print(“\033[48;5;236m\033[38;5;231mStack \033[38;5;208mAbuse\033[0;0m”)
```

运算结果（图pcs.4-3）：

图pcs.4-3　字体颜色显示示例1

14

描述：

color_lst为提取的ANSI code格式颜色数据列表，每一元组值对应text styles（字体类型，包括normal/0、bold/1、light/2、italicized/3、underlined/4、blink/5），foreground（Text）color（字

体颜色，包括black/30、red/31、green/32、yellow/33、blue/34、purple/35、cyan/36、white/37计8个颜色），及background color（字体的背景色，颜色同字体色，但是编号为40-47）。每次循环配置打印颜色值，并以颜色值为打印的字符串（图pcs.4-4）。

需要注意对于循环语句，通常包括多个值，甚至千万个待循环值，因此在书写代码时需要增加终止循环的代码if i==5:break，当变量i每次循环自增1～5时，调用break终止语句，跳出循环。待调试一次或几次循环无误后，再循环所有的值，避免等待运算时间，尤其需要花费10min以上，甚至多到几个小时或几天才能运算完的代码段。

下述示例代码保留了调试代码，除变量i和终止条件语句行外，调试时，要不断用print()函数查看变量值，确定变量值是否正确，以及确认变量值结构，从而知晓后续代码行应用该变量的方式；或者，通过后续要求的数据结构来处理数据为后续所用结构的类型（通常是用后者的方式判断和书写代码）。例如，示例中通过print(color)来查看通过color=';'.join([str(i) for i in c])语句编写满足ANSI code要求的颜色格式，例如0;30;47，注意这里的数字为使用str()转换数字为字符串，满足使用%s格式化符号的要求；也可以不转换为字符串，而是使用%d的方式直接格式化。

代码段：

```
color_lst = [label.color for label in labels]
print(color_lst)

# i=0 # 调试用
for c in color_lst:
    color = ';'.join([str(i) for i in c])
    # print(color) # 调试用
    s = '\x1b[%sm %s \x1b[0m' % (color, color)
    print(s)
    # if i==5:break # 调试用
    # i+=1 # 调试用
```

图pcs.4-4 字体颜色显示示例2

运算结果：

```
[(0, 30, 47), (0, 31, 46), (0, 32, 45), (0, 33, 44), (0, 34, 43), (1, 35, 42),
(1, 36, 41), (1, 37, 40), (1, 30, 41), (1, 31, 42), (2, 32, 43), (2, 33, 44), (2,
34, 45), (2, 35, 46), (2, 36, 47), (3, 37, 40), (3, 30, 42), (3, 31, 43), (3,
32, 44), (3, 33, 45), (4, 34, 46), (4, 35, 47), (4, 36, 40), (4, 37, 41), (4,
30, 43), (5, 31, 44), (5, 32, 45), (5, 33, 46), (5, 34, 47), (5, 35, 42), (5,
36, 41), (0, 37, 40), (1, 30, 44), (2, 32, 45), (3, 33, 47)]
```

15

描述：

enumerate(iterable, start=0)同时成对返回计数值和列表值，为一个枚举对象（return an enumerate object）。对列表执行enumerate()之后，在使用循环语句时可以将成对的计数值和元素值分别赋予给两个变量，如idx和c。如果并不在for循环中直接序列解包，赋予了一个变量，如i，则其如(0, (0, 30, 47))，仍然需要索引方式或序列解包方式（idx,c=i）提取值。

在打印字符时，如果不换行，可以增加参数end=''来避免起新行，结果如图pcs.4-5所示。

代码段：

```
for idx, c in enumerate(color_lst):
    color = ';'.join([str(i) for i in c])
    s = '\x1b[%sm %s \x1b[0m' % (color, idx)
    print(s, end='')
print('\n', "_"*50, '\n')
for i in enumerate(color_lst):
    # print(i)
    idx, c = i
    color = ';'.join([str(i) for i in c])
    s = '\x1b[%sm %s \x1b[0m' % (color, idx)
    print(s, end='')
```

运算结果：

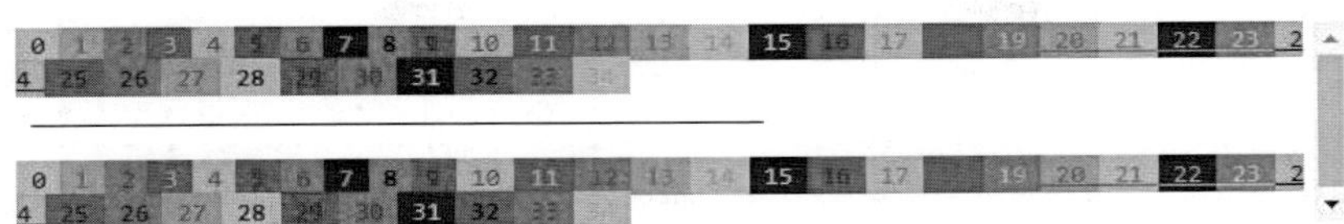

图pcs.4-5 字体颜色显示示例3

16

描述：

- 循环字典——键值对形式

以键值对形式循环字典是经常使用到的一种方式，可以很方便地同时提取键名和元素值，并在每一次循环中同时处理键名和元素值，再成对输出。例如，建立了一个空字典name2NewColor_dict，在每次成对循环原有字典值时，修改了颜色值（+1），并按键名和新颜色值成对存储在新建的字典中。

对于新键字典也在终端打印具有色彩的字符串（图pcs.4-6），因为ANSI code格式颜色有值域，如果超出范围，则可以看到对应部分不会发生颜色变化。

这里在调试代码时，直接使用了break语句。没有结合条件语句，因此该循环在调试时只执行一次循环。

代码段：

```
name2color_dict = {label.name: label.color for label in labels}
print(name2color_dict)
name2NewColor_dict = {}
for name, color in name2color_dict.items():
    c = ‘;’.join([str(i) for i in color])
    s = ‘\x1b[%sm %s \x1b[0m’ % (c, name)
    print(s, end=‘’)
    name2NewColor_dict[name] = (i+1 for i in color)
    # break
print(‘\n’, “_”*50, ‘\n’)
for name, color in name2NewColor_dict.items():
    c = ‘;’.join([str(i) for i in color])
    s = ‘\x1b[%sm %s \x1b[0m’ % (c, name)
    print(s, end=‘’)
```

运算结果：

```
{‘unlabeled’: (0, 30, 47), ‘ego_vehicle’: (0, 31, 46), ‘rectification_border’:
(0, 32, 45), ‘out_of_roi’: (0, 33, 44), ‘static’: (0, 34, 43), ‘dynamic’: (1,
35, 42), ‘ground’: (1, 36, 41), ‘road’: (1, 37, 40), ‘sidewalk’: (1, 30, 41),
‘parking’: (1, 31, 42), ‘rail_track’: (2, 32, 43), ‘building’: (2, 33, 44),
‘wall’: (2, 34, 45), ‘fence’: (2, 35, 46), ‘guard_rail’: (2, 36, 47), ‘bridge’:
(3, 37, 40), ‘tunnel’: (3, 30, 42), ‘pole’: (3, 31, 43), ‘polegroup’: (3, 32,
44), ‘traffic_light’: (3, 33, 45), ‘traffic_sign’: (4, 34, 46), ‘vegetation’:
(4, 35, 47), ‘terrain’: (4, 36, 40), ‘sky’: (4, 37, 41), ‘person’: (4, 30, 43),
‘rider’: (5, 31, 44), ‘car’: (5, 32, 45), ‘truck’: (5, 33, 46), ‘bus’: (5,
34, 47), ‘caravan’: (5, 35, 42), ‘trailer’: (5, 36, 41), ‘train’: (0, 37, 40),
‘motorcycle’: (1, 30, 44), ‘bicycle’: (2, 32, 45), ‘license_plate’: (3, 33, 47)}
```

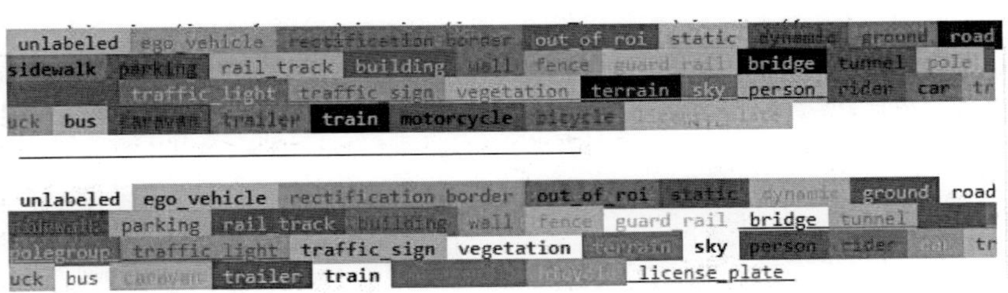

图pcs.4-6 字体颜色显示示例4

17

描述：

● zip()和map()

zip(*iterables)将多个列表（序列）返回为成对的值，for Loops可以逐个成对循环。下述示例使用了ANSI code格式颜色256模式，用catId作为背景颜色，未配置字体颜色；同时，字符串之间增加了一个空格，断开名称，结果如图pcs.4-7所示。

代码段：

```
name_lst = [label.name for label in labels]
catId_lst = [label.catId for label in labels]
for name, catId in zip(name_lst, catId_lst):
    s = '\033[48;5;%dm%s\033[0;0m ' % (catId, name)
    print(s, end='')
    # break
```

运算结果：

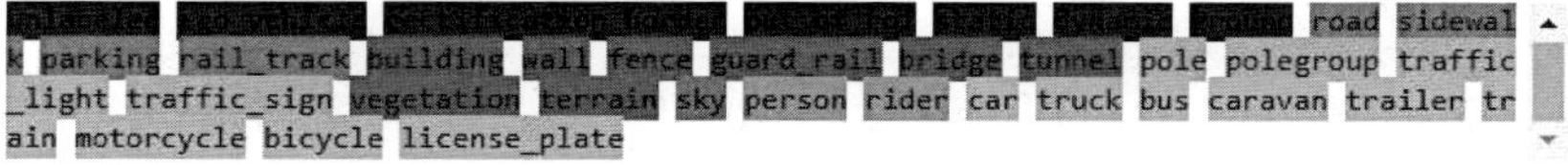

图pcs.4-7 字体颜色显示示例5

18

描述：

map(func, *iterables)输入参数func，自定义为lambda（'lamdə）函数，并给了两个输入参数x和n，返回一个元组，其中一个值保持不变（即name名称），另一个值加1（即用作颜色值的catId加1）。用for Loops可以逐个循环给定列表中的值经过map()中func参数函数的计算的返回值，结果如图pcs.4-8所示。

代码段：

```
xAdd33_Wname = lambda x,n:(n,x+33)
for name, catId_new in map(xAdd33_Wname, catId_lst, name_lst):
    # print( name,color)
    s = '\033[48;5;%dm%s\033[0;0m ' % (catId_new, name)
    print(s, end='')
    # break
```

运算结果：

图pcs.4-8 字体颜色显示示例6

19

描述：

● 循环range(len())

在组织列表中数值之间的运算模式时，经常通过列表的不同索引组合规律相加、相乘，或者任何更为复杂的计算来获取新的符合某一规律的列表值。例如，regularAdd_1为逐个计算列表中相邻两个值之和（即循环时为当前索引对应值和其后索引对应值之和）；regularAdd_2为循环时，计算当前索引之前所有值之和；而regularAdd_3则结合slicing，实现间隔相加的结果。

代码段：

```
import random
```

```
for idx in range(len(name_lst)):
    print(‘%d-%s;’ % (idx, name_lst[idx]), end=’’)
lst = [random.randint(10, 30) for i in range(10)]
print(‘\n’, “_”*50)
print(lst)
regularAdd_1 = []
regularAdd_2 = []
for i in range(len(lst)-1):
    regularAdd_1.append(lst[i]+lst[i+1])
    regularAdd_2.append(sum([lst[i] for i in range(i+1)]))
regularAdd_3 = []
for i in range(1, len(lst)-1, 2):
    print(i)
    regularAdd_3.append(lst[i]+lst[i+2])
print(regularAdd_1, ‘\n’, regularAdd_2, ‘\n’, regularAdd_3)
```

运算结果：

```
0-unlabeled;1-ego_vehicle;2-rectification_border;3-out_of_roi;4-static;5-
dynamic;6-ground;7-road;8-sidewalk;9-parking;10-rail_track;11-building;12-
wall;13-fence;14-guard_rail;15-bridge;16-tunnel;17-pole;18-polegroup;19-traffic_
light;20-traffic_sign;21-vegetation;22-terrain;23-sky;24-person;25-rider;26-
car;27-truck;28-bus;29-caravan;30-trailer;31-train;32-motorcycle;33-bicycle;34-
license_plate;
 __________________________________________________________________________________
[28, 24, 16, 19, 29, 20, 18, 11, 26, 21]
1
3
5
7
[52, 40, 35, 48, 49, 38, 29, 37, 47]
 [28, 52, 68, 87, 116, 136, 154, 165, 191]
 [43, 39, 31, 32]
```

20

描述：

- 嵌套循环（nested for loops）

如果要获取嵌套列表、嵌套字典或这任何包含多个嵌套关系的数据结构，或者需要多个序列值运算处理，通常需要用嵌套循环逐层的拆解或嵌套循环不同序列运算。下述示例对同一个列表执行嵌套循环，并相加。为了方便观察数据关系，使用列表推导式切分为嵌套列表，并循逐行环打印，可以看到构建了一个列表中每一个元素与该列表全部值相加的矩阵，这可用于计算多个点列表、两两点之间距离的成本矩阵（起点-目的地 (OD) 成本矩阵）、用于城市交通等分析。

代码段：

```
lst = [random.randint(10, 30) for i in range(10)]
print(lst)
regularAdd_4 = []
for i in range(len(lst)):
    for j in range(len(lst)):
        regularAdd_4.append(lst[i]+lst[j])
n = len(lst)
regularAdd_4_chunks = [regularAdd_4[i:i+n]
                       for i in range(0, len(regularAdd_4), n)]
for sub_lst in regularAdd_4_chunks:
    print(sub_lst)
```

运算结果：

```
[12, 18, 28, 24, 20, 14, 27, 25, 18, 23]
[24, 30, 40, 36, 32, 26, 39, 37, 30, 35]
[30, 36, 46, 42, 38, 32, 45, 43, 36, 41]
[40, 46, 56, 52, 48, 42, 55, 53, 46, 51]
[36, 42, 52, 48, 44, 38, 51, 49, 42, 47]
[32, 38, 48, 44, 40, 34, 47, 45, 38, 43]
[26, 32, 42, 38, 34, 28, 41, 39, 32, 37]
[39, 45, 55, 51, 47, 41, 54, 52, 45, 50]
[37, 43, 53, 49, 45, 39, 52, 50, 43, 48]
[30, 36, 46, 42, 38, 32, 45, 43, 36, 41]
[35, 41, 51, 47, 43, 37, 50, 48, 41, 46]
```

21

描述：

- for Loops中的星号*（asterisk）

python中*和**，除了作为运算符或者字符串中的特殊字符外，通常作为前缀运算符（prefix operators），即在变量之前使用*和**运算符。*和**运算符具有丰富的用法。下述应用到循环中的示例是用*来收集多个元素值。

代码段：

```
for a, *b, c in [(1, 2, 3, 4, 5), (5, 6, 7, 8)]:
    print(a, b, c)
```

运算结果：

```
1 [2, 3, 4] 5
5 [6, 7] 8
```

知识点：

4.4 循环语句_while 模式

●●●22

描述：

基于语法为：

```
while test:
   statements
else:
   statements
Copy to clipboardErrorCopied
```

或

```
while test:
   statements
   if test:break
   if test: continue
else:
   statements
```

while的关键是处理循环停止的条件，一种是在while之后给出停止条件，例如while x<=10:，只要不满足条件就会停止循环。这时，给出的条件变量通常在while代码块中参与执行相关的运算并会因为变化而不满足给出的条件后跳出循环，例如x+=1；或者在while代码块内给出条件语句，配合使用break停止循环，类似于for循环中的break。

代码段：

```
x = 1
while x <= 10:
    print(x, ';', end='')
    x += 1
print('\n', "_"*50)
x = 1
while True:
    print(x, ';', end='')
    x += 1
    if x > 10:
        break
print('\n', "_"*50)
x = 1
for i in range(1000):
    print(x, ';', end='')
    x += 1
    if x > 10:
        break
```

运算结果：

```
1 ;2 ;3 ;4 ;5 ;6 ;7 ;8 ;9 ;10 ;
 __________________________________________________
1 ;2 ;3 ;4 ;5 ;6 ;7 ;8 ;9 ;10 ;
```

```
 ----------------------------------------------------------------------------
1 ;2 ;3 ;4 ;5 ;6 ;7 ;8 ;9 ;10 ;
```

23

描述：

将while True/break用于input()函数，结果如图pcs.4-9所示。

代码段：

```
while True:
    value = input(
        'To calculate the square root. Enter a number:(enter "stop" to stop the
operation)')
    if value == 'stop':
        break
    import math
    print("The square root of \033[1;31;40m%s\033[0m is \033[1;31;40m%.2f\033[0m."
% (
        value, math.sqrt(float(value))))
```

运算结果：

```
To calculate the square root. Enter a number:(enter "stop" to stop the operation) 5
The square root of [illegible] is [illegible].
To calculate the square root. Enter a number:(enter "stop" to stop the operation) 6
The square root of [illegible] is [illegible].
To calculate the square root. Enter a number:(enter "stop" to stop the operation) 7
The square root of [illegible] is [illegible].
To calculate the square root. Enter a number:(enter "stop" to stop the operation) stop
```

图pcs.4-9　字体颜色显示示例7

24

描述：

可以给定不同的要求，使用多个elif的方式分别计算。不过，对于这种在终端交互的方式通常是使用既有的交互图表库。

代码段：

```
while True:
    import math
    command = input('sqrt, power,sum, or stop:')
    if command == 'sqrt':
        number = input('Enter 1 number:')
        print('square root=%.2f' % math.sqrt(float(number)))
    elif command == 'power':
        number = input('Enter 1 number:')
        print('power=%.2f' % math.pow(float(number), 2))
    elif command == 'sum':
        numbers = input('Enter 2 numbers, separated by commas:')
        print('sum=%.2f' % sum([float(v) for v in numbers.split(",")]))
    elif command == 'stop':
        break
```

运算结果：

```
sqrt, power,sum, or stop: sqrt
Enter 1 number: 5
square root=2.24
sqrt, power,sum, or stop: power
Enter 1 number: 5
power=25.00
sqrt, power,sum, or stop: sum
Enter 2 numbers, separated by commas: 5,6
sum=11.00
sqrt, power,sum, or stop: stop
```

25

描述：

将while True/break用于文件读取。注意在书写代码时，是将poi_lst.append(POI._

make(line.split(",")))语句写于if not line:break中断语句之后。先判断是否已到行末尾（读取的是否为空行），否则会提示运行错误，因为空行不能执行line.split(",")运算。

代码段：

```
xian_poi_fn = './data/xian_poi.csv'  # 存储有西安 POI, point of interesting 兴趣点数据
f = open(xian_poi_fn, 'r', encoding="utf-8")
POI = namedtuple('POI', ['idx', 'unknown', 'name', 'lat',
                 'lon', 'category', 'score', 'price', 'x', 'y'])
poi_lst = []
while True:
    line = f.readline()
    if not line:
        break
    poi_lst.append(POI._make(line.split(",")))
    # break # 调试用
f.close()
print('The length of the poi list is %d.' % len(poi_lst))
print("_"*50)
for i in poi_lst[:3]:
    print(i, '\n')
```

运算结果：

```
The length of the poi list is 13732.
__________________________________________________________________________________
POI(idx='1', unknown='0101000020897F000008D599CB26E312418530CECF16EB4C41',
name='美香源', lat='34.23709808337344', lon='108.93100212046282', category='美食;
中餐厅', score='0', price='', x='309449.6988290106', y='3790381.623479905\n')

POI(idx='2', unknown='0101000020897F0000B038F4CF9BE31241389AD606BFEC4C41', name='
雷记澄城水盆羊肉(红樱路店)', lat='34.244750060429915', lon='108.93113243785623',
category='美食;中餐厅', score='3.9', price='20.0', x='309478.95308006834',
y='3791230.053424146\n')

POI(idx='3', unknown='0101000020897F00005C24B17F97E3124160F86A81E5EC4C41', name='
段府农家菠菜面(红缨路店)', lat='34.245443456204875', lon='108.93110375582032',
category='美食;中餐厅', score='4', price='', x='309477.8746991807',
y='3791307.011076972\n')
```

知识点：

4.5 列表推导式（comprehension）

26

描述：

因为使用for loops会占据多行，而只要不是复杂的循环，使用列表推导式是优先选择。不仅会简化代码行，而且书写起来相对要便捷得多。其语法为newlist=[expression for item in iterable if condition == True]，或者newlist=[expression1 if condition == True else expression2 for item in iterable]。

代码段：

```
lst = [9, 8, 7, 6, 5, 4, 3]
for i in range(len(lst)):
    lst[i] += 10
print(lst)
lst = [9, 8, 7, 6, 5, 4, 3]
lst_A = [i+10 for i in lst]
print(lst_A)
```

运算结果：

```
[19, 18, 17, 16, 15, 14, 13]
[19, 18, 17, 16, 15, 14, 13]
```

27

描述：

列表推导式也可以转换具有条件语句的for loops。

代码段：

```
lst = [9, 8, 7, 6, 5, 4, 3]
for i in range(len(lst)):
    if i % 2 == 0:
        lst[i] += 10
    else:
        lst[i] += 100
print(lst)
lst = [9, 8, 7, 6, 5, 4, 3]
lst_B = [lst[i]+10 if i % 2 == 0 else lst[i]+100 for i in range(len(lst))]
print(lst_B)
```

运算结果：

```
[19, 108, 17, 106, 15, 104, 13]
[19, 108, 17, 106, 15, 104, 13]
```

28

描述：

将嵌套循环转换为列表推导式计算。

代码段：

```
monogram_lst = []
for i in ‘abcd’:
    for j in ‘hijk’:
        monogram_lst.append(‘%s-%s’ % (i, j))
print(monogram_lst)
monogram_lst_A = [‘%s-%s’ % (i, j) for i in ‘abcd’ for j in ‘hijk’]
print(monogram_lst_A)
```

运算结果：

```
['a-h', 'a-i', 'a-j', 'a-k', 'b-h', 'b-i', 'b-j', 'b-k', 'c-h', 'c-i', 'c-j',
'c-k', 'd-h', 'd-i', 'd-j', 'd-k']
['a-h', 'a-i', 'a-j', 'a-k', 'b-h', 'b-i', 'b-j', 'b-k', 'c-h', 'c-i', 'c-j',
'c-k', 'd-h', 'd-i', 'd-j', 'd-k']
```

29

描述：

列表推导式用于dict数据结构。将hasInstances字段的布尔值转换为0或1。

代码段：

```
name2hasInstances_dict = {label.name: label.hasInstances for label in labels}
print(name2hasInstances_dict)
print({k: int(v) for k, v in name2hasInstances_dict.items()})
```

运算结果：

```
{‘unlabeled’: False, ‘ego_vehicle’: False, ‘rectification_border’: False, ‘out_
of_roi’: False, ‘static’: False, ‘dynamic’: False, ‘ground’: False, ‘road’:
False, ‘sidewalk’: False, ‘parking’: False, ‘rail_track’: False, ‘building’:
False, ‘wall’: False, ‘fence’: False, ‘guard_rail’: False, ‘bridge’: False,
‘tunnel’: False, ‘pole’: False, ‘polegroup’: False, ‘traffic_light’: False,
‘traffic_sign’: False, ‘vegetation’: False, ‘terrain’: False, ‘sky’: False,
‘person’: True, ‘rider’: True, ‘car’: True, ‘truck’: True, ‘bus’: True, ‘caravan’:
True, ‘trailer’: True, ‘train’: True, ‘motorcycle’: True, ‘bicycle’: True,
‘license_plate’: False}
{‘unlabeled’: 0, ‘ego_vehicle’: 0, ‘rectification_border’: 0, ‘out_of_roi’: 0,
‘static’: 0, ‘dynamic’: 0, ‘ground’: 0, ‘road’: 0, ‘sidewalk’: 0, ‘parking’: 0,
‘rail_track’: 0, ‘building’: 0, ‘wall’: 0, ‘fence’: 0, ‘guard_rail’: 0, ‘bridge’:
0, ‘tunnel’: 0, ‘pole’: 0, ‘polegroup’: 0, ‘traffic_light’: 0, ‘traffic_sign’:
0, ‘vegetation’: 0, ‘terrain’: 0, ‘sky’: 0, ‘person’: 1, ‘rider’: 1, ‘car’: 1,
‘truck’: 1, ‘bus’: 1, ‘caravan’: 1, ‘trailer’: 1, ‘train’: 1, ‘motorcycle’: 1,
‘bicycle’: 1, ‘license_plate’: 0}
```

参考文献（References）：

[1]（日）高桥 信著，陈刚译．株式会社TREND-PRO漫画制作．漫画统计学[M].科学出版社．北京，2019．8.

PCS-5. 函数（function）、作用域（scope）与命名空间（namespace）、参数（arguments）

PCS-5（　　）：5.1 定义函数；5.2 作用域（scope）和命名空间（namespace）；5.3 工厂函数（Factory Functions）；5.4 函数的输入参数（Arguments）；5.5 函数定义综合实验-自定义箱形图打印样式

知识点：

5.1 定义函数

••• 1

描述：

不定义函数，按行执行代码可以完成一个任务。但是，如果要用该组代码完成不止一次同样的计算任务，重复地复制代码。或者调整位于不同位置行的输入参数，则异常繁琐，并容易发生错误；如果所要处理的任务较为复杂，例如完成遗传算法或蚁群算法等复杂的任务，采取逐行执行代码的方式不是很现实。为了让代码书写更流畅、精简、易读，方便调试与减少出错率，以及代码的迁移，重复调用，函数定义必不可免。

函数定义的基本语法如下：

```
def name(arg1,arg2,...,argN):
    statemetns
    return value # 可以返回值，也可以移除该行，则返回值默认为空（None）
```

下述定义了两个小函数，通过获取当前时间和代码运行后时间与其差值，用于较大计算量代码运行时间长度的计算。`start_time()`函数没有输入参数，定义函数名为`start_time`，在函数内部，调入了一个时间模块datetime，使用`datetime.datetime.now()`方法获取当前时间，并赋值给变量`start_time`。打印当前时间，以及`return start_time`返回当前时间变量值。调用执行该函数时，如果需要接收函数的返回值，则赋值给一个变量。

第2个函数`duration(start_time)`有一个输入参数，为时间格式的值。函数块内部，同样计算了当前时间并赋值给变量`end_time`，由`(end_time-start_time).seconds/60`方法计算时间差，并将时间格式转换为易读的分钟形式。这个函数没有提供返回值。

注：是在函数内，还是在函数外调用模块（module，库的部分），需要根据具体情况衡量利弊。例如，如果模块调入在文件开始，则只需要调入一次；但是如果并不是所有函数均调用该模块，而该模块可能会与其他模块（库的安装）发生冲突，或者模块自身体量庞大，而使用者只使用该模块中的其他方法时，则函数调入在特定的函数内执行可能会合理。但是，需要注意，通常为在模块开始一次性全部调入所用到的库，而不是重复调用。

代码段：

```python
def start_time():
    import datetime
    start_time = datetime.datetime.now()
    print("start time:", start_time)
    return start_time
def duration(start_time):
    import datetime
    end_time = datetime.datetime.now()
    print("end time:", end_time)
    duration = (end_time-start_time).seconds/60
    print("Total time spend:%.2f minutes" % duration)
```

--- 2

代码段：

```python
s_t = start_time()
print(type(s_t))
```

运算结果：

```
start time: 2022-08-07 19:19:12.115484
<class 'datetime.datetime'>
```

3

代码段：

```
duration(s_t)
```

运算结果：

```
end time: 2022-08-07 19:19:18.036481
Total time spend:0.08 minutes
```

4

描述：

下例为应用上述组合函数的一个场景，计算代码`for i in range(10**8):value=i`运行的时间。

代码段：

```
s_t = start_time()
for i in range(10**8):
    value = i
duration(s_t)
```

运算结果：

```
start time: 2022-08-07 20:25:19.170276
end time: 2022-08-07 20:25:29.551021
Total time spend:0.17 minutes
```

5

描述：

- 多态性（polymorphism）—— 数据类型类

一些运算不仅对于数值起作用，同样对其他类型的数据起作用。例如，下述案例中定义了一个乘积函数`times(x,y)`，含两个输入参数，返回两者之积。python中，并不会在赋值变量或输入参数定义时，定义变量或输入参数的数据类型。python会自动判断数据类型，并根据提供的运算返回计算结果。这样的处理方式可以减轻程序员思考的负担，也使得语言精简并富有弹性。

代码段：

```
def times(x, y):
    print('- - '*3, 'X={};y={}'.format(x, y))
    return (x*y)
print(times(5, 7))
print(times([5], 3))
print(times('polymorphism_', 3))
```

运算结果：

```
- - - - - -  X=5;y=7
35
- - - - - -  X=[5];y=3
[5, 5, 5]
- - - - - -  X=polymorphism_;y=3
polymorphism_polymorphism_polymorphism_
```

6

描述：

如果不区分参数输入的数据类型，常常不容易直接判断参数类型是什么，因此可以增加形如`def func(arg1:type,arg2:type)->type`的函数注释。其中，`:type`为参数类型，`->type`为返回值类型。注意，Python解释器并不会因为这些注解而提供额外的校验，没有任何的类型检查工作。这些类型注解加不加，对代码来说没有任何影响。

代码段：

```
def times_type(x: float, y: float) -> float:
    print('- - '*3, 'X={};y={}'.format(x, y))
    return (x*y)
```

```
print(times_type(5, 7))
print(times_type([5], 3))
print(times_type('polymorphism_', 3))
```

运算结果：

```
- - - - - - -  X=5;y=7
35
- - - - - - -  X=[5];y=3
[5, 5, 5]
- - - - - - -  X=polymorphism_;y=3
polymorphism_polymorphism_polymorphism_
```

7

描述：

● 函数定义的诸多考量——定义描述性统计函数

这是一个略微复杂些的例子，用于对给定的一组数据做描述性统计分析。下述示例结果是书写调试，增加对函数功能、输入和输出参数说明后的最终定稿代码段。实际编写代码过程是一个反复修改、调试的过程，这不仅包括函数功能实现的内容的修改、增补或削减；也包括功能实现过程中，结构逻辑的调整，例如是否用字典的形式先计算所有的统计量。如果给定了measure的方法，这样的逻辑设计将会增加无关的计算量；再者，选用何种的方式计算这些统计量，不调用库，而自行根据公式编写计算流程，或调入哪个库计算，这可以用math、statistics、NumPy、pandas、SciPy等[①]任何方法。如果需要调入多个库，往往需要综合考虑，尽量减少库的调入数量；而返回值的形式也需要认真考虑。如果需要返回值参与到其他计算中，则字符串表述形式的返回形式是不合适的，需要考虑直接返回具体的数值。对于函数的定义，在实际的数据分析时，考虑的内容会因为所要解决问题的不同而存在差异，需要具体情况具体分析。

① math（https://docs.python.org/3/library/math.html）；statistics（https://docs.python.org/3/library/statistics.html）；NumPy（https://numpy.org/）；pandas（https://pandas.pydata.org/）；SciPy（https://scipy.org/）。

在PCS_4中，列举的函数是将一段代码调整为函数定义，这个过程需要注意对变量名的重新定义和重复代码段的调整。而该处的函数定义则直接定义函数，不涉及代码的转换。变量名的定义时，就已经考量到命名的一般性，以及上述所考量的内容。因此，在实际代码书写时，非必要单行时，则直接以函数形式定义，建议避免由逐行再转换为函数形式，因为转换过程会耗费不必要的精力。

代码段：

```
def descriptive_statistics(data, measure=None, decimals=2):
    '''
    计算给定数值列表的描述性统计值，包括数量、均值、标准差、方差、中位数、众数、最小值和最大值。

    Parameters
    ----------
    data : list(numerical)
        待统计的数值列表 .
    measure : str, optional
        包括：'count', 'mean', 'std', 'variance', 'median', 'mode', 'min', 'max'. The default is None.
    decimals : int, optional
        小数位数 . The default is 2.

    Returns
    -------
    dict
        如果不给定参数 measure，则以字典形式返回所有值；否则，返回给定 measure 对应值的表述字符串 .

    '''
    import statistics
    d_s = {
        'count': len(data),  #样本数
        'mean': round(statistics.mean(data), decimals),  #均值
        'std': round(statistics.stdev(data), decimals),  #标准差
        'variance': round(statistics.variance(data), decimals),  #方差
        'median': statistics.median(data),  #中位数
        'mode': statistics.mode(data),  #众数
        'min': min(data),  #最小值
        'max': max(data),  #最大值
    }
    if measure:
```

```
        return '{}={}'.format(measure, d_s[measure])
    else:
        return d_s
ramen_price_lst = [700, 850, 600, 650, 980, 750, 500, 890, 880,
                   700, 890, 720, 680, 650, 790, 670, 680, 900,
                   880, 720, 850, 700, 780, 850, 750, 780, 590,
                   650, 580, 750, 800, 550, 750, 700, 600, 800,
                   800, 880, 790, 790, 780, 600, 690, 680, 650,
                   890, 930, 650, 777, 700]
d_s_1 = descriptive_statistics(ramen_price_lst)
print(d_s_1)
print('--'*30)
d_s_2 = descriptive_statistics(ramen_price_lst, 'std')
print(d_s_2)
d_s_3 = descriptive_statistics(ramen_price_lst, measure='mean', decimals=1)
print(d_s_3)
```

运算结果：

```
{'count': 50, 'mean': 743.34, 'std': 108.26, 'variance': 11720.64, 'median':
750.0, 'mode': 700, 'min': 500, 'max': 980}
------------------------------------------------------------
std=108.26
mean=743.3
```

知识点：
5.2 作用域（scope）和命名空间（namespace）

8

描述：

● 对作用域的描述

Python中，变量的访问权限取决于该变量赋值的位置，这个位置所在的代码块称为该变量所属的作用域。首先，需要明确作用域是一个嵌套的关系，如图pcs.5-1所示；同时，需要明确这作用域的嵌套关系作用于一个文件（或模块，module）。如果要使用另一个文件内定义的方法（函数），或属性（变量），则需要使用import方法调入模块。整个模块（或文件）即为内置作用域，包含Python自身内置函数所包括的各类函数或方法名称，例如示例中的print()函数，以及open()、range()等，可以通过import builtins;print(dir(builtins))查看所有内容。内置作用域可以访问函数外定义的变量，但是无法访问函数内定义的变量。除非使用global关键字，将在函数内（局部作用域）定义的变量声明为全局变量，从而函数外也可以访问。闭包局部作用域是函数内作用域，只有函数内对象可以访问的变量。但是，如果含有嵌套函数，则嵌套函数形成一个局部作用域，嵌套函数外的对象无法访问并以此类推，除非使用nonlocal关键字。当将嵌套局部作用域外定义的变量在嵌套局部作用域内更新时，嵌套局部作用域外的变量值也会发生对应改变。从上述描述可以注意到，作用域是从外层作用域中剥离内层作用域的过程，外层作用域无法访问内层作用域变量，而内层作用域可以访问外层作用域变量。如果外层作用域要访问内层作用域，则需要是使用global或nonlocal关键字。其中，global可以在任何内层作用域中使用，将变量声明为全局变量；但nonlocal仅作用于函数内使用，并需要在外层存在有该变量名，只是在内层操作更新该变量时，外层的变量对应更新。

通常将这4个作用域Built-in Scope、Global Scope、Enclosed Local Scope和Nested Local Scope缩写为B、G、E、L。

● 命名空间

命名空间是为了防止项目中命名冲突的一种机制。如果代码量较大，项目内定义的变量名较多，必然容易发生重复命名的事件。而命名空间与作用域对应，既然为命名空间，不同作用域之间变量名没有关联，可以用同样的名称在不同作用域中定义，但是需要注意不能与Python内置函数名称和关键字名称同。可以将命名空间对应作用域分为内置名称（Built-tin Names）、全局名称（Global Names）和局部名称（Local Names）。命名空间的查找顺序为局部名称到全局名称到内置名称。如果找不到变量，则会引发NameError异常。

内层作用域要访问外层作用域的变量，最好是通过函数的输入参数调入，而不是直接使用。避免在局部代码迁移时发生变量名未找到的错误，也能够更好、更清晰地组织代码结构，避免内

```
G_1=3.5
G_2=6.0

def outer():
    L_E_1=5.3

    def inner():
        L_N_1=6.7
        print('G_1={};\nG_2={};\nL_E_1={};\nL_N_1={}'.format(G_1,G_2,L_E_1,L_N_1))

    inner()

outer()

print('--'*30)
print('G_1={};G_2={};'.format(G_1,G_2))
```

Nested Local Scope（嵌套局部作用域）

Enclosed Local Scope（闭包局部作用域）

Global Scope（全局作用域）

Built-in Scope（内置作用域）

```
G_1=3.5
G_2=6.0

def outer():
    L_E_1=5.3

    global L_E_2
    L_E_2=7.9

    L_E_N_3=78

    def inner():
        L_N_1=6.7

        global L_N_2
        L_N_2=5.5

        nonlocal L_E_N_3
        L_E_N_3+=1

        print('Nested Scope:\nG_1={};\nG_2={};\nL_E_1={};\nL_E_2={};\nL_N_1
            ={}\nL_N_2={};\nL_N_3={}'.format(G_1,G_2,L_E_1,L_E_2,L_N_1,L_N_2,L_E_N_3
            ))

    print("_"*50)
    print('L_E_N_3={} in Enclosd Scope.'.format(L_E_N_3))
    inner()

outer()

print('--'*30)
print('Global Scope:\nG_1={};\nG_2={};\nL_E_2={};\nL_N_2={}'.format(G_1,G_2,L_E_2
    ,L_N_2))
```

Nested Local Scope（嵌套局部作用域）

Enclosed Local Scope（闭包局部作用域）

Global Scope（全局作用域）

Built-in Scope（内置作用域）

图pcs.5-1 LEGB 规则

外层命名混乱，削弱了代码的易读性。

代码段：

```
G_1 = 3.5
G_2 = 6.0
def outer():
    L_E_1 = 5.3
    global L_E_2
    L_E_2 = 7.9
    L_E_N_3 = 78
    def inner():
        L_N_1 = 6.7
        global L_N_2
        L_N_2 = 5.5
        nonlocal L_E_N_3
        L_E_N_3 += 1
        print('Nested Scope:\nG_1={};\nG_2={};\nL_E_1={};\nL_E_2={};\nL_N_1={}\
nL_N_2={};\nL_N_3={}'.format(
            G_1, G_2, L_E_1, L_E_2, L_N_1, L_N_2, L_E_N_3))
    print("_"*50)
    print('L_E_N_3={} in Enclosd Scope.'.format(L_E_N_3))
    inner()
outer()
print('--'*30)
print('Global Scope:\nG_1={};\nG_2={};\nL_E_2={};\nL_N_2={}'.format(
    G_1, G_2, L_E_2, L_N_2))
```

运算结果：

```
--------------------------------------------------------------------------------
L_E_N_3=78 in Enclosd Scope.
Nested Scope:
G_1=3.5;
G_2=6.0;
L_E_1=5.3;
L_E_2=7.9;
L_N_1=6.7
L_N_2=5.5;
L_N_3=79
--------------------------------------------------------------------------------
Global Scope:
G_1=3.5;
G_2=6.0;
L_E_2=7.9;
L_N_2=5.5
```

9

描述：

外层作用域无法访问内层作用域中定义的变量

代码段：

```
print(L_E_1)
```

运算结果：

```
--------------------------------------------------------------------------------

NameError                                 Traceback (most recent call last)
Input In [30], in <cell line: 1>()
----> 1 print(L_E_1)
NameError: name 'L_E_1' is not defined
```

10

代码段：

```
import builtins
print(dir(builtins))
```

运算结果：

```
['ArithmeticError', 'AssertionError', 'AttributeError', 'BaseException',
'BlockingIOError', 'BrokenPipeError', 'BufferError', 'BytesWarning',
```

```
'ChildProcessError', 'ConnectionAbortedError', 'ConnectionError',
'ConnectionRefusedError', 'ConnectionResetError', 'DeprecationWarning', 'EOFError',
'Ellipsis', 'EnvironmentError', 'Exception', 'False', 'FileExistsError',
'FileNotFoundError', 'FloatingPointError', 'FutureWarning', 'GeneratorExit', 'IOError',
'ImportError', 'ImportWarning', 'IndentationError', 'IndexError', 'InterruptedError',
'IsADirectoryError', 'KeyError', 'KeyboardInterrupt', 'LookupError', 'MemoryError',
'ModuleNotFoundError', 'NameError', 'None', 'NotADirectoryError', 'NotImplemented',
'NotImplementedError', 'OSError', 'OverflowError', 'PendingDeprecationWarning',
'PermissionError', 'ProcessLookupError', 'RecursionError', 'ReferenceError',
'ResourceWarning', 'RuntimeError', 'RuntimeWarning', 'StopAsyncIteration',
'StopIteration', 'SyntaxError', 'SyntaxWarning', 'SystemError', 'SystemExit', 'TabError',
'TimeoutError', 'True', 'TypeError', 'UnboundLocalError', 'UnicodeDecodeError',
'UnicodeEncodeError', 'UnicodeError', 'UnicodeTranslateError', 'UnicodeWarning',
'UserWarning', 'ValueError', 'Warning', 'WindowsError', 'ZeroDivisionError', '__
IPYTHON__', '__build_class__', '__debug__', '__doc__', '__import__', '__loader__',
'__name__', '__package__', '__spec__', 'abs', 'all', 'any', 'ascii', 'bin', 'bool',
'breakpoint', 'bytearray', 'bytes', 'callable', 'chr', 'classmethod', 'compile',
'complex', 'copyright', 'credits', 'delattr', 'dict', 'dir', 'display', 'divmod',
'enumerate', 'eval', 'exec', 'execfile', 'filter', 'float', 'format', 'frozenset', 'get_
ipython', 'getattr', 'globals', 'hasattr', 'hash', 'help', 'hex', 'id', 'input',
'int', 'isinstance', 'issubclass', 'iter', 'len', 'license', 'list', 'locals', 'map', 'max',
'memoryview', 'min', 'next', 'object', 'oct', 'open', 'ord', 'pow', 'print', 'property',
'range', 'repr', 'reversed', 'round', 'runfile', 'set', 'setattr', 'slice', 'sorted',
'staticmethod', 'str', 'sum', 'super', 'tuple', 'type', 'vars', 'zip']
```

11

描述:

- 局部作用域对全局作用域变量的更新

在配置参数时，参数值往往需要配置为不同值观察比较计算结果。例如，如果参数值为全局变量，可以定义下述示例函数来更新该变量，使得代码易读、不容易发生混淆，尤其避免不容易查找到的错误出现。

代码段:

```
g_var = 5927
def setGvar(new_Gvar):
    global g_var
    g_var = new_Gvar
setGvar(9527)
print(g_var)
```

运算结果:

```
9527
```

知识点:

5.3 工厂函数（Factory Functions）

12

描述:

工厂函数类似于类方法（Class，称为工厂方法）的本质，可以实例化外层函数，再显式地调用嵌套函数。只是只能返回一个嵌套函数，而不能并行多个内层函数。整个过程为当调用外层函数，并将其赋值给一个变量ds，即实例化，该变量称为实例化对象。此时，运行到嵌套函数时只是完成对嵌套函数的定义，并不执行该函数；当执行实例化对象时ds(ramen_price_lst)，将会完成对内层函数的调用。

代码段:

```
def descriptive_statistics_factory(decimals=2):
    def std(data):
        import statistics
        return round(statistics.stdev(data), decimals)
    return std
ramen_price_lst = [700, 850, 600, 650, 980, 750, 500, 890, 880,
                   700, 890, 720, 680, 650, 790, 670, 680, 900,
                   880, 720, 850, 700, 780, 850, 750, 780, 590,
                   650, 580, 750, 800, 550, 750, 700, 600, 800,
                   800, 880, 790, 790, 780, 600, 690, 680, 650,
```

```
                     890, 930, 650, 777, 700]
ds_a = descriptive_statistics_factory(5)
print(ds_a)
ds_a(ramen_price_lst)
```

运算结果：

```
<function descriptive_statistics_factory.<locals>.std at 0x0000026CD4E8DAF0>
108.26189
```

---------- 13

描述：

可以实例化多个对象，例如下述实例化为输入参数decimals为3的实例对象ds_b，并多次调用该实例化对象，计算不同列表值的标准差。

代码段：

```
ds_b = descriptive_statistics_factory(3)
print(ds_b(ramen_price_lst))
course_grade_lst = [90, 81, 73, 97, 85]
print(ds_b(course_grade_lst))
```

运算结果：

```
108.262
9.066
```

知识点：

5.4 函数的输入参数（Arguments）

•••••••••• 14

描述：

● mutable（可变）和immutable（不可变）数据结构作为输入参数

数据结构含有mutable（可变）和immutable（不可变）两种类型，对应到函数的输入参数则为不可变参数（immutable arguments）和可变参数（mutable arguments）。对于不可变参数，诸如整数（int）、字符串（string），是按值传递（by value）。虽然通过引用（reference）而非复制（copy）来传递参数值，但不可变对象无法原地更改，因此效果同复制；对于可变参数，诸如列表和字典，则是通过指针传递（by pointer），类似于C语言的指针传递方式，可变参数可以就地更改，因此函数内部传入可变参数，改变引用的可变参数，对应的全局变量也会发生变化。为防止改变全局变量，通常通过复制的方法copy()避免此类更改。

下述定义的3个函数，第一个直接使用全局变量；第二个传入参数，直接引用全局变量；第3个传入参数，复制引用的全局变量。可以发现前2个定义的函数都更新了全局变量，但第3个因为复制而没有更新的全局变量。

代码段：

```
biology_score_dict = {“Mason”: 59, “Reece”: 73, ‘A’: 47, ‘B’: 38, ‘C’: 63, ‘D’:
56, ‘E’: 75,
                      ‘F’: 53, ‘G’: 80, ‘H’: 50, ‘I’: 41, ‘J’: 62, ‘K’: 44, ‘L’:
26, ‘M’: 91, ‘N’: 35, ‘O’: 53, ‘P’: 68}
name, new_score = ‘Reece’, 100
print(biology_score_dict)
print(“_”*50)
def biology_score_update_A(name, new_score):
    biology_score_dict[name] = new_score
score_update_A(name, new_score)
print(biology_score_dict)
print(“--”*30)
def biology_score_update_B(score_dict, name, new_score):
    score_dict[name] = new_score
    return score_dict
biology_score_updated_B = biology_score_update_B(
    biology_score_dict, ‘Mason’, 100)
print(biology_score_dict, ‘\n’, biology_score_updated_B)
print(“--”*30)
def biology_score_update_C(score_dict, name, new_score):
    import copy
```

```
    score_dict_copy = copy.copy(score_dict)
    score_dict_copy[name] = new_score
    return score_dict_copy
biology_score_updated_C = biology_score_update_C(biology_score_dict, ‘A’, 100)
print(biology_score_dict, ‘\n’, biology_score_updated_C)
```

运算结果：

```
{‘Mason’: 59, ‘Reece’: 73, ‘A’: 47, ‘B’: 38, ‘C’: 63, ‘D’: 56, ‘E’: 75, ‘F’: 53, ‘G’:
80, ‘H’: 50, ‘I’: 41, ‘J’: 62, ‘K’: 44, ‘L’: 26, ‘M’: 91, ‘N’: 35, ‘O’: 53, ‘P’: 68}
----------------------------------------------------------------------------
{‘Mason’: 59, ‘Reece’: 100, ‘A’: 47, ‘B’: 38, ‘C’: 63, ‘D’: 56, ‘E’: 75, ‘F’:
53, ‘G’: 80, ‘H’: 50, ‘I’: 41, ‘J’: 62, ‘K’: 44, ‘L’: 26, ‘M’: 91, ‘N’: 35, ‘O’:
53, ‘P’: 68}
----------------------------------------------------------------------------
{‘Mason’: 100, ‘Reece’: 100, ‘A’: 47, ‘B’: 38, ‘C’: 63, ‘D’: 56, ‘E’: 75, ‘F’:
53, ‘G’: 80, ‘H’: 50, ‘I’: 41, ‘J’: 62, ‘K’: 44, ‘L’: 26, ‘M’: 91, ‘N’: 35, ‘O’:
53, ‘P’: 68}
 {‘Mason’: 100, ‘Reece’: 100, ‘A’: 47, ‘B’: 38, ‘C’: 63, ‘D’: 56, ‘E’: 75, ‘F’:
53, ‘G’: 80, ‘H’: 50, ‘I’: 41, ‘J’: 62, ‘K’: 44, ‘L’: 26, ‘M’: 91, ‘N’: 35, ‘O’:
53, ‘P’: 68}
----------------------------------------------------------------------------
{‘Mason’: 100, ‘Reece’: 100, ‘A’: 47, ‘B’: 38, ‘C’: 63, ‘D’: 56, ‘E’: 75, ‘F’:
53, ‘G’: 80, ‘H’: 50, ‘I’: 41, ‘J’: 62, ‘K’: 44, ‘L’: 26, ‘M’: 91, ‘N’: 35, ‘O’:
53, ‘P’: 68}
 {‘Mason’: 100, ‘Reece’: 100, ‘A’: 100, ‘B’: 38, ‘C’: 63, ‘D’: 56, ‘E’: 75, ‘F’:
53, ‘G’: 80, ‘H’: 50, ‘I’: 41, ‘J’: 62, ‘K’: 44, ‘L’: 26, ‘M’: 91, ‘N’: 35, ‘O’:
53, ‘P’: 68}
```

15

描述：

对于复制需要注意，包含浅复制copy.copy()和深复制copy.deepcopy()。如下述案例，对于嵌套字典或列表为参数值传递，修改嵌套部分的值时，对于浅复制，全局变量值仍会发生改变；而深复制，则可以避免嵌套字典或列表对全局对应变量的更改。

代码段：

```
import copy
test_score_dic = {“English”: {“Mason”: 90, “Reece”: 81, ‘A’: 73, ‘B’: 97, ‘C’:
85, ‘D’: 60, ‘E’: 74, ‘F’: 64, ‘G’: 72, ‘H’: 67, ‘I’: 87, ‘J’: 78, ‘K’: 85, ‘L’:
96, ‘M’: 77, ‘N’: 100, ‘O’: 92, ‘P’: 86},
                  “Chinese”: {“Mason”: 71, “Reece”: 90, ‘A’: 79, ‘B’: 70, ‘C’:
67, ‘D’: 66, ‘E’: 60, ‘F’: 83, ‘G’: 57, ‘H’: 85, ‘I’: 93, ‘J’: 89, ‘K’: 78, ‘L’:
74, ‘M’: 65, ‘N’: 78, ‘O’: 53, ‘P’: 80},
                  “history”: {“Mason”: 73, “Reece”: 61, ‘A’: 74, ‘B’: 47, ‘C’:
49, ‘D’: 87, ‘E’: 69, ‘F’: 65, ‘G’: 36, ‘H’: 7, ‘I’: 53, ‘J’: 100, ‘K’: 57, ‘L’:
45, ‘M’: 56, ‘N’: 34, ‘O’: 37, ‘P’: 70},
                  “biology”: {“Mason”: 59, “Reece”: 73, ‘A’: 47, ‘B’: 38, ‘C’:
63, ‘D’: 56, ‘E’: 75, ‘F’: 53, ‘G’: 80, ‘H’: 50, ‘I’: 41, ‘J’: 62, ‘K’: 44, ‘L’:
26, ‘M’: 91, ‘N’: 35, ‘O’: 53, ‘P’: 68},
                  }
print(test_score_dic)
print(“_”*50)
def test_score_update_A(score_dict, subject, name, new_score):
    import copy
    score_dict_copy = copy.copy(score_dict)
    score_dict_copy[subject][name] = new_score
    return score_dict_copy
score_dict, subject, name, new_score = test_score_dic, ‘biology’, ‘Reece’, ‘100’
test_score_updated_A = test_score_update_A(
    score_dict, subject, name, new_score)
print(test_score_dic, ‘\n’, test_score_updated_A)
print(“--”*30)
def test_score_update_B(score_dict, subject, name, new_score):
    import copy
    score_dict_copy = copy.deepcopy(score_dict)
    score_dict_copy[subject][name] = new_score
    return score_dict_copy
score_dict, subject, name, new_score = test_score_dic, ‘biology’, ‘Mason’, ‘100’
test_score_updated_B = test_score_update_B(
    score_dict, subject, name, new_score)
print(test_score_dic, ‘\n’, test_score_updated_B)
```

运算结果：

```
{'English': {'Mason': 90, 'Reece': 81, 'A': 73, 'B': 97, 'C': 85, 'D': 60, 'E': 74,
'F': 64, 'G': 72, 'H': 67, 'I': 87, 'J': 78, 'K': 85, 'L': 96, 'M': 77, 'N': 100,
'O': 92, 'P': 86}, 'Chinese': {'Mason': 71, 'Reece': 90, 'A': 79, 'B': 70, 'C': 67,
'D': 66, 'E': 60, 'F': 83, 'G': 57, 'H': 85, 'I': 93, 'J': 89, 'K': 78, 'L': 74,
'M': 65, 'N': 78, 'O': 53, 'P': 80}, 'history': {'Mason': 73, 'Reece': 61, 'A': 74,
'B': 47, 'C': 49, 'D': 87, 'E': 69, 'F': 65, 'G': 36, 'H': 7, 'I': 53, 'J': 100, 'K':
57, 'L': 45, 'M': 56, 'N': 34, 'O': 37, 'P': 70}, 'biology': {'Mason': 59, 'Reece':
73, 'A': 47, 'B': 38, 'C': 63, 'D': 56, 'E': 75, 'F': 53, 'G': 80, 'H': 50, 'I':
41, 'J': 62, 'K': 44, 'L': 26, 'M': 91, 'N': 35, 'O': 53, 'P': 68}}
--------------------------------------------------------------------------------
{'English': {'Mason': 90, 'Reece': 81, 'A': 73, 'B': 97, 'C': 85, 'D': 60, 'E':
74, 'F': 64, 'G': 72, 'H': 67, 'I': 87, 'J': 78, 'K': 85, 'L': 96, 'M': 77, 'N':
100, 'O': 92, 'P': 86}, 'Chinese': {'Mason': 71, 'Reece': 90, 'A': 79, 'B': 70,
'C': 67, 'D': 66, 'E': 60, 'F': 83, 'G': 57, 'H': 85, 'I': 93, 'J': 89, 'K': 78,
'L': 74, 'M': 65, 'N': 78, 'O': 53, 'P': 80}, 'history': {'Mason': 73, 'Reece':
61, 'A': 74, 'B': 47, 'C': 49, 'D': 87, 'E': 69, 'F': 65, 'G': 36, 'H': 7, 'I':
53, 'J': 100, 'K': 57, 'L': 45, 'M': 56, 'N': 34, 'O': 37, 'P': 70}, 'biology':
{'Mason': 59, 'Reece': '100', 'A': 47, 'B': 38, 'C': 63, 'D': 56, 'E': 75, 'F':
53, 'G': 80, 'H': 50, 'I': 41, 'J': 62, 'K': 44, 'L': 26, 'M': 91, 'N': 35, 'O':
53, 'P': 68}}
 {'English': {'Mason': 90, 'Reece': 81, 'A': 73, 'B': 97, 'C': 85, 'D': 60, 'E':
74, 'F': 64, 'G': 72, 'H': 67, 'I': 87, 'J': 78, 'K': 85, 'L': 96, 'M': 77, 'N':
100, 'O': 92, 'P': 86}, 'Chinese': {'Mason': 71, 'Reece': 90, 'A': 79, 'B': 70,
'C': 67, 'D': 66, 'E': 60, 'F': 83, 'G': 57, 'H': 85, 'I': 93, 'J': 89, 'K': 78,
'L': 74, 'M': 65, 'N': 78, 'O': 53, 'P': 80}, 'history': {'Mason': 73, 'Reece':
61, 'A': 74, 'B': 47, 'C': 49, 'D': 87, 'E': 69, 'F': 65, 'G': 36, 'H': 7, 'I':
53, 'J': 100, 'K': 57, 'L': 45, 'M': 56, 'N': 34, 'O': 37, 'P': 70}, 'biology':
{'Mason': 59, 'Reece': '100', 'A': 47, 'B': 38, 'C': 63, 'D': 56, 'E': 75, 'F':
53, 'G': 80, 'H': 50, 'I': 41, 'J': 62, 'K': 44, 'L': 26, 'M': 91, 'N': 35, 'O':
53, 'P': 68}}
--------------------------------------------------------------------------------
{'English': {'Mason': 90, 'Reece': 81, 'A': 73, 'B': 97, 'C': 85, 'D': 60, 'E':
74, 'F': 64, 'G': 72, 'H': 67, 'I': 87, 'J': 78, 'K': 85, 'L': 96, 'M': 77, 'N':
100, 'O': 92, 'P': 86}, 'Chinese': {'Mason': 71, 'Reece': 90, 'A': 79, 'B': 70,
'C': 67, 'D': 66, 'E': 60, 'F': 83, 'G': 57, 'H': 85, 'I': 93, 'J': 89, 'K': 78,
'L': 74, 'M': 65, 'N': 78, 'O': 53, 'P': 80}, 'history': {'Mason': 73, 'Reece':
61, 'A': 74, 'B': 47, 'C': 49, 'D': 87, 'E': 69, 'F': 65, 'G': 36, 'H': 7, 'I':
53, 'J': 100, 'K': 57, 'L': 45, 'M': 56, 'N': 34, 'O': 37, 'P': 70}, 'biology':
{'Mason': 59, 'Reece': '100', 'A': 47, 'B': 38, 'C': 63, 'D': 56, 'E': 75, 'F':
53, 'G': 80, 'H': 50, 'I': 41, 'J': 62, 'K': 44, 'L': 26, 'M': 91, 'N': 35, 'O':
53, 'P': 68}}
 {'English': {'Mason': 90, 'Reece': 81, 'A': 73, 'B': 97, 'C': 85, 'D': 60, 'E':
74, 'F': 64, 'G': 72, 'H': 67, 'I': 87, 'J': 78, 'K': 85, 'L': 96, 'M': 77, 'N':
100, 'O': 92, 'P': 86}, 'Chinese': {'Mason': 71, 'Reece': 90, 'A': 79, 'B': 70,
'C': 67, 'D': 66, 'E': 60, 'F': 83, 'G': 57, 'H': 85, 'I': 93, 'J': 89, 'K': 78,
'L': 74, 'M': 65, 'N': 78, 'O': 53, 'P': 80}, 'history': {'Mason': 73, 'Reece': 61,
'A': 74, 'B': 47, 'C': 49, 'D': 87, 'E': 69, 'F': 65, 'G': 36, 'H': 7, 'I': 53, 'J':
100, 'K': 57, 'L': 45, 'M': 56, 'N': 34, 'O': 37, 'P': 70}, 'biology': {'Mason':
'100', 'Reece': '100', 'A': 47, 'B': 38, 'C': 63, 'D': 56, 'E': 75, 'F': 53, 'G': 80,
'H': 50, 'I': 41, 'J': 62, 'K': 44, 'L': 26, 'M': 91, 'N': 35, 'O': 53, 'P': 68}}
```

16

描述：

- 参数匹配语法（Argument Matching Syntax）

函数的参数匹配包括两个位置，一个是定义函数时的传入参数；再者为调用时的传入参数，如表pcs.5-1所示。常规模式为位置参数，按照顺序从左到右对应参数；调用时，可以给定关键字参数，不受位置参数顺序的影响，但是需要将关键字参数放置于位置参数之后。如果是在定义函数时，给定关键字参数，则为为该参数指定默认值。当调用时，不传递该参数值，则以提供的默认值替代；收集参数（Varargs collecting）包括只有一个星号*的元组形式收集模式，和包括有两个星号**的字典形式收集模式。

不同的匹配语法可以根据需要自由组合。但是，排序通常为：一般模式在前，再跟元组收集，最后跟字典收集。如果位置不对，会引发异常提示，根据提示修改位置，直至满足要求。

表pcs.5-1 参数匹配语法

语法（Syntax）	位置（Location）	解释（Interpretation）
func(value)	调用（Caller）	常规参数（位置参数）：按位顺序匹配（从左至右）
func(name=value)	调用（Caller）	关键字参数：按名称匹配
func(*iterable)	调用（Caller）	将可迭代对象作为单独的位置参数传入：按位顺序匹配
func(**dict)	调用（Caller）	将字典键值对作为关键字参数传入：按键名匹配
def func(name)	函数（Function）	常规参数（位置参数）：按位置或名称匹配任何传递值
def func(name=value)	函数（Function）	配置函数默认参数值，如果没有在调用中传递值（配置参数默认值 default value）
def func(*name)	函数（Function）	以元组形式匹配并收集剩余的位置参数：收集参数（Varargs collecting）-位置参数
def func(**name)	函数（Function）	以字典的形式匹配并收集剩余的关键字参数：收集参数（Varargs collecting）-关键字参数
def func(*other, name)	函数（Function）	只能在调用中通过关键字传递的参数
def func(*, name=value)	函数（Function）	只能在调用中通过关键字传递的参数

代码段：

```
x = 2
y = 3
z = 5
xyz_lst = [2, 3, 5]
xyz_dict = {'X': 2, 'Y': 3, 'Z': 5}

# 常规参数
def xyz_normal(X, Y, Z):
    print(X, Y, Z)
xyz_normal(x, y, z)
xyz_normal(X=x, Y=y, Z=z)
xyz_normal(x, y, Z=z)
xyz_normal(*xyz_lst)
xyz_normal(**xyz_dict)
print("--"*30)
# 配置参数默认值
def xyz_default(X, Y=7, Z=9):
    print(X, Y, Z)
xyz_default(x, y, z)
xyz_default(x, y)
xyz_default(x)
print("--"*30)
# 收集参数 - 位置参数
def xyz_collect_positional(*args):
    print(args)
xyz_collect_positional(x, y, z)
xyz_collect_positional(xyz_lst)
print("--"*30)
# 收集参数 - 位置参数 - 变化组合
def xyz_collect_positional_alter(X, *args, Z):
    a, b, c = args
    print(X, args, Z)
    print(a, b, c)
xyz_collect_positional_alter(x, 12, 13, 15, Z=z)
xyz_collect_positional_alter(11, 12, 13, 15, Z=z)
print("--"*30)
# 收集参数 - 关键字参数
def xyz_collect_keyword(**args):
    print(args)
xyz_collect_keyword(x=2, y=3, z=5)
xyz_collect_keyword(**xyz_dict)
print("--"*30)
# 只能由关键字传递参数
def xyz_keyword(X, *, Y, Z):
    print(X, Y, Z)
xyz_keyword(x, Y=3, Z=5)
print("--"*30)
# 组合匹配
```

```
def xyz_normal_collect(X_n, Y_d=97, *pargs, **kargs):
    print(X_n, Y_d, pargs, kargs)
xyz_normal_collect(x, y, *xyz_lst, **xyz_dict)
```

运算结果：

```
2 3 5
2 3 5
2 3 5
2 3 5
2 3 5
----------------------------------------------------------------------
2 3 5
2 3 9
2 7 9
----------------------------------------------------------------------
(2, 3, 5)
([2, 3, 5],)
----------------------------------------------------------------------
2 (12, 13, 15) 5
12 13 15
11 (12, 13, 15) 5
12 13 15
----------------------------------------------------------------------
{'x': 2, 'y': 3, 'z': 5}
{'X': 2, 'Y': 3, 'Z': 5}
----------------------------------------------------------------------
2 3 5
----------------------------------------------------------------------
2 3 (2, 3, 5) {'X': 2, 'Y': 3, 'Z': 5}
```

知识点：

5.5 函数定义综合实验-自定义箱形图打印样式

●●17

描述：

数据分析，需要图表辅助观察数据变化关系或数据之间的差异，这使得难以理解的数据，在统计图表下变得易读、易于理解。matplotlib[①]图表库，提供了丰富的图表形式。如果不作为最终论文发表或报告，默认的参数配置或者提供的案例代码足可以用于数据分析；但是，如果要发表研究内容，佐证研究结果，对图表的样式则提出了一些较高的要求。清晰地表达图表并尽量美观，会让读者更容易尝试去理解你的研究。下述定义的函数boxplot_custom(data_dict,**args)，依托matplotlib库实现自定箱形图的样式。代码书写过程是先确定输入数据参数data_dict的数据类型，这里使用了字典的数据格式（多数图表库通常支持使用Pandas库的DataFrame格式数据），并给定了一个简单的数据样例test_score_dic；因为要调整默认的图表样式，因此函数内建立了一个字典paras用于初始化需要配置的样式参数，并以关键字参数**args的方式更新字典。这样，可以让使用者在不输入样式参数时，快速地打印一个默认样式箱形图，快速地查看数据关系。而不必要一开始就配置每一个参数，过于烦琐而放弃使用；在确定数据结果无误后，如果希望用于正式的论文图表，则再进一步根据需要有选择性地配置参数。

① matplotlib，是一个综合库，用于在Python中创建静态、动画和交互式数据可视化（https://matplotlib.org/）。

对于matplotlib图表库的样式配置直接搜索或从官方文档说明中获取，无须记忆各个参数名。根据需要配置完所需的样式参数，确认代码逻辑设计合理。调试无误后，在函数开头书写函数功能和参数说明。说明的文件格式直接由scipy解释器生成。

从下述箱形图pcs.5-2中很容易发现，英语成绩整体都较高；其次为中文成绩，并且两者的成绩相对比较集中，即每一得分对排名的影响很大；而历史和生物整体得分相对低，并且成绩分散，即每一得分对排名影响相对较弱，同时可以观察到历史有成绩很高的少数得分，也有一个最低的异常值，小于分数刻度线20。经核验，该得分为7。从箱形图中，可以观察出很多数据的关系，而各类图表对于不同研究内容都是很重要的分析工具，这尤其体现在数据分析领域。

注：即使完成了一个函数定义的所有内容，但是往往在后续调用时会出现这样那样的问题，需要不断调整代码。这是正常的代码编写过程。即使一开始认为完全无误，无须调试，也可能会出现意想不到的异常，因此很必要保持不断调整代码的心态。

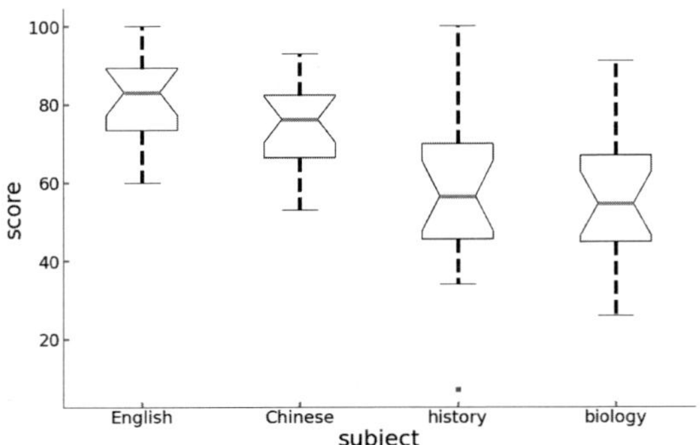

图pcs.5-2 自定义箱形图样式试验1

代码段:

```
def boxplot_custom(data_dict, **args):
    '''
    根据 matplotlib 库的箱型图打印方法，自定义箱型图可调整的打印样式。

    Parameters
    ----------
    data_dict : dict(list,numerical)
        字典结构形式的数据，键为横坐分类数据，值为数值列表 .
    **args : keyword arguments
        可调整的箱型图样式参数包括 ['figsize', 'fontsize', 'frameOn', 'xlabel',
'ylabel', 'labelsize', 'tick_length', 'tick_width', 'tick_color', 'tick_
direction', 'notch', 'sym', 'whisker_linestyle', 'whisker_linewidth', 'median_
linewidth', 'median_capstyle'].
    Returns
    -------
    paras : dict
        样式更新后的参数值 .

    '''
    import matplotlib.pyplot as plt

    # 计算值提取
    data_keys = list(data_dict.keys())
    data_values = list(data_dict.values())

    # 配置与更新参数
    paras = {'figsize': (10, 10),
             'fontsize': 15,
             'frameOn': ['top', 'right', 'bottom', 'left'],
             'xlabel': None,
             'ylabel': None,
             'labelsize': 15,
             'tick_length': 7,
             'tick_width': 3,
             'tick_color': 'b',
             'tick_direction': 'in',
             'notch': 0,
             'sym': 'b+',
             'whisker_linestyle': None,
             'whisker_linewidth': None,
             'median_linewidth': None,
             'median_capstyle': 'butt'}
    print(paras)
    paras.update(args)
    print(paras)

    # 根据参数调整打印图表样式
    plt.rcParams.update({'font.size': paras['fontsize']})
    frameOff = set(['top', 'right', 'bottom', 'left'])-set(paras['frameOn'])

    # 图表打印
    fig, ax = plt.subplots(figsize=paras['figsize'])
```

```
    ax.boxplot(data_values,
               notch=paras['notch'],
               sym=paras['sym'],
               whiskerprops=dict(
                   linestyle=paras['whisker_linestyle'],
linewidth=paras['whisker_linewidth']),
               medianprops={"linewidth": paras['median_linewidth'], "solid_
capstyle": paras['median_capstyle']})
    ax.set_xticklabels(data_keys)  #配置X轴刻度标签
    for f in frameOff:
        ax.spines[f].set_visible(False)  #配置边框是否显示

    #配置X和Y轴标签
    ax.set_xlabel(paras['xlabel'])
    ax.set_ylabel(paras['ylabel'])

    #配置X和Y轴标签字体大小
    ax.xaxis.label.set_size(paras['labelsize'])
    ax.yaxis.label.set_size(paras['labelsize'])

    #配置轴刻度样式
  ax.tick_params(length=paras['tick_length'],
                   width=paras['tick_width'],
                   color=paras['tick_color'],
                   direction=paras['tick_direction'])
    plt.show()
    return paras
test_score_dic = {"English": {"Mason": 90, "Reece": 81, 'A': 73, 'B': 97, 'C':
85, 'D': 60, 'E': 74, 'F': 64, 'G': 72, 'H': 67, 'I': 87, 'J': 78, 'K': 85, 'L':
96, 'M': 77, 'N': 100, 'O': 92, 'P': 86},
                  "Chinese": {"Mason": 71, "Reece": 90, 'A': 79, 'B': 70, 'C':
67, 'D': 66, 'E': 60, 'F': 83, 'G': 57, 'H': 85, 'I': 93, 'J': 89, 'K': 78, 'L':
74, 'M': 65, 'N': 78, 'O': 53, 'P': 80},
                  "history": {"Mason": 73, "Reece": 61, 'A': 74, 'B': 47, 'C':
49, 'D': 87, 'E': 69, 'F': 65, 'G': 36, 'H': 7, 'I': 53, 'J': 100, 'K': 57, 'L':
45, 'M': 56, 'N': 34, 'O': 37, 'P': 70},
                  "biology": {"Mason": 59, "Reece": 73, 'A': 47, 'B': 38, 'C':
63, 'D': 56, 'E': 75, 'F': 53, 'G': 80, 'H': 50, 'I': 41, 'J': 62, 'K': 44, 'L':
26, 'M': 91, 'N': 35, 'O': 53, 'P': 68},
                  }
test_score_lst_dic = {subject: list(v_subject.values())
                      for subject, v_subject in test_score_dic.items()}
print(test_score_lst_dic)
print("--"*30)
_ = boxplot_custom(test_score_lst_dic,
                   figsize=(15, 10),
                   fontsize=23,
                   frameOn=['bottom', 'left'],
                   xlabel='subject',
                   ylabel='score',
                   labelsize='30',
                   tick_color='r',
                   notch=1,
                   sym='rs',
                   whisker_linestyle='--',
                   whisker_linewidth=5,
                   median_linewidth=5
                   )
```

运算结果：

```
{'English': [90, 81, 73, 97, 85, 60, 74, 64, 72, 67, 87, 78, 85, 96, 77, 100,
92, 86], 'Chinese': [71, 90, 79, 70, 67, 66, 60, 83, 57, 85, 93, 89, 78, 74, 65,
78, 53, 80], 'history': [73, 61, 74, 47, 49, 87, 69, 65, 36, 7, 53, 100, 57, 45,
56, 34, 37, 70], 'biology': [59, 73, 47, 38, 63, 56, 75, 53, 80, 50, 41, 62, 44,
26, 91, 35, 53, 68]}
------------------------------------------------------------
{'figsize': (10, 10), 'fontsize': 15, 'frameOn': ['top', 'right', 'bottom',
'left'], 'xlabel': None, 'ylabel': None, 'labelsize': 15, 'tick_length': 7, 'tick_
width': 3, 'tick_color': 'b', 'tick_direction': 'in', 'notch': 0, 'sym': 'b+',
'whisker_linestyle': None, 'whisker_linewidth': None, 'median_linewidth': None,
'median_capstyle': 'butt'}
{'figsize': (15, 10), 'fontsize': 23, 'frameOn': ['bottom', 'left'], 'xlabel':
'subject', 'ylabel': 'score', 'labelsize': '30', 'tick_length': 7, 'tick_width':
```

```
3, 'tick_color': 'r', 'tick_direction': 'in', 'notch': 1, 'sym': 'rs', 'whisker_
linestyle': '--', 'whisker_linewidth': 5, 'median_linewidth': 5, 'median_
capstyle': 'butt'}
```

18

描述：

自动生成了新的一组数据random_val_dict，调用该函数执行箱型图打印（图pcs.5-3），输入参数仅随意配置了边框显示与中位数横线的线型宽度。

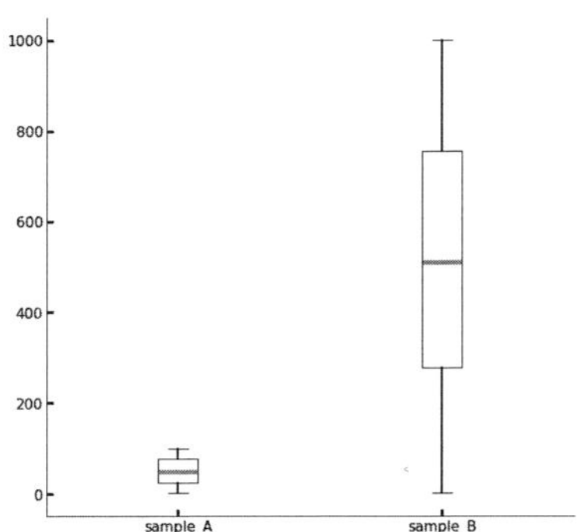

图pcs.5-3　自定义箱型图样式试验2

代码段：

```
import numpy as np
random_val_dict = {'sample_A': np.random.randint(low=1, high=100, size=1000),
'sample_B': np.random.randint(low=1, high=1000, size=1000)
                   }
boxplot_custom(random_val_dict, frameOn=['bottom', 'left'], median_linewidth=5)
```

运算结果：

```
{'figsize': (10, 10), 'fontsize': 15, 'frameOn': ['top', 'right', 'bottom',
'left'], 'xlabel': None, 'ylabel': None, 'labelsize': 15, 'tick_length': 7, 'tick_
width': 3, 'tick_color': 'b', 'tick_direction': 'in', 'notch': 0, 'sym': 'b+',
'whisker_linestyle': None, 'whisker_linewidth': None, 'median_linewidth': None,
'median_capstyle': 'butt'}
{'figsize': (10, 10), 'fontsize': 15, 'frameOn': ['bottom', 'left'], 'xlabel':
None, 'ylabel': None, 'labelsize': 15, 'tick_length': 7, 'tick_width': 3,
'tick_color': 'b', 'tick_direction': 'in', 'notch': 0, 'sym': 'b+', 'whisker_
linestyle': None, 'whisker_linewidth': None, 'median_linewidth': 5, 'median_
capstyle': 'butt'}
{'figsize': (10, 10),
 'fontsize': 15,
 'frameOn': ['bottom', 'left'],
 'xlabel': None,
 'ylabel': None,
 'labelsize': 15,
 'tick_length': 7,
 'tick_width': 3,
 'tick_color': 'b',
 'tick_direction': 'in',
 'notch': 0,
 'sym': 'b+',
 'whisker_linestyle': None,
 'whisker_linewidth': None,
 'median_linewidth': 5,
 'median_capstyle': 'butt'}
```

PCS-6. recursion（递归）、lambda（Anonymous Function，匿名函数）、generator（生成器）

PCS-6（ ）：6.1 递归；6.2 匿名函数（lambda）；6.3 生成器（generator）

知识点：

6.1 递归

1

描述：

通过简单的案例查看每一个循环，满足条件代码块运行变量值和返回值的变化，是理解递归最好的途径。递归的核心是调用自身，重复同一个代码逻辑过程，实现返回值满足某一变化特征的叠加；并且，需要有结束条件，以结束递归，即结束调用自身的循环；同时，给结束条件一个返回值，为最后一次调用自身的返回值。

下述两个案例一个为递归求和，一个为递归阶乘，通常用此来解释递归函数的方法。

● 递归求和

第1个解释案例是递归求和，其过程如图pcs.6-1所示。用具体的语句和值解释上一段话为：开始输入值（开始值）为`lst=[1,2,3,4,5]`，不满足结束条件`if not lst`，执行`else`语句块，返回值语句为`return lst[0]+recursive_sum(lst[1:])`，将值代入语句（表达式），为`1+recursive_sum([2,3,4,5])`。其中，包含调用自身语句`recursive_sum(lst[1:]`，因此从函数名定义行重新开始执行该函数；因为输入参数值`[2,3,4,5]`不满足结束条件，再次执行`else`语句块。将值代入语句（表达式），为`2+recursive_sum([3,4,5])`，继续调用自身，此时的输入值为`[3,4,5]`；以此类推，不断迭代，直至执行到`5+recursive_sum([])`时，可以发现调用递归的输入值为`[ ]`，满足结束条件。因此，执行`if not lst`语句块，返回值为0，结束递归迭代。当结束递归后，将每一次的返回值代入上一次迭代的返回值，计算获得最终结果为`1+2+3+4+5+0=15`。

从上述递归过程的描述中，可以确定递归的基本逻辑结构，无外乎调用自身即返回值计算逻辑和该逻辑能够产生结束条件，因此具有递归属性的问题都可以借由此点切入设计代码。

代码段：

```python
def recursive_sum(lst):
    if not lst:
        print("if not lst:{}".format(lst))
        return 0
    else:
        print(lst[1:])
        return lst[0] + recursive_sum(lst[1:])
lst = [1, 2, 3, 4, 5]
lst_sum = recursive_sum(lst)
```

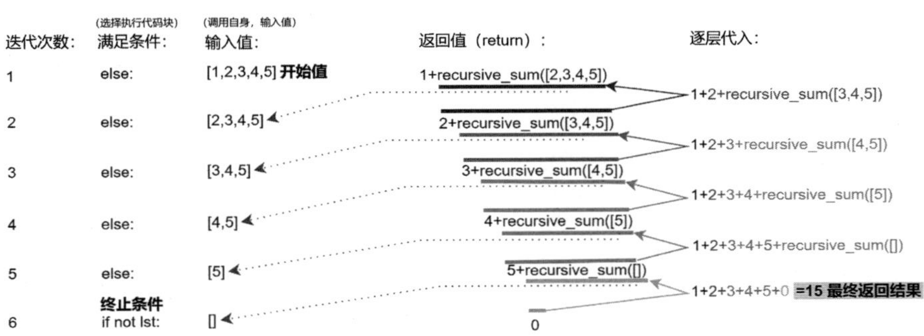

图pcs.6-1 递归求和图解

```
print(“--” * 30)
print(lst_sum)
```

运算结果：

```
[2, 3, 4, 5]
[3, 4, 5]
[4, 5]
[5]
[]
if not lst:[]
------------------------------------------------------------
15
```

2

描述：

可以通过小的语句核实确定某些语句的运行结果，辅助理解程序。

代码段：

```
print(not [])
print(not [5])
print([5][1:])
```

运算结果：

```
True
False
[]
```

3

描述：

返回值是需要代入到上次迭代调用自身的部分，因此只要返回值满足计算要求，可以根据需要返回任何值。下述代码将结束条件返回值配置为入1000时，计算结果为1015。

代码段：

```
def recursive_sum(lst):
    if not lst:
        print(“if not lst:{}”.format(lst))
        return 1000
    else:
        print(lst[1:])
        return lst[0] + recursive_sum(lst[1:])
lst = [1, 2, 3, 4, 5]
lst_sum = recursive_sum(lst)
print(“--” * 30)
print(lst_sum)
```

运算结果：

```
[2, 3, 4, 5]
[3, 4, 5]
[4, 5]
[5]
[]
if not lst:[]
------------------------------------------------------------
1015
```

4

描述：

- 递归阶乘

直接可以替换递归求和图中对应的值，来表达递归阶乘的过程，如图pcs.6-2所示。可以解释为：开始输入值（开始值）为6，不满足结束条件`if n==1`，执行`else`语句块，返回值语句为`return n*recursive_factorial(n-1)`。将值代入语句（表达式），为6×`recursive_factorial(5)`。其中，包含调用自身语句`recursive_factorial(5)`，因此从函数名定义行重新开始执行该函数；因为输入参数值5不满足结束条件，再次执行`else`语句块，将值代入语句（表达式），为5×`recursive_factorial(4)`，继续调用自身，此时的输入值为4；以此类推，不断迭代，直至执行到2×`recursive_factorial(1)`时，可以发现调用递归的输入值为

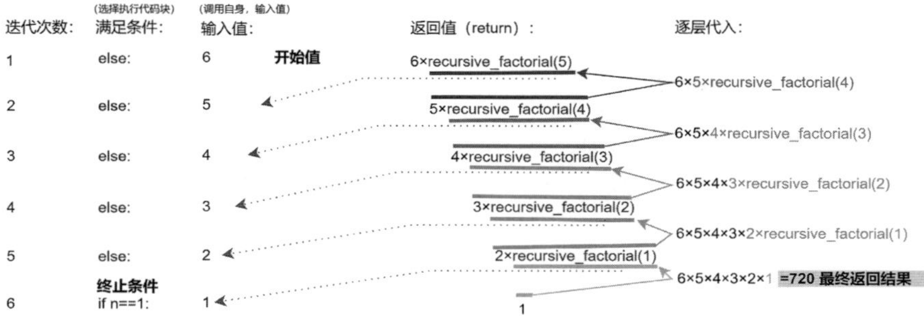

图pcs.6-2 递归阶乘 图解

1，满足结束条件，因此执行if n==1语句块，返回值为1，结束递归迭代。当结束递归后，将每一次的返回值代入上一次迭代的返回值，计算获得最终结果为6 × 5 × 4 × 3 × 2 × 1=720。

代码段：

```
def recursive_factorial(n):
    if n == 1:
        return 1
    else:
        return n * recursive_factorial(n - 1)
fatorial_result = recursive_factorial(6)
print(fatorial_result)
```

运算结果：

```
720
```

5

描述：

● 用循环语句替换递归的方法

递归的方法可以用循环的方式解决，例如下述代码改写的递归求和及阶乘。虽然循环的方式更加易读，但是递归的方法似乎也符合python的宗旨，保持简单，除非它必须复杂！因此，究竟是使用递归还是循环，可以由书写代码的作者自己决定。

代码段：

```
lst = [1, 2, 3, 4, 5]
lst_sum = 0
while lst:
    lst_sum += lst[0]
    lst = lst[1:]
print(lst_sum)
n = 6
fatorial_result = 1
while n:
    if n == 1:
        break
    else:
        fatorial_result *= n
        n = n - 1
print(fatorial_result)
```

运算结果：

```
15
720
```

6

描述：

● 递归列表展平

下述递归嵌套列表展平的代码，虽然代码行数不多，但是相对递归求和和阶乘要复杂。这里需要注意两个问题：一个是存在有三个条件，if是结束条件。elif和else分别针对两种不同情况采取的措施。结束条件不是像递归求和和阶乘最后执行一次结束，是对elif和else

下不断调用自身，最终达到各个分枝下满足结束条件的输入参数，逐个结束各个分枝；另一个是，返回值中存在return flatten_lst(lst[0])+flatten_lst(lst[1:])，为调用自身的函数求和，需要从左到右一次计算，先返回 flatten_lst(lst[0])部分，待这部分递归完毕后，再执行flatten_lst(lst[1:])部分；而各自部分同样有满足elif isinstance(lst[0],list)条件的对象，因此还会产生调用自身两个函数相加的情况。依旧从左到右依次计算，至递归结束。具体的过程可以从图pcs.6-3中观察，一个是正序过程，由k-i,j的方式标注执行顺序；粉色部分则是逆序逐步代入值的过程。比较清晰地表述了该递归的整个过程。

注：当很难通过代码直接理解算法时，必然需要通过 print()，打印各个变化值查看整个过程。如果打印的对象比较多，尤其类似递归这样要不断循环的算法，最好给出些辅助标识，定位每一轮次迭代。方便查看每次迭代对应变量的变化，从而找出规律，理解算法。

代码段：

```
i = 0
j = 0
k = 0
def flatten_lst(lst):
    global k, i, j
    print("#" * 30, '%d-%d-%d' % (k, i, j))
    print("lst:", lst)
    if lst == []:
        print('_' * 30, 'if')
        k += 1
        return lst
    elif isinstance(lst[0], list):
        print('_' * 30, "elif")
        print(lst[0], ';', lst[1:])
        i += 1
        return flatten_lst(lst[0]) + flatten_lst(lst[1:])
    else:
        print('_' * 30, 'else')
        print(lst[:1], ";", lst[1:])
        j += 1
        return lst[:1] + flatten_lst(lst[1:])
nested_lst = [['A', 'B', ['C', 'D'], 'E'], [6, [7, 8, [9]]]]
```

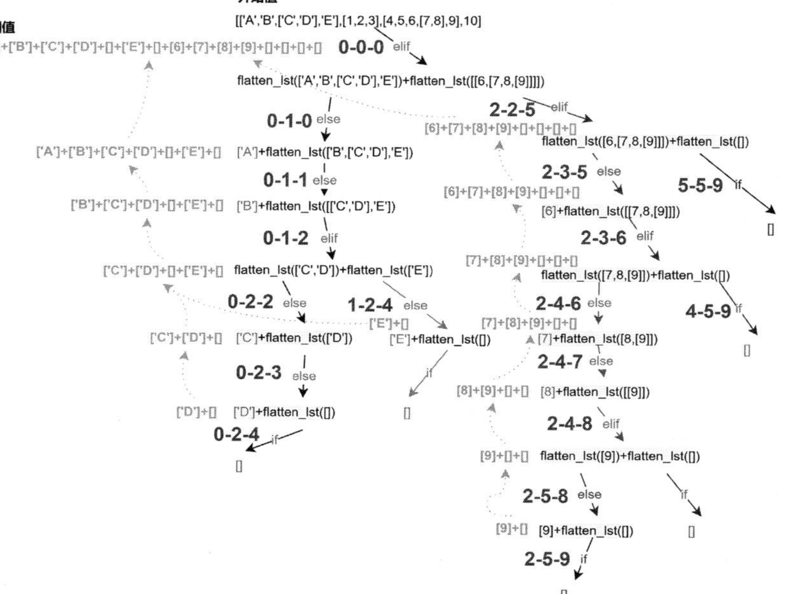

图pcs.6-3 递归列表展平图解

```
flatten_nested_lst = flatten_lst(nested_lst)
print("--" * 30)
print(flatten_nested_lst)
```

运算结果：

```
################################ 0-0-0
lst: [['A', 'B', ['C', 'D'], 'E'], [6, [7, 8, [9]]]]
________________________________ elif
['A', 'B', ['C', 'D'], 'E'] ; [[6, [7, 8, [9]]]]
################################ 0-1-0
lst: ['A', 'B', ['C', 'D'], 'E']
________________________________ else
['A'] ; ['B', ['C', 'D'], 'E']
################################ 0-1-1
lst: ['B', ['C', 'D'], 'E']
________________________________ else
['B'] ; [['C', 'D'], 'E']
################################ 0-1-2
lst: [['C', 'D'], 'E']
________________________________ elif
['C', 'D'] ; ['E']
################################ 0-2-2
lst: ['C', 'D']
________________________________ else
['C'] ; ['D']
################################ 0-2-3
lst: ['D']
________________________________ else
['D'] ; []
################################ 0-2-4
lst: []
________________________________ if
################################ 1-2-4
lst: ['E']
________________________________ else
['E'] ; []
################################ 1-2-5
lst: []
________________________________ if
################################ 2-2-5
lst: [[6, [7, 8, [9]]]]
________________________________ elif
[6, [7, 8, [9]]] ; []
################################ 2-3-5
lst: [6, [7, 8, [9]]]
________________________________ else
[6] ; [[7, 8, [9]]]
################################ 2-3-6
lst: [[7, 8, [9]]]
________________________________ elif
[7, 8, [9]] ; []
################################ 2-4-6
lst: [7, 8, [9]]
________________________________ else
[7] ; [8, [9]]
################################ 2-4-7
lst: [8, [9]]
________________________________ else
[8] ; [[9]]
################################ 2-4-8
lst: [[9]]
________________________________ elif
[9] ; []
################################ 2-5-8
lst: [9]
________________________________ else
[9] ; []
################################ 2-5-9
lst: []
________________________________ if
################################ 3-5-9
lst: []
________________________________ if
```

```
############################### 4-5-9
lst: []
_______________________________ if
############################### 5-5-9
lst: []
_______________________________ if
----------------------------------------------------------------------------------
['A', 'B', 'C', 'D', 'E', 6, 7, 8, 9]
```

知识点：

6.2 匿名函数（lambda）

7

描述：

lambda的基本语法为lambda argument1,argument2,argument3,...,argumentN: expression using arguments。需要注意的是，lambda使用的是表达式（expression）不是语句（statement），返回一个函数表达式，不需要定义函数名（因此称为匿名函数）。lambda的表达式中可以加入if条件语句，语法为lambda <arguments>:<return value if condition is True> if <condition> else <return value if condition is False>；如果包含类似elif的结构，则语法为lambda <args>:<return value> if <condition> else (return value if <condition> else <return value>)。

对于较为简单的函数定义，使用lambda为内联函数（inline function），可以简化代码，而def方法则会稍显烦琐。

代码段：

```
def sum_func(x, y, z):
    return x + y + z
lst = [1, 2, 3]
print(sum_func(*lst))
sum_lambda = lambda x, y, z: x + y + z
print(sum_lambda(*lst))
print("-" * 60)
comparison_operation_A = lambda x: True if x > 10 else False #只有一个条件
print(comparison_operation_A(20))
CompOpera_B = lambda x: True if (x > 10 and x < 20) else False  #包含多个条件，需要括起
print(CompOpera_B(18))

CompOpera_C = lambda x: x > 10 and x < 20  #不使用 if else 的方式
print(CompOpera_C(18))
print("-" * 60)
CompOpera_D = lambda x: 1 if x < 10 else (2 if x < 20 else 0)  #类似 elif
print(CompOpera_D(7))
print(CompOpera_D(15))
print(CompOpera_D(30))
```

运算结果：

```
6
6
----------------------------------------------------------------------------------
True
True
True
----------------------------------------------------------------------------------
1
2
0
```

8

描述：

- 嵌套的匿名函数

与嵌套函数类似。

代码段：

```
nested_sum = lambda x: lambda y: x + y
```

```
sum_instance = nested_sum(10)  #输入值对应参数 x
print(sum_instance(20))  #输入值对应参数 y
print(sum_instance(50))
```

运算结果：

```
30
60
```

9

描述：

- lambda 与filter(),map(),reduce(),sorted

lambda经常配合其他以函数作为输入参数的函数，可以非常便捷地以嵌入函数（inline function）方式行内完成代码。通过help()非常方便地确定filter()、map()、reduce()、sorted输入参数，包含使用函数作为参数的部分。

代码段：

```
from functools import reduce
print(help(filter))
print(“+” * 60)
print(help(map))
print(“+” * 60)
print(help(reduce))
print(“+” * 60)
print(help(sorted))
```

运算结果：

```
Help on class filter in module builtins:
class filter(object)
 |  filter(function or None, iterable) --> filter object
 |
 |  Return an iterator yielding those items of iterable for which function(item)
 |  is true. If function is None, return the items that are true.
 |
 |  Methods defined here:
 |
 |  __getattribute__(self, name, /)
 |      Return getattr(self, name).
 |
 |  __iter__(self, /)
 |      Implement iter(self).
 |
 |  __next__(self, /)
 |      Implement next(self).
 |
 |  __reduce__(...)
 |      Return state information for pickling.
 |
 |  ----------------------------------------------------------------------------
 |  Static methods defined here:
 |
 |  __new__(*args, **kwargs) from builtins.type
 |      Create and return a new object.  See help(type) for accurate signature.
None
++++++++++++++++++++++++++++++++++++++++++++++++++++++++++++++++++++++++++++++++
Help on class map in module builtins:
class map(object)
 |  map(func, *iterables) --> map object
 |
 |  Make an iterator that computes the function using arguments from
 |  each of the iterables.  Stops when the shortest iterable is exhausted.
 |
 |  Methods defined here:
 |
 |  __getattribute__(self, name, /)
 |      Return getattr(self, name).
 |
 |  __iter__(self, /)
 |      Implement iter(self).
 |
 |  __next__(self, /)
```

```
 |      Implement next(self).
 |
 |  __reduce__(...)
 |      Return state information for pickling.
 |
 |  ----------------------------------------------------------------------
 |  Static methods defined here:
 |
 |  __new__(*args, **kwargs) from builtins.type
 |      Create and return a new object.  See help(type) for accurate signature.
None
++++++++++++++++++++++++++++++++++++++++++++++++++++++++++++++++++++++++++++
Help on built-in function reduce in module _functools:
reduce(...)
    reduce(function, sequence[, initial]) -> value
    Apply a function of two arguments cumulatively to the items of a sequence,
    from left to right, so as to reduce the sequence to a single value.
    For example, reduce(lambda x, y: x+y, [1, 2, 3, 4, 5]) calculates
    ((((1+2)+3)+4)+5).  If initial is present, it is placed before the items
    of the sequence in the calculation, and serves as a default when the
    sequence is empty.
None
++++++++++++++++++++++++++++++++++++++++++++++++++++++++++++++++++++++++++++
Help on built-in function sorted in module builtins:
sorted(iterable, /, *, key=None, reverse=False)
    Return a new list containing all items from the iterable in ascending order.
    A custom key function can be supplied to customize the sort order, and the
    reverse flag can be set to request the result in descending order.
None
```

10

描述：

如果某一函数或者方法的输入参数有函数参数，则首先考虑使用lambda的方式。这经常用于Pandas下配合方法apply，根据行数据增加新的列数据。

下述案例将以字典保存的保龄球大赛得分转换为DataFrame数据格式，并用apply方法，以lambda方式计算标准计分（Standard Score，z_score，代表原始数值和平均值之间的距离，并以标准差为单位计算，即z-score是从感兴趣的点到均值之间有多少个标准差，这样就可以在不同组数据间比较某一数值的重要程度。）

注：浅试 Pandas 的 DataFrame 数据结构。Pandas 数据处理方法异常丰富，是数据处理的核心应用库，可以查看 Pandas 手册或相关说明。

代码段：

```
vals = [2, 3, 4, 10, 15, 17, 19, 30, 50]
filtered_vals = filter(lambda x: x > 10, vals)  #返回的是一个迭代器（iterator）
print(filtered_vals)
print(list(filtered_vals))
print("-" * 60)
mapped_vals = map(lambda x: x**2, vals)
print(list(mapped_vals))
print("-" * 60)
reduced_vals = reduce(lambda x, y: x + y, vals)  #相当于求和
print(reduced_vals)
print("-" * 60)
scores = [('Mason', 90), ('Reece', 81), ('A', 73), ('B', 97), ('C', 85)]
sorted_vals = sorted(scores, key=lambda score: score[1])
print(sorted_vals)
```

运算结果：

```
<filter object at 0x0000020982DF1730>
[15, 17, 19, 30, 50]
------------------------------------------------------------
[4, 9, 16, 100, 225, 289, 361, 900, 2500]
------------------------------------------------------------
150
------------------------------------------------------------
[('A', 73), ('Reece', 81), ('C', 85), ('Mason', 90), ('B', 97)]
```

11

代码段：

```
import pandas as pd
bowlingContest_scores_dic = {‘Barney’: 86, ‘Harold’: 73, ‘Chris’: 124, ‘Neil’:
111, ‘Tony’: 90, ‘Simon’: 38,
                                ‘Jo’: 84, ‘Dina’: 71, ‘Graham’: 103, ‘Joe’: 85,
‘Alan’: 90, ‘Billy’: 89,
                                ‘Gordon’: 229, ‘Wade’: 77, ‘Cliff’: 59, ‘Arthur’:
95, ‘David’: 70, ‘Charles’: 88}
print(bowlingContest_scores_dic)
bowlingContest_scores_df = pd.DataFrame.from_dict(
    bowlingContest_scores_dic, orient=’index’, columns=[‘score’])
print(bowlingContest_scores_df)
scores_mean = bowlingContest_scores_df.score.mean()
scores_std = bowlingContest_scores_df.score.std()
print(“-” * 60)
print(‘scores_mean=%.2f,scores_std=%.2f’ % (scores_mean, scores_std))
bowlingContest_scores_df[‘z_score’] = bowlingContest_scores_df.score.apply(
    lambda score: round((score - scores_mean) / scores_std, 3))
print(bowlingContest_scores_df)
```

运算结果：

```
{‘Barney’: 86, ‘Harold’: 73, ‘Chris’: 124, ‘Neil’: 111, ‘Tony’: 90, ‘Simon’:
38, ‘Jo’: 84, ‘Dina’: 71, ‘Graham’: 103, ‘Joe’: 85, ‘Alan’: 90, ‘Billy’: 89,
‘Gordon’: 229, ‘Wade’: 77, ‘Cliff’: 59, ‘Arthur’: 95, ‘David’: 70, ‘Charles’:
88}
         score
Barney      86
Harold      73
Chris      124
Neil       111
Tony        90
Simon       38
Jo          84
Dina        71
Graham     103
Joe         85
Alan        90
Billy       89
Gordon     229
Wade        77
Cliff       59
Arthur      95
David       70
Charles     88
------------------------------------------------------------
scores_mean=92.33,scores_std=39.09
         score  z_score
Barney      86   -0.162
Harold      73   -0.495
Chris      124    0.810
Neil       111    0.477
Tony        90   -0.060
Simon       38   -1.390
Jo          84   -0.213
Dina        71   -0.546
Graham     103    0.273
Joe         85   -0.188
Alan        90   -0.060
Billy       89   -0.085
Gordon     229    3.496
Wade        77   -0.392
Cliff       59   -0.853
Arthur      95    0.068
David       70   -0.571
Charles     88   -0.111
```

12

描述：

- 用lambda定义递归展平嵌套列表

前述递归嵌套列表是根据索引为0的项值是否为列表给出elif和else的处理路径，如果列表为空则执行if，其为终止条件。此次给出的方法是使用lambda函数，lambda函数组合了列表推导式的方法，递归lambda自身完成嵌套列表的展平。直接看包括列表推导式和递归的lambda函数理解递归的过程，因为无法print()变量，很难查看整个计算流程，因此需要将lambda转换为def的形式，并将列表推导式转换为for循环。

图pcs.6-4解释了整个计算流程。对循环的子列表执行递归操作，通过判断该列表是否为列表，执行if或者else语句块。else语句块为终止条件，为列表项值不再是子列表，而是单个值时的情况。则返回包含该一个值的列表，通过extend的方法，将其追加到变量lst_collection中。注意，追加的过程，是根据循环从左到右、从外到内（子列表）迭代过程的逆序。例如，先['A','B',['C','D'],'E']子列表，然后[6,[7,8,[9]]]子列表；前者子列表仍旧从左到右，先A，再B，在调用自身位置追加到列表为['A','B']；到['C','D']时，先执行嵌套子列表，从左到右，在调用到自身位置处追加到列表为['C','D']；在A,B和['C','D']齐平的位置合并列表为['A','B','C','D','E']，依次类推。

终止条件的返回值是返回递归调用自身的位置，例如满足终止条件的A，返回值为['A']，该值对应到递归调用的位置lst_collection.extend(flatten_lst_loop(n_lst))，而被追加到 lst_collection列表中，因为返回的是含一个值的列表，因此使用extend方法。

思考：如果将else终止条件返回值改为lst，而不是[lst]，并将调用位置语句改为lst_collection.append(flatten_lst_loop(n_lst))，即将extend改为append，为什么不可以？

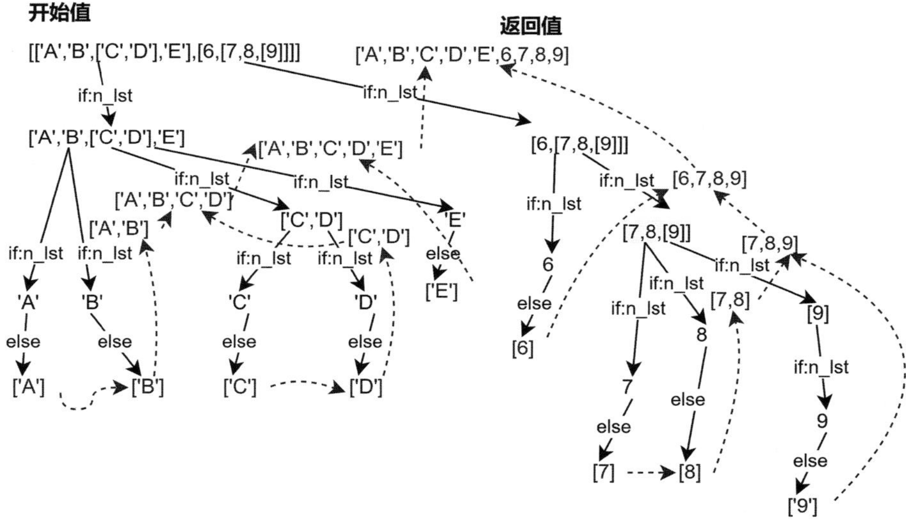

图pcs.6-4 递归展平嵌套列表图解

代码段：

```
flatten_lst = lambda lst: [m for n_lst in lst for m in flatten_lst(
    n_lst)] if type(lst) is list else [lst]
nested_lst = [['A', 'B', ['C', 'D'], 'E'], [6, [7, 8, [9]]]]
print(flatten_lst(nested_lst))
```

运算结果：

```
['A', 'B', 'C', 'D', 'E', 6, 7, 8, 9]
```

13

描述：

将lambda转换为def定义的形式，并拆解列表推导式，方便通过print方式查看变量值的变化。

代码段：

```
i = 0
def flatten_lst_loop(lst):
```

```
    global i
    print("-" * 50, i)
    lst_collection = []
    if type(lst) is list:
        i += 1
        print('**', lst)
        for n_lst in lst:
            print("##", n_lst)
            lst_collection.extend(flatten_lst_loop(n_lst))
            print(':', lst_collection)
    else:
        i += 1
        print('++', lst)
        return [lst]
    return lst_collection
flattened_nestedLst = flatten_lst_loop(nested_lst)
print("-" * 60)
print(flattened_nestedLst)
```

运算结果：

```
-------------------------------------------------- 0
** [['A', 'B', ['C', 'D'], 'E'], [6, [7, 8, [9]]]]
## ['A', 'B', ['C', 'D'], 'E']
-------------------------------------------------- 1
** ['A', 'B', ['C', 'D'], 'E']
## A
-------------------------------------------------- 2
++ A
: ['A']
## B
-------------------------------------------------- 3
++ B
: ['A', 'B']
## ['C', 'D']
-------------------------------------------------- 4
** ['C', 'D']
## C
-------------------------------------------------- 5
++ C
: ['C']
## D
-------------------------------------------------- 6
++ D
: ['C', 'D']
: ['A', 'B', 'C', 'D']
## E
-------------------------------------------------- 7
++ E
: ['A', 'B', 'C', 'D', 'E']
: ['A', 'B', 'C', 'D', 'E']
## [6, [7, 8, [9]]]
-------------------------------------------------- 8
** [6, [7, 8, [9]]]
## 6
-------------------------------------------------- 9
++ 6
: [6]
## [7, 8, [9]]
-------------------------------------------------- 10
** [7, 8, [9]]
## 7
-------------------------------------------------- 11
++ 7
: [7]
## 8
-------------------------------------------------- 12
++ 8
: [7, 8]
## [9]
-------------------------------------------------- 13
** [9]
## 9
-------------------------------------------------- 14
```

```
++ 9
: [9]
: [7, 8, 9]
: [6, 7, 8, 9]
: ['A', 'B', 'C', 'D', 'E', 6, 7, 8, 9]
------------------------------------------------------------------------------
['A', 'B', 'C', 'D', 'E', 6, 7, 8, 9]
```

知识点：

6.3 生成器（generator）

14

描述：

生成器包括生成器函数和生成器表达式。生成器不会一次性地计算所有结果，例如以列表形式返回所有计算值（如果为海量数据，列表形式会非常耗内存）；而是根据需要提取数值，这样可以增加计算的效率，节约内存空间。例如，filter()、map()、zip()等函数的返回值均为迭代器（iterator）(可迭代对象)，即生成器返回迭代器（对象）。

生成器函数，就是在def定义函数时，返回值以yield的方式一次返回一个结果；生成器表达式，就是将列表推导式的[ ]改为( )即可；或者，使用iter()函数将可迭代对象转换为迭代器（iterator）。

通过help()查看生成器说明文件，有__next__方法，即next(iterable)实现逐个读取生成器返回的迭代对象。

● 生成器函数

代码段：

```
def squared_generator(range_start, range_stop, range_step=1):
    for i in range(range_start, range_stop, range_step):
        yield i**2
squared_iterable = squared_generator(1, 5)
print(squared_iterable)
for i in squared_iterable:
    print(i)
print("-" * 60)
for i in squared_iterable:
    print(i)  #当迭代完毕后，在执行则为空，需重新调用生成器函数
```

运算结果：

```
<generator object squared_generator at 0x000001E013AF4430>
1
4
9
16
------------------------------------------------------------------------------
```

15

代码段：

```
squared_iterable = squared_generator(1, 5)
for i in squared_iterable:
    print(i)
```

运算结果：

```
1
4
9
16
```

16

代码段：

```
def computing_generator(x):
    x += 1
    print('Performed addition')
    yield x
    x *= 2
```

```
    print('Performed multiplication')
    yield x
computing_iterable = computing_generator(5)
print(next(computing_iterable))
print(next(computing_iterable))
```

运算结果：

```
Performed addition
6
Performed multiplication
12
```

17

代码段：

```
def infinite_generator():
    i = 0
    while True:
        i += 1
        yield i
infinite_iterable = infinite()
print(next(infinite_iterable))
print(next(infinite_iterable))
print(next(infinite_iterable))
help(infinite_iterable)
```

运算结果：

```
1
2
3
Help on generator object:
infinite = class generator(object)
 |  Methods defined here:
 |
 |  __del__(...)
 |
 |  __getattribute__(self, name, /)
 |      Return getattr(self, name).
 |
 |  __iter__(self, /)
 |      Implement iter(self).
 |
 |  __next__(self, /)
 |      Implement next(self).
 |
 |  __repr__(self, /)
 |      Return repr(self).
 |
 |  close(...)
 |      close() -> raise GeneratorExit inside generator.
 |
 |  send(...)
 |      send(arg) -> send 'arg' into generator,
 |      return next yielded value or raise StopIteration.
 |
 |  throw(...)
 |      throw(typ[,val[,tb]]) -> raise exception in generator,
 |      return next yielded value or raise StopIteration.
 |
 |  ----------------------------------------------------------------------------
 |  Data descriptors defined here:
 |
 |  gi_code
 |
 |  gi_frame
 |
 |  gi_running
 |
 |  gi_yieldfrom
 |      object being iterated by yield from, or None
```

18

描述：

- 迭代对象的send()方法

send()方法可以向yield表达式传递参数，作为yield表达式的值。

代码段：

```
def double_inputs():
    while True:
        x = yield
        print("-" * 10, x)
        yield x * 2
di_gen = double_inputs()
print(next(di_gen))
print(di_gen.send(5))
next(di_gen)
print(di_gen.send(7))
```

运算结果：

```
None
---------- 5
10
---------- 7
14
```

19

代码段：

```
def squared_generator():
    for i in range(1, 10):
        print("-" * 10, i)
        yield i**2
sg = squared_generator()
print(next(sg))
print(next(sg))
print(next(sg))
print("-" * 60)
print(sg.send(7))
print(sg.send(9))
```

运算结果：

```
---------- 1
1
---------- 2
4
---------- 3
9
------------------------------------------------------------
---------- 4
16
---------- 5
25
```

20

描述：

需要注意的是，下述计数求和代码，需要先运行next()，然后执行send()；否则，提示TypeError: can't send non-None value to a just-started generator错误。send()将值传递给的是yield表达式，因此可以看到下述代码，在send(1)之后，先运行delta=yield countNtotal_nt(count,total)语句，此时delta对象的值为1，执行if语句块；然后，依次迭代。

代码段：

```
from collections import namedtuple
countNtotal_nt = namedtuple('countNsum', ['count', 'total'])
def countNtotal(count=0, total=0):
    i = 0
    while True:
        print('-' * 10, i)
        delta = yield countNtotal_nt(count, total)
        print("###", delta)
```

```
        if delta:
            print('+' * 10, i)
            count += 1
            total += delta
            i += 1
vals = [1, 2, 3, 4, None, 7, 8, 9]
countNtotal_gen = countNtotal()
print(next(countNtotal_gen))
print("-" * 60)
for v in vals:
    print(countNtotal_gen.send(v))
```

运算结果：

```
---------- 0
countNsum(count=0, total=0)
------------------------------------------------------------
### 1
++++++++++ 0
---------- 1
countNsum(count=1, total=1)
### 2
++++++++++ 1
---------- 2
countNsum(count=2, total=3)
### 3
++++++++++ 2
---------- 3
countNsum(count=3, total=6)
### 4
++++++++++ 3
---------- 4
countNsum(count=4, total=10)
### None
---------- 4
countNsum(count=4, total=10)
### 7
++++++++++ 4
---------- 5
countNsum(count=5, total=17)
### 8
++++++++++ 5
---------- 6
countNsum(count=6, total=25)
### 9
++++++++++ 6
---------- 7
countNsum(count=7, total=34)
```

21

描述：

- 生成器表达式

代码段：

```
plus10_generator = (i + 10 for i in [1, 2, 3, 4, 5])
print(plus10_generator)
print(list(plus10_generator))
print("-" * 60)
lst_iter = iter([1, 2, 3, 4, 5])
print(lst_iter)
print(next(lst_iter))
print(next(lst_iter))
```

运算结果：

```
<generator object <genexpr> at 0x000001E013AE7430>
[11, 12, 13, 14, 15]
------------------------------------------------------------
<list_iterator object at 0x000001E013BB9670>
1
2
```

22

描述：

● 用生成器递归展平列表

该方法逻辑同用lambda定义递归展平嵌套列表，只是这里有些处理方法的替换，1是，将`if type(lst) is list`:判断条件用`try`,`except`异常处理替换，当`lst`变量不为列表时，就会引发异常，从而执行`except`语句块，为结束条件；2是，用`yield`替换`lst_collection.extend(flatten_lst_loop(n_lst))`值追加到列表中，而是直接构建生成器，产生迭代对象。

代码段：

```
def flatten_lst_generator(lst):
    try:  # 使用语句 try/except 捕捉异常
        for n_lst in lst:
            for m in flatten_lst_generator(n_lst):
                yield m
    except:
        yield lst
nested_lst = [['A', 'B', ['C', 'D'], 'E'], [6, [7, 8, [9]]]]
flatten_nestedLst = flatten_lst_generator(nested_lst)
print(list(flatten_nestedLst))
```

运算结果：

```
['A', 'B', 'C', 'D', 'E', 6, 7, 8, 9]
```

23

描述：

itertools[①]库可以实现一系列迭代器（iterator），可以简洁而高效地完成相关的算法，而避免自行重新编写相关的工具。理解与善用itertools库，可以提高代码书写效率，以及高效利用内存。对itertools可以直接查看官网。下述摘抄了官网的说明表格，记录如下，方便查阅。通过表格给出的参数，结果和示例很容易推断工具的使用方法和算法目的。

① itertools — Functions creating iterators for efficient looping（https://docs.python.org/3/library/itertools.html）。

● Infinite iterators（无穷迭代器），如表pcs.6-1所示。

表pcs.6-1 无穷迭代器

Iterator（迭代器）	Arguments（参数）	Results（结果）	Example（示例）
count()	start, [step]	start, start+step, start+2*step, ...	count(10) --> 10 11 12 13 14 ...
cycle()	p	p0, p1, ... plast, p0, p1, ...	cycle('ABCD') --> A B C D A B C D ...
repeat()	elem [,n]	elem, elem, elem, ... endlessly or up to n times	repeat(10, 3) --> 10 10 10

● Iterators terminating on the shortest input sequence（根据最短输入序列长度停止的迭代器），如表pcs.6-2所示。

表pcs.6-2 根据最短输入序列长度停止的迭代器

accumulate()	**p [,func]**	**p0, p0+p1, p0+p1+p2, ...**	**accumulate([1,2,3,4,5]) --> 1 3 6 10 15**
chain()	p, q, ...	p0, p1, ... plast, q0, q1, ...	chain('ABC', 'DEF') --> A B C D E F
chain.from_iterable()	iterable	p0, p1, ... plast, q0, q1, ...	chain.from_iterable(['ABC', 'DEF']) --> A B C D E F
compress()	data, selectors	(d[0] if s[0]), (d[1] if s[1]), ...	compress('ABCDEF', [1,0,1,0,1,1]) --> A C E F
dropwhile()	pred, seq	seq[n], seq[n+1], starting when pred fails	dropwhile(lambda x: x<5, [1,4,6,4,1]) --> 6 4 1
filterfalse()	pred, seq	elements of seq where pred(elem) is false	filterfalse(lambda x: x%2, range(10)) --> 0 2 4 6 8
groupby()	iterable[, key]	sub-iterators grouped by value of key(v)	
islice()	seq, [start,] stop [, step]	elements from seq[start:stop:step]	islice('ABCDEFG', 2, None) --> C D E F G
pairwise()	iterable	(p[0], p[1]), (p[1], p[2])	pairwise('ABCDEFG') --> AB BC CD DE EF FG

续表

accumulate()	**p [,func]**	**p0, p0+p1, p0+p1+p2, ...**	**accumulate([1,2,3,4,5]) --> 1 3 6 10 15**
starmap()	func, seq	func(*seq[0]), func(*seq[1]), ...	starmap(pow, [(2,5), (3,2), (10,3)]) --> 32 9 1000
takewhile()	pred, seq	seq[0], seq[1], until pred fails	takewhile(lambda x: x<5, [1,4,6,4,1]) --> 1 4
tee()	it, n	it1, it2, ... itn splits one iterator into n	
zip_longest()	p, q, ...	(p[0], q[0]), (p[1], q[1]), ...	zip_longest(‘ABCD’ , ‘xy’ , fillvalue=’ -’) --> Ax By C- D-

- Combinatoric iterators（排列组合迭代器），如表pcs.6-3所示。

表pcs.6-3 排列组合迭代器

Iterator（迭代器）	**Arguments**（参数）	**Results**（示例）
product()	p, q, … [repeat=1]	cartesian product, equivalent to a nested for-loop
permutations()	p[, r]	r-length tuples, all possible orderings, no repeated elements
combinations()	p, r	r-length tuples, in sorted order, no repeated elements
combinations_with_replacement()	p, r	r-length tuples, in sorted order, with repeated elements

PCS-7. 模块与包（module and package），及PyPI发布（distribution）

PCS-7（ ）：模块与包

PCS7 之前必读参考：*29. Packages(python-course.eu)*[①]

知识点：

模块与包

1

描述：

一个单独的.py文件即为一个模块（module），包含定义的函数、类或变量；当模块的数量不断增加，最好将其分类于不同的文件夹中，这种有效组织模块的方式就是创建一个包（package）。包通常为一个含有__init__.py文件的文件夹，可以包含子模块或子包(子目录)。通常把一个项目组织为一个包，然后可以使用setuptools[②]工具创建和分发包，从而像标准库或扩展库一样安装包和调用包，使用包内的工具（即定义的函数或类等）。

下述为一个包的文件夹结构。整个项目位于PCS_7_package_example文件夹下，其中包放置于src文件夹下。toolkit4beginner文件夹是创建的包，包括两个子包graph和utility文件夹，其中graph子包组织图表绘制类的工具，例如自定义样式的打印箱形图和四象限图（4 quadrant diagram），包含在graph.py文件中；utility子包组织通用的工具，例如展平列表，计算时间长度等都位于general_tools.py文件中，而描述性统计和获取最近值的函数都位于stats_tools.py文件中。test文件夹包含用于测试包的模块，确保运行正确。如果出现错误，则需要进行调试。

在包中，都有一个__init__.py文件，确保包能够正确导入及用setuptools打包时，可以找到包的位置，该文件通常为空。也可以像常规的模块写入代码，代码会在python第1次导入一个目录时自动运行，因此主要作为执行（软件）包所需初始化步骤的钩子（hook），控制定义包级别的变量。因为python文件都是按目录（子包）组织模块，python会通过搜索该目录下的文件导入相关模块。而不是搜索所有目录，只搜索目录下包含有__init__.py文件的目录。此时，这个目录当作一个包目录，进而搜索包内的模块。如果未含有__init__.py文件，调用该目录下的模块时会引发未找到等错误提示。

如果在python 3.3及以上的版本，如果目录中不含有__init__.py文件，也可以作为单目录命名空间包实现（single-directory namespace packages），正确导入模块。但是也意味着，没有自定义初始化运行时加载控制变量的代码。

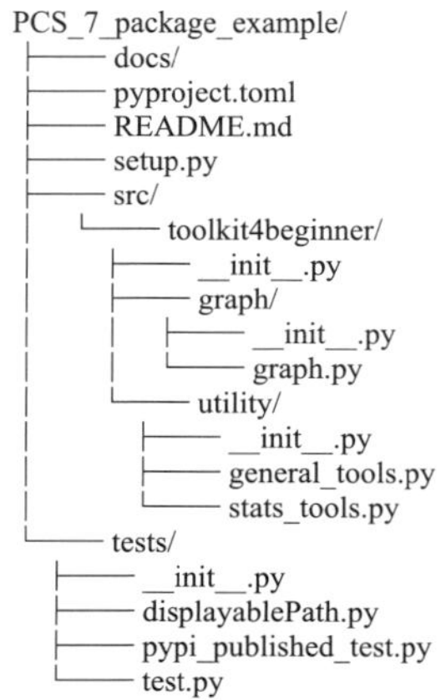

```
PCS_7_package_example/
├──── docs/
├──── pyproject.toml
├──── README.md
├──── setup.py
├──── src/
│   └──── toolkit4beginner/
│       ├──── __init__.py
│       ├──── graph/
│       │   ├──── __init__.py
│       │   └──── graph.py
│       └──── utility/
│           ├──── __init__.py
│           ├──── general_tools.py
│           └──── stats_tools.py
└──── tests/
    ├──── __init__.py
    ├──── displayablePath.py
    ├──── pypi_published_test.py
    └──── test.py
```

上述文件夹目录结构中，setup.py文件是setuptools创建和分发包的配置文件；README.md

① 29. Packages（python-course.eu），（https://python-course.eu/python-tutorial/packages.php）。

② setuptools，（https://pypi.org/project/setuptools/）。

文件是在setup.py文件中读取，用于配置long_description参数，方便编辑；pyproject.toml文件是从PEP 621[①]开始，Python 社区选择 pyproject.toml 作为指定项目元数据的标准方式。Setuptools 采用了此标准，使用此文件中包含的信息作为构建过程的输入（与setup.py作用同，但格式不同，此包中可以移除该文件）。创建分发包有多种打包方式，例如在anaconda下所建立环境的终端（terminal）执行`python setup.py sdist`，为打包成源码包（即压缩包，.zip，.tar，.gz等，如果不指定参数`--formats=gztar`，按当前平台默认格式；执行`python setup.py bdist_wheel`，为打包成二进制包（需要先安装wheel[②]）。打包完成后，会在文件夹下生成相应的文件，目录结构变化为：

① PEP 621 – Storing project metadata in pyproject.toml，（https://peps.python.org/pep-0621/）。

② wheel，（https://pypi.org/project/wheel/）。

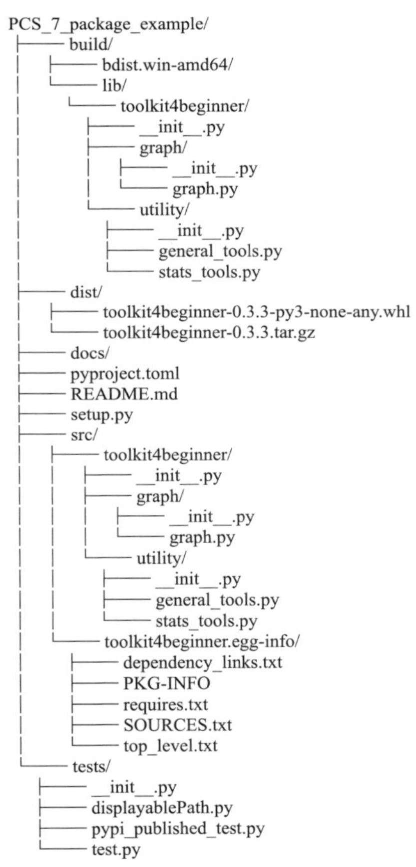

```
PCS_7_package_example/
├── build/
│   ├── bdist.win-amd64/
│   └── lib/
│       └── toolkit4beginner/
│           ├── __init__.py
│           ├── graph/
│           │   ├── __init__.py
│           │   └── graph.py
│           └── utility/
│               ├── __init__.py
│               ├── general_tools.py
│               └── stats_tools.py
├── dist/
│   ├── toolkit4beginner-0.3.3-py3-none-any.whl
│   └── toolkit4beginner-0.3.3.tar.gz
├── docs/
├── pyproject.toml
├── README.md
├── setup.py
├── src/
│   ├── toolkit4beginner/
│   │   ├── __init__.py
│   │   ├── graph/
│   │   │   ├── __init__.py
│   │   │   └── graph.py
│   │   └── utility/
│   │       ├── __init__.py
│   │       ├── general_tools.py
│   │       └── stats_tools.py
│   └── toolkit4beginner.egg-info/
│       ├── dependency_links.txt
│       ├── PKG-INFO
│       ├── requires.txt
│       ├── SOURCES.txt
│       └── top_level.txt
└── tests/
    ├── __init__.py
    ├── displayablePath.py
    ├── pypi_published_test.py
    └── test.py
```

打包的二进制文件或源码文件为新生成的dist文件夹下的toolkit4beginner-0.3.3-py3-none-any.whl和toolkit4beginner-0.3.3.tar.gz。安装分发包同一般库的安装，可以在终端定位到该文件夹后，直接执行`pip install XXX.whl`或源码包执行`pip install xxx.tar.gz`，完成安装。如果要在PyPI[③]上发布，可以提供给更多人使用，则可以注册该网站，将生成的源码或者二进制文件通过`twine upload dist/...`方式推送至自己的注册地址（需要安装twine[④]工具，推送时提示输入pypi的注册用户名和密码）。上传成功后（pypi可能不支持代理上传），则可以在pypi网站注册地址的your projects下找到发布的包（因为推送的包不能有重名，如果名字已存在，需要修改后，重新打包推送），在view[⑤]下查看安装代码为`pip install toolkit4beginner==0.3.3`（或存在之前版本，则可以用`pip install toolkit4beginner -U`方法更新）。此时，该包同任何第三方库一样，可以直接在终端根目录安装，并且不需要源文件，可以为任何人安装使用。

建立一个项目，如果希望最终发布为一个分发包提供给更多人使用，则在建立项目时最好就按照包的文件目录结构布局，根据项目要求初步构思好基本的分类，然后在书写代码时，将函数、类和变量按类置于不同的模块下，并将模块置于对应的分类目录下。如果开始粗略的分类

③ PyPI.Find, install and publish Python packages with the Python Package Index，（https://pypi.org/）。

④ twine，用于在PyPI上发布Python包的实用程序（https://pypi.org/project/twine/）。

⑤ toolkit4beginner，（https://pypi.org/project/toolkit4beginner/0.3.3/）。

不满足要求，则实时调整目录结构，只要保持一个包合理的目录结构即可。最终配置setup.py文件，打包推送发布。

本节的案例是将之前PCS1-6中部分函数分类置于不同的模块和目录下，并增加了一个新的函数boxplot_custom(data_dict,**args)，位于graph/graph.py模块下。

每一模块都使用if __name__=="__main__":语句进行模块内的测试。如果构建包，在直接使用模块做数据分析时，通常也将函数、类和全局性的变量置于该语句之前；而在该语句之下（缩进），执行调用计算的代码。该语句的目的是避免其他模块调用该模块内的工具时，执行了测试或者用于直接计算部分的代码。

2

描述：

- toolkit/setup.py

该配置文件不在交互式解释器中运行，需要在终端中定位到该文件目录后，执行python setup.py sdist或python setup.py bdist_wheel，完成打包。通常确定包名name之后，每次模块发生更新时，修改version参数后，重新打包上传至pypi。pypi会将包置于同一个项目下，其下包含每次包更新的版本。参数配置时主要包括两大部分，一部分是项目(版权、管理、版本)信息，例如name，version,license，author,author_email,description,long_description,long_description_content_type和url等；另一部分是打包要求、依赖和包位置等胚子信息，定位包所在目录的packages参数，对python版本有要求的python_requires参数，安装依赖库的install_requires参数，是否包含额外文件的include_package_data参数等。更多参数可以查看setuptools documentation①，其中包括pyproject.toml②配置说明。

① setuptools documentation，（https://setuptools.pypa.io/en/latest/）。
② Configuring setuptools using pyproject.toml files，（https://setuptools.pypa.io/en/latest/userguide/pyproject_config.html）。

代码段：

```
# -*- coding: utf-8 -*-
"""
Created on Sun Aug 21 14:57:02 2022

@author: Richie Bao
"""
from setuptools import setup, find_packages, find_namespace_packages
with open("README.md", "r", encoding="utf-8") as fh:
    long_description = fh.read()

setup(name='toolkit4beginner',  #应用名，即包名
      version='0.3.3',  #版本号
      license="MIT",  #版权声明，BSD,MIT
      author='Richie Bao',  #作者名
      author_email='richiebao@outlook.com',  #作者邮箱
      description='模块、包和分发文件目录组织结构说明',  #描述
      long_description=long_description,
      long_description_content_type="text/markdown",
      url='https://richiebao.github.io/USDA_CH_final',  #项目主页
      #--------------------------------------------------------------
      package_dir={"": "src"},
      #包括安装包内的python包；find_namespace_packages()，和find_packages()['toolkit4beginner']
      packages=find_packages(where='src'),
      python_requires='>=3.6',  #pyton版本控制
      platforms='any',
      install_requires=['matplotlib', 'statistics', 'numpy']  #自动安装依赖包（库）
      #include_package_data=True,#如果配置有MANIFEST.in，包含数据文件或额外其他文，该参数配置为
True，则一同打包
      )
```

3

描述：

- toolkit4beginner/graph/graph.py

自定义样式图表打印模块，测试结果如图pcs.7-1和图pcs.7-2所示。

代码段：

```
# -*- coding: utf-8 -*-
"""
```

```
Created on Sun Aug 21 15:08:21 2022
@author: Richie Bao
"""
import matplotlib.pyplot as plt
import matplotlib as mpl
from statistics import mean, median
def boxplot_custom(data_dict, **args):
    '''
    根据 matplotliB 库的箱形图打印方法，自定义箱形图可调整的打印样式。

    Parameters
    ----------
    data_dict : dict(list,numerical)
        字典结构形式的数据，键为横坐分类数据，值为数值列表 .
    **args : keyword arguments
        可调整的箱形图样式参数包括 ['figsize', 'fontsize', 'frameOn', 'xlabel',
'ylabel', 'labelsize', 'tick_length', 'tick_width', 'tick_color', 'tick_
direction', 'notch', 'sym', 'whisker_linestyle', 'whisker_linewidth',
'median_linewidth', 'median_capstyle'].
    Returns
    -------
    paras : dict
        样式更新后的参数值 .

    '''

    # 计算值提取
    data_keys = list(data_dict.keys())
    data_values = list(data_dict.values())

    # 配置与更新参数
    paras = {'figsize': (10, 10),
             'fontsize': 15,
             'frameOn': ['top', 'right', 'bottom', 'left'],
             'xlabel': None,
             'ylabel': None,
             'labelsize': 15,
             'tick_length': 7,
             'tick_width': 3,
             'tick_color': 'b',
             'tick_direction': 'in',
             'notch': 0,
             'sym': 'b+',
             'whisker_linestyle': None,
             'whisker_linewidth': None,
             'median_linewidth': None,
             'median_capstyle': 'butt'}

    # print(paras)
    paras.update(args)
    # print(paras)

    # 根据参数调整打印图表样式
    plt.rcParams.update({'font.size': paras['fontsize']})
    frameOff = set(['top', 'right', 'bottom', 'left']) - set(paras['frameOn'])

    # 图表打印
    fig, ax = plt.subplots(figsize=paras['figsize'])
    ax.boxplot(data_values,
               notch=paras['notch'],
               sym=paras['sym'],
               whiskerprops=dict(
                   linestyle=paras['whisker_linestyle'],
linewidth=paras['whisker_linewidth']),
               medianprops={"linewidth": paras['median_linewidth'], "solid_
capstyle": paras['median_capstyle']})

    ax.set_xticklabels(data_keys)  # 配置 X 轴刻度标签
    for f in frameOff:
        ax.spines[f].set_visible(False)  # 配置边框是否显示

    # 配置 X 和 Y 轴标签
    ax.set_xlabel(paras['xlabel'])
    ax.set_ylabel(paras['ylabel'])
```

```
    # 配置 X 和 Y 轴标签字体大小
    ax.xaxis.label.set_size(paras['labelsize'])
    ax.yaxis.label.set_size(paras['labelsize'])

    # 配置轴刻度样式
    ax.tick_params(length=paras['tick_length'],
                   width=paras['tick_width'],
                   color=paras['tick_color'],
                   direction=paras['tick_direction'])
    plt.show()
    return paras
def four_quadrant_diagram(data_dict, method='mean', **args):
    '''
    绘制四象限图。

    Parameters
    ----------
    data_dict : dict(list)
        字典格式数据，包括两组键值 . 其中键名将用作轴标签
    method : string, optional
        按照均值 (mean) 或中位数 (median) 方式划分四象限 . The default is 'mean'.
    **args : keyword arguments
        可调整的图表样式参数包括：['figsize', 'fontsize', 'frameOn', 'crosshair_
color', 'crosshair_linstyle', 'crosshair_linewidth', 'labelsize', 'tick_length',
'tick_width', 'tick_color', 'tick_direction', 'dot_color', 'dot_size',
'annotation_position_finetune', 'annotation_fontsize', 'annotation_lable',
'annotation_color'].
    Returns
    -------
    None.
    '''

    # 解决中文乱字符的问题
    mpl.rcParams['font.sans-serif'] = ['SimHei']
    mpl.rcParams['font.serif'] = ['SimHei']

    # 计算值提取
    (key_x, x), (key_y, y) = data_dict.items()

    # 配置与更新参数
    paras = {'figsize': (10, 10),
             'fontsize': 15,
             'frameOn': ['top', 'right', 'bottom', 'left'],
             'crosshair_color': 'red',
             'crosshair_linstyle': '--',
             'crosshair_linewidth': 3,
             'labelsize': 15,
             'tick_length': 7,
             'tick_width': 3,
             'tick_color': 'b',
             'tick_direction': 'in',
             'dot_color': 'k',
             'dot_size': 50,
             'annotation_position_finetune': 0.5,
             'annotation_fontsize': 10,
             'annotation_lable': None,
             'annotation_color': 'gray',
             }
    paras.update(args)

    # 根据参数调整打印图表样式
    plt.rcParams.update({'font.size': paras['fontsize']})
    frameOff = set(['top', 'right', 'bottom', 'left']) - set(paras['frameOn'])

    # 图表打印
    fig, ax = plt.subplots(figsize=paras['figsize'])
    ax.scatter(x, y, c=paras['dot_color'], s=paras['dot_size'])
    crosshair_color = paras['crosshair_color']
    crosshair_linestyle = paras['crosshair_linstyle']
    crosshair_linewidth = paras['crosshair_linewidth']
    if method == 'mean':
        ax.axhline(y=mean(y), color=crosshair_color,
                   linestyle=crosshair_linestyle, linewidth=crosshair_linewidth)
        ax.axvline(x=mean(x), color=crosshair_color,
```

```
                    linestyle=crosshair_linestyle, linewidth=crosshair_linewidth)
    elif method == 'median':
        plt.axhline(y=median(y), color=crosshair_color,
                    linestyle=crosshair_linestyle, linewidth=crosshair_linewidth)
        plt.axvline(x=median(x), color=crosshair_color,
                    linestyle=crosshair_linestyle, linewidth=crosshair_linewidth)
    for f in frameOff:
        ax.spines[f].set_visible(False)  #配置边框是否显示

    #标注点
    if paras['annotation_lable']:
        annotations = paras['annotation_lable']
    else:
        annotations = list(range(len(x)))
    annotation_position_finetune = paras['annotation_position_finetune']
    for label, x, y in zip(annotations, x, y):
        ax.annotate(label, (x + annotation_position_finetune, y + annotation_
position_finetune),
                    fontsize=paras['annotation_fontsize'],
                    color=paras['annotation_color'])

    #配置 X 和 Y 轴标签
    ax.set_xlabel(key_x)
    ax.set_ylabel(key_y)

    #配置 X 和 Y 轴标签字体大小
    ax.xaxis.label.set_size(paras['labelsize'])
    ax.yaxis.label.set_size(paras['labelsize'])

    #配置轴刻度样式
    ax.tick_params(length=paras['tick_length'],
                   width=paras['tick_width'],
                   color=paras['tick_color'],
                   direction=paras['tick_direction'])
    plt.tight_layout()
    plt.show()
# four_quadrant_diagram(test_EC_scores,
#             # method=' median' ,
#             figsize=(15,15),
#             fontsize=23,
#             frameOn=[ 'bottom' ,' left' ],
#             labelsize=' 30' ,
#             dot_size=200,
#             annotation_fontsize=30,
#             annotation_lable=test_names,
#             )
if __name__ == "__main__":
    #模块内测试
    #A_ 测试 -boxplot_custom(data_dict,**args)
    #%%
    test_score_lst_dic = {'English': [90, 81, 73, 97, 85, 60, 74, 64, 72, 67,
87, 78, 85, 96, 77, 100, 92, 86],
                          'Chinese': [71, 90, 79, 70, 67, 66, 60, 83, 57, 85,
93, 89, 78, 74, 65, 78, 53, 80],
                          'history': [73, 61, 74, 47, 49, 87, 69, 65, 36, 7, 53,
100, 57, 45, 56, 34, 37, 70],
                          'biology': [59, 73, 47, 38, 63, 56, 75, 53, 80, 50,
41, 62, 44, 26, 91, 35, 53, 68]}
    _ = boxplot_custom(test_score_lst_dic,
                       figsize=(15, 10),
                       fontsize=23,
                       frameOn=['bottom', 'left'],
                       xlabel='subject',
                       ylabel='score',
                       labelsize='30',
                       tick_color='r',
                       notch=1,
                       sym='rs',
                       whisker_linestyle='--',
                       whisker_linewidth=5,
                       median_linewidth=5
                       )

    #B_ 测试 -four_quadrant_diagram(data_dict,method=' mean' ,**args)
```

```
    test_EC_scores = {'English': [90, 81, 73, 97, 85, 60, 74, 64, 72, 67, 87,
78, 85, 96, 77, 100, 92, 86],
                      'Chinese': [71, 90, 79, 70, 67, 66, 60, 83, 57, 85, 93,
89, 78, 74, 65, 78, 53, 80], }
    test_names = ['Mason', 'Reece', 'A', 'B', 'C', 'D', 'E',
                 'F', 'G', 'H', 'I', 'J', 'K', 'L', 'M', 'N', 'O', 'P']
    four_quadrant_diagram(test_EC_scores,
                          # method=' median' ,
                          figsize=(15, 15),
                          fontsize=23,
                          frameOn=['bottom', 'left'],
                          labelsize='30',
                          dot_size=200,
                          annotation_fontsize=30,
                          annotation_lable=test_names,
                          )
```

运算结果：

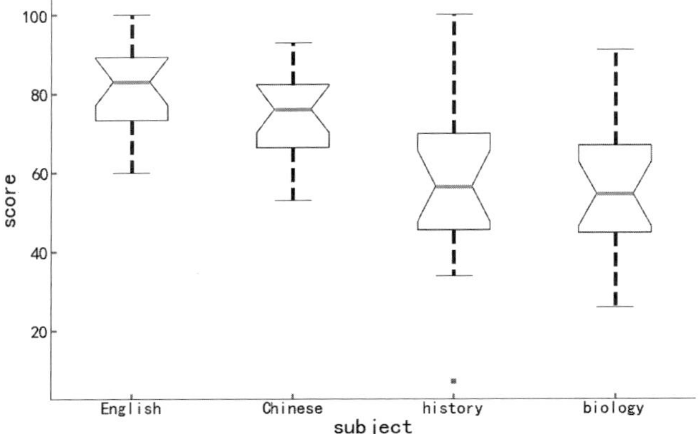

图pcs.7-1　自定义样式图表打印模块 boxplot_custom 函数测试

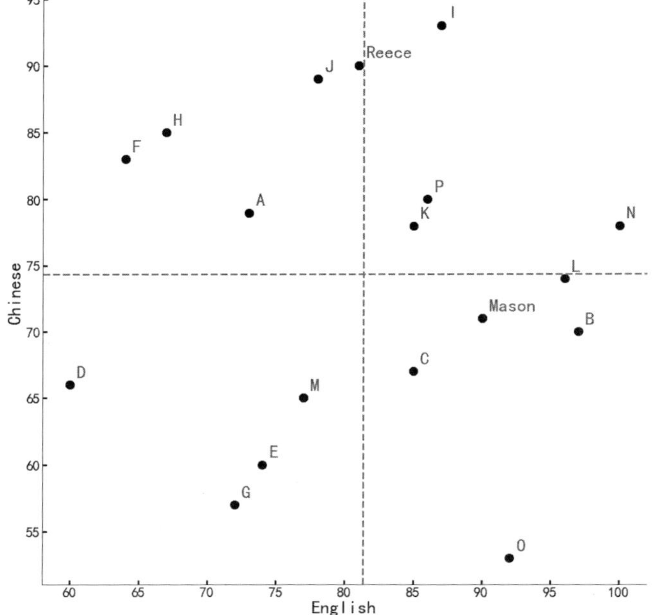

图pcs.7-2　自定义样式图表打印模块 four_quadrant_diagram 函数测试

描述：

- toolkit4beginner/utility/general_tools.py

一般常用工具模块。

代码段：

```
# -*- coding: utf-8 -*-
"""
Created on Sun Aug 21 18:35:24 2022

@author: Richie Bao
"""
import datetime
import datetime

# 展平嵌套列表，返回列表
flatten_lst = lambda lst: [m for n_lst in lst for m in flatten_lst(
    n_lst)] if type(lst) is list else [lst]
def flatten_lst_generator(lst):
    '''
    展平嵌套列表，返回迭代对象。

    Parameters
    ----------
    lst : list
        嵌套列表 .

    Yields
    ------
    iterable object
        嵌套列表展平后的迭代对象 .

    '''
    try:  # 使用语句 try/except 捕捉异常
        for n_lst in lst:
            for m in flatten_lst_generator(n_lst):
                yield m
    except:
        yield lst
def infinite_generator():
    '''
    无穷整数生成器。

    Yields
    ------
    i : int
        整数值 .

    '''
    i = 0
    while True:
        i += 1
        yield i
def recursive_factorial(n):
    '''
    阶乘计算。

    Parameters
    ----------
    n : int
        阶乘计算的末尾值 .

    Returns
    -------
    int
        阶乘计算结果 .

    '''
    if n == 1:
        return 1
    else:
        return n * recursive_factorial(n - 1)
```

```
def start_time():
    '''
    获取当前时间

    Returns
    -------
    start_time : datetime
        返回当前时间 .

    '''
    start_time = datetime.datetime.now()
    print("start time:", start_time)
    return start_time
def duration(start_time):
    '''
    配合 start_time() 使用。计算时间长度。

    Parameters
    ----------
    start_time : datetime
        用于计算时间长度的开会时间 .

    Returns
    -------
    None.
    '''
    end_time = datetime.datetime.now()
    print("end time:", end_time)
    duration = (end_time - start_time).seconds / 60
    print("Total time spend:%.2f minutes" % duration)
if __name__ == "__main__":
    # 模块内测试
    # A_ 测试 -flatten_lst
    nested_lst = [['A', 'B', ['C', 'D'], 'E'], [6, [7, 8, [9]]]]
    print(flatten_lst(nested_lst))

    # B_ 测试 -flatten_lst_generator(lst)
    fl = flatten_lst_generator(nested_lst)
    print(list(fl))

    # C_ 测试 -infinite_generator()
    inf = infinite_generator()
    print(next(inf))
    print(next(inf))

    # D_ 测试 -recursive_factorial(n)
    print(recursive_factorial(6))

    # E_ 测试 -start_time();duration(start_time)
    s_t = start_time()
    for i in range(10**8):
        value = i
    duration(s_t)
```

运算结果：

```
['A', 'B', 'C', 'D', 'E', 6, 7, 8, 9]
['A', 'B', 'C', 'D', 'E', 6, 7, 8, 9]
1
2
720
start time: 2022-08-21 23:16:51.631636
end time: 2022-08-21 23:17:04.243942
Total time spend:0.20 minutes
```

5

描述：

- toolkit4beginner/utility/stats_tools.py

统计类工具模块。

代码段：

```
# -*- coding: utf-8 -*-
"""
```

```
Created on Sun Aug 21 18:56:45 2022

@author: Richie Bao
"""
import statistics
import numpy as np
def descriptive_statistics(data, measure=None, decimals=2):
    '''
    计算给定数值列表的描述性统计值，包括数量、均值、标准差、方差、中位数、众数、最小值和最大值。

    Parameters
    ----------
    data : list(numerical)
        待统计的数值列表 .
    measure : str, optional
        包括：'count', 'mean', 'std', 'variance', 'median', 'mode', 'min', 'max'.
The default is None.
    decimals : int, optional
        小数位数 . The default is 2.

    Returns
    -------
    dict
        如果不给定参数 measure，则以字典形式返回所有值；否则，返回给定 measure 对应值的表述字
符串 .

    '''
    d_s = {
        'count': len(data),  # 样本数
        'mean': round(statistics.mean(data), decimals),  # 均值
        'std': round(statistics.stdev(data), decimals),  # 标准差
        'variance': round(statistics.variance(data), decimals),  # 方差
        'median': statistics.median(data),  # 中位数
        'mode': statistics.mode(data),  # 众数
        'min': min(data),  # 最小值
        'max': max(data),  # 最大值
    }
    if measure:
        return '{}={}'.format(measure, d_s[measure])
    else:
        return d_s
def value2values_comparison(x, lst):
    '''
    判断给定的一个值是否在一个列表中，如果在，则返回给定值；如果不在，则返回列表中与其差值绝对值
最小的值。

    Parameters
    ----------
    x : numerical
        数值 .
    lst : list(numerical)
        用于寻找给定值的参考列表 .

    Returns
    -------
    numerical
        列表中最接近给定值的值 .

    '''
    lst_mean = np.mean(lst)
    abs_difference_func = lambda value: abs(x)
    comparisonOF2values = lambda v1, v2: print('x is higher than the average %s
of the list.' % lst_mean) if v1 > v2 else print(
        'x is lower than the average %s of the list.' % lst_mean)
    if x in lst:
        print("%s in the given list." % x)
        comparisonOF2values(x, lst_mean)
        return x
    else:
        print('%.3f is not in the list.' % x)
        closest_value = min(lst, key=abs_difference_func)
        print('the nearest value to %s is %s.' % (x, closest_value))
        print("_" * 50)
        comparisonOF2values(closest_value, lst_mean)
```

```
        return closest_value
if __name__ == '__main__':
    # A_ 测试 -descriptive_statistics(data,measure=None,decimals=2)
    ramen_price_lst = [700, 850, 600, 650, 980, 750, 500, 890, 880,
                        700, 890, 720, 680, 650, 790, 670, 680, 900,
                        880, 720, 850, 700, 780, 850, 750, 780, 590,
                        650, 580, 750, 800, 550, 750, 700, 600, 800,
                        800, 880, 790, 790, 780, 600, 690, 680, 650,
                        890, 930, 650, 777, 700]
    d_s_1 = descriptive_statistics(ramen_price_lst)
    print(d_s_1)
    print('--' * 30)
    d_s_2 = descriptive_statistics(ramen_price_lst, 'std')
    print(d_s_2)
    d_s_3 = descriptive_statistics(
        ranmen_price_lst, measure='mean', decimals=1)
    print(d_s_3)

    # B_ 测试 -value2values_comparison(x,lst)
    ramen_price_lst = [700, 850, 600, 650, 980, 750, 500, 890, 880, 700, 890,
720, 680, 650, 790, 670, 680, 900, 880, 720, 850, 700, 780, 850, 750,
                        80, 590, 650, 580, 750, 800, 550, 750, 700, 600, 800,
800, 880, 790, 790, 780, 600, 690, 680, 650, 890, 930, 650, 777, 700]
    price_x = 200.68
    price_x_closestValue = value2values_comparison(price_x, ramen_price_lst)
    print(price_x_closestValue)
```

运算结果：

```
{'count': 50, 'mean': 743.34, 'std': 108.26, 'variance': 11720.64, 'median':
750.0, 'mode': 700, 'min': 500, 'max': 980}
--------------------------------------------------------------------------------
std=108.26
mean=743.3
200.680 is not in the list.
the nearest value to 200.68 is 700.
--------------------------------------------------------------------------------
x is lower than the average 729.34 of the list.
700
```

6

描述：

- toolkit4beginner/test/test.py

如果打包推送至PyPI上发布，并通过pip install toolkit4beginner==0.3.3或pip install toolkit4beginner -U安装，则可以使用该包的工具。首先，是通过一行调入语句import toolkit4beginner进行简单的测试，是否包已经正确安装。如果没有提示错误，则已正确安装并调入；否则，需要确认之前的安装步骤是否正确，或者源码和打包的整个流程中是否有疏漏，重新安装或者调试后重新打包，以新版本（配置version参数）发布后再安装或更新包。

代码段：

```
import toolkit4beginner
```

7

描述：

下述测试代码是对具体模块的测试。在模块调入方式上，选择了不同的调入方式，包括from toolkit4beginner.graph import graph，从子包（子目录）中调入一个模块；from toolkit4beginner.utility.general_tools import flatten_lst，从子包的模块中调入一个方法（函数）；from toolkit4beginner.utility.stats_tools import *，一次性调入一个模块中的所有方法。也可以使用import toolkit4beginner as t4b，或者from toolkit4beginner.utility.general_tools import flatten_lst as FL等方式为调入的对象重新命名，这尤其适合于有同名的文件或方法，避免同名冲突。测试结果如图pcs.7-3和图pcs.7-4所示。

代码段：

```
# -*- coding: utf-8 -*-
"""
Created on Sun Aug 21 15:23:51 2022
```

```
@author: Richie Bao
"""
# 测试

from toolkit4beginner.utility.stats_tools import *
from toolkit4beginner.utility.general_tools import flatten_lst
from toolkit4beginner.graph import graph
test_score_lst_dic = {'English': [90, 81, 73, 97, 85, 60, 74, 64, 72, 67, 87,
78, 85, 96, 77, 100, 92, 86],
                      'Chinese': [71, 90, 79, 70, 67, 66, 60, 83, 57, 85, 93,
89, 78, 74, 65, 78, 53, 80],
                      'history': [73, 61, 74, 47, 49, 87, 69, 65, 36, 7, 53,
100, 57, 45, 56, 34, 37, 70],
                      'biology': [59, 73, 47, 38, 63, 56, 75, 53, 80, 50, 41,
62, 44, 26, 91, 35, 53, 68]}
_ = graph.boxplot_custom(test_score_lst_dic,
                         figsize=(15, 10),
                         fontsize=23,
                         frameOn=['bottom', 'left'],
                         xlabel='subject',
                         ylabel='score',
                         labelsize='30',
                         tick_color='r',
                         notch=1,
                         sym='rs',
                         whisker_linestyle='--',
                         whisker_linewidth=5,
                         median_linewidth=5
                         )
test_EC_scores = {'English': [90, 81, 73, 97, 85, 60, 74, 64, 72, 67, 87, 78,
85, 96, 77, 100, 92, 86],
                  'Chinese': [71, 90, 79, 70, 67, 66, 60, 83, 57, 85, 93, 89,
78, 74, 65, 78, 53, 80], }
test_names = ['Mason', 'Reece', 'A', 'B', 'C', 'D', 'E',
              'F', 'G', 'H', 'I', 'J', 'K', 'L', 'M', 'N', 'O', 'P']
graph.four_quadrant_diagram(test_EC_scores,
                            # method='median',
                            figsize=(15, 15),
                            fontsize=23,
                            frameOn=['bottom', 'left'],
                            labelsize='30',
                            dot_size=200,
                            annotation_fontsize=30,
                            annotation_lable=test_names,
                            )
nested_lst = [['A', 'B', ['C', 'D'], 'E'], [6, [7, 8, [9]]]]
print(flatten_lst(nested_lst))
ramen_price_lst = [700, 850, 600, 650, 980, 750, 500, 890, 880, 700, 890, 720,
680, 650, 790, 670, 680, 900, 880, 720, 850, 700, 780, 850, 750,
                   80, 590, 650, 580, 750, 800, 550, 750, 700, 600, 800, 800,
880, 790, 790, 780, 600, 690, 680, 650, 890, 930, 650, 777, 700]
price_x = 200.68
price_x_closestValue = value2values_comparison(price_x, ramen_price_lst)
print(price_x_closestValue)
# ---------------------------------------------------------------------------
ramen_price_lst = [700, 850, 600, 650, 980, 750, 500, 890, 880,
                   700, 890, 720, 680, 650, 790, 670, 680, 900,
                   880, 720, 850, 700, 780, 850, 750, 780, 590,
                   650, 580, 750, 800, 550, 750, 700, 600, 800,
                   800, 880, 790, 790, 780, 600, 690, 680, 650,
                   890, 930, 650, 777, 700]
d_s_1 = descriptive_statistics(ramen_price_lst)
print(d_s_1)
print('--' * 30)
d_s_2 = descriptive_statistics(ramen_price_lst, 'std')
print(d_s_2)
d_s_3 = descriptive_statistics(ramen_price_lst, measure='mean', decimals=1)
print(d_s_3)
```

运算结果：

```
['A', 'B', 'C', 'D', 'E', 6, 7, 8, 9]
200.680 is not in the list.
the nearest value to 200.68 is 700.
```

```
--------------------------------------------------------------------------------
x is lower than the average 729.34 of the list.
700
{'count': 50, 'mean': 743.34, 'std': 108.26, 'variance': 11720.64, 'median':
750.0, 'mode': 700, 'min': 500, 'max': 980}
--------------------------------------------------------------------------------
std=108.26
mean=743.3
```

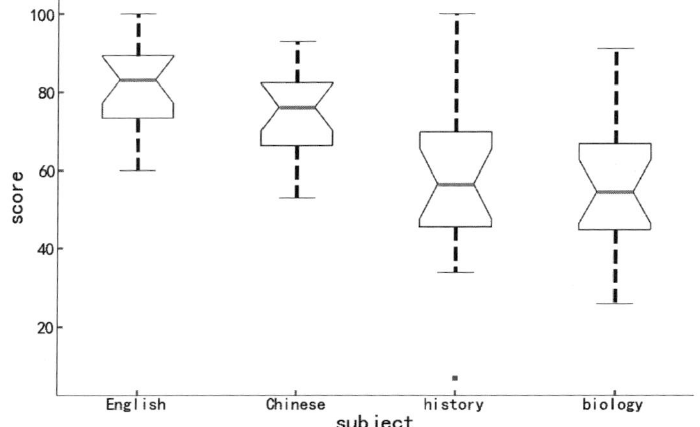

图pcs.7-3 安装后，自定义样式图表打印模块 boxplot_custom 函数测试

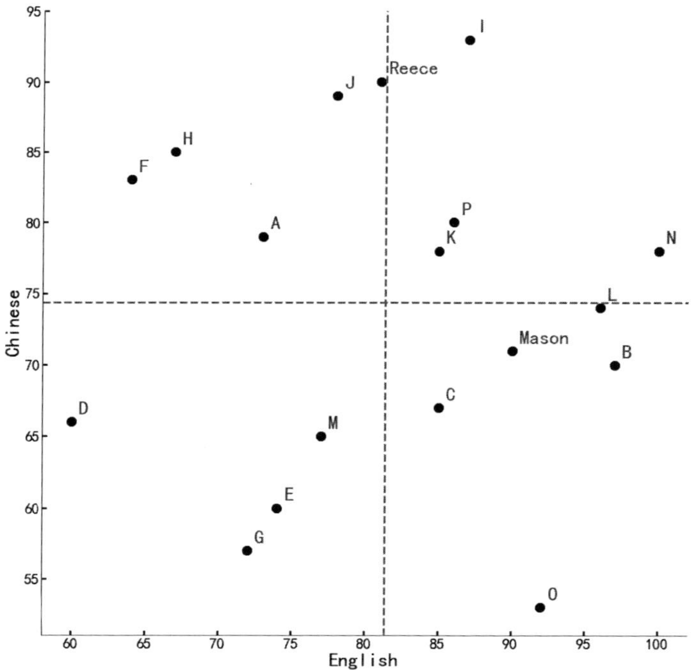

图pcs.7-4 安装后，自定义样式图表打印模块 four_quadrant_diagram 函数测试

PCS-8. 类（OOP）Classes_定义，继承，__init_()构造方法，私有变量/方法

PCS-8（ ）：8.1 OOP(Object-oriented programming)_面向对象编程，及Classes定义和__init__()构造方法；8.2 Inheritance（继承）——superclasses（父类）和subclasses（子类）；8.3 private variables(私有变量/成员)和private methods(私有方法)；8.4 Operator Overloading（运算符/操作符重载）与类属性格式"字典"

知识点：

8.1 OOP(Object-oriented programming)_面向对象编程，及Classes定义和__init__()构造方法

●●1

描述：

OOP与定义类

面向对象编程（Object-oriented programming，OOP）是一种计算机编程模型，是围绕数据或对象而不是功能和逻辑来组织软件设计，更专注于对象与对象之间的交互。对象涉及的方法（methods）和属性(attributes)都在对象内部。其中，类（Classes）是相同种类对象的抽象，是该类对象的公共属性；将该类实例化一个或多个对象（instance objects），即为该对象的实例。例如，定义一个类为鸟类，所有的鸟类都具有觅食的行为，可以将该行为定义为类的函数（成员函数，member functions），称为类的方法；可以将鸟类的飞翔动作定义为类的变量（成员变量），称为类的属性。鸟类通常会根据鸟类特征，划分很多子类，例如雨燕目、鸽形目、雁形目等，不同子类的羽毛颜色、鸟鸣声都会不同，因此可以基于已有鸟类（父类，superclass），定义继承父类属性方法的子类（subclass）。

python类提供了面向对象编程的所有标准特性：类的继承机制（inheritance mechanism）允许多个基类（basse classes/superclasses）；一个派生类（derived class/subclass）可以覆盖基类或任何类的方法；一个方法可以调用（call/invoke）其基类的方法；对象可以包含任意数量和种类的数据；类具有动态特性（dynamic nature），可以在运行时添加或者删除属性方法等。

类定义的语法，可以简单表示为：

```
class ClassName:
    <statement-1>
    .
    .
    .
    <statement-N>
```

通常，类的定义名首字母大写（共识非标准），其中语句一般为函数定义（允许其他类型的语句，例如定义变量赋值语句等）。下面，引入Python官网的几个小案例[①]，稍许调整，辅助理解类定义和调用的方式。

① 关于Class类Python官网的几个小案例（https://docs.python.org/3/tutorial/classes.html）。

下述类定义的类名称为SayHello。该类之下的开始行，函数定义之前，定义了一行赋值语句name_default='Who'；定义的函数helloWorld(self,name)，其中self参数为类方法定义的第一个必要参数（本身的意思），在方法调用时指向类的实例。同时，该函数也传入了另一个参数name，指定默认值为变量name_default。需要注意，通过该函数内print(SayHello.name_default)语句可以发现，通过Class.attribute的方式可以调用函数外类内的变量。x=SayHello()即实例化类对象语句，将类实例化为变量x，则该实例x具有了该类的所有属性和方法。x.name_default为调用类属性的方式；x.helloWorld()和x.helloWorld('Swift')为调用类方法（函数）的方式。

类可以使用内置方法__doc__读取类下的注释语句。

代码段：

```
class SayHello:
    """A simple example class"""
    name_default = 'Who'
    def helloWorld(self, name=name_default):
        print(SayHello.name_default)  #可以通过 Class.attribute 的方式获取函数外的变量值
        return '%s says hello, world!' % name
x = SayHello()
print(x.name_default)
print(x.helloWorld())
print(x.helloWorld('Swift'))
```

运算结果：

```
Who
Who
Who says hello, world!
Who
Swift says hello, world!
--------------------------------------------------------------------------------
A simple example class
```

2

描述：

通过help()可以查看定义类的内容，这包括开始行的注释语句、方法（Methods）和属性（attributes），以及定义的数据描述（data description defined）等。其中，__dict__是存储对象属性的一个字典，键为属性名，而值为属性值。通过__dict__可以查看所有类的属性和方法。

代码段：

```
print('-' * 50)
print(SayHello.__doc__)
print(help(SayHello))
print(SayHello.__dict__)
print('-' * 50)
print(SayHello.__dict__['name_default'])
print(x.__dict__)
```

运算结果：

```
Help on class SayHello in module __main__:
class SayHello(builtins.object)
 |  A simple example class
 |
 |  Methods defined here:
 |
 |  helloWorld(self, name='Who')
 |
 |  ------------------------------------------------------------------------------
 |  Data descriptors defined here:
 |
 |  __dict__
 |      dictionary for instance variables (if defined)
 |
 |  __weakref__
 |      list of weak references to the object (if defined)
 |
 |  ------------------------------------------------------------------------------
 |  Data and other attributes defined here:
 |
 |  name_default = 'Who'
None
{'__module__': '__main__', '__doc__': 'A simple example class', 'name_default':
'Who', 'helloWorld': <function SayHello.helloWorld at 0x00000152FBEC9CA0>, '__
dict__': <attribute '__dict__' of 'SayHello' objects>, '__weakref__': <attribute
'__weakref__' of 'SayHello' objects>}
--------------------------------------------------------------------------------
Who
{}
```

3

描述：

● '__init__()'构造方法与可变对象

如果类实例化时要传入参数（初始化参数），则可以定义类时，定义一个名为__init__()的特殊方法（constructor method，构造方法），下述案例为构造方法传入了一个name参数。在实例化时，如果该参数未指定默认值，则需要传入该参数值，如d = Dog('Fido')。类的实例化会自动调用__init__()来处理新创建的类的实例，通过 self.name = name，初始化对应的参数，则self.name变量的作用域为整个类，内部的所有函数均可以调用self.name变量，外部实例化对象也可以通过d.name访问。也可以在该构造函数内书写其他代码，完成相关的任务，例如定义新的变量或相关计算。

下述引用官网的案例在说明__init__()构造方法初始化参数值和相关运算，也阐述在类下函数外定义的变量，如果为可变对象（mutable objects）例如tricks = []，变量tricks为一个列表，是可变对象，则实例化多个对象。执行类的add_trick()方法时，新的值都会被追加到tricks列表中，即各个实例化对象的tricks属性值相同，为共同追加的值列表。如果要将其分开，则需要在初始化时，通过self.tricks = []方法定义可变变量，self指向各个实例化对象自身，而不是类本身。因此，在各各个实例化对象（d和e）执行add_trick()方法时，仅会将值追加在各自的self.tricks列表中。

在SayHello案例中，因为没有以self引入的变量，因此print(x.__dict__)时为空。在Dog案例中，因为存在变量self.name，因此print(d.__dict__)时，会返回{'name': 'Fido'}。

代码段：

```
class Dog:

    tricks = []             # 错误地使用类变量

    def __init__(self, name):
        self.name = name
    def add_trick(self, trick):
        self.tricks.append(trick)
d = Dog('Fido')
print(d.name)
e = Dog('Buddy')
d.add_trick('roll over')
e.add_trick('play dead')
print(d.tricks)
print(e.tricks)
print("-" * 50)
print(d.__dict__)
```

运算结果：

```
Fido
['roll over', 'play dead']
['roll over', 'play dead']
--------------------------------------------------
{'name': 'Fido'}
```

4

代码段：

```
class Dog:
    def __init__(self, name):
        self.name = name
        self.tricks = []    # 为每只狗创建一个新的空列表

    def add_trick(self, trick):
        self.tricks.append(trick)
d = Dog('Fido')
e = Dog('Buddy')
d.add_trick('roll over')
e.add_trick('play dead')
print(d.tricks)
print(e.tricks)
```

运算结果:

```
['roll over']
['play dead']
```

知识点:

8.2 Inheritance（继承）——superclasses（父类）和subclasses（子类）

5

描述:

继承主要作用是实现代码的复用，使子类拥有父类的方法和属性，在已有类的继承上扩展子类，可以大幅度降低代码书写量，并避免重复；同时，可以更加方便地理清类之间、子类父类之间的关系，这使得代码的逻辑更加清晰；配置类的组合关系，则使得代码书写具有更好的弹性。

继承的语法为:

```
class DerivedClassName(BaseClassName):
    <statement-1>
    .
    .
    .
    <statement-N>
```

如果父类和子类不在同一个模块（module）中（即不在同一个文件），则可以用下述方法:

```
class DerivedClassName(modname.BaseClassName):
```

python的类支持多重继承（multiple inheritance），即可以同时继承多个父类，语法如下:

```
class DerivedClassName(Base1, Base2, Base3):
<statement-1>
.
.
.
<statement-N>
```

- 继承搜索

具有继承关系类的属性和方法搜索，基本路径是自下而上、自左而右。例如，上述多重继承，首先在子类自身DerivedClassName中搜索。如果搜索不到，则向上在父类中搜索。首先，在Base1中搜索，如果仍搜索不到，则继续搜索Base2和Base3，直至锁定；或者找不到属性和方法，则引发异常。

- 继承的第1个案例——1个父类+1个子类

下述案例定义了父类Bird，拥有的属性有fly和hungry，拥有的方法有eat()。定义的子类Apodidae(Bird)，继承父类Bird的所有属性和方法，同时增加了自身的属性有sound；方法有sing()。通过super(Apodidae,self).__init__(hungry)语句，为调用父类的初始化方法来初始化子类，也可以在子类中重新定义初始化的方法，来覆盖父类初始化的属性结果。

代码段:

```
class Bird:
    fly = 'Whirring'  # 美 /wɜːr/

    def __init__(self, hungry=True):
        self.hungry = hungry
    def eat(self):
        if self.hungry:
            print('Aaaah...')
            self.hungry = False
        else:
            print('No.Thanks!')

class Apodidae(Bird):  # 美 /'æpədədiː/
    def __init__(self, sound='Squawk!', hungry=True):
        super(Apodidae, self).__init__(hungry)
        self.sound = sound  # 美 /skwɔːk/

    def sing(self):
```

```
        print(self.sound)
```

6

描述：

将子类实例化为两个实例对象，为swift和blackswift，实例化时，传入的初始化参数不同，对于swift使用了默认参数值，即(Squawk,True)，而blackswift传入的参数为('hooting',False)。子类具有自身定义的属性（sound）和方法(sing())，也继承了父类的属性（hungry,fly）和方法(eat())。

代码段：

```
swift = Apodidae()
print("swift.fly->{}".format(swift.fly))
print("swift.hungry->{}".format(swift.hungry))
swift.eat()
print("swift.hungry->{}".format(swift.hungry))
swift.eat()
print("swift.sing->{}".format(swift.sound))
swift.sing()
print("-" * 50)
blackswift = Apodidae('hooting', False)
print("blackswift.fly->{}".format(blackswift.fly))
print("blackswift.hungry->{}".format(blackswift.hungry))
blackswift.eat()
print("blackswift.hungry->{}".format(blackswift.hungry))
print("blackswift.sing->{}".format(blackswift.sound))
blackswift.sing()
```

运算结果：

```
swift.fly->Whirring
swift.hungry->True
Aaaah...
swift.hungry->False
No.Thanks!
swift.sing->Squawk!
Squawk!
--------------------------------------------------------------------------------
blackswift.fly->Whirring
blackswift.hungry->False
No.Thanks!
blackswift.hungry->False
blackswift.sing->hooting
hooting
```

7

描述：

如果将父类Bird实例化为对象crow，则crow仅具有Bird类的属性和方法，而不具有子类的属性和方法。

代码段：

```
crow = Bird()
print("crow.fly->{}".format(crow.fly))
print("crow.hungry->{}".format(crow.hungry))
crow.eat()
print("crow.hungry->{}".format(crow.hungry))
print("-" * 50)
crow.sing()
```

运算结果：

```
crow.fly->Whirring
crow.hungry->True
Aaaah...
crow.hungry->False
--------------------------------------------------------------------------------
AttributeError                          Traceback (most recent call last)
Input In [20], in <cell line: 7>()
      5 print("crow.hungry->{}".format(crow.hungry))
      6 print("-"*50)
----> 7 crow.sing()
AttributeError: 'Bird' object has no attribute 'sing'
```

8

描述：

- 继承的第2个案例——连续继承与`__repr__()`重写（overload）

该案例引自*Mark Lutz. (2013). Learning python(5th Edition).O'Reilly Media, Inc, USA*中关于类阐述的章节。其中，`AttriDisplay`是`TopTest`的父类，`TopTest`是`SubTest`的父类。因为`__dict__`可以获取类的所有属性和方法的键值对，而`getattr()`内置函数用于一个对象的属性值，因此`AttrDisplay`方法`gatherAttrs()`提取了`self`（实例化对象）具有的属性，并存储在`attrs`列表中。

`repr()`通过`__repr__()`这个特殊方法来得到一个对象的字符串表示形式，方便查看。如果对象没有实现`__repr__()`，返回的字符串则如下述的`<__main__.Example object at 0x00000152FBE9E190>`。在`AttriDisplay`类定义中，则定义了`__repr__()`方法，该方法返回值必须是字符串对象。因此，当打印实例化对象时，则会执行`__repr__()`方法，返回对应格式化后的字符串。

代码段：

```
string = 'example'
print(repr(string))
class Example:
    pass
print(repr(Example()))
```

运算结果：

```
<__main__.Example object at 0x00000152FBE9E190>
'example'
```

9

代码段：

```
class AttrDisplay:
    """
    提供一个可继承显示重载的方法，该方法显示具有类名的实例，以及存储在实例本身上的每个属性的名称=值对（但不包括从其类继承的属性）。可以混合到任何类中，并且在任何实例上工作。
    """
    def gatherAttrs(self):
        attrs = []
        for key in sorted(self.__dict__):
            attrs.append('%s=%s' % (key, getattr(self, key)))
        return ', '.join(attrs)
    def __repr__(self):
        return '[%s: %s]' % (self.__class__.__name__, self.gatherAttrs())
class TopTest(AttrDisplay):
    count = 0
    def __init__(self):
        self.attr1 = TopTest.count
        self.attr2 = TopTest.count + 1
        TopTest.count += 2
class SubTest(TopTest):
    pass

print(AttrDisplay())  #父类 AttriDisplay 没有 self 引导的属性
X, Y = TopTest(), SubTest()
print(X)
print(Y)
print("-" * 50)
print(getattr(X, 'count'))
```

运算结果：

```
[AttrDisplay: ]
[TopTest: attr1=0, attr2=1]
[SubTest: attr1=2, attr2=3]
--------------------------------------------------------------------------------
4
```

10

描述：

● 继承的第3个案例——shelve[①]数据录入、更新、删除与读取

shelve，是python数据储存的一种方式，类似key-value数据库存储形式，便于保存python对象。shelve只有一个open()函数，用来打开指定的文件（字典），返回一个类似字典的对象。为了方便shelve存储方式的操作，编写下述DB_Shelve类，实现读、写、更新和删除4类基本操作。同时，继承了AttrDisplay类，方便查看属性对象。当调试完后，可以移除该父类。

写入(write())与更新(update)数据的方法，使用了关键字(**kwargs)的形式。同时，为了方便查看操作结果，增加了feedback()方法。可以在读写更新等方法中调用，反馈操作结果等信息。

① shelve — Python object persistence，（https://docs.python.org/3/library/shelve.html）。

代码段：

```
import shelve
class DB_Shelve(AttrDisplay):
    '''
    为方便 shelve 库方式的数据存储，编写数据写入、读取、更新和删除等操作，方便调用。
    '''
    def __init__(self, db_fn, flag='c'):
        '''
        初始化读写 shelve 数据库（存储文件）的基本参数

        Parameters
        ----------
        db_fn : string
            存储文件路径名 .
        flag : string, optional
            读写方式。包括 r- 只读；
            w- 可读写 ;
            n- 每次调用 open() 都重新创建一个空的文件，可读写 .
            The default is 'c'- 如果数据文件不存在，就创建，允许读写 .

        Returns
        -------
        None.
        '''
        self.db_fn = db_fn
        self.flag = flag
    def feedback(self, process=None, msg='OK.', data=None):
        '''
        用于反馈读写信息

        Parameters
        ----------
        process : string, optional
            标识操作，例如 write, read,update 或 trash 等 . The default is None.
        msg : string, optional
            返回信息，表述运行是否成功等 . The default is 'OK.'.
        data : any python built-in types, optional
            打印数据 . The default is None.

        Returns
        -------
        dict
            反馈操作结果信息 .

        '''
        if data:
            return {"process": process, 'msg': msg, 'data': data}
        else:
            return {"process": process, 'msg': msg}
    def write(self, **kwargs):
        '''
        向存储文件中写入数据

        Parameters
        ----------
        **kwargs : dict/kwargs
            以字典形式（加 **dict）或关键字参数方式输入数据 .

        Returns
        -------
```

```
        dict
            反馈操作结果 .

        '''
        with shelve.open(filename=self.db_fn, flag=self.flag) as db:
            for k, v in kwargs.items():
                db[k] = v
        return self.feedback('write')
    def read(self, keys_selection=[]):
        '''
        从存储文件中读取数据

        Parameters
        ----------
        keys_selection : list, optional
            待读取存储文件数据的键 . The default is [].

        Returns
        -------
        dict
            反馈操作结果 .

        '''
        with shelve.open(filename=self.db_fn, flag=self.flag) as db:
            keys = list(db.keys())
            data = {}
            if keys_selection:
                for k in keys_selection:
                    data[k] = db[k]
            else:
                for k in keys:
                    data[k] = db[k]
        return self.feedback('read', data=data)
    def trash(self, trash_keys):
        '''
        根据指定的键，删除存储文件对应键值数据

        Parameters
        ----------
        trash_keys : list
            待删除的存储文件键列表 .

        Returns
        -------
        dict
            反馈操作结果 .

        '''
        with shelve.open(filename=self.db_fn, flag=self.flag) as db:
            keys = set(db.keys())
            if set(trash_keys).issubset(keys):
                for t_k in trash_keys:
                    del db[t_k]
                return self.feedback('trash')
            else:
                return self.feedback('trash', 'failed, %s not exist.' % trash_
keys)
    def update(self, clear_all=False, **kwargs):
        '''
        给定键，更新存储文件对应键的值

        Parameters
        ----------
        clear_all : bool, optional
            是否清空存储文件所有数据 . The default is False.
        **kwargs : dict/kwargs
            以字典形式（加 **dict）或关键字参数方式输入更新数据 .

        Returns
        -------
        dict
            反馈操作结果 .

        '''
```

```
        with shelve.open(filename=self.db_fn, flag=self.flag) as db:
            if clear_all:
                keys = list(db.keys())
                for k in keys:
                    del db[k]
            if len(kwargs) > 0:
                for k, v in kwargs.items():
                    db[k] = v
            return self.feedback('update')
db_fn = './database/shelve_db'
db = DB_Shelve(db_fn)
print(db)
print("-" * 50)
fb_w = db.write(**{'commodity_name': 'muskmelon', 'sales': 1100})
print(fb_w)
fb_r_all = db.read()
print(fb_r_all)
fb_r_selection = db.read(['sales'])
print(fb_r_selection)
db.write(**{'exporting_country_name': 'peru'})
print("-" * 50)
print(db.read())
fb_t = db.trash(['sales'])
print(fb_t)
print(db.read())
print("-" * 50)
db.update(sales=1100)
db.update(commodity_code=101, idx=1101)
```

运算结果：

```
[DB_Shelve: db_fn=./database/shelve_db, flag=c]
--------------------------------------------------------------------------------
{'process': 'write', 'msg': 'OK.'}
{'process': 'read', 'msg': 'OK.', 'data': {'commodity_name': 'muskmelon',
'exporting_country_name': 'peru', 'commodity_code': 101, 'idx': 1101, 'commodity_
dic': {'commodity_code': 103, 'commodity_name': 'muskmelon'}, 'sales': 1100}}
{'process': 'read', 'msg': 'OK.', 'data': {'sales': 1100}}
--------------------------------------------------------------------------------
{'process': 'read', 'msg': 'OK.', 'data': {'commodity_name': 'muskmelon',
'exporting_country_name': 'peru', 'commodity_code': 101, 'idx': 1101, 'commodity_
dic': {'commodity_code': 103, 'commodity_name': 'muskmelon'}, 'sales': 1100}}
{'process': 'trash', 'msg': 'OK.'}
{'process': 'read', 'msg': 'OK.', 'data': {'commodity_name': 'muskmelon',
'exporting_country_name': 'peru', 'commodity_code': 101, 'idx': 1101, 'commodity_
dic': {'commodity_code': 103, 'commodity_name': 'muskmelon'}}}
--------------------------------------------------------------------------------
{'process': 'update', 'msg': 'OK.'}
```

11

描述：

DB_Shelve类是对shelve数据存储的基本操作。如果要拓展列表追加或字典更新等值的存储类型，可以以子类的方式增加功能。同时，也可以对原有父类的某些方法重新。这里，仅增加了update_lstExtend()和update_dict两种方法，未重写父类方法。注意，要列表追加或更新字典，需要配置shelve.open(file_name, flag='', writeback=True\False)中的writeback为True。

代码段：

```
class DB_Shelve_Extension(DB_Shelve):
    '''
    拓展 DB_Shelve 操作 shelve 数据存储的方式，包括值类型的，列表形式追加和字典更新
    '''
    def __init__(self, db_fn, flag='c', writeback=True):
        '''
        初始化读写 shelve 数据库（存储文件）的基本参数，较之 DB_Shelve 类，增加了 writeback
参数

        Parameters
        ----------
        db_fn : string
            存储文件路径名 .
        flag : string, optional
```

```
                读写方式。包括 r- 只读；
                w- 可读写；
                n- 每次调用 open() 都重新创建一个空的文件，可读写 .
                The default is 'c'- 如果数据文件不存在，就创建，允许读写 .
            writeback : bool, optional
                当设置为 True 以后，shelve 对象为所有访问过的条目保留缓存并在 close() 或 sync()
时将他们写回到 DB. The default is True.

            Returns
            -------
            None.
            '''
            super(DB_Shelve_Extension, self).__init__(db_fn, flag)
            self.writeback = writeback
        def update_lstExtend(self, **kwargs):
            '''
            对值为列表形式的数据，追加新的数据

            Parameters
            ----------
            **kwargs : dict/kwargs
                 以字典形式（加 **dict）或关键字参数方式输入更新数据，值为列表 .

            Returns
            -------
            dict
                反馈操作结果 .

            '''
            with shelve.open(filename=self.db_fn, flag=self.flag, writeback=self.
writeback) as db:
                if len(kwargs) > 0:
                    for k, v in kwargs.items():
                        # print('-'*30,k,v)
                        # print(db[k])
                        db[k].extend(v)
                        # print(db[k])
            return self.feedback('update_lstExtend')
        def update_dict(self, **kwargs):
            '''
            对值为字典形式的数据，根据新的字典数据更新值

            Parameters
            ----------
            **kwargs : dict/kwargs
                 以字典形式（加 **dict）或关键字参数方式输入更新数据，值为字典 .

            Returns
            -------
            dict
                反馈操作结果 .

            '''
            with shelve.open(filename=self.db_fn, flag=self.flag, writeback=self.
writeback) as db:
                if len(kwargs) > 0:
                    for k, v in kwargs.items():
                        # print(db[k])
                        db[k].update(v)
                        # print(db[k])
            return self.feedback('update_dict')
db_fn = './database/shelve_db'
db_ex = DB_Shelve_Extension(db_fn)
print(db_ex)
print("-" * 50)
print(db_ex.read())
print("-" * 50)
db_ex.update(commodity_name=['muskmelon', 'strawberry', 'apple'])
print(db_ex.read())
db_ex.update_lstExtend(commodity_name=['lemon'])
print(db_ex.read(keys_selection=['commodity_name']))
print("-" * 50)
db_ex.write(commodity_dic={'commodity_code': 101,
            'commodity_name': 'muskmelon'})
```

```
print(db_ex.read())
print("-" * 50)
db_ex.update_dict(commodity_dic={'commodity_code': 103})
print(db_ex.read())
```

运算结果：

```
[DB_Shelve_Extension: db_fn=./database/shelve_db, flag=c, writeback=True]
--------------------------------------------------------------------------
{'process': 'read', 'msg': 'OK.', 'data': {'commodity_name': ['muskmelon',
'strawberry', 'apple', 'lemon'], 'exporting_country_name': 'peru', 'commodity_
code': 101, 'idx': 1101, 'commodity_dic': {'commodity_code': 103, 'commodity_
name': 'muskmelon'}, 'sales': 1100}}
--------------------------------------------------------------------------
{'process': 'read', 'msg': 'OK.', 'data': {'commodity_name': ['muskmelon',
'strawberry', 'apple'], 'exporting_country_name': 'peru', 'commodity_code':
101, 'idx': 1101, 'commodity_dic': {'commodity_code': 103, 'commodity_name':
'muskmelon'}, 'sales': 1100}}
{'process': 'read', 'msg': 'OK.', 'data': {'commodity_name': ['muskmelon',
'strawberry', 'apple', 'lemon']}}
--------------------------------------------------------------------------
{'process': 'read', 'msg': 'OK.', 'data': {'commodity_name': ['muskmelon',
'strawberry', 'apple', 'lemon'], 'exporting_country_name': 'peru', 'commodity_
code': 101, 'idx': 1101, 'commodity_dic': {'commodity_code': 101, 'commodity_
name': 'muskmelon'}, 'sales': 1100}}
--------------------------------------------------------------------------
{'process': 'read', 'msg': 'OK.', 'data': {'commodity_name': ['muskmelon',
'strawberry', 'apple', 'lemon'], 'exporting_country_name': 'peru', 'commodity_
code': 101, 'idx': 1101, 'commodity_dic': {'commodity_code': 103, 'commodity_
name': 'muskmelon'}, 'sales': 1100}}
```

知识点：

8.3 private variables(私有变量/成员)和private methods(私有方法)

12

描述：

python实际上没有类似public、private等关键字修饰的成员变量和方法，通常在变量名或方法名前增加__两个下画线，将公有变量和方法转换为私有变量和方法。实际上，私有变量和方法是可以在外部访问的，只须将例如__data的私有变量，__say()或__subPrintData()的私有方法，替换为_PA__data，_PA__say()和_PB__subPrintData()（类名之前一个下画线，私有变量和方法前两个下画线），因此python中并没有真正的私有变量和方法。私有变量和方法仅在类内访问，其子类和父类无法直接访问。

代码段：

```
class PA:
    def __init__(self):
        self.__data = []  #私有变量，实际被替换为：self._PA__data=[]
        self.__name = 'Guido van Rossum'  #私有变量，实际被替换为：self._PA__name=' Guido
van Rossum'
    def add(self, item):
        self.__data.append(item)  #实际被替换为：self._PA__data.append(item)

    def printData(self):
        print(self.__data)  #实际被替换为：self._PA__data

    def __say(self):  #私有方法，实际被替换为：def _PA__say（self）
        print("%s created python." % self.__name)  #实际被替换为：self._PA__name

class PB(PA):
    def printSuperclassAttri(self):
        print(self._PA__data)  #self.__data 无法访问，属于父类的私有变量，需按替换语句操作

    def __subPrintData(self):
        print('This is a private method of the subclass.')
instance = PB()
for i in ['c', 'c#', 'C++', 'python']:
    instance.add(i)
instance.printData()
print(instance._PA__data)  #访问私有变量，需要按替换语句操作
```

```
print(instance._PA__say())  # 调用私有方法，需要按替换语句操作

print("-" * 50)
instance.printSuperclassAttri()
instance._PB__subPrintData()
```

运算结果：

```
['c', 'c#', 'C++', 'python']
['c', 'c#', 'C++', 'python']
Guido van Rossum created python.
None
--------------------------------------------------------------------------------
['c', 'c#', 'C++', 'python']
This is a private method of the subclass.
```

知识点：

8.4 Operator Overloading（运算符/操作符重载）

13

描述：

Operator Overloading（操作符重载）

操作符重载理解为，如果在类的内部重写了python内置操作（build-in operations），类的实例化时会自动调用重写的方法，由重写的方法计算的结果作为返回值。对于操作符重载：即重写的方法拦截了python正常的内置操作方法；类可以重载所有的python表达式操作；类也可以重载python内置操作，例如打印（printing）、函数调用（function calls）和属性访问（attribute access）等；重载使类的实例表现得更像内置类型；重载是通过在一个类中提供特别命名的方法来实现的。

例如，下述案例在类里重写了python的运算符加号+，即__add__方法。为了区别正常的加法运算，这里在两个数相加后，再加10。下述类重写加法运算符直接调用了重写的方法运算方式，同时调用类进行计算时类似于常规的内置类型运算方式，直接使用+进行计算，而不必按照类调用方法的途径实现。

代码段：

```
class OO_Add:
    def __init__(self, data):
        self.data = data
    def __add__(self, y):
        return OO_Add(self.data + y + 10)
x = OO_Add(3)
y = x + 9
print(y.data)
print("-" * 50)
class OO_Add_:
    def __init__(self, data):
        self.data = data
    def __add__(self, y):
        return self.data + y + 10
print(OO_Add_(3) + 9)
```

运算结果：

```
22
--------------------------------------------------------------------------------
22
```

14

描述：

常见可用于操作符重载方法的内置运算符，如表pcs.8-1所示。

表pcs.8-1 操作符重载方法的内置运算符

Method	Implements	Called for
__init__	Constructor 构造函数	Object creation: X=Class(args)
__del__	Destructor 析构函数	Object deletion of X

续表

Method	Implements	Called for
__add__	Operator + 加法	X+Y, X+=Y if no __iadd__
__or__	Operator |(bitwise OR) 运算符|	X|Y, X|=Y if no __ior__
__repr__, __str__	Printing, conversions 打印 / 转换	print(X), repr(X),str(X)
__call__	Function calls 函数调用	X(*args, **kwargs)
__getattr__	Attribute fetch 属性引用	X.undefined
__setattr__	Attribute assignment 属性赋值	X.attribute=value
__delattr__	Attribute deletion 属性删除	del X.attribute
__getattribute__	Attribute fetch 属性获取	X.any
__getitem__	Indexing, slicing, iteration 索引运算	X[Key], X[i:j], for loops and other iterations if no __iter__
__setitem__	Index and slice assignment 索引赋值	X[Key]=value, X[i:j]=iterable
__delitem__	Index and slice deletion 索引和分片删除	del X[Key],del X[i:j]
__len__	Length 长度	len(X), truth tests if no __bool__
__bool__	Boolean tests 布尔测试	bool(X), truth tests(named __nonzero__ in 2.X)
__lt__, __gt__, __le__, __ge__, __eq__, __ne__	Comparisons 特定的比较	X<Y, X>Y, X<=Y, X>=Y, X==Y, X!=Y
__radd__	Right-side operators 右侧加法	Other+X
__iadd__	In-place augmented operators 实地（增强的）加法	X+=Y (or else __add__)
__iter__, __next__	Iteration contexts 迭代	I=iter(X), next(I); for loops, in if no __contains__, all comprehensions, map(F,X), others(__next__ is named next in 2.X)
__contains__	Membership test 成员关系测试	item in X(any iterable)
__index__	Integer value 整数值	hex(X), bin(X), oct(X), O[X],O[X:]
__enter__, __exit__	Context manager 环境管理器	with obj as var:
__get__, __set__, __delete__	Descriptor attributes 描述符属性	X.attr, X.attr=value, del X.attr
__new__	Creation 创建	Object creation, before __init__

知识点：

8.5 类属性格式"字典"

15

描述：

如果以dict.key获取类属性的形式替代字典dict[key]的形式，在代码书写上相对简单。要达到这样的目的，可以建立一个空的类，通过赋予属性的方式实现，例如下述案例中的class Sales:pass；Sales类方法在赋值时需要逐一赋值，也不能够将已有的字典转换为属性字典的形式，因此定义类class AttrDict_simple(dict)，该种方法继承了父类dict内置字典类，并以父类的方式super(AttrDict_simple,self).__init__(*args,**kwargs)初始化参数值，并通过self.__dict__=self将实例化对象本身赋予__dict__对象（可以获取类的所有属性和方法的键值对），因此可以通过属性方式访问字典成员；AttrDict_simple类方式虽然可以直接将字典转换为属性字典的格式，但是缺少编辑功能，例如增减成员，因此定义class AttriDict(dict)类，重写（overload）了部分父类dict的方法，实现增加键值对（属性值）的方法。

类似 __xx__ 或 __xx__() 前后都有2个下画线的变量或方法，通常是python中内置的特殊变量属性或方法的标识，在书写代码时应尽量避免用该类方式自定义变量或方法。

代码段:

```
help(dict)
```

运算结果:

```
Help on class dict in module builtins:
class dict(object)
 |  dict() -> new empty dictionary
 |  dict(mapping) -> new dictionary initialized from a mapping object's
 |      (key, value) pairs
 |  dict(iterable) -> new dictionary initialized as if via:
 |      d = {}
 |      for k, v in iterable:
 |          d[k] = v
 |  dict(**kwargs) -> new dictionary initialized with the name=value pairs
 |      in the keyword argument list.  For example:  dict(one=1, two=2)
 |
 |  Built-in subclasses:
 |      StgDict
 |
 |  Methods defined here:
 |
 |  __contains__(self, key, /)
 |      True if the dictionary has the specified key, else False.
 |
 |  __delitem__(self, key, /)
 |      Delete self[key].
 |
 |  __eq__(self, value, /)
 |      Return self==value.
 |
 |  __ge__(self, value, /)
 |      Return self>=value.
 |
 |  __getattribute__(self, name, /)
 |      Return getattr(self, name).
 |
 |  __getitem__(...)
 |      x.__getitem__(y) <==> x[y]
 |
 |  __gt__(self, value, /)
 |      Return self>value.
 |
 |  __init__(self, /, *args, **kwargs)
 |      Initialize self.  See help(type(self)) for accurate signature.
 |
 |  __iter__(self, /)
 |      Implement iter(self).
 |
 |  __le__(self, value, /)
 |      Return self<=value.
 |
 |  __len__(self, /)
 |      Return len(self).
 |
 |  __lt__(self, value, /)
 |      Return self<value.
 |
 |  __ne__(self, value, /)
 |      Return self!=value.
 |
 |  __repr__(self, /)
 |      Return repr(self).
 |
 |  __reversed__(self, /)
 |      Return a reverse iterator over the dict keys.
 |
 |  __setitem__(self, key, value, /)
 |      Set self[key] to value.
 |
 |  __sizeof__(...)
 |      D.__sizeof__() -> size of D in memory, in bytes
 |
```

```
 |  clear(...)
 |      D.clear() -> None.  Remove all items from D.
 |
 |  copy(...)
 |      D.copy() -> a shallow copy of D
 |
 |  get(self, key, default=None, /)
 |      Return the value for key if key is in the dictionary, else default.
 |
 |  items(...)
 |      D.items() -> a set-like object providing a view on D's items
 |
 |  keys(...)
 |      D.keys() -> a set-like object providing a view on D's keys
 |
 |  pop(...)
 |      D.pop(k[,d]) -> v, remove specified key and return the corresponding
value.
 |      If key is not found, d is returned if given, otherwise KeyError is
raised
 |
 |  popitem(self, /)
 |      Remove and return a (key, value) pair as a 2-tuple.
 |
 |      Pairs are returned in LIFO (last-in, first-out) order.
 |      Raises KeyError if the dict is empty.
 |
 |  setdefault(self, key, default=None, /)
 |      Insert key with a value of default if key is not in the dictionary.
 |
 |      Return the value for key if key is in the dictionary, else default.
 |
 |  update(...)
 |      D.update([E, ]**F) -> None.  Update D from dict/iterable E and F.
 |      If E is present and has a .keys() method, then does:  for k in E: D[k] =
E[k]
 |      If E is present and lacks a .keys() method, then does:  for k, v in E:
D[k] = v
 |      In either case, this is followed by: for k in F:  D[k] = F[k]
 |
 |  values(...)
 |      D.values() -> an object providing a view on D's values
 |
 |  ----------------------------------------------------------------------
 |  Class methods defined here:
 |
 |  fromkeys(iterable, value=None, /) from builtins.type
 |      Create a new dictionary with keys from iterable and values set to value.
 |
 |  ----------------------------------------------------------------------
 |  Static methods defined here:
 |
 |  __new__(*args, **kwargs) from builtins.type
 |      Create and return a new object.  See help(type) for accurate signature.
 |
 |  ----------------------------------------------------------------------
 |  Data and other attributes defined here:
 |
 |  __hash__ = None
```

16

代码段：

```
from datetime import datetime
date = datetime.strptime('2020-3-30', '%Y-%m-%d').date()
sales_dict = {'idx': 1101, 'date': date, 'exporting_country': 'peru',
              'sales': 2500, 'commodity_name': 'lemon'}
print(sales_dict['commodity_name'])
print("-" * 50)
class Sales:
    pass
sales_rec = Sales()
sales_rec.idx = 1101
sales_rec.date = date
```

```
sales_rec.exporting_country = 'peru'
sales_rec.sales = 2500
sales_rec.commodity_name = 'lemon'
print(sales_rec.commodity_name)
print("-" * 50)
class AttrDict_simple(dict):
    def __init__(self, *args, **kwargs):
        super(AttrDict_simple, self).__init__(*args, **kwargs)
        self.__dict__ = self
sales_attrDict_simple = AttrDict_simple(sales_dict)
print(sales_attrDict_simple.commodity_name)
```

运算结果：

```
lemon
--------------------------------------------------
lemon
--------------------------------------------------
lemon
```

17

代码段：

```
class AttriDict(dict):
    def __init__(self, *args, **kwargs):
        super(AttriDict, self).__init__(*args, **kwargs)
        for arg in args:
            if isinstance(arg, dict):
                for k, v in arg.items():
                    self[k] = v
        if kwargs:
            for k, v in kwargs.items():
                self[k] = v
    def __getattr__(self, attr):
        return self.get(attr)
    def __setattr__(self, key, value):
        self.__setitem__(key, value)
    def __setitem__(self, key, value):
        super(AttriDict, self).__setitem__(key, value)
        self.__dict__.update({key: value})
    def __delattr__(self, item):
        self.__delitem__(item)
    def __delitem__(self, key):
        super(AttriDict, self).__delitem__(key)
        del self.__dict__[key]
sales_attrDict = AttriDict(sales_dict)
print(sales_attrDict)
print(sales_attrDict.commodity_name)
sales_attrDict.commodity_code = 102
print(sales_attrDict)
sales_attrDict['exporting_country_ID'] = 25
print(sales_attrDict)
del sales_attrDict.idx  # 或用：del sales_attrDict['idx']
print(sales_attrDict)
```

运算结果：

```
{'idx': 1101, 'date': datetime.date(2020, 3, 30), 'exporting_country': 'peru',
'sales': 2500, 'commodity_name': 'lemon'}
lemon
{'idx': 1101, 'date': datetime.date(2020, 3, 30), 'exporting_country': 'peru',
'sales': 2500, 'commodity_name': 'lemon', 'commodity_code': 102}
{'idx': 1101, 'date': datetime.date(2020, 3, 30), 'exporting_country': 'peru',
'sales': 2500, 'commodity_name': 'lemon', 'commodity_code': 102, 'exporting_
country_ID': 25}
{'date': datetime.date(2020, 3, 30), 'exporting_country': 'peru', 'sales': 2500,
'commodity_name': 'lemon', 'commodity_code': 102, 'exporting_country_ID': 25}
```

PCS-9.（OOP）_Classes_Decorators（装饰器）_Slots

PCS-9（　）：9.1 装饰器-函数；9.2 装饰器-类；9.3 __slots__

知识点：

9.1 装饰器-函数

1

描述：

1．函数调用另一个函数（函数作为参数）

这里定义了3个函数，say_hello(name)和be_awesome(name)，传入的为常规参数（不是以函数作为参数），并进行了不同方式字符串格式化。而对于greet_bob(greeter_func)，从 greeter_func("Bob")语句可以判断出函数参数greeter_func为一个函数。将say_hello(name)和be_awesome(name)函数作为参数传入到函数greet_bob(greeter_func)，可以对应将参数替换为参数函数思考代码运行机制，比较方便理解。

代码段：

```
def say_hello(name):
    # f-string 字符串格式化方法，Literal String Interpolation（文字字符串插值）
    return f"Hello {name}"
def be_awesome(name):
    return f"Yo {name}, together we are the awesomest!"
def greet_bob(greeter_func):
    return greeter_func("Bob")
print(greet_bob(say_hello))
print(greet_bob(be_awesome))
```

运算结果：

```
Hello Bob
Yo Bob, together we are the awesomest!
```

2

描述：

2．内置函数（inner functions）

如果函数内部存在多个内置函数，函数定义的前后位置并不重要，主要由执行语句的顺序确定。同时，内部函数在调用父函数之前不会被定义，属于父函数parent()的局部作用域，仅在parent()内部作为局部变量存在。

代码段：

```
def parent():
    print("Printing from the parent() function")
    def first_child():
        print("Printing from the first_child() function")
    def second_child():
        print("Printing from the second_child() function")
    second_child()
    first_child()
parent()
```

运算结果：

```
Printing from the parent() function
Printing from the second_child() function
Printing from the first_child() function
```

3

描述：

3．函数返回值为一个函数

python允许使用函数作为返回值，下述parent()函数返回了一个内置函数。需要注意，返回函数时为return first_child，是一个没有给()的函数名，意味返回first_child函数的引用。如果给了()，则是返回first_child函数的一个结果。当函数返回值为函数，则返回值（通常赋予于新的变量名）可以像普通函数一样调用（使用）。

代码段:

```
def parent(num):
    def first_child():
        return "Hi, I am Emma"
    def second_child():
        return "Call me Liam"
    if num == 1:
        return first_child
    else:
        return second_child
first = parent(1)
second = parent(2)
print(first)
print(first())
print(second())
```

运算结果:

```
<function parent.<locals>.first_child at 0x000001B52F2933A0>
Hi, I am Emma
Call me Liam
```

4

描述:

4. 简单的装饰器

say_whee = my_decorator(say_whee)返回my_decorator(func)父函数内置函数wrapper()的引用，为return wrapper。该内置函数包含父类传入的一个函数参数func，并执行func()，即执行参数函数的计算结果。执行say_whee()时，即执行父类my_decorator(func)内的wrapper()内置函数，只是此时，该函数已经独立于父函数my_decorator(func)，并包含有执行wrapper()函数所需的所有参数，这里为参数函数func。因此，say_whee = my_decorator(say_whee)中的say_whee为一个闭包（Closure或Lexical Closure），为一个结构体，存储了一个函数和与其关联的环境参数。

因为内置函数wrapper()，实际上对传入的参数函数say_whee()的功能进行了增加，即"装饰"。所以，可以简单地说，装饰器就是对一个函数进行包装，修改已有的功能。

代码段:

```
def my_decorator(func):
    def wrapper():
        print("Something is happening before the function is called.")
        func()
        print("Something is happening after the function is called.")
    return wrapper
def say_whee():
    print("Whee!")
say_whee = my_decorator(say_whee)
say_whee()
```

运算结果:

```
Something is happening before the function is called.
Whee!
Something is happening after the function is called.
```

5

描述:

say_whee = not_during_the_night(say_whee)装饰，则根据条件判断执行不同的操作。如果满足7 <= datetime.now().hour < 22，则执行外部函数say_whee()；否则，什么都不发生。

代码段:

```
from datetime import datetime
def not_during_the_night(func):
    def wrapper():
```

```
        if 7 <= datetime.now().hour < 22:
            func()
        else:
            pass
    return wrapper
def say_whee():
    print("Whee!")
say_whee = not_during_the_night(say_whee)
say_whee()
```

运算结果：

```
Whee!
```

6

描述：

5. 语法糖（Syntactic Sugar）

上面的装饰器方法笨拙，为了简化代码过程，python允许用@symbol方式使用装饰器，有时称为pie语法。下述案例与上述结果一致，但是通过@my_decorator pie方法，替代了say_whee = not_during_the_night(say_whee)代码，简化操作。

代码段：

```
def my_decorator(func):
    def wrapper():
        print("Something is happening before the function is called.")
        func()
        print("Something is happening after the function is called.")
    return wrapper
@my_decorator
def say_whee():
    print("Whee!")
say_whee()
```

运算结果：

```
Something is happening before the function is called.
Whee!
Something is happening after the function is called.
```

7

描述：

6. 带参数的装饰器

在wrapper_do_twice(*args, **kwargs)内置函数传入参数为*args、**kwargs，接受任意数量的位置参数和关键字参数；并将其传入参数函数。

代码段：

```
def do_twice(func):
    def wrapper_do_twice(*args, **kwargs):
        func(*args, **kwargs)
        func(*args, **kwargs)
    return wrapper_do_twice
@do_twice
def greet(name):
    print(f"Hello {name}")
greet("World")
```

运算结果：

```
Hello World
Hello World
```

8

描述：

7. 装饰器的返回值

如果装饰器要返回值，do_twice(func)内置函数 wrapper_do_twice(*args, **kwargs)在调用参数函数func时，需要执行return func(*args, **kwargs) 返回参数函数的返回值。

代码段：

```
def do_twice(func):
    def wrapper_do_twice(*args, **kwargs):
        func(*args, **kwargs)
        return func(*args, **kwargs)
    return wrapper_do_twice
@do_twice
def return_greeting(name):
    print("Creating greeting")
    return f"Hi {name}"
hi_adam = return_greeting("Adam")
print("-" * 50)
print(hi_adam)
print("-" * 50)
print(return_greeting)
print(return_greeting.__name__)
print("-" * 50)
print(help(return_greeting))
```

运算结果：

```
Creating greeting
Creating greeting
--------------------------------------------------
Hi Adam
--------------------------------------------------
<function do_twice.<locals>.wrapper_do_twice at 0x000001B52F293EE0>
wrapper_do_twice
--------------------------------------------------
Help on function wrapper_do_twice in module __main__:
wrapper_do_twice(*args, **kwargs)
None
```

9

描述：

8. 保留原始函数的信息-自省（introspection）调整

自省是指一个对象在运行时了解自己的属性的能力。例如，一个函数知道它自己的名字和文档。在上述示例中，通过return_greeting.__name__，help(return_greeting)等方式可以查看函数对象的相关属性；但是，发现给出的是wrapper_do_twice的内置函数，而不是return_greeting函数。因此，可以通过functools[①] 的 @functools.wraps(func)方法解决这个问题，保留原始函数的信息。

① functools，用于高阶函数：作用于或返回其他函数的函数，通常，任何可调用对象都可以视为函数，（https://docs.python.org/3/library/functools.html）。

代码段：

```
import functools
def do_twice(func):
    @functools.wraps(func)
    def wrapper_do_twice(*args, **kwargs):
        func(*args, **kwargs)
        return func(*args, **kwargs)
    return wrapper_do_twice
@do_twice
def return_greeting(name):
    print("Creating greeting")
    return f"Hi {name}"
hi_adam = return_greeting("Adam")
print("-" * 50)
print(hi_adam)
print("-" * 50)
print(return_greeting)
print(return_greeting.__name__)
print("-" * 50)
print(help(return_greeting))
```

运算结果：

```
Creating greeting
Creating greeting
--------------------------------------------------
Hi Adam
--------------------------------------------------
```

```
<function return_greeting at 0x000001B530D988B0>
return_greeting
------------------------------------------------------------------------------
Help on function return_greeting in module __main__:
return_greeting(name)
None
```

10

描述：

9．带参数的装饰器

装饰器中可以带参数，例如@repeat(num_times=3)中num_times=3。此时，对装饰器函数做了调整，增加了一层嵌套内置函数，传递装饰器参数。

代码段：

```
def repeat(num_times):
    def decorator_repeat(func):
        @functools.wraps(func)
        def wrapper_repeat(*args, **kwargs):
            for _ in range(num_times):
                value = func(*args, **kwargs)
            return value
        return wrapper_repeat
    return decorator_repeat
@repeat(num_times=3)
def greet(name):
    print(f"Hello {name}")
greet("Galaxy")
```

运算结果：

```
Hello Galaxy
Hello Galaxy
Hello Galaxy
```

11

描述：

10．多个装饰器装饰一个函数

可以将多个装饰器堆叠在一起，应用在一个函数上。此时，执行的装饰器执行的顺序是从内到外，例如示例先执行@decor，返回值为20；而后，再执行@decor1，返回值为400。

代码段：

```
#用于测试装饰器链的代码
def decor1(func):
    def inner():
        x = func()
        return x * x
    return inner
def decor(func):
    def inner():
        x = func()
        return 2 * x
    return inner
@decor1
@decor
def num():
    return 10
print(num())
```

运算结果：

```
400
```

12

描述：

11．decorator模块简化装饰器

使用decorator模块[①]库的@decorator装饰器装饰‘装饰函数’，可以简化装饰器定义。例如，下述代码取消了内置函数，将原始函数和输入参数都在do_print(func,*args, **kwargs)，装饰函数中一起输入。

① decorator，模块的目标是使定义保留签名（signature-preserving）的函数装饰器和装饰器工厂（factories）变得容易（https://pypi.org/project/decorator/）。

代码段：

```
from decorator import decorator
@decorator
def do_print(func, *args, **kwargs):
    print('Hi {}!'.format(*args, **kwargs))
    return func(*args, **kwargs)
@do_print
def greet(name):
    print(f"Hello {name}!")
greet("World")
```

运算结果：

```
Hi World!
Hello World!
```

13

描述：

12. 示例

● 执行时间长度

这个装饰器存储函数开始运行前的时间start_time = time.perf_counter()和函数结束后的时间end_time = time.perf_counter()；然后，计算运行函数的时间，run_time = end_time - start_time并打印。

代码段：

```
import time
def timer(func):
    """ 打印被装饰函数的运行时间 """
    @functools.wraps(func)
    def wrapper_timer(*args, **kwargs):
        start_time = time.perf_counter()    # 1
        value = func(*args, **kwargs)
        end_time = time.perf_counter()      # 2
        run_time = end_time - start_time    # 3
        print(f"Finished {func.__name__!r} in {run_time:.4f} secs")
        return value
    return wrapper_timer
@timer
def waste_some_time(num_times):
    for _ in range(num_times):
        sum([i**2 for i in range(10000)])
waste_some_time(999)
```

运算结果：

```
Finished 'waste_some_time' in 4.1239 secs
```

14

描述：

● 减缓运行

对执行的函数进行运行速度的限制。

代码段：

```
def slow_down(func):
    """ 在调用函数之前休眠 1 秒 """
    @functools.wraps(func)
    def wrapper_slow_down(*args, **kwargs):
        time.sleep(1)
        return func(*args, **kwargs)
    return wrapper_slow_down
@slow_down
def countdown(from_number):
    if from_number < 1:
        print("Liftoff!")
    else:
        print(from_number)
        countdown(from_number - 1)
countdown(3)
```

运算结果：

```
3
2
1
Liftoff!
```

知识点：

9.2 装饰器-类

15

描述：

1. @property

@property 内置装饰器可以将类的方法，转换为只能读取的属性。例如，使用andy.password类属性操作模式，而不是andy.password()类方法操作模式。如果要修改或者删除属性，则需要重新实现属性的setter、getter和deleter方法，例如@password.setter和@password.deleter装饰器。

代码段：

```
class Bank_acount:
    def __init__(self):
        self._password = ‘preset password: 0000’
    @property
    def password(self):
        return self._password
    @password.setter
    def password(self, value):
        self._password = value
    @password.deleter
    def password(self):
        del self._password
        print(‘del complete’)
andy = Bank_acount()
print(andy.password)  # getter
andy.password = ‘1q2w3e’  # setter
print(andy.password)
del andy.password  # deleter
```

运算结果：

```
preset password: 0000
1q2w3e
del complete
```

16

描述：

2. @classmethod和@staticmethod

类方法@classmethod和静态方法@staticmethod，都可以直接通过Class/Instance.method()调用，可以不用实例化对象，直接由类直接调用，例如类方法的Person.fromBirthYear('mayank', 1996)和静态方法的Person.isAdult(22)。对于类方法，需要将self参数转换为cls；对于静态方法，则不需要self等任何参数。

代码段：

```
from datetime import date
# 演示 Python 程序
# 使用类方法和静态方法 .
class Person:
    def __init__(self, name, age):
        self.name = name
        self.age = age

    # 一个按出生年份创建 Person 对象的类方法 .
    @classmethod
    def fromBirthYear(cls, name, year):
        return cls(name, date.today().year - year)
```

```
    # 静态方法来检查一个人是否成人.
    @staticmethod
    def isAdult(age):
        return age > 18
person1 = Person('mayank', 21)
person2 = Person.fromBirthYear('mayank', 1996)
print(person1.age)
print(person2.age)

# 打印结果
print(Person.isAdult(22))
print(person1.isAdult(18))
```

运算结果：

```
21
26
True
False
```

17

描述：

3. @abstractmethod

标准库abc[①]提供有@abstractmethod抽象方法。当所在的类继承了abc.ABC，并给需要抽象的实例方法添加装饰器@abstractmethod后，这个类就成为了抽象类，不能够被直接实例化，例如示例的Animal类，抽象方法为info()。如果要使用抽象类，必须继承该类并实现该类的所有抽象方法，例如Bird子类继承了抽象类Animal，并在子类info()中实现父类抽象类的info()方法。

代码段：

```
from abc import ABC, abstractmethod
class Animal(ABC):
    @abstractmethod
    def info(self):
        print("Animal")
class Bird(Animal):
    # 实现抽象方法
    def info(self):
        # 调用基类方法（即抽象方法）
        super().info()
        print("Bird")
animal = Animal()
```

运算结果：

```
---------------------------------------------------------------------------------
TypeError                                   Traceback (most recent call last)
Input In [65], in <cell line: 1>()
----> 1 animal = Animal()
TypeError: Can't instantiate abstract class Animal with abstract methods info
```

18

代码段：

```
bird = Bird()
bird.info()
```

运算结果：

```
Animal
Bird
```

19

描述：

4. 装饰整个类

装饰器接收的是一个类，而不是一个函数。

代码段：

```
def timer(func):
```

① abc，提供了在Python中定义抽象基类(ABC)的基础结构（https://docs.python.org/3/library/abc.html）。

```
    """ 打印被装饰函数的运行时间 """
    @functools.wraps(func)
    def wrapper_timer(*args, **kwargs):
        start_time = time.perf_counter()    # 1
        value = func(*args, **kwargs)
        end_time = time.perf_counter()      # 2
        run_time = end_time - start_time    # 3
        print(f"Finished {func.__name__!r} in {run_time:.4f} secs")
        return value
    return wrapper_timer
@timer
class TimeWaster:
    def __init__(self, max_num):
        self.max_num = max_num
    def waste_time(self, num_times):
        for _ in range(num_times):
            sum([i**2 for i in range(self.max_num)])
tw = TimeWaster(1000)
tw.waste_time(999)
```

运算结果：

```
Finished 'TimeWaster' in 0.0000 secs
```

---20

描述：

5. 示例

记录状态的装饰器

使用类作为装饰器，实现__init__()和__call__方法，完成函数运行状态的记录。

代码段：

```
class CountCalls:
    def __init__(self, func):
        functools.update_wrapper(self, func)
        self.func = func
        self.num_calls = 0
    def __call__(self, *args, **kwargs):
        self.num_calls += 1
        print(f"Call {self.num_calls} of {self.func.__name__!r}")
        return self.func(*args, **kwargs)
@CountCalls
def say_whee():
    print("Whee!")
say_whee()
say_whee()
say_whee()
```

运算结果：

```
Call 1 of 'say_whee'
Whee!
Call 2 of 'say_whee'
Whee!
Call 3 of 'say_whee'
Whee!
```

知识点：

9.3 __slots__

••••••21

描述：

通过__slots__类属性分配一连串的字符串属性名称进行属性声明，从而限制类实例对象将拥有的合法属性集，达到优化内存、提高程序运行速度的作用。当为__slots__ 分配一串字符串名称，则只有__slots__ 列表中的那些名称可以被分配为实例属性。并且在实例化时，阻止了为实例分配__dict__对象，除非在__slots__ 中包含该对象。

下述案例类IceTeaSales中配置 __slots__对象的属性名称包括['temperature', 'iceTeaSales']，因此当配置非该列表中所列的属性名，例如iceTea.price时，就会引发异常。

代码段：

```
class IceTeaSales:
    __slots__ = [‘temperature’, ‘iceTeaSales’]
    def __init__(self):
        self.temperature = 0
        self.iceTeaSales = 0
iceTea = IceTeaSales()
print(iceTea.temperature)
iceTea.temperature = 29
setattr(iceTea, ‘iceTeaSales’, 77)
print(iceTea.iceTeaSales, iceTea.temperature)
print(getattr(iceTea, ‘temperature’))
iceTea.price
```

运算结果：

```
0
77 29
29
---------------------------------------------------------------------------
AttributeError                            Traceback (most recent call last)
Input In [96], in <cell line: 13>()
     11 print(iceTea.iceTeaSales,iceTea.temperature)
     12 print(getattr(iceTea,'temperature’))
---> 13 iceTea.price
AttributeError: ‘IceTeaSales’ object has no attribute ‘price’
```

22

描述：

__slots__阻止了__dict__对象分配给实例，因此iceTea.__dict__会引发异常，提示实例化对象没有属性__dict_。

代码段：

```
iceTea.__dict__
```

运算结果：

```
---------------------------------------------------------------------------
AttributeError                            Traceback (most recent call last)
Input In [90], in <cell line: 1>()
----> 1 iceTea.__dict__
AttributeError: ‘IceTeaSales’ object has no attribute ‘__dict__’
```

23

描述：

dir()收集整个类树中所有继承的名称。

代码段：

```
print(dir(iceTea))
print(‘temperature’ in dir(iceTea))
```

运算结果：

```
[‘__class__’, ‘__delattr__’, ‘__dir__’, ‘__doc__’, ‘__eq__’, ‘__format__’, ‘__
ge__’, ‘__getattribute__’, ‘__gt__’, ‘__hash__’, ‘__init__’, ‘__init_subclass__’,
‘__le__’, ‘__lt__’, ‘__module__’, ‘__ne__’, ‘__new__’, ‘__reduce__’, ‘__reduce_
ex__’, ‘__repr__’, ‘__setattr__’, ‘__sizeof__’, ‘__slots__’, ‘__str__’, ‘__
subclasshook__’, ‘iceTeaSales’, ‘temperature’]
True
```

24

描述：

__init__构造方法初始化参数，如果参数名不在__slots__列表中，也会引发异常。

代码段：

```
class IceTeaSales:
    __slots__ = [‘temperature’, ‘iceTeaSales’]
    def __init__(self):
        self.temperature = 0
```

```
        self.iceTeaSales = 0
        self.price = 0
iceTea = IceTeaSales()
```

运算结果：

```
---------------------------------------------------------------------------

AttributeError                            Traceback (most recent call last)
Input In [97], in <cell line: 7>()
      5         self.iceTeaSales=0
      6         self.price=0
----> 7 iceTea=IceTeaSales()
Input In [97], in IceTeaSales.__init__(self)
      4 self.temperature=0
      5 self.iceTeaSales=0
----> 6 self.price=0
AttributeError: 'IceTeaSales' object has no attribute 'price'
```

25

描述：

如果在__slots__列表中包含__dict__，则可以增加新的属性名，__dict__则会包含非__slots__列表中新增加的属性名键值对。

代码段：

```
class IceTeaSales:
    __slots__ = ['temperature', 'iceTeaSales', '__dict__']
    def __init__(self):
        self.temperature = 0
        self.iceTeaSales = 0
        self.price = 0
iceTea = IceTeaSales()
print(iceTea.price)
iceTea.name = 'flower tea'
print(iceTea.name)
print(iceTea.__slots__)
print(iceTea.__dict__)
```

运算结果：

```
0
flower tea
['temperature', 'iceTeaSales', '__dict__']
{'price': 0, 'name': 'flower tea'}
```

26

描述：

- Slot应用规则：

如果存在子类，在用__slots__时则需要注意：

（1）子类中有__slots__，但父类中未配置__slots__，则实例对象总可以访问__dict__属性，因此没有意义。父类中有__slots__，而子类没有，同上，也没有意义。

（2）子类定义了与父类相同的__slots__，只能从父类中的__slots__获取定义的属性名。

代码段：

```
class C:
    pass
class D(C):
    __slots__ = ['a']
X = D()
X.a = 1
X.b = 2
print(X.__dict__)
print(D.__dict__.keys())
```

运算结果：

```
{'b': 2}
dict_keys(['__module__', '__slots__', 'a', '__doc__'])
```

27

描述：

- 内存使用量测试

使用memory-profiler[①]库，测量代码内存的使用率。该模块对python程序的内存消耗进行逐行分析，从而监控一个进程的内存消耗，该模块依赖psutil[②]库。

从计算结果来看，未使用__slots__，内存变化为16.7MB；使用__slots__，内存变化为5.8MB，因此使用__slots__可以有效节约内存空间。

JupyterLab中无法执行，需要在Spyder中运行（保存为模块）

未使用__slots__:

代码段：

```
from memory_profiler import profile
class A(object):
    def __init__(self, x):
        self.x = x
@profile
def main():
    f = [A(523825) for i in range(100000)]
if __name__ == '__main__':
    main()
```

运算结果：

```
7    142.2 MiB    142.2 MiB           1   @profile
8                                         def main():
9    158.9 MiB     16.7 MiB      100003       f=[A(523825) for i in
range(100000)]
```

28

描述：

使用__slots__:

代码段：

```
class A(object):
    __slots__ = ('x')
    def __init__(self, x):
        self.x = x
@profile
def main():
    f = [A(523825) for i in range(100000)]
if __name__ == '__main__':
    main()
```

运算结果：

```
12    142.1 MiB    142.1 MiB           1   @profile
13                                         def main():
14    147.9 MiB      5.8 MiB      100003
f=[A(523825) for i in range(100000)]
```

① memory-profiler，用于监视进程的内存消耗及对Python程序内存消耗进行逐行分析。注意，该库已不再维护（https://pypi.org/project/memory-profiler/）。

② psutil，用于在Python中检索有关正在运行的进程和系统利用率（CPU、内存、磁盘、网络、传感器）的信息。它主要用于系统监视、分析和限制进程资源及运行进程的管理（https://pypi.org/project/psutil/）。

参考文献（References）：

[1] Primer on Python Decorators, https://realpython.com/primer-on-python-decorators/.
[2] classmethod() in Python, https://www.geeksforgeeks.org/classmethod-in-python/.

PCS-10.异常-Errors and Exceptions

PCS-10（　　）：10.1 内置异常（Built-in Exceptions）；10.2 处理异常的方式

知识点：

10.1 内置异常（Built-in Exceptions）

●●1

描述：

for i in range(10) print(i)代码缺少了:，为语法/句法错误（syntax error），会引发内置异常SyntaxError错误，并通常会给出错误的详细原因，例如invalid syntax等。反馈的异常信息中，通常会标识行号，并用^等符号标示错误位置，方便快速定位修改。

代码段：

```
for i in range(10) print(i)
```

运算结果：

```
Input In [14]
    for i in range(10) print(i)
                       ^
SyntaxError: invalid syntax
```

———2

描述：

又例如，索引值5超出了lst列表索引数，引发IndexError异常。

代码段：

```
lst = [1,2,3,4,5]
element = lst[5]
```

运算结果：

```
---------------------------------------------------------------------------
IndexError                                Traceback (most recent call last)
Input In [16], in <cell line: 2>()
      1 lst=[1,2,3,4,5]
----> 2 element=lst[5]
IndexError: list index out of range
```

———3

描述：

内置异常的层次结构如下：

```
BaseException 所有内置异常的基类
 +-- SystemExit 此异常由 sys.exit() 函数引发
 +-- KeyboardInterrupt 当用户按下中断键 (通常为 Ctrl-C 或 Delete 键) 时将被引发
 +-- GeneratorExit 当一个 generator 或 coroutine 被关闭时将被引发
 +-- Exception 所有内置的非系统退出类异常都派生自此类
      +-- StopIteration 由内置函数 next() 和 iterator 的 __next__() 方法所引发，用来表
示该迭代器不能产生下一项
      +-- StopAsyncIteration 必须由一个 asynchronous iterator 对象的 __anext__() 方
法来引发以停止迭代操作
      +-- ArithmeticError 此基类用于派生针对各种算术类错误而引发的内置异常
      |    +-- FloatingPointError 目前未被使用
      |    +-- OverflowError 当算术运算的结果大到无法表示时将被引发
      |    +-- ZeroDivisionError 当除法或取余运算的第二个参数为零时将被引发
      +-- AssertionError 当 assert 语句失败时将被引发
      +-- AttributeError 当属性引用或赋值失败时将被引发
      +-- BufferError 当与 缓冲区 (buffer) 相关的操作无法执行时将被引发
      +-- EOFError 当 input() 函数未读取任何数据即达到文件结束条件 (EOF) 时将被引发
      +-- ImportError 当 import 语句尝试加载模块遇到麻烦时将被引发
      |    +-- ModuleNotFoundError ImportError 的子类，当一个模块无法被定位时将由
import 引发
      +-- LookupError 此基类用于派生当映射或序列所使用的键或索引无效时引发的异常
```

```
 |      +-- IndexError 当序列抽取超出范围时将被引发
 |      +-- KeyError 当在现有键集合中找不到指定的映射（字典）键时将被引发
 +-- MemoryError 当一个操作耗尽内存但情况仍可（通过删除一些对象）进行挽救时将被引发
 +-- NameError 当某个局部或全局名称未找到时将被引发
 |      +-- UnboundLocalError 当在函数或方法中对某个局部变量进行引用，但该变量并未绑定
任何值时将被引发
 +-- OSError 此异常在一个系统函数返回系统相关的错误时将被引发，此类错误包括 I/O 操作失
败例如 " 文件未找到 " 或 " 磁盘已满 " 等（不包括非法参数类型或其他偶然性错误）
 |      +-- BlockingIOError 当一个操作将会在设置为非阻塞操作的对象（例如套接字）上发生
阻塞时将被引发
 |      +-- ChildProcessError 当一个子进程上的操作失败时将被引发
 |      +-- ConnectionError 与连接相关问题的基类
 |      |      +-- BrokenPipeError ConnectionError 的子类，当试图写入一个管道而该管道
的另一端已关闭，或者试图写入一个套接字而该套接字已关闭写入时将被引发
 |      |      +-- ConnectionAbortedError ConnectionError 的子类，当一个连接尝试被对
端中止时将被引发
 |      |      +-- ConnectionRefusedError ConnectionError 的子类，当一个连接尝试被对
端拒绝时将被引发
 |      |      +-- ConnectionResetError ConnectionError 的子类，当一个连接尝试被对端
重置时将被引发
 |      +-- FileExistsError 当试图创建一个已存在的文件或目录时将被引发
 |      +-- FileNotFoundError 将所请求的文件或目录不存在时将被引发
 |      +-- InterruptedError 当系统调用被输入信号中断时将被引发
 |      +-- IsADirectoryError 当请求对一个目录执行文件操作 （例如 os.remove()） 时将
被引发
 |      +-- NotADirectoryError 当请求对一个非目录执行目录操作 （例如 os.listdir()）
时将被引发
 |      +-- PermissionError 当在没有足够访问权限的情况下试图执行某个操作时将被引发 --
例如文件系统权限
 |      +-- ProcessLookupError 当给定的进程不存在时将被引发
 |      +-- TimeoutError 当一个系统函数在系统层级发生超时的情况下将被引发
 +-- ReferenceError 此异常将在使用 weakref.proxy() 函数所创建的弱引用来访问该引用的
某个已被作为垃圾回收的属性时被引发
 +-- RuntimeError 当检测到一个不归属于任何其他类别的错误时将被引发
 |      +-- NotImplementedError 此异常派生自 RuntimeError。在用户自定义的基类中，抽
象方法应当在其要求所派生类重载该方法，或是在其要求所开发的类提示具体实现尚待添加时引发此异常
 |      +-- RecursionError 此异常派生自 RuntimeError。 它会在解释器检测发现超过最大递
归深度时被引发
 +-- SyntaxError 当解析器遇到语法错误时引发
 |      +-- IndentationError 与不正确的缩进相关的语法错误的基类
 |             +-- TabError 当缩进包含对制表符和空格符不一致地使用时将被引发
 +-- SystemError 当解释器发现内部错误，但情况看起来尚未严重到要放弃所有希望时将被引发
 +-- TypeError 当一个操作或函数被应用于类型不适当的对象时将被引发
 +-- ValueError 当操作或函数接收到具有正确类型但值不适合的参数，并且情况不能用更精确的
异常例如 IndexError 来描述时将被引发
 |      +-- UnicodeError 当发生与 Unicode 相关的编码或解码错误时将被引
 |             +-- UnicodeDecodeError 当在解码过程中发生与 Unicode 相关的错误时将被引发
 |             +-- UnicodeEncodeError 当在编码过程中发生与 Unicode 相关的错误时将被引发
 |             +-- UnicodeTranslateError 在转写过程中发生与 Unicode 相关的错误时将被
引发
 +-- Warning 警告类别的基类
        +-- DeprecationWarning 如果所发出的警告是针对其他 Python 开发者的，则以此作为
与已弃用特性相关警告的基类
        +-- PendingDeprecationWarning 对于已过时并预计在未来弃用，但目前尚未弃用的特
性相关警告的基类
        +-- RuntimeWarning 与模糊的运行时行为相关的警告的基类
        +-- SyntaxWarning 与模糊的语法相关的警告的基类
        +-- UserWarning 用户代码所产生警告的基类
        +-- FutureWarning 如果所发出的警告是针对以 Python 所编写应用的最终用户的，则以
此作为与已弃用特性相关警告的基类
        +-- ImportWarning 与在模块导入中可能的错误相关的警告的基类
        +-- UnicodeWarning 与 Unicode 相关的警告的基类
        +-- BytesWarning 与 bytes 和 bytearray 相关的警告的基类
        +-- EncodingWarning 与编码格式相关的警告的基类
        +-- ResourceWarning 资源使用相关警告的基类
```

知识点：

10.2 处理异常的方式

4

描述：

1.

```
try:
    statements
except [built-in exception/(exceptions)]:
    statements
```

首先，执行try代码块；如果没有触发异常，则跳过except代码块，执行完try代码块。如果在执行try代码块时发生了异常，则跳过代码块中剩余的部分。如果触发的异常与except关键字后指定的异常相匹配，则会执行except代码块，然后跳到try/except代码块之后执行；如果触发的异常与except语句中指定的异常不匹配，则它会被传递到外部的try语句中。如果没有找到处理程序，则是一个未处理异常且终止程序。

代码段：

```
while True:
    try:
        x = int(input('Please enter a number:'))
        break
    except ValueError:
        print('Oops! That was no valid number. Try again...')
```

运算结果：

```
Please enter a number: d
Oops! That was no valid number. Try again...
Please enter a number: 3
```

5

代码段：

```
while True:
    try:
        x = 9/int(input('Please enter a number:'))
        break
    except ValueError:
        print('Oops! That was no valid number. Try again...')
```

运算结果：

```
Please enter a number: 0
---------------------------------------------------------------------------
ZeroDivisionError                         Traceback (most recent call last)
Input In [18], in <cell line: 2>()
      1 while True:
      2     try:
----> 3         x=9/int(input('Please enter a number:'))
      4         break
      5     except ValueError:
ZeroDivisionError: division by zero
```

6

描述：

except后可以用()追加多个异常，只要满足其中一个，就执行except代码块。

代码段：

```
while True:
    try:
        x = 9/int(input('Please enter a number:'))
        break
    except (ValueError, ZeroDivisionError): #或者使用 ArithmeticError
        print('Oops! That was no valid number or 0. Try again...')
```

运算结果：

```
Please enter a number: 0
Oops! That was no valid number or 0. Try again...
Please enter a number: 7
```

7

描述：

如果except后不指定异常，则触发任何存在的异常。

代码段：

```
while True:
    try:
        x = 9/int(input('Please enter a number:'))
        break
    except:
        print('Oops! That was no valid number or 0. Try again...')
```

运算结果：

```
Please enter a number: d
Oops! That was no valid number or 0. Try again...
Please enter a number: 0
Oops! That was no valid number or 0. Try again...
Please enter a number: 3
```

8

描述：

2.

```
try:
    statements
except exception as alias:
    statements
except exception(s):
    statements
except:
    statements
...
```

可以有多个except，根据异常不同执行不同的代码块。可以用except exception as alias:方式为异常定义别名（变量）。该变量绑定到一个异常实例，并将参数存储在instance.args中。为了能够直接调入存储的参数而不必引用.args，该异常实例定义了__str__()，从而可以直接用定义的变量读取参数值。

代码段：

```
import sys
try:
    f = open('myfile.txt')
    s = f.readline()
    i = int(s.strip())
except OSError as err:
    print("OS error: {0}".format(err))
except ValueError:
    print("Could not convert data to an integer.")
except BaseException as err:
    print(f"Unexpected {err=}, {type(err)=}")
    raise
```

运算结果：

```
1.5
3
```

9

描述：

3.

```
try:
    statements
except:
```

```
    statements
else:
    statements
```

try...except后可以跟随else。当try代码块没有引发异常，但又必须执行的代码，可以放置在else代码块中。如果将必须执行的代码块放置于try中，则可能会意外捕捉到try...except语句保护的代码触发的异常。

代码段：

```
def fetcher(obj,index):
     print(obj[index])
try:
    obj=[9/i for i in range(1,10)]
    index=5
    fetcher(obj,index)
except IndexError:
    print('ndexError:{}'.format(IndexError))
else:
    fetcher(list(range(10)),3)
```

运算结果：

```
1.5
3
```

10

描述：

4. 触发异常（Raising Exceptions）

raise可以强制触发指定的异常。raise唯一的参数就是触发的异常。

代码段：

```
raise NameError('HiThere')
```

运算结果：

```
---------------------------------------------------------------------------
NameError                                 Traceback (most recent call last)
Input In [44], in <cell line: 1>()
----> 1 raise NameError('HiThere')
NameError: HiThere
```

11

描述：

通过raise触发异常，并在except下打印该异常实例的参数。

代码段：

```
try:
    raise Exception('spam', 'eggs')
except Exception as inst:
    print(type(inst))      # 异常实例 异常实例
    print(inst.args)       # 存储在 .args 中的参数
    print(inst)            # __str__ 允许直接打印参数，但是可以在异常子类中被重写

    x, y = inst.args       # 序列解包
    print('x =', x)
    print('y =', y)
```

运算结果：

```
<class 'Exception'> ('spam', 'eggs') ('spam', 'eggs') x = spam y = eggs
```

12

描述：

如果只想判断是否触发了异常，但并不打算处理该异常，则可以使用更简单的raise语句重新触发异常。

代码段：

```
try:
    raise NameError('HiThere')
```

```
except NameError:
    print('An exception flew by!')
    raise
```

运算结果：

```
An exception flew by!
---------------------------------------------------------------------------
NameError                                 Traceback (most recent call last)
Input In [45], in <cell line: 1>()
      1 try:
----> 2     raise NameError('HiThere')
      3 except NameError:
      4     print('An exception flew by!')
NameError: HiThere
```

13

描述：

5. 异常链（Exception Chaining）

raise 语句支持可选的from 子句，该子句用于启用链式异常。

代码段：

```
def func():
    raise ConnectionError
try:
    func()
except ConnectionError as exc:
    raise RuntimeError('Failed to open database') from exc
```

运算结果：

```
---------------------------------------------------------------------------
ConnectionError                           Traceback (most recent call last)
Input In [46], in <cell line: 4>()
      4 try:
----> 5     func()
      6 except ConnectionError as exc:
Input In [46], in func()
      1 def func():
----> 2     raise ConnectionError
ConnectionError:
The above exception was the direct cause of the following exception:
RuntimeError                              Traceback (most recent call last)
Input In [46], in <cell line: 4>()
      5     func()
      6 except ConnectionError as exc:
----> 7     raise RuntimeError('Failed to open database') from exc
RuntimeError: Failed to open database
```

14

描述：

6. 自定义异常（User-defined Exceptions）

通过定义内置异常类，通常为Exception的子类来自定义异常。通常，异常命名以Error结尾，类似标准异常的命名。同时，可以定义__str__()类，附加状态信息或者方法。

代码段：

```
class AlreadyGotOneError(Exception):
    def __str__(self):return 'So you got an exception...'
    pass  #自定义异常

def grail():
    raise AlreadyGotOneError #引发自定义异常

try:
    grail()
except AlreadyGotOneError as ago_e:
    print(ago_e)
    print('got exception!')
```

运算结果：

```
So you got an exception...
got exception!
```

15

描述：

7. 定义清理操作（Defining Clean-up Actions）

```
try:
    statements
finally:
    statements
```

如果存在finally子句，则finally子句是try语句结束前执行的最后一项任务。不论try语句是否触发异常，都会执行finally子句。以下内容介绍了几种比较复杂的触发异常情景：

- 如果执行try子句期间触发了某个异常，则某个except子句应处理该异常。如果该异常没有except子句处理，在finally子句执行后会被重新触发。
- except 或else子句执行期间也会触发异常。同样，该异常会在finally子句执行之后被重新触发。
- finally子句中包含break、continue或return等语句，异常将不会被重新引发。
- 如果执行try语句时遇到break、continue或return语句，则finally子句在执行break、continue或return语句之前执行。
- 如果finally子句中包含return语句，则返回值来自finally子句的某个return语句的返回值，而不是来自try子句的return语句的返回值。

代码段：

```
try:
    raise KeyboardInterrupt
finally:
    print('Goodbye, world!')
```

运算结果：

```
Goodbye, world!
---------------------------------------------------------------------------
KeyboardInterrupt                         Traceback (most recent call last)
Input In [58], in <cell line: 1>()
      1 try:
----> 2     raise KeyboardInterrupt
      3 finally:
      4     print('Goodbye, world!')
KeyboardInterrupt:
```

16

代码段：

```
def bool_return():
    try:
        return True
    finally:
        return False
bool_return()
```

运算结果：

```
False
```

17

描述：

从下述案例可以看出，不管是不是触发了异常，finally代码块都会执行。

代码段：

```
def divide(x, y):
    try:
        result = x / y
    except ZeroDivisionError:
        print("division by zero!")
```

```
    else:
        print("result is", result)
    finally:
        print("executing finally clause")
divide(2, 1)
```

运算结果：

```
result is 2.0
executing finally clause
```

18

代码段：

```
divide(2, 0)
```

运算结果：

```
division by zero!
executing finally clause
```

19

代码段：

```
divide("2", "1")
```

运算结果：

```
executing finally clause
--------------------------------------------------------------------------------
TypeError                                 Traceback (most recent call last)
Input In [62], in <cell line: 1>()
----> 1 divide("2", "1")
Input In [60], in divide(x, y)
      1 def divide(x, y):
      2     try:
----> 3         result = x / y
      4     except ZeroDivisionError:
      5         print("division by zero!")
TypeError: unsupported operand type(s) for /: 'str' and 'str'
```

20

描述：

8. 预定义的清理操作（Predefined Clean-up Actions）

某些对象定义了不需要该对象时要执行的标准清理操作。无论使用该对象的操作是否成功，都会执行清理操作。

例如，下述案例在语句执行完毕后，即使在处理时遇到问题，也都会关闭文件f。

```
with open("myfile.txt") as f:
    for line in f:
        print(line, end="")
```

参考文献（References）：

[1] Errors and Exceptions, https://docs.python.org/3/tutorial/errors.html.
[2] Built-in Exceptions, https://docs.python.org/3/library/exceptions.html.